Industrial Applications of the Mössbauer Effect

Industrial Applications of the Mössbauer Effect

Edited by

Gary J. Long
University of Missouri–Rolla
Rolla, Missouri

and

John G. Stevens
University of North Carolina at Asheville
Asheville, North Carolina

Plenum Press • New York and London

Library of Congress Cataloging in Publication Data

Industrial applications of the Mössbauer effect.

"Based on a symposium on Industrial Applications of the Mössbauer Effect, held December 16–21, 1984, at the International Chemical Congress of Pacific Basin Societies, Honolulu, Hawaii"—T.p. verso.
Includes bibliographical references and indexes.
1. Mössbauer spectroscopy—Congresses. 2. Mössbauer effect—Congresses. I. Long, Gary J., 1941- . II. International Chemical Congress of Pacific Basin Societies (1984: Honolulu, Hawaii)
QD96.M6I53 1986 543′.08586 86-22669
ISBN 0-306-42463-0

Based on a symposium on Industrial Applications of the Mössbauer Effect, held December 16–21, 1984, at the International Chemical Congress of Pacific Basin Societies, Honolulu, Hawaii

A Division of Plenum Publishing Corporation
233 Spring Street, New York, N.Y. 10013

Printed in the United States of America

PREFACE

As is often the case, the preface is the last task to be finished during the preparation of a large volume such as you are now holding. The first task, obtaining approval for a symposium on the industrial applications, now seems a long time ago. The idea orginated with John Stevens, probably in 1982, from his observation of papers dealing with industrial applications of the Mössbauer effect appearing in the Mössbauer Effect Reference and Data Journal. His initial suggestion for a symposium entitled "Industrial Applications of the Mössbauer Effect" to be held at a national meeting of the American Chemical Society eventually led to the symposium at the International Chemical Congress of Pacific Basin Societies which met in Honolulu, Hawaii in December 1984.

This volume is the result of the symposium at the above mentioned Congress, but is not actually the 'proceedings' of the symposium because this volume does not contain all of the over one hundred Mössbauer effect papers that were presented at the symposium. Rather it contains a selection of papers that the organizing committee for the symposium deemed most appropriate for a volume devoted to industrial applications of the Mössbauer effect. The final volume also contains six chapters that were not a part of the symposium but which are closely related to the topic. There is another difference from many proceedings. Each chapter in this volume has been carefully reviewed by at least two referees and the contents critically evaluated, edited, and revised when necessary. Although this process led to the loss of some chapters and to some delay in its appearance, the editors believe that this process has yielded a more valuable contribution to the Mössbauer literature.

It seems very appropriate that the symposium was part of an International Congress because, as this volume indicates, the Mössbauer effect has become a truly international science. This volume contains contributions from six continents and from twenty countries, ranging in size from Brazil to Belgium, and in place from India to China. There are chapters from universities, private companies, and government laboratories covering a broad range of applications.

In a project of this type there are many people who deserve thanks in making it a success. First we would like to thank the people who paid the bills. The symposium was financially supported by the Division of Inorganic Chemistry of the American Chemical Society, the organizing committee of the International Chemical Congress, the Mössbauer Effect Data Center, Austin Science Associates, Dupont-New England Nuclear, Elscint, Monsanto, Nuclear Data, Oxford Instruments, Ranger Scientific, Signal-UOP, The Nucleus, and Wissel. Their financial support is greatly appreciated.

Finally we would like to thank the people who made it all possible. These certainly must include all the unnamed but hard working referees who devoted many hours to reading and commenting on the manuscripts. The assistance of the staff of the Mössbauer Effect Data Center, and especially Ms. Virginia Stevens, is greatly appreciated for making the meeting such a pleasant success both scientifically and socially. Ms. Phyllis Johnson of the University of

Missouri-Rolla receives our thanks for typing the entire camera ready text. The assistance of Ms. Ellen Raynor-Enco, Ms. Linda Piccinino, and Mr. John Matzka, of the Plenum editorial staff has been a great help in the production of this volume.

Gary J. Long
January 1987

CONTENTS

Section 1
INTRODUCTION

Chapter 1. The Dozen Uniquenesses of the Mössbauer Effect 1
U. Gonser

Chapter 2. Metal Research of Industrial Significance by Mössbauer Spectroscopy 7
T. E. Cranshaw

Section 2
TECHNIQUES AND GENERAL APPLICATIONS

Chapter 3. Mössbauer Spectroscopy in Physical Metallurgy 25
U. Gonser, M. Ackermann, H.-J. Bauer, N. Blaes, S. M. Fries, R. Gaa, and H.-G. Wagner

Chapter 4. Industrial Applications of Mössbauer Spectroscopy to Microcrystals 63
Steen Mørup

Chapter 5. Conversion Electron Mössbauer Scattering. A Technique for Structural Microanalysis of Thin Films and Subsurface Regions in Materials 83
J. A. Sawicki

Chapter 6. Surface Phase Analysis by Depth-Selective Conversion Electron Mössbauer Spectroscopy 121
K. Saneyoshi, K. Debusmann, W. Keune, R. A. Brand, and D. Liljequist

Chapter 7. Rapid Mössbauer-Based Methods for Applications in the Coal and Steel Industries 137
F. E. Huggins, G. P. Huffman, and G. R. Dunmyre

Section 3
APPLICATIONS TO STEEL AND STEEL ALLOYS

Chapter 8. Mössbauer Spectroscopy in Steel Industry 153
K. R. P. M. Rao

Chapter 9. Mössbauer Investigation of High-Manganese Steels 171
G. P. Huffman and Chiaki Ouchi

Chapter 10. The Application of Mössbauer Spectroscopy to Iron-Aluminum Alloys and Industrial Aluminum Samples 189
Thomas Birchall, Dave Hodgson, and Harish D. Merchant

Chapter 11. Mössbauer Investigation of AlFe and AlFeSi Phases in Specially Produced Aluminum Alloys and in Industrial Aluminum 201
Sàndor Nagy, Attila Vèrtes, Zoltàn Homonnay Lajos Murgàs, Tibor Turmezey, Agnes Griger, and Jòzsef Lakner

Chapter 12. Sulfidation-Induced Changes in Iron-Aluminum and Iron-Silicon Alloys 217
S. M. Dubiel, Z. Żurek, and M. Przybylski

Chapter 13. Hyperfine Fields in Fe-Ni-X Alloys and Their Applications to a Study of Tempering of 9Ni Steel 237
B. Fultz and J. W. Morris, Jr.

Chapter 14. Surface Characterization of Nitrogen Implanted Carbon- and Alloy-Steels: A Review and Recent Results 267
G. Principi, S. Lo Russo, and C. Tosello

Chapter 15. Studies of the Phosphatization of Steel and Its Corrosion Products 295
W. Meisel

Chapter 16. Conversion Electron Mössbauer Studies of the Corrosion Products Formed on the Steel Surface in Atmospheres Containing HCl, H_2S, and SO_2 315
Yusuke Ujihira, Arimichi Handa, and Masanori Fujinami

Section 4
APPLICATIONS TO AMORPHOUS ALLOYS AND GLASSES

Chapter 17. Transition Metal Substitution in $Fe_{75}Tm_5B_{20}$ Amorphous Alloys. Effect on Structure, Thermal Stability, and Structural Relaxation 329
P. Matteazzi, G. Cocco, G. Le Caer, and G. Riontino

Chapter 18. Mössbauer Investigations of Magnetic Interactions and Crystallization Behavior of the Amorphous Alloy $Fe_{72}Ni_4Mo_6B_{16}Si_2$ 355
R. Jagannathan, T. G. N. Babu, B. B. Prasad, A. K. Bhatnagar, and V. R. V. Ramanan

Chapter 19. The Fitting Method for Mössbauer Spectra of Amorphous Alloys and an Investigation of Their Properties and Structures 371
Zuxiong Xu, Ruzhang Ma, and Jueyun Ping

Chapter 20. Mössbauer Effect Study of the Cluster Distribution in the Spin Glass Θ-$Me_{0.35}V_2O_5$ (Me=Fe, Al) 399
H. Lass and J. Pebler

Chapter 21. Application of Mössbauer Effect on the Crystallization of Several Glasses 409
Tetsuaki Nishida and Yoshimasa Takashima

Chapter 22. Application of Iron-57 Mössbauer Spectroscopy to the Characterization of Silicate Glasses Used for High Temperature Incinerated Low-Level Radioactive Waste Products 423
E. De Grave and P. Van Iseghem

Section 5
APPLICATIONS TO COAL, MINERALS, AND MINERAL PROCESSING

Chapter 23. Mössbauer Analysis of Q-Basic Oxygen Process Slags 447
F. E. Huggins and G. P. Huffman

Chapter 24. Slag Attack of Firebrick in a Tin Smelting Furnace 467
Pamela McNamara, John D. Cashion, and Mary S. J. Gani

Chapter 25. Mössbauer Effect Study of Decreptation Phenomena in Iron Ores 479
Joice Terra, Elisa B. Saitovitch, Jacques Danon, and Nilce G. Souza

Chapter 26. Coal Treatment and Desulfurization 491
H. Pollak and J. G. Stevens

Chapter 27. Iron Phases in Brown Coal Ashes 513
J. D. Cashion, L. Joan Brown, and A. L. Ottrey

Chapter 28. Mössbauer Spectroscopy as a Tool for Studying Hydrometallurgical Treatments of Copper-Iron-Sulfides 523
Stuart A. Fysh

Chapter 29. Mössbauer Effect Study of the Thermal Reactions of Phosphate Minerals from Christmas Island Fertilizer Rocks 533
I. W. M. Brown and J. E. Davies

Chapter 30. Mössbauer Studies of Thermal Reaction Mechanisms in Hydrous Iron Layer Silicates 547
K. J. D. MacKenzie, M. E. Bowden, and R. M. Berezowski

Chapter 31. A Study of the Initially-Formed Hydrolysis Species and Intermediate Polymers and Their Role in Determining the Product Iron Oxides Formed in the Weathering of Iron 565
James H. Johnston and David G. Lewis

Chapter 32. The Oxidation of Hydroxycarbonate Green Rusts 585
Enver Murad and Reginald M. Taylor

Chapter 33. Metal Adsorption by Activated Carbon 595
J. D. Cashion, D. J. Cookson, L. J. Brown, and D. G. Howard

Section 6
APPLICATIONS TO CATALYTIC MATERIALS

Chapter 34. Mössbauer Spectroscopy of Iron and Iron Alloy Fischer-Tropsch Catalysts 609
A. M. van der Kraan and J. W. Niemantsverdriet

Chapter 35. Mössbauer Effect Characterization of Silica Supported Rhodium-Iron Catalysts 635
Yoshitaka Minai, Takeshi Tominaga, Takakazu Fukushima, and Masaru Ichikawa

Chapter 36. The Application of Mössbauer Spectroscopy to Metallic, Bimetallic, and Mixed Metal Oxide Catalysts 649
F. J. Berry

Chapter 37. Influence of Promoters on the Physical Properties of Iron Zeolite Catalysts 667
Friedrich Schmidt

Chapter 38. Application of Mössbauer Emission Spectroscopy to the Study of Cobalt-Molybdenum Hydrodesulfurization Catalysts 683
Steen Mørup, Bjerne S. Clausen, and Henrik Topsøe

Chapter 39. Applications of the Mössbauer Effect in Electrolysis . . 697
A. Vèrtes, I. Czakò-Nagy, and M. L. Varsànyi

Section 7
MISCELLANEOUS APPLICATIONS

Chapter 40. Role of Tin Complexes in Bonding Silver to Glass: Mössbauer Studies 715
T. M. Thomas, D. L. Williamson, and D. P. Rainville

Chapter 41. Use of Mössbauer Spectroscopy in the Study of IV-VI Semiconductors 725
Ida Ortalli and Vincenzo Fano

Chapter 42. The Mössbauer Characterization of Multilayered Films with Artificial Superstructures 739
Teruya Shinjo and Toshio Takada

Author Index 751

Subject Index 769

THE DOZEN UNIQUENESSES OF THE MÖSSBAUER EFFECT

Uli Gonser

Angewandte Physik
Universitat des Saarlandes
D-6600 Saarbrucken
Federal Republic of Germany

The Mössbauer effect is a unique method. To prove this point, its one dozen extraordinary features are presented:

1. The Methods

Scientific methods might be distinguished formally as:

(a) Descriptive methods: among these are the old and classical methods such as electrical and thermal conductivity, calorimetry, specific heat, magnetization, strength, fracture, etc.

(b) Abbreviated methods: these methods have become numerous and nowadays one needs a guide to find one's way through the jungle of abbreviations: APS, BIS, EDPD, ENDOR, EPR, ESCA, EXAFS, FIM, HEAD, IAS, IMPACT, LEED, LEIS, NMR, NQR, PAC, PSD, SEXAFS, SHEED, SIMS, STEM, STM, SXAPS, TDPAC, TEM, UPS, XANES, XPS...........

(c) Name-tag methods: they are usually named after their discoverers and, as a result, they have the flair of nobility. Examples of name-tag methods are Mössbauer spectroscopy, Auger spectroscopy, Rutherford scattering, Hall effect, de Haas - van Alphen effect, etc. In some cases hybrids are used where the name-tag is incorporated into the abbreviation. For instance conversion electron Mössbauer spectroscopy, CEMS, or Rayleigh scattering of Mössbauer radiation, RSMR.

If we compare the types of methods to wine, we can associate the first two categories to ordinary barrel wines or country wines, whereas the third is a quality vintage with a label. Focussing our attention on the Mössbauer effect, this comparison seems quite appropriate because in German vineyards one sees everywhere the sign Mössbauer since 1856. One is tempted to assume that Mössbauer's discovery was missed by just one century. But behind this sign is a company involved in wine production. One could imagine that a combined conference would be quite productive if the Mössbauer company provided the wine and the Mössbauer scientists provided some superfine spectra. Maybe the 150 year old wine unit of Ochsle could be improved.

100 Jahre
Mössbauer
seit 1856

2. The Discovery

In general, methods originate by advancing, refining, and combining certain aspects of theory, experimental sophistication, and engineering skill. In most cases a number of scientists from different laboratories have their share in the development. Often it is difficult to trace all of the individual contributions leading to the final success. Sometimes theoreticians have made predictions and have announced their priority by a Gedanken-experiment. In contrast, Mössbauer's discovery has all the attributes of a wonderful and unexpected revelation, starting with his stumbling upon a strange phenomenon, his finding the correct cause and interpretation, until finally a new name-tag spectroscopy developed.

Mössbauer's discovery was the realization that nuclear γ-ray emission and absorption processes can occur in a recoil-free fashion and thus resonance can easily be achieved. Recoil-free is the catch. One might think that there was one moment when this idea sparked, but instead, as Mössbauer recalled, it was a long struggle of belief and disbelief, until finally belief was the winner. Besides the discovery itself, Mössbauer made use of the Doppler shift in order to modulate the γ-ray energy, thus developing his effect into a spectroscopy. Finally he gave his effect the correct theoretical interpretation.

The Mössbauer effect may perhaps be regarded as the last great fundamental discovery in the Quantum age. The delay in its discovery is embarrassing for theoreticians as well as for experimentalists and, moreover, in retrospect, the disbelief which occurred when the Mössbauer effect was first announced seems strange because the underlying laws governing the effect had been discovered much earlier.

3. The Discoverer

At the time of his great discovery in the nineteen fifties, Mössbauer was a graduate student struggling towards his Ph.D. In addition to the usual scientific and experimental difficulties, Mössbauer faced problems during the postwar years caused by some rather strict rules imposed upon Germany dealing with the possession of radioactive material. Only through some dark channels did he succeed in obtaining the desired radioactive isotope.

4. The History

The early history of the Mössbauer effect was appropriately told by Harry Lipkin: in the Prehistoric Times (before 1958) the effect might have been discovered, but was not. In the Early-Iridium Age (1958) the effect was discovered, but not noticed. In the Middle-Iridium Age (1958/59) the effect was noticed but not believed. In the Late-Iridium Age (1959) the effect was believed but not considered interesting. Then nature was so kind as to provide iron-57. With the advent of the Iron Age all doubts were swept away and a fierce battle began for the remaining glamour and priorities, even to the extent that as early as April 1960 the editor of the Physical Review Letters had to intervene and wrote: 'We believe that the time has come to put a damper on the influx of Letters on the Mössbauer Effect.'

In a simplified way the succeeding decades can be described as follows:

The Fifties - the discovery of the Mössbauer effect and of the hyperfine interaction parameters.

The Sixties - expansion into all fields of natural science, recognition of the potential applications, mainly qualitative measurements.

The Seventies - quantitative evaluations, computer fitting procedures.

The Eighties - practical applications.

The Nineties - ?

5. The Price Tag

The Mössbauer effect is a cheap method. Mössbauer himself needed only a toy to produce the Doppler velocity. With the addition of a source and a γ-ray counter he was in business.

6. The Honor

In the scientific world the highest honour is the Noble Prize. Uniquely, Mössbauer won the Nobel Prize in 1961 with his very first publication. Only three years separated the first announcement of his effect and the award. The prediction he made at the end of his speech of acceptance has been fully realized:

> "We may therefore hope that this young branch of physics stands only at its threshold, and that it will be developed in the future, not only to extend the application of existing knowledge but to make possible new advances in the exciting world of unknown phenomena and effects."

7. The Significance in Science and Education

The Mössbauer effect has gained easy entrance into all fields of natural science from nuclear physics to industrial applications. Much important interdisciplinary information has resulted.

Some of the fascinating and exotic applications are: measuring the apparent weight of photons in the earth's gravitational field, resonances in isotopes having theoretical natural line widths on the order of 10^{-17}eV, identification of the material composing the biological compass, aerosols, moon samples, applications in art, archaeology, and medicine. The fact that the nuclear hyperfine interaction of one specific isotope can mirror macroscopic diseases in medicine is very impressive.

In addition, the effect is ideally suited to science education, in a broad sense: nuclear physics, solid state physics, magnetism, physical metallurgy, chemistry, biology, and geology, to mention just a few fields, have benefited.

8. The Leader

We have a leader in Rudi Mössbauer. This gives the effect a personal note. He has remained faithful to his effect. Actually he uses it in the most sophisticated experiments. With the strongest of sources and a special counter with high energy and spatial resolution, he is probing the structure and dynamics of biological macromolecules.

9. The Mössbauer Effect Data Center

We have the Mössbauer Effect Data Center operated by John and Ginnie Stevens. They are conscientious bookkeepers of everything new in the field and the Mössbauer community is provided with excellent data and information.

10. Conferences

In most cases conferences are focused on specific fields such as magnetism, low temperature physics, semiconductors, amorphous metals, defects, etc. Method conferences are extremely rare. We are fortunate in having regular international and national conferences on the various applications of

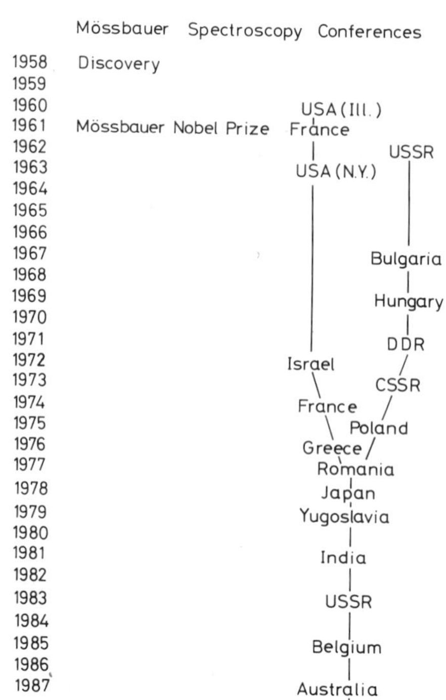

Figure 1. Mössbauer Spectroscopy Conferences over the years.

the Mössbauer effect (see Figure 1). They integrate all scientific fields and are therefore never dull.

11. Our Vocabulary.

The Mössbauer effect has a personal note of a psychological nature. The following contrast is significant: instead of dealing with tough words such as: fracture, cracks, shock waves, corrosion, stress, and other typical features and expressions of physics and chemistry, the effect seems to provide many kind and tender words for our daily work, such as: recoil-free processes, lots of interactions, finest interactions, and resonances. An outsider who is listening and hears again and again about hyperfine interactions and hyperfine spectra waits in vain for an 'ugly' interaction. I believe that the observation of resonances in nature leads us to search for resonances among ourselves.

12. The Future True Fairy-Tale

My last point is addressed especially to Rudi Mössbauer. One element in the story is still missing. The story as a whole has to be recorded in a book which will read like a fairy-tale:

> "Once upon a time there was a young scientist confronted and frustrated by a tough problem •••••."

Of course, this tale can only be written by the master himself. Coming generations will be impressed by the romantic air of the story, and it will be hard to convince them that it is really true.

METAL RESEARCH OF INDUSTRIAL SIGNIFICANCE BY MÖSSBAUER SPECTROSCOPY

T. E. Cranshaw

Nuclear Physics Division
AERE Harwell
England

INTRODUCTION

In 1970 Lyle Schwartz [1] read a paper to the 10th Annual Symposium of Physics and Non-Destructive Testing in Chicago. In it he claimed that Mössbauer spectroscopy was coming of age, and identified three areas in which it could make contributions to industry

1. as a research tool,
2. in quality control,
3. for in-service evaluation.

He clearly expected that in future there would be plenty to be said about the second and third areas, but at that date he had to be satisfied with giving a review of the use of Mössbauer spectroscopy as a research tool, and ending with an account of his own most recent research.

Unfortunately, his expectations have not been fulfilled. To the best of the author's knowledge, there is still no example of a Mössbauer instrument in use for quality control, or for in-service evaluation of materials or surfaces. Most of the other papers in this volume come into the first category, Mössbauer spectroscopy as a research tool. Papers on instrumentation for industrial use, such as the paper by Huggins, Huffman and Dunmyre on a pyritemeter, or ferritemeter are few and far between. The plan of this paper is therefore the same as that of Schwartz, fourteen years ago. I will describe first some near misses in the attempt to get a Mössbauer instrument onto the factory floor, and then describe some recent work at Harwell using Mössbauer spectroscopy to make a contribution to the solution of industrial problems.

THE NEAREST APPROACH TO THE USE OF MÖSSBAUER SPECTROSCOPY FOR QUALITY CONTROL, OR IN-SERVICE EVALUATION.

In areas 2 and 3 noted by Schwartz, it is a striking fact that several physicists have found Mössbauer spectroscopy to be the best way of making particular measurements required in industrial processes. For example Huffman [2] has stated that Mössbauer spectroscopy is the best way of monitoring iron-phase distribution in pellets during direct reduction of beneficiated ores. Karl and Zuckerman [3] have said that the Mössbauer technique has important advantages over current methods for analyzing iron-bearing minerals in fossil fuels. R. G. Campany [4] at the limits of lubrication conference, 1977, said that backscatter Mössbauer spectroscopy is the most successful technique for examining the structure and composition of wear layers. Why then, do we not find Mössbauer instruments in service in industry? Perhaps these are the drawbacks. Ideally, a device to be used in an industrial environment must be able to function without the attention of an expert Mössbauer spectroscopist to interpret its responses. Furthermore, it must be reliable, in the sense that it is not more likely to produce false results than other similar measuring instruments in use in industry. These restrictions reduce the range very considerably, perhaps even to the unique case of the ferritemeter or austenite meter. Here the very sharp distinction between single line spectrum of austenite and the six line spectrum of ferrite or martensite gives one hope that an instrument could be developed which would give a digital reading of percent austenite. Jaggi and Rao [5] gave some consideration to methods of producing the answer quickly, i.e., in a time measured in seconds rather than hours, by using sources in which the matrix is the material under investigation, so that resonance is obtained at zero velocity, and is destroyed by the application of a high velocity.

If these restrictions are relaxed somewhat, then other measurements may become possible candidates for industrial use, and it is worthwhile to look for such candidates. One may suppose that if one discovers, or thinks he has discovered, a process which might have applications in industry, he would try to patent it. A search of the patent literature turned up about thirty specifications in which the word Mössbauer appeared. Of course, this does not mean that there are no other patents relating to Mössbauer methods which do not contain the word Mössbauer in their titles, but of these thirty, only two concerned applications in industry. One was a method of monitoring vibrations, and was really the inverse of a calibration, which is familiar to all Mössbauer spectroscopists. The other was a method measuring the thickness of a plated film during the plating process by taking advantage of the fact that the Mössbauer isotope in the plating solution shows no resonance absorption, but the plated layer does. Huffman and Podgurski [7] have described a method of determining the thickness of haematite and magnetite layers on oxidized iron by conversion electron Mössbauer spectroscopy, the layers in question having a thickness of about 0.3μm. I will now describe some measurements by x-ray Mössbauer spectroscopy of the thickness of nitride layers on iron, where the thickness of the layers is about 30 μm.

THE MEASUREMENT OF THE THICKNESS OF NITRIDE LAYERS BY MÖSSBAUER METHODS

One widely used process for carbonitriding an iron component is the immersion of the component in a molten salt bath containing, for example, cyanides, and other compounds. The salts are broken down by interaction with the iron, and the carbon and nitrogen from a hard layer of iron nitrides and carbides. The thickness of the layer depends on the combination of the time of immersion of the component and the temperature. However, it appears that this cannot always be relied on to give a desired thickness of layer, and it is common practice to extract a few components from a batch, to cut them up and mount them and make a metallographic inspection with a microscope. This is quite a costly procedure.

Now in general, as one goes down in the nitrided layer, the concentration of nitrogen or carbon falls off, and between the outer nitrided layer and the iron base, there is a layer of carbon or nitrogen austenite, as shown in Figure 1. The principle of the method is once again to exploit the power of a Mössbauer spectrum to discriminate between the martensite, the austenite and the nitride layers. Figure 2 shows Mössbauer spectra taken by detecting back-scattered x-rays, of the surfaces of components treated by immersion in a commercial salt mixture for times between 2 and 30 minutes.

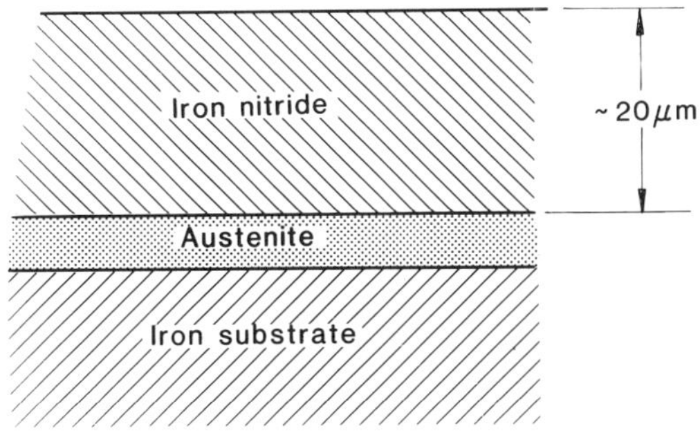

Figure 1. A representation of the layers on a nitrided iron component.

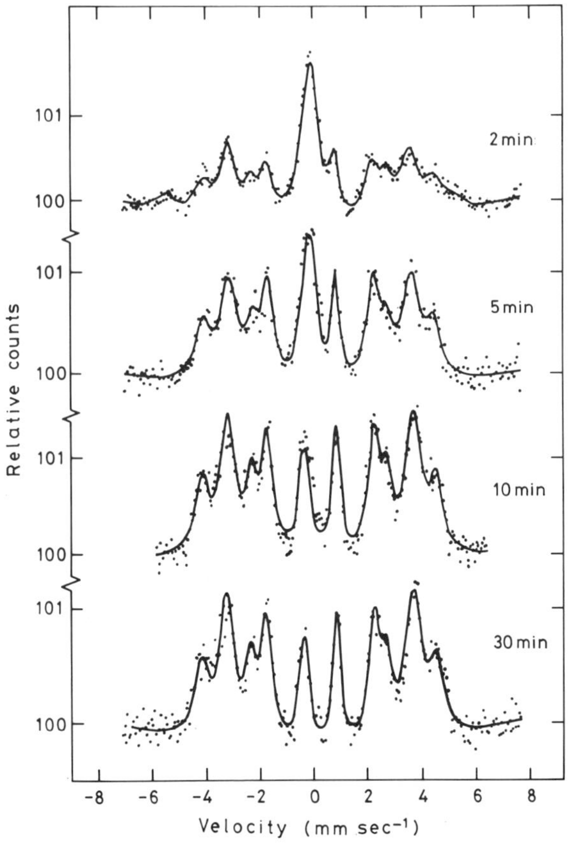

Figure 2. Mössbauer spectra by back-scattered x-rays of an iron component treated for different times in a commercial nitriding salt bath.

The interpretation of the spectra is quite straightforward. In the top spectrum, a trace of martensite or ferrite from the underlying steel is visible. The central peak near zero velocity comes from the layer of nitrogen austenite between the nitrided layer and the underlying steel. The remainder of the spectrum comes from hexagonal epsilon nitride, εFe_2N_{1-x} with x ca. 0.3, in which the iron site field may take two values depending on the number of nitrogen neighbors. As the immmersion time is increased, the martensite component disappears, and then after it, the austenite peak, until only the ε-nitride is seen. It is clear that the signal from the austenite layer is attenuated by the iron nitride above it, and it only remains to make a calibration of the intensity of the austenitic line relative to the nitride spectrum to be able to measure the thickness of the nitride layer. The absorption lengths of the 14keV γ-ray and the 7keV x-ray in iron nitride are similar, namely about 25μ. The absorption of the austenite signal therefore occurs with an absorption length of about 12μ, and measurements of the thickness should be possible up to about 30μ. This is often the region of interest.

THE STRUCTURE AND COMPOSITION OF WEAR SURFACES BY MÖSSBAUER SPECTROSCOPY

Another example where the components of the Mössbauer spectra are simple, so that it is imaginable that a computer program could extract the data occurs in the study of wear surfaces. When lubricated sliding surfaces are loaded to the point of failure, subsequent microscopic examination of the surfaces reveals white areas of transformed material, and the characterization of these surfaces became known as the "white layer problem". The work to be described here was an investigation of white layers in collaboration with R. G. Campany [4] at Thornton Research Centre, Shell U.K. Research Ltd.

Specimens representing wear surfaces were produced in a David Brown Two Disc machine, in which lubricated steel blanks could be run against each other with a predetermined slip velocity. The load on the surface was increased at five minute intervals until breakdown of the surface was detected by a sharp rise in friction and disc temperature, usually accompanied by noise and vibration. The disc after removal from the machine showed regions of "scuffing", defined as "a gross damage characterized by the forming of local welds between the sliding surfaces". Transmission electron microscopy and x-ray diffraction suggested that these were composed of a heavily deformed mixture of austenite and martensite.

The discs also showed regions of high polish, and such regions were common on discs removed before the occurrence of breakdown. Under a microscope, these surfaces, called "scored surfaces" show many streaks in the direction of sliding, and sections through the surface show a thin transformed layer. This is identified with the material called WI by Rogers [8], whereas the scuffed regions he called WII.

The ability of Mössbauer back-scatter spectroscopy to reveal the presence of martensite, austenite and iron carbide make it a powerful method of investigating these surfaces. Figure 3 shows two examples of spectra of wear surfaces. The first spectrum is of a scuffed surface, and shows the presence of martensite and about 13% of austenite. The second, of a polished and scored surface shows the presence of martensite, about 20% of austenite and 32% of iron carbide. In cases of rather uniform polished surface about 50% of the surface was in the form of iron carbide. It is believed that in the course of breakdown of the surfaces a hard layer of iron carbide is produced first, which subsequently breaks up, possibly by a fatigue mechanism. It is these particles that then cause scuffing and catastrophic breakdown of the surface. It may be thought that testing for iron carbide in

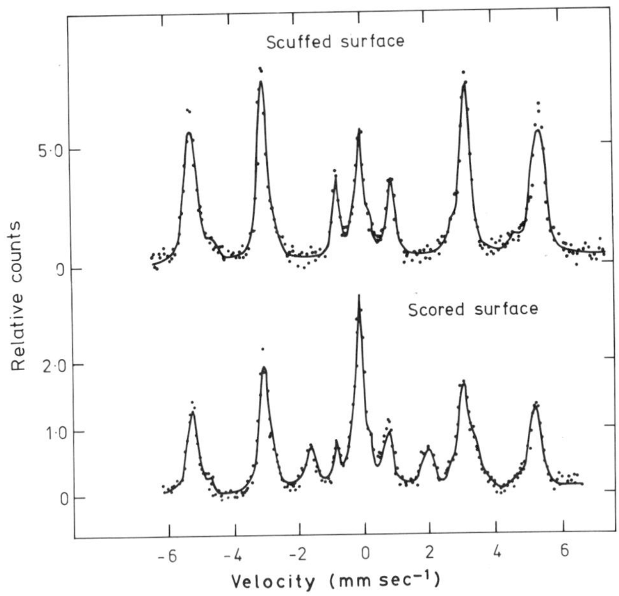

Figure 3. Spectra of scuffed and scored surfaces produced in a two disc wear machine.

a surface could be a useful method of monitoring wear in a bearing surface. It would, of course, be non-destructive, but would require a degree of dismantling of the machinery under scrutiny.

MÖSSBAUER SPECTROSCOPY AS A RESEARCH TOOL IN THE METALS INDUSTRY

1. Temper embrittlement, and the behavior of impurity elements.

The work to be described first, rose out of an investigation of the phenomenon of "reversible temper embrittlement", which may be defined as follows. When some steels are held at a temperature in the region of 450 degrees C for long periods, the material is found to become brittle, and failure under load occurs at grain boundaries. A short heating at higher temperatures will restore the former properties. It has been established that a prerequisite for the phenomenon to occur is the presence of a major alloying element, for example, nickel, and a minor alloying element, possibly a tramp element in concentrations of only parts per million, for example, tin or antimony. The combination of alloying elements is far more damaging than the sum of the elements separately.

Although the metallurgical damage evidently occurs at grain boundaries, whereas Mössbauer spectra in general relate to the bulk of the material, and although the concentrations of metallurgical importance may be well below the levels which are accessible to Mössbauer spectroscopy an investigation of tin-119 in iron-nickel alloys proved to be very rewarding. In Figure 4 are shown tin spectra of iron-nickel alloys containing 2% tin-119 after different heat treatments. In the embrittled state, it is very clear that an interaction between nickel and tin has taken place, leading to a non-magnetic precipitate in the bulk material, and it is plausible that similar interactions occur at grain boundaries and result in weakening of the boundary. From the Mössbauer spectra, it is impossible to determine whether the

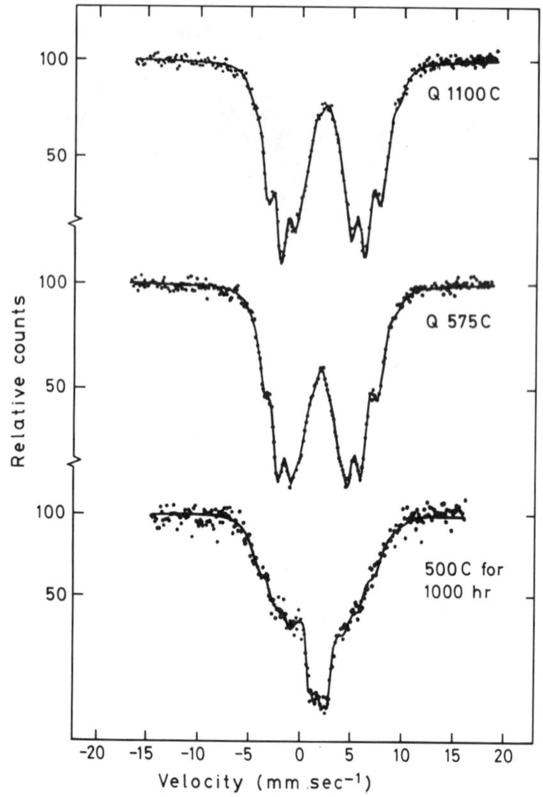

Figure 4. Mössbauer spectra of tin-119 in iron nickel alloys with different treatments.

precipitate is Ni_3Sn_2 or Ni_3Sn_4. Subsequent work by electron microscopy has shown that both occur under appropriate conditions.

A later investigation of the spectra of tin-119 in solution in alloys of iron and nickel showed that there were two components in the spectrum in addition to the undisturbed component due to tin atoms far from any impurity atom, one with a smaller field and one with a larger field than the undisturbed component of -80kOe. The attribution of these components to tin atoms with nickel in the first and second neighbor shells respectively is not yet certain, but so far all the evidence is consistent with it. It was found that the number of nickel neighbors in the first shell was substantially greater than the expectation for a random distribution, while the number in the second shell was close to the expected value. With suitable heat treatment, large number of nickel atoms could be found in the first shell, giving confused spectra with broad distributions of field.

We represent the interaction potential by a potential well, as in Figure 5. Then in the approximation of a low concentration of probe tin atoms, we may use the law of mass action to find for the concentration of nickel atoms in the i-th neighbor shell

$$C_i = C_\infty \exp(-B_i/kT)$$

where C_∞s is the concentration of nickel atoms far from a tin atom, B_i is the attractive potential in the i-th neighbor shell, and T is the temperature.

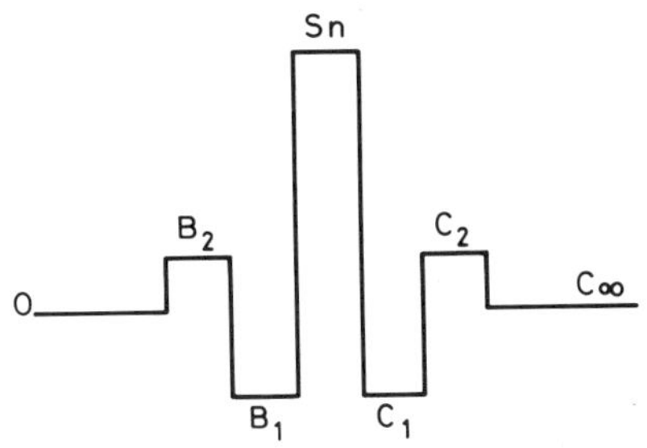

$$C_i = C_\infty \exp(-B_i/kT)$$

Figure 5. Diagram of potential wells assumed to exist in the first and second neighbor shells of a tin atom. B represents the depth of the well, and C the concentration of impurity atoms in that shell.

By quenching from a range of temperatures, and employing a range of values of C_∞, one can deduce values for B. The value found for the interaction between tin and nickel is about -100meV, where the negative value indicates an attractive potential.

In similar experiments, using Mössbauer spectroscopy, NMR and x-ray diffraction, Cadeville, et al. [9] showed that interactions of about the same magnitude exist between antimony atoms and titanium and nickel dissolved in iron. They were able to deduce solubility limits for antimony at a range of temperatures in nickel iron alloy, and deduced values of -50meV and -10meV for the attractive interaction between antimony and nickel in first and third neighbor shells.

In an attempt to learn more about the potential existing between the tin and nickel atoms, spectra were taken of tin in iron-palladium and iron-platinum alloys [10]. It was found that the attractive potential between tin and palladium dissolved in iron was about -280meV even greater than between tin and nickel, whereas between platinum and nickel, the potential was near to zero, i.e. the number of platinum atoms in the neighbor shells of the tin atom was close to the expectation for a random solution. It was also observed that no component could be found in the spectrum of tin-119 in iron-palladium alloys which might correspond to tin with palladium in its second neighbor shell, indicating that either a palladium atom in the second neighbor shell makes no difference to the hyperfine field at the tin atom, or that a strong repulsive potential exists at this distance.

A step forward in the interpretation of these potentials can be taken from the work of Krolas [11] on the correlation between impurity binding energies and the heat of formation of alloys. Krolas pointed out that when a pair of impurity atoms become first neighbors to each other, a bond with a host atom is broken for both impurity atoms, while a bond between impurities is created instead. Thus the binding energy, E_B, may be written

$$E_B = \Delta E_{BC} - \Delta E_{AB} - \Delta E_{AC}$$

where ΔE_{BC}, ΔE_{AB}, and ΔE_{AC} represent the energy of a single bond between BC, AB, and AC atoms respectively.

By plotting the observed values of E_B, determined by PAC, Mössbauer or NMR methods against $(H_{BC}-H_{AB}-H_{AC})$, where $\bar{H}$ is the enthalpy of formation of the binary alloy, Krolas showed that there was indeed a correlation, and that the constant in the correlation was Z, the coordination number. When the enthalpy of formation of the alloy was not know from experiment, Krolas used the semi-empirical methods of Miedema [12] to calculate them.

For the heat of solution H_{AB} of metal B in a transition metal A, Miedema writes

$$H_{AB} = \frac{2V_B^{2/3}}{n_A^{-1/3}-n_B^{-1/3}} \{-P(\Delta\phi)^2 + Q\ (\Delta n^{1/3})^2-R\}$$

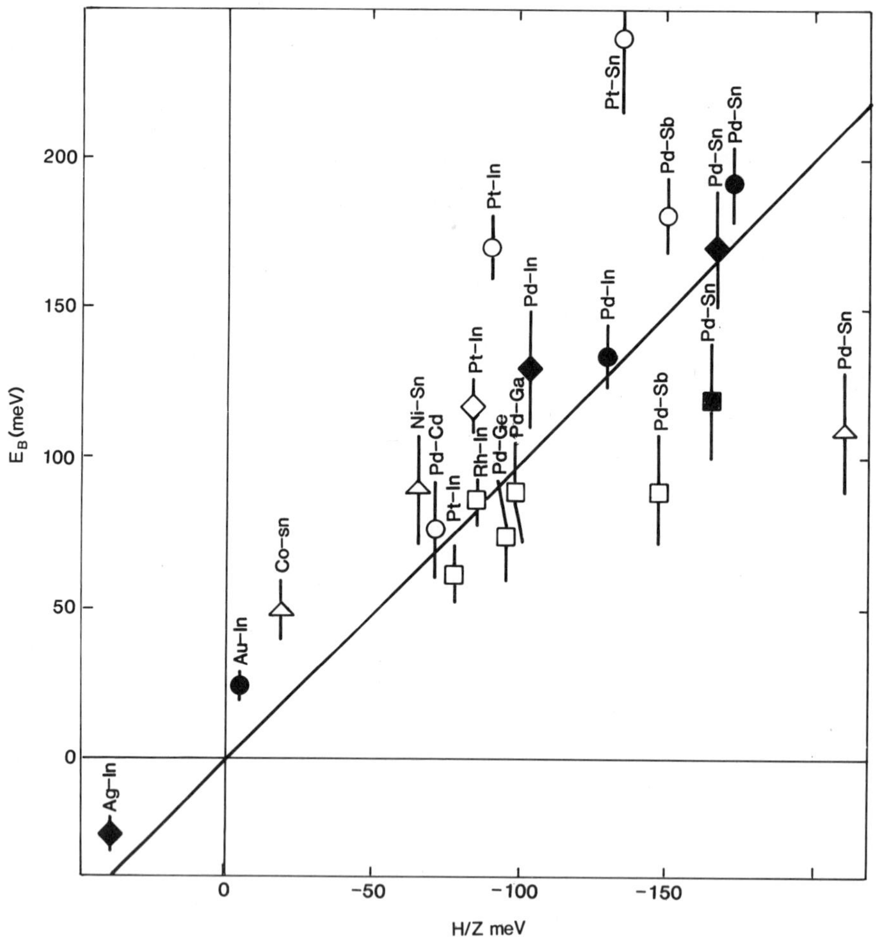

Figure 6. A plot of measured binding energies against H/Z where H is defined in the text, and Z is the coordination number. Points drawn Δ are for iron host, for copper, O for silver, and □ for gold. Solid symbols indicate calculation from heats of formation, open symbols indicate use of Miedema prediction.

In this equation, $\Delta\phi = \phi_A - \phi_B$, is the electronegativity difference between A and B, $\Delta n^{1/3}$ is the difference of electron density to the one third power at the boundary of bulk Wigner-Seitz cells, V_B is the atomic volume of metal B, and P, Q, and R are empirical constants. All the parameters needed in the expression for the heat of solution are given in the publications of Miedema. The difference in electronegativity results in electron transfer which favors alloy formation. The smoothing of the disjcontinuity in electron density leads to an increase in energy opposing alloy formation. Finally, R is a term representing hybridization of p electrons of an sp impurity with d electrons of a transition metal host. The success of this procedure as shown in Figure 6 suggests that it would be a reasonable step to discard the experimental values of heats of formation of alloys, and compare the measured pair interaction potentials with the prediction of the Miedema-Krolas model.

The experimental evidence which has been considered in the work of Krolas relates to any host, and any pair of impurity atoms. It seemed that a good test of the Miedema relations might be made if the host and one of the impurity pair atoms were kept constant, and the interaction with a series of other atoms measured. It turns out that this can be done if the host is iron, the constant probe atom is tin-119, and the impurity atom one of the 4th period atoms from cobalt to germanium.

In all cases, the specimens consist of iron containing about 3% of the alloying element and 0.2% tin-119. In the rolled state, the spectra usually show a magnetically split component with a field near 82kOe, corresponding to the field at tin-119 in iron, which can be attributed to tin atoms far from an atom of the alloying element, and two other components, one with a field smaller than 82kOe, and one with a field larger than 82kOe. These are attributed to tin atoms which have an atom of the alloying element in their first or second neighbor shell respectively. Although this attribution is

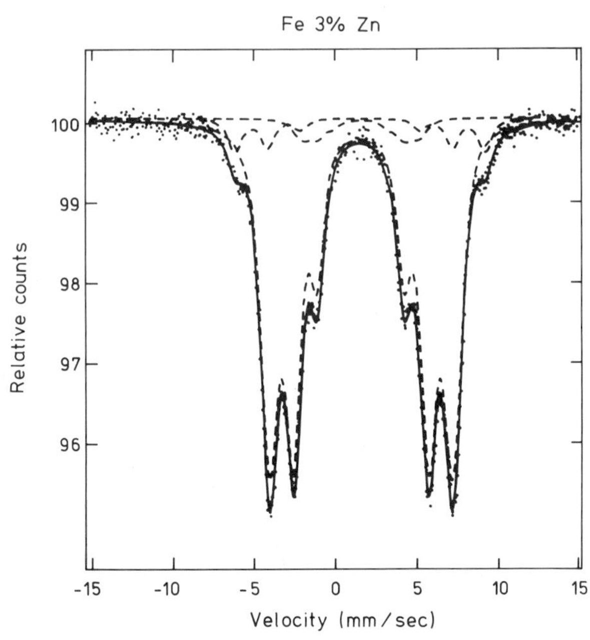

Figure 7. Spectrum of tin-119, concentration 0.2% in an alloy of Fe 3% Zn.

not beyond doubt, it is consistent with the work on tin-119 in iron-nickel alloys, and also with theories of the disturbance of conduction electron polarization based on Friedels' model [13]. An example of such a spectrum is shown in Figure 7 for an annealed alloy of iron and zinc.

Figure 8 shows a similar spectrum for a specimen of iron-gallium, in the as-rolled state. Components due to gallium atoms in the first and second neighbor shells of the tin atom are again seen in the spectrum. Figure 9 shows the same specimen after annealing at 400 degrees C for 2 hours, and it is clear that the strength of the neighbor components has been greatly reduced.

This phenomenon is attributed to a repulsion between the atoms of tin and the alloying element, so that at a temperature at which diffusion can occur, the atoms move apart. An increase in intensity may then be brought about by quenching the specimen from a high temperature.

It must be admitted that the values of C_i found for different values of T, taken as the quenching temperature are not very concordant, but this may well be due to accidents of the quenching process. The point is that to find B_i, we have to take logarithms of the concentrations, so that bismuth is not too sensitive to changes in C_i.

The results of the analysis are shown in Figure 10a, and b. Along the abscissa are the alloying elements, with Z increasing by one at each step. In Figure 10 a, the measured values of the interaction potential at first neighbor distance with tin atoms are displayed, together with the predictions of the Miedema-Krolas model. It is clear that there is a very good

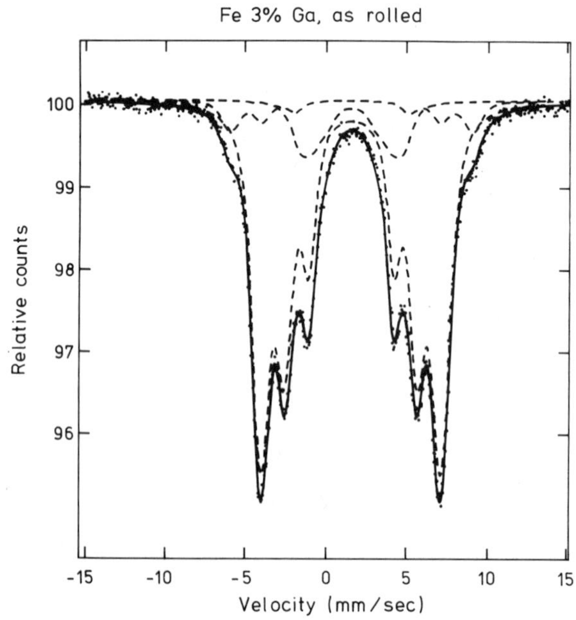

Figure 8. Spectrum of tin-119, concentration 0.2% in Fe 3% Ga, in the as-rolled state.

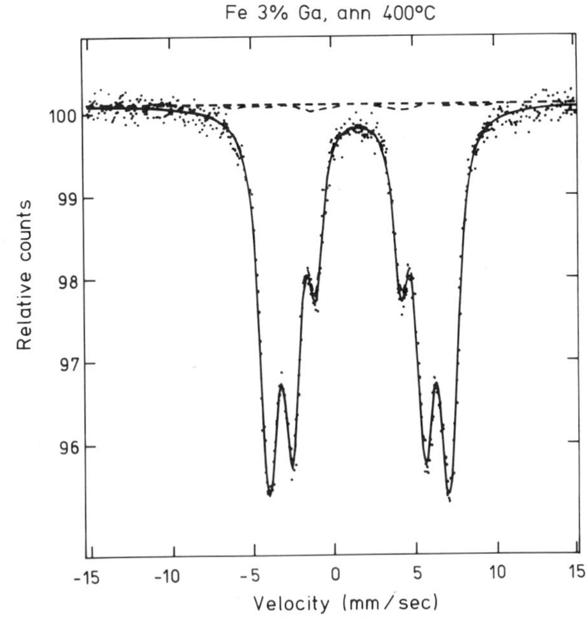

Figure 9. Spectrum of tin-119, concentration 0.2% in Fe 3% Ga after annealing for 2 hrs. at 400°C.

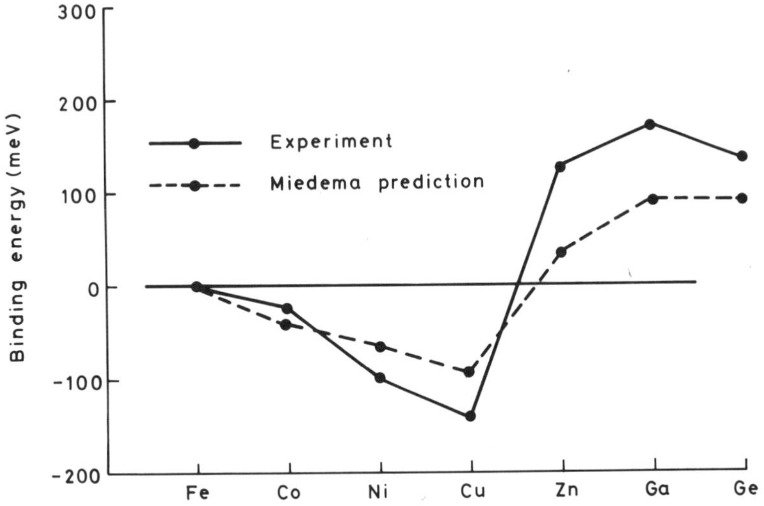

Figure 10a. The measured binding energies between Sn and the elements Co, Ni, Cu, Zn, Ga, and Ge relates to first neighbors.

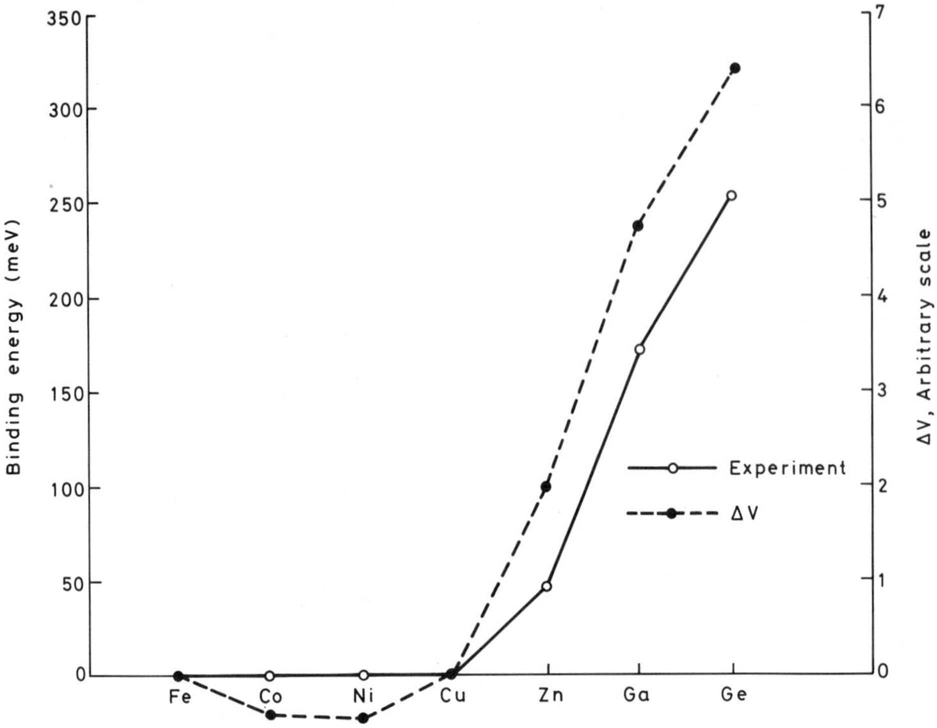

Figure 10b. The measured binding energies between Sn and the elements Co, Ni, Cu, Zn, Ga and Ge relates to second neighbors.

correlation. In particular, the change from attraction at copper to repulsion at zinc is well reproduced, and the apparent "saturation" at gallium and germanium. In Figure 10b, are the measured potentials at second neighbor distances. It is clear that there is very little correlation with the Miedema prediction. Perhaps this should be expected, because the Miedema model depends on a contact interaction. Instead, there is a good correlation with the quantity ΔV, the difference between the atomic volume of the host atom and the alloying atom. Alonso [14] and others have emphasized the importance of an elastic energy term arising from size mismatch. Alonso has suggested that H^{size} should be proportional to the product $(V_A-V_B)(V_A-V_C)$ where V_A is the atomic volume of the host material and V_B and V_C the atomic volumes of the two impurity atoms. In our case, (V_A-V_B) will be constant, and H will be proportional to $\Delta V=V_A-V_C$.

It is interesting to note the surprising fact that the potentials at first and second neighbor distances are of the same order of magnitude, in fact, in the case of gallium and germanium, the second neighbor repulsion appears to be stronger than the first neighbor repulsion. This may be a peculiarity of the BCC lattice, which is not close packed. The second neighbor distance is only 15% greater than the first neighbor distance, and this may result in first neighbors being able to accommodate a size mismatch more easily than the second neighbors.

We may conclude that some progress has been made in obtaining a semi-empirical understanding of the problem of pair interactions in solution in metals, a problem which seems almost impossibly complicated from first principles.

2. The use of single crystals in alloys research

I would now like to draw attention to the increase of information which can be obtained from magnetic alloys when the specimens are single crystal rather than polycrystalline. A good example of the specific information which can be obtained from the use of single crystals is the work of Van Diepen [15] on magnetite, which showed that the broadening of the line commonly attributed to electron hopping was in fact due to broadening by dipolar components of the hyperfine field. The new information in the case of alloys arises because in general, even when the alloy is based on a cubic lattice, the cubic symmetry at the probe site is destroyed by the presence of alloying atoms, allowing the existence of dipolar magnetic fields and electric field gradients. The superimposition of these anisotropic fields on the usual hyperfine fields broadens or splits the spectral lines by an amount which is dependent on the magnetization direction. Thus by taking spectra with the magnetization lying along certain chosen crystalline axes, and noting the differences between them, by, for example, subtracting one spectrum from another, one can reveal the anisotropic interactions. Apart from the intrinsic interest of the anisotropic components, this procedure can be helpful in identifying the components of an alloy spectrum, because the relative magnitude of the broadening produced by the anisotropic interactions as a function of magnetization direction is predictable on geometric grounds alone, and does not depend upon assumptions about the line intensities or other preconceptions. For example, the spectrum of a polycrystalline specimen of iron containing a few percent of nickel is a six line spectrum with a field slightly larger than the natural iron field, and with slightly broadened lines. Figure 11 shows two spectra of an iron-nickel single

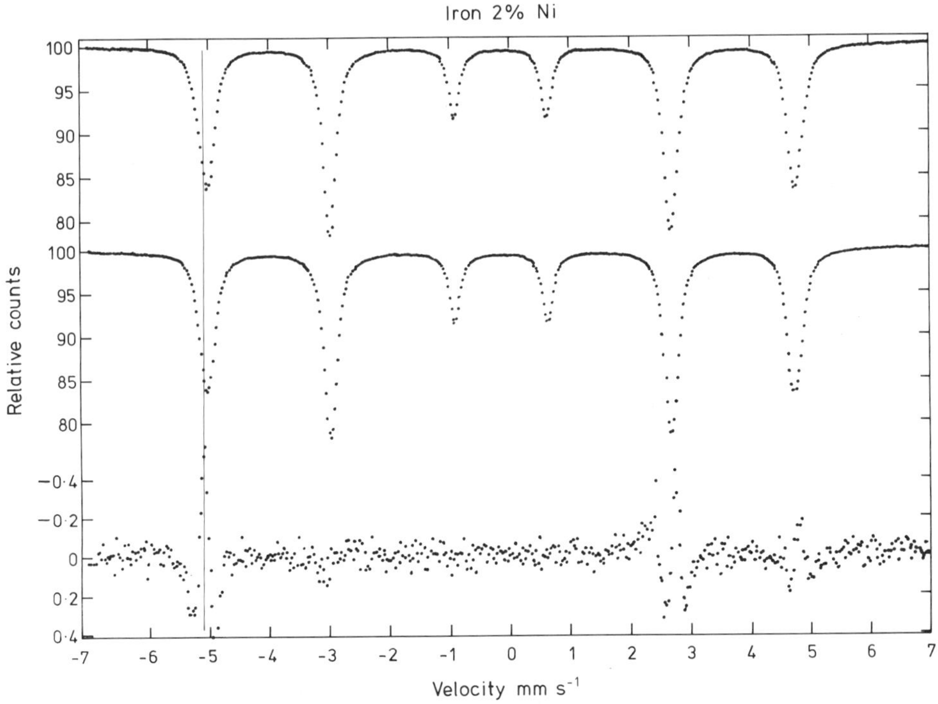

Figure 11. Spectra of an iron crystal containing 3% nickel, taken with the magnetization along the 100 axis, (first spectrum) and along the 111 axis (second spectrum). The third spectrum shows the difference.

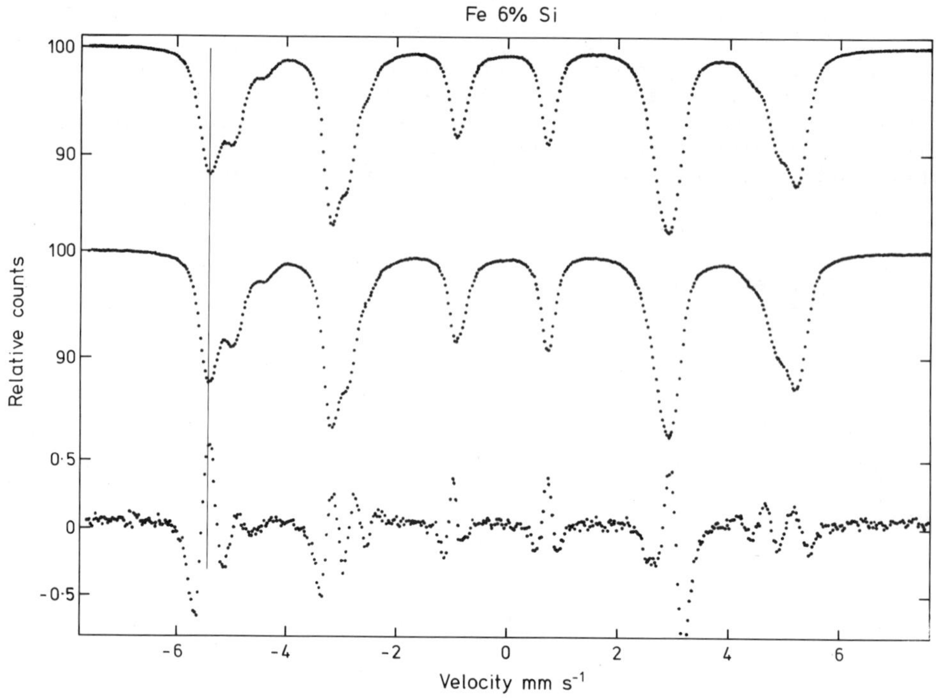

Figure 12. Spectra of an iron crystal containing 6% silicon, taken with the magnetization lying along the 200 axis (first spectrum), and the 111 axis (second spectrum). The third spectrum shows the difference.

crystal, taken with the magnetization along the 100 axis and the 111 axis. Below them is the result of subtracting the spectra, and the sign of the peaks in this spectrum enables one to determine unambiguously that they are due to first neighbor interactions. A computer fit shows that δH_1, the change of hyperfine field due to the presence of a nickel atom in the first neighbor shell of an iron atom is 5.6kOe, a quantity which is much smaller than the line width. Again Figure 12 shows two spectra of an iron silicon single crystal taken under the same conditions. In this case, the sign of the spike marked x shows unambiguously that it is due to second neighbor anisotropic interactions, and we can deduce that δH_2 is very close to zero.

The example I wish to show now is that of ordered Ni_3Fe. Drijver [16] et al. have carried out extensive work on the ordering process of this material. In the course of their work they fitted their spectra on the assumption that the field at an iron atom depended only on the number of iron atoms in the first and second neighbor shells. They found δH_1=-11kOe and δH_2=-2.5kOe, the sign of these two quantities indicating that the presence of an iron atom in the first or second neighbor shell of another iron atom increases the absolute value of the hyperfine field, whose magnitude is about -272kOe. The work on single crystals show that the true situation is quite different.

Figure 13 shows the structure of Ni_3Fe. The solid circles are iron sites, and the open circles nickel sites. In practice, perfect order cannot be achieved, and there are therefore two kinds of defects in the crystal,

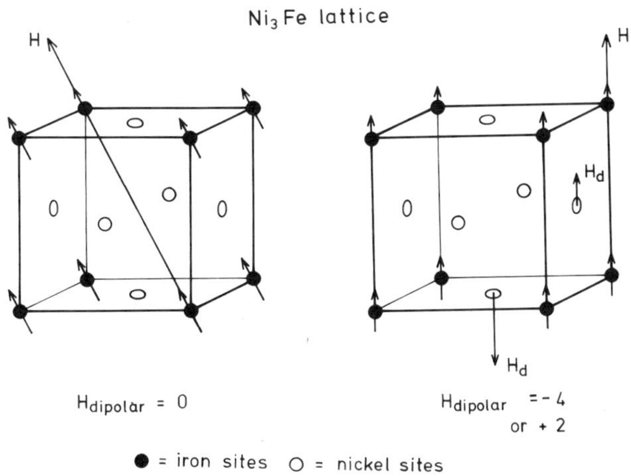

Figure 13. The crystal structure of Ni_3Fe, showing the dipolar field at nickel sites when the magnetization axis lies along a 111 axis or a 100 axis.

iron atoms on nickel sites, and nickel atoms on iron sites, which we will call defects of type 1 and type 2 respectively. Let us consider the first of these in more detail.

It can be seen in the figure that iron atoms on nickel sites, defects of type 1, have four iron neighbors forming a square. When the magnetization lies along the 111 axis, the values of $3\cos^2\theta-1$ for the four neighbors are 1, 1, -1, and -1, so that the component of the dipolar field along the 111 axis is zero. On the other hand, Figure 13b shows that when the magnetization is along a 100 axis, there are two possible defect sites, according to whether the plane of the square is normal to the magnetization or contains the magnetization direction. The component of the dipolar field is then proportional to -4 or +2. For each site with $3\cos^2\theta-1=-4$, there are two with $3\cos^2\theta-1=2$, so that overall, the anisotropic terms cancel, consistent with cubic symmetry.

Two single crystal specimens with Ni_3Fe order were prepared with compositions $Ni_{74}Fe_{26}$ and $Ni_{76}Fe_{24}$. Clearly the specimen with the iron excess will contain more defects of type 1 than the specimen which is iron deficient. The spectra of the specimen with iron excess is shown in Figure 14. The top spectrum is taken with an external magnetic field of 1kOe directed along the 100 axis. The second is taken with the field along the 111 axis. This is the "easy" axis of magnetization, so that this spectrum corresponds to the spectrum of polycrystalline Ni_3Fe. The two spectra are strikingly different, showing immediately that there are strong anisotropic interactions.

Let us assume that the differences are mainly due to the defects of the type 1. A few trials with fitting programs leads to the model shown between the two spectra. When the magnetization lies along the 100 axis anisotropic interactions split the contribution to the spectrum due to iron atoms on nickel sites by a total of 1.1mm/s on line 1, and by 0.3mm/s on line 6. The magnetic dipolar contribution is 42kOe, larger than any of the shifts of mean hyperfine field produced by neighbor effects acting through electron polarization effects. The presence of dipolar fields with strengths of this order should occasion no surprise, because Gros and Pauleve [17] found a

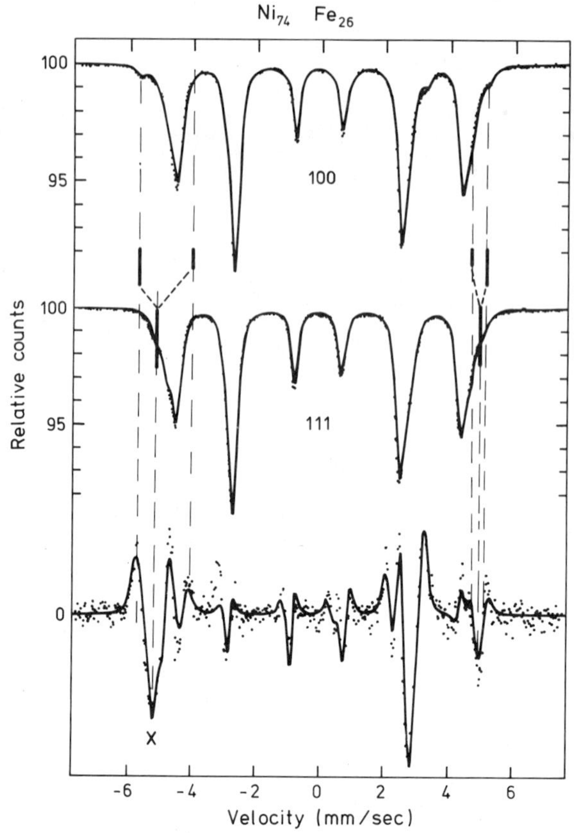

Figure 14. Spectra of $Ni_{74}Fe_{26}$, taken with the magnetization lying along the 100 axis and the 111 axis. The third spectrum is the difference. The solid line is a fitted curve according to the model discussed in the text.

splitting of the mean hyperfine field in ordered nickel-iron of 39kOe by dipolar fields. This is a very relevant result, because the structure of nickel-iron is obtained if all the nickel sites on alternate planes of Ni_3Fe are occupied by iron atoms. Also, in iron-57 doped Ni_3Al, Drijver and Van Der Woude found dipolar fields of 22kOe.

Just as in the case of the iron-nickel and iron-silicon, we can make a precise determination of the mean field at the iron atoms on nickel sites, although the resolution is insufficient to make such a determination if only polycrystalline specimens are available. The result is that the field H_{Ni}=-317kOe, whereas H_{Fe}=-274kOe. We can also find the shift of field at iron on iron sites caused by the presence of defect iron atoms on nickel sites, and find $\delta H_1 Fe$=-6kOe. This compares with the value found by Drijver et al. of -11kOe.

It will be noticed that if we were to attribute the difference in field for iron at nickel sites and iron at iron sites entirely to the fact that it has four iron first neighbors, we would have H(4)-H(0)=-24kOe (where H(n) means an iron atom with n iron first neighbors) whereas the experimental value is -43kOe. We might attribute this discrepancy to the fact that an iron atom on an iron site has 6 iron second neighbors, whereas iron on nickel

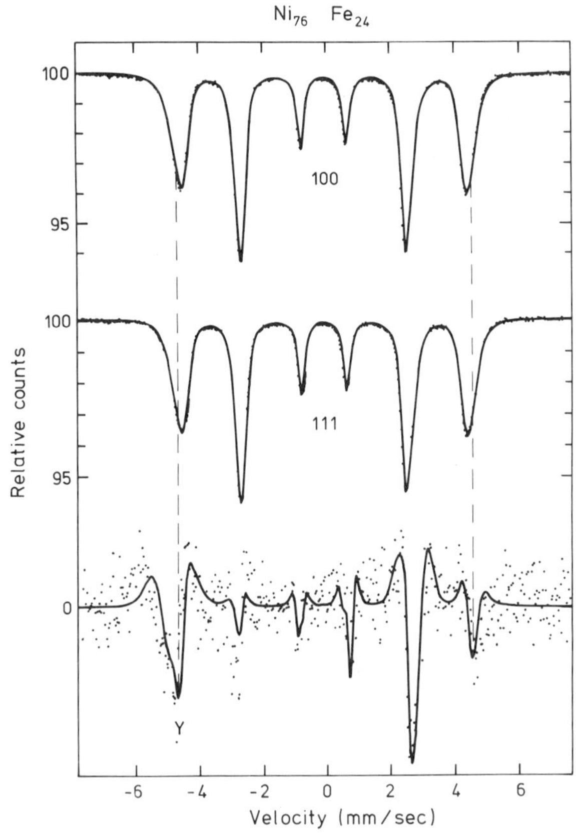

Figure 15. Spectra of $Ni_{76}Fe_{24}$, under the the same conditions as Figure 14.

sites has all nickel second neighbors. The second specimen, having the composition $Ni_{76}Fe_{24}$, will have many defects of the type nickel-on-iron sites. Each of these nickel atoms is second neighbor to six iron atoms on iron sites, and so should enable us to determine δH_2Fe.

Figure 15 shows the spectra of $Ni_{76}Fe_{24}$, taken under the same conditions as Figure 14. It will be seen that the components due to the type 1 defects are almost absent in the spectra with magnetization along 100 or 111 axes, and there is just a remnant of the line x in the difference spectrum. But now, a stronger component is the line marked Y. The peak in the difference spectrum has the expected sign for iron atoms which are second neighbors to type 2 defects, and its position indicates that the effect of exchanging an iron atom in the second neighbor shell of iron atoms on iron sites for a nickel atom is to increase the negative field. This indicates that ΔH_2Fe is positive, and we find $\Delta H_2Fe=+12kOe$.

If we were now to call the field on iron atoms on iron sites H(0,6) and the field on iron atoms on nickel sites as H(4,0), we would have H(4,0)-H(0,6)=-96kOe, whereas the experimental value is -43kOe. In fact, by including the effect of second neighbors, the discrepancy has become greater.

The discrepancy is probably to be explained by a difference of moment between iron atoms on iron sites and iron atoms on nickel sites. The saturation magnetization changes from $4.71\mu_B$ to $4.94\mu_B$ per unit cell on ordering. This suggests that the disordered iron atoms, i.e. iron atoms on nickel sites may carry a smaller moment than iron atoms on nickel sites, accounting for the difference of 53kOe in the hyperfine field.

This work is continuing with a nickel-iron crystal doped with 2% chromium. It is hoped that it will be possible to determine whether chromium occupies mainly iron or nickel sites.

REFERENCES

1. Schwartz, L. H., International Journal of Nondestructive Testing 1, 353 (1970).

2. G. P. Huffman and F. E. Huggins, Mössbauer Spectroscopy and Its Chemical Applications, p. 265, Am. Chem. Soc., Washington (1981).

3. R. E. Karl and J. J. Zuckerman, Mössbauer Spectroscopy and Its Chemical Applications, p. 221, Am. Chem. Soc., Washington (1981).

4. R. G. Campany, Thornton Research Centre Paper 2198.R, Thornton Research Centre, PO Box 1, Chester CH1 3SH, England.

5. N. K. Jaggi and K. R. P. M. Rao, J. App. Phys. 55, 1404 (1984).

6. F. E. Huggins and G. P. Huffman, J. App. Phys. 55, 1404 (1984).

7. G. P. Huffman and H. H. Podgurski, Oxidation of Metals 15, 323 (1981).

8. M. D. Rogers, Tribology 2, 123 (1969).

9. M. C. Cadeville, M. Maurer, and J. M. Friedt, Sci. Eng. 51, 147 (1981).

10. T. E. Cranshaw, J. Phys. F. Mat. Phys. 10, 1323 (1980).

11. K. Krolas, Phys. Lett. 85A, 107 (1981).

12. A. R. Miedema, P. F. de Chatel, and F. R. de Boer, Physica, 100B, 1 (1980); A. R. Miedema and A. K. Niessen, Calphad 7, 27 (1983).

13. I. A. Campbell and A. Blandin, J. Mag. Mag. Matt. 1, 1 (1975).

14. J. A. Alonso and S. Simozar, Phys. Rev. B22, 5583 (1980).

15. A. M. van Diepen, Phys. Lett. 57A, 354 (1976).

16. J. W. Drijver, F. van der Woude, and S. Radellaar, Phys. Rev. B 16, 985 (1977).

17. Y. Gros and J. Pauleve, J. de Physique 31, 459 (1970).

MÖSSBAUER SPECTROSCOPY IN PHYSICAL METALLURGY

U. Gonser, M. Ackermann, H.-J. Bauer, N. Blaes,
S. M. Fries, R. Gaa, and H.-G. Wagner

Angewandte Physik
Universitat des Saarlandes
D-6600 Saarbrücken
Federal Republic of Germany

I. INTRODUCTION

In materials science, three discoveries had a great impact on our society, that is, on its understanding of science and technology. The first of these, dislocations, has just had its 50th anniversary. Dislocations were discovered by Orowan [1], Polanyi [2], and Taylor [3]. With the concept of dislocations we are able to get a better insight into the nature of metals, in particular their mechanical properties and their plastic deformation behavior. Second,in the early fifties Welker [4] discovered the electrical properties of III-V compounds. They play a dominating role in the semiconductor and microelectronic industries. Third, also in the middle of the century, amorphous metals were discovered and, in the following years more and more technological applications were found for them. Although the great significance of these three discoveries has been well recognized, they have not been honored by the Nobel Prize, the highest award for achievement in natural science. In our opinion the reason for this unfortunate omission is the fact that, for the chemist, these discoveries invoke too much physics and, for the physicist, the materials are too dirty.

When a new method or field develops, the scientists are usually quick to do experiments of a qualitative nature or even "Gedankenexperimente". At the universities there are many professors and students just waiting for a chance to try out a new method. In contrast, it normally takes a new method quite a long time to reach the practical state. Very careful quantitative analyses and evaluations are necessary and both the method and its technical parameters have to match properly before useful results can finally be obtained. The Mössbauer effect is a good example. Soon after the Mössbauer effect was observed in iron-57 all the important Mössbauer parameters had been found and published or presented at numerous Mössbauer conferences: isomer shift, quadrupole splitting, magnetic interaction, relativistic effects, etc. Of course, at these meetings remarks could be heard about the practical importance, but only a few people really dug in. After 28 years we finally have a Mössbauer conference on technological applications, even if the Hawaiian site is anything but technological.

Physical metallurgy is a rather wide field and in this chapter it is only possible to report on certain selected topics. Furthermore, a number of other contributions deal with problems of physical metallurgy such as steel, surfaces, alloying, phase analysis, and others.

Because iron is the star performer in physical metallurgy and iron-57 is the star performer in Mössbauer spectroscopy, we have an ideal combination for match-making.

The following three subjects will be discussed: lattice defects, hydrogen in metals, and amorphous metals. In particular, we want to outline some of the successes of the method and give some examples of its limitations.

II. LATTICE DEFECTS

Lattice defects are of great importance in physical metallurgy. The following distinction is often made: according to their dimensionality we divide the defects into point defects (of zero dimension), line defects (of one dimension), interfaces (of two dimensions), and disorder (of three dimensions). Soon after the Mössbauer effect was discovered, its various applications in connection with defects were described and, in the meantime, many experiments have been performed [5]. Focussing on the local environment (of atomic scale) one can say that the following property changes occur in the vicinity of the defect, when compared to the ideal crystal: symmetry, effective mass, force constant, elastic strain, charge and spin density, texture, disorder, etc.

Let us consider an impurity (I) being a resonance atom and occupying substitutionally--as a monomer--a lattice site. Its spectrum will reflect some of the properties of the matrix and the change in the electronic environment which it causes. When such a resonance atom is in close proximity to a defect it is quite obvious that the association of defect and impurity will influence the Mössbauer parameters. In particular, in the common face-centered cubic (fcc) and body-centered cubic (bcc) metals, the local cubic environment is removed. This gives rise to quadrupole interactions and changes in local electron distribution which are made evident by characteristic isomer shifts. Two extreme cases can be distinguished.

In the first case, the association of the defect with the resonance atom is unique. This situation is to a great extent verified in monomers, dimers (pairs of impurity atoms in a metallic matrix) and when an impurity (I) atom traps an interstitial atom [6] or a vacancy [7]. This is schematically shown in Figure 1(a). Such associations have been investigated in detail in various matrices. When interstitials are produced by irradiation or vacancies are produced by plastic deformation or fast quenching techniques, the resonance probe makes it possible to follow the processes by which, at elevated temperatures, these defects become mobile and are trapped by the impurity resonance atom, I. At even higher temperatures one can observe the defects, either interstitials or vacancies, dislodge themselves from the traps, the impurities, and move to deeper traps like edge dislocations or grain boundaries. As a result the spectrum of the impurity monomer is once again observed.

In the second situation there exists no unique correlation between impurity and defect. Instead, for the resonance atom, a large number of lattice sites in the proximity of the defect are available. This is the case for the lattice sites in the vicinity of dislocations and grain boundaries. In Figure 1(b) an edge dislocation is simulated and represented by a bubble model. The dislocation line is located at the center and runs perpendicular to the figure. Resonance atoms in the vicinity of a dislocation exhibit Mössbauer parameters for each site which are slightly different from neighboring sites. The situation becomes worse in amorphous metals. This is the reason why, despite all efforts, we have so far failed to elucidate the atomic structure of such metals.

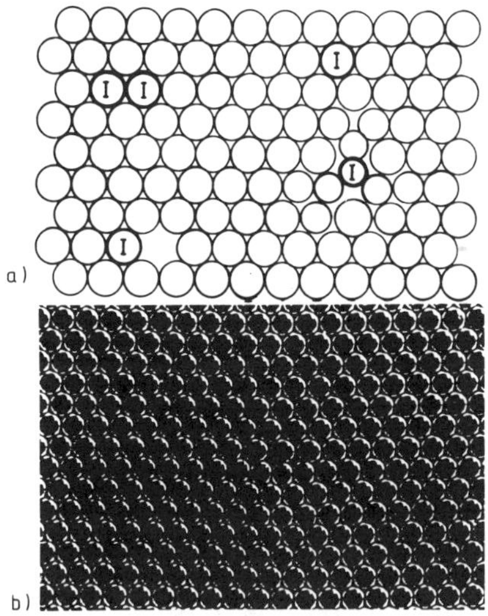

Figure 1. (a) Point defects and their associations with impurity resonance isotopes, I. (b) Cut through an edge dislocation realized by a bubble model

III. HYDROGEN IN METALS

Mössbauer spectroscopy has proven a powerful tool in the investigation of hydrogen storage materials [8,9]. When the resonance spectrum of the host metal is known, the influence of the hydrogen on the environment of Mössbauer isotopes can easily be detected. The resulting spectrum yields information on hydrogen solubility and hydride formation, the location of hydrogen atoms, the electronic and magnetic properties, hydrogen induced stresses, surface processes, as well as hydrogen diffusion and effusion.

In this section we will treat some aspects of Mössbauer spectroscopy in titanium-iron.

Hydrogen in TiFe may form three different phases [10]:

I. The α-phase with a solid solution of hydrogen. In this phase the bcc-structure of TiFe is retained and the hydrogen is distributed over certain interstitial sites [11].

II. The monohydride of the β-phase. Here the host lattice shows a small orthorhombic distortion and the hydrogen is stored in certain octahedral interstitial sites only [12,13,14].

III. The dihydride of the γ-phase. In this case the lattice is monoclinic and the hydrogen again occupies certain octahedral interstitial sites [15,16].

As shown in Figure 2, these phases could be resolved in the iron-57 Mössbauer spectrum [17,18]. The α-phase is represented by a single line of isomer shift -0.135mm/s relative to α-iron at room temperature. This line is identical to the TiFe resonance line. The β-phase shows a doublet with an isomer shift of 0.03mm/s and a splitting of 0.23mm/s. Because of the small deviation of the host lattice from cubic symmetry, the occurence of an electric field gradient (EFG) at the iron nucleus may be attributed to the

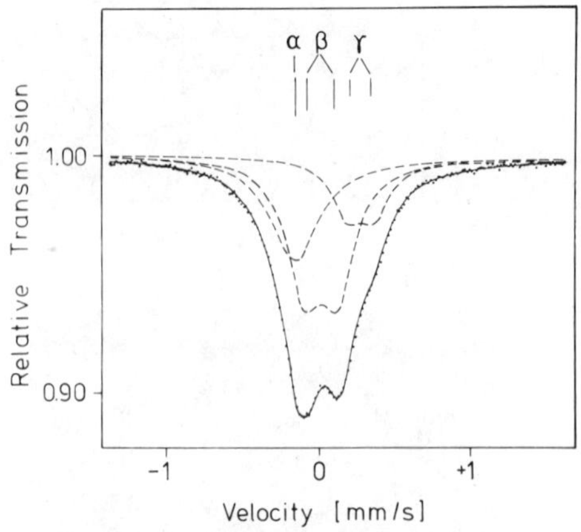

Figure 2. Mössbauer transmission spectrum of an inhomogeneously hydrided TiFe sample. All three phases α, β, and γ can be identified.

asymmetric arrangement of hydrogen atoms around the iron site. Also, the absorption due to the γ-phase is split by 0.18mm/s and shifted to 0.2mm/s. As has been observed for most metal hydride systems, the values of the isomer shifts increase with increasing hydrogen content. This behavior may be traced to volume expansion and sometimes to a charge transfer [13].

Once the phases are identified it becomes possible to determine the phase boundaries in the hydrogen phase diagram [18]. If, on the other hand, the hydrogen concentrations of the various phases are known, one can estimate the total hydrogen content of a sample from the Mössbauer spectrum by using the areas under the corresponding resonance absorption lines. In most amorphous materials, such as $Zr_{76}Fe_{24}$, no hydrogen phases can be detected in the Mössbauer spectrum and thus the hydrogen concentration must be estimated from the isomer shift alone [19]. The identification of specific hydrogen phases also allows the investigation of the properties of samples containing several of these hydrogen phases. This is an important advantage over most other methods, particularly for such systems where the production of single phase samples is difficult. In many ordinary investigations of hydrogen diffusion, the occurrence of different hydrogen phases can often falsify the result.

To date, in the field of Mössbauer spectroscopy, three different methods have been applied to the determination of hydrogen diffusion.

I. Time-dependent isomer shift variation at tantalum-181 atoms [20,21].

II. Hydrogen induced displacements of iron-57 atoms [22,23].

III. Hydrogen induced reorientation of the principal axes of electric field gradients at iron-57 atoms [24,25].

In this chapter we will confine ourselves to the observation of hydrogen diffusion by the EFG reorientation at iron-57 in TiFe-H.

The introduction of hydrogen in cubic lattices may destroy this symmetry, leading to the occurrence of quadrupole split lines in the Mössbauer spectrum, as shown above for the β- and γ-phase of TiFe-H. If the hydrogen atoms are stored in a non-cubic arrangement in the immediate vicinity of the

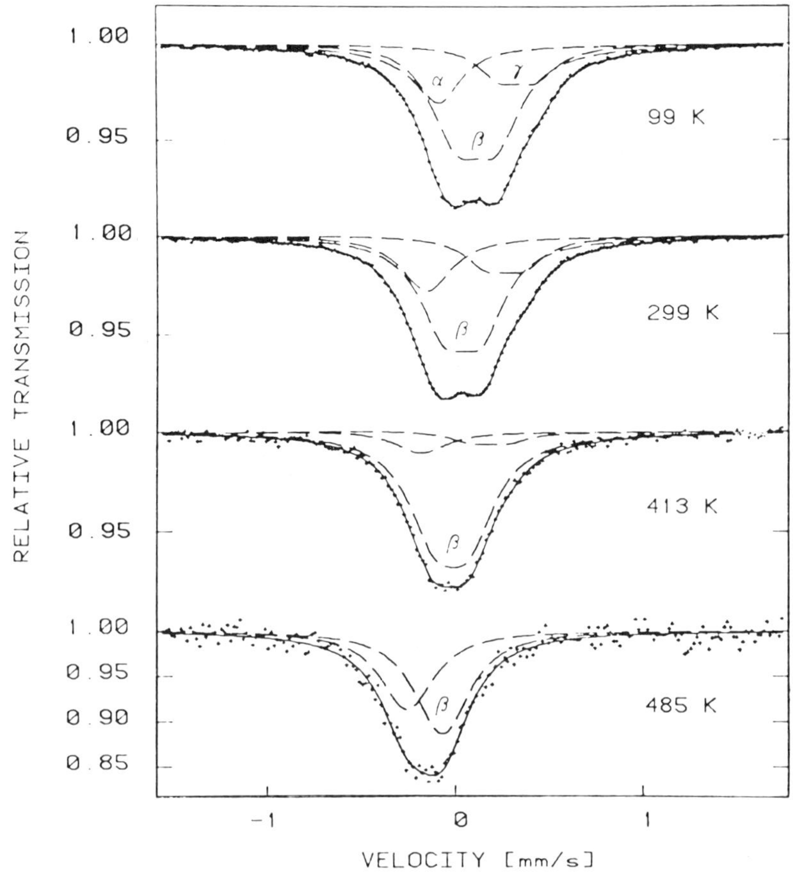

Figure 3. Mössbauer transmission spectra of TiFe-H at various temperatures. The zero of the velocity scale corresponds to the centroid of the α-Fe spectrum at room temperature.

Mössbauer atom, and other host atoms are situated farther away in a cubic manner, one can assume that the EFG is produced directly by the hydrogen atoms as stated above for the β-phase of TiFe-H. The interaction of the EFG with the nuclear quadrupole moment results in a splitting of the nuclear states, where in the classical model, the splitting energy corresponds to the precession frequency of the nuclear moment.

Because hydrogen normally occupies interstitial sites in its host lattice, there are only a finite number of different hydrogen arrangements around the Mössbauer atom. This in turn leads to a finite number of different EFGs. For a simple model we assume that these arrangements differ only in the direction of the principal component of the EFG, but produce the same quadrupole splitting. For powder samples the spectrum then consists of a single quadrupole split doublet. As a result of hydrogen mobility the arrangements of the hydrogen atoms may change at a temperature-dependent rate. In the static case of low hydrogen diffusion, the nuclear precession is not affected by these changes of the environment and the quadrupole splitting observed in the Mössbauer spectrum remains unchanged. If the jump rate of the hydrogen atom is comparable with the precession frequency, the precession is strongly disturbed and produces relaxation effects in the resonance line. Finally, for very high hydrogen mobility one can observe only the average over all hydrogen arrangements. This leads to a collapse of the quadrupole splitting.

Figure 3 shows the Mössbauer spectra of TiFe-H in the temperature range between 99 and 485K. At low temperatures the spectrum consists of the α-, β- and γ-line. With increasing temperature the γ-line vanishes and the splitting of the β-line decreases strongly above 400K, as is also shown in Figure 4. We have explained this reduction by an EFG reorientation, as described above [24]. In order to simplify the interpretation of these spectra, we used a well-known simple model in which only three orientations of the EFG

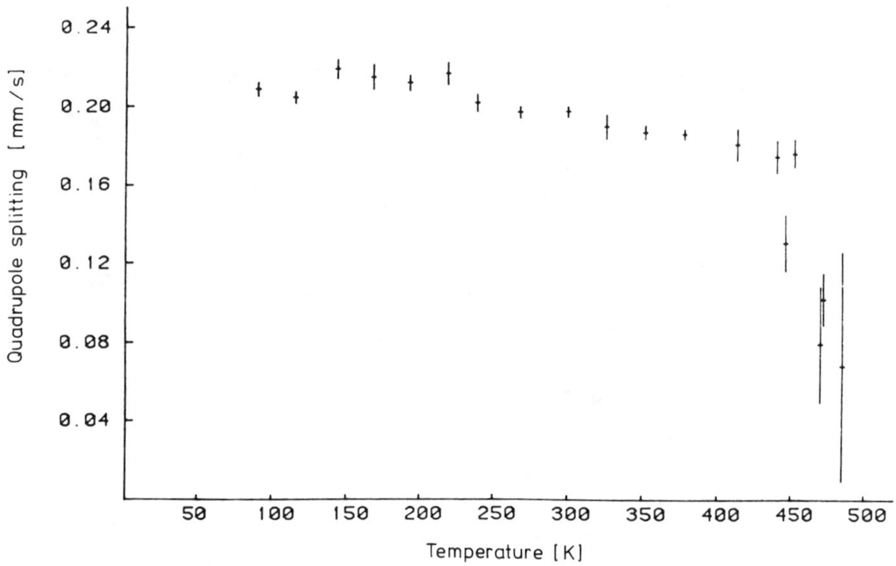

Figure 4. Temperature dependence of the quadrupole splitting in the β-phase of TiFe-H

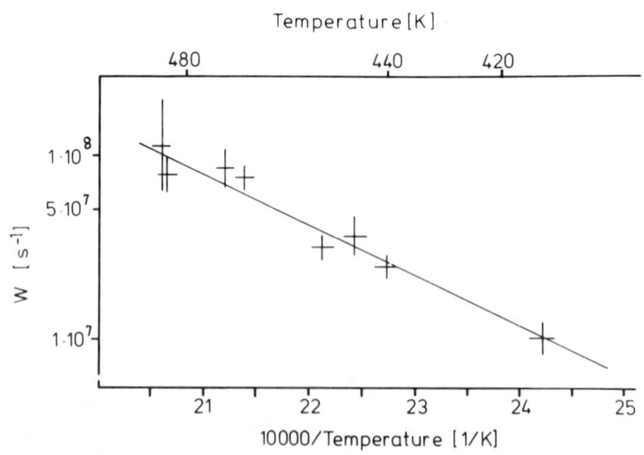

Figure 5. Probability per unit time for the relaxation of the EFG tensor between its three orientations, W, vs. 1/T in the temperature range of 400-486K.

principal axes are possible and each of these orientations occurs with the same probability. For such a model Leider and Pipkorn [26] and Tjon and Blume [27] have calculated the Mössbauer spectrum. Our spectra have been fitted in terms of their model. As a fitting parameter we used the transition probability per unit time, W, for the interchange of the EFG tensor principal axes between the three orientations. This corresponds to the mean dwell time of a hydrogen atom at an interstitial site. In the temperature range of strong relaxation this probability obeys an Arrhenius law with an activation energy of about 0.54eV (see Figure 5). Although the model is simple, the diffusion coefficient could also be estimated. These results are in quite good agreement with other methods, as cited in Ref. 28, where it should be noted that we have to take into account that most of the other methods integrated over at least two hydrogen phases.

IV. AMORPHOUS METALS

Physicists prefer the pure, ideal state, which in practice means they like to work with perfect single crystals. The motivitation for this preference is best expressed by Kittel in what is probably the most widely read book in physics, "Introduction to Solid State Physics". He states 'Measurements on single crystals are always much more significant and informative than are measurements on polycrystalline specimens.' In recent years the amorphous state has gained great importance from a scientific as well as a technological point of view, and one might reverse the above statement to read 'any fool can handle the crystalline state but it takes a genius for the amorphous state.'

There was never any doubt that materials with covalent, ionic, or van der Waals bonding exist in the amorphous state, but it was not until the middle of this century that it was realized that metals can be produced in the amorphous state. In the years that followed, this field experienced a development of dramatic dimensions. Its short history is presented schematically in a plot of the effort measured in publications per year vs. time (Figure 6). In the fifties a number of techniques were applied to produce amorphous metals. This was accomplished by electrodeposition by Brenner et al. [29], by vapor condensation by Buckel et al [30] and by irradiation by Gonser et al. [31,32]. At the beginning of the sixties the Bernal model [33] was developed and applied to the liquid state as a topological assembly of random dense packed hard spheres. It often served as a starting point in the interpretation of the atomic arrangements in amorphous metals. Also at this time Duwez et al. [34,35] started systematic investigations in the field. They invented a number of new fast quenching techniques, and, in addition, produced a large number of various new types of amorphous alloys. In this work, they discovered the $T_{80}M_{20}$ alloys where T is a transition metal such as Fe, Co, Ni, or Pd, and M is a metalloid such as B, C, N, P, or Si. These alloys play a significant role in technological applications due to their ferromagnetic ordering. In the mid-sixties Mössbauer spectroscopy was used to study these metals [36]. At the beginning of the seventies the meltspinning technique was developed and it then became possible to produce large quantities of amorphous metals in a continuous fashion [37-39], first only as narrow ribbons. More recently it has become possible to manufacture sheets of these amorphous metals in widths of about one meter [40]. Parallel to this development great efforts were devoted to the understanding of amorphous alloys from a theoretical point of view. Numerous models were suggested to describe their short range order. Judging these efforts in terms of the number of publications over ten-year periods, we note that there were about ten at the end of the fifties, one hundred at the end of the sixties, and, by the end of the seventies, a level of one thousand was reached. It is doubtful whether the ten thousand mark will be reached by the end of the eighties, but several thousands of papers are now being published annually. It seems

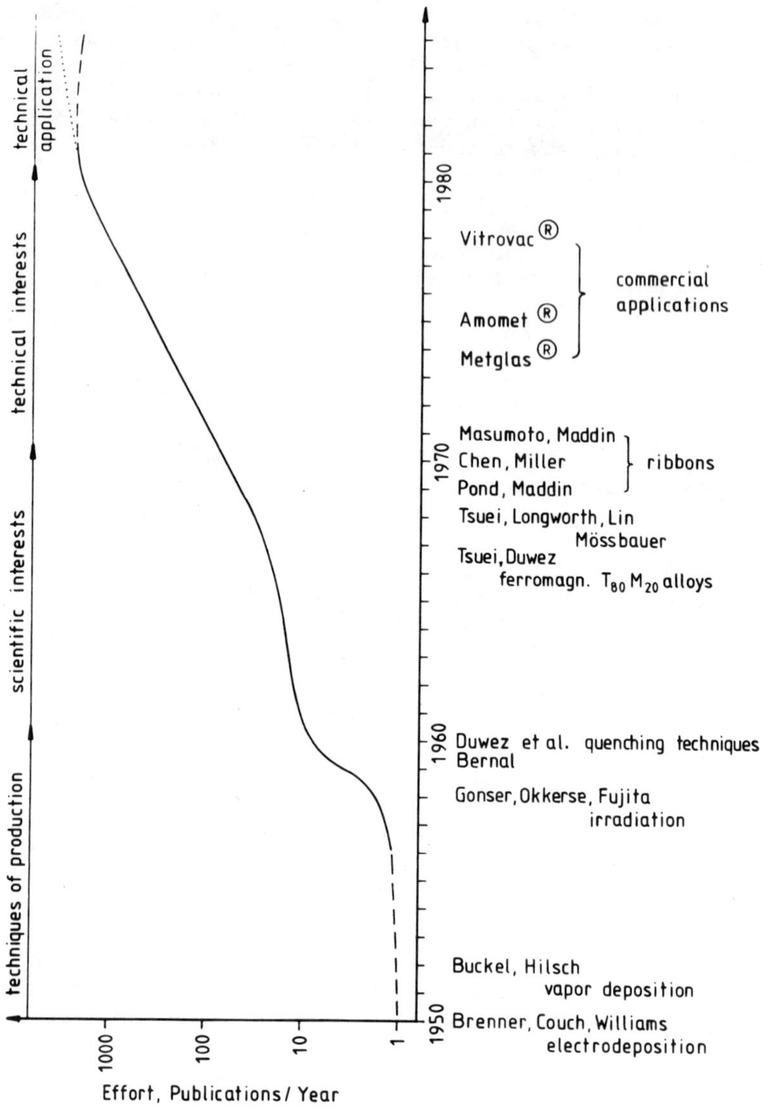

Figure 6. Schematic representation of the history of amorphous metals.

significant that the emphasis is shifting from the purely scientific side to the technological side. The industrial interest is also documented by such trade names as Melglas ®, Amomet ®, and Vitrovac ®.

Amorphous metals exhibit rather interesting magnetic, electrical, and mechanical properties. Their general features are relatively easy to understand. Due to lack of magnetocrystalline anisotropy they are usually magnetically soft. The mean free path of electrons is short, therefore their electrical resistivity is high. They lack distortions with well defined glide planes and defined Burger vectors, which result in a relatively high hardness.

Because many of the amorphous metals, in particular the alloys important for technological application, contain iron and also because these metals are usually produced as foils or ribbons with thicknesses ideal for Mössbauer spectroscopy, it was quite natural that Mössbauer spectroscopists were eager to try out their method on these materials. In fact, it seems there is no other field where this resonance method plays such a dominant role. Initially, it was thought that Mössbauer spectroscopy would be particularly useful in establishing the local atomic arrangement or short range order in these alloys. However, it has turned out that this task is extremely difficult. Just as for the many other techniques which have been used to try to obtain the atomic topology in amorphous metals, Mössbauer spectroscopy has also, at least so far, been destined to fail because of the infinite number of atomic configurations present and its lack of resolution.

Data Analysis

The analysis of Mössbauer spectra of ferromagnetic amorphous metals is extremely difficult. It has been the subject of much controversy and remains very much an open question. One glance at some typical spectra (Figure 7) illustrates the formidable task at hand. Instead of narrow lines corresponding to well defined nuclear sites in crystalline materials, the spectra of amorphous ferromagnets show broad overlapping lines reflecting distributions in the magnetic hyperfine fields, H, the EFG parameters, splitting Δ and asymmetry parameter η, and the isomer shifts, IS. The spectra may be further complicated by thickness effects and texture. The distributions in the hyperfine parameters are caused by the large number of inequivalent nuclear sites in the amorphous state. It is virtually impossible to extract all these distributions from the experimental data, especially because they may be partially correlated. Usually one tries to obtain just one distribution, while making relatively crude approximations about the others. The details of the distribution functions extracted from this approach are very often unreliable. The experimental spectrum is given by

$$C(v) = C_\infty - \int_{H_1}^{H_2} p(H)F(v,H)dH$$

where p(H) is the probability density for H, F(v,H) is the elementary magnetic sextet for velocity v and field H, C_∞ is the background, and C(v) are the counts recorded for relative velocity v. This problem is ill defined

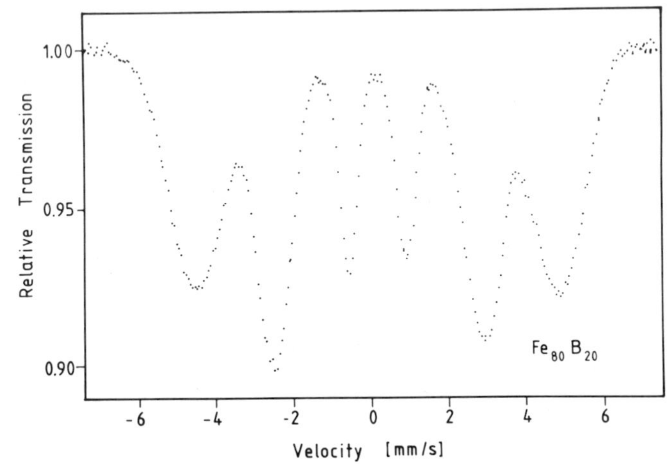

Figure 7. Mössbauer spectrum of amorphous $Fe_{80}B_{20}$ recorded at room temperature.

in that it has no unique solution for p(H). One can select classes of solutions by imposing quantitative and/or qualitative restrictions on p(H) such as $p(H) \geq 0$ for all allowed values of H.

Many different methods have been used to find p(H) (see e.g. ref. 41) and various claims have been made about the relative merits of the different methods. When F(v,H) is known and p(H) is a Gaussian distribution centered at high average fields, $\bar{H}$, with small width, σ_H, all methods produce reliable results. However these conditions are not fulfilled for amorphous metals. In general we do not know F(v,H),

$$F(v,H) = \int g(z|H)f(v,H,z)dz,$$

where g is the conditional probability density for z given H. Here z includes all other parameters (V_{zz}, η, IS, etc.). Also, for amorphous alloys, p(H) is not necessarily Gaussian and may extend to relatively low values of H. The approximations commonly used for F(v,H) and the methods used to extract p(H) will influence the calculated p(H) and may distort it considerably. In particular, the use of first order perturbation theory depends upon the assumption that the quadrupole interaction is small compared to the magnetic dipole interaction. In amorphous transition metal-metalloid alloys average quadrupole splittings of 0.3-0.8mm/s are found. This can be shown directly [42,43] by inducing the radio frequency (rf) collapse of the magnetic splitting in the Mössbauer spectra (Figure 8). As this figure illustrates, sidebands are produced which, however, give no further information on the basic Mössbauer parameters. However, the collapsed central part of the spectrum can be measured with higher resolution and a distribution of quadrupole splittings can be determined (see Figure 9). The corresponding distribution is shown in Figure 10. Because of the magnitude of the observed quadrupole splittings the application of first order perturbation theory is only justified for field distributions with no low

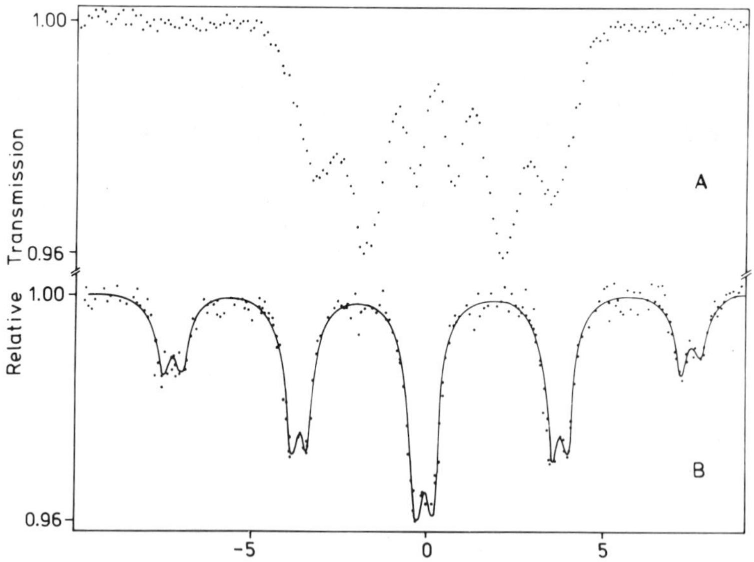

Figure 8. Room temperature Mössbauer spectrum of amorphous $Fe_{40}Ni_{40}P_{14}B_6$ without (A) and with a radio frequency field (B). The rf field causes the collapse of the magnetic splitting and the appearance of 2 orders of sidebands.

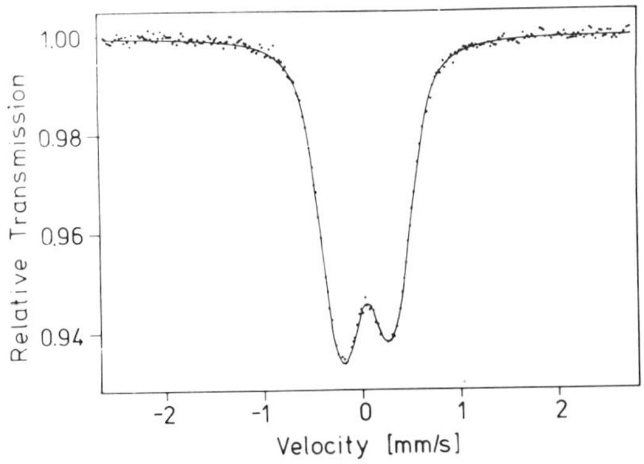

Figure 9. Central doublet of the collapsed spectrum of Figure 8. The continuous line is the result of a fit to a distribution of quad rupole splittings.

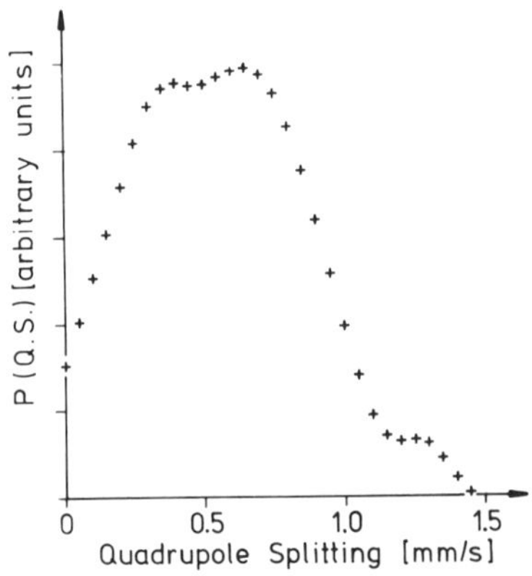

Figure 10. Distribution of quadrupole splittings found in the collapsed spectrum shown in Figure 9.

field component. Approximations are also necessary for the line shape of the elementary sextet [41]. Le Caer and Dubois [44] have shown that in general it is necessary to use relatively complex line shapes.

To check the influence of some of these approximations on the p(H), calculated from fits of experimental spectra, spectra were simulated by using the full Hamiltonian for mixed interaction with the field direction either parallel or perpendicular to the direction of the γ-radiation. The electric field gradient axes were assumed to be randomly distributed relative to the direction of the hyperfine field, H. The quadrupole splitting, Δ, was ± 0.5mm/s, the asymmetry parameter was taken to be zero and p(H) was a normal distribution with the mean, $\bar{H}$, varying from 40 to 280kOe with a fixed standard deviation, σ_H, of 24kOe. The simulated spectra were then fitted by using the constrained Hesse-Rubartsch method [44]. Figure 11 shows that for large H the calculated p(H) is close to the true distribution. When H decreases extra peaks appear. Their position depends on the angle between the magnetic hyperfine field and the direction of observation of the γ-radiation. If the magnetic hyperfine field is perpendicular to the direction of observation the extra peak is located at 0.48 $\bar{H}$. With increasing overlap

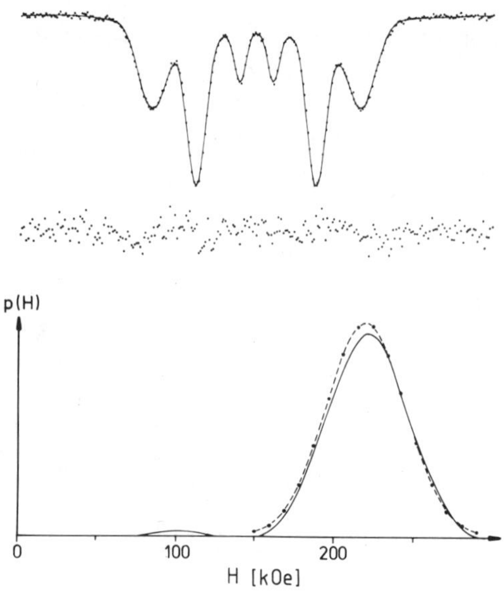

Figure 11. Fit of simulated spectrum, deviations of fit, and comparison of calculated (continuous line) and true distribution p(H) (dashed line) for small overlap and small perturbation.

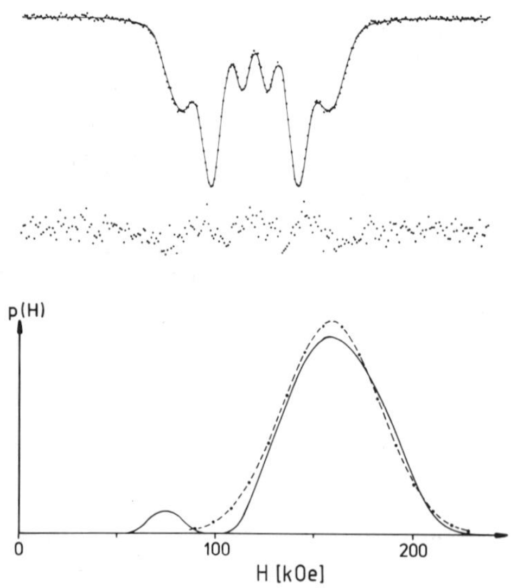

Figure 12. Fit of simulated spectrum, deviations of fit, and comparison of calculated (continuous line) and true distribution p(H) (dashed line) for medium overlap and medium perturbation.

the fits improve although the calculated p(H) becomes very different from the true distribution (Figures 12 and 13). The results are summarized in a "validity diagram" (see Figure 14). The ratio of the mean quadrupole splitting, $\bar{\Delta}$, to the mean hyperfine field, $\bar{H}$, is plotted along the vertical axis. The horizontal axis gives the ratio of the standard deviation, σ_H, to the mean hyperfine field, H. The letters O and P stand for overlap and quadrupole perturbations respectively and their relative size indicates the errors they introduce in the extraction of p(H).

The calculated p(H) is reliable within the hatched region. Outside this region there may be problems with spurious peaks and distortion in p(H).

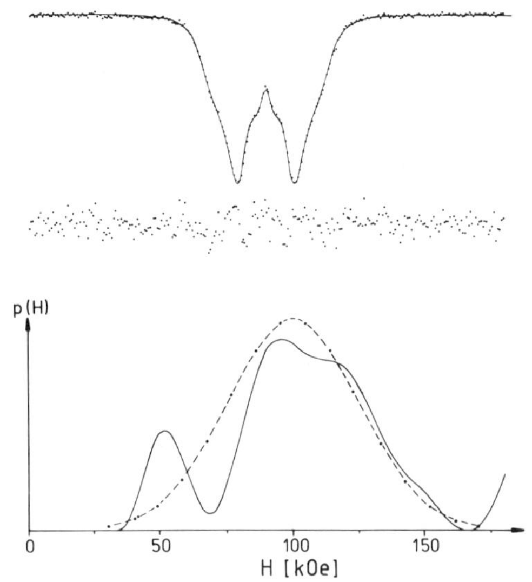

Figure 13. Fit of simulated spectrum, deviations of fit, and comparison of calculated (continuous line) and true distribution p(H) (dashed line) for large overlap and large perturbation.

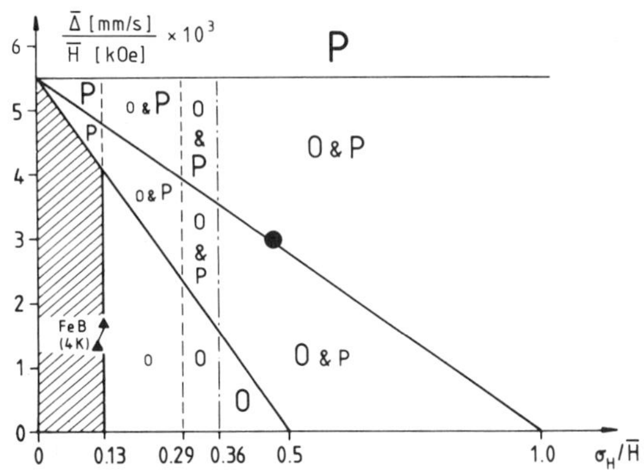

Figure 14. Validity diagram for calculated p(H). The ratio of the mean quadrupole splitting, $\bar{\Delta}$, to the mean hyperfine field, $\bar{H}$, is plotted along the vertical axis. The horizontal axis gives the ratio of the standard deviation, σ_H, to the mean hyper fine field, $\bar{H}$. O and P stand for overlap and quadrupole perturbation. Spurious peaks and distortions of p(H) increase with increasing letter size.

Unfortunately very few ferromagnetic amorphous metals fall within the hatched region. Therefore it seems doubtful whether any conclusions about the validity of structural models can be drawn from the shape of p(H). Even if one has reliable distribution functions, p(H), the problem remains of how to compare these with model structures which are given in terms of atom coordinates [45]. A rigorous treatment would require the determination of Mössbauer parameters from an accurate calculation of the electronic states of model structures. With the exception of simple metals this cannot be done with sufficient accuracy. Hence it is again necessary to make relatively crude assumptions. Some of these problems can be avoided by looking only at average hyperfine parameters which, at least in the case of the magnetic hyperfine field, can be determined quite well without complicated distribution fitting procedures. Examples of this are discussed in the next section. A rather ingeneous method of analysis introduced by Lines and Eibschutz [46] allows the calculation of the averages, the standard deviations and the degree of correlation of the different distribution functions. However, this technique will only work when the overlap in the spectra is small.

Metal-Metalloid Alloys

Amorphous metals are of great interest from a technological point of view. Of the various types of amorphous metals, the most promising technologically are the alloys consisting of transition metals (T) and metalloids (M) with a composition of about $T_{80}M_{20}$. Many are now commercially available. The composition ranges of high relative stability coincide with pronounced eutectics in the phase diagrams. They are of considerable interest in view of the potential application of these materials which combine high mechanical strength and magnetic softness. A number of studies [42-54] have demonstrated that Mössbauer spectroscopy can play an important part in the investigation of these properties because of its sensitivity to the hyperfine interaction parameters.

It is remarkable that ferromagnetic amorphous alloys of $T_{80}M_{20}$ with different constituents exhibit similarities in their X-ray diffraction pattern as well as in their Mössbauer spectra. The Mössbauer lines are typically broad and the relative strength of the Δm equals zero lines indicates a preferred orientation of the spins in the plane of the ribbon. As mentioned in the previous section, instead of the unique Mössbauer parameters typical of crystalline materials, distributions are present in amorphous alloys and significant information can only be deduced from the average hyperfine parameters. Some research groups in the field agree that these similarities, independent of the individual properties of the various possible constituents, may reflect similarities in the structure in the structure.

Scientists apply various models of the structure to interpret the Mössbauer spectra of these ferromagnetic amorphous metals in terms of local variations in the environment of the iron atoms. Three extreme view-points can be distinguished:

I. Microcrystallites (quasi-crystallites).
II. Elementary cells (building blocks).
III. Dense random packing of hard spheres (Bernal structure).

In general, one has to realize that each of these starting models represents a rather idealized concept and that in reality a variety of concepts ought to be taken into account. The real structure of these amorphous alloys might best be represented by points somewhere within the interrelationship of these three models [55].

Several papers [56-58] have investigated the effect of alloying elements on the crystallization temperature of $T_{80}M_{20}$metals. Substitution of one species of metalloid for another in amorphous $T_{80}M_{20}$ produces certain changes in the Mössbauer spectrum. Systematic studies of these effects have been made on the alloy system [59], $Fe_{80}B_xM_{20-x}$ where M is Ge, Si, C, or P and $Fe_{80}P_xM_{20-x}$, where M is Si, C, or B. Some results are shown in Figures 15 and 16 where the average hyperfine field is plotted versus the concentration of the substituted metalloid. In Figures 15 and 16 linear changes of the mean hyperfine field as a function of the metalloid concentration can be observed. Other correlations have been found by considering the influence of the radius, electronegativity, concentration and other properties of the metalloid atoms on the isomer shift and internal field. However, due to the problem of resolution we are still far from a full understanding of these correlations. In particular, we are not able to predict or deduce the macroscopic behavior of the material from the measured microscopic parameters.

Considerable work has been focused on the alloy system, $Fe_{80-x}T_xB_{20}$, where T is Ti, V, Cr, Mn, Co, or Ni. The results of a systematic study [60] of the variation of the transition metal atoms are shown in Figure 17. The mean hyperfine field sharply decreases with an increase of the Ti, V, Cr, and Mn concentration. The behavior of the magnetic moment per transition metal atom is similar to that found for crystalline alloys with a general trend toward smaller moments in the case of the amorphous alloys [61,62]. Assuming that the 3d subbands of the metal atoms in the amorphous alloys were filled by electrons of the metalloid atoms leads to good agreement with the Slater-Pauling diagram [63-65]. However, the application of the rigid band model to this problem is questionable as it is not clear to what extent the shape of bands is changed by alloying [66.67].

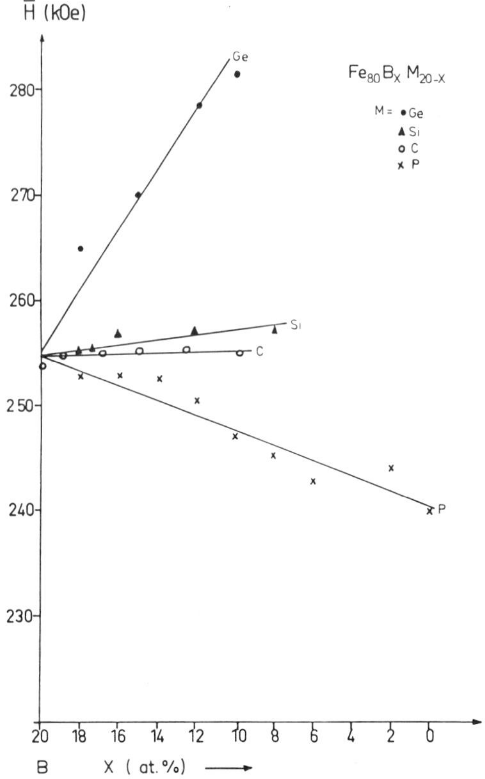

Figure 15. Average hyperfine fields vs. metalloid type and concentration of $Fe_{80}B_xM_{20-x}$ alloys.

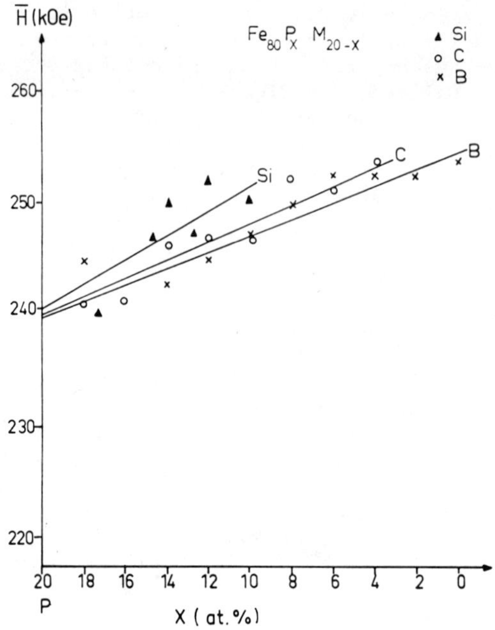

Figure 16. Average hyperfine fields vs. metalloid type and concentration of $Fe_{80}P_xM_{20-x}$ alloys.

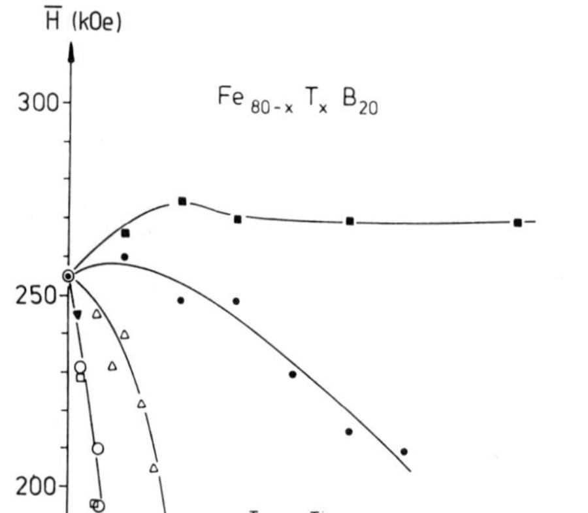

Figure 17. Average hyperfine fields of amorphous $Fe_{80-x}T_xB_{20}$ alloys as a function of transition metal and concentration.

Surface versus Bulk Crystallization

Mössbauer spectroscopy is a very sensitive method for studying the crystallization of amorphous metals. It is of particular interest to observe the onset of crystallization. Conversion electron emission Mössbauer spectroscopy (CEMS) when combined with γ-transmission Mössbauer spectroscopy [68] makes it possible to measure simultaneously the effects of heat treatments on surface layers of ca. 1000Å and on the bulk of the sample. The experimental setup is shown in Figure 18. The sample is placed inside an electron counter. The electrons which are emitted after the corresponding nuclei have experienced nuclear resonances are registered in one section of a multichannel analyzer, while γ-radiation which passes from the source through the electron counter and the sample is detected with a γ-ray counter and registered in another section of the multichannel analyzer.

It was found that for amorphous metals containing boron, crystallization starts at the surface [69-71] and that boron depletion occurs during annealing. This can happen either through sublimation and/or selective oxidation of the boron. Thus the atmosphere in which annealing treatments are performed plays a crucial role in the crystallization of amorphous metals. By way of example, the crystallization of binary amorphous FeB alloys will be discussed. The surfaces of amorphous ribbons produced by single roller quenching are quite different. The surface in contact with the roller during the quench is relatively rough and appears dull whereas the free surface is smooth and shiny.

With $Fe_{100-x}B_x$ alloys ($15 \leq x < 22$) heat treated in dynamic or in static vacuum, crystallization starts at the surface of the sample at temperatures well below the crystallization temperature determined by differential scanning calorimeter (DSC) measurements. No traces of the crystalline phases are observed in the bulk (Figure 19). This is still true for samples after repeated heat treatment and removal of previous crystalline surface layers. The new narrow lines in the conversion electron Mössbauer spectra indicated by the stick diagram, can be attributed to crystalline α-iron. We are thus dealing with primary crystallization although the sample composition would suggest eutectic crystallization with α-iron and Fe_3B as products. From the known boron concentration dependence of the crystallization of amorphous FeB it can be concluded that during the heat treatment a considerable amount of boron leaves the sample. In the case of a dynamic vacuum this occurs by sublimation, thereby changing the sample composition particularly in the surface regions, with a consequent reduction of the crystallization temperature.

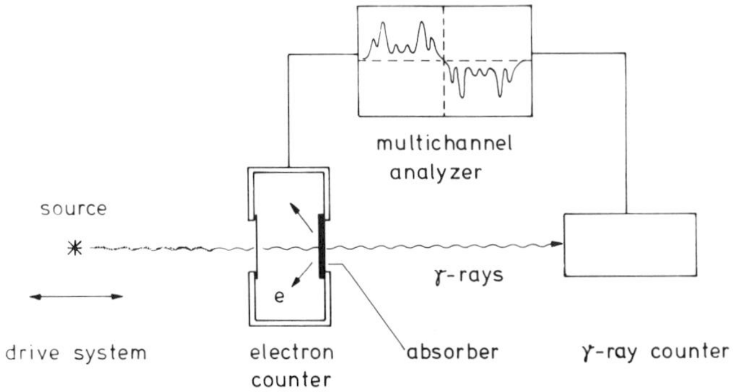

Figure 18. Experimental setup for simultaneous recording of conconversion electron and γ-ray transmission Mössbauer spectra.

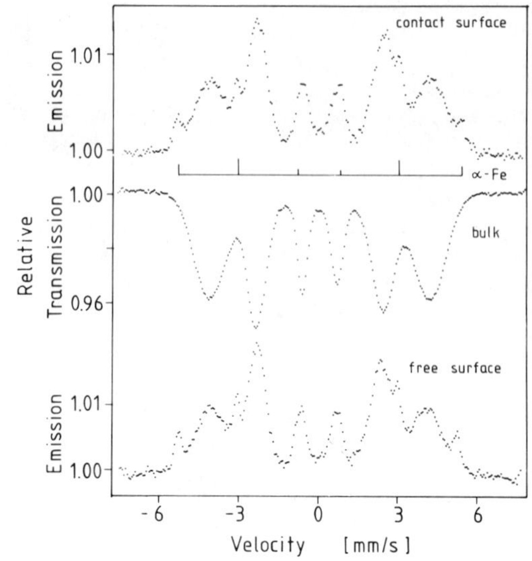

Figure 19. γ-ray transmission and conversion electron Mössbauer spectra of amorphous $Fe_{80}B_{20}$ after annealing at 613K for two hours in a dynamic vacuum.

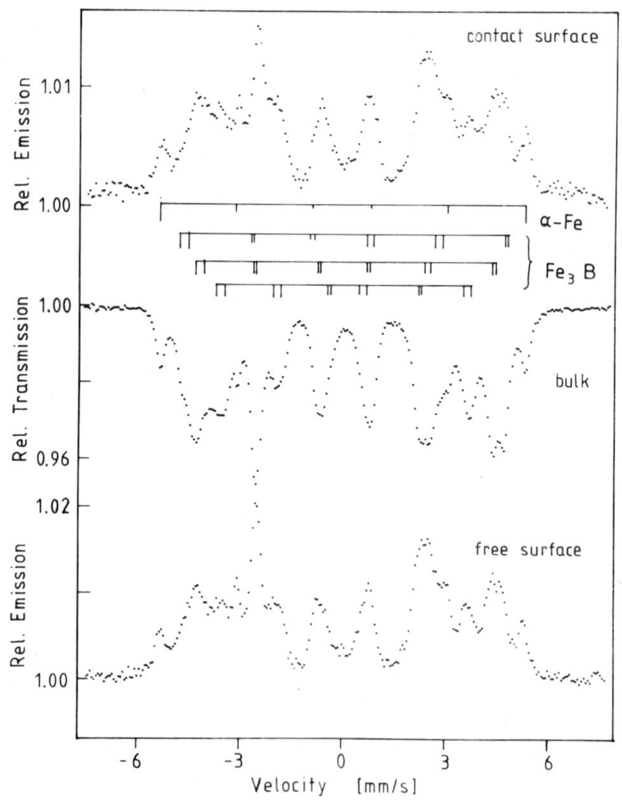

Figure 20. γ-ray transmission and conversion electron Mössbauer spectra of amorphous $Fe_{77}B_{23}$ after annealing at 613K for one hour in a dynamic vacuum.

A different crystallization behavior depending on the condition of heat treatment is shown by the $Fe_{100-x}B_x$ alloys with $22 \leq x \leq 24$. After annealing of an $Fe_{77}B_{23}$ sample at 613K for one hour in a dynamic vacuum, new lines are found corresponding to crystalline α-iron and tetragonal Fe_3B in both the surface and the bulk (Figure 20). This indicates simultaneous eutectic crystallization of the surfaces and the bulk material. In contrast to this, the heat treatment of the $Fe_{77}B_{23}$ alloy in static vacuum for the same length of time, produces only the crystalline α-iron phase on both surfaces. There is no indication of crystallization in the bulk. The decrease of bulk boron concentration by sublimation is now reduced. Now the surface spectra indicate a characteristic decrease in the boron content of the iron bearing phases most probably due to the selective oxidation of boron.

Figure 21 shows the conversion electron and γ-ray transmission Mössbauer spectra of $Fe_{75}B_{25}$ annealed in a dynamic vacuum. Both the surface and the bulk spectra show only one phase, Fe_3B, resulting from polymorphous crystallization. This indicates that, in this case, no substantial loss of boron occurred. In contrast, the surface spectrum of the $Fe_{75}B_{25}$ sample, heat treated in a static vacuum, shows α-iron and Fe_3B (eutectic crystallization); therefore a reduction in the relative amount of boron must have taken place. Additional X-ray photoelectron spectroscopic measurements carried out on this sample showed that selective oxidation of the boron is responsible for the decrease of the boron content and not, as first suggested, the existence of boron precipitates at the surface.

In order to clarify the onset of the crystallization process, further investigations were carried out with ternary, $Fe_{70}B_{20}Zr_{10}$, alloys. In contrast to the $Fe_{100-x}B_x$ alloys, which were produced by continuous techniques such as melt spinning, these amorphous $Fe_{70}B_{20}Zr_{10}$ samples were obtained by a single foil "splat quenching" technique which involved higher cooling rates. Because of the uniform heat transfer from both sides of the sample during cooling, there are no obvious differences in the magnetic properties or the structure of the two surfaces obtained by this method. Heat treatment for short periods of time at ca. 67 per cent of the crystallization temperature (one hour at 650K in a static vacuum of 10Pa) led to primary crystallization

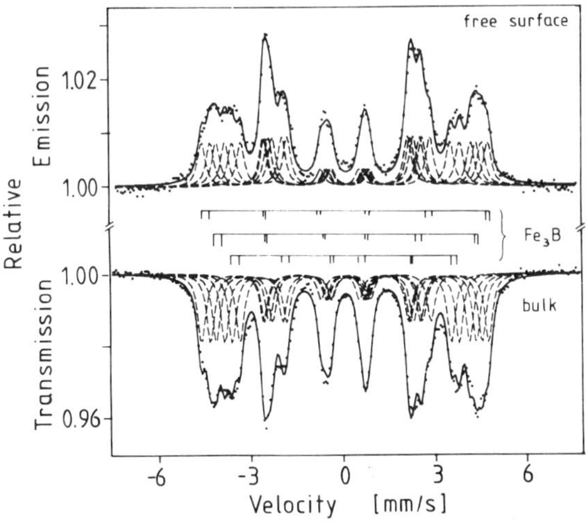

Figure 21. γ-ray transmission and conversion electron Mössbauer spectra of amorphous $Fe_{75}B_{25}$ after annealing at 660K for two hours in a dynamic vacuum.

of α-iron precipitates on both surfaces. Because with this production method the quenching rate is lowest in the center of the bulk it was possible to make a sample with residual crystalline sections inside the sample. By additional polishing of the surfaces we were able to obtain splats with a larger fraction of crystalline material in the bulk than in the surface. As shown by the Mössbauer spectra in Figure 22(a), we have a splat with a surface which is amorphous, whereas crystalline precipitates which were identified as α-iron, are present in the bulk. The question arose as to whether the crystallization process occurs at the surfaces or at the crystalline seeds of the bulk. Figure 22(b) shows the Mössbauer spectra of this sample which was heat treated under the same conditions as before (one hour at 650K in a static vacuum of 10Pa). The primary crystallization of α-iron observed at the surface indicates that, in this case also, crystallization and precipitation processes proceed from the surface.

Zirconium-Iron Alloys

Whereas amorphous transition metal-metalloid alloys have been of great scientific and technological interest ever since the discovery of their astonishing magnetic properties, relatively little is known about amorphous intertransition metal alloys. The two methods most generally used to produce these amorphous alloys are the liquid-quench and the vapor deposition techniques. While the liquid-quench method is in general only successful for compositions near deep eutectics in the phase diagram, the vapor deposition method is capable of producing amorphous alloys over a much wider range of compositions. As an example for an amorphous metal-metal system we discuss

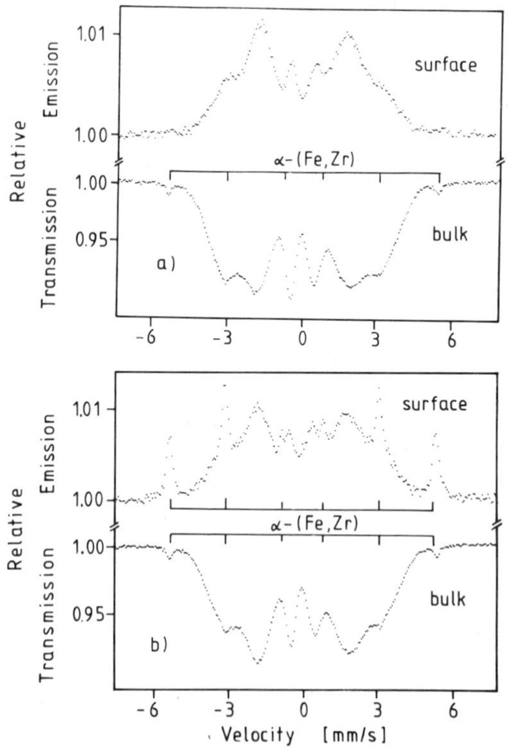

Figure 22. γ-ray transmission and conversion electron Mössbauer spectra of splat quenched $Fe_{70}B_{20}Zr_{10}$ as prepared (a) and spectra of the same sample after heat treatment described in the text (b).

the amorphous zirconium-iron alloys. The zirconium-iron binary phase diagram, as revised by Aubertin et al. [72] is shown in Figure 23. The existence of two eutectics in the phase diagram has made it possible to produce amorphous $Zr_{100-x}Fe_x$ alloys near the compositions with x of 90 and 24. Recently there has been success in producing amorphous zirconium-iron alloys on the zirconium rich side by meltspinning over a composition range between 20 and 40 atomic percent iron [73,74]. The iron-rich alloys could be produced in a composition range of 87 to 93 atomic percent iron. Many investigations of these liquid-quenched alloys have shown that the iron-rich and zirconium-rich alloys exhibit drastically different properties. The iron-rich samples for example show a magnetic behavior which has not been

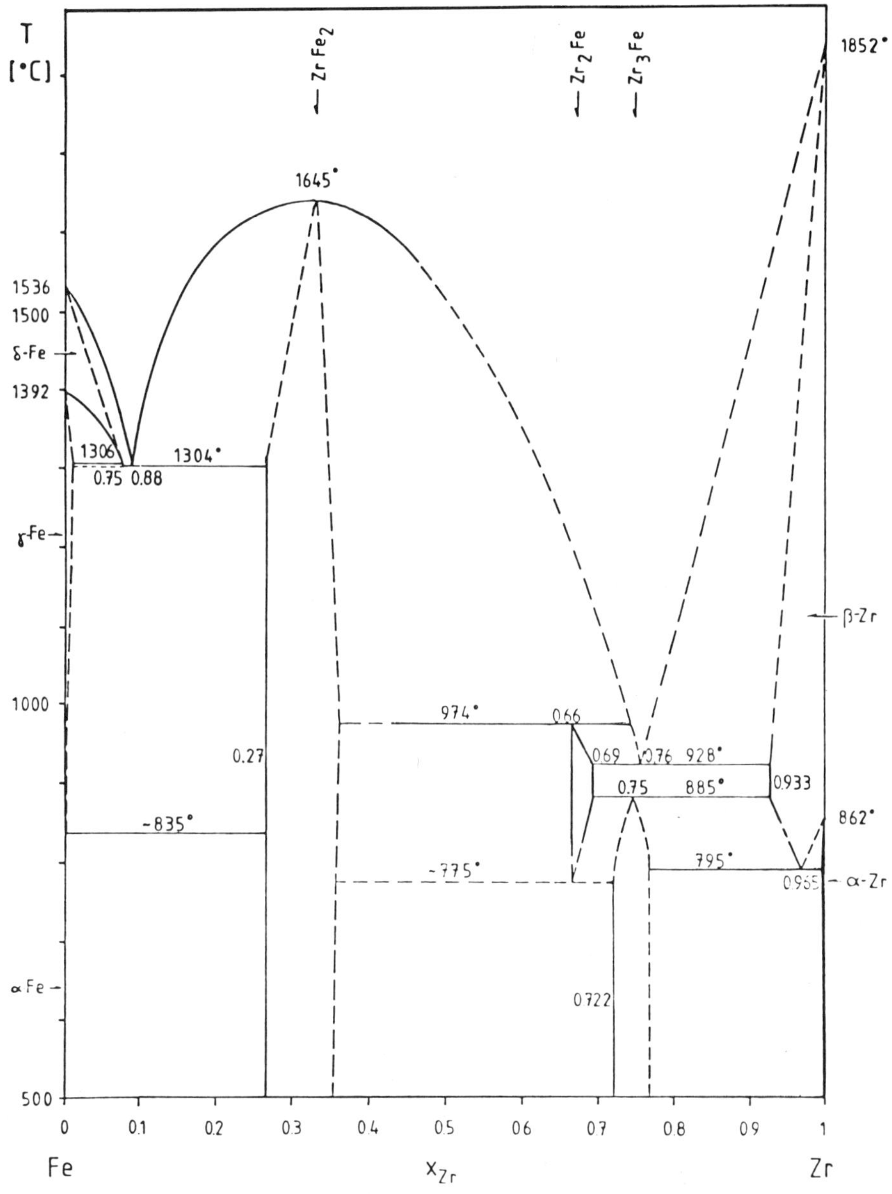

Figure 23. The binary phase diagram of the zirconium-iron system as given by reference 72.

fully understood until recently, while the iron-poor samples are superconducting [75-82]. The vapor deposition method has been used in several instances to prepare amorphous samples of non-eutectic composition [83-86].

Egami and Waseda [87] have reported a model in which the maximum and minimum composition of a binary amorphous alloy can be determined. Their approximation applied to $Zr_{100-x}Fe_x$ leads to a minimum of 19 and a maximum of 91 atomic percent of iron by using atomic radii of 1.24Å for iron and 1.60Å for zirconium [88]. Recently Unruh and Chien presented an extended combined Mössbauer spectroscopy and magnetization study of amorphous $Zr_{100-x}Fe_x$ films, deposited onto liquid-nitrogen cooled substrates at sputtering rates of about 0.05μm/min [89]. The composition range, where the vapor deposited samples were amorphous, was in good agreement with the predictions of Egami and Waseda. Samples with an iron content of 20 to 93 percent iron were found to be amorphous, whereas samples with higher or lower iron concentrations were found to be partially or completely crystalline. The instability of elemental pure amorphous metals is indicated by the precipitous drop in the crystallization temperature as the elemental endpoints in an amorphous metallic binary system are approached [90] and leads to the restrictions on the available composition range of amorphous samples.

Magnetic Behavior

In many amorphous alloys, formed by the combination of iron with a nonmagnetic metal, $M_{100-x}Fe_x$, ferromagnetism occurs above a critical concentration of x of ca. 40. These binary systems normally show an increase of the Curie temperature, T_c, and the average moment per iron atom with x beyond x_c. An exception is the zirconium-iron system. To date the magnetic behavior of the iron rich amorphous Zr-Fe alloys has not been fully understood. Some investigations show the transformation of the magnetic order from colinear ferromagnetism to asperomagnetism [76,77,82]. The different behavior apparently is caused by inhomogeneities from the sample preparation [60] and magnetically heterogeneous behavior is observed in the magnetic hyperfine field distributions observed by Mössbauer spectroscopy [79]. The magnetic ordering temperatures of the amorphous Zr-Fe samples, as a function of the iron concentrations, are shown on the right hand side of Figure 24. There is a clear maximum in T_c at an iron concentration of ca. 85. Above this value of x there is a clear decrease in T_c and an extrapolation to an x of 100 yields a value of T_c in pure amorphous Fe of ca. 200K. For iron concentrations below an x of ca. 85, T_c decreases monotonically to the critical

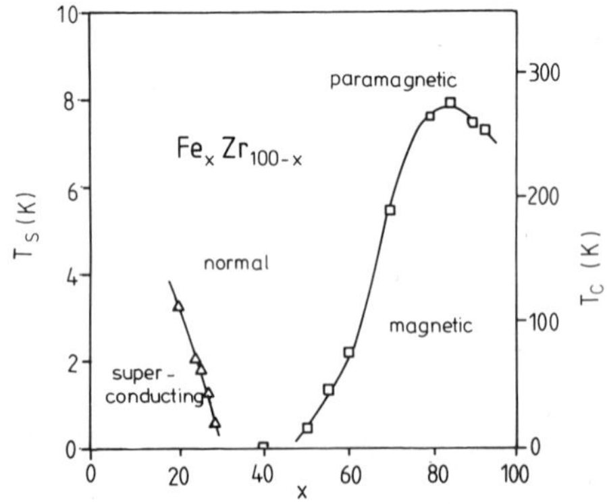

Figure 24. Phase diagram of amorphous Fe Zr_{100-x} alloy system. The magnetic ordering temperatures, T_c, and the superconducting transition temperatures, T_s, are denoted as squares and triangles respectively [89].

composition for magnetic ordering which occurs in the vicinity of x_c of ca. 45. Magnetization measurements from Buschow and Smit [84] exhibit a finite moment for samples above x_c and show Pauli-paramagnetism for samples below x_c. This means that the magnetic threshold at a critical composition, x_c, is due to the loss of the iron moment.

A maximum in T_c can be qualitatively understood, from either a localized moment or itinerant magnetic model point of view. In the localized moment approximation, where T_c measures the net magnetic exchange interaction, a decrease in the strength of ferromagnetic exchange could be due to a decrease in iron-iron spacing or the emergence of an antiferromagnetic exchange component. A decrease in the coordination number is unlikely [91]. The itinerant model postulates an evolution from weak (holes in both 3d subbands) to strong (holes in only the minority subband) and back to weak ferromagnetism with increasing iron concentration [92]. The iron moment will also decrease at sufficiently large iron concentrations. Whereas magnetization measurements in applied fields of up to 15kOe have indicated a decrease in the iron moments [84,93,94], recent magnetization measurements in a very large applied field of 110kOe indicate a monotonically increasing iron moment [95]. From these points of view, the emergence of an antiferromagnetic exchange component could explain the observed composition dependence of T_c. This point has not been fully understood until now, particularly as an intensive Mössbauer investigation on amorphous meltspun $Fe_{91}Zr_9$ suggests that no antiferromagnetic component is present [75].

Superconductivity and Magnetic Order

In the amorphous zirconium-iron system a very interesting feature is the existence of both magnetic order and superconductivity. Therefore this system is well suited to study the processes involved in the transformation from superconductivity to magnetism. Whereas the iron rich alloys show magnetic order as mentioned above, the zirconium rich alloys are superconducting with superconducting transition temperatures, T_s, up to above 3K for $Zr_{80}Fe_{20}$ [89]. This transition occurs despite the large amounts of iron present in these samples. In comparison the transition temperature for pure crystalline zirconium is only ca. 0.6K. The concentration dependence of T_s for amorphous sputtered zirconium rich zirconium-iron alloys is shown on the left hand side of Figure 24. The decrease in T_s with increasing iron content has been explained by Altounian and Strom-Olsen [78] in terms of the influence of spin fluctuations on T_s. The strength of the spin fluctuations increases with the iron concentration and produces an increase in the susceptibility of these alloys.

Magnetic Hyperfine Interactions

The Mössbauer spectra of amorphous metallic alloys in general exhibit broadened spectral lines, which are due to a distribution of the corresponding hyperfine interactions present in these alloys. The results of a temperature dependent Mössbauer investigation of meltspun $Fe_{89}Zr_{11}$ is shown in Figure 25. The Mössbauer spectra obtained at 220K and below consist of broad asymmetric patterns of Zeeman split lines. The spectrum obtained at 290K, which is correlated to a pure electric quadrupole interaction, will be discussed later. The lines through the measured data represent the fits obtained by using a constrained Hesse-Rubartsch method [96] developed by Le Caer and Dubois [44]. The magnetic hyperfine field distributions obtained from the fits, assuming as a first approximation a linear correlation between magnetic field and isomer shift, are shown alongside the corresponding spectra in Figure 25. The spectra clearly show the magnetic phase transition taking place between room temperature and 220K and the continuous increase of the average magnetic field at the iron-57 site from 101kOe at 220K to 226kOe at 4.2K. The magnetic hyperfine field distributions consist of a low field

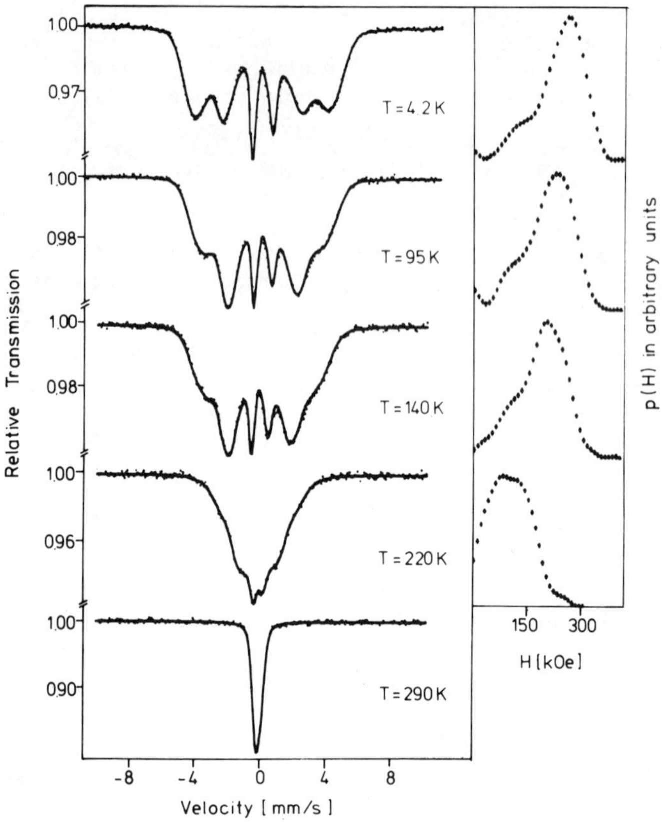

Figure 25. Mössbauer spectra of amorphous meltspun $Fe_{89}Zr_{11}$ as a function of temperature. The corresponding magnetic hyperfine field distributions are given on the right.

component as well as a large peak as found by Yamamoto et al. [79] for meltspun $Fe_{92}Zr_8$ and $Fe_{90}Zr_{10}$. One part of the spins was found to be strongly ferromagnetically correlated, whereas at least a part of the spins, corresponding to the low field component was claimed to be weakly coupled to the host.

A comparison of the Mössbauer spectra for amorphous $Fe_{89}Zr_{11}$ at 4.2K with and without an applied magnetic field is shown in Figure 26. An external magnetic field of 200Oe perpendicular to the γ-ray direction turns nearly all the spins into the ribbon plane, without any strong influence on the hyperfine field distribution. The intensity ratio, I_2/I_1, changes from 0.44 to 1.03. This change in the intensity ratio indicates that the magnetic moments originally had an average orientation pointing out of the plane. The application of a small external magnetic field leads to an increased spin alignment in the sample plane. This behavior is consistent with the properties of soft ferromagnetic materials. The results for the Mössbauer parameters of the spectra shown in Figures 25 and 26 are summarized in Table I.

Unruh and Chien [89] reported similar behavior for iron rich amorphous vapor-deposited zirconium-iron films. Mössbauer spectra taken at 4.2K of amorphous $Zr_{100-x}Fe_x$ samples with x ranging from 93 to 50, show the typical broadened spectral lines resulting from the distribution of the internal magnetic fields. They also found the soft ferromagnetic behavior of these

Table I. Mössbauer data obtained from the hyperfine distribution fitting procedure

Alloy Composition	H_{ext}, Oe	T, K	$\bar{H}$, kOe	I_2/I_1	$\overline{IS}$, mm/s
$Fe_{89}Zr_{11}$	0	4.2	226	0.44	0.03
		95	197	0.82	0.08
		140	182	0.90	-0.03
		220	101	0.96	-0.07
$Fe_{89}Zr_{11}$	200	4.2	232	1.03	0.07

materials, as indicated by an increased spin alignment in the sample plane when a small external magnetic field was applied perpendicular to the γ-ray. The average splitting and the average effective hyperfine field decreases with decreasing iron concentration. The hyperfine field distributions are found to be single maxima functions of H with an indication of a low field component in the p(H). In the case of amorphous $Zr_{50}Fe_{50}$, the determination of the average effective hyperfine field was not possible due to the predominant electric quadrupole interaction existing at 4.2K in this sample. As a result, the spectrum appears as a broadened quadrupole doublet. Figure 27

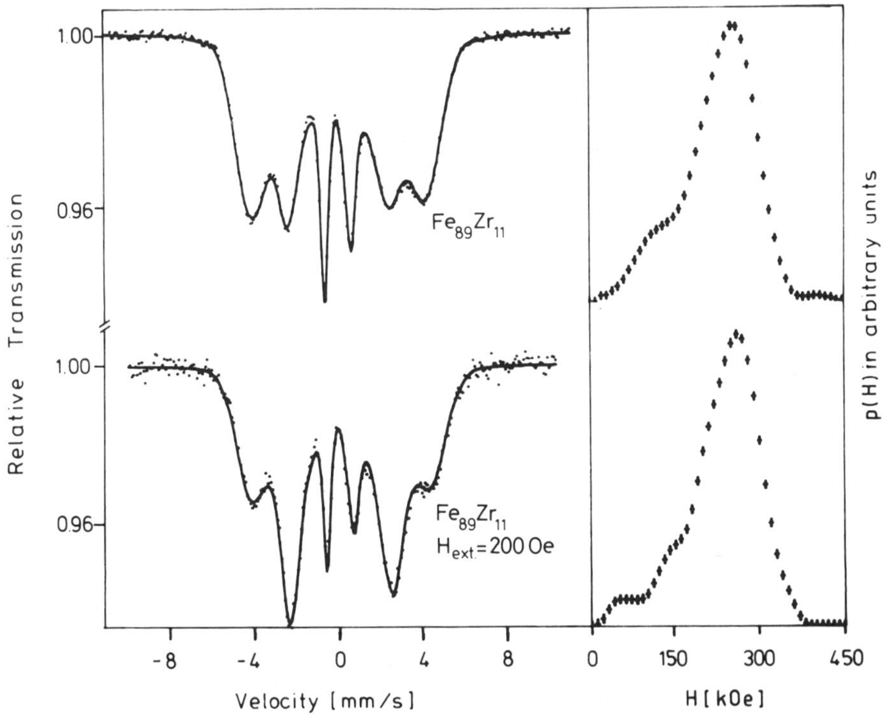

Figure 26. The 4.2K Mössbauer spectra of amorphous meltspun $Fe_{89}Zr_{11}$ with and without an external magnetic field together with the corresponding magnetic hyperfine field distributions. The orientation of the magnetic field is perpendicular to the γ-ray propagation direction.

shows the concentration dependence of the average effective hyperfine field obtained from Mössbauer measurements at 4.2K. Note that $\bar{H}$ is increasing monotonically with increasing iron concentration. No maximum is found as for the T_c data. From the proportionality between H and the magnetic moment on iron, one can conclude that the moment should also increase monotonically with x [97]. Several magnetization measurements indicated an unexpected decrease in the magnetization for x values above 88 [84,93,94]. Complete alignment of the iron moments would lead to a corresponding decrease in the effective moment. However, if complete spin alignment is not achieved by the external applied magnetic field, the corresponding μ_{Fe} would be underestimated. The results of magnetization measurements in an external field of 110kOe [95] are shown in Figure 28. The solid circles illustrate the data obtained in the external field of 10kOe. The magnetic moments decrease for x above 88 and extrapolation to pure amorphous iron would lead to a μ_{Fe} of zero. The μ_{Fe} values, obtained from the high field magnetization measurements, indicated as the open circles in Figure 28 completely reverse the behavior exhibited by the low field data. Here the magnetic moment increases monotonically without anomaly as concluded from the H values.

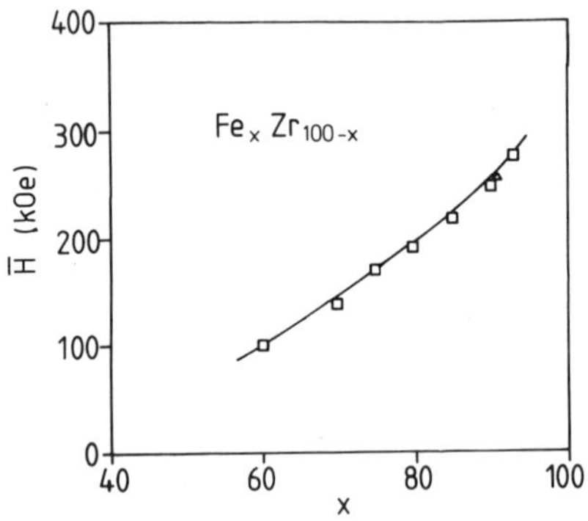

Figure 27. Average hyperfine field $\bar{H}$ as obtained from Mössbauer measurements of amorphous zirco nium iron alloys. Except for one liquid-quenched sample (triangle), all values are taken from sputtered samples [89].

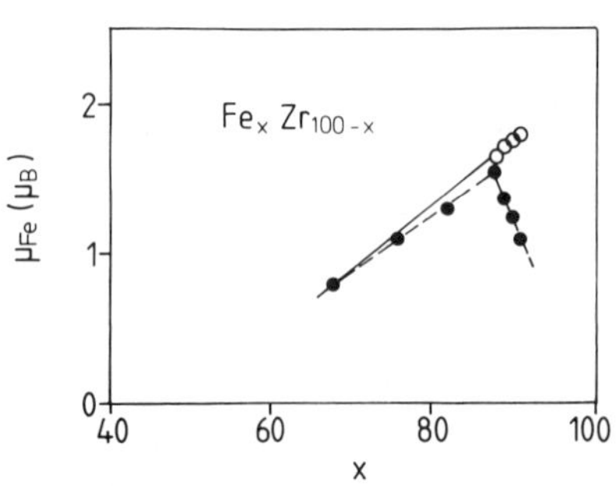

Figure 28. The iron magnetic moment deduced from magnetization of amorphous zirconium-iron alloys. The high field magnetiza tion measurements are denoted as the open circles [95] and the low field data as the closed circles [94].

Isomer Shift and Electric Quadrupole Splitting

Mössbauer spectra of the amorphous zirconium-iron alloys, taken at room temperature, show in all cases an asymmetric broadened doublet. The broad lines, which are non-Lorentzian, are likely to be caused by different electric field gradients existing at the iron-57 sites in the amorphous state. Because there are no texture effects present in the samples, and, to a first approximation there is, in general, no lattice vibrational anisotropy (Goldanskii-Karyagin effect) in amorphous materials, the asymmetry is due to a correlation between the distributions of quadrupole splittings and isomer shifts. To obtain a satisfactory fit it was necessary to introduce, as originally suggested by Vincze, et al. [98], a linear correlation between the isomer shifts, IS, and the quadrupole splittings, QS, such that

$$IS(QS) = a(QS-QS_{min}) + b.$$

The distribution of quadrupole splittings derived from this approach, along with the resultant fits for three selected compositions, are shown in Figure 29. The advantage of this fitting procedure is that it provides an accurate

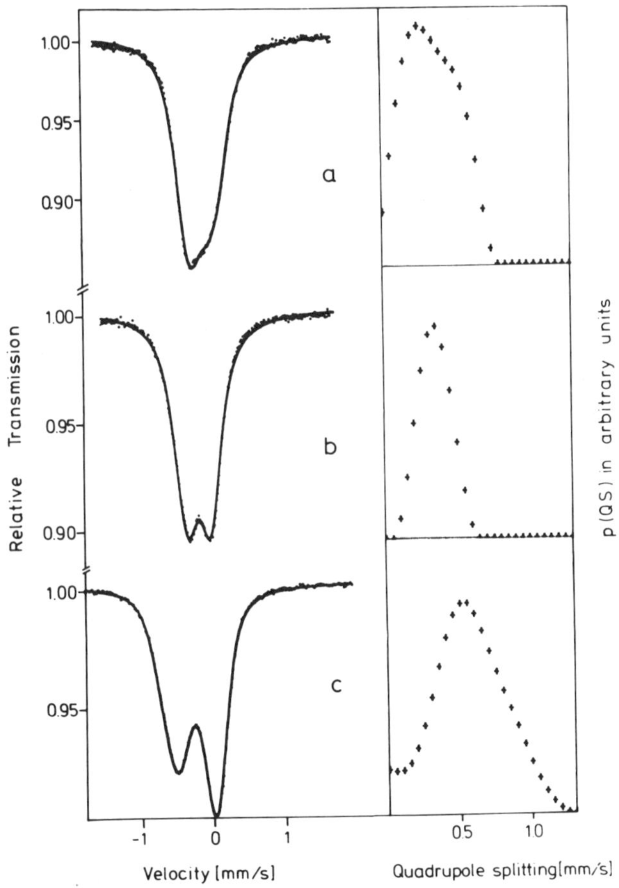

Figure 29. Mössbauer spectra of amorphous $Zr_{100-x}Fe_x$ samples. Whereas the samples with an x of 89 (a) and an x of 24 (c) were prepared by meltspinning, the sample with x of 50 (b) was prepared by sputtering. The full curves through the data show the fits obtained on the basis of the distributions of quadrupole splittings p(QS) also shown in the figure.

Table II. Average Mössbauer spectral parameters for $Zr_{100-x}Fe_x$ [a]

x	a	b	$\overline{QS}$ mm/s	$\overline{IS}$ mm/s	Preparation
24	-0.116	-0.192	0.55	-0.256	meltspinning
50	0.000	-0.168	0.35	-0.168	vapor deposition
89	+0.104	-0.144	0.36	-0.107	meltspinning

[a] Parameters obtained for the fits to the spectra shown in Figure 29.

description of the spectra, while allowing changes produced by different iron concentrations to be monitored readily by the parameters a,b, the average quadrupole splitting, $\overline{QS}$, and the isomer shift $\overline{IS}$ (see Table II). The spectra show clearly that the left component of the asymmetric doublet becomes deeper, while the right component decreases in intensity as the iron concentration increases. Also there is a general trend for the isomer shift to become more positive with increasing iron content.

Measured room temperature isomer shifts of amorphous zirconium-iron films by Unruh and Chien [89] are shown as a function of the iron concentration in Figure 30. Whereas the isomer shifts are all negative (typical for amorphous iron-early transition metals), the quantitative dependence of the isomer shift on composition is different from that found in amorphous Nb-Fe and Ti-Fe [99,100], where the isomer shift is characterized by a rapid drop to more

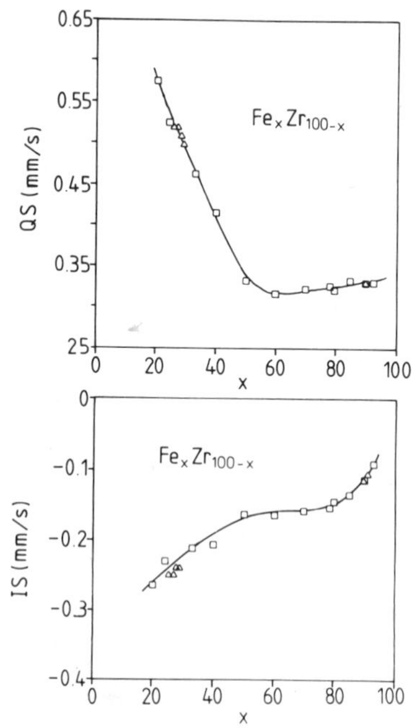

Figure 30. Quadrupole splittings and isomer shifts of amorphous zirconium-iron alloys at room temperature. The squares and triangles are measured from sputtered and liquid-quenched samples respectively [89].

negative values with decreasing iron concentration. This drop is followed by a very shallow minimum. For the Zr-Fe system the isomer shift versus x curve is very similar to those of other iron-early transition metal systems in the iron-rich region, whereas for the zirconium-rich region the isomer shift falls more rapidly with no indication of a minimum. Applying Miedema's model [101] of the concentration dependence of the isomer shift to the Zr-Fe system, the detailed features found in the data are not reproduced, but the qualitative agreement is very satisfactory, despite the simplicity of this model.

The quadrupole splitting (Figure 30) exhibits different behavior for iron-rich and zirconium-rich alloys. Whereas the iron-rich samples ($x>50$) show a weak compositional dependence, similar to Ti-Fe and Nb-Fe [99,100] alloys, the zirconium-rich samples ($x<50$) show an unexpected increase in the quadrupole splitting with a much stronger compositional dependence. Such a dependence is not observed for Ti-Fe or Nb-Fe. The anomalous behavior of both the isomer shift and quadrupole splitting suggests that there are subtle changes in the structure of amorphous iron-rich and zirconium-rich alloys.

As can be seen from Figures 27 and 30 no appreciable differences in the average values of the isomer shift, quadrupole splitting or internal field can be observed between the sputter-deposited and meltspun samples despite very dissimilar quenching rates. Because of the very sensitive dependence of these Mössbauer parameters to subtle changes in the local iron-57 environment, one can assume that the structures resulting from vapor -deposited and liquid-quenched alloys are very similar.

Structure of Meltspun Amorphous $Zr_{76}Fe_{24}$

As seen from Figure 29 the asymmetric broadened doublet of amorphous $Zr_{76}Fe_{24}$ consists of a quadrupole splitting distribution correlated with an isomer shift distribution. The distribution of quadrupole splittings occurs as the result of two distributions, that of the electric field gradient (EFG) and that of the asymmetry parameter, η, where

$$QS = \frac{1}{2}eQV_{zz}(1+\frac{1}{3}\eta^2)^{1/2}$$

$$\eta = (V_{xx}-V_{yy})/V_{zz}$$

$$\mathrm{Trv} = 0$$

Here V_{zz}, V_{yy}, and V_{xx} are the eigenvalues of the EFG tensor in order of decreasing magnitude, e is the elementary charge, and Q is the nuclear quadrupole moment [102]. The spectrum of amorphous $Zr_{76}Fe_{24}$ (Figure 29) provides information on the magnitude of the splitting, QS, but not on the sign of the main component of the EFG, V_{zz}, or the asymmetry parameter, η. This information can be obtained by measurements in an external magnetic field, if the magnetic interaction is of the same magnitude as the electric quadrupole interaction. The strength of the mixing between both hyperfine interactions is given by

$$R = \left|\frac{eQV_z}{2g_{3/2}\mu_N H}\right|$$

where $g_{3/2}$ is the nuclear g-factor, μ_N is the nuclear magneton, and H is the magnetic field at the nuclear site.

We present the results of a Mössbauer measurement of amorphous $Zr_{76}Fe_{24}$ in an external magnetic field of 46.5kOe parallel to the γ-ray propagation direction. In this case the value of R for an average quadrupole splitting of 0.55mm/s is determined to be between 1.74 and 1.51 depending upon whether η is zero or one respectively.

In the cases of R>>1 or R<<1 the eigenvalues and eigenvectors of the hyperfine Hamiltonian could be calculated conventionally by first order perturbation theories. In the case of R ca. one, where the spectrum depends on the strength of the interactions as well as the relative orientation of the EFG principal axes, the magnetic field, and the direction of observation, one usually calculates the line shape by solving the secular equation of the full nuclear excited state Hamiltonian [103] in order to get the eigenvalues and eigenvectors. In general the eigenstates of the excited state are not pure angular momentum states, thus the selection rules allow eight lines, whose intensities are given by the transition probabilities determined by the transition matrix elements between the corresponding eigenstates. Assuming, as in the case of powder samples, that the principal axes of the EFG in amorphous $Zr_{76}Fe_{24}$ are randomly oriented with respect to the external field, we have to average over the angles between both. Conventional calculation of such spectra by using an analytical solution of the secular equation [104], together with an improved method for spherical angular averages [105], requires long computing times, especially as we have a distribution of quadrupole splittings and in addition expect a distribution of η for each splitting. The computer simulations of the applied field spectrum have therefore been calculated by using a superoperator formalism to yield analytically closed form expressions for the Mössbauer line shape in the presence of a magnetic dipole and an electric quadrupole hyperfine interaction [106].

The best result of the computer simulations is shown in Figure 31 and was obtained for a predominant (90%) positive sign of V_{zz}, an average η value of 0.7, and a Gaussian probability distribution for η. More detailed features about the sign determination in amorphous $Zr_{76}Fe_{24}$ and the computer simulations are given by Fries et al. [19] and Le Caër et al. [107].

The results for the Mössbauer parameters of amorphous $Zr_{76}Fe_{24}$ in an external magnetic field support earlier conclusions about the local structure

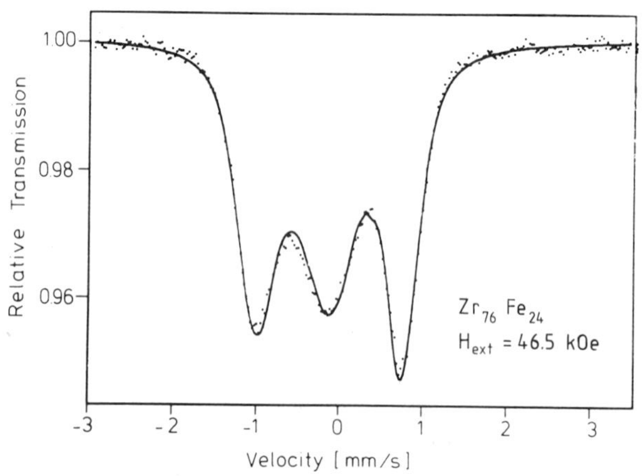

Figure 31. Mössbauer spectrum of amorphous $Zr_{76}Fe_{24}$ in an external field of 46.5kOe parallel to the γ-ray beam. The full curve is the simulated spectrum calculated in the manner described in the text.

around iron in this alloy [108]. The predominant positive sign of V_{zz} confirms that there is a well defined iron environment, a trigonal prism of zirconium atoms as suggested by Dubois and Le Caer [109]. A dense random packing model as proposed by Czjzek et al. [110] would lead to almost equal contributions of environments with positive and negative V_{zz}. More detailed information about the structure of amorphous $Zr_{76}Fe_{24}$ and in general of amorphous zirconium-iron alloys should be obtained in the future by means of EXAFS and neutron-scattering experiments in order to clarify suggested structure changes indicated by the anomalous behavior of the isomer shift and quadrupole splitting in this system.

XPS Measurements on Meltspun Amorphous $Zr_{76}Fe_{24}$

In order to get information about the surface properties of amorphous $Zr_{76}Fe_{24}$ the free surface (not in contact with the wheel) of an amorphous $Zr_{76}Fe_{24}$ ribbon produced by meltspinning under an inert atmosphere has been investigated by x-ray photoelectron spectroscopy, XPS. The XPS spectrum of the as quenched ribbon is shown in Figure 32. The spectrum clearly shows zirconium, carbon, and oxygen but exhibits no sign of the iron-2p lines which would appear above 700eV. A better resolved investigation of zirconium-3d lines in the energy range between 175 and 195eV shows lines which could be attributed to a ZrO_2 layer [111] and no indication of metallic zirconium was found.

All contaminations disappeared after prolonged sputtering with 6keV argon ions. The resultant spectrum of the sputtered surface is shown in Figure 33. Because of the long recording-time of the spectrum there appear again carbon and oxygen lines due to surface contamination caused by their partial pressures in the vacuum. The spectrum clearly shows the iron-2p doublet and in addition argon lines due to the sputtering with argon ions. The Fe-$2p_{3/2}$ line of the doublet appears at 707eV and the Zr-3d doublet at 178.6eV with a spin-orbit splitting of 2.4eV indicating metallic behavior in both cases. From the intensities of the zirconium and iron lines one calculates the zirconium/iron ratio to be 4.2 for $Zr_{76}Fe_{24}$ compared with the expected value

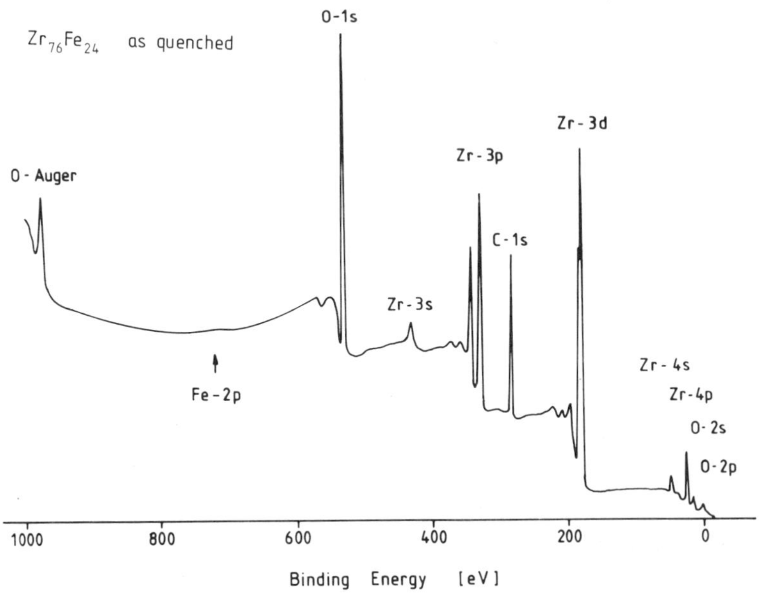

Figure 32. The x-ray photoelectron spectrum of amorphous meltspun $Zr_{76}Fe_{24}$ as quenched.

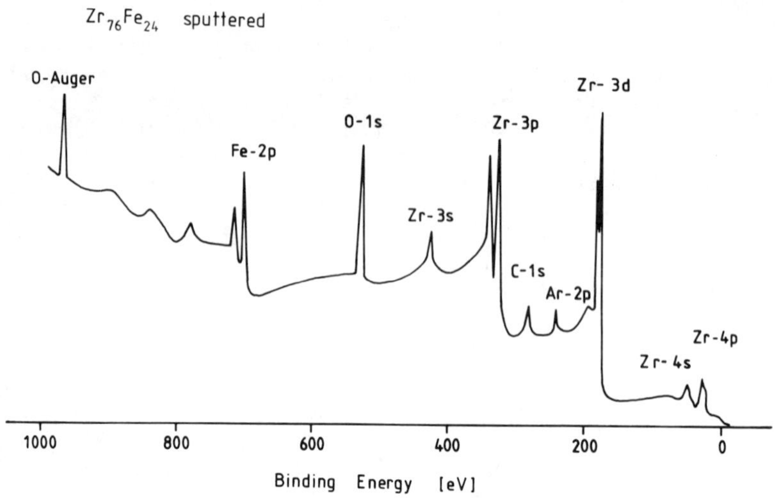

Figure 33. The x-ray photoelectron spectrum of amorphous meltspun $Zr_{76}Fe_{24}$ after prolonged sputtering with 6keV argon ions

of 3.2. This discrepancy is probably related to the small sampling depth of only a few angstrom associated with the XPS technique. The measured intensity ratio represents the composition of the initial atomic layers which might be subject to selective sputtering of the different components, thereby causing a deviation from the bulk value of 3.2 for the Zr/Fe ratio. Intentional oxidation of the cleaned surface under an oxygen atmosphere of 10^{-4} Pa leads to a highly preferred oxidation of zirconium, whereas the iron is only weakly affected by the oxygen present.

CONCLUSIONS

Mössbauer spectroscopy has become a widely used method in physical metallurgy with iron as the most convenient Mössbauer isotope still predominant. Many different problems have been successfully dealt with. By way of example this paper showed applications to defects, phase analysis, and diffusion in metal; hydrides, and to amorphous metals. Amorphous metals serve as model systems for both the advantages and limitations of Mössbauer spectroscopy. In fields such as the investigation of magnetic properties and of crystallization and hydrogenation processes, important contributions were made by Mössbauer spectroscopy. Unfortunately, the question of short range order could not be resolved by Mössbauer spectroscopy.

ACKNOWLEDGEMENT

Part of this work was supported by the Deutsche Forschungsgemeinschaft. We are grateful for the financial support of the Vereinigung der Freunde der Universitat des Saarlandes.

REFERENCES

1. E. Orowan, Z. Physik <u>89</u>, 634 (1934).
2. M. Polanyi, Z. Physik <u>89</u> 660 (1934).
3. G. I. Taylor, Proc. Roy. Soc. (London) <u>A145</u> (1934).

4. H. Welker, Z. Naturforsch. 7a, 744 (1952).

5. U. Gonser and H. Wiedersich, Proc. Int. Conf. on Crystal Lattice Defects, J. Phys. Soc. Japan 18, Suppl. 11 (1963).

6. G. Vogel, W. Mansel, and P. H. Dederichs, Phys. Rev. Lett. 36, 1497 (1976).

7. R. S. Preston, S. Nasu, and U. Gonser, J. Physique C2, 244 (1979).

8. G. Wortmann, J. Physique C6, 333 (1976).

9. F. E. Wagner and G. Wortmann, in Topics in Applied Physics, Vol. 28, Ed. G. Alefeld, J. Volkl, Springer (1978).

10. J. J. Reilly and R. H. Wisall, Inorg. Chem. 13, 218 (1974).

11. P. Thompson, F. Reidinger, J. J. Reilly, L. M. Corliss, and J. M. Hastings, J. Phys. F10, L57 (1980).

12. P. Fischer, W. Halg, L. Schlapbach, F. Stucki, and A. F. Andresen, Mat. Res. Bull. 13, 931 (1978).

13. E. Lebsanft, Berichte der KFA Jülich-1547 (1978).

14. P. Thompson, M. A. Pick, F. Reidinger, L. M. Corliss, J. M. Hastings, and J. J. Reilly, J. Phys. F8, L75 (1978).

15. P. Thompson, J. J. Reilly, F. Reidinger, J. M. Hastings, and L. M. Corliss, J. Phys. F9, L61 (1979).

16. J. Schefer, P. Fischer, W. Hälg, F.Stucki, L.Schlapbach, and A. F. Andresen, Mat. Res. Bull. 14, 1281 (1979).

17. L. J. Swartzendruber, L. H. Bennet, and R. E. Watson, J. Phys. F6, L331 (1976).

18. W. Schafer, E. Lebsanft, and A. Blasius, Z. Phys. Chem. NF 115, 201 (1979).

19. S. M. Fries, H.-G. Wagner, S. J. Campbell, U. Gonser, N. Blaes, and P. Steiner, J. Phys. F5, 79 (1985).

20. A. Heidemann, G. Kaindl, D. Salomon, H. Wipf, and G. Wortmann, Phys. Rev. Lett. 36, 213 (1976).

21. A. Heidemann, H. Wipf, and G. Wortmann, Hyperfine Interactions 4, 844 (1978).

22. A. Blasius, R. S. Preston, and U. Gonser, Z. Phys. Chem. NF 115, 187 (1979).

23. F. E. Wagner, R. Wordel, and M. Zelger, J. Phys. F14, 535 (1984).

24. N. Blaes, A. Blasius, U. Gonser, and R. S. Preston, Proc. Int. Conf. Appl. Mössbauer Effect, Jaipur, 819 (1981).

25. U. Gonser, N. Blaes, and R. Preston, Z. Metallkde. 73, 605 (1982).

26. H. R. Leider and D. N. Pipkorn, Phys. Rev. 165, 494 (1968).

27. J. A. Tjon and M. Blume, Phys. Rev. 165, 456 (1968).

28. G. Arnold, and J.-M. Welter, Metallurgical Transactions 14A, 1573 (1983).

29. A. Brenner, D. E. Couch, and E. K. Williams, J. Res. Natl. Bur. Stand. 44, 109 (1950).

30. W. Buckel and R. Hilsch, Z. Phys. 132, 420 (1952).

31. U. Gonser and B. Obkerse, J. Phys. Chem. Solids 7, 55 (1958).

32. F. E. Fujita and U. Gonser, J. Phys. Soc. Japan 13, 1068 (1958).

33. J. D. Bernal, Nature (London) 188, 908 (1960).

34. P. Duwez, R. H. Willens, and W. Klement, J. Appl. Phys. 31, 1136 (1960).

35. P. Duwez and S. C. H. Lin, J. Appl. Phys. 38, 4096 (1967).

36. C. C. Tsuei, G. Longworth, and S. C. H. Lin, Phys. Rev. 170, 603 (1968).

37. R. Pond and R. Maddin, Trans. Metall. Soc. AIME (Am. Inst. Min., Metall., Pet. Eng.) 245, 2475 (1969).

38. H. S. Chen and C. E. Miller, Rev. Sci. Instrum. 41, 1237 (1970).

39. T. Masumoto and R. Maddin, Acta Metall. 19, 725 (1971).

40. H. J. Guntherodt, priv. communication.

41. G. Le Caer, J. M. Dubois, H. Fischer, U. Gonser, and H.-G. Wagner, Nucl. Instr. Meth. B5, 25 (1984).

42. M. Kopcewicz, H.-G. Wagner, and U. Gonser, Sol. State Commun. 48, 531 (1983).

43. M. Kopcewicz, H.-G. Wagner, and U. Gonser, Hyperfine Interactions 15/16, 729 (1983).

44. G. Le Caer and J. M. Dubois, J. Phys. E12, 1083 (1979).

45. H.-G. Wagner, Proc. Second Seeheim Workshop, ed. P. Gutlich and G. M. Kalvius, Mainz (1983).

46. M. E. Lines and M. Eibschutz, Sol. State Commun. 45, 435 (1983).

47. C. C. Tsuei, H. Lilienthal, Phys. Rev. B13, 4899 (1976).

48. C. L. Chien and R. Hasegawa, J. Appl. Phys. 47, 2234 (1976).

49. P. Mangin, G. Marchal, M. Piecuch, and C. Janot, J. Phys. E9, 1101 (1976).

50. F. E. Fujita, T. Masumoto, M. Kitaguchi, and M. Ura, Jap. J. Appl. Phys. 10, 1731 (1977).

51. C. L. Chien and R. Hasegawa, Phys. Rev. B16, 3024 (1977).

52. A. M. van Diepen, and F. J. A. den Breoder, J. Appl. Phys. 48, 3165 (1977).

53. R. Hasegawa and C. L. Chien, Solid State Commun. 18, 913 (1976).

54. U. Gonser, M. Ghafari, and H.-G. Wagner, Proc. Int. Conf. Mössbauer Spectroscopy, Bucharest, Romania, p. 237 (1977).

55. U. Gonser and R. S. Preston, in Topics in Applied Physics, Vol. 53, Glassy Metals II, ed. H. Beck and H.-J. Guntherodt, Springer, Berlin (1983).

56. F. E. Luborsky, Mat. Sci. Eng. 28, 139 (1977).

57. M. Naka, K. Hashimoto, and T. Masumoto, J. Non-Crystalline Solids 34, 257 (1979).

58. H. J. V. Nielsen, J. Magn. Mat. 13, 187 (1979).

59. H.-G. Wagner, M. Ghafari, U. Gonser, and M. Naka, J. Physique C8, 199 (1980).

60. M. Ghafari, Dissertation, Universitat des Saarlandes (1982).

61. R. Hasegawa, R. C. O'Handley, and L. I. Mendelsohn, AIP Conf. Proc. 34, 298 (1976).

62. P. Haasen, Phys. Bl. 34, 573 (1978).

63. K. Yamauchi and T. Mizoguchi, J. Phys. Soc. Japan 39, 541 (1975).

64. T. Mizoguchi, AIP Conf. Proc. 34, 286 (1976).

65. R. C. O'Handley, R. Ray, and C. P. Chou, Appl. Phys. Lett. 29, 330 (1976).

66. R. Alben, J. I. Budnick, and G. S. Cargill III, in Metallic Glasses, American Society for Metals, Metals Park, Ohio, p. 304 (1978).

67. E. M. Gyorgy, in Metallic Glasses, American Society for Metals, Metals Park, Ohio, p. 275 (1978).

68. S. Nasu and U. Gonser, J. Physique C8, 690 (1980).

69. H.-P. Klein, M. Ghafari, M. Ackermann, U. Gonser, and H.-G. Wagner, Nucl. Instr. Methods 199, 159 (1982).

70. H.-G. Wagner, M. Ackermann, and U. Gonser, J. Non-Cryst. Sol. 61/62, 847 (1984).

71. H.-G. Wagner, M. Ackermann, R. Gaa, and U. Gonser, Proc. 5th Int. Conf., Rapidly Quenched Metals, in press (1984).

72. F. Aubertin, U. Gonser, S. J. Campbell, and H.-G. Wagner, Z. f. Metallk. 76, 237 (1985).

73. H. U. Krebs, C. Michaelsen, J. Reichelt, H. A. Wagner, J. Wecker, Q. R. Zhang, and H. C. Freyhardt, J. Non-Cryst. Sol. 61/62, 463 (1984).

74. H. A. Wagner, C. Michaelsen, M. Meyer, and H. C. Freyhardt, Proc. 5th Int. Conf., Rapidly Quenched Metals, in press (1984).

75. M. Ghafari, U. Gonser, H.-G. Wagner, and M. Naka, Nucl. Instr. Meth. 199, 197 (1982).

76. H. Hiroyoshi and K. Fukamichi, Phys. Lett. **85A**, 242 (1981).

77. H. Hiroyoshi and K. Fukamichi, J. Appl. Phys. **53**, 2226 (1982).

78. Z. Altounian and J. O. Ström-Olsen, Phys. Rev. **B27**, 4149 (1982).

79. H. Yamamoto, H. Onodera, K. Hosoyama, T. Masumoto, and H. Yamauchi, J. Magn. Magn. Mat. **31-34**, 1579 (1983).

80. T. Masumoto, S. Ohnuma, K. Shirakawa, M. Nose, and K. Kobayashi, J. Physique **C8**, 686 (1980).

81. S. M. Fries, H.-G. Wagner, U. Gonser, L. Schlapbach, and R. Montiel-Montoya, J. Magn. Magn. Mat. **45**, 331 (1984).

82. Y. Boliang, D. H. Ryan, J. M. D. Coey, Z. Altounian, J. O. Strom-Olsen, and F. Razavi, J. Phys. **F13**, L217 (1983).

83. H. Heiman and N. Kazama, Phys. Rev. **B19**, 1623 (1978).

84. K. H. J. Buschow and P. H. Smit, J. Magn. Magn. Mat. **23**, 85 (1981).

85. K. M. Unruh and C. L. Chien, J. Magn. Magn. Mat. **31-34**, 1578 (1983).

86. C. L. Chien and K. M. Unruh, Phys. Rev. **B25**, 5709 (1982).

87. T. Egami and Y. Waseda, J. Non-Cryst. Sol. (1984).

88. Lange's Handbook of Chemistry, 12th Edition, edited by J. A. Dean,McGraw-Hill Book Co., New York, pp. 3-120 (1979).

89. K. M. Unruh and C. L. Chien, Phys. Rev. **B30**, 4968 (1984).

90. K. H. J. Buschow, J. Less-Common Metals **79**, 243 (1981).

91. H. S. Chen, K.T. Aust, and Y. Waseda, J. Non-Cryst. Sol. **46**, 307 (1981).

92. E. P. Wohlfarth, in Amorphous Metallic Alloys, edited by F. E. Luborsky, Butterworth Monographs in Materials, London, pp. 283-299 (1983).

93. K. Fukamichi and R. J. Gambino, IEEE Trans. Mag. **17**, 3059 (1981).

94. S. Ohnuma, K. Shirakawa, M. Nose, and T. Masumoto, IEEE Trans. Mag. **16**, 1129 (1980).

95. H. Hiroyoshi and K. Fukamichi, in High Field Magnetism, edited by M. Date, North-Holland, Amsterdam (1983).

96. J. Hesse and A. Rubartsch, J. Phys. **E7**, 526 (1974).

97. P. Panissod, J. Durand, and J. I. Budnick, Nucl. Instr. Meth. **199**, 99 (1982).

98. I. Vincze, F. van der Woude, and M. G. Scott, Sol. State Commun. **37**, 567 (1981).

99. S. H. Liou and C. L. Chien, J. Appl. Phys. **55**, 1820 (1984).

100. C. L. Chien, K. M. Unruh, and S.H. Liou, J. Appl. Phys. **53**, 7756 (1982).

101. A. R. Miedema and F. van der Woude, Physics 100B, 145 (1980).

102. Mössbauer Spectroscopy, Topics in Applied Physics, Vol. 5, edited by U. Gonser, Springer, Berlin (1975).

103. P. M. Parker, J. Chem. Phys. 24, 1096 (1956).

104. L. Haggstrom, Report UUIP-851, University of Uppsala Sweden, SW ISSN 0042-0263, (1974).

105. K. M. Hasselbach, H. Spiering, Nucl. Instr. Meth. 176, 537 (1980).

106. N. Blaes, H. Fischer, and U. Gonser, Nucl. Instr. Meth. B9, 20 (1985).

107. G. Le Caër, J. M. Cadogan, R. A. Brand, J. M. Dubois, H. J. Güntherodt, J. Phys. F14, L73 (1984).

108. H.-G. Wagner, M. Ghafari, H.-P. Klein, and U. Gonser, J. Non-Cryst. Sol. 61/62, 427 (1984).

109. J. M. Dubois and G. Le Caer, Acta Met. 32, 2101 (1984).

110. G. Czjzek, J. Fink, F. Gotz, H. Schmitt, J. M. D. Coey, J. P. Rebouillat, and A. Lienard, Phys. Rev. B23, 2513 (1981).

111. C. P. Wagner, W. M. Riggs, and L. E. Davis, in Handbook of X-Ray Photoelectron Spectroscopy, edited by J. F. Maulder and S. E. Mailenberg, Perkin Elmer Corp., Eden Prairie (1979).

INDUSTRIAL APPLICATIONS OF MÖSSBAUER SPECTROSCOPY TO MICROCRYSTALS

Steen Mørup

Laboratory of Applied Physics II
Technical University of Denmark
DK-2800 Lyngby
Denmark

INTRODUCTION

Microcrystalline materials with crystallite dimensions in the range 5-100nm are very common on the earth. For example, sediments often contain microcrystals of clay minerals, iron oxides and iron oxyhydroxides. Moreover, rapidly cooled lavas may also contain microcrystalline minerals. Even in biological samples microcrystals are common. The iron storage proteins of animals contain microcrystals of iron oxyhydroxides. Because clays are commonly used for manufacturing ceramics and building materials, such samples may also be microcrystalline. Fly ash, which is another example of a microcrystalline material is used, for example, as an additive to concrete.

In the chemical industry microcrystals play an important role in catalysts. In paints microcrystals of, for example, Fe_2O_3 and FeOOH are used as pigments. Suspensions of magnetic microcrystals in water or organic liquids are applied as ferrofluids which have found many interesting applications. One of the most important technogical applications of magnetic microcrystals is in magnetic tapes and other magnetic memory systems.

Because of their small size, microcrystals are often difficult to characterize by the conventional techniques of solid state physics, such as x-ray diffraction. Mössbauer spectroscopy is, however, very useful for studies of microcrystalline materials. In particular the magnetic properties of microcrystals can be studied in great detail by this technique. Such studies permit the determination of the magnetic anisotropy, the alignment of magnetic domains, the particle size, the magnetic interaction among microcrystals, and surface effects. Several reviews on the application of Mössbauer spectroscopy to studies of magnetic microcrystals have recently been published [1-4].

SUPERPARAMAGNETIC RELAXATION

Magnetic microcrystals with dimensions below a certain critical size are single-domain particles. The critical diameter is normally of the order of 50nm. The easy directions of magnetization for a single-domain particle are determined by the total magnetic anisotropy energy which generally is made up by several contributions, such as magnetocrystalline anisotropy, shape anisotropy, stress anisotropy, and surface anisotropy [1]. In studies of macroscopic crystals the surface anisotropy is normally negligible. However, as will be discussed in Section 5, this contribution to the magnetic anisotropy energy constant may be substantial in small particles.

The magnetic energy of a microcrystal with uniaxial anisotropy may be expressed by

$$E(\theta) = KV\sin^2\theta \tag{1}$$

where K is the magnetic anisotropy energy constant, V is the volume, and θ is the angle between the magnetization vector and the easy direction of magnetization. Because the magnetic energy is proportional to the volume, it may, for very small particles, be comparable to the thermal energy, even well below room temperature. Under these circumstances the direction of the magnetization vector is not fixed along one of the easy directions as in large crystals, but fluctuates. The effect of such fluctuations on Mössbauer spectra depends on the fluctuation frequency compared to the time scale of Mössbauer spectroscopy, which is approximately given by the Larmor precession time, τ_L, of the nuclear magnetic moment in the magnetic hyperfine field. In the absence of an applied magnetic field the magnetization vector may fluctuate among the easy directions of magnetization with a relaxation time τ. This process is the so-called superparamagnetic relaxation. Roughly speaking, one observes for $\tau \gg \tau_L$ a magnetically split Mössbauer spectrum, with six lines in the case of iron-57 Mössbauer spectroscopy. For $\tau \ll \tau_L$ a paramagnetic spectrum typically with one or two lines is found. In the intermediate range ($\tau \approx \tau_L$) complex spectra with broadened lines can be observed. In iron-57 Mössbauer spectroscopy τ_L is of the order 10^{-3}-10^{-9} s.

For a microcrystal with magnetic energy given by (1) the superparamagnetic relaxation time for KV is less than or ca. equal to kT is given by

$$\tau \approx \frac{M\pi^{1/2}}{K\gamma_0} \left(\frac{KV}{kT}\right)^{-1/2} \exp\left\{\frac{KV}{kT}\right\} \tag{2}$$

where M is the magnetization, γ_0 is the gyromagnetic ratio, k is Boltzmann's constant, and T is the temperature [1,3,5-8].

For microcrystals with other forms of magnetic anisotropy energy (e.g., cubic) the superparamagnetic relaxation time is given by similar expressions and one may in general assume that the relaxation time is given by

$$\tau \approx \tau_o \exp\left\{\frac{K_eV}{kT}\right\} \tag{3}$$

where K_e is an effective magnetic anisotropy energy constant.

The superparamagnetic blocking temperature, T_B, is defined as the temperature below which the superparamagnetic relaxation is slow compared to the time scale of the experimental technique used for the study of the magnetic properties of the small particles (i.e., about 10^{-8} to 10^{-9} s for the iron-57 Mössbauer spectroscopy). It is seen from equation (2) that τ depends critically on the magnitude of the energy barrier, KV, separating the two easy directions. Therefore, Mössbauer studies in the vicinity of the blocking temperature give information about this parameter. In practice, a sample of small particles always exhibits a particle size distribution. At a given temperature some particles may therefore be below whereas other may be above the blocking temperature. Studies of the temperature dependence of the spectra thus give information on the particle size distribution [1,3,9] or, strictly speaking, on the distribution in KV because, as discussed in section 5, K may itself be a function of particle size.

Figure 1 shows a typical example of Mössbauer spectra of a microcrystalline sample which exhibits superparamagnetic behavior. The sample is a

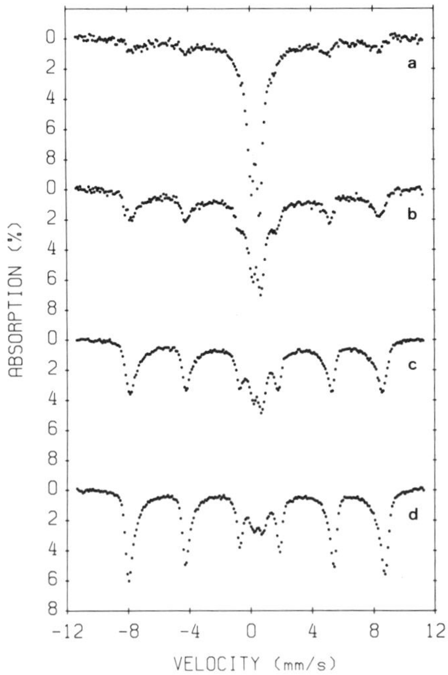

Figure 1. Mössbauer spectra of a sample of tropical red soil from the island Oahu, Hawaii, obtained at various temperatures. a: 295K; b: 186K; c: 130K; d: 80K.

tropical red soil, collected by the author on the island Oahu, Hawaii, in December 1984 during the symposium on Industrial Applications of the Mössbauer Effect. At 80K the main component in the spectrum is magnetically split. The Mössbauer parameters show that this component is due to hematite. The minor component is a quadrupole doublet with an isomer shift of 0.43mm/s relative to α-iron at 300K and quadrupole splitting of 0.47mm/s. At higher temperatures the relative area of the quadrupole doublet gradually increases at the expense of the six-line component. This illustrates that the hematite particles are superparamagnetic with a distribution in superparamagnetic blocking temperatures that is related to the particle size distribution.

MAGNETIC HYPERFINE SPLITTING BELOW THE BLOCKING TEMPERATURE

Below the superparamagnetic blocking temperature the Mössbauer spectra of microcrystals are magnetically split. However, although the probability that the magnetization vector will surmount the energy barrier between the easy directions is negligible, the spectra may still be affected by fluctuations of the magnetization vector. For a microcrystal with a magnetic energy given by equation 1 the probability that the magnetization vector forms an angle between θ and $\theta + d\theta$ with the easy direction is given by [1,10]

$$p(\theta)d\theta = \frac{\exp[-E(\theta)/kT]\sin\theta d\theta}{\int_0^{\pi/2}\exp[-E(\theta)/kT]\sin\theta d\theta} \tag{4}$$

Thus, the magnetization vector may fluctuate in directions close to an easy direction of magnetization, unless the parameter KV is infinitely large compared to the thermal energy. This kind of fluctuations (collective magnetic excitations), which are illustrated in Figure 2, are fast compared to the time scale of Mössbauer spectroscopy. Therefore, the observed magnetic hyperfine field is the average hyperfine field given by [1,3,10-12]

$$B_{obs} = B_o \langle\cos\theta\rangle \tag{5}$$

where B_o is the hyperfine field in the absence of fluctuations and $\langle\cos\theta\rangle$ is given by

$$\langle\cos\theta\rangle = \int_o^{\pi/2} \cos\theta\, p(\theta)\, d\theta \tag{6}$$

For a particle with magnetic energy given by equation 1, one finds [10-12]

$$\langle\cos\theta\rangle = \frac{1}{2}\beta^{1/2} \frac{\exp(\beta^{-1})-1}{\int_o^{\beta^{1/2}} \exp(x^2)\,dx}, \tag{7}$$

where β is KV/kT. For $\beta << 1$ one may use the approximation [12]

$$\langle\cos\theta\rangle \approx 1 - \frac{1}{2}\beta - \frac{1}{2}\beta^2 - \frac{5}{4}\beta^3 - \frac{37}{8}\beta^4. \tag{8}$$

The factor $M\pi^{1/2}/(K\gamma_o)$ in equation 2 is typically of the order of 10^{-9}-10^{-11} s. Therefore values of KV/kT below 5-10 result in relaxation times shorter than about 10^{-8} s. Consequently, the expression for the reduction in the magnetic hyperfine field due to collective magnetic excitations, equation 7, cannot be applied for values of KV/kT below about 5-10 because this results in a superparamagnetic relaxation time which is so short that it significantly affects the spectra [12].

The temperature dependence of the magnetic hyperfine field of microcrystals is schematically illustrated in Figure 3. The maximum reduction in B_{obs} due to collective magnetic excitations is of the order of 5-15 percent [12]. Additional increase in the temperature results in a collapse of the magnetic hyperfine splitting because of fast superparamagnetic relaxation.

The dependence of the magnetic energy of a microcrystal on the magnetization direction may be much more complicated than that given in equation 1 and, therefore, instead of using equations 7 and 8, one may derive expressions which are also valid for other forms of the magnetic energy. The

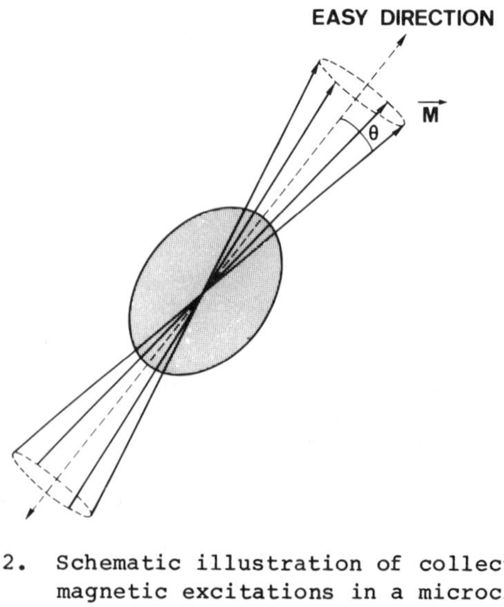

Figure 2. Schematic illustration of collective magnetic excitations in a microcrystal.

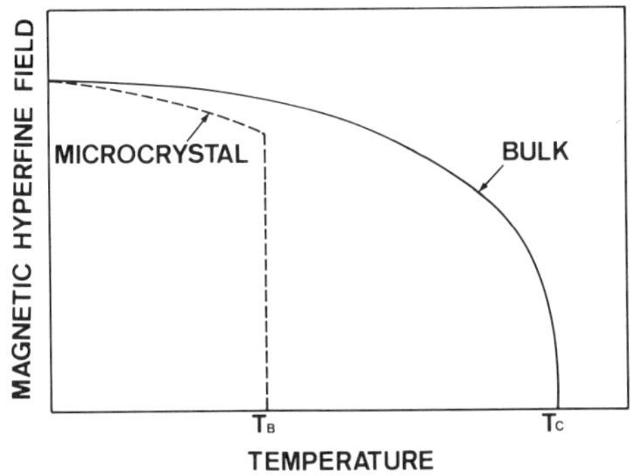

Figure 3. Schematic illustration of the temperature dependence of the magnetic hyperfine splitting in a microcrystal.

magnetic energy of a small particle may in general be expressed as a function of the direction cosines, u_x, u_y, u_z, of the magnetization vector [12], i.e.,

$$E = E(u_x, u_y, u_z). \tag{9}$$

Well below the superparamagnetic blocking temperature the magnetization vector remains near one of the easy directions (which we define as the z-direction) with a finite probability that it forms a small angle with the easy direction. Because the correlation time of the fluctuations of the magnetization vector around an easy direction is short compared to the time-scale of Mössbauer spectroscopy, the observed magnetic splitting in a spectrum is proportional to the average value of the hyperfine field [12],

$$B_{obs} = B_o \langle u_z \rangle. \tag{10}$$

Well below T_B, i.e., when the thermal energy is small compared to the magnetic energy one finds the first order approximation [12]

$$B_{obs}/B_o \approx 1 - \frac{1}{2}kT\left[\left(\frac{\partial^2 E}{\partial u_x^2}\right)_o^{-1} + \left(\frac{\partial^2 E}{\partial u_y^2}\right)_o^{-1}\right]. \tag{11}$$

Equation 11 may be written [12]

$$B_{obs}/B_o \approx 1 - \frac{kT}{2\kappa V} \tag{12}$$

where κ is an effective anisotropy constant. For particles with magnetic energy given by equation 1, equation 12 reduces to

$$B_{obs}/B_o \approx 1 - \frac{kT}{2KV}. \tag{13}$$

Because the reduction in the magnetic hyperfine splitting depends on the particle volume, a particle size distribution will result in a line broadening especially at temperatures just below T_B. This is illustrated in the spectra shown in Figure 1. At 80K the line width of lines 1 and 6 is about 0.7mm/s, but at higher temperatures it increases, and for example at 186K it is about 1.11mm/s.

Apart from the magnetic fluctuations, other effects may also affect the magnetic splitting in the Mössbauer spectra of microcrystals. The magnitude of the total magnetic field at the nucleus is given by

$$\underline{B}_o = \underline{B}_h + \underline{B}_L + \underline{B}_D. \tag{14}$$

Here $\underline{B}_h$ is the field due to the electrons surrounding the nucleus, $\underline{B}_L$ is given by

$$\underline{B}_L = \frac{1}{3}\mu_o\underline{M}, \tag{15}$$

and is the Lorentz field due to the neighboring atoms, and $\underline{B}_D$ is given by

$$\underline{B}_D = -\mu_o N\underline{M}, \tag{16}$$

and is the demagnetizing field. Here $\underline{M}$ is the magnetization, μ_o is the vacuum permeability and N is the demagnetization coefficient which depends on the particle shape and the direction of the magnetization vector. In antiferromagnetic crystals one finds that $\underline{M}$ is zero and therefore

$$\underline{B}_L = \underline{B}_D = 0 \tag{17}$$

Because ferro- and ferrimagnetic crystallites with dimensions below about 50nm are single-domain particles, the demagnetizing field contributes to the total magnetic field at a nucleus in a microcrystal.

To illustrate the influence of $\underline{B}_D$ on the value of $\underline{B}_o$, we consider a spherical particle of metallic iron at room temperature. In this case $\underline{B}_h$ is -33.7T and $\underline{B}_L$ is +0.7T. These two terms together give the hyperfine field, $\underline{B}_o$, of -33.0T in multi-domain particles, because the domains are arranged such that the demagnetizing field is negligible. However, in spherical single-domain particles, N is one-third and $\underline{B}_D$ is -0.7T. The total magnetic field at the nucleus is therefore $\underline{B}_o$ = -33.7T, i.e., the magnetic hyperfine splitting in a single-domain particle is larger than that of a multidomain particle [1,3,13]. Particles with dimensions larger than about 100nm essentially exhibit the bulk behavior.

APPLIED MAGNETIC FIELDS

In the presence of an external magnetic field, $\underline{B}_{ext}$, the magnetic energy, $E(u_x,u_y,u_z)$, of a ferromagnetic or ferrimagnetic particle depends not only on the magnetic anisotropy but also on the size and direction of the external field. Moreover, the applied field also contributes directly to the magnetic splitting. Therefore equation 11 must be replaced by

$$\underline{B}_{obs} \sim \underline{B}_o[1 - \frac{1}{2}kT((\frac{\partial^2 E}{\partial u_x^2})_o^{-1} + (\frac{\partial^2 E}{\partial u_y^2})_o^{-1})] + \underline{B}_{ext}. \tag{18}$$

The contribution to $E(u_x,u_y,u_z)$, which is due to the external field, is given by

$$E = -\underline{\mu} \cdot \underline{B}_{ext}, \tag{19}$$

where $\underline{\mu}$ is magnetic moment of the particle. For a particle with dimensions of the order of 5 to 10nm, $\underline{\mu}$ is typically in the range of 10^3 to 10^5 Bohr magnetons. Therefore, even in a moderate external field, the contribution to the magnetic energy given by equation 19 is often predominant. The average magnetization of the particle is then given by [1,3]

$$\langle M\rangle \approx M\, L\, \{\frac{\mu B_{ext}}{kT}\}, \tag{20}$$

where $L\,\{\frac{\mu B_{ext}}{kT}\}$ is the Langevin function.

If the superparamagnetic relaxation time is shorter than about 10^{-9} s, the observed magnetic hyperfine field in the Mössbauer spectrum is given by [1,3]

$$\underline{B}_{obs} \approx \underline{B}_o L \left\{\frac{\mu B_{ext}}{kT}\right\} + \underline{B}_{ext}. \tag{21}$$

For $\mu B_{ext}/kT$ greater than ca. 4, we may use the high field expansion of the Langevin function and find [14]

$$\underline{B}_{obs} \approx \underline{B}_o (1 - \frac{kT}{\mu B_{ext}}) + \underline{B}_{ext}. \tag{22}$$

Thus a plot of $|\underline{B}_{obs} - \underline{B}_{ext}|$ as a function of $1/B_{ext}$ gives a straight line with the slope $B_o kT/\mu$ and therefore V, which is μ/M, can be determined [1,3,14].

Figure 4 shows spectra of microcrystals of Fe_3O_4 obtained at 260K in various applied magnetic fields. The iron atoms at the octahedral and the tetrahedral sites in Fe_3O_4 have different hyperfine fields with opposite directions, and the corresponding Mössbauer lines are not completely resolved when the external magnetic field is small. However, an average splitting of the inner lines, Δ_{34}, could be obtained from the spectra [14]. Figure 5 shows a plot of Δ_{34} as a function of inverse applied field. The experimental

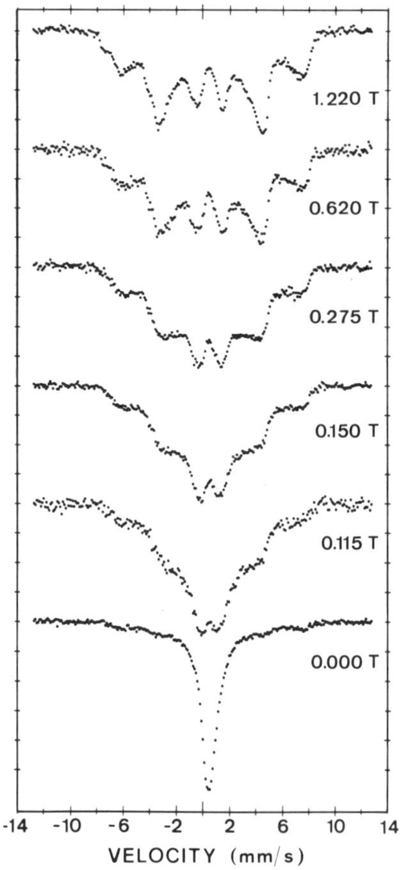

Figure 4. Mössbauer spectra of 6nm particles of Fe_3O_4 at 260K in various applied magnetic fields (adapted from [14]).

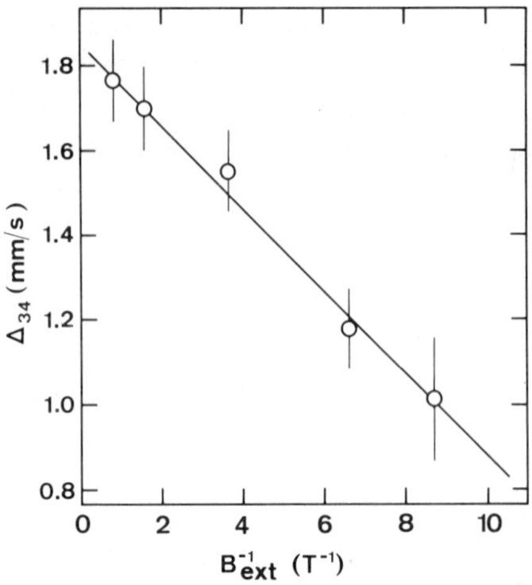

Figure 5. Magnetic splitting of lines 3 and 4 in the spectra shown in Figure 4 plotted as a function of the reciprocal applied magnetic field (adapted from [14]).

points are fitted with a straight line, from the slope of which an average particle size of 6.4nm was determined. This particle size is in excellent agreement with x-ray diffraction and electron microscopy results [14]. The line broadening in the high field spectra is mainly due to the particle size distribution. At lower fields the anisotropy energy cannot be considered negligible and this gives rise to additional line broadening.

MAGNETIC ANISOTROPY IN MICROCRYSTALS

As discussed in the previous sections, the particle volume, V, can be determined from the magnetic field dependence of Mössbauer spectra of microcrystals above T_B. Moreover, the parameter KV can be determined from studies of superparamagnetic relaxation and collective magnetic excitations. Therefore the magnetic anisotropy energy constant of microcrystals can be calculated on the basis of Mössbauer results. Such investigations have been performed in studies of ferrofluids and catalysts.

Ferrofluids are colloidal suspensions of single-domain particles in a nonmagnetic carrier fluid [15]. The particles are coated by surfactant molecules which prevent agglomeration of the particles due to van der Waals forces and magnetic dipole forces. Ferrofluids are used in several types of devices [15]. Mössbauer studies of suspensions of 6.0nm particles of Fe_3O_4 in acetone have shown that the superparamagnetic relaxation time is affected by adsorption of the surfactant molecules. For example, exposure of the particles to oleic acid was found to result in a significant increase in the relaxation time. This result was interpreted in terms of an increase in the surface contribution to the magnetic anisotropy energy constant when molecules of oleic acid were chemisorbed on the surface of the particles [11,16].

Small particles of metallic iron, cobalt and nickel are used as catalysts in the chemical industry. Such particles can be prepared by impregnation of high surface area silica supports with solutions of the nitrates of the metals, followed by drying and reduction in hydrogen [3]. The surface of these particles can be cleaned by heating them in hydrogen in

Table I. Magnitude of magnetic anisotropy energy constants in J/m^3.

Metal	Particle size (nm)	Small Particles, K: In Vacuum	With H_2	With CO	Macroscopic Crystals, K_e
Fe	6.3	$0.8\ 10^5$	$1.2\ 10^5$	$1.3\ 10^5$	$0.1\ 10^5$
Co	7.2	$0.35\ 10^5$	$0.7\ 10^5$	$0.5\ 10^5$	$0.08\ 10^5$
Ni	4.0	$0.6\ 10^5$	$0.3\ 10^5$		$0.006\ 10^5$

The uncertainties in the results for the small particles are mainly given by the uncertainty in the determinations of the volume (about ± 25 percent. However, the uncertainty in the ratio between the anisotropies for a given sample but under different surface conditions is much smaller (about ± 10 percent).

an in situ cell [17]. This treatment leads to removal of, for example, surface oxide layers. Particles free of chemisorbed molecules may then by prepared by subsequent pumping to ultra-high vacuum at high temperature which results in desorption of chemisorbed hydrogen. Such particles have been investigated by means of Mössbauer spectroscopy [3]. The particle size was determined from the field dependence of the spectra, and the magnetic anisotropy energy constants were then determined from the temperature dependence of superparamagnetic relaxation and the reduction in the magnetic hyperfine splitting due to collective magnetic excitations as discussed above. Results were obtained both for samples free of chemisorbed molecules and with H_2 or CO chemisorbed. The magnetic anisotropy energy constants are given in Table I. It is interesting to compare these values with the magnetocrystalline anisotropy of large crystals. For microcrystals with cubic anisotropy the magnetic energy is given by

$$E = K_1(u_x^2u_y^2 + u_x^2u_z^2 + u_y^2u_z^2) + K_2u_x^2u_y^2u_z^2 \tag{23}$$

The effective magnetic anisotropy energy constant K_e in equation 3 for a crystal with cubic anisotropy is given [18] by

$$K_e = K_1/4 \text{ for } K_1 > 0 \tag{24}$$

and

$$K_e = K_1/12 \text{ for } K_1 < 0. \tag{25}$$

Calculated values of K_e for iron, cobalt, and nickel are included in Table I.

It is seen that the magnetic anisotropy of the small particles is considerably larger than the magnetocrystalline anisotropy of macroscopic crystals. There may be several reasons for this [1]. For example, if the particle shape is not spherical, the shape anisotropy may be significant for ferromagnetic and ferrimagnetic microcrystals. Moreover, the interaction with the silica support may give rise to stress in the particles, resulting in stress anisotropy. Furthermore, in all the samples studied here the average particle size is smaller than 10nm and therefore the surface is expected to give a significant contribution to the anisotropy energy constant because of the large relative number of surface atoms having a lower symmetry than bulk atoms [19]. The sensitivity of the magnetic anisotropy to chemisorption suggests that such surface contributions to the magnetic anisotropy are quite important since the surface anisotropy is expected to be sensitive to chemisorption of gases.

It is remarkable that chemisorption of H_2 on nickel particles leads to a decrease in the anisotropy energy constant, whereas chemisorption of H_2 or CO on iron and cobalt particles leads to an increase in the ansiotropy energy constants. This difference may be explained by differences in the chemisorption bonds at the surface of iron, cobalt, and nickel. There are, however, several other possible explanations of the different behavior of iron, cobalt, and nickel particles. It has been shown by Neel [19] that the surface anisotropy energy depends on the particle shape and also on the types of exposed crystallographic planes. Therefore, morphological differences may also explain the different effect of chemisorption on the magnetic anisotropy energy constants.

MAGNETIC HYPERFINE FIELD OF ATOMS AT THE SURFACE AND IN THE INTERIOR OF MICROCRYSTALS

Mössbauer spectroscopy has been extensively used for studies of the magnetic properties of surface and interface atoms. Most of these studies have been carried out with samples of ultrathin films. The hyperfine field of iron-57 surface atoms in thin films of metallic iron covered by various metals or insulators have been studied extensively. Reviews of the results have been given by Shinjo [20], Topsøe et al. [21] and Walker [22]. The experiments have shown that a distribution of hyperfine fields is generally present near the surface, and it has also been found that only a few surface layers are affected by the presence of the surface. The average value of the hyperfine field near the surface is only slightly different from the bulk value. The actual distribution depends on the material covering the surface.

The use of small particles for studies of magnetic surface effects has several advantages compared to the use of thin films covered with various metals. For example, the particles may be stable even at quite high temperatures and this greatly facilitates the preparation of clean surfaces. A complication in studies of surface atoms of small particles is that the Mössbauer spectra may, as discussed above, be affected by magnetic fluctuation phenomena, i.e., superparamagnetic relaxation and collective magnetic excitations. However, when a large magnetic field is applied at low temperatures, these fluctuation effects are negligible for samples of ferromagnetic and ferrimagnetic particles, and the Mössbauer spectra then give information on the hyperfine fields of atoms at the surface and atoms in the interior of the particles.

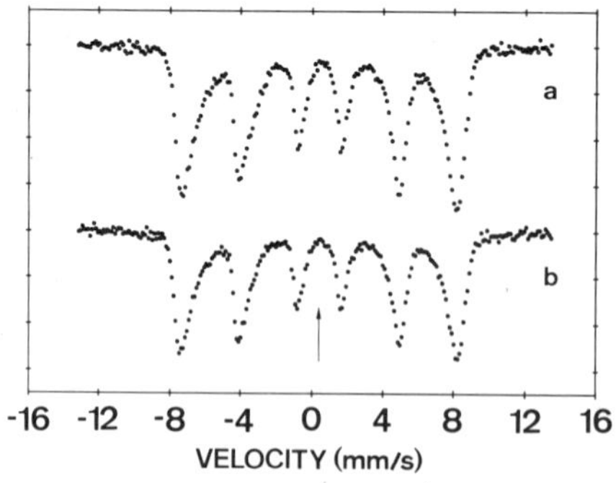

Figure 6. Mössbauer spectra of 6.0nm Fe_3O_4 particles at 78K in acetone (a) and in oleic acid (b). The arrow indicates the expected position of a line due to a 'magnetically dead layer' (adapted from [3]).

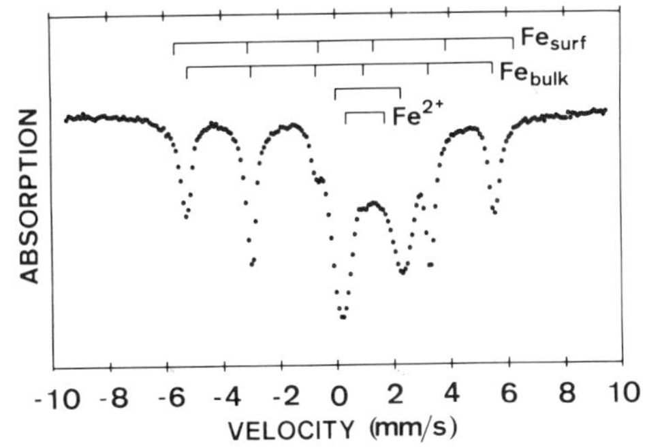

Figure 7. Mössbauer spectrum of silica supported α-iron particles at 78K in an applied magnetic field of 1.24T. The bar diagrams indicate the decomposition of the spectrum in terms of a 'bulk' and a 'surface' magnetic α-iron component and two iron(II) components due to nonreduced iron (adapted from [3]).

In order to investigate the presence of surface effects, Mössbauer studies have been made on suspensions of 6.0nm Fe_3O_4 microcrystals in acetone and in a mixture of acetone and oleic acid [3,11,16]. One of the aims of the studies was to test the proposal by Kaiser and Miskolczy [23] that chemisorption of oleic acid and other polar molecules on magnetite surfaces result in the formation of a 'magnetically dead layer', i.e., a nonmagnetic surface layer with a thickness of about 0.8nm. This proposal was based on magnetization measurements on microcrystals which showed a decrease in the magnetic moment of the particles upon chemisorption of oleic acid. Below the blocking temperature, the presence of a 'magnetically deal layer' should result in a central absorption peak in the Mössbauer spectrum (at about 0.6nm/s) with an intensity corresponding to more than 30percent of the total spectral area. Figure 6 shows spectra of the two samples at 78K. It is clearly seen that no such central absorption line is present. Therefore, it can be concluded that no 'magnetically dead layer' is present. The small magnetization, observed by Kaiser and Mizkolczy, may be due to, for example, a noncollinear spin arrangement near the surface. Such affects have been observed in microcrystals of $NiFe_2O_4$ [24,25] and γ-Fe_2O_3 [26,27].

The magnetic properties of surface atoms of α-iron particles have also been studied [28,29]. Figure 7 shows a spectrum of small particles of α-iron with an average size of 6.3nm. The spectrum was obtained in vacuum at 78K in an applied magnetic field of 1.24T and shows the presence of a ferromagnetic (sextet) component arising from the metallic iron particles. In addition, paramagnetic iron(II) components due to nonreduced iron, are also present [28]. The lines of the six-line component are clearly asymmetric. This indicates the presence of more than one magnetic hyperfine field. The bar diagram in the figure indicates a fit of this component with two magnetically split components with hyperfine fields of 33.3 ± 0.1T and 36.7 ± 0.3T, respectively. The component with the smaller hyperfine field has Mössbauer parameters which, as discussed below, correspond to those of 'bulk' α-iron [28]. The second component was found to ch nge upon chemisorption of H_2 or CO and its relative area (about 15 percent of the total area of the magnetic components) corresponds to the relative number of surface atoms. Therefore, this component could be attributed to the surface atoms of the metallic iron particles. The value of the surface magnetic hyperfine field agrees qualitatively with theoretical calculations by Wang and Freeman [30,31], predicting the magnetic moment of surface iron atoms to be about 20 percent larger than the bulk value.

When CO or H_2 is chemisorbed on the iron particles the magnetic hyperfine field and the isomer shift of the 'surface' component increase, but the isomer shift is still characteristic of a metallic state. The 'bulk' component is unaffected by the chemisorption. This result indicates that the chemisorption bonds do not change the properties of all the iron atoms in the metallic iron particles but only the properties of the surface atoms. The localized character of the chemisorption bonds is in accordance with a molecular orbital model for the chemisorption of CO and H_2 [28]. Recently the surface component of α-iron particles has also been identified in studies of particles prepared by other methods [32,33]. In these studies an enhancement of the magnetic hyperfine field of surface atoms by about 10 percent was also observed.

Apart from the magnetic surface effects, there may also be other size effects in microcrystals which may result in magnetic hyperfine fields which are different from those of large crystals. In order to investigate if such effects are present, the intense magnetic component with the smaller magnetic hyperfine field was analyzed in detail. This component can be attributed to atoms in the interior of the iron particles. Since $\mu B_{ext} \gg KV$ and $\mu B_{ext}/kT \gg 1$ the value of B_{oo} can be estimated on the basis of the observed magnetic splitting of this component. By using equation 22, we find

$$B_o = \frac{B_{obs} + B_{ext}}{1-kT/\mu B_{ext}} = 34.7 \pm 0.1T \tag{26}$$

Before comparing this value with the magnetic hyperfine field of macroscopic iron crystals it must be taken into account that the magnetic splitting of the spectra of small particles has a contribution from the demagnetization field B_D, because the particles are single-domain particles [13]. This contribution is not present in multi-domain particles (see above). For an assembly of small particles of α-iron of different particle shapes and random orientations with respect to the applied magnetic field the average value of the demagnetizing field is $B_D \approx 0.7T$ and B_D is parallel to the magnetic hyperfine field. Thus after correction for the influence of B_D we find a hyperfine field of $34.0 \pm 0.1T$ which is essentially identical to the bulk value at 78K ($33.9 \pm 0.1T$). We can therefore conclude that B_o in small particles of α-iron at 78K is essentially equal to the bulk value.

SURFACE ENRICHMENT IN ALLOY PARTICLES

A special type of surface effect has recently been found in studies of catalysts made of iron-cobalt alloy particles [34,35]. The microcrystals were doped with cobalt-57. In this way it is possible to study the cobalt atoms by use of emission Mössbauer spectroscopy whereas the iron atoms can be studied by absorption Mössbauer spectroscopy. Spectra obtained in external magnetic fields show slightly different hyperfine fields in the emission and absorption spectra. This indicates that the surroundings of the iron and the cobalt atoms are not identical. The emisssion and absorption spectra obtained in zero applied magnetic field were significantly influenced by superparamagnetic relaxation and were quite similar. These results show that the iron and cobalt atoms are present in the same particles and not in separate iron-rich and cobalt-rich particles [34]. Studies of the samples after exposure to different gases showed that the iron atoms react much more easily than the cobalt atoms. Figure 8 shows absorption and emission Mössbauer spectra of a catalyst with an iron to cobalt ratio of 3 to 1, obtained after carburization in a gas mixture of CO and H_2 [35]. The spectra were obtained at 300K with an applied magnetic field of 1.0T in order to suppress relaxation effects. The absorption spectrum (a) clearly shows the presence of both an iron-cobalt alloy and a carbide phase whereas the emission spectrum (b) contains an alloy component but only weak absorption lines due to carbides. This result and the different hyperfine fields in the

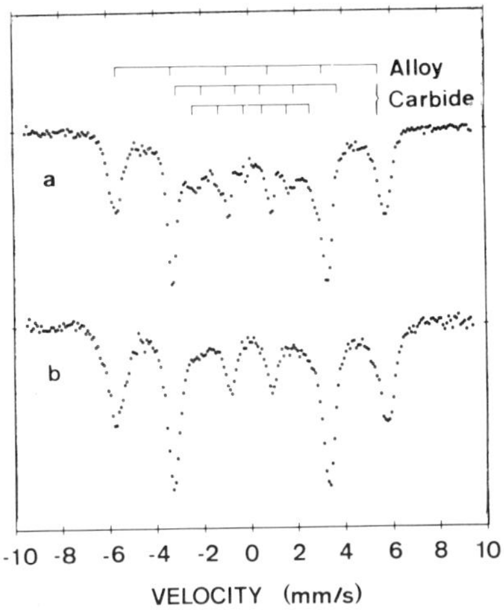

Figure 8. Room temperature Mössbauer spectra of a carburized iron-cobalt catalyst with an atomic ratio of 3 to 1 in an applied magnetic field of 1.0T where a is an absorption spectrum and b is an emission spectrum (adapted from [35].

emission and absorption spectra show that the iron and cobalt atoms are separated in iron-rich and cobalt-rich regions. It is likely that the surface may be enriched with one of the elements. The greater reactivity of the iron atoms suggests that the cobalt atoms are mainly present in the interior of the particles.

MAGNETIC ORDER IN SMALL PARTICLES

The magnetic order in a microcrystal may differ from that of a bulk sample of the same material. For example, the Curie or Neel temperature may be particle-size dependent. There are, however, several results which indicate that small particles have magnetic ordering temperatures close to the values found in large crystals. For example, magnetization measurements on small particles of cobalt, nickel and γ-Fe_2O_3 have shown that the Curie temperature is little affected by the particle size [36]. Other types of magnetic transitions may also be affected by particle size. In α-Fe_2O_3 the spin orientation changes at the Morin transition temperature, T_M, of ca. 260K. Below T_M, the spins are parallel to the (001) direction, and above T_M the spins are perpendicular to this direction (with a small canting angle). This transition gives rise to changes in the sign and magnitude of the quadrupole shift and in the size of the hyperfine field and is therefore easily observed by using Mössbauer spectroscopy. The Morin temperature has been observed to decrease when the particle size decreases [9,37]. Schroeer and Nininger [38] found a Morin temperature of 166±10K in particles with an average diameter of 57.5nm. These same authors later published a detailed study of the particle size dependence of the Morin transition [39].

Large crystals of Fe_3O_4 show a magnetic transition at the so-called Verwey transition temperature, T_V, of ca. 119K. This transition has also been studied in small particles by using Mössbauer spectroscopy. An investigation of the Verwey transition in 10nm particles yielded a transition temperature of about 100K, indicating a decrease in T_V with decreasing particle size [40]. Recently, the Verwey transition of 6.0nm Fe_3O_4 microcrystals has been investigated [41]. Figure 9 shows the spectra of bulk

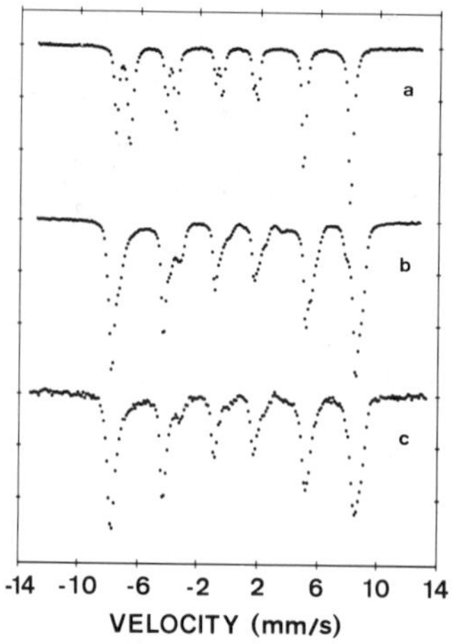

Figure 9. Mössbauer spectra of Fe_3O_4. (a) Bulk at 300K, (b) bulk at 80K, and (c) 6.0nm particles at 4.2K (adapted from [41]).

Fe_3O_4 at 300K (a), and at 80K, (b). The spectrum c was obtained from a sample of 6.0nm Fe_3O_4 particles at 4.2K. It is clearly seen that the very complex structure of spectrum (b), with several low-intensity lines reflecting the complex electronic and magnetic structure below T_V, is also found in the spectrum (c) of the microcrystalline sample. The area ratios of the lines are essentially identical in the two spectra. Because the lattice constant of Fe_3O_4 is about 0.84nm, more than 50 percent of the atoms in the 6.0nm particles are in the first surface layer. The similarity of the spectra of bulk and microcrystals therefore shows that even in the first surface layer a Verwey transition has taken place.

MAGNETIC INTERACTIONS AMONG MICROCRYSTALS

In most of the work that has been published on Mössbauer studies of magnetic microcrystals the particles have been assumed to be isolated, i.e., the magnetic interaction among neighboring particles has been neglected. In practice, however, such magnetic interaction may be quite important. Consider, for example, the two magnetic microcrystals, shown in Figure 10. For arbitrary directions of the magnetic moments $\underline{\mu}_1$ and $\underline{\mu}_2$ the magnetic dipole interaction energy is given by

$$E = \frac{\mu_o}{4\pi}\left\{\frac{\underline{\mu}_1\cdot\underline{\mu}_2}{r_{ij}^3}\right\}-\left\{\frac{3(\underline{\mu}_1\cdot\underline{r}_{ij})\,(\underline{\mu}_2\cdot\underline{r}_{ij})}{r_{ij}^5}\right\}, \qquad (27)$$

where $\underline{r}_{ij}$ is the vector connecting the centers of the two particles. The difference in interaction energy between the two states shown in Figure 10a and b, i.e., the energy required to reverse the magnetization of one of the particles, is given by

$$\Delta E = \frac{\mu_o}{\pi}\,\frac{\mu^2}{r_{ij}^3} \qquad (28)$$

For two particles of metallic iron with diameters of 5nm and separated by 10nm one finds that ΔE is $5.1\ 10^{21}$J. The magnetic anisotropy energy constants of such α-iron particles is of the order of $10^5 J/m^3$ (see Table I). Thus one finds that the energy barrier for an isolated particle is given by

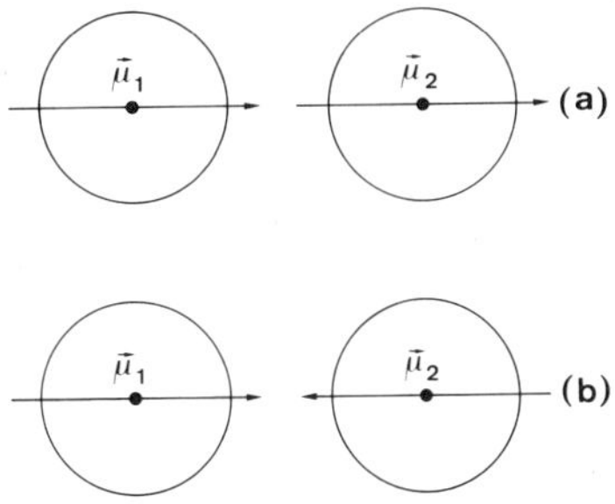

Figure 10. Schematic illustration of two interacting magnetic microcrystals with parallel (a) and antiparallel (b) magnetizations.

KV of ca. 6.5 10^{21}J, i.e., approximately the same magnitude as ΔE. This example illustrates that the magnetic dipole interaction may play a significant role for the magnetic properties of microcrystalline samples.

In samples of closely packed magnetic microcrystals, another type of magnetic interaction may also be present. In such samples the exchange interaction between surface atoms belonging to two neighboring crystals may be significant and this may lead to a large contribution to the magnetic interaction energy. Because the above discussion of relaxation effects was based on models for isolated particles the result derived in these sections may not apply to samples of interacting particles.

Recently, a detailed Mössbauer study of magnetic interaction effects in samples of microcrystalline α-FeOOH has been carried out [42,43]. Figure 11 shows some typical spectra of one of the samples. These spectra are significantly different from those shown in Figure 1. For example, the lines are much broader and there is no paramagnetic component below 290K, although there is a significant reduction in the hyperfine field. Between 300 and 320K the sextet rapidly collapses to a quadrupole doublet [42]. The distributions in magnetic hyperfine fields were determined by use of a computer program developed by Wivel and Mørup [44] and the temperature dependence of the average hyperfine field was determined. It was found that the average hyperfine field at high temperatures was less than 60 percent of the bulk value. As discussed above, the reduction in the observed field in microcrystals due to collective magnetic excitations is not expected to exceed 5-15 percent. Thus, the results are not in accordance with the theories for collective magnetic excitations and superparamagnetic relaxation in isolated microcrystals.

Electron micrographs of the α-FeOOH samples showed that the crystallites are in very close contact [42]. Therefore, it is likely that the results may be explained by magnetic interactions among the crystallites. For a crystallite in an assembly of interacting crystallites the magnetic energy may be written (neglecting the magnetic anisotropy) [12,42] as

$$E_i = \sum_j K_{ex}^{ij} \underline{M}_i \underline{M}_j . \tag{29}$$

where the summation is taken over the neighboring crystallites j. K_{ex} is the exchange coupling constant for the interaction between the crystallites i and j and depends on the exchange coupling between pairs of atoms belonging to the two crystallites. $\underline{M}_i$ and $\underline{M}_j$ are the sublattice magnetization of the crystallites i and j, respectively. The magnetic hyperfine splitting in such

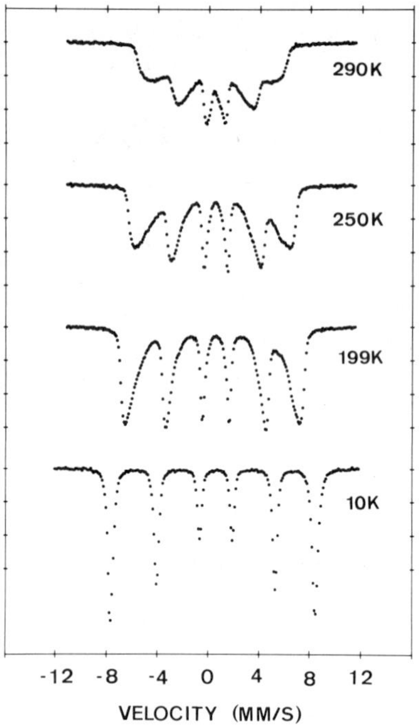

Figure 11. Mössbauer spectra of microcrystalline α-FeOOH at various temperatures (adapted from [43]).

a system has been calculated by use of a modified Weiss mean field theory [42]. According to this model the normalized hyperfine field,

$$b(T) = B_{obs}(T)/B_o(T) \tag{30}$$

can be obtained from the expression [42]

$$b(T) = L\left\{\frac{3T_p}{T}\left\{\frac{B_o(T)}{B_o(T_p)}\right\}^2 b(T)\right\} \tag{31}$$

where $B_o(T)$ is the hyperfine field of bulk α-FeOOH and T_p is the magnetic transition temperature which can be found from the equation [12,42]

$$T_p = \frac{M^2(T_p)\Sigma_j K_{ex}^{ij}}{3k}. \tag{32}$$

Here $M(T_p)$ is the sublattice magnetization in bulk α-FeOOH at T_p.

Figure 12 shows the experimental values of $\langle B_{obs}(T)\rangle/B_o(T)$ for the sample of microcrystalline α-FeOOH. The solid line represents a fit with the modified Weiss mean field theory (equation 31) with a T_p of 316K. The theoretical curves fit the experimental results remarkably well and therefore the magnetic interaction among the crystallites seems to play a significant role in their magnetic behavior. A similar model has been used to explain the results obtained in studies of corrosion products and passivation layers [33].

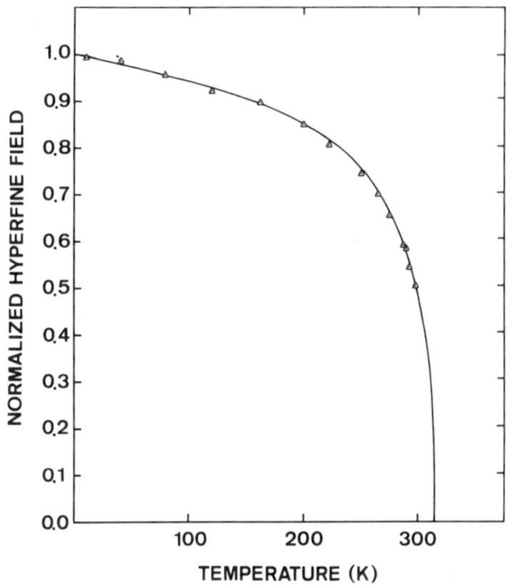

Figure 12. The experimental values of $<B_{obs}(T)>/B_o(T)$ as a function of temperature for closely packed α-FeOOH microcrystals.The solid line represents a fit of the results with the modified Weiss mean field theory discussed in the text (adapted from [43]).

Such microcrystalline systems in which the magnetic coupling between neighboring crystallites is significant and hinders superparamagnetic relaxation have been described as "superferromagnetic" [42].

CONCLUSIONS

The examples discussed in this paper illustrate that Mössbauer spectroscopy is a very useful technique for studies of magnetism in microcrystals. The technique gives information on important parameters such as particle size, magnetic anisotropy, magnetic phase transitions, surface magnetism, and magnetic interactions among microcrystals. In several respects Mössbauer spectroscopy seems to be superior to other experimental techniques that are used to characterize microcrystalline materials. Because many industrially important materials contain microcrystals it seems likely that the application of Mössbauer spectroscopy to studies of such materials will increase in the future.

ACKNOWLEDGMENTS

Support from the Danish Council for Scientific and Industrial Research is gratefully acknowledged.

REFERENCES

1. S. Mørup, J. A. Dumesic and H. Topsøe, in Applications of Mössbauer Spectroscopy, Vol. II, ed. R. L. Cohen (Academic Press, New York), p. 1 (1980).

2. J. L. Dormann, Rev. Phys. Appl. 16, 275 (1981).

3. S. Mørup, H. Topsøe and B. S. Clausen, Phys. Scripta 25, 713 (1982).

4. A. H. Morrish and K. Haneda, J. Magn. Magn. Mat. 35, 105 (1983).

5. W. F. Brown, Jr., J. Appl. Phys. 30, suppl. 130S (1959).

6. W. F. Brown, Jr., J. Appl. Phys. 34, 1319 (1963).

7. W. F. Brown, Jr., Phys. Rev. 130, 1677 (1963).

8. A. Aharoni, Phys. Rev. 135, A447 (1964).

9. W. Kündig, H. Bömmel, G. Constabaris, and R. H. Lindquist, Phys. Rev. 142, 327 (1966).

10. S. Mørup and H. Topsøe, Appl. Phys. 11, 63 (1976).

11. S. Mørup, H. Topsøe, and J. Lipka, J. de Phys. Colloq. 37, C6-287 (1976).

12. S. Mørup, J. Magn. Magn. Mat. 37, 39 (1983).

13. J. E. Knudsen and S. Mørup, J. de Phys. Colloq. 41, C1-155 (1980).

14. S. Mørup and H. Topsøe, Proc. Internat. Conf. Mössbauer Spectroscopy, Bucharest, Romania (Edited by D. Barb and D. Tarina), p. 229 (1977).

15. S. W. Charles and J. Popplewell, in Ferromagnetic Materials, Vol. 2 (Ed. E. P. Wohlfarth), North Holland, p. 509 (1980).

16. S. Mørup, J. Magn. Magn. Mat. 39, 45 (1983).

17. B. S. Clausen, S. Mørup, P. Nielsen, N. Thrane, and H. Topsøe, J. Phys. E., Sci. Instrum. 12, 439 (1979).

18. K. Krop, J. Korecki, J. Zukrowski, and W. Laras, Int. J. Magn. 6, 19 (1974).

19. L. Neel, J. Phys. Radium 15, 225 (1954).

20. T. Shinjo, J. de Phys. Colloq. 40, C2-63 (1979).

21. H. Topsøe, J. A. Dumesic, and S. Mørup, in Applications of Mössbauer Spectroscopy (Edited by R. L. Cohen), Vol. II, p. 55 Academic Press, New York (1980).

22. J. C. Walker, Proceedings of the Indian National Science Academy, International Conference on the Applications of the Mössbauer Effect, Jaipur, India, 1981, (New Delhi, 1982). p. 21.

23. R. Kaiser and G. J. Miskolczy, J. Appl. Phys. 41, 1064 (1970).

24. A. E. Berkowitz, J. A. Lahut, I. S. Jacobs, L. M. Levinson, and D. W. Forester, Phys. Rev. Lett. 34, 594 (1975).

25. A. H. Morrish and K. Haneda, J. Appl. Phys. 52, 2496 (1981).

26. J. M. D. Coey, Phys. Rev. Lett. 27, 1140 (1971).

27. A. H. Morrish, K. Haneda, and P. J. Schurer, J. de Phys. Colloq. 37, C6-301 (1976).

28. B. S. Clausen, S. Mørup and H. Topsøe, Surface Sci. 106, 438 (1981).

29. B. S. Clausen, H. Topsøe and S. Mørup, International Conference on the Applications of the Mössbauer Effect, Jaipur, India (Indian National Science Academy, New Delhi), p. 423 (1982).

30. A. J. Freeman, C. S. Wang, H. Krakauer, and M. Pasternak, J. de Physique Colloq. 41, C1-39 (1980).

31. C. S. Wang and A. J. Freeman, J. Mag.Magn. Mater. 15-18, 869 (1980).

32. R. Birringer, H. Gleiter, H.-P. Klein, and P. Marquard, Phys. Lett. 202A, 365 (1984).

33. I. Tamura and M. Hayashi, Surface Sci. 146, 501 (1984).

34. P. H. Christensen, S. Mørup, B. S. Clausen, and H. Topsøe, Proc. 8th Int. Congress on Catalysis, Berlin (West), Vol. II, (Verlag Chemie), p. 545 (1984).

35. P. H. Christensen, S. Mørup, B. S. Clausen, and H. Topsøe, Proceedings of the International Conference on the Applications of the Mössbauer Effect, Alma-Ata, USSR, in press (1983).

36. I. S. Jacobs and C. P. Bean, in Magnetism" (Eds., G. T. Rado and H. Suhl), Vol. III, Academic Press, New York (1963).

37. A. M. van der Kraan, Phys. Status Solidi (a) 18, 215 (1973).

38. D. Schroeer and R. C. Nininger, Phys. Rev. Lett. 19, 632 (1967).

39. R. C. Nininger and D. Schroeer, J. Phys. Chem. Solids 39, 137 (1978).

40. H. Topsøe and S. Mørup, Proc. Int. Conf. Mössbauer Spectroscopy, Krakow, Poland, (Eds., A. Z. Hrynkiewicz and J. A. Sawicki), Vol. 1, p. 321 (1975).

41. S. Mørup and H. Topsøe, J. Magn. Magn. Mat. 31-34, 953 (1983).

42. S. Mørup, M. B. Madsen, J. Franck, J. Villadsen, and C. J. W. Koch, J. Magn. Magn. Mat. 40, 163 (1983).

43. M. B. Madsen, S. Mørup, J. Franck, C. Wivel, J. Villadsen, and C. J. W. Koch, Proc. Internat. Conf. Mössbauer Effect, Alma Ata, USSR, in press (1983).

44. C. Wivel and S. Mørup, J. Phys. E., Sci. Instrum. 14, 605 (1981).

CONVERSION ELECTRON MÖSSBAUER SCATTERING. A TECHNIQUE FOR STRUCTURAL MICROANALYSIS OF THIN FILMS AND SUBSURFACE REGIONS IN MATERIALS

J. A. Sawicki*

Institute of Physics
Jagiellonian University
Cracow, Poland

and

Department of Physics
Technical University of Munich
Federal Republic of Germany

INTRODUCTION

Within the past few years, conversion electron Mössbauer spectroscopy, CEMS, has been used increasingly for structural analysis of thin films and subsurface regions in solids. Because of its attractive probing depth, ranging from first atomic layers on the surface to about 1μm below the surface, and because of its high sensitivity, which in most favorable cases enables studies of films as thin as 1 to 100ng/cm^2, i.e., 10^{13}-10^{15}at/cm^2, CEMS has established itself as a sensitive depth-selective method for materials characterization in different areas of high-technology and related basic research. Various problems currently investigated with the help of CEMS include surface physics and chemistry, studies of thin films and interfaces, magnetism, metallurgy, catalysis, materials fatigue, corrosion, ion implantation, radiation damage, and plasma-wall interaction. Experimental aspects of CEMS and its applications were discussed in several earlier reviews [1-7] and a complete list of references until November 1982 can be found in the "Conversion Electron Mössbauer Spectroscopy Handbook" [8].

In this chapter recent advances in the methodology of the technique and some of its applications are surveyed. Techniques for CEMS measurements at temperatures from 4 to 1200K and at ultra-high vacuum are described and an extended list of Mössbauer resonances useful in CEMS is presented. The emphasis is placed on studies of thin films and overlayers, materials fatique, near-surface magnetization phenomena, ion implantation in metals and ionic crystals, and the characterization of plasma induced effects in the walls of controlled thermonuclear devices. Measurements for the 77.3keV transition in gold-197 with the sensitivity of 5μg/cm^2 of gold are reported for the first time and applied to analysis of very thin gold films.

*Atomic Energy of Canada Limited, Chalk River Nuclear Laboratories, Chalk River, Ontario, Canada, KOJ 1J0

The nucleus, resonantly excited by the recoil-free absorption of a gamma ray, can discharge its surplus energy in various ways, involving either the re-emission of gamma rays or internal conversion electrons. Because the low energy isomeric states in the nuclei are usually characterized by large electron conversion coefficients (1-100), the resonant Mössbauer scattering of gamma rays is dominated by electron conversion processes.

A total intensity of conversion electrons is determined by the internal conversion coefficient for the given gamma ray transition, defined as the ratio between the probability of electron conversion and gamma ray emission, such that

$$\alpha = N_e/N_\gamma = \alpha_K + \alpha_L + \alpha_M + \cdots\cdot. \qquad (1)$$

Internal electron conversion is accompanied by Auger electrons, X-rays, as well as by numerous secondary electrons and low energy photons.

Figure 1 presents the scheme of the de-excitation processes after the resonant absorption of 14.4keV gamma rays in iron-57 nuclei. The energies and relative intensities of the main electron transitions in this case are: conversion electrons - K(7.3keV, 79%), L(13.6keV, 8%) and M(14.3keV, 1%); Auger electrons - KLL(5.4keV, 60%), LMM(0.5keV, 60%). In the case of the 23.9keV γ-transition in tin-119, the values are L(19.6keV, 83%), M(23.0keV, 13%), LMM(2.8keV, 74%). For the 77.3keV γ-transition in gold-197, the electron transitions are L(63.0-65.4keV, 74%), M(73.9-74.6keV, 19%), N,O(76.6-77.3keV, 7%); LMM(7-10keV, 47%); MNN(1-2keV, 18%). The relative intensities of the conversion and Auger lines are given in percent per one resonant absorption.

All radiation emitted after the resonant absorption can be employed in recording Mössbauer scattering spectra. The techniques which are based on detection of electrons are commonly denoted a conversion electron Mössbauer spectroscopy, CEMS, irrespective of the origin of the electrons being detected.

Conversion electron Mössbauer spectra were first observed in 1961 by Kankeleit [9] who employed a magnetic wedge spectrometer for detecting the 30.5keV K-shell conversion electrons generated by the 100.1keV γ-transition in tungsten-182. The potential of CEMS in materials analysis was recognized

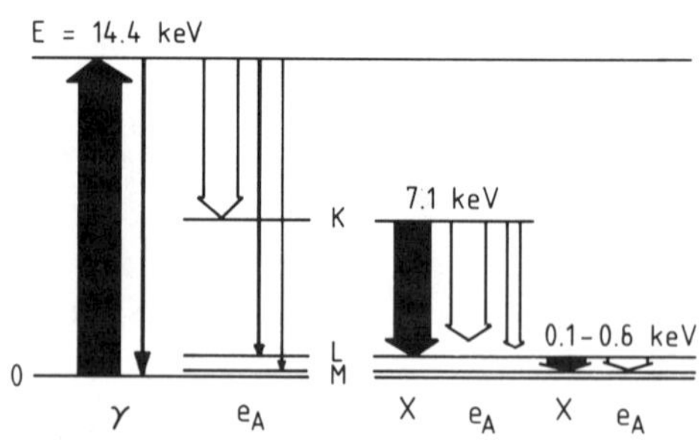

Figure 1. The resonant absorption of 14.4keV γ-rays by iron-57 is followed either by the emission of 14.4keV γ-rays or the internal conversion electrons, e_c, which is accompanied by X-rays and Auger electrons, e_A. The relative intensities of the occurring transitions are represented in the scheme by the widths of the corresponding arrows.

only in the late sixties, starting from work by Bonchev et al. [10] and by Swanson and Spijkerman [11]. The first group demonstrated the possibility of using energy-resolved conversion electron spectra for the depth selective analysis of materials, DSCEMS. Experiments were performed on tin-119 with the magnetic solenoid β-spectrometer. The second group showed the excellent performance of a helium-flow backscatterrer proportional counter for the sensitive nondestructive investigation of bulk specimens using the iron-57 resonance. A list of Mössbauer isotopes for which the CEMS spectra have been observed is given in Table I.

PROBING DEPTH OF CONVERSION ELECTRON MÖSSBAUER SCATTERING

The probing depth of CEMS is determined by the escape range of the conversion, Auger and secondary electrons in the absorber. This depth is usually by two to three orders of magnitude smaller than the depth probed by Mössbauer transmission spectroscopy or by γ-ray and x-ray scattering techniques. An estimate of the electron penetration range can be given on the basis of the formula derived from the analysis of electron transmission measurements and confirmed by recent Monte Carlo simulations [12]. For electrons with kinetic energies, E, between 5 and 50keV, the flux of electrons after transversing a distance x in a material with density, d, given in g/cm^3, is attenuated exponentially,

$$I = I_o e^{-x/\tau}, \tag{2}$$

where the mean penetration depth τ given in nm^{-1} is given by

$$\tau = A\frac{E^n}{d}, \tag{3}$$

where A is 3 to $4*10^{-2} mg/cm^2$ and n is 1.4 to 1.6. A and n are material dependent parameters. Characteristic probing depths for different Mössbauer transitions are listed in Table I. As seen, the depths probed differ from about 10nm in the case of tantalum-181, through 100nm for iron-57 and tin-119 to about 1μm in the case of high energy transitions with energetic conversion electrons.

DEPTH RESOLUTION

Two high resolution electron spectra for iron-57 are shown in Figure 2. The spectra were measured by Toriyama et al. [13,14] by using the ellipsoidal cylindrical mirror analyzer and a very thin 30μCi cobalt-57 source electroplated on stainless steel rod. The B spectrum was obtained after the source was covered with a 6.2nm layer of iron; this spectrum illustrates the character of the energy losses due to inelastic scattering of electrons in the film.

The energy losses of electrons in the sample are related to the depth of their origin below the surface. This fact can be used in the depth selective iron-57 CEMS for depth profiling of iron specimens with the resolution of about 5-10nm. A thorough discussion of depth-selective iron-57 CEMS was recently presented by Liljequist et al. [15] and by Keune et al. [16], in this volume. The characteristics of the electron spectrometers used at present in depth selective CEMS are given in Table II.

A considerable improvement in the energy resolution for the electron spectra of iron-57 was reported very recently by Briancon et al. [17], who constructed a new double electrostatic cylindrical mirror analyzer with a precision approaching one eV. For the cobalt-57 source with the weight of

Table I. Characteristics of Mössbauer Transitions Used in CEMS

Isotope	Abun. %	Parent Isotope	Half-Life	Trans. Energy keV	Internal Conversion Factor, α	Cross Section Ratio σ_o/σ_p	Electron Energies, keV	Mean Electron Range nm e_c, e_A	Ab.	Abs. Temp. K	Maximum Effect Observed η,%	Detection Limit atoms/cm^2
^{57}Fe	2.1	^{57}Co	270d	14.4	8.2	400	K 7.3, KLL 5.4	57, 36	Fe	300	3600	10^{14}
^{119}Sn	8.6	^{119m}Sn	210d	23.8	5.1	470	L19.6, LMM 2.8	270, 8	SnO_2	4	2000	10^{15}
^{125}Te	7.0	^{125}Sb	2.8y	35.5	13.3	100	K 3.7, KLL22.3	26,380	Te	4	1	10^{18}
^{145}Pm		^{145}Nd	11.1d	61.2		50	K16.0, KLL30.3	206,538	Pm_2O_3	10	0.3	
^{151}Eu	47.8	^{151}Sm	93y	21.6	28.6	25	L13.6	215	Eu_2O_3	80	60	10^{16}
^{161}Dy	18.9	^{161}Tb	6.9d	25.7	2.9	125	L16.2, LMM 4.9	170, 29	DyF_3	300	10	10^{18}
^{169}Tm	100	^{169}Er	9.4d	8.4	325	200	M 6.7, MNN 1.4	42, 4	Tm_2P_3	300	1.5	
^{170}Yb	3.0	^{170}Tm	130d	84.3	6.7	30	K23.0, KLL41.3	357,860	Yb	4	1.5	10^{20}
^{180}Hf	35.2	^{180}Ta	8.1h	93.3	4.2	150	K28.0, KLL42.8	256,483	HfO_2	4	100	10^{17}
^{181}Ta	100	^{181}W	140d	6.2	82	20	M 3.5, MNN 1.6	10, 3	Ta	300	1	10^{16}
^{180}W	0.14	^{180}Ta	8.1d	103.7	3.4	160	K34.2, KLL42.8	234,328	W	10	6	
^{182}W	26.4	^{182}Ta	115d	100.1	3.8	200	K30.5, KLL42.8	197,328	W	22	27	10^{18}
^{186}W	28.4	^{186}Re	90h	122.5	1.6	280	K50.7, KLL42.8	423,328	W	10	2	
^{186}Os	1.6	^{186}Re	89h	137.2	1.3	360	K62.3, KLL47.9	508,334	Os	10	25	
^{188}Os	13.3	^{188}Re	16.7h	155.0	0.8	500	K81.1, KLL47.9	737,334	Os	10	10	
^{193}Ir	62.7	^{193}Os	31h	73.0	6.8	40	L59.1, LMM 7.4	456, 20	I	4	5	
^{197}Pt	100	^{197}Et	18h	77.3	4.3	50	L62.9, LMM 7.5	584, 24	Au	4	40	$1.5 \cdot 10^{16}$

Table II. Comparison of Electron Spectrometers for Depth Selective Conversion Electron Mössbauer Spectroscopy [13]

Location	Type	% Energy Resolution at 7.3keV	Acceptance Angle	Detection Efficiency %	Absorber Dimension mmϕ	S/N Ratio[a]
Stockholm University	Cylindrical mirror	2.5	38°-48°	6[b]	5	1
Stevens University	Cylindrical mirror	0.67	56°-60°	2	1	3.6
Rutgers University	Spherical condenser	2.0	75°-90°	13	10	8
Duisburg Institute	150° spherical sector	2.0	9°-21°	0.27	10x4	3
Freie Universität Berlin	Ellipsoidal mirror retarding grid	9	10°-90°	35	5	1
Tokyo Institute of Technology	Spherical retarding grid	2 1 0.3	30°-90° 30°-60° 30°-60°	5.5 2.1 1.2	4 4 3	3 1
Institute of Nuclear Study, Tokyo University	Air core magnetic	2	±15%	1.2	5	10

[a]Obtained with an iron-57 enriched absorber. [b]Transmission detection efficiency.

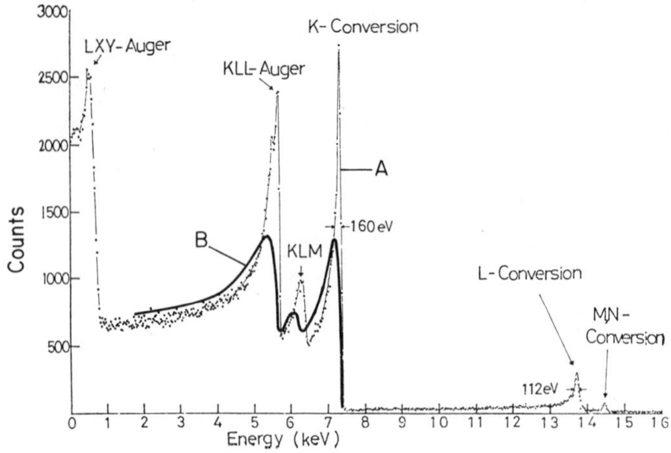

Figure 2. Electron spectra associated with the decay of cobalt-57 to iron-57 measured with an ellipsoidal mirror analyzer [13, 14]. (A), the spectrum obtained for the 30μCi cobalt-57 source electroplated on the surface of stainless steel rod; (B), the spectrum obtained after covering the source with 6.2nm layer of iron.

the order of 1μg/cm^2, electrodeposited onto aluminum, they obtained the line-width 7eV for the 7.3keV K-shell line, which is about one order of magnitude better than the former best measurements.

Our group [18] have studied the low-energy secondary electrons, coming from the resonant iron samples as the result of multiple scattering and energy degradation processes. The spectrum obtained by using a cylindrical mirror analyzer, as shown in Figure 3, clearly indicates a high yield of

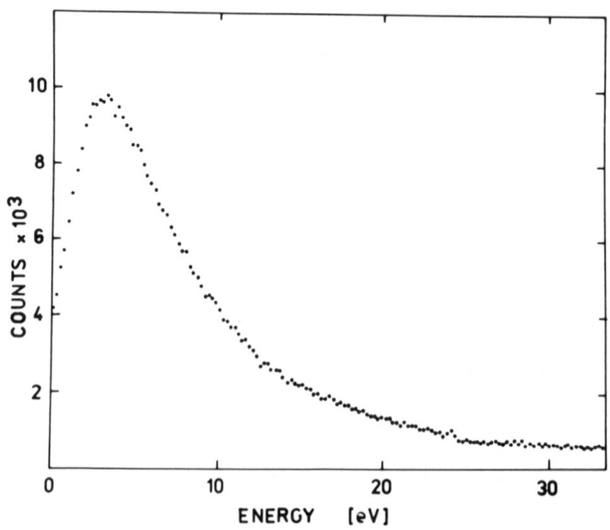

Figure 3. The spectrum of secondary electrons associated with the resonant absorption of 14.4keV gamma rays in a metallic iron-57 absorber, measured with a cylindrical mirror analyzer [18].

secondary electrons with the energies below ca. 10eV. Consequently it was shown, that these electrons can be efficiently detected by using channel electron multipliers and that, by detecting them, one can considerably improve the efficiency of CEMS measurements (see below).

SENSITIVITY OF CONVERSION ELECTRON MÖSSBAUER SPECTROSCOPY

The high sensitivity of CEMS is another very attractive feature of this method. In several favorable cases, such as iron-57, tin-119, europium-151, and gold-197, the required quantity of resonant isotope in the sample can be 10 to 1000 times smaller than that necessary for the usual transmission measurements. The lowest quantities measured so far by CEMS for the nuclides listed above are 10^{13}-10^{14} atoms/cm^2 of iron-57, 10^{14}-10^{15} atoms/cm^2 of tin-119, 10^{15}-10^{16} atoms/cm^2 of europium-151, and 10^{16}-10^{17}atoms/cm^2 of gold-197 (Table I). Thus, the sensitivity of CEMS is generally placed in between the sensitivity ranges for Mössbauer emission experiments using low-activity sources and that for Mössbauer transmission experiments. For these reasons, CEMS is especially suitable in the studies of thin films, coverages and near surface regions in solid materials.

The sensitivity of CEMS is determined by many factors, depending on nuclear, atomic, and solid state parameters of the source and absorber. For a very thin single-line resonant scatterer, the maximum magnitude of the resonant effect is approximately given by the expression

$$\eta_{CEMS} = \frac{N_o - N_\infty}{N_\infty} = \frac{1}{2} f_s f_a a \frac{\alpha}{1+\alpha} \frac{\sigma_o}{\sigma_p}, \quad (4)$$

where N_o and N_∞ are the count rates at resonance and far off resonance, respectively, f_s and f_a are the recoilless fractions, respectively, for the source and the absorber, α is the internal conversion coefficient, σ_o is the resonant absorption cross section, σ_p is the photoeffect cross section, and a is the abundance of the resonant isotope. This can be compared with the magnitude of the resonant effect for the transmission experiment,

$$\eta_T = \frac{1}{2} f_s f_a \frac{1}{1+\alpha} \sigma_o n, \quad (5)$$

where n is the number of resonant nuclei per cm^2. Because the cross section ratio σ_o/σ_p is often as large as 100 or 400 (see Table I), one can, in principle, obtain much larger effects in the CEMS mode than in transmission mode, especially in the case of very thin absorbers. In this case the gain in the effect is given by

$$\frac{\eta_{CEMS}}{\eta_T} = \frac{\alpha a}{\sigma_p n}. \quad (6)$$

Hence, the use of CEMS is especially advantageous as compared to the transmission technique when, (1), the γ-ray transition is highly converted, (2), the photoeffect cross sections of the elements in the sample are small, and (3), when the number of resonant nuclei per cm^2 is small.

The observed effective magnitude of the effect in CEMS is usually significantly smaller than the maximum effect predicted by equation 4, such that

$$\eta = \eta_{max} \frac{1}{1+B} \quad (7)$$

where B expresses the total background contribution.

The nonresonant background is mainly caused by scattering of gamma rays and X-rays on the detector itself and on other parts of the equipment. Especially unfavorable are cases where the resonant γ-ray transition is accompanied by intensive nonresonant γ-ray transitions and X-rays, as well as high-energy γ-rays. In addition, all sources of spurious pulses in the detector and the noise in the electronics badly diminish the effect and should, therefore, be carefully reduced.

The usefulness of a Mössbauer resonance is largely determined by the statistical quality of the spectra [19] which can be defined as

$$Q = \sqrt{N_\infty}\,\frac{\eta}{\sqrt{2+\eta}}. \tag{8}$$

The high value of Q, as generated by a high η-value, is reduced by the generally small count rate in CEMS measurements. In order to count the electrons more effectively one has to use detectors with large solid angle and as high as possible efficiency. Of course, the use of very strong sources is also most helpful.

For more detailed analysis of various factors affecting the statistical quality of CEMS measurements, the reader can be referred to the papers of Bokemeyer, et al. [19] and Oswald and Ohring [20].

The statistical quality of Mössbauer transmission, X-rays scattering and CEMS measurements depends on the thickness of the sampled iron foil as shown in Figure 4 [20]. As seen, the detection of electrons is especially advantageous for iron samples thinner than about 100nm. Thicker iron samples can be usually more efficiently investigated by X-ray scattering (0.1 to 1μm) or γ-ray transmission for thicknesses of less than 1μm.

BASIC EXPERIMENTAL EQUIPMENT

The experimental techniques and the applications of various detectors in CEMS through 1982 has been surveyed [6]. Therefore, only the basic experimental arrangement will be presented here. The techniques developed more recently will be described in more detail below.

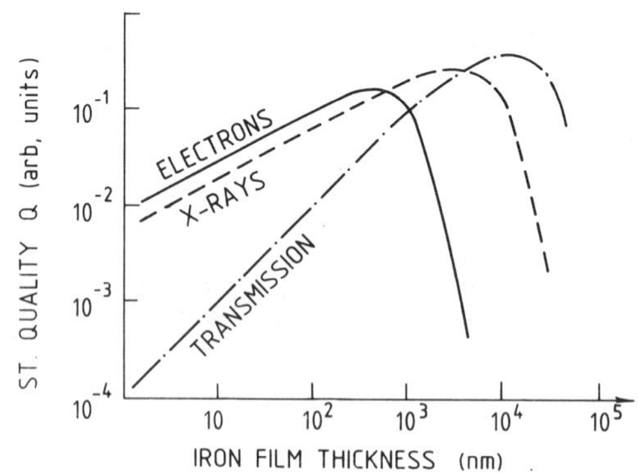

Figure 4. Comparison of the statistical quality Q of the iron-57 Mössbauer spectra for iron measured with with the detection of γ-rays, X-ray, and in the the electron scattering CEMS mode, with Q values calculated from the formulas derived in [20].

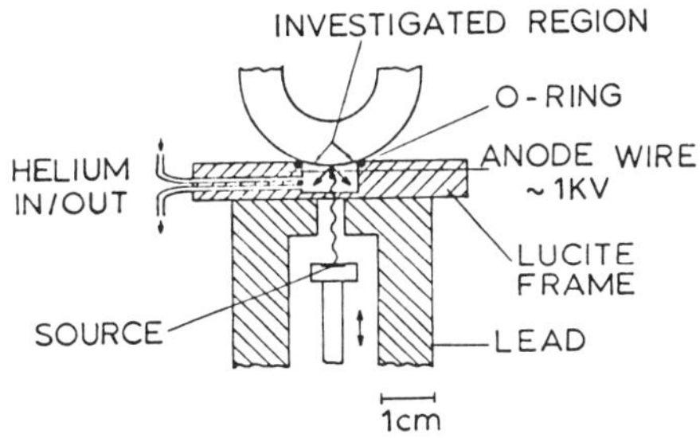

Figure 5. The configuration for a conversion electron Mössbauer analysis of near-surface regions in bulk objects using a helium-flow proportional counter [7].

CEMS spectra are usually measured in backscattering geometry and with the help of helium-flow proportional counters. Such geometry ensures the detection of electrons at a large solid angle of ca. 2π and allows for efficient surface analysis of bulk samples irrespective of their size and dimensions. The efficiency of a helium counter for low energy electrons of ca. 10keV is close to 100 percent, while they are virtually insensitive to incident and backscattered γ-rays and X-rays. Counters of this type can be easily manufactured in various configurations, depending on the requirement of the particular experiment. Typical construction materials are lucite sheets for the box frame, 30μm tungsten wire for the anode, aluminized tape for the electrical screen and windows. As working gases, the 90 percent helium-10 percent methane mixture or pure helium with a flow of about 1 to $3cm^3/min$ are typical. Anode potentials range from 500 to 1500V.

Figure 5 presents a simple set-up for CEMS measurements at room temperature which can be used both for thin films and bulk samples analysis. The counter of this type can be mounted on the surface of big objects, and it can also be operated in remote locations, for instance inside of tubes, chambers, etc. To test the performance of the detectors, one uses iron-57 enriched metallic iron foil or SnO_2 films enriched with tin-119. At room temperature the observed effect should be as high as 1200 percent for the maximum line in metallic iron and about 700 percent for the SnO_2.

Two examples of the measurements using the above described set-up are shown in Figure 6 and 7. Figure 6 presents the CEMS and transmission spectrum of an amorphous 100nm $Dy_{0.23}Co_{0.77}$ film coevaporated on a thin 0.1mm cooper foil. Due to the small thickness of the substrate, the transmission spectrum could be measured, but the time of measurement was about 10 times longer than in the case of CEMS spectrum. Figure 7 presents the CEMS spectra recorded on a low-alloy steel sample, the outer raceway of a roller bearing, taken before and after plastic deformation [21]. The fraction of the retained-austenite, represented by a single line at zero velocity, decreases with the increase in fatigue. A contribution of the carbide phase, represented by a quadrupole-split doublet, could also be observed in the spectra. Numerous other applications of the helium counters in conversion electron Mössbauer spectroscopy at room temperature are presented in the literature.

NEW EXPERIMENTAL TECHNIQUES

In this section several new techniques developed recently, primarily by the author and his colleagues are briefly presented. For the details the reader is referred to the original publications.

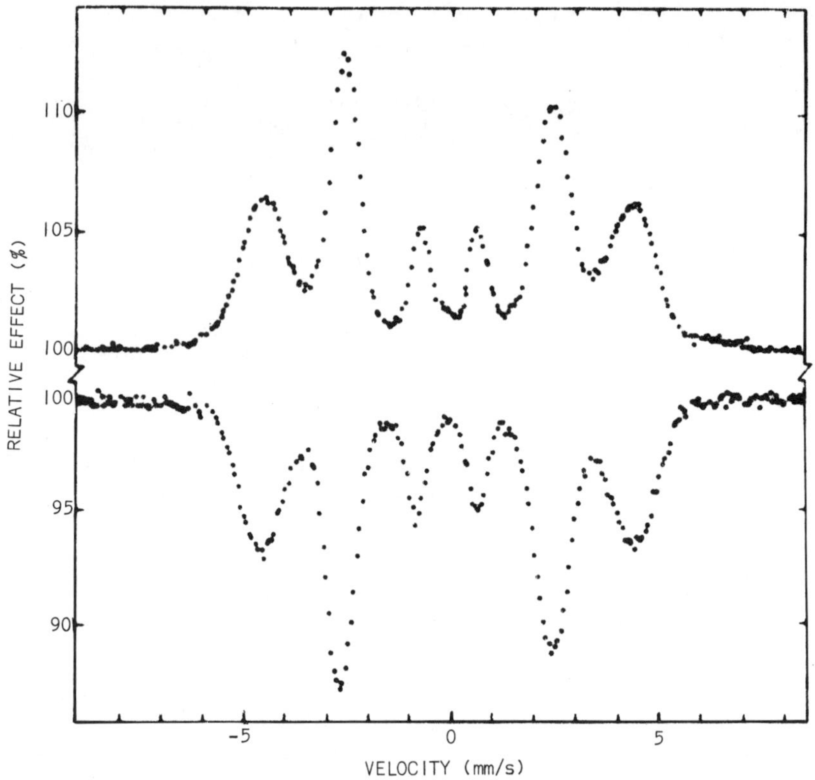

Figure 6. The Mössbauer spectra of $Dy_{0.23}Co_{0.77}$ alloy measured in conversion electron mode and in transmission mode, upper and lower, respectively.

MEASUREMENTS AT LOW TEMPERATURES

Because measurements at low temperatures are very important in Mössbauer spectroscopy, the development of experimental techniques for CEMS in the liquid helium temperature range is of particular interest. The first CEMS experiments at liquid helium temperature were carried out by using complex magnetic wedge [9], orange [19], or solenoid type [22], β-spectrometers which are available in nuclear spectroscopy laboratories. Unfortunately, these spectrometers turned out to be quite inefficient for material studies because of their complexity and low counting efficiency. The various high-resolution electron spectrometers listed in Table II, built for depth-selective CEMS, are not usually easily adapted to cryogenic conditions. The attempts to overcome efficiency limitations by using proportional helium counters proved to be unsuccessful at temperatures below 30K [23], mainly because of a drastic decrease in the gas-amplification at low temperatures.

In a series of experiments [24-30], we have been searching for more convenient methods of measuring conversion electron Mössbauer spectra at low temperatures. The technique which proved to be most practical is based on the observation that channel electron multipliers can be used for efficient detecting of electrons even at liquid helium temperature [24]. The principle of measurements and the details of the detectors used most often, are shown

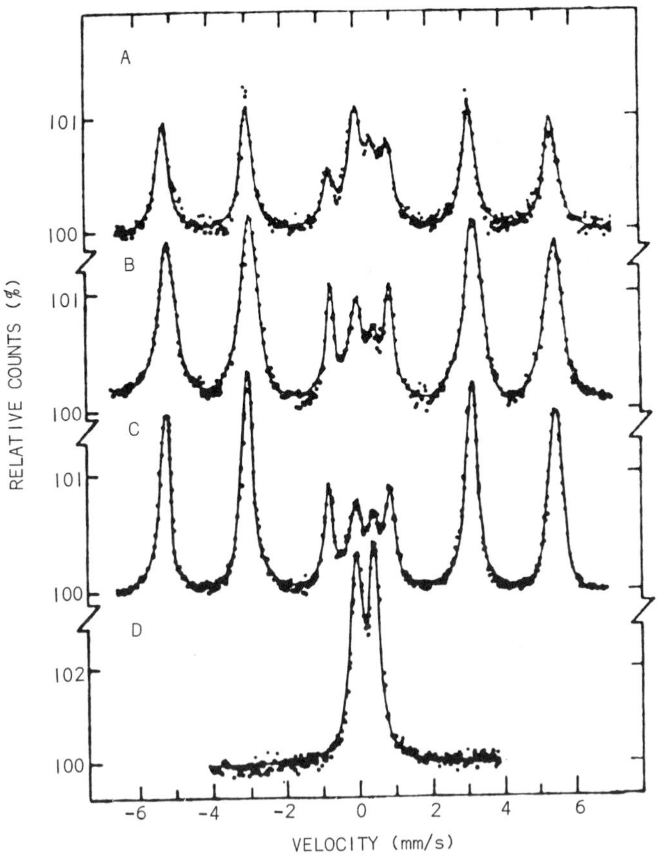

Figure 7. Conversion electron Mössbauer spectra of a ball-bearing steel sample; (A), before plastic deformation; (B), after stressing the sample at 5900MPa for 10 cycles; (C), after 40 cycles at 5900MPa stress; (D), spectrum of the alloyed carbide, $(Fe_{0.85}Cr_{0.15})_3C$, electrochemically extracted from the investigated steel.

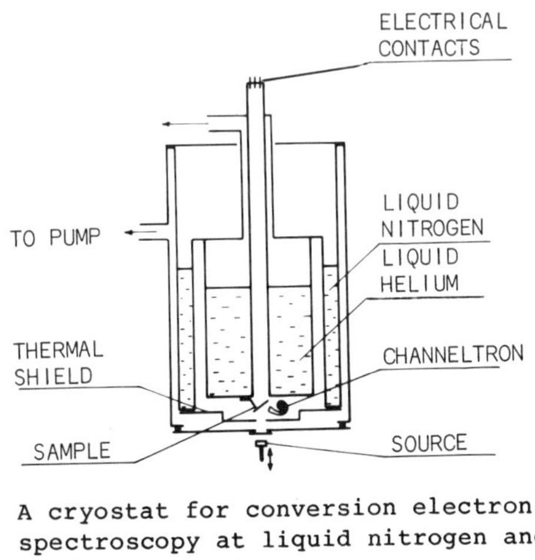

Figure 8. A cryostat for conversion electron Mössbauer spectroscopy at liquid nitrogen and liquid helium temperature.

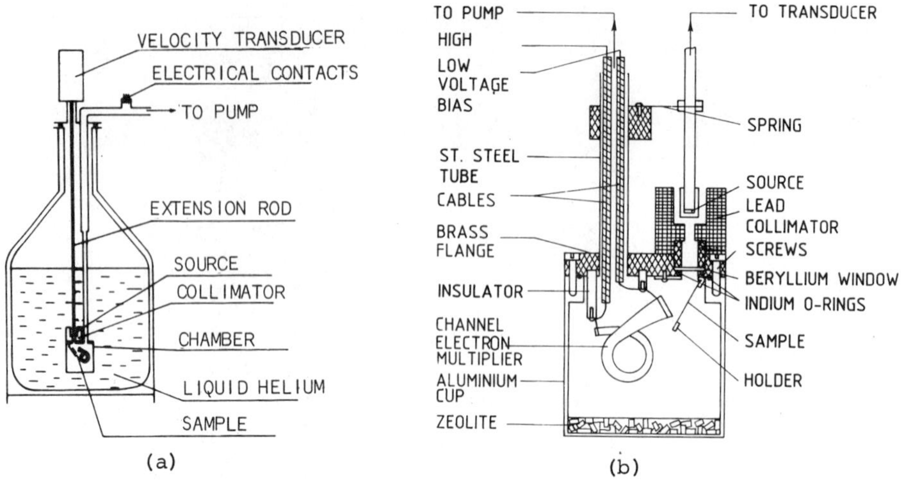

Figure 9. A set-up for conversion electron Mössbauer spectroscopy with both the source, absorber and channel electron multiplier cooled to 78 or 4.2K by immersion into a cryostat. (A), general lay-out; (B), detector chamber.

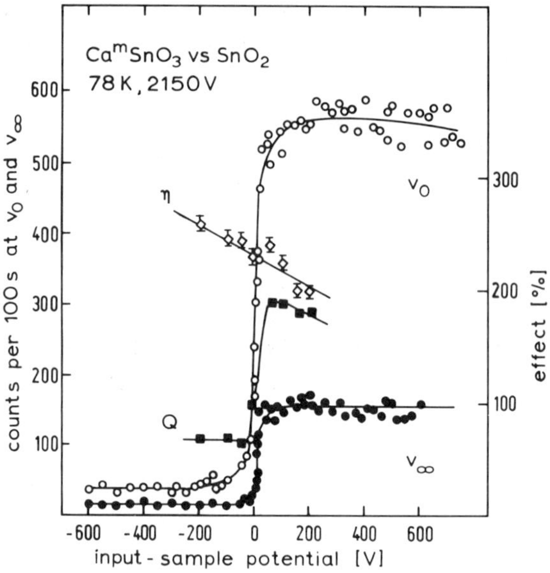

Figure 10. The count rate, the magnitude of the resonant effect, η, and the statistical quality of the spectra, Q, in the CEMS measurements with SnO_2 scatterer, as a function of the input potential on the channel electron multiplier [29].

in Figures 8 and 9. Typically the sample and the channel electron multiplier are installed in a sideways scattering geometry and cooled together either by using a conventional Dewar (Figure 8) or by inserting a detector chamber into a liquid helium or liquid nitrogen container (Figure 9). Because of the small distances between the source, sample, and the detector, this configuration yields high count rates which are comparable with those obtained for gas counters. Further, by applying a positive potential of 50 to 100V to the channel input, one can extract and collect more efficiently secondary electrons coming from the sample.

The characteristics of the detector, i.e., the count rate, the resonant effect and the statistical quality of the spectra measured vs. the applied input potential are shown in Figure 10. A drastic rise in both the count rate and the quality of the spectra is observed with a positive input potential. The magnitude of the resonant effect, η, in the configuration described above depends very much on the level of the nonresonant electron scattering from the window and walls of the chamber and the channel multiplier itself. Therefore, the geometry of the experiment has to be carefully optimized. In some cases, it is possible to reduce the background substantially by using a properly shaped and graded collimator and by filtering out X-rays and γ-rays with lower energies, e.g., with 2mm of lucite for iron-57, 25μm of palladium for tin-119 or 2mm of copper for gold-197. By using this method, it has been possible to record the conversion electron Mössbauer spectra of iron-57, tin-119, tellurium-125, europium-151, ytterbium-170, tungsten-182, iridium-193, and gold-197. The full list of isotopes for which the CEMS measurements have been made and values of the effect measured are given in Table I. Several examples of the spectra measured recently by using 73.0keV iridium-193 for thin iridium metal and iridium oxide films are presented in Figure 11 [30]. The applications of the technique in the case of the 77.3keV resonance in gold-197 will be discussed in the last paragraph of this chapter.

A configuration presented in Figure 12 was designed for CEMS measurements as a function of temperature. In this case, the insert containing the

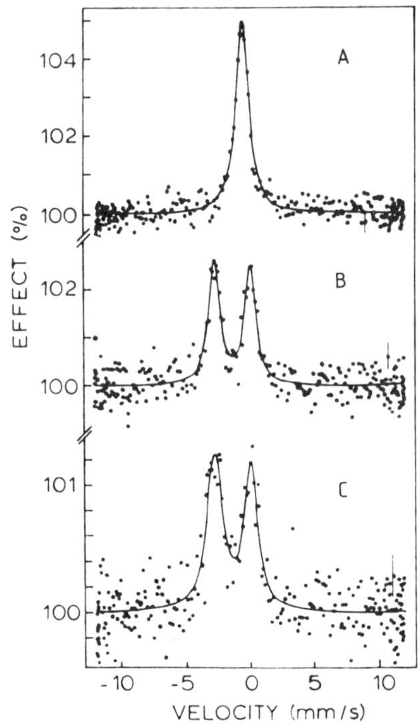

Figure 11. Conversion electron Mössbauer spectra of the 73.0keV gamma-ray transition in iridium-193 measured at 4.2K; (A), spectrum of $5mg/cm^2$ of iridium metal; (B), spectrum of $5mg/cm^2$ of IrO_2; (C), spectrum of $1mg/cm^2$ of IrO_2 [30].

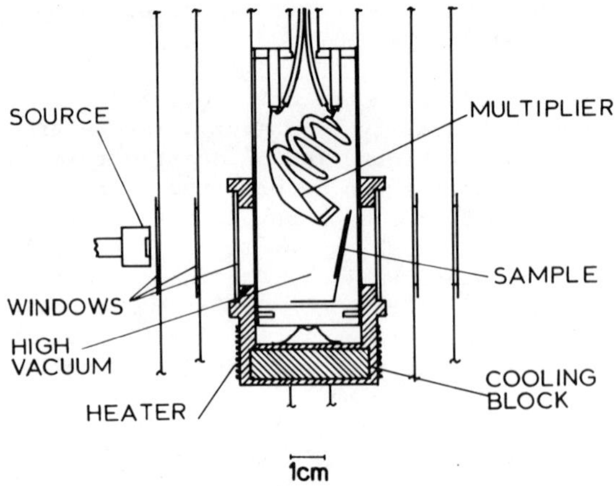

Figure 12. A configuration for the measurements of CEMS spectra between 4 and 400K [29].

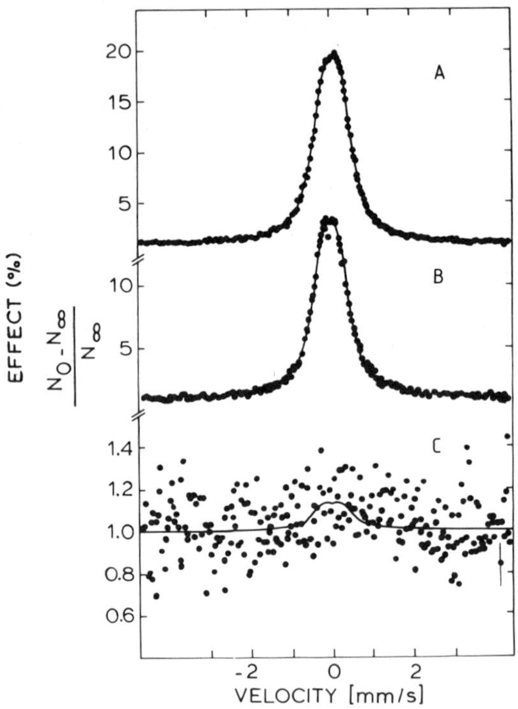

Figure 13. CEMS spectra for $^{119}SnO_2$ measured at 4K (A), and 100K (B). Bottom spectrum (C) was obtained with a negative potential at the detector input [29].

sample and detector is top-loaded into the inner vacuum space of a continuous flow helium or nitrogen cryostat, so that any temperature between 4 and 400K can be quickly attained and maintained with the precision of ±0.1K. Due to the favorable geometry and low atomic number materials used, this detector makes it possible to get a very low nonresonant background and a high percent effect. To give an example, we obtained a resonant effect of more than 2000 percent for SnO_2 containing tin-119 at 4K with a $CaSnO_3$ source at 300K, which is the largest percent effect ever observed in Mössbauer spectroscopy. Several spectra measured with this detector are shown in Figure 13.

A configuration shown in Figure 12 can easily be used for simultaneous measurements in transmission geometry. It can also be coupled to the beam port of the accelerator which permits in-situ beam studies. The sample can be very quickly installed and cooled to 4K within about 30 minutes. Further, it can be rotated while cold with respect to the γ-ray beam direction. The described detector could be used as a resonant counter in experiments with isotope separator implanted short-lived radioactive nuclei, such as indium-119 which has a halflife of 2.3min or antimony-119 with a halflife of 38.5h. Each decays to tin-119.

A few comments are worthwhile about the excellent detecting capabilities of the channel electron multipliers at low temperatures. The detecting capabilities of these detectors at low temperatures have not been described in the general literature, or in the data sheets and technical specifications published by the manufacturers, Galileo Electrooptics Corp. and Philips. Our measurements at 4K [24-26] were made with homemade detectors and later the commercial ones were used successfully. The basic characteristics of the channel electron multipliers, manufactured at Technical University of Gdansk, Poland, were an operating voltage of 2 to 3.4keV, a gain of 10^8, a resistance of 10^9ohm, a standard count rate of 10^3 counts/sec, an input diameter of 10mm, a diameter-to-length ratio of 50, a size of 25x15mm, an operating pressure below $2{*}10^{-4}$torr, a bake temperature of 100°C, and a prize of ca. $500. Aging effects were observed only after prolonged exposure to radiation under a vacuum, and were mostly caused by contamination and cracking of hydrocarbons on the glass surface. Usually, the operating voltage at 4K had to be increased by 200 to 500V, as compared to the voltage at room temperature. More detailed description of the low-temperature properties of channel electron multipliers is presented elsewhere [29].

MEASUREMENTS AT HIGH TEMPERATURES

It is often desirable to investigate the behavior of a material at high temperature, in a vacuum or in various different atmospheres. Because the Mössbauer effect in iron-57 exists up to high temperatures, iron-57 CEMS is especially suitable for such studies. CEMS measurements at temperatures close to 1000K and above are, however, rather difficult and have only recently been accomplished [31,32].

The maximum temperature at which a gas counter can work is limited by the thermal emission of electrons and by the decrease of the electrical resistivity between the cathode and the anode. To avoid the latter effect, Isozumi, et al., [31] have built a special proportional counter in which the insulator between the anode and the cathode is kept far away from the heated volume. It permitted operating temperatures of up to 1100K. The main disadvantage, when using a gas counter, is associated with the possible influence of the working gas, and the impurities contained therein, on the surface of the investigated sample. This influence can be greatly enhanced at high temperatures. On the other hand, the presence of the gas in the detector provides us with the possibility for convenient in situ observation of surface phenomena and the gas induced changes in the samples heated at various atmospheres. The influence of helium, methane, carbon dioxide, hydrogen, nitrogen, and oxygen, either as working gases or as quenching admixtures, on the iron surface in CEMS measurements at high temperatures has been studied [31].

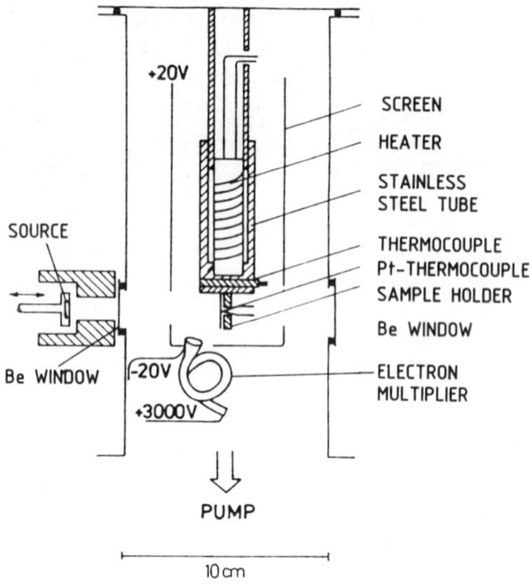

Figure 14. A configuration for conversion electron Mössbauer spectroscopy at high temperatures using a channel electron multiplier [32].

Another type of apparatus for CEMS measurements at high temperatures was built by Kowalski, et al., [32], and is shown in Figure 14. In this case the electrons leaving the sample are detected with a channel electron multiplier at a vacuum of 10^{-5} torr. The background due to thermally activated electrons and ions was suppresssed by setting the detector input at a negative potential of -20V and the screening grid at a positive potential of +20V. Measurements with iron-57 enriched metallic iron films were possible at temperatures up to 1200K, above the Curie point of iron, and the thermochemical changes in an ionic crystal of LiF implanted with $2x10^{16}$ions/cm^2 of iron-57 were followed at temperatures of up to 700K [32]. Because of the effective thermal coupling to the walls, via mica plates, the temperature of the electron multiplier during the measurements did not exceed 400K.

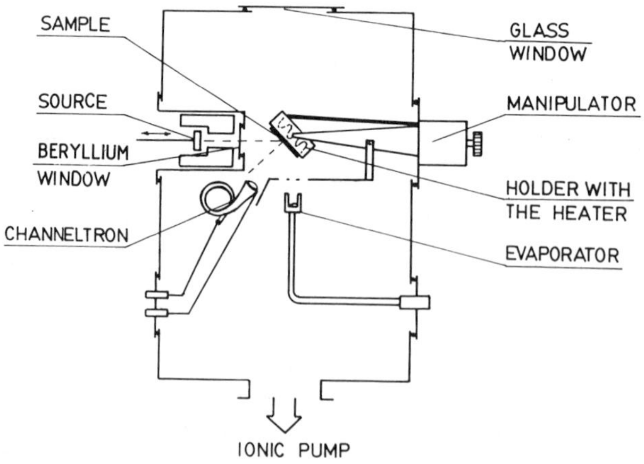

Figure 15. Equipment for conversion electron Mössbauer spectroscopy in ultra-high vacuum combined with a thin film evaporation facility [33].

MEASUREMENTS AT ULTRA-HIGH VACUUM

The channel electron multiplier detectors permit the measurements of CEMS spectra in ultra-high vacuum. Several simple instruments of this type have recently been constructed [33,34]. Figure 15 shows the CEMS configuration combined with a thin film evaporation and heat treatment facility, which permits a relatively clean deposition of iron-57 and for an in-situ CEMS analysis at temperatures up to 800K and at pressures in the range from 10^{-7} to 10^{-9} torr [33]. In more advanced equipment [34], CEMS was supplemented by low-energy electron diffraction, Auger electron spectroscopy, and secondary ion analysis. Because of the efficient recording of secondary electrons and the low background, sensitivity sufficient for the investigation of iron films of a monoatomic layer, 10^{15} atoms/cm^2, was achieved. By using this technique, films of iron-57 as thin as a 3, 10, 15, and 30Å, evaporated onto silicon and nickel substrates and heated to 700K were investigated and interface processes between the films and the substrate were followed up to about 1000K [34]. Recently, using a similar technique Korecki and Gradmann [35] have managed to characterize first two monolayers of Fe on Fe(110) surface.

CEMS IN EXTERNAL MAGNETIC FIELDS

Mössbauer spectroscopy is an important technique in the studies of magnetic properties of materials. In particular, it is capable of determining the magnitudes and directions of the local magnetic fields, at the resonant nuclei, and hence the orientation of magnetic domains near the surface of specimens. The experimental method for CEMS measurements at high magnetic fields of up to 10T was described earlier [4]. Here, a technique, which has been developed recently be de Waard et al. [36], is presented for CEMS measurements at low magnetic field.

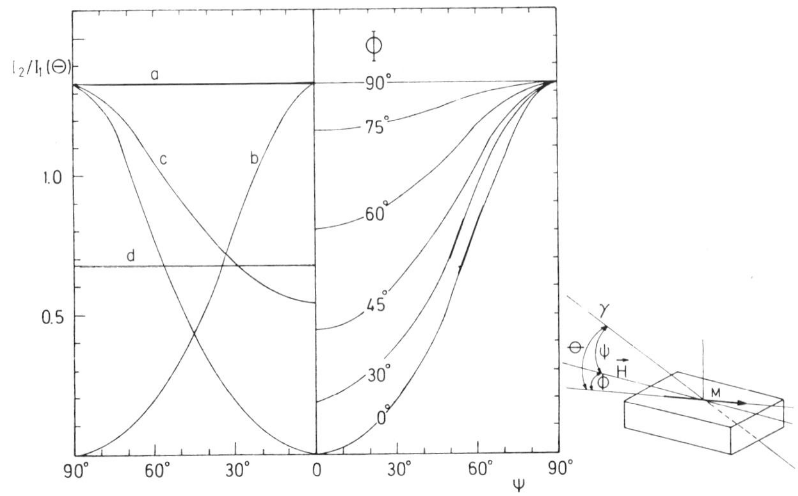

Figure 16. The intensity ratio of the second and first lines in the magnetic spectrum of iron, calculated for various configurations of magnetic moments; magnetization perpendicular to the direction of γ-rays; a, normal to the sample plane; b, for the moments randomly oriented in the sample plane; c, and for an isotropic distribution of moments; d, [36].

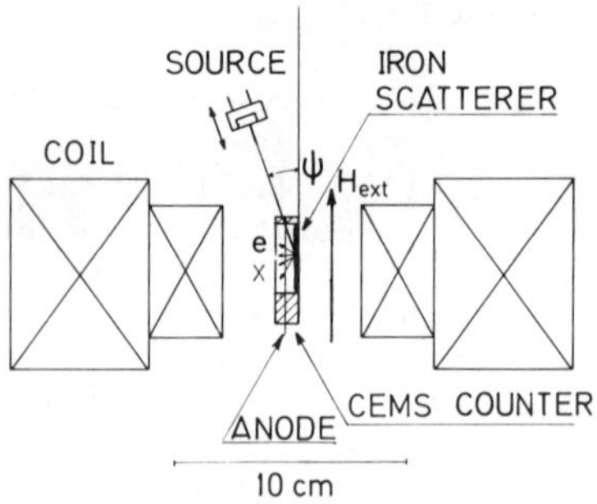

Figure 17. A configuration for near-surface magnetometry with iron-57 CEMS.

As is well known, the relative line intensities, R, in the magnetically-split Mössbauer spectrum of iron-57 depend on the angle, θ, between the local magnetic field and the γ-ray propagation direction and is given by

$$R = \frac{I_2+I_5}{I_1+I_6} = \frac{4\sin^2\theta}{1+\cos^2\theta} \qquad (9)$$

where 1, 2,•••6 stand for consecutive lines in iron-57 sextuplet. By measuring the line intensity ratio, R, as the function of the magnitude of the external magnetic field, one can find the average orientation of the domains and the way in which they rotate towards the direction of the applied field. The magnitude of the ratio, R, as a function of the angle θ, for different configurations of the domains, e.g., axial, normal to the surface plane, planar, or isotropic, is plotted in Figure 16.

The arrangement of the domains lying in the sample plane in a longitudinally applied magnetic field can most easily be traced by measuring the spectra with glancing incidence of γ-rays. However, because equation 9 is even with regard to the angle θ, the measurement of the ratio, R, cannot distinguish between the parallel and antiparallel directions, and, therefore one cannot, for instance, detect the shifts of the walls between two oppositely oriented domains nor the change in the orientation of domains by 180°. Shifts of the walls between nonparallel domains, e.g., shifts of 90° walls, and rotations of the magnetization vector to the direction of the applied field, both reduce the magnitude of the line intensity ratio, so that, at saturation, the ratio is reduced to 0.153 at the incidence angle of 75°. A more complete description of the domain rearrangements, including 180° inversions, can be obtained from Mössbauer spectroscopy with polarized γ-rays, or by perturbed angular correlations of γ-rays, PAC.

The arrangement for magnetometric CEMS measurements is shown in Figure 17. A specially designed proportional counter with the side window makes it to perform the measurements at glancing γ-ray incidence. By using this equipment magnetization curves of closure domains in various samples of iron and in some other iron-containing materials were measured. Several examples are presented in Figure 18.

In thin iron foils and films, as is seen in Figure 18A-C, the magnetization starts to increase as soon as the external field is applied, and the intensity ratio decreases until the value of the saturation magnetization is reached. Various surface treatment processes, such as oxidation, carbonization, or implantation of various ion species, can drastically modify the magnetic properties of the material, particularly by affecting the pattern of closure domains and their response to an external magnetic field. In Figure 18A,B it may be observed that the approach to magnetic saturation in

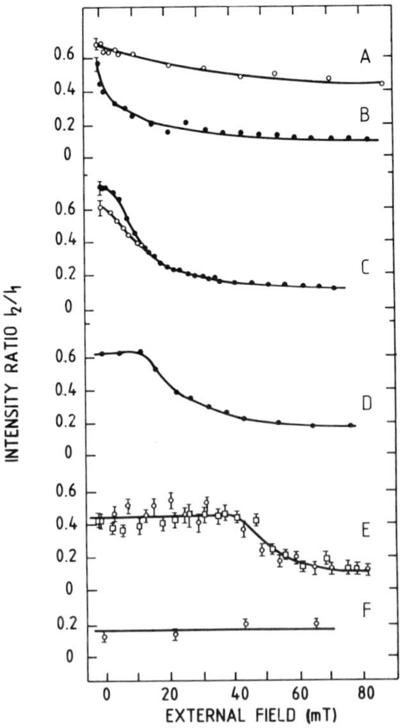

Figure 18. Near-surface magnetization curves measured by means of conversion electron Mössbauer spectroscopy for (A), an iron foil after carborization; (B), the same foil before carbonization; (C), an iron film evaporated on a nickel substrate; o first run, • second run; (D), an iron polycrystal 25x5x1mm^3; (E), an iron monocrystal (110), 15x15x1mm^3 and 4° misorientation; (F), the same but with 1° misorientation.

the carbonized iron foil (A) is much slower than in pure foil (B). Thorough investigation of such effects by CEMS, particularly in ion-implanted materials and in magnetic recording materials, would be of considerable interest.

In iron plates with seizable thickness the situation is different. As is shown in Figure 18D, the magnetization does not change until the external field reaches a certain threshold value. This field corresponds roughly to the demagnetizing field for the specimen, $H_{st}=DM_{sat}$. In addition, in single iron crystals, we observed the magnetization curve to vary very much from sample to sample, depending on the arrangement of the closure domains with respect to the applied field direction. The data in Figure 18E and F for two crystals cut in the (100) orientation, but misaligned within ±1° and ±4°, clearly demonstrated this effect. The sudden growth of R for a misaligned crystal is attributed to the formation of a "tree-like" pattern of closure domains. For more detailed discussion of this subject, the reader is referred to reference 36.

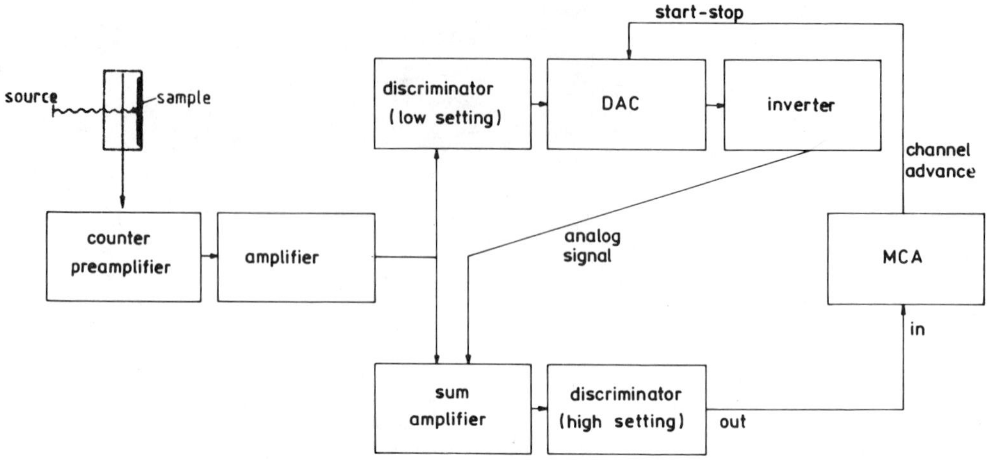

Figure 19. Schematic for the electronics for the rapid, active, acquisition of CEMS spectra using a helium electron counter [38].

RAPID ACCUMULATION OF CEMS SPECTRA

The long Mössbauer spectral count times, typically of the order of several hours per spectrum are frequently a serious obstacle in the use of CEMS, especially when industrial applications are concerned. Therefore, the development of techniques for rapid accumulation of the spectra presents a considerable challenge. The rapid accumulation of transmission Mössbauer spectra can be accomplished by using very strong sources combined with fast counting or current integration technique. However, because of the low rate of scattering events, the fast acquisition or current integration techniques cannot be used in scattering experiments.

An alternative technique for the rapid accumulation of a Mössbauer spectrum, called active acquisition, was suggested by Sawicki and Stanek [37]. Recently we investigated the application of this concept in CEMS and a schematic of the apparatus used in such measurements is shown in Figure 19. The measurements are performed with the use of an ordinary helium-flow proportional counter. However, in addition to the normal "high-level" pulse discriminator, a second, "low-level" discriminator is used and plays a significant part in the recording process. The low-level discriminator is set such that it passes a considerable amount of the noise pulses. Because the amplification of the preamplifier depends on the input load, i.e., on the instantaneous conductance of the counter, the noise amplitude and the noise pulse count rate decreases whenever there are more electrons detected in the counter. Thus, at the velocities (or, time intervals) corresponding to the resonance lines, there are more counts in the high discrimination, normal, branch of the circuit and fewer counts in the low discrimination, auxiliary branch. The signal from the auxiliary branch, after digital-to-analog conversion and inversion, is used to enhance the intensity of the Mössbauer lines in the CEMS spectrum.

This technique permits the measurement of a CEMS in a period of time shorter by an order of magnitude than that required for normal measurements as is illustrated in Figure 20. However, this technique introduces distortions in the line shapes and affects the relative intensities of lines. Also, the statistical significance of the data points in the spectra measured in such a way is not well defined. The described method can be useful,

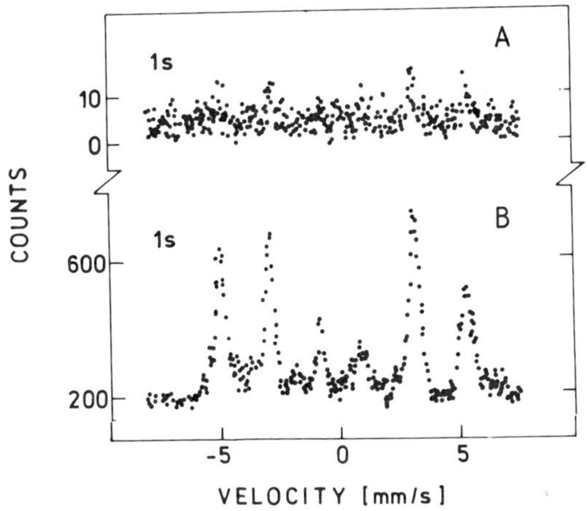

Figure 20. CEMS spectra of iron foil enriched in iron-57 measured simultaneously for one second; (A), by using the conventional counting technique and (B) in the active mode. The source was 10mCi of cobalt-57 in chromium.

especially in cases in which the qualitative analysis of the spectra has to be performed on a short time. Typical applications are to velocity calibration and the fast identification of various iron states in industrial samples.

ION IMPLANTATION STUDIES

Ion implantation is proving to be both a valuable research tool and a tool for the controlled modification of the properties of materials. Current uses of ion implantation extend from studies of metastable phase formation, to the production line fabrication of integrated circuits [39]. The controlled introduction of impurity atoms and defects near the surface of materials provides a valuable method for exploring kinetics and equilibria in metal alloying reactions, for investigation of the formation of new phases, and for the simulation of radiation effects relevant to a nuclear reactor environment.

Conversion electron Mössbauer spectroscopy has already found many applications in studies of ion implantation and modifications of various materials by ion beams. Research in this field was initiated in Cracow about 10 years ago and was published in several survey papers [1,4,7,40,41]. Most of the studies were performed by using iron-57 as a probe, but the applicability of other nuclei, such as tin-119, europium-151, and gold-197, was also recently demonstrated. Several general aspects of these studies and some examples of applications will be briefly discussed below.

IRON-57 IN ELEMENTAL MATRICES

The implantation of iron-57 ions was performed mostly at ion energies of from 20 to 100keV and at doses from 10^{15} to 10^{16}ions/cm^2, which corresponds to the average implantation range of about 10-50nm and an average concentration of iron of about 0.1 to 10 atomic percent. A variety of sites, created both during the implantation and in the subsequent annealing or other post-implantation processes, can be studied. As an example, Figure 21 shows the CEMS spectra for iron-57 implanted in copper, first for the as-implanted

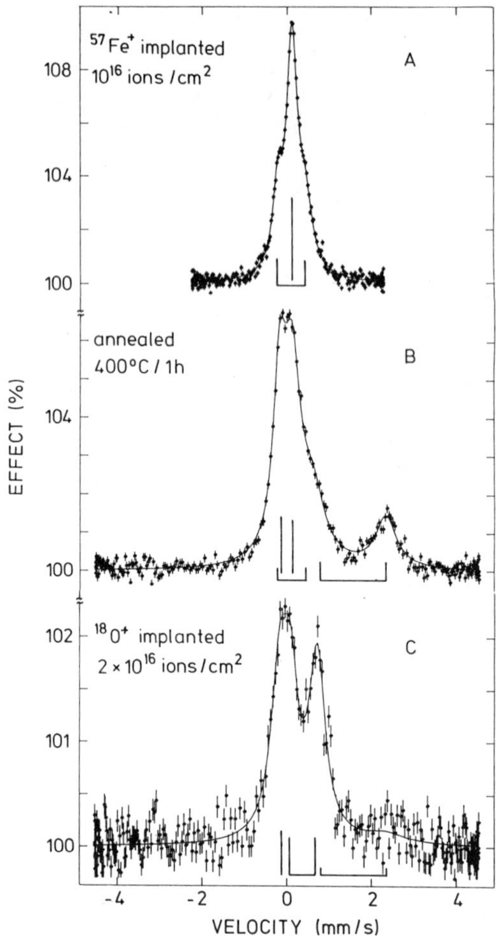

Figure 21. CEMS spectra of iron-57 implanted in copper. The energy of implantation was 30keV with 10^{16} ions/cm^2. (A), as-implanted sample; (B), after annealing at 400°C for 1 hour in the presence of oxygen; (C), after additional implantation with 2×10^{16} ions/cm^2 of O_2^+ at an energy of 10-20keV [42].

sample (A), then annealed in the presence of oxygen (B), and last for the sample post-implanted with O_2^+ (C) [42]. The spectrum for the as-implanted sample consists of a singlet and a doublet corresponding to isolated substitutional iron atoms and isolated iron dimers, respectively, in agreement with the statistical distribution of iron in copper at the given iron concentration. Annealing in the presence of oxygen caused the oxidation of iron to the iron(II) and iron(III) state. The subsequent implantation of O_2^+ ions caused the change in the iron oxidation state to mainly iron(III), which points to an increased oxidation potential within the implanted layer.

From the CEMS spectra of iron-57 implanted in elemental targets, one can extract information on the final impurity location and electronic configuration, the character of the lattice damage and amorphous condition, the nature of the impurity-impurity and impurity-host interactions, the aggregation processes, the presence of metastable phases and the thermal annealing

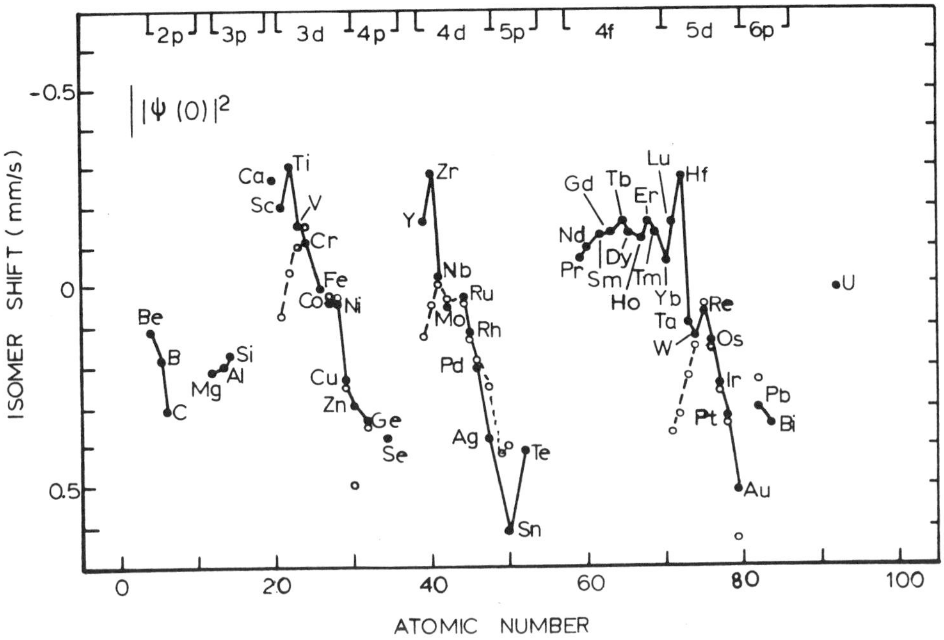

Figure 22. Systematics of the isomer shifts for the one to three atomic percent iron alloys obtained by implantation of iron-57 ions (50-100keV) in various matrices - black points. The open points are for the corresponding alloys obtained by diffusion or alloying at thermal equilibrium.

behavior. In particular, it was found that the site localization and aggregation trends of implanted iron impurities, in various metal matrices, can be described in terms of the Miedema model of alloying, whereas the amorphous components observed in some metals were found to be correlated with the size and electronic configuration mismatch between the iron and host atoms. For the details the reader is referred to references 7 and 41, and to the references contained therein.

Figure 22 presents the recently extended systematics of the isomer shift data for iron impurities implanted into various elemental matrices. Implantations were performed at room temperature with doses of 5×10^{15} and 10^{16} ions/cm^2, and an implantation energy of 50 to 100keV, so that the concentration of iron in the samples was ca. one to three atomic percent. The isomer shift data in Figure 22 represent the dominant component in the spectra, if two or more components were present. In many matrices the isomer shift for the implanted iron-57 differs significantly from the isomer shift for the dilute iron alloy obtained by thermal equilibrium process such as diffusion or melting. In these cases the Mössbauer spectra of the implanted samples closely resemble the spectra of similar samples obtained by athermal methods such as spin-melting, sputtering, or coevaporation. Typical examples of matrices are boron, carbon, silicon, germanium, selenium, tellurium, zirconium, and rare earth matrices. The CEMS spectra of iron in all these matrices display a pronounced quadrupole splitting, indicating a high degree of atomic disorder in the matrix. In particular, it was shown that, for rare earth matrices, a good description of the isomer shift and quadrupole splitting can be obtained by assuming a strain-free packing of atoms in an amorphous closely packed network [43], whereas in carbon, silicon, and germanium, a tetrahedral random network had to be assumed [44].

Significant progress has recently been made in understanding the structural and chemical changes caused by high-dose ion bombardment in compound matrices. Ionic crystals such as LiF and other alkali halides [45-47], and refractory oxides such as MgO [48], TiO_2 [49], garnets [50,51] and olivines [52] were most thoroughly investigated. In addition to the iron-57 CEMS studies, the samples were characterized by optical absorption, electrical microscopy, Rutherford backscattering, and ion channeling measurements. A review of the data was presented by Perez [53]. Several aspects of these studies are worth mentioning here.

The solid phase which is created as a result of ion implantation incorporates impurities in various charge states and electronic configurations, e.g., iron(III), iron(II), and metallic iron(0). Also, various impurity-defect complexes are formed to fulfill the requirement of electric charge equilibrium, e.g., an iron(III)-vacancy in MgO or vacancy-iron(III)-vacancy and iron(II)-vacancy in LiF. Fortunately, because of the distinctly different nature of the various contributing charge states, iron-57 Mössbauer spectra permit the various processes and transformations occurring in the implanted layer to be observed with good resolution.

It was found that the occurrence of various charge states in the as-implanted samples depends on the concentration of the implanted atoms and that it can be well described by a simple statistical model, at least for the cases of MgO and LiF which have been investigated to date. The variation in the occurrence of different charge states of iron in MgO and LiF with dose and hence concentration, as determined from the conversion electron Mössbauer spectra, is presented in Figure 23. The lines represent the probabilities

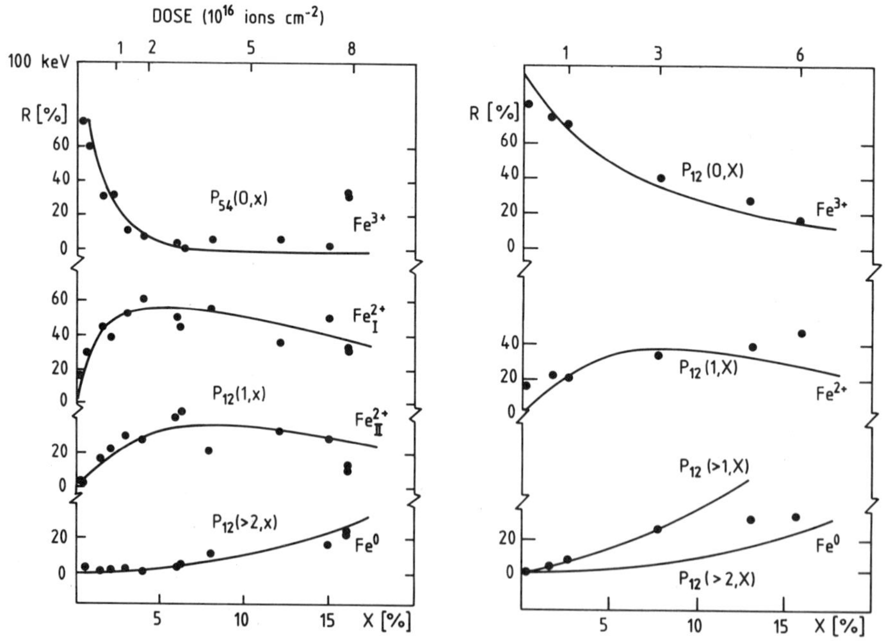

Figure 23. Fractional contribution of various states of iron in MgO (left) and LiF (right) crystals versus the concentration (dose) of the implanted iron ions. Curves represent the probability functions calculated for the binomial distribution [46,48].

calculated by assuming a binomial distribution for the various arrangements of iron atoms such that n sites among N are occupied by iron atoms, i.e.,

$$P_N(n,x) = \frac{N!}{n!(N-n)!}x^n(1-x)^{N-n}, \tag{10}$$

where x is the iron concentration.

Thermal annealing usually removes a great deal of damage and leads, via diffusion and agglomeration, to the growth of various equilibrium states. Some of these can be predicted on the basis of the equilibrium phase diagrams and were correspondingly identified in the implanted specimens. For instance, upon thermal annealing, small precipitates of magnesiowustite, $Mg_{1-x}Fe_xO$, and magnesioferrite, $MgFe_2O_4$ were found in (Fe,Mg)O [48], or pseudobrookite, $FeTi_2O_5$ precipitates in $(Fe,Ti)O_2$ [49]. Because the chemical processes in the implanted layer, and the nature of precipitates, depend critically on the annealing atmosphere and temperature, one can control and stimulate the process, so that the desired properties of the top layer can be

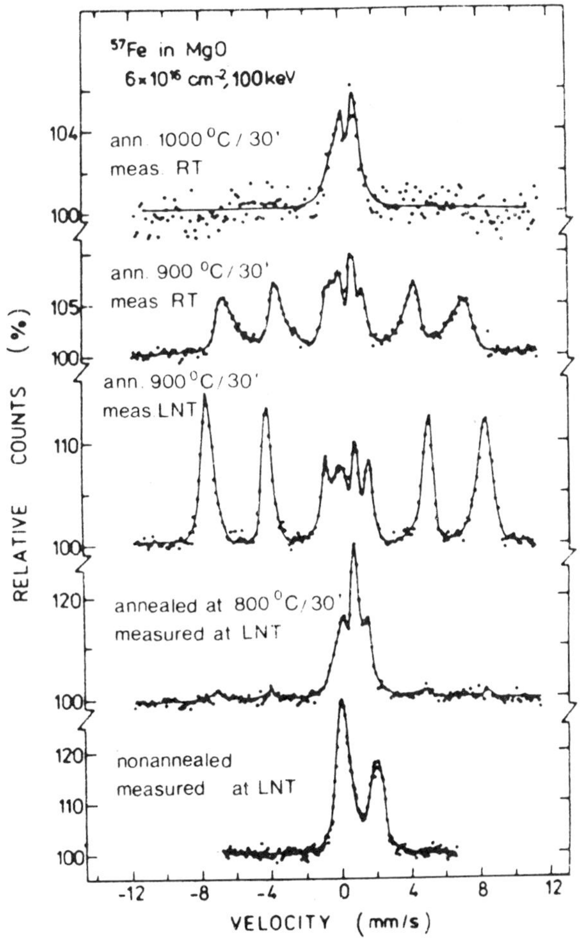

Figure 24. CEMS spectra for the (Fe,Mg)O sample implanted with $6*10^{16}$ ions/cm^2, 100keV, indicating the precipitation and growth of the superparamagnetic particles of magnesioferrite, $MgFe_2O_4$. The annealing and measurement temperatures are indicated. At 1000°C, one observes the dissolution of precipitates in the matrix [48].

obtained. One can also control the reduction-oxidation potential of the top layer by appropriate doping. In addition, it is possible to create well defined crystalline ferromagnetic particles in paramagnetic substrates, for instance, $MgFe_2O_4$ in MgO, or to modify the magnetic properties of the top layer by hydrogen or nitrogen ion implantation in garnets [51]. As an example, several spectra representing various steps of iron precipitation in an MgO matrix, are shown in Figure 24.

IMPLANTS OTHER THAN IRON-57

Only a few CEMS studies have been performed by using implants other than iron-57. Baumvoll, et al. [54,55] used tin-119 CEMS to investigate the influence of tin ion implantation on the oxidation and tribological characteristics of pure iron samples. Binczycka, et al. [56] studied tin-119 implanted into aluminum at 70keV with a dose of $2*10^{16}$ions/cm^2 at temperatures of 4, 78, and 300K. The experiments showed that the sensitivity of tin-119 CEMS is sufficient for measuring the implanted tin-119 at the level of $5*10^{15}$ to 10^{16}ions/cm^2.

The feasibility of CEMS in studying the implantation of europium-151 ions was also demonstrated [26]. In the reported experiment the 70keV Eu(I) ions of europium-151 were implanted into the (100) face of an oriented silicon single crystal, up to the dose of $6*10^{16}$ions/cm^2 of the europium. The spectrum obtained at 78K showed a superposition of europium valence states with the principal line corresponding to europium(II) and a small fraction of ca 5 percent of the europium(III). The measurements were carried out by using a 200mCi SmF_2 source containing samarium-151 and the configuration presented in Figure 9. The first CEMS measurements of implanted gold-197 atoms [57] will be discussed below.

MATERIALS STUDIES FOR CONTROLLED THERMONUCLEAR FUSION

CEMS is well suited for studying materials with practical significance in nuclear power technology. Irradiation and corrosion effects in austenitic and ferritic steels, zircaloys, and other iron alloys, as well as plasma-wall interaction effects and properties of solid breeder ceramics can be investigated. Some of these studies, initiated recently at the Technical University of Munich and the Max Planck Institute for Plasma Physics will be mentioned here.

PLASMA-WALL INTERACTION IN A TOKAMAK

The first problem under study concerns the influence of a hydrogen plasma on the properties of stainless steel. Austenitic stainless steels are widely used in nuclear fission reactors and are also considered as primary materials for the first walls and other structural elements in thermonuclear fusion devices. The high-flux irradiations cause changes in local structure of steel which result in deletorious effects like swelling and the deterioration of its mechanical properties. Irradiation induced phase changes in the first wall material may also have a dramatic influence on the trapping and retention of hydrogen and, therefore, may affect hydrogen isotope recycling and, in particular, the tritium inventory in a deuterium-tritium burning thermonuclear reactor.

In order to explain the nature of structural modifications in austenitic steels under high-flux hydrogen bombardment, we have recently begun CEMS studies of various steel specimens [58] implanted with hydrogen. The AISI-304 stainless steel used as the main material for walls and other components exposed to the plasma in the ASDEX Tokamak was examined. The irradiation

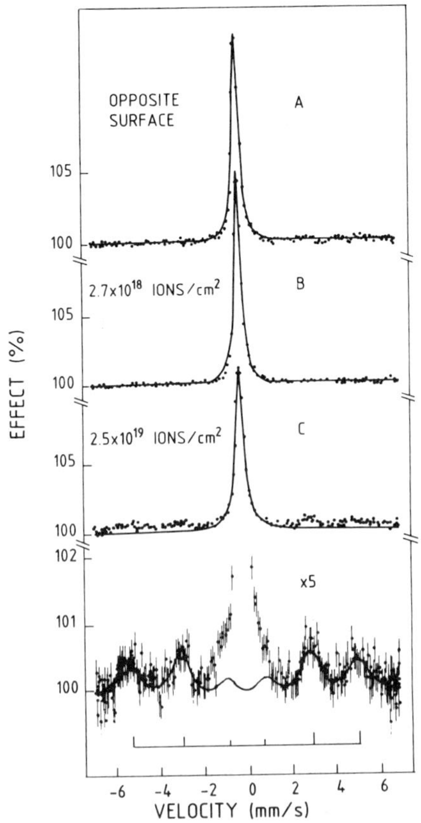

Figure 25. CEMS spectra of AISI-304 stainless steel measured as a function of the fluence of 7keV deuterium ions [58]. (A), the virgin sample; (B), after bombardment with $2.7*10^{18}$ ions/cm^2; (C), after bombardment with $2.5*10^{19}$ ions/cm^2.

energy of 7keV deuterium ions was chosen so that it was comparable with the thermal energy of the plasma near the ignition conditions. It was found that, after bombardment with up to $2.5*10^{19}$ ions/cm^2, an approximate 30 percent fraction of the austenitic γ-phase, represented by a single Mössbauer line, has been converted to the martensitic α-phase. This new phase is ferromagnetic and is represented by the magnetic sextet seen in Figure 25. It is to be noted, that similar phase changes were observed by CEMS in 304 and 316 stainless steels after bombardment at 200°C with 40keV helium(I) ions [59]. In this case, however, it was found that the $\gamma \rightarrow \alpha$ transformation occurred at lower doses, i.e., in the case of 316 stainless steel after the irradiation to a dose of $3*10^{18}$ ions/cm^2.

PLASMA INDUCED EROSION AND REDEPOSITION OF WALL MATERIAL

The second problem under study was connected with the deposition of metal on the insulating components inside of the ASDEX Tokamak. The presence of metal ions in the plasma is one of the most troublesome problems in current controlled fusion experiments, because it increases the effective charge of plasma and causes considerable cooling effects due to emitted radiation. In particular, the presence of 100ppm of iron impurities in the hydrogen plasma will completely prevent the plasma ignition and thus self-sustaining fusion. Thus, investigation of the migration of iron ions in Tokamak presents a subject of considerable significance in fusion technology.

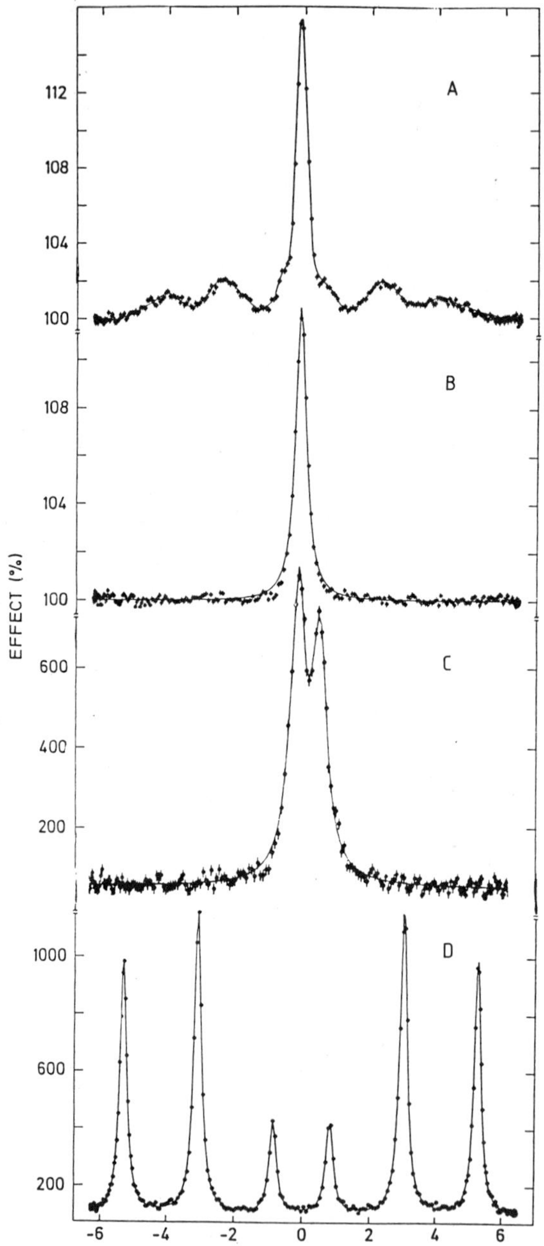

Figure 26. CEMS spectra of some iron containing materials in the ASDEX Tokamak [61]. (A,B), spectra of 304 stainless steel taken before (as machined) and after aging at 900°C, respectively; this steel is a main construction material in the ASDEX; (C), spectrum for iron containing droplets deposited on the graphite tiles of the plasma limiters in the course of plasma discharges; (D), α-iron reference spectrum.

Behrisch et al. [60] investigated the graphite protective tiles from the plasma edge limiters in the ASDEX Tokamak. The scanning electron micrographs taken after 150 plasma discharges showed a considerable number of spheroidal droplets on the surface of the graphite. Their diameter was in the range of 0.1÷1μm. Elemental analysis using proton induced X-ray emission, PIXE, indicated the presence of iron, nickel, and chromium in the composition corresponding to that of 304 stainless steel. However, according to the iron-57 CEMS spectra measured by Sawicka et al. [61], the droplets are not composed of the stainless steel but consist of the alloyed carbide, $(Fe,Cr,Ni)_3C$. A detailed analysis of the CEMS spectra allowed us to make conclusions about the mechanism of wall erosion (mostly sputtering), migration and deposition of the material, as well as about the thermal conditions of the carbides formation in the Tokamak. It was concluded that, because of the high stability of carbides, metal impurities in the plasma can be effectively trapped by using graphite coatings. Some of the CEMS spectra measured in this work are shown in Figure 26.

SOLID BREEDERS

Another area of possible application of CEMS in fusion studies is the characterization of the solid breeder materials. The tritium breeding reaction

$$^{6}Li + n \rightarrow {}^{3}T + {}^{4}He$$

is considered to be a crucial process for ensuring the economical operation of thermonuclear power stations. Mössbauer spectroscopy could answer many questions concerning the behavior of solid lithium compounds, doped with resonant nuclei, such as iron-57 and tin-119, particularly at high radiation damage levels and at high temperatures. Simulation of the interface processes between the breeding substance and the containment vessel can be performed, for instance, by low energy implantation of iron-57 [45-48]. From the Mössbauer measurements, one can obtain useful data about lattice defects and changes in the reduction-oxidation characteristics of these materials, changes which might occur both due to ion bombardment and during thermal treatment at various temperatures and purge gas pressures and flow rates. High temperature corrosion of the containment vessel walls, and in particular at the lithium-iron interface could be investigated by CEMS even at temperatures as high as 1200K, the conditions at which breeders are designed to operate.

CONVERSION ELECTRON MÖSSBAUER SPECTROSCOPY OF GOLD-197

Gold is the only noble metal which displays the Mössbauer effect and the 77.3keV transition in monoisotopic gold-197 provides nearly equal value for structural and chemical analysis as does the iron-57 resonance. The Mössbauer studies of gold-containing specimens have in the past been carried out exclusively in transmission geometry. It has recently been shown that CEMS can be used as a particularly powerful method for gold-197 Mössbauer spectroscopy [27,28]. CEMS spectra for gold films as thin as 5 to 10 monolayers of gold (ca. 5μgAu/cm^2), can now be measured. This should be compared with the thickness of gold absorbers typically required for transmission measurements, which is about 50 to 100mg/cm^2. Thus, gold-197 CEMS has considerable capability in the studying gold-containing microsamples as well as gold-based technological processes of particular significance in molecular electronics and semiconductor technology.

The platinum-197 source with a half-life of 18 hours is produced by thermal neutron irradiation in a reactor of about 20mg of platinum-196 enriched metallic platinum. The activity of about 100mCi is usually obtained

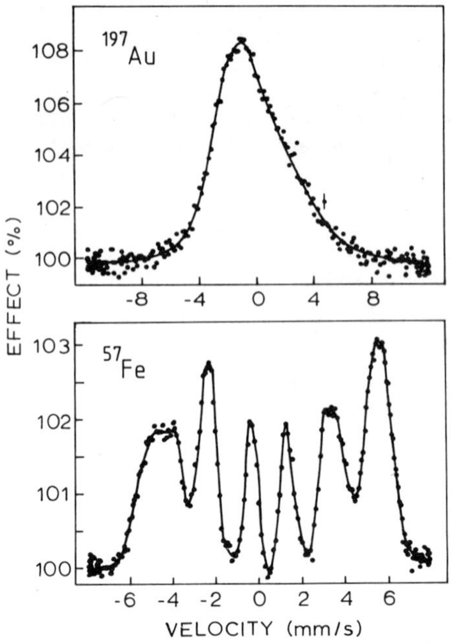

Figure 27. CEMS spectra of gold-197 and iron-57 in spin-glass gold-20 atomic percent iron measured at 4.2K.

and the source can be used without any further radiochemical processing. Because of the low recoil-free fraction, the Mössbauer spectra of gold-197 must be measured at liquid helium temperature. Gold-197 CEMS spectra of various thin-film gold alloys, compounds, and ion-implanted materials have recently been investigated by the author. All measurements were performed by using the channeletron electron detecting device described in Figure 9. Several examples of the experimental data presented below will indicate additional applications of the method.

One of the most intensively studied spin-glass alloys is AuFe [62]. Figure 27 shows the conversion electron Mössbaeur spectra measured by the author for both iron-57 and gold-197 in the alloy with the composition $Au_{0.8}Fe_{0.2}$. The sample was 25μm thick and the iron was enriched with up to 66 percent iron-57. The spectra were measured with the 50mCi cobalt-57 in rhodium source and a 30mCi platinum-197 source for 5 hours in each case. The count rates were 3000 and 1000 counts/sec, respectively. The spectra show rather broad lines with a distribution of hyperfine fields, a quadrupole interaction and an isomer shift characteristic of disordered alloys. It should be noticed, that CEMS spectra of gold-197, representing only 0.1 to 1μm beneath the sample surface, are identical to the transmission spectra of this spectrum which has a sample thickness of 25μm. This proves that the local atomic structure and the spin arrangement in the investigated spin-glass is the same near the surface as in the bulk. As was discussed above, such identity usually does not occur in ordered ferromagnetic materials, because of the presence of closure domains near the surface, but is characteristic of magnetic materials with a spin-glass structure.

Another application of gold-197 CEMS is shown in Figure 28, which presents the spectra of thin films of gold, $Au_{0.25}Sc_{0.75}$, and $Au_{0.1}Cr_{0.9}$ which were vapor deposited on a self-supporting aluminum substrate of 200μg/cm². The composition and homogeneity of the films were examined by Rutherford

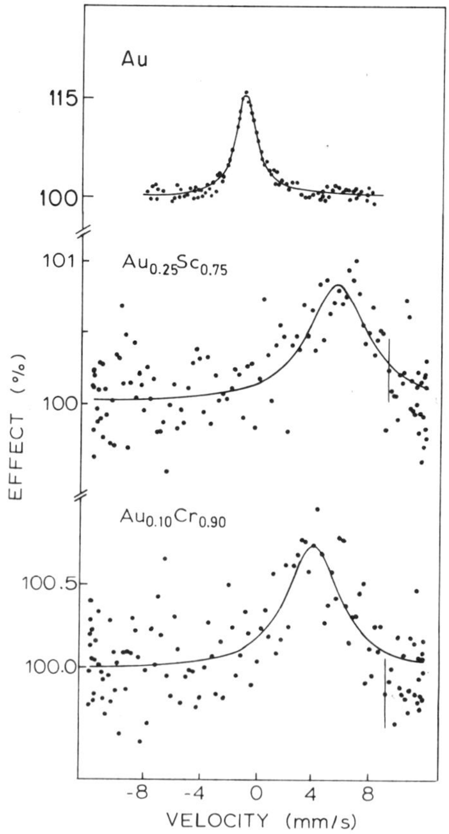

Figure 28. CEMS spectra of gold-197 for gold alloys thin films measured at 4.2K [64].

backscattering of 2MeV α-particles. It was found that the content of gold in both alloy films varies with film depth; it decreases by about 20 percent when going from the top surface of the film to the aluminum substrate. This variation in the gold concentration results in a considerable broadening of the lines in the CEMS spectra of both alloys, as compared with the natural line width observed in gold film. The high values of the gold-197 isomer shift of 5.55(25)mm/sec in $Au_{0.25}Sc_{0.75}$ and of 3.96(21)mm/sec in $Au_{0.1}Cr_{0.9}$, are associated with the high s-electron density on the gold nuclei in these alloys as compared to metallic gold. This is caused by the electron charge transfer from the less electronegative scandium or chromium to the more electronegative gold. The experimental values of the measured isomer shifts are in very good agreement with the values calculated according to the model of Miedema and van der Woude [63] for gold alloys.

The analyzed films have been neutron activated in the high-flux ILL-reactor in Grenoble to produce well defined sources of gold-199 for the precise determinations of the internal conversion coefficients for M, N, and O electron shells in gold by using the on-line β-spectrometer [64]. Combining the values of these coefficients with the isomer shift data will help to calibrate the isomer shift scale for gold-197 and to evaluate the change in the gold-197 nuclear radii during the 77.3keV Mössbauer transition. This information is necessary for a precise determination of the electron density at the gold-197 nuclei in various materials.

From a practical point of view, the two examples discussed above demonstrate the potential of CEMS in studying metallurgical processes in very thin films of new, metastable, gold alloys. Materials of this type usually cannot be obtained by conventional melting but can be produced by using athermal techniques such as spin-melting, coevaporation, sputtering, ion beam mixing, or ion implantation. The characterization of such materials, sometimes produced in the form of thin overlayers on bulk substrates, opens a challenging field of activity for conversion electron Mössbauer spectroscopy.

Another region of a considerable practical significance, which can be successfully pursued with gold-197 CEMS, is associated with the interaction of the gold coverage with the semiconductor substrate, which is crucial for the manufacturing and functioning of many semiconductor devices [65]. By using gold-197 CEMS we have recently begun the investigation of gold-silicon interfaces [28]. The samples were obtained by vapor deposition of metallic gold on the (100) surface of a single crystal wafer, in the form of thin films with thicknesses of 100, 50, 20, 10, and 5μg/cm^2. The CEMS spectra show reasonable resolution even for a sample as thin as 5μg/cm^2 which corresponds to $1.5*10^{16}$atoms/cm^2 of gold (10 monolayers of gold). In fact, the amount of gold in this sample, as seen by the γ-ray beam, was lowered by the beam collimation and inner aperture to ca. 2μg or only $6*10^{15}$atoms of gold. The last number represents the sensitivity attained so far for gold-197 CEMS, but it is still not the ultimate detection limit of the method. By using a platinum-197 source with an activity of 500mCi to 1Ci, one could improve the sensitivity limit to about 10^{15}atoms/cm^2 of gold and approach the detecting capability of one monolayer. The observed dependence of the resonant effect on the thickness of the gold film is shown in Figure 29.

The Mössbauer spectra given in Figure 30 point out that although the presence of the gold-silicon interface causes only minor changes in the values of the isomer shift and linewidth for the as-evaporated films (A), the resultant modifications after laser-treatment (B) or after heating of the samples (C) can be much more significant. Hence, by measuring CEMS spectra one can follow the formation of gold silicides or the diffusion of the gold

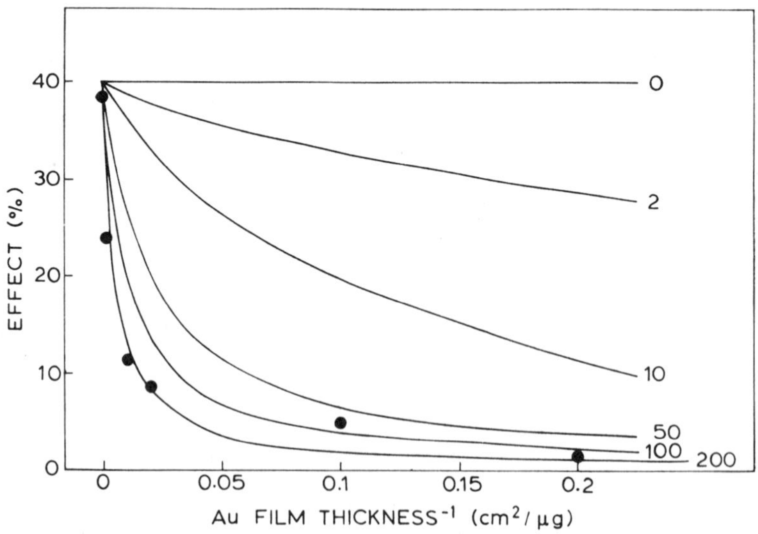

Figure 29. The dependence of the magnitude of the resonant effect in gold-197 CEMS measurements on the thickness of the gold film. Lines represent the influence of the non-resonant background according to equation 7.

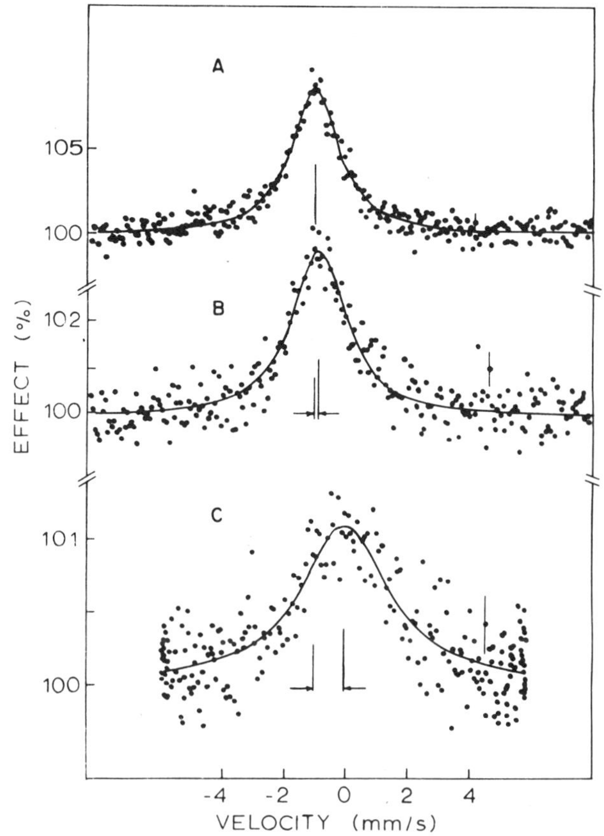

Figure 30. CEMS spectra of 77.3keV gamma-ray transition in gold-197 measured for a 50µg/cm^2 gold film evaporated on the (100) silicon substrate. (A), as-evaporated; (B), after laser irradiation; (C), after heating for 10 minutes at 900°C in vacuum.

into the silicon. These processes are the basis of gold metallization in semiconductor technology.

Figure 31 presents the first applications of gold-197 CEMS in ion implantation studies [57]. The MgO single crystal was implanted with $8*10^{16}$ ions/cm^2 of gold at the energy of 180keV, which corresponds to the mean depth of 100nm and an average concentration of 10 atomic percent. CEMS spectra were measured at 4.2K, for the as-implanted sample and for the sample annealed at 900°C for 33h. The spectrum of the as-implanted sample presents a broad line with the width of 7.6(1.4)mm/s. The observed isomer shift of -1.09(17)mm/s is close to that for metallic gold. It indicates that, although the gold atoms in the as-implanted sample occupy non-unique, defect-associated sites, they do have electronic structure characteristic for metallic gold. In the annealed sample one observes a single line with an isomer shift -1.30(6)mm/s and a linewidth of 2.48(23)mm/s, corresponding to the already well-defined metallic gold precipitates. This finding helped to answer a question about the nature of the gold precipitates in MgO [66]. Electron microscopy, optical, and X-ray studies of the same sample indicate that metallic particles with a face centered cubic structure, which could be ascribed either to metallic gold or to some AuMg alloy precipitates. The value of the isomer shift for the AuMg alloy rises with the magnesium content from -1.23mm/s for pure metallic gold to 3.74mm/s for AuMg and up to 6.3mm/s

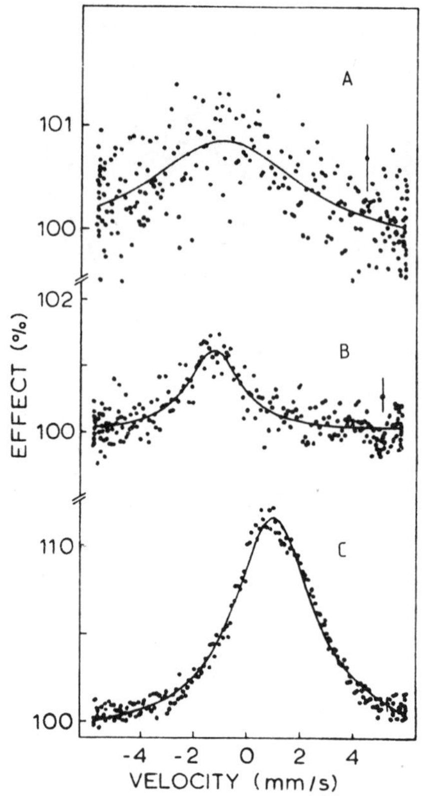

Figure 31. CEMS spectra of gold-197 implanted in MgO crystals measured at 4.2K. (A), the as implanted sample; (B), sample annealed at 900°C for 33 hours in air; (C), spectrum of gold oxide, Au_2O_3.

for gold diluted in magnesium. Thus, CEMS measurements proved that the precipitates are composed of mostly metallic gold, while magnesium is not included. Additional information obtained from CEMS spectra is that the gold atoms have no tendency to oxidize in MgO matrix, because the gold oxide line is typically found at 1.0mm/s (see Figure 31C).

The examples presented above indicate that the gold-197 CEMS studies just initiated, have, similarly to iron-57 and tin-119 CEMS, a considerable potential in studying various microsystems. Further improvement of the experimental technique of gold-197 CEMS is possible and is in progress.

CONCLUDING REMARKS

It can be expected that in the near future the conversion electron Mössbauer spectroscopy will expand in the following directions. Iron-57 CEMS will become a standard tool for near-surface characterization of iron-containing materials, routinely applied in the materials science of phase analysis, identification of impurity states, studies of thermochemical effects, nondestructive and remote testing, corrosion, ion implantation and beam mixing, amorphous thin films, near-surface magnetism, etc. The development of experimental techniques for CEMS will continue and improved spectrometers dedicated both for depth integral and depth selective measurements will be built. The use of multichannel array detectors and advances in rapid spectra acquisition will be very helpful. Resonances other than iron-57 and tin-119 will be used more often. CEMS analysis will be combined more closely with other techniques, such as electron microscopy, low energy electron diffraction, and energy loss spectrometry, Auger electron spectroscopy, ESCA, X-ray diffraction methods, EXAFS, proton-induced X-ray emission, Rutherford backscattering, ion channeling, and nuclear reaction analysis. Advanced ultra-high vacuum systems for combined measurements with these methods and CEMS will become available.

ACKNOWLEDGMENTS

Many results described in this chapter originated in the years 1975 to 1982, at the Jagiellonian University in Cracow. The work has been considerably extended during the author's visits at the University of Groningen in 1983 and at the Technical University of Munich in 1984. The collaboration with many colleagues and in particular with Drs. B. D. Sawicka, J. Stanek, T. Tyliszczak, G. Marest, A. Perez, H. de Waard, G. M. Kalvius, F. E. Wagner, R. Behrisch, J. Auleytner, O. Gzowski and last, but not least, A. Z. Hrynkiewicz was most fruitful and is greatly appreciated.

REFERENCES

1. B. D. Sawicka and J. A. Sawicki, Proc. Int. Conf. on Mössbauer Spectroscopy, Vol. 2, p. 35, eds., D. Barb and D. Tarina, Bucharest, 1977.

2. W. Jones, J. M. Thomas, R. K. Thorpe, and M. J. Tricker, Appl. Surf. Sci. 1, 338 (1978).

3. M. J. Tricker, Proc. Conf. on Chemical Applications of Mössbauer Spectroscopy, eds. J. G. Stevens and G. Shenoy, Advances in Chemistry Series 491, 63 (1980).

4. B. D. Sawicka and J. A. Sawicki, in Topics in Current Physics, Vol. 25, ed. U. Gonser, Springer Verlag, p. 139 (1981).

5. D. Liljequist, Scanning Electron Microscopy III, 997 (1983).

6. J. A. Sawicki and B. D. Sawicka, Hyperfine Interactions 13, 199 (1983).

7. J. A. Sawicki, Materials Science and Engineering 69, 501 (1985).

8. Conversion Electron Mössbauer Spectroscopy Handbook, eds. J. G. Stevens and V. E. Stevens, Mössbauer Effect Data Center (1982).

9. E. Kankeleit, Z. Phys. 164, 442 (1961).

10. Zw. Bonchev. A. Jordanov, and A. Minkova, Nucl. Instr. Meth. 70, 36 (1970).

11. K. R. Swanson and J. J. Spijkerman, J. Appl. Phys. 41, 3155 (1969).

12. S. Valkealahti and R. M. Nieminen, Appl. Phys. A32, 95 (1983).

13. T. Toriyama, K. Asano, K.Saneyoshi, and K. Hisatake, Nucl. Instr. Methods, 232, 170 (1984).

14. J. Itoh, T. Toriyama, K. Saneyoshi, and K. Hisatake, Nucl. Instr. Meth. 205, 279 (1983).

15. D. Liljequist, M. Ismail, K. Saneyoshi, K. Debusmann, W. Keune, R. A. Brand, and W. Kiauko, Phys. Rev. B31, 4137 (1985).

16. W. Keune, K. Saneyoshi, K. Debusmann, R. A. Brand, and D. Liljequist, Chapter 6, this volume

17. C. Briancon, B. Legrand, R. J. Wallen, Ts. Vylov, A. Minkova, and A. Inoyatov, Nucl. Instr. Methods 221, 547 (1984).

18. J. A. Sawicki and T. Tyliszczak, Nucl. Instr. Methods 216, 501 (1983).

19. H. Bokemeyer, K. Wohlfahrt, E. Kankeleit, and D. Eckardt, Z. Phys. A274, 305 (1975).

20. R. Oswald and M. Ohring, J. Vac. Sci. Tech. 13. 40 (1976).

21. S. Skrzypek, E. Kalawa, J. A. Sawicki, and T. Tyliszczak, Material Science and Engineering 66, 145 (1984).

22. O. Massenet, Nucl. Instr. Methods 144, 225 (1977).

23. J. A. Sawicki, Nucl. Instr. Methods 152, 577 (1978).

24. J. A. Sawicki, T. Tyliszczak, and O. Gzowski, Nucl. Instr. Methods 190 (1981).

25. J. A. Sawicki and T. Tyliszczak, Proc. Inter. Conf. on the Applications of the Mössbauer Effect, Jaipur, India, Vol. 2, p. 800 (1981).

26. J. A. Sawicki, T. Tyliszczak, B. D. Sawicka, and J. Kowalski, Physics Letters A91, 414 (1982).

27. J. A. Sawicki, T. Tyliszczak, J. Stanek, B. D. Sawicka, and J. Kowalski, Nucl. Instr. Methods 215, 567 (1983).

28. J. A. Sawicki and B. D. Sawicka, Hyperfine Interactions 15/16, 1013 (1983).

29. J. A. Sawicki, Nucl. Instr. Methods, B (1986), in press.

30. J. A. Sawicki, B. D. Sawicka, and F. E. Wagner, to be published.

31. Y. Isozumi, M. Kurakado, and M. Katano, Rev. Sci. Instr. 180, 413 (1981).

32. J. Kowalski, J. Stanek, T. Tyliszczak, and J. A. Sawicki, Nucl. Instr. Methods 216, 299 (1983).

33. T. Tyliszczak, A. Budkowski, and J. A. Sawicki, Phys. Stat. Solidi (a) 96 (1986), in press.

34. T. Tyliszczak, J. A. Sawicki, and W. Wilk, Hyperfine Interactions 15/16, 1001 (1983).

35. J. Korecki and U. Gradmann, Phys. Rev. Letters, in press.

36. H. de Waard, J. A. Sawicki, B. D. Sawicka, F. Pleiter, and K. Frentz, J. Magn. Magnetic Materials, 49, 50 (1985).

37. J. A. Sawicki and J. Stanek, J. Physique 40, C-258 (1979).

38. J. Stanek and J. A. Sawicki, to be published.

39. S. T. Picraux, Ann. Rev. Mater. Sci. 14, 355 (1984); Physics Today 37, 38 (1984).

40. B. D. Sawicka, Nucl. Instr. Methods 182/183, 1039 (1981).

41. B. D. Sawicka and J. A. Sawicki, Nucl. Instr. Metods 209/210, 799 (1983).

42. W. Gries, B. D. Sawicka, and J. A. Sawicki, Nucl. Instr. Methods, to be published.

43. B. D. Sawicka, E. Letocha, J. A. Sawicki, T. Tyliszczak, and H. Binczycka, Physics Letters 100A, 45 (1984).

44. B. D. Sawicka and J. A. Sawicki, Nucl. Instr. Methods 194, 465 (1982).

45. J. Stanek, J. Kowalski, B. D. Sawicka, and J. A. Sawicki, Nucl. Instr. Methods 182/183, 499 (1981).

46. J. Kowalski, G. Marest, A. Perez, B. D. Sawicka, J. A. Sawicki, J. Stanek, and T. Tyliszczak, Nucl. Instr. Methods 209/210, 1145 (1983).

47. J. Kowalski, J. Stanek, T. Tyliszczak, B. D. Sawicka, J. A. Sawicki, and M. Drwiega, Hyperfine Interactions 15/16, 499 (1983).

48. A. Perez, G. Marest, B. D. Sawicka, J. A. Sawicki, and T. Tyliszczak, Phys. Rev. B28, 1227 (1983).

49. M. Guermazi, G. Marest, A. Perez, B. D. Sawicka, J. A. Sawicki, P. Thevenard, and T. Tyliszczak, Nucl. Instr. Methods 209/210, 1145 (1983).

50. P. J. Picone and A. H. Morrish, J. Appl. Phys. 53, 247 (1982).

51. A. Perez, G. Marest, P. Gerard, M. Madare, and P. Martin, Nucl. Instr. Methods 209/210, 1179 (1983).

52. S. Massouh, A. Perez, and J. Serughetti, Phys.Rev. B., in press.

53. A. Perez, Nucl. Instr. Methods B1, 621 (1984).

54. I. J. R. Baumvol, J. Appl. Phys. 52, 4583 (1981).

55. I. J. R. Baumvol, G. Longworth, I. W. Backer, and R. E. J. Watkins, Hyperfine Interactions 10, 1123 (1981).

56. H. Binczycka, M. Drwiega, M. Miczko, J. A. Sawicki, J. Stanek, and T. Tyliszczak, Materials Science and Engineering 69, 519 (1985).

57. J. A. Sawicki, G. Abouchacra, J. Serughetti, and A. Perez, Nucl. Instr. Methods, B16, 355 (1986).

58. P. Børgesen, B. D. Sawicka, J. A. Sawicki, and B. M. U. Scherzer, J. Appl. Phys., to be published.

59. H. Hayashi, I. Sakamoto, and T. Takahashi, J. Nucl. Materials 128/129, 756 (1985).

60. R. Behrisch, P. Børgesen, J. Ehrenberg, B. M. U. Scherzer, B. D. Sawicka, J. A.Sawicki, and ASDEX-team, J. Nucl. Materials 128/129, 470 (1985).

61. B. D. Sawicka, J. A. Sawicki, and R. Behrisch, Materials Science and Engineering 69, 45 (1985).

62. R. J. Borg and D. N. Pipkorn, J. Appl. Phys. 40, 1483 (1969).

63. A. R. Miedema and F. van der Woude, Physica 100B, 145 (1980).

64. K. Schreckenbach, J. A. Sawicki, P. Maier-Komor, G. M. Kalvius, and F. E. Wagner, Hyperfine Interactions, 27, 701 (1986).

65. V. G. Keramides, Thin Solid Films 96, 347 (1982).

66. G. Abouchacra, A. Chassagne, and J. Serughetti, Rad. Effects 64, 189 (1982).

SURFACE PHASE ANALYSIS BY DEPTH-SELECTIVE CONVERSION ELECTRON MÖSSBAUER SPECTROSCOPY

K. Saneyoshi*, K. Debusmann, W. Keune and R. A. Brand

Laboratorium für Angewandte Physik
Universität Duisburg
Lotharstr. 1, D-4100 Duisburg
Federal Republic of Germany

D. Liljequist

Department of Physics
University of Stockholm
Vanadisvägen 9, S-11346 Stockholm, Sweden

Dedicated to Professor Edgar Lüscher at the occasion of his 60th anniversary

INTRODUCTION

Depth-selective iron-57 conversion electron Mössbauer spectroscopy, DCEMS, is a rapidly developing technique for the nondestructive investigation of iron-57 containing surface layers of materials [1]. The potential of DCEMS lies in the unique possibility of studying, to a certain extent and resolution, local physical properties as a function of depth below the surface, without relying on destructive and possibly disturbing methods such as etching or sputtering off surface layers. In DCEMS, electrons emerging from the surface of a Mössbauer absorber are detected by means of an electron spectrometer of several percent energy resolution, and Mössbauer electron backscattering spectra are obtained at various selected electron energies. Mono-energetic conversion electrons are emitted after the deexcitation of Mössbauer nuclei (e.g., 7.3keV K-conversion electrons in the case of iron-57), and their energy loss is related to their depth of origin below the surface of the Mössbauer absorber. In this way, information on the depth dependence of local physical parameters can be obtained, DCEMS can, in principle, be applied to any sample of fundamental or technological interest which contains a sufficient amount of iron-57.

It has been demonstrated experimentally and theoretically that depth-selective surface layer studies in a region of 0-1000Å are feasible by using iron-57 K-conversion electrons [2,3]. In these investigations [2,3] it has been shown that measurements at different electron emission angles relative

*Present Address: Department of Energy Science, Tokyo Institute of Technology, 4259 Nagatsuta-cho, Midori-ku, Yokohama 227, Japan

to the absorber surface normal (i.e., angular resolved DCEMS) can give additional information on depth-dependent properties. Thus DCEMS measurements at extreme glancing angles ($\theta \sim 80^0$-90^0) show enhanced surface sensitivity, and selective observation of a surface layer with a mean depth of less than 10Å appears to be possible [2]. A description and a comparison of various electron spectrometers currently being used for DCEMS have been given recently [4]. So far DCEMS has found application in various fields of materials science, such as oxidation and corrosion [5-10], ion implantation [11-13], thin film interdiffusion [14], and thin film and surface magnetism [15-20].

We have used the DCEMS technique to analyze surface layers of bcc iron ion-implanted with nitrogen ions of relatively low incident ion energy. Ion implantation of N_2^+ ions has been shown to provide a new method for improving mechanical surface properties, i.e., the wear resistance and fatique behavior of engineering components [21]. Usually ion-implanation is performed with high-energy ions above 10keV, and the existing theory [22] for the implantation range calculations has been shown to be applicable above 10keV [23]. We wish to determine the interaction of low-energy (less than 10keV) light ions with surfaces. The present work therefore is part of a study of the depth distribution of low-energy nitrogen ions (4keV in polycrystalline bcc iron. It is a unique feature of DCEMS that, in principle, the depth distribution of iron nitride phases (which might be formed during the implantation process) can be determined nondestructively in a finger-print type manner via their different Mössbauer spectral parameters.

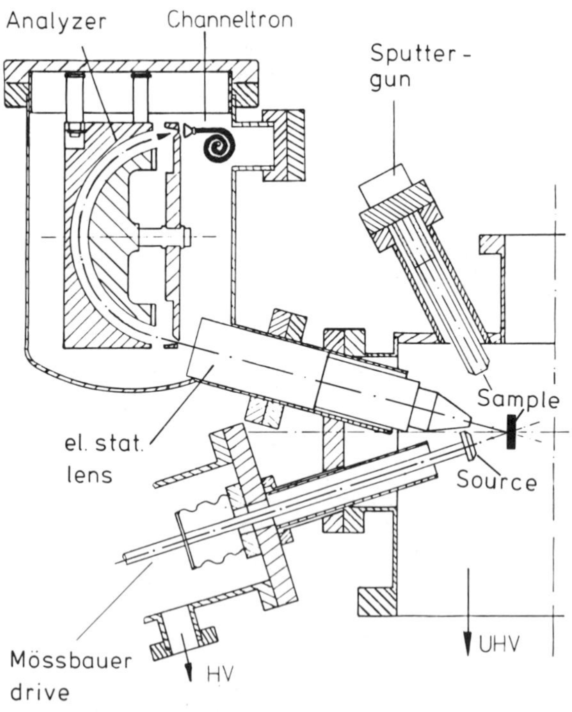

Figure 1. Schematic view of the electron spectrometer for DCEMS studies showing the 150° spherical-sector electron analyzer, electrostatic inlet lens, Mössbauer absorber (sample), cobalt-57 Mössbauer source and ion sputter gun, all parts being placed in an ultra high vacuum.

In this paper the notations used in references 2 and 3 will be followed.

EXPERIMENTAL

Apparatus

The DCEMS measurements were carried out with a modified commercial electrostatic electron analyzer (150° spherical sector analyzer) combined with conventional Mössbauer instrumentation. The apparatus has been described in more detail elsewhere [24]. Figure 1 and Figure 2 show the experimental arrangement for the present measurements. It consists of an ca. 80mCi initial activity cobalt-57 in rhodium Mössbauer source and a Mössbauer absorber which can be tilted in ultra-high vacuum so as to make different angles, θ, relative to the entrance of the electron spectrometer and hence different angles relative to the γ-ray direction, all parts being placed in an ultrahigh vacuum with a base pressure 5×10^{-11} torr. The width Δθ of the angular interval selected by the electron analyzer is approximately ±6°. The correspondingly small solid angle results in very low electron count rates which are typically several counts per hour per channel when 512 channels are used. However, because the electron analyzer acts as a filter for background electrons and because the noise count-rate of the channeltron detector is extremely low and very stable over periods of several months in an ultra high vacuum, DCEM spectra can be measured with reasonable statistics within a period of one to several days per spectrum if samples 95 percent enriched in iron-57 are used.

The spectrometer profile S(V,E), i.e, the spectral shape on the spectrometer energy setting (V-) axis for monoenergetic electrons of energy E, was determined by a calibration procedure as described in reference 3, and was found to be of triangular shape with a FWHM energy resolution of 2.7 percent. The calibration constant V/E (nominal spectrometer energy setting/electron energy) was found to be 1.007. Both quantities, S(V,E) and V/E, must be known accurately in the practical DCEMS analysis.

Sample Preparation

The target for the ion implantation was a 12μm thick polycrystalline α-iron foil 95.1 percent enriched in iron-57. The foil was mounted to the sample holder in the ultrahigh vacuum chamber, and its surface was cleaned of oxides and other contaminants by argon ion sputtering at oblique incidence (45° relative to the sample normal at 5keV for 30min) prior to the nitrogen implantation. The subsequent N_2^+ implantation was carried out by using the same ion sputter gun as for the surface cleaning procedure, but by flooding

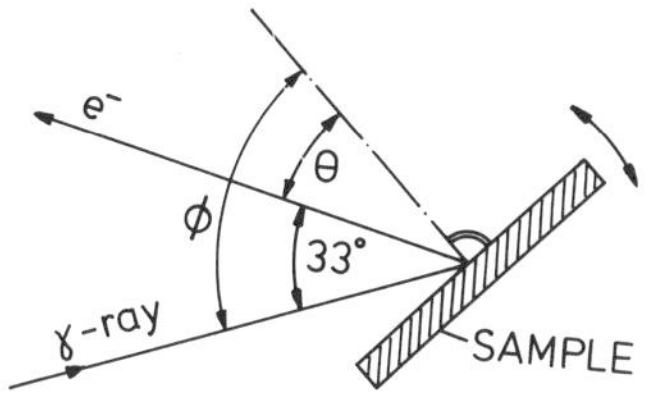

Figure 2. Experimental geometry chosen for the present investigation. θ is the angle between electron emission direction and sample surface normal. ϕ is the angle between γ-ray direction and sample surface normal.

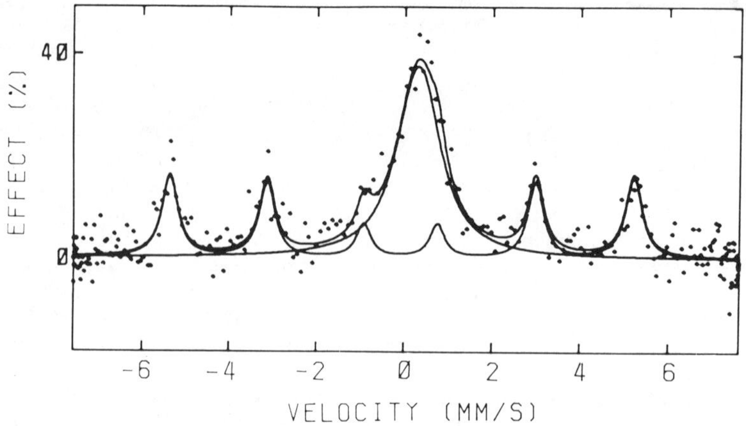

Figure 3. DCEM spectrum at room temperature for electron energy setting V = 585eV and almost perpendicular electron emission (θ=6°) obtained from α-iron-foil implanted with a nominal dose of $3.2x10^{17}$ N_2^+ ions/cm^2 (implantation energy 4keV).

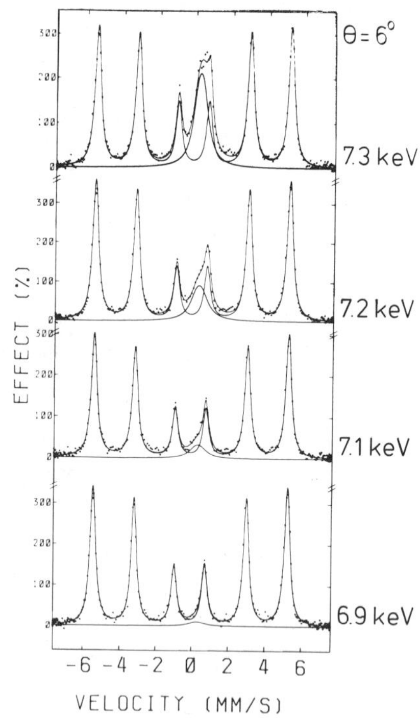

Figure 4. DCEM spectra at room temperature at θ=6° (almost perpendicular electron emission) for electron energy settings V of 7.3, 7.2, 7.1 and 6.9keV, respectively, (from top), obtained from an α-iron-57 foil implanted with a nominal dose of $3.2x10^{17}$ N_2^+ ions/cm^2 (implantation energy 4keV).

the ultrahigh vacuum chamber with dry nitrogen gas (5N purity) at $8x10^{-5}$ torr instead of with argon gas. The N_2^+ ion implantation energy was 4keV (current density $14\mu A/cm^2$) at almost perpendicular ion incidence (6^0 relative to the surface normal). The nominal dose chosen was $3.2x10^{17}$ N_2^+ ions/cm^2. Argon ion surface cleaning, nitrogen implantation, and all subsequent DCEMS measurements were performed in situ. The pressure during the DCEMS experiments was $2x10^{-10}$ torr. All DCEMS measurements were performed with the sample at room temperature.

EXPERIMENTAL RESULTS

Figure 3 shows a DCEM spectrum of the nitrogen implanted sample measured at perpendicular electron emission (i.e., $\theta=6^0$) with a selected electron energy setting of only 585eV. At this energy setting, we observed one of the LMM-Auger electron peaks of iron in the photoelectron spectrum emitted from our sample. The typical escape depth of these low-energy Auger electrons of iron is about 10-20Å [25]. Therefore, the DCEM spectrum shown in Figure 3 partially consists in an Auger electron Mössbauer spectrum (AEMS) which gives average information upon a surface layer of ca. 20Å depth (due to LMM-Auger electrons). We identify this surface spectral contribution with the (broadened) central emission line visible in Figure 3. However, because the discriminated LMM-Auger line of iron (at 585eV) was found to be superimposed upon a background tail of inelastically scattered K-conversion electrons and KLL-Auger electrons (of 5.4keV original energy) which both originate from

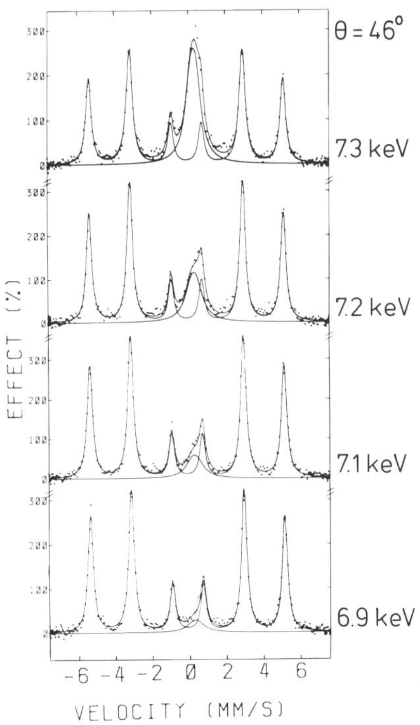

Figure 5. DCEM spectra at room temperature at $\theta=46^0$ for electron energy settings V of 7.3, 7.2, 7.1 and 6.9keV, respectively (from top), obtained from an α-iron foil implanted with a nominal dose of $3.2x10^{17}$ N_2^+ ions/cm^2 (implantation energy 4keV).

very large depths below the surface and consequently have experienced very large energy losses, the AEM spectral contribution in Figure 3 is superimposed upon a substrate Mössbauer spectrum (visible in Figure 3 as a six-line spectrum) which is caused by these highly inelastically scattered electrons. We attribute the central line in Figure 3 to some paramagnetic nitride phase existing at the surface and in the subsurface region.

A series of DCEM spectra were measured at angles $\theta=6°$, $46°$, $60°$ and $72°$, and in each case, at spectrometer settings V = 7.3, 7.2, 7.1, 6.9 and (in 2 cases) 6.6keV (nominal energy). Thus the measurements involve a total of 18 Mössbauer spectra. The measuring time per spectrum ranged between 20 and 67 hours, depending on the spectrum. Typical sequences of DCEM spectra obtained at $\theta=6°$ (almost perpendicular electron emission) and $\theta=46°$ for energy settings between 7.3 and 6.9keV are shown in Figure 4 and Figure 5, respectively. Each DCEM spectrum (including the AEM spectrum of Figure 3) was least-squares fitted to a Lorentzian six-line spectrum due to α-iron superimposed upon a central Lorentzian quadrupole-split line (with small quadrupole splitting) due to paramagnetic iron nitride. Clear experimental evidence for depth selectivity is found from the spectra in Figure 4 and Figure 5. Thus the central nitride peak is clearly enhanced relative to the six-line α-iron spectrum if electrons of high energy (7.3keV) are selected (which originate from regions closer to the surface), while decreasing the electron energy reduces the central nitride signal and enhances the α-iron spectral intensity. It is remarkable that the nitride signal intensity has almost vanished

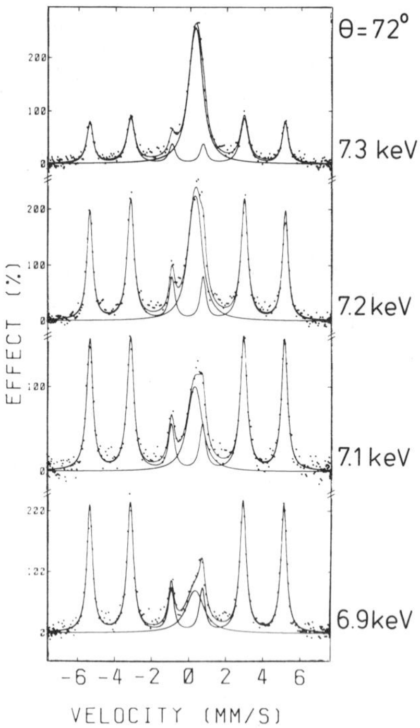

Figure 6. DCEM spectra at room temperature at $\theta=72°$ (near glancing angle electron emission) for electron energy settings V of 7.3, 7.2, 7.1 and 6.9keV, respectively (from top), obtained from an α-iron-57 foil implanted with a nominal dose of $3.2x10^{17}$ N_2 ions/cm^2 (implantation energy 4keV).

Table I. Mössbauer Spectral Parameters for Paramagnetic Iron Nitrides

Nitride	I.S.[a] (mm/s)	ΔE_Q[b] (mm/s)	Γ[c] (mm/s)	Reference
ζ-Fe_2N	0.41	0.28	0.29	26
ε-Fe_2N	0.40	0.27	~0.3	27
Surface Nitride	0.39	0.28	0.7-0.9	Present work

[a]Isomer shift relative to α-iron at 295K. [b]Quadrupole splitting, $1/2e^2qQ$. [c]Line width (FWHM) of Lorentzian lines.

at 6.9keV in Figure 4 (almost perpendicular electron emission), indicating qualitatively that the nitride is located within a thin surface layer region, in agreement with the interpretation of the Auger electron Mössbauer spectrum of Figure 3. Another sequence of DCEM spectra, but measured at θ=72°, i.e,, near glancing electron emission, is shown in Figure 6. Figure 7 shows a

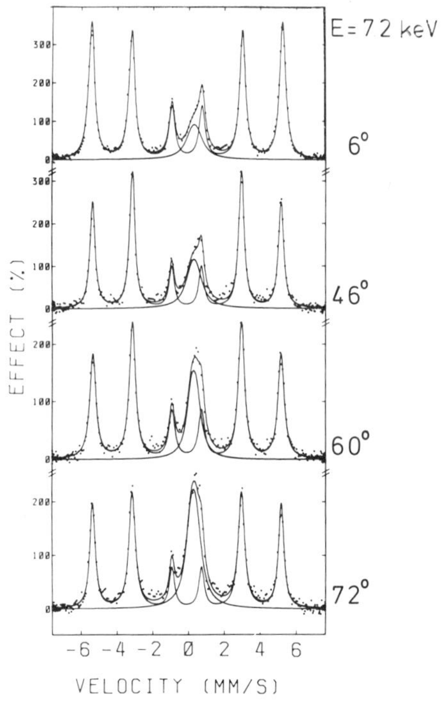

Figure 7. DCEM spectra at room temperature at electron energy setting, V = 7.2keV, for electron emission angles θ of 6°, 46°, 60°, and 72°, respectively (from top), obtained from an α-iron-57 foil implanted with a nominal dose of 3.2×10^{17} N_2^+ ions/cm^2 (implantation energy 4keV).

Table II. Experimental DCEMS Signals a (in Arbitrary Units) for Iron Nitride (Fe_2N) and α-Iron.

θ	V	$I^{Nitr.}$	I^{Fe}	θ	V	$I^{Nitr.}$	I^{Fe}
6°	7.3	33.1(0.6)	116.3(14.3)	46°	7.3	32.5(0.2)	62.2(2.8)
	7.2	19.2(0.2)	139.3(3.7)		7.2	17.5(1.0)	74.9(3.8)
	7.1	8.4(0.2)	133.3(3.6)		7.1	9.5(0.2)	87.0(14.7)
	6.9	2.1(0.25)	113.8(10.6)		6.9	4.4(1.0)	75.9(4.4)
	6.6	1.3(0.1)	89.6(1.3)				
60°	7.3	37.1(0.7)	56.5(3.6)	72°	7.3	34.5(0.5)	28.4(3.1)
	7.2	19.3(0.7)	68.8(6.9)		7.2	21.7(0.4)	44.1(1.8)
	7.1	12.3((0.2)	73.1(3.1)		7.1	13.2(0.5)	43.1(1.9)
	6.9	5.1(0.1)	68.2(2.5)		6.9	8.4(0.7)	40.7(2.6)
					6.6	3.4(0.6)	35.5(2.3)

[a] Normalized to the corresponding measuring time and Mössbauer source activity, and corrected for nonresonant background, and obtained at different angles θ and electron energy settings V, given in keV. The errors given in parenthesis are statistical errors

sequence of DCEM spectra obtained at V=7.2keV for different angles θ. It is clear from Figures 6 and 7 that the nitride signal is drastically enhanced relative to the α-iron spectral intensity if measurements at glancing electron emission (θ=72°) are performed. Also it is worthwhile to mention that an integral CEMS measurement performed by removing our sample from the UHV system after the DCEMS experiments and placing it into a conventional $He\text{-}CH_4$-proportional counter resulted in a CEM spectrum similar to that shown in Figure 5 for 6.9keV and θ=46°. DCEMS, therefore, is much more surface sensitive than integral CEMS, if high electron energies and glancing emission angles are selected.

The computer analysis of our spectra mentioned above gave consistent results and fairly satisfactory least-squares fits for all measured spectra. In Table I the Mössbauer parameters for the nitride phase as obtained from the fits are given, together with literature values [26,27] for paramagnetic iron nitrides. Our parameters are in very good agreement with those of bulk ζ-Fe_2N or ε-Fe_2N, indicating that an iron nitride phase very close to the Fe_2N composition (i.e., an iron nitride with the highest nitrogen content) has been formed during implantation. Table I shows, however, that the line width of our Fe_2N phase is a factor of two to three larger than the literature values for bulk Fe_2N. A possible explanation for this difference in line width could be strong structural damage in our Fe_2N caused by the heavy dose of implanted ions which could even lead to amorphization of Fe_2N during the implantation process. Fe_2N has been discovered also after N_2^+ implantation into iron and steels at higher implantation energies than ours [11,21,28,29].

From each DCEM spectrum the signals (spectral areas),

$$I^{Nitr}_{\theta,V} \text{ and } I^{Fe}_{\theta,V}$$

for iron nitride (Fe_2N) and α-iron, respectively, at different angles θ and energy settings V, were determined by least-squares fitting of the spectra. The results are summarized in Table II and also plotted in Figure 8, together with theoretical values calculated as described below. The typical experimental line width (FWHM) for the α-iron lines was 0.4-0.45mm/s.

ANALYSIS AND RESULTS

Method

In order to obtain information on the depth dependence of the nitride (Fe_2N-) concentration, we have performed a data analysis similar to that described in reference 3.

Theoretical DCEM signals (spectral areas) from the iron nitride phase and α-iron phase have, respectively, been calculated by the expressions

$$I^{Nitr}_{\theta,V}(\text{theor}) = C_\theta \int_0^\infty n^{Nitr}_{eff} \bullet \text{DISTR}(x) \bullet T_{\theta,V}(x)dx \tag{1a}$$

$$I^{Fe}_{\theta,V}(\text{theor}) = C_\theta \int_0^\infty n^{Fe}_{eff} \bullet [1-\text{DISTR}(x)] \bullet \gamma(x) \bullet T_{\theta,V}(x)dx \tag{1b}$$

where the experimentally obtained signals, denoted by exp are subject to the condition that the ratio,

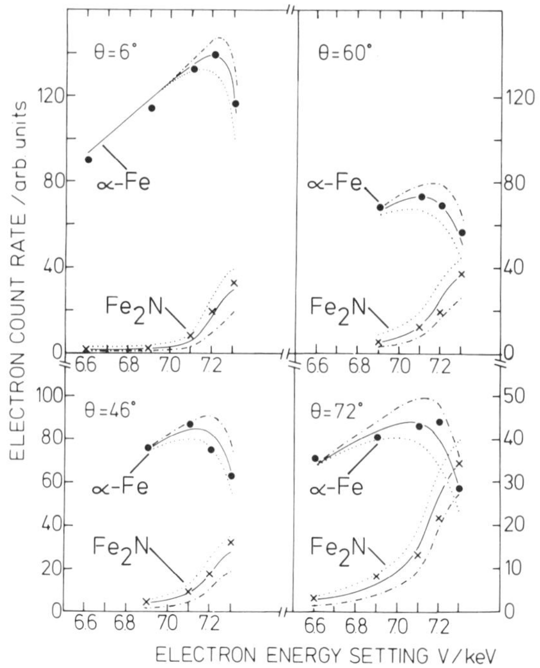

Figure 8. Comparison between experimental and theoretical DCEM signals $I_{\theta,V}$ (electron count rate) for the same Mössbauer absorber as in Figures 3-7. Experimental signals (Table 2) from the α-iron and Fe_2N are shown by (•) and (x), respectively. Full curves show the DCEM signals theoretically calculated by equation 1a-c from Monte Carlo simulated data on electron scattering and energy loss, assuming a Gaussian depth distribution for the nitride phase with B=75FeÅ. Dashed-dotted curve: theoretical curve for B=50FeÅ; dotted curve: theoretical curve for B=100FeÅ.

$$\frac{\sum_V [I^{Nitr}_{\theta,V}(exp) + I^{Fe}_{\theta,V}(exp)]}{\sum_V [I^{Nitr}_{\theta,V}(theor) + I^{Fe}_{\theta,V}(theor)]} = 1 \quad (1c)$$

This condition determines the scaling factor, C_θ, for a selected angle θ. C_θ should ideally have the same value in all measurements; the dependence on θ observed in our results is due to a variation of the effective spectrometer luminosity [3] with θ. The other notations have the following meaning: n^{Nitr}_{eff} is the effective iron-57 density of the nitride phase, relative to that of α-iron. n^{Fe}_{eff} is the effective iron-57 density of α-iron-Fe, $n^{Fe}_{eff} = 1$. DISTR(x) is the distribution of the nitride phase as a function of depth, x, [1-DISTR(x)] is the corresponding depth distribution of the α-iron phase, $\gamma(x)$ is the γ-ray attenuation factor, and $T_{\theta,V}(x)$ is the weight function at energy setting V and angle θ.

The depth x is to be measured [2] as mass/area. We use the unit $1FeÅ=0,0786\mu g/cm^2$. The weight functions $T_{\theta,V}(x)$ take into account K-conversion electrons as well as corrections for L-conversion electrons and for the GCE contribution (secondary Mössbauer absorption-cf, reference 2). Figure 9 shows Monte Carlo simulated K-conversion weight functions for the angles selected with our spectrometer, convoluted with the triangular spectrometer profile described above. The factor $\gamma(x)$ takes account of the resonant attenuation of the incident gamma radiation in the absorber, and has been calculated as described in reference 3. The resonant gamma attenuation has to be taken into account for the L-conversion and GCE contribution, as it derives from the deep bulk phase (α-iron) of the absorber. Non-resonant

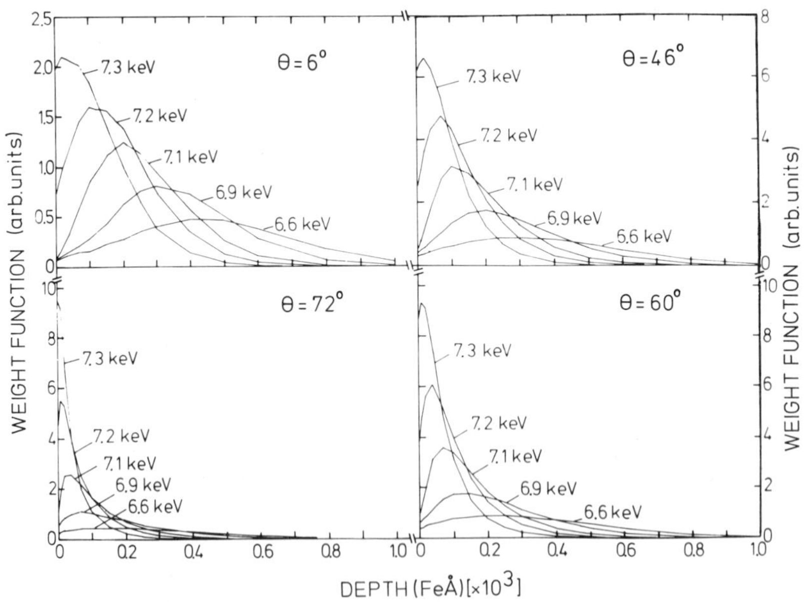

Figure 9. Some typical theoretical K-conversion weight functions $T^K_{\theta,V}(x)$ calculated for the present spectrometer (2.7% FWHM energy resolution). Calculated curves are labeled by electron energy setting (V) value, corresponding to energy E by V/E=1.007. The selected angular intervals are $\theta=6°\pm6°$, $46°\pm6°$, $60°\pm6°$ and $72°\pm6°$, respectively. Depth x is measured in $FeÅ=0.0786\mu g/cm^2$.

attenuation may be entirely neglected. It should be noted that the L-conversion and the GCE contributions, as well as the actual effect of the gamma attenuation factors on the DCEM signals, $I_{\theta,V}$, are small corrections. The gamma attenuation is appreciable only at larger depths from which the K-conversion electron transmission to the surface is always small. For the signal from the thin nitride layer, the effect of gamma attenuation has been neglected.

The effective iron-57 density of nitride (Fe_2N) relative to that of α-iron (per unit mass thickness) is given by

$$n_{eff}^{Nitr} = \frac{f^{Nitr}}{f^{Fe}} \bullet N^{Nitr}/(\frac{M^{Nitr}}{M^{Fe}}) \qquad (2)$$

$$= \frac{f^{Nitr}}{f^{Fe}} \bullet 0.89$$

with the notations which have the following meanings: f^{Nitr}, f^{Fe} are the f-factor of nitride and α-iron, respectively, N^{Nitr} is the number of iron atoms per formular unit of nitride, and M^{Nitr}, M^{Fe} are the molecular weight of nitride and α-iron, respectively.

Taking the values f^{Fe}=0.7 for α-iron and f^{Nitr} ca. 0.5 for Fe_2N (reference 26), we obtain

$$n_{eff}^{Nitr} \sim 0.65.$$

The correctness of this value has been checked by using the sum rule of DCEMS [3] which in the present case states that the value of

$$[\frac{I_{\theta,V}^{Fe}(exp)}{n_{eff}^{Fe}} + \frac{I_{\theta,V}^{Nitr}(exp)}{n_{eff}^{Nitr}}] \bullet [\int_o^\infty T_{\theta,V}(x)dx]^{-1} \qquad (3)$$

should be independent of v (at a given angle θ). Taking n_{eff}^{Fe} = 1, and using the experimental data from Table II, the best fit to the sum rule was found for

$$n_{eff}^{Nitr} \sim 0.6\text{-}0.75,$$

in agreement with the value of ~ 0.65 mentioned above.

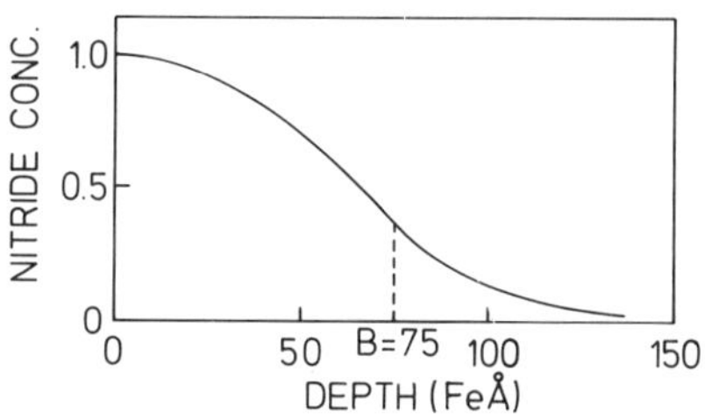

Figure 10. Distribution of nitride phase (Fe_2N) versus depth (Gaussian) as obtained from the present analysis. At the surface 100% nitride phase is present and no α-iron. The average depth B is (75 ± 10) FeÅ.

Table III. Shape parameter, R, and mean depth, B obtained from least-squares fit of theoretical and experimental DCEM signals independently for each angle θ.

θ	R	B(FeÅ)
6°	0.53	68
46°	0.63	80
60°	0.69	66
72°	1.02	80

For the analysis the following absorber model has been used. The depth distribution of the (Fe_2N) nitride phase, DISTR(x), was assumed to be of the form

$$\mathrm{DISTR}(x) = A \cdot \exp\left[-\left(\frac{x-\bar{x}}{B}\right)^{1/R}\right] \tag{4}$$

B is the mean depth of this phase (measured in mass/area), $\bar{x}$ is the peak position, and R is a parameter giving the shape of the distribution. For a Gaussian shape R is 0.5. It became clear from our analysis described below that a good fit to experimental data can only be obtained with A=1 (or very close to 1) and $\bar{x}$=0, i.e., the surface is covered to 100 percent by nitride, and the distribution peak is at the surface. This is also reasonable in view of the large implantation dose used here.

RESULTS

Our main results are shown in Figure 8 and Figure 10. In Figure 8 the full drawn curves give the theoretical DCEM signals $I_{\theta,V}$(theor) for the Fe_2N and α-iron, respectively, calculated with the model parameters B=75FeÅ and R=0.5 (Gaussian distribution, Figure 10). The agreement between the theoretical curves and the experimental data points is rather good at all selected angles, θ. In order to demonstrate the sensitivity to the parameter B, i.e., the mean thickness of the nitride layer, B has been varied slightly (B=50, 75 and 100FeÅ, respectively). The theoretical DCEM signals computed with R=0.5 (Gaussian) and B=50FeÅ and B=100FeÅ also are shown in Figure 8. A comparison of these curves with the experimental points allows us to estimate the uncertainty of B to be about ±10FeÅ.

We also attempted to determine the shape parameter R (which was set to R=0.5 in the above cases). This has been done independently for each angle θ by searching the best least-squares fit values of R (shape) and B (thickness) simultaneously by using

$$n_{eff}^{Nitr} = 0.65$$

as before. The result is shown in Table III. Apart from the case of θ=72°, all R-values tend to find the best least-squares fit near R = 0.6±0.1 near Gaussian with corresponding B-values in the range between 66 and 80 FeÅ. For θ - 72°, the R-value obtained is ca. 1, corresponding to an exponential nitride distribution. It must be pointed out, however, that although the relative nitride signal intensity is very large at nearly glancing electron emission (θ=72°), the effect of the <u>shape</u> of the nitride depth distribution on the DCEM signal is rather small, since there is practically vanishing depth resolution at glancing angles (as demonstrated in Figure 9 for θ=72°).

Thus the R-value obtained at $\theta=72°$ (exponential) should be considered with somewhat less confidence.

The analysis described so far points to a Gaussian or near Gaussian nitride depth distribution. We also have made attempts to fit our experimental results (Figure 8) with other nitride distribution models (exponential (R=1) or step function (R=0) distributions) for each selected angle θ, leaving the B-parameter as a variable. The least-squares fit χ^2-values obtained favor in general a Gaussian or near Gaussian distribution which also appears to be physically reasonable. On the other hand, the mean depth B is well determined by all models used, being within the range of B-values given in Table III. In addition it became evident from our least-squares fits that, regardless of the distribution model (R-value) used the parameter A and $\underline{x}$ in equation (4) have to be ca. one and zero, respectively, because, otherwise remarkable discrepancies between experimental and theoretical points were found in our fits. This result undoubtfully indicates that the surface is entirely covered by nitride (Fe_2N).

DISCUSSION

Our experimental and theoretical DCEM investigation of N_2^+ implanted at 4keV into a α-iron at a high nominal dose (3.2×10^{17} ions/cm^2) indicates that a surface and subsurface nitride phase similar to Fe_2N is formed. Our results show for the first time that very thin surface layers (of a few tens of FeÅ thickness) of iron-57 containing metallurgical phases (in our case Fe_2N) can be analyzed depth-selectively with sufficient sensitivity of detection if energy-selective and angular selective DCEMS is performed. Particularly DCEMS at glancing electron emission shows good surface layer sensitivity. Quantitatively the mean depth B of the Fe_2N depth distribution is very well determined to be (75 ± 10)FeÅ corresponding to an area density of $(5.9\pm0.8)\mu g/cm^2$. Furthermore, it appears certain from our analysis that the surface consists of 100 percent of Fe_2N (i.e., no α-iron phase at the surface), and that the peak of the nitride depth distribution is located at the surface.

In addition we may conclude from our analysis that the depth profile for the nitride (Fe_2N) concentration has a Gaussian or nearly Gaussian shape. Thus our data analysis favors a nitride depth distribution of the form

$$\text{DISTR}(x) = \exp\left[-\left(\frac{x}{B}\right)^{1/R}\right],$$

with R of ca. 0.6. The uncertainty in the determination of the exact shape of the depth profile in the present study is related to the rather small implantation depth chosen in our experiment. Since in the present case the average implantation depth, B, is much smaller than the typical width of the relevant weight functions, particularly at small angles Θ, where depth resolution is best, according to Figure 8, variations in the shape of the nitride depth distribution will give only small effects in the energy dependence of the DCEM signals, i.e., in $I_{\theta,V}$. It is clear that the depth profile can be obtained only with little accuracy in this case. On the other hand, if we had a thicker implantation layer of ca. 200 to 400FeÅ mean thickness (but less than ca. 1000FeÅ), in principle, the shape of the depth profile could be determined with much better accuracy.

The fact that the nitride distribution peak is located at the surface is very likely the result of strong sputtering effects due to the high implantation dose in our experiment. It is known that sputtering affects the depth distribution of implanted ions such that the distribution peak shifts towards the surface [30]. Such a distribution peak at the surface has been observed also for N_2^+ implanted at low energies (1-5keV) with doses up to 5×10^{16} ions/cm^2 into copper and silver [31]. In addition, inspection of the range distributions of nitrogen ions in copper as given in reference 31 shows that our value for the mean depth, B, of nitrogen in iron is of reasonable magnitude.

That sputtering effects are important in our experiment can be concluded by a comparison of the nominal dose with the actually retained dose, i.e., the surface layer of nitrogen as found from the DCEM analysis. The number of nitrogen atoms per unit area after implantation should be about

$$(B/M^{Nitr})\bullet N_A$$

where N_A is Avogadro's number. Evaluating this with $B=5.9\mu g/cm^2$ for Fe_2N, one finds a value of 2.8×10^{16} nitrogen atoms/cm^2 for the retained dose. Because the nominal dose is 6.4×10^{17} nitrogen atoms/cm^2, only about 4 percent of the incident nitrogen atoms are actually retained in the surface layer.

SUMMARY

The present investigation shows that energy and angular dependent depth-selective iron-57 conversion electron Mössbauer spectroscopy (DCEMS) is a very useful technique for in-situ metallurgical phase analysis of surface layers of a few tens of Å thickness. The method has much larger surface sensitivity than conventional integral CEMS. Our DCEM spectra show that N_2^+ ion-implantation at 4keV with a high nominal dose of ca. 10^{17} N_2^+ions/cm^2 into a α-iron yields a surface and subsurface nitride phase similar to Fe_2N. From the energy and angular dependence of the DCEM signal intensity (spectral area) we were able to draw conclusions on the depth distribution of the nitride phase. The analysis shows that the peak of the Fe_2N depth profile is located at the surface, and that only Fe_2N (and no α-iron) is present at the surface. Quantitatively the mean depth of the nitride distribution has been determined with reasonable accuracy as 75 ± 10FeÅ or $5.9\pm0.8\mu g/cm^2$. In addition our data analysis points of a Gaussian (or nearly Gaussian) nitride depth profile which appears plausible because of physical reasons, however, the uncertainty in the profile shape is large due to the fairly small implantation depth chosen in our experiment. In principle the shape of the depth profile could be obtained with much better accuracy with somewhat thicker implantation layers of ca. 200-400FeÅ.

ACKNOWLEDGEMENTS

The technical assistance of U. von Hörsten and W. Weyers is gratefully acknowledged. This work was supported financially by the Deutsche Forschungsgemeinschaft.

REFERENCES

1. For a recent review see D. Liljequist, in Scanning Electron Microscopy III, SEM Inc., AMF O'Hare, Chicago, p. 997 (1983).

2. D. Liljequist and M. Ismail, Phys. Rev. B 31, 4131 (1985).

3. D. Liljequist, M. Ismail, K. Saneyoshi, K. Debusmann, W. Keune, R. A. Brand, and W. Kiauko, Phys. Rev. B31 4137 (1985).

4. T. Toriyama, K. Asano, K. Saneyoshi, and K. Hisatake, Nucl. Instr. Methods B4, 232, 170 (1984).

5. B. Bodlund-Ringström, U. Bäverstam, and C. Bohm, J. Vac. Sci. Technol. 16, 1013 (1979).

6. C. Bohm, U. Bäverstam, B. Bodlund-Ringström, and D. Liljequist, USIP Report 79-07, Uniersity of Stockholm, Sweden (1979).

7. G. Belozerskii, C. Bohm, T. Ekdahl, and D. Liljequist, Corrosion Sci. 22, 831 (1982).

8. G. N. Belozerskii, C. Bohm, T. Ekdahl, and D. Liljequist, Nucl. Instr. Methods 192, 539 (1982).

9. M. Domke, B. Kyvelos, and G. Kaindl, Surface Sci. 126, 727 (1983).

10. M. Domke and B. Kyvelos, Corrosion Sci. 23, 921 (1983).

11. M. Carbucicchio, L. Bardani, and S. Tosto, J. Appl. Phys. 52, 4589 (1981).

12. T. Shigematsu, S. Staniek, R. A. Brand, W. Keune, R. H. Nussbaum, H.-D. Pfannes, D. Liljequist, G. Longworth, and R. Atkinson, Hyperfine Interactions 15/16, 383 (1983).

13. J. Itoh, Y. Yonekura, T. Toriyama, H. Miyasaka, and K. Hisatake, Hyperfine Interactions 15/16, 383 (1983).

14. T. Shigematsu, S. Staniek, W. Keune, H.-D. Pfannes, and D. Liljequist, Proceed. Intern. Conf. on the Mössbauer Effect, Alama Ata, USSR, 1983, Gordon and Breach (in press).

15. T. Yang, A. Krishnan, N. Benczer-Koller, and G. Bayreuther, Phys. Rev. Lett. 48, 1292 (1982).

16. K. Saneyoshi, T. Toriyama, J. Itoh, K. Hisatake, and S. Chikazumi, J. Magn. Mag. Mater. 31-34, 705 (1983).

17. J. Itoh, Y. Yonekura, K. Saneyoshi, T. Toriyama, and K. Hisatake, J. Magn. Mag. Mater. 35, 340 (1983).

18. S. Staniek, T. Shigematsu, W. Keune, and H.-D. Pfannes, J. Magn. Mag. Mater. 35, 347 (1983).

19. J. Korecki and U. Gradmann, Phys. Rev. Letters, in press; Hyperfine Interactions, in press (1985).

20. W. Keune, Hyperfine Interactions, in press (1985).

21. For a recent review see: G. Principi, S. LoRusso, and C. Tosello, these proceedings.

22. J. Lindhard, M. Scharff, and H. E. Schiott, Mat. Fys. Medd. Dan. Vidensk. Selsk. 33, 14 (1963).

23. H. E. Schiott, Can. J. Phys. 46, 449 (1968).

24. T. Shigematsu, H.-D. Pfannes, and W. Keune, in Mössbauer Spectroscopy and Its Chemical Applications, edited by J. G. Stevens and G. K. Shenoy, Advances in Chemistry, Series 194, Am. Chem. Soc., Washington, DC p. 125 (1981).

25. H. Ibach, in Electron Spectroscopy for Surface Analysis, edited by H. Ibach, Springer, Berlin, p. 1 (1977).

26. J. Bainbridge, D. A. Channing, W. H. Whitlow, and R. E. Pendlebury, J. Phys. Chem. Solids 34, 1579 (1973).

27. M. Chabanel, C. Janot, and J. P. Motte, C. R. Acad. Sci. 266, B419 (1986).

28. G. Longworth and N. E. W. Hartley, Thin Solid Films 48, 95 (1978).

29. C. A. dos Santos, B. A. S. de Barros, Jr., J. P. de Souza, and I. J. R. Baumvol, Appl. Phys. Lett. 41, 237 (1982).

30. G. Longworth and R. Atkinson, in Mössbauer Spectroscopy and Its Chemical Applications, edited by J. G. Stevens and G. K. Shenoy, Adv. in Chemistry Series 194, Am. Chem. Soc., Washington, DC, p. 101 (1981).

31. A. Zomorrodian, S. Tougaard, and A. Ignatiev, Phys. Rev. B30, 3124 (1984).

RAPID MÖSSBAUER-BASED METHODS FOR APPLICATIONS IN THE COAL AND STEEL INDUSTRIES

F. E. Huggins*, G. P. Huffman*, and G. R. Dunmyre*

U.S. Steel Technical Center
Monroeville, PA 15146

INTRODUCTION

Conventional Mössbauer spectroscopy is not suited for routine quality-control applications in industry because of the long sample turn-around times and relatively sophisticated nature of the technique and its interpretation. Moreover, because quality-control applications tend to involve repetitive determinations of the same quantity, the full power of the technique is generally not necessary. It appears that Mössbauer spectroscopy is perhaps 'overqualified' for industrial applications on the shop floor. As a result, simpler and more rapid Mössbauer techniques have been proposed and developed for such applications.

The most promising of such variations of the Mössbauer technique for specific industrial applications is the 'on-off' resonance method. Basically, this method reduces the conventional Mössbauer spectrum of typically 400 to 1024 data points to two data points. One of these two points corresponds to the background, off-resonance radiation, $\bar{I}$, and is measured with the source oscillating strongly with respect to the sample to minimize the resonant absorption or scattering between the source and sample. The other data point corresponds to the on-resonance radiation, $\bar{I}_o$, and includes contributions from the resonant absorption or scattering between source and sample and from the background, nonresonant radiation. Generally, this latter measurement is made with the source and sample at rest. The difference in signal between the two data points corresponds therefore to the resonant absorption or scattering between source and sample. For absorption geometry, it follows that:

$$R = 100(\bar{I}-\bar{I}_o)/\bar{I} \qquad (1a)$$

whereas for scattering geometry:

$$R = 100(\bar{I}_o-\bar{I})/\bar{I} \qquad (1b)$$

where R is the percent resonance effect.

This concept was first utilized in a device for measuring cassiterite (SnO_2) contents of rocks and ores [1] and became the basis for a geophysical

*Current address: BCR National Laboratory, Monroe, PA 15146

prospecting instrument for tin [2,3]. In this case, the Mössbauer isotope was tin-119. More recently, a number of similar methods have been proposed for iron-bearing materials based on the Mössbauer effect in iron-57. Such methods include the measurement of the austenite [4,5] and ferrite [5] contents of steel, the pyrite contents of coals [6,7], and the metallic iron contents of basic oxygen furnace slags [7].

In the current study, we will principally discuss the austenite and pyrite methods because the former involves scattering geometry, the latter absorption geometry, and both applications involve the presence of additional iron-bearing phases that complicate the analysis. Moreover, both methods show potential for practical industrial application, particularly as these methods can be further extended by combining measurement of the nonresonant absorption or scattering with the resonant effect measurement. However, before discussing the austenite and pyrite methods in detail, it will be necessary to describe briefly how such methods are simulated and experimentally tested.

SIMULATION AND EXPERIMENTAL VERIFICATION

For developmental work, conventional Mössbauer spectroscopy equipment operating in the pulse-height-analysis (PHA) mode can be readily utilized for the design and testing of a rapid Mössbauer-based method. However, much can also be learned from simulation of the rapid method, including how the method is altered by changing the source matrix and how important are secondary phases that also contain the Mössbauer isotope. In this regard, we have developed theory [7] and a computer routine to calculate, from Mössbauer parametric data for the source and phase or phases of interest, the time- or velocity-averaged resonant signal (equivalent to count-rate) as a function of the velocity amplitude of the vibrating source. In the general case, the time-averaged resonant signal for the source at rest is given by:

$$\bar{I}_o = \left(\frac{\Gamma}{2}\right)^2 \sum_{i,j} \frac{a_i b_j}{(v_i-v_j)^2+(\Gamma/2)^2} \tag{2}$$

and for the source oscillating in the constant acceleration mode with maximum velocity amplitude, V_m, is given by:

$$\bar{I} = \frac{\Gamma}{4V_m} \sum_{i,j} \left\{ a_i b_j \left[\tan^{-1}\left(\frac{2[V_m-(v_i-v_j)]}{\Gamma}\right) + \tan^{-1}\left(\frac{2[V_m+(v_i-v_j)]}{\Gamma}\right)\right]\right\} \tag{3}$$

where Γ is the full width at half maximum of the absorption peaks, a_i and b_j are the relative transition probabilities of the various absorber lines and source emission lines, respectively, and v_i and v_j are the resonant velocities of the absorber lines and source emission lines, respectively.

With these equations, all possible combinations of source and absorber materials can be readily assessed for a given application. From such calculations the optimal rapid method can be specified in terms of source matrix and minimization of secondary phase influences. For example, simulation techniques showed that a cobalt-57 rhodium source and a cobalt-57 stainless-steel source would be the best metal-matrix sources for the pyrite and austenite determination, respectively [7].

Although simulation will reveal the relative merits of different variations of the same method, experimental verification is necessary to confirm that the method is indeed feasible. To verify a given method, counts are measured with the source at rest and with the source in rapid oscillation (V_m > 10mm/s) for typically 60s PHA live time. To guard against the possibility of electronic drift and other spurious effects, each measurement is repeated alternately a number of times. The values of I_o and $\bar{I}$ are then averaged in order to calculate the percentage resonant effect, R, according to the appropriate form of Equation 1. The variation of R with known quantities of the measured phase in calibration samples can then be used to estimate the total counts that should be accumulated to determine the phase of interest in an unknown sample with a desired precision. As with conventional Mössbauer spectroscopy, the precision improves proportionately with the square root of the counts or the measurement time.

By using a combination of such simulation and experimental verification techniques, methods were developed and optimized for the determination of austenite in steel and of pyrite in coals. These methods are described in more detail in the next sections.

DETERMINATION OF AUSTENITE IN STEEL

The austenite determination is made in the backscattering geometry and is based on the detection of 6.3keV X-rays that are emitted from a steel surface after absorption of 14.4keV iron-57 Mössbauer γ-rays from a cobalt-57 stainless-steel source. As shown in Figure 1, the 6.3keV X-ray peak is the only region in the PHA spectrum of an austenite-rich steel that shows significant variation in intensity (count rate) between when the source is stationary and when it is in rapid oscillation. As indicated in Figure 2, for a steel that is a mixture of austenite and martensite and/or ferrite, only austenite contributes to the resonant scattering of X-rays with the stainless-steel source at rest. Hence, the magnitude of the difference in X-ray count-rate between the source at rest and the source in rapid oscillation

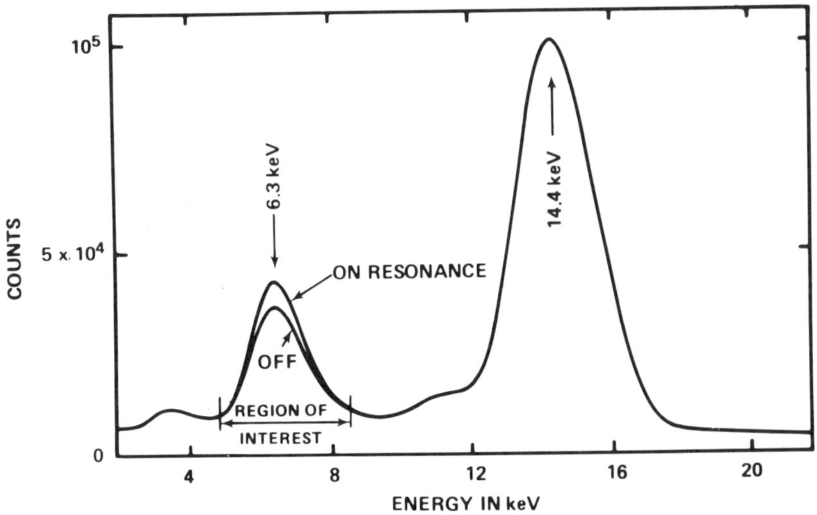

Figure 1. PHA spectra from an austenite-rich sample recorded with the drive at rest (on resonance) and vibrating with an amplitude of 10mm/s (off resonance). Both spectra were recorded for exactly 1000s and virtually superimpose upon each other, except in the 6.3keV region of interest.

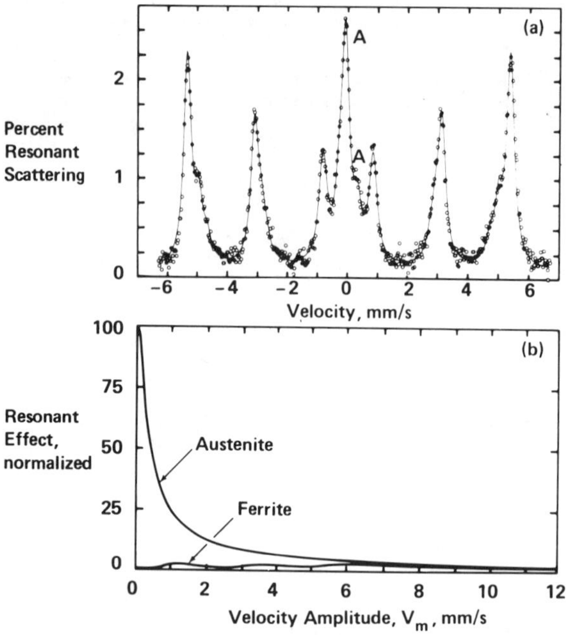

Figure 2. (a) Mössbauer spectrum of roll surface showing central peaks, marked with A, due to austenite and peaks due to ferrite-martensite and (b) calculated variation of Mössbauer resonant signal for austenite and ferritic iron as function of velocity amplitude of oscillating cobalt-57 stainless-steel source. Note negligible resonance for ferrite at V_m equals zero.

(V_m > 10mm/s) is directly proportional to the percentage of iron in the form of austenite in approximately the top 15μm of the steel. For low-alloy steels (Fe > 95%), this value has been shown [8] to be a close approximation to the volume fraction of austenite measured by X-ray diffraction or metallography. For high-alloy steels, however, because of the presence of significant amounts of chromium, nickel, manganese, etc., the resonant signal from the iron is effectively diminished and it is more precise to equate the resonant signal in the rapid Mössbauer method to the weight of iron in the steel in the form of austenite. This parameter, WFA, can be calculated as a percentage from conventional Mössbauer and compositional data as:

$$\text{WFA}(\%) = (\%\text{Fe as austenite} \times \text{wt.\% Fe in steel})/100 \quad (4)$$

To test the rapid austenite method, nine calibration samples, varying in austenite content from 0 to 100 volume percent and in alloy content from 1 to 47 weight percent, were prepared from eight different steels. The compositions of the steels, determined by X-ray fluorescence (XRF), are shown in Table I. Total iron was estimated by difference. Conventional Mössbauer spectra were obtained to determine percent iron as austenite in the samples. One sample, RW 1639, from a spalled roll, gave a significantly lower value for percent iron as austenite after grinding of its surface for the XRF determination, and therefore there were two calibration samples available for this steel. Values for the weight percent iron in the steel in the form of austenite, WFA, were calculated according to Equation 4 from the conventional Mössbauer and XRF data.

The calibration specimens, which were cut to dimensions of no less than 1 inch by 1.5 inches, were taped over the thin (3mil) beryllium exit window of a proportional counter filled to one atmosphere with a standard P-10 gas

Table I. Weight Percentage Compositions of Steel Samples as Determined by XRF

Steel	Cr	Ni	Mn	Si	Mo	V	C (nom.)	Fe (bal.)
Carbon	0.02	0.01	0.33				0.6	99.0
CA-1 Alloy			4.25	0.26	0.23			95.2
CA-5 Alloy			4.28	0.25	0.23			95.2
RW 1639	1.65	0.17	0.45	0.40	0.18	0.17	0.8	96.2
430 SS	16.23	0.30	0.37	0.36	0.15		<0.1	82.5
304 SS	18.23	8.42	1.35	0.48	0.25		<0.1	71.2
310 SS	24.88	19.15	1.92	0.47	0.24		0.2	53.2
316 SS	17.17	10.81	1.77	0 32	2.08		<0.1	67.8

Table II. Data from Conventional and Rapid Mössbauer Methods

Steel Sample	%Fe as Austenite from Conventional Mössbauer	WFA, %[a]	Rapid Mössbauer Method $\bar{I}_o$ [b]	$\bar{I}$ [b]	R, %
Carbon	0	0	402277	403030	-0.2
CA-1	2	2	391173	390502	0.2
CA-5	11	11	399750	392539	1.8
RW 1639 gr.[c]	19	18	404435	390701	3.5
RW 1639	24	23	402883	386100	4.3
430 SS	0	0	340436	341584	-0.3
304 SS	100	71	375015	334495	12.1
310 SS	100	53	350370	320141	9.4
316 SS	100	68	369374	329687	12.0
Std. Error	±1	±1	±730	±730	±0.4

[a]WFA--weight percent iron in form of austenite in steel. [b]Counts/60s average of three determinations: I_o with source stationary, I with source in rapid oscillation. [c]Surface of sample was ground.

mixture (90% argon-10% methane). Gamma rays from a stainless-steel source impinged on the sample after passing through the proportional counter in a standard backscattering geometry [8]. The counting times for I_o and I were set to 60s live time (approximately 87s real time) and each parameter was determined, alternately, three times. Data were recorded on a printer connected serially to the multichannel analyzer. The value of R for each sample was calculated according to Equation 1b, above, from the average values of I_o and I. Data from both conventional and rapid Mössbauer methods are summarized in Table II.

Selected data from Tables I and II were then subjected to correlation and regression analysis. The correlation matrix is shown in Table III. The best correlation, with a coefficient, r, of 0.99, was between R and WFA, the weight percent iron in the steel in the form of austenite. The regression analysis gave the following equation for the relationship between these variables:

$$WFA = 5.656R + 0.4 \tag{5}$$

Table III. Correlation Matrix (r values)

Variable	$\bar{I}_o$	$\bar{I}$	R	%Fe	WFA	%Cr	%Ni
$\bar{I}_o$	1.0						
$\bar{I}$	0.871	1.0					
R	-0.361	-0.740	1.0				
%Fe	0.769	<u>0.948</u>	-0.790	1.0			
WFA	-0.333	-0.751	<u>0.999</u>	-0.795	1.0		
%Cr	-0.881	<u>-0.987</u>	0.704	<u>-0.958</u>	0.715	1.0	
%Ni	-0.592	-0.826	0.802	<u>-0.960</u>	0.800	0.851	1.0

$\bar{I}_o$ - Average counts/60s, buzzer off - column 4, Table II
$\bar{I}$ - Average counts/60s, buzzer on - column 5, Table II
%Fe - Wt% Fe in steel - column 9, Table I
WFA - Wt% Austenitic iron in steel - column 3, Table II
%Cr - Wt% Cr in steel - column 2, Table I
%Ni - Wt% Ni in steel - column 3, Table I

for which the standard error of an estimate of WFA is ±1.5 percent. The calibration curve is shown in Figure 3. The correlation between the two variables is surprisingly good, given the range in both austenite content and composition, and is an excellent validation of the rapid austenite method. However, the determined parameter, WFA, has still to be related to the volume percentage of austenite in the steel, which is usually the parameter of interest to metallographers and metallurgists. In the general case, this relationship can be approximated as:

$$\text{Vol \% Austenite} = 100(\text{WFA}/\text{wt\%Fe}) \tag{6}$$

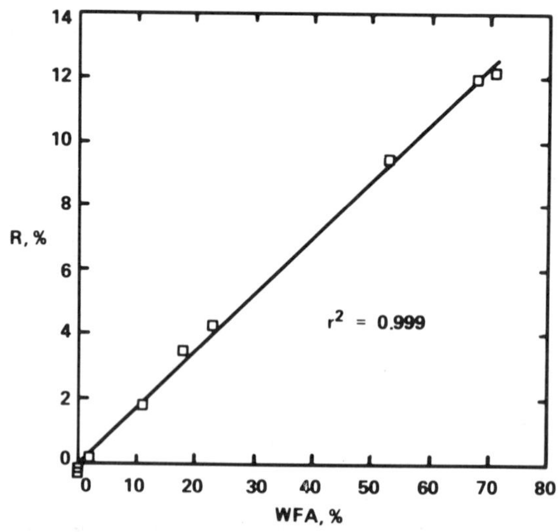

Figure 3. Calibration of austenite determination showing high degree of correlation between R, the measured austenite resonant effect, and WFA, the weight percentage of iron as austenite in the steel. The solid line is the least-squares-fit relationship between the two variables derived from regression analysis.

This equation is an approximation because it implies that the iron contents of the austenite and the martensite or ferrite phases in the steel are the same. However, there are very few situations in which this assumption will give rise to significant bias in the determination of the volume percentage of austenite.

It should be noted that there are a number of other highly correlated variables indicated by the correlation matrix in Table III. Such correlated variables include $\bar{I}$ and percent iron ($r = 0.95$), $\bar{I}$ and percent chromium ($r = -0.99$), percent iron and percent chromium ($r = -0.96$), percent iron and percent nickel ($r = -0.96$). The last two correlations arise because iron, chromium and nickel are the three principal compositional variables. The strong negative correlation between total counts, $\bar{I}$, and percent chromium, however, arises because of the close similarity of the energy of the X-rays emitted in the Mössbauer process to the K-shell absorption edge of chromium, which makes this element particularly effective in absorbing X-rays of this energy and reducing the number of counts triggered in the detector. This correlation may therefore be used for estimating the chromium content of a steel. The following equation was derived from regression analysis:

$$\%Cr = 119 - 3.0 \times 10^{-4} I \tag{7}$$

for which the standard error of an estimate of percent chromium is ± 1.75 percent. This relationship adds an extra dimension to the rapid Mössbauer austenite technique.

For low-alloy steels, a prime application would be the determination of retained austenite in a steel resulting from thermal processing. For example, rapid quenching of forged rolls causes retained austenite contents of up to 25 volume percent near the surface of the rolls [9]. Although desirable for toughness and strength, the austenite may also give rise to internal stresses that can cause spontaneous spalling. Retained austenite in dual-phase steels, on the other hand, is beneficial because it transforms to martensite during plastic straining, thereby enhancing the ductility of such steels [10]. In both cases, the amount of retained austenite importantly affects properties of the steel, and rapid determinations of the austenite content of the steel in the plant would provide a means for quality control.

For austenitic stainless steels (iron-chromium-nickel alloys), the conversion of austenite to ferrite or martensite due to deformation during pressing, cold working, or other processing could be estimated from the combination of compositional information and the rapid Mössbauer determination of WFA. Alternatively, for completely austenitic stainless steels, the technique could be used to monitor the approximate composition of the steel. In such situations, the measurement of R becomes, in effect, a measure of the weight percent iron in the steel. As percent chromium can be estimated from Equation 7, the remaining alloying elements, principally nickel, plus minor manganese, silicon, etc. can be estimated as:

$$\%Ni \text{ (+ other elements)} = 100 - \%Fe - \%Cr. \tag{8}$$

Any molybdenum would be measured as chromium, as this element also strongly absorbs the scattered Mössbauer X-rays. This compositional information may be sufficient to infer the type of stainless steel.

Applications involving magnetic ferritic and martensitic stainless steels would be limited to determinations of chromium content. However, such steels usually contain only chromium and iron as major alloying elements and therefore the austenite determinations would not only confirm the lack of austenite but also would indicate the approximate composition of the alloy and possibly the type of steel.

A prototype Mössbauer austenitemeter, designed and constructed at the U. S. Steel Technical Center, is shown in Figure 4. This instrument consists of two major components. The experimental end consists of a lead-lined aluminum housing in which the cobalt-57 stainless-steel source and drive are located. The drive is no more than a direct current buzzer. Attached to the housing are a proportional counter and a preamplifier. The experimental end is interfaced to a Tracor Northern 7200 multichannel analyzer (MCA) modified to amplify correctly the proportional counter signal and to alternately switch the buzzer on and off after each acquisition cycle to make use of an autorepeat feature. Much of the operation of the TN 7200 MCA is programmable, which makes it very easy to run once it is set up. In addition, options are available to send either the whole spectrum or one or more regions of interest to a printer for a hard-copy record or to a minicomputer for electronic storage and further data reduction. Except for such optional peripherals, the austenitemeter is powered by a single 110V AC standard supply and can be readily carried by a single person. It can therefore double both as a laboratory instrument for steel samples and as a portable device for use in plants directly on steel products. The austenitemeter has been extensively tested; indeed, the data shown in Table II and Figure 3 were obtained with this instrument.

With the above configuration and a source strength of about 40mCi, an austenite determination can be made in about 5 minutes with a precision of ±2 percent. The technique is nondestructive and about the only requirements are that the steel surface be clean and reasonably flat. The surface need not be ground or highly polished. In fact, such preparation can cause strain-induced transformation of unstable austenite and should be avoided.

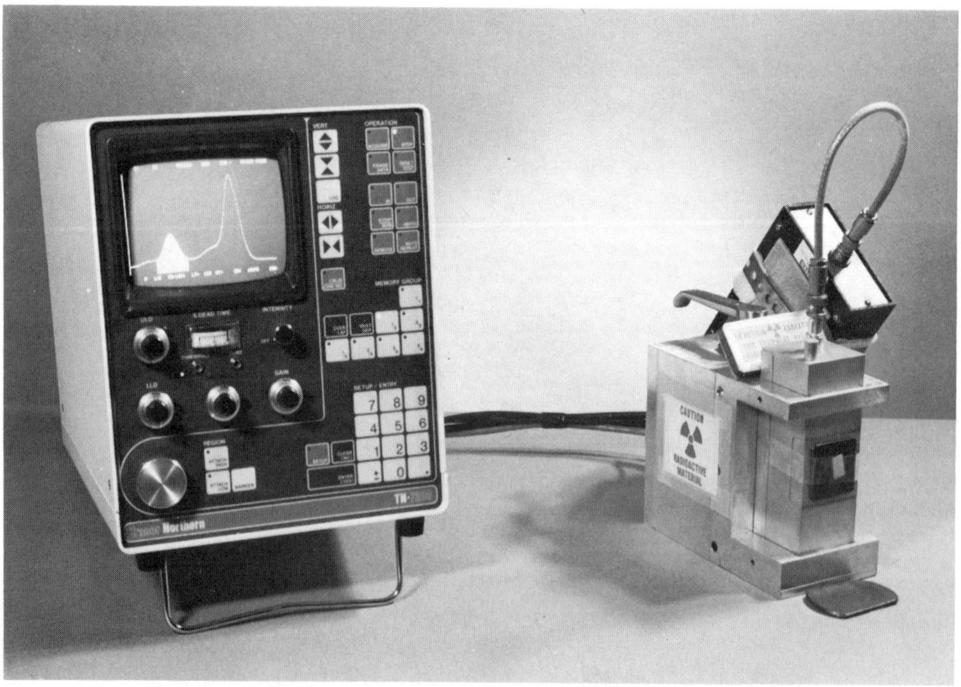

Figure 4. Prototype austenitemeter showing the experimental end (right) and the programmable Tracor Northern 7200 MCA for signal acquisition and control.

Table IV. Analytical Data for Mixtures of Raw and Washed Coals, Harlan Seam, KY and Pittsburgh Seam, PA

Mixture % Raw	% Washed	Harlan Seams, KY % Ash	% S	Pittsburgh Seam, PA % Ash	% S
100	0	31.4	0.92	15.8	2.18
90	10	28.9	0.93	15.1	2.23
80	20	26.3	0.93	14.3	2.11
70	30	24.0	0.92	12.4[a]	2.11[a]
60	40	21.3	0.83	12.5	2.11
50	50	18.8	0.91	11.5	2.00
40	60	16.5	0.91	10.7	1.84
30	70	14.0	0.89	9.9	1.87
20	80	11.5	0.89	9.1	1.76
10	90	9.2	0.81	8.0	1.73
0	100	6.7	0.91	7.3	1.69

[a]The similarity of these data for the Pittsburgh 70-30 mixture to the 60-40 mixture indicated that the 70-30 mixture had been improperly weighed out according to the 60-40 mixture weights.

DETERMINATION OF PYRITE IN COAL

A rapid method for the determination of pyrite in coal using a cobalt-57 doped pyrite source was originally proposed by Jaggi and Rao [6] but, as far as we are aware, was never tested experimentally by them, presumably because of the difficulty of preparing the source. As discussed elsewhere [7], the hypothetical pyrite source can be replaced for this application with a conventional source with little or no loss in effectiveness. Standard sources can be used because their recoilless fractions are significantly higher than that of pyrite and because significant improvement (ca. 20%) can be obtained for a cobalt-57 palladium source by measuring I_o with the source oscillating at a small velocity amplitude (V_m ca. 0.3mm/s) rather than at rest.

For this application, the absorption geometry is best because coal is quite transparent to the 14.4keV Mössbauer γ-rays and the pyrite is generally dilute, rarely exceeding five percent by weight of the coal. Sample holders were used that presented a constant cross-sectional area of sample to the beam, but permitted a variable thickness to accommodate the variation of volume of the same weight of different coals.

A number of different tests have been performed on the rapid pyrite determination; these have varied from simply correlating the resonant absorption with pyrite added to a coal to more general tests involving 20 different coals [7]. Each test, however, was made with a cobalt-57 source and with conventional constant acceleration Mössbauer apparatus.

A particularly illustrative test is documented in Tables IV, V, and VI. For this test, raw and washed (cleaned) coals were obtained from two coal preparation plants; one plant processed coal from the Harlan and associated seams in eastern Kentucky, the other, coal from the Pittsburgh seam in southwestern Pennsylvania. Nine mixtures of weighed amounts of raw and washed coals were then prepared for both the Harlan and Pittsburgh coals. Total ash and sulfur contents were measured for all samples (Table IV) and conventional Mössbauer data, including pyritic-sulfur values [11], were obtained for the end-member coals (Table V). The percent resonance effect, R, was measured for each mixture and a further parameter, A, was also determined. The parameter,

Table V. Conventional Mössbauer Data for Raw and Washed Harlan and Pittsburgh Coals

Coal	% Fe in					Wt%[b]
	Pyrite	Clay	Siderite	Jarosite	Magnetite[a]	Pyr S
Harlan, Raw	37	51	8	4	--	0.47
Harlan, Washed	60	32	5	4	--	0.23
Pittsburgh, Raw	92	4	4	-	--	1.25
Pittsburgh, Washed	86	4	-	-	10	0.70

[a]Magnetite is picked up during coal washing. [b]From Mössbauer data, by method described elsewhere [11].

Table VI. Reduced Count-Rate Data for the Mixtures of Table V.

% Raw	% Washed	Harlan Seams, KY		Pittsburgh Seam, PA	
		R(%)[a]	A[b]	R(%)[a]	A[b]
100	0	4.0	1.423	6.2	1.453
90	10	3.6	1.364	6.0	1.428
80	20	3.5	1.294	5.5	1.428
70	30	3.4	1.220	--[c]	----[c]
60	40	3.1	1.142	5.3	1.310
50	50	3.0	1.070	5.2	1.266
40	60	2.9	1.003	4.9	1.217
30	70	2.3	0.927	4.7	1.177
20	80	2.3	0.859	4.4	1.135
10	90	2.0	0.779	4.2	1.093
0	100	1.6	0.701	3.7	1.040
Standard Error		±0.2	±0.002	±0.2	+0.002

[a]Calculated from the average values of five 18s determinations of $\bar{I}_o$ and $\bar{I}$ according to Equation 1a. [b]Calculated according to Equation 9.
[c]Data combined with Pittsburgh 60-40 mixture for the reason indicated in Table IV.

A, is defined as the off-resonance count-rate, $\bar{I}$, for the coal sample relative to that for the empty sample holder. For absorption processes, it follows that A is best defined as:

$$A = \ln(\bar{I}_E/\bar{I}), \tag{9}$$

where $\bar{I}_E$ is the count-rate measured for the empty sample holder. Data for both A and R are listed in Table VI.

As indicated in Figure 5a, R is a linear function of mixture composition for both coals, within the precision of the measurement. R is also a linear function of the pyritic-sulfur values interpolated between the end-member values (Figure 5b), but the trends of these two coals are quite different. Moreover, this difference is enhanced when the Pittsburgh data are normalized to the same weight as the Harlan data (dashed line in Figure 5b). The explanation for this difference lies in the different iron-bearing minerals present in the coals (Table V). Whereas pyrite in both raw and washed Pittsburgh coals constitutes over 85 percent of the iron, pyrite varies from only 37 percent of the iron in the raw to 60 percent in the washed Harlan coal. As documented by conventional Mössbauer spectroscopy, much of the remaining iron is in the form of iron-bearing clay in this coal. As shown by calculation [7], iron in the form of clay is about 28 percent as effective as pyritic iron in contributing to the value of R, measured with a cobalt-57 palladium source at rest. Hence, for Harlan mixtures, particularly those rich in the raw coal, the resonant signal includes appreciable contribution from nonpyritic iron. If the contributions from clays and other minor iron-bearing minerals are calculated and subtracted from the measured value of R, the Harlan trend would lie almost exactly on the dashed trend for Pittsburgh coal in Figure 5b.

This result shows that the rapid Mössbauer technique for pyrite must be applied with considerable care because of uncertainty regarding other iron-bearing minerals. However, if pyrite is the dominant iron-bearing mineral in the coal (iron in pyrite > 80% of the iron), as is the case for nearly all lignites and subbituminous coals and many high-volatile bituminous coals, then such uncertainty is minimal. The technique therefore could find wide application in coal preparation plant and testing laboratories, where continuous, repeated monitoring of the pyritic-sulfur content of such a coal or blend of such coals is needed. In terms of ease of operation and speed (ca. 10 minutes per determination), it compares favorably with the standard ASTM test (D2492), which is an involved wet chemical procedure that takes a few hours to perform [12]. The precision of both techniques is comparable, typically ±0.05 weight percent pyritic sulfur.

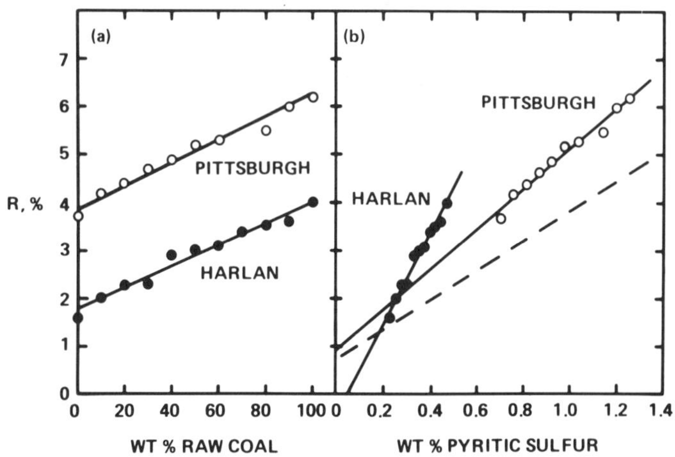

Figure 5. (a) Variation of resonant effect, R, with composition for Pittsburgh and Harlan coal mixtures and (b) variation of R with interpolated pyritic-sulfur contents for the same mixtures. Dashed line in (b) is the calculated trend for Pittsburgh coal normalized to the same weight of Harlan coal.

Furthermore, there is additional information to be obtained from the rapid Mössbauer determination that would appear to have application in coal quality control. We have noted that the energy of the Mössbauer γ-rays, 14.4keV, is similar to that (6 to 17keV) employed in low-energy X-ray devices for monitoring the ash content of coal, such as the Sortex and Cendrex devices [13]. The major problem with these devices, however, is that they are unable to compensate for any significant variation in ash composition, particularly in the iron content. Hence, because the rapid Mössbauer device can be used to measure separately the resonant absorption, which reflects the iron content, and the nonresonant absorption of 14.4keV γ-rays, the two measurements can be combined to yield a more precise determination of percent ash than can be obtained from the nonresonant absorption alone.

As shown in Figure 6, which is a plot of the nonresonant absorption, A, against percent ash content for the Harlan and Pittsburgh coals, the nonresonant absorption by itself, could be a highly precise means to determine weight percent ash. However, implicit in the test documented in Tables IV and VI, is a linear relationship between weight percent ash and ash composition that necessarily constrains the trends in Figure 6 to be linear. In practice, there is no such relationship and more general tests show a considerable scatter between A and weight percent ash because of the variation in ash composition. For example, Figure 7a shows the correlation (or lack of it) found for 18 coals. The standard error of an estimate of percent ash from a measurement of A from this correlation is ± 5 weight percent, which is unacceptably high. Inclusion of a term involving the resonant absorption does improve the relationship (Figure 7b), but the improvement only reduces the standard error of an estimate of percent ash to ± 3.6 weight percent. Once again, the variation in ash composition of 18 different coals is still too great to be adequately represented by two variables.

A test involving a single coal, to which measured amounts of pyrite (FeS_2) and quartz (SiO_2) were added, indicates that percent ash could be determined adequately in restricted circumstances from knowledge of R and A

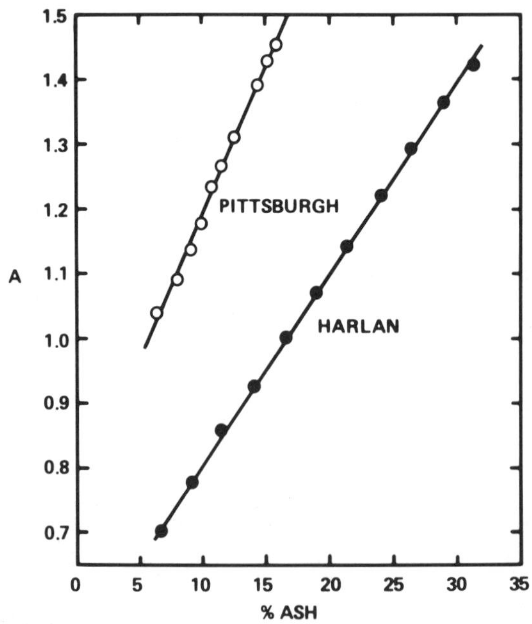

Figure 6. Correlation between A, the nonresonant absorption by the coal, and the ash content of the Pittsburgh and Harlan coal mixtures.

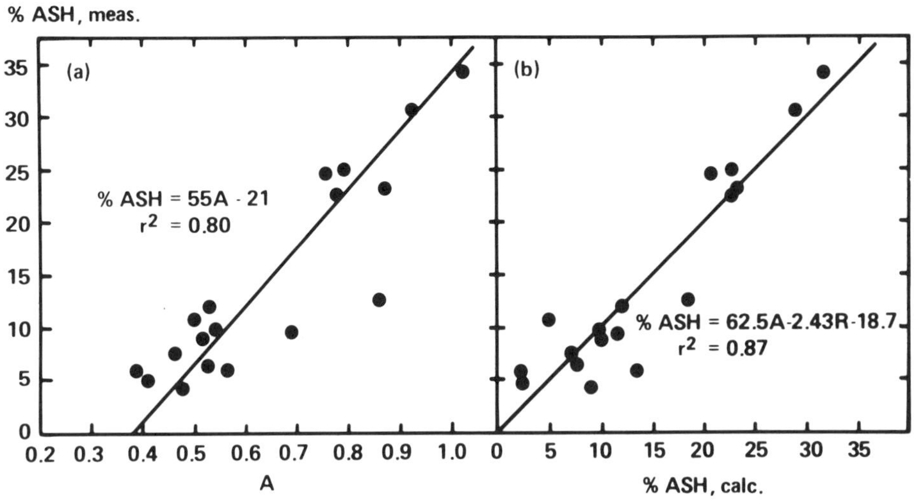

Figure 7. (a) Relationship between measured ash content and nonresonant absorption, A, for 18 coals and (b) comparison between measured ash content and ash content calculated from the regression equation incorporating terms for both A and R.

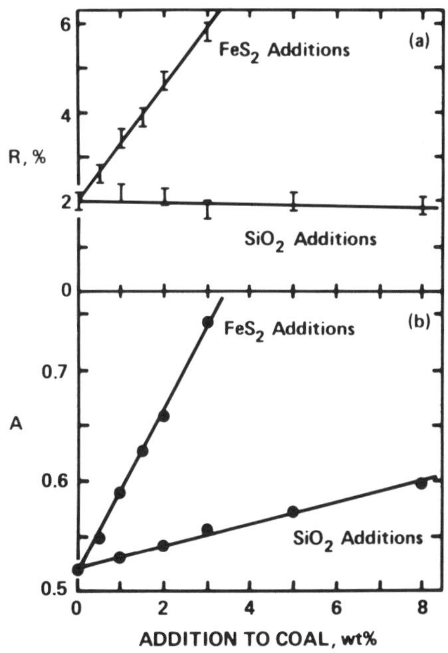

Figure 8. (a) Relationship between R, the resonant effect, and additions of known weights of pyrite (FeS_2) and quartz (SiO_2) to a sample of Pittsburgh coal and (b) relationship between A, the nonresonant absorption, and the same additions of FeS_2 and SiO_2.

measured in the rapid Mössbauer technique. The variation of A and R for additions of quartz (representing noniron minerals) and pyrite are documented in Figure 8. As can be seen, R is independent of the addition of quartz, whereas the additions of both pyrite and quartz increase A, albeit at quite different slopes. These data are combined into a single equation by multiple linear regression that represents the addition (percent ash) in terms of A and R. The calculated percent ash is shown plotted against the measured addition in Figure 9. By use of this relation, the percent ash can be estimated from R and A with a standard error of ±0.7 weight percent, which is acceptable for many applications. Most of the error arises from the large relative uncertainty in R (up to ±10%) and efforts to determine this parameter more precisely would significantly improve the ash determination. Such efforts might include replacing the cobalt-57 (palladium) source with a cobalt-57 (rhodium) source to take advantage of its higher recoilless fraction and better zero velocity resonance with pyrite. Also, better experimental design to improve the signal-to-noise ratio could be implemented [5,6] because the small solid angle geometry required by Mössbauer spectroscopy is not needed for the rapid Mössbauer measurement as the source is usually at rest (V_m = 0.0mm/s) or vibrating strongly (V_m > 10mm/s) so that geometry broadening is not important. Hence, much larger solid angles and smaller source-to-detector distances could be employed to enhance count-rates and signal-to-noise ratios.

CONCLUSIONS

In this work, two rapid Mössbauer-based techniques that have significant potential for industrial monitoring applications have been described and discussed in detail. These techniques employ the Mössbauer effect, but not conventional spectroscopy, to determine specific iron-bearing phases in materials. One technique, based on the emission of 6.3keV X-rays from a steel following Mössbauer absorption, determines austenite volume fractions in approximately the top 15μm of the steel with a precision of ±2% in less than ten minutes. Interference from the other major iron-bearing phases, martensite and/or ferrite, is negligible. Equipment has been designed for this technique that can be used in the laboratory or directly on steel products in a steel mill. The other technique, based on the absorption of 14.4keV γ-rays, is capable of determining pyritic-sulfur contents of coals

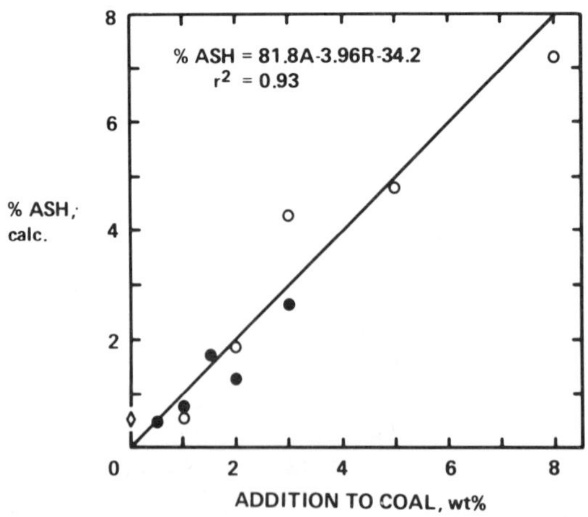

Figure 9. Comparison of 'ash' contents calculated from regression equation and measured additions of FeS_2 (closed circles) and SiO_2 (open circles) to Pittsburgh coal.

with a precision of ± 0.05 weight percent. However, because of the influence of other iron-bearing minerals in the coal on the resonant signal, this latter technique is best suited for coals that are known to contain relatively small amounts of other iron-bearing minerals. Alternatively, it should be possible to design features into the technique, such as additional sources or measurements at different velocity amplitude settings, so that the other minerals could be estimated. The simulation techniques described elsewhere [7] should be most useful in this regard.

As demonstrated in this paper, it is also possible to combine the measurements of resonant signal with the nonresonant signal to extend the range of application of such rapid Mössbauer techniques. For the austenite and pyrite determination, such additional information as the chromium content of the steel and the ash content of the coal may also be determined from the rapid technique under certain circumstances.

Rapid Mössbauer-based techniques appear to offer the best opportunity to find significant application for the Mössbauer effect in industrial shop-floor environments. These techniques can be designed for specific problems, are considerably more rapid than the conventional technique, are easy to operate and interpret, and the required instrumentation is rugged and relatively inexpensive. All these requirements are necessary for use in quality-control applications. Although we have discussed only two such applications, similar techniques could be designed and implemented for other iron-bearing materials or for materials containing other Mössbauer elements.

REFERENCES

1. A. V. Dolenko, B. G. Egiazarov, L. V. Isakov, V. I. Goldanskii, E. F. Makarov, and V. A. Trukhtanov, U.S. Patent No. 3,597,615, August 3, 1971.

2. I. Ya. Garzanov, V. I. Kardashuk, V. A. Kotelnikov, V. G. Labushkin, and E. F. Makarov, Proceedings, 5th International Conference on Mössbauer Spectroscopy, (in Russian), Bratislava, Czechoslovakia, Vol. 3, 660 (1975).

3. I. Ya. Garzanov, V. A. Kulikov, V. G. Labushkin, and E. F. Mararov, Meas. Tech. 24, 159 (1981).

4. T. Zemcik and S. Havlicek, Proceedings of the International Conference on Mössbauer Spectroscopy, Bucharest, Romania, Vol. 1, 405 (1977).

5. N. K. Jaggi and K. R. P. M. Rao, NDT International, 281 (1978). N. K. Jaggi and K. R. P. M. Rao, Proceedings of the Nuclear Physics and Solid State Physics Symposium, 21st, Bombay, India, Vol. 21C, 804 (1978).

6. N. K. Jaggi and K. R. P. M. Rao, Fuel 58, 688 (1979).

7. F. E. Huggins and G. P. Huffman, J. Appl. Phys. 55, 1404 (1979).

8. G. P. Huffman and F. E. Huggins, AIP Conference Proceedings No. 84, Physics in the Steel Industry, Bethlehemn, PA, 150 (1982).

9. F. E. Huggins and G. P. Huffman, unpublished data.

10. G. R. Speich, A. J. Schwoeble, and G. P. Huffman, Met. Trans. 14A, 1079 (1983).

11. G. P. Huffman and F. E. Huggins, Fuel 57, 592 (1978).

12. ASTM Standard Test D2492, 1976 Annual Book of Standards, Part 26, 319 (1976).

13. P. Cammack and A. Balint, Trans. SME 260, 363 (1976).

MÖSSBAUER SPECTROSCOPY IN STEEL INDUSTRY

K. R. P. M. Rao
Nuclear Physics Division
Bhabha Atomic Research Centre, Trombay
Bombay-400 085, India

INTRODUCTION

This review article is not intended to be an alternative to the numerous excellent articles written by various research workers and in particular to these written by Huffman and Huggins, who have pioneered the utilization of the Mössbauer spectroscopy to the steel industry. It is rather intended to draw the attention of research workers in this field and related fields to the exciting and most practical commercial utilization of the technique in the industry by consolidating a number of studies related to the steel industry reported in the literature. It is very likely that the survey of the literature is not complete. Only a few examples from the published work have been chosen more because of the biased interest of the author.

Although Mössbauer spectroscopy is over 25 years old, it has not been utilized initially in the field of commercially important steel industry. It is only in the past five to eight years that both transmission and scattering geometry techniques have been profitably used in the steel industry. This has occurred after the realization of the potential of backscatter and conversion electron Mössbauer spectroscopy as a non-destructive testing tool for routine monitoring purposes as well as for investigative studies in the industry. The utilization of Möesbauer spectroscopy in the steel industry may broadly be classified as indicated in Table I.

STEEL MAKING PROCESSES

It has amply been demonstrated by Huffman et al. [1,2] that Mössbauer spectroscopy holds a good future for its extensive application in the steel industry both as a routine monitoring tool and as a powerful experimental technique for the investigation of important metallurgical problems. The work carried out at U.S. Steel Corporation by Huffman et al. [1] vindicates the role of Mössbauer spectroscopy at every stage in the steel industry, viz., in the selection of the raw iron ore that goes into steel making, in characterizing the coals that go into the blast furnaces, in investigating the ash and slag deposits that are formed during combustion of coals, in quantitative estimation of the phase distribution in the final steel products, in the study of surface treated steels, and corrosion phenomena.

Iron bearing ores which normally contain 25-50 percent iron are concentrated by various methods such as magnetic separation, washing and

Table I. Mössbauer Spectroscopy in Steel Industry

Steel Making Process	Steel Surface Process	Steel Corrosion Studies	Metallurgical Studies
Ore selection and Analysis	Nitrided Surface	Bulk Corrosion	Composition
		Surface Corrosion	Phase Transformation
Sinter Monitoring	Carbided Surface	Protective Layers	Retained Austenite in Ferritic Steels
Direct Ore Reduction	Borided Surface	Corrosion Products	
Coal Characterization	Galvanized Surface	Corrosion Process	Ferrite in Austenitic Welds and Castings
Monitoring of Pyrite Content	Aluminized Surface	Inhibitor Influence	Precipitation Hardening
Boiler Ash and Slag Studies	Coated Steels	Ion Implanted Surfaces	Wear-Resistance
End Product Phase Analysis	Wear and Hardness	Alloyed Surfaces	Tempering
	Implanted Steels		Cold Working
	Case Hardened		Aging
	Surface Stresses		Heat Treatment

floating. The particular method suitable for a particular ore depends upon the type and distribution of iron phases in the ore. The characterization and distribution of the different iron phases present in a given ore can be conveniently studied with Mössbauer spectroscopy. For example in the case of the ore containing magnetite as its principal iron bearing constituent, a magnetic separation method is used for the beneficiation of the ore. After beneficiation of the ore to obtain the desired iron bearing phase, it is subjected to a sintering process to obtain a compact material--a sinter, which can be fed into the blast furnace. The sinter should be porous for good reducibility in the blast furnace. The quality of a sinter is related inversely to its glossy content apart from other factors. The quality of a sinter can easily be tested by quantitative estimation of its glossy content. Huffman et al. [1] studied two sinters, one of which gave good performance in the blast furnace and the other poor. They found that the poor sinter contained approximately twice as much glossy material as the good sinter.

Some beneficated ores are reduced to metallic iron and are used directly in electric furnaces in the place of expensive scrap, as what is called 'burden' material in the blast furnace. This leads to saving of significant amounts of coke over what is normally used. Reduction of the beneficiated ores to metallic iron under various processes can routinely be monitored.

The iron bearing FeS_2 phase and the pyritic sulfur content of coals that are used in steel making can be conveniently and accurately determined, as has been demonstrated by Levinson et al. [3] and Huggins et al. [4]. The damaging ash and slag deposits that are formed during direct combustion of coals can be studied to minimize the tendency of the coals to form such harmful deposits. It is well known that mechanical properties of steels like tensile strength, fracture toughness and ductility depend to some extent upon the relative amounts of the various phases present. This phase distribution in a finished end product of steel can be monitored and quantitatively determined.

FERRITE IN AUSTENITIC WELDS AND CASTINGS

The determination of the austenite to martensite ratio [5], the amount of austenite in ferritic steels, and ferrite in austenitic steels is of practical importance in the steel industry.

The first backscatter 14.4keV spectra from a series of bulk stainless steel welds and castings were reported by Schwartzendruber et al. [6] in 1973. The ferrite content in the microstructure of predominantly austenitic steel alloys (viz., Fe-Cr-Ni) significantly influences [7] the mechanical properties of these steels. It has been shown that partially ferritic stainless steel welds are more resistant to fissuring or cracking during or immediately after welding of heavy sections [6]. In addition, they exhibit increased strength, substantially greater resistance to stress corrosion cracking, and improved welding characteristics. On the other hand, it has been shown that too much ferrite can decrease corrosion resistance. Hence it is of practical importance to be able to estimate quantitatively the ferrite content in austenitic steel welds and castings. It has been demonstrated by Schwartzenruber et al. [6] that Mössbauer spectroscopy can be utilized to measure the ferrite content fairly accurately without the problems that are usually connected with shape, size and orientation of ferrite particles in the case of magnetic methods.

Whenever the different phases present in a sample give well-resolved Mössbauer spectra, the relative areas of the respective Mössbauer components give a quantitative estimate of the ratio of the distinct phases present in the sample. However before one gets the actual numbers, in some cases, one has to make corrections for finite sample thickness, phase composition, and the different recoil-free fractions for the different phases.

AUSTENITE IN FERRITIC STEELS

Chow and Bogner [8] measured retained austenite in ferritic steels, and through comparison with x-ray measurements, showed that Mössbauer spectroscopy gives results in agreement with x-ray data to within two percent. Retained austenite in ferritic steel is known to improve the fracture toughness. Fracture toughness of nickel bearing ferritic steels was found to increase dramatically when some austenite was retained after tempering in the austenite-ferrite two phase field. Kim and Schartz [9] carried out a study of the influence of retained austenite in Fe-9Ni-0.1C steel on the fracture toughness of the steel. They showed that the toughness increases by fourfold at room temperature and tenfold at 77°C. Their results indicate that the retained austenite increases the fracture toughness by an amount proportional to the volume fraction of austenite. They studied the fractured surface of the steel by Mössbauer spectroscopy and showed that part of the austenite has been converted into a martensitic phase. They proposed that the improved fracture toughness of the steel could be due to the energy dissipated in the mechanically induced austenite-martensite phase transition. Obviously estimation of retained austenite in ferritic steels is important in understanding the fracture toughness and other physical properties of steels. The fractured surface can be conveniently studied by backscattering Mössbauer spectroscopy whereas x-ray diffraction becomes ineffective because of the roughness of the surface. In the case of Mössbauer scattering studies, the surface irregularities do not effect the results, on the other hand, in the case of x-ray diffraction methods, there is a loss of low angle scattering intensity due to surface roughness. The surface roughness can also broaden the Bragg scattering peaks. Both the effects make the quantitiative estimates of retained austenite uncertain in the x-ray case.

ZERO VELOCITY METHODS FOR RAPID QUANTITATIVE ESTIMATION OF PHASES

For any practical shop floor routine application of Mössbauer spectroscopy in the steel industry, the technique should be simple and quick in providing the information sought. Even a lay person should be able to use the method which does not require complicated analysis of the data to get meaningful results. In this context both the conventional transmission geometry and scattering geometry Mössbauer spectroscopy are not suitable for direct utilization in the steel industry. The transmission geometry technique suffers from the requirement of a thin sample, while in industry, in most cases of practical importance and interest, one should be able to monitor or investigate bulk samples without subjecting them to any mechanical slicing or grinding. The scattering geometry method, although it can monitor bulk samples directly without the requirement of a thin sample, takes too much time to collect the data. Second, and most important, the data has to be corrected for a number of factors that affect the scattered intensity. The analysis of the data in both the above methods is complex and takes time as one has to take recourse to a computer. This is probably the reason for the confinement of Mössbauer spectroscopy to the research laboratories and its not being taken seriously by the steel industry, at least on the shop floor.

Jaggi and Rao [10] proposed a very simple, non-destructive, rapid and inexpensive method, based on just the Mössbauer effect for shop floor utilization in the steel industry, for specific tasks such as monitoring the ratio of ferrite to austenite in steels or for monitoring the pyrite content in coals [11]. This is possible under certain favourable conditions, viz., when there is no overlap of the resonant line or lines belonging to the phase being measured with those of the remaining phases in the sample. The essential principle of this method lies in utilizing a chemically matched source corresponding to the phase to be determined and measuring the scattered intensity when the source is at rest and when the source is vigorously vibrated with an infinite velocity. The change in the two scattered intensities is shown to be proportional to the content of the phase in question. The proportionality constant has to be measured experimentally by calibration procedures. Because only zero and infinite Doppler velocities are involved, the 'cos θ' broadening is of no consequence and hence large solid angles can be utilized. This leads to a large reduction in the counting time. As can be easily realized, this method enables one to do away with the need for a sophisticated drive, multichannel analyser, strong Mössbauer source, long counting times and the need of a computer for data analysis. However, this method calls for utilization of a specialized source for each individual application.

The proposal of Jaggi and Rao [11] for the measurement of pyrite content in coals has been examined in detail by Huggins and Huffman [12] both theoretically and experimentally. They have estimated theortically the resonance effect for a pyrite absorber as a function of the velocity amplitude of a conventional Mössbauer drive for three Mössbauer sources: cobalt-57 doped into pyrite, cobalt-57 in rhodium and cobalt-57 in palladuim. It has been shown that cobalt-57 in pyrite is nearly 100 percent more efficient than the paladuim source and about 40 percent more efficient than the rhodium source for measuring the resonance effect in pyrite in coal, provided the f-factors are assumed to be the same in all the three cases. However, in practice the f-factors for cobalt-57 in palladium, rhodium, and pyrite are 0.66, 0.76, and 0.5 respectively. So, when the f-factors are taken into account, it has been shown that the cobalt-57 in rhodium in effect is 15 percent more effective and cobalt-57 in palladium is about 20 percent less effective than the cobalt-57 doped pyrite source for measuring the pyrite content in coals. So it appears that cobalt-57 in rhodium is a good choice for the quantitative estimation of pyrite in coals rather than the cobalt-57 in pyrite source.

In order to test the concept of the pyritemeter experimentally, Huggins and Huffman, by using the rapid transmission method [12] measured the content of pyrite quantitatively in 20 coals of known pyrite content. They used a cobalt-57 in palladium source moving with a velocity of 0.3mm/sec to maximize the resonance effect and with a velocity of 12mm/sec to obtain the non-resonance background. The difference in the counts between the two measurements was a measure of the pyrite content. They find only a fair correlation between the pyrite content of the coal and the measured resonant-absorption signal while a better correlation between the total iron content of the coal and the resonant absorption signal was obtained. Iron in coals is present not only in the form of pyrite, but also in other iron bearing phases like siderite ($FeCO_3$) or clay, the two other phases found routinely in unweathered coals. It has been pointed out by Huggins and Huffman [12] that because the experimentally observed resonant-absorption signal is better correlated with the total iron content in coals rather than the pyrite phase, the palladium is nearly as sensitive to iron in siderite or clay as to the iron in the pyrite phase. Relative Mössbauer resonant absorption effects for various source combinations are calculated by Huggins and Huffman. They argue that, although a pyrite source may be less sensitive to other iron bearing minerals compared to the pyrite phase, it is still sensitive enough to give resonance effects with siderite and iron bearing clays. This leads to an erroneous estimate for the pyrite content of coals. The source used should not be sensitive to other iron bearing phases under ideal conditions in order that the pyritemeter is useful for routine determination of a given phase in coals. As it is not possible to obtain an ideal source which fulfills all the above conditions, Huggins and Huffman point out that the pyritemeter is useful only in the cases where the coals under consideration are known to contain only pyrite or pyrite and relatively minor amounts of other iron bearing phases. It is also useful when one wants to monitor repeatedly the relative pyrite content in a given coal sample.

The correlation obtained [12] between the resonance effect and various known amounts of pyrite added to Pittsburgh seam coal obtained by the rapid method is shown in Fig. 1. The correlation is linear as expected.

Theoretical calculations for resonance effects as a function of Doppler velocity amplitude for an austenite absorber were done for various sources vis., in austenitic phase, chromium, rhodium, and palledium matrices. It was demonstrated that a chromium source could be an acceptable alternative to

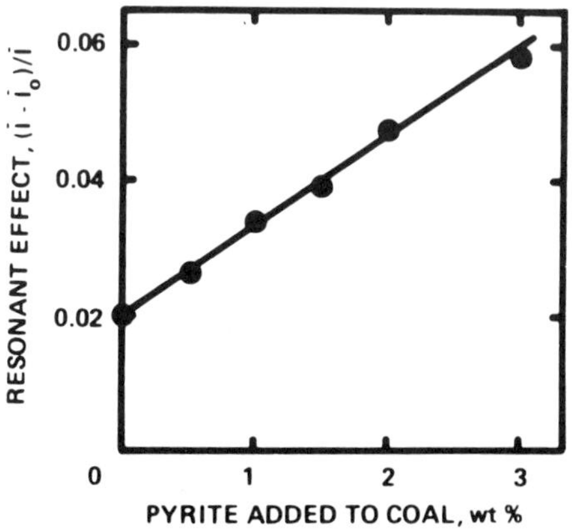

Figure 1. Correlation of measured Mössbauer resonant aborption with additions of pyrite to a sample of Pittsburgh seam coal (Ref. 12).

cobalt-57 in an austenitic phase for the estimation of the austenitic phase in martensitic steels.

Huggins and Huffman [12] have shown that the metallic iron content in a basic oxygen furnace slag can be determined to within two weight percent by using the rapid Mössbauer method and x-ray fluorescence. The slag contains a complex wustite phases. $(Fe,Ca,Mn)_{1-x}O$, substituted dicalcium ferrite, $Ca_2Fe_2O_5$, iron bearing silicates or glass, and iron-metal. It is seen that the iron-metal absorption lines do not overlap with the lines from other phases and this observation by Huggins and Huffman lead them to use a cobalt-57 doped ferrite source for estimating the iron-metal content in the slag. However, due to a slight overlap of the iron-metal lines with the other lines, it is expected theoretically that a ferrite resonance effect may be obtained even when the iron-metal content in the sample is zero. The difficulty has been overcome ingeniously by Huggins and Huffman by using what they called a 'determinative grid' method in which the resonance effect is plotted against the total iron in the slag as determined by x-ray fluorescence for various percentage values of the iron present as metal in the slag. A typical determinative grid is shown in Figure 2.

Once this 'determinative grid' is prepared, it is trivial to estimate the percentage iron present as metal in the slag. It has been pointed out by Huffman, et al. [12] that a quick estimation of iron in the slag would be extremely useful to the efficient operation of basic oxygen furnaces.

A WORD OF CAUTION FOR PHASE ESTIMATION FROM SCATTERING MEASUREMENTS

Mössbauer spectrometry has, over years, evolved into an excellent analytical tool for both identification and quantitative estimation of different phases that are present in a given sample. In particular, it is of great commercial significance in the steel industry to measure quantitatively

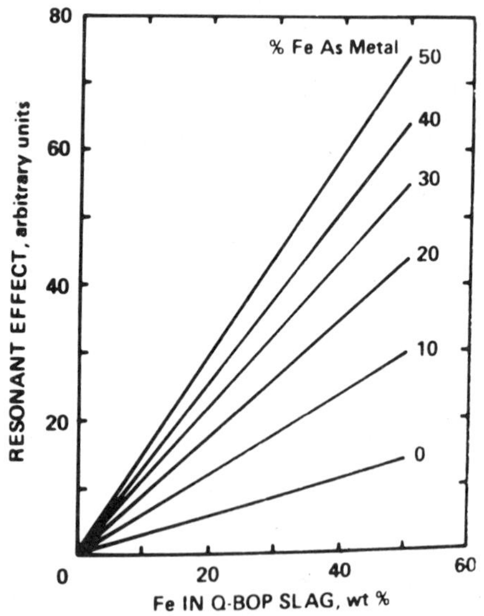

Figure 2. Determinative grid for estimating metallic iron of basic oxygen furnace slag from measurements of the total iron content and the Mössbauer resonant effect. The source is cobalt-57 in ferritic iron (Ref. 12).

the retained austenite [5] in ferritic steel welds and castings and the various metallurgical phases that are present in a given steel sample. The mechanical properties of the steels depend, to a certain extent, on the different phases and their relative proportions present. It was first suggested by Marcus, Schwartz, and Fine [13,14] that Mössbauer spectroscopy could be used for the determination of retained austenite. Subsequently Chow and Bogner [8] have determined the retained austenite both by Mössbauer spectroscopy and x-ray analysis and showed that the two results agree to within one percent. Abe and Schwartz [15], by using transmission Mössbauer spectroscopy showed that the paramagnetic fraction V13., austenite in a two phase ferromagnetic and paramagnetic duplex system, can be fairly accurately measured by just measuring the austenitic peak resonance area, provided the total sample thickness (iron atoms/cm^2) is known. Schwartzenruber [6] was the first to obtain scattering Mössbauer spectra of a series of stainless steel welds and casting and estimate the ferrite content in them.

Although it is quite feasible to get fairly accurate estimates of different phases from Mössbauer transmission experiments, it is not always as straight forward in the case of Mössbauer scattering experiments. Bara, et al. [16] have shown theoretically and experimentally that, in a scattering geometry experiment, under certain conditions, one would actually record an absorption line rather than a scattering line, contrary to the conventional belief that one always observes a scattering line in a scattering geometry experiment. The theoretical and experimental spectra obtained by Bara et al. [16] in the case of iron impurities in thick beryllium plates are shown respectively in Figure 3 and Figure 4. Obviously, in such a case where one gets an absorption line instead of a scattering line, the concept of quantitative estimation of a given phase from the corresponding resonance

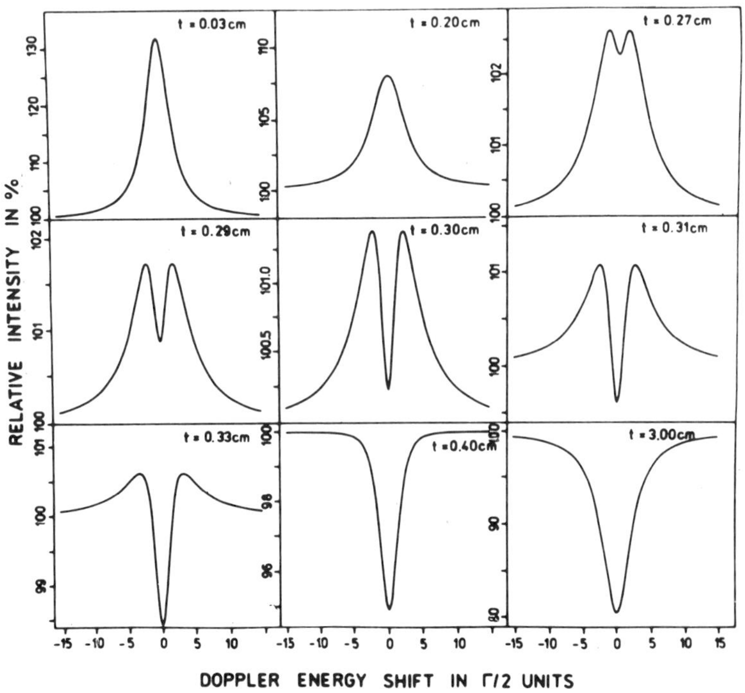

Figure 3. Scattering lines calculated for hypothetical beryllium scatterer with 0.1 at percent of nonenriched iron impurities. Scatters of of various thicknesses t were considered in the calculations [16].

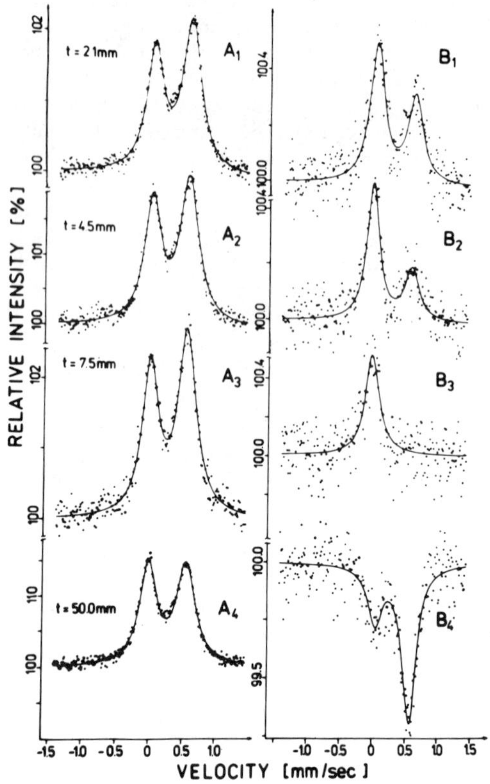

Figure 4. Mössbauer x-ray (A1,A2,A3,A4) and gamma ray (B1,B2,B3,B4) scattering spectra measured at a scattering angle of 130° for iron impurities. The A1,A2,A3,A4, and corresponding B1,B2,B3,B4, spectra were recorded simultaneously in the same geometry and by the same counter. (Ref. 16)

area of the peak loses its meaning. As has been pointed out by Jaggi [17] this kind of an anomoly arises because, in the case of transmission geometry, the Doppler velocity dependent contribution to the absorption line is only from the resonant nuclear recoilless absorption, which is linearly proportional to the number of Mössbauer nuclei for a thin sample. In the case of a scattering geometry experiment, as enumerated by Bara et al. [16] and Jaggi et al. [17] there are three types of Doppler velocity dependent contributions, viz., resonant nuclear, Rayleigh and Compton scattering. It has been theoretically shown by Bara et al. [16] and Jaggi [17] that the contributions from the resonant nuclear scattering is positive while those from Rayleigh and Compton scattering are negative. The theoretical calculations of Bara are shown in Figure 5.

The net intensity depends upon which process is dominating in a particular sample. It has been pointed out by Bara [16] that the unusual shape of an absorption line in a scattering geometry will arise from Mössbauer isotopes with large values of internal conversion coefficient contained in samples with small photoelectric cross-sections compared to those of Rayleigh and Compton scattering processes.

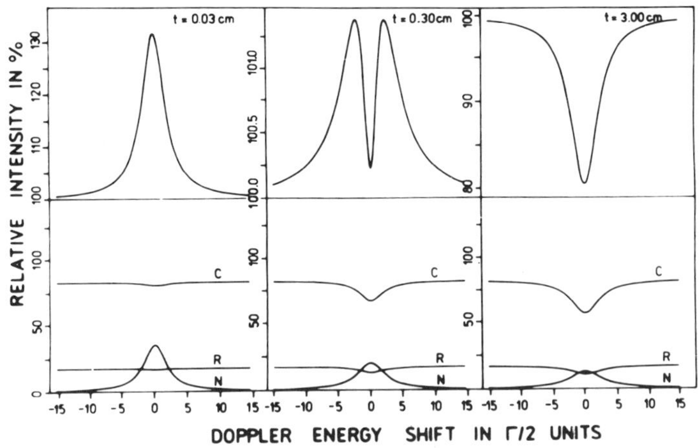

Figure 5. Scattering lines and their nuclear resonance (N), Rayleigh (R), and Compton (C) contributions calculated for 0.1 at percent of non-enriched iron impurities in 0.03 cm, 0.3 cm, and 3.0 cm thick hypothetical beryllium scatterers. (Ref. 16)

It is usual to assume equivalent f-factors for all the different phases in a sample for quantitative estimation of the phases. However, Jaggi [17] has shown that the scattered intensity is sensitively and non-linearly dependent upon the f-factor to the extent that the shape of the spectrum itself varies with f-factor in the case of magnetically split spectra. In this case, the area under a sextet can not simply represent quantitatively the phase present even if a linear scaling for the differences in the f-factors for the different phases is taken into account. Jaggi [18] has shown through his theoretical calculations that the net scattered intensity varies in a highly non-linear fashion with thickness, composition and f-factors of the sample. He cautions that quantitative estimation of the different phases is obviously not straightforward. It is emphasized that for quantitative work one must do complete calculations for the scattered intensity or use elaborately determined calibration curves.

BORIDED STEELS

It is now common knowledge that boriding the surfaces of unalloyed and alloyed steels, in general, gives rise to better surface wear resistance characteristics [19,20]. Carbiding and nitriding the surfaces also lead to improved surface wear protection. It is of great practical importance [21] to know what phases of borides, carbides and nitrides are formed in a particular process and to correlate the wear resistance or toughness with the phase formed. If there is any particular phase which maximizes the wear protection or toughness, then it would obviously be desirable to develop a suitable process for depositing that particular phase on the steel surface. It may not be possible to deposit a particular required phase in a general process of boriding, carbiding, or nitriding; so it becomes necessary to monitor the phases that are formed while different heat treatments or methods are tried. Mössbauer spectroscopy is ideal for such studies.

It has been shown by Habig et al. [19] that boriding alloyed steels leads to some interesting complexities. It was shown that the alloying elements like chromium, molybdenum, and vanadium are dissolved into the boride layers and in some cases concentrated under the surface layer. In both cases, the phase composition, morphology, and the growth of the layer are modified further. It has been shown that the dissolved elements in the boride layer alter the wear behavior of the boride layers. Again, it is of great practical importance to follow the changes in the phases formed and the corresponding changes in the wear resistance behavior of the borided layers.

It has been shown that a single phase FeB layer wears out due to adhesion while a two phase viz., a FeB and Fe_2B layer, offers better wear resistance. It was inferred that Fe_2B plays a positive role in adhesion. It was shown that borided alloy steels wear faster compared to hardened steels whenever the main wear mechanism is due to adhesion. It was further shown that the borided steels wear less compared to hardened or nitrided layers if the wear mechanism is due to tribooxidation. In this case the different boride phases do not seem to affect the wear properties. If the mechanism of wear is due to abrasion, the borided steels show superior wear resistance compared to hardened or nitrited steels. It is observed that the alloying elements in steels, when dissolved in the boride layers on the steel, lead to a marked influence in the wear behavior. Future Mössbauer spectroscopic study of the borided layers on steels to establish the phases formed, their thickness, and the changes that occur in them when the steels are subjected to various kinds of wear causing mechanisms and heat treatments, can profitably be undertaken.

Rao et al. [22] carried out conversion electron Mössbauer spectroscopic (CEMS) studies of borided mild steels with a view to characterizing the different phases that are formed as the thickness of the boride layer is increased and to study the effect of laser treatment on the borided surface. Borided mild steel samples with different thickenss of the borided layers ranging from 10μm to 45μm were prepared by the pack cementation method. The CEMS spectra of the borided steels are presented in Figure 6. The spectra clearly indicate that, upto a thickness of 40μm, mainly the Fe_2B phase is formed. Subsequently when the thickness of the borided phase is further increased to 45 μm mainly the FeB phase is formed. Laser induced rapid melt quenching of borided steels, which have a glass forming composition, is likely to produce an amorphous surface layer, intrinsic with the underlaying crystalline substrate. It is known that such amorphous layers could substantially enhance the wear and corrosion resistance. With a view to check whether the top crystalline FeB phase on the steel sample with 45μm thick boride layer is converted into an amorphous phase on laser treatment, the entire surface of the steel sample was subjected to laser treatment with a $3J/cm^2$ and 3mm diameter spacially homogenized overlapping ruby laser beam of 25μsec. duration. The calculations of Jain et al. [23] indicated surface melting up to 0.4μm deep and rapid quenching at cooling rates greater than 10^9K/sec. The spectrum of the laser treated sample shown in Figure 6 indicates at least partial formation of the amorphous FeB phase.

SURFACE TREATED STEELS

Steel surfaces, in some cases, are galvanized, aluminized, painted with an oxide, or allowed to form some protective coating to prevent or reduce the corrosion of the substrate. Scattering Mössbauer spectroscopy can be profitably utilized to characterize the surfaces and to understand their role in the reduction of the corrosion process. One example of this kind of study [24] is the work done at U.S. Steel Corp. on COR-TEN, a popular structural steel which develops a protective oxide layer when exposed to the atmosphere. It is known that this steel has superior corrosion resistance compared to normal structural steels and at the same time, the surface does not require any additional painting as it develops an attractive colour on atmospheric exposure. Huffman [24] analyzed the surface utilizing both electron and x-ray scattering Mössbauer spectroscopy and identified the oxide layer on the

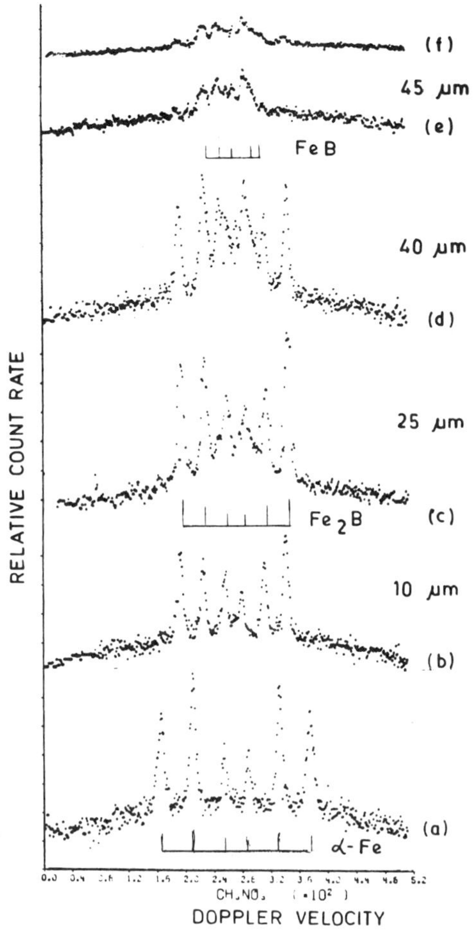

Figure 6. Mössbauer spectra of borided mild steel samples with different thicknesses of the borided layers: (a) pure ferrite (b) 10μm, (c) 25μm, (d) 40μm, (e) 45μm, (f) 45μm laser treated.

COR-TEN steel as consisting of a mixture of γ-FeOOH and β-FeOOH. They find the thickness of the oxide layer to be 13μm after the steel is exposed to the atmosphere for a period of three months.

Galvanized and aluminized steels have been studied by Jones and Denner [25]. They studied the kinetics of the hot dip galvanizing and aluminizing reactions. The intermetallic compounds formed on the surface have been investigated. It is inferred that the phases formed in the case of aluminized steels are governed by a parabolic law.

It is of practical importance to investigate the oxides that are formed after prolonged atmospheric exposure of steels coated with corrosion protective material. An analysis of the products formed might lead to an understanding of the corrosion process on such surfaces and the role played by the protective material coated. A detailed review of the Mössbauer spectroscopic studies on surface treated steels is available in the literature [24].

MEASUREMENT OF RESIDUAL STRESS

Residual surface stresses in steels are caused by a number of factors like shaping, cutting, welding, finishing operation, and the geometrical configuration in the final assembly. It is well known that tensile stresses locked-in on the surface of steel eventually lead to stress-corrosion cracking and failure of the product, if the stress in service is also tensile in nature. Hence identification of the nature of the stress and quantitative measurement of the surface residual stresses in structural steels is of great commercial importance. So far, x-ray diffraction is the only non-destructive method for the determination of the residual stresses. A large number of efforts have been made to make the x-ray diffraction method portable for the shop-floor utilization of the technique. Ruud [26] has recently developed a portable x-ray stress analysis instrument utilizing a position sensitive scintillation detector based on fiber optics which could achieve an accuracy of $\pm$1.5ksi in measuring the residual stresses on austenitic steels. However this technique is quite sophisticated and expensive. It is worthwhile to look into the possibility of developing an alternative, cheap, simple, rapid, and non-destructive method for the measurement of residual stresses. In this context, Mössbauer spectroscopy seems to be promising, although to date, the technique has not been developed well enough to compete with the x-ray method, although a few attempts [27,28,29] have been made in the past.

Measurement of residual stresses by conventional Mössbauer spectroscopy is almost impossible because applying as much as 100ksi stress in any ferrous metal will cause only a very small shift [26] of the Mössbauer line, viz., a shift of about 0.005mm/sec, which is only two percent of the natural linewidth. The normal stresses that are encountered in any real situation of technological interest are expected to produce only one percent change in the linewidth.

An elegant method for the measurement of the residual stresses has been adopted by Collins [28] and Baer et al. [27]. Consider the Mössbauer absorption line shown in Figure 7.

It has been calculated that the line shifts by about 5×10^{-5}mm/sec/ksi. Suppose one shifts the energy of the emission line such that one 'sits' at

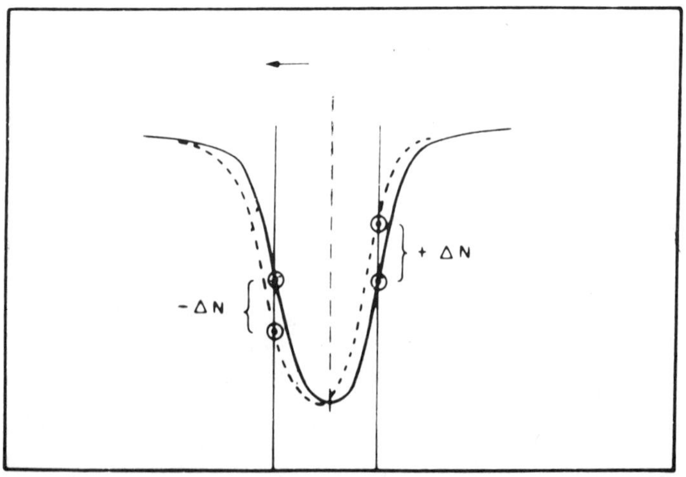

Figure 7. The effect of stress on a Mössbauer absorption line.

one of the points of inflexion of the line, i.e., at a point on the curve with maximum slope. Then any small shift in the line position in either direction, due to any stress or compression, will lead to a large change in the count rate. Similarly when one shifts the emission line position to the other point of inflexion of the absorption line, the change in the count rate will be in the opposite direction. By counting the changes in the count rates at these two points, one can double the sensitivity of the measurement. Baer et al. [27] used a thermal method to, in effect, shift an emission line near to the two points of inflexion of a central absorption line by maintaining the source respectively at +100°C and -80°C, in transmission geometry and at +80°C and -55°C in the scattering geometry measurements of stress in a ferritic steel. Obviously different calibration curves have to be obtained for different steels before the method could be used in any practical application. It was demonstrated by Baer et al. [27] that the stress measurements can be made with an accuracy of $\pm$15ksi in 30 minutes by using their thermal scanning technique. In fact, the ideal temperature [27] by which the source has to be heated in order shift the source line position to the point of inflexion would be 144°C and such temperature changes would improve further the accuracy of the measurement provided the f-factor is not drastically reduced at that temperature.

Jaggi and Rao [30] used the method of chemically matched sources to shift the emission line to the points of inflexion. In the case of a stainless steel scatterer, the desirable sources would be cobalt-57 in tantalium and cobalt-57 in vanadium, which shift the emission line to the two points of inflexion. Preliminary trials to make these sources proved to be disappointing as the sources prepared were found to be highly brittle and came apart as powder. In the absence of the ideal sources, they utilized cobalt-57 in chromium which shifts the emission line from near the center of the stainless steel absorption line half way to one of the points of inflexion. Intergerated differences in the count rates of the scattered 14.4kev gamma rays were recorded with the source at rest and vigorously vibrating. The change in the differences in the integrated count rates with and without a 7ksi external tensile force was just above the statistical fluctuations of the count rates. Utilization of the ideal chemically matched sources to shift to the points of inflexion is expected to improve the sensitivity of the technique. There appears to be scope for further improvement of the sensitivity of the method.

MÖSSBAUER STUDY OF COALS

The depleting supply of gasoline and petroleum products in the world has awakened renewed interest in coal as a major source of electricity and synthetic fuels. Because of the importance of coal as a source energy and the associated possible hazardous environmental pollution from coal has led to considerable research to understand fully the different compounds present in coals and how the compounds transform during processing and utilization in the steel industry. The suitability of coal for a particular industrial application, in the steel industry depends on the constituents of the coal. Thus characterization of different coals is an important step. Mössbauer spectroscopy is an ideal and probably indispensible and unique analytical tool for the identification of iron-bearing minerals in coal. Particularly, detection and quantitative estimation of pyrite, FeS_2, by Mössbauer spectroscopy is extremely relevant for the steel industry. The three main areas where coal is utilized are: 1) direct use as a combustible material for steam production, 2) in the steel industry, and 3) for synthetic fuel production.

Here we confine ourselves to the steel industry. The coal that is used in the steel industry is first converted into coke. The conversion of coal into coke is a chemical and physical process. In this case, an appropriate variety of coal, containing minimum permissible amounts of pyrite, is exposed to sufficiently high temperature for a long period of time in the absence of air. First volatile gases and tars are expelled from the coal and carbon and ash are produced. Not all coals are suitable for coking. It has been shown that heavily weathered coal is not suitable for coking. Huggins and Huffman [4] studied several cokes produced from different coals and identified the various phases in the coals.

STEEL CORROSION

Corrosion is essentially a surface phenomena. It is possible to prevent or minimize the corrosion reaction electrochemically by stimulating the anodic reaction which promotes a protective film formation on the surface (the so called anodic protection) or the cathodic reaction (the so called cathodic protection) with a corresponding lowering of the metal oxidation rate. It is also possible to reduce corrosion by surface alloying which gives adequate corrosion resistance to a metal with intrinsically inferior corrosion resistance but otherwise having all the required mechanical properties. However, because the solid solubility of the alloying elements in the base metal is restricted by the equilibrium phase diagram, this method has limitations and can be used only in certain favorable cases. Ion inplantation provides an alternative method to surface alloying to produce corrosion-resistant exotic surface alloys even in the case where the equilibrium phase diagrams are unfavorable for the required concentrations of the alloying elements in the base metal.

Backscatter Mössbauer spectroscopy utilizing the 14.4kev gamma rays scattered from the bulk near the surface and the 6.4kev conversion electrons scattered from the thin top layer of the surface, offers an excellent method to understand the process of surface corrosion of steels in all the above mentioned preventive methods of corrosion. Excellent reviews of the use of Mössbauer spectroscopy in corrosion studies are available [31,32].

To learn how to prevent corrosion of steel in sea-water is obviously of commercial importance. Such a study will lead to enormous saving of money, material, and manual labor. Low alloy steels are known [33] to form two layers of iron rust during corrosion in sea water. The first layer, which is of the order of 2000Å is called the barrier layer. On top of this layer another iron rust layer is formed. It is important to know the properties of the barrier layer viz., the nature of its composition, structure, and thickness. One example of this kind of study is that carried out by Chen et al. [33]. They corroded low alloyed steels containing chromium and nickel and another set of low alloy steels not containing any chromium or nickel, in sea water for one year. They determined the deposited layer (top layer) composition. Then they scrapped off the rust on the top and by using transmission geometry Mössbauer spectroscopy, they showed that the rust of nickel and chromium bearing steels consists of 65 percent β-FeOOH, 20 percent Fe_3O_4 and 15 percent of γ- and α-FeOOH. The barrier layer rust was found to consist of 80 percent β-FeOOH and 20 percent γ-FeOOH for both the alloys.

Jaggi et al. [34] carried out CEMS studies on plain carbon steels corroded in alkaline water at 310°C and a pH of 10.5 and containing a low dissolved oxygen content of 0.05ppm, simulating the typical operating conditions of a pressurized water nuclear reactor. Corrosion studies have also been done under the above conditions but with dissolved 'high' oxygen (3.5ppm) conditions. Analysis of the spectra obtained in the case of a 'low' oxygen sample showed that a protective film of nonstoichiometric ferrous, cation deficient magnetite was first formed and that any further corrosion leads to the growth of the film at the oxide-water interface with the composition approaching stoichiometry as the film grows thicker. In the case of 'high' oxygen sample, it was shown that the oxide layer consisted of about 85 percent hematite and 15 percent Fe_3O_4. It was further inferred that roughly the top 4000A is pure hematite and that the non-stoichiometric magnetite was adjacent to the base metal.

Their study has shown that even large amounts of hematite form a continuous, adherent, and protective thick film on an underlying thin Fe_3O_4 film which in turn is adherent to the base metal. This is contrary to the popular belief that hematite does not form an adherent and protective film.

Although it may be difficult to stop corrosion process completely, it is possible to decrease the corrosion rate by adding known corrosion inhibitors. It is of interest in such cases to investigate the corrosion process with inhibitors. Choudhury et al. [35] carried out corrosion studies on mild steels in NaCl and HCl solutions both without and with corrosion inhibitors, viz., thiourea and a surfactant known as N-laurlyβ-aminopropionic acid. They find that the corrosion products formed are mainly γ-FeOOH with small quantities of α-FeOOH and α-Fe_2O_3. The corrosion rate without the inhibitor was formed to follow a parabolic law. The resonance area under the corrosion product is shown as a function of time in Figure 8. Choudhuri et al. [35] obtained a corrosion rate constant of $0.14\text{x}10^{12} cm^2$/sec for the corrosion of mild steel in a N/10 HCl solution. They showed that thiourea drastically reduced the corrosion process, while the surfactant could control and prevent the corrosion process only during the first two hours. It was found that the surfactant loses its inhibiting property progressively on continued use beyond two hours.

One of the commercially important problems is the corrosion of widely used cast iron boilers in domestic central heating systems. It was shown by McEnaney et al. [36], through chromatographic analysis, that the most significant corrosion process in the system is due to the formation of Fe_3O_4 and H_2 gas. The suspension of magnetic particles is known to cause blockage and seizure of water pipes. It was observed that a substantial amount of these particles in the system circulate as a sludge formed from high proportions of ferrous corrosion products, $FeOH^+$ ions and $Fe(OH)_2$. They are converted to Fe_3O_4 via the Schikorr reaction. Copper, nickel, cobalt, and platinium, or their ions in solution, are believed to aid the formation of Fe_3O_4 through the Schikorr reaction. However, McEnaney et al. [36] did not find any evidence for the catalytic action of copper in the Schikorr reaction. It was inferred that Fe_3O_4 forms a protective boiler scale under low dissolved oxygen conditions and the corrosion rate controlling process is the transport of $FeOH^+$ ions through the pores in the Fe_3O_4 film. Consequently the particle size of the magnetite formed plays a significant role in inhibiting or promoting the corrosion process. Hence it is probably worth investigating the formation of magnetite under various conditions of oxygen levels, water

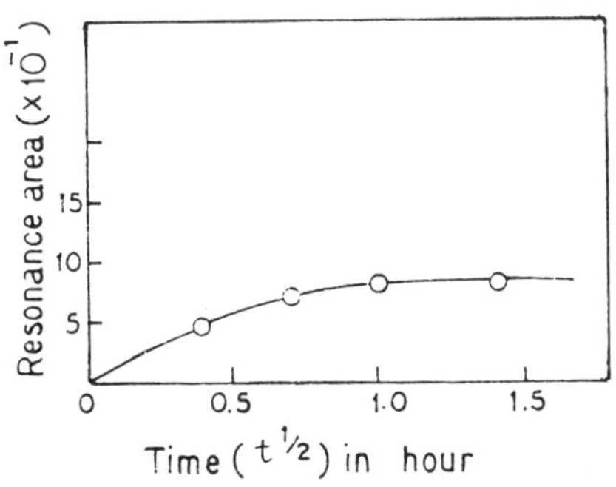

Figure 8. Resonance area of corrosion products formed as a function of corrosion time. (Ref. 35)

chemistry, temperature of operation, impurity elements, and in the presence of corrosion inhibitors to understand the particle size of Fe_3O_4 and the intermediate products that are formed.

It is known that the presence of alkalies in the water used in steam generating equipment made of mild steel, for instance in nuclear reactors, enhances the protective character of the Fe_3O_4 layer formed on the surface of the steel. The underlying mechanism for this improved characteristic is understood through Mössbauer backscatter measurements [37]. It is observed that when a steel specimen is heated in plain water an intermediate corrosion product, $Fe(OH)_2$ is formed at the interface between the initial corrosion product (Fe_3O_4) layer and water due to the diffusion of iron(II) ions from the steel through the porous Fe_3O_4 corrosion product to water. $Fe(OH)_2$ is observed to decompose to fine particles of Fe_3O_4 which exhibited superparamagnetism. However, when the water is made alkaline, it has been observed that the $Fe(OH)_2$ decomposes to coarse Fe_3O_4 particles. These coarse particles block the porous Fe_3O_4 layer and prevent any further movement of iron(II) ions from steel to water and thus lead to a drastic decrease in the corrosion rate of the steel.

CONCLUSION

We see that Mössbauer spectroscopy is being utilized in many aspects of the steel industry. The examples chosen clearly point out that the technique is useful both as a routine monitoring tool as well as an investigative tool in the industry.

ACKNOWLEDGEMENTS

The author is grateful to Prof. J. J. Bara and Dr. G. P. Huffman for their kind permission to use their published data and figures. The Mössbauer scattering measurements and theoretical calculations carried out at Trombay and referred to in this article were performed by Dr. N. K. Jaggi (now at Northeastern University, Boston, U.S.A.) as a part of his doctoral thesis work. The author has greatly been benefited from his association and wishes sincerely to acknowledge the same. The author is grateful to Dr. P. K. Iyengar, Director, Bhabha Atomic Research Centre, Trombay for constant encouragement.

REFERENCES

1. G. P. Huffman and F. E. Huggins in 'Mössbauer Spectroscopy and its Chemical Applications' Eds. J. G. Stevens and Gopal K. Shenoy, Advances in Chemistry Series 194, 265 (1981).

2. G. P. Huffman and F. E. Huggins, AIP Conference Proceedings, 84, (1982), 149 (1982).

3. L. M. Levinson and I. S. Jacobs, Fuel, 56, (1977), 453.

4. F. E. Huggins and G. P. Huffman , 'Analytical Methods for Coal and Coal Products', Ed. Clarence Karr, Jr., Academic Press, vol. 3, pp 371-423 (1979).

5. T. E. Cranshaw in 'Advances in Mössbauer Spectroscopy', Eds. P. K. Iyengar, B. V. Thosar, S. C. Bhargava and J. K. Srivastava, Elsevier Sci. Pub. Co., (1983), pp 256-257.

6. L. J. Schwartzendruber, L. H. Bennet, E. A. Schoefer, W. T. Delong, and H. C. Cambell, Welding Journal, Jan. 1974.

7. G. Brouwer, Philips Welding Reporter No. 3, 1978.

8. H. K. Chow and R. L. Bogner, 'Developments in Appl. Spec.' Vol. 8, Ed. E. L. Growe, Plenum Press, 229 (1970).

9. K. J. Kim and L. H. Schwartz, J. de Physique, Colloque C6-405, (1976).

10. N. K. Jaggi and K. R. P. M. Rao, NDT International, 281 (1978).

11. N. K. Jaggi and K. R. P. M. Rao, Fuel 58, 688 (1979).

12. F. E. Huggins and J. P. Huffman, J. Appl. Phys. 55, 1404 (1984).

13. H. L. Marcus, L. H. Schwartz, and M. E. Fine, Trans. Am. Soc. Metals 59, 468 (1966).

14. B. W. Christ and P. M. Giles, Trans. Met. Soc. of AIME 242, 1915 (1965).

15. N. Abe and LO. H. Schwartz, Mat. Sci. and Eng. 14, 239 (1974).

16. Jozef A. Bara and F. Bogacz, Nucl. Instr. Methods 186, 561 (1981).

17. N. K. Jaggi and K. R. P. M. Rao, 'Intr. Conf. on the Appl. of Mössbauer Effect', Jaipur 1981, Proc. Indian National Science Academy, New Delhi, 268 (1982).

18. N. K. Jaggi, Ph.D. thesis, Bombay University, India (1982).

19. K. H. Habig and R. Chattergee-Fischert, Tribology International, 209 (1981).

20. M. Carbucichio, L. Bardani, and G. Palombarini, J. Materials Science 15, 711 (1980).

21. A. Handa and Y. Ujihira, 'Intr. Conf. on the Appl. of Mössbauer Effect' Jaipur 1981, Proc. Indian National Science Academy, New Delhi, 306 (1982).

22. K. R. P. M. Rao, D. K. Sood, Animesh K. Jain, and J. Subrahamanyam, Nucl. Phys. and Solid State Phys. (India) 25C 523 (1982).

23. K. Jain, V. N. Kulkarni, and D. K. Sood, Appl. Phys. 25, 127 (1981).

24. G. P. Huffman, in 'Applications of Mössbauer Spectroscopy', vol. II, Ed. R. L. Cohen, Academic Press, 202 (1980).

25. R. D. Jones and S. G. Denner, Scripta Met. 8, 175 (1974).

26. C. O. Ruud, J. of Metals 31, 10 (1979).

27. W. H. Baer, R. L. Collins, R. A. Mazak, and C. M. Yagnik, Proc. 9th Symposium on NDE, San Antonio, (1973).

28. R. L. Collins, R. A. Mazak, and C. M. Yagnik, Mössbauer Effect Methodology Vol. 8, Eds., I. J. Gruverman and C. W. Seidel, Plenum Press 191 (1973). R. L. Collins, Intr. J. Nondestructive Testing 4, 77 (1972).

29. J. H. Terrel, Intr. J. of Nondestructive Testing 2, 267 (1976).

30. N. K. Jaggi and K. R. P. M. Rao, Unpublished work (1979).

31. G. W. Simmons and H. Leidheiser, J. of Applications of Mössbauer Spectroscopy, Vol. I, Ed. R. L. Cohen, Academic Press 85 (1971).

32. H. Leidheiser, Jr., G. W. Simmons, and E. Kellerman, Croatica Chemica Acta 45 257 (1973).

33. J. Chen, Z. Cai, Zhijie Wang, Chihu Wang and H. Zhang, Intr. Conf. Appl. of Mössbauer Effect, Jaipur 1981, Proc. Indian National Science Academy, New Delhi 261 (1982).

34. N. K. Jaggi, K. R. P. M. Rao, Y. D. Dande, P. K. Chauhan, and H. S. Gadiyar, Intr. Conf. Appl. of Mössbauer Effect, Jaipur 1981, Proc. Indian National Science Academy, New Delhi (1982) 313-315.

35. K. Choudhury, D. Das, C. K. Majumdar, K. R. P. M. Rao, and M. Adhikari, Phys. Stata Solidi 91, to be published (1985).

36. B. McEnaney and D. C. Smith, Corrosion Science 18, 591 (1978)..

37. A. M. Pritchard, J. R. Haddon, and G. N. Walton, Corrosion Sci. 11, 11 (1971).

MÖSSBAUER INVESTIGATION OF HIGH-MANGANESE STEELS

G. P. Huffman

U.S. Steel Corporation
Technical Center
Monroeville, PA 15146

Chiaki Ouchi

Nippon Kokan K. K.
Technical Research Center
Kawasaki, Japan

ABSTRACT

Austenitic high-manganese steels containing from 16 to 36 weight percent manganese and from 0.01 to 0.75 weight percent carbon were investigated by Mössbauer spectroscopy in both hot-rolled and cold-rolled conditions. Spectra were obtained over the temperature range 10 to 295K in the transmission mode and at room temperature in the X-ray re-emission mode. In the hot-rolled condition, the steels were primarily antiferromagnetic γ-phase austenite, and exhibited Neel temperatures ranging from 234 to 460K. After cold reduction, varying degrees of transformation to the nonmagnetic hcp ε-phase were observed, the resistance of the austenite to mechanically induced transformation increased with increasing manganese and carbon content. Steels that exhibited relatively small amounts of the ε-phase after cold reduction showed continuous work hardening with increasing cold reduction. Steels that exhibited large amounts of ε-phase formation during cold reduction exhibited rapid work hardening followed by saturation. No magnetic α-phase material was observed by Mössbauer spectroscopy, in agreement with the very small changes observed in magnetic susceptibility.

INTRODUCTION

The best known high-manganese steel is Hadfield steel, invented by Hadfield in 1882. Hadfield steel is an austenitic steel that typically contains 1.0 to 1.4 percent carbon and 10 to 14 percent manganese*, and is primarily used as an abrasion-resistant steel. In recent years, new applications of high-manganese austenitic steel have been investigated. Advanced technologies such as the plasma fusion reactor or the high-speed repulsive-levitation railway will require a nonmagnetic structural steel. A high magnetic field generated by means of superconductive devices is utilized as an energy source in these technologies, and the use of nonmagnetic structural steel is important to minimize magnetic energy loss.

*All compositions given as weight percentage unless otherwise noted.

Recently, the greatest amount of research and development activity for high-manganese austenitic steel has been carried out in Japan [2,3]. As a result, several new types of high-manganese steel have been developed. In general, these new steels have featured lower carbon (~ 0.2 to 0.8%) and higher manganese contents (~15 to 30%) than Hadfield steels. Chromium additions (~2%) are also made to improve corrosion resistance. Traditionally, high manganese steels have had carbon contents of 1.0% or larger because the strength of these steels in the solution-treated condition decreases markedly with decreasing carbon content. However, application of thermomechanical treatment has allowed the carbon content to be decreased while still maintaining or increasing the yield strength and simultaneously increasing ductility and machinability. Additionally, the combination of high manganese and moderate carbon contents has produced an increase in the stability of nonmagnetic properties, a decrease in thermal expansion coefficients to values similar to those of ferritic steels, and an increase in low-temperature toughness. These alterations in properties are all highly desirable for the applications mentioned earlier, and also for numerous other applications including: submarine hulls requiring high-strength nonmagnetic steel, piling steels requiring good abrasion resistance and high strength against buckling, steels for use in cryogenic applications, and line-pipe steel requiring superior low-temperature toughness and good corrosion resistance.

As well documented through studies of Hadfield steel, the austenite of high manganese steel can transform to ε-phase (hcp structure) or α martensite by cold working and also by change of temperature, depending on the chemical composition. Additionally, both the austenite and ε-phases of high manganese steel exhibit a magnetic transition from paramagnetism to antiferromagnetism with decreasing temperature. Although changes of both the structure and the magnetic state significantly influence the mechanical and physical properties of high-manganese steel, very few systematic evaluations of the phase mixture and magnetic state have been conducted on high-manganese steels with different combinations of carbon and manganese contents. In this investigation, four steel compositions were investigated in annealed, hot-rolled, and various cold-rolled conditions. Mössbauer spectroscopy was used to determine the percentages and magnetic transitions of the different phases present. The Mössbauer results were correlated with the results of various mechanical and physical property tests.

EXPERIMENTAL PROCEDURES

The chemical compositions of the steels investigated are shown in Table I. Steels HA and HM were mill-production heats hot-rolled as H-shapes. Steels 50B and 51B were melted in a 50-kg induction-heating furnace, and hot-rolled to 12mm thick plates. Hot rolling in both the H-shape mill and the laboratory rolling mill was carried out at the reheating temperature of 1200°C and at the finishing rolling temperature of 850°C. The specimens for tensile testing were 10mm in diameter and 50mm in gage length, and were machined from the mid-thickness parallel to the rolling direction of the plate or the flange in the H-shape.

The steels were investigated in basically three conditions: as hot-rolled, solution-treated, and cold-rolled. The solution treatment was carried out by reheating at 1050°C for 30 minutes, followed by water quenching. For cold rolling, sheets 0.5mm or 0.8mm thick were machined. The 0.5mm thick sheets were cold-rolled to thickness reductions of 5, 10, and 20 percent, and the 0.8mm thick sheets were cold-reduced 50 percent. From these thin sheets, specimens for Mössbauer spectroscopy and other measurements were prepared.

Table 1. Chemical Composition of Hi-Mn Steels

Steel	C	Si	Mn	P	S	Cr
HA	0.74	0.40	15.90	0.0240	0.005	1.82
HM	0.27	0.44	24.40	0.026	0.008	1.65
50B	0.009	0.25	26.68	0.011	0.0061	----
51B	0.018	0.24	35.58	0.012	0.0088	----

Mössbauer spectroscopy was used to investigate the magnetic properties and mechanically induced phase transformations of the steels. The Mössbauer spectrometer used was a constant acceleration spectrometer of standard design, described in more detail elsewhere [4]. Room-temperature Mössbauer spectra were obtained from thick specimens in the X-ray re-emission or backscatter mode. X-ray re-emission Mössbauer (XRM) spectra are obtained by detecting the 6.4keV X-rays emitted following the internal-conversion process that follows resonant absorption of the 14.4keV Mössbauer gamma ray by iron-57. The attenuation coefficients of the 6.4keV X-rays are such that the XRM technique probes the sample to a depth of 15 to 20mm [4,5].

Low temperature spectra were obtained in the transmission mode. The samples were thinned chemically to 25 to 30mm thick foils, and cooled to temperatures in the range from 10 to 250K in a Displex cryogenic system [6].

Neel temperatures and magnetic permeabilities were measured by using a vibrating sample magnetometer. The specimens, prepared from hot-rolled plates, were heated from 50 to 300°C under a magnetic field of 10 kilogauss.

Various additional measurements, including X-ray diffraction, transmission electron microscopy, abrasion resistance measurements, and dynamic strain aging tests, were also made. The current paper will concentration primarily on the relationship of the Mössbauer spectroscopy results to the mechanical and magnetic properties of the steels. A more complete report will be published elsewhere [7].

EXPERIMENTAL RESULTS

Mechanical and Magnetic Properties

Tensile properties in the as-hot-rolled condition are listed in Table II. It was found that the strength of high-manganese steels depends primarily on carbon content, but relatively little on manganese content. The nominal stress-strain curves of the steels are shown in Figure 1. The work-hardening rate increased with increasing carbon content, except for steel 50B, which exhibited a high work-hardening rate despite its very low carbon content. As discussed later, this is probably due to a strain-induced transformation from γ austenite to ε martensite.

Figure 2 shows the change of hardness with cold-rolling reduction. The hardness of steel HA (16Mn-0.74C) increased almost lineraly with rolling reduction up to 50 percent, and the hardness value at a cold-reduction of 50 percent reached 700 in Vicker's hardness value. The hardness variations of steels HM (24Mn-0.27C) and 50B (27Mn-0.01C) were very similar. The hardness of these steels increased rapidly with relatively small cold reductions, and

Table II. Tensile Properties of the Steels

Steel	0.2% PS kg/mn^2	TS kg/mn^2	EL %	RA %
HA	58.8	112.7	29.8	26
HM	51.3	92.7	43.8	60
50B	22.0	74.7	39.3	52
51B	19.1	52.0	50.3	79

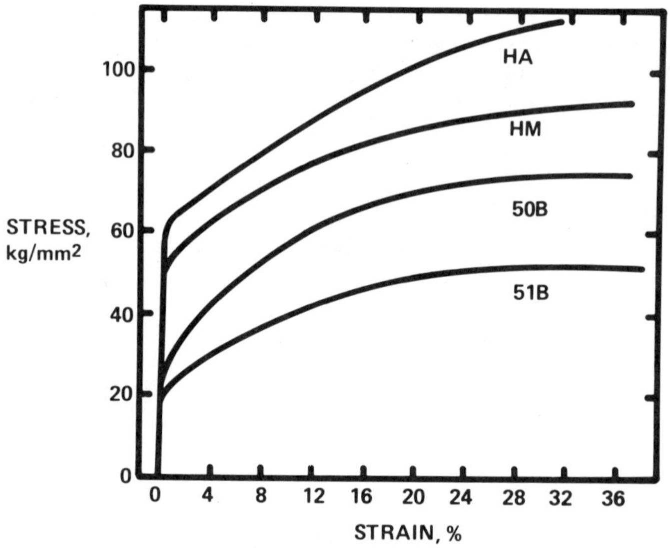

Figure 1. Stress-strain behavior of austenitic high-manganese steels.

then tended to saturate at reductions above approximately 20 percent. Steel 51B (36Mn-0.02C) also exhibited a continuous increase of hardness with reduction. The hardening rate, however, was lower than that of steel HA.

The stability of the nonmagnetic properties of the steels is illustrated by Figure 3. It is seen that the increase of the magnetic permeability of the high-manganese steels with increasing cold reduction is very small compared to that of a typical stainless steel or that of a Hadfield steel. Indeed, with the exception of steel HA (16Mn-0.75C), the variation of the magnetic permeabilities of the high-manganese steels with cold reduction is essentially negligible.

As an example of one application of hot-rolled, high manganese steel, steel HA was used as an H-pile. A newly developed piling method, called gunpiling, was specially designed to be suitable for piling in very hard, rocky soil. As a primary property of the steel pile used in this method, very high abrasion resistance as well as high strength against buckling was required. Figure 4 shows the outer appearance of the top end of the H-pile after striking to 4 m depth. The suitability of this steel for this application was confirmed by the extremely small abrasion wear that occurred in the sharp knife edge at the top end of H-pile. Hardness distributions of this portion were measured as a function of depth from the surface as shown in Figure 5. The steep hardness distribution extended 0.4 or 0.5mm into the subsurface layer.

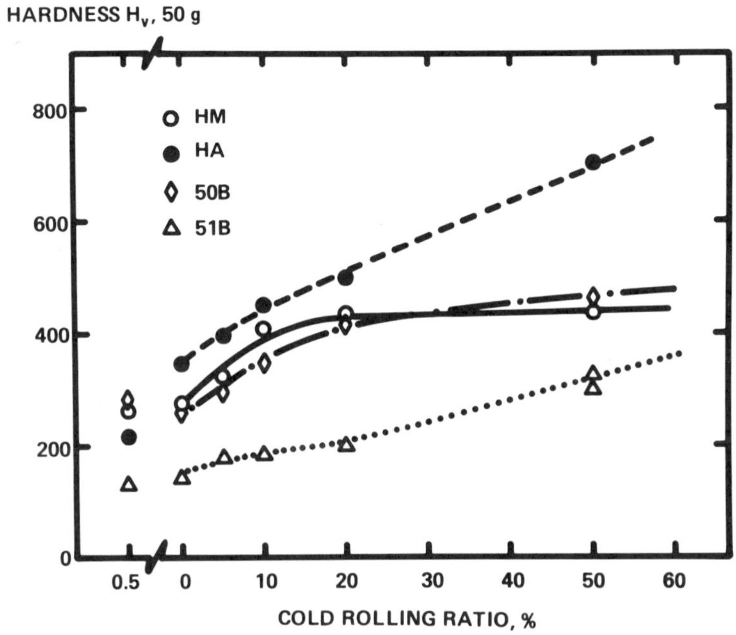

Figure 2. Changes in hardness values with increasing cold-rolling ratio.

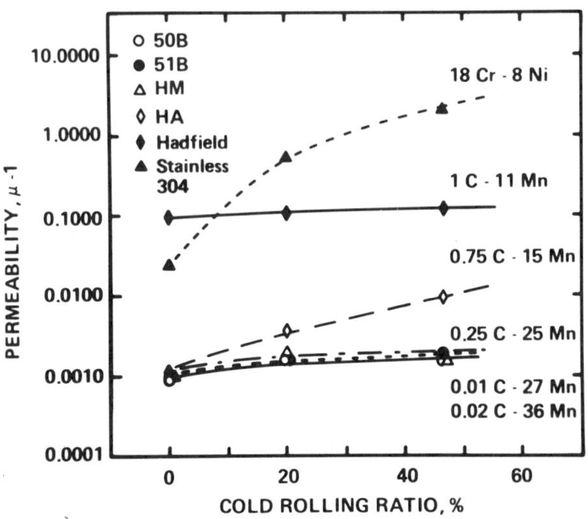

Figure 3. Change of magnetic permeability with increasing cold-rolling ratio for high-manganese austenitic steels, stainless steel, and Hadfield steel.

Figure 4. End of an H-pile made from steel HA after striking to a 4 m depth.

Room Temperature Mössbauer Data

Typical XRM spectra obtained from several high-manganese steels at room temperature are shown in Figure 6 and 7. In either the solution-annealed state (1050°C for 30 min., followed by quenching) or the hot-rolled condition, the steels contained primarily the fcc γ-phase (austenite). During cold rolling, varying amounts of the steels were transformed to the hcp ε-phase. No α-martensite was observed in any of the samples investigated. For most of the samples, the γ phase was antiferromagnetic with a Neel temperature above room temperature, while the ε-phase was paramagnetic at room temperature and transformed to an antiferromagnetic state only at lower temperatures ($T < 200K$). As illustrated by Figure 6, the antiferromagnetic γ-phase contributed to a six-peak magnetic hyperfine spectrum, while the paramagnetic ε-phase contributed a single peak. Because the magnetic

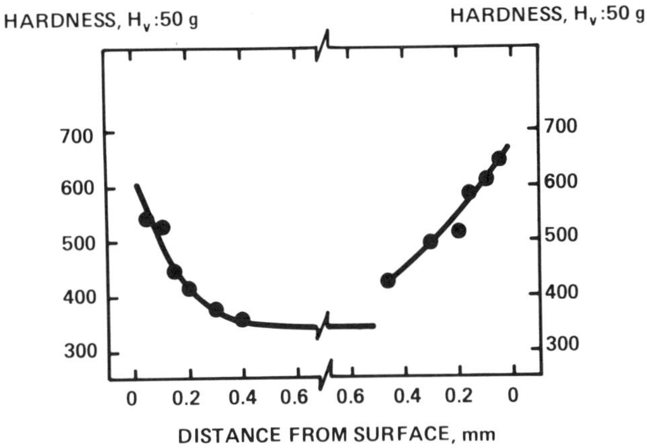

Figure 5. Hardness profile in the surface layer of the knife-edge portion of the H-pile.

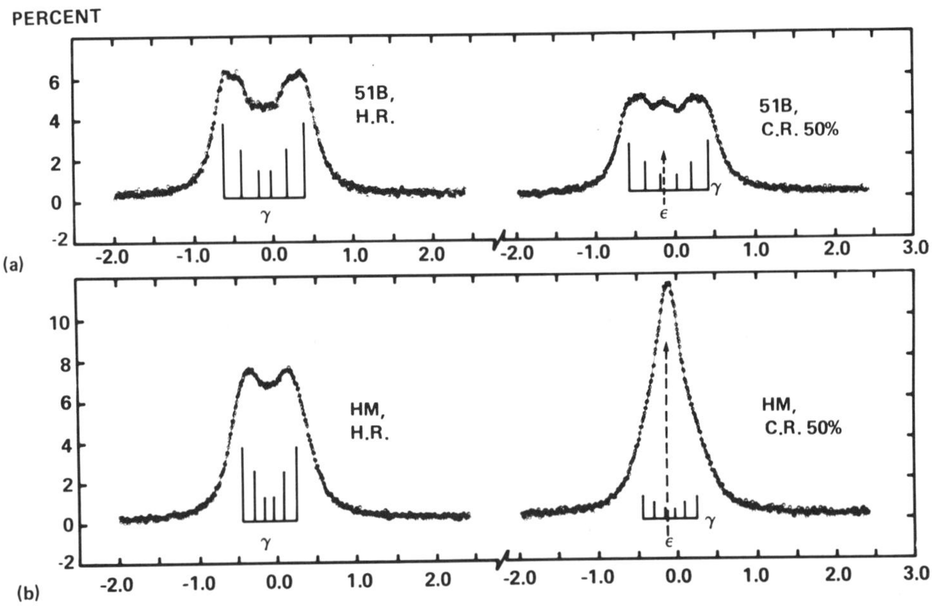

Figure 6. Room temperature XRM spectra of high manganese steels: (a) steel 51B as hot-rolled (left) and after 50% cold-rolling reduction (right), (b) steel HM as hot-rolled (left) and after 50% cold-rolling reduction (right). Approximate locations and intensities of the peaks contributed by the γ and ε phases are indicated.

hyperfine field of the γ-phase is relatively small (~20 to 40 kilogauss), there is considerable overlap between the paramagnetic ε-phase peak and the magnetic γ-phase spectrum. Nevertheless, the spectral contributions of the two phases were usually resolvable by computer analysis.

For the steel of lowest manganese and highest carbon content (HA, 16 percent manganese-0.75 percent carbon), the γ-phase Neel point is

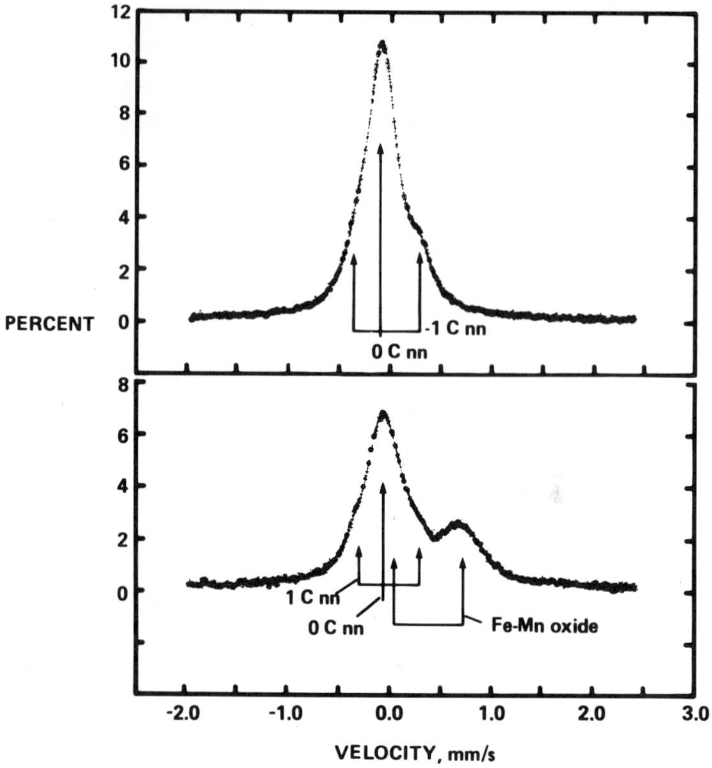

Figure 7. XRM spectra of steel HA: top - after 50% cold-rolling reduction, bottom - surface section removed from the knife-edge of the H-shape pile.

substantially below room temperature. Consequently, the γ- and ε-phases in this steel were not readily distinguishable at room temperature. The top XRM spectrum of Figure 7, obtained from a sample subjected to 50 percent cold reduction, is essentially identical to that obtained from the solution-annealed and hot-rolled samples of the same steel. The spectrum consists of a single peak arising from iron atoms with no carbon nearest neighbors (nn), and a quadrupole doublet arising from iron atoms having a carbon nn in an octahedral interstice. For a random distribution of carbon in austenite, the percentage of iron atoms having one carbon nn would be

$$P_1 = 6\ a_c\ (1 - a_c)^5 \times 100$$

where a_c is the atomic fraction of carbon. For this 0.75 wt. % carbon steel, a_c = 0.034, and P_1 from Equation (1) is 17 percent. The observed values of P_1 for the annealed, hot-rolled, and cold-rolled samples of this steel are all in reasonable agreement with this value.

A sample of the same composition was also investigated after the relatively severe deformation resulting from application as a pile driver. The XRM spectrum obtained from a surface section removed from the knife edge of the pile shown in Figure 4 is shown in the bottom half of Figure 7. In addition to the original single peak and quadrupole doublet, an intense new ferric quadrupole doublet (I.S. = 0.35, Q.S. = 0.70) was observed.

As discussed below, this doublet may be derived from a mixed iron-manganese oxide, presumably formed by rapid surface heating during the pile-driving application.

Low-Temperature Mössbauer Data

A number of samples were thinned to foils and examined by transmission Mössbauer spectroscopy in the temperature range from 10 to 250K. This phase of the investigation had several objectives: (1) to obtain better resolution of the γ- and ε-phase spectra, particularly at low manganese content, (2) to make more conclusive identification of the oxide phase formed on the pile knife-edge, (3) to better define the Neel points of some of the steels, and (4) to identify any minor phases that might be obscured in the room temperature spectra but detectable at low temperatures.

Spectra obtained near liquid-helium temperature of steels containing 24% manganese-0.27 percent carbon and 27 percent manganese-0.01 percent carbon (HM and 50B) are shown in Figure 8. As anticipated, the magnetic hyperfine field of the γ-phase is significantly increased, as may be seen by comparing the top spectrum of Figure 8 with the spectrum at the lower left in Figure 6. The middle and lower spectra of Figure 8 illustrate the improved resolution of the γ and ε phases at low temperatures. For example, in the case of the 27% manganese 0.01 percent carbon steel (50B) cold-reduced 50 percent, the γ phase constitutes 17 percent of the sample, but was not resolved at room temperature because of its small magnetic hyperfine splitting.

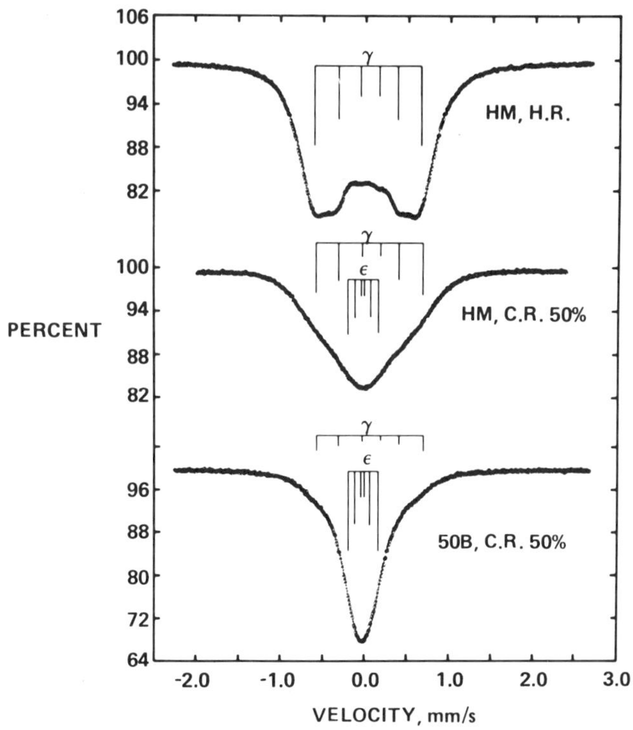

Figure 8. Low temperature (approximately 12K) transmission Mössbauer spectra obtained from high-manganese steels in hot-rolled and cold-rolled conditions.

It should be noted that the ε-phase is also antiferromagnetic at these temperatures. The magnetic hyperfine field is so small (H = 8 to 13 kilogauss), however, that the ε-contribution to the spectra near liquid-helium temperature has the appearance of a single broadened peak. Nevertheless, the value of chi-squared was noticeably improved by fitting the ε-phase spectrum as a magnetic component rather than a single broad peak. This conclusion is in agreement with earlier data on iron-manganese alloys which indicates Neel points of the order of 200K and very small magnetic hyperfine fields for the ε-phase [8].

In the case of the steel HA, both the γ and ε-phases are antiferromagnetic near liquid-helium temperature and can now be resolved as illustrated by Figure 9. The difference spectrum obtained by subtracting the spectrum of the hot-rolled sample from that of the 50 percent cold-reduced sample clearly shows a broad central component due to the ε-phase. Least squares analysis of the upper spectrum in Figure 9 indicated that the low-magnetic field ε-phase component constituted 17 percent of steel HA after 50 percent cold reduction.

It may also be noted that the low-temperature spectra of steels HM and HA (Figures 8 and 9) exhibit a slight asymmetry. This asymmetry can be approximately accounted for by adding a weak quadrupole doublet (~ 1-2% of the iron) in the least squares analysis with Mössbauer parameters (see Table III) qualitatively similar to those of some iron-manganese carbides [9].

In order to examine the knife-edge specimen at low temperatures, a section of the edge was thinned by mechanical polishing from the interior towards the surface. Although the resulting specimen was relatively thick and nonuniform (~ 30 to 80mm), it gave adequate low-temperature spectra, as illustrated in Figure 10. It appears from this spectrum that the ferric compound observed in the room temperature XRM spectrum is a mixed oxide phase

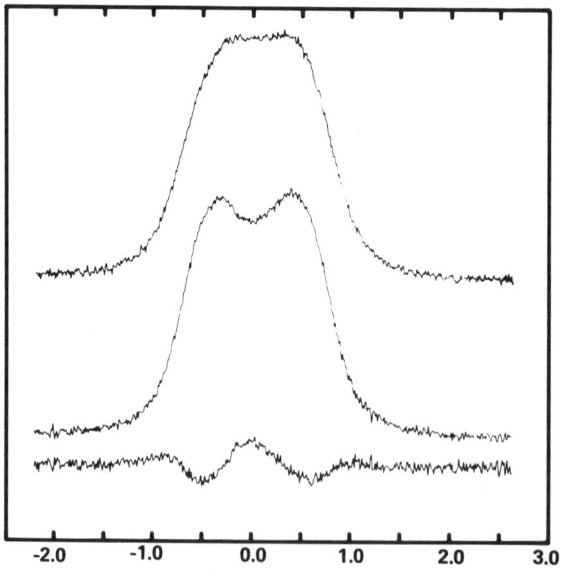

Figure 9. Transmission spectra (inverted) of steel HA at 12K, top - 50% cold-rolling reduction, middle - as hot-rolled, bottom - difference spectrum.

Table III. Mössbauer Parameters Measured at Room Temperature and Near Liquid Helium Temperature

Sample	Composition (wt%)	Temperature K	Spectrum Component and Phase Identification	Mössbauer Parameters I.S.(mm/s)	Q.S.(mm/s)	H(kG)	T_N' (K)
51B	36Mn-0.02C	295[a]	Magnetic-γ	-0.097	-	31.8	450[c]
			Single-ε	-0.12	-	-	
		15.5	Magnetic 1-γ	0.015	-	44.7	
			Magnetic 2-γ	0.010	-	37.2	
			Magnetic average-γ	<0.012>	-	<40.1>	
50B	27Mn-0.01C	295	Magnetic-γ	-0.096	-	27.8	403[c]
			Single-ε	-0.13	-	-	
		10.5	Magnetic 1-γ	0.016	-	35.9	
			Magnetic 2-ε	-0.019	-	7.8	
HM	24Mn-0.27C-1.75Cr	295	Magnetic-γ	-0.093	-	25.4	328[d]
			Single-ε	-0.12	-	-	337[c]
		12.0K	Magnetic 1-γ	0.011	-	39.6	
			Magnetic 2-ε	0.006	-	10.8	
			Quadrupole (weak)	0.30	0.35	-	234[d]
HA	16Mn-0.75C-1.7Cr	295	Single-γ[b], 0 C nn	-0.089	-	-	234[d]
			Quadrupole-γ, 1 C nn	-0.014	0.598	-	
		10.7K	Magnetic 1-γ	0.034	-	39.5	
			Magnetic 2-γ	0.037	-	26.9	
			Magnetic average-γ	<0.035>	-	35.1	
			Magnetic 3-ε	0.048	-	12.5	
			Quadrupole (weak)	0.35	0.36		
HA	Knife edge on pile, 16Mn-0.75-1.7Cr, oxide parameters only	295K	Quadrupole-Fe-Mn oxide	0.35	0.70	-	
		12.0K	Magnetic 1-Fe-Mn oxide	0.51	-0.004	492	
			Magnetic 2-Fe-Mn oxide	0.47	0.04	449	

[a]Room temperature. [b]At room temperature, the ε-phase could not be distinguished from the γ-phase for this steel. [c]Determined from variation of the magnetic susceptibility with temperature. [d]Mössbauer value, determined from variation of the hyperfine field with temperature.

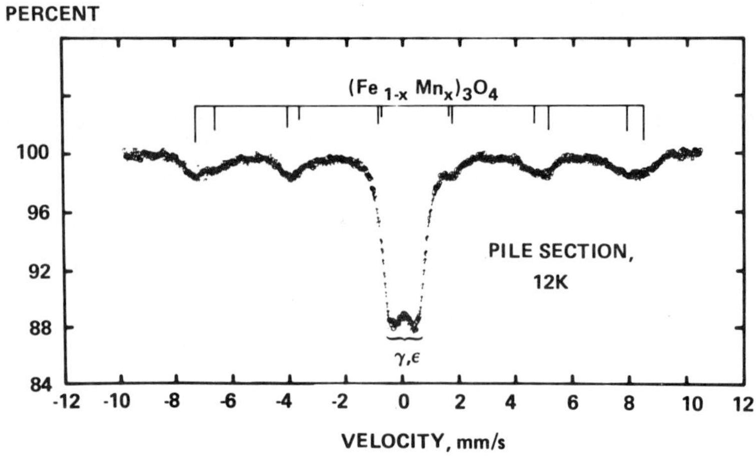

Figure 10. Low-temperature transmission spectrum of surface section removed from the knife-edge of the H-shape pile.

containing iron, manganese, and possibly chromium. The spectrum is similar to that exhibited by M_3O_4 types of oxide; however, there appears to be very little ferrous iron present. The variation of the magnetic hyperfine pattern with temperature indicates that the oxide phase is superparamagnetic.

Finally, the γ-phase Neel points of the 24% manganese-0.25 percent carbon (HM) and the 16 percent manganese-0.75 percent carbon (HA) steels were approximately determined by measuring the hyperfine field as a function of temperature and extrapolating it to zero.

SUMMARY AND DISCUSSION

All Mössbauer data are summarized in Tables III and IV. The Mössbauer parameters observed for all compositions at room temperature and near liquid-helium temperature are summarized in Table III, together with the Neel points of the γ-phase. Although the ε-phase is weakly antiferromagnetic near liquid-helium temperatures, no attempt was made to determine the Neel points of this phase because of the very small magnetic hyperfine field that it exhibited. For most of the entries in Table III, the γ-phase Mössbauer parameters were obtained from spectra of the annealed or hot-rolled samples, while the ε-phase parameters were determined from the 50 percent cold-reduced samples. In the case of the γ-phase, a better fit was sometimes obtained by using two overlapping magnetic components rather than a single magnetic component. Presumably this behavior arises from alloying effects. For these samples, an averaged value of the Mössbauer parameters is also given.

The iron phase percentages* determined by Mössbauer spectroscopy are summarized in Table IV, together with the sample treatments and hardness values. For these samples, the percentages of the total iron contained in the γ- and ε-phases should be close to the volume percentages of those phases. The increase in the percentage of the ε-phase as a function of cold reduction (% CR) is shown in Figure 11. It is apparent that the stability of the austenite phase increases significantly with increasing carbon and manganese contents. This is illustrated in a somewhat different fashion in

*Percentages of the total iron contained in the different iron-bearing phases identified.

Table IV. Iron Percentages[a] and Hardness Values

Sample	Composition (wt%)	Sample Treatment[b]	% Fe in γ Phase	% Fe in ε Phase	Hardness H_v (50g)	% Fe in Other Phases
51G	36Mn-0.0-2C	Ann.	100	0	135	
		HR	100	0	145	
		CR-20%	99	1	195	
		CR-50%	93	7	325	
50B	27Mn-0.01C	Ann.	74	26	285	
		CR-5%	56	44	305	
		CR-10%	29	71	355	
		CR-20%	22	78	410	
		CR-50%	17	83	465	
HM	24Mn-0.28C-1.75Cr	Ann.	99		270	1[c]
		HR	95	<5	280	
		CR-5%	87	13	325	
		CR-20%	52	48	440	
		CR-50%	43	57	440	
HA	16Mn-0.75C-1.75Cr	Ann.	100		220	
		HR	98		350	2[c]
		CR (50%)	83	17	705	
HA	Knife edge on pile	XRM results	50	e		50[d]
	16Mn-0.75C-1.7Cr	Low T results	61	7		32[d]

[a]Percentages of the total iron in different phases. [b]Ann. - annealed at 2050°C for one hour and quenched. CR - cold rolled to the indicated percentage reduction in thickness. HR - hot rolled. [c]Possible Fe-Mn carbide. [d]Fe-Mn oxide phase [e]Not distinguished from γ at r.t.

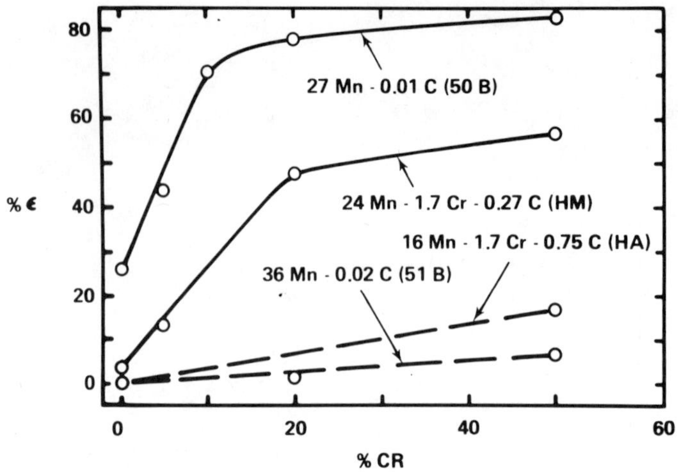

Figure 11. Percentage of ε-phase as a function of the percentage of cold reduction in high-manganese steels

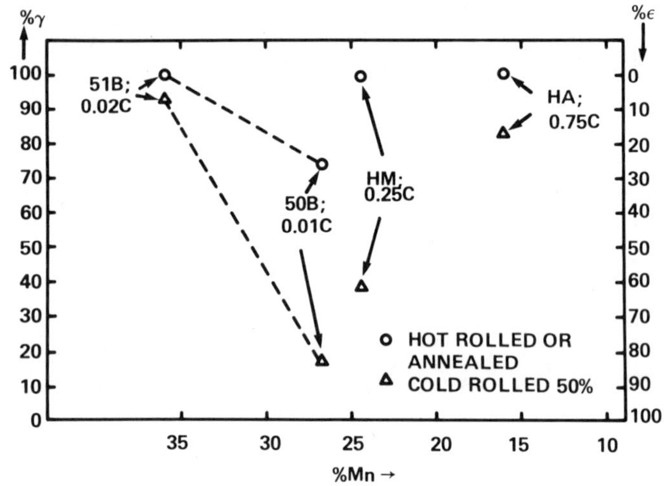

Figure 12. Percentages of the γ- and ε-phases observed by Mössbauer spectroscopy in hot- and cold-rolled high manganese steels as a function of manganese and carbon content.

Figure 12, where the phase percentages of the four steels in the hot-rolled and 50% CR states are shown as a function of manganese and carbon content. In Figure 13, the hardness values are plotted as a function of the percentage of the ε phase in the hot-rolled and cold-rolled steels. The hardness increases more rapidly with increasing percentages of the ε-phase for the two steels having the most stable austenite phase. Thus, for the 35% manganese 0.02% C (51B) and the 16 percent manganese-0.75 percent carbon (HA) steels, 50 percent cold-reduction transforms relatively small percentages of γ-austenite to the ε-phase, but causes the hardness values to increase by factors of approximately 2 and 3, respectively. The remaining two steels exhibit much larger amounts of the ε-phase after 50 percent cold-reduction, but

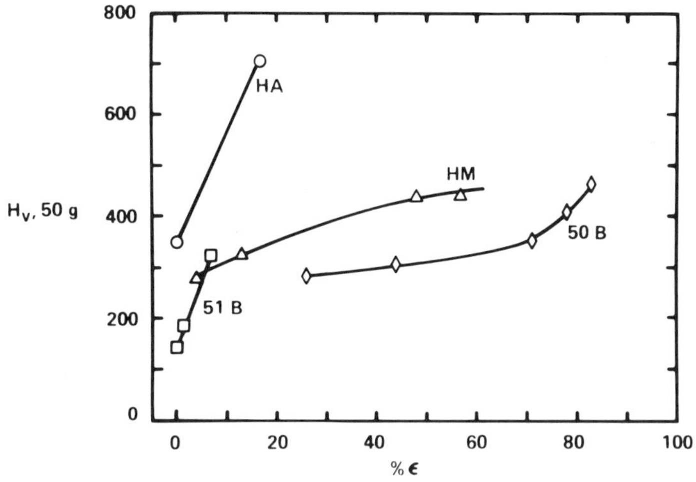

Figure 13. Hardness values as a function of the percentage of ε-phase in high-manganese steels.

the hardness increase is only about 60 percent. Clearly, the mechanical properties of these steels are influenced not only by the amount of ε-phase, but also by other factors, such as the manner in which the ε-phase is dispersed within the γ-phase and the dislocation distribution induced by deformation. As noted earlier, there is also a significant increase in the strength of these steels with increasing carbon content. Tentative identification of a minor carbide phase was made for the two steels of highest carbon content.

With regard to magnetic properties, the Mössbauer data indicate that the γ- to ε-transformation causes little or no change in the magnetic

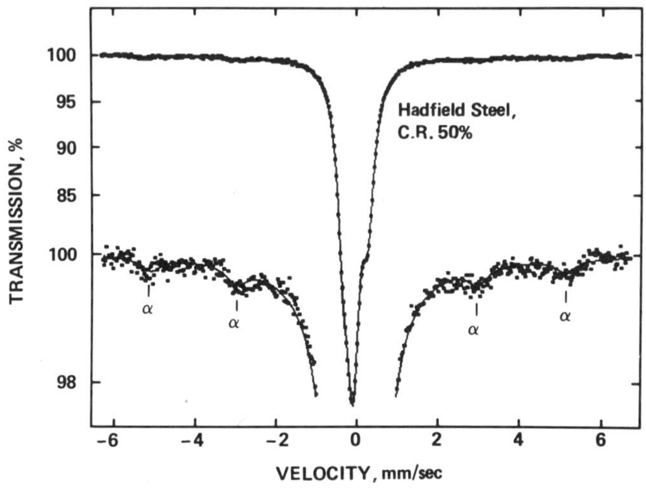

Figure 14. Room-temperature Mössbauer spectrum of a Hadfield steel (1 percent carbon, 11 percent manganese) cold reduced 50%. The α-martensite peaks can be seen in the blow-up.

permeability (Figure 3). No α-martensite was detected by Mössbauer spectroscopy in any of the high manganese steels. In contrast, the Hadfield steel from which the magnetic permeability data in Figure 3 were obtained exhibited approximately two percent of ferromagnetic α-martensite after 50 percent cold reduction (see Figure 14). The Neel temperatures of the austenite phase in these steels increased with manganese content and decreased with carbon content. It varied from 234 to 450K (-39°C to 177°C) among the four steels. The ε-phase was paramagnetic at room temperature and antiferromagnetic below approximately 200K. Consequently, all the steels exhibited very low magnetic permeabilities, even in a heavily cold-worked condition.

Finally, the room-temperature XRM spectrum of the H-pile section indicates that a mixed ferric oxide constitutes approximately half the material in the top 20mm of the sample. The low-temperature Mössbauer data indicate that the oxide is a superparamagnetic mixed oxide, possibly of the form $(Fe_{1-x}Mn_x)_3O_4$, and that it accounts for approximately 1/3 of the material in a somewhat thicker surface layer (~ 30 to 80mm). Additionally, a small amount of ε-phase (7%) was detected in low-temperature transmission spectra obtained at small velocity spans. On the basis of Figure 13, this would yield a hardness value of approximately 500, which is in accord with the increase in hardness observed on moving from the interior of the piece (> 300mm deep) to the surface (< 50mm) [Figure 5]. Whether or not the mixed oxide phase makes any contribution to the abrasion resistance of the steel is not clear at this time.

CONCLUSIONS

Several high-manganese steels containing from 16 to 36 percent manganese and from 0.01 to 0.75 percent carbon have been investigated by Mössbauer spectroscopy and by a variety of mechanical and physical property tests. Based on the phase identification, percentages, and magnetic states determined by Mössbauer spectroscopy, the following conclusions may be reached.

1. During cold-rolling, the austenite phase in these high-manganese steels transforms to the hcp ε-phase. No α-martensite was observed. The stability of the austenite phase increased with increasing carbon and manganese contents.

2. The steels with the most stable austenite (HA, 16 manganese-0.75 carbon and 51B, 36 manganese-0.02 carbon) exhibited continuous work hardening with increasing cold-rolling reduction up to 50 percent. The work hardening rate was higher for the steel with higher carbon content (HA). On the other hand, the steels that exhibited significant amounts of ε-phase formation during cold-reduction (HM, 24 manganese-0.27 carbon and 50B, 27 manganese-0.01C) exhibited rapid work hardening in a relatively small strain, and then tended to saturate at higher strain.

3. Both the γ- and ε-phases in the current suite of steels exhibited antiferromagnetism. The γ-phase Neel point ranged from 234 to 450K, while the ε-phase was antiferromagnetic below approximately 200K. This accounts for the low magnetic permeability of these steels even in a heavily cold-worked condition.

ACKNOWLEDGMENT

We are grateful to G. R. Dunmyre and J. W. Conroy for their assistance with the Mössbauer measurements, and to F. C. Schwerer and C. E. Spangler for magnetic susceptibility measurements.

REFERENCES

1. T. T. Hagane, 'Symposium on High-Manganese Austenitic Steel' 67, No. 2, A81 (1981).

2. The Specialty Steels (Japan) 28, No. 5 (1979), special issue devoted to high-manganese steels.

3. Y. N. Dastur and W. C. Leslie, Met. Trans. A 12A, 749 (1981).

4. G. P. Huffman and F. E. Huggins, The Use of Mössbauer Spectroscopy in the Analysis of the Raw Materials and Products of the Steel Industry, Proc. of APS/AISI Conference on 'Physics in the Steel Industry', Lehigh University, 1981, Ed., Fred C. Schwerer, AIP Conf. Proc. No. 84, Series Ed., Hugh C. Wolfe, American Institute of Physics, New York, 1982, pp. 149-198. ibid, Applications of Mössbauer Spectroscopy in the Steel Industry, pp. 265-301 in 'Mössbauer Spectroscopy and its Chemical Applications', Eds., J. G. Stevens and G. K. Shenoy, Advances in Chemistry Series 194, Amer. Chem. Soc., Washington, D. D., 1981.

5. K. R. Swanson and J. J. Spijkerman, J. Appl. Phys. 41, 3155 (1970).

6. Manufactured by Air Products and Chemicals, Inc., Allentown, PA.

7. C. Ouchi and G. P. Huffman, paper to be submitted to the J. Japanese Iron and Steel Institute.

8. Y. Endoh and Y. Ishikawa, J. Phys. Soc. Japan 30, 1614 (1971).

9. G. P. Huffman, P. R. Errington, and R. M. Fisher , Phys. Stat. Sol. 22, 473 (1967).

THE APPLICATION OF MÖSSBAUER SPECTROSCOPY TO IRON-ALUMINUM ALLOYS AND INDUSTRIAL ALUMINUM SAMPLES

Thomas Birchall and Dave Hodgson

Department of Chemistry and the
Institute for Material Research
McMaster University
1280 Main Street West
Hamilton, Ontario
L8S 4M1, Canada

Harish D. Merchant

Continental Can Company
Stamford, CT 06904

INTRODUCTION

The use of aluminum in manufacturing is expanding and in many areas there is a move to replace iron based alloys by aluminum based alloys. The merits of iron versus aluminum based alloys in packaging will not be commented on here. Rather the discussion will center on the application of iron-57 Mössbauer spectroscopy of the study of iron-aluminum alloys and as a probe for investigating the nature of the aluminum used in the manufacturing industry, because all commercial aluminum contains iron as an impurity. The form in which the iron is incorporated into the aluminum depends upon the nature of the other contaminants and, in some cases, upon the treatment that the metal received in processing.

The iron-aluminum system has been well studied by Mössbauer spectroscopy by using both iron-57 and implanted cobalt-57, and continues to attract the attention of research scientists. An excellent summary, by Gonser and Ron [1], covers the work up to 1980. However this earlier review is concerned with iron-aluminum phases which have, for the most part, been prepared from the pure elements. In cases where the iron content was low, iron enriched in the iron-57 isotope was used [2-7]. The various phases and defects of iron in aluminum, that have been identified by Mössbauer spectroscopy, have been summarized in an article by Nasu et al. [8].

This review will be concerned principally with studies carried out since 1979 on the iron-aluminum system, though earlier work will be included where relevant to the discussion. Unfortunately, the number of studies on commercial alloys remains small and these will be discussed in a separate section.

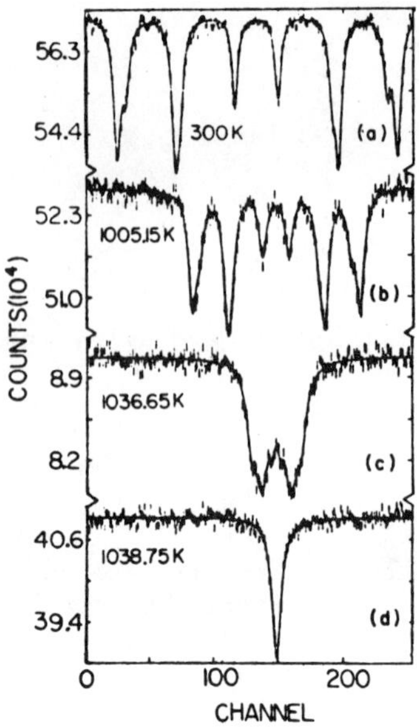

Figure 1. Iron-57 Mössbauer spectra of $Fe_{0.96}Al_{0.04}$ at temperatures below (a,b,c) and above (d) T_c. [Reproduction from reference 9]

IRON RICH ALLOYS

In spite of the difficulties involved in interpreting the Mössbauer data of iron-rich iron-aluminum alloys, that have been enumerated by Gonser and Ron [1], this region of the iron-aluminum phase diagram continues to attract attention. Iron containing increasing amounts of aluminum, up to 15 atomic percent produced spectra in which the hyperfine field, at iron, decreased from 330 to 300kG [9]. This decrease in hyperfine field was accompanied by an increasing distribution of fields -- a not unexpected result. The spectral changes were interpreted in terms of spin and charge density changes. In a subsequent report [10] these workers conclude that substitution by aluminum in body centered cubic iron causes changes in the solution redistribution which are due to s-d electron conversion. Other s-p solutes have a similar effect. Collins et al. [11] have examined $Fe_{1-x}Al_x$ alloys, for $0 < x < 0.04$. AT $x = 0.04$ the Mössbauer spectrum, Figure 1, shows two distinct fields below Tc, one due to iron with no aluminum near neighbors and a smaller field arising from iron having one aluminum near neighbor. A model was developed to assist in the interpretation of critical behavior in the presence of a small distribution of critical temperatures. In the expression

$$H_{(T,T_c)} = B\left(1-\frac{T}{T_c}\right)^{\beta}$$

where $T < T_c$, the disorder exponent β was found to have a value of 0.366.

Sidorenko and coworkers have examined iron-aluminum alloys containing up to 20 atomic percent aluminum [12-15]. Analysis of the Mössbauer spectra of these alloys was claimed to provide a means for investigating the dependencies of the hyperfine structure parameters on the composition. An estimate of the probability of various configurations of the environment of the resonant nuclei was made, as well as the correct selection of the physical model used in the later stages of data analysis [13]. These data, in conjunction with NMR results, were used to study short range order in these alloys and to obtain correlation moments [15]. This paper criticizes the use of a binomial distribution to calculate probabilities of atomic configurations in ordered alloys.

A distribution of near-neighbor configurations was obtained by using a Gaussian distribution of hyperfine fields in iron-aluminum alloys in the 25-35 atomic percent aluminum range [16]. The DO_3 alloy was nonmagnetic, or only weakly magnetic, but became magnetic on filing, the filed alloy being different from the thermally disordered state. This alloy reordered very slowly during annealling. The ordered, nonmagnetic, alloy with B2 symmetry became partially disordered on filing and was weakly magnetic. This alloy promptly reverted to its initial state with low temperature annealling. Iron-aluminum-silicon alloys, of composition Fe_3Al/Si, were found to separate into Fe_3Al and Fe_3Si components [17]. The magnetic, electrical and mean hyperfine field did not change significantly with composition, nor did annealling result in an increased degree of DO_3 order and had only a minor effect on the physical properties. The properties are governed by the composition.

Eymery and coworkers [18,19] have implanted aluminum atoms into an alloy of composition $Fe_{59.5}Al_{40.5}$. The starting alloy, at room temperature, gave a spectrum consisting of a single line (δ = 0.15mm/s) which showed, on cooling to 4.2K, magnetic satellite lines whose separation corresponded to a field of 180kG. This spectrum was interpreted as arising from superparamagnetic Fe_3Al. Implantation of Al at doses of 10^{14} ions/cm^2 gave a broad line when examined by conversion electron Mössbauer spectroscopy. Doses of 10^{15} ions/cm^2 resulted in the observation of a central resonance and satellite

Table I. Iron-57 Mössbauer Parameters for Aluminum Rich Iron Alloys

Composition	Isomer	Shift	Quadrupole	Line Width	Reference
Al-Fe	0.0104[a]	(0.1954)	0		20
		0.36	0		21
Al_5Fe_2	-0.058[b]	0.289	0.462	0.350	4
	~0.44[c]	(~0.18)	0.5		22
$Al_{13}Fe_4$		0.010	0	0.265	3
		0.202	0	0.265	
		0.392	0	0.265	
	-0.337[b]	(0.010)	0	0.245	4
	-0.137[b]	(0.210)	0	0.285	
	+0.068[b]	(0.415)	0	0.245	23
Al_6Fe	-0.124[b]	(0.223)	0.288	0.266	4
		0.22	0.26		23
Al-Fe (solid solution)		0.43	0		24
		0.440	0	0.252	6
		0.422	0	0.242	3
	0.082[b]	(0.429)	0	0.229	4

[a] shift versus cobalt-57/Pd, [b] versus cobalt-57/Pt, [c] versus sodium nitroprusside. Isomer shifts in parentheses have been adjusted to be relative to α-iron foil by using the conversion given in N. N.Greenwood and T.C. Gibb, 'Mössbauer Spectroscopy', Chapman and Hall, London, 1971, p. 90.

lines which corresponded to a magnetic field of 205kG. Surface damage recovers after 2 hours at 600K.

ALUMINUM RICH ALLOYS

This section reviews the data available on aluminum-iron alloys containing less than 50 percent iron. As in the previous section, the early work has been summarized by Gonser and Ron [1]. In Table I are collected the literature data for Al-Fe, Al_5Fe_2, $Al_{13}Fe_4(Al_3Fe)$, Al_6Fe and Al-Fe solid solution, the Mössbauer parameters of which are important for subsequent discussion. It is generally agreed that the Mössbauer parameters for these species are well established. Of these compositions only Al-Fe and the solid solutions of iron in aluminum give rise to single unsplit resonances as expected. The solubility of iron in aluminum is only 0.005 weight percent at 450°C [25]. Because of the low solubility, the spectrum of this species was obtained by using iron enriched in iron-57. Even so a pure solid solution spectrum was not obtained [2,4], see Figure 2. The additional resonances in these spectra have variously been attributed to precipitated iron [2], as $Al_{13}Fe_4$ [3], or as dimers [1]. The second minor component in the spectra obtained from low concentration of iron in aluminum is difficult to identify unambiguously because it appears only as a weak shoulder on the main peak, Figure 2(a). The careful study by Preston and Gerlach [3] would seem to favor $Al_{13}Fe_4$ as the most likely contaminant.

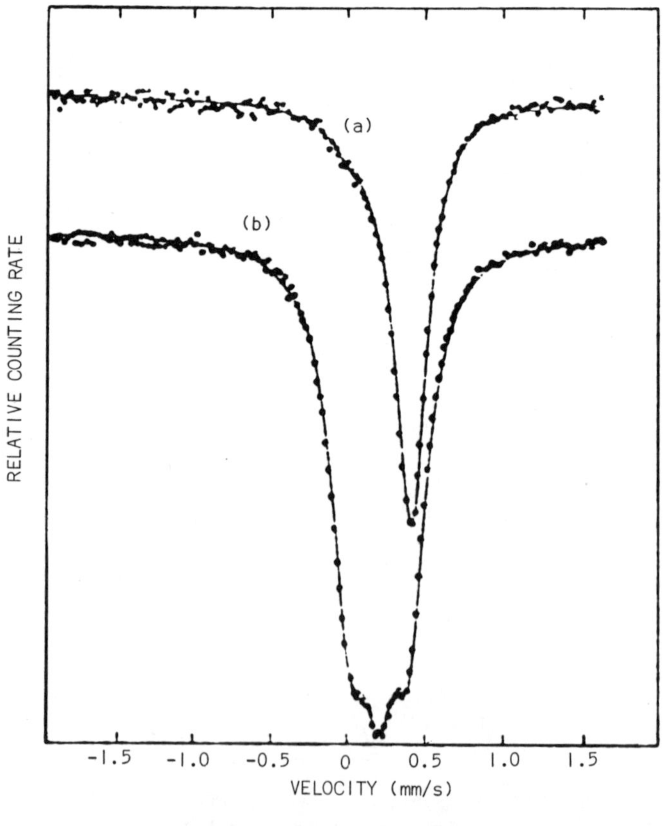

Figure 2. Iron-57 Mössbauer spectra of (a) 0.02% iron in aluminum; (b) $Al_{13}Fe_4$. [Reproduced from reference 3.]

The orthorhombic phases, Al_5Fe_2 and Al_6Fe, both give quadrupole split absorptions indicating distorted iron environments. The Al_5Fe_2 phase is stable with one basic iron environment and the spectrum resembles that of the metastable Al_6Fe phase. However its isomer shift is more positive and the quadrupole splitting is much larger than the corresponding Al_6Fe parameters. It is also of interest to note the significantly larger line width, suggesting a number of slightly different environments due to a variety of near neighbor iron environments. The linewidths for the Al_6Fe spectrum are indicative of only one environment. The transformation of this phase to the stable $Al_{13}Fe_4$ phase can be followed by Mössbauer spectroscopy. The pure $Al_{13}Fe_4$ phase has the three-line spectrum shown in Figure 2(b). X-ray data for this composition indicates that there are five different iron sites but the Mössbauer spectrum is best fitted to three lines [3,4] which are usually interpreted as arising from a quadrupole doublet and a single central resonance [8]. It is not clear how five iron environments give rise to a three-line spectrum, nevertheless the spectrum is characteristic of $Al_{13}Fe_4$.

Nasu et al. [8] have included in their summary data, supporting the existence of dimers [26], vacancies [5,17,27], interstitials [28,29] and clusters [26]. However, these species have been identified, for the most part, from an analysis of a weak shoulder on a strong solid solution single resonance. Examination of these spectra leads one to the conclusion that the isomer shifts and quadrupole splitting are not as certain as is claimed, and hence the nature of the site is open to question. For example, in a series of splat quenched alloys [26], the 0.5 percent iron alloy showed a weak satellite line which was interpreted as a doublet, δ = 0.12mm/s, Δ = 0.39mm/s, arising from dimers. The 5 percent iron alloy was fitted to a double, δ = 0.15mm/s, Δ = 0.32mm/s, and a single line δ = 0.424mm/s, the former being attributed to clusters. These assignments were justified on the basis of probabilities of dimers, trimers, etc. existing in these alloys. However the authors [26] appear to have overlooked their predicted high probability of dimers in the 5 percent alloy, and the parameters quoted for these species are in fact close to those for Al_6Fe (Table I). As pointed out earlier, the use of probability distributions, to estimate the species present, has been criticized [15]. Nasu and coworkers have also attributed satellite lines to the association of iron atoms with vacancies [7,30]. Here again the preparative procedures used result in samples which give spectra that are invariably dominated by the iron solid solution line, and this makes the assignment of the weak satellite line(s) open to question.

Several groups have used conversion electron Mössbauer spectroscopy, CEMS, to investigate iron-57 implanted into aluminum [31,33]. Sassa et al [31] obtained similar results to those of Nasu and coworkers [26], but these are explained as being due to the formation of an ordered phase similar to Al_6Fe rather than a randomly distributed iron atom. Laser irradiated films of iron-57 on aluminum have been examined by CEMS and the spectra that arose were interpreted in terms of the formation of α-iron, iron clusters, and $Al_{13}Fe_4$, though this latter conclusion is not convincing [32]. CEM spectra, obtained from iron implanted aluminum surfaces, give symmetric doublet spectra having δ = 0.201(4)mm/s and quite constant, while the quadrupole splittings varied from Δ = 0.501 to 0.521(6)mm/s. These spectra were similar to those recorded by Sawicka et al. [34] and assigned to iron atoms associated with another iron atom as nearest neighbor. The closer the iron is to the surface the more distorted is its environment, which could be the result of higher radiation damage or, to a decrease in symmetry as the surface is approached. An analysis of the fractional intensities of spectral components arising from iron implanted in aluminum was interpreted as arising from iron aggregates rather than to a random distribution [35]. The conclusions were justified on the basis of dosage and annealling dependence studies. Theoretical studies of the Debye-Waller factor of iron-57 trapped in aluminum have been made by Tewari et al [36].

Implantation of cobalt-57 into aluminum has attracted the attention of several groups to investigate vacancies produced by quenching. Sassa and coworkers [27,37,38] interpret their spectra in terms of iron-57 atoms associated with vacancies which have isomer shifts 0.33mm/s less positive than the solid solution peak [37]. Spectral changes were used to calculate the Debye temperatures for the solid solution and vacancy peak as 235 and 300K respectively. This result does not agree with an earlier result by this group [5] and it is clear that further study is necessary. Implantation of cobalt-57 at 4.2K leads to spectra which are less well resolved than in the previous cases, but these spectra have been interpreted as showing evidence for the presence of trapped vacancies and interstitials [39].

Petry and coworkers [40-43] have also studied the trapping of cobalt-57 (iron-57) in single crystals of aluminum. At 4.2K the f-factor (87%) is found to be isotopic, but as the temperature is raised to 25.5K it becomes anisotropic. The iron atoms, occupying the eight cage sites, are displaced in the [III] direction by 0.52Å [40]. Pauling [44] has analyzed this behavior in terms of a valence bond model. Figure 3 shows the central part of the magnetic Mössbauer spectrum arising from cobalt-57 implanted into aluminum at 4.2K when measured against an iron foil [45]. Two distinct sites are observed in an area ratio of 46:54 which are assigned to substitutional and interstitial cobalt (iron) atoms. The latter has a much broader line profile (0.47mm/s) and this is taken as indicative of a small quadrupole splitting (0.16mm/s).

COMMERCIAL ALUMINUM ALLOYS

Many of the publications cited here, and in the earlier reviews [1], have originated in industrial laboratories, yet hardly any of these studies have been concerned with commercial quality aluminum, which always contains iron. This is surprising since the main phases in the aluminum-iron system

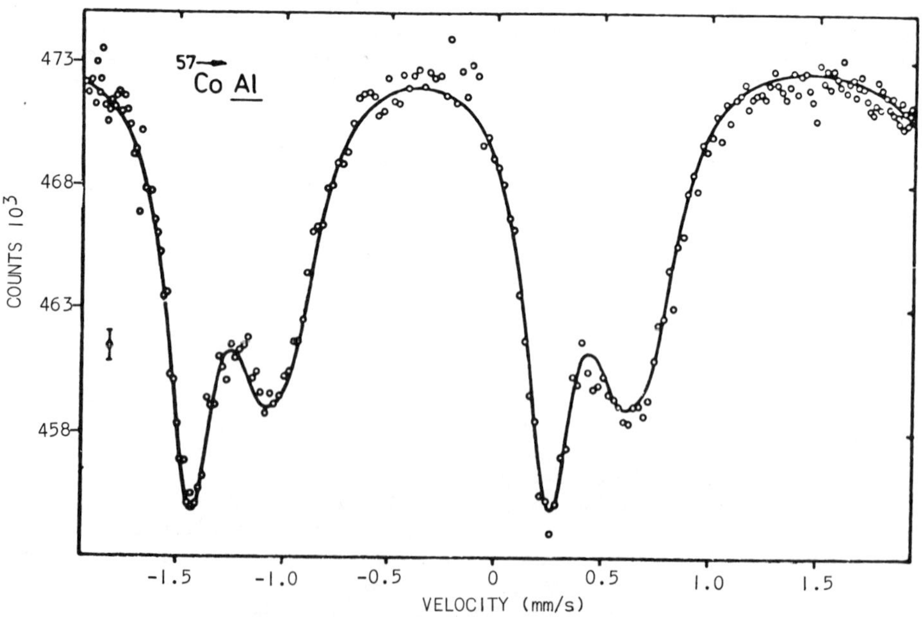

Figure 3. Mössbauer spectrum of cobalt-57 in aluminum at 4.2K, the absorber is an iron-foil and the central lines are shown [reference 45].

are now so well characterized (Table I). It is of course possible that many results, on such systems, have been collected but remain unpublished.

Vertes and coworkers [46] have recognized the possibilities for using iron-57 Mössbauer spectroscopy as a tool to study commercial alloys during the industrial processing steps. Although their alloys were synthetically prepared, these alloys contained concentrations of iron commonly encountered in industrial samples. For those alloys containing low concentrations of iron, 0.02-0.05 weight percent, the spectra consisted of asymmetric doublets which were made up of spectra from solid solution, iron in aluminum, clusters and $Al_{13}Fe_4$. In these cases annealling resulted in the latter two species being converted to solid solution (monomer). At higher iron concentrations the spectra appear to consist of monomer and Al_6Fe, though it is not clear why the metastable Al_6Fe would not revert to $Al_{13}Fe_4$. Addition of silicon to these alloys does have an influence on the spectra and this results in much of the resolution being lost, Figure 4b. This was attributed to the preference of silicon for sites that are close to iron.

Birchall et al. [47] appear to be the only group to have examined alloys used in commercial can production. Table II summarizes some of their results. A sample of hot rolled 3004 alloy, containing 0.45 percent iron, gave a Mössbauer spectrum which appears as a symmetric doublet. The parameters for this speciesare very close to those for the metastable Al_6Fe phase (Table I). Presumably this Al_6Fe phase is stabilized by the presence of 1.10 percent manganese. Al_6Mn is isostructural with Al_6Fe. Cold rolling this sample into a sheet results in a broadening of the lines but otherwise no significant change in the parameters. An ice-brine quench from 610°C induces an

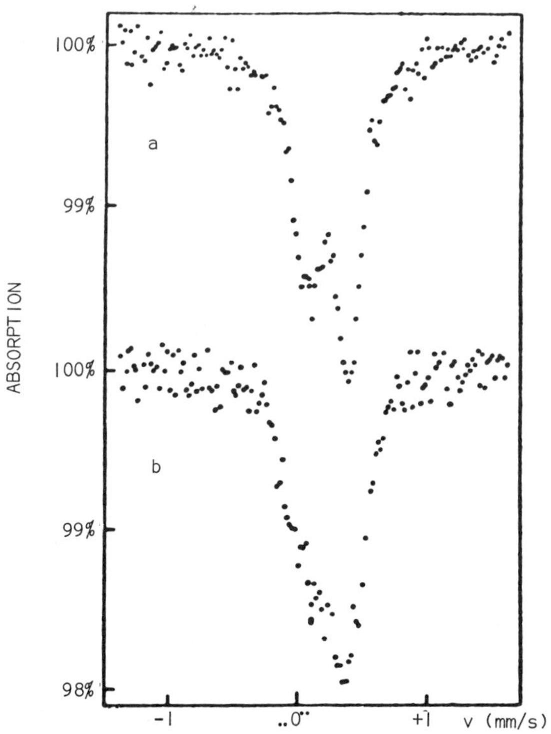

Figure 4. Iron-57 Mössbauer spectra of aluminum samples containing(a) 0.26 weight percent iron, (b) 0.25 weight percent iron, 0.21 weight percent silicon [reference 46].

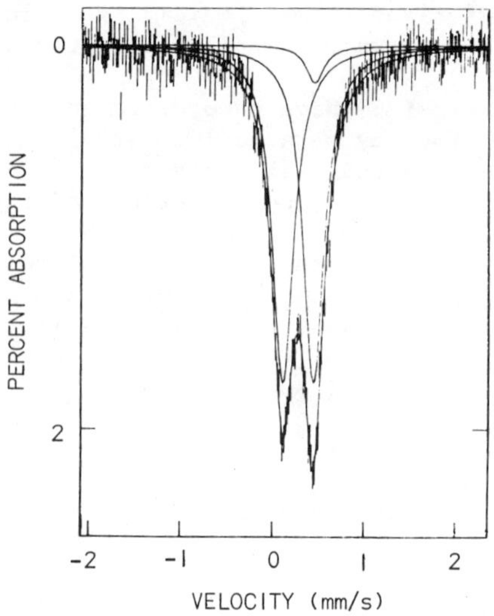

Figure 5. Iron-57 Mössbauer spectrum of 3004 aluminum cold rolled sheet, ice brine quenched from 610°C

asymmetry to the spectrum, Figure 5, and significant changes in the physical properties of the alloy. This spectrum was interpreted as an Al_6Fe doublet plus a solid solution of iron in aluminum, the latter amounting to 4.5 percent of the total or 0.0203 percent iron in solution. Slow cooling, whether at a controlled rate or not, results in a lower amount of iron retained in solid solution (0.0106 percent) and results in a less brittle alloy. Addition of 0.26 per cent chromium to this alloy gave a spectrum which was analyzed in terms of a doublet and a singlet, sample 2, Table II. A lower concentration of 0.27 percent iron, sample 3 gives a spectrum which resembles that reported by Vertes et al. [46], Figure 4b: a higher percentage of the total iron being retained in a monomer like environment. Samples 4 and 5 have a higher iron content (ca. one percent) than the other samples, but differ in the content of other elements (sample 4 has magnesium, 0.86, manganese, 0.95, silicon, 0.45, chromium, 0.0). The spectrum of sample 4 has a simple doublet structure while that of sample 5 is asymmetric. It is not clear which of the composition changes is responsible for the spectral changes at this stage. The parameters for the doublets and single lines are somewhat different from those for Al_6Fe and solid solution monomer, and are likely due to the additional impurity elements present. For example, manganese would likely be present as Al_6Mn which is isostructural with Al_6Fe and undoubtedly influences its spectrum.

In samples 6 to 8 an attempt was made to monitor the effect of increasing copper content on a 1251 alloy: the percent copper increased from 0.09 through 0.27 to 0.41 percent. The spectra were all asymmetric, Figure 6, and contained 84 percent doublet and 16 percent singlet. The variation in copper content appears to have had no influence on the environment of the iron containing components.

CONCLUSIONS

It is clear that there is much more information to be gained from Mössbauer studies of this type of sample, but the number of variables, such as composition and heat treatment, are large. However the technique does present a viable means of determining the form that the iron takes in these

Table II. Iron-57 Mössbauer Spectra of Continental Can Samples at 298K

	Sample		δ	Δ	Γ	% Area
1	3004 Hot rolled	1mm thinned from 3.0mm	0.23	0.31	0.29	100
	3004 Cold rolled	Sheet	0.23	0.29	0.33	100
		Ice-brine quench	0.25 0.42	0.34 ----	0.28 0.24	95.6 4.5
		Slow cooled	0.25 0.41	0.33 ----	0.28 0.31	97.4 2.6
		Controlled cooling	0.22 0.41	0.29 ----	0.25 0.24	97.9 2.1
2	3004 + Cr		0.24 0.30	0.36 ----	0.27 0.27	86.7 13.3
3	3004 + Cr		0.22 0.30	0.37 ----	0.28 0.28	78.4 21.6
4	3004 + Cr/Fe		0.20	0.32	0.32	100
5	3004/HiFe		0.23 0.30	0.37 ----	0.31 0.31	84.5 15.5
6	1251		0.22 0.25	0.31 ----	0.27 0.27	83.8 16.2
7	1251/Cu		0.23 0.26	0.31 ----	0.26 0.26	83.2 16.8
8	1251/Cu		0.23 0.26	0.32 ----	0.26 0.26	83.4 16.6

[a]Data in mm/s relative to α-iron foil.

alloys and in estimating the relative amount. This type of study will continue to attract the attention of Mössbauer spectroscopists.

ACKNOWLEDGEMENTS

The Natural Sciences and Engineering Research Council of Canada are thanked for financial support.

REFERENCES

1. U. Gonser and M. Ron, 'Applications of Mössbauer Spectroscopy', Vol. II, ed. R. L. Cohen, Academic Press, New York (1980), p. 299-305 and refs. therein.

2. R. H. Bush, C. A. Stickels, and L. W. Hobbs, Scripta Met. 1, 75 (1967).

3. R. S. Preston and G. Gerlach, Phys. Rev. B. Solid State 3, 1519 (1971).

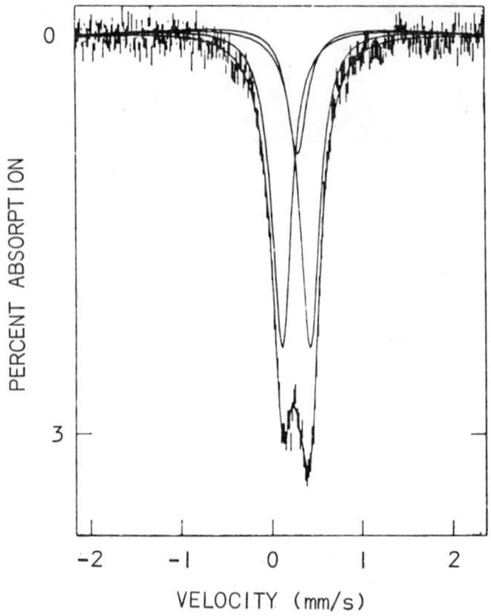

Figure 6. Iron-57 Mössbauer spectrum of 1251 aluminum alloy containing 0.41 weight percent copper.

4. C. A. Stickels and R. H. Bush, Met. Trans. 2, 2031 (1971).

5. H. Ichinose, K. Sassa, Y. Ishida, and M. Kato, Phil. Mag. 36, 1367 (1977)

6. C. Janot and H. Georges-Gilbert, J. Phys. F: Met. Phys. 7, 231 (1977).

7. R. S. Preston, S. Nasu, and U. Gonser, J. de Phys. Colloque, C2, 40, 564 (1979).

8. S. Nasu, U. Gonser, and R. S. Preston, J. de Phys. C1, 41, 385 (1980).

9. S. M. Dubiel and W. Zinn, Phys. Rev. B. 26, 4574 (1982).

10. S. M. Dubiel and W. Zinn, Phys. Rev. B. 29, 2279 (1984).

11. G. S. Collins, A. R. Chowdhury, and C. Hokenemser, Phys. Rev. B. 26, 4997 (1982).

12. F. A. Sidorenko, Sov. Phys. Solid State, 23, 2046 (1981).

13. V. V. Ovchinnchov, N. A. Pervukhin, B. V. Ryzhcnko, F. A. Sidorenko, and P. V. Gel'd, Sov. Phys. Solid State, 23, 350 (1981).

14. F. A. Sidorenko, P. V. Gel'd, V. Ya. El'ner, and B. V. Ryzhenko, J. Phys. Chem. Solids, 43, 297 (1982).

15. F. A. Sidorenko, N. A. Pervukhin, B. V. Ryzhenko, and P. V. Gel'd, Sov. Phys. Solid State, 25, 1297 (1983).

16. K. Ohi, S. Towata, M. Tamiya, and T. Egachi, J. Japan Inst. Metals, 45, 354 (1981).

17. O. Schneeweiss, T. Zemcik, T. Zak, and S. Mager, Phys. State Sol. 79, 125 (1983).

18. J. P. Eymery, A. Fnidiki, and J. P. Riviere, Nucl. Inst. Methods, 309/210, 919 (1983).

19. J. P. Eymery, A. Fnidiki, and J. Deleford, J. Mag. Mag. Met. 40, 130 (1983).

20. G. K. Wertheim and J. H. Wernick, Acta Met. 15, 297 (1967).

21. K. Ono, Y. Ishikawa, and A. Ito, J. Phys. Soc. Japan, 17, 1747 (1962).

22. V. V. Nemoshkalenko, O. N. Rasumov, and V. V. Gorskii, Phys. State Sol., 29, 45 (1968).

23. S. Nasu and U. Gonser, Proceedings Mössbauer Conference, Bratislava, 311 (1973).

24. S. Nasu and Y. Murakami, Phys. State Sol. (B), 46, 711 (1971).

25. M. Nishio, S. Nasu, and Y. Murakami, Nippon-Kimzoky-Gokkai-Shi, 34, 1173 (1970), quoted in Ref. 1.

26. S. Nasu, U. Gonser, P. H. Shingu, and Y. Murakami, J. Phys. F. Metal Phys., 4, L24 (1974).

27. K. Sassa, H. Goto, Y. Ishida, and M. Kato, Scripta Met. 11, 1029 (1977).

28. W. Mansel and G. Vogel, J. Phys. F. Metal Phys. 7, 253 (1977).

29. G. Vogel, W. Mansel, and P. H. Dederichs, Phys. Rev. Letters, 36, 1497 (1976).

30. S. Nasu, U. Gonser, R. S. Preston, and G. Ibe, Proc. Indian Nat. Sci. Acad., Phys. Sci. Special Vol. C-58, 777 (1982).

31. K. Sassa, Y. Ishida, and K. Kaneoko, J. de Physique C2, 40, 556 (1979).

32. K. G. Prasad, Proc. Indian Nat. Sci. Acad., Phys. Sci. Special Vol. MA-41, 561 (1982).

33. W. Jones, M. J. Tricker, and G. A. Gord, J. Mat. Sci. Letters, 14, 751 (1979).

34. B. D. Sawicka, J. A. Sawicki, and J. Stanek, Phys. Letters, 59A, 59 (1975).

35. B. D. Sawicka and J. A. Sawicki, Nucl. Inst. Methods, 209-210, 799 (1983).

36. A. K. Tewari, S. K. Roy, and D. L. Bhattacharya, Phys. State Sol. B98, 323 (1980).

37. K. Sassa, H. Goto, and Y. Ishida, J. de Physique C2, 40, 559 (1979).

38. K. Sassa, S. Umeyama, Y. Ishida, and H. Yashida, J. Phys. F. Metal Phys. 11, L163 (1981).

39. E. Verbiest, H. Pattyn, and J. Odeurs, J. de Physique Cl, 41, 431 (1980).

40. W. Petry, G. Vogel, and W. Mansel, Phys. Rev. Letters, 45, 1862 (1980).

41. W. Petry, G. Vogel, and W. Mansel, Z. für Phys. B, 46, 319 (1982).

42. S. Mantl, W. Petry, K. Schroeder, and G. Vogel, Phys. Rev. B, 27, 5313 (1983).

43. W. Petry, M. Brussler, V. Groger, H. G. Muller, and G. Vogel, Hyperfine Int. 15/16, 371 (1983).

44. L. Pauling, J. Solid State Chem. 40, 266 (1981).

45. J. Odeurs, H. Pattyn, E. Verbiest, and R. S. Reintsoma, Nucl. Inst. Methods, 170, 373 (1980).

46. A. Vèrtes, S. Nagy, M. Zoki-Awad, J. Lakner, E. Kovacs-Crselonyi, and G. Groma, Scripta Met. 16, 145 (1982).

47. T. Birchall, D. Hodgson, and H. D. Merchant, J. Mat. Sci. Letters, in press (1986).

MÖSSBAUER INVESTIGATION OF AlFe AND AlFeSi PHASES IN SPECIALLY PRODUCED ALUMINUM ALLOYS AND IN INDUSTRIAL ALUMINUM

Sàndor Nagy, Attila Vèrtes, and Zoltàn Homonnay

Laboratory of Nuclear Chemistry
Eötvös University
H-1088 Budapest, Puskin u. 11-13
Hungary

Lajos Murgàs, Tibor Turmezey, Agnes Griger, and Jòzsef Lakner

Aluterv-fki
Engineering and Development Centre Hungalu
H-1389 Budapest
Hungary

INTRODUCTION

This paper is a summary of the Mössbauer investigations done in the framework of a project supported by the Ministry of Culture and Education of Hungary and by the ALUTERV-FKI. The aim of the project is to study industrial aluminum with respect to the precipitation and transformation processes of the various phases formed by iron, an unwanted alloying element which, unfortunately, is a constant impurity in aluminum.

In our investigations we followed a step-by-step strategy. Starting with specially prepared binary phases and alloys, then through more complicated ternary systems containing silicon besides iron and aluminum. Then we turned our attention towards industrially produced samples. Throughout this paper the adjective "industrial" is used in a narrow sense, referring to alloys in which the iron and silicon content is about the same as in commercial aluminum alloys but the concentration of other impurities is very low. Although the reason for this strategy is fairly obvious, we mention that according to a recent review article [1], in 1983-84 there were about 90 impurities in aluminum, most of them written about specially, rather than industrially, prepared binary aluminum-iron systems but none of them dealt with ternary AlFeSi phases. The first work on hexagonal α-(AlFeSi) was reported by us recently [2].

We should mention that, besides Mössbauer investigations positron lifetime measurements, x-ray analytical and diffraction studies, differential thermal analysis, electrical resistivity measurements, various mechanical tests as well as optical and electron microscopy studies were also carried out but, here we mention only such measurements which were closely related to our Mössbauer results.

Mössbauer spectra were measured at room temperature with a cobalt-57 on palladium source. All the isomer shifts refer to α-iron. The x-ray diffraction measurements were carried out with a Guinier-Hägg high precision vacuum camera.

ALUMINUM-IRON BINARY ALLOYS

Although there are reliable data on the Mössbauer parameters of binary phases containing aluminum and iron [3-8], for the present study, Mössbauer parameters of the aluminum(iron) solid solution as well as those of the Al_3Fe and Al_6Fe phases were verified by the measurement of special alloys. Samples containing 0.02, 0.5, and 39.0 mass percent iron, the first one enriched in iron-57 were prepared by melting 99.99 percent pure aluminum with iron of the same purity in a vacuum intermediate-frequency induction furnace. For the Mössbauer measurements sheet specimens of 1mm thickness and 10mm diameter were cut from the cast material.

The aluminum 0.02 mass percent iron alloy was annealed for 24 hours at 640°C. The Mössbauer spectrum of this sample consisted of a singlet corresponding to iron dissolved in aluminum. The measured isomer shift was 0.40mm/s. In order to obtain a sample containing only the phase Al_3Fe, the aluminum 39.0 mass percent iron alloy was heat treated for 5 hours at 1000°C. The sample was checked by x-ray diffraction [9] to ensure that the annealed sample did not contain any aluminum or other phases besides Al_3Fe. The spectrum of this phase could be fit with three equidistant lines with the same amplitude and width as found by Preston and Gerlach [3]. The isomer shifts were 0.00, 0.19, and 0.38mm/s.

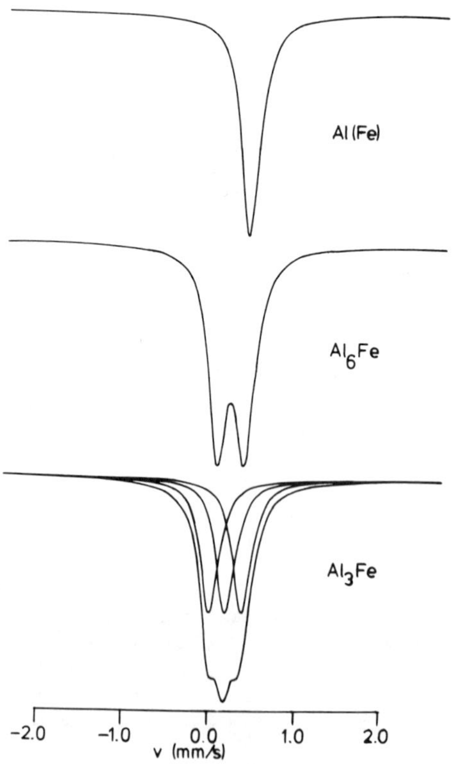

Figure 1. Computer simulated spectra of binary aluminum-iron phases formed in aluminum.

X-ray diffraction investigations showed that the as cast aluminum 0.5 mass percent iron sample contained only Al_6Fe besides aluminum. The Mössbauer spectrum of this specimen could be interpreted as the superposition of a single line and a symmetric quadrupole split component arising from iron dissolved in aluminum and Al_6Fe phase, respectively. The doublet has an isomer shift of 0.06mm/s and a quadrupole splitting of 0.38mm/s. Figure 1 shows the computer simulated spectra of the above phases. The Mössbauer parameters of the investigated phases agreed fairly well with published data [3-8].

An industrial direct chill cast billet of 90mm diameter was prepared through semicontinuous DC casting from an aluminum 0.58 mass percent iron alloy. The melt temperature was 685°C, and the casting speed was 80mm/min. For the Mössbauer measurements, sheet samples were cut from the billet at the middle of its radius with their planes at right angles to the direction of casting. In order to study the Al_6Fe to Al_3Fe phase transformation, heat treatments were carried out at 450, 530, and 620°C for 10, 30, 90, 300, 600, and 3000 minutes.

POSSIBLE PROBLEMS IN PHASE ANALYSIS

Texture. In order to determine the distribution of iron between the different phases in the aluminum-iron binary alloys, one has to decompose the Mössbauer spectrum into the above mentioned singlet, doublet, and triplet attributed to dissolved iron, Al_6Fe, and Al_3Fe, respectively. However, the results presented here demonstrate that relative concentrations obtained from the line areas of sheet-like samples must be taken with reservations, because of the possible texture [10] of the sample caused by the anisotropic features of the technological process used for the production of aluminum. As we can see from Figure 2a, there was no apparent texture in sample 1. In accordance with this, its spectrum (Figure 2b) is a symmetric doublet which was attributed to randomly oriented Al_6Fe. The contribution of dissolved iron, the line of which would show up round the second line of the doublet, can be neglected owing to the fairly high iron content and to the low solubility of iron in aluminum.

The spectra of samples 2 and 3 indicate that there is a correlation between the texture of the aluminum-matrix, which can be observed in the center of the billet where the dendrites run parallel with the direction of casting, and the orientation of the minor component Al_6Fe. From Figure 2b we can see that the crystals of Al_6Fe are partly oriented in the center, i.e., the electric field gradients, EFG, show perpendicular to the axis of the billet.

Even if the exact EFG distribution is unknown, the extent of orientation of a partly oriented sample can be characterized quantitatively by the following expression,

$$P = \frac{(S^- - R) + (S^+ - R - M)}{S^- + S^+ - M} 100 = (1 - \frac{2R}{100-M}) 100\%,$$

where S^- and S^+ are the measured normalized, $S^- + S^+ = 100$, areas of the left and right-hand line, respectively, and M is the area contribution of the monomer. The partial line area, R, of the fraction of crystals assumed as perfectly random can be calculated from the following expression [11,12],

$$\frac{S^- - R}{S^+ - R - M} = \frac{5-3\cos^2\theta}{3+3\cos^2\theta} = \delta(\theta),$$

where θ is the angle between the direction of the γ-radiation and the principal axis of the EFG of the oriented fraction (see Figure 2b). If the

Table I. Percentage of Orientation (P) Calculated for Samples 2 and 3 Assuming that the Contribution of Monomer Iron is Negligible.

Sample	θ	δ(θ)	%S^-	%S^+	%P
3	0°	1/3	42.3	57.7	30.8
2	90°	5/3	54.4	45.6	35.2

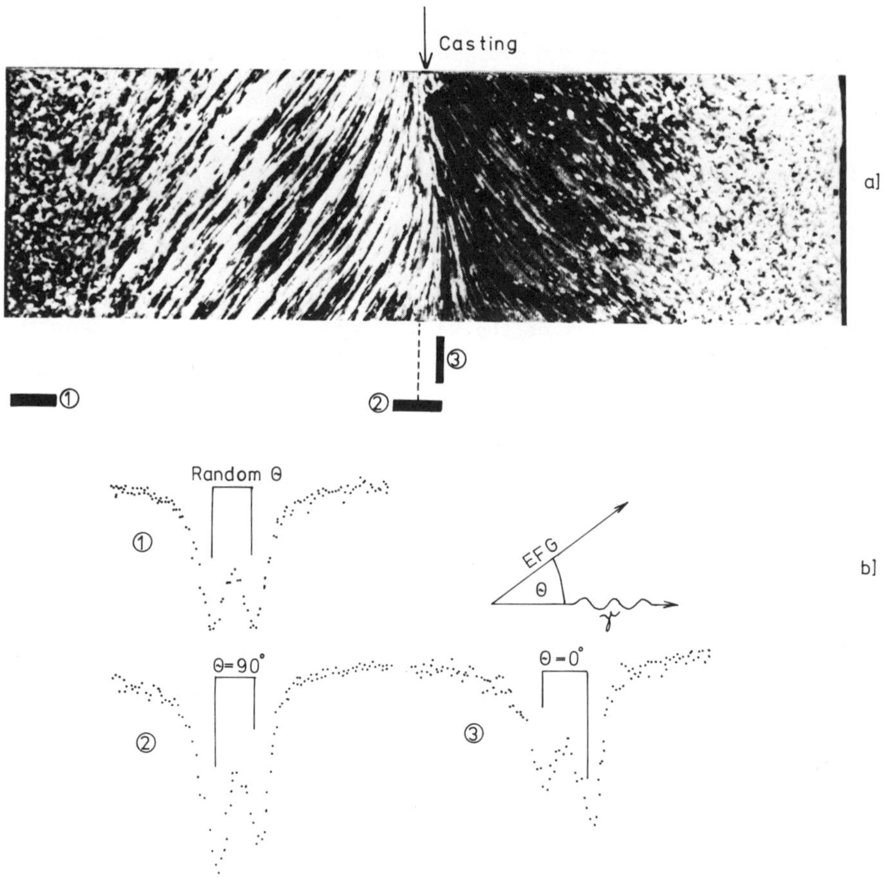

Figure 2. (a) Axial cross section of the aluminum-iron billet etched with NaOH showing the origin of the Mössbauer absorbers used for the study of the texture of Al_6Fe. (b) Mössbauer spectra of the above samples and the interpretation of the angle θ. The vertical lines in the spectra represent the area ratios calculated for perfectly random and totally ordered Al_6Fe. Note that the asymmetry of spectrum 2 caused by texture is the opposite of that expected from the contributions of dissolved iron (Figure 1).

oriented EFG are parallel with one another, two measurements at a θ of 0° and 90° are sufficient for the simultaneous determination of R and M.

Table I shows the calculated percentage of orientation, P, for samples 2 and 3 assuming M is zero. The data are consistent with the assumption that there is a 30-36 percent radial orientation of the EFG in the center of the billet.

Deformation. The anisotropic mechanical stress field arising in the billet during solidification could lead to an effect similar to that of texture. Janot and Delcroix [13] investigated the influence of mechanical stress on the Mössbauer spectrum of a undirectionally solidified Al-Al_3Fe composite material. The direction of the applied elongation stress coincided with that of the crystal growth and with the [001] axis of the Al_3Fe crystallites. In that particular work the spectrum of Al_3Fe was evaluated as an asymmetric doublet. The isomer shift, quadrupole splitting, and intensity ratio changed significantly owing to the deformation. Taking these results into consideration, it cannot be precluded that the above discussed asymmetry (Figure 2b) was caused partly by the anisotropic mechanical stress field in the billet. The EFG of the Al_6Fe crystallites could be partially ordered by the stress field.

SPECTRUM EVALUATION

In order to fit the Mössbauer spectrum of a sheet sample taken from one particular site of the billet one should know how the intensity-ratio of the Al_6Fe doublet is affected by texture and/or deformation. The problem was solved by comparing the spectrum of the as cast sheet sample with that of fine filings of random orientation. In the latter case, the effect of

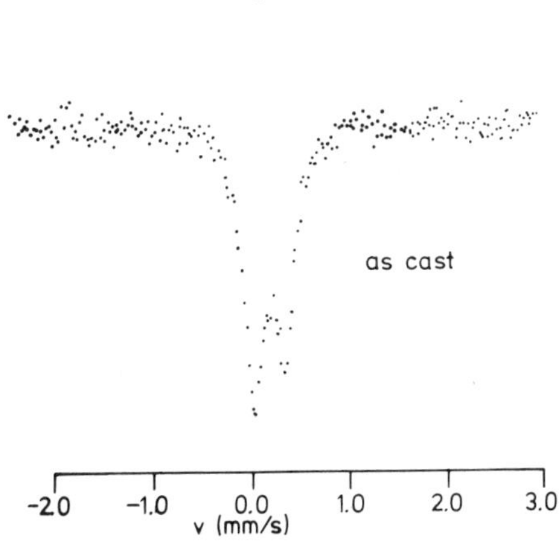

Figure 3. Mössbauer spectra of two industrial aluminum-iron samples. Although it is not obvious, the spectral lines of the heat treated samples were 10 to 25 percent broader than those of the as cast samples.

texture and deformation could be neglected. With samples cut out from the same site and with the same plane direction, we could fix the intensity-ratio of the Al_6Fe doublet because transmission electron microscopic investigations revealed that the orientation of the Al_6Fe crystallites did not change during the heat treatments [14].

The best fits were obtained under the following conditions. The positions of all lines were fixed. All lines were constrained to have the same variable width. The three lines of Al_3Fe were characterized by the same but variable total intensity.

The relative chi-square of the accepted fits varied between 0.93 and 1.26. We have not given the individual values, however, as they are more adequate for the comparison of different evaluations of the same spectrum than for that of the best evaluation of a series of spectra.

RESULTS AND DISCUSSION

The spectrum of the as cast sample could be well fit by the singlet of the aluminum-iron solid solution and the doublet of the Al_6Fe phase, but the assumption of Al_3Fe in this sample led to unacceptable results. Three percent of the iron content of the sample was in the solid solution and 97 percent was in the Al_6Fe phase.

Al_3Fe could not be found in the samples heat treated at 450 or 530°C. Assuming only Al(Fe) solid solution and Al_6Fe, the fit was quite

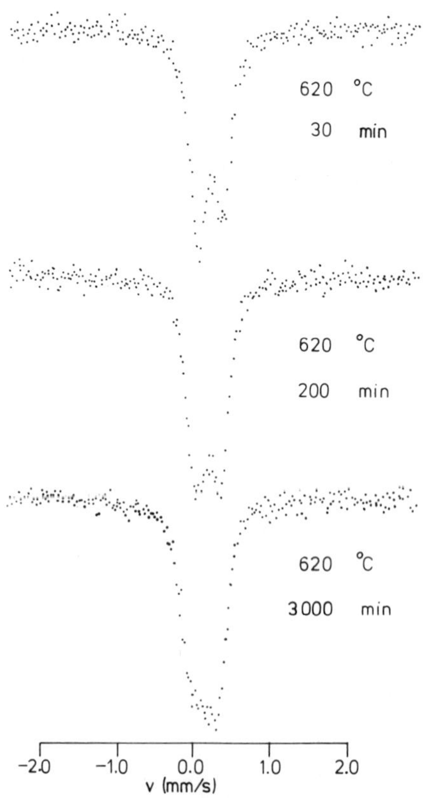

Figure 4. Mössbauer spectra obtained at room temperature of industrial aluminum-iron samples heat treated at 620°C.

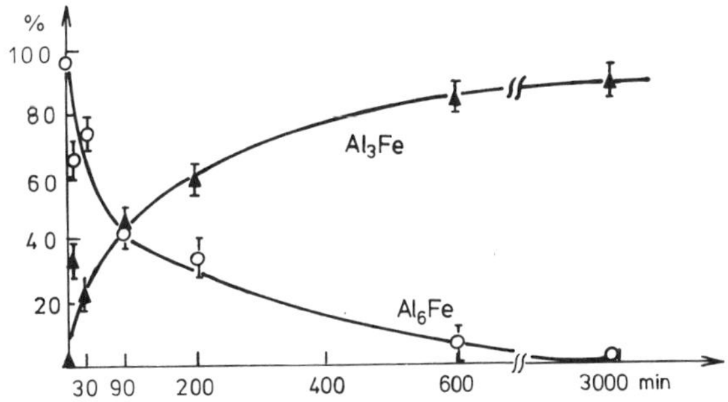

Figure 5. Distribution of iron between Al_6Fe and Al_3Fe as a function of the heat treatment at 620°C. The data were obtained from the Mössbauer spectra shown in Figure 4.

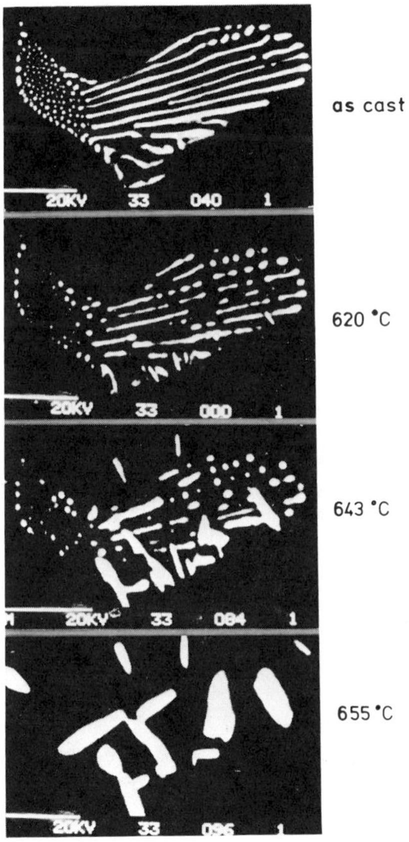

Figure 6. Scanning electron microscopic pictures taken from the same spot of a heat treated industrial aluminum-iron sample. The pictures clearly indicate the dissolution of Al_6Fe and the formation of bulky Al_3Fe particles. (Courtesy of Dr. Ferenc Tranta, Technical University of Miskolc).

satisfactory, but the spectral lines were 10 to 25 percent broader than those of the as cast sample (Figure 3). These changes took place within 30 minutes at both the heat treating temperatures. The significant line broadening might arise from the formation of some kind of a transient phase. The possibility of this was supported by transmission electron microscopic investigations [14] which indicated that during the heat treatments at 450 and 530°C for 10 hours a coherent phase was formed consisting of needle-shaped particles, 0.1μm in length, with no characteristic diffraction pattern of their own. Because of the difficulty caused by the appearance of these phases, the distribution of iron between the Al(Fe) solid solution and Al_6Fe phase could not be determined accurately.

The spectra of the samples annealed at 620°C and shown in Figure 4 could be fit very well on the assumption that they consisted of Al(Fe) solid solution, Al_6Fe and Al_3Fe phases. As it can be seen from Figure 5, the quantity the phase Al_3Fe increased and that of the phase Al_6Fe decreased with duration of heat treatment. The quantitative evaluation indicated that the phase Al_6Fe which was the dominant phase in the as cast sample, completely disappeared after 3000 minutes at 620°C and the majority or iron was present in the stable Al_3Fe phase. The phase transformation is illustrated in Figure 6 by showing a series of scanning electron microscopic pictures taken from the same spot of a polished aluminum surface after various heat treatments. The sample was heated up to the different temperatures shown in the figure at the rate of 10°/min. After each heat treatment the sample was quenched to room temperature at which point the pictures were taken.

TERNARY ALLOYS

The samples used in this series of measurements originated from two semicontinuously cast 90mm diameter extrusion aluminum billets. Billet number one contained 0.58 mass percent iron and 0.21 mass percent silicon, while billet number 2 contained 0.54 mass percent iron and 0.95 mass percent silicon. The melt temperature was 670°C and the casting speed was 80mm/min. In order to study the phase transformation in the ternary alloys, heat treatments were carried out at 450, 530, 575, and 620°C from 10 to 3000 minutes.

The number and type of the aluminum-iron and aluminum-iron-silicon phases occurring in the investigated alloys depends first on the concentration of iron and silicon and on the cooling rate. On the basis of the phase diagram, we concluded that the most likely components to show up in our

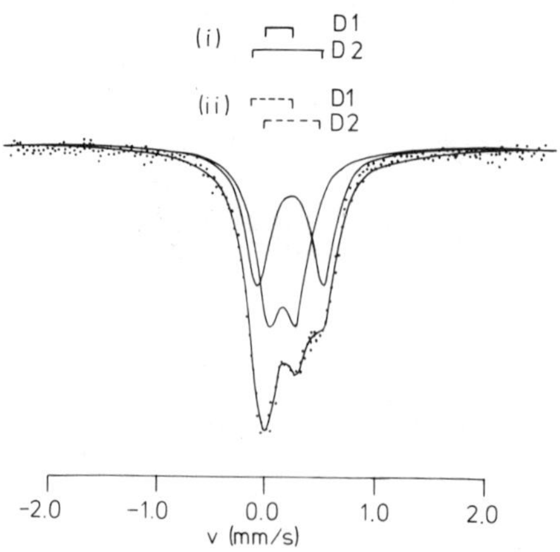

Figure 7. Mössbauer spectrum of hexagonal α-(AlFeSi). The two doublets drawn by the solid line, variant i, were chosen for the purpose of phase analysis.

Table II. Mössbauer Parameters of the Hexagonal α-(AlFeSi)[a]

Fit	δ	D1	% Area	δ	ΔE_Q	% Area	Γ	χ^2
(i)	0.175	0.257	52.2(7)	0.248	0.599	47.8(7)	0.287	1.108
(ii)	0.129	0.350	52.2(7)	0.299	0.496	47.8(7)	0.288	1.104

[a]All data in mm/s obtained at room temperature relative to α-iron.

spectra are dissolved iron, Al_6Fe, Al_3Fe, hexagonal α-(AlFeSi), cubic α-(AlFeSi), β-(AlFeSi) and γ-(AlFeSi). As the Mössbauer parameters of the aluminum-iron-silicon phases were unknown, the first task was to prepare and measure pure monophasic samples.

Hexagonal α-(AlFeSi). This alloy was prepared from 99.99 percent aluminum to which 32.5 mass percent iron and 9.5 mass percent silicon was added. The degassed melt was cast in an iron mold, 30mm in diameter and 130mm in length. The estimated solidification rate was 2000 to 3000°/min. The alloy was heat treated at 600°C for 500 hours, cooled in air, and ground to a fine powder. The x-ray powder diffraction [16] indicated hexagonal α-(AlFeSi). Its Mössbauer spectrum, measured at room temperature and illustrated in Figure 7, could be interpreted as the superposition of two symmetric doublets of equal linewidths. The spectrum could be fit in two different ways as we can see from Figure 7 and Table II. Although the relative χ^2 is somewhat less in the second case and the quadrupole splitting of D2 is also smaller, the first fit is more consistent with the metallic character of hexagonal α-(AlFeSi), because the isomer shifts of the D1 and D2 sites are closer to each other. It is important to note that the area percentages obtained for the two components were the same in either fit.

There are 46 iron atoms in the unit cell of hexagonal α-(AlFeSi) occupying five different sites [17], the ratio of Fe(1):Fe(2):Fe(3):Fe(4):Fe(5) is 12:12:12:6:4. If we assume that the Mössbauer-Lamb factors are the same for each site and note that (12+6+4)/45 is 0.478, we have a strong temptation to attribute D2 to the atoms Fe(4) and Fe(5) as well as to one of the remaining three sites. We have several reasons to believe that the right assignment is D1:Fe(1), Fe(2) and D2:Fe(3), Fe(4), Fe(5). First the Fe(1) and Fe(2) polyhedra [17] form one, type A, sheet and the Fe(3) polyhedra form a second, type B sheet. Further the short iron-aluminum distance of 0.229nm is found in the Fe(3) polyhedra which would explain the higher isomer shift and quadrupole splitting of D2.

The cubic α-(AlFeSi) phase is a metastable phase. It forms only at high cooling rates in alloys of low iron and silicon concentration, in this case 0.5 mass percent iron and 0.2 mass percent silicon, and therefore it cannot be produced in its pure state. According to the x-ray diffraction measurements [15], our preparation contained only one to two mass percent of the intermetallic compound. Unfortunately the contribution of the always present dissolved iron of probably less than 10 percent, could not be determined in the spectra because of the overlapping spectral lines as shown in Figure 8. The results obtained for a symmetric two-doublet fit, in which the contribution of dissolved iron was neglected, can be seen in Figure 9 and Table III. Naturally this kind of simplification must affect the interpretation of all the spectra in which cubic α-(AlFeSi) appeared.

β-(AlFeSi). The sample, prepared from aluminum 17.8 mass percent iron, 9.7 mass percent silicon was heat treated at 580°C for 330 hours to reach equilibrium. Microprobe and x-ray diffraction measurements [15] indicated

Table III. Mössbauer Data for Cubic α-(AlFeSi), β-(AlFeSi), and γ-(AlFeSi)[a]

Phase	δ	ΔE_Q	% Area	Γ
Cubic α-(AlFeSi)	0.30	0.20	34.6	0.29
	0.18	0.30	65.4	0.29
Cubic β-(AlFeSi)	0.12	0.42	43.6	0.27
	0.23	0.40	49.2	0.27
	0.34	--	7.2	0.27
Cubic γ-(AlFeSi)	0.16	0.24	46.1	0.31
	0.16	0.56	53.9	0.31

[a]All data in mm/s obtained at room temperature relative to α-iron.

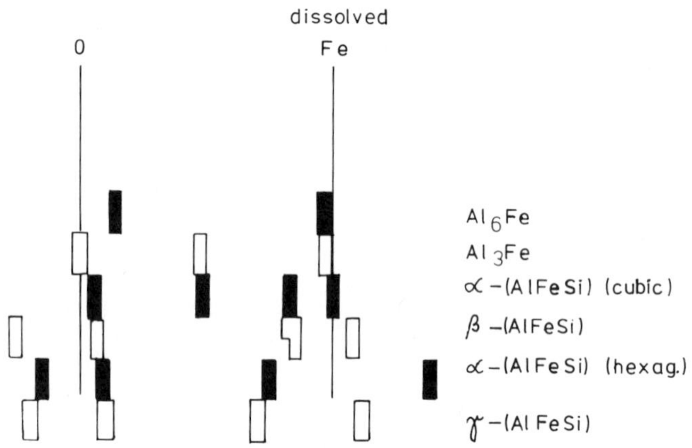

Figure 8. The possible spectral lines observed in the Mössbauer spectrum of a ternary aluminum-iron-silicon industrial alloy.

that the annealed sample contained about 66 volume percent β-(AlFeSi) besides aluminum. The contribution of dissolved iron is still negligible. The best fit for the Mössbauer spectrum (Figure 10) was based on two doublets and one singlet, the parameters of which are given in Table III. Because of the lack of reliable data on the atomic positions in the unit cell of β-(AlFeSi), we could not decide whether the fit had any physical meaning. On the other hand, with respect to the χ^2, the fit was excellent and therefore it could be used for the purpose of phase analysis. It would be interesting to compare these results with those of a single crystal x-ray diffraction study.

γ-(AlFeSi). Although this phase could not be found in our industrial samples, we have measured its Mössbauer spectrum which is shown in Figure 11. The spectral parameters are given in Table III.

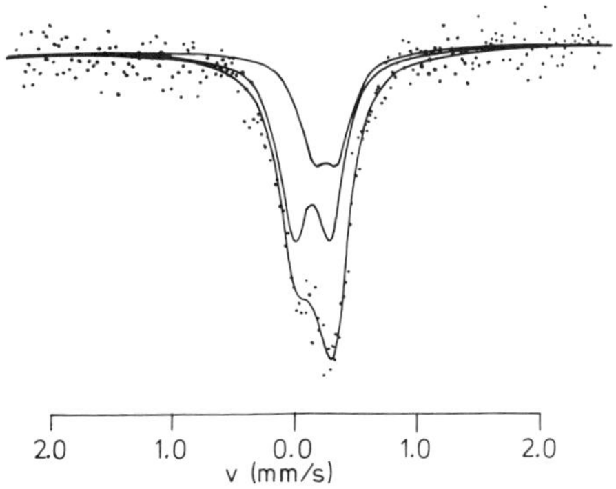

Figure 9. The room temperature Mössbauer spectrum of cubic α-(AlFeSi).

The γ-(AlFeSi) sample was prepared from aluminum, 34.9 mass percent iron, 12.3 mass percent silicon. The alloy was annealed for 32 hours at 600°C, and then ground to a fine powder. According to the microprobe and x-ray diffraction measurements the preparation contained 90 percent γ-(AlFeSi) besides aluminum.

INDUSTRIAL AlFeSi ALLOYS

For the Mössbauer measurements on the aluminum-iron-silicon alloys, fine filings were distributed in paraffin to avoid problems arising from texture or stress. The annealings were performed in the same manner as in the case of the aluminum-iron alloys. The samples were cut out at the half radius of the billets. The more complicated spectra of the industrial alloys were fit by linear combinations of the basis spectra shown in Figures 1, 7, 9, 10. It was also assumed that the isomer shift of dissolved iron was unaffected by the silicon content of the alloy. The spectra were evaluated on a formal basis independently of the results of the x-ray measurements [15]. However, there was a good agreement between the results obtained by the two methods proving the applicability of Mössbauer spectroscopy for the phase analysis of such alloys.

Because each spectrum could be interpreted in terms of one or two major components, we had only n+2 free variables in the computer evaluations, where n is the number of the actually used basis spectra. The parameters were baseline, line width and n amplitude parameters, one for each basis spectrum.

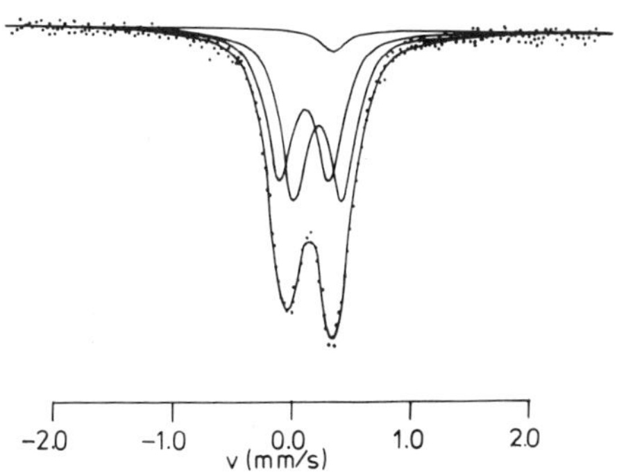

Figure 10. The room temperature Mössbauer spectrum of β-(AlFeSi)

Table IV. Qualitative Characterization of the Spectral Evaluations Connected with Billet Number 1. The Notation is Defined in the Text.

Basis	Basis Spectra Used							
Al(Fe)	Yes	Yes	Yes	Yes	Yes	Yes	Yes	Yes
Al_6Fe	Yes	No	No	Yes	Yes	No	No	No
Al_3Fe	No	Yes	No	No	Yes	Yes	No	Yes
Cubic α-(AlFeSi)	Yes	Yes	Yes	Yes	No	No	Yes	Yes
Hexagonal α-(AlFeSi)	No	No	No	Yes	No	Yes	Yes	Yes
Heat Tr. T,°C Hours	Characterization							
As Cast	2+	2	1	0	1	0	0	0
450 10	2+	2	1	-	-	0	1	1
530 0.5	2+	2	-	-	-	-	-	-
530 1.5	2+	2	1	-	-	-	-	-
530 5	2+	2	1	-	-	-	-	-
530 10	2+	2	1	-	-	0	0	0
530 16	2+	-	-	-	-	-	-	-
575 16	2	2+	1	-	-	-	-	-
620 0.5	0	0	0	0	-	2+	0	0
620 1.5	0	0	0	0	1	2+	1	0
620 5	0	0	0	0	1	2+	0	0
620 10	0	0	0	0	1	2+	0	0
620 50	0	0	0	0	1	2+	1	0

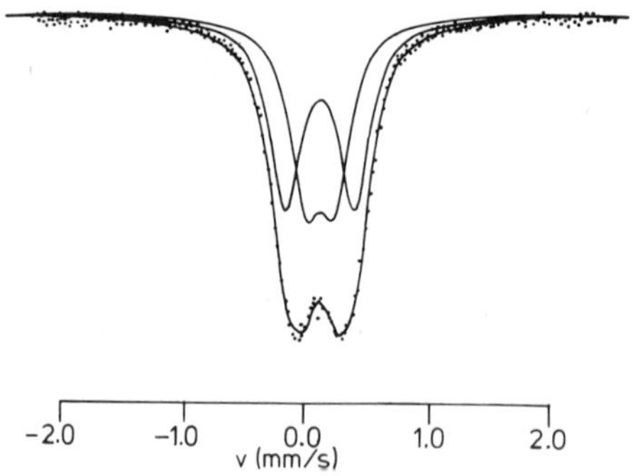

Figure 11. The room temperature Mössbauer spectrum of γ-(AlFeSi). This phase was not found in our industrial samples.

Thus, the number of fitted parameters could usually be kept below six which was good considering the similar spectral lines (Figure 8). Only the distinction between Al_3Fe and cubic α-(AlFeSi) was somewhat difficult.

Table IV gives a qualitative characterization of the various trials to fit the spectra of the samples cut from billet number 1. The quality of the evaluations was characterized by the following numbers. A zero indicates the fit was totally unacceptable. A 1 indicates from a formal point of view the fit was good, but that its physical meaning was doubtful because of negative amplitudes or of too large linewidths. A 2 indicates the fit seems to be reasonable in every respect. The fits we finally accepted are indicated by plus signs. Besides the above combinations two more bases were also tried which included β-(AlFeSi), but the fits were unacceptable.

For the as cast sample, the following conclusions can be drawn from Table IV. First, whenever the fitting basis included the spectrum of hexgonal α-(AlFeSi), the fits were very poor. Therefore, apparently this component is unlikely to form while casting. Second, if the basis included the spectrum of cubic α-(AlFeSi), the fits were usually very good. These conclusions are in good agreement with the general observation that, at the cooling rate of casting, the thermodynamically metastable phases form rather than the stable phases. Best fits were obtained when Al_6Fe or Al_3Fe was looked for besides Al(Fe) and cubic α-(AlFeSi). Attempts to use all the four components at the same time were unsuccessful because of the big correlation between the three minor components. The final choice was

$$Al(Fe) + Al_6Fe + \text{cubic } \alpha\text{-}(AlFeSi)$$

as this selection resulted in the lowest χ^2 value. Naturally the presence of some Al_3Fe cannot be precluded because of the above mentioned correlation. One also has to consider that some uncertainty also arises from our having neglected the contribution of dissolved iron when fitting the reference spectrum of cubic α-(AlFeSi).

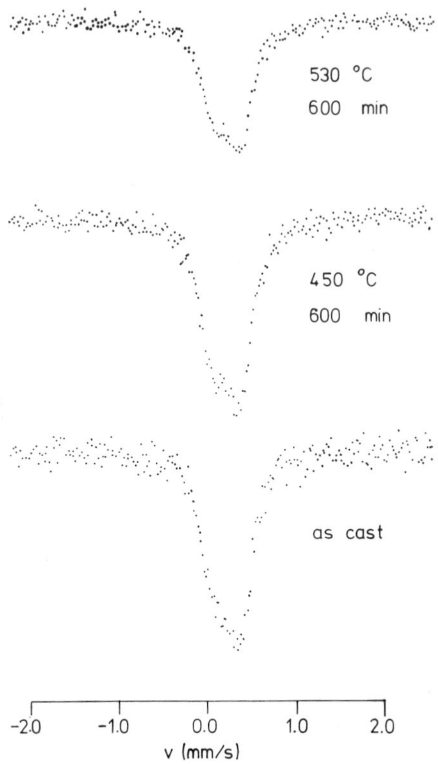

Figure 12. Heat treatments up to 530°C did not affect the room temperature Mössbauer spectrum of the ternary aluminum-iron-silicon samples cut from billet number 1.

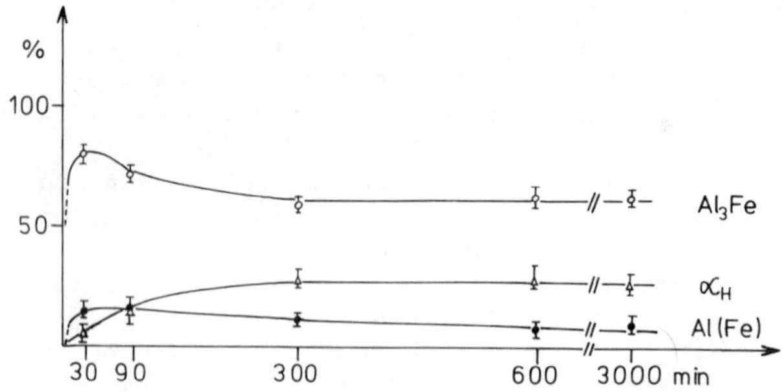

Figure 13. Distribution of iron between Al_3Fe, hexagonal α-(AlFeSi) and dissolved iron after heat treatments performed on billet number 1 at 620°C. The data were calculated from the Mössbauer spectra shown in Figure 14.

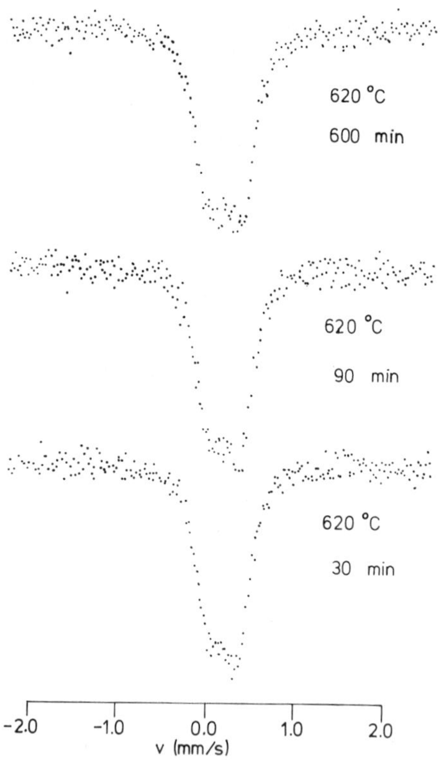

Figure 14. The room temperature Mössbauer spectra of billet number 1 after heat treatment at 620°C.

The above selection of basis spectra proved to be the best also in the case of the samples heat treated at 450 and 530°C. However, at 575°C the χ^2 was lower, if the Al_6Fe was replaced with Al_3Fe, in agreement with the expectation that the metastable Al_6Fe would transform to the stable Al_3Fe. Some of the measured spectra are shown in Figure 12.

In the case of the samples heat treated at 620°C the above set of basis spectra turned out to be a bad choice. Two important conclusions can be drawn from the eight different basis selections we tried. First, the evaluations including cubic α-(AlFeSi) usually resulted in unsatisfactory fits. Second, if the basis selection included hexagonal α-(AlFeSi), the fits obtained were usually good. Most satisfactory results were obtained with the selection:

$$Al(Fe) + Al_3Fe + \text{hexagonal } \alpha\text{-}(AlFeSi)$$

shown in Figure 13 and 14. Assuming that these evaluations have a real physical basis, the following conclusions can be drawn regarding the phase transformations at 620°C. First, cubic α-(AlFeSi) phase suffers fast decomposition within 30 minutes to give Al_3Fe and Si. Second, after subsequent heat treatments the formation of α_H hexagonal α-(AlFeSi) could also be observed, probably as a consequence of the reaction

$$Al_3Fe + Si \rightarrow \text{hexagonal } \alpha\text{-}(AlFeSi).$$

The above reactions could serve with an explanation for the maximum on the diagram for Al_3Fe seen in Figure 13. In order to support this idea, we are planning further heat treatments to study the first reaction step in the temperature range between 575 and 620°C.

The samples cut from billet number 2 were heat treated at 450, 530, and 575°C. The spectra were evaluated on eight different bases combined from the six reference spectra. The spectra could be fairly well fit by the parameters of β-(AlFeSi) along. The problem with the basis of Al(Fe) and β-(AlFeSi) was that the line areas calculated for dissolved iron were always negative by a few percent. The possible explanation for this is the inhomogeneity of the β-(AlFeSi) particles. X-ray microprobe investigations

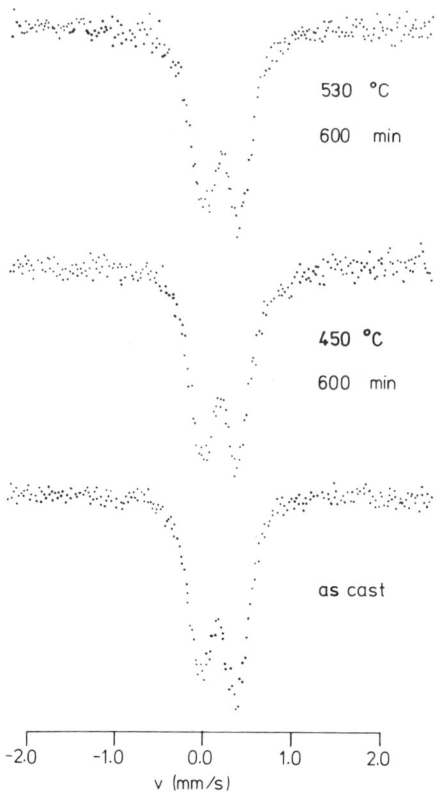

Figure 15. The room temperature Mössbauer spectra of billet number 2 indicate mainly the presence of β-(AlFeSi).

revealed that the stoichiometric ratio of iron and silicon was varying even within one crystallite [15]. It is possible that the stoichiometry of β-(AlFeSi) has an effect on the area ratio of the two doublets with which we tried to fit the spectra. Some spectra are shown in Figure 15.

ACKNOWLEDGEMENT

The authors thank Dr. Ferenc Tranta of the Technical University of Miskolc, Hungary for his permission to publish the electron microscopic pictures shown in Figure 6.

REFERENCES

1. S. Nagy, C. Szeles, and A. Vertes, Akademiai Kiado, in press (1985).
2. S. Nagy, A. Vertes, A. Griger, E. Hidvegi, and T. Turmezey, Proceedings of the ICAME, Alma-Ata, 1985, to be published.
3. R. S. Preston and R. Gerlach, Phys. Rev. B 3, 1519 (1971).
4. C. A. Stickels and R. H. Bush, Metal. Trans. 2, 2031 (1971).
5. R. S. Preston, Metal. Trans. 3, 1931 (1972).
6. S. Nasu and V. Gonser, 5th Int. Conf. Mössbauer Sp., Proc. 2, 311, Bratislava (1973).
7. S. Nasu, V. Gonser, P. H. Shingu, and Y. Murakami, J. Phys. F. 4, L24 (1974).
8. S. Nasu, V. Gonser, and R. S. Preston, J. Phys. Colloq. 41, C1-385 (1985).
9. A. Griger, V. Stefaniay, and T. Turmezey, to be published (1985).
10. S.Nagy, A. Vertes, E. Hidvegi, A. Griger, and T. Turmezey, Proceedings of the ICAME, Alma Ata, 1983, to be published.
11. V. I. Goldanskii and R. H. Herber, in Chemical Application of Mössbauer Spectroscopy, Academic Press, New York and London (1968).
12. H. Wegener, in Der Mössbauereffekt und seine Anwendung in Physik und Chemie Bibliographisches Institute, Mannheim (1966).
13. C. Janot and P. Delcroix, Acta Met. 24, 197 (1976).
14. A. Vertes, S. Nagy, Z. Homonnay, L. Murgas, A. Cziraki, B. Fogarassy, and I. Szabo, Eighth European Congress on Electron Microscopy, Proc., Vol. 1, 747, Budapest (1984).
15. V. Stefaniay, A. Griger, T. Turmezey, S. Nagy, Z. Homonnay, and A. Vertes, 1985, to be published.
16. A. Griger, 1985, to be published
17. R. N. Corby and P. J. Black, Acta Cryst. B33, 3468 (1977).

SULFIDATION-INDUCED CHANGES IN IRON-ALUMINUM AND IRON-SILICON ALLOYS

S. M. Dubiel

Department of Solid State Physics
Institute of Metallurgy
Academy of Mining and Metallurgy
30-050 Krakòw, Poland

Laboratoire de Mètallurgie associè au CNRS U.A. 159
Ècole des Mines
Parc de Saurupt
54042 Nancy Cedex, France

Z. Żurek

Institute of Inorganic Chemistry and Technology
Technical University
31-155 Krakòw, Poland

M. Przybylski

Department of Solid State Physics
Institute of Metallurgy
Academy of Mining and Metallurgy
30-050 Krakòw, Poland

ABSTRACT

By using X-ray phase and microanalysis, optical and scanning electron microscopy and Mössbauer spectroscopy, we investigated the influence of sulfidation of iron-aluminum alloys containing 0.8 and 3 weight percent aluminum and a iron-3 weight percent silicon. We find that the near-surface zone of the metallic core is enriched in the solute element and in the case of iron-silicon alloy it contains an admixture of the Fe_3Si phase. A mechanism is proposed in order to account for the enrichment.

INTRODUCTION

Corrosion of iron and its alloys is certainly one of the most common and technologically important problems. Corrosion which occurs in an atmosphere containing sulfur or its compounds, i.e., sulfidation, proceeds at elevated temperatures with a rate much larger than the corrosion caused by oxygen, i.e., oxidation. A knowledge of how to prevent sulfur corrosion has been growing as it is almost impossible to get rid of sulfur from hot gaseous atmospheres present often in modern technical installations. In particular, sulfur corrosion protection is one of the major problems to be solved in gas turbines which are used, for instance, in the energy and aviation industry. It is also a major problem in the rapidly developing technology of coal gasification.

In order to know how to protect materials against the corrosion or, at least to minimize its effects, one has to know the mechanism involved. Intensive investigations of this mechanism have been carried out in various laboratories by means of different experimental techniques. In particular, the sulfidation of iron and its alloys has been a subject of great interest. From the investigations carried out to date [1-6], we know that (1) the process of the sulfidation of these materials proceeds, in general, according to a parabolic law, and that (2) the resulting scale that forms on a metal surface consists of a ferrous sulfide, $Fe_{1-y}S$, and grows as a result of an out-of-the core diffusion of the metal.

Although the existing experimental data has helped to explain or at least helped to better understand some related phenomena, many other problems have been posed and remain unexplained. Hence a further investigation of the problem seems to be justified. Results on the kinetics of the sulfidation of iron and its alloys, and on the phenomena accompanying and occurring within the near-surface zone of the metallic core show that the temperature and pressure of the sulfidation are very important parameters.

The relatively large disagreement between the results obtained by different authors in their experiments on the sulfidation may result because such experiments are more difficult to carry out than similar experiments on oxidation. Further complications associated with the former are due to its much higher rate. Rapidly proceeding sulfidation causes not only geometrical changes in a sample, but also promotes phase transformations, especially in the near-surface zone of the core. These transformations must be taken into account in order to properly describe and understand the sulfidation. These changes result from a selective, preferential inner sulfidation that occurs in the near surface zone as well as from an oversaturation in the vacancies being generated on the core surface. The kind of phenomenon that occurs during a sulfidation of Fe-Al and Fe-Si alloys has been described in our previous papers [7,8]. In this paper we present and discuss further these phenomena based on extended experiments carried out by means of X-ray phase and microanalysis, optical and scanning electron microscopy (SEM), as well as by the Mössbauer effect spectroscopy.

II. EXPERIMENTAL

Measurement of the Corrosion Kinetics

Measurement of the oxidation or sulfidation rate of metals plays a key role in the investigation of the corrosion mechanism. Such studies were initiated by Tammann [9] and by Pilling and Bedworth [10]. Several empirical laws, known as the kinetic laws of metal oxidation or sulfidation, have been presented to describe the process of metal and alloy oxidation or sulfidation. At elevated temperatures, for instance, the law has, in general, a parabolic form and it can be expressed by the following equation

$$\frac{dx}{dt} = \frac{k_p}{x} \qquad (1)$$

where x is the thickness of the reaction product after a time t and k_p is the parabolic constant of the reaction rate. After integration, equation 1 transforms into

$$x^2 = 2\,k_p t + C \qquad (1a)$$

where C is an integration constant. Practically, this equation is rather seldom used to measure the rate of metal oxidation or sulfidation because measuring the thickness of the scale formed as a function of time is neither an easy nor a precise task.

If the oxidation or sulfidation proceeds homogeneously on the entire surface of the sample, i.e., edge effects can be neglected, the scale thickness in then proportional to an increase of the sample mass, Δm. In this case the relationship between the two quantities can be described by

$$dx = \frac{V_{Me_rX_s}}{\frac{s}{2}M_{X_2}A_{Me}} d\Delta m \tag{2}$$

$V_{Me_rX_s}$ is the molar volume of the compound forming the scale,

M_{X_2} is the molecule mass of the oxidizer, X,

A_{Me} is the area through which the diffusion of reacting elements proceeds,

and r and s are the stoichiometric coefficients of the chemical reaction which forms the scale

$$rMe + \frac{s}{2}X_2 = Me_rX_s.$$

After integration, equation (2) has the following form

$$x = \frac{V_{Me_rX_s}}{\frac{s}{2}M_{X_2}} \int_o^{\Delta m} \frac{d\Delta m}{A_{Me}}. \tag{2a}$$

By putting this result into equation (1a), one obtains the following

$$\left(\int_o^{\Delta m} \frac{d\Delta m}{A_{Me}}\right)^2 = 2\left(\frac{\frac{s}{2}M_{X_2}}{V_{Me_rX_s}}\right)^2 k_p t + C. \tag{3}$$

By taking into account the change of the surface of the oxidized or sulfidized sample and by calling

$$2\left(\frac{s}{2} M_{X_2}/V_{Me_rX_s}\right)^2 k_p = k'_p.$$

the equation (3) transforms into

$$\left(\frac{\Delta m}{A}\right)^2 = k'_p t + C, \tag{4}$$

where A describes a change of the sample area in terms of an increase of the sample mass.

The parabolic law describing the process of oxidation or sulfidation indicates that the slowest elemental process which determines the rate of a product formation is the diffusion of the reacting elements within the layer of the scale. With this law one can describe the oxidation or sulfidation of many metals and alloys over a wide range of temperatures [11].

Experimental Equipment

Among the numerous experimental methods used in the investigation of the kinetics of oxidation or sulfidation, the thermogravimetric method has been perhaps most frequently applied, mainly because of its practical advantages. A review of different thermogravimetric equipment is given, for instance, by Mrowec [11]. In the present study we used a rather simple setup which enabled us to study the kinetics of the sulfidation of iron-aluminum and iron-silicon alloys in a sulfur vapor at atmospheric pressure. Its schematic layout is shown in Figure 1. It consists of a vertical quartz tube, 6cm in diameter and 100cm long, with a closed lower end. The tube is covered with a tantalum foil coated with an asbestos layer. The tube has two zones. The upper is the zone in which sulfidation takes place and the lower contains liquid sulfur. The two zones have independent heating systems. Within the lower one the sulphur is kept at its boiling point which ensures a constant pressure of a sulfur vapor inside the upper zone. The heating system in the upper zone enables us to maintain a desired temperature for the sulfidation which is automatically regulated to within $\pm$ 3°C.

After leaving the upper zone, the sulfur vapor is burned and thereby produces the so-called flame barrier which prevents air from penetrating into the zone. The quartz tube is placed inside a chamber from which SO_2 and the sulfur vapor are continuously evacuated during the experiment. The sample to be sulfidized is suspended in the middle of the upper zone with a molybdenum wire which is connected to a balance and permits the measurement of an increase in the sample mass to an accuracy of about 10^{-4} g.

Alloy and Sample Preparation

The alloys investigated were iron-aluminum containing 0.8 and 2.96 weight percent aluminum and an iron-silicon alloy with 3.12 weight percent silicon.

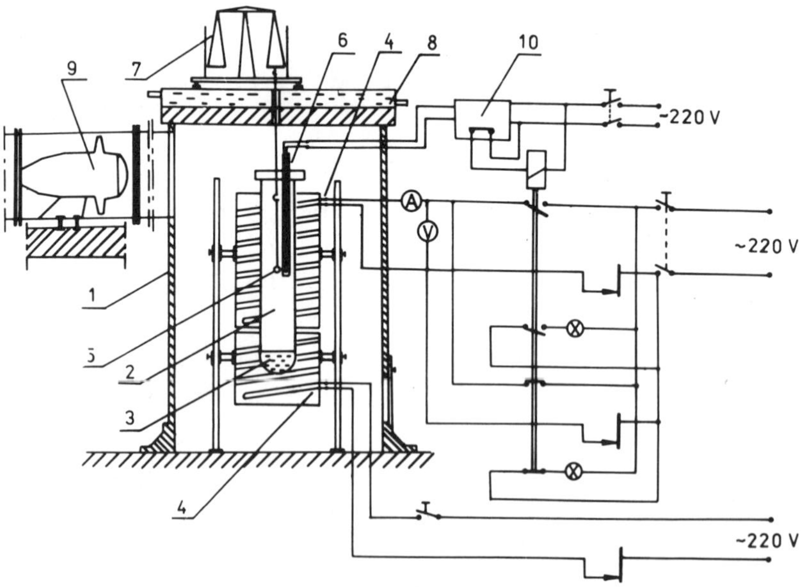

Figure 1. A schematic of the equipment used to sulfidize iron-aluminum and iron-silicon alloys. 1-chamber, 2-quartz tube, 3-liquid sulfur, 4-heating element, 5-sample, 6-thermocouples, 7-balance, 8-thermoisolation, 9-fan, 10-power supply and temperature regulating unit.

They were prepared, by melting in an induction furnace under a high vacuum, ARMCO iron with aluminum or silicon, of 4N purity. From the ingots slabs of 30x20x2.0mm were cut for the iron-aluminum alloys and of 30x20x1.5mm for the iron-silicon alloy. Directly before the sulfidation, the surface of each sample was hand-ground with glasspapers of a decreasing grain down to 5/0, and then washed, first with alcohol, and then with acetone. The sulfidation process was carried out as described above. The iron-aluminum samples were sulfidized for 15min at 550 and 650°C and those of iron-silicon at 600, 700 and 800°C for 35 to 45min depending on the temperature of the sulfidation.

For the iron-aluminum alloys the kinetics of the sulfidation was not studied, we were mainly interested in the phenomena occurring in the near-surface zone of the metallic core. For the iron-silicon alloy, the kinetic behavior is illustrated in Figure 2. The resulting values of the parabolic constant, k_p, are presented in Table I together with the corresponding results for a pure iron.

In iron-aluminum alloys, the near-surface zone was studied by optical microscopy, X-ray microanalyzer, and by Mössbauer spectroscopy. The first method failed to detect any changes after the sulfidation. In order to measure the distribution of iron, aluminum and sulfur, we used a JOEL JXA-3A X-ray microanalyzer for the 3 percent iron-aluminum. After the sulfidation process was completed, the sample was sectioned perpendicular to its surface, ground, and finally polished. The polished cross-sections were coated directly before the analysis with graphite in order to increase their surface conductivity.

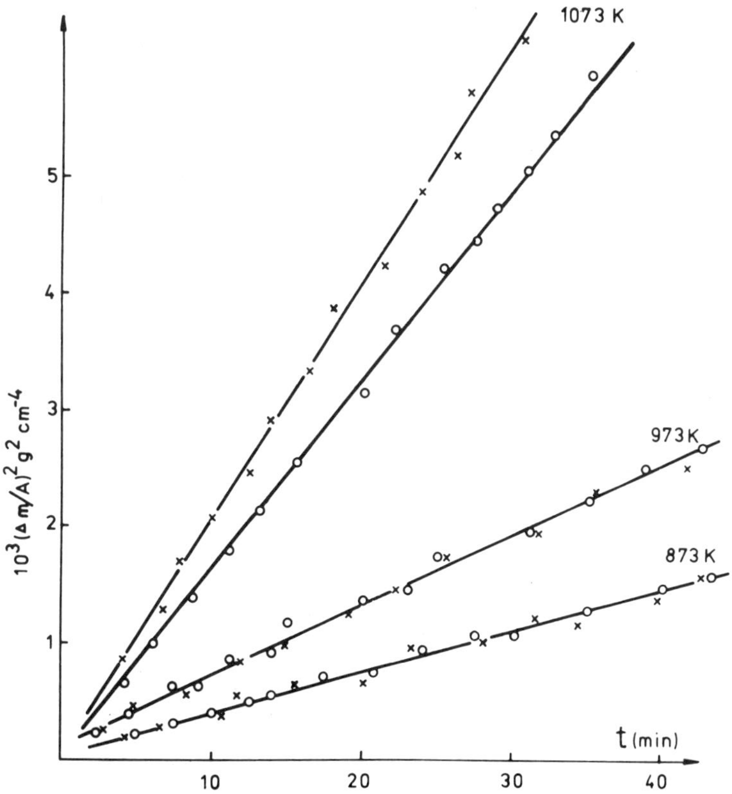

Figure 2. Sulfidation lines for an iron (x) and the iron-3 percent silicon alloy (o) in the parabolic plane for different temperatures.

Table I. Parabolic Constants[a], k'_p, for pure iron and for iron-silicon alloy sulfidized

T_S(°C)	Iron	Fe-3% Si
600	0.57	0.59
700	1.08	1.02
800	4.26	2.73

[a]Parabolic constants given in $10^6 g^2/cm^4 s$

For the Mössbauer measurements on the iron-aluminum alloys, samples were prepared from (a) the non-sulfidized alloys, (b) the center of the core of the sulfidized alloys, (c) the near-surface zone of the core of the sulfidized alloys, and (d) the surface-zone of the core of the sulfidized alloys, i.e., the region having a direct contact with the scale. For the iron-silicon alloy, samples (a), (b), and (d) were prepared. Each sample had a size of ca. 15x10mm and it was 20-30μm thick. The Mössbauer spectra were measured at room temperature in a transmission geometry with a standard spectrometer, which utilized a cobalt-57 source in a chromium matrix and was calibrated with iron foil. Its activity was ca. 20mCi and a spectrum could be collected with good statistics within 24 hours. Examples of the resulting spectra are shown in Figures 3 and 4.

More detailed investigations of the physical and chemical properties of the scales and of particular layers of interest were carried out for the iron-silicon alloy. The morphology of the scale, the near-surface zone and the inner zone of the core were studied on microsections obtained by cutting

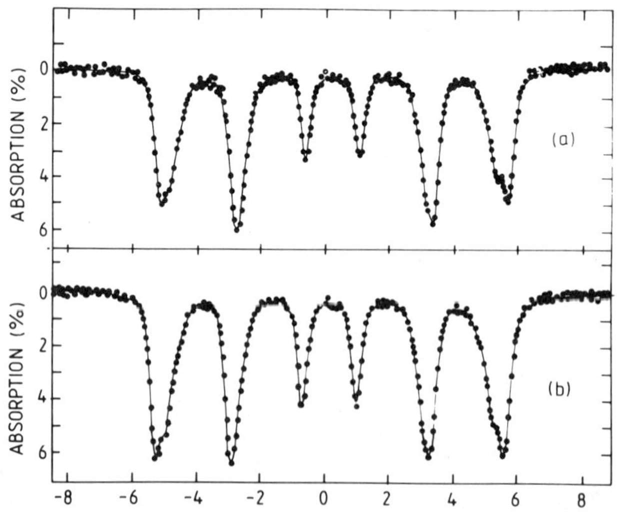

Figure 3. Mössbauer spectra of the iron-3 weight percent aluminum alloy after 15min of sulfidation at 650°C, for the (a) near-surface zone of the core and (b) the center of the core.

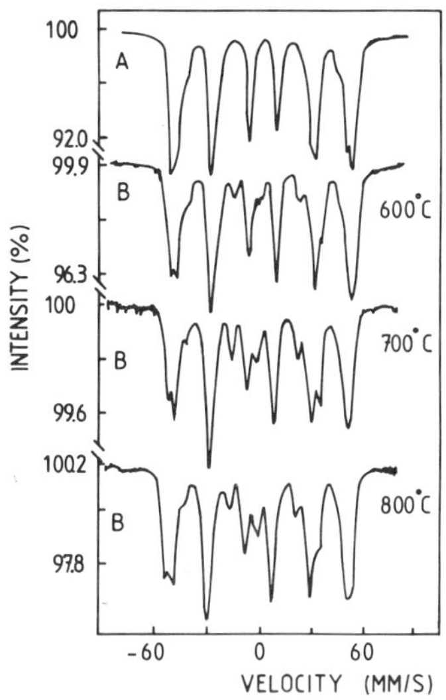

Figure 4. Mössbauer spectra of the iron-3 weight percent silicon alloy after sulfidation at different temperatures for (A) the sample from the center of the core and (B) the sample from the surface zone of the core.

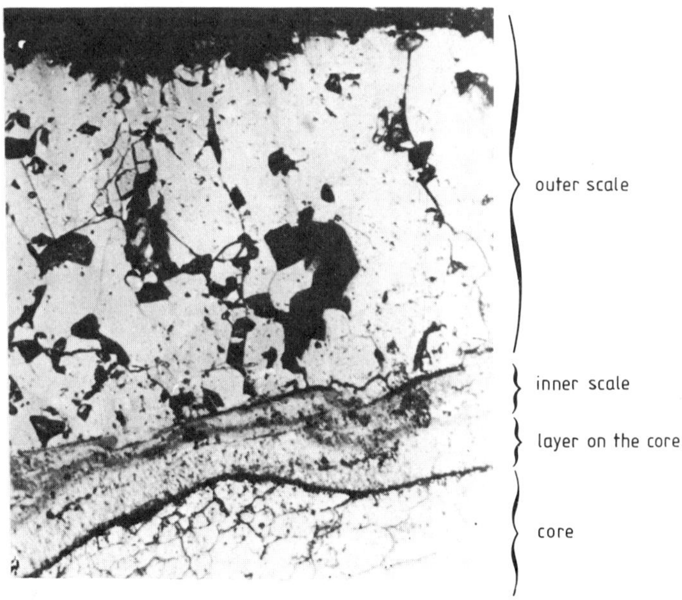

Figure 5. The microstructure of the iron-3 weight percent silicon alloy sulfidized at 600°C for 45min, x100.

the sample perpendicular to its surface. In order to make grains more visible, the microsections were etched in an alcoholic solution of nitrous acid.

Figures 5 and 6 show the microstructure of the layers of interest for the sample sulfidized at 600°C and 800°C. For the analysis by scanning electron microscopy, the samples were fractured perpendicular to the sulfidized surface, and the electron beam was perpendicular to the area of the fracture. Figures 7 to 9 illustrate the microstructure of the particular layers of interest for the sample sulfidized at 800°C.

The distribution of the elements in the iron-silicon samples was obtained with a SMQ ARL X-ray microanalyzer. FeS was used as a standard to determine iron and sulfur and metallic silicon was used to determine silicon. The analysis of the phase composition of the layers was carried out by means of X-ray diffraction. The outer layer of the scale was removed prior to this analysis. The resulting reflections were interpreted on the basis of the ASTM tables [12]. The diffraction was carried out on subsequent layers starting with the inner scale, followed by a layer of the core and ending with a layer taken from the center of the core. A particular layer was made accessible by grinding the preceding one. This procedure was continued until the diffraction pattern obtained was identical to that characteristic of the primary, non-sulfidized sample. Figure 10 shows the phases as deduced from

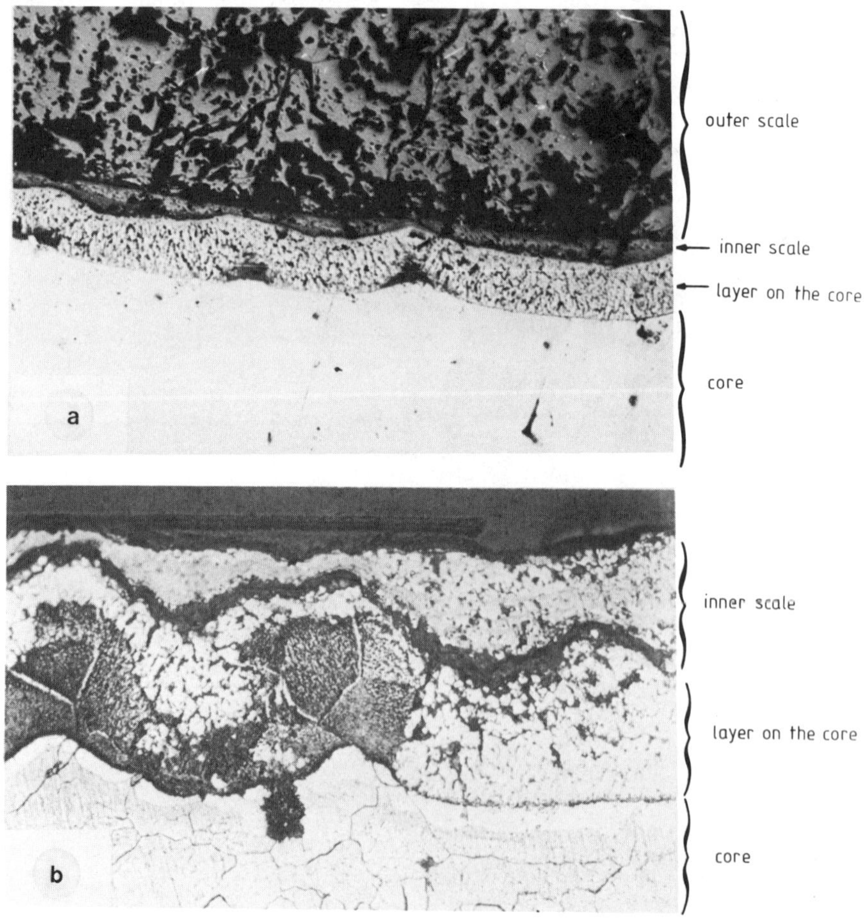

Figure 6. The microstructure of the iron-3 weight percent alloy sulfidized at 800°C for 35min, (a) x100, (b) x250

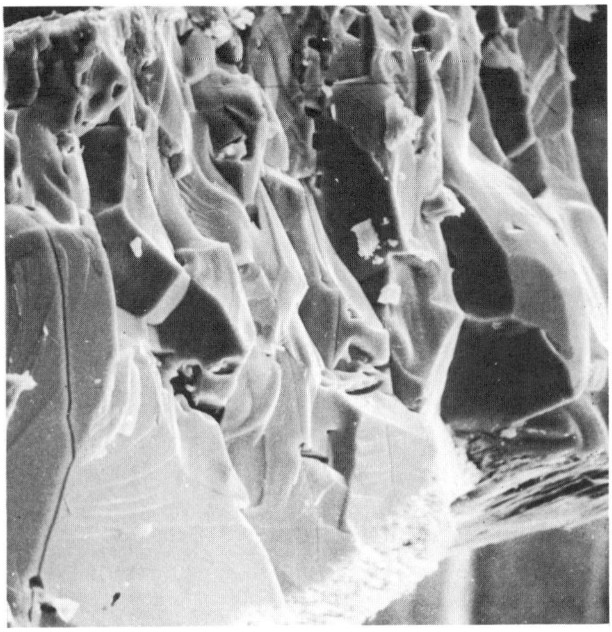

Figure 7. The microstructure of the outer scale formed on the surface of the iron-3 weight percent silicon alloy sulfidized at 800°C. Note the well-formed columnar structure.

Figure 8. The microstructure of the layer on the core of the iron-3 weight percent silicon alloy sulfidized at 800°C x6000. Note the irregular spheres and thin plates.

this study. The kind and relative abundance of the particular phases occurring in the four layers can be distinguished.

Chemical Composition of the Layers

Figure 11 shows the distribution of iron, aluminum and sulfur along the cross-section of the iron-3 weight percent aluminum sample sulfidized at 650°C. We note that within the near-surface zone of the core there is some increase in the aluminum concentration. The depth of this aluminum-enriched

Figure 9. The microstructure of the near-surface zone of the iron-3 weight percent silicon alloy sulfidized at 800°C, x5200. Note the small, detached, sphere-like elements.

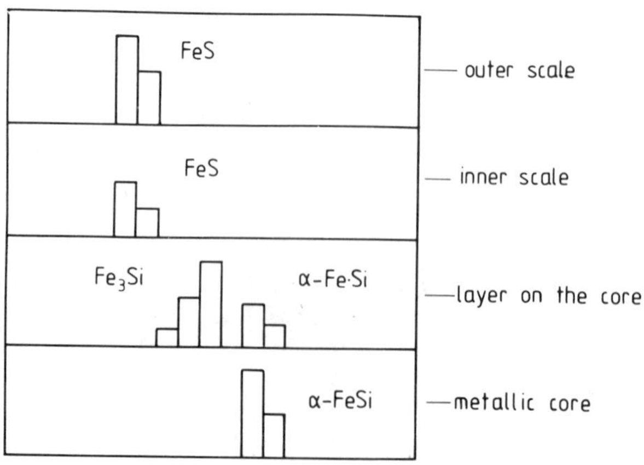

Figure 10. Phase distribution as obtained for the iron-3 weight percent silicon alloy sulfidized at

zone is rather small and amounts to, at most, 4μm. As illustrated in Figure 12 and 13 for the iron-silicon alloy, the distribution of elements within the metallic core is rather homogeneous. In the layer on the core, however, one observes a rather sharp increase in silicon content and a decrease in iron concentration. The sulfur content in this layer increases smoothly on approaching the scale, but its concentration is less than 0.5-1.5%. There is more silicon present in the inner scale and the layer on the core than there is within the core. The sulfur content is rather high except in regions where a silicon is present where its concentration is significantly smaller. In the outer scale silicon is absent and the concentration of iron and sulfur corresponds to that of FeS. Because of the morphology of the layers, these results should be treated only qualitatively.

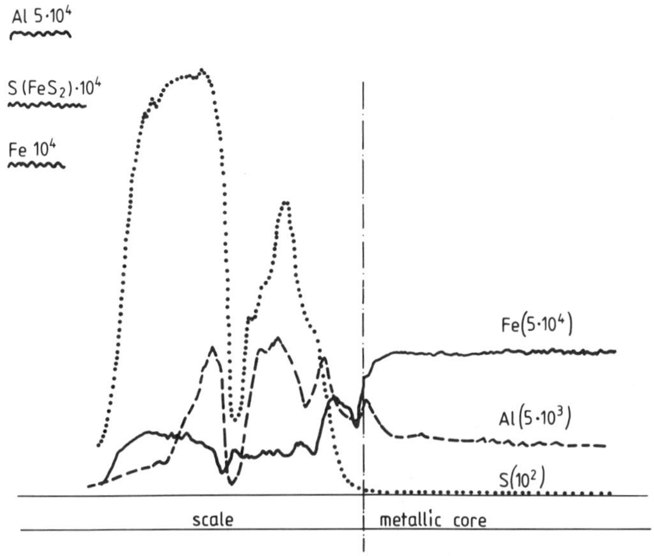

Figure 11. Distribution of elements for the iron-3 weight percent aluminum sample sulfidized at 650°C (relative scale factors in brackets).

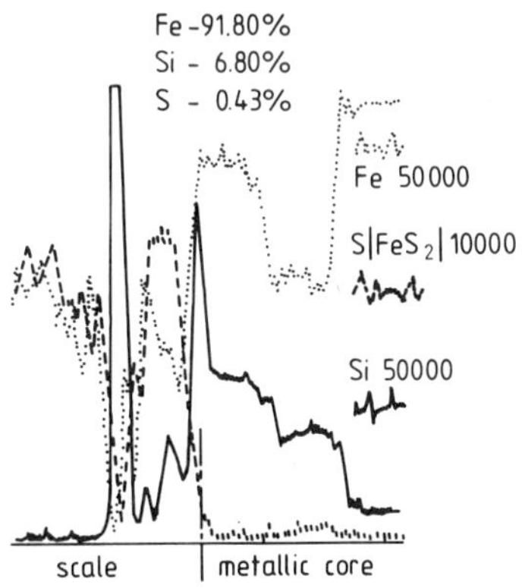

Figure 12. Distribution of elements for the iron-3 weight percent silicon sample sulfidized at 600°C.

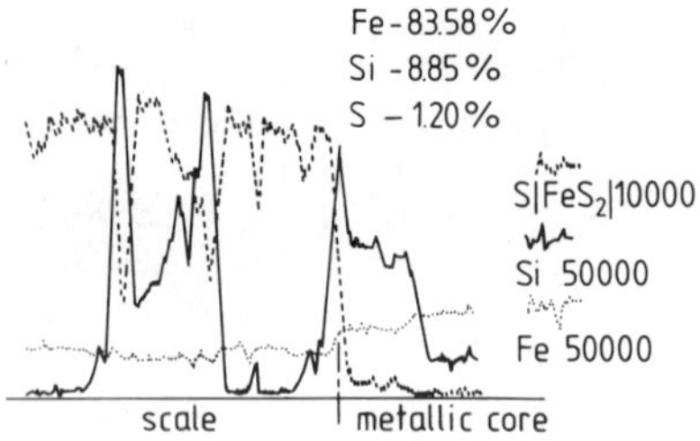

Figure 13. Distribution of elements for the iron-3 weight percent silicon sample sulfidized at 800°C.

Morphology of the Layers

The morphology of the layers of interest was studied for the iron-silicon alloy by means of metallographic methods and by observing the fractures be scanning electron microscopy. In all cases we could distinguish (a) an outer layer of the scale, (c) an inner layer of the scale, and (c) a layer on the metallic core. The outer layer, which is rather distinct from the other two, can be described as containing numerous holes, cracks, and pores which mainly originate from the preparation of the microsections. The inner scale is very thin and appears more compact and homogeneous and is less gray than the outer scale. The third layer has a consistency similar to the second layer and has the characteristic metallic luster. This layer, which is firmly bound to its predecessor, can be relatively easily detached from the metallic core.

The morphology of the layers (see Figure 7) indicates that the outer scale is built from well-formed columnar crystals. No holes, cracks, or pores have been detected which is evidence to support earlier remark that such effects could result from the preparation of the microsections. We could not observe the morphology of the thin inner scale by scanning electron microscopy, probably because it is similar to the morphology of the much thicker layer occurring beneath it, i.e., the layer on the core. The layer on the core is composed of irregular spheres and thin plates and has numerous defects such as pores and discontinuities (see Figure 8). A surface of the core, which was investigated after removal of all the previous layers, consists of small detached sphere-like elements (see Figure 9).

Mössbauer Measurements

All the Mössbauer spectra were fitted by a least-squares computer iteration procedure. Each spectrum was assumed to be composed of a number of sextets with characteristic hyperfine fields, H, and isomer shifts, δ. Each such subspectrum was ascribed to a certain atomic configuration (m,n) for a probe iron-57 with m aluminum or silicon atoms in the first (NN) shell and n impurity atoms in the second (NNN) neighbor shell. It was further assumed that the influence of the impurity atoms on the hyperfine parameters is additive such that

$$H(m,n) = H(0,0) + m\Delta H + n\Delta H_2 \quad (5)$$

$$\delta(m,n) = \delta(0,0) + m\Delta\delta_1 + n\Delta\delta_2. \quad (6)$$

In these expressions, H(m,n) is the hyperfine field at the iron-57 which is in the (m,n) atomic configuration, and $\Delta H_{1,2}$ are the hyperfine field changes caused by one impurity atom residing in the NN, index 1, or in NNN, index 2, shells. The meaning of the $\delta(m,n)$ and $\Delta\delta_{1,2}$ symbols is similar. The spectra were computer-decomposed into a number of subspectra whose statistical weights, $P(m,n)$, were variable parameters. The number of subspectra was calculated from the binomial distribution by taking into account those with the largest $P(m,n)$ values such that $\Sigma P(m,n) \geq 0.98$.

By using the above fitting procedure and assumptions, we could satisfactorily fit all the spectra for the iron-aluminum samples. The resulting parameters are given in Table II, along with the average hyperfine fields, $\bar{H}$, calculated according to the formula

$$\bar{H} = \Sigma H(m,n)P(m,n)/\Sigma P(m,n). \tag{7}$$

In order to reveal the influence of the sulfidation on the distribution of aluminum atoms in the iron matrix, the average number of aluminum atoms within the NN-NNN volume, $\bar{N}$, was calculated from the expression

$$\bar{N} = \Sigma(m+n)P(m,n)/\Sigma P(m,n). \tag{8}$$

The resulting values are presented in Table II along with the corresponding values, $\bar{N}_o$, expected for a random distribution. One can readily note that, for the less concentrated iron-aluminum alloy, no difference between $\bar{N}$ and $\bar{N}_o$ was found. For the more concentrated iron-aluminum alloys, however, $\bar{N}$ is greater than $\bar{N}_o$ for the sulfidized samples. It is clear that for the sample sulfidized at 650°C there is a significant increase in $\bar{N}$ from 1.18 to 1.33 on going from the center of the core to its surface. This increase in the concentration of aluminum is believed to be a direct result of the sulfidation.

Iron-aluminum alloys are known to order crystallographically. The ordered structure, with the lowest content of aluminum is Fe_3Al also depicted as DO_3. One may, therefore, expect that, within the aluminum enriched zone there is a tendency toward such ordering. In order to test this expectation, we introduced the order parameter, Ω, where

$$\Omega(\%) = \frac{\bar{N}_o - \bar{N}}{\bar{N}_o - 4.67} 100. \tag{9}$$

Ω is equal to zero for a random distribution of atoms and to 100 for the Fe_3Al, DO_3 structure. The Ω-values obtained for the iron-aluminum alloys are included in Table II. We note that, for the iron-0.8 weight percent aluminum samples as well as for the nonsulfidized iron-3 weight percent aluminum alloys, Ω is zero to within the error limits. On the other hand, for the sulfidized sample of the more concentrated alloys Ω is greater than zero with a maximum value of 13 percent for the sample taken from the near-surface zone of the core of the alloy sulfidized at 650°C. For another indication of departure from the random distribution, one can study the values of the hyperfine fields, $H(0,0)$ and $\bar{H}$. Both fields have enhanced values for the more ordered samples in agreement with the observation of Yamamoto [13].

For the iron-silicon alloy we could adequately fit with the above described procedure and assumptions, only the spectrum obtained for the nonsulfidized sample (spectrum A in Figure 4). The spectra obtained for the layers of interest in the sulfidized samples show additional lines which could only be satisfactorily accounted for by adding two additional sextet subspectra. These are possibly related to iron atoms occupying two crystallographically different sites in the DO_3 structure. It is well known that in this structure there are two nonequivalent iron sites, I and II, which the following NN, NNN environments: Fe_I = (4 Fe 4 Si, 6 Fe) and Fe_{II} = (8 Fe, 6 Si). Their relative abundance is 2:1. Further, it appears (see spectrum B in Figure 4) that the central part of the spectrum contains a poorly resolved line which could be well fitted as a doublet.

Table II. Mössbauer Spectral Parameters for the Iron-Aluminum Alloy

Sample	Weight Percent	T_s °C	H(0,0), kOe	$\bar{H}$ kOe	$\bar{N}_o$	$\bar{N}$	Ω(%) Percent
Unsulfidized	0.83	---	330.4(1)	325.4	0.23	0.24(10)	0
Near Surface Zone	0.83	550	330.5(4)	326.2	0.23	0.23(10)	0
Unsulfidized	2.96	---	330.7(3)	315.3	0.825	0.77(14)	1
Core Zone	2.96	650	332.1(3)	314.0	0.825	1.05(15)	6
Surface Zone	2.96	650	336.9(4)	318.4	0.825	1.33(12)	13
Near Surface Zone	2.96	650	331.7(2)	318.0	0.825	1.18(18)	9
Near Surface Zone	2.96	550	332.0	317.6	0.825	1.18(10)	9

The best fit parameters obtained for the iron-silicon samples are given in Table III We note that the values of the hyperfine parameters, H(0,0), $\Delta H_{1,2}$, $\delta(0,0)$ and $\Delta\delta_{1,2}$ agree with the corresponding parameters of the nonsulfidized sample. The values of the parameters describing the two additional subspectra $H_{I,II}$ and $\delta_{I,II}$ are in accord with those characteristic of the iron 3-silicon structure [14,15]. Finally, the parameters characterizing the doublet, δ and Q, the quadrupole splitting, are in agreement with those of FeS_2 [16].

As for the iron-aluminum alloys, we have evaluated the order parameter, Ω, for the iron-silicon samples in order to obtain a more quantitative insight into the problem of the development of the DO_3 structure. Ω was evaluated for the entire magnetic portion of the spectra, i.e., only the doublet was excluded, as well as for the nonordered magnetic part, i.e., the doublet and the two additional subspectra related to the ordered phase were excluded. In the latter case the order parameter was designated Ω'. One can readily see from the results presented in Table IV that Ω' is not zero. This means that in fact silicon atoms are distributed nonrandomly within the nonordered phase.

Finally, the relative abundance of the iron atoms in the DO_3 structure, $P(DO_3)$, and the FeS_2 phase, $P(FeS_2)$ was determined and is given in Table IV which also contains the values of P(0,0), the probabilities of (0,0) atomic configuration as obtained from the fitting procedure. We note that the decrease in the P(0,0) values is correlated with an enhanced degree of order, Ω, with its maximum value at about 41 percent for the sample sulfidized at 700°C.

DISCUSSION AND INTERPRETATION OF THE RESULTS

It is interesting that for the iron-silicon alloy the rate of the sulfidation is similar to that of pure iron over the temperature range studied. This is in contrast with the rates obtained for the iron-aluminum alloys within which the rate of the sulfidation was found to depend upon the concentration of aluminum [17]. This occurs even though the scale has a similar bilayer structure with a heterogeneous and porous inner part. The difference in the rate of the reaction results perhaps because, for the iron-silicon alloy, the scale proceeds toward the zone of the metal consumption. This does not occur for the iron-aluminum alloy. A similar independence of the rate of the sulfidation upon the content of silicon in copper-silicon alloys has been observed [18].

Table III. Mössbauer Spectra Parameters for Iron-3 Weight Percent Silicon Alloy

Sample	T_s °C	H(O,O) kOe	ΔH_1 kOe	ΔH_2 kOe	$\delta(O,O)$ mm/s	$\Delta\delta_1$ mm/s	$\Delta\delta_2$ mm/s	H_I kOe	H_{II} kOe	δ_I mm/s	δ_{II} mm/s	Q mm/s	δ mm/s
Unsulfidized	----	337.1	-29.6	- 9.2	-0.008	0.04	0.00	---	---	----	----	----	----
Surface zone	600°	341.0	-32.0	-10.0	0.000	0.04	0.00	315	201	0.07	0.24	0.58	0.26
Surface zone	700°	336.0	-32.5	-10.7	0.003	0.04	0.01	311	199	0.08	0.24	0.57	0.23
Surface zone	800°	338.0	-31.0	-11.0	0.000	0.05	0.01	314	202	0.06	0.25	0.59	0.29

Table IV. Mössbauer Fit Parameters for the Iron-3 Weight Percent Silicon Alloy

Sample	T_s, °C	Ω, %	Ω', %	P(DO_3)	P(FeS_2)	P(0,0)
Unsulfidized	---	11.4	11.4	-----	-----	0.196
Surface zone	600	30.2	12.4	0.202	0.024	0.130
Surface zone	700	40.8	11.3	0.332	0.046	0.110
Surface zone	800	35.5	15.3	0.313	0.079	0.090

For the near-surface zone of the core of the iron-aluminum alloys, the Mössbauer measurements on a sample cut from this region of the alloy containing 3 weight percent aluminum revealed an increase in aluminum content due to the sulfidation. An X-ray microanalysis of this zone confirmed this enrichment, but indicated that the depth of this aluminum enriched zone does not exceed 4μm. An X-ray analysis applied to the iron-silicon alloy revealed that sulfidation at 600-800°C results in a formation of three layers on the metallic core. (1) The first outermost layer, called the outer scale, has the composition of FeS_2. (2) The second layer, called the inner scale, is much thinner, and its composition could not be identified. When this layer is detached from the outer scale, it has a smell characteristic of H_2S and an X-ray powder diffraction pattern typical of iron sulfide. (3) The third layer, which is on the core, is a mixture of the ordered, Fe_3Si, DO_3, structure with the nonordered, but nonrandom α-phase of the iron-silicon alloy. No reflections characteristic of FeS or FeS_2 were detected for this layer.

The analysis of the Mössbauer spectra of the third layer revealed that its actual composition depends upon the temperature of sulfidation, T_S. For a T_S of 700°C the contribution of the DO_3 structure, P(DO_3), has a maximum value of about 41 percent and Ω' is at a minimum. At T_S of 800°C, P(DO_3) has decreased, perhaps because the abundance of the FeS_2 phase has increased. The abundance of FeS_2 steadily increases with increasing T_S (see Table IV). The distribution of elements show that the concentration of silicon in this layer has doubled in comparison with its value inside the core, whereas the concentration of iron has decreased. However, there is no simple correlation between the concentration of silicon and the sulfidation temperature. The content of sulfur does not exceed two percent although it increases with T_S. For a given T_S it increases from the surface of the core toward the inner scale.

A X-ray microanalysis of the second layer, the inner scale, reveals peaks for silicon. They may be due to SiO_2 which, like H_2S, is a product of a disintegration of SiS_2 in moist air. This was also indicated by the smell of H_2S after removal of the outer scale.

For the first layer, the outer scale, the concentration of iron and sulfur are constant over the entire cross section of the layer and correspond to those of iron sulfide. No evidence of silicon was detected in this layer.

We believe that both a thermodynamic and a kinetic mechanism are responsible for the enrichment of the near-surface zone upon sulfidation of the iron-aluminum or the iron-silicon alloys. In the iron-aluminum alloy, the concentration of aluminum in the near-surface zone should be diminished by the sulfidation because aluminum has a high affinity for sulfur than does iron. However, the present results indicate that the effect is just the opposite. A similar enrichment of aluminum on a sulfidation was discovered for copper-aluminum alloys [19], in which Cu_9Al_4 was found to be present in the near-surfaced zone of the core after sulfidation. Therefore, it seems reasonable to assure that a common mechanism is responsible for the aluminum enrichment in both systems. We suggest that, for the iron-aluminum alloys, the primary scale cannot move toward the zone of the metal consumption because of the polyphasic inner scale which forms. Instead, the surface of the metallic core retreats after reacting with sulfur. Therefore, on the surface of the core, a process of nucleation of the sulfidation products, FeS and Al_2S_3, which do not form a solid solution with each other, must occur continuously. We believe that both a place for nucleation and its probability are crucial parameters for the nucleation of these two phases. Let us consider an elementary process for the nucleation of FeS and Al_2S_3. In order to form a nucleus of FeS, which consists of at least several atoms, the iron atoms, due to their concentration do not need to move away from their primary positions to meet sulfur atoms. An increase of the FeS nucleus is favored, not only by a large concentration of iron in the iron-aluminum alloy, but also by a larger mobility of iron ions in FeS [6]. We can, therefore, assume that the nucleation of FeS occurs at any place on the core surface and freely increases in size. In contrast, the nucleation of Al_2S_3 is more difficult, due to the lower concentration of aluminum atoms, which consequently have to travel further in order to meet sulfur atoms. In other words, less aluminum than iron within the near-surface zone can react with sulfur leading to the observed enrichment in aluminum in this zone. Further, the relatively low temperature of sulfidation does not favor the nucleation of Al_2S_3 [20].

The near-surface zone of the core is not only enriched in aluminum but it is also more than saturated with vacancies which cause a distortion of the lattice. Hence, it is in a nonequilibrium thermodynamic state. In order to restore equilibrium, there must be a rearrangement of atoms leading in turn to a partly ordered structure. This is indicated by the Ω-$\underline{v}$alues as well as by the enhanced values of the hyperfine fields, H(O,O) and $\bar{H}$.

A mechanism responsible for the formation of the scale on the iron-silicon alloy should be similar to that driving the formation of the scale on the iron-aluminum and copper-aluminum alloys due to their sulfidation [19,21]. This follows because the scale formed on the former has a similar structure as that on the latter. Moreover, as a result of the sulfidation of copper-aluminum, precipitates of Cu_9Al_4 were found within the near-surface zone. This has its corresponding phenomenon in the iron-silicon alloys in the form of the Fe_3Si phase detected in the similar zone. Another similarity is the rate of the sulfidation which, for the both systems, does not depend on the concentration of silicon. This has already been explained by the authors of reference 18.

We shall try to explain the mechanism for the formation of the Fe_3Si phase on the analogy of the formation of the η'-phase in copper-silicon [18] and the γ_2-phase in copper-aluminum [22]. The standard enthalpy needed to form FeS is smaller than that needed to form SiS_2. In addition, the activity of silicon in the alloy is diminished by its low concentration. The sandard euthalpies, G of the following reactions at T_S of 600-800°C are found [23] to be

$$G(Si + S_2 \rightarrow SiS_2) = -50.3 \text{ to } -43.2\text{kcal/mol} \tag{10}$$

$$G(2Fe + S_2 \rightarrow 2FeS) = -77.7 \text{ to } -50.5\text{kcal/mol.} \tag{11}$$

According to the parabolic law for sulfidation, the parabolic constants for pure iron is for the studied range of T_S, 0.6 to $4.6 \times 10^{-6} g^2/cm^4 s$ and for the iron-silicon alloy is 0.6 to $2.8 \times 10^{-6} g^2/cm^4 s$. If, in addition, we note that the rate of sulfidation of pure silicon is a few orders of magnitude lower than that of iron, we must conclude that the rate of sulfidation of the alloy is mainly determined by the reaction of iron with sulfur. Consequently, from the very beginning of the sulfidation, FeS forms on the surface of the alloy blocking further access of sulfur to the metallic surface. As a result of this blocking, SiS_2 cannot be formed and the near-surface zone of the core gets enriched in silicon. The increase of the concentration of silicon in an iron-silicon alloy leads, in general, to an ordering of silicon atoms in the iron matrix. An iron-silicon alloy has the bcc structure whose atomic configuration can be described in terms of four sublattices, I, II, III, and IV, each having the fcc structure as illustrated in Figure 14. Atoms occupying the I and II sites are the nearest neighbors of those residing on the III and IV sites, and vice versa. In contrast, atoms occupying sites I and III are the next-nearest neighbors to those located on sites II and IV. A distribution of atoms over these four sites can be described [24] in terms of three parameters, x, y, and z, such that

$$x = 0.25\ (P_{Fe}^{I} + P_{Fe}^{II} - P_{Fe}^{III} - P_{Fe}^{IV})$$

$$y = 0.5\ (P_{Fe}^{III} - P_{Fe}^{IV}) \tag{12}$$

$$z = 0.5\ (P_{Fe}^{I} - P_{Fe}^{II}),$$

where P_{Fe}^{i} is the probability of finding an iron atom in the i-th sublattice (i = I, II, III, IV). If x = y = z = 0, the lattice is called A2. If $x \neq 0$ and y = z = 0, the iron atoms preferentially occupy sites I and II when x is greater than zero, or sites III and IV when x is less than zero, and then the structure is called B2. Consequently, silicon atoms may occupy sites III and IV or I and II, respectively. As a result of this selective site occupation the number of iron-silicon near-neighbor pairs increases in comparison with the situation in the A2. If $y \neq$ or $z \neq 0$, the structure is called DO_3 and its characteristic feature is a larger number of the next-nearest iron-silicon pairs than are found for the B2 structure. Sites III and IV are occupied by iron atoms, site II by iron atoms and site I by silicon atoms.

The present study has revealed that sulfidation of the iron-silicon alloy results in the formation of the DO_3 superstructure in the near-surface zone of the core. There will be a tendency for silicon atoms to "bind" neighboring iron atoms, because of the high affinity between silicon and iron which results from the interaction between the 3s and 3p electrons of silicon and the 3d electrons of iron. In fact, strong interactions between iron and silicon atoms have been reported to occur [25,26]. Consequently, only those iron atoms which do not participate in the formation of the superstructures will react with sulfur atoms forming FeS and thus causing a relative increase in silicon content. Its maximum occurs at the metallic core/layer interface

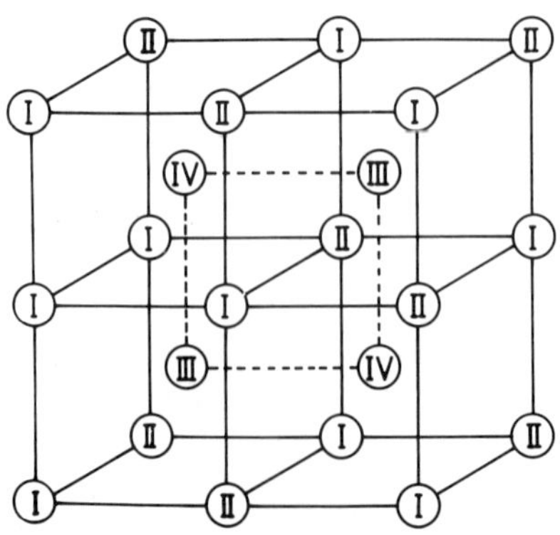

Figure 14. Spatial arrangement of the sublattice sites I, II, III, and IV for the DO_3 structure.

on the core. At this interface the concentration of sulfur is sufficient to react with iron atoms not participating in the formation of the superstructures. A flux of vacancies resulting from a transfer of iron atoms from the core into the scale and in the opposite direction may also favor the formation of the superstructures by an enhancement of the mobility of the atoms. This in turn makes the formation of a new configuration easier.

The actual abundance of the DO_3 structure, $P(DO_3)$, is determined by at least two competing phenomena, its formation and its decay. From the corresponding $P(DO_3)$, values displayed in Table IV, it follows first that for a T_S of 600-700°C the rate of the formation, R_f, is larger than the decay rate, R_d, and second that for T_S of 700-800°C, $R_f < R_d$.

The decay of the DO_3 structure occurring at the interface between the inner scale and the layer on the core is due both to the activity of sulfur in this region as well as to a high mobility of iron atoms in FeS and of silicon atoms in SiS_2. Iron atoms in FeS diffuse faster (3.7×10^{-7} cm/s [27]) than those in the α-phase of iron-silicon (3.3×10^{-11} cm/s [28]), and probably those in the DO_3 structure. As a result of this diffusion, iron atoms quickly reach the outer scale, where, either after having reacted with sulfur they increase the thickness of the outer scale, or they react with sulfur originating from a dissociation of the outer scale. Finally, they form FeS which remains on the inner scale. The activity of sulfur at the metallic core/scale interface is, in the equilibrium state, equal to the decay pressure of the iron sulfide. In the presence of superstructures this activity is even larger because Fe_3Si decays into FeS and SiS_2 is more easily formed from Fe_3Si than from the α-phase of the iron-silicon alloy because the chemical potential of silicon in the former is larger than that in the latter.

CONCLUSIONS

Based on the results presented in the previous sections we can draw the following conclusions:

(1) The rate of sulfidation of the iron-silicon alloy is similar to that of pure iron for the temperature range studied.

(2) The sulfidation of the iron 3 weight percent aluminum and 3 weight percent silicon alloys results in an enrichment of the near-surface zone in aluminum or silicon, respectively, and consequently in the formation of the DO_3 structure in that zone for the iron-silicon alloy.

(3) The main source of the aluminum or silicon enrichment is a selective sulfidation of iron atoms which occurs on the surface of the core.

(4) The Fe_3Si structure formed at the inner scale/metallic core interface decays into FeS and SiS_2.

ACKNOWLEDGMENTS

We are grateful to Professor T. Werber for the stimulating interest in this work and to Dr. Ing. S. Bzowski for an illumination discussion of the superstructures.

REFERENCES

1. K. Hauffe and A. Rahmel, Z. Phys. Chem. 199, 152 (1952).

2. R. A. Meussner and C. E. Birchenall, Corrosion 13, 677 (1957).

3. A. Sterten and S. Hangen, Oxid. Met. 3, 545 (1971).

4. T. Narita and K. Nishida, Trans. Jpn. Inst. Met., 15, 314 (1971).

5. D. J. Young and W. W. Smaltzer, J. Electrochem. Soc. 123, 299 (1976).

6. M. Danielewski, S. Mrowec, A. Stoklosa and J. Przybyszewska, Bull. Pol. Acad. Sci. (Ser. Sci. Chim.) 26, 175 (1978).

7. S. M. Dubiel, T. Werber and Z. Żurek, Proc. 8th Int. Congress on Met. Corrosion, p. 81, ed. DECHEMA, Frankfurt/M,(1981).

8. S. M. Dubiel and Z. Zurek, High. Temp.-High. Press 16, 409 (1984).

9. G. Tammann, Z.Anorg. Chem. 111, 78 (1920).

10. N. Pilling and R. Bedworth, J. Inst. Met. 29, 529 (1923).

11. S. Mrowec, in Zarys Teorii Utleniania Metali, ed. Śląsk, Katowice, (1971).

12. ASTM-Data, ed. Joint Committee on Powder Diffraction Standards, Pennsylvania (1979).

13. H. Yamamoto, Jpn. J. Appl. Phys. 3, 745 (1964).

14. M. B. Stearns, Phys. Rev. 168, 588 (1968).

15. M. C. Lin, R. G. Barnes, and D. R. Torgesen, Phys. Rev. B, 24, 3712 (1981).

16. A. A. Temperlay and H. W. Lefevre, J. Phys. Chem. Solids 27, 85 (1966).

17. K. N. Strafford and R. Manifold, Corrosion Sci. 9, 489 (1969).

18. K. Szawlowska-Wallisch and T. Werber, Arch. Hutnictwa 27, 381 (1982).

19. T. Werber and Z. Zurek, Rev. Inst. Hautes Temp. Refract. 14, 281 (1977).

20. G. V. Samsonov and S. V. Drozdova, in Sulfidy, Moskva (1972).

21. S. Mrowec and T. Werber, Werkstoffe and Korrosion 18, 116 (1967).

22. T. Werber and Z. Zurek, to be published.

23. J. Barin and O. Knacke, in Thermochemical properties of inorganic substances, pp. 297 and 692, ed. Springer Verlag, New York (1973).

24. G. Inden and W. Pitsch, Z. Metallkunde 62, 627 (1971).

25. M. B. Stearns, Phys. Rev. 147, 439 (1966).

26. F. Korber and W. Oelsen, Mitt. Kaiser Wilhelm Inst. fur Eisenforsch. 18, 109 (1936).

27. E. Fryt, W. W. Smaltzer, and J. S. Kirkaidy, J. Electrochemical Soc. 126, 673 (1979).

28. C. J. Smithells, in Metals Reference Book, p. 897, ed. Butherworths, London-Boston (1978).

HYPERFINE FIELDS IN Fe-Ni-X ALLOYS AND THEIR APPLICATION TO A STUDY OF TEMPERING OF 9Ni STEEL

B. Fultz* and J. W. Morris, Jr.

Department of Materials Science and Mineral Engineering
University of California, Berkeley, and the
Materials and Molecular Research Division
Lawrence Berkeley Laboratory
Berkeley, CA 94720

ABSTRACT

Hyperfine magnetic field perturbations due to the solutes in Fe-Ni-X alloys were systematized, and interpreted with a model of linear response of hyperfine magnetic fields to magnetic moments. The effects of solutes on the iron-57 hyperfine magnetic field were used for chemical analysis of the austenite formed in 9Ni steel during tempering. Diffusion kinetics of the nickel and X solutes were found to play an important role in the formation of the austenite particles.

PHYSICAL METALLURGY OF 9Ni STEEL

9Ni steel is used in large industrial structures for the transportation, storage, and processing of cryogenic liquids. For cryogenic service, 9Ni steel is heat-treated to develop a primarily martensitic (bcc) microstructure. From a cryogenic structural engineering standpoint, such martensitic steels have three salient features. The first positive attribute of martensitic steels is their high strength, which becomes even greater at lower temperatures. Second, martensitic steels have low levels of alloy additions and are rather inexpensive. The general negative attribute of martensitic steels is the temperature dependence of their resistance to crack propagation. At "high" temperatures near room temperature, their toughness is usually good. However, as the temperature is reduced to a temperature characteristic of of each material, its toughness deteriorates rapidly. The toughness may decrease by a factor of ten over only a few tens of degrees C, and this ductile-to-brittle transition is the central problem in the use of martensitic steels for cryogenic service.

In the development of 9Ni steel it was found that a tempering at 570°C was particularly beneficial; it suppressed the ductile-to-brittle transition temperature by more than 100°C [1]. Since the microstructures of all 2-9 percent nickel steels are otherwise similar, early workers attributed these

*Present Address: Keck Laboratory, California Institute of Technology, Pasadena, California 91125.

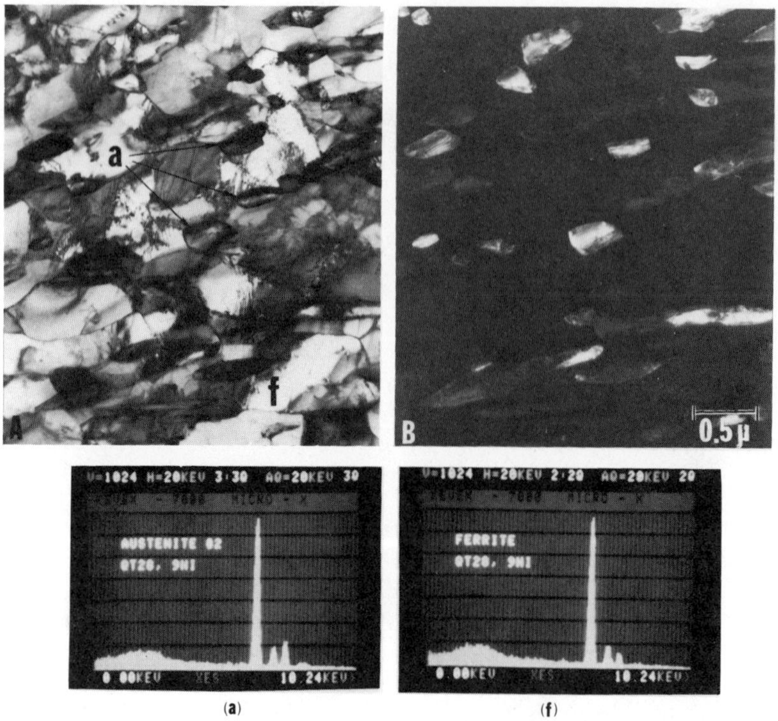

Figure 1. A. The bright field transmission electron micrograph of commercial 9Ni steel after tempering for 81 hours at 600°C. B. Corresponding dark field micrograph taken with (111) austenite reflection and (110) zone axis. a. X-ray fluorescence spectrum from austenite particle (strong peak is Fe Kα) f. X-ray fluorescence spectrum from tempered martensite.

different tempering responses to the austenite (fcc) phase that is formed in 9Ni steel [2]. Some of this austenite is retained at room temperature as a metastable phase, and can be seen along martensite lath boundaries in the transmission electron micrograph of Figure 1. Austenite is a high temperature phase, and the austenite in 9Ni steel may transform martensitically when exposed to low temperatures. Early workers [3] found that the austenite must be thermally stable against the martensitic transformation if good cryogenic toughness is to be obtained. Tempering for less than 10 hours at temperatures below 600°C was found to be especially effective in producing stable austenite. Higher temperatures and longer tempering times were found to be deleterious to both austenite stability and the ductile-to-brittle transition.

Many mechanisms have been proposed to account for the effects of austenite on mechanical behavior. Mössbauer spectrometry has played a decisive role in eliminating one class of mechanisms. The poor mechanical properties of steels with unstable austenite lead to several hypotheses that the austenite must remain untransformed during deformation if good toughness is to be obtained [4-7]. In backscatter Mössbauer spectrometry studies [8-10] no austenite was detected near the fracture surfaces of 6Ni or 9Ni steels subjected to a wide variety of heat treatments and testing conditions. These observations showed that the austenite transforms in front of an advancing crack, so these early hypotheses were discarded.

Nevertheless, there is no doubt that the stability of the austenite is somehow important to the fracture toughness, and the solute concentration of the austenite particles was believed to be a major factor controlling their stability. The present research was organized to measure the chemical composition of the austenite. Because the austenite particles are so small, when we began this work there was no reliable data on their solute concentration. We have found only two experimental techniques which are capable of measuring the chemical composition of the austenite; analytical scanning transmission electron microscopy [11,12], and Mössbauer difference spectrometry [13]. The information provided by these two techniques is largely complementary. In Mössbauer spectrometry the chemical composition information comes from the presence of nearest neighbor solute atoms around the iron-57 nucleus. However, this information is averaged over all the material. Although the broadening of the incident electron beam causes the chemical composition information from analytical scanning transmission electron microscopy to be of a less local origin, it is not globally averaged. The scanning transmission electron microscopy measurements provided an important check on our Mössbauer spectrometry results.

EXPERIMENTAL PROCEDURES AND DATA PROCESSING

Commercial 9Ni steel plate was kindly supplied by the Nippon Kokan Company with a chemical composition of 9.1 weight percent nickel, 0.50 manganese, 0.20 silicon, 0.17 chromium, 0.06 carbon, 0.004 phosphorous, 0.004 sulfur, and the remainder iron. High purity specimen materials of controlled composition were prepared from 9.995 percent pure iron and 99.95 percent pure nickel by melting measured amounts of these starting materials in new alumina crucibles under a backpressure of helium gas. To ensure chemical homogeneity of the ingots, the melt was held at 1650°C for 2h and cooled at 50°C/min to room temperature. Weight losses after melting were negligible, and discoloration was not observed on the ingot surfaces, so the alloy compositions were based on the weight of the starting materials. These compositions were checked with x-ray fluorescence spectrometry. To further homogenize the ingots, and to erase the starting microstructure of the commercial material, the alloys were heated in evacuated quartz ampules at 1150°C for 3h. Thin specimens were then prepared from the bulk material with a wafering saw, or by cold rolling. These foils were austenitized in evacuated quartz ampules. A fully martensitic microstructure with a uniform 10μm grain size was produced during the water quench which terminated the austenitizing heat treatment. Finally the foils were chemically polished to 5-8μm thickness in fresh solutions of 3ml HF in 100ml of H_2O_2 of 30 percent concentration. Tempering at temperatures in the fcc-bcc two-phase region of the phase diagram was performed by heating the specimen in evacuated quartz ampules or in a vacuum furnace on the Mössbauer spectrometer.

All Mössbauer spectra were obtained in transmission geometry with an Austin Science Associates S-600 constant acceleration spectrometer. Thin specimens of pure iron at 18°C had a linewidth of 0.24mm/sec for the ±3/2 to ±1/2 peaks. Spectra at elevated temperatures were obtained with the specimen foils sandwiched between beryllium disks in the bore of an evacuated tube furnace. Spectrum collection alternated between the Fe-Ni alloys and pure iron foils. From the pure iron spectra small long-term drifts of the Doppler velocity were identified and then used to determine corrections to the experimental spectra. The positions of the corrected ±3/2 to ±1/2 peaks are believed accurate to ±0.002mm/sec.

Determining the actual shape of the hyperfine magnetic field distribution requires corrections of the experimental spectra for broadening due to the linewidth of the incident γ-rays and the natural linewidth of the iron-57 in the specimen. These corrections were performed in two ways. In the first method a Lorentzian function of width determined from a pure iron spectrum was deconvolved from the spectrum by using the deconvolution theorem of harmonic analysis. It is necessary to attenuate the deconvolution at large values of the transformed variable in order to prevent the statistical scatter from causing the deconvolution to diverge. In our deconvolutions a Gaussian filter function was used to attenuate the deconvolution at large values of the transformed variable. The threshhold value for this Gaussian rolloff and the steepness of this rolloff were chosen by examining the noise in the Fourier transform of the data. Because of the Gaussian rolloff, our deconvolution procedure did not deconvolve a true Lorentzian function (with Fourier transform of approximately $\exp(-k|x|)$ from the experimental data, but rather a function whose Fourier transform is $\exp(-k|x|)$ divided by the Gaussian filter function. Consequently, near the center of the deconvolved peaks the width is larger than expected from the deconvolution of a pure Lorentzian function, and small oscillations appear in the tails of the peak.

The second method of extracting hyperfine magnetic field distributions from experimental spectra involved direct fittings of Lorentzian functions to the experimental data. Unconstrained fitting procedures will occasionally fit small Lorentzian functions to fluctuations in the background regions between the six main absorption peaks. Such small peaks represent large hyperfine magnetic field perturbations, and will cause the high moments of the hyperfine magnetic field distribution to depend strongly on the fluctuations in the background. Our method of constrained fitting involved three parameters, corresponding to the position, width, and asymmetry of the hyperfine magnetic field distribution. These three parameters defined a peaked function that was the product of a Gaussian function and a linear function whose positive or negative slope provided the positive or negative skewness. This peaked function was convolved with a Lorentzian function with the characteristic width of pure iron peaks. The three parameters were varied until the minimum root-mean-squared deviation between the calculated peak and the experimental peak was found. This constrained curve fitting procedure was performed for all four outer peaks of the experimental spectrum because the tails of the second and fifth peaks overlapped with the first and sixth peaks. Data scatter affected the outcome of the curve fitting procedure by occasionally causing the fitting program to get trapped into a local minimum of the root-mean-squared deviation. It was necessary to start the fitting process with different initial parameters, and compare the final parameters and root-mean-square deviations found from these different processes. The authors' discretion was occasionally required when these different processes gave different results which had similar root-mean-square deviations.

We used a difference spectrum method to provide a direct and sensitive means of quantifying small changes in Mössbauer spectra. In this approach, Mössbauer spectra from materials with different heat treatments were obtained under experimental conditions as similar as possible. Intensity differences between the two Mössbauer spectra were revealed by digitally subtracting the data points of the second spectrum from the data points of the first spectrum. Systematic experimental errors common to both starting spectra, such as the parabolic intensity correction associated with the constant acceleration mode, were largely cancelled out in this way. The most important part of our differencing process was the normalization, before differencing, of the two starting martensite sextets. Four distinctly different methods for normalizing the starting spectra were used, and all proved at least semiquantitatively useful for chemical composition analysis. The normalization criterion preferred by the authors is a peak height normalization. Peak number one from the two spectra are normalized so that their dips below the background count are equal, and then the two spectra are differenced. This normalization procedure will accurately discriminate between nickel and X (where X is manganese, chromium, silicon, and carbon) concentration differences if their effects are confined to opposite sides of peak number one.

Our Doppler drive occasionally underwent small changes in its velocity scanning range of about 0.005mm/sec. An ongoing record of the positions of pure iron peaks numbers one and six versus time was maintained, and it was found that these positions changed symmetrically about the center of the spectrum. A procedure to correct for these small shifts was developed in which an entire spectrum was expanded or contracted by replacing the contents of a data channel by a fraction of its contents plus a fraction of the contents of an adjacent channel. Adding data channels together in pairs has the undesirable effect of slightly broadening spectral features. Differences between a spectrum corrected for a shift of the Doppler drive and an uncorrected spectrum show artificial features due to the broadening from the shift correction procedure. To minimize this problem, shift corrections of opposing sign were simultaneously performed on the two spectra that were differenced.

Thickness distortion is a saturation effect in which a thicker absorber is unable to produce proportionately more resonant absorption. For a specimen with a single Lorentzian absorption peak, the area, A, under the observed peak has been evaluated [14-16], and we used these results for correcting peak areas of our experimental data. The most difficult and least accurate part of our thickness distortion correction procedure was the determination of the thickness. Since the mean resonant scattering length of the material was mostly the same for all specimens, the ratio of peak number one depth to background count was used to determine the specimen thickness. Martensite peak number one depth to background count ratios were collected for numerous specimens in each experimental configuration, and were correlated to the specimen thicknesses as measured by a micrometer. These calibrations for specimen thickness per percent dip were then used for thickness determinations. We found it easiest to first perform the differencing of two spectra, and then correct the difference for thickness distortion effects. Data from the difference spectrum was multiplied by a thickness distortion correction factor. Correction factors for difference spectrum intensitites were determined from the dips of the starting spectra at the velocities of maximum difference intensity.

RESPONSE OF IRON-57 HYPERFINE MAGNETIC FIELDS TO MAGNETIC MOMENTS

Early investigators recognized that in alloys of iron with 3d transition metal solutes, the iron-57 hyperfine magnetic field is due to a mechanism of core polarization and a mechanism of conduction electron polarization [17]. This general picture has since developed into a detailed model of how the iron-57 hyperfine magnetic field depends on the magnetic moments at the different lattice sites in an alloy [18-22]. Here we describe our implementation of these ideas. This model is developed further in the two following sections in order to estimate the effects of nickel and other (X) solutes on Mössbauer spectra of 9Ni steel. These estimates support our use of the difference spectrum method for measuring changes in the nickel and X composition of the martensite phase.

The important contributions to the iron-57 hyperfine magnetic field in iron-based alloys arise from the Fermi contact interaction; the spins of electrons that interpenetrate the iron-57 nucleus interact with the nuclear spin to perturb the nuclear energy levels. By exchange polarization, an increased fraction of 3d spin-up electrons reduces the Coulomb repulsion between 3d spin-up electrons and the 1s and 2s core spin-up electrons, allowing these core electrons to expand outwards from the iron-57 nucleus to reduce their kinetic energy. A slight excess of 1s and 2s spin-down electrons is left at the iron-57 nucleus, and this makes a negative contribution to the iron-57 hyperfine magnetic field [22-25]. Exchange interactions between the unpaired 3d spin-up electrons and the more outlying 3s and 4s spin-up electrons cause these electrons to move closer to the iron-57 nucleus and make a positive contribution to the iron-57 hyperfine magnetic field. The total response of the iron-57 hyperfine magnetic field to changes in the magnetic moment local to our iron-57 atom at the origin, $\Delta\mu(0)$, is given by

$$\Delta H_L = (\alpha_{CP} + \alpha_{CEP})\Delta\mu(0),$$

and

$$\Delta H_L = -90 \frac{kG}{\mu_B} \Delta\mu(0). \tag{3}$$

The constants α_{CP} and α_{CEP} relate changes in the iron-57 hyperfine magnetic field to changes in $\Delta\mu(0)$ through the mechanisms of core polarization and conduction electron polarization, respectively.

The spin-polarization at the iron-57 nucleus due to the nonlocal 4s electrons is also sensitive to magnetic moments at neighboring atoms. For solutes with atomic volumes similar to that of iron, experimental systematics show that these changes in 4s contributions to the iron-57 hyperfine magnetic field upon alloying are proportional to the difference between the magnetic moment of the neighboring atom, $\mu(\mathbf{r})$, and the magnetic moment of an iron in pure iron, μ_{Fe} [18,21]. The change in the iron-57 hyperfine magnetic field due to neighboring (non-local) magnetic moments in the alloy, ΔH_{NL}, is given by

$$\Delta H_{NL} = \alpha_{CEP} \sum_{\mathbf{r}\neq 0} f(r)\,[\mu(\mathbf{r})-\mu_{Fe}], \tag{4}$$

where f(r) is the fraction of 4s conduction electron polarization at the iron-57 nucleus produced by the magnetic moment at $|\mathbf{r}|$, with respect to the amount of 4s conduction electron polarization produced by the same magnetic moment at the iron-57 atom itself.

For further analysis it is convenient to express ΔH_{NL} as a sum of two terms. The first term is the contribution from those neighboring sites occupied by solute atoms, ΔH_{DNL} (direct non-local), and the second term is the contribution from those neighboring lattice sites occupied by iron atoms, ΔH_{INL} (indirect non-local); then,

$$\Delta H_{NL} = \Delta H_{DNL} + \Delta H_{INL}, \tag{5}$$

$$\Delta H_{DNL} = \alpha_{CEP} \sum_{\mathbf{r}\neq 0} f(r)\delta(\mathbf{r})\Big[\mu_x + \sum_{\mathbf{r}'\neq 0} \delta(\mathbf{r}')g_x^x(\mathbf{r}'-\mathbf{r})-\mu_{Fe}\Big], \tag{6a}$$

and

$$\Delta H_{INL} = \alpha_{CEP} \sum_{\mathbf{r}\neq 0} f(r)(1-\delta(\mathbf{r}))\sum_{\mathbf{r}'\neq 0} \delta(\mathbf{r}')g_X^{Fe}(\mathbf{r}'-\mathbf{r}) . \tag{6b}$$

The function, $\delta(r)$, is one if the site is occupied by a solute atom, and is zero if it is occupied by an iron atom. We use the variable $g_X^Y(r_j)$ to express the change in magnetic moment of a Y atom when it has an X solute atom in its r_j^{th} coordination sphere. All perturbations of iron magnetic moments are referenced to the magnetic moment of an iron atom in pure iron. The additivity of the $g_X^{Fe}(r)$ parameters is expected only for dilute Fe-X alloys. Below we will discuss the saturation of the iron magnetic moments in non-dilute Fe-Ni alloys. It is common practice to use the same $\{f(r_j)\}$ parameters in both Equations 6a and 6b. This implicitly assumes one unpaired spin is equally effective in spin-polarizing the 4s electrons if it is at an iron atom or at a solute atom. This assumption is acceptable for solutes with no magnetic moment, i.e., silicon, germanium, or aluminum, because they cause minimal 4s spin polarization.

In a non-dilute alloy there will be many solute environments contributing to a distribution of iron-57 hyperfine magnetic fields. It is often adequate [26,27] to parameterize the hyperfine magnetic field at a iron-57 nucleus in terms of the numbers of solute atoms in its various nearest-neighbor shells, $\{n_j\}$. When the solutes are randomly distributed with an average concentration, c, there will be a probability $P(\{n_j'\})$, of a specific set of nearest-neighbor shell occupancies $\{n'(r_1), n'(r_2), n'(r_3), \cdots\}$ equal to the product of the binomial probabilities associated with the occupancy of each nearest neighbor shell, given by

$$P(\{n_j'\},c) = \prod_{j>0} P(N_j,n_j',c), \tag{7a}$$

and

$$P(N_j,n_j',c) = \frac{N_j!}{n_j'!(N_j-n_j')!}\, c^{n_j'}\,(1-c)^{N_j-n_j'} . \tag{7b}$$

For the bcc structure the total number of sites in consecutive nearest neighbor shells form the ordered set, $\{N_j\}$, which begins: $\{8, 6, 12, 24, 8, 6\}$.

With knowledge of the arrangements of the solute atoms (Equation 7) and the systematics of how they perturb the iron-57 hyperfine magnetic field (Equations 5 and 6), it is possible to calculate the hyperfine magnetic field distribution. Equally importantly, it is possible to evaluate the constants in Equations 5 and 6 by comparing measured Mössbauer spectra with the results of such calculations. For such "calibration" purposes it is convenient to study alloys of Fe-Si, Fe-Ge, or Fe-Al. These and other sp-series solutes develop no magnetic moment in bcc iron, so the parameters $g_X^X(r)$ and $(\mu_X-\mu_{Fe})$ in Equation 6a are known to be zero and $2.2\mu_B$, respectively. Furthermore, these solutes do not perturb the magnetic moments at neighboring iron atoms [28], so ΔH_L and ΔH_{INL} will conveniently be zero. This simplifies the contribution to the iron-57 hyperfine magnetic field, and the Mössbauer spectra from dilute Fe-Si alloys show well-defined sub-peaks which can be associated with the effects of solutes in individual nearest-neighbor shells around the iron-57 atom. We use the parameters

$$\alpha_{CEP} f(r_1) = -11.5\text{kG}/\mu_B,$$

$$\alpha_{CEP} f(r_2) = -3.5\text{kG}/\mu_B,$$

and

$$\alpha_{CEP} f(r_3) = +2.5\text{kG}/\mu_B.$$

because they were obtained from our own data from Fe-Si alloys, and are consistent with previously reported values [18,20,21].

EFFECTS OF NICKEL SOLUTES ON IRON-57 HYPERFINE MAGNETIC FIELDS

With nickel solutes it is necessary to accurately account for ΔH_{INL} in order to calculate the hyperfine magnetic field distribution. This was done in detail by Stearns with NMR spectra of dilute Fe-Ni alloys obtained by Budnick, et al. [19,29]. It was found that the largest hyperfine magnetic field perturbation came from a nickel solute in the third nearest neighbor shell about the iron-57 nucleus; this non-intuitive situation is, however, consistent with Equations 3 and 6. A good fit of the calculated spectra to the experimental spectrum was found for precise values of μ_{Ni} and $\{g_{Ni}^{Fe}(r)\}$.

Mössbauer peaks from Fe-Ni alloys at 18°C are shown in Figure 2. In agreement with previous studies [21,22,27,31,32], we find that the mean iron-57 hyperfine magnetic field in Fe-Ni is larger than in pure iron at room temperature, and the hyperfine magnetic field distribution is broadened. Materials with greater nickel concentrations give increased absorption intensity on the high velocity sides, i.e., the sides away from the zero Doppler shift, of the martensite peaks at 18°C. The difference spectrum intensity on the high velocity sides of the martensite peaks was linearly related to the difference in nickel concentration. An example of difference spectra from binary Fe-Ni alloys used for the calibration for nickel concentration

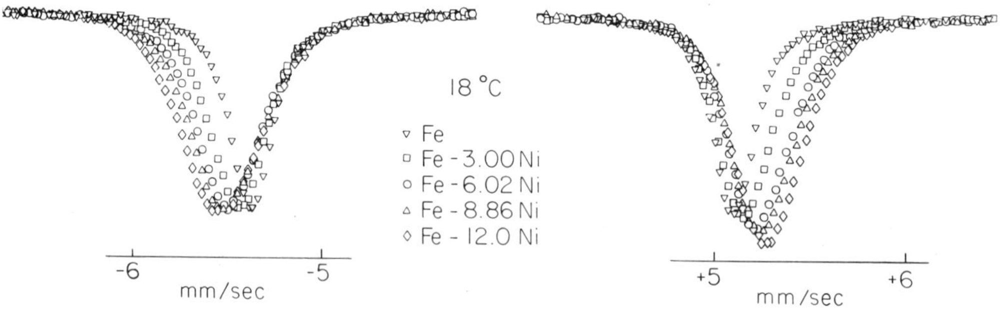

Figure 2. Mössbauer peaks number 1 and 6 from Fe-Ni alloys at 18°C.

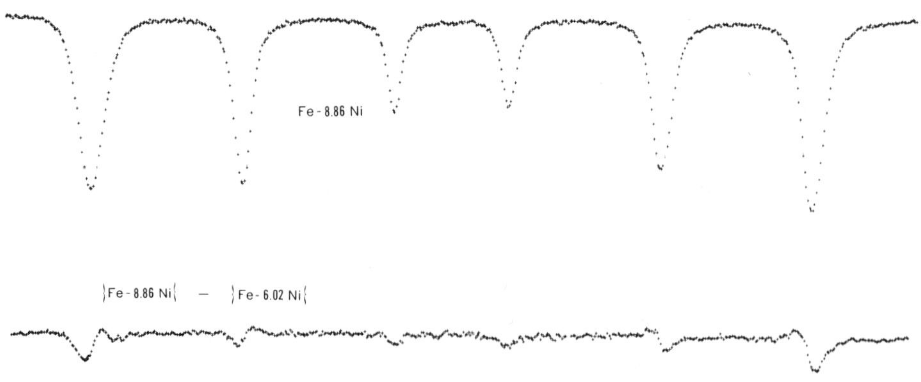

Figure 3. Calibration for effects of nickel concentration changes on changes on difference spectra at 18°C. Spectra were normalized for height of peak number 1.

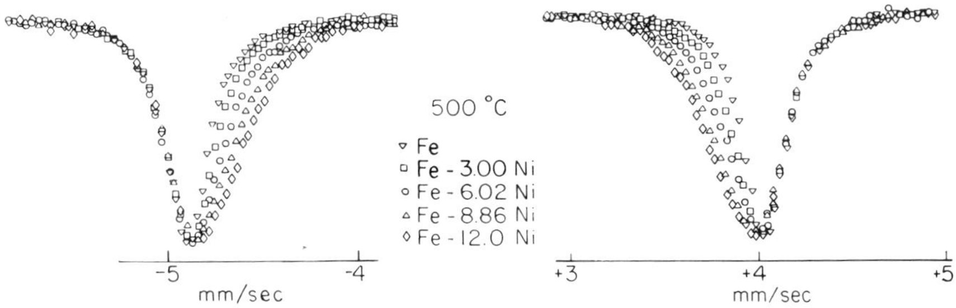

Figure 4. Mössbauer peaks numbers 1 and 6 from Ni-Fe alloys at 500°C.

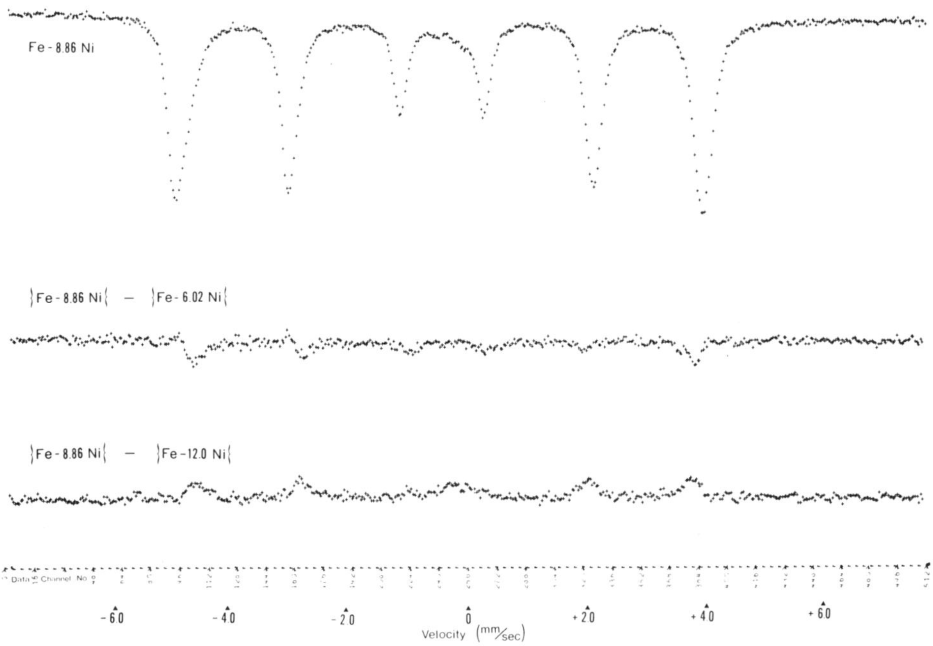

Figure 5. Calibration for effects of nickel concentration changes on difference spectra at 500°C. Spectra were normalized for height of peak number 1.

analysis is shown in Figure 3. Mössbauer peaks from Fe-Ni alloys at 500°C are shown in Figure 4, and calibration spectra for nickel analysis by the difference method at 500°C are shown in Figure 5. Unfortunately, in spectra taken at 500°C the difference spectrum intensities due to changes in nickel and X concentrations overlapped on the low velocity sides of the absorption peaks, so independent determinations of the nickel and X concentrations were not possible at this temperature.

The distinct hyperfine magnetic field structure observed in high resolution NMR spectra of dilute Fe-Ni alloys [29,31] is not evident in Mössbauer spectra of non-dilute Fe-Ni alloys. It is therefore appropriate to use the model of linear response of hyperfine magnetic fields to magnetic moments to calculate the mean, variance and skewness of the hyperfine magnetic field distribution. The theorem in the Appendix shows that when the iron-57 hyperfine magnetic field distribution can be expressed in terms of the set $\{n_j\}$, exact calculations of the hyperfine magnetic field distribution are possible. It is convenient to express ΔH_L and ΔH_{DNL} in terms of $\{n_j\}$. Unfortunately, the term ΔH_{INL} does not only depend on $\{n_j\}$. Because

$$\{g_{Ni}^{Fe}(r_j)\} \neq \{g_{Ni}^{Ni}(r_j)\}$$

ΔH_{INL} will depend on the specific arrangement of nickel atoms, which requires more information than is supplied by the set $\{n_j\}$. Parameterizing the iron-57 hyperfine magnetic field distribution in terms of the numbers of solutes around the iron-57 atom together with the number of solutes around each solute atom is analytically difficult because these numbers are not independent, and they depend on details of the solute configuration. We have been forced to consider an average value of ΔH_{INL}. Since we replace ΔH_{INL} with a single average value, we expect the calculated width and variance of the hyperfine magnetic field distribution to be too small. With respect to the hyperfine magnetic field of pure iron metal, the hyperfine magnetic field perturbation caused by $\{n_j\}$ solutes when the average nickel concentration is c is given by

$$\begin{aligned}\Delta H &= \sum_{r_j \neq 0} \Big[(\alpha_{CP} + \alpha_{CEP}) n_j g_{Ni}^{Fe}(r_j) \\ &\quad + \alpha_{CEP} f(r_j) n_j (\mu_{Ni} - \mu_{Fe} + c\, M_{Ni}^{Ni}) + \alpha_{CEP} f(r_j)(N_j - n_j) c\, M_{Ni}^{Fe} \Big] \\ &= \Delta H_L + \Delta H_{DNL} + \Delta H_{INL}, \end{aligned} \tag{8a}$$

where

$$M_{Ni}^{Fe} = \sum_{r_j \neq r} N_j g_{Ni}^{Fe}(r_j) \tag{8b}$$

is the total disturbance of iron magnetic moments around each nickel solute.

We now use Equation 8 to calculate the various moments, and then the mean, variance, and skewness of the iron-57 hyperfine magnetic field distribution. With the definitions

$$A_j = (\alpha_{CP} + \alpha_{CEP}) n_j g_{Ni}^{Fe}(R_j) + \alpha_{CEP} f(r_j) [\mu_{Ni} - \mu_{Fe} + c(M_{Ni}^{Ni} - M_{Ni}^{Fe})]$$

and

$$B_j = \alpha_{CEP} f(r_j) c M_{Ni}^{Fe},$$

we find, after calculating moments of the binomial distribution as described in the Appendix,

$$\langle \Delta H \rangle = c \sum_j N_j (A_j + B_j) \tag{9a}$$

$$\langle (\Delta H)^2 \rangle = c(1-c) \sum_j N_j A_j^2 + c^2 \Big(\sum_j N_j A_j\Big)^2 + 2c \sum_j N_j A_j \sum_k N_k B_k + \Big(\sum_j N_j B_j\Big)^2 \tag{9b}$$

$$\begin{aligned}\langle (\Delta H)^3 \rangle &= c^3 \sum_j N_j^3 A_j^3 + 3c(1-c) \sum_j N_j A_j^2 \Big(\sum_j N_j A_j + \sum_j N_j B_j\Big) \\ &\quad - 3c^2 \sum_j N_j A_j^2 + c(1+c^2) \sum_j N_j A_j^3 \end{aligned} \tag{9c}$$

$$+ c^3(\sum_j N_j A_j)^3 + 3c^2(\sum_j N_j A_j)^2 \sum_{j} N_j B_j + 3c\sum_j N_j A_j (\sum_j N_j B_j)^2 + (\sum_j N_j B_j)^3.$$

We have evaluated these moments of the hyperfine magnetic field distribution by using the following sets of parameters appropriate for 18°C [18-21,30-35]:

$$\{N_j\} = \{8, 6, 12, 24, 8\} \tag{10a}$$

$$\alpha_{CEP}\{f_i\} = \{-11,5, -3.5, +2.5\} \frac{kG}{\mu_B} \tag{10b}$$

$$\alpha_{CP} + \alpha_{CEP} = -90 \frac{kG}{\mu_B} \tag{10c}$$

$$\mu_{Ni} - \mu_{Fe} = -0.8\mu_b \tag{10d}$$

$$\{g^{Fe}_{Ni}\} = \{0.065, 0.043, 0.015, 0.009, 0.008\} \tag{10e}$$

$$\{g^{Ni}_{Ni}\} = \{0\}. \tag{10f}$$

The mean, square root of the variance and cube root of the skewness that we obtain from this calculation are -14.0, +5.8, and -14.8kG, respectively, when the nickel concentration is 0.1, and -7.3, +3.9, and -12.0kG when the nickel concentration is 0.05.

The measured mean, square root of the variance, and cube root of the skewness of the iron-57 hyperfine magnetic field distribution at 18°C are -8, +8, and +1kG, respectively, when the nickel concentration is 0.09, and are -6.5, +7, and +1kG when the nickel concentration is 0.06. Our calculation seems most accurate for the lower nickel concentration data, although the calculated variance and skewness are too small and too large, respectively. The lack of quantitative agreement between the calculated and observed hyperfine magnetic field distribution of Fe-Ni results in part from a nickel dependence of the set $\{g^{Fe}_{Ni}(r_j)\}$, as shown by magnetization data of non-dilute Fe-Ni alloys [36-38] and by neutron scattering data on iron moments in Fe-Ni alloys [39,40]. The reduction in $\{g^{Fe}_{Ni}(r_j)\}$ with nickel concentration causes ΔH_{INL} to saturate with nickel concentration. This effect reduces the mean of the hyperfine magnetic field distribution. It also strongly reduces the negative hyperfine magnetic field perturbation for the nickel-rich configurations, and therefore increases the positive skewness of the iron-57 hyperfine magnetic field distribution. The suppressed variance of the hyperfine magnetic field distribution is primarily due to our use of a single average value for ΔH_{INL}.

Some previous workers have been tempted to interpret the iron-57 hyperfine magnetic field in Fe-Ni alloys with a simple model based on the assumption that each nearest neighbor configuration produces a hyperfine magnetic field perturbation of

$$\Delta H(\{n'_j\}) = \sum_j n'_j \Delta H_j.$$

This simple model provides no means for determining the set of $\{\Delta H_j\}$ parameters, except by comparison with experimental data while applying a general guideline of $\Delta H_j > \Delta H_i$ if $i > j$. For Fe-Ni it is tempting to choose $\Delta H_1 = \Delta H_2 = -10$kG, because with these parameters the simple model reproduces the correct mean of the iron-57 hyperfine magnetic field distribution. By using the theorem in the Appendix, the mean, square root of the variance, and cube root of the skewness are -14, +11.3, and -10.3kG, respectively, when c_{Ni} is 0.1 and -7, +8.2, and 8.,5kG when c_{Ni} is 0.05. It is interesting that the simple model with its phenomenological hyperfine magnetic field perturbation parameter is capable of providing as good agreement with measured data as does our calculation with the model of linear response. However, the process of choosing the $\{\Delta H_j\}$ parameters for the simple model amounted to curve fitting, whereas the model of linear response uses measured magnetic moment and hyperfine magnetic field response data. Furthermore, the choice of $\Delta H_1 = \Delta H_2 = -10$kG is physically incorrect. The simple model attributes the largest hyperfine magnetic field perturbations to those iron-57 atoms with the closest nickel atoms, but this is in disagreement with Equation 8 and detailed NMR data [29,30]. Although a more physically realistic choice of $\{\Delta H_j\}$ parameters could be used with the simple phenomenological model, this choice of parameters would be determined from Equation 8. An additional problem arises with the simple model when c_{Ni} is on the order of 0.1. Equations 6a,b show that iron-57 hyperfine magnetic field perturbations in Fe-Ni cannot be parameterized simply in terms of the set $\{n_j\}$. The iron-57 hyperfine magnetic field perturbation due to a specified nickel solute atom will not only depend on its position about the iron-57 atom, but also on the positions of other nickel solutes around this specific nickel solute atom. Our approximations for ΔH_{INL} in Equation 8 also did not fully account for this effect. However, with computerized bookkeeping it is possible to include the various environments of the solute atoms in the model of linear response. No new parameters are involved in such a treatment, but the resulting variance and skewness of the calculated hyperfine magnetic field distribution are in much better agreement with experiment [41].

EFFECTS OF MANGANESE, CHROMIUM, SILICON, AND CARBON ON IRON-57 HYPERFINE MAGNETIC FIELDS

Here we use the model of linear response of hyperfine magnetic fields to magnetic moments from above to estimate the effects of X solutes on the Mössbauer spectra of Fe-9Ni-1X ternary alloys, where X is manganese, chromium, silicon, and carbon. Our goal is to understand possible changes in the effects of X solutes when the iron host is alloyed with nickel.

The average iron magnetic moment depends on the concentration of both nickel and X solutes is given by

$$\mu^{Fe}_{Fe\text{-}Ne\text{-}X}(c_{Ni}c_X) = \sum_{i_1}^{8}\sum_{j_2}^{8}\sum_{k_1}^{12}\sum_{i_2}^{8}\sum_{j_2}^{6}\sum_{k_2}^{12} P(8,i_1,c_{Ni})P(6,j_1,c_{Ni})P(12,k_1,c_{Ni}) \tag{11}$$

$$\times P(8,i_2,c_X)P(6,j_2,c_X)P(12,k_2,c_X)$$

$$\times \left[\mu^0_{Fe} + i_1 g^{Fe}_{Ni}(r_1) + j_1 g^{Fe}_{Ni}(r_2) + k_1 g^{Fe}_{Ni}(r_3)\right.$$

$$\left. + i_2 g^{Fe}_X(r_1) + j_2 g^{Fe}_X(r_2) + k_2 g^{Fe}_X(r_3)\right].$$

Note that this treatment involves the independence of the magnetic perturbations around nickel and X atoms, and does not even prevent both nickel and X atoms from simultaneously occupying the same lattice site. This could be a problem for more concentrated alloys, but no difficulties are expected for alloys with X solute concentrations of less than one percent. We identify first moments of the binomial distribution and find all average magnetic moments. By using the notation of Equation 8b, we find

$$\mu^{Fe}_{Fe\text{-}Ni\text{-}X}(c_{Ni},c_X) = \mu^0_{Fe} + c_{Ni}M^{Fe}_{Ni} + c_X M^{Fe}_X , \tag{12a}$$

$$\mu_{\substack{Ni\\Fe-Ni-X}}(c_{Ni},c_X) = \mu^0_{Ni} + c_{Ni}M^{Ni}_{Ni} + c_X M^{Ni}_X, \tag{12b}$$

$$\mu_{\substack{X\\Fe-Ni-X}}(c_{Ni},c_X) = \mu^0_X + c_{Ni}M^X_{Ni} + c_X M^X_X. \tag{12c}$$

We now substitute one X atom for one j-th nearest neighbor iron atom about many iron-57 nuclei. Although we could conceivably substitute the new X atoms for nickel atoms as well as for iron atoms, this would confuse the effect of nickel on $\Delta H^X_j(c_{Ni},c_X)$ by altering the nickel concentration of the lattice. This increases the concentration of X from c_X to $c_X+\Delta c_X$. The iron-57 hyperfine magnetic field is changed by the amount

$$\begin{aligned}\Delta H^X_j(c_{Ni},c_X) = {} & (\alpha_{CP}+\alpha_{CEP})1g^{Fe}_X(r_j) \\ & + \alpha_{CEP}(1-c_{Ni}-c_X) \sum_{0<r\neq r_j} f(r)\Delta c_X M^{Fe}_X \\ & + \alpha_{CEP}c_{Ni} \sum_{0<r\neq r_j} f(r)\Delta c_X M^{Ni}_X \\ & + \alpha_{CEP}c_X \sum_{0<r\neq r_j} f(r)\Delta c_X M^X_X \\ & + \alpha_{CEP}f(r_j)\left[\mu^0_X + c_{Ni}M^X_{Ni} + c_X M^X_X\right] \\ & - \alpha_{CEP}f(r_j)\left[\mu^0_{Fe} + c_{Ni}M^{Fe}_{Ni} + c_X M^{Fe}_X\right].\end{aligned} \tag{13}$$

After rearranging and performing lattice sums, we obtain

$$\begin{aligned}\Delta H^X_{\substack{j\\Fe-Ni-X}}(c_{Ni},c_X) = {} & 1g^{Fe}_X(r_j)(\alpha_{CP} + \alpha_{CEP}) \\ & + \alpha_{CEP}[8f(r_1) + 6f(r_2) + 12f(r_3) + \cdots](\Delta c_X) \\ & \times \left[(1-c_{Ni}-c_X)M^{Fe}_X + c_{Ni}M^{Ni}_X + c_X M^X_X\right] \\ & + \alpha_{CEP}f(r_j)\left[\left(\mu^0_X + c_{Ni}M^X_{Ni} + c_X M^X_X\right) - \left(\mu^0_{Fe} + c_{Ni}M^{Fe}_{Ni} + c_X M^{Fe}_X\right)\right]\end{aligned} \tag{14}$$

The first term of Equation 14 is the local hyperfine magnetic field perturbation, ΔH_L. The second term, linear in Δc_X, is similar to the indirect nonlocal term, ΔH_{INL}, although here it arises from the magnetic moment perturbations due to X solutes on X and nickel moments as well as on iron moments. The third term is essentially the direct nonlocal response, ΔH_{DNL}, of the conduction electrons at the iron-57 nucleus caused by the substitution of an X atom for an iron atom at the j-th nearest neighbor site.

In order to estimate the hyperfine magnetic field perturbation associated with a specific set $\{n_j\}$, we are forced to make further approximation for ΔH_{INL}. We average over the first three nearest neighbor shells, such that

$$\begin{aligned}\Delta H^X(n_1,n_2,n_3,c_{Ni},c_X) = {} & \Delta c_X(\alpha_{CP} + \alpha_{CEP}) \\ & \times \left(n_1 g^{Fe}_X(r_1) + n_2 g^{Fe}_X(r_2) + n_3 g^{Fe}_X(r_3)\right)\end{aligned} \tag{15}$$

$$+ \alpha_{CEP}(8f(r_1) + 6f(r_2) + 12f(r_3))[\frac{n_1+n_2+n_3}{8+6+12}]$$

$$x \left[(1-c_{Ni}-c_X)M_X^{Fe} + c_{Ni}M_X^{Ni} + c_X M_X^X \right]$$

$$+ \alpha_{CEP}(n_1 f(r_1) + n_2 f(r_2) + n_3 f(r_3))$$

$$x \left[(\mu_X^0 + c_{Ni}M_{Ni}^X + c_X M_X^X) - (\mu_{Fe}^0 + C_{Ni}M_{Ni}^{Fe} + c_X M_X^{Fe}) \right]$$

From Equation 14 we see that the substitution of an X atom for an iron atom causes a different iron-57 hyperfine magnetic field perturbation in an Fe-9Ni-1X alloy than in an Fe-1X alloy for four reasons: (1) In Fe-Ni-X there is a perturbation of nickel moments by X atoms. (2) In Fe-Ni-X there is a perturbation of X moments by nickel atoms. (3) In Fe-Ni-X the increase of nickel concentration is at the expense of iron atoms and consequently, there for fewer iron magnetic moments to be perturbed by X atoms. (4) in Fe-Ni-X the X atom replaces an iron atom whose magnetic moment was enhanced by nickel. All four of these effects increase in proportion of c_{Ni}, and all four also become more important in proportion to Δc_X.

The position of the "X satellite" with respect to the main absorption peak can be unchanged when the iron host is alloyed with nickel only if the X and the nickel atoms perturb the iron, nickel and X magnetic moments equally, that is if

$$M_X^X = M_X^{Ni} = M_X^{Fe} \text{ and } M_{Ni}^X = M_{Ni}^{Ni} = M_{Ni}^{Fe}. \tag{16}$$

When these conditions are not met, a different satellite peak separation is expected in Fe-9Ni-1X than in Fe-1X alloys.

With a knowledge of the constants α_{CP}, α_{CEP}, $\{f(r_j)\}$, and the necessary sets $\{g\}$, effects of nickel on the "X satellite" peak can be predicted. Because $c_X << c_{Ni}$, we need only consider the sets

$$\{g_{Ni}^{Fe}\}, \{g_{Ni}^{Ni}\}, \text{ and } \{g_{Ni}^X\}.$$

For a specific example, consider the effect that alloying the iron host with nickel has on the total "chromium satellite" revealed by the difference spectrum procedure. We can use Equation 15 to predict that there will still be a significant hyperfine magnetic field perturbation associated with ΔH_1^{Cr} and ΔH_2^{Cr} so long as M_{Cr}^{Ni} and M_{Ni}^{Cr} are less than approximately $+30\mu_B$. This would seem a safe assumption. For a second example ΔH_3^{Cr} is experimentally insignificant for Fe-X alloys [21]. With Equation 15 we predict that ΔH_3^{Cr} will still be experimentally insignificant in a Fe-9Ni-1Cr alloy, provided that M_{Cr}^{Ni} is less than approximately $10\mu_B$.

Our data as listed in Table I show that a manganese solute in the second nearest neighbor shell of the iron-57 atom causes an experimentally insignificant hyperfine magnetic field perturbation in Fe-Mn alloys, but causes a significant perturbation in Fe-Ni-Mn alloys. If we assume that

$$g_{Ni}^{Mn}(1) = g_{Ni}^{Mn}(2) = \frac{1}{2}\mu_B,$$

then Equation 15 predicts a change in ΔH_2^{Mn} of 4.2kG. Thus, through a large effect on the manganese magnetic moment, nickel atoms could have a qualitative effect on the "manganese satellite" by transforming the insignificant ΔH_2^{Mn} of a Fe-1Mn alloy into a significant ΔH_2^{Mn} in a Fe-9Ni-1Mn alloy.

Since a silicon atom has no magnetic moment and causes no magnetic moment perturbations at iron and nickel atoms around it, there are no

Table I. N_S^X for Fe-Ni-X Alloys at 18°C

X	Fe	Fe-3Ni	Fe-6Ni	Fe-9Ni	Fe-12Ni
Mn	7(2)	11(4)	13(3)	13(2)	-
Cr	13(2)	14(3)	15(3)	14(2)	15(4)
Si	13(3)	-	-	13(3)	-
C	6	-	-	4(2)	-

Error limits are given in parenthesis.

unknowns preventing us from making concrete predictions with Equation 15. The change in hyperfine magnetic field at iron-57 nuclei with j-th nearest neighbor silicon atom is given by

$$\Delta H^{Si}(i,j,k,.09,c_{Si}) - \Delta H^{Si}(i,j,k,0,c_{Si}) = \alpha_{CEP} f(r_j) M_{Ni}^{Fe}(.09), \qquad (17)$$

when c_{Ni} changes from zero to 0.09. This gives changes of -1.3, -0.4, and +0.3kG for ΔH_1^{Si}, ΔH_2^{Si}, and ΔH_3^{Si}, respectively. Adding Ni to the alloy will therefore have an insignificant effect on the "silicon satellite", and our value of N_S^{Si} (in Equation 18) for Fe-Si alloys should be unchanged for Fe-9Ni-Si alloys, at least so long as the response parameters $\{f(r_j)\}$ are unchanged with c_{Ni}. Our observations that the "silicon satellite" in Fe-Ni-Si is similar to the "silicon satellite" in Fe-Si is strong evidence that the $\{f(r_j)\}$ parameters do not change with c_{Ni} [41].

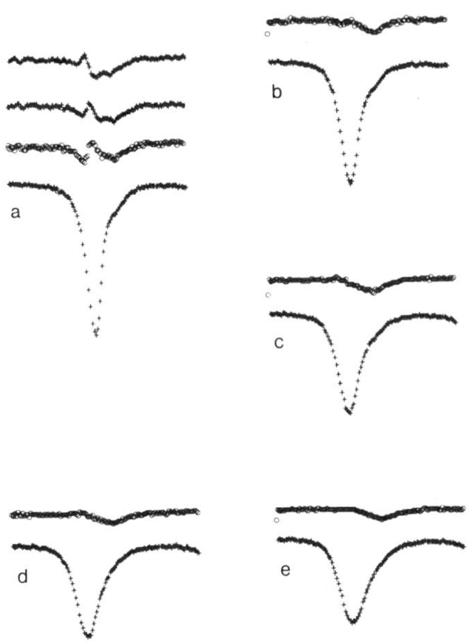

Figure 6. Calibration for effects of chromium "satellites" on difference spectra of peak Number 1 at 18°C. a. Fe-0.75Cr top; excess shift of V_{max} = +0.3 channels; middle: proper proper shift of V_{max}; bottom: excess shift of V_{max} = -0.3 channels. b. Fe-3.0Ni-0.75Cr. c. Fe-6.0Ni-0.75Cr. d. Fe-8.9Ni-0.75Cr. e. Fe-12.0Ni-0.75Cr.

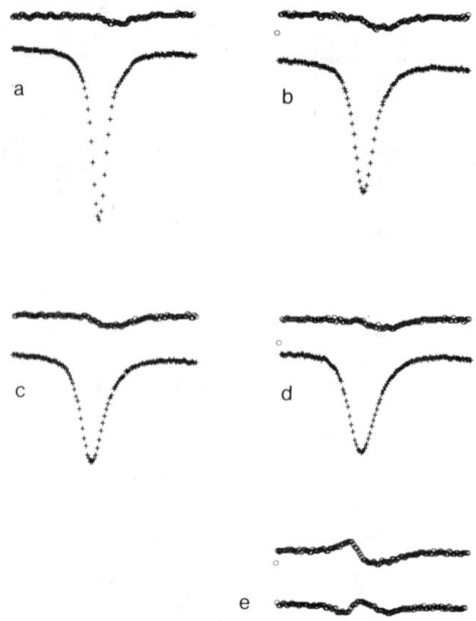

Figure 7. Calibration for effects of manganese "satellites" on difference spectra of peak number 1 at 18°C. a. Fe-0.75Mn. b. Fe-3.0Ni-0.75Mn. c. Fe-6.0Ni-0.75Mn. d. Fe-8.9Ni-0.75Mn. e. other differences between spectra of 7d., but with shifts of +1 and -1 channels with respect to 7d.

The difference spectrum intensity associated with effects of X solutes was revealed by taking differences between Fe-Ni and Fe-Ni-X spectra from alloys with the same nickel concentration. Examples of difference spectrum intensities around peak number one of Fe-Ni-.75Cr and Fe-Ni-.75Mn alloys are shown in Figures 6 and 7. For convenience we call this total difference spectrum intensity the "X satellite" peak. At low concentrations of X solutes, the integrated intensity of the "X satellite" peak, I_S, with respect to the integrated intensity of the full Fe-Ni-X peak, I_T, is proportional to the solute concentration times the number of neighboring sites, N_S^X, at which a solute atom can cause an experimentally significant iron-57 hyperfine magnetic field perturbation, thus

$$c_X N_S^X = \frac{I_S}{I_T}. \qquad (18)$$

Our calculations indicate that adding nickel to the alloy will not affect the values of N_S^X unless very large magnetic changes are associated with Ni-X interactions. By using Fe-Ni-X alloys with X concentrations of 0.75 and 1.3 percent, we experimentally determined N_S^X, and these data are listed in Table I. Additionally, for a Fe-9Ni host at 500°C, we found a N_S^{Mn} value of 12(4), and a N_S^{Cr} value of 16(4). The hyperfine magnetic field perturbations due to carbon in Fe-Ni-C alloys were difficult to quantify because carbon contributes so little intensity to the "X satellite". Our data from low carbon concentrations in Fe-Ni-C alloys suggest that N_S^C is 4, but in the present work N_S^C was not measured with accuracy. We believe that N_3^C is 6 for Fe-9Ni-C alloys because this is consistent with what is found for binary Fe-C alloys [42-44], and the value of N_S^C of 6 in Table I is given for this reason. Two experiments with a Fe-3Ni-0.75Mn ingot gave values for N_S^{Mn} of 9 or 10. Nevertheless, we are confident that N_S^{Mn} changes from about 8 to 14 as nickel

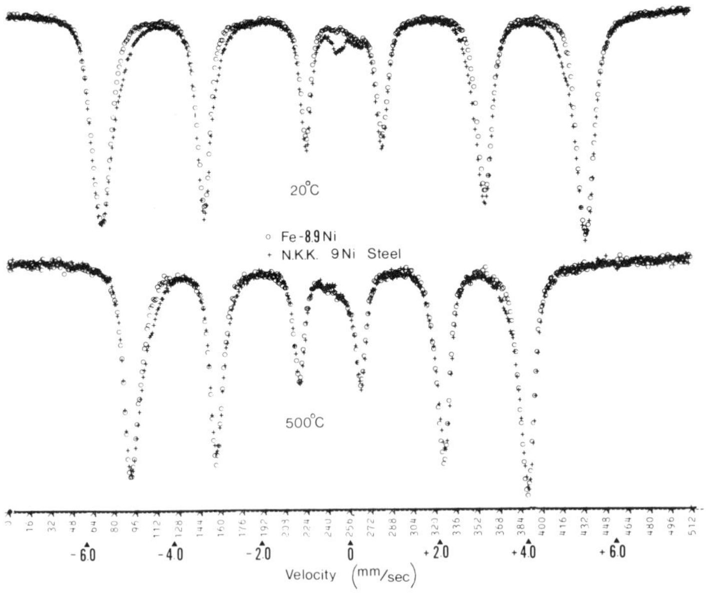

Figure 8. Comparison of Mössbauer spectra of Fe-8.9Ni with 9Ni steel at 18°C (top), and at 500°C (bottom).

is added to the host, although the precise composition at which this occurs is somewhat uncertain, and may depend upon the manganese concentration in the alloy.

Intensity changes caused by the other "X" solutes in commercial 9Ni steel can be seen in Figure 8 for spectra taken at 18°C and 500°C. At both temperatures, the "X" solutes cause some iron-57 nuclei in the martensite to absorb γ-rays on the low velocity sides of the six peaks. The difference of the spectra in Figure 8 gives an integral intensity consistent with the manganese, chromium, and silicon concentrations given in the experimental section and an N_S^X of 14.

MÖSSBAUER SPECTRA OF TEMPERING SEQUENCES

Phase Analysis

The fraction of austenite present in our materials was determined by taking the ratio of integrated intensity of the austenite peak to the total intensity of the Mössbauer spectrum [45-47]. Changes in the intensity of the austenite peak during tempering were clearly shown in the difference spectra of tempering sequences. The amount of austenite formed at 600°C was obtained primarily from Mössbauer spectra taken at 500°C. At 500°C the austenite is stable against the fcc to bcc martensitic transformation, and the amount of austenite does not change during the collection of the Mössbauer spectrum. For materials with thermally stable austenite, including commercial 9Ni steel tempered at 500 and 600°C, the fraction of austenite at 18°C was equal to the actual fraction of austenite that formed during tempering. However, in several high purity alloys, such as Fe-9Ni, Fe-9Ni-1Cr, Fe-9Ni-0.4Si, and Fe-9Ni-0.3C, much of the austenite had retransformed to martensite upon cooling to 18°C.

In order to convert data on the fraction of iron-57 nuclei in the austenite into the actual volume fraction of austenite, it is necessary to know the chemical compositions and the densities of the austenite and martensite.

For austenite formed at 600°C in 9Ni steel, this conversion factor was about 1.15. We have neglected the difference in recoil-free fraction of the austenite and martensite phases because we believe that it is small. By using the Debye model with Debye temperatures of 420°K for martensite and 450°K for austenite [48,49], the recoil-free fractions of austenite and martensite should differ by about 1 percent.

ANISOTROPIC HYPERFINE MAGNETIC FIELD PERTURBATIONS AROUND NICKEL ATOMS AT 18°C

The difference spectra shown in Figure 9 were obtained at 18°C between temperings at 600°C. Scales for the abscissa and the ordinate of the difference spectra are the same as for the spectrum from untempered material at the top of the figure. The Q-QT1/3 and Q-QT27 difference spectra show a disproportionately large residual intensity of peaks number 2 and 5 simultaneously with large difference spectrum intensities around peaks number 1 and 6. On the other hand, the Q-QT3 difference spectrum shows nearly zero residual intensity of peaks number 2 and 5, and uncharacteristically small difference spectrum intensities around peaks numbers 1 and 6. This same correlation between the residual intensities of peaks numbers 2 and 5 and the difference spectrum intensities around peaks numbers 1 and 6 was also observed in several tempering sequences of commercial 9Ni steel.

This correlation between the residual intensity of peaks numbers 2 and 5 and the difference spectrum intensity around peaks 1 and 6 originates with changes in the anisotropic or "pseudo-dipolar" contributions to the iron hyperfine magnetic field [50-52] that occur as the magnetization direction

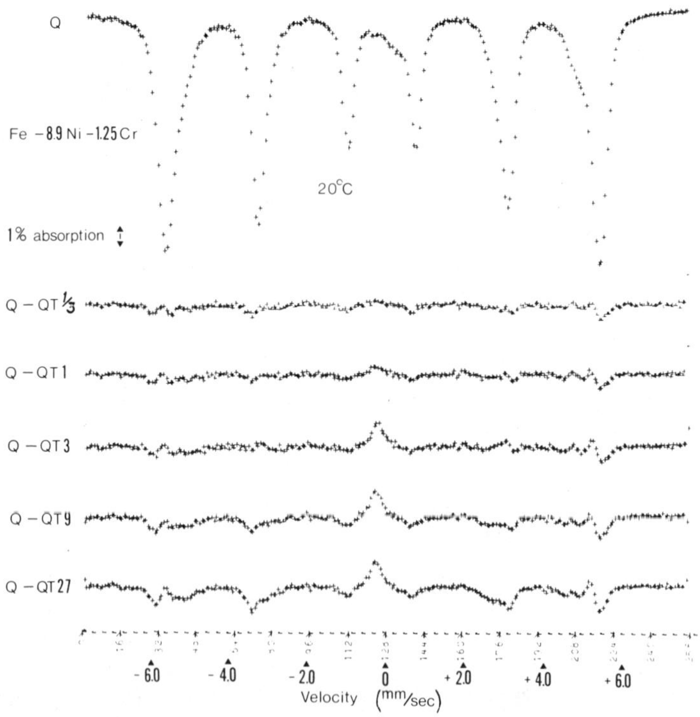

Figure 9. Differences of Mössbauer spectra of Fe-8.9Ni-1.25Cr alloy taken between temperings at 600°C from 20 minutes to 27 hours. Peak number one height normalization was used.

changes after tempering. There is no unique geometrical relationship between the angle formed by the hyperfine magnetic field direction and the direction from the iron-57 nucleus to a neighboring solute (which controls the magnitude of pseudo-dipolar interactions), and the angle formed by the hyperfine magnetic field direction and the incident γ-ray direction (which controls the residual intensity of peaks numbers 2 and 5 when peak number one height normalization is used). Nevertheless, both the pseudo-dipolar interactions and the ratio of peak number one to peak number two are controlled by the lattice magnetization, and they will both change as the direction of the lattice magnetization changes. Because we expect the largest pseudo-dipolar contributions from first near neighbor sites, we expect the smallest pseudo-dipolar contributions when the magnetization is along [100], i.e., when $3\cos^2(54.7°)-1$ is zero. The magnitudes of the pseudo-dipolar contributions and their diversity should increase when an applied magnetic field aligns the specimen magnetization perpendicular to the incident γ-ray direction and away from the [100] easy axes of magnetization. A difference spectrum of a binary Fe-8.9Ni specimen with different states of magnetization is shown in Figure 10. As in Figure 9, we see that as the specimen magnetization lies more in plane of the foil, the breadth of peaks number 1 and 6 is increased, and negative difference spectrum intensity appears around peaks number 1 and 6.

To minimize the empirical corrections for the effects of changing pseudo-dipolar contributions, Mössbauer spectra from a sequence of temperings were obtained with the specimen mounted in a reproducible way in a 2.2kG magnetic field. "Locking" the hyperfine magnetic field directions with this applied magnetic field significantly reduced variations in the chemical composition data. Although the experimenter does not have the control over the crystallographic direction of magnetization as for single crystals [50-52], the "locking" magnetic field is recommended for polycrystalline metals in order to obtain a reproducible change in iron-57 hyperfine magnetic field for a given change in solute concentration. We reproduced our chemical composition calibration work in the 2.2kG magnetic field, and found that N_S^X did not depend on the presence of the magnetic field.

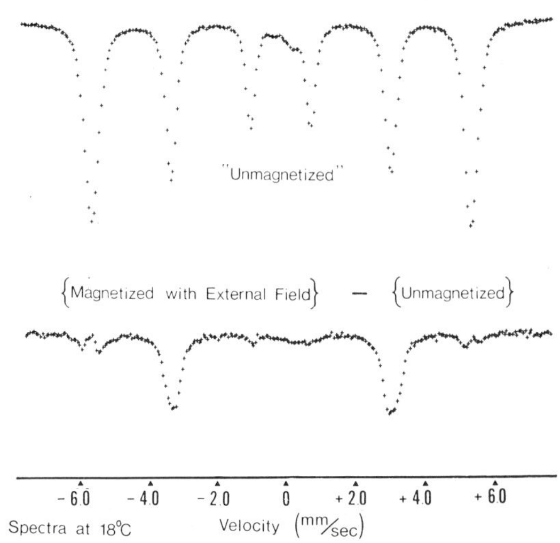

Figure 10. Magnetic orientation and anisotropic hyperfine magnetic field perturbation difference spectra of Fe-8.9Ni alloy taken with and without an applied 2.2kG magnetic field. For explanation, see text.

An improved set of difference spectra of 9Ni steel taken during a sequence of temperings is shown in Figure 11. Note that these difference spectra show features around peaks numbers 2 and 5 which are qualitatively similar to the features around peaks numbers 1 and 6. This similarity is promoted by the use of the saturating applied magnetic field, since it maintains a constant intensity ratio of peak 1 to peak 2 for both spectra.

CHEMICAL COMPOSITION OF THE AUSTENITE

The austenite which formed after different times at a given temperature appears to have a steady chemical composition. This is suggested by the constant linewidth of the paramagnetic austenite peak. Unfortunately, the isomer shifts and the electric quadrupole effects which broaden the austenite peak are small. Much larger hyperfine magnetic field effects due to solute concentration changes are seen in the peaks of the bcc martensite phase.

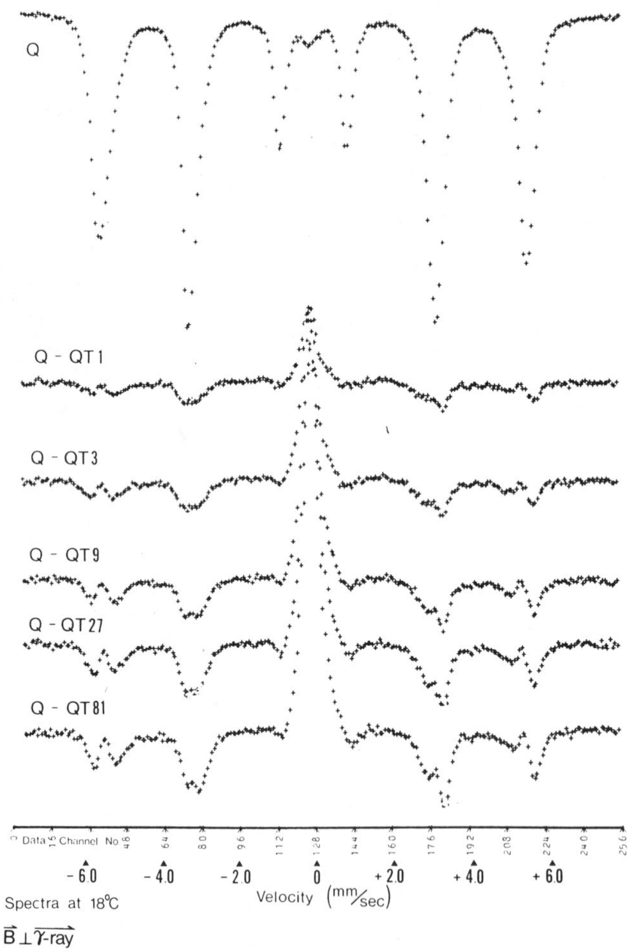

Figure 11. Differences of 18°C Mössbauer spectra of commercial 9Ni steel taken between temperings at 600°C from 1 to 81 hrs. Peak number 1 height normalization was used. Saturating magnetic field was applied to specimen.

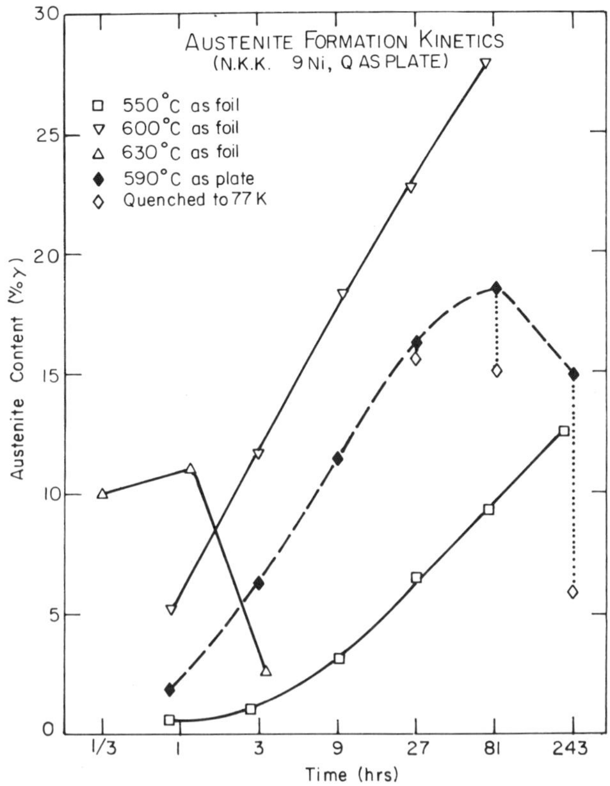

Figure 12. The volume fraction of austenite which formed in commercial 9Ni steel as a function of tempering time at four tempering temperatures.

Since austenite and martensite were the only two phases present in the material, the changes in the martensite composition in conjunction with the amount of austenite present (Figure 12) can be used to deduce the composition of the austenite. The data of Figures 13 and 14 were obtained in this way, through the use of difference spectra such as shown in Figure 11. Accuracy is poorest for the shorter tempering times when small amounts of austenite had formed, and only small solute depletions of the martensite had occurred. After more austenite had formed, the accuracy and precision of the austenite solute concentration data obtained by Mössbauer spectrometry was comparable to that obtained by the analytical scanning transmission electron microscopy.

We found the austenite formed at 600°C in the binary Fe-9Ni alloy to have a nickel content of 23 atomic percent. This is more than the 20 atomic percent indicated by the equilibrium Fe-Ni phase diagram, and we suspect that this discrepancy is due to systematic errors in our analysis method. However, the nickel content of the austenite in both the Fe-9Ni-1.25Mn alloy and Nippon Kokan Company 9Ni steel is only about 15 atomic percent, and we are confident that these data show an austenite nickel concentration that is lower than predicted by the Fe-Ni phase diagram. Although leaner in nickel, the austenite in the Fe-9Ni-1.25Mn alloy had a manganese concentration of about 8 atomic percent, and the austenite in commercial 9Ni steel had a combined concentration of manganese, chromium, silicon, and carbon of about 4 atomic percent.

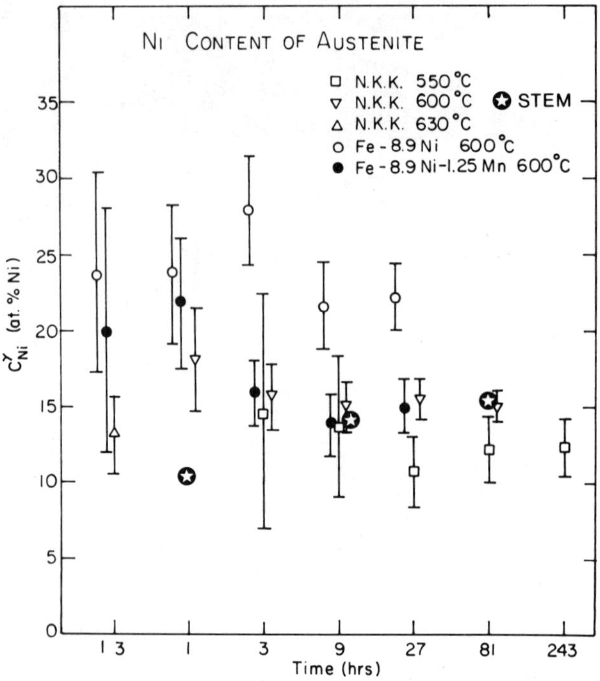

Figure 13. The concentration of nickel in the austenite of 3 alloys as a function of tempering time at several temperatures.

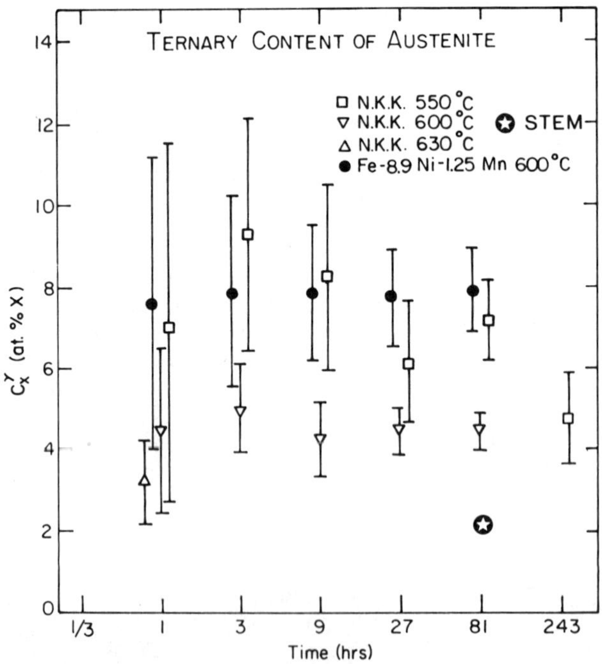

Figure 14. The net concentration of manganese, chromium, silicon, and carbon in the austenite of two alloys as a function of tempering time at several temperatures.

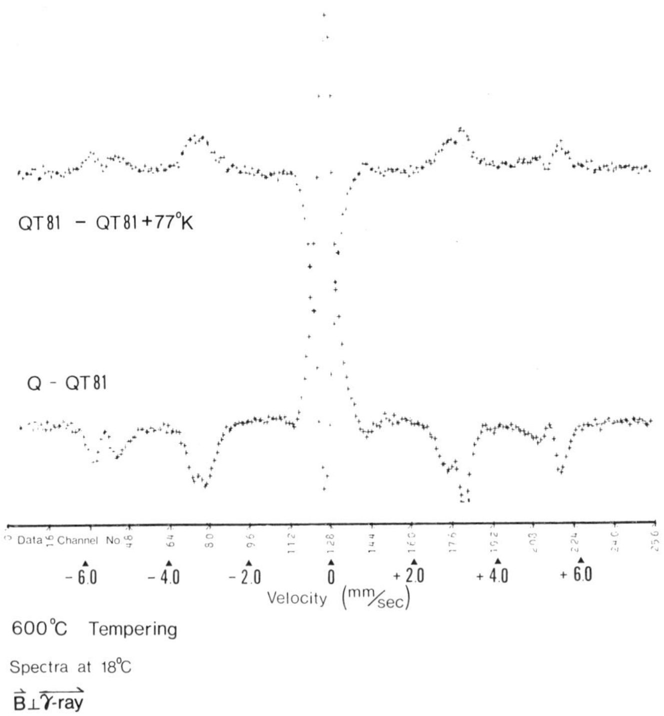

Figure 15. Retransformation of austenite and martensite chemistry changes. Bottom: difference of spectra of 9Ni steel before and after tempering at 600°C for 81hrs. Top: difference of spectra of 9Ni steel tempered at 600°C for 81hrs before and after hammering the specimen at 77K.

We found a small demonstration of the validity of the difference spectrum method for chemical and phase analysis. At the bottom of Figure 15 is the difference between Mössbauer spectra obtained before and after tempering a specimen for 81 hours at 600°C. The Q-QT81 difference spectrum shows the formation of a large amount of austenite and shows a large loss of nickel and X solutes from the martensite. After 81 h of tempering, the austenite is thermo-mechanically unstable; lightly hammering the specimen at 77K caused most of the austenite to transform martensitically. This transformation is seen in the top difference spectrum of Figure 15, which is the difference between the spectra of the QT81 specimen before and after hammering at 77K. The previously solute-rich austenite particles are now solute-rich martensite particles. In addition to the loss of austenite, this top difference spectrum also shows a return of nickel and X solutes to the martensite phase after this cryogenic treatment. The difference spectrum intensities corresponding to nickel and X concentration changes in the martensite are not exactly the reverse of the difference spectrum intensities at the bottom of Figure 15. This is because some austenite still remains in the cold-hammered specimen, and the chemical composition of the martensite in the cold-hammered material is not homogeneous.

METALLURGICAL DISCUSSION

Kinetics of Austenite Formation

Most of the formation of austenite during tempering occurs by diffusion of solute elements to growing austenite particles. The alternative process of a diffusionless transformation of bcc to fcc phase is inconsistent with the strong segregation of solute elements from the martensite to the austenite during tempering (see Figures 13 and 14). The Mössbauer difference spectrum technique is not especially sensitive for chemical analysis of the austenite when little austenite has formed, but even with this uncertainty we can set an upper bound of two percent for the amount of austenite which may have formed by a diffusionless process in 9Ni steel at 600°C.

The kinetics of austenite formation depend on both the kinetics of nucleation of new austenite particles and on the diffusion-limited growth of these particles. Both the processes of nucleation and growth are temperature-dependent, and together they account for the temperature-dependence of the overall kinetics of austenite formation. It is traditional to express the overall rate of austenite formation as a product of a nucleation factor and a growth factor. The growth rate of the austenite particles depends in the diffusive mobilities of the solute elements, which must move through the bcc phase to reach the growing austenite particles. The diffusive mobilities of the nickel and the other "X" solutes become greater with temperature as described by Arrhenius relationships given by

$$D_{Ni}(T) = D_{ONi} \exp - \left(\frac{Q_{Ni}}{RT}\right)$$

and

$$D_X(T) = D_{OX} \exp - \left(\frac{Q_X}{RT}\right). \tag{19}$$

The temperature dependence of the nucleation factor is less well understood. However, since the formation of austenite is more thermodynamically favorable at higher tempering temperatures, the austenite particles can nucleate at sites which are less energetically favorable. At the lower tempering temperatures there are fewer usable sites for austenite nucleation. This temperature dependence of the number of nucleation sites causes the activation energy for the overall kinetics of austenite formation to be much larger than the activation energy for solute diffusion in the bcc phase. At high temperatures, however, all of the suitable nucleation sites for austenite formation are used, so the nucleation factor is not expected to contribute any further to the temperature dependence of austenite formation. At higher temperatures the overall activation energy for austenite formation approaches the activation energy for solute diffusion (Q_{Ni} and Q_X of Equation 19), which is about 50-60kcal/mole for nickel, manganese, chromium, and silicon in the bcc martensite phase.

Activation energies for the formation of austenite taken from the data of Figure 12 support this general picture of the temperature dependence of nucleation and growth. The activation energy determined from the two lower tempering temperatures, 550°C and 600°C, is 160kcal/mole. This decreases to 90kcal/mole for our higher pair of tempering temperatures, 600°C and 630°C. We therefore expect that more nucleation sites are still coming into use at temperatures above 600°C. The importance of austenite nucleation kinetics in the overall process of austenite formation suggests a modification to the tempering of 9Ni steel. It is believed that the best mechanical properties are associated with fine scale dispersions of austenite particles in the microstructure. Our data suggest that carrying out the nucleation of austenite particles at higher temperatures could help to produce such a microstructure.

Solute Diffusion to the Austenite

We have compared our data on austenite composition to equilibrium phase diagrams for iron-nickel-manganese and iron-nickel-chromium. At our highest tempering temperature, 630°C, it may be possible that the austenite is forming with a composition consistent with tie-lines across the two-phase field of the phase diagram. However, if the austenite were forming with its equilibrium composition at 550°C, the large concentration of manganese, chromium, and silicon in the austenite would cause the equilibrium bcc phase to have a negative concentration of these elements! This suggests that the austenite does not form as an equilibrium precipitate at the lower tempering temperatures. Furthermore, equilibrium phase diagrams show that the nickel concentration of the austenite should increase with decreasing tempering temperatures. Although this is consistent with data for the nickel concentration at 630 and 600°C, our measurements show a higher nickel concentration at 600 than at 550°C.

Perhaps these inconsistencies in the chemical composition of the austenite are associated with the morphology or small size of the austenite particles. However, here we explore an alternative kinetic basis for the chemical composition of austenite formed at low tempering temperatures. We try to describe the formation of austenite by a model which is entirely based on diffusion kinetics of nickel, manganese, chromium, and silicon solutes. In this kinetic model the austenite forms with a composition determined by which solutes reach it first. From Equation 19 we find that the ratio of X diffusivity to nickel diffusivity is temperature-dependent and given by

$$\frac{D_X(T)}{D_{Ni}(T)} = \frac{D_{OX}}{D_{ONi}}\, e^{\frac{Q_{Ni}-Q_X}{RT}} \qquad (20)$$

In the kinetic model we expect this ratio of diffusivities to control the ratio of concentration of X to the concentration of nickel, c_X/c_{Ni} in the austenite. In order to actually relate our data on the ratio c_X/c_{Ni} to the ratio $D_X(T)/D_{Ni}(T)$ of Equation 20, it is necessary to make some geometric assumptions about how the solutes diffuse to the austenite. These assumptions can be developed into detailed models of austenite growth [53], but here we seek merely to determine the geometrical dimensionality of the solute diffusion to the austenite. For example, if the austenite were to form in the center of an isotropic solid, it would be equally probable that solutes would arrive at the surface of the austenite from all radial directions. This is an example of a three-dimensional model. On the other hand, a one-dimensional model can be envisioned in which the solutes need not diffuse radially inwards to the growing austenite, but must diffuse only to the nearest martensite/martensite boundary. Once the solutes reach this boundary, they are rapidly transported by a grain boundary diffusion mechanism to a growing austenite particle. This "collector plate mechanism" [53,54] could be appropriate for the formation of austenite in 9Ni steel because the austenite particles are known to form on the boundaries of martensite laths.

By considering the amount of nickel and X solutes which are able to diffuse through one, two, and three-dimensions to the surface of a small austenite particle, we find that the ratio of c_X to c_{Ni} is related to the ratio of D_X to D_{Ni} by

$$\frac{c_X}{c_{Ni}} = \left[\frac{D_X(T)}{D_{Ni}(T)}\right]^{n/2} \qquad (21)$$

where n is the geometrical dimensionality of the diffusion process. By using our experimental data for c_X/c_{Ni} in Equation 21 at the pairs of temperatures

550, 600°C and 600, 630°C, we are able to eliminate the pre-exponential factor in Equation 20 and obtain $Q_{Ni}-Q_X$. For the lower temperature pair we find that $Q_{Ni}-Q_X$ is +18, 9, and 4.5kcal/mole for one, two, and three-dimensional models, respectively. For the higher temperature pair this difference is +44, 22, and 11kcal/mole for one, two, and three-dimensional models, respectively. It is disappointing that our data disagree for the low temperature and high temperature pairs by more than a factor of two. However, even with this inaccuracy it is apparent that the difference $Q_{Ni}-Q_X$ obtained with the one-dimensional model is too large; from diffusion data we know that $Q_{Ni}-Q_X$ should be +5 to 10kcal/mole. The chemical composition of the austenite is most consistent with a three-dimensional, or possibly a two-dimensional, kinetic model of solute diffusion to the austenite.

Although high temperatures are preferred for the nucleation of the austenite, the growth of austenite particles at lower tempering temperatures can be used to optimize their solute concentration. A two-step tempering treatment is suggested, a brief higher temperature treatment at approximately 630°C is desirable for dense austenite nucleation, followed by a lower temperature tempering at approximately 580°C which is desirable for the best solute concentration. We point out that a denser nucleation of austenite particles reduces the required solute diffusion distances,so in the second step the austenite will grow more rapidly than shown by the data of Figure 12.

SUMMARY AND CONCLUSIONS

The fracture toughness of 9Ni steel at cryogenic temperatures can be controlled by the amount and stability of small fcc austenite particles. By using Mössbauer spectrometry, we undertook a metallographic study of how this austenite is formed during tempering. Phase analysis was straightforward, but measuring the chemical concentration changes required an investigation of the systematics of iron-57 hyperfine magnetic fields in Fe-Ni-X alloys. A model of linear response of the iron-57 hyperfine magnetic field to magnetic moments can account, at least qualitatively, for the systematics of how changes in nickel and X solute concentrations affect the iron-57 hyperfine magnetic field distribution in the bcc martensite phase. Some of the lack of quantitative agreement is due to approximations in our calculations, and not necessarily to deficiencies in the model. To avoid variations in the "pseudo-dipolar" contributions to the iron-57 hyperfine magnetic field, it was necessary to "lock" the lattice magnetizations with an applied magnetic field during spectrum collection.

We suggest that the chemical composition of the austenite is determined in part by diffusion kinetics, especially at the lower tempering temperatures. The relative rates at which the X and nickel solutes are able to diffuse in two or three dimensions to an austenite particle controls its composition. As a means of forming a fine dispersion of stable austenite particles with a single tempering temperature, the commercial "QT" treatment (one hour at 600°C) is well-chosen. However, the present work suggests potential advantages of a two-step heat treatment consisting of a brief high temperature nucleation step followed by a longer low temperature solute enrichment step.

ACKNOWLEDGMENTS

The authors are grateful to Dr. J. I. Kim of the IBM Thomas J. Watson Research Center for important discussions and help with the scanning transmission electron microscopy work. This work was supported by the Director, Office of Energy Research, Office of Basic Energy Science, Materials Science

Division of the U.S. Department of Energy under Contract number De-AC03-76SF00098.

APPENDIX

Moments Theorem. If the hyperfine magnetic field at an iron-57 nucleus can be expressed as a polynomial function of the variables $\{n_j\}$, then all moments of the iron-57 hyperfine magnetic field distribution from a random alloy can be evaluated.

Proof. (1) Statistical independence of the variables $\{n_j\}$ causes the average of products to equal the product of averages, thus

$$\langle n_{j1}^{X}\, n_{j2}^{Y}\rangle = \langle n_{j1}^{X}\rangle\langle n_{j2}^{Y}\rangle .$$

The moments $\langle n_j^X\rangle$ can then be individually evaluated, as shown in the next step. (2) The Nth moment of the binomial distribution for n trials can be determined from the N-1th, N-2th,•••• moments of the binomial distribution for n-1 trials. As an example, we develop the relation between the third moment of the binomial distribution for 14 trials in terms of the second and first moments of the binomial distribution for 13 trials:

$$\langle x^3\rangle_{14} = \sum_{i=1}^{14} i^3 \frac{14!}{i!(14-i)!} c^i (1-c)^{14-i} .$$

Note that the i equals zero term vanishes and

$$\langle x^3\rangle_{14} = 14c \sum_{i=1}^{14} i^2 \frac{13!}{(i-1)!\,(13-(i-1))!} c^{i-1}(1-c)^{13-(i-1)} .$$

Now define j = i - 1, so

$$\langle x^3\rangle_{14} = 14c \sum_{j=0}^{13} (j+1)^2 \frac{13!}{j!(13-j)!} c^j(1-c)^{13-j}$$

and define

$$\langle x^N\rangle_{13} = \sum_{j=0}^{13} j^N \frac{13!}{j!(13-j)!} c^j (1-c)^{13-j}$$

so that

$$\langle x^3\rangle_{14} = 14c[\langle x^2\rangle_{13} + 2\langle x\rangle_{13} + 1].$$

With care and patience, higher moments of the binomial distribution may be evaluated by repeated applications of this procedure. Finally, the zeroth moment is evaluated in the next step. (3) The binomial coefficients are normalized so that:

$$\sum_{n=0}^{N_j} P(N_j, n_j, c) = 1$$

Corollary. The mean, variance and skewness of the hyperfine magnetic field distribution of a random alloy can be always be evaluated when the iron-57 hyperfine magnetic field can be expressed in terms of the set $\{n_j\}$, where

$$\text{mean} = \langle \Delta H \rangle$$

$$\text{variance} = \langle \Delta H^2 \rangle - \langle \Delta H \rangle^2$$

$$\text{skewness} = \langle \Delta H^3 \rangle - 3\langle \Delta H^2 \rangle \langle \Delta H \rangle + 2\langle \Delta H \rangle^3.$$

REFERENCES

1. A. W. Pense and R. D. Stout, Welding Research Council Bulletin No. 205, 1975.

2. G. R. Brophy and A. J. Miller, Trans. ASM 41, 1185 (1949); T. N. Armstrong and G. R. Brophy, The American Society of Mechanical Engineers, The International Nickel Company, New York, Oct. 1947.

3. C. W. Marschall, R. F. Hehemann, and A. R. Troiano, Trans. ASM 55, 135 (1962); C. W. Marschall, Ph.D. Thesis, Case Inst. of Tech., Cleveland, OH (1961).

4. T. Ōoka and K. Sugino, J. Japan Inst. Metals 30, 435 (1966); T. Ōoka, H. Mimura, S. Yano, K.Sugino, and Y. Toizumi, J. Japan Inst. Metals 30, 442 (1966).

5. M. Kron, A. Constant, A. Clerc, J. Plateau, G. Henry, M. Robert, and C. Crussard, Mem. Sci. Rev. Metal. 58, 901 (1961).

6. S. Yano, H. Sakurai, H. Mimura, N. Wakita, T. Ozawa, and K. Aoki, Iron and Steel Inst. Japan 13, 133 (1973).

7. H. Haga, Trans. Iron and Steel Inst. Japan 13, 141 (1973).

8. K. J. Kim and L. H. Schwartz, Mater. Sci. Eng. 33, 5 (1978).

9. C. K. Syn, B. Fultz, and J. W. Morris, Jr., Met. Trans. A9, 1635 (1978).

10. B. Fultz and J. W. Morris, Jr., Met. Trans. A16, 173 (1985); B. Fultz, M.Sc. Thesis, University of California, Berkeley, LBL-7671 (1978).

11. A. D. Romig, Jr. and J. I. Goldstein, Met. Trans. A11, 1151 (1980); A. D. Romig, Jr. and J. I. Goldstein, Met. Trans. A12, 243 (1981).

12. J. I. Kim and J. W. Morris, Jr., Met. Trans. A12, 1957 (1981).

13. B. Fultz and J. W. Morris, Jr., in Nuclear and Electron Resonance Spectroscopies Applied to Materials Science, E. N. Kaufmann and G. K. Shenoy, eds., North Holland, New York, p. 377 (1980); B. Fultz, Ph.D. Thesis, Univ. of Calif., Berkeley (1982).

14. G. A. Bykov and P. J. Hien, Sov. Phys. JETP 16, 646 (1963).

15. D. P. Johnson and J. G. Dash, Phys. Rev. 172, 983 (1968).

16. J. Heberle, Nucl. Instr. and Methods 58, 90 (1968).

17. D. A. Shirley and G. A. Westenbarger, Phys. Rev. 138, A170 (1965).

18. M. B. Stearns, Phys. Rev. B4, 4069 (1971).; M. B. Stearns, Phys. Rev. B4 4081 (1971).

19. M. B. Stearns, Phys. Rev. B9, 2311 (1974); M. B. Stearns, Phys. Rev. B13, 1183 (1976).

20. I. Vincze and G. Grüner, Phys. Rev. Lett. 28, 178 (1972).

21. I. Vincze and I. A. Campbell, J. Phys. F: Metal Phys. 3, 647 (1973).

22. F. van der Woude and G. A. Sawatzky, Phys. Repts. 12, 335 (1974).

23. R. E. Watson and A. J. Freeman, Phys. Rev. 123, 2027 (1961); R. E. Watson and A. J. Freeman, in Hyperfine Interactions, A.J. Freeman and R. B. Frankel, eds., Academic Press, New York, Chapter 2 (1967).

24. K. J. Duff and T. P. Das, Phys. Rev. B12, 3870 (1975).

25. T. Yang, A. Kreshnan, and N. Benczer-Koller, Phys. Res. B30, 2438 (1984).

26. P. A. Flinn and S. L. Ruby, Phys. Rev. 124, 34 (1961).

27. G. K. Wertheim, V. Jaccarino, J. H. Wernick, and D. N. E. Buchanan, Phys. Rev. Lett. 12, 24 (1964).

28. T. M. Holden, J. B. Comly, and G. G. Low, Proc. Phys. Soc. 92, 726 (1967).

29. J. I. Budnick, T. J. Burch, S. Skalski, and K. Raj, Phys. Rev. Lett. 24, 511 (1970).

30. M. B. Stearns and L.A. Feldkamp, Phys. Rev. B13, 1198 (1976).

31. P. C. Riedi, J. Phys. F: Metal Phys. 8, L201 (1978).

32. I. Vincze, I. A. Campbell, and A. J. Meyer, Solid State Comm. 15, 1495 (1974).

33. M. F. Collins and G. G. Low, Proc. Phys. Soc. 86, 535 (1965).

34. I. A. Campbell, Proc. Phys. Soc. 89, 71 (1966).

35. H. R. Child and J. W. Cable, Phys.Rev. B13, 227 (1976).

36. A. T. Pickles and W. Sucksmith, Proc. Royal Soc. London 175, 331 (1940).

37. J. Crangle and G. C. Hallam, Proc. Soc. London A272, 119 (1963).

38. D. I. Bardos, J. L. Beeby, A. T. Aldred, Phys. Rev. 177, 878 (1969).

39. C. G. Shull and M. K. Wilkinson, Phys. Rev. 97, 304 (1955).

40. M. F. Collins, R. V. Jones, and R. D. Lowde, J. Phys. Soc. Japan 17 Suppl. BIII, 19 (1962).

41. B. Fultz and J. W. Morris, Jr., Phys. Rev. B, in press (1986).

42. N. DeCristofaro, R. Kaplow, and W. S. Owen, Met. Trans. A9, 821 (1978).

43. V. N. Gridnev, G. G. Gavrilyuk, V. V. Nemoshkalenko, Y. A. Polushkin, and O. N. Razumov, Fiz. Metal. Metalloved 32, 582 (1977).

44. T. Moriya, S. Sumitomo, H. Ino, and F. E. Fujita, J. Phys. Soc. Japan 35, 1378 (1973).

45. H. L. Marcus, L. H. Schwartz, and M. E. Fine, Trans. ASM 59, 468 (1966).

46. H. K. Chow, R. F. Weise, and P. A. Flinn, Report to USAEC Contract No. AT-(30-1)-4023; USAEC Rept. NSEC-4023-1 (1969).

47. L. H. Schwartz, in Applications of Mössbauer Spectroscopy, Vol. 1, Chapter 2, R. L. Cohen, ed., Academic Press, New York (1976).

48. J. de Launay, in Solid State Physics, Vol. 2, F. Seitz and D. Turnbull, eds., Academic Press, New York (1956).

49. Y. Tanji, J. Phys. Soc. Japan 30, 133 (1971).

50. T. E. Cranshaw, C. E. Johnson, and M. S. Ridout, Phys. Lett. 21, 481 (1966).

51. T. E. Cranshaw, J. Phys. F: Metal Phys. 2, 615 (1972).

52. A. Asano, Ph.D. Thesis, Northwestern Univ. (1974).

53. H. I. Aaronson, J. K. Lee, and K. C. Russell, in Precipitation Processes in Solids, K. C. Russell and H. I. Aaronson, eds., TMS-AIME, Warrendale, PA, Chapter 2 (1978).

54. H. B. Aaron and H. I. Aaronson, Acta Met. 16, 789 (1968).

SURFACE CHARACTERIZATION OF NITROGEN IMPLANTED CARBON- AND ALLOY-STEELS: A REVIEW AND RECENT RESULTS

G. Principi

Istituto di Chimica Industriale
Facoltà di Ingegneria
Università di Padova
Via Marzolo 9
I-35100 Padova, Italy

and

Istituto per la Ricerca Scientifica e Tecnologica
I-38050 Povo, Trento, Italy

S. Lo Russo

Dipartimento di Fisica
Facoltà di Scienze
Università di Padova
Via Marzolo 8
I-35100 Padova, Italy

and

Istituto per la Ricerca Scientifica e Tecnologica
I-38050 Povo, Trento, Italy

C. Tosello

Dipartimento di Fisica
Facoltà di Scienze
Uniersità di Trento
I-38050 Povo, Trento, Italy

and

Istituto per la Ricerca Scientifica e Tecnologica
I-38050 Povo, Trento, Italy

1. INTRODUCTION

Ion implantation, consisting of the direct injection of energetic ions into the subsurface region of materials, has been utilized in recent years to improve the technological properties of metallic surfaces. Many experimental studies as well as industrial applications have established in particular the specific advantages of nitrogen implantation with respect to conventional nitriding methods. These include, in particular, the possibility of avoiding high process temperatures, and of exceeding the solubility limit of impurities in the matrix, prevent dimensional changes of treated specimens, and ensure clean and well controlled process conditions. Furthermore, it has been found that thermally nitrided steels may be improved in their surface properties by additional nitrogen implantation. The subject has been recently reviewed [1-11].

Particularly interesting is the effect of nitrogen implantation on steels to increase both the wear resistance and the fatigue lifetime. Much experimental work has been carried out to clarify the mechanisms responsible for these effects and to correlate the technological improvements with the structural modifications induced on the treated surfaces. A variety of materials (iron and steels) have been implanted under different conditions and studied with several characterization techniques. Particularly useful are nuclear reaction analysis, NRA, to measure concentration and depth profile of the implanted species, and conversion electron Mössbauer scattering, CEMS, to analyze the surface structural changes.

We present here an outline of the current literature on nitrogen implantation on steels as well as some recent new results. The second section will deal with the technological improvements of nitrogen implanted surfaces and with the proposed mechanisms to account for such improvement. In the third section, the complementary information derived from various characterization techniques is briefly discussed. The influence of matrix composition and structure, of and implantation parameters, in light of the most recent results, are considered in the fourth section.

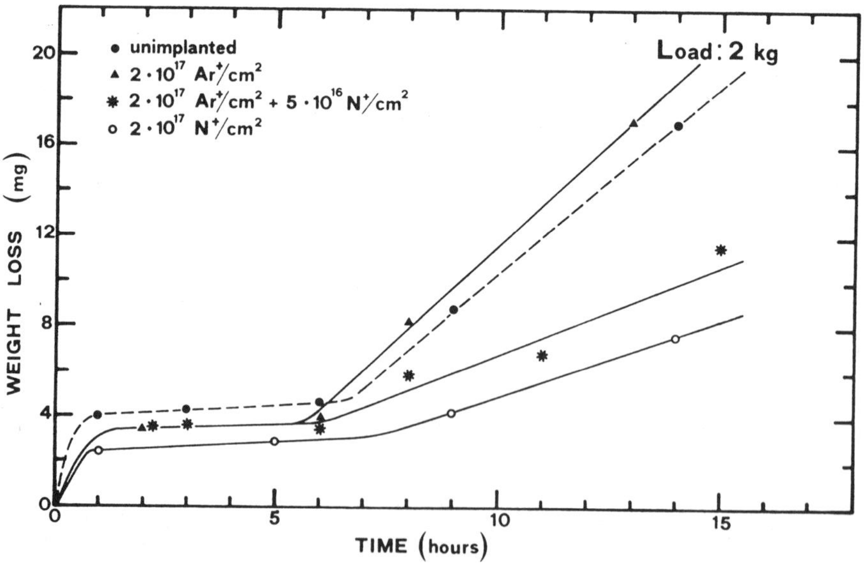

Figure 1. Weight loss following various periods of unlubricated reciprocating wear test for implanted and unimplanted conditions. Argon ion implantation (radiation damage) alone was not effective, but enhanced the effect of a subsequent implantation of nitrogen at a dose lower than that usually employed [17].

Table I. Improvements in Wear Lifetime Obtained by Nitrogen Implantation on Steel [15].

Application	Material	Lifetime Increase or Additional Benefits
Scoring die for aluminum beverage can lids	D2 tool steel	3X
Forming die for aluminum beverage can bottoms	D2 tool steel	Lowered wear and material pickup
Finishing rolls for copper rod	H-13	3X; improved surface finish
Paper slitters	1.6% Cr; 1% C	2X
Taps for phenolic resin	M2 high speed	Up to 5X
Tool inserts	4% Ni; 1% Cr corrosion	Reduced tool
Forming tools	12% Cr; 2% C	Greatly reduced adhesive wear
Injection molding nozzle, mold, screws, gate pads	Tool and hard Cr plated steel	4X-6X
Thermally nitrided steel molds	Tool steel	Combination better than either process alone
Fuel injectors and metering pump	Tool steel	>100X in engine tests
Plastic calibrator die	Nitrided H-13	2X
Profile hot die for plastic extrusion	P-20 tool steel	4X

2. TECHNICAL IMPROVEMENTS OF NITROGEN IMPLANTATED ON STEELS

2.1. Wear Reduction

Table I gives some significant examples of tribological effects of nitrogen implantation under different conditions on a variety of steels. Hartley [12-14] reported that the volumetric wear rate for lubricated sliding wear was reduced by a factor of 20-30 in a 50keV $2{*}10^{17}$ ions/cm^2 implanted nitriding steel. Similar experiments were carried out by Hirvonen [15] on AISI 416 steel specimens, which gave a volumetric wear rate 25-50 times less than the unimplanted specimens. No improvements were found for argon implantation under the same conditions. Because argon is a chemically inert species, this demonstrates that the effect is not due solely to the radiation damage nor to the compressive stresses that are produced by the implantation process. Moreover, it appeared that the beneficial effect due to nitrogen implantation persisted even after removal by wear of amounts of material exceeding the penetration depth of the implanted ions. Apparently, the mobile-interstitial-implanted species can decorate dislocations produced during wear, impeding their motion, and producing a hard skin. Sliding wear reduction for an implanted medium-carbon low-alloy steel was also detected by Lo Russo et al. [16] under unlubricated conditions. In this case, however, the treatment was less effective (about three times wear reduction), probably due to the more severe wear conditions. Double implantation of argon and nitrogen was accomplished by Lo Russo et al. [17]: it was demonstrated that radiation damage alone (argon implantation) did not produce any beneficial effect, as illustrated in Figure 1. Moreover, it has been verified that double implantation of Ar^+ and N^+ reduces wear even at low N^+ dose, indicating that a small amount of N^+ is effective only if high defect concentration is previously created in the matrix. In the same work additional evidence was obtained (Figure 2) of nitrogen presence after prolonged wear, as noticed also by Hirvonen [15]. The nitrogen dose most effective to reduce wear is typically about $2*10^{17}$ ions/cm^2, as confirmed also by Varjoranta et al. [18] on the relative wear reduction behavior of a low-carbon steel under lubricated and unlubricated conditions. In the last case, however, the authors attributed the beneficial effect to the surface stresses induced by the process.

The dependence of the wear behavior on operating conditions was studied by Goode and Baumvol [19]. They measured by nuclear reaction analysis the effective N dose retained in iron, high-carbon low-alloy steel, and austenitic stainless steel, after implants at 90keV and with a nominal dose of $4*10^{17}$N ions/cm^2, but with different sample temperatures and beam current-densities, of up to 25μA/cm^2. Whereas the stainless steel, whatever the current-density used, retained in the subsuperfical region all the implanted nitrogen, in the tool steel the amount of retained nitrogen was only 60 percent. Sliding wear tests indicated that a more lasting improvement was obtained for stainless steel implanted at low current-density and high temperature (5μA/cm^2, 330°C), and for tool steel implanted at low current-density and low temperaure (5μA/cm^2, 46°C). These conditions corresponded to the highest level of iron nitrides, mainly Fe_2N, formed in the subsuperficial

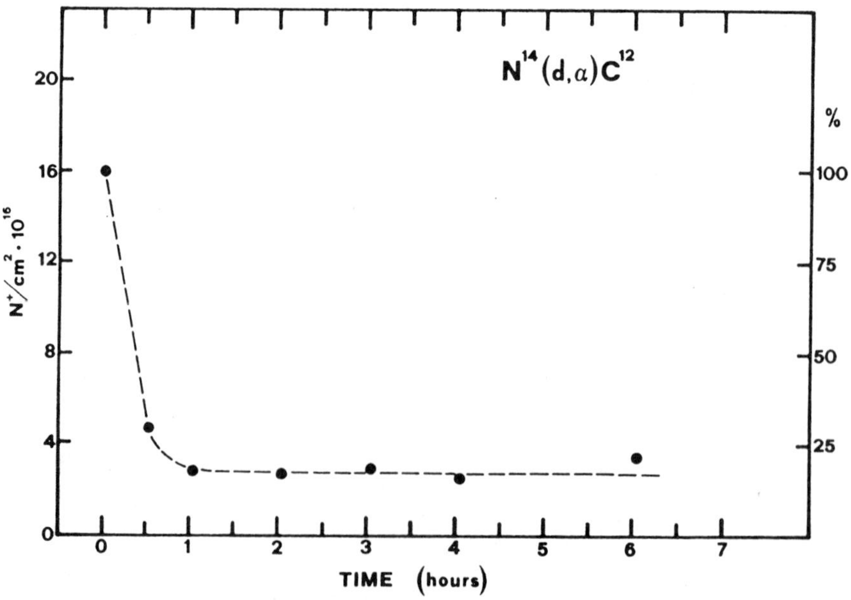

Figure 2. Yield of N-14(d,α)C-12 nuclear reaction indicating the retention of nitrogen following wear test [17]

region, as revealed by CEMS. However, a link between iron nitride formation and wear properties was not definitely established, because alloying elements like chromium rather than iron, may tie up the implanted nitrogen. Indeed, chemical correlation between nitrogen and chromium was detected in N implanted austenitic stainless steel by AES and XPS measurements [20]. That the wear rate reduction should be dependent on the steel composition at given operating implantation conditions, was also noticed by Kun et al. [21].

Another interesting feature is the synergistic effect of nitrogen implantation on previously thermally nitrided or carbided steels [22]. Refinement or dispersion of large precipitates formed by thermal treatment, as well as enrichment of nitrogen concentration near the surface, have been considered the main result of 40keV-$5*10^{17}$ions/cm^2-250μA/cm^2 nitrogen implantation, which increased the wear resistance by a reproducible factor of two.

The relatively good wear resistance to abrasion, noticed for carbon and austenitic stainless steels [23-25] was initially believed to be a measure of surface hardness, but was eventually related, for ductile steels, to the work-hardening rates during wear [7,8]. Singer's conclusions on nitrogen implantated and worn steels (partially in contradiction with his early reports) were (a) the friction coefficient did not decrease significantly in any of the treated steels, (b) large increases in wear resistance were obtained in ductile steels, but no changes were found in the examined carbon and low-chromium martensitic steels, (c) the wear resistance was attributed to changes in the wear mode and not simply to a "slowing down" of the wear rate, and (d) because wear resistance persisted up to loads that reach the load-carrying capacity of the steels, implantation must affect the way the steel surface responds to deformations induced by the stress transmitted during sliding.

In a very recent paper, Dimigen et al. [26] have reported more detailed results. Ball-on-disk wear tests under oscillating conditions without lubrication led to the conclusion that: (a) wear resistance increased up to three orders of magnitude in both ferritic and austenitic stainless steels with high chromium content for a nitrogen implanted nominal dose greater than $3*10^{17}$ions/cm^2 at 100keV, with no effect being detected at lower dose, (b) poor or no wear reduction was found for bearing and heat-treated low-chromium steels, whatever the dose, (c) increase of temperature during implanation led to a greater loss of nitrogen in low-chromium steels than in high-chromium steels, and (d) increased wear resistance persisted in high-chromium steels even for track depths of about one order of magnitude greater than the implantation depth (part of the implanted nitrogen was found below a wear track depth of about 2μm). The observed behavior of nitrogen implanted high-chromium steels was attributed to the formation, above a certain dose, of special phases within which part of the implanted nitrogen was tightly bound to the chromium atoms. These phases were believed responsible for, among other things, the high compressive stresses in the implanted layer. This agrees, in particular, with the observations by Yost et al. [27] of Cr-N and (Cr,Fe)-N precipitates in the nitrogen implanted layers of chromium steels, but is in contrast with the results of Feller et al. [28] who reported on metallic chromium.

Not all authors agree with the motion of nitrogen ahead of the wear front into the substrate, which is supported also by Fayeulle et al. [29]. Strong criticisms have been reported for example by Singer [7]. In a recent work, Doyle et al. [30] measured by nuclear reaction analysis the amount of nitrogen remaining in the tracks of a worn nitrogen implanted 15-5 PH chromium steel and found significant fractions of nitrogen remaining only at wear depths lower than the original implanted layer thickness. On the other hand, in an even more recent paper, Marest et al. [31] related NRA and CEMS detection of nitrogen after wear, at about 250nm below the original surface, in a 40keV nitrogen implanted medium-carbon low-alloy steel.

The literature on the wear behavior of nitrogen implanted steels often provide doubtful and sometimes contradictory results. Factors such as the wear test device utilized (pin-on-disc, crossed cylinders, alternative motion), applied load, and wear conditions (lubricated or unlubricated) should be carefully considered, when one attempts to compare results of different laboratories. Moreover, steels of different composition, structure and surface finish respond quite differently to the implantation process.

2.2. Fatigue Failure Reduction

Fatigue failure is generally thought to be a surface controlled mechanism, the crack initiation occurring at or near the surface of the specimen. Experiments have been carried out to see whether and how the surface modifications induced by ion implantation might improve the fatigue failure resistance of steels.

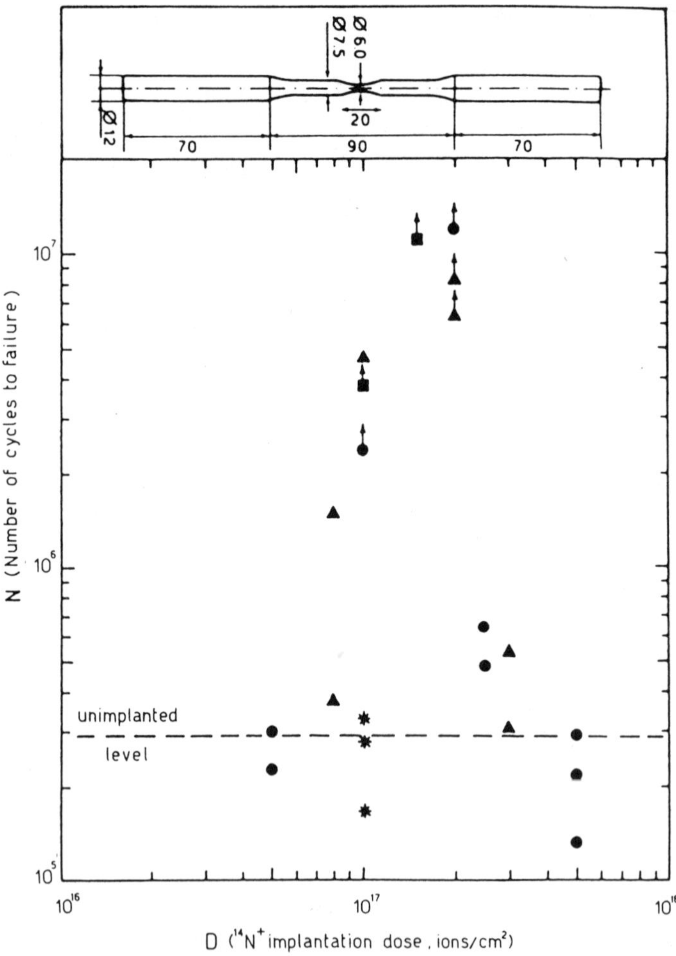

Figure 3. Lifetime versus nitrogen concentration for fatigue samples implanted at 30keV for different values of current density: stars 50, circles 100, triangles 140, and squares 200μA/cm^2. Arrows indicate no failure. In the insert the sample geometry is indicated [35].

Work at Harwell [14,32] showed significant extension of fatigue lifetime for nitrogen implanted low carbon steel and stainless steels. Hu, et al. [33,34] observed an increased effect due to nitrogen implantation of AISI 1018 steel specimens after natural or artificial aging. TEM studies showed the presence at the surface of $Fe_{16}N_2$ precipitates in both the as implanted and aged samples. However, the role played by nitrides seemed to be less important than that of nitrogen itself. The nitrogen would interact with defects (either preexisting or irradiation induced) in such a way as to inhibit the motion of dislocations. The interaction would be favored by aging.

Improvements in the fatigue lifetime of nitrogen implanted steel fatigue samples, without the need of aging, have been reported by Lo Russo et al. [35]. The difference with the Hu et al. [33,34] results was tentatively attributed to the higher beam current-density used, which would produce an aging effect during implantation and thermal diffusion of the implanted ions. Lifetime data vs. nitrogen dose showed a maximum near a dose of $2*10^{17}$ ions/cm^2 as illustrated in Figure 3.

Although the role of residual stresses are known to have an important influence on fatigue resistance (shot peening is one of the classical methods), implantation-induced stresses have been considered to have little influence in fatigue reduction [5].

3. ANALYSIS OF IMPLANTED SURFACES

A variety of surface analytical techniques, besides the already mentioned CEMS and NRA, have been utilized to examine the composition and microstructure of implanted materials. These include transmission electron microscopy, TEM, Auger electron spectroscopy, AES, X-ray photoelectron spectroscopy, XPS, Rutheford backscattering spectgroscopy, RBS, secondary ion mass spectrometry, SIMS, glancing angle X-ray diffraction, XRD, and others [8,10,11]. Complementary information on the same specimens is often essential to obtain a better knowledge of the materials under study. Some salient features and applications will be given in the next section.

3.1. Conversion Electron Mössbauer Scattering (CEMS)

Conversion electron Mössbauer scattering is the technique best suited to investigate nitrogen bonding to iron in nitrogen implanted layers, because scattered electrons escape mainly from a depth of about 0.2μm [36]. Such an analyzed depth is well suited to the examination of steel surfaces implanted with nitrogen ions with energy of the order of 100keV. An integral scattered spectrum, as obtained by a helium methane flowing gas proportional detector with a 2π collection geometry [37], provides information on the layers directly below the scattered surface without depth resolution. Many studies have been carried out by this method on the surface iron-nitrogen phases induced by the implantation process (see below).

Integral scattered spectra may also be obtained with channeltron devices which, working in vacuum, allow measurements at low and high temperature and, moreover, are suitable to extend CEMS investigation to Mössbauer isotopes other than iron-57, tin-119, and europium-151 [38]. The use of electrostatic and magnetic spectrometers provides for the selection of a narrow range of electron energies. Although electrons having the same energy do not necessarily originate at the same scattering depth, spectra measured at several electron energies may be used to obtain depth selective information. A depth selective qualitative picture of the nitrogen implanted layer of a low-alloy steel was obtained by Carbucicchio et al. [39]. The observed sequence of iron-epsilon-nitrides showed a maximum in nitrogen content below an outer surface layer, whose structure was interpreted as nitrogen containing austenite. By using Monte Carlo simulated weight functions for electron escape and detection, a depth profiling of the implantation induced phases was reported by Saneyoshi et al. [40] for specimens of 4keV nitrogen implanted iron. The gaussian shaped profile gave 100 percent nitride at the surface. Extensive reviews on the CEMS techniques are available [41].

3.2. Nuclear Reaction Analysis (NRA)

Nuclear reaction analysis is a powerful nondestructive technique to measure selectively the concentration and depth profile of light ions implanted in a heavy matrix [42,43] and is particularly useful in the case of nitrogen implanted on steels. For this purpose, three different nuclear reactions are currently utilized: N-14(d,α)C-12, N-14(d,p)N-15, and N-15(p,αγ)C-12, each one having peculiar characteristics.

The N-14(d,α)C-12 reaction is usually utilized at a deuteron incident energy higher than one MeV, so that the analyzed depth in steels is of the order of several microns below the surface. The emitted alpha particles are very energetic (~10MeV) and completely free from undesirable interferences. This reaction has been used by the authors to determinate the nitrogen concentration in nitrogen implanted steel after different times of sliding wear (Figure 2), as well as by Goode and Baumvol [19], for retained nitrogen measurements after nitrogen implantation on steels in different conditions. Very recently, Dimigen et al. [26] used this technique to evaluate the differences in the nitrogen retained dose and profile shape in implanted low-alloy and chromium steels.

Using the N-14(d,p_5)N-15 reaction at a deuteron incident energy of 610keV, the analyzed depth in steel is about 200nm below the surface and is comparable with the theoretically expected range plus a range straggling of 100keV nitrogen implanted ions [44]. As already noticed in the previous section, this depth is of the same order as that analyzed by the CEMS technique. The emitted protons have an energy of 1.56MeV and care must be taken to avoid interferences with reaction products from other elements. This reaction has been carefully studied for nitrogen microanalysis by Berti and Drigo [45]. They verified that an incident deuteron energy of 610keV permits the microanalysis of nitrogen simultaneously with that of oxygen, the latter through the O-16(d,p_0)O-17 reaction. This possibility is very useful in the examination of nitrogen implanted steels, where the surface oxygen content contribute in determining the surface mechanical properties.

For the described reactions the emitted particles are recorded by a silicon surface barrier detector, typically positioned at 150 degrees from the collimated deuteron incident beam. A mylar absorber placed in front of the detector prevents detection of the backscattered deuterons.

An excellent method for obtaining the implanted nitrogen concentration profile in a nondestructive way is to use the N-15(p,αγ)C-12 resonant nuclear reaction. This reaction presents a narrow resonance cross-section at the proton energy 429keV, so that it is possible to analyze increasing depths below the surface by increasing, step by step, the proton incident energy. The use of this reaction requires the implantation of the nitrogen-15 isotope instead of nitrogen-14. As reaction product, the γ-rays are usually detected by means of a NaI(Tl) detector. The depth resolution in steels of this profiling method is about 4nm, corresponding to the resonance width of 0.9keV. As examples of the utilization of this reaction in the study of nitrogen-15 implanted steels, we mention the works by Anttild and Hirvonen [46], Marest et al. [47], Moncoffre et al. [48], and Barnavon et al. [49].

3.3. Other Techniques

This section will be limited to a brief mention of other current methods which provide peculiar characteristics and convenience in the analysis of implanted surfaces.

TEM has been successfully utilized in detailed microstructural analysis of nitrogen implanted iron and steels by means of selected area X-ray diffraction patterns [27,50,51,52].

AES and XPS (the latter has lower spatial resolution and longer data acquisition times but is easier to use) provide compositional and chemical analyses and, in conjuction with ion sputtering, depth profiling of the elements in the implanted layer. The reliability of chemical and compositional information obtained from ion milled surfaces is associated with problems like differential sputtering, knock-out, cascade mixing, and ion enhanced diffusion, which may alter the surface. Moreover, depth profiling obtained in this way is destructive. Nonetheless, useful measurements have been performed on the bonding of implanted nitrogen with chromium atoms in steels [20,50] and on nitrogen distribution in implanted iron and steels [19,28,50].

RBS, which detects elastically backscattered protons or alpha particles, has a small sensitivity to light elements in heavy matrices and is most efficiently applied to study heavy atoms implanted into light matrices. However, profiles of nitrogen concentration vs. depth may be obtained for implanted iron and steel, on the basis of the reduction in the iron scattering yield from the surface region of the steel sample containing implanted nitrogen [18,26,27].

SIMS has a greater atomic sensitivity (ppm for many elements) than AES but is highly influenced by matrix effects, especially in the case of high dose implantation. The high sensitivity is however very useful to determine the amount of implanted nitrogen remaining in wear tracks [26,28,53].

A signal from nitrides more specific than AES may be provided by glancing angle X-ray diffraction. For example, the clear presence of Cr_2N in the nitrogen implanted layer of a single crystal of martensite has been noticed by Whitton et al. [54].

4. MATRIX CHARACTERISTICS AND OPERATING PARAMETERS

The considerable interest in studying nitrogen implanted iron and steel surfaces in a variety of conditions has been established in the previous sections. The composition and the structure of the implanted layers have often been found to depend strongly on the matrix characteristics and on the implantation parameters.

4.1. Pure Iron

In the first combined CEMS and NRA measurements on nitrogen ion implanted pure iron, Longworth and Hartley [55] reported that iron-nitrogen phases were produced by implantation at doses ranging from $2*10^{17}$ to $8*10^{17}$ ions/cm^2. Initially the γ'-Fe_4N phase was formed and, with increasing the dose, ε-$Fe_{2+x}N$ became predominant. Reference characterizations [56-60] on pure phases were utilized to recognize the surface compounds produced by nitrogen implantation (see Tables IIA and B). The annealing in vacuum of implanted samples at successive temperatures up to 500°C showed progressive destabilization of the nitrides, which disappeared completely when heated

Table IIA. Mössbauer Spectral Parameters for the Iron-Nitrogen-Carbon Phases Obtained Thermally (δ and QS in mm/s, H_{int} in kG).

Process	Phase	δ	QS	H_{int}	Ref.
Iron, gas nitriding	α''-$Fe_{16}N_2$	0.05		297	59
		0.18		407	
		0.18		314	
Iron, gas nitriding	γ'-Fe_4N	0.24		341	57
		0.52		216	
		-0.15		219	
Iron, gas nitriding	ε-$Fe_{2+x}N$ x=1.2	0.24		298	58
		0.33		238	
Iron, gas nitriding	x=0	0.40	0.27		56
Iron, gas nitriding and carbo-nitriding	x~1.2	0.25		290/300	68
	0.3<x<2.7	~0.30		0/235	
	0.3<x<1.2	0.40/0.60		0/145	
	x=0	0.31/0.49	0.28/0.86		

above 500°C. This was accompanied by the migration of the implanted nitrogen into the bulk and by the formation of α"-nitrogen-martensite.

Drako and Gumanskij [51] utilized high-voltage TEM to study the formation of precipitates at the surface of massive iron samples implanted with 60keV M^+ and C^+ at doses from 1 to $3*10^{17}$ ions/cm^2. Nitrogen and carbon irradiation under these conditions led to the formation of highly dispersed hexagonal ε-nitrides (40-70nm) and uniformly distributed Θ-carbides (15-20nm) respectively, with a close structural resemblance. The nitrides were mainly located at some depth under the surface and proved to be stable when heated up to 600°C, whereas the carbides, which did not penetrate deeply, were transformed into cementite at 200°C.

In agreement with observations on nitrogen implanted steels (see below), where the final surface composition is influenced by the structure previous to the treatment, dos Santos, et al. [61] showed that the double implantation on pure iron of carbon and nitrogen led to the formation of surface compounds, carbonitrides, which differed according to the order of implantation. In iron samples doubly implanted and successively annealed in a temperature interval up to 450°C, the concentration of carbonitrides was very similar and similar also to that of a nitrogen implanted low-carbon steel.

4.2. Steels

Nitrogen ion implantation on high-carbon martensite has been found to originate only the iron ε-phase in the subsuperfical region [62]. The preexistence of interstitial carbon, indeed, would support the growth of the ε-phase, where carbon and nitrogen are mutually replaced [63]. This interpretation has been confirmed by dos Santos et al. [64]. Frattini et al. [65] reported the formation of some γ-nitride in nitrogen implanted medium-carbon tempered martensite and attributed this formation to the presence of ferrite islands in the matrix before implantation. In a systematic study on the surface compound formation in implanted steels with

Table IIB. Mössbauer Spectral Parameters for the Iron Nitrogen-Carbon Phases Obtained by Ion Implantation (δ and QS in mm/s, H_{int} in kG).

Process	Phase	δ	QS	H_{int}	Ref.
Iron, nitrogen ion implant	γ'-Fe_4N	-0.30/0.34	-0.13/0.07	191/224	55
	ε-$Fe_{2+x}N$	0.35/0.34	0.29/0.48	240/244	
	ε-Fe_2N	0.39/0.44			
High-carbon martensite, nitrogen ion implant	α''-$Fe_{16}N_2$	0.06		288	62
	ε-$Fe_{2+x}(N,C)$	0.21		239	
	ε-$Fe_2(N,C)$	0.06/0.42	0.42/0.73		
Medium-carbon steel, nitrogen ion implant	ε-$Fe_{2+x}(N,C)$	0.09/0.12		299/300	65
		0.29/0.34		212/300	
		0.35/0.80		92/131	
	ε-$Fe_2(C,N)$	0.41/0.45	0.30/0.63		
Low-alloy steel, nitrogen ion implant	α''-$Fe_{16}N_2$	0.27		299	47
		0.10		393	
	γ'-Fe_4N	0.29/0.50	-0.24/0	197/231	
		0.39/0.50		328/368	
	ε-Fe_3N	0.26/0.49		260/323	
	ε-Fe_2N	0.23/0.43	0.23/0.46		
Carbon steels, nitrogen ion implant	θ-$Fe_3(C,N)$	0.40	0.63	203	62
		0.23	-0.73	200	
	γ'-Fe_4N	0.32	0.17	216	
	ε-$Fe_{3.2}(N,C)$	0.29		243	
	ε-$Fe_{2+x}(N,C)$	0.34	0.48		
		0.24	0.06	225	
		0.27	0.06	165	
	α''-$Fe_{16}N_2$	-0.01	-0.04/-0.06	300/314	
Iron, nitrogen and carbon ions implant	θ-$Fe_3(N,C)$	0.28/0.58	0.08/0.68	194/203	63
		0.21	-0.67	196	
		0.16/0.44	0.63/0.69		
	ε-$Fe_3(N,C)$	0.25/0.30	0.30	240	

different carbon content, dos Santos et al. [66], suggested that the pre-existent carbon in the steel matrix and the dose of nitrogen implanted may be two competitive factors. Their combinations would determine the total amount of iron nitrides formed. Experimental data seemed to indicate that, at high-carbon concentrations, the precipitation of nitrides was inhibited, because all the interstitial sites in the steel matrix were already occupied. The role of chromium in the steel as a nitrogen trap to form chromium nitrides was considered less important than that of carbon. All these results were obtained by CEMS analysis.

Frattini et al. [67] showed by CEMS that nitrogen implantation on a high-carbon low-chromium steel induced surface modifications similar to those found in carbon-steels. The modifications consisted principally in the formation of ε-iron carbonitride, and were independent on the structure before implantation. Retained austenite, present in a large amount in quenched and in tempered samples, was destabilized at the highest implantation doses. The nonmagnetic spectral contribution of carbides in tempered and annealed samples seemed to vanish by increasing the H dose in tempered samples, but not in the annealed samples, probably due to the larger particle size.

Medium-carbon and high-carbon low-alloy steels implanted with 40keV-10μA/cm^2 nitrogen-15 ions at various doses have been studied by Marest et al. [47]. NRA profiles of the nitrogen-15 distribution and CEMS spectra obtained on air annealed samples indicated: (a) saturation of retained nitrogen above the $2*10^{17}$ion/cm^2 dose, (b) no thermal broadening of nitrogen distribution profiles, suggesting that the evolution of the nitrogen content into the implanted zone is controlled by the transformation kinetics of the nitrides and not by the elementary nitrogen diffusion process, (c) an increase of the total amount of the implantation-induced iron-nitrides (but with a simultaneous decrease of the γ'-Fe_4N relative contribution) by increasing the carbon content in the matrix. The last point is evidently in contrast with the results of dos Santos et al. [66].

Whereas experimental evidence indicates that heating after implantation leads to progressive dissolution of the surface nitrides and/or carbonitrides, the effect of temperature during implantation is not yet well understood, although several studies have been carried out on the subject. As mentioned above, Goode and Baumvol [19] noticed some influence of implantation temperature on the nitrogen retention in steels. Ramous et al. [44] analysed a high-carbon steel which was nitrogen implanted at a fixed dose in the beam current-density range of 50 to 200μA/cm^2. NRA measurements showed that the retained dose decreased with the beam current-density and was systematically lower in quenched steel than in annealed steel (Figure 4). Analysis of CEMS spectra of samples implanted at 50 and 200μA/cm^2 indicated that the amount of induced surface nitrides was considerably lower in quenched than in annealed samples (Figures 5 and 6). These findings have

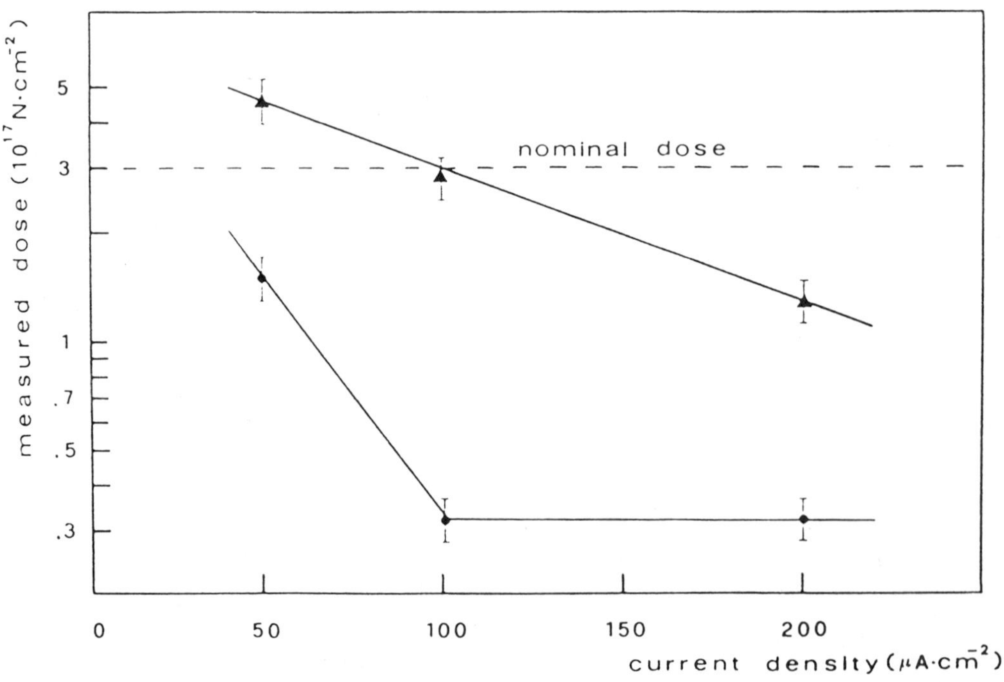

Figure 4. Retained dose measured by NRA versus current density for annealed (triangles) and quenched (circles) nitrogen implanted high-carbon steel [44].

been interpreted as due to the easier migration of interstitials in a quenched structure which is more defective than an annealed structure. Moreover, the increase in sample temperature during implantation at high current-density, as indicated by microhardness measurements, produced an undesirable tempering effect on the quenched structures.

NRA and CEMS complementary information on nitrogen implanted steels with varius carbon and chromium content have recently been reported by Barnavon et al. [49]. Data on samples heated during implantation and annealed after implantation in vacuum or in air confirmed that temperature is one of the most important parameters which determines the chemical state and distribution of implanted nitrogen. The temperature influence was different during and after implantation and for annealing performed in air or vacuum. The thinning of the nitrogen distribution, observed as a consequence of heating, was associated with the carbon contamination at the sample surface. Nitrogen diffusing towards the surface would combine with this carbon to form carbonitrides more stable than the thermally dissolved nitrides.

The nonstandardized conditions of implantation and analysis make it rather difficult to obtain from the mentioned works, whose results are often

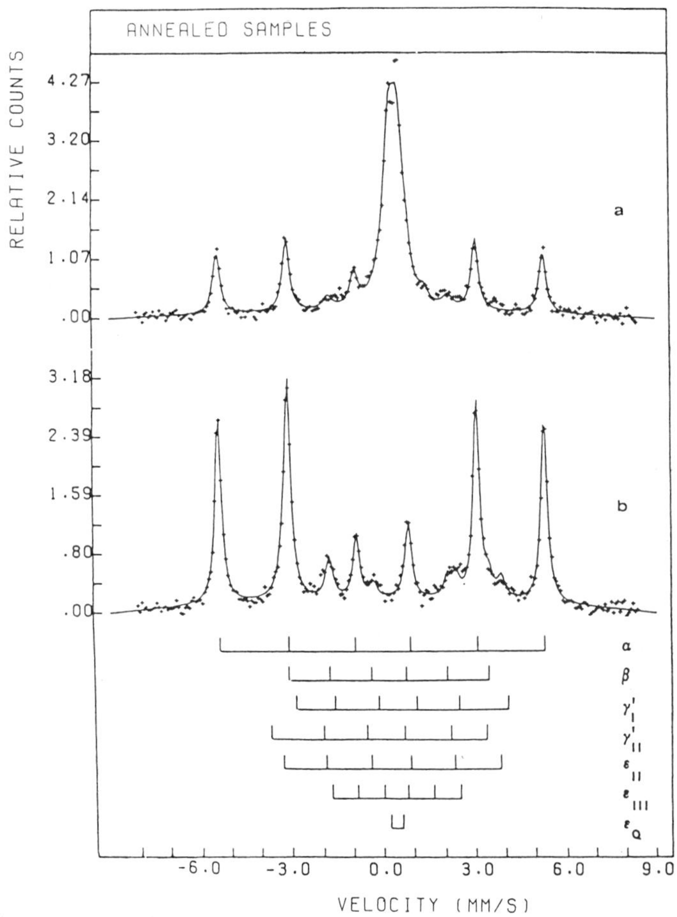

Figure 5. CEMS spectra of annealed samples of Figure 4 implanted at current densities of 50μA/cm² (a) and 200μA/cm², (b). The recognized components are α-ferrite; β-cementite; γ'_{I}-Fe_4N; γ'_{II}-Fe_4N; ε_{II}-$Fe_{2+x}(N,C)$; ε_{III}-$Fe_{2+x}(N,C)$; and ε_{Q}-$Fe_{2+x}(N,C)$ [44].

in contrast, a definite picture of the state of the art. We may notice, in particular, that CEMS data considered alone may be interpreted too subjectively. The wide spread of evaluated hyperfine parameters in the samples reported in Table IIB is meaningful. It is enough to consider that the values pertaining to ε-iron nitride and carbonitride phases cover the ranges reported by Firrao et al [68] and reproduced in Figures 7, 8 and 9. The inhomogeneity in depth of implanted nitrogen concentration and of the

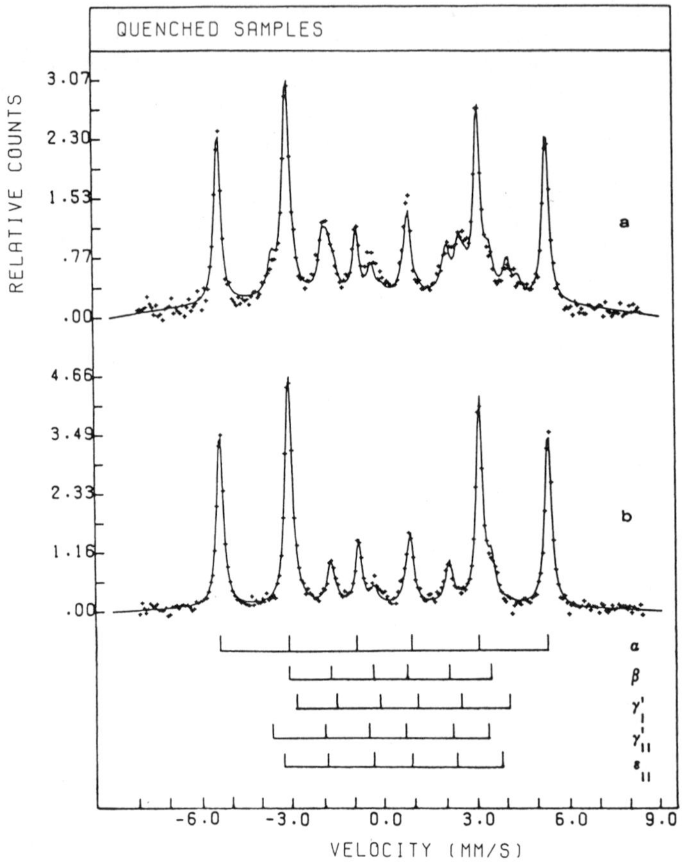

Figure 6. CEMS spectra of quenched samples of Figure 4 implanted at current densities of 50μA/cm² (a) and 200μA/cm² (b). The components are as in Figure 5 [44].

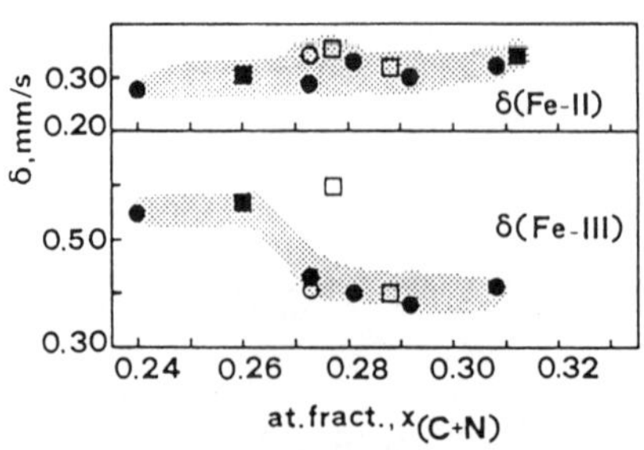

Figure 7. Isomer shift of the Mössbauer spectral components for the iron-epsilon phases of Figure 9 versus interstitial atom content [68].

structure of the implanted layer is an additional factor which complicates the spectral profiles and makes their interpretation more difficult. Hence many questions remain unanswered.

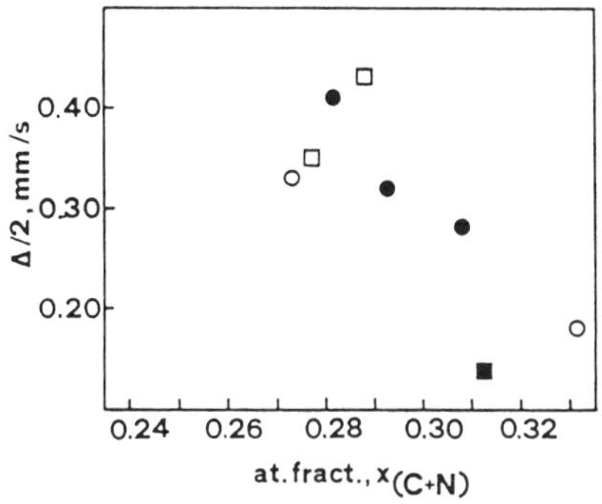

Figure 8. Quadrupole splitting of the Mössbauer spectral components for the iron-epsilon phases of Figure 9 versus interstitial atom content [68].

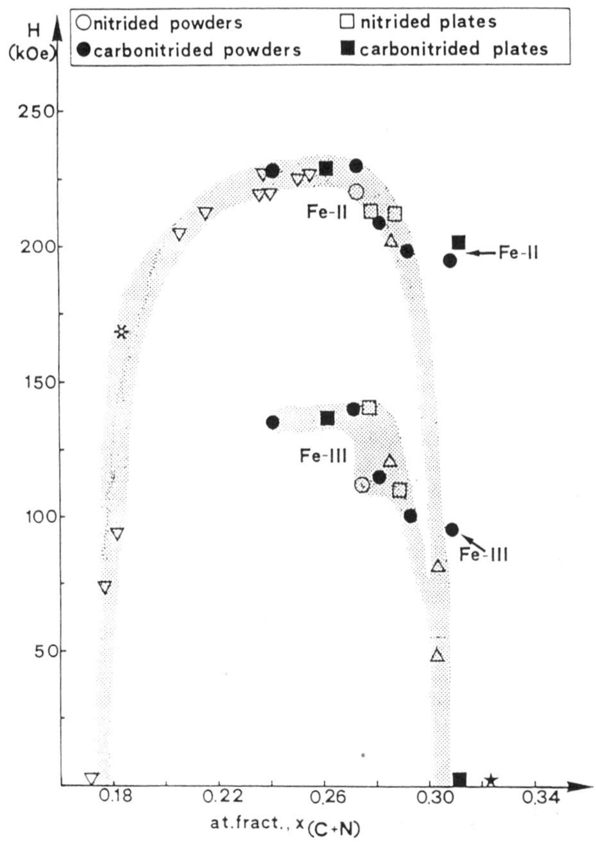

Figure 9. The room-temperature hyperfine magnetic field of the Mössbauer spectral components for the iron-epsilon phase versus interstitial atom content [68]. Fe-II and Fe-III are the ε_{II} and ε_{III} components indicated in Figures 5 and 6.

Table III. Nominal Weight Percent Composition of the Treated Steels

Steel	C	Mn	Si	Cr	S
100Cr6	1.00	0.50	<0.40	1.50	
X30Cr13	0.32	<1.00	<1.00	13.00	
X10Cr17	<0.12	<1.00	<1.00	17.00	0.25

5. SOME RECENT RESULTS

A systematic study on nitrogen-implanted alloy-steels has been carried out with the aim of correlating the chemical state of the modified surface with the steel composition and structure and with experimental implantation parameters. A previous report has been presented by Principi et al. [70].

5.1. Materials and Experimental Conditions

As received (annealed) and after austenitizing, oil quenched, samples of UNI100Cr6 and X30Cr13 steels and as received (annealed) samples of X10Cr17 steel whose compositions are given in Table III, were implanted, after mechanical polishing of the surfaces, by means of the 120kV IRST (Trento) ion implanter at the fixed accelerating voltage of 100kV and with different current-densities in the range of 20-140μA/cm^2. The rectangular shaped nitrogen beam was held fixed and the target holder containing a number of samples was moved back and forth under the beam. Due to the absence of mass analysis, a mixture of N^+ and N_2^+ was present in the beam and affected the implantation profile. The profile and the minor amount of impurities are not expected to affect appreciably the correlation between nitride formation and the nitrogen retained dose in the subsuperficial region. The N-14(d,p_5)N-15 nuclear reaction induced by a 610keV deuteron beam was used to measure the amount of nitrogen retained in the subsuperficial region. CEMS spectra were obtained with a flowing gas proportional counter on a standard spectrometer.

5.2. Experimental Data

Figures 10, 11, and 12 show the CEMS spectra of the three annealed steels before and after implantation at different current densities. The spectral profiles of untreated samples are rather different. Besides the weak contribution to the central zone due to carbides like M_3C, M_7C_3, and $M_{23}C_6$ [71], the contribution of magnetic components becomes more complex with increasing the chromium content of the steel. It is known (see, for example, Janot [72]) that chromium, as well as aluminum, silicon, manganese, vanadium, etc., when dissolved as an impurity in a ferritic matrix produces a depression of the internal field which increases by increasing concentration.

The samples implanted at 50μA/cm^2 current-density display a drastically modified CEMS profile. It is particularly evident the paramagnetic contribution of ε-iron-carbonitride, with the highest interstitial content, which overlaps the doublet with lower isomer shift and greater quadrupole shift due to the superparamagnetic oxide, Fe_2O_3. The more considerable effects of implantation at the highest current density (140μA/cm^2) on the CEMS profiles are the very lowspectral contribution of iron nitrides and the simplification of the magnetic contribution.

Despite of their complexity, the CEMS spectra have been interpreted on the basis of the nitrides and/or carbonitrides characterization given in Table II, by assuming the more reasonable fitting models. A good guide are the values of the mean nitrogen concentration supplied by NRA. The combined CEMS and NRA results are given in Figure 13, where the vertical scale measures the ratio, ρ, of the sum of spectral areas of the iron-nitrogen components and the sum of the spectral areas of the matrix components, a variety of sextets corresponding to the different iron-chromium coordination. The evaluation of ρ and its uncertainty (the upper limit is shown in Figure 13) was accomplished by means of several computer runs of a standard minimization program.

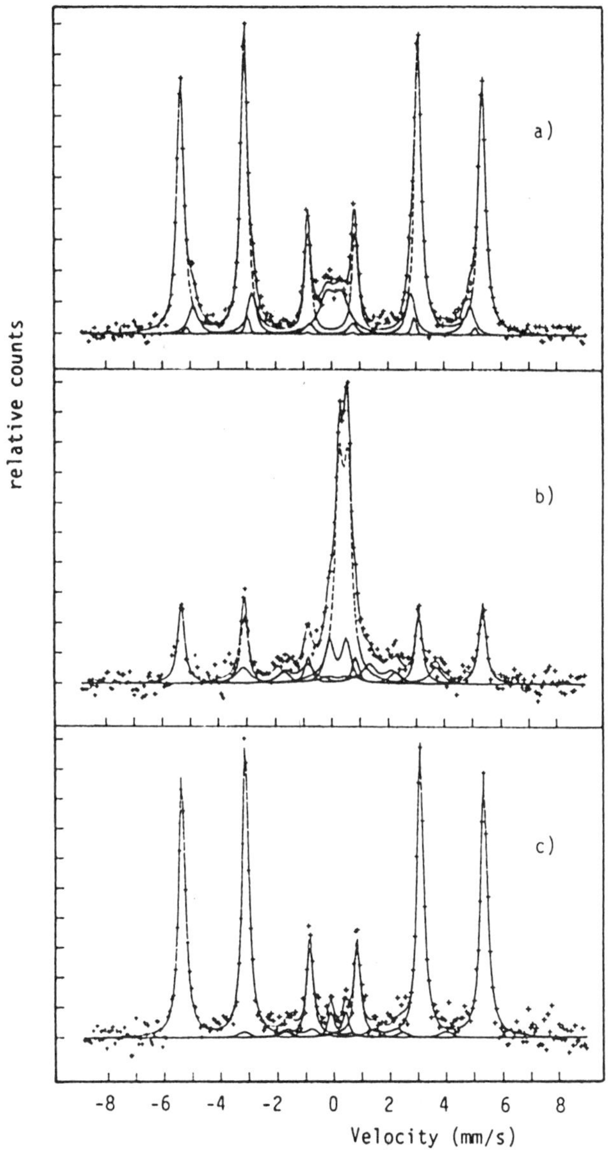

Figure 10. CEMS spectra of 100Cr6 samples unimplanted (a) and nitrogen implanted at 50μA/cm^2 (b) and 140μA/cm^2 (c) current density.

5.3. Interpretation

Figure 13 displays the unexpected but regular behavior of ρ versus the retained dose. ρ increases linearly with the nitrogen dose for each steel composition and structure, whatever the current density. The slope of the straight lines approximating each set of data is practically the same. The intercept of each of these lines, which represents the threshold concentration of nitrogen retained in the near surface region to form iron-carbonitrides, increases with chromium content in the matrix and, at a given matrix composition, with the defect density (greater for quenched than for annealed structures). The dashed curves A and B, which cross the points corresponding to the annealed samples of the three types of steel implanted simultaneously (in two batches) at a beam current-density of 50μA/cm^2 and of 140μA/cm^2, respectively, suggest some effect due to the highest current-

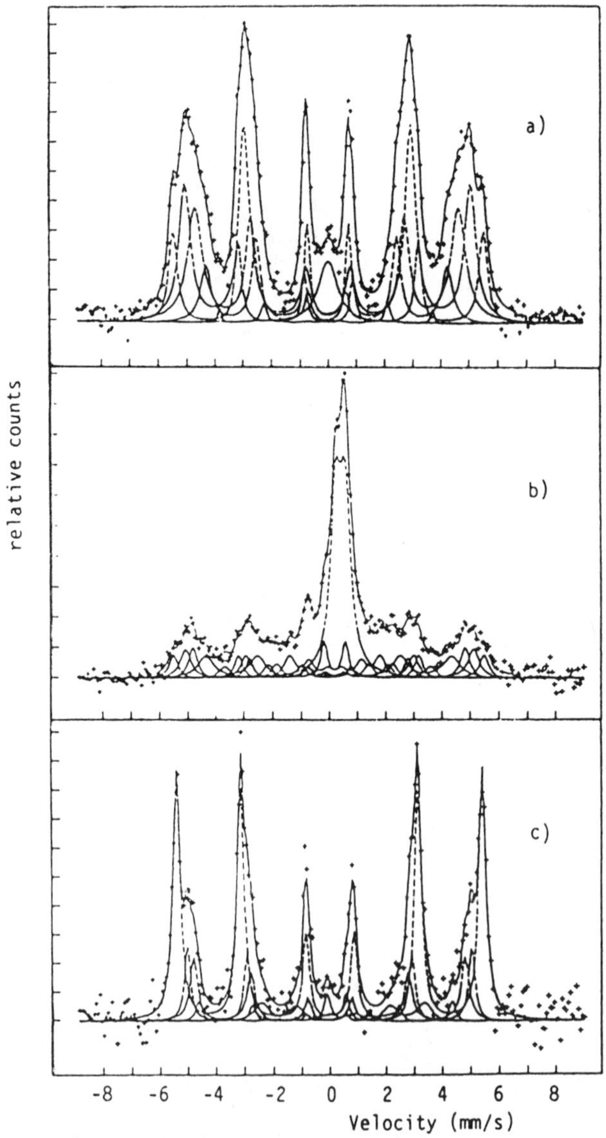

Figure 11. CEMS spectra of X30Cr13 samples unimplanted (a) and nitrogen implanted at 50μA/cm^2 (b) and 140μA/cm^2 and (c) current density.

density, such as diffusion of nitrogen into the bulk, as is supposed to occur for high-carbon steels [44]. Reconsideration of previous data on implanted high-carbon steel yields the full (annealed structure) and the open (quenched structure) small circles reported in the figure. It seems that the relationship between ρ and dose for high-carbon steels is insensitive to the structure, but more data are necessary to confirm this point.

The data points relevant to the annealed samples belonging to the curves A and B of Figure 13 are reported in Figure 14, and to depict the influence of the chromium content in the matrix on the amount of iron-nitrogen compounds (a) and on the near-surface nitrogen retained dose (b) for the two indicated values of the nitrogen beam current density. The nitrogen retained

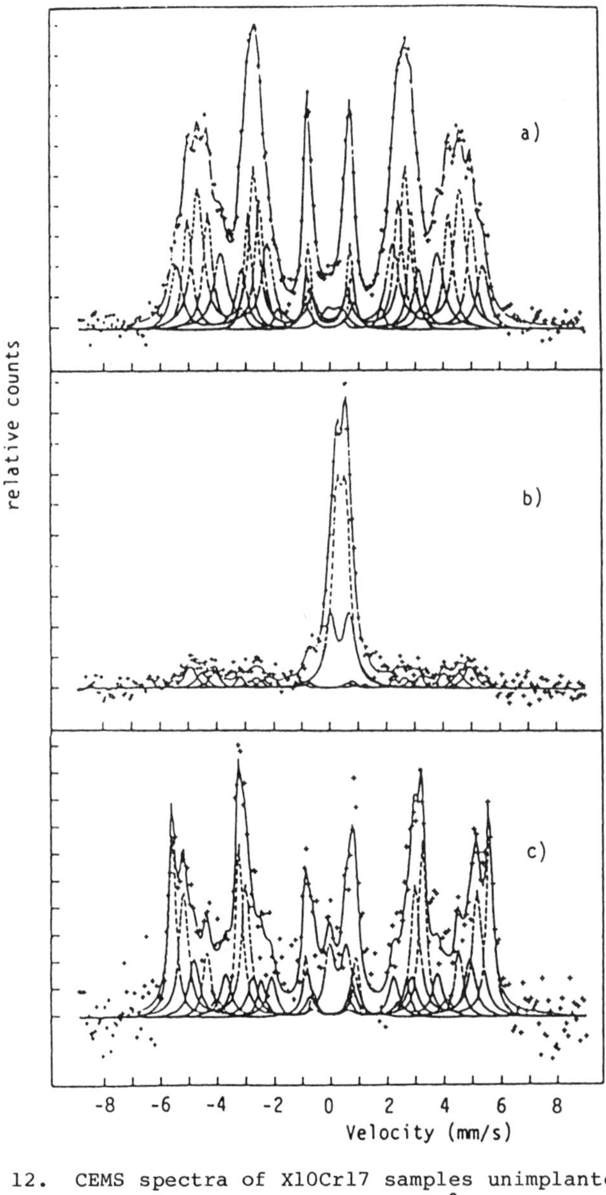

Figure 12. CEMS spectra of X10Cr17 samples unimplanted (a) and nitrogen implanted at 50μA/cm^2 (b) and 140μA/cm^2 (c) current density.

dose increases and ρ decreases as the chromium concentration increases. At 140μA/cm², both retained dose and ρ are significantly lower than the values at 50μA/cm², but with opposite trends with respect to chromium concentration. This last effect goes beyond the possible different molecular content in the unanalyzed nitrogen beam. This is also confirmed by the dose measured on

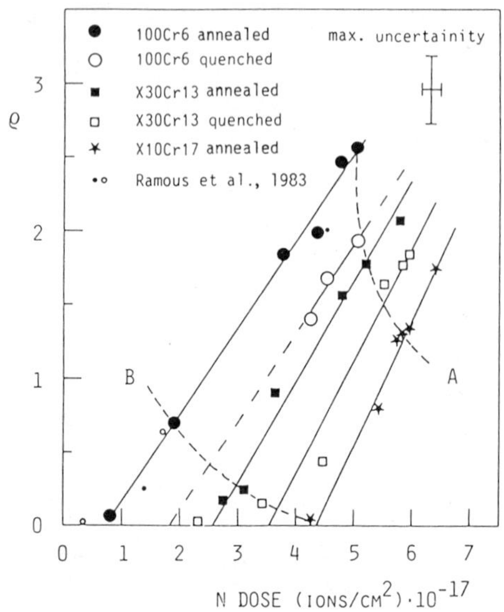

Figure 13. Behavior of ρ (see text) versus nitrogen retained dose in the near-surface region of nitrogen-implanted chromium-steels. Curves A and B cross data points corresponding to implants of annealed samples at 50 and at 140μA/cm² current density, respectively. The small open and full circles have been deduced from ref. 44.

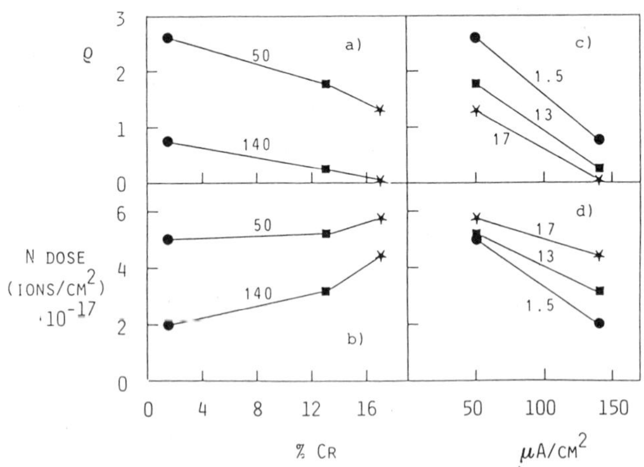

Figure 14. Behavior of ρ (a) and of nitrogen retained dose (b) versus chromium content in two sets of steel samples nitrogen implanted in the same batch at a beam current density of 50 and of 140μA/cm², respectively. The same data are reported versus current density in (c) and versus weight percent chromium content in (d).

18Cr-10Ni austenitic steel samples, included in the same batches which was about 5.8ions/cm^2 for both current-densities.

It is then possible to conclude the following. At a given beam current-density and dose, the nitrogen retained in the near-surface region increases with chromium content in the steel. The influence of composition on nitrogen retention was earlier noticed by Hartley [14]. In the already mentioned work by Goode and Baumvol [19] (see above) chromium is assumed to act as a nitrogen trap. This is confirmed also by AES and TEM observations on the presence of chromium nitrides in the surface layers of nitrogen-implanted high-chromium steels [20,26,50,52]. Our experimental evidence confirms the above assumptions. When the amount of implanted nitrogen exceeds the solubility limit of the matrix, the formation of chromium-nitrides appears to precede that of iron-carbonitrides. The latter are more favored by reducing the current-density and, obviously, the chromium content (Figure 14). The composition of the matrix coexisting with nitrogen in the implanted layer must necessarily undergo important variations, like a decrease of chromium concentration, the chromium-nitrogen compounds precipitation being more favored than that of the iron-nitrogen compounds, as expected by thermodynamic considerations [9]. The threshold value of nitrogen retained dose necessary for the formation of iron-carbonitrides appears to increase with the chromium content in the matrix and with its defect concentration (Figure 14).

5.4. Iron-Chromium Coordination in the Implanted Layer

In terms of the matrix modifications in the implanted layer, another approach to the spectral analysis is possible on the basis of a model of the iron-chromium coordination proposed by Dubiel and coworkers [73-79]. They studied the iron-chromium coordination of alloys with chromium content up to 45 atomic percent and, following the work of Schwartz and Chandra [80], considered a random distribution of chromium in the first two coordination spheres. Assuming an additivity effect of first, m, and second, n, neighbors on the hyperfine field at iron, the resulting spectrum is then a superposition of 63 sextets with hyperfine fields,

$$H(m,n,c) = 330 - m\Delta H_1 - n\Delta H_2 + f(c),$$

where c is the chromium concentration, ΔH_1 and ΔH_2 are the contributions of each of the first and second neighbors, and f(c) is a suitable concentration factor. The probability of each component is given by the binomial distribution,

$$P(m,n,c) = \binom{8}{m}\binom{6}{n}c^{m+n}(1+c)^{14-m-n}.$$

The results of Dubiel on the application of this model, where an additivity effect of first and second neighbors on the isomer shift is taken into account, may be summarized by several points. The model corresponds satisfactorily to the experimental data for the overall considered range of chromium concentration. The values of H(0,0), H(0,1) and H(1,0) increase with increasing chromium content, from 330, 307, and 298kG to about 340, 318, and 308kG, respectively, at about 10-12 atomic percent chromium and then progressively decrease. The mean value of H, in the interval 1-17 atomic percent chromium decreases almost linearly from 330 to 270kG. H(0,0) increases linearly with IS(0,0). The mean values of H and isomer shift follow a decreasing linear relationship. These changes in H and the isomer shift are interpreted in terms of spin and charge transfer between chromium and iron atoms. Annealing for different times and up to 1100°C does not change appreciably the hyperfine parameters of the spectra for chromium concentrations of up to 15 atomic percent.

Table IV. A Comparison Between M, the Iron-Chromium Coordination Probability from the Dubiel Model and E, the Mössbauer Spectral Areas for a Nitrogen Implanted 13 Weight Percent Chromium Ferritic Steel

Sample	Weight Percent Chromium	Coordination				
		0,0	0,1 1,0	0,2 1,1 2,0	0,3 1,2 2,1 3,0	0,4 1,3 2,2 3,1 4,0
Virgin	M 13	12	28	29	19	12
	E ?	14	46	19	18	3
Implanted at $20\mu A/cm^2$	M 10	21	37	29	13	
	E ~10	21	43	22	14	
Implanted at $100\mu A/cm^2$	M 3.5	59.5	32.5	8		
	E ~4	61	33	6		

In order to learn more about the ferritic structure coexistent with the implanted species, the Dubiel model has been applied to the spectra of the nitrogen implanted X30Cr13 steel (13 weight percent chromium), with annealing structure previous to implantation. The hyperfine field distribution, obtained by a modified Hesse-Rubartsch method [81] has been compared with the occupancy probability of the different coordination sites deduced from the model, as shown in Figure 15, for virgin, $20\mu A/cm^2$, and $100\mu A/cm^2$ implanted samples. Only the region of high hyperfine field is considered. Because the observed hyperfine field distributions are very different, rough comparison is possible with occupancy probability only for very different values of the chromium concentration. In Table IV, the percent probabilities from the model (M), for each set k=m+n of iron-chromium coordination sites, disregarding the values for $k \geq 5$, are compared with the corresponding subspectral areas as evaluated by the standard fitting procedure (E). The chromium concentrations utilized in the Dubiel model and the deduced actual values are indicated in the second column.

In the case of the virgin sample, the sum of the theoretical occupancy probabilities for k=2 and 4 are greater than the corresponding sum of experimental areas. For k=1 they are lower. Moreover, we have verified that the value of H(0,0) is slightly greater than that deduced from the model. This could mean that the chromium distribution in the steel is not completely random as assumed in the model and, in addition, that the effective chromium content in the ferritic structure can be less than the nominal 13 weight percent. We are dealing, indeed, with a system alloying both carbon and chromium, which is characterized in the annealed state by the presence of precipitates of the type $M_{23}C_6$, where M is either iron or chromium. If all the available carbon atoms participate in the precipitation of the $Cr_{23}C_6$ carbide, the chromium content in the ferritic structure would be reduced by about 5 weight percent. Obviously, this is not the actual case, as is indicated by the weak paramagnetic component recognizable in the spectrum and attributable to the mixed carbide, $(Fe,Cr)_{23}C_6$. Therefore, part of the chromium is associated with carbon and the chromium content in the ferritic structure is less than the nominal 13 weight percent. The exact value is not easily obtained due to the discrepancies between the model and the experimental evidence.

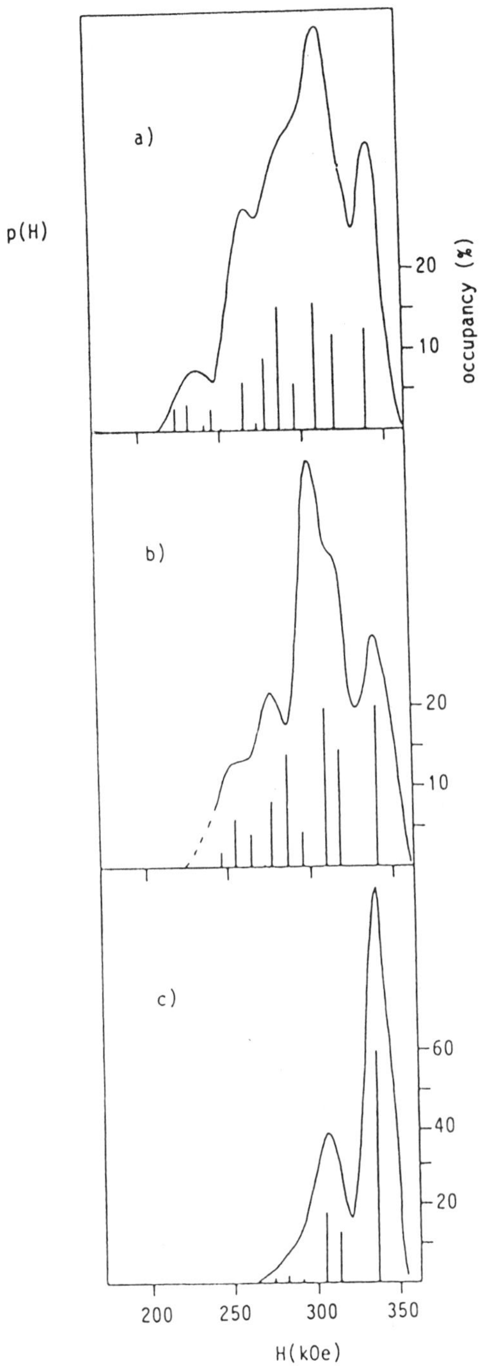

Figure 15. Comparison between iron-chromium coordination site occupancy (vertical segments) based on the Dubiel model and the HFD computed from the Mössbauer spectra of X30Cr13 samples, virgin (a), and nitrogen implanted at $20\mu A/cm^2$ (b) and at $100\mu A/cm^2$ (c) current densities. The shown iron-chromium site occupancies correspond to a weight percent chromium content of 13 in (a) of 10 in (b) and of 3.5 in (c).

As noticed above, the magnetic contribution due to the ferrite phase, which coexists with the implantation induced compounds, is simplified with respect to the unimplanted sample. Agreement of the hyperfine field distributions in Figures 15b, and c, with the predictions of the Dubiel model is excellent for chromium concentrations, considerably lower than the nominal value (see Table IV). The only apparent disagreement is the peak in the distribution located at about 300kOe in Figure 15b, which is attributable, as confirmed by standard fitting analysis, to the α''-$Fe_{16}N_2$ nitride, and is not associated with any iron-chromium coordination.

The implantation process has, therefore, on the considered steel, not only the effect of producing new surface compounds, but also of modifying the coexisting matrix by lowering its chromium content. This occurs more consistently for the implantation at a very high current-density. This analysis confirms the above interpretation of Figures 13 and 14, according to which the high sample temperature, during implantation at the highest current-densities, leads more easily to the formation of chromium nitrides at the expense of iron nitrides, besides a noticeably lower nitrogen retention.

6. CONCLUSIONS

To draw firm conclusions from the sometimes contradictory literature on nitrogen-implanted steels is rather difficult. It seems that nitrogen implants at high doses are generally more effective for ductile than for hard steels to improve the surface technological properties, . This is mainly attributable to the large number of alloying elements like chromium present, for example, in stainless steels, which would favor the formation of special phases. It may be significant that, in high-chromium steels, the dose threshold leading to dramatic increase in wear resistance roughly corresponds to the dose threshold for the formation of iron-containing nitrides.

The data on carbon and low-alloy steels are still controversial and require additional work. Very interesting is the possibility of further improving, by nitrogen implantation, the resistant surfaces previously treated by conventional methods.

The optimization of implantation parameters is yet to be established for each class of steels. Energy, dose, current density, and temperature strongly influence the effect of treatment.

With regard to the surface analysis techniques, we underline the convenience of using complementary data to enrich our knowledge of the surface composition and structural changes produced on steels by nitrogen implantation and subsequent tribological tests. The combination of CEMS, NRA, TEM, AES, and XPS is surely a very valuable experimental basis.

The particular ability of Mössbauer spectroscopy in providing information on the iron-chromium coordination in the implanted matrices has been exploited herein.

ACKNOWLEDGMENTS

Thanks are due to G. Le Caer and J. M. Dubois for the use of thier program to compute the hyperfine field distributions. This work has been financially supported by Consiglio Naziomale delle Ricerche, Progetto Finalizzato Metallurgia, Sottoprogetto B.

REFERENCES

1. C. A. dos Santos and I. T. R. Baumvol, in "Ion Implantation: Equipment and Techniques, eds., H. Ryssel and H. Gloswishing, Springer, Berlin, pp. 347-356 (1983).

2. G. Dearnaley, in "Ion Implantation: Equipment and Techniques", eds. H. Ryssel and H. Glaswishing, Springer, Berlin, pp. 332-346 (1983).

3. G. Dearnaley, in "Surface Engineering", eds. R. Kossowsky and S. C. Singal, NATO ASI Series, Series E: Applied Sciences n. 85, pp. 125-147 (1984).

4. J. K. Hirvonen, Mat. Res. Soc. Symp. Proc. 27, 621 (1984).

5. J. K. Hirvonen and C. R. Clayton, in "Surface Modification and Alloying", eds. J. M. Poate, G. Foti, and D. C. Jacobson, Plenum Press, NY, pp. 323-384 (1984).

6. I. J. R. Baumvol, in "Ion Implantation Science and Technology", ed. J. F. Zugler, Academic Press, NY, pp. 261-320 (1984).

7. I. L. Singer, Mat. Res. Soc. Symp. Proc. 27, 585 (1984).

8. I. L. Singer, Appl. Surf. Sci. 18, 28 (1984).

9. I. L. Singer, Vacuum 34, 853 (1984).

10. U. K. Chaturvedi and S. Lo Russo, Proc. Int. Conf. SM2IB, Heidelberg, Sept. 1984, Mater. Sci. Eng. 69, 233 (1985).

11. U. K. Chaturvedi and S. Lo Russo, unpublished work (1985).

12. N. E. W. Hartley, Tribology 8, 65 (1975).

13. N. E. W. Hartley, Wear 34, 427 (1975).

14. N. E. W. Hartley, in "Applications of Ion Beams to Materials", Conf. Series N. 28, Institute of of Physics, London, pp. 210-223 (1976).

15. J. K. Hirvonen, J. Vac. Sci. Technol. 15 1662 (1978).

16. S. Lo Russo, P. Mazzoldi, I. Scotoni, C. Tosello, and S. Tosto, Appl. Phys. Lett. 34, 627 (1979).

17. S. Lo Russo, P. Mazzoldi, I. Scotoni, B. Tiveron, C. Tosello, G. Wolf, and G. L. Zhang, Atti VII Congr. Naz. Sci. Tecnol. Vusto, Bressanone,Italy, A.I.V. and CLEUP, Perdak, pp. 273-278 (1981). Results also reported in reference 5.

18. T. Varjoranta, J. K. Hirvonen, and A. Anttila, Thin Solid Films 75, 241 (1981).

19. P. D. Goode and I. J. R. Baumvol, Nucl. Instr. Meth. 189, 161 (1981).

20. I. L. Singer and J. S. Murday, J. Vac. Sci. Technol. 17, 327 (1980).

21. Y. Kun, H. Li, X. Zhang, and J. Tian, Nucl. Instr. Meth. 209, 1063 (1983).

22. A. H. Deutchman and R. J. Partyka, in "Surface Engineering", eds. H. Ryssel and H.Glaswishing, Springer, Berline, pp. 332-346 (1984).

23. I. L. Singer and R. N. Bolster, in "Ion Implantation Metallurgy", eds. C. M. Preece and J. K. Hirvonen, Metallurgical Soc. of AIME, Warrendale, PA, p. 116 (1980).

24. R. N. Bolster and I. L. Singer, Appl. Phys. Lett. 36, 208 (1980).

25. R. N. Bolster and I. L. Singer, ASLE Trans. 24, 526 (1981).

26. H. Dimigen, K. Kobs, R. Leutenecker, H. Ryssel, and P. Eichinger, Proc. Int. Conf. SM2IB, Heidelberg, Sept. 1984, Mater. Sci. Eng. 69, 181 (1985).

27. F. G. Yost, S. T. Picraux, D. M. Follstaedt, L. E. Pope, and J. A. Knapp, Thin Solid Films 107, 285 (1983).

28. H. G. Feller, R. Klinger, and W. Benecke, Proc. Int. Conf. SM2IB, Heidelberg, Sept. 1984, Mater. Sci. Eng. 69, 173 (1985).

29. S. Fayeulle, D. Treheux, P. Guiraldenq, T. Barnavon, J. Tousset, and M. Robelet, Scr. Metall. 17, 459 (1983).

30. F. G. Doyle, F. G. Yost, S. T. Picraux, D. M.Follstaedt, L. E. Pope, and J. A. Knapp, Nucl. Instr. Meth. in Phys. Res. B7/8, 166 (1985).

31. G. Merest, M. Moncoffre, and S. Fayeulle, Appl. Surf. Sci. 20, 205 (1985).

32. N. E. W. Hartley, Thin Solid Films 64, 427 (1975).

33. W. W. Hu, C. R. Clayton, H. Herman, and J. K. Hirvonen, Scr. Metall. 12, 697 (1978).

34. W. W. Hu, H. Herman, C. R. Clayton, J. Kozubowski, R. A. Kant, J. K. Hirvonen, and R. K. Maccrone, in "Ion Implantation Metallurgy", eds. C. M. Preece and J. K. Hirvonen, Metallurical Soc. of AIME, Warrendale, PA, pp. 92-102 (1980).

35. S. Lo Russo, P. Mazzoldi, I. Scotoni, C. Tosello, and S. Tosto, Appl. Phys. Lett. 36, 822 (1980).

36. G. Longworth and R. Atkinson, in Mössbauer Spectroscopy and Its Chemical Applications", eds. J. G. Stevens and G. K. Shenoy, Adv. in Chemistry Series 194, Amer. Chem. Soc., Washington, DC, pp. 101-105 (1981).

37. J. J. Spijkerman, in "Mössbauer Effect Methodology", vol. 7, ed. I. J. Gruverman, Plenum Press, NY, pp. 85-96 (1971).

38. J. A. Sawicki, these proceedings.

39. M. Carbucicchio, L. Bardani, and S. Tosto, J. Appl. Phys. 52, 4589 (1981).

40. K. Kaneyoshi, K. Debusmann, W. Keune, and R. A. Brand, these proceedings.

41. M. J. Tricker, in "Mössbauer Spectroscopy and Its Chemical Applications", eds. J. G.Stevens and G. K. Shenony, Adv. in Chemistry Series 194, Amer. Chem. Soc., Washington, DC, pp. 63-100 (1981).

42. G. Amsel, J. P. Nadai, E. D'Artemare, D. David, E. Girard, and J. Moulin, Nucl. Instr. Meth. 92, 481 (1971).

43. G. Amsel and W. A. Lanford, Ann. Rev. Nucl. Part. Sci. 34, 435 (1984).

44. E. Ramous, G. Principi, L. Giordano, S. Lo Russo, and C. Tosello, Thin Solid Films 102, 97 (1983).

45. M. Berti and A. V. Drigo, Nucl. Instr. Meth. 201, 473 (1982).

46. A. Anttila and J. K. Hirvonen, Appl. Phys. Lett. 33 394 (1978).

47. G. Marest, C. Skoutarides, T. Barnavon, J. Tousset, S. Fayeulle, and M. Robelet, Nucl. Instr. Meth. 209/210, 259 (1983).

48. N. Moncoffre, T. Barnavon, J. Tousset, S. Fayeulle, P. Guiraldenq, D. Treheux, and R. Robelet, in "Surface Engineering", eds. R. Kossowsky and S. C. Singal, NATO ASI Series, Series E: Applied Sciences n. 85, pp. 148-157 (1984).

49. T. Barnavon, H. Jaffrezic, G. Marest, N. Moncoffre, J. Tousset, and S. Fayeulle, Proc. Int. Conf. SM21B, Heidelberg, Sept. 1984, Mater. Sci. Eng. 69, 531 (1985).

50. M. Baron, A. L. Chang, J. Schreurs, and R. Kossowsky, Nucl. Instr. Meth. 182/183, 531 (1981).

51. V. M. Drako and G. A. Gumanskij, Rad. Eff. 66, 101 (1982).

52. S. Fayeulle, D. Treheux, P. Guiraldenq, T. Barnavon, J. Tousset, and M. Robelet, Scr. Metall. 17, 459 (1983).

53. F. Z. Cui, H. D. Li, and X. Z. Zhang, Nucl. Instr. Meth. 209/210, 881 (1983).

54. J. L. Whitton, G. T.Ewan, M. M. Ferguson, T. Laursen, I. V. Mitchell, H. H. Plattner, M. L. Swanson, A. V. Drigo, G. Celotti, M. Servidori, and W. A. Grant, Mater. Sci. Eng., in press (1985).

55. G. Longworth and N. E. W. Hartley, Thin Solid Silms 48, 95 (1978).

56. M. Chabanel, C. Janot, and P. Motte, C. R. Acad. Sci. 266, B419 (1968).

57. A. J. Nozik, J. C. Wood, and G. Haacke, Solid State Comm. 7, 1677 (1969).

58. K. H. Eickel and W. Pitsch, Phys. Status Solidi 39, 121 (1970).

59. T. Moriya, Y. Sumitomo, H. Ino, F. E. Fujita, and Y. Maeda, J. Phys. Soc. Jpn. 35, 1378 (1973).

60. J. Bainbridge, D. A. Channing, W. H. Whitlow, and R. E. Pendlebury, J. Chem. Phys. Solids 34, 1579 (1973).

61. C. A. dos Santos, I. J. R. Baumvol, E. A. Garcia, and M. Behar, J. Phys. D17, 969 (1984).

62. G. Principi, P. Matteazzi, E. Ramous, and G. Longworth, J. Mater. Sci. 15, 2665 (1980).

63. D. H. Jack and K. H. Jack, Mater. Sci. Eng. 11, 1 (1973).

64. C. A. dos Santos, B. A. S. de Barros, J. P. de Souza, and I. J. R. Baumvol, Appl. Phys. Lett. 41, 237 (1982).

65. R. Frattini, G. Principi, S. Lo Russo, B. Tiveron, and C. Tosello, J. Mater. Sci. 17, 1683 (1982).

66. C. A. dos Santos, M. Behar, J. P. de Souza, and I. J. R. Baumvol, Nucl. Instr. Meth. 209/210, 907 (1983).

67. R. Frattini, L. Giordano, S. Lo Russo, G. Principi, and C. Tosello, Proc. ICAME, Jaipur, India, Indian Nat. Sci. Ac. New Delhi, pp. 378-382 (1982).

68. D. Firrao, M. Rosso, G. Principi, and R. Frattini, J. Mater. Sci. 17, 1773 (1982).

69. C. A. dos Santos, M. Behar, and I. J. R. Baumvol, J. Phys. D17, 551 (1984).

70. G. Principi, S. Lo Russo, and C. Tosello, T. Phys. F: Metal Physics 15, L207 (1985).

71. G. Principi, R. Frattini, and M. Magrini, Gazz Chim. Ital. 113, 281 (1983).

72. C. Janot, Pure and Appl. Chem. 48, 53 (1976).

73. S. M. Dubiel and K. Krop, J. Phys. Coll. C6, 459 (1974).

74. S. M. Dubiel and J. Zukrowski, Acta Phys. Pol. A48, 315 (1975).

75. S. M. Dubiel, Acta Phys. Pol. A49, 619 (1976).

76. S. M. Dubiel, T. Wrzeniewska, and S. Gorczyca, J. Phys. F9, 494 (1979).

77. S. M. Dubiel and J. Zukrowski, Acta Phys. Pol. A58,37 (1980).

78. S. M. Dubiel and J. Zukrowski, J. Magn. Magn. Mater. 23, 214 (1981).

79. S. M. Dubiel, Proc. 5th Riso Int. Symp. on Metallurgy and Material Science, Roskilde, DK, Sept. 84, ed. Riso Nat. Lab., Roskilde, DK, pp. 231-237 (1984).

80. L. H. Schwartz and D. Chandra, Phys. Status Sol. (b) 45, 201 (1971).

81. G. Le Caer and J. M. Dubois, J. Phys. E12, 1083 (1979).

STUDIES OF THE PHOSPHATIZATION OF STEEL AND ITS CORROSION PRODUCTS

W. Meisel

University of Mainz
Institute of Inorganic and
Analytical Chemistry
D-6500 Mainz
Federal Republic of Germany

INTRODUCTION

The reaction of phosphoric acid with steel is a basic process in many technological procedures. A typical example is the pretreatment of steel surfaces for the application of coatings. The aim of the treatment is to clean the surface, to produce a homogeneous, tight, and smooth cover and to provide a reliable adhesion between the steel substrate and the coating. This example points out that the phase composition of the layer formed is of great interest, and that reactions with residues on the surface, such as oxides and corrosion products, have to be considered, too. The same holds for phosphate layers formed in order to modify the mechanical or corrosion behavior of steel surfaces. Because of the electrolytic character of these processes, in addition to the usual reaction parameters, such as temperature, concentration, and time, the electrochemical potential of the sample has to be controlled during the reaction. In these examples, diluted phosphoric acid, phosphate solutions, and mixed solutions are used. Solutions of higher concentration are used as "rust transformers". An example for the use of concentrated phosphoric acid is its application for the removal of rust from vessels, tubes, etc. If the corrosion products to be removed are contaminated, as for instance in nuclear power plants, a minimum quantity of the reagent needed for the treatment is desired.

These few examples show that the study of surface reactions with both diluted and concentrated phosphoric acid is of great practical interest. There are a large number of publications dealing with such studies, most of them in the fields of electrochemistry and optical spectroscopy. Mösssbauer spectroscopy can contribute additional and important information, because it allows a direct phase analysis of the iron containing products.

The majority of relevant publications is based on experiments with pure iron, sometimes with single crystals, or with pure alloys. Such basic experiments are very important to our understanding of the fundamental processes. On the other hand, the surface of a technical steel differs very much in its properties from those of such pure materials. There are the effects of strains and texture, and of the chemical inhomogeneities in the microscopic size range. The latter effects are very important. One has to consider the reactions of the environmental medium with the metal of the grains, with the material of the grain boundaries, with carbide inclusions, etc. Even neighboring grains may differ in structure and impurity concentration. The technical phosphatization is a superposition of all these partial reactions. Therefore, studies by use of technical grade materials are difficult, but they are necessary.

EXPERIMENTAL

Two Mössbauer spectrometers were employed, which were made from commercial components and were calibrated with metallic iron foil. The sources were cobalt-57 on rhodium. For transmission and gamma scattering experiments scintillation detectors were used. Gamma scattering was performed in a symmetrical 90° geometry. For low temperature gamma scattering experiments, a simple styrofoam cryostat with a 20 liter tank for liquid nitrogen was constructed. Conversion electron Mössbauer spectra (CEMS) were recorded with a gas-flow counter, the sample acting as the back electrode.

For the investigation of layers formed in electrolytes at a defined potential, in-situ measurements are of particular interest. Only in this case can the influence of the atmosphere or other gases and of handling in vacuo be excluded. Therefore, for gamma scattering experiments a special cell was constructed, which is shown schematically in Figure 1. The thickness of the electrolyte layer to be penetrated by the gamma radiation is variable and can be reduced to about one millimeter. Pulse height spectra recorded in this arrangement are shown in Figure 2. The 14.4keV radiation appears around channel 60. For this cell and for ex-situ preparations, a $PbSO_4$ reference electrode inside a capillary was used. For the control of the electrode potential a potentiostat was constructed and its most reliable version is shown in Figure 3. The samples were made of commercial steel, 0.5mm thick, the quality corresponding to the standard DIN 1623.05 (maximum weight percentage: 0.2 C, 0.4 Cu, 0.3 Ni, 0.5 Si, 0.3 Cr, 1.6 Mn). The orthophosphoric acid used was 85 percent H_3PO_4, and the electrolyte used for

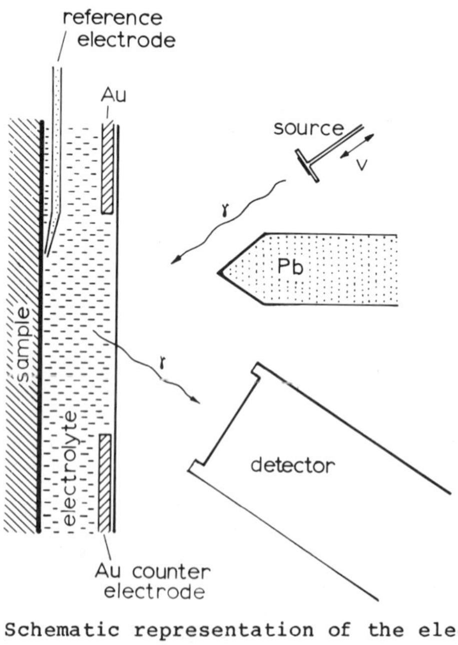

Figure 1. Schematic representation of the electrolyic cell for in-situ gamma scattering Mössbauer spectroscopy.

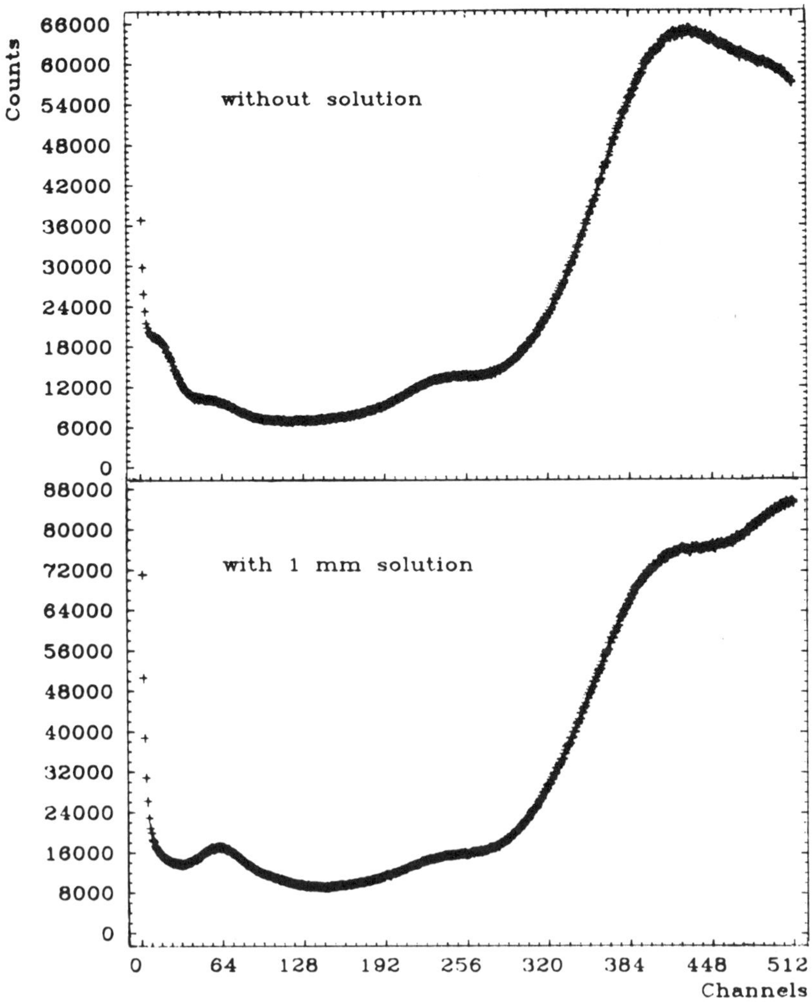

Figure 2. Pulse height spectra of gamma quanta scattered from a steel sample in the electrolyic cell with and without electrolyte (0.1mm NaI(Tl)).

sample preparations and in-situ measurements was a phosphate buffer of pH 3.0 (7.52 pt. 0.15m NaH_2PO_4 + 1 pt. 0.15m H_3PO_4).

Surface analysis under ultrahigh vacuum conditions was performed by a commercial scanning Auger microprobe allowing depth profiling of up to six multiplexed elements under 2keV argon ion sputtering. The sputtered depth was calibrated under the assumption that the sputter yield is the same for the treated samples as for Ta_2O_5 on tantalum.

STUDIES AT LOW CONCENTRATION

In Figure 4, the current density vs. electrode potential curve for a steel sample in an electrolyte as described above is shown. The voltage axis was measured with respect to the $PbSO_4$ reference electrode. Reference to the normal hydrogen electrode (NHE) would shift the voltage scale to the right by 0.276V. In Figure 4 the immunity range, corrosion potential U_{corr} and

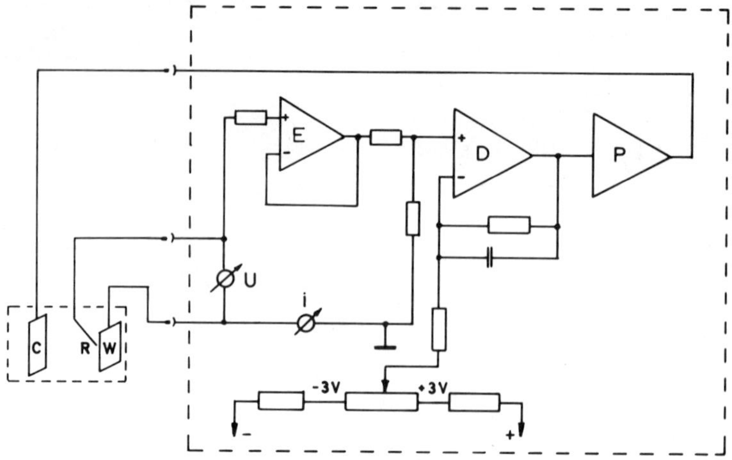

Figure 3. Mode of operation of the potentiostat controlling the electrode potential. E - electrometer amplifier, C - counter electrode, D - difference amplifier, R - reference electrode, P - Power amplifier, W - working electrode "sample".

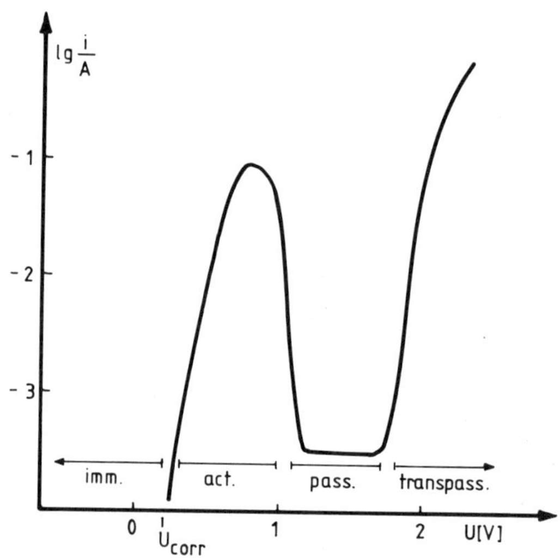

Figure 4. Current density vs. potential of a steel sample in phosphate buffer solution at a pH of 3.0.

active, passive, and transpassive potential ranges are indicated. The curve should be compared with that of pure iron [1]. The steel sample exhibits a shorter range of passivity, and the current drop from the active to the passive region is somewhat smaller than in the case of pure iron. This can be understood to result from the macroscopic superposition of different curves due to differences in the surface material on a microscopical scale, as described above. Nevertheless, the current decreases by more than two orders of magnitude in going from the active to the passive region.

In Figure 5, a gamma scattering Mössbauer spectrum is shown, measured in-situ and with a potential in the active range applied during the recording time which required several days. As found by computational fitting, it

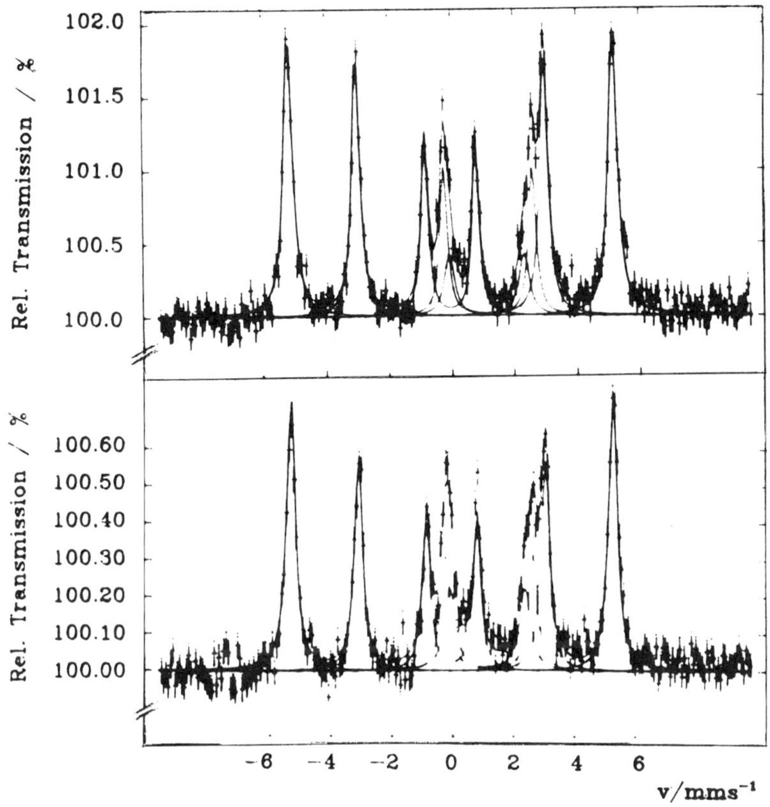

Figure 5. In-situ gamma scattering Mössbauer spectra recorded in the active potential range (upper spectrum) and at corrosion potential.

consists of the sextet from the steel substrate and of two doublets with isomer shifts corresponding to iron(II). The doublets can be ascribed to vivianite, the normal divalent iron phosphate, $Fe_3(PO_4)_2 \cdot 8H_2O$ whose parameters are given in Table I. No significant amount of iron(III) can be observed. The lower part of Figure 5 shows an in-situ spectrum obtained from a sample without an applied potential. In this case (open circuit), the floating or corrosion potential of the sample was found to be -20mV with respect to the normal hydrogen electrode. The spectrum is very similar to the former, demonstrating that the layers formed on the surface are practically equal in both cases. For comparison, in Figure 6, the transmission spectrum of a mixture of normal iron(II) and iron(III) phosphate is reproduced, exhibiting the two subspectra with a large quadrupole splitting characteristic for the iron(II) phosphate, and the narrower doublets for the iron(III) phosphate.

While in-situ gamma scattering was useful to detect surface layers formed in the active potential region and at corrosion potential, it was not successful in the potential regions of immunity, passivity, and transpassivity. In these regions the spectra showed only the sextet of the metal demonstrating that the layers formed in the electrolyte at these potential ranges are too thin, i.e., less than about 1μm, to be detected by gamma scattering. In these cases, CEMS has to be used because its information depth is much less, namely about 30 to 3000Å.

Table I. Fit Parameters of the Doublet Subspectra of Divalent Iron in Figures 5 and 6 and in Pure $Fe_3(PO_4)_2 \cdot 8H_2O$.[a]

	Figure	δ	ΔE_Q
Active potential (upper spectrum)	5	1.18 ± 0.05 1.22 ± 0.02	2.5 ± 0.2 2.91 ± 0.05
Corrosion potential (lower spectrum)	5	1.17 ± 0.05 1.22 ± 0.04	2.5 ± 0.2 2.9 ± 0.1
Mixed phosphates	6	1.23 ± 0.06 1.20 ± 0.02	2.3 ± 0.1 2.99 ± 0.04
Fe(II)-phosphate (vivianite)	-	1.23 ± 0.02 1.22 ± 0.2	2.42 ± 0.03 2.94 ± 0.03

[a]Isomer shift in mm/s with respect to metallic iron. Quadrupole splitting in mm/s.

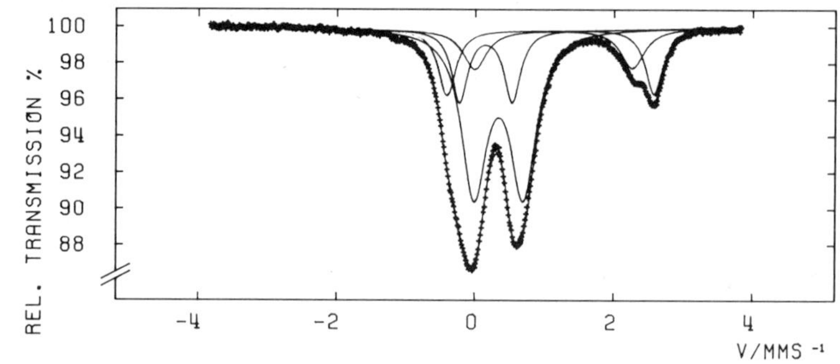

Figure 6. Transmission spectrum of a mixture of normal Fe(II) and Fe(III) phosphate (reproduced with permission from [4]).

CEMS is an ex-situ method. The samples were prepared in the electrolyte at a given potentiostatic voltage for 30 minutes or longer, then quickly removed from the electrolyte, rinsed with water and acetone, dried in air and inserted into the CEMS detector, a helium-methane flow counter. The lower part of Figure 7 presents a spectrum measured in this way after preparation at the active potential. There is no qualitative difference as compared with the in-situ spectrum shown in Figure 5. If the flow rate of the electrolyte through the cell during the preparation of the sample is drastically increased from 0.1 to 0.5mm/s at the surface of the sample, the iron(II) phosphate on the surface does not appear, and only a small iron(III) signal can be seen in the upper spectrum in Figure 7. Thus, the iron(II) phosphate seems to be merely a deposit on the surface due to an increased concentration of the solution close to the surface, exceeding the solubility constant at relatively high current densities.

The spectra reproduced in Figure 8 show that only iron(III), and no iron(II), is formed in the other potential regions. The layers are very

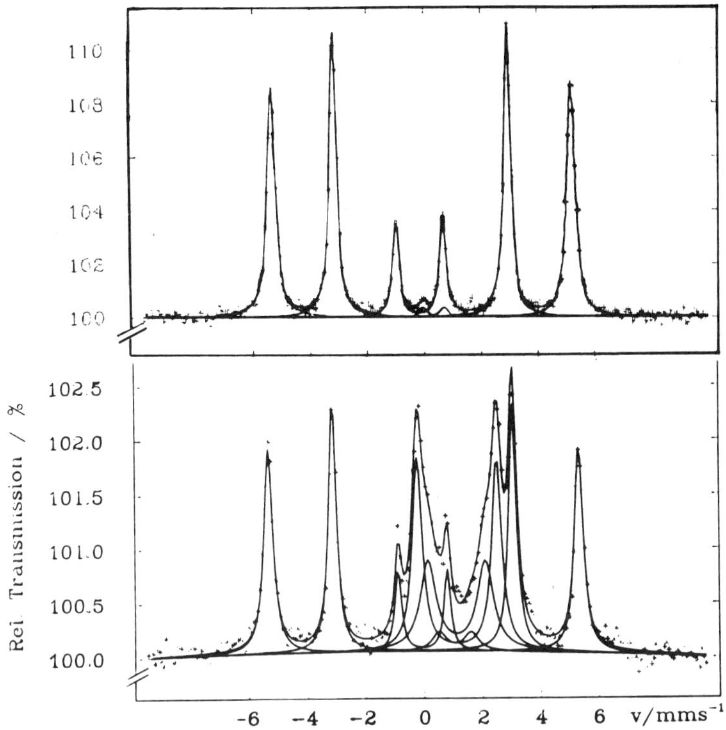

Figure 7. CEM spectra of steel samples prepared at active potential at high flow (upper spectrum) and at low flow of electrolyte.

thin, and those found for the immunity and transpassive conditions may even result from atmospheric oxidation. In the passive region, a thicker layer is formed, but no iron(II) is observed. Thus, the formation of a comparatively thick layer of Fe_3O_4, as it is stated in many electrochemical publications and textbooks, for example [1], is not verified by these spectra. This is also demonstrated in Figure 9. These CEM spectra were recorded by use of an iron-57 film electroplated onto a gold foil and treated at passive potential. The lower spectrum shows the absence of iron(II) and of any magnetically ordered phases such as α-Fe_2O_3, γ-Fe_2O_3, or Fe_3O_4 with a much higher sensitivity.

In Figure 10 concentration profiles of the elements iron, oxygen, phosphorus, and carbon are shown as derived from Auger intensities by applying the sputter technique. Each curve has to be multiplied by a different factor to reflect the real concentration, but some conclusions are possible without such an intensity correction. The steel sample treated in the passive region at a potential of 903mV exhibits a surface layer with a half-width thickness of approximately 500Å. The layer does not consist of a pure oxide but also contains some phosphate. The layer does not consist of one or a few well defined oxidic phases such as γ-Fe_2O_3 or Fe_3O_4, but exhibits a nonstoichiometric composition changing smoothly with the depth. Calibration by a sample of Ta_2O_5 on tantalum metal (Figure 11) shows that the concentration profiles are not smeared out by the experimental technique. The other curves in Figure 11 are oxygen profiles which illustrate the different layer thicknesses formed after treatment in the active, passive, and transpassive potential regions.

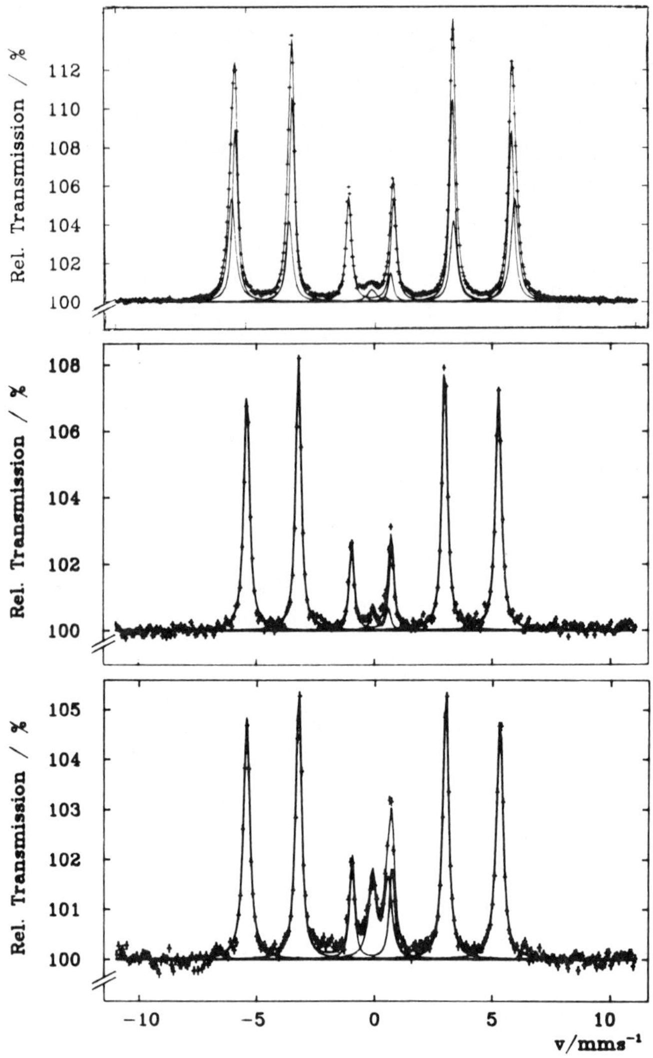

Figure 8. CEM spectra of steel samples prepared at immunity potential (upper), transpassive potential U(NHE)=1774mV (middle), and passive potential U(NHE)=1224mV (lower spectrum).

It should be noted that, because of the low (technical grade) purity of the materials used, the reproducibility of analytical results as described above is relatively low. Thus the analytical results reported have, to some extent, a preliminary character, and more quantitative data are needed to substantiate a high statistical reliability.

STUDIES AT HIGH CONCENTRATION

The so-called rust transformers are liquids commercially available and sometimes applied to rusty surfaces before the application of a coating. Their influence on corroded steel has been described [2]. The majority of these rust transformers contains a high concentration of phosphoric acid as the most reactive component. By use of the liquid nitrogen cryostat

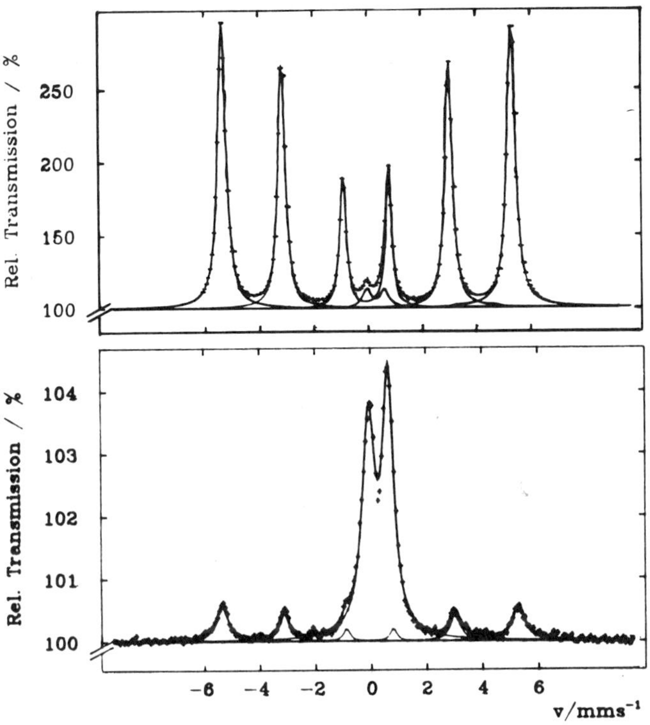

Figure 9. CEM spectra of a iron-57 film electroplated onto a gold foil, before (upper) and after passivation (lower spectrum).

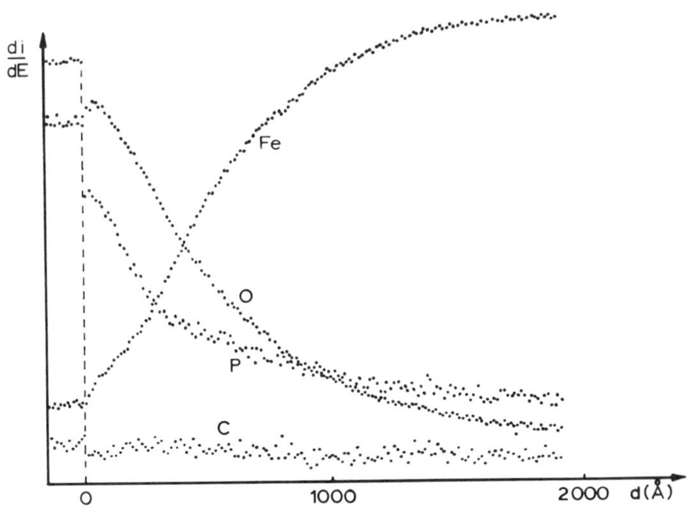

Figure 10. Auger concentration profiles (uncorrected) of a steel surface after passivation at U(NHE)=903mV

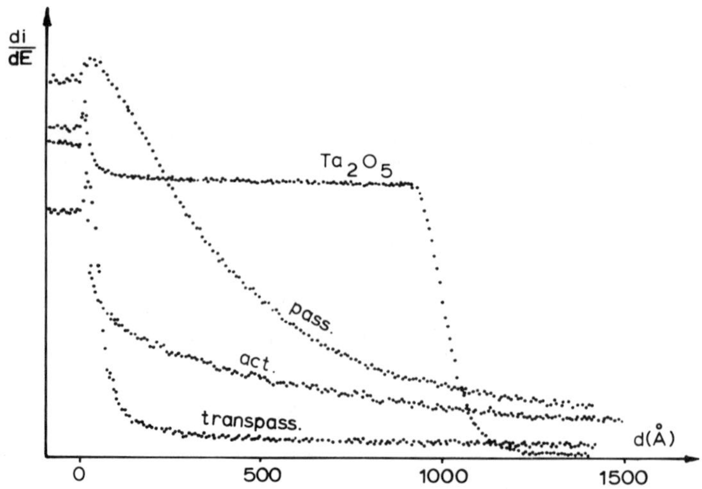

Figure 11. Auger oxygen profiles of a calibration sample and of steel samples after treatment in different potential regions

mentioned above, a rusty steel plate was analyzed by Mössbauer gamma scattering. A spectrum recorded before treatment is reproduced in Figure 12a. Figure 12b shows the same spectrum after fitting with standard spectra and after subtraction of the sextet subspectrum corresponding to the metallic substrate. From this spectrum it follows that the corrosion product consists of γ-FeOOH, β-FeOOH, and $FeCl_2 \cdot 4H_2O$. This composition is expected, because the sample had been exposed to an aggressive, chloride containing, atmosphere. The spectrum in Figure 12c was recorded after application of a rust transformer which was known to contain phosphoric acid as the active component. Again the substrate sextet has been subtracted from the measured spectrum. A successful fitting of this spectrum was possible only under the assumption that, in addition to some doublets, a singlet exists as a subspectrum. From a comparison with Figure 6 it is evident that the doublets but not the singlet can be explained by the formation of normal phosphates. Thus, an iron phosphate was formed, the singlet spectrum of which was not previously known. Furthermore, it seems that the formation of such a compound from the corrosion products was found here for the first time. In the following it is shown that this compound may appear generally with the application of concentrated phosphoric acid.

Detailed investigations with iron oxides and some salts has lead [3] to the following conclusion. The substance exhibiting the single line Mössbauer spectrum will be formed if concentrated orthophosphoric acid reacts with iron compounds containing only iron(III), otherwise the normal iron phosphates are formed. An example is given in Figure 13. Both maghemtite and magnetite have the same spinel structure, but magnetite contains both iron(II) and iron(III) ions. After the reaction of phosphoric acid with maghemite the single line phase is formed, while the reaction with magnetite results in the formation of the normal phosphates, indicated by doublets.

The best way to synthesize the compound in question was found to be the reaction of phosphoric acid with ferric orthophosphate. According to the phase diagram [5], the compound in question seems to be the acid ferric phosphate $FeH_3(PO_4)_2 \cdot 2.5H_2O$. Although this is not yet fully conclusive, in the following text the abbreviation "AFP" will be used for the compound which exhibits the single line Mössbauer spectrum.

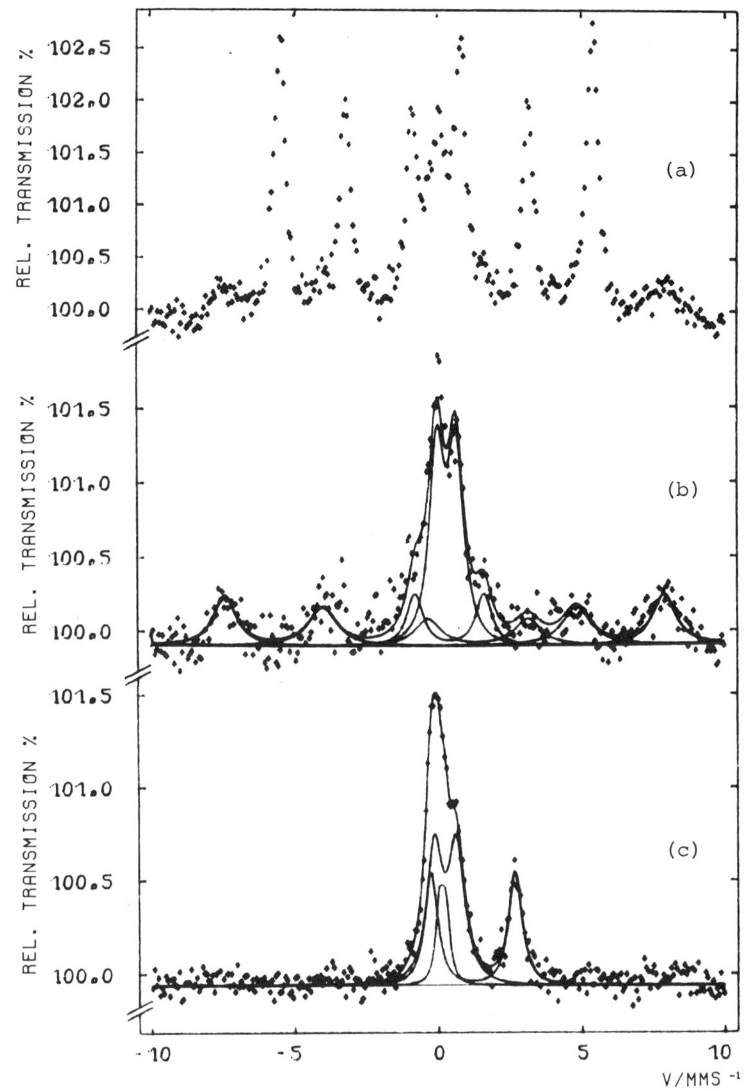

Figure 12. Gamma scattering Mössbauer spectra of a rusty steel plate before treatment (a), after subtraction of the sextet belonging to the metal substrate (b), and after treatment with a commercial rust transformer and subtraction of the metal sextet (c), measured at 80K (reproduced with permission from [3]).

Numerous attempts to synthesize pure AFP have been performed. Apparently the phase diagram [5] is misleading. The reaction product does not depend only on the concentrations of the reagents, but also depends on other parameters such as temperature, reaction time, and stirring. In principle, a very broad variety of phosphates, orthophosphates, pyrophosphates, polyphosphates, metaphosphates, and polymetaphosphates may be formed, and in fact, it was impossible to prepare AFP with less than 8 percent impurities. In Figure 14 the room temperature Mössbauer spectrum of an AFP sample is shown exhibiting a small doublet contribution from an impurity. Even analytical grade ferric orthophosphate always contains a remarkable percentage

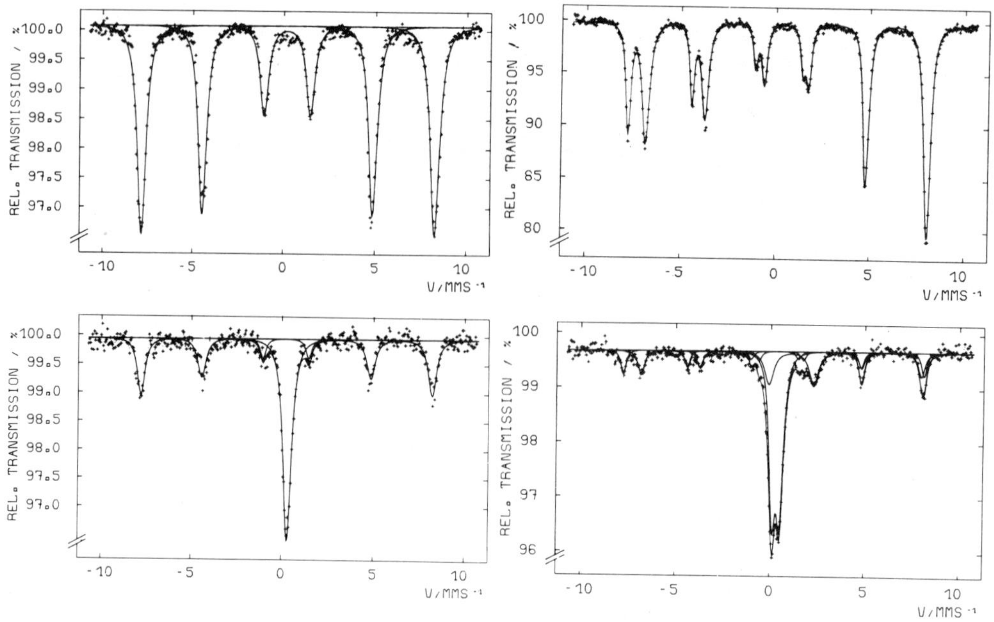

Figure 13. Transmission Mössbauer spectra of maghemite, γ-Fe_2O_3, (left) and magnetite Fe_3O_4, (right) before (upper) and after (lower) treatment with concentrated phosphoric acid (reproduced with permission from [4]).

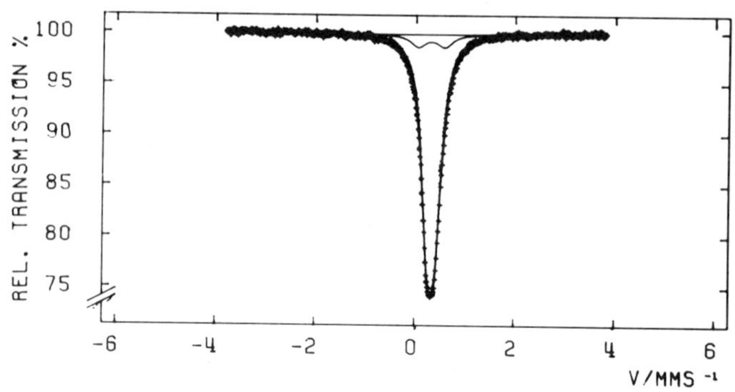

Figure 14. Room temperature Mössbauer spectrum of an AFP sample containing a small impurity doublet.

of other phosphates, mainly pyrophosphate, as analyzed by Mössbauer spectroscopy. The best results in synthesizing AFP, a pale pink powder, were found by the use of 0.2 to 0.3g of ferric orthophosphate per millimeter of 85 percent phosphoric acid, the ferric orthophosphate is slowly added to the acid at room temperature. After stirring for several days, the precipitate was washed with water and vacuum dried. The slow addition of ferric phosphate to boiling phosphoric acid until precipitation and formation of a highly viscous product occurred, followed by long stirring, was also successful. The elemental analyses of AFP samples produced in this way showed iron to phosphorus to hydrogen ratios close to that of $FeH_3(PO_4)_2 \cdot 2.5H_2O$, although

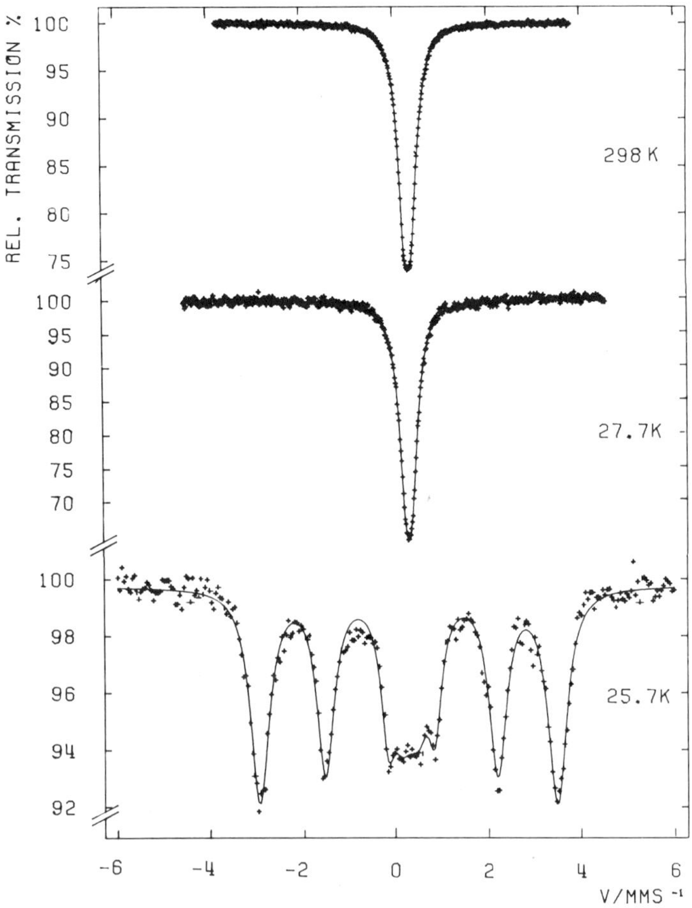

Figure 15. Mössbauer spectra of an AFP sample showing the transition to magnetic order (Reproduced with permission from [4]).

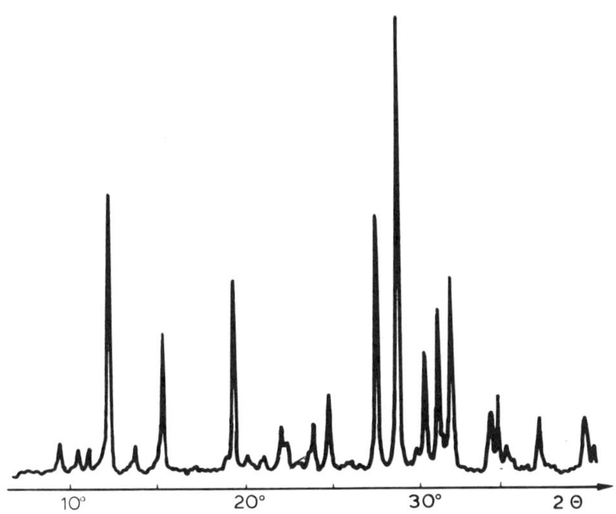

Figure 16. X-ray diffraction pattern of an AFP sample.

sometimes with some hydrogen deficit. However, this ratio is not a good criterion, because it is not very sensitive to other related moieties. On the base of quantitative Mössbauer phase analyses and elemental analyses of the iron, phosphorus, and hydrogen contents, for one of the synthesized products (AFPII), the reaction can be described formally as

$$FePO_4 + 0.185\ Fe_4(P_2O_7)_3 + 1.5\ H_3PO_4 + 4.5\ H_2O$$

$$\rightarrow 1.5\ FeH_3(PO_4)_2 \bullet 2.5\ H_2O + 0.3\ FeHP_2O_7 \bullet 1.5\ H_2O.$$

The Mössbauer spectrum of AFP exhibits a single line which is best fit to a doublet with a very small splitting. The quadrupole splitting increases slightly with decreasing temperature, but even at 30K the splitting is less than the linewidth as illustrated in Figure 15. The isomer shift relative to iron metal is 0.42 ± 0.03mm/s at 300K and 0.49 ± 0.03mm/s at 80K. The splitting does not show a relaxation behavior. Below 27K the single line splits abruptly into a magnetic sextet as seen in Figure 15.

The powders prepared were of good crystallinity with rectangular crystallites up to 20μm length. Unfortunately, monocrystals large enough for an x-ray structure determination have not yet been found. An x-ray powder diffraction diagram is shown in Figure 16. The symmetry seems to be rather low; an indexing is not yet possible. The symmetry around the iron(III) ion, however, should be close to cubic, as indicated by the small quadrupole splitting.

At low temperatures, two different sets of spectra were observed, depending on the preparation technique of the samples. Thus, at least two different compounds exhibit a single line Mössbauer spectrum, referred to as AFPI and AFPII. In Figure 17 the upper spectra correspond to AFPI, the lower to AFPII. The middle spectra clearly show that the preparation of this sample lead to a mixture of both of these moieties. The spectra on the right were recorded with an external applied magnetic field of 50kOe parallel to

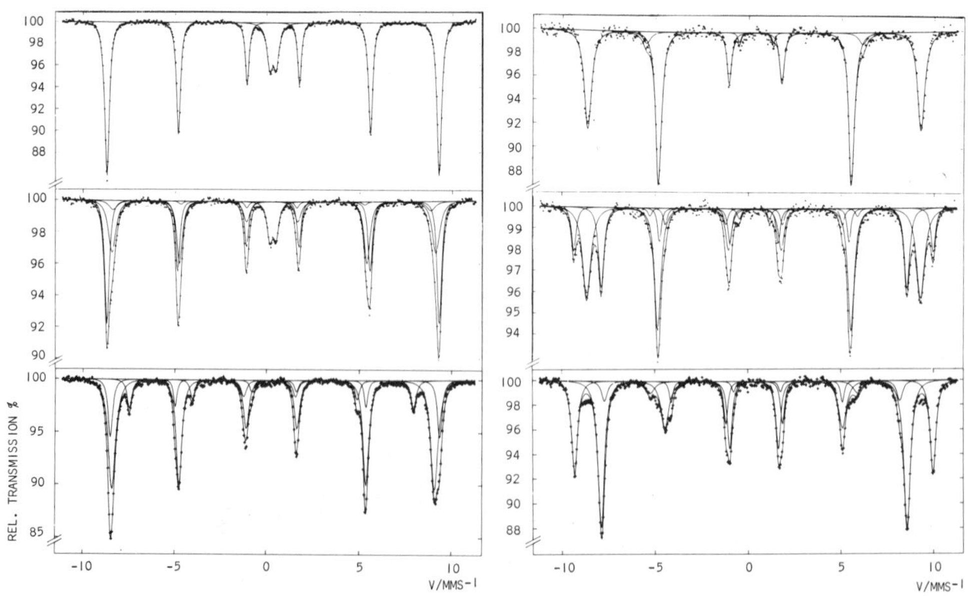

Figure 17. Mössbauer spectra at 4.2K of AFPI, AFP(I+II), and AFPII without (left) and with an external field of 50kOe.

the gamma radiation. In this case the differences between the two species are more obvious. A small sextet of an impurity with a magnetic ordering temperature of about 8K is always observed. While AFPII is ascribed to $FeH_3(PO_4)_2 \cdot 2.5H_2O$, AFPI seems to differ from it only in the reduced amount of water.

As is seen in the upper spectra in Figure 17, an external field has little influence on the spectrum of AFPI. The effective field for the sextet is slightly reduced and the ratio of the sextet line intensities, 3:x:1:1:x:3, is close to an x of 4, indicating that the angle of the effective field direction with the gamma ray direction is close to 90°. This suggests antiferromagnetic order with canted spins. The magnetic moments are canted by an angle α, as displayed in Figure 18, and add up to an effective moment which, at saturation, is directed parallel to the external field. Hyperfine fields are known to be antiparallel to the atomic moments and, therefore, the external field reduces the original field, H_o, to the measured effective field according to Figure 18. By measurements of x and the effective field at 4.2K, a canting angle of $\alpha = 7 \pm 2°$ was determined. Furthermore, from the experiments it is concluded that the easy axis of magnetization is oriented parallel to the electric field gradient and that the anisotropy field is relatively small. The magnetization curve in Figure 19 is derived

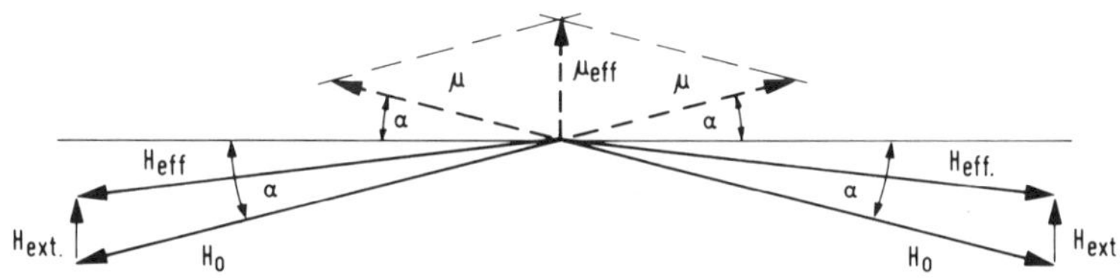

Figure 18. Canted antiferromagnetic order of AFPI at 4.2K.

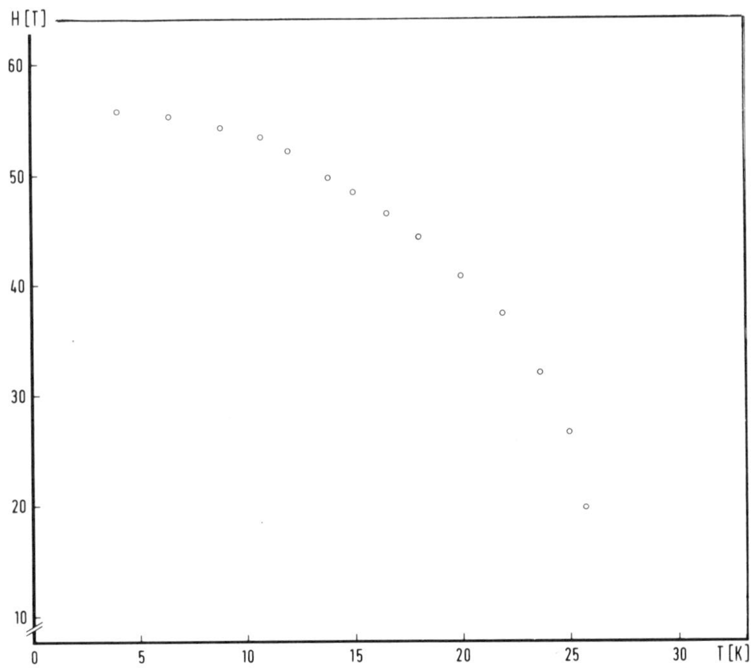

Figure 19. Hyperfine field vs. temperature for AFPI.

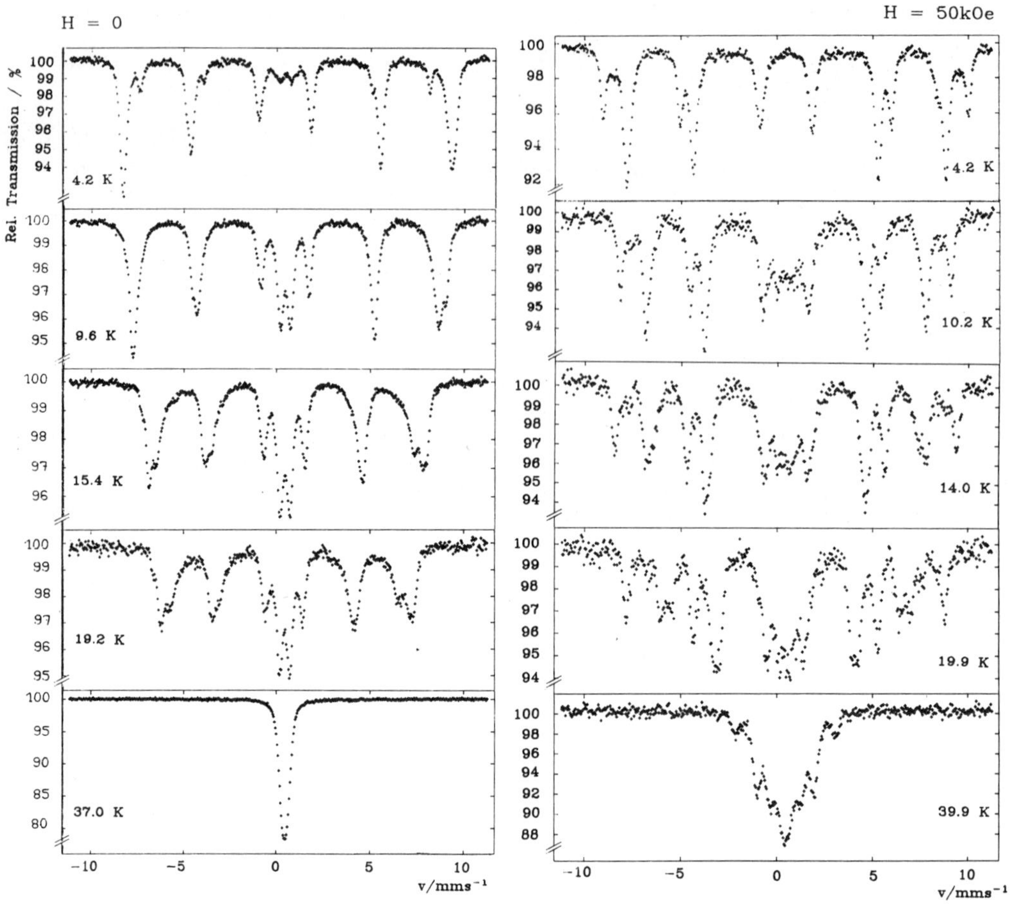

Figure 20. Spectra of AFPII at different temperatures without and with an external field parallel to the gamma ray direction.

from Mössbauer spectral results. It shows a magnetic ordering temperature of about 27K and a bend at about 13K indicating a phase transition.

Mössbauer spectra of a sample of AFPII at low temperatures without and with an external field are shown in Figure 20. The temperature dependence of their two main sextets is plotted in Figure 21. Again the order temperature is about 27K, and the bend around 14K may be interpreted as the indication of a phase transition. Close to this transition temperature, the spectra consist of four sextets, indicating that the transition for different particles or domains takes place at slightly different temperatures. At 25K the curves drop to zero more rapidly than predicted by a Brillouin function [6]. This suggests superparamagnetic behavior of small particles. In Figure 20, the 4.2K spectra suggest simple ferrimagnetic ordering. In a pure ferrimagnet, the effective fields as measured from the Mössbauer sextets should follow the dashed lines in Figure 22, where H_{ext} is the external applied field in Tesla. The experimental points, however, do not follow these lines. Therefore, a more complicated spin structure must be present. In Figure 22, the numbers on the experimental points are the measured ratio x of the sextet intensities, 3:x:1:1:x:3, and A and B refer to the two magnetic sublattices. The B-sextet is always about twice as intense as the A-sextet. The values of x show that, at low applied fields, H_B is aligned nearly perpendicularly to the applied field, and H_A is nearly parallel, while at

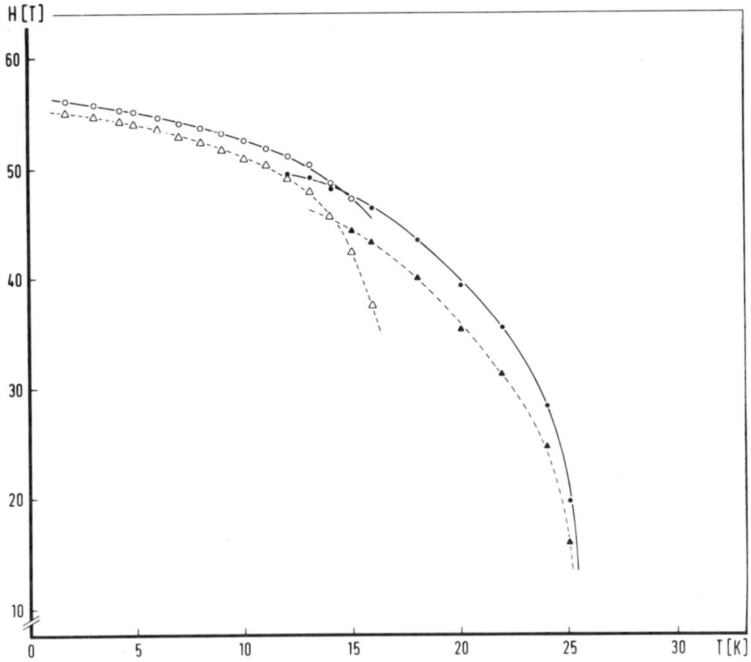

Figure 21. Magnetization curves of AFPII as derived from Mössbauer spectra.

large applied fields, H_A as well as H_B align more and more parallel to the applied field. The simplest model for the hyperfine field orientations in accordance with the experimental results is shown in Figure 23. H_A is larger than H_B, and the corresponding four magnetic moments couple to a net magnetization M_B that is larger than that for the A-sublattice. The sublattice magnetizations, M_A and M_B, have opposite directions, and the net magnetization is parallel to the applied field. Therefore, the spin model is a ferrimagnetic structure with spin canting in both sublattices. H_A points along the direction of an easy axis of magnetization.

CONCLUSIONS

Mössbauer in-situ gamma scattering and conventional CEMS techniques as well as Auger spectroscopy were used to study the layers formed on steel surfaces by immersion in diluted phosphoric acid under potentiostatic conditions. The immune, active, passive, and transpassive potential regions were investigated, and differences were found in the layer thicknesses as well as in the chemical species formed. The product formed under floating potential is the same as that formed in the active region. Iron(II) was formed in the active region as a deposit of vivianite, if the electrolyte flow is slow. If the flow is fast and in the other potential regions only iron(III) was found. In no case was a magnetically ordered oxide such as Fe_3O_4 or Fe_2O_3 found. The surface layers contain some phosphate and exhibit a nonstoichiometric oxide composition. Therefore, these experiments do not confirm the common assumption that these electrochemical processes lead to layers of defined magnetite and maghemite on steel surfaces.

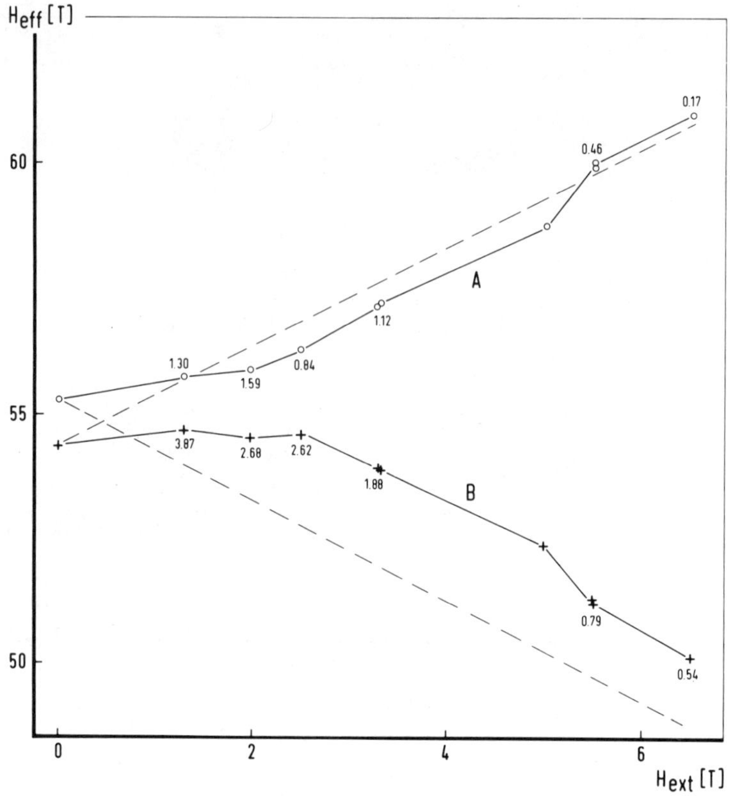

Figure 22. Effective hyperfine fields of sublattices A and B vs. external field for AFPII. The numbers are the x-values observed for the sextet intensity ratios 3:x:1:1:x:3.

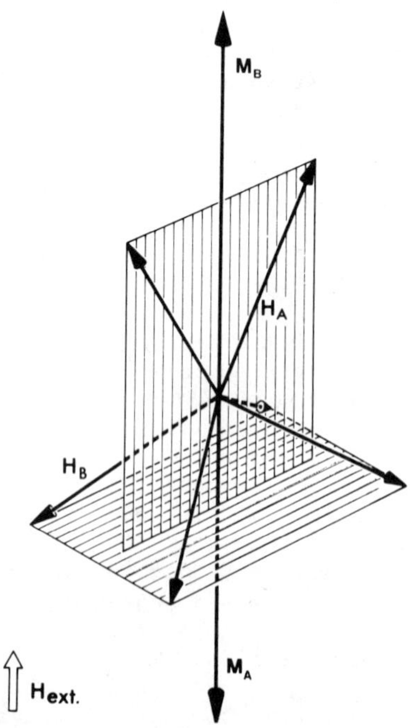

Figure 23. A model for the canted ferrimagnetic order in AFPII at 4.2K.

Gamma scattering and transmission Mössbauer techniques were used to study the reaction of concentrated phosphoric acid with pure and corroded steel. Besides the normal iron phosphates, two other products (probably acid ferric phosphates) showing a single Mössbauer line were found. The conditions for their formation were studied in detail. Furthermore, models for the spin structure of these substances at low temperatures were derived from Mössbauer investigations.

The experimental results described above point out that the phosphatization of steel, a process of technical importance, is not yet completely understood, and that modern analytical techniques such as Mössbauer and Auger spectroscopy may add new information to our understanding of this process. Similar studies are to be continued and extended to other processes such as sulfatization and other different types of corrosion.

ACKNOWLEDGEMENT

A great deal of the experimental work was done by U.Stumm and H.-J. Guttmann, Mainz, Federal Republic of Germany, and A.Swolfs, E. Janssen, and J. Mintjens, Nijmegen, The Netherlands. Their valuable contributions are gratefully acknowledged. Furthermore, the author is very much indebted to Professors P. Gütlich of Mainz, and E. de Boer and J. Fuggle of Nijmegen, for their interest in this project, and to the Deutsche Forschungsgemeinschaft for financial support for a part of this work.

REFERENCES

1. N. Sato and G. Okamoto, in "Comprehensive Treatise of Electrochemistry", v. 4, J. O'M. Bockris, ed., Plenum Press, New York (1981).

2. W. Meisel and P. Gütlich, Werkstoffe und Korrosion 32, 296 (1981), and Fresenius Z.Anal. Chem. 314, 324 (1983).

3. W. Meisel, H.-J. Guttmann and P. Gütlich, Corr. Sci. 23,1373 (1983).

4. W. Meisel, A.Swolfs, and P. Gütlich, Fresenius Z. Anal. Chem. 319, 735 (1984).

5. R. F. Jameson and J. E.Salmon, J. Chem. Soc. 28 (1954).

6. W. Meisel, Hyperfine Interactions 15/16, 941 (1983).

CONVERSION ELECTRON MÖSSBAUER STUDIES OF THE CORROSION PRODUCTS FORMED ON THE STEEL SURFACE IN ATMOSPHERES CONTAINING HCl, H_2S, AND SO_2

Yusuke Ujihira, Arimichi Handa, and Masanori Fujinami

Department of Industrial Chemistry
Faculty of Engineering
The University of Tokyo
Hongo, Bunkyo-ku, Tokyo, Japan

ABSTRACT

Conversion electron Mössbauer spectrometry, CEMS, was applied to the analyses of the steel surface corroded and the corrosion products formed in HCl, H_2S, and SO_2 environments. In an HCl environment, $FeCl_2 \cdot 2H_2O$ and $FeCl_2 \cdot 4H_2O$ was formed at the initial stage and transformed into γ-FeOOH if the relative humidity was high and β-FeOOH if it was low. The exposure to an H_2S atmosphere resulted in the formation of paramagnetic FeS_{1-x}, as fine particles at the initial stage, coarse ferromagnetic FeS_{1-x} particles at the intermediate stage, and finally the surface was sulfided to Fe_3S_4. In a combined SO_2 air atmosphere, the formation of $FeSO_4 \cdot H_2O$ and $FeSO_4 \cdot 4H_2O$ as the initial and γ-FeOOH as the final corrosion product could be confirmed. The utility of CEMS for the corrosion study was demonstrated.

INTRODUCTION

The aim of this paper is to describe the results obtained by the application of conversion electron Mössbauer spectrometry, CEMS, to the analyses of corrosion products formed on steel exposed to an atmosphere that contained small amounts of HCl, H_2S, or SO_2. These gases are typical environment pollutants that cause materials to corrode. Therefore, the elucidation of their corrosion process and transformation of corrosion products with weathering are very important. Although X-ray diffractometry is a typical analytical technique to be applicable for the identification of the corrosion products, it becomes impotent if the corrosion layer is very thin as is the case of the initial corrosion stage and if the corrosion products are in an amorphous state. The application of surface analytical methods, such as ESCA, sometimes may induce chemical changes of the corrosion products because the specimen is introduced into an ultra-high vacuum. Moreover, the observation depth of these surface analytical techniques is restricted to several Ångstrom. In contrast, iron-57 CEMS, which collects electrons associated with the 7.3keV conversion and the 5.5keV Auger electrons reemitted after resonance absorption of 14.4keV γ-rays in iron-57 nuclei, provides useful information on corrosion products grown to the thickness of more than 10nm nondestructively. It is concluded that CEMS is one of the most suitable techniques for the investigation of the corrosion phenomena [1,2]. Hitherto, CEMS has been applied not only to the oxidation of iron [3-5] and its corrosion in various solutions [6-9], but also to the estimation of the physical and chemical state of iron in modified surface layer which have been boronized [10,11], carburized [12], nitride [13,14], galvanized [15], tin coated [16], and ion implanted [17,18]. Also studies have been extended to the analyses of coatings produced by the immersion of steels in zinc phosphate [19], manganese phosphate [20], iron phosphate [21], sodium oxalate [22], and alkaline [23] solutions. CEMS has been used to analyze the crystallization of amorphous iron [24]. CEMS studies of tin-119 [25,26], europium-151 [27], tantalum-181 [28], and gold-197 [29] have also been carried out.

EXPERIMENTAL

Room temperature CEM spectra were taken with a back scatter type gas-flow proportional counter, fabricated in our laboratory and operated at 1500V. The counting gas of 90 percent helium and 10 percent methane had a flow rate of 20ml/min. Liquid nitrogen temperature CEM spectra were obtained by detecting electrons with an electron multiplier as shown in Figure 1 [30,31]. A small heater was attached to a sample holder in order to avoid the condensation of residual gases in the chamber on the sample surface on cooling to 80K. The heater was turned off after the vacuum reached 10^{-5} Torr and then the sample was cooled to 80K. The counting rate was about 300 counts per second. The Doppler velocity was calibrated by using α-iron at room temperature.

CORROSION OF A STEEL IN AN HCl ATMOSPHERE

The study of the formation of β-FeOOH, akaganeite, as a corrosion product of a steel near the sea or in a fluoride or chloride brine environment has been the subject of many corrosion studies [32-34]. A preparatory CEMS study of the corrosion of SM-50 steel, with carbon < 0.2 percent, sulfur < 0.04 percent, silicon < 0.5 percent, manganese < 1.5 percent, and phosphorus < 0.04 percent, exposed to atmospheres containing HCl gas diluted with nitrogen or dry air, proved that water vapor was required for the initiation of corrosion and that both β-FeOOH and γ-FeOOH were produced in the HCl and air environment. To clarify the conditions of β-FeOOH and γ-FeOOH formation, CEMS observations of steel surfaces corroded by successive exposure to air containing HCl vapor and an atmosphere of various humidity levels were carried out at room temperature and 80K. CEM spectra of the steel corroded for 3 minutes at room temperature in water vapor saturated air containing HCl at partial pressure of 40 to 80Torr revealed that the initial corrosion product was $FeCl_2 \cdot 4H_2O$, because the observed isomer shift and quadrupole splitting

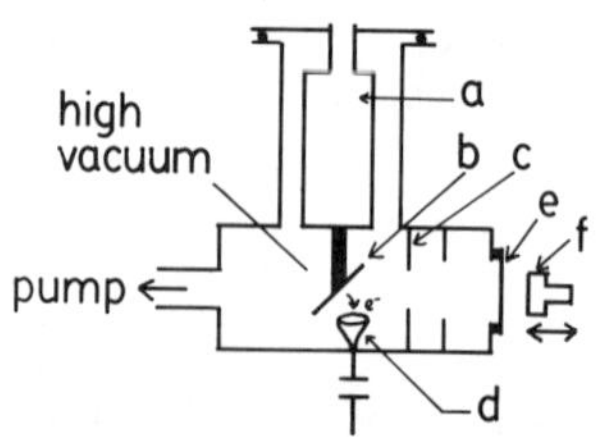

Figure 1. A schematic diagram of the apparatus for CEMS at 80K with an electron multiplier.(a) liquid nitrogen, (b) sample, (c) lead shield, (d) electron multiplier, (e) window, and (f) source.

were 1.29 and 2.59mm/s, respectively. CEMS study also indicated that $FeCl_2 \cdot 2H_2O$ did not change in dry air, but it was transformed into $FeCl_2 \cdot 4H_2O$ if the steel was held for 8 minutes in air saturated with water vapor containing 42Torr HCl, as shown in Figure 2.

Although the white layer of $FeCl_2 \cdot 2H_2O$ and $FeCl_2 \cdot 4H_2O$ formed at the surface of the SPCC-D steel, with carbon < 0.12 percent, manganese < 0.5 percent, phosphorus < 0.04 percent, sulfur < 0.45 percent, did not show any change in color when kept in completely dry air for 7 days, the color of the surface gradually changed when kept in air with a relative humidity of less than 10 percent for 10 days. The CEM spectra of the steel surface after the color change were measured at room temperature and 80K, and shown in Figure 3. Because only a limited counting rate was afforded by the electron multiplier equipped CEM spectrometer due to the geometrical handicap of around 1 in 50, when compared with a 2π gas flow proportional counter, enough counts could not be accumulated to resolve the doublet peaks for either $FeCl_2 \cdot 2H_2O$ or $FeCl_2 \cdot 4H_2O$ within a practical measurement period of a week or so.

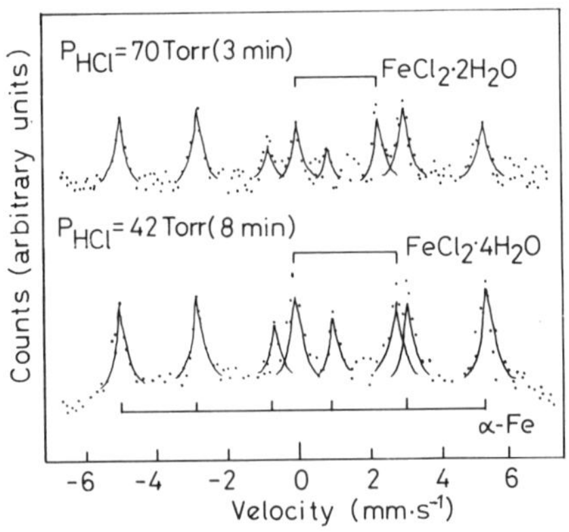

Figure 2. Room temperature CEM spectra for the steels exposed to moist air containing HCl.

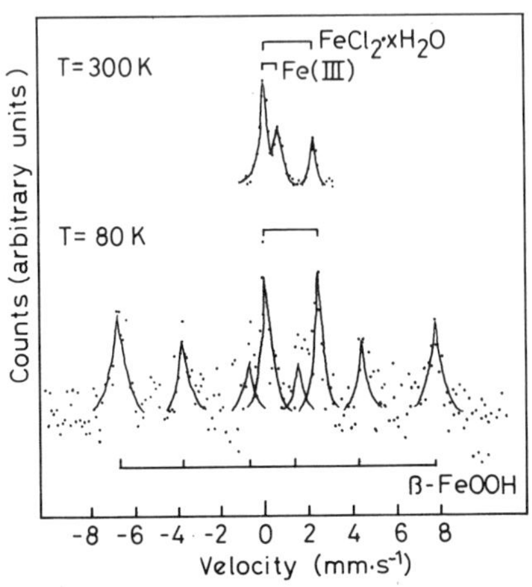

Figure 3. CEM spectra for steel exposed to moist air containing HCl for 3 minutes followed by the exposure to dry air for 20 days.

Therefore, the corrosion product was designated as $FeCl_2 \cdot xH_2O$. In the upper spectrum shown in Figure 3, doublets due to $FeCl_2 \cdot xH_2O$ and an iron(III) compound are observed. Considerations based on wet chemical knowledge suggested that FeOOH was the most probable iron(III) compound [35]. By the CEM measurement at 80K, the FeOOH can be identified as either β-FeOOH or γ-FeOOH, because β-FeOOH has a hyperfine field of 48.0T at 80K. The lower spectrum shown in Figure 3, taken at 80K, consisted of a sextet with the hyperfine field of 48T and a quadrupole split doublet due to $FeCl_2 \cdot xH_2O$. A combined consideration of both the spectra led us to conclude that the iron(III) oxide was β-FeOOH. These results indicated that in a relatively dry atmosphere the $FeCl_2 \cdot 2H_2O$ and $FeCl_2 \cdot 4H_2O$ layer, produced by the exposure to a HCl atmosphere, was oxidized by atmospheric oxygen to $FeCl_3 \cdot xH_2O$, which was then hydrolyzed to form β-FeOOH.

The CEM spectra of steel corroded in HCl atmosphere for 3 minutes, and subsequently exposed to air with a 65 percent relative humidity for 7 days, are shown in Figure 4. The doublet due to FeOOH which appeared in the CEM spectrum at room temperature was transformed into a sextet attributed to β-FeOOH and a doublet attributed to γ-FeOOH at 80K. Thus, both β-FeOOH and γ-FeOOH were produced by keeping the HCl gas corroded steel in a moist atmosphere.

CEM spectra of the steel, prepared by keeping the $FeCl_2 \cdot xH_2O$ covered steel, obtained by the exposure of the steel to a HCl atmosphere for 3 minutes and then exposed to moist air with 100 percent relative humidity for 15 minutes, are shown in Figure 5. Because the doublet due to FeOOH hardly changed at 80K, the FeOOH was identified as γ-FeOOH, which has a Nèel temperature of 50K.

In conclusion the corrosion of steel in an HCl atmosphere begins with the production of $FeCl_2 \cdot 2H_2O$ and $FeCl_2 \cdot 4H_2O$. The atmospheric humidity in which the $FeCl_2 \cdot xH_2O$ surface is kept has a decisive influence on whether β-FeOOH or γ-FeOOH forms. Lower humidity favors the formation of β-FeOOH. A mechanism for the formation of β-FeOOH or γ-FeOOH is proposed as Figure 6. The left route is followed for the β-FeOOH formation, in which, in the less moist atmosphere, $FeCl_2 \cdot xH_2O$ may be oxidized with oxygen in air to form an

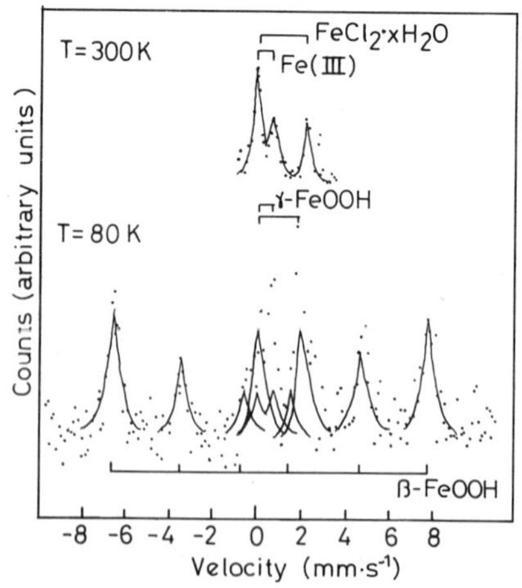

Figure 4. CEM spectra for steel exposed to moist air containing HCl for 3 minutes followed by the exposure to air of 65 percent relative humidity for 7 days.

iron(III) species of chloride, which is dissolved in adsorbed water and hydrolyzed to form the hydroxyl compound from which β-FeOOH is precipitated. The right route is followed for the γ-FeOOH formation, where, in a highly moist atmosphere, $FeCl_2 \cdot xH_2O$ may be dissolved in water adsorbed on the surface to become aqueous iron(II) hexaaquoiron(II), which is oxidized by the dissolved oxygen in a solution to hexaaquoiron(III), and polymerized through hydroxylation to precipitate γ-FeOOH.

CORROSION OF A STEEL IN H_2S ATMOSPHERE

Hydrogen sulfide is a pollutant of atmosphere found in the area of spas, oil refining plants, and chemical industries, and causes the corrosion to metallic materials or the stress cracking in the construction materials due to hydrogen embrittlement. The variation of CEM spectra for the steel plate, JIS G3141 with carbon < 0.12 percent, manganese < 0.50 percent, phosphorus < 0.04 percent, and sulfur < 0.45 percent, exposed to 0.86 volume percent H_2S

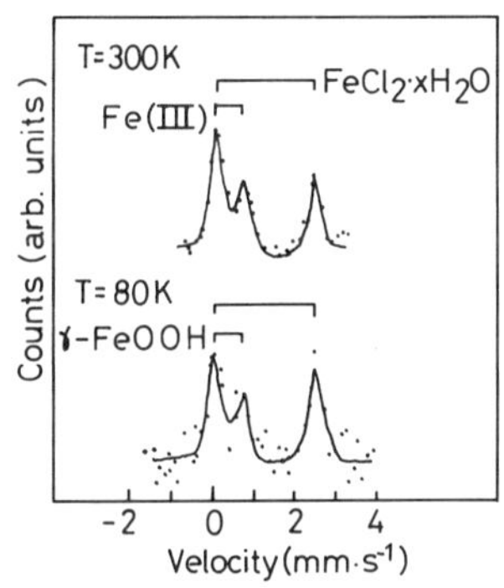

Figure 5. CEM spectra for steel exposed to moist air containing HCl for 3 minutes followed by the exposure to moist air of 100 percent humidity for 15 minutes

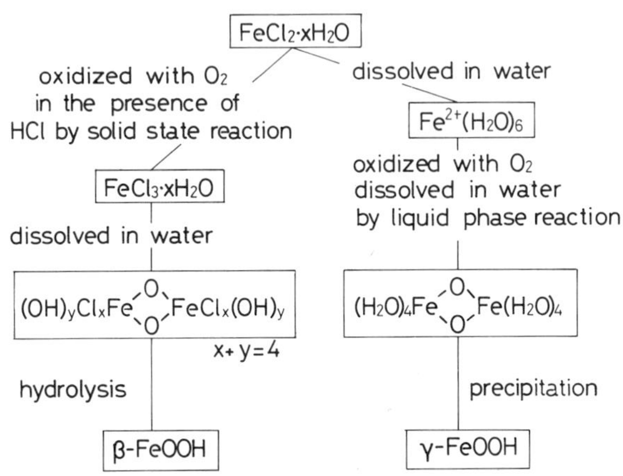

Figure 6. Proposed mechanism for the formation of β-FeOOH and γ-FeOOH in an HCl environment.

in a nitrogen moist atmosphere for different periods is shown in Figure 7. After 30 minutes exposure, a paramagnetic singlet appeared in addition to the sextet due to the substrate. The singlet has an isomer shift of 0.42mm/s, close to that of mackinawite, FeS_{1-x}, which was produced by the reaction between iron and aqueous hydrogen sulfides [36,37]. The peak intensity ratios of the singlet to sextet increased with the exposure time and only the singlet was observed after 2 days exposure. The thickness of the layer was estimated to grow to more than 300nm. After 2.5 days exposure, additional magnetically split peaks were observed in the CEM spectrum and their Mössbauer parameters corresponded to those of ferromagnetic mackinawite which may be synthesized by the reaction of iron(II) sulfate and sodium sulfide in a solution at pH 7.5 [38]. Scanning electron microscopic observation of the corroded surface revealed that the paramagnetic mackinawite was composed of fine-grained particles and that ferromagnetic mackinawite was composed of large-grained particles which were produced near highly localized corrosion spots such as cracks in the fine-grained mackinawite film [39]. After 3 days exposure, the peaks due to ferromagnetic mackinawite vanished and a new sextet and doublet appeared in the CEM spectrum. The magnetically split peaks

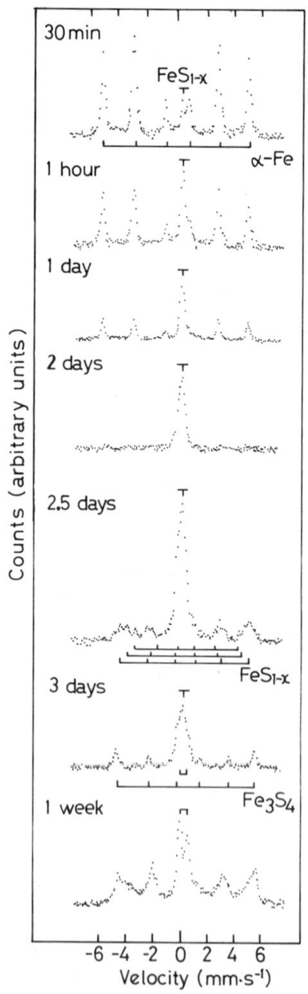

Figure 7. Room temperature CEM spectra for steel exposed to 0.86 volume percent H_2S in a nitrogen environment.

were attributed to greigite, Fe_3S_4, by reference to its Mössbauer spectral parameters 0.38mm/s for the isomer shift and 29.7T for the hyperfine field. The quadrupole split peaks were assigned to superparamagnetic fine particles of greigite [37]. Greigite has an inverse spinel structure, Fe(III)[Fe(II), Fe(III)]S_4, and is ferromagnetic. After a week, the surface layer was composed of only Fe_3S_4. These results confirmed that mackinawite produced on a steel could be subsequently sulfided to greigite. It was likely that the sulfidation of mackinawite was initiated heterogeneously at the points where very small water drops coagulated.

Figure 8 shows the CEM spectra for the steel exposed to 0.86 volume percent H_2S in a moist atmosphere for different periods of time, followed by further exposure to moist air without H_2S for one day. In all of the CEM spectra, the doublet peaks due to γ-FeOOH with an isomer shift of 0.38mm/s and a quadrupole splitting of 0.69mm/s, could be observed as well as the singlet of mackinawite. It was considered that γ-FeOOH was produced by the oxidation of FeS_{1-x} in moist air. The peak intensity ratios of γ-FeOOH to FeS_{1-x} tended to decrease with the increase of corrosion time in the H_2S and N_2 environment. The results indicated that the oxidation of mackinawite to γ-FeOOH became more difficult as the exposure time was lengthened. This tendency might result because the easily oxidizable fine mackinawite decreased with the prolonged sulfidation. That is, large particle mackinawite increased and was transformed into greigite.

Figure 9 shows the variation of the CEM spectra for the steels corroded in H_2S 0.86 volume percent in a moist air atmosphere for different periods.

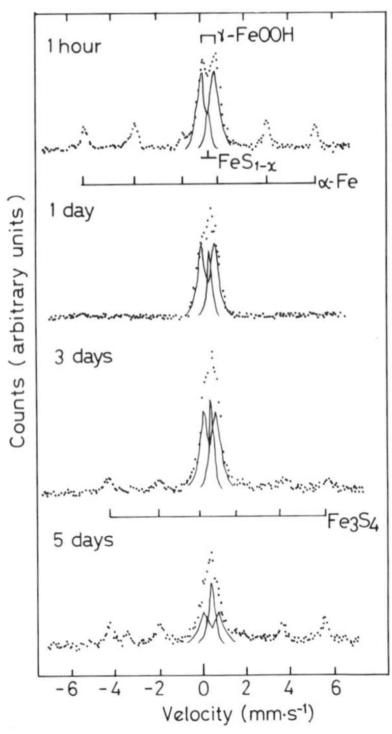

Figure 8. Room temperature CEM spectra for steel exposed to 0.86 volume percent H_2S in nitrogen for different periods followed by the exposure to moist air of 100 percent relative humidity without H_2S for 1 day.

After an hour and a day of exposure, mackinawite was the sole product. Comparison of the growth rate of the peak intensity ratios of mackinawite to substrate during the corrosion in the H_2S and air atmosphere (Figure 9) and that in H_2S and N_2 (Figure 7) revealed that the corrosion rate was faster in the presence of oxygen. An ESCA study of a steel corrosion under an atmosphere of several Torr of dry H_2S at 275°C suggested that the production of iron sulfide was promoted by the presence of iron oxide at the surface [40]. After 3 days and a week of exposure, only the doublet of γ-FeOOH was observed. Mackinawite was oxidized to γ-FeOOH in the presence of oxygen before it changed to greigite.

It was confirmed that, in an oxygen free atmosphere containing H_2S, the initial corrosion product was paramagnetic mackinawite, FeS_{1-x}, and that an intermediate compound was ferromagnetic coarse mackinawite, which was transformed into greigite, Fe_2S_3. The presence of oxygen in the H_2S and N_2 atmosphere promoted the sulfidation reaction of the steel. Mackinawite produced in the initial stage was gradually oxidized to γ-FeOOH after several days exposure.

CORROSION OF STEEL IN SO_2 ATMOSPHERE

Sulfur dioxide, which is one of the most prevalent air pollutants in industrial districts, corrodes steels to a considerable extent. The corrosion process of a steel in moist air containing 0.16 volume percent of SO_2 was studied by CEMS as a function of the exposure duration and condition [41].

Figure 10 shows the variation of the CEM spectra for JIS G3141 steel exposed to moist air containing 0.16 volume percent of SO_2 for different periods of time. The CEM spectrum of the steel exposed for 15 minutes consisted of a broadened quadrupole doublet (1) corresponding to a high-spin

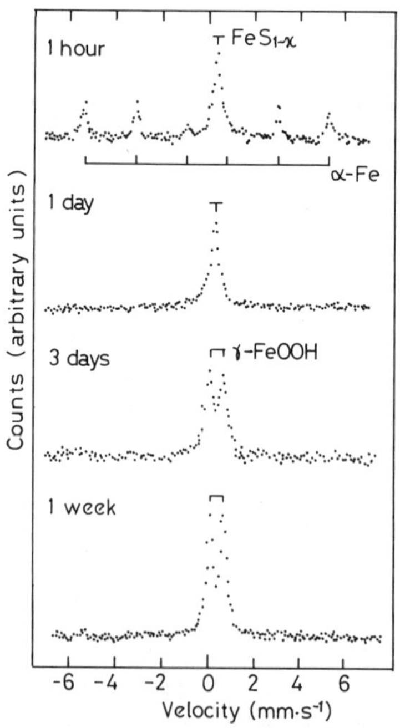

Figure 9. Room temperature CEM spectra for steel exposed to 0.86 volume percent H_2S in air of 100 percent relative humidity.

iron(II) species and a sextet due to the substrate. Because the isomer shift and quadrupole splitting of doublet (1) were 1.20mm/s and 2.68mm/s, respectively, it was attributed to $FeSO_4 \cdot H_2O$ [42]. The $FeSO_4 \cdot H_2O$ film could be produced through the reaction between iron and sulfuric acid, which was produced by the oxidation of SO_2 to SO_3 and the subsequent absorption into the atmospheric water vapor at the initial stage of the corrosion process. The application of X-ray diffraction to the same specimen was not possible because the corroded layer was too thin to give diffraction peaks. The increase of the peak intensity radio of quadrupole doublet 1, or $FeSO_4 \cdot H_2O$, to substrate suggested that the $FeSO_4 \cdot H_2O$ layer had grown to a thickness of several tens of nanometer.

When the specimen was corroded for a day, three doublets due to a high-spin iron(II) species doublets (1 and 2) and a high-spin iron(III) species doublet (3) appeared. Doublet (2) was assigned to $FeSO_4 \cdot 4H_2O$ on the basis of its isomer shift of 1.24mm/s, and its quadrupole splitting of 3.18mm/s [43]. Doublet 3 had an isomer shift of 0.36mm/s and a quadrupole splitting of 0.62mm/s which are characteristic of FeOOH. After 3 days of exposure, the peak intensity ratio of the iron(II) to iron(III) species was almost the same as that observed after one day. After 5 days exposure, $FeSO_4 \cdot H_2O$ and FeOOH were dominant and $FeSO_4 \cdot 4H_2O$ had disappeared. After a week of exposure, the sextet due to substrate vanished, indicating that the surface layer had grown to a thickness of several hundred nanometers. The X-ray diffraction pattern

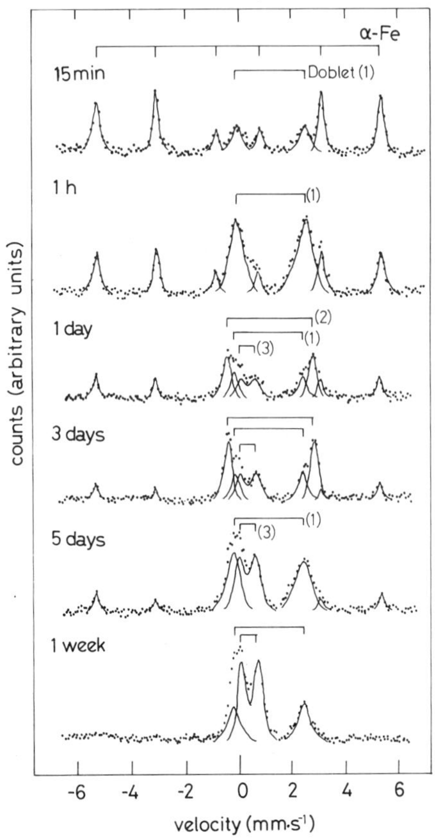

Figure 10. Room temperature CEM spectra for steels exposed to moist air containing 0.16 volume percent of SO_2

of this thick layer indicated the presence of $FeSO_4 \cdot H_2O$ and γ-FeOOH as indicated by CEM study.

Figure 11 shows the CEM spectrum for the steel, first exposed to a SO_2 atmosphere for 15 minutes when only $FeSO_4 \cdot H_2O$ had formed, followed by exposure to moist air without SO_2 for an hour and twelve hours. It shows that the exposure to moist air without SO_2 rapidly transformed the initially formed $FeSO_4 \cdot H_2O$ into FeOOH. This result revealed that the presence of excessive sulfur dioxide suppressed the main reaction of iron to iron(II) because the oxidation of sulfite to sulfate was a competitive reaction [44]. When a CEM spectrum was taken at 80K, a similar quadrupole doublet with an isomer shift of 0.44mm/s and a quadrupole splitting of 0.68mm/s was observed in the spectrum. Hence doublet (3) was assigned to γ-FeOOH.

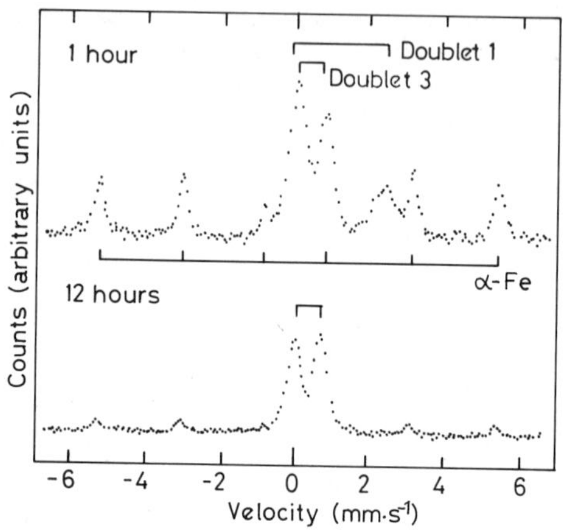

Figure 11. Room temperature CEM spectra for steel exposed to moist air containing 0.16 volume percent of SO_2 for 15 minutes followed by the exposure to moist air without SO_2 for 1 and 12 hours.

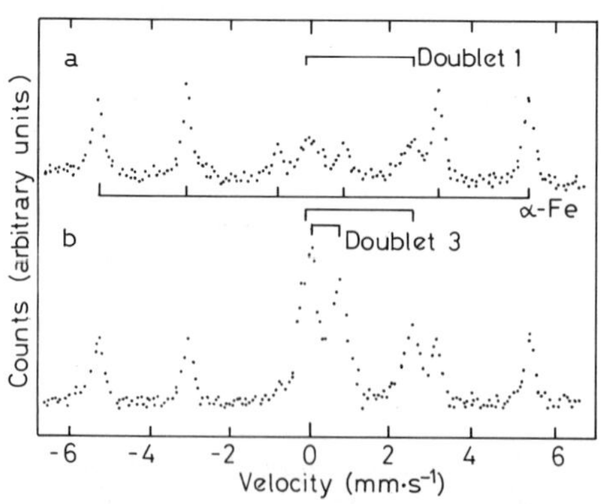

Figure 12. Room temperature CEM spectra for the steels exposed to moist air containing 0.16 volume percent of SO_2 for 15 minutes (a) followed by the exposure to dry air without SO_2 for without SO_2 for 1 day (b).

Figure 12 shows the CEM spectra for the steels corroded in a SO_2 atmosphere for 15 minutes and subsequently exposed to dry air without SO_2 for one day. The growth of the layer was slower compared with exposure to moist air at the second stage. However, the oxidation rate of $FeSO_4 \cdot H_2O$ to γ-FeOOH was faster in air without SO_2 than in air containing SO_2. This result is the basis of the above-mentioned suppressive effect of SO_2.

CONCLUSIONS

The chemical states of the corrosion products formed on a steel under HCl, H_2S and SO_2 environments could be identified, and the utility of CEMS to the task has been demonstrated. The improvement of the in situ CEMS and depth selective CEMS, and the establishment of the quantitative interpretation of the Mössbauer spectrum will afford us more detailed and useful information on the study of corrosion.

REFERENCES

1. Y. Ujihira and K. Nomura, Bunseki 1978, 555 (1978).
2. Y. Ujihira, Rev. Anal. Chem. 8, 125 (1985).
3. M. Domke and B. Kyuelos, Corrosion Sci. 23, 921 (1983).
4. G. W. Simmons, E. Kellerman, and H. Leihdeiser, Jr., Corrosion 29, 227 (1973).
5. M. J. Tricker, J. M. Thomas, and A. P. Winterbottom, Surface Sci. 45, 601 (1974).
6. P. Gütlich, W. Meisel, and E. Mohs, Werkstoffe Korrosion 33, 35 (1982).
7. G. Belozerskii, C. Bohna, T. Ekdahl, and D. Liljequist, Corrosion Sci. 22, 831 (1982).
8. Y. Ujihira and K. Nomura, Bunseki Kagaku 27, 782 (1978).
9. J. Ensling, J. Fleisch, R. Grimm, J. Gruber, and P.Gütlich, Corrosion Sci. 18, 797 (1978).
10. M. Carbucicchio, L. Bardani, and G. Palombarini, J. Mater. Sci. 15, 711 (1980).
11. A. Handa and Y.Ujihira, ibid 18, 1887 (1983).
12. A. Handa, I.Okabe, and Y. Ujihira, ibid 16, 1999 (1981).
13. Y. Ujihira and A. Handa, J.Physique 40, C2-586 (1979).
14. Y. Ujihira, A. Handa, Y. Abe, and I. Okabe, Nippon Kagakukaishi 1979, 626 (1979).
15. M. J.Graham, P. E. Beaubien, and F. I. Sproule, J. Mater. Sci. 15, 626 (1980).
16. M. Fujinami, K. Yoshida, Y. Ujihira, and M. Terasaka, ibid, 20, 4099 (1985).
17. J. A. Sawicki, Mater. Sci. Eng. 69, 501 (1985).

18. B. M. Ullrich, eds., Nucl. Instrum Methods B7/8, 1 (1985).

19. K. Nomura, Y. Ujihira, Y. Matsushima, R. Kojima, and Y. Sugawara, Nippon Kagakukaishi 1980, 1372 (1980).

20. K. Nomura and Y. Ujihira, ibid 1982, 1352 (1982).

21. Y. Ujihira and K. Nomura, J. Physique 41, Cl-391 (1980).

22. K. Nomura and Y. Ujihira, J. Mater. Sci. 18, 1751 (1983).

23. K. Nomura and Y. Ujihira, ibid 19, 2664 (1984).

24. M. Fujinami and Y. Ujihira, J. Non-crystal. Solids 69, 361 (1985).

25. A. Kato, K. Endo, and H. Sano, Bunseki Kagaku 29, 708 (1980).

26. G. P. Huffman and G. R. Dunmyre, J. Electrochem. Soc. 125, 1652 (1978).

27. J. A. Sawicki, T. Tyliszczak, B. D. Sawicka, and J. Kowalski, Phys. Lett. A91, 414 (1982).

28. D.Salomon, P. J. West, and G. Weyer, Hyperfine Interaction 5, 61 (1977).

29. J. A. Sawicki, T. Tyliszczak, J. Stanek, B. D. Sawicka, and J. Kowalsky, Nucl. Instrum. Methods 215, 567 (1983).

30. H. Nakagawa, K. Nomura, and Y. Ujihira, Radioisotopes 31, 94 (1982).

31. A. Handa, J. Kobayashi, and Y. Ujihira, Appl. Surface Sci. 20, 581 (1985).

32. M. Ohyabu and Y. Ujihira, J. Inorg. Nucl. chem. 43, 3125 (1981).

33. M. Ohyabu and Y. Ujihira, Nippon Kagakukaishi 1979, 624 (1979).

34. K. Kaneko and K. Inouye, Corrosion Sci. 27, 639 (1981).

35. N. N. Greenwood and T. C. Gibb, in "Mössbauer Spectroscopy", Chapman and Hall Ltd., London (1971).

36. M. Fujinami and Y. Ujihira, Appl. Surface Sci. 17, 276 (1984).

37. D. J. Vaughan and M. S. Ridout, J. Inorg. Nucl. Chem. 33, 741 (1971).

38. J. A. Morice, L. V. C. Rees, and D. T. Rickard, J. Inorg. Nucl. Chem. 31, 3797 (1969).

39. D. W. Shoesmith, P. Taylor, M. G. Bailey, and D. G. Owen, J. Electrochem. Soc. 127, 1007 (1980).

40. P. B. V. Narayan, J. W. Anderegg, and C. W. Chen, J. Electron Spectrosc. Related Phenomena 27, 233 (1982).

41. M. Fujinami and Y. Ujihira, Appl. Surface Sci. 17 265 (1984).

42. A Vèrtes, T. Szekely, and T. Tarnoczy, Magy. Kem. Folyoirat 75, 172 (1969).

43. G. W. Simmons and H. Leidheiser, Jr., in "Applications of Mössbauer Spectroscopy", Vol. 1, R. L. Cohen, ed., Academic Press, New York, p. 85 (1976).

44. U. R. Evans and C. A. J. Taylor, Corrosion Sci. 12, 227 (1972).

TRANSITION METAL SUBSTITUTION IN $Fe_{75}TM_5B_{20}$ AMORPHOUS ALLOYS. EFFECT ON STRUCTURE, THERMAL STABILITY, AND STRUCTURAL RELAXATION

P. Matteazzi

Istituto di Chimica Industriale
Università di Padova, V. Marzolo 9
I-35100 Padova, Italy

G. Cocco

Dipartimento di Chimica-Fisica
Università di Venezia
Venezia, Italy

G. Le Caer

Laboratoire de Metallurgie
Ecole des Mines
Nancy, France

G. Riontino

Istituto di Chimica Generale ed Inorganica
Facoltà di Farmacia
Università di Torino
Torino, Italy

ABSTRACT

A series of amorphous alloys of $Fe_{80}B_{20}$, with a constant amount of transition metal substitution and the nominal composition $Fe_{75}TM_5B_{20}$, have been obtained by the "melt-spinning" technique. Differential scanning calorimetry, resistivity, small angle x-ray scattering measurements, and Mössbauer spectroscopy have been utilized to study the substitution-induced modification of the behavior, in both the amorphous and annealed states, of the temperature and enthalpy of crystallization, the activation energy for crystallization, the electrical resistivity, the local magnetic and electronic properties, and the structural homogeneity. The effects of substitution on the electronic structure and hyperfine field at the iron atoms are also discussed.

1. INTRODUCTION

The effect of the addition of metallic elements on the various properties of iron-boron based metallic glasses has been extensively studied [1-5], because of the technological importance of these materials, and their useful magnetic characteristics.

Generally, the onset of crystallization destroys these useful properties, so it is important to shift to higher temperatures or longer times the beginning of the cooperative ordering processes which lead to the formation of the more stable phases and/or to complete crystallization. In this respect, it is often sufficient to control the composition of the alloy in order to prevent the primary crystallization as, for example, in the iron-boron amorphous alloys, for which [6] slightly ipereutectic boron compositions are found to correspond to the highest crystallization temperatures. The same objective may be reached by the addition of high-melting metals, which retard the diffusion processes upon which these transformations are based. This topic has been exensively discussed over the past few years, however, the phenomena involved are far from being fully understood. For example, there is much yet to be learned about the correlation between the local atomic arrangement and the properties of a glassy alloy which undergoes structural relaxation induced by annealing at various working temperatures.

The aim of this work is to study the influence that small amounts of substitutional elements of the fourth period have on the thermal stability of the $Fe_{80}B_{20}$ amorphous alloy. These influences have been characterized by differential scanning calorimetry and resistivity analysis. On the atomic scale, some information on the iron environment has been gained by Mössbauer spectroscopy. Beyond the structural and compositional aspects on the atomic level, and their possible relationship to thermally induced effects, further questions arise concerning the structural changes brought about on a larger scale by the substitutional iron neighbors. Thus subsidiary information has been obtained by small angle x-ray scattering analysis which makes possible the discovery and measurement of submicroscopic inhomogeneities within the spatial network of the glass.

2. EXPERIMENTAL

A master alloy of composition near $Fe_{80}B_{20}$ has been prepared by melting high-purity iron and boron in a purified argon atmosphere. By successive dilution with the addition of high purity transition metals of the fourth period, alloys of nominal composition $Fe_{75}TM_5B_{20}$ were obtained.

A melt-spinning technique has been employed in air to produce ribbons of about 30 microns thickness and about one to two mm width. A copper-berylium wheel has been used, rotating at about 40m/sec, as the cooling substrate. The amorphousness of the samples was checked both by x-ray diffraction and by differential scanning calorimetric analysis. The composition of each alloy ribbon was determined by spectrographic emission for boron and atomic absorption for the transition metal. The results of the analysis are reported in Table I in terms of the stoichiometric composition of each alloy (the titanium-containing alloy was $Fe_{76}Ti_4B_{20}$).

A Du Pont 1090 thermal analyzer with a 910 differential scanning calorimeter cell and a 1091 disk memory was used to measure the crystallization temperature, and the enthalpy and activation energy for the crystallization processes. Where necessary, in order to stop the crystallization process the sample was quenched by a quick admission of liquid nitrogen onto the measuring cell.

Electrical resistance measurements were performed with a d.c. four-probe ESI ohmeter, of resolution of 10^{-4} ohms, on samples one to two cm long by using pressure contact and platinum wires. The relative variation of the resistivity has been followed with increasing temperature at the rate of

Table I. Chemical Composition and Mössbauer Spectral Parameters for the Quenched Alloys[a]

Alloy	$\bar{H}$	H_{max}	σ_H	Γ	Δ	δ	$\sigma_H/\bar{H}$	M	dM
$Fe_{75}V_4B_{21}$	191	198	51	0.275	0.43	0.050	0.27	0.060	0.007
$Fe_{75}V_4B_{21}$[b]	189	207	50	0.280	0.41	0.060	0.26	0.030	0.008
$Fe_{77}Cr_5B_{18}$	189	203	53	0.280	0.42	0.046	0.28	0.050	0.005
$Fe_{77}Cr_5B_{18}$[b]	184	194	51	0.280	0.39	0.053	0.28	0.060	0.008
$Fe_{77}Mn_4B_{19}$	208	225	49	0.260	0.41	0.074	0.23	0.050	0.003
$Fe_{77}Mn_4B_{19}$[b]	209	225	46	0.270	0.42	0.079	0.22	0.060	0.008
$Fe_{81}B_{19}$	251	257	38	0.260	0.44	0.077	0.15	0.038	0.003
$Fe_{76}Co_4B_{20}$	261	265	37	0.260	0.45	0.081	0.14	0.020	0.005
$Fe_{76}Ni_5B_{19}$	253	257	39	0.260	0.49	0.085	0.15	0.030	0.007

[a]The various terms are defined in the text. [b]Data from the spectra recorded at the "magic angle" [8].

2°C/min, up to the beginning of crystallization and at constant temperatures, in a protective argon flux. In the latter case thermal equilibrium of the sample was obtained about 15 minutes after the introduction of the sample holder into a preheated horizontal furnace.

The small angle x-ray study, SAXS, was carried out with an air-free Kratky camera. Depending on the samples studied, filtered radiation was employed. The infinite beam approximation was met and a minimal accessible scattering angle was set to achieve an h value of about $7.6*10^{-3}$/Å which corresponds to a maximum value for the inhomogeneities of about 800Å. Here h is $4\pi\sin\theta/\lambda$ which is the modulus of the scattering vector with the 2θ scattering angle and λ is the x-ray radiation wavelength. The observed intensities were desmeared from the beam height effect and the SAXS analysis was accomplished according to routine procedures [7].

Mössbauer spectroscopy absorbers were made by placing several ribbons parallel to each other and had an area of about $1cm^2$. The spectra were obtained in transmission geometry in a commercial spectrometer with constant acceleration. A cobalt-57 source in a rhodium matrix, with an activity of about 10mCi, was used. The spectra of three as quenched alloys containing vanadium, chromium, and manganese were recorded with the method described in [8], at the magic angle, which allowed a texture-free evaluation of the spectra.

The distribution of magnetic hyperfine fields has been calculated with a modified Hesse-Rubartsch method [9] as described in [10]. The calculations were performed utilizing 70 magnetic components between 0 and 400kG. The results of the fitting of the experimental spectra, were evaluated with the parameters [11] Misfit M and dM. The mean isomer shift, mean quadrupole splitting, line width, anisotropic field, and line intensity ratios (except for the magic-angle spectra) were all evaluated through an iterative refining of the values, to obtain the best fit in terms of M and dM [11] by using the calculation procedure for the distribution of hyperfine fields described above.

3. RESULTS

Crystallization Temperature

Both differential scanning calorimetry, DSC, and resistivity measurements, at increasing temperatures, were used to determine the crystallization temperatures. In Figure 1 typical DSC traces are illustrated. The temperature of the onset of the exothermal emission peaks relevant to the eutectic transformation (Figure 1) was taken as the crystallization temperature, T_x. A similar trend in T_x values, against the transition metal atomic number is observed at various scanning rates, as shown in Figure 2. The relative variation in resistivity, $\Delta\rho$, with respect to the room temperature value, ρ_o, is shown in Figure 3. Here the onset of crystallization is taken as the point of maximum of $\Delta\rho/\rho_o$ curves. The resulting values are reported in Figure 4 as a function of the transition metal atomic number.

Crystallization Enthalpy

Because of the various temperature ranges and scanning rates involved, calibration procedures were adopted to convert the heat flow during crystallization into the crystallization enthalpy, H_x. Figure 5 gives the results obtained at a scanning rate of 30°C/min, a rate at which the overheating of the samples during crystallization can be considered to be negligible, because of the small quantity of the specimen (about 1.5mg).

Activation Energy for Crystallization

The activation energy, E_c, for the crystallization process has been estimated, under non-isothermal conditions, according to the Ozawa method [12]. By plotting the logarithm of the heating rate, ϕ, as a function of the inverse absolute temperature of the crystallization peak, T_m, at its maximum, one obtains a straight line whose slope gives E_c. In Figure 6 the results are reported for the as-quenched completely amorphous samples.

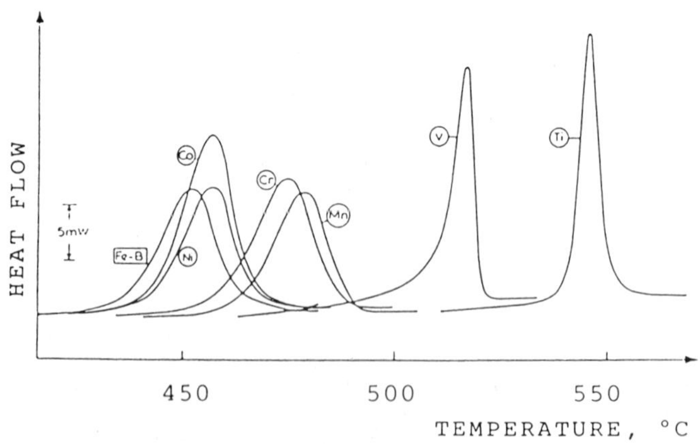

Figure 1. Calorimetric traces of the crystallization exotherm for the transition metal containing $Fe_{80}B_{20}$ amorphous alloys. The scanning rate is 30°C/min.

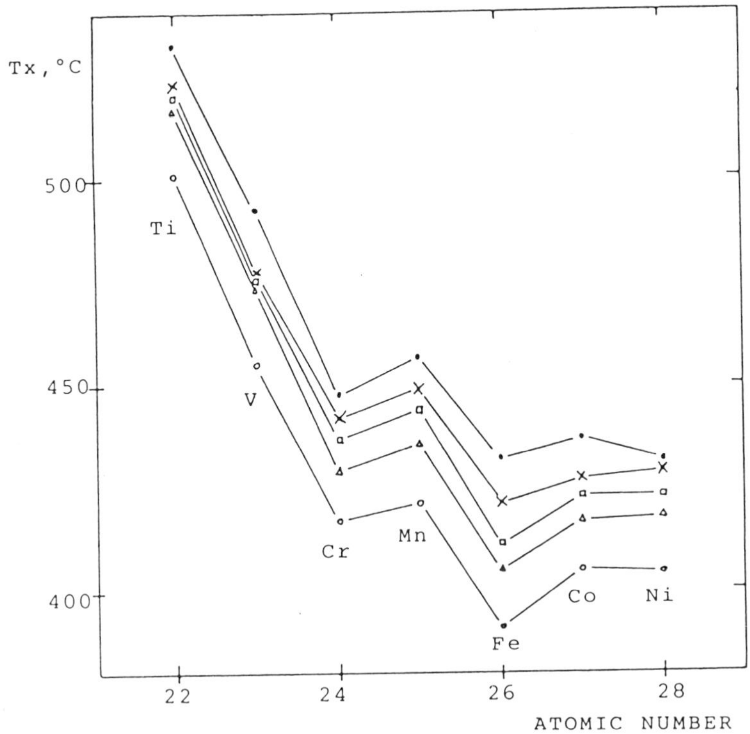

Figure 2. Crystallization temperatures T_x at the indicated scanning rates for the transition metal containing $Fe_{80}B_{20}$ amorphous alloys (●, 5°C/min; Δ, 10°C/min; □, 20°C/min; X, 30°C/min.; o, 50°C/min).

The same procedure has been adopted for partially crystallized specimens. The overall results are summarized in Figure 7, by reporting directly the E_c values as a function of the α-transformed fractions. These latter have been obtained by annealing the as-quenched sample at a suitable temperature for short times. The crystallized amount, α, was evaluated from the energy needed to complete the residual transformation, compared to the total energy emission of the corresponding completely amorphous sample.

Electrical Resistivity

Annealing at low temperature has been performed to investigate the stability of the glassy alloys against structural relaxation. This relaxation is commonly attributed to short range chemical ordering and/or short range topological ordering of the atomic species constituting the alloy. Electrical resistivity measurements under isothermal conditions, starting from temperatures as high as 60 percent of the relevant crystallization temperature, give the results reported in Figure 8. In this case, a decreasing trend is observed for cobalt and nickel, whereas the opposite behavior is found for the remaining elements.

In Figure 9 the complete results are shown, in terms of the $\Delta\rho/\rho_o$ values obtained at a constant annealing time of 200min, plotted as a function of the reduced annealing temperature, T/T_x. The initial trend of the electrical resistivity variation with respect to the room temperature value, is reported in Figure 10, together with the slope of the resulting straight lines, i.e., the temperature coefficient of resistivity.

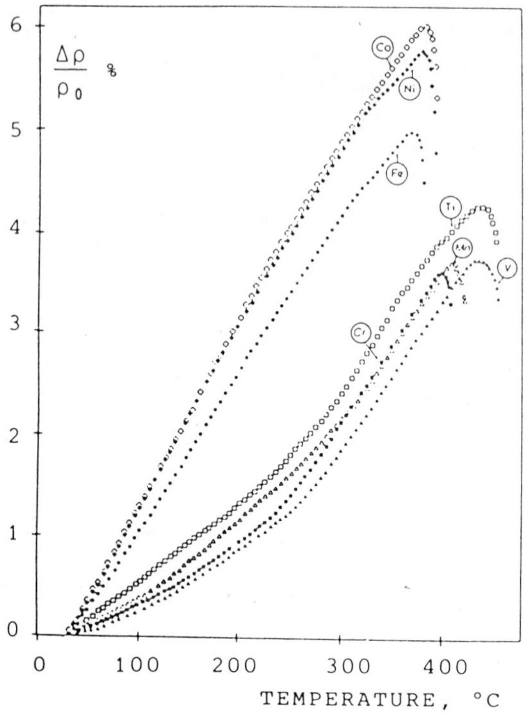

Figure 3. The relative variation of resistivity with respect to its room temperature value, as a function of temperatures.

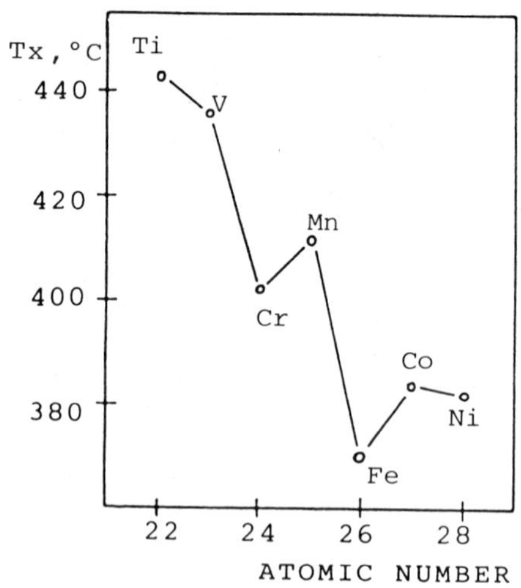

Figure 4. Crystallization temperatures, T_x, determined from the point of inflection of the resistivity curves shown in Figure 3.

Small Angle X-ray Scattering Measurements

The normalized small angle x-ray scattering, SAXS, intensities, obtained on the as-quenched alloys are shown in Figure 11. Two distinct trends are observable through the scattering range measured. A steeply descending

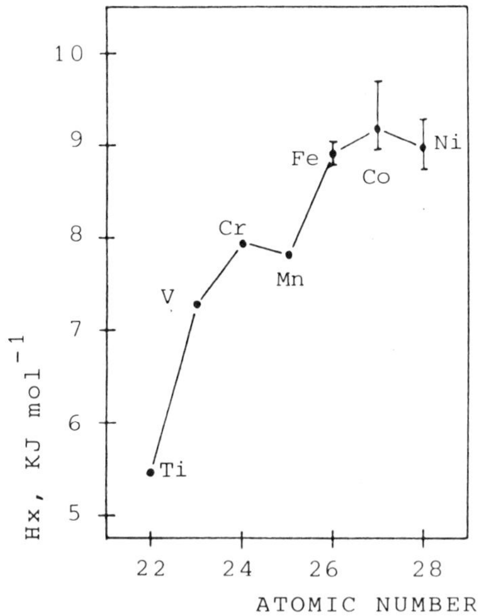

Figure 5. Crystallization enthalpy, H_x, evaluated at 30°C/min, as a function of the atomic number of the substituting elements.

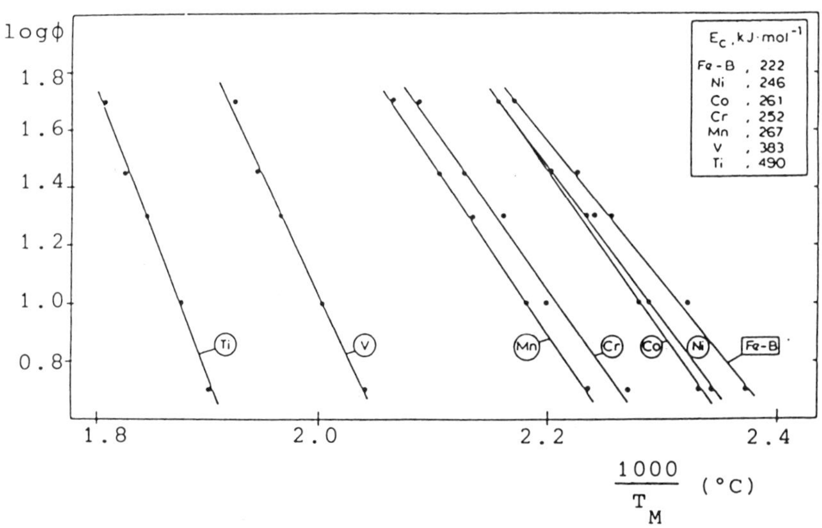

Figure 6. Logarithm of heating rate, ϕ, as a function inverse temperature at the maximum deflection of the differential scanning calorimetry signal. The activation energies for the various amorphous alloys are indicated.

behavior common to all samples can be recognized at the lowest diffraction angles. However, the intensities found for $Fe_{80}B_{20}$, as well as for the cobalt and nickel containing alloys show the peculiar "wings" caused by small

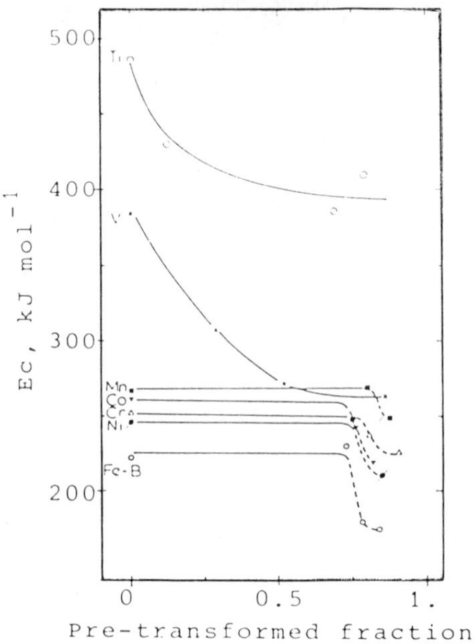

Figure 7. Activation energies, E_c, for samples in which an α pre-transformed fraction has been previously attained.

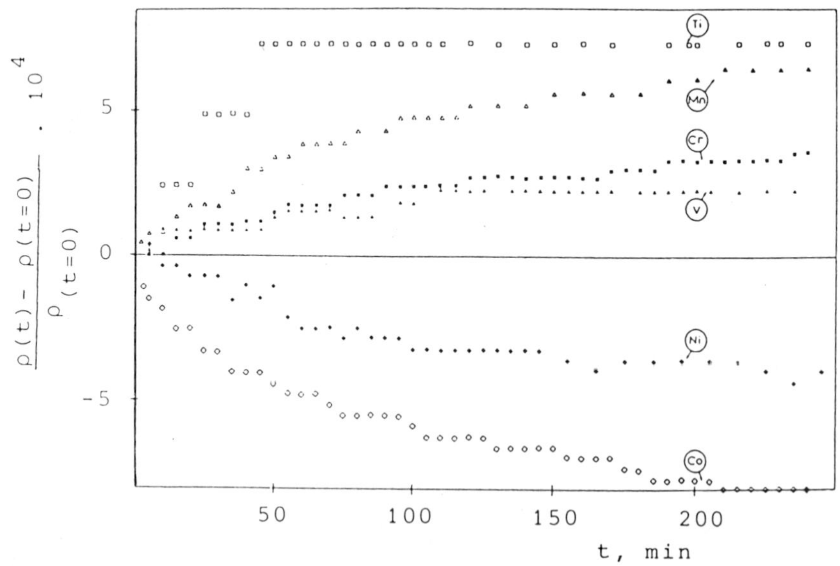

Figure 8. Relative variation of resistivity with respect to the value at zero time of annealing, as a function of time. For all the samples, the annealing temperature is about 0.7 of the respective crystallization temperature.

electron fluctuation domains. In the remaining samples, the sharp descent of the scattering curves fades out to a featureless background level. Annealing treatments up to 377°C do not appreciably change these experimental trends.

It is worth noting that surface scattering effects have a small influence at the very low angles, but are quite negligible in the outer portion of the scattering curves, as was established by further experiments on different variously thinned and polished ribbons. Thus, the essential features mentioned above, with special reference to the broad shoulders in the iron, cobalt, and nickel alloy patterns, appear to be linked to the internal structural inhomogeneities, which belong to the whole volume of the glasses.

Although the theoretical requirements to invert the experimental data were not rigorously fulfilled, an attempt was made to obtain some real-space knowledge of the structural heterogeneities in the samples. A significative picture of the different spatial electron distributions in the samples was obtained and is shown in Figure 12. We report here the electron density-density correlation functions obtained by Fourier-Bessel transformation of the intensities, obtained from specimens annealed for five hours at 207°C.

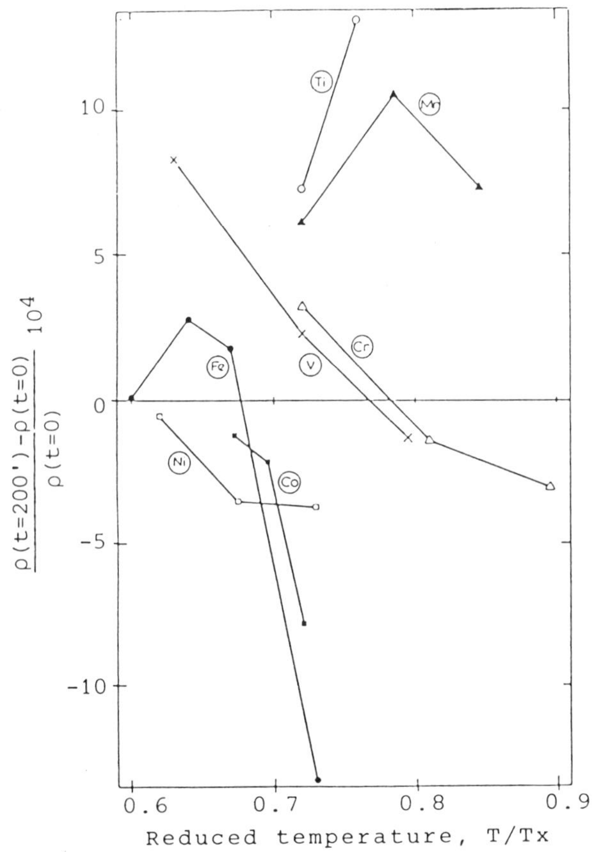

Figure 9. The fractional value of the resistivity variation, with respect to zero time of annealing, taken at 200min, as a function of the reduced annealing temperature, T/T_x

As expected from the intensity increase at the lowest angles, large inhomogeneous regions, extending up to 400Å in size, are present in all the samples studied. However, at short distances, a strong electron density-density correlation of about 20Å is distinctive of iron, cobalt, and nickel alloys, whereas the titanium and vanadium containing samples appear to be homogeneous on the same structural scale. An intermediate structural condition can be inferred for the chromium and manganese glassy alloys.

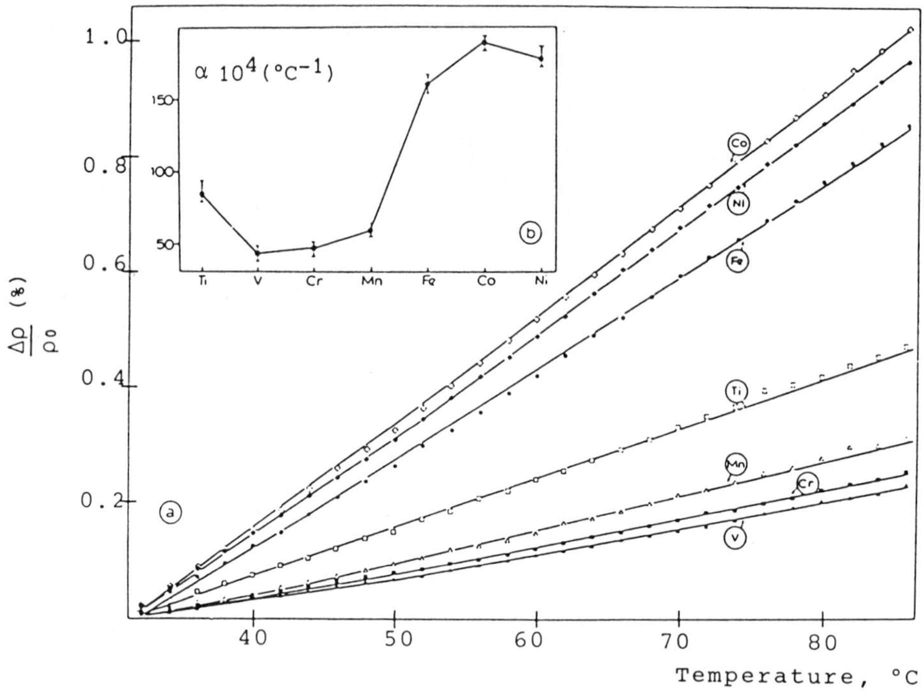

Figure 10. Relative variation of resistivity with respect to the room temperature value, as a function of temperature for the as-quenched samples. In the insert the slopes of the straight lines are indicated.

MÖSSBAUR SPECTROSCOPY

The experimental Mössbauer spectra for the as-quenched and annealed alloys, together with the fits obtained by the calculation of the distribution of hyperfine fields by the method described above, are shown in Figure 13. Figure 14 shows the Mössbauer spectra recorded at the "magic-angle" [8], for the vanadium, chromium, and manganese substituted, as quenched, alloy. The hyperfine field distributions obtained are shown in Figure 15 for the as quenched alloys and in Figure 16 for the annealed alloys. The various distributions have been plotted on the same graph for comparison. In Table I and II, the Mössbauer spectral parameters, determined by the fitting procedure described above, and some numerical results of the field distribution are presented. In these tables, $\bar{H}_{max}$ is the field of maximum intensity in the distribution, σ_H is the standard deviation in the field distribution, Δ is the mean quadrupole splitting, M and dM indicate the goodness of the fit [11] and the other symbols have their usual meaning.

Table II. Mössbauer Spectral Parameter for the Alloys Annealed at Temperature, T, for Three Hours

Alloy	T	$\bar{H}$	H_{max}	σ_H	Γ	Δ	δ	$\sigma_H/\bar{H}$	M	dM
$Fe_{75}V_4B_{21}$	230.0	188	198	52	0.280	0.36	0.045	0.28	0.030	0.008
$Fe_{77}Cr_5B_{18}$	233.0	184	203	54	0.285	0.43	0.047	0.29	0.040	0.006
$Fe_{77}Mn_4B_{19}$	233.0	209	221	48	0.270	0.40	0.078	0.23	0.030	0.007
$Fe_{81}B_{19}$	208.5	253	257	37	0.270	0.44	0.077	0.15	0.050	0.006
$Fe_{76}Co_4B_{20}$	208.5	261	270	40	0.270	0.45	0.081	0.15	0.007	0.007
$Fe_{76}Ni_5B_{19}$	208.0	256	261	37	0.260	0.47	0.087	0.14	0.016	0.006

[a]The various terms are defined in the text.

Table III. The Influence of Substitution on the Field Distribution

Substitution Element	$(\sigma^2-\sigma_b^2)/(\bar{H}-\bar{H}b)^2$
V	0.32(6)
Cr	0.30(6)
Mn	0.47(7)

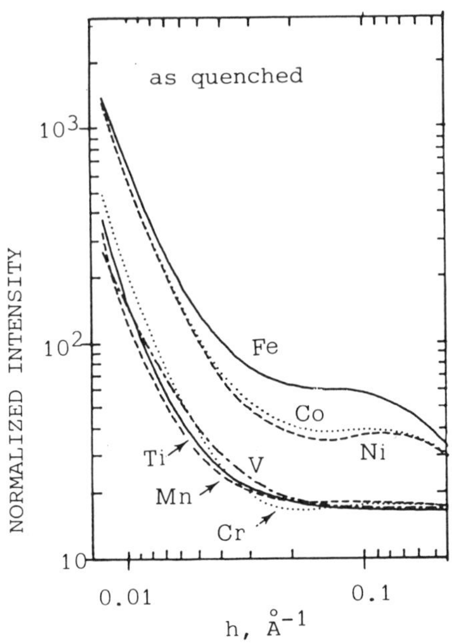

Figure 11. Small angle x-ray scattering curves for the transition metal substituted alloys.

Figure 17 shows a comparison of the field distribution obtained from the spectra recorded at the magic angle and from the spectra recorded with the gamma rays normal to the absorber. Very good agreement between the two distributions is observed. Moreover, the fitted values of line intensity (A-2,5/A1,6) ratio are close to the theoretically [8] expected ratio of 3:2:1:1:2:3 for the magic angle spectra. It should also be noted that the measurements have been performed in two different laboratories.

DISCUSSION

Crystallization Temperature

The Specific displacement, $\Delta T_x/\Delta_c$, of the crystallization temperature, produced by the unitary atomic addition of a transition element in the $Fe_{81}B_{19}$ master alloy, is illustrated in Figure 18 as a function of the atomic number. The presence of the iron-substituting element, in each case increases the crystallization temperature of the master alloy, titanium being the most effective. The same holds for the crystallization temperatures, determined from the beginning of the rapid decrease of the relative resistivity variation as a function of temperature (Figure 4). The observed trend is similar to the one reported by Naka, et al. [4], who also found an almost linear increase in the $Fe_{80}B_{20}$ crystallization temperature with increasing levels of substitution up to 5 to 10 atomic percent.

In a search for a possible correlation, it has been observed [5,13] that, in some cases, the sublimation enthalpy of the elements, a measure of their cohesive energy, may give a possible explanation for the observed behavior (Figure 18 and 19). Alternatively, the atomic radius difference between solute and iron atom has also been considered [3]. However, by

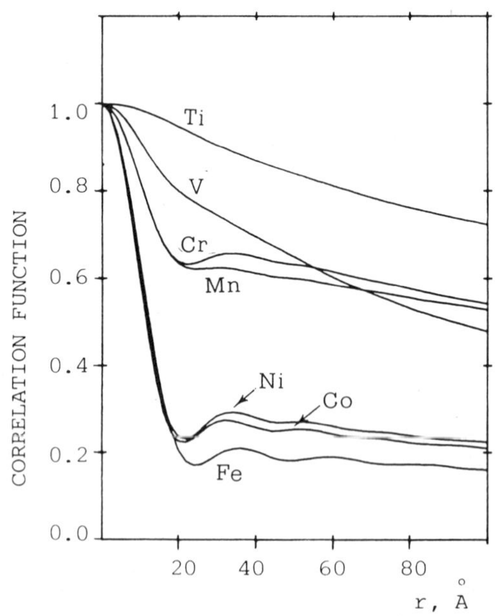

Figure 12. The normalized correlation functions obtained by the Fourier-Bessel transformation of the intensities for the transition metal substituted alloys indicated.

taking in account the nearly constant boron content in our alloys, a better correlation is found by considering the steric effect displayed, by the substitutional element, in the corresponding iron solid solution. According to [15], a volume size factor, Ω_{sf}, can be defined as

$$\Omega_{sf} = \frac{\Omega^*_B - \Omega_A}{\Omega_A},$$

in which Ω^*_B is the effective atomic volume of a solute B in a solvent A, and may be obtained by a linear extrapolation of the measured solid solution volumes up to 100 percent solute. In Figure 20, the Ω_{sf} values are plotted against the atomic number of the transition element. Good agreement is observed with the corresponding trend of the crystallization temperature variation illustrated in Figure 18. A similar dependence on the atomic number of the same added element, has been previously [16] found in the case of a recrystallization study of cold-worked iron solid solutions. This analogous

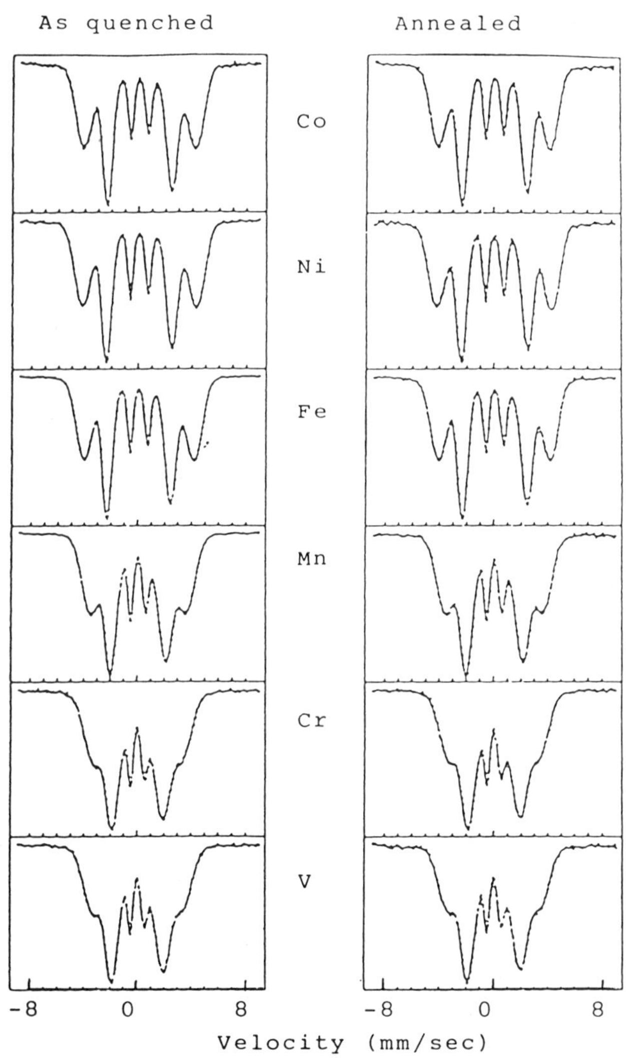

Figure 13. Experimental Mössbauer spectra for the as-quenched and annealed alloys listed in Table I and II.

behavior suggest that a steric effect is the determining factor for the mobility of atoms in defective systems and, in particular, for the thermal stability of glassy alloys.

The correlation of the specific displacement, $\Delta T_x/\Delta c$ found for the substituting elements with the crystallization enthalpy, H_x, of the corresponding alloy yields the trend shown in Figure 21. As a rough approximation, it can be observed that the effectiveness of the element in retarding

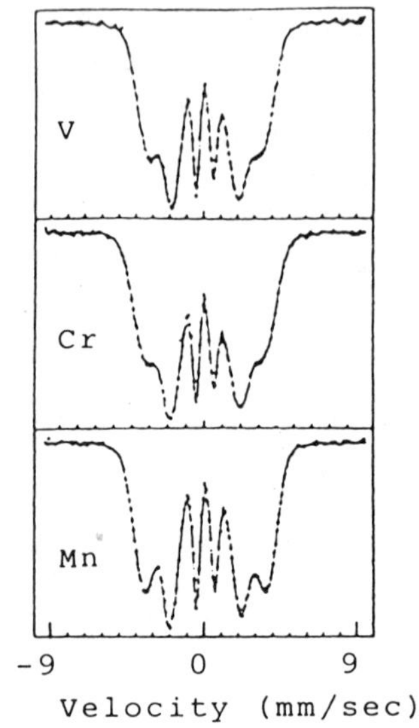

Figure 14. Experimental Mössbauer spectra recorded with the magic-angle method for the as-quenched alloys indicated.

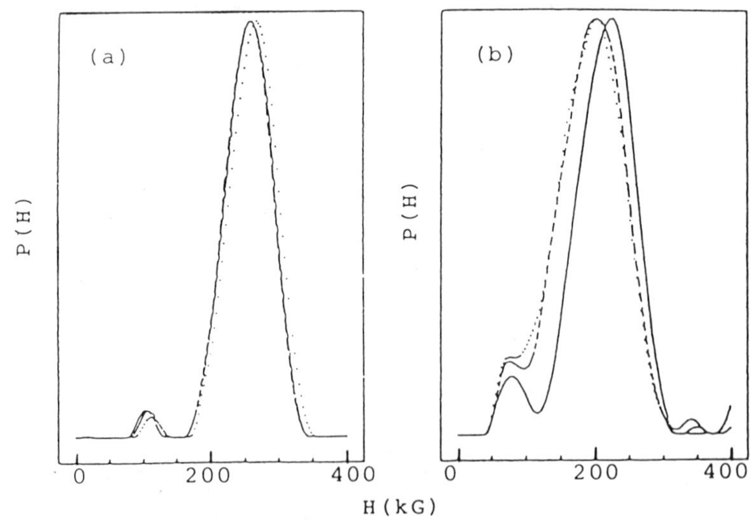

Figure 15. Hyperfine field distributions for the as-quenched alloys calculated from the spectra shown in Figure 13. The substituting transition metals are (a) Ni, ————, Co, •••••••, Fe, -------; (b) Mn————, Cr, •••••••, V, -------.

the crystallization increases as the enthalpy of the process decreases. It should be noted that cobalt and nickel, as compared with the other substituting elements, act in different manners on the structural relaxation of the iron-boron master alloy, as is discussed below. However, other aspects, such as the presence of quenched-in nuclei, strongly affect the overall energy content, of the amorphous material. Thus, additional data will be necessary to separate the contributions of relaxation and crystallization. Further insight into these topics will be presented below.

ACTIVATION ENERGY FOR CRYSTALLIZATION

The thermal stability as crystallization occurrence of amorphous alloys is related to kinetic factors. The energy barrier that must be overcome in the thermally activated crystallization process plays a fundamental role in the determination, for example, of the best service properties. To evaluate the apparent activation energy, E_c for crystallization, we have adopted the Ozawa method [12], whose validity has been previously [16-18] demonstrated. From the slopes of the straight lines reported in Figure 6, we obtain the E_c values shown in Figure 22 as a function of the atomic number of the substituting elements. The trend is quite similar to the one shown in Figure 18 for the crystallization temperatures, indicating, to a first approximation, a common basis for the crystallization event. It is commonly accepted that the crystallization of glassy alloys proceeds through nucleation and crystal growth. In some cases, [19-21] indirect evidence for the presence of quenched-in nuclei during melt-spinning has been presented.

Some authors [19,20] have advanced this hypothesis on the basis of the differential scanning calorimetry, DSC, behavior displayed by $Fe_{80}B_{20}$ alloy. The presence of a shoulder in the high-temperature side of the DSC peak, has been shown to pertain to the nuclei-free fraction of the ribbons.

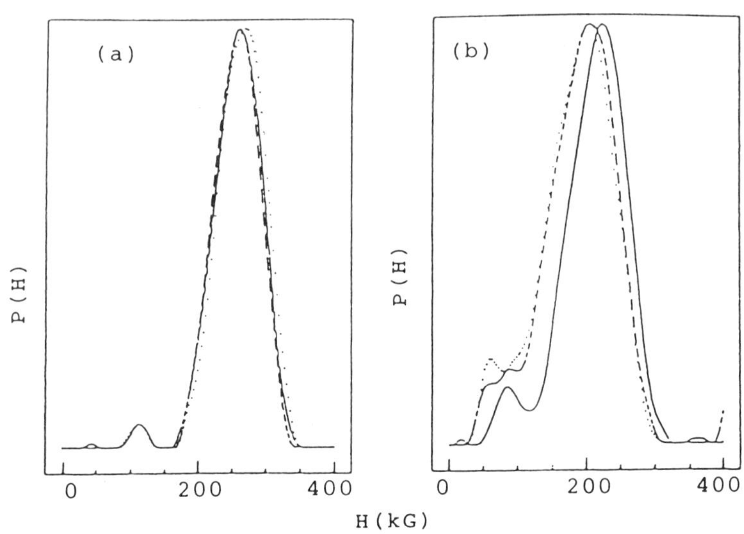

Figure 16. Hyperfine field distributions for the annealed alloys calculated from the spectra shown in Figure 13. The substituting transition metals are (a) Ni, ———, Co, •••••••, Fe, -------; (b) Mn, ———, Cr, •••••••, V, -------.

In order to separate the contribution to the overall activation energy, E_c, by the nucleation energy, E_n, and by the growth energy, E_g, some preliminary thermal treatments have been performed to promote, in all the samples, a partial crystallization. This procedure will mainly produce nucleation or growth, depending on the initial concentration of the nuclei. Because the partial area under a DSC peak is proportional to the amount of the crystalline fraction [22], the α-transformed fraction was evaluated as the ratio between the area under the DSC peaks of the heat treated and as-quenched samples. In Figure 7 we report the E_c data so obtained.

Two different trends are observed in the activation energy. For the titanium and vanadium containing metallic glasses, a substantial reduction of E_c is observed, even at low α values. For the other alloys the activation energy remains practically constant up to an α of 75 percent, beyond which a small decrease is observed. This behavior has been explained through an analysis, in non-isothermal conditions [23], of the n-exponent of the Avrami law, governing the crystallization process. According to the equation proposed [24] for the contribution of E_n and E_g to the total measured

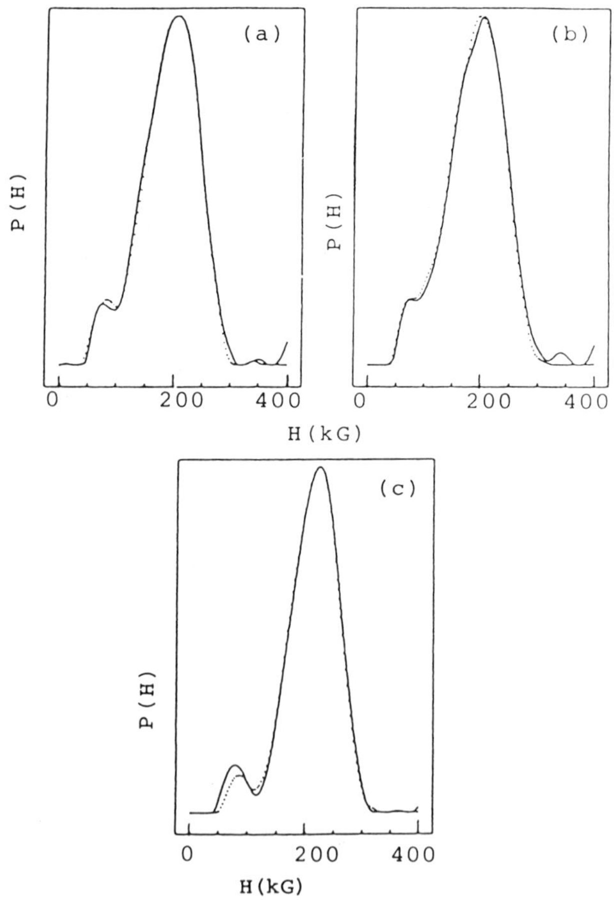

Figure 17. A comparison of the hyperfine field distribution calculated from the spectra recorded with a standard geometry, ————, and at the [8] magic angle, •••••••. The alloys are (a) $Fe_{75}V_4B_{21}$, (b) $Fe_{77}Cr_5B_{18}$, and (c) $Fe_{77}Mn_4B_{19}$.

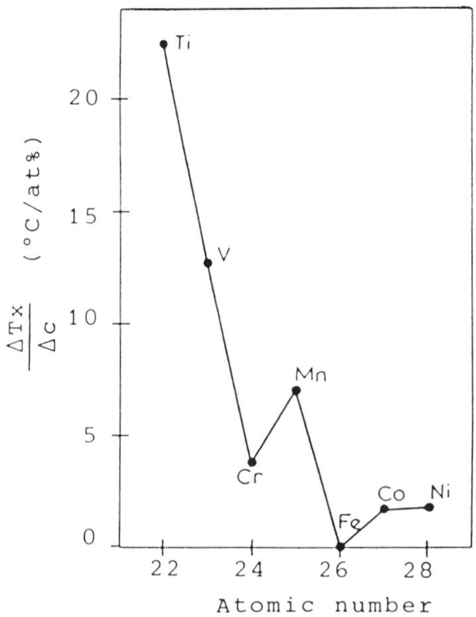

Figure 18. The specific displacement of the crystallization temperature produced by a unitary atomic percent addition of the elements.

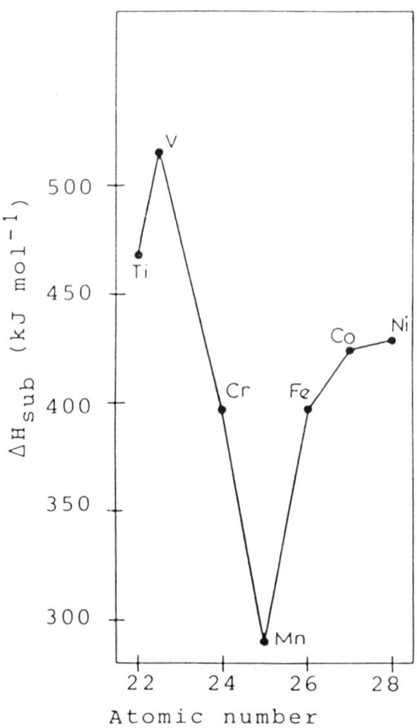

Figure 19. The sublimation enthalpies of the substituting elements [14].

activation energy, E_c, in the case of titanium and vanadium substitution, a large part of E_c results from E_n; there is a progressive lowering of E_n at high α-fractions, and only crystal growth is significant. This can explain the relatively high stability against crystallization found for these alloys [25]. For the other alloys, because a constant E_c value is observed, crystal growth is significant. This is the case as can be seen in Figure 7. The decrease in E_c when the highest pretransformed fractions are reached, can be attributed to the enhanced mobility of the crystallization fronts towards the initial nuclei-free portion of the ribbon.

STRUCTURAL RELAXATION

Thermal aging may induce some modification of the physical properties, [26,28] which reflect the local rearrangements occurring in the amorphous structure. Chemical and/or topological short range ordering are usually claimed to occur, with different modalities, during the structural relaxation. In this regard, some insight can be gained from isothermal behaviors of electrical resistivity, at different temperatures, because this property is very sensitive to the small modifications of an amorphous structure taking place during the first stages of transformation [29]. Some typical results have already been reported in Figure 8, which shows the two different trends of the electrical resistivity variations with respect to the zero time of annealing. The results may be interpreted in terms of a higher probability for clustering, or chemical short range ordering, induced by the elements at the left of iron in the periodic table, in analogy with the analogous behavior observed [29] in FeNiMoB amorphous alloys. The clustering may be due to a local redistribution of the transition metal and boron atoms, which

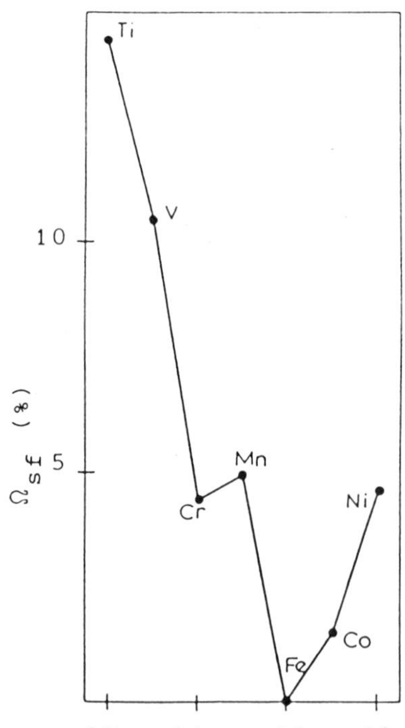

Figure 20. The volume size factor for the substituting elements (see text for explanation).

precede the mixed borides of the iron and transition metal found [4,30] during eutectic crystallization of the corresponding alloys. For the other alloys, a decrease of the resistivity with time is observed, due to the reduction of the free volume frozen-in during quenching, as predicted by a recent theory [29].

Because the free volume reduction must be a common path towards crystallization, one can conclude that elements like titanium, vanadium, chromium, and manganese, for which an increase in resistivity is observed, act as stabilizing factors for the amorphous structure during relaxation. In other words, they shift to higher temperatures or times the beginning of the more cooperative processes, which lead to the free-volume annihilation and, subsequently, to crystallization.

The question still remains as to whether the transition metal forms preferential bonds with iron atoms or boron atoms, because, during crystallization, a solid solution iron-transition metal is also observed [30]. In nickel-iron-boron amorphous alloys, theoretical [31] and experimental [32] results suggest a preferential nickel-boron interaction.

Because the initial relative structural conditions of the amorphous with respect to the crystalline state, are rate determining for structural relaxation, a more detailed investigation has been made at temperatures slightly higher than room temperature, at which no irreversible structural relaxation takes place. Several scans at 2°C/min were performed on different samples of the same alloy. The results are summarized in Figure 10, in which the slopes for each composition are also reported. It was previously shown [33] that the initial slope of the resistivity-temperature curves can be related to the quenched-in free volume content. For the same composition, but at different free-volume contents, lower initial resistivity slopes have been found to be related to higher free volume content.

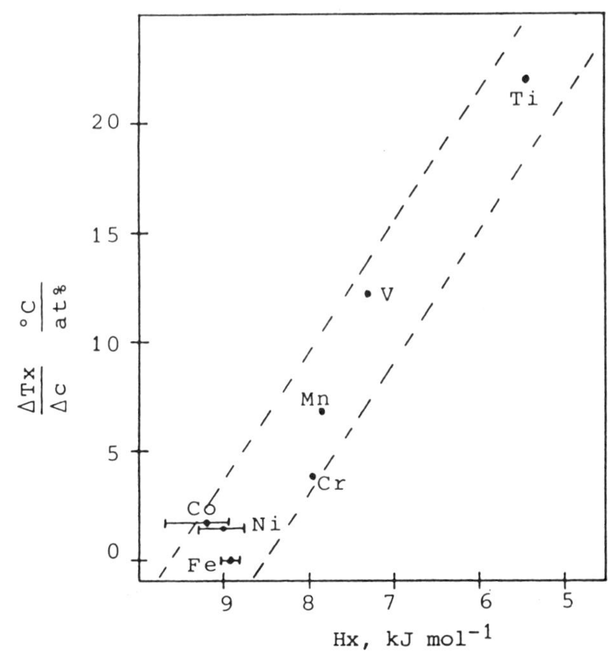

Figure 21. The specific displacement of the crysallization temperature as a function of the crystallization enthalpy of the corresponding amorphous alloy.

In the case of the present alloys, although other parameters, such as the electronic structure cannot be neglected, the trend of the initial resistivity slope as a function of the atomic number of the substituting elements seems to indicate that a lower quenched-in free volume content exists in the iron-boron and cobalt and nickel containing alloys. In terms of its annihilation, a less stable condition pertains in these alloys, as deduced from Figures 8 and 9.

LOCAL MAGNETIC AND ELECTRONIC PROPERTIES

In Figure 23 the dependence of $\bar{H}$, σ_H, and the ratio $\sigma H/\bar{H}$ upon the substituting metal is plotted. As can be seen, the addition causes first a change in the hyperfine fields, which may be related to a change of the spin density, and second a change in the isomer shift (Figure 24), which may be related to a change of the charge density. In agreement with some previous [34] results on amorphous alloys, the addition of elements to the left of iron in the periodic tables causes a decrease in the hyperfine field, whereas the elements to the right increases slightly the hyperfine field. According to Walker et al. [35] the electronic charge density at the iron nuclei decreases with increasing isomer shift. Thus the charge density in these alloys (Figure 24) increases from nickel to vanadium substitution. Moreover, the behaviors of both $\bar{H}$ and the isomer shift, as a function of atomic number, are similar to those observed in crystalline alloys by some authors [36]. This seems to indicate that similar effects, related to valence changes, are present in both cases. Apparently structural effects do not have obvious consequences. Cser et al. [36] have found that the hyperfine field and isomer shift vary as a function of the atomic number of the transition metal

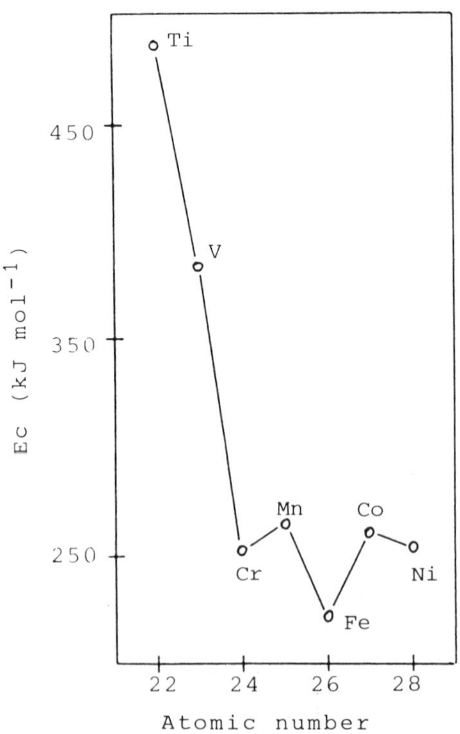

Figure 22. Total activation energy for crystallization.

in a similar manner. They found an average ratio $\Delta\bar{H}/\Delta\delta$ of $0.13T/\mu_s$. In the amorphous alloys of the present investigation the ratio is about $0.2T/\mu_s$ higher than in the crystalline alloys. According to Cser et al. [36], increasing charge density leads to a positive change of the hyperfine field and, for iron, to lower values of the hyperfine field.

The change of charge density (of one sign) is not necessarily correlated with a change of the spin (up or down) density. But, in these amorphous

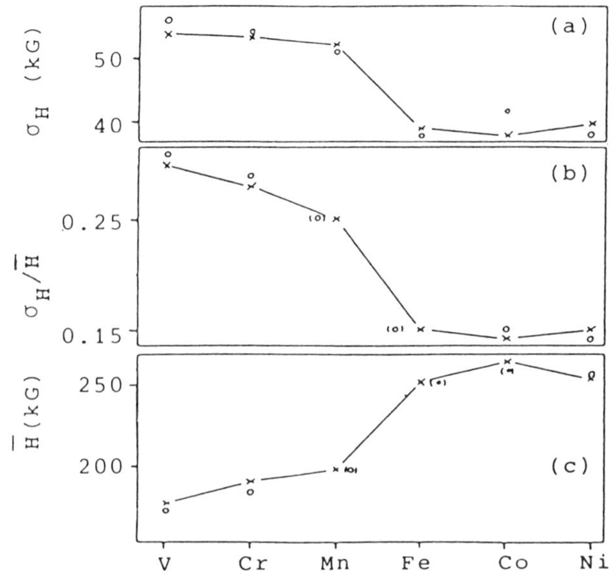

Figure 23. Behavior of various quantities calculated from the hyperfine field distribution, corrected for a 5 atomic percent substitution, for the as-quenched (x) and annealed (o) alloys.

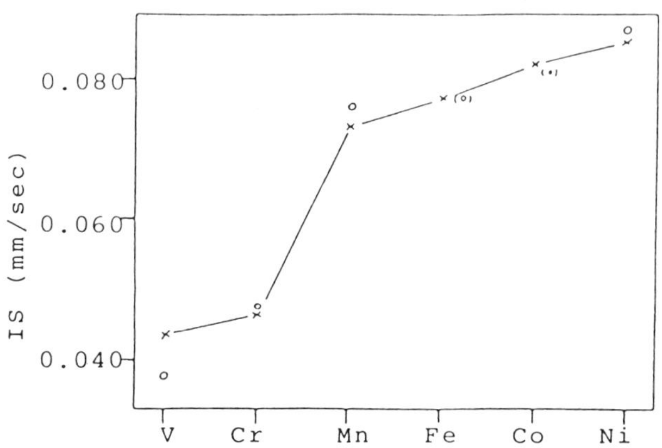

Figure 24. The behavior of the isomer shift corrected to a 5 atomic percent substitution, for the as-quenched (x) and annealed (o) alloys.

alloys, as in the crystalline case, the charge density and the spin density seems to vary together. The simplest explanation of this correlation [37] is that only spin-up or only spin-down electrons are involved in the transfer. Indeed for chromium and vanadium substitution in crystalline alloys, it has been found [38] that the decrease in $\bar{H}$ must be due to an increase of the spin-up electron density.

The effect of the substitution has been also interpreted on the basis of the virtual bound state concept [39] or by assuming that the addition of elements from vanadium to manganese have magnetic moments which are antiferromagnetically coupled to the matrix. This causes the decrease in the average magnetization [40].

The width of the hyperfine field distribution as indicated by σ_H, and the ratio between this width and the mean field, $\sigma_H/\bar{H}$, varies as a function of the substituting transition metal, decreasing with increasing atomic number as shown in Figure 23.

In the as-quenched and annealed alloys, there is an obvious correlation between the decrease in the mean field and the increase in the standard deviation of the distribution. In a simple model, in which we neglect the influence of structural changes due to the substitution of iron by the other transition metal atoms, which yields almost no change in coordination number and only small changes in the metal-metal distances, and if we assume that additivity holds in $Fe_{80-x}TM_xB_{20}$, when x is small, we would have

$$\bar{H} = \bar{H}_b + x\sum_i N_i \Delta H_i$$

and

$$\sigma^2 = \sigma_b^2 + x(1-x)\sum_i N_i \Delta H_i^2.$$

This corresponds to a random substitution of iron by the transition metal, where ΔH_i is the field change due to one atom in the ith coordination shell with a mean coordination, N_i, and $\bar{H}_b$ is the corresponding field in $Fe_{80}B_{20}$. If ΔH_i, for i greater than one, are neglected, we get

$$\sigma^2 = \sigma_b^2 + \frac{(1-x)(\bar{H}-\bar{H}_b)^2}{N_1 x}.$$

If we take N_1 at 12.4 in $Fe_{80}B_{20}$ for bonds between 2.25 and 3.45Å [41] or N_2 at 18.1 for bonds between 3.5 and 4.6Å, we deduce that for x between 0.04 and 0.05, $(1-x)/N_1x$ is 1.5 to 2.4; if we use two sheels with the assumption that ΔH_1 equals ΔH_2, we would obtain 0.6 to 0.8 and numbers in between for $\Delta H_2 < \Delta H_1$. If we use the experimental results in Table I and II for vanadium, chromium and manganese (for cobalt and nickel, the precision of the calculation too low) we obtain, for the ratio $(\sigma^2-\sigma_b^2)/(\bar{H}-\bar{H}_b)^2$, the values given in Table III. The experimental values are of the correct order of magnitude, but lower than the values expected for the very simple model used. Nevertheless, the trends are correctly accounted for by the model.

The samples annealed at the temperature indicated in Table II for three hours show a behavior of the various parameters similar to those of the as-quenched samples. Figure 25 shows the comparison of the hyperfine field distribution for the as-quenched and annealed samples. The distribution of the annealed samples appear in general 'smoothed' with respect to the as-quenched samples, but the differences in the shape of the distributions are very small for the cobalt and nickel substitutions. From Table II an increase in the linewidth, Γ, in the annealed samples can be observed. The isomer shift (Figure 24) is observed to increase on annealing corresponding to a reduction of electronic charge density, an exception occurs for vanadium substitution. The hyperfine fields decrease on annealing for substitution with metals to the left of iron.

From these various effects we may conclude that the annealing generally introduces a decrease in the electron charge density and that the changes in the hyperfine field, depend on the atomic number of the substituting transition metal.

STRUCTURAL HOMOGENEITY FROM SMALL ANGLE X-RAY SCATTERING

Because the experimental conditions were maintained strictly constant during the preparation process, small angle x-ray scattering results show the presence of distinctive features in the underlying structure of the alloys. Large scale heterogeneities are common to the examined samples, and it is probably best to say that there are regions of extended defects because of the non-uniformity of the solidification process. However, the heterogeneties cannot be regarded as representative of the structural effects induced by the transition metals with which we are dealing. Conversely, density-density correlations extending a few atomic diameters in size are characteristics of iron-boron and nickel and cobalt containing alloys, and allow a structural distinction as a function of the substituting elements. This

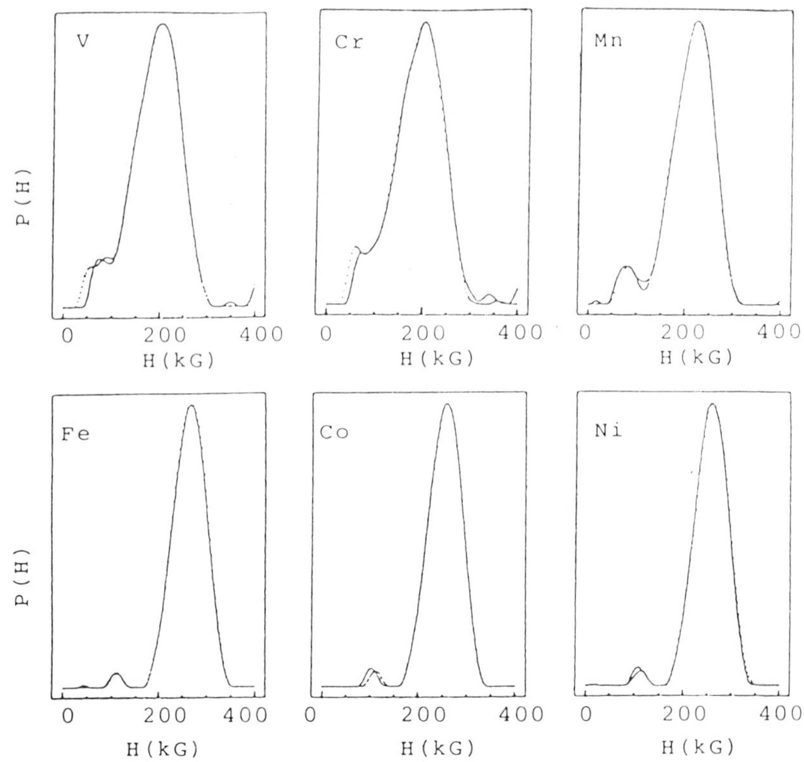

Figure 25. A comparison of the hyperfine field distributions in the as-quenched, ————, and annealed, •••••••, alloys for the substituting transition metals indicated.

real-space structural pattern is in keeping with and supplements the overall results reported above. Particularly their correlation with the activation energies for crystallization appears to be straightforward. Except for the titanium and vanadium containing glasses, quenched-in nuclei were inferred from the E_c trends relevant to the occurrence of the crystallization process in the preheated samples. Thus the demixing phenomena found by small angle x-ray scattering both in the as-quenched and annealed conditions, appears properly associated with the final phase separation required for the establishment lone-range order.

A troubling problem concerns the extent to which these structural inhomogeneities depend on the quenching operations. However, it must be noted that a growing body of experimental evidence indicates the occurrence of a microscale separation phenomena in iron-boron based glasses. Small angle scattering [42,43] and transmission electron microscopy [44] are noticeable examples. Details accounts on this topic are presented elsewhere [45]. The role of the separating events in determining the basic magnetic properties of the alloys has been reviewed [46].

By turning back to the general characterization resulting from our data, we can see that even the differentiation among the samples resulting from both the resistivity and Mössbauer results fit in the small angle x-ray scattering structural framework. Thus, a correspondence can be recognized between the inhomogeneous structures and the decrease in resistivity observed with iron, cobalt, and nickel. This correspondence is also seen in the small variations in the hyperfine field and isomer shift for these alloys, as opposed to the remaining alloys. This correspondence is also in agreement with the conclusions recently reported [47] on similar amorphous systems.

CONCLUSIONS

Mössbauer spectroscopy appears to give answers to the basic structural questions associated with these amorphous alloys. Based on the strong charge transfer and increased spin observed, it appears reasonable to assume a random substitution of iron by the elements on its left in the periodic table. A different quenched structural condition can be inferred for cobalt and nickel substitution. On this basis, preferential interactions lead to an inhomogeneous structural material probably due to a dual structure as is indicated by small angle x-ray scattering and confirmed by differential scanning calorimetry. Moreover, the evolution of the Mössbauer parameters during the thermal annealing, indicates the occurrence of structural relaxation as evidenced by resistivity changes, primarily due to the irreversible reduction of the quenched-in free volume in the cobalt and nickel substituted alloys. In contrast, clustering phenomena prevail during this stage, particularly for titanium and vanadium containing alloys, for which a more homogeneous glassy network may be proposed. Finally, it should be noted that the effects of substitution on the local magnetic and electronic structure, are similar in the amorphous and crystalline alloys.

REFERENCES

1. L. Granasy, A. Lovas, L. Kiss, T. Kemeny, and E. Kisdi-Koszo, J. Magn. Magn. Mater. 26, 109 (1982).

2. A. Lovas, L. Granasy, K. Zambo-Balla, and J. Kiraly, in Proceedings Conference on "Metallic Glasses: Science and Technology", CRIP Budapest, vol. II, p. 291 (1980).

3. J. L. Walter, Mater. Sci. Eng. 50, 137 (1981).

4. M. Naka, U. Gonser, C. Gorlitz, and H. Ruppersberg, J. Non-Cryst. Solids 45, 99 (1981).

5. I. W. Donald and M. A. Davies, in "Proc. Third Int. Conf. on Rapidly Quenched Metals", B. Cantor, ed., vol. I, The Metals Society, London, p. 273 (1978).

6. R. Ray, R. Hasegawa, C. P. Chou, and L. A. Davis, Scripta Met. 11, 973 (1977).

7. G. Cocco, S. Enzo, C. Antonione, G. Riontino, G. Venturello, Solid State Comm. 51, 771 (1984).

8. J. M. Greneche and F. Varret, J. Phys. C 15, 5333 (1982).

9. J. Hesse and A. Rubartsch, J. Phys. E 7, 526 (1974).

10. G. Le Caer and J. M. Dubois, J. Phys. E 12, 1083 (1979).

11. S. L. Ruby, Mössbauer Effect Meth. 8, 263 (1973).

12. T. Ozawa, J. Therm. Anal. 2, 301 (1970).

13. M. Naka, J. Non-Cryst. Solids 41, 71 (1980).

14. R. Hultgren, R. L. Orr, P. D. Anderson, and K. K. Kelley, in "Selected Values of Thermodynamic Properties of Metals and Alloys", J. Wiley and Sons, N. Y., 1963.

15. H. W. King, J. Mater. Sci. 1, 79 (1966).

16. C. Antonione, G. Della Gatta, A. Lucci, G. Riontino, and G. Venturello, Acta Met. 18, 1169 (1970).

17. C. Antonione, L. Battezzati, A. Lucci, G. Riontino, and G. Venturello, Scripta Met. 12, 1011 (1978).

18. L. Battezzati, A. Lucci, and G. Riontino, Thermochem. Acta 23, 213 (1978).

19. A. L. Greer, Acta Met. 30, 171 (1982).

20. A. S. Schaafsma, H. Snijders, F. van der Woude, J. W. Drijver, and S. Radelaar, Phys. Rev. B 20, 4423 (1979).

21. F. E. Luborsky and H. H. Liebermann, Appl. Phys. Lett. 33, 233 (1978).

22. J. Borchardt and F. Daniels, J. Am. Chem. Soc. 79, 41 (1957).

23. J. Colmenero, J. Ilarraz, and J. M. Barandiaran, Thermochim. Acta 35, 381 (1980).

24. S. Ranganathan and M. Von Heimendhal, J. Mat. Sci. 16, 2401 (1981).

25. A. Lucci, G. Riontino, F. Marino, C. Antonione, and G. Cocco, Materials Chemistry and Physics 10, 549 (1984).

26. R. W. Cahn, Contemp. Phys. 21, 43 (1980).

27. T. Egami, K. Maeda, and V. Vitek, Phil. Mag. 41, 883 (1980).

28. T. Egami, Ann. N. Y. Acad. Sci. 371, 236 (1981).

29. P. Allia, D. Andreone, R. Sato Turtelli, F. Vinai, and G. Riontino, J. Appl. Phys. 53, 8798 (1982).

30. L. Battezzati and G. Riontino, to be published.

31. R. P. Messmer, in "Amorphous Metallic Alloys", F. E. Luborsky, eds., Butterworths, London p. 114 (1983).

32. I. Vincze, F. Van Der Woude, T. Kemeny, and A. S. Schaafsma, J. Magn. Magn. Mater. 15-18, 1336 (1980).

33. P. Allia, R. Sato Turtelli, F. Vinai, and G. Riontino, Solid State Commun. 43, 821 (1982).

34. L. Takacs, A. Vertes, A. Lovas, E. Kovacs, J. Farkas, and L. Kiss, Nuclear Instruments and Methods 199, 281 (1982).

35. L. R. Walker, G. K. Wertheim, and V. Jaccarino, Phys. Rev. Letters 6, 3 (1961).

36. L. Cser, I. Gladkin, and C. Hargitai, Phys. Stat. Sol. (b) 124, 271 (1984).

37. S. M. Dubiel and J. Zukrowski, J. Magn. Magn. Mater. 23, 214 (1981).

38. S. M. Dubiel and W. Zinn, J. Magn. Magn. Mater. 45, 298 (1984).

39. R. C. O'Handley, Solid State Comm. 38, 703 (1981).

40. T. Kemeny, B. Fogarassy, I. Vincze, I. W. Donald, M. J. Besnus, and H. A. Davies, Proc. 4th Int. Conf. on Rapidly Quenched Metals, T. Masumoto and K. Suzuki, eds., The Japan Institute of Metals, Sendai, Japan, p. 851 (1982).

41. E. Nold, P. Lamparter, H. Olbrich, A. Rainier-Harbach, and S.Steeb, Z. Naturforsch. 36a, 1032 (1981).

42. K. Osamura, K. Shibue, R. Suzuki, and Y. Murakami, J. Mat. Sci. 16, 957 (1981).

43. E. Nold, S. Steeb, and P. Lamparter, Z. Naturforsch 35a, 610 (1980).

44. J. L. Walter, S. F. Bartram, and I. Mello, Mater. Sci. Eng. 36, 193 (1978).

45. J. M. Dubois and G. Le Caer, in "The Structure of Non-Crystalline Materials", P. H. Gaskell, ed., Taylor & Francis, London, p. 206 (1982).

46. J. Durand, in "Glassy Metals: Magnetic, Chemical and Structural Properties", R. Hasegawa, ed., CRC Press, Inc., Florida (USA), p. 109 (1983).

47. J. L. Walter and A. E. Berkowitz, in "Rapidly Quenched Metals, S. Steeb and H. Warlimont, eds., Elsevier Science Publishers B.V., New York, p. 1303 (1985).

MÖSSBAUER INVESTIGATIONS OF MAGNETIC INTERACTIONS AND CRYSTALLIZATION BEHAVIOR OF THE AMORPHOUS ALLOY $Fe_{72}Ni_4Mo_6B_{16}Si_2$

R. Jagannathan and T. G. N. Babu

School of Chemistry
University of Hyderabad
Hyderabad 500 134, India

B. B. Prasad and A. K. Bhatnagar

School of Physics
University of Hyderabad
Hyderabad 500 134, India

V. R. V. Ramanan

Allied Corporation
Morristown, NJ 07960

INTRODUCTION

Metallic glasses, produced by rapid quenching of metallic melts have emerged as an important class of materials with unusual combinations of desired properties. Further, many of the unique characteristics arise due to their amorphous structure. Study of the amorphous state is of great current interest as evidenced by the publications of books and the proceedings of international conferences on this subject [1-3]. One of the features of these alloys is that they can be prepared with a variety of compositions with a number of elements, providing a method to alter or modify their magnetic, mechanical, electrical, and corrosion behavior. The alloys constitute a large range of material containing from two to as many as five elements. They behave as very soft magnetic materials, exhibit exceptional hardness, high tensile strengths, and almost zero thermal expansion in certain instances. Some of them show marked resistance to corrosion. Such combinations of properties and in particular the existence of ferromagnetism in these alloys, a property earlier thought to be impossible in non-crystalline materials, have made them potential and competitive candidates in a wide range of applications. Recent developments include their suitability in large and electronic-size transformers, in motors, in mechanical applications such as brazing filler materials, as transducers and as potential catalysts.

Among technologically important magnetic metallic glasses, which fall into three categories, viz. transition metal-metalloid, rare earth-transition metal and transition metal-(zirconium, hafnium) alloys, the first mentioned has attracted significant attention, particularly by way of understanding the variations in properties due to composition change and thermal and magnetic annealing effects. They typically contain about 80 atomic percent iron, cobalt and/or nickel and 20 atomic percent metalloid atoms such as carbon silicon, boron, phosphorus, or aluminum. Compositions which include small percentages of other transition metals, such as titanium, vanadium, chromium, molybdenum, tungsten, copper, etc., are known to influence resistance to corrosion significantly [4]. Compositional change also brings about significant differences in the active sites associated with metallic glasses used as hydrogenation catalysts [5]. In addition, they bring about large changes in the Curie, T_c, and amorphous to crystalline transformation, T_x, temperatures and produce differences in their thermal and magnetic annealing behavior. Most of the alloys which are rich in iron content lend themselves to in-depth studies by Mössbauer spectroscopy as they give rise to several spectral features characteristic of their amorphous nature and give information about T_c, T_x, thermal stability, magnetic anisotrophy, hyperfine field distribution and the nature of amorphous to crystalline state transformation and crystalline products. The present work deals with the ferromagnetic metallic glass of nominal composition $Fe_{72}Ni_4Mo_6B_{16}Si_2$ in continuation of our earlier published [6] work on a closely related molybdenum containing amorphous alloy, $Fe_{39}Ni_{39}Mo_4Si_6B_{12}$. The presence of molybdenum in these alloys is known to improve their corrosion properties. Further, the amorphous alloys containing these constituent elements in different compositions have been shown to lead to a minimum in alternating current core-loss, after annealing treatments, when the nickel content is rather low at ca. 5 atomic percent [7].

We present here a detailed Mössbauer study of the metallic glass $Fe_{72}Ni_4Mo_6B_{16}Si_2$ in the temperature range 80-900K. Its Curie, T_c, and crystallization, T_x, temperatures have been determined. The variation of reduced magnetic hyperfine field with reduced temperature, which shows a significant difference, characteristic of amorphous alloys, is dealt with in the light of existing theories. By using the variation of the hyperfine field, H_{eff}, with temperature, spin-wave stiffness coefficients were obtained and compared with values for crystalline systems. The value of β, the critical exponent has been determined. A Fourier analysis of the hyperfine field distribution obtained for the Mössbauer spectra measured below T_c is presented. A comparison of T_c and T_x values with related systems containing various amounts of molybdenum has been made. From the values of the hyperfine fields of the components of the room temperature Mössbauer spectrum of the crystallized alloy and x-ray data, the products of crystallization have been deduced.

EXPERIMENTAL

The metallic glass, $Fe_{72}Ni_4Mo_6B_{16}Si_2$, was received from Allied Corporation (USA) in a ribbon form 0.47cm wide and ~25μm thick. The Mössbauer spectra at different temperatures from 80 to 900K were recorded by using an Elscint Mössbauer spectrometer in the constant acceleration mode. A 60mci cobalt-57 rhodium source matrix is used. Mössbauer measurements were performed on an absorber containing pieces of the ribbons kept parallel. The velocity scale was calibrated with a thin natural iron foil as reference. A laboratory made cryostat was used for the low temperature measurements. An Elscint vacuum furnace was used for high temperature measurements, the temperature of which had an accuracy of ±0.5K. All the spectra were analyzed by fitting the data with a least square fit computer program which assumed Lorentzian line shapes. The Curie temperature of the sample was determined both from the temperature dependence of the internal magnetic field as well as by the thermal scan method [8].

MÖSSBAUER SPECTRA AND MAGNETIC INTERACTIONS

Representative Mössbauer spectra of the "as received" glassy alloy, $Fe_{72}Ni_4Mo_6B_{16}Si_2$, recorded in the temperature range of 80 to 900K are shown in Figures 1 and 2. The six-line pattern indicative of the ferromagnetic state of the sample is observed up to 443K. The absorption lines are characterized by rather large linewidths of ca. 0.6 to 1.6mm/sec compared to the 0.28mm/sec found for a thin natural iron-foil. The observed broad lines are typical or iron-rich metallic glasses, and arise from a random distribution of iron atoms with varying coordinations. From 420 to 460K the Mössbauer spectra are unresolved with large line widths of the order of ca. 1.8mm/sec which clearly indicate the presence of magnetic interactions. The spectrum at 490K is a narrow doublet with a quadrupole splitting of 0.46(2)mm/sec.

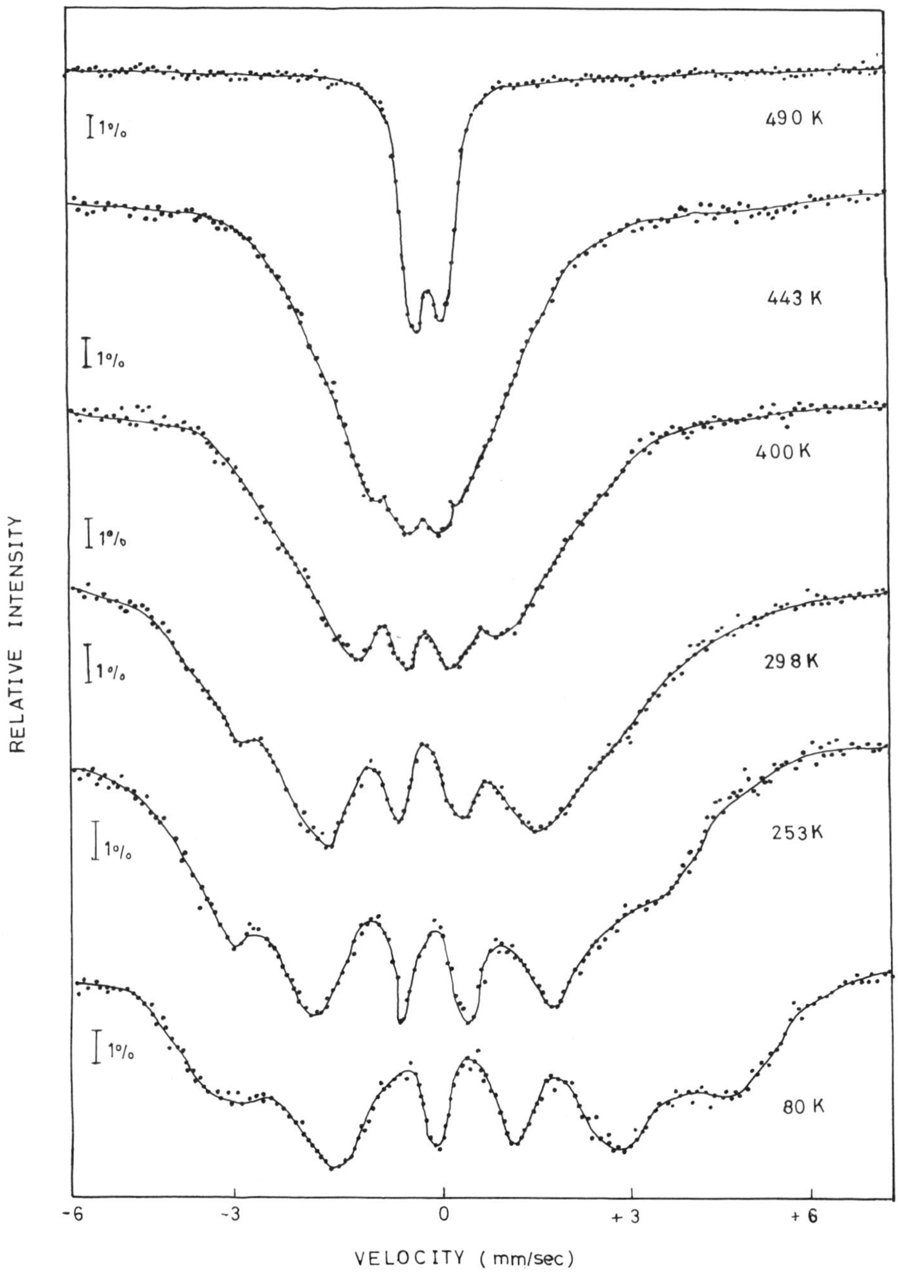

Figure 1. Mössbauer spectra of amorphous $Fe_{72}Ni_4Mo_6B_{16}Si_2$ from 80K to 490K.

The paramagnetic profile persists up to 850K. At ca. 880K and above, the spectra show complex profiles due to the crystallization of the glassy sample. Thus the Curie temperature, T_c, and the crystallization temperature, T_x, for this sample are determined to be 490(5)K and 880(5)K, respectively. A least square fit analysis of the fairly well resolved Zeeman split spectrum recorded at 80K gave rise to relative intensities for the 2 and 5 lines of ca. 1.7 indicating that the average magnetization direction lies along the ribbon plane. This is found to be the case for the amorphous glass, $Fe_{40}Ni_{38}Mo_4B_{18}$ (METGLAS 2826 MB), [9], and $Fe_{72}Ni_8Mo_2B_{16}Si_2$, [10], as well. In the former case thermal treatment and compression effects have been found to tilt the magnetization direction away from the plane. The poor resolution of the lines of the Mössbauer spectra recorded at higher temperatures does not permit any definite conclusion regarding the temperature dependence of orientation in the present case; but the absence of any change in the intensities of the second and fifth lines indicates the absence of change in the magnetization orientation.

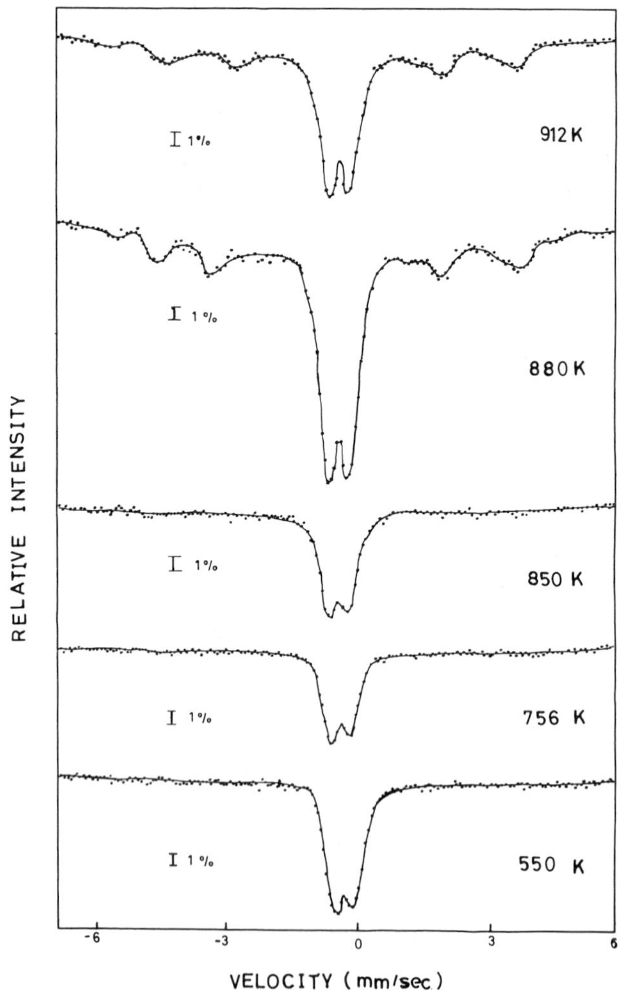

Figure 2. Mössbauer spectra of amorphous $Fe_{72}Ni_4Mo_6B_{16}Si_2$ from 550K to 912K.

Figure 3 shows normalized average line widths of peaks 1 and 6, 2 and 5, and 3 and 4. It is observed that the normalized average linewidths decrease as a function of temperature indicating that the distribution of magnetic hyperfine fields becomes narrower with an increase in temperature. This is further confirmed by the decrease in the full width at half maximum of the hyperfine field distribution curves with increasing temperature as discussed later.

CURIE AND CRYSTALLIZATION TEMPERATURE

The Curie temperature, T_c, of amorphous $Fe_{72}Ni_4Mo_6B_{16}Si_2$ was deduced by both the extrapolation of the $H_{eff}(T)$ versus temperature curves and by the thermal scan method. From the $H_{eff}(T)$ versus temperature plot, the vanishing magnetic hyperfine field, corresponding to T_c, is found to occur at 490(5)K and the saturation magnetic hyperfine field $H_{eff}(0)$ is found to be 248(5)kOe. A more accurate determination of T_c was accomplished by using the thermal scan method and was found to be 490(2)K as shown in Figure 4. This method also showed that the amorphous ferromagnet crystallizes at ca. 880K. The crystallization of the sample at temperature below 880K cannot be ruled out because bulk Mössbauer measurements indicate crystallization at a higher temperature as compared to the value given by surface Mössbauer measurements [11].

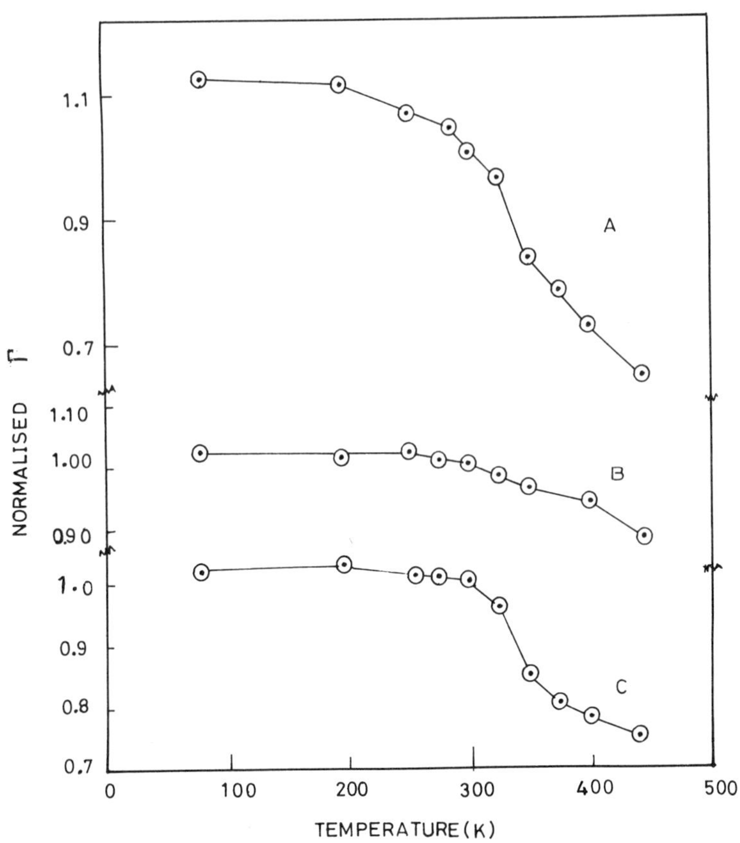

Figure 3. Normalized average linewidths of lines 1 and 6(A), 2 and 5(B), and 3 and 4(C) of amorphous $Fe_{72}Ni_4Mo_6B_{16}Si_2$ as a function of temperature.

Table I. Comparison of Curie Temperature, T_c, and crystallization temperature, T_x, of related metallic glasses

Metallic Glass	T_c, °K	T_x, °K	Reference
$Fe_{40}Ni_{40}B_{20}$	695	700	20
$Fe_{40}Ni_{38}Mo_4B_{18}$	608	813	a
$Fe_{59}Ni_{23}Mo_4B_{14}$	655	698	b
$Fe_{109}Ni_{45}Mo_{17}B^{29}$	570	706	b
$Fe_{39}Ni_{39}Mo_4B_{12}Si_6$	575	725	6
$Fe_{72}Ni_4Mo_6B_{16}Si_2$	490	880	this work

[a] R. Chandra and G. Chandra, Proc. Int. Conf. on Mössbauer Effect, Jaipur, India, Indian National Science Academy, New Delhi, p. 357, 1982. [b] P. Allia, D. Andreone, R. Sato Turtelli, F. Vinai, and Giontino, J. Magn. Magn. Mat. 26, 139 (1982).

It is worthwhile at this point to look into the variations of T_c and T_x brought about by compositional change. It is seen in Table I that the introduction of molybdenum reduces down T_c. The large reduction in T_c in $Fe_{72}Ni_4Mo_6B_{16}Si_2$ is brought about both by lowering the nickel content as well as by increasing the molybdenum content. Change in the nickel to iron ratio exceeding 60 to 40 is known to reduce T_c significantly [12]. On the other hand, T_x is influenced considerably by changes in the composition of the metalloid atoms. Replacing boron by silicon also helps in enhancing T_x and simultaneously decreasing the critical cooling rate, making it easier to produce the alloy. A comparison of T_c and T_x of various amorphous alloys, as shown in Table I, however, shows that a small increase in molybdenum or a large decrease in nickel content also results in a significant increase in T_x. This is in agreement with the observations of Walter [13], Donald et

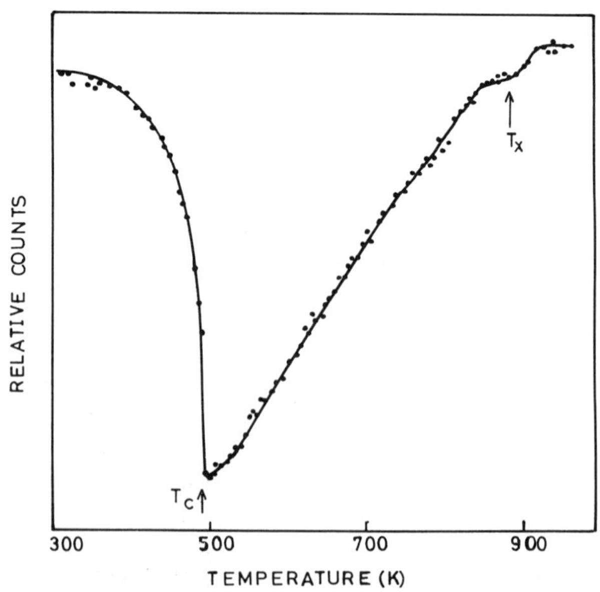

Figure 4. Count rate at Zero Doppler velocity as a function of temperature for amorphous $Fe_{72}Ni_4Mo_6B_{16}Si_2$.

al. [14], and Lovas et al. [15] that, in glasses of the type (Fe,Ni)-M-B-Si where the atomic size of M is larger than that of iron or nickel, T_x is enhanced as the free volume and diffusion coefficient render the alloy more resistant to crystallization. Replacement of iron by up to 2 to 3 atomic percent nickel initially increases T_x, but subsequently brings it down rapidly [16]. An increase of molybdenum content no doubt favors T_x but drastically decreases T_c [17]. An increase in boron increases both T_c as well as T_x. Naka et al. [18] have shown that elements on the right of iron in the periodic table, such as cobalt or nickel enhance T_x only marginally, whereas elements on the left of iron such as titanium, vanadium, and chromium increase T_x significantly. Larger, T_x for the present sample (Table 1) arises from the molybdenum which apparently resembles elements such as those lying on the left of iron in the periodic table, the addition of which even in small quantities of up to 10 atomic percent enhances T_x quite appreciably [18].

TEMPERATURE DEPENDENCE OF MAGNETIC HYPERFINE FIELD

The temperature dependence of the reduced averaged magnetic hyperfine field, $H_{eff}(T)/H_{eff}(O)$ vs. T/T_c is plotted in Figure 5 along with the Brillouin curve for S equal to one. For any given reduced temperature, $H_{eff}(T)$, is always smaller than the corresponding value for α-iron due to the presence of the nonmagnetic metalloid atoms in the alloy. Experimental points lie considerably below the Brillouin curve as observed for other metallic glasses, indicating that $H_{eff}(T)$ decreases more rapidly with temperature than in crystalline ferromagnets. This rapid decrease in the hyperfine fields, typical of amorphous ferromagnets can be explained by the

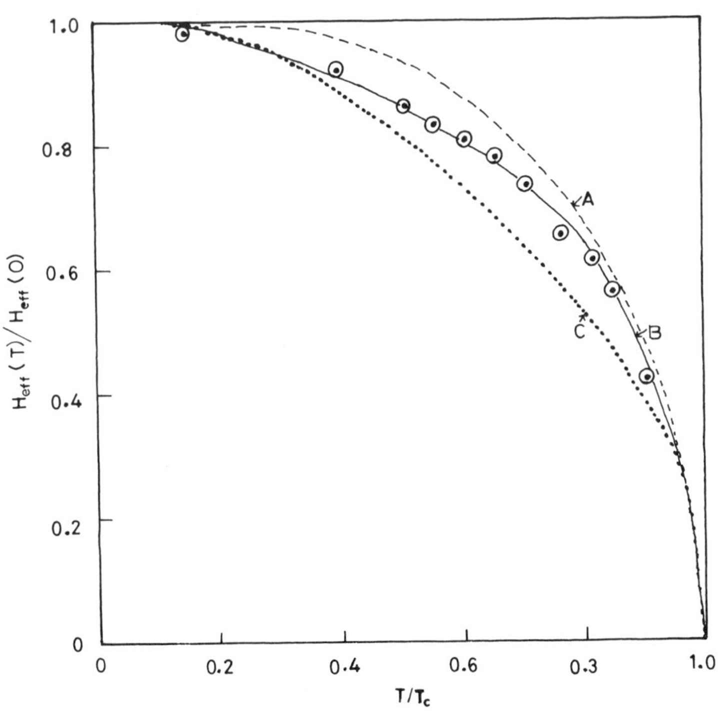

Figure 5. Reduced hyperfine field vs. reduced temperature of amorphous $Fe_{72}Ni_4Mo_6B_{16}Si_2$. Curves (A), (B), and (C) represent the theoretical curves of equation (1) for δ values of 0, $0.55(1-t^2)$, and 0.5 respectively, where S is 1, and t is T/T_c.

Handrich model [19]. The expression for reduced hyperfine fields in amorphous ferromagnets, according to this model may be written as

$$m(T) = \frac{M(T)}{M(0)} = \frac{1}{2}B_s[(1+\delta)x] + \frac{1}{2}B_s[(1-\delta)x] \qquad (1)$$

where B is the Brillouin function for the spin S, and

$$x = \frac{3S}{(S+1)}\frac{m}{t}$$

where t is T/T_c. The parameter δ is a measure of random fluctuations in the exchange interaction and its value lies between zero and one. Equation (1) reduces to the case of crystalline ferromagnets when δ is zero. Equation (1) gives a faster decrease in the reduced magnetic hyperfine fields with T/T_c when δ is not equal to zero. However, it has been observed that even with δ equal to 0.5 or higher there is a serious quantitative disagreement between experimental data and the Handrich model in the low and high temperature regions. If δ is assumed to be temperature dependent and an empirical relationship,

$$\delta = \delta_0(1-t^2)$$

where t is T/T_c is substituted into equation (1) it was found to give a good quantitative agreement for δ_0 of 0.55 as shown in Figure 5. This supposition has been found to be valid for many other iron-rich metallic glasses reported from our laboratory [8,11,20].

Yet another theoretical approach which explains the flattening of magnetization curve is based on the Monte Carlo method as illustrated for systems with S of 1/2 by Fahnle [21] who took into account fluctuations of the exchange interactions (bond disorder) and of the magnetic moments (site disorder). This model does show the importance of correlations between the exchange fluctuations (correlated bond disorder) but still lacks agreement with experimental data at high temperatures. Our good empirical fit obtained by using the temperature dependence of δ and the poor agreement between the Monte Carlo calculations and the experimental points at high temperatures suggest that thermal correlations between spin fluctuations play a significant role.

HYPERFINE FIELD DISTRIBUTION

The distribution in the hyperfine field in the ferromagnetic glasses arises from the presence of a number of inequivalent iron sites in the glassy sample. Several methods have been developed to analyze the distribution in H as a function of the probability of the field, P(H). The method due to Window [22] employing a Fourier series has been shown to have the merit of being model independent and giving a distribution information different from those based on assumptions such as binomial distribution in the types of coordination [23]. The field distribution, P(H), in this method is given by

$$P(H) = \sum_{n=1}^{M} a_n[\cos(n\pi H/H_o)-(-1)^n],$$

where H_o is a large value of H at which P(H) is zero and a_n are the fourier coefficients. In this analysis it is found that the smoothness of the desired solution as judged by the χ^2 value depends on the number of terms, M, used and the relative line intensities [24]. In the present case M equals 9 has been found to give lower χ^2 value for the relative intensities 3:b:1:1:b:3. Addition of further terms is found to influence the χ^2 value very little but introduces more oscillations in the distributions. The P(H) distribution curves thus obtained, for temperatures from 80 to 400K are given

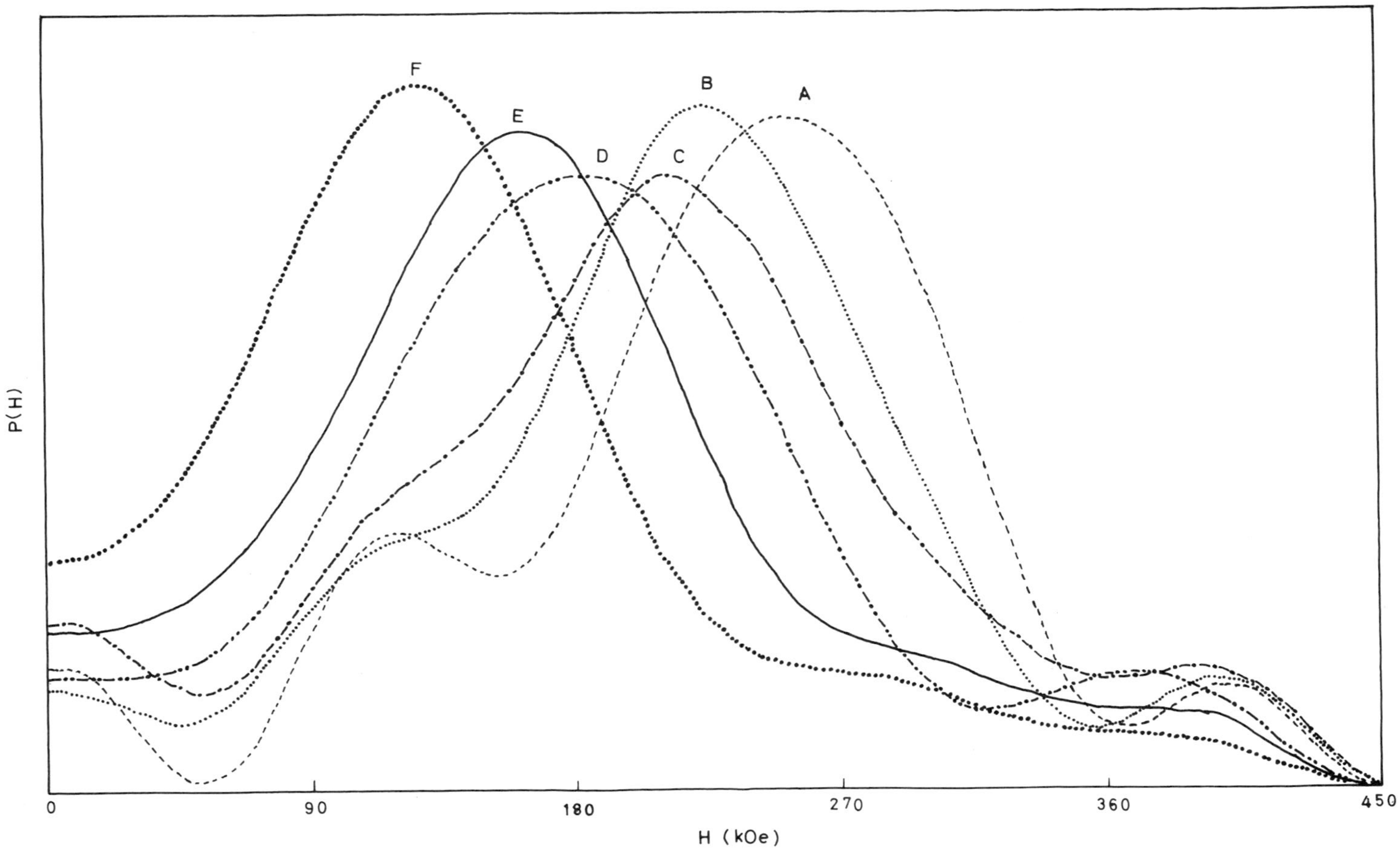

Figure 6. P(H) distribution for the amorphous $Fe_{72}Ni_4Mo_6B_{16}Si_2$ at (A) 80K (B) 196K, (C) 253K, (D) 298 K, (E) 350K, and (F) 400K.

Table II. Most Probable Field, H_p, Mean Field, $\bar{H}$, Average Magnetic Hyperfine Field, $H_{eff}(T)$, and Field Distribution Width, for Amorphous $Fe_{72}Ni_4Mo_6B_{16}Si_2$

T, °K	$\bar{H}$,kOe	$H_{eff}(T)$,kOe	H_p,kOe	ΔH,kOe
80	227.2	240.0	252.0	140.5
196	210.5	230.2	224.5	130.5
253	201.8	214.4	209.0	131.0
298	184.5	200.6	184.5	133.0
350	169.3	182.3	162.0	112.5
400	139.5	152.0	136.0	103.0

in Figure 6. It was found that the data could be well fit with a relative intensity of b by 2.6 to 2.8 for all the spectra below T_c.

From the resultant P(H) curves, the mean hyperfine field, $\bar{H}$, the most probable field, H_p, and the full width at half maximum ΔH, can be obtained. These values obtained from the analyses of Mössbauer measurements of amorphous $Fe_{72}Ni_4Mo_6B_{16}Si_2$ at different temperatures are given in Table II. The temperature dependence of $\bar{H}$, H_p, ΔH is shown in Figure 7. It is of interest to compare the results based on the P(H) analysis with similar analysis of other ferromagnetic glasses. In many instances only one dominant peak has been observed [24,25] with ΔH of about 90kOe. In the present case ΔH is 140kOe. The significantly broader and asymmetric distribution may be attributed to the presence of molybdenum leading to an unresolved bimodal

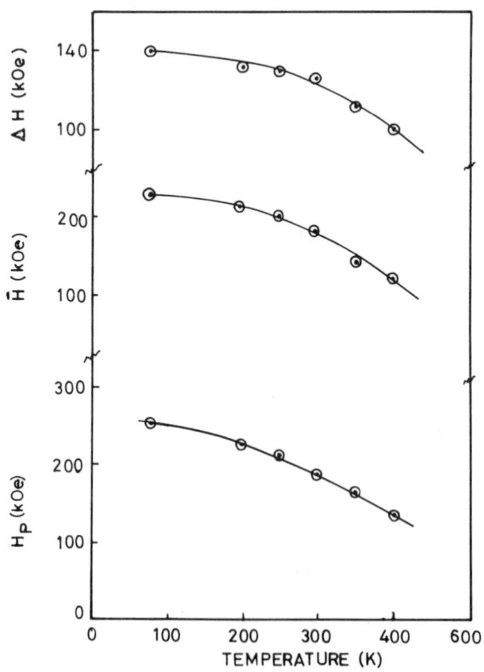

Figure 7. Temperature dependence of most the probable field, H_p, the mean field, $\bar{H}$, and ΔH, the distribution width for amorphous $Fe_{72}Ni_4Mo_6B_{16}Si_2$.

distribution of the hyperfine field as in the case of the chromium containing alloy, $Fe_{32}Ni_{36}Cr_{14}P_{12}B_6$, where a well resolved two component distribution has been observed [23]. The minor component may be attributed to iron sites containing molybdenum near neighbors. The lower field of the minor component arises due to the antiferromagnetic interaction of the iron and molybdenum atoms as suggested in the literature [23,26]. The dominant contribution arises from the iron-iron and iron-nickel ferromagnetic interaction. A decrease in ΔH with increasing temperature as shown in Figure 7 indicates that the distribution of hyperfine fields becomes narrower with increase in temperature and that long range interactions in the amorphous alloys are important [24,27]

SPIN WAVE EXCITATIONS

Early measurements of low temperature $T^{3/2}$ demagnetization behavior suggested the existence of spin wave excitations in amorphous alloys [28,29]. Subsequently inelastic neutron scattering confirmed the existence of well defined spin wave excitations. In the Mössbauer studies the relevant spin wave stiffness coefficients B and C may be obtained by using the expression [30],

$$\frac{H_{eff}(T)}{H_{eff}(O)} = 1 - BT^{3/2} - C\,T^{5/2}. \qquad (2)$$

This expression is valid at low temperature and the $T^{3/2}$ dependence comes from the excitation of long wavelength spin waves. Equation (2) may be rewritten as

$$\frac{\Delta H_{eff}(T)}{H_{eff}(O)} = -B_{3/2}(T/T_c)^{3/2} - C_{5/2}(T/T_c)^{5/2} \qquad (3)$$

where

$$\Delta H_{eff}(T) = H_{eff}(T) - H_{eff}(O).$$

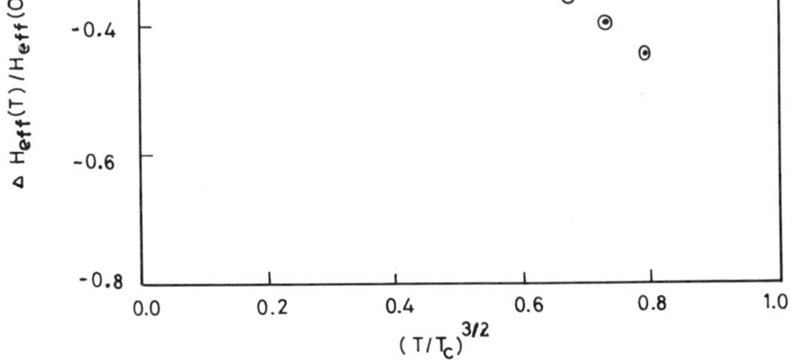

Figure 8. Fractional change of the hyperfine magnetic field, for the amorphous glass, $Fe_{72}Ni_4Mo_6B_{16}Si_2$.

Table III. Spin Wave Parameters for Amorphous $Fe_{72}Ni_4Mo_6B_{16}Si_2$ Resulting from Different Temperature Intervals

T/T_c	$B, K^{-3/2}$	$C, K^{-5/2}$	$B_{3/2}$	$C_{5/2}$
0.516	$47.20(8)*10^{-6}$	$6.61(4)*10^{-8}$	0.51(2)	0.19(2)
0.557	38.64(7)	3.07(9)	0.41(9)	0.03(2)
0.608	36.50(1)	1.54(5)	0.39(5)	0.08(2)
Crystalline Fe[a]	3.4	0.1	0.11(4)	0.04
Crystalline Ni[a]	7.5	1.5	0.11(7)	0.15

[a] B. E. Argyle, S. H. Charap, and E. W. Pugh, Phys. Rev. 132, 2051 (1963).

Figure 8 shows the fractional change of the reduced magnetic hyperfine field as a function of $(T/T_c)^{3/2}$. It is found that data points do not show a strictly linear behavior. A least square fit of equation (2) and (3) to the data gave the values listed in Table III. These values may be compared with those obtained for other ferromagnetic metallic glasses such as $Fe_{80}B_{20}$, $Fe_{40}Ni_{38}Mo_4B_{18}$, and $Fe_{81}B_{27}Si_7C_2$. The observed $B_{3/2}$ and $C_{5/2}$ values for the present case and the other metallic glasses are significantly larger than in crystalline ferromagnets such as iron and nickel. This has been attributed to excitation of spin-waves having long wavelengths in metallic glasses compared to those in crystalline ferromagnets [31].

CRITICAL EXPONENT

The value of the critical exponent, β, may be deduced from the temperature dependence of the internal magnetic field at different temperatures by using the relation

$$\frac{H_{eff}(T)}{H_{eff}(0)} = D[1-T/T_c]^{\beta} \quad (4)$$

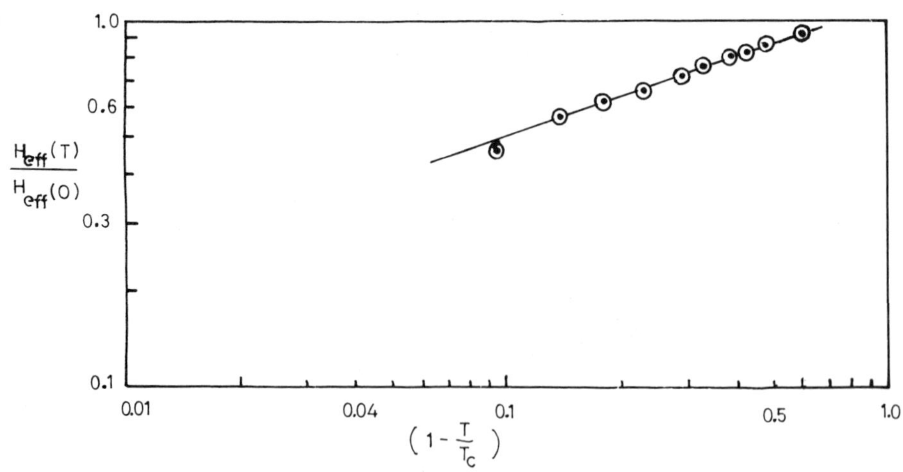

Figure 9. The reduced field vs. (1-T/TC) for amorphous $Fe_{72}Ni_4Mo_6B_{16}Si_2$.

Table IV. Values of the critical exponents D and β for the amorphous $Fe_{72}Ni_4Mo_6B_{16}Si_2$

T/T_c Interval	D[a]	β[a]
0.7142	1.319	0.35
0.6630	1.417	0.37
0.6081	1.385	0.37
0.5571	1.338	0.36
0.5163	1.321	0.35
0.4000	1.302	0.35
0.1632	1.205	0.33

[a] Error limits of ±0.03.

where D is a constant. It may be seen from Figure 9 that the reduced hyperfine field plotted against $\ln(1-T/T_c)$, follows a fairly linear relationship in the region 0.16 < 0.72. The experimental points closer to T_c could not be obtained due to the uncertainty in the hyperfine field values as the spectra were poorly resolved. A least square fit of the data to expression (4) yielded a β of 0.35(4) and a D of 1.31(2). The values of β and D calculated from different temperature regions are given in Table IV. Closer to T_c the trend reflects a β value greater than 0.35. The addition of molybdenum to the amorphous alloys strongly reduces the transition temperature as discussed earlier and greatly modifies the critical behavior [17,32]. It is known that a β value significantly greater than the typical value of 0.33 for three dimensional Heisenberg ferromagnets, is observed in molybdenum containing metallic glasses, the deviation arising from a random anisotropy due to the difference in the spin-orbit coupling constants of iron and molybdenum [32].

CRYSTALLIZATION BEHAVIOR

The sample heated at 850K showed a simple paramagnetic spectrum. However, the Mössbauer spectrum measured at 882(2)K showed the presence of α-iron. The absence of any other phase in significant amounts in the early stage of crystallization indicates that the crystalline products discussed below are the result of a "primary" crystallization as in the case of related systems [33]. Room temperature Mössbauer spectra of the metallic glass, $Fe_{72}Ni_4Mo_6B_{16}Si_2$, heated in vacuum above T_x at 912 and 1273K for about 2 hours are shown in Figure 10. The complex spectra reveal the presence of several phases resulting from crystallization. An attempt to identify the phases corresponding to the three observed magnetic components, and a paramagnetic phase with a quadrupole splitting of 0.44(2)mm/s, showed that they correspond to an α(Fe-Mo) alloy of composition $Fe_{95}Mo_5$ [33] with a room temperature internal field of 335(4)kOe, α-iron, Fe_2B, with a field of 235kOe, and Fe_2Mo, respectively. The presence of these phases are also supported by the X-ray diffraction data. The sample heated to 1273K and cooled to room temperature showed additional features. The presence of an additional paramagnetic phase is discernible by an absorption peak from -1 to 1mm/sec which could be attributed to the formation of an invar alloy with a composition [35] of $Fe_{1-x}Ni_x$. This additional phase is thus precipitated only at temperature above 912K. Interestingly, the crystalline products of amorphous $Fe_{40}Ni_{40}B_{20}$, with T_x of 700K, when heated to 900K and cooled to room temperature is found to give rise to significantly different products such as a magnetic Fe-Ni alloy phase and borides [36] of the type $(Fe_xNi_{1-x})_3B$.

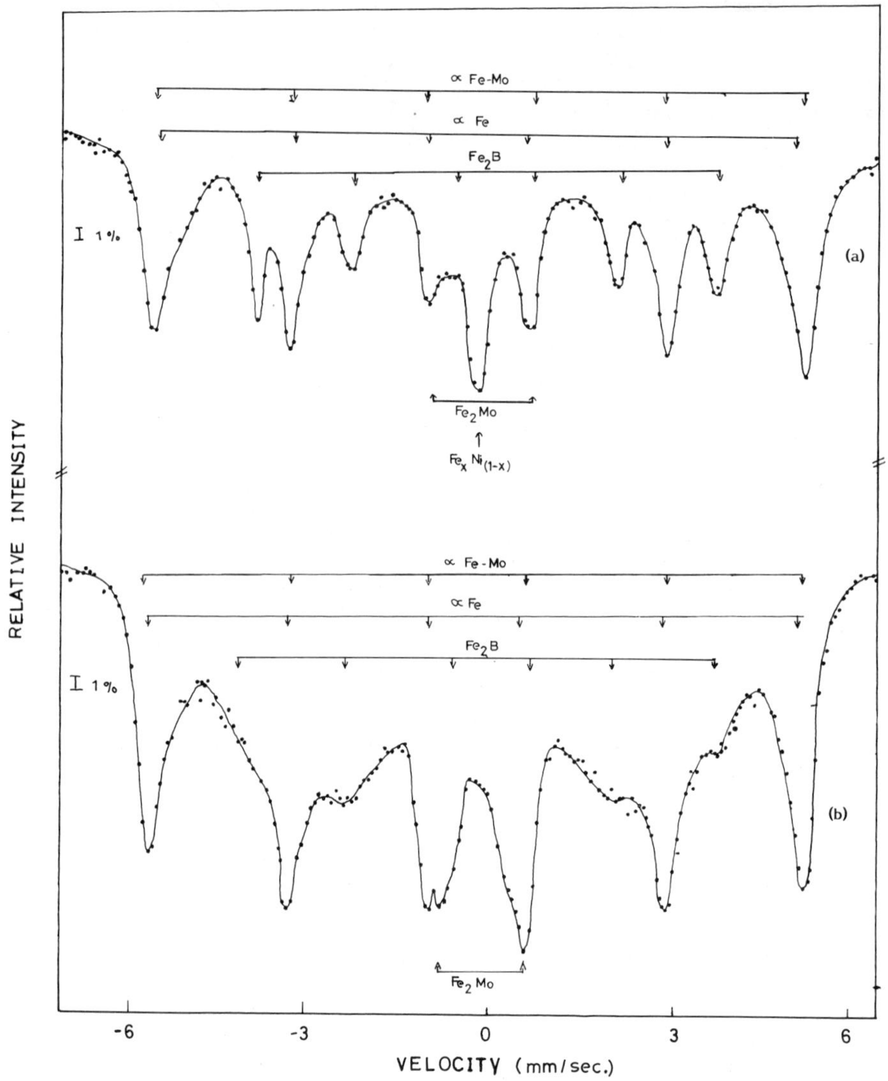

Figure 10. Room temperature Mössbauer spectra of the crystalized $Fe_{72}Ni_4Mo_6B_{16}Si_2$ sample heated at (a) 912K and (b) 1273K

CONCLUSIONS

Mössbauer spectroscopy enables in-depth studies of iron-rich ferromagnetic metallic glasses. A detailed study of the amorphous alloy, $Fe_{72}Ni_4Mo_6B_{16}Si_2$, is presented here as an illustrative example. Temperature dependence of Mössbauer measurements show that the sample has a Curie temperature, T_c, of 490K and a crystallization temperature, T_x, of 880K. Comparison of T_c and T_x with those of related systems show that molybdenum, like the transition metals to the right of iron in the periodic table decreases T_c but enhances T_x. The temperature dependence of reduced hyperfine field versus the reduced temperature is found to deviate significantly from the theoretical Brillouin curves for a spin of 1/2 or 1. The assumption of a temperature dependence in the term representing the fluctuation in the exchange interactions in the Handrich model [19], of the form $\delta = \delta_o(1-t^2)$, where t is T/T_{cx}, is found to lead to very close agreement with experimental data for many of the metallic glasses. The model independent method of analyzing the hyperfine field distribution due to Window [22] shows a bimodel distribution similar to the case of chromium containing metallic glasses and is attributed to the presence of molybdenum near neighbors with iron-molybdenum antiferromagnetic interactions. The temperature dependence of width of the distribution curves further showed that long range interactions are important in these alloys. Analysis of the temperature dependence of the hyperfine field data at low temperature led to a $B_{3/2}$ value of 0.16 and a $C_{5/2}$ value of 0.37 for the spin-wave stiffness constants. Analysis of the data close to T_c led to values of the critical constants, D of 1.37(5) and β = 0.36(1). Mössbauer measurement at 880K revealed the precipitation of α-iron indicating a "primary" crystallization process. Room temperature measurement of vacuum heated sample at a temperature above T_x at 912K showed the presence of $Fe_{95}Mo_5$, Fe_2B, and Fe_2Mo in addition to α-iron. When heated to a still higher temperature, the presence of an additional paramagnetic phase, which may be attributed to invar alloy formation with composition of the type $Fe_{1-x}Ni_x$, is indicated. The products were found to be significantly different from nickel rich metallic glasses such as $Fe_{40}Ni_{40}B_{20}$.

ACKNOWLEDGMENTS

The work has been sponsored by the Science and Engineering Research Council, Department of Science and Technology, New Delhi, India, through Grant No. 21(11)/82-STP.II. T. G. Narendra Babu would like to thank the Council of Scientific and Industrial Research, India, for a predoctoral fellowship.

REFERENCES

1. Metallic Glasses: Production, Properties and Applications, T. R. Anantharaman, Ed., Trans. Tech. Pub., Switzerland (1984).

2. Amorphous Metallic Alloys, F. E. Luborsky, Ed., Butterworth Co., Ltd., London (1983).

3. Proc. 4th Int. Conf. on Rapidly Quenched Metals, T. Masumoto and K. Suzuki, Eds., The Japan Institute of Metals, Sendai (1982).

4. M. Naka, K. Hashimoto, A. Injoue, and T. Masumoto, J. Non-cryst. Solids 31, 347 (1979).

5. R. B. Diegle, Proc. 5th Int. Conf. on Liquid and Amorphous Metals, C. N. J. Wagner and W. I. Johnson, Eds, p. 601 (1983).

6. A. K. Bhatnagar, B. B. Prasad, and N. R. M. Rathnam, J. Non-Cryst. Solids 61 & 62, 1201 (1984).

7. Ref. 3, p. 971.

8. B. Bhanu Prasad, A. K. Bhatnagar, and R. Jagannathan, J. Appl. Phys. 54, 2019 (1983).

9. P. J. Schurer and A. H. Morrish, J. Magn. Magn. 15-18, 577 (1980).

10. T. G. N. Babu, R. Jagannathan, and A. K. Bhatnagar, to be published.

11. H. G. Wagner, M. Ackermann, and U. Gonser, J. Non-cryst. Solids, 61 & 62, 847 (1984).

12. F. E. Luborsky, J. Appl. Phys. 51, 2808 (1980).

13. J. L. Walter, Mater. Sci. Eng. 50, 137 (1981).

14. Ref. 3, p. 597.

15. A. Lovas, L. Granasy, K. Zambo Balla, and J. Kiraley, Proc. Conf. on Metallic Glasses, Science and Technology, C. Hargitai, I.Bakonyi, and T. Kemeny, Eds., vol. II, p. 291, Kultura, Budapest (1981).

16. S. Chattopadhyay, J. Appl. Phys. 53, 2291 (1982).

17. C. L. Chien and H. S. Chien, J. Appl. Phys. 50, 1574 (1979).

18. M. Naka, U. Gonser, C. Gorlity, and H. Ruppersberg, J. Non-Cryst. Solids 45, 99 (1981).

19. K. Handrich, Phys. Stat. Solidi 33, K55 (1969).

20. A. K. Bhatnagar, B. Bhanu Prasad, N. Ravi, R. Jagannathan, and T. R. Anantharaman, Solid State Commun. 44, 905 (1982).

21. M. Fähnle, Solid State Commun. 49, 391 (1984).

22. B. Window, J. Phys. E4, 401 (1971).

23. A. Narayanaswamy, T. Nagarajan, P. Muthukumarasamy, and T. S. Radhakrishnan, J. Phys. F. Metal Phys. 9, 2261 (1979), and ref. 24.

24. A. K. Bhatnagar, B. B. Prasad, and R. Jagannathan, Phys. Rev. B29, 4896 (1984).

25. G. L. Whittle and A. M. Steward, Phys. Stat. Solidi (a) 80, 101 (1983).

26. C. L. Chien, Phys. Rev. B19, 81 (1979).

27. C. L. Chien and H. S. Chien, J. Appl. Phys. 50, 1574 (1979).

28. R. W. Cochrane and G. S. Cargil, Phys. Rev. Lett. 32, 476 (1974).

29. N. S. Nazama, T. Masumoto, and H. Watanabe, J. Phys. Soc. Japan 37, 1171 (1974).

30. M. Eibschutz and M. E. Lines, Phys. Rev. B7, 4907 (1973).

31. A. K. Bhatnagar and N. Ravi, Phys. Rev. B28, 359 (1983).

32. L. S. Barton and M. B. Salamon, Phys. Rev. B25, 2030 (1982).

33. F. L. Cumbrera, M. Millan, A. Conde, R. Marquez, and P. Vigier, J. Mater. Sci. 17, 861 (1962).

34. H. L. Marcus and L. H. Schwartz, Phys. Rev. 162, 259 (1967).

35. Y. Nakamura, M. Shiga, and N. Shikazone, J. Phys. Soc. Japan 19, 1177 (1964).

36. H. Franke and M. Rosenberg, IEEE Trans. Magn. Mag. 14, 952 (1978).

THE FITTING METHOD FOR MÖSSBAUER SPECTRA OF AMORPHOUS ALLOYS AND AN INVESTIGATION OF THEIR PROPERTIES AND STRUCTURE

Zuxiong Xu, Ruzhang Ma, and Jueyun Ping

Beijing University of Iron and Steel Technology
Beijing, People's Republic of China

INTRODUCTION

Over the past several years, the amorphous alloys have been of scientific and technological interest and have been extensively investigated. In particular, much attention has been devoted to ferromagnetic metallic glasses. For investigation of the magnetic anisotropy or the structural relaxation and crystallization during annealing Mössbauer spectroscopy is known to be a suitable widely used technique. A number of results have reported that, for many amorphous alloys, the structural relaxation below the crystallization temperature increases the average magnetic hyperfine field [1-3] and that the crystallization pattern above the crystallization temperature varies for amorphous alloys, such as Fe-P [4], Fe-B [5,6], and Fe-B-Si [2,7], and Fe-B-Si-C [3,8,9]. Another interesting Mössbauer result has reported that, despite the large demagnetizing effect which would favor the moments lying in the ribbon plane, out-of-plane anisotropy is observed for various amorphous alloys when measured at low temperature [10,11] or when the alloys are annealed at higher temperature [1,12-16].

In this paper, in view of the current fitting methods for Mössbauer spectra of amorphous alloys, we suggest a modified fitting method. The magnetic texture, the structural relaxation, and crystallization during annealing, as well as the change of the magnetic anisotropy in amorphous alloys are investigated with Mössbauer spectroscopy.

ASYMMETRY IN MÖSSBAUER SPECTRA AND A MODIFIED FITTING METHOD FOR SPECTRA

At present, there are two problems to be resolved to improve the goodness of fit for iron-57 Mössbauer spectra of ferromagnetic amorphous alloys. The first is the problem of the asymmetry of peak amplitudes and peak widths in the spectrum about the spectral center and the second is the method for correcting parameters, such as the peak area ratios, peak widths in the subspectra, and the average isomer shift, when fitting spectra with the methods of Hesse [17] or Window [18]. In this section the asymmetry in spetra of amorphous alloys is analyzed and a modified method for fitting spectra is proposed.

There is usually some asymmetry in the iron-57 Mössbauer spectral peaks for ferromagnetic amorphous alloys. According to the apparent peak width, $\Gamma_1>\Gamma_6$, $\Gamma_2>\Gamma_5$, and $\Gamma_3>\Gamma_4$, as is found for the amorphous alloys $(Fe_{0.2}Co_{0.8})_{80}P_{13}Al_3$ [19] and Fe-P [20]. In the second, $\Gamma_1>\Gamma_6$, $\Gamma_2<\Gamma_5$, and $\Gamma_3<_4$, as is found in the amorphous alloy $Fe_{80}B_{20}$ [21]. It has been suggested that the linear relationship between the hyperfine field, H_{hf}, and the isomer shift, δ, [19,20], or the linear correlation between the hyperfine field and the quadrupole splitting, ΔE_Q, [21] or the dipolar magnetic field, H_d, [22] are the possible causes of these types of asymmetry.

Our experience in fitting spectra shows that it is unsatisfactory to explain the asymmetry in spectra of most amorphous alloys by one of the above-mentioned suggestions. All three hyperfine parameters H_{hf}, ΔE_Q, and δ depend on the nearest neighbors around the iron atom, so it is reasonable to suggest that there are simultaneously the following approximate linear relations between three hyperfine parameters H_{hf}, ΔE_Q, and δ,

$$\Delta E_Q = aH_{hf} + c$$

$$\delta = b_{hf} + d. \tag{1}$$

If a or b is equal to zero, there is no correlation between H_{hf} and ΔE_Q or between H_{hf} and δ. The constants c and d influence the positions of six broadened peaks and are independent on their asymmetry. The magnitude and the signs of the constants a and b, which are called asymmetry parameters, determine the type and amplitude of the asymmetry in the spectrum. For two typical amorphous alloys, $Fe_{80}B_{20}$ and $Fe_{82}P_{18}$, by simultaneously assuming the two linear relations, H_{hf}-δ and H_{hf}-ΔE_Q, we can obtain good quality fits as shown in Figure 1. However, when we fit these spectra to only one of the

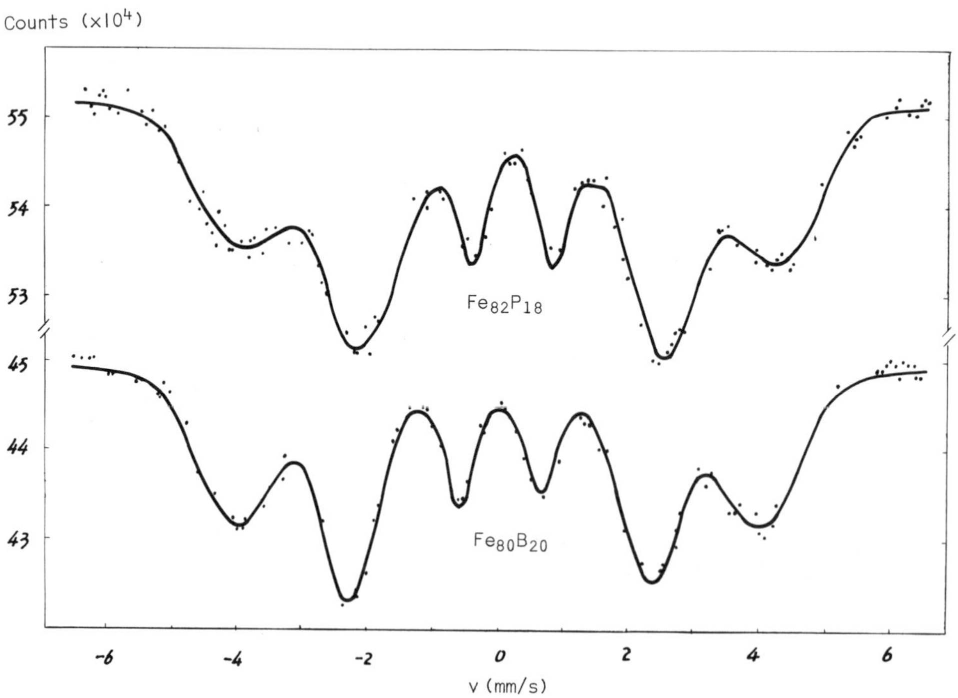

Figure 1. The asymmetric spectra of $Fe_{80}B_{20}$ and $Fe_{82}P_{18}$

Table I. Assymmetry Parameters for Several Amorphous Alloys

Alloy	a	b
$Fe_{40}Ni_{40}P_{14}B_6$	$-0.86x10^{-3}$	$-0.43x10^{-3}$
$Fe_{82}P_{18}$	$-1.0x10^{-3}$	$-0.6x10^{-3}$
$Fe_{80}B_{20}$	$-1.0x10^{-3}$	$0.4x10^{-3}$
$Fe_{78}Si_{10}B_{12}$	$-1.1x10^{-3}$	$0.5x10^{-3}$
$Fe_{81}B_{13.5}Si_{3.5}C_2$	$-0.9x10^{-3}$	$0.5x10^{-3}$

linear relationships, the amplitudes of the fourth and the fifth broadened peaks in the calculated spectra are larger than that in the experimental spectrum, and the amplitudes of the second and third peaks are smaller than that in the experimental spectrum, if the quality of the fit for the first and sixth broadened peaks is satisfactory. Our experience of fitting such spectra shows that a reasonable selection of parameters a and b may efficiently improve qualities of fits to asymmetric spectra. Asymmetry parameters a and b for spectra of several common amorphous alloys with the optimum fits are listed in Table I.

The quality of spectral fits for amorphous alloys depends on the reasonableness of the derived parameters, such as the intensity ratios, linewidths, and average isomer shift, which are adjusted in the methods of Hesse [17] or Window [18]. Often it is difficult to estimate accurately these parameters. To obtain good quality fits, it is often necessary to trial-fit repeatedly. Moreover, these two methods do not deal with the asymmetry in the spectrum, and a complex spectrum which consists of several crystalline and amorphous components often cannot be fit. To solve these problems, we proposed a modified method for fitting spectra on the basis of the above mentioned analysis of the asymmetry in spectra.

MATHEMATICAL MODEL

With the view of fitting general spectra, we assume that the total spectrum comprises spectra for the crystalline component and the amorphous component with a continuous distribution of hyperfine fields. Thus we note that the crystalline spectrum consists of M_1'' subspectra and the amorphous spectrum is divided into M_1' subspectra where the total of all subspectra M_1 is $M_1'+M_1''$. We also assume that all M parameters to be fit constitute a parameter vector,

$$q=(q_1,q_2,\cdots q_M)^T.$$

These M parameters are divided into two groups. The first consists of a (M_1+3)-dimensional linear vector, q_I, which comprises the intensities of the M_1' subspectra in the amorphous spectrum, $q_i(i=1,2,\cdots M_1')$, the intensities of M_1'' subspectra in the crystalline spectrum, q_i $(i=M_1'+1, M_1'+2,\cdots M_1)$, and three parabolic background parameters E, F, and G called q_{M_1+1}, q_{M_1+2} and q_{M_1+3}, respectively. The second consists of a $M_2$0-dimensional non-linear parameter vector,

$$q_{II} = (q_{M_1+4}, q_{M_1+5},\cdots q_M)^T,$$

which consists of the a, b, c, and d parameters in the amorphous spectrum, the normalized intensities, a_K, and widths Γ_{ik}, of six peaks in each subspectrum, and the hyperfine parameters, H_{hf}, ΔE_Q, and δ, to be evaluated in the crystalline subspectrum.

The counts of the theoretical spectrum at the n-th channel are given by

$$Y_t(n)=E+F\bullet X_n+G\bullet X_n^2 - \sum_{i=1}^{M_1} S_i\bullet \sum_{k=1}^{6} \frac{a_k}{1+4(X_n-P_{ik})^2/\Gamma_{ik}^2}, \tag{2}$$

where S_i is the intensity of the i-th subspectrum, P_{ik} is the position of the k-th peak in the i-th subspectrum, and X_n is the velocity of the n-th channel. The modified method for fitting spectra adopts the objective function

$$X_m^2= \sum_{n=1}^{N} [Y_e(n)-Y_t(n)]^2/Y_e(n)+g \sum_{i=2}^{M_1'-1} (\Delta q_{i-1}-2\Delta q_i+\Delta q_{i+1})^2$$
$$+ \lambda \sum_{i=1}^{M} \sum_{n=1}^{N} (\frac{\partial Y_t(n)}{\partial q_i} \bullet\Delta q_i)^2/Y_e(n), \tag{3}$$

where $Y_e(n)$ is the counts of the experimental spectrum at the n-th channel, N is the number of channels, g is a smoothing factor, and λ is the damping factor. The parameter increment vector,

$$\Delta q = (\Delta q_I, \Delta q_{II})^T = (\Delta q_1, \Delta q_2,\bullet\bullet\bullet\Delta q_M)^T,$$

is determined by the condition which minimizes X^2. By linearizing the condition, we can obtain the iterating equation in matrix form,

$$(A^{(1)} + g\bullet D + \lambda^{(1)}\bullet C^{(1)})\Delta q^{(1)} = B^{(1)}, \tag{4}$$

where the smooth matrix is

$$D = \begin{pmatrix} D_{II} & 0 \\ 0 & 0 \end{pmatrix},$$

$$D_{II} = \begin{pmatrix} 1 & -2 & -1 & & & & & & \\ -2 & 5 & -4 & 1 & & & & & \\ 1 & -4 & 6 & -4 & 1 & & & & \\ & 1 & -4 & 6 & -4 & 1 & & & \\ & & \bullet & & & \bullet & & & \\ & & & \bullet & & & \bullet & & \\ & & & & \bullet & & & \bullet & \\ & & & & 1 & -4 & 6 & -4 & 1 \\ & & & & & 1 & -4 & 5 & -2 \\ & & & & & & 1 & -2 & 1 \end{pmatrix} \tag{5}$$

and the elements of other matrixes are

$$A_{ij}^{(l)} = \sum_{n=1}^{N} \frac{\partial Y_t(n)}{\partial q_i}^{(l)} \cdot \frac{\partial Y_t(n)}{\partial q_j}^{(l)} / Y_e(n), \tag{6}$$

$$C_{ij}^{(l)} = \begin{cases} 0 & (i \neq j), \\ A_{ii}^{(l)} & (i=j), \end{cases} \tag{7}$$

$$B_i^{(l)} = \sum_{n=1}^{N} \frac{\partial Y_t(n)}{\partial q_i}^{(l)} \cdot \left[Y_e(n) - Y_t^{(l)}(n)\right] / Y_e(n), \tag{8}$$

for i, j=1,2,•••M.

DETERMINING LINEAR PARAMETERS

In the beginning of iteration, i.e., when l=0 we put

$\lambda^{(0)}=0$, $q_I^{(0)}=0$, and $q_{II}^{(0)}$

with a set of suitable initial values. In this condition all partial differential quotients of the $Y_t(n)$ with respect to the nonlinear parameter are equal to zero. Therefore, equation (5) becomes

$$\begin{pmatrix} A_{II}^{(o)} & 0 \\ 0 & 0 \end{pmatrix} \begin{pmatrix} \Delta q_I^{(o)} \\ \Delta q_{II}^{(o)} \end{pmatrix} + g \begin{pmatrix} D_{II} & 0 \\ 0 & 0 \end{pmatrix} \begin{pmatrix} \Delta q_I^{(o)} \\ \Delta q_{II}^{(o)} \end{pmatrix} = \begin{pmatrix} B_I^{(o)} \\ 0 \end{pmatrix} . \tag{9}$$

Consequently, we obtain reasonable values for the parameter increments in the o-th iteration,

$$\Delta q_I^{(o)} = (A_{II}^{(o)} + g \bullet D_{II})^{-1} B_I^{(o)},$$

$$\Delta q_{II}^{(o)} = 0, \tag{10}$$

and new values of parameters

$$q_I^{(1)} = q_I^{(o)} + \Delta q_I^{(o)} = (A_{II}^{(o)} + g \bullet D_{II})^{-1} B_I^{(o)},$$

$$q_{II}^{(1)} = q_{II}^{(o)} + \Delta q_{II}^{(o)} = q_{II}^{(o)} . \tag{11}$$

By comparing equation 11 with the solution of the method of Hesse, [17], it is found that, except for the weight and the three parameters, E,F, and G, which may be determined from equation 11, these methods for fitting spectra are essentially identical. Therefore, the above-mentioned method in the o-th iteration is a modified method by Hesse.

REVISED ITERATION PARAMETERS

When l=1, the revision of the parameters begins. In order to improve the convergence of iteration, we set $\lambda^{(1)}$ to a positive value. If $q^{(1)}$ is used as the new initial values of these parameters, we can determine the parameter increment, $\Delta q^{(1)}$ from equation 4. Then the new values of these parameters are

$$q^{(2)} = q^{(1)} + \Delta q^{(1)}.$$

We repeat this program

$$q^{(l+1)} = q^{(l)} + \Delta q^{(l)},$$

until the value of X^2 is satisfactory. Therefore, when $l \neq 0$, this method is essentially identical with the method of nonlinear least-square minimization.

DISCUSSION OF THE MODIFIED METHOD

From the above analysis, we find that the modified method is essentially the combination of the modified method of Hesse [17] with the common nonlinear least-square minimization. This method has the following characteristics. (1) It does not require initial values of linear parameters, so it is suitable for fitting complex and overlapping spectra. (2) It possesses a procedure for revising the parameters and improves the efficiency of fitting spectra of amorphous alloys. (3) It has wide application, e.g., it can be applied to either the spectra of the amorphous alloys or to mixed spectra which contain crystalline and amorphous components. Hence, it is a universal method for fitting spectra. (4) For mixed spectra which consist of spectra of known and unknown phases, it possesses the function of mixed calculation with stripping and fitting.

EXAMPLES WHICH APPLY THE MODIFIED METHOD

Three spectra of the quenched and annealed amorphous alloy, $Fe_{78}Si_{10}B_{12}$, were measured at room temperature and fitted with the modified method as shown in Figure 2. The quality of fit is satisfactory and the fitted parameters are listed in Table II. The fitted result shows that the asymmetry parameters, a and b, of these spectra become smaller at higher annealing temperatures and that the asymmetry in the spectrum depends not only on the composition of the alloy but also on the state of thermal treatment.

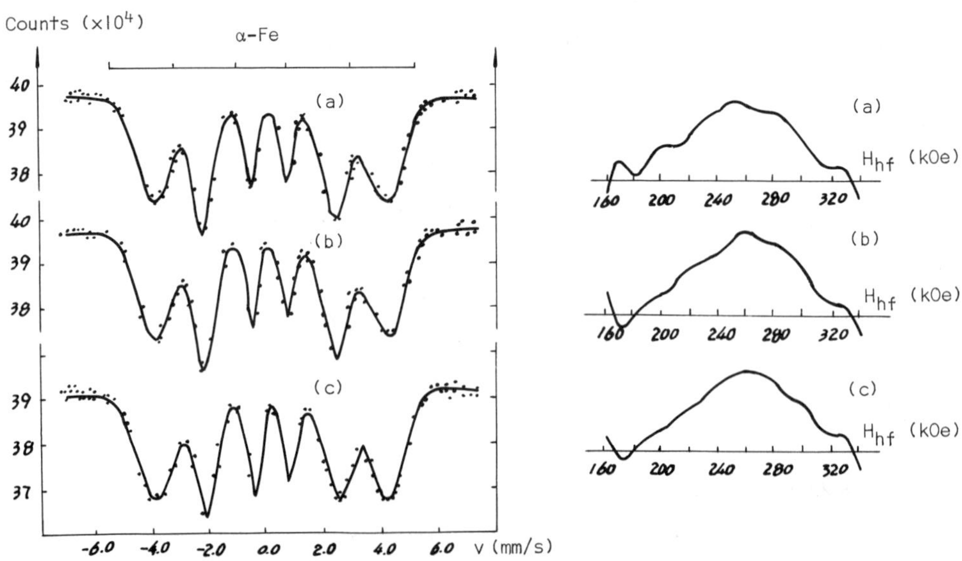

Figure 2. The Mössbauer spectra and the distribution of hyperfine fields for $Fe_{78}Si_{10}B_{12}$ as-quenched, (a) and annealed at 385°C (b) and 440°C (c).

Table II. Mössbauer Spectral Parameters for the Amorphous Alloy, $Fe_{78}Si_{10}B_{12}$

Parameter	Alloy As-Quenched	Alloy Annealed at 385°C	440°C
$a(*10^{-3})$	-1.06	-1.16	-1.30
$b(*10^{-3})$	0.48	0.40	0.26
c, mm/s	0.26	0.30	0.32
d, mm/s	-0.01	0.02	0.05
Ave H_{hf}, kOe	251.5	255.5	257.8
Ave δ, mm/s	0.11	0.12	0.12
Ave ΔE_Q, mm/s	-0.01	0.01	-0.01

Moreover, the experimental results show that, during annealing, the average hyperfine field increases, the average isomer shift, remains constant and the average quadrupole splitting remains nearly zero.

The room temperature spectrum of partly crystallized amorphous alloy, $Fe_{162}B_{27}Si_{27}C_4$, annealed at 450°C for one hour has been fit as shown in Figure 3. The result shows that the sample comprises crystalline components Fe-4 atomic percent Si and $Fe_2(B,C)$ as well as amorphous components which are approximately 50 percent of the sample. The result of fitting the room temperature spectrum of the completely crystallized alloy, $Fe_{80}P_{17.6}Si_{1.6}C_{0.8}$, is shown in Figure 4.

It can be concluded that, because of simultaneous correlations between the three hyperfine parameters, H_{hf}, ΔE_Q, and δ, the proposed method for fitting the spectra can effectively improve the quality of fits for amorphous alloys. This modified method has some advantages, such as more functions for fitting the spectra, wide application, and convenience in application.

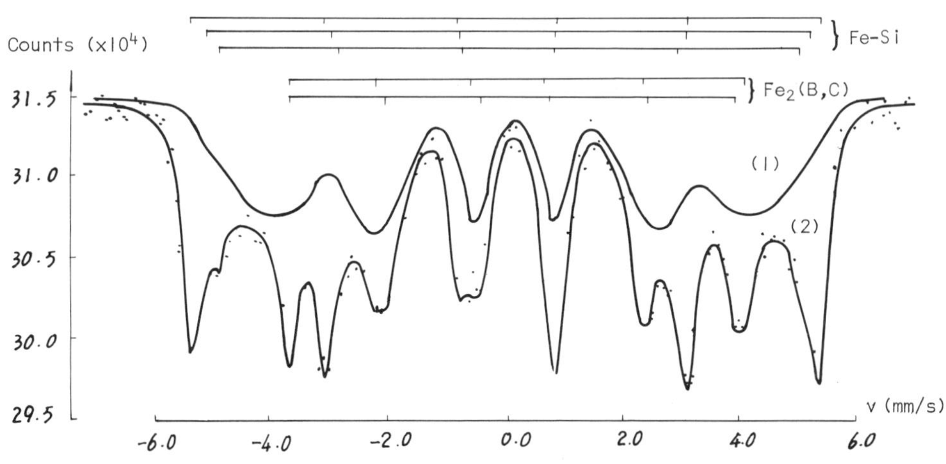

Figure 3. The spectrum of the partly crystallized amorphous alloy, $Fe_{162}B_{27}Si_7C_4$, annealed at 450°C where (1) is the amorphous subspectrum and (2) is the fitted mixed spectrum.

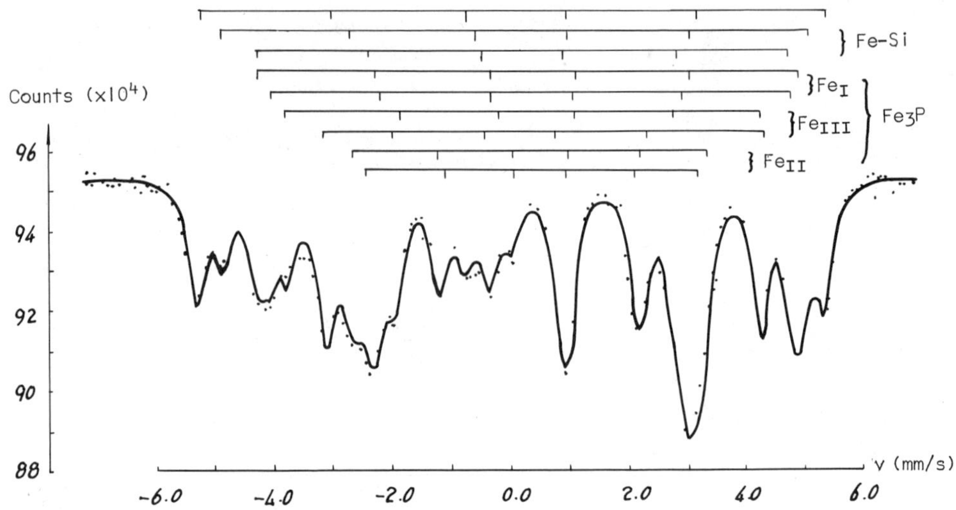

Figure 4. The spectrum of the completely crystallized alloy, $Fe_{80}P_{17.6}Si_{1.6}C_{0.8}$

INVESTIGATION OF MAGNETIC TEXTURE IN FERROMAGNETIC AMORPHOUS ALLOY

Because of the thermal treatment under a magnetic field or an external stress, or rolling deformation, ferromagnetic amorphous alloys may have a structure with a preferential orientation of the moments and the magnetic anisotropy. Investigations of the magnetic texture and anisotropy are important for improving the application of amorphous alloys. The study of the magnetic texture and anisotropy without an external field is a typical application of Mössbauer spectroscopy.

The angular factors of iron-57 Mössbauer spectra for ferromagnets provide information about the orientation of the hyperfine field at the iron nucleus. For most ferromagnetic alloys with a predominant contact field, H_c, the orientation of the magnetic moment is identical with that of the hyperfine field, but their directions are reverse. Therefore, we can study the magnetic texture with Mössbauer spectroscopy. Pfannes [23] has discussed thoroughly the mathematics relationship between texture and the Mössbauer spectrum and suggested a probability distribution function of hyperfine fields, or magnetic moments $D(\theta,\phi)$, developed as a spherical harmonic series, to express this relationship. However, the function $D(\theta,\phi)$, sometimes has negative values and it seems unreasonable as an approximation for describing the distribution of magnetic moments, and a clear physical basis is lacking for this description. Greneche [24] improved partly upon this description, but their basic ideas are identical. In this paper a statistical description for the distribution of the moments is used to represent the magnetic texture of several ferromagnetic amorphous alloys.

STATISTICAL DESCRIPTION OF MAGNETIC TEXTURE

The bulk magnetic properties of a complex magnetic system with texture in a unit vector direction $\vec{D}$ usually depends upon two statistical quantities,

$$\Sigma p_i \cdot \cos^2\psi_i$$

and

$$\Sigma p_i \cdot |\cos\psi_i|,$$

where p_i is the distribution probability in the i-th direction of the magnetic moment, $\vec{M}_i$ and ψ_i is the angle between the vectors $\vec{M}_i$ and $\vec{D}$. We define two kinds of preferences for the distribution of magnetic moments as follows. First, the degree of the direction preference for the magnetic moments of a magnetic system with texture in a given direction $\vec{D}$ can be defined as

$$P_d(\vec{D}) = \Sigma p_i \bullet \cos\psi_i,$$

for

$$-1 \leq P_d(\vec{D}) \leq 1. \tag{12}$$

It is obvious that $P_d(\vec{D}) = -P_d(-\vec{D})$. We shall abbreviate $P_d(\vec{D})$ as P_d hereafter. It can be seen from (12) that the greater P_d, the greater the distribution probability of magnetic moments in or near the direction $\vec{D}$. If P_d is 1, then all the magnetic moments point exactly in $\vec{D}$ direction. If P_d = -1, then they point in the opposite direction. Secondly, the degree of orientation preference for magnetic moments in a given direction, $\vec{D}$, can be defined as

$$P_o(\vec{D}) = \Sigma p_i \bullet \cos^2\psi_i,$$

for

$$0 \leq P_o(\vec{D}) \leq 1. \tag{13}$$

It is obvious that $P_o(\vec{D}) = P_o(-\vec{D})$. We shall abbreviate $P_o(\vec{D})$ as P_o hereafter. It can be seen from (13) that the greater P_o, the greater the distribution probability of magnetic moments in or near the orientation of $\vec{D}$. If P_o is 1, then all the magnetic moments are parallel to $\vec{D}$. If P_o is zero, then they point perpendicular to $\vec{D}$. Here we define the degree of the orientation preference as the expression (13) rather than

$$\Sigma p_i \bullet |\cos\psi_i|,$$

because (13) is easy to analyze mathematically and can be measured by Mössbauer spectroscopy.

Mathematical analysis reveals that the degrees of the direction preference and orientation preference for the magnetic texture in a given direction, $\vec{D}$, can be described with several independent statistical parameters [25]. If the coordinates of the magnetic moment, $\vec{M}_i$ are (x_i, y_i, z_i), the direction $\vec{D}$ is (l, m, n) in the reference frame of sample XYZ. The vector $\vec{R}$ is $(P'_x P'_y P'_z$, where

$$
\begin{aligned}
P'_x &= \Sigma p_i \bullet x_i, & -1 &\leq P'_x \leq 1,\\
P'_y &= \Sigma p_i \bullet y_i, & -1 &\leq P'_y \leq 1,\\
P'_z &= \Sigma p_i \bullet z_i, & -1 &\leq P'_z \leq 1.
\end{aligned}
\tag{14}
$$

The matrix B is

$$B = \begin{pmatrix} P_x \eta_z \eta_y \\ \eta_z P_y \eta_x \\ \eta_y \eta_x P_z \end{pmatrix},$$

where

$$P_x = \Sigma p_i \bullet x_i^2, \qquad 0 \leq P_x \leq 1,$$
$$P_y = \Sigma p_i \bullet y_i^2, \qquad 0 \leq P_y \leq 1, \tag{15}$$
$$P_z = \Sigma p_i \bullet z_i^2, \qquad 0 \leq P_z \leq 1,$$
$$P_x + P_y + P_z = 1, \qquad P_y = 1 - P_x - P_z; \tag{16}$$

and

$$\eta_x = \Sigma p_i \bullet y_i \bullet z_i, \qquad -1/2 \leq \eta_x \leq 1/2,$$
$$\eta_y = \Sigma p_i \bullet z_i \bullet x_i, \qquad -1/2 \leq \eta_y \leq 1/2, \tag{17}$$
$$\eta_z = \Sigma p_i \bullet x_i \bullet y_i, \qquad -1/2 \leq \eta_z \leq 1/2.$$

It may be shown that the degrees of the direction preference and orientation preference for the magnetic texture can be expressed respectively as [25]

$$P_d = \vec{R} \bullet \vec{D} = P'_x \bullet l + P'_y \bullet m + P'_z \bullet n, \tag{18}$$

$$P_o = \begin{pmatrix} l \\ m \\ n \end{pmatrix}^T B \begin{pmatrix} l \\ m \\ n \end{pmatrix} = \begin{pmatrix} l \\ m \\ n \end{pmatrix}^T \begin{pmatrix} P_x & \eta_z & \eta_y \\ \eta_z & P_y & \eta_x \\ \eta_y & \eta_x & P_z \end{pmatrix} \begin{pmatrix} l \\ m \\ n \end{pmatrix}. \tag{19}$$

From (14) it is seen that P'_x, P'_y, P'_z are the degrees of the direction preference in three coordinate axes respectively. From (15) it is seen that P_x, P_y, P_z are the degrees of the orientation preference in three coordinate axes respectively, which are called the axial parameters of the orientation preference and there are only two independent parameters among them. The other three parameters, η_x, η_y, and η_z, which describe the statistical asymmetry of the distribution of magnetic moments, are called the deviation parameters. We can perform a linear transformation of the coordinates, so that all the deviation parameters will vanish in the main coordinate system. Therefore, the following expression is valid in the main coordinate system, say $P_1 \geq P_2 \geq P_3$,

$$P_o = \begin{pmatrix} l' \\ m' \\ n' \end{pmatrix}^T \begin{pmatrix} P_1 & 0 & 0 \\ 0 & P_2 & 0 \\ 0 & 0 & P_3 \end{pmatrix} \begin{pmatrix} l' \\ m' \\ n' \end{pmatrix} = P_1 \bullet l'^2 + P_2 \bullet m'^2 + P_3 \bullet n'^2, \tag{20}$$

where (l', m', n') are the coordinates of $\vec{D}$ in the main coordinate system. Moreover, there are three different independent parameters, which describe the orientation of the main coordinate system with respect to the reference frame XYZ of sample, the three Euler angles θ, Φ, and Ψ.

From the above analysis it is found that, with regard to the degree of orientation preference, any distribution of magnetic texture is equivalent to a simple one in which the magnetic moments distribute only along the three main axes and the distribution probabilities are P_1, P_2, and P_3, respectively. By realizing that

$$l'^2 + m'^2 + n'^2 = 1$$

equation (20) can be rewritten as

$$P_o = P_1 - (P_1 - P_2)m'^2 - (P_1 - P_3)n'^2$$

$$= (P_1-P_3)l'^2-(P_2-P_3)m'^2-P_3. \tag{21}$$

Hence for $P_1 \leq P_0 \leq P_3$, the maximum degree of the orientation preference is P_1 in the 1-1 main axis, while the minimum one is P_3 in the 3-3 main axis. Moreover, $|\vec{P}_d|$ has a maximum parallel with R and a minimum in the plane normal to R.

As mentioned above, the 3-dimensional vector $\vec{R}$, and the matrix B, or the simple distribution model of magnetic moments, can describe the preference for the magnetic texture in space, and also represent the main characteristics of the moment distribution for the magnetic texture. This statistical description has a clear physical meaning and is easy to express, which is an advantage. It is a weakness of this statistical description, that the dispersion information about the distribution of magnetic moments is missing from the statistical parameters.

MAGNETIC TEXTURE AND MÖSSBAUER SPECTROSCOPY

In this paper, we deal only with the case in which the magnetic interaction is stronger than quadrupole interaction, and the recoil-free fraction is isotropic. A theoretical analysis of the Mössbauer spectra for ferromagnetic alloys shows that the angular factors provide information about the direction preference and orientation preference of magnetic moments in the magnetic texture.

POLARIZED γ-RAY MÖSSBAUER SPECTROSCOPY AND MAGNETIC TEXTURE

The laboratory coordinate system is X'Y'Z', where the Z'-axis is the incident direction of the Mössbauer γ-ray, the X'-axis is in the plane determined by the Z'-axis and the internal field, $\vec{H}_s$, of the polarized γ-ray source. The coordinate system XYZ is identical with X'Y'Z' in the beginning. The angle between the source internal field, $\vec{H}_s$, and Z'-axis is θ_s and the coordinates of $\vec{M}_i$ in X'Y'Z' are $(x_i'y_i'z_i')$. Their correlation is shown in Figure 5. Accordingly, the five angular factors, S_i, of the polarized γ-ray Mössbauer spectrum are expressed as follows [26]:

$$\begin{aligned}
S_1 &= \cos^2\theta_s - 2\cos\theta_5(\Sigma P_i \bullet z_i') + \Sigma P_i \bullet z_i'^2 + \sin^2\theta_s \bullet (\Sigma P_i \bullet x_i'^2),\\
S_2 &= \cos^2\theta_s + 2\cos\theta_s \bullet (\Sigma P_i \bullet z_i') + \Sigma P_i \bullet z_i'^2 + \sin^2\theta_s \bullet (\Sigma P_i \bullet x_i'^2),\\
S_3 &= 1 - \Sigma P_i \bullet z_i'^2 - \sin^2\theta_s(\Sigma P_i \bullet x_i'^2),\\
S_4 &= \sin^2\theta_s \bullet (1-\Sigma P_i \bullet x_i'^2),\\
S_5 &= \sin^2\theta_s \bullet (\Sigma P_i \bullet x_i'^2).
\end{aligned} \tag{22}$$

It is known that the degree of the direction preference for the magnetic texture along Z'-axis, $\Sigma P_i \bullet z_i'$, the degrees of the orientation preference in Z'-axis and X'-axis, $\Sigma P_i \bullet z_i'^2$ and $\Sigma P_i \bullet x_i'^2$, can be obtained by measurement of the five angular factors in the spectrum. The degree of the orientation preference in the Y'-axis is

$$\Sigma P_i \bullet y_i'^2 = 1 - \Sigma P_i \bullet x_i'^2 - \Sigma P_i \bullet z_i'.$$

Rotating the sample to change the incident orientation of the γ-ray and orienting the X'-axis and Z'-axis at any given direction in the sample, we can measure the direction preference and orientation preference for the magnetic texture in any given direction.

From (22) it is known that, when the γ-ray is linearly polarized, i.e., θ_s is 90°, the five angular factors do not provide the information about $\Sigma \vec{P}_i \cdot z_i'$, so they imply only the orientation preference for the magnetic texture.

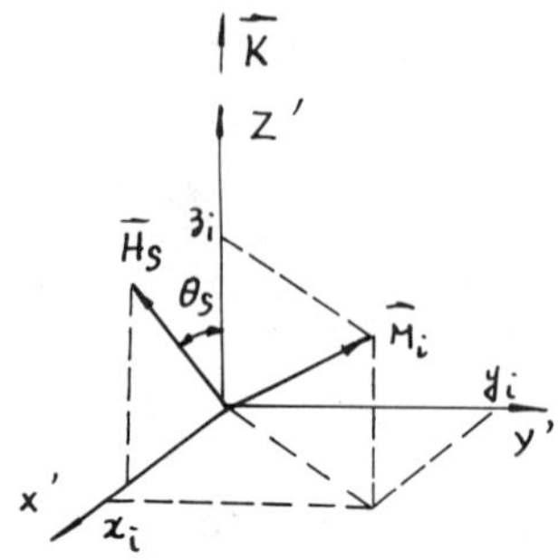

Figure 5. The coordinate system relationship between the incident direction, $\vec{K}$, of the γ-ray, the internal field, $\vec{H}_s$, and the magnetic moment, $\vec{M}_i$.

NON-POLARIZED X-RAY MÖSSBAUER SPECTROSCOPY AND MAGNETIC TEXTURE

There are six magnetic peaks in a iron-57 Mössbauer spectrum of ferromagnetic alloys in the case of a non-polarized γ-ray source. When the incident γ-ray is in a direction $\vec{D}$, the magnetic texture and the angular factor, A, expressing the ratio of the second and fifth peak area to the third and fourth peak area in the spectrum are

$$A = \frac{A_{2,5}}{A_{3,4}} = 4 \bullet \frac{\Sigma P_i \bullet \sin^2 \chi_i}{\Sigma P_i (1+\cos^2 \chi_i)} = 4 \bullet \frac{1-P_o}{1+P_o} \tag{23}$$

$$P_o = \frac{4 - A}{4 + A}. \tag{24}$$

Therefore, the angular factor of the non-polarized γ-ray Mössbauer spectrum implies only the orientation preference, P_o, for the magnetic texture in the incident direction of the γ-ray. By rotating the sample, the orientation preference for the magnetic texture in any given direction can be measured.

In order to find the independent statistical parameters describing the magnetic texture, we should measure several values of the degrees of the direction and orientation preference in various directions by rotating the sample and substituting the results into (18) and (19) obtaining two sets of simultaneous linear equations of rank 3 and 5, respectively.

In summary, the angular factors of a Mössbauer spectrum imply the statistical preference of the distribution for magnetic moments in the magnetic texture. The non-polarized or linearly polarized γ-ray Mössbauer spectrum provides only the information about the orientation preference for the magnetic texture, and the elliptically polarized γ-ray Mössbauer spectrum implies the information about both the direction and orientation preferences for the magnetic texture. By rotating the sample we can determine the independent statistical parameters characterizing the two preferences of the distribution for magnetic moments within the magnetic texture.

Table III. The Angular Factors, A, for the As-Quenched $Fe_{40}Ni_{40}P_{14}B_6$

Angle ϕ	θ	Measured A	Calculated A
0°	0°	2.95	2.97
0°	15°	2.89	2.87
0°	30°	2.58	2.58
0°	45°	2.22	2.22
90°	30°	2.63	2.62
90°	45°	2.26	2.27
45°	45°	2.19	2.19

MAGNETIC TEXTURE IN QUENCHED AMORPHOUS ALLOYS RIBBONS

The non-polarized γ-ray Mössbauer spectrum provides less information than that of the polarized spectrum, and it requires more measurements in order to determine the statistical parameters describing the magnetic texture. However, the non-polarized spectrum is easier to fit and does not require a polarized γ-ray source. The energies of various magnetic anisotropies are related to the orientation preference for the magnetic moments. Therefore, the non-polarized γ-ray Mössbauer spectrum are used widely in study of the magnetic texture for amorphous alloys.

MAGNETIC TEXTURE IN THE AS-QUENCHED AMORPHOUS ALLOYS

We have measured the magnetic texture of the as-quenched sample of amorphous $Fe_{40}Ni_{40}P_{14}B_6$ in a ribbon 12mm wide and 20μm thick with the transmission Mössbauer spectra of non-polarized γ-rays. Suppose the major axis, the minor axis of the ribbon, and the normal of the ribbon plane are the X-, Y-, and Z-axis respectively. The measurements are carried out in the arrangement shown in Figure 6. The measured angular factors, fitted with the modified method for fitting spectra, are listed in Table III. The parameters of the magnetic texture for $Fe_{40}Ni_{40}P_{14}B_6$, calculated with the data in Table III are as follows:

$$P_x = 0.436,\ P_y = 0.416,\ P_z = 0.148,$$

$$2\eta_x = 1.42 \times 10^{-2},\ 2\eta_y = -1.19 \times 10^{-2},\ 2\eta_z = 5.82 \times 10^{-2}.$$

The θ dependences of the measured and calculated angular factors, A, are plotted in Figure 7 where it is seen that the theoretically calculated magnitudes of A coincide well with those measured in several orientations. Thus, the above mentioned theoretical analysis is valid. The experiment shows that all the deviation parameters are approximately zero and the distribution of magnetic moments is nearly symmetric, which is consistent with the observation of the magnetic domains. therefore, the three X-, Y-, and Z-axis of the as-quenched ribbon of $Fe_{40}Ni_{40}P_{14}B_6$ are considered to be approximately the main axes. From its axial parameters P_x, P_y, and P_z, it is known that the magnetic moments are preferentially oriented in the ribbon plane, and that they have the greatest degree of the orientation preference in the major X-axis and the smallest one in the normal Z-axis, which thus shows a zig-zag domain. Thus, we have a quantitative description of the orientation preference for magnetic moments in the sample. For the

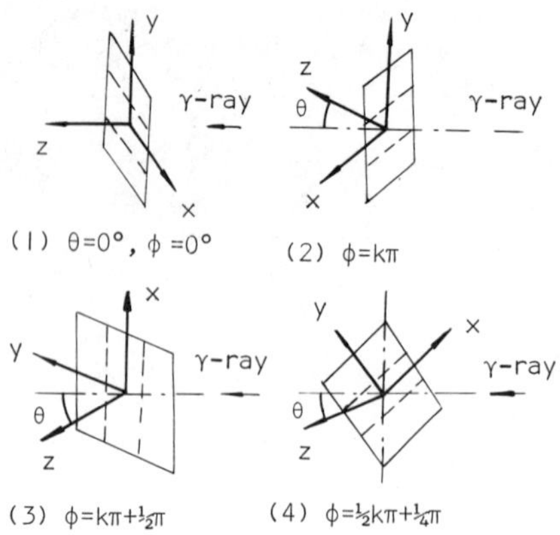

Figure 6. Four usual arrangements of the measured sample. θ and φ are the polar and azimuthal angles of γ-ray in the coordinate systems, respectively.

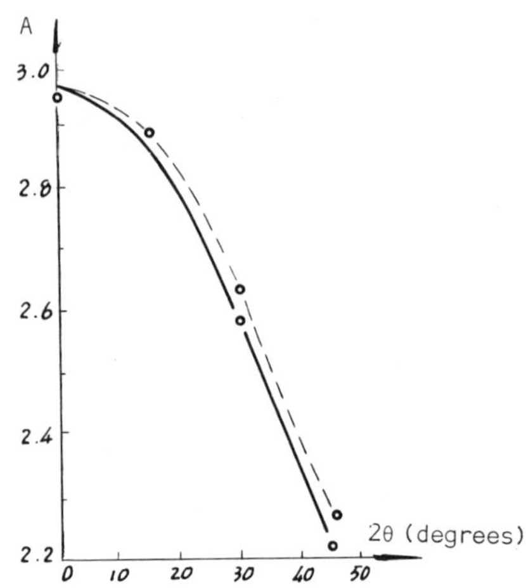

Figure 7. The angle dependence of the measured and calculated angular factors, A, for $Fe_{40}Ni_{40}P_{14}B_6$ • measured for $\phi = 90°$; o measured for $\phi = 0°$; ---- calculated for $\phi = 90°$; —— calculated for $\phi = 0°$.

as-quenched ribbon of $Fe_{62}B_{27}Si_7C_4$, similar results are obtained as shown below.

From the above result it is known that the main coordinate system for the magnetic texture in amorphous alloys, can in fact be determined by analyzing the symmetry of the distribution of magnetic moments. In this case we need two angular factors in any orientation to find out the three axial parameters in the main coordinate system. If the three main axes are X-, Y-, and Z-axis respectively, we can usually measure the angular factors A_1 and A_2 in the arrangements in Figure 6(1) and 6(2), i.e., with the ribbon plane perpendicular to γ-ray and by rotating the sample by θ around the Y-axis, where the angle θ is usually between 30° and 45°. The three axial parameters, P_x, P_y, and P_z, in the main coordinate system are determined according to the following equations:

$$\frac{4 - A_1}{4 + A_1} = P_z,$$

$$\frac{4 - A_2}{4 + A_2} = P_x \cdot \sin^2\theta + P_z \cdot \cos^2\theta, \tag{25}$$

$$P_y = 1 - P_x - P_z.$$

Thus the required number of measurements decrease. With this method, we have studied the magnetic texture for several amorphous alloys.

INFLUENCE OF THINNING ON THE MAGNETIC ANISOTROPY OF $Fe_{162}B_{27}Si_7C_4$

The amorphous $Fe_{162}B_{27}Si_7C_4$ alloy in ribbons 16mm wide and 20μm thick was obtained from Central Iron and Steel Research Institute, where they were prepared with the wheel-quenching technique. Suppose the major axis, X, minor axis, Y, of the ribbon and the normal, Z, of ribbon plane, are the three main axes. The spectra with the ribbon plane perpendicular to the γ-ray, and with the ribbon at angle θ of 30° around Y-axis are collected for the as-quenched and thinned samples at room temperature. The thickness dependence of the obtained axial parameters, P_x, P_y, and P_z, in the main coordinate system for the as-quenched and thinned samples is plotted in Figure 8. In Figure 8 is shown the thickness dependence of the average hyperfine field, $\bar{H}_{hf}$, calculated with the distribution of hyperfine fields, $P(H_{hf})$, independent on the thickness of the thinned sample, which shows that the composition and structure throughout the cross section are homogeneous in macroscopic statistics. This result is identical with that obtained by Becker [27] by magnetic measurement.

It is interesting that the three axial parameters, P_x, P_y, and P_z, in the main coordinate system are unchanged upon eliminating the surface layers and thinning, i.e., the distribution of magnetic moments for the as-quenched samples is also independent. Therefore, we conclude that the quenched internal stresses in the as-quenched sample are nearly constant throughout its cross section; the distribution of magnetic moments and main magnetic domains in the sample are dependent mainly on the internal stress and the surface unevenness has little influence on those. The above experimental results coincide with that obtained by Becker [27] with magnetic measurements and differ from those obtained by Takahashi [28]. This difference is probably attributed to greater magnetic anisotropy near the surfaces of the as-quenched ribbon as obtained with the measurement of the torque curve and is related with the surface unevenness.

Therefore, above-mentioned experimental results indicate that the predominant domain structures, deduced from a study with the Kerr effect or Bitter method, on the as-quenched ribbon eliminated, with original surface, can represent the original predominant domain structures in the as-quenched ribbon. It is obvious that, for the annealed sample with the bulk compressive stress resulting from the surface crystallization [12] or oxidation [16], the original predominant domain structures cannot be represented by those studied with the Kerr effect or Bitter method after eliminating the original surface.

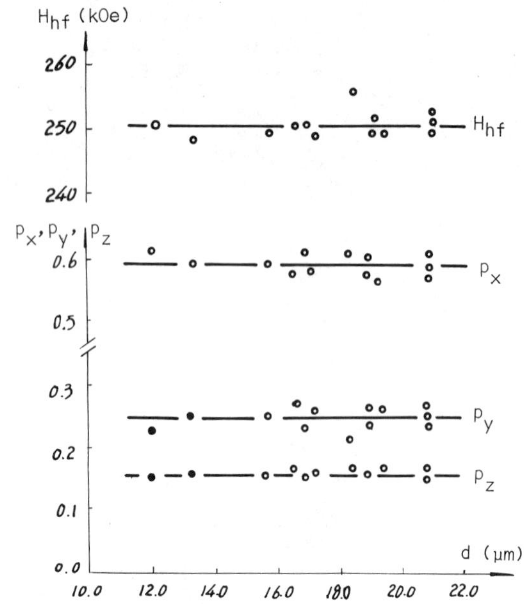

Figure 8. The dependence of the magnetic anisotropy and H_{hf} on the thickness of the thinned samples for $Fe_{62}B_{27}Si_7C_4$.

ABNORMAL MAGNETIC ANISOTROPY OF $Fe_{71}Ni_{10}B_{13}Si_4C_2$ AFTER MAGNETIC ANNEALING

The amorphous $Fe_{71}Ni_{10}B_{13}Si_4C_2$ alloy in ribbons 10mm wide and 20μm thick was obtained from Beijing Metallurgical Research Institute. The samples were annealed in an argon atmosphere under external magnetic field orientated along the major axis, X, of the ribbon. Suppose the major axis, X, minor axis, Y, and the normal, Z, of the ribbon plane are the main axes for the as-quenched and annealed samples. The two angular factor A_1 and A_2 with the ribbon perpendicular to γ-ray and with rotating the ribbon at a θ of 30° around the Y-axis are measured respectively. The experimental results are listed in Table V. The experiment shows that the magnetic properties, H_c and B_{50}, of the annealed samples are improved. However, the magnetic anisotropy of the annealed sample varies abnormally. The zig-zag domains, with magnetic moments perpendicular to the ribbon plane increase, and there are a lot of the horizontal domains with moments perpendicular to the major X-axis. These results differ from the usual views and the Mössbauer result [29] of amorphous $Fe_{75}P_{15}C_{10}$ annealed under the magnetic field.

In order to understand the above results, we have measured the X-ray diffraction patterns of the as-quenched and annealed samples as shown in Figure 9 where it is seen that the as-quenched sample shows the typical broadened peak of the amorphous alloy and the annealed samples show a sharp peak of the crystalline iron-based alloy on top of the broadened peak of the amorphous alloy. In contrast, all the Mössbauer spectra of the as-quenched and annealed samples are the typical broadened sextet of amorphous alloy which has partly crystallized on the surfaces of the ribbon during magnetic annealing. therefore, in addition to the magnetic anisotropy induced during magnetic annealing, there are the bulk compressive stresses which resulted from the surface crystallization and have their magnetic anisotropy perpendicular to the ribbon plane [12] in the samples after magnetic annealing. The two kinds of the magnetic anisotropies together influence and determine the distribution of magnetic moments in the sample after magnetic annealing, which is probably the main cause of above-mentioned abnormal magnetic anisotropy in these samples after magnetic annealing.

Table V. The Magnetic Properties and Anisotropy of $Fe_{71}Ni_{10}B_{13}Si_4C_2$ [a]

Sample	$B_{50}(G_s)$	$Hc(O_e)$	P_x	P_y	P_z
As-quenched	1.2×10^4	0.20	0.382	0.527	0.091
Annealed at 430°C	1.57×10^4	0.056	0.511	0.312	0.177
Annealed at 410°C	1.6×10^4	0.032	0.246	0.492	0.262

[a] T_{Cr} = 480°C (warming rate of 20°C/min), T_c = 460°C

INVESTIGATION OF ANNEALING PROCESS IN AMORPHOUS $Fe_{62}B_{27}Si_7C_4$

Among the many magnetic glasses recently developed for the utilization in low frequency power transformers, it has been reported that Metglas 2605 SC, having a nominal composition of $Fe_{62}B_{27}Si_7C_4$, is superior to many other commercially produced metallic glasses. In the present paper Mössbauer and X-ray techniques are used to study the structure, magnetic relaxation, the amorphous-to-crystalline transformation, and the change of the magnetic anisotropy during annealing for this alloy.

SAMPLE PREPARATION AND EXPERIMENTAL RESULTS

The amorphous $Fe_{62}B_{27}Si_7C_4$ alloy, in the form of a ribbon 16mm wide and 20μm thick was produced by single-wheel quenching technique at the Central Iron and Steel Research Institute. Samples were annealed in a static air atmosphere at temperatures of 245, 300, 358, 400, 420, 440, and 450°C for an hour, respectively. During annealing the samples were not clamped, and were laid on glass plates to avoid any influence of external stresses. There were

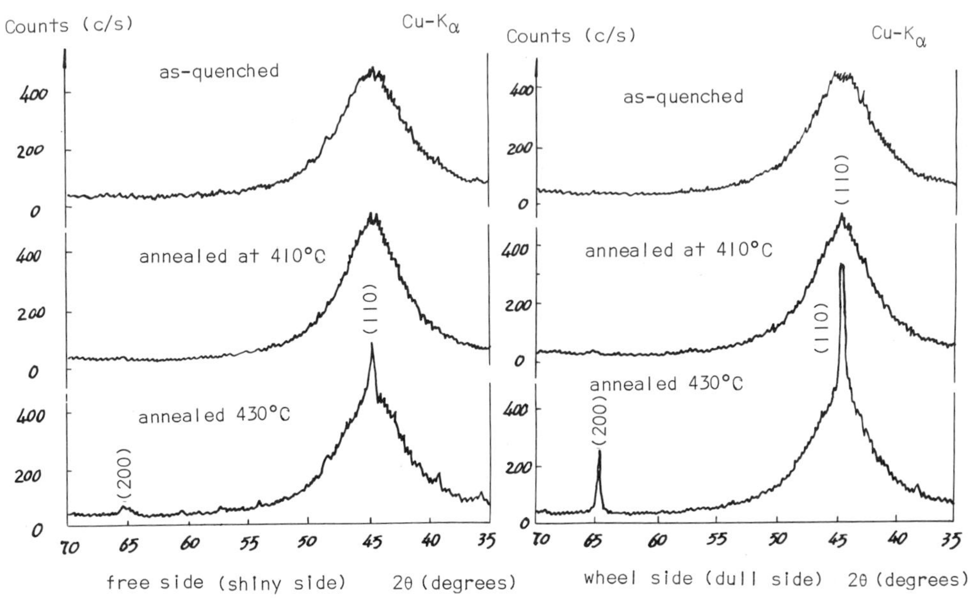

Figure 9. The X-ray diffractograms of $Fe_{71}Ni_{10}B_{13}Si_4C_2$ (40kV, 100mA).

orange or tangerine-colored oxide films on the surfaces of all the treated samples.

The crystallization temperature, 482°C, which approximates the 487°C value found for warming rates of 5°C/min [8], was measured by differential thermal analysis for a warming rate of 4°C/min.

Mössbauer spectra of the as-quenched and annealed samples (Figure 10) with the γ-ray perpendicular to the ribbon plane were collected at room temperature. The X-ray diffraction patterns, obtained with CuKα radiation, on both the free (shiny) and wheel (dull) sides of the samples were obtained with an APD-10 diffractometer (Figure 11). Moreover, profiles of the samples annealed at temperatures of 420°C and 450°C were observed with a S250-II scanning electron microscope.

AMORPHOUS-TO-CRYSTALLINE TRANSFORMATION

From Figure 10 it is seen that Mössbauer spectra of samples annealed at temperatures below 440°C do not show the appearance of a crystalline subspectra and their hyperfine field distribution $P(H_{hf})$, do not have sharp peaks corresponding to the crystalline phases. The spectrum of the sample annealed at a temperature of 450°C has an extensive crystalline subspectra overlapping with amorphous subspectrum. This mixed spectrum is fitted according to the crystalline subspectra, which comprises the alloy Fe-Si and two subspectra of $Fe_2(B,C)$ [30] (refer to Figure 3). The fitted results are listed in Table V, and there is no trace of the crystalline phase Fe_3B. The crystallization of amorphous $Fe_{162}B_{27}Si_7C_4$ seems to be following the one-step reaction

$$Fe_{162}B_{27}Si_7C_4 \rightarrow Fe\text{-}Si + Fe_2(B,C).$$

This pattern of crystallization differs from the two-step pattern of crystallization [5,6,8,31]. Prasad [8] shows that the first step of crystallization precipitated at the alloy, Fe-Si, which was close to the second step where the remaining amorphous phase decomposes to Fe-Si and Fe_2B. The two steps overlap partly. Therefore, the difference in the above result from the reported results are probably because the steps of crystallization overlap and turn out to be a one-step pattern.

The appearance of the brittle fractures for the samples annealed at 420°C and 450°C were observed with a S250-II scanning electron microscope.

Table V. The Mössbauer Parameters of Crystalline Subspectra in Amorphous $Fe_{162}B_{27}Si_7C_4$ Annealed at 450°C[a]

	H_{hf}, kOe	δ, mm/s	ΔE_Q, mm/s	% Intensity
Fe-Si	334.2	0.035	-0.037	20.3 (61.3)
	322.5	0.051	-0.025	5.2 (15.7)
	308.0	0.079	0.000	7.6 (23.0)
$Fe_2(B,C)$	241.0	0.142	0.178	6.7
	234.0	0.175	-0.067	6.6

[a]The values in brackets are the relative intensity ratios of three subspectra for alloy Fe-Si.

No apparent difference in the structure throughout the fracture, for the sample annealed at 420°C is observed. Many crossed-dendritic crystallites (Figure 12), which are analogous to those observed in transmission-electron microscopy [31], are found to be distributed homogeneously in an amorphous matrix throughout the fracture of the sample annealed at 450°C.

Figure 11 shows that partly crystallization takes place in samples annealed at temperatures from 358 to 440°C and that the sharp diffraction peaks of the bcc alloy, Fe-Si overlap with the broadened diffraction peak of the amorphous phase. In the light of above analysis of the Mössbauer results, samples annealed at temperatures from 358 to 440°C are considered to

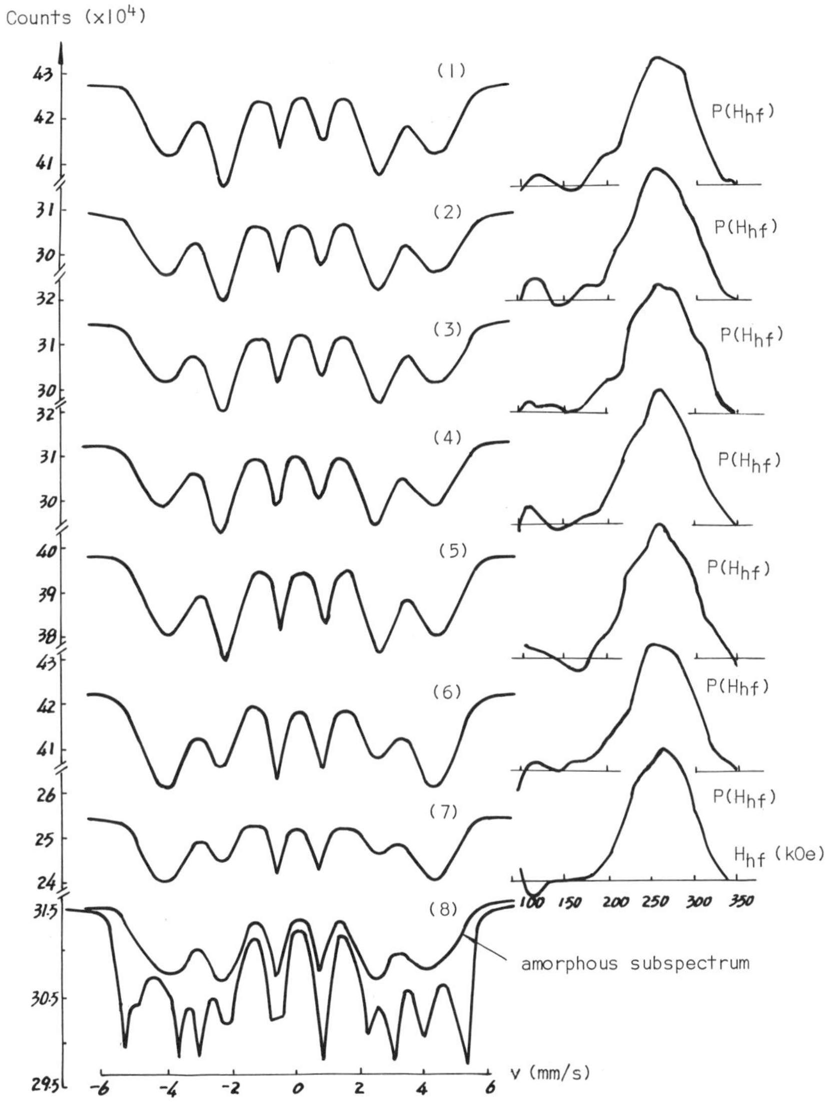

Figure 10. Mössbauer spectra of the as-quenched and annealed samples for the amorphous $Fe_{62}B_{27}Si_7C_4$ obtained at room temperature for the (1) as-quenched; and (2)-(8) annealed at 245, 300, 358, 400, 420, 440, and 450°C sample.

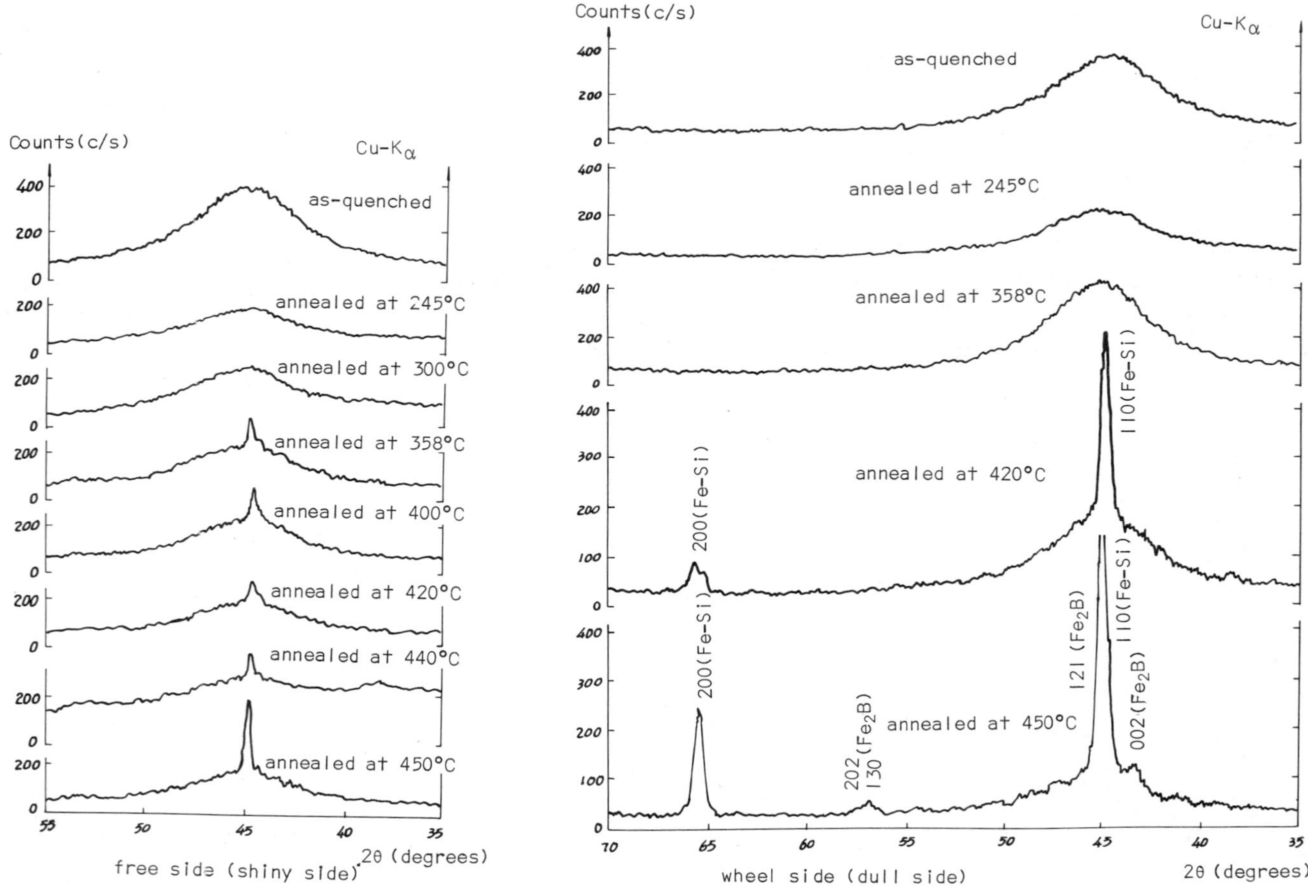

Figure. 11. X-ray diffraction patterns of both surfaces of the as-quenched and and annealed sample of amorphous $Fe_{162}Si_7C_2$ at room temperature.

crystallize only on their surfaces, because only surface 1μm deep contributes to the diffraction intensity [6]. The X-ray diffraction pattern of the partly crystallized sample annealed at 450°C has sharp peaks corresponding to the alloy Fe-Si, and to the compound $Fe_2(B,C)$, which coincides with the Mössabuer result. It is reported that from X-ray diffraction results, the crystallized precipitate from amorphous $Fe_{162}B_{27}Si_7C_4$ are considered to be α-iron and Fe_3B [31]. However, all investigations [5-9] of the crystallization of amorphous Fe-B-Si or Fe-B-Si-C by Mössbauer spectroscopy show that there is only Fe_2B or $Fe_2(B,C)$, and no trace of Fe_3B in the crystallization products. From Figure 11 it is difficult to identify Fe-B which has weak peaks and overlaps with the diffraction pattern of the iron-based alloy. In contrast, in the spectra of the iron-based alloys, Fe_2B and Fe_3B differ from each other and are easy to identify, so that the Mössbauer sepctral interpretation is reliable.

By comparing the diffraction patterns of both the free and the wheel surfaces shown in Figure 11, it is known that both surfaces of the amorphous ribbon have different abilities to crystallize; the free surface being easier to crystallize than the wheel surface. The diffraction pattern of the free surface for the sample annealed at 358°C has the sharp peaks of the crystalline alloy Fe-Si, and the diffraction pattern of the wheel surface has a broadened amorphous pattern similar to that of the as-quenched sample. This result is opposite to that reported for $Fe_{82}B_{12}Si_6$ [12] and $Fe_{162}B_{27}Si_7C_4$ [31] which exhibit primary crystallization on the wheel surface, but is identical with that reported for $Fe_{80}B_{20}$, [6] which exhibits primary crystallization on the free surface, Furthermore, it has been reported [31] that amorphous $Fe_{162}B_{27}Si_7C_4$ begins surface crystallization after annealing for 3 hours at 390°C, which is a longer time and higher temperature than in our experiment. The above results show that the surface of ribbon is easier to crystallize and the initial temperature of crystallization on the surface differ from that reported. It seems reasonable that the wheel surface of the single-wheel-quenched ribbon is more stable than the free surface, because it is quenched more rapidly. However, it has been reported that, even for the as-quenched samples of $Fe_{82}B_{12}Si_6$ [32], $Fe_{67}Co_{18}B_{14}Si_1$ [33] and $Fe_{75}B_{15}$ [34], there is a surface-crystallization layer on the wheel surface. Davis [32] attributes the surface crystallization to the oxidation of boron by air trapped in the "valleys" of the irregular wheel surface. It is reported that the surface oxidation of many amorphous ribbons may bring about an advance in their crystallization. Probably, the oxidation depletes the metaloids of metglass-former on the surface layer and the amorphous state under the oxidized layer becomes unstable. Therefore, the reported differences in surface crystallization for amorphous alloys are probably related to the

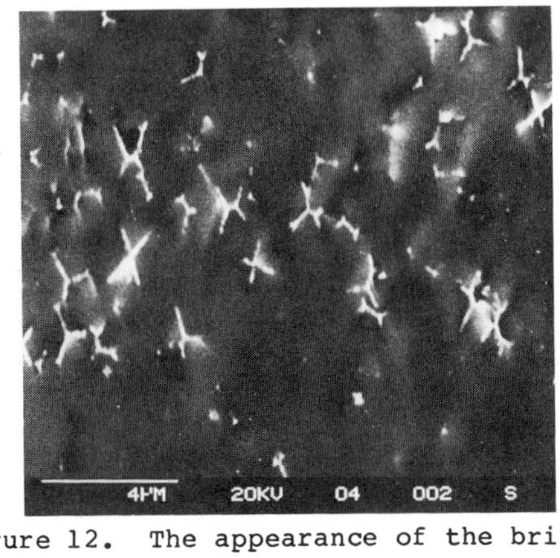

Figure 12. The appearance of the brittle fracture for the $Fe_{162}B_{27}Si_7C_4$ annealed at 420°C.

preparative conditions and their subsequent treatment. The characteristic of surface crystallization in our experiment is probably related to the surface oxidation of the experimental samples in the air.

COMPOSITIONS OF THE CRYSTALLINE AND AMORPHOUS PHASES

From the Mössbauer data given in Table V, the compositions of the crystalline phase and remaining amorphous phase of the sample after annealling at 450°C are identified as follows. One iron atom in the bcc alloy, Fe-Si has eight nearest neighbors and six next nearest neighbors. Suppose that the silicon atoms are randomly distributed and that the probability p(m,n), for the iron-atom having n nearest and m next nearest neighbors of Si-atoms [26] is given by

$$p(m,n) = C_8^m C_6^m c^{m+n} (1-c)^{14-m-n},$$

where c is the concentration of silicon in atomic percent. Then the probabilities for several iron-silicon alloys may be calculated and are listed in Table VI.

The relative ratios of the occupation probabilities for the iron-atoms with only the three main configurations are listed in Table VII. By comparing the data in Tables V and VII, we can conclude that the iron-silicon crystallization product comprises 4 atomic percent silicon. From the experimental Mössbauer spectra data for iron-silicon alloys [35] it may also be concluded that the iron-silicon precipitation in the partly crystallized sample in present experimental consists of 3.5 to 4.9 atomic percent silicon which is consistent with the above analysis. It is difficult to identify the compositions of the iron-silicon alloy from its diffraction data in Figure 11, so the measurement of the exact lattice constant with X-ray diffraction should be undertaken.

Table VI. Normalized Relative Occupation Probabilities in Percent for Iron Atoms With Various Configurations

c	0.030	0.035	0.040	0.050
P(0,0)	65.3	60.7	56.5	48.8
P(0,1)	16.2	17.6	18.8	20.5
P(0,2)	1.70	2.2	2.7	3.8
P(1,0)	12.1	13.2	14.1	15.4
P(1,1)	3.0	3.8	4.7	6.8

Table VII. Relative Ratios of Occupation Probabilities for Iron-Atoms with Three Main Configurations

c	0.030	0.035	0.040	0.050
P(0,0)	69.8	66.3	63.2	57.6
P(0,1)	17.3	19.2	21.0	24.2
P(1,0)	12.9	14.4	15.8	18.2

The composition of the remaining amorphous phase, for the partly crystallized sample annealed at 450°C, may be determined from the following reaction,

$$Fe_{162}B_{27}Si_7C_4 \rightarrow x(Fe_{96}Si_4) + y(Fe_2B_{27}C_4)_2$$
$$+ Fe_{(81-96x-2y)}B_{(13.5 - \frac{13.5}{15.5}y)}Si_{(3.5-4x)}C_{(2 - \frac{2}{15.5}y)}.$$

By assuming that the recoil-free factors for the iron-silicon alloy, the boride $Fe_2(B,C)$, and the remaining amorphous phase are f_1, f_2, and f_3, respectively, with the intensities given in Table V, we can obtain

$$96x : 2y : (81-96x-2y) = 33.2f_1 : 13.3f_2 : 53.6f_3.$$

Then if $f_1=f_2=f_3$, we obtain x=0.279, y=5.39, and the composition of the remaining amorphous phase is

$$Fe : B : Si : C = 43.43 : 8.81 : 2.38 : 1.30$$

$$= 77.7 : 15.8 : 4.3 : 2.3.$$

Accordingly the atomic ratio of iron and metaloids in the remaining amorphous phase is

$$Fe : (B+Si+C) = 78 : 22,$$

which approximates the ratio for Fe_3B. In fact, because the f factors of the crystalline phases are 15-39 percent larger than that of the amorphous phase [2,6], the phase fraction of the remaining amorphous phase should be larger than 53.6 percent its composition should also be revised corresponding to this larger phase and fraction.

AVERAGE HYPERFINE FIELD, $\bar{H}_{hf}$

Below the crystallization temperature the amorphous alloy exhibits a structural relaxation process, which influences the fine structure, width and centers of the distributions of the hyperfine fields, $P(H_{hf})$, as shown in Figure 10. The annealing temperature, T_a, vs. the average hyperfine field, $\bar{H}_{hf}$, which is the center of the hyperfine fields distributions is plotted in Figure 13. The average hyperfine field is calculated with the following formula,

$$\bar{H}_{hf} = \int_{H_{min}}^{H_{max}} P(H_{hf})H_{hf}dH_{hf}.$$

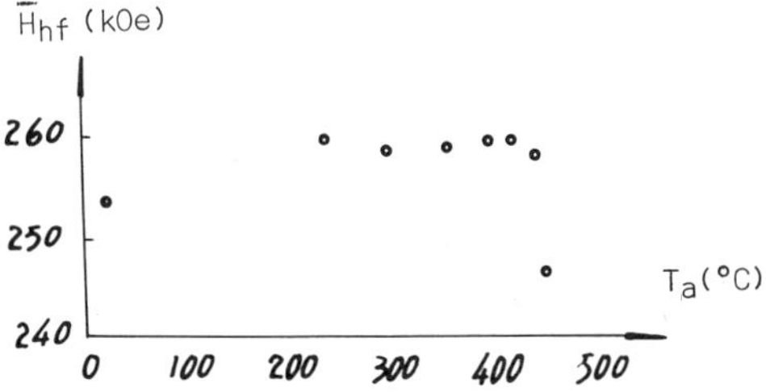

Figure 13. Average hyperfine field vs. T_a.

The increase of the average hyperfine field, $\bar{H}_{hf}$, implies a change in the short-range order around the iron atom during the relaxation process. From Figure 13 it is found that the average hyperfine field, $\bar{H}_{hf}$, increases markedly below 245°C and remains nearly constant with increasing annealing temperature, until 440°C. It is interesting that the increase of the average hyperfine field, $\bar{H}_{hf}$, occurs predominantly at the stage of low temperature relaxation. Moreover, the average hyperfine field of the remaining amorphous phase, $Fe_{78}(B,Si,C)_{22}$, for the partly crystallized sample annealed at 450°C, decreases markedly, because of the decrease in the iron content for the remaining amorphous phase [2,3,13].

ASYMMETRY IN THE MÖSSBAUER SPECTRA

The Mössbauer spectra at room temperature for samples annealed at various temperatures, as shown in Figure 10, are fit to six broadened Gaussian peaks. The differences in the fitted width of the three couples of broadened peaks, $\Delta\Gamma_{1,6}$, $\Delta\Gamma_{2,5}$, and $\Delta\Gamma_{3,4}$, are plotted in Figure 14. This figure shows that annealing influences the spectral asymmetry of amorphous alloys and all the curves of the width differences, $\Delta\Gamma_{1,6}$, $\Delta\Gamma_{2,5}$, and $\Delta\Gamma_{3,4}$, exhibit a sharp change in the region of the annealing temperatures of 300-400°C. This change in the spectral asymmetry probably indicates two different structural relaxation processes before crystallization, one of which corresponds to structural relaxation at low temperature and the other to preliminary crystallization at high temperature.

MAGNETIC ANISOTROPY

The Mössbauer spectra of the as-quenched sample of $Fe_{62}B_{27}Si_7C_4$ were measured at five orientations. The fitted angular factors are listed in Table VIII. The six parameters describing the orientation preference for the magnetic texture are

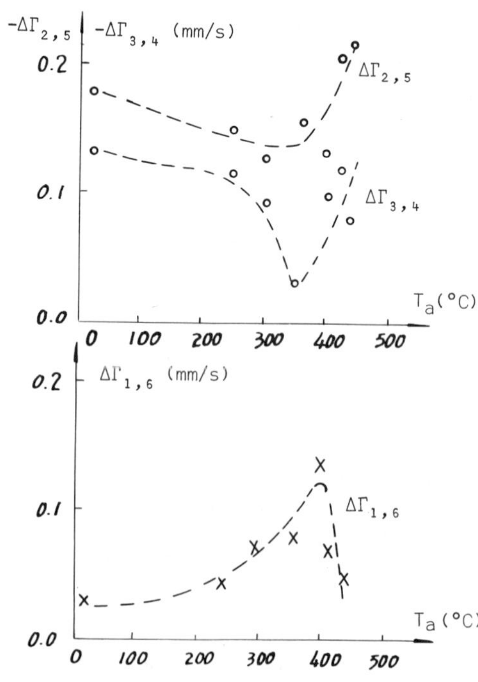

Figure 14. The spectrum asymmetry vs. T_a

Table VIII. The Angular Factors A at Five Orientation

Arrangement	θ = 0° φ = 0°	θ = 45° φ = 180°	θ = 45° φ = -90°	θ = 45° φ = 90°	θ = 45° φ = 135°
A	2.75	2.09	2.30	2.27	2.14

$$P_x = 0.453, \quad P_y = 0.363 \quad P_z = 0.184$$

$$2\eta_x = 0.53 \times 10^{-2}, \ 2\eta_y = -0.85 \times 10^{-2}, \ 2\eta_z = -0.73 \times 1\text{-}^{-2}.$$

These parameters show that there are few zig-zag domains; the magnetic moments are predominantly distributed in the ribbon plane and the preferential order toward the X-axis in the ribbon plane is the largest.

The dependence of the annealing temperature, T_a on the angular factor, A, of the Mössbauer spectrum obtained at room temperature with γ-ray perpendicular to the ribbon plane, is plotted in Figure 15. It shows that the angular factor, A_{25}/A_{34} for the sample annealed below 420°C decreases slowly. Apparently the stress relaxation during annealing at these low temperatures has little influence on the preferential ordering of magnetic moments. In contrast, the angular factors of the samples annealed at 420 and 440°C decrease sharply and the zig-zag domains are nearly 50 percent of the domains in the samples. The angular factor, A, for the sample annealed at 450°C increases again and is approximately 2.0, which indicates a random distribution of magnetic moments. This result is similar to that reported earlier [12-14].

In the light of the analyses of the Mössbauer spectra and the X-ray diffraction patterns, it is known that the origin and change in the magnetic anosotropy perpendicular to the ribbon plane during annealing are related to the origin of the surface and bulk crustallization. As previously reported [12-14], for many amorphous alloys annealed in an inert-gas or vacuum, the perpendicular magnetic anisotropy has a similar origin and relationship with the surface and bulk crystallization. Therefore, it is concluded that the sharp increase in the perpendicular magnetic anisotropy during annealing at 420 to 440°C is attributed mainly to the bulk compressive stress which results from surface crystallization [12] rather than surface oxidation [16,36]. It is obvious that the disappearance of the perpendicular magnetic

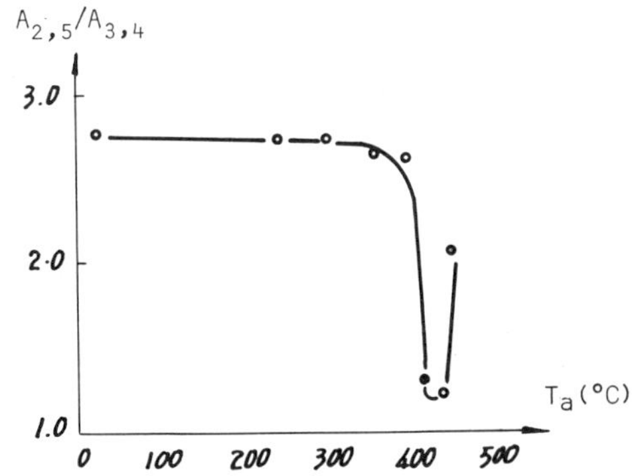

Figure 15. The angular factor A vs. T_a

anisotropy for the sample annealed at 450°C is related to the development of bulk crystallization and the elimination of the bulk compressive stress which resulted from the surface crystallization.

The bulk compressive stress in samples annealed at 420 to 440°C results from the difference between the volume changes of the surface layers and the core of the ribbon. This change is attributed to the differences in the densities which resulted from the transformation,or their different contractions when cooling to room temperature. Based upon density measurements, Ok [12] concludes that the bulk compressive stress results from the smaller density of the amorphous phase as compared with that of crystalline phases. In fact, the cause of the bulk compressive stress should still be determined by comparing the in situ spectra during isothermal annealing with the corresponding spectra at room temperature. In Figure 16 the angular factors for the abovementioned Mössbauer spectra reported by Schurer [13] are compared. This figure shows that the dependences of the angular factor, A, on annealing temperatures for the two kinds of Mössbauer spectra are close and the perpendicular magnetic anisotropy results mainly from the non-reversible volume change during isothermal annealing. However, in Figure 16 there is still some difference between the two. This difference results from the different contractions of the surface layers and the core upon cooling and are considered one of the causes of the perpendicular anisotropy.

MAGNETIC ANOSOTROPY IN RIBBON PLANE FOR THE ANNEALED SAMPLE

For the sample annealed at 420°C, Mössbauer spectra at room temperature are measured either with the ribbon plane perpendicular to the γ-ray or by rotating the sample at 45° around the minor Y-axis. The fitted angular factors are 2.09 and 1.32 respectively. By supposing the major X-axis, the minor Y-axis and the normal Z-axis of the ribbon are the three main axes, we can obtain three parameters in the main coordinate system, P_x=0.123, P_y=0.373, and P_z=0.504. By comparing these parameters with those of the as-quenched sample, we can conclude that, in addition to the intense perpendicular magnetic anisotropy, the distribution of magnetic moments in the ribbon plane for the sample annealed at 420°C also varied, i.e., in the ribbon plane the magnetic moments are distributed preferentially around the Y-axis. This change in the magnetic anisotry in the ribbon plane shows that the bulk compressive stress resulted from the crystallized surface layer is inhomogeneous and larger in the direction of X-axis than of the Y-axis. Ito [15] has reported in a Mössbauer study on the amorphous alloy, Co-Fe-Si-B,

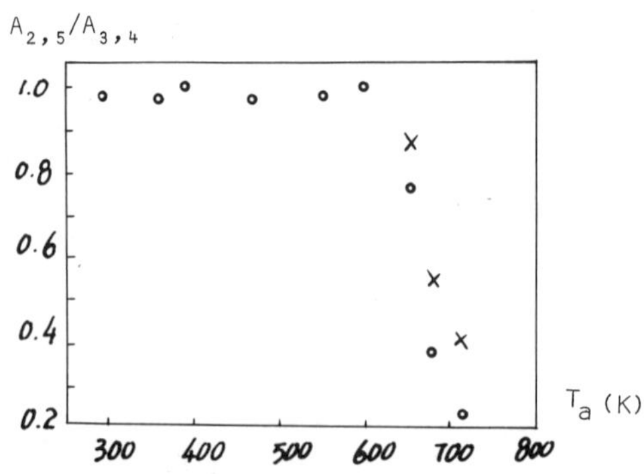

Figure 16. The dependences of the angular factors A on annealing temperature for unclamped-annealed samples of $Fe_{78}B_{12}Si_{10}$ [13] obtained at room temperature after annealing, o, and during isothermal annealing, x.

that annealing at 753K for 5 minutes eliminated the preference of the magnetic moments for the major axis of the ribbon plane in the as-quenched sample. Moreover, investigations [37] of the domain structure in annealed amorphous alloys show that, in the sample annealed at high temperature before crystallization, the magnetic domains with moments oriented along X-axis vanish and the magnetic domains with moments oriented perpendicular to X-axis increase. This is attributed to the elimination of the as-quenched internal stress and to the effect of intrinsic structure anisotropy. Therefore, there is a divergences of views on the explanation of the change of magnetic anisotropy in ribbon plane.

REFERENCES

1. P. J. Schurer and A. H. Morrish, Phys. Rev. B20 4660 (1979).

2. H. N. Ok and A. H. Morrish, Phys. Rev. B22, 3471 (1980).

3. R. E. Vandenberghe and D. Gruffroy, J. Magn. Magn. Mater. 36, 259 (1980).

4. D. Musser, C. L. Chien, and H. S. Chien, J. Appl. Phys. 50, 7659 (1979).

5. T. Kemeny, I. Vincze, and B. Fogarassy, Phys. Rev. B20, 476 (1979).

6. A. S. Shaafma, H. Snijders, and F. van der Woude, Phys. Rev. B20, 4423 (1979).

7. Y. P. Yelsukov, A. A. Glazer, V. R. Galakhov, A. P. Potapov, and Ye Ye Yurchikov, Metal Physics and Metallography 57, 578, in Russian (1984).

8. B. B. Prasad and A. K. Bhatnagar, and R. Jagannathan, J. Appl. Phys. 54, 2019 (1983).

9. N. Saegusa and A. H. Morrish, Phys. Rev. B27, 4027 (1983).

10. A. M. Van Diepen and F. J. A. den Broeder, J. Appl. Phys. 48, 3165 (1977).

11. C. L. Chien and R. Hasegawa, J. Appl. Phys. 47, 2234 (1976).

12. H. N. Ok and A. H. Morrish, Phys. Rev. B23, 2257 (1981).

13. P. J. Schurer, A. H. Morrish, and M. J. Stavan, Phys. Stat. Sol., (a) 64, 343 (1981).

14. N. Saegusa and A. H. Morrish, J. Magn. Magn. Mater. 31-34, 1555 (1983).

15. Atsuko I., Japan. J. Appl. Phys. 20, 797 (1981).

16. U. Gonser, M. Ackermann, and H. G. Wagner, J. Magn. Magn. Mater. 31-34, 1605 (1983).

17. J. Hesse and A. Rubartsch, J. Phys. E7, 526 (1974).

18. B. Window, J. Phys. E4, 401 (1971).

19. C. L. Chien and H. S. Chien, J. de Phys. Colloq. C2-118 (1979).

20. J. Logan and E. Sun, J. Non-Cryst. Solids 20, 285 (1976).

21. D. C. Price, S. J. Campbell, and P. J. Back, J. de Phys. Colloq. C1-263 (1980).

22. G. Le Caer and J. M. Dubois, Phys. Stat. Sol. (a) 64, 275 (1981).

23. H. L. Pfannes and H. Fischer, Appl. Phys. 13, 317 (1977).

24. J. M. Greneche and F. Varret, J. Phys. C15, 5333 (1982).

25. Z. Xu, R. Ma, and J. Ping, Scientia Sinica (Series A), in press.

26. U. Gonser, ed., Mössbauer Spectroscopy, Springer, p. 121 (1975).

27. J. J. Becker, J. Appl. Phys. 52, 1095 (1981).

28. M. Takahashi, Japan. J. Appl. Phys. 15, 1821 (1976).

29. D. K. Brown, I. Nowik, and D. I. Paul, J. Magn. Magn. Mater. 7, 182 (1978).

30. I. D. Weisman, L. J. Swartzendruber, and L. H. Benett, Phys. Rev. 177, 465 (1969).

31. J. C. Swartz, R. Kossowsky, J. J. Haugh, and R. F. Krause, J. Appl. Phys. 52, 3324 (1981).

32. L. A. Davis, N. DeCristofara, and C. H. Smith, Proceedings of the Conference on Mettallic Glasses: Science and Technology, Budapest, Hungary (1980).

33. J. C. Ford, W. A. Hines, J. I. Budnick, A. Paoluzi, D. M. Pease, L. T. Kabacoff, and C. U. Modzelewski, J. Appl. Phys. 53, 2288 (1982).

34. Y. V. Baldokhin, O. S. Kozlova, V. A. Makarov, and V. P. Ovcharov, Metal Physics and Metallography 53, 1027, in Russian (1982).

35. L. Haggstrom, L. Granas, R. Wappling, and S. Dvanarayanan, Physica Scripta 7, 125 (1973).

36. R. Gemperle, L. Kraus, F. Kroupa, and J. Schneider, Phys. Stat. Sol. (1) 60, 265 (1980).

37. B. Groger, and H. Kronmuller, J. Magn. Magn. Mater. 9, 203 (1978).

MÖSSBAUER EFFECT STUDY OF THE CLUSTER DISTRIBUTION IN THE SPIN GLASS θ-$Me_{0.35}V_2O_5$ (Me=Fe, Al)

H. Lass and J. Pebler

Department of Chemistry
University of Marburg und Sonderforschungsbereich
127 "Kristallstruktur und Chemische Bindung"
Hans-Meerwein-Str.
D-3550 Marburg/Lahn
Federal Republic of Germany

ABSTRACT

Mössbauer data for the vanadium oxide bronzes, θ-$Fe_{0.35}V_2O_5$ and θ-$Al_{0.26}Fe_{0.09}V_2O_5$ reveal a randomly oriented spin arrangement with only weak indirect exchange interaction possibly stimulating a spin-glass like order. Below 18K the onset of magnetic order shows up as a broadened magnetic hyperfine patterns. Two sextets are observed below 6K. The Mössbauer spectra can be fit assuming superparamagnetic relaxation of a cluster magnetization. Our relaxation model is appropriate for a flipping of the hyperfine field into a reversed direction. From the resolved spectra, it is apparent that clusters with different relaxation frequencies of 4.0 and 0.9MHz and different hyperfine fields of 45.0 and 38.5 Tesla are present. The cluster distribution function was derived from the temperature dependence of the relative intensity of the doublets.

INTRODUCTION

The observation of a more or less sharp maximum in the temperature dependence of the magnetic susceptibility of spin glasses has aroused great experimental and theoretical interest. Two groups of models are used to describe spin glasses. The phase transition concept [1] belongs to the first group. In the second group the approximation of the spin glass effects arises from the clustering of the magnetic moments [2]. These clusters are blocked into stable states as the temperature is decreased. At higher temperatures, the thermal excitations prevent the intercluster correlation. When the temperature is reduced towards the spin glass freezing temperature, the growing clusters of various sizes and shapes begin to be blocked with their preferred axis distributed randomly over the samples. Hence, one expects a competition between the ferromagnetic interaction and the antiferromagnetic interaction together with random anisotropy. The original ideal of Neel [3] of that the superparamagnetism in such clusters was frozen into a more stable ferromagnetic state at a blocking temperature was proposed by Beck [4], Tholence and Tournier [2], Wohlfarth [5], Herber [6], and Guy [7] as being the essential interpretation of spin glass behavior. The interest [8] in θ-$Fe_{0.35}V_2O_5$ was stimulated by the suggestion of a spin glass type phase arising below 20K. In this temperature region θ-$Fe_{0.35}V_2O_5$ is an insulator and the RKKY interactions which are present in the metallic spin glass are excluded. On the other hand, θ-$Fe_{0.35}V_2O_5$ has a well defined crystalline structure, and is thus different from amorphous spin glasses. It seems most probable, in view of crystallographic and other chemical considerations, that the appearance of the spin glass phase in θ-$Fe_{0.35}V_2O_5$ is due to a frustrated antiferromagnetic interactions.

EXPERIMENTAL

$Fe_{0.35}V_2O_5$ and $Fe_{0.09}Al_{0.26}V_2O_5$ enriched in iron-57 were synthesized by using the solid state reaction method of Galy [9]. The mixed iron-aluminum sample was additionally tempered for 200 hours at 640°C in vacuo and then rapidly quenched. The x-ray powder diffraction patterns revealed presence of only the θ-phase. Magnetic susceptibility measurements were performed on powdered samples from 4.2 to 300K with the Foner method. The measured values were corrected for diamagnetic contributions. The Mössbauer spectra were obtained with a conventional spectrometer. The liquid helium of cobalt-57 temperature source consisted of about 10mCi diffused into metallic rhodium. The absorber temperature accuracy was about 0.5K and the temperature stability was somewhat better. The velocity transducer was metallic iron and gave a linearity for the velocity range of better than one percent.

CRYSTAL STRUCTURE

Monoclinic θ-$Fe_xV_2O_5$ crystallizes in the C2/m space group with six formula units per unit cell and has a homogenity range for x of 0.33 to 0.38.

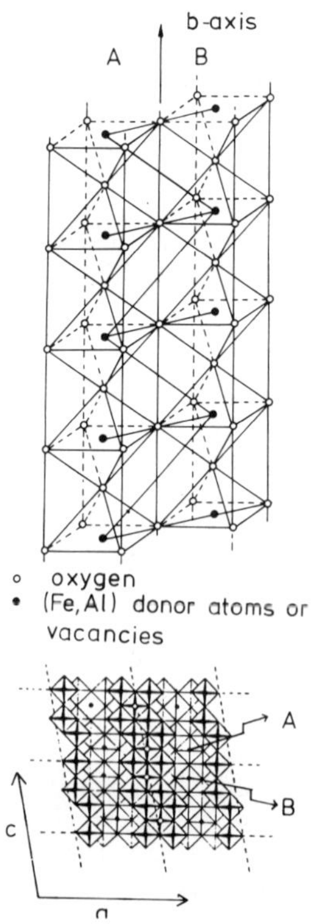

Figure 1. A perspective view of the structure of $Fe_xAl_{0.35-x}V_2O_5$.

Both on the basis of its structure and its electrical and magnetic properties it belongs to the class of compounds known as vanadium oxide bronzes (VOB). The high-spin iron(III) occupy a 3x/2 fraction of the accessible interstitial lattice positions as is illustrated in Figure 1. Charge neutrality for the compound

$$Fe^{3+}_{0.35}V^{4+}_{1.05}V^{5+}_{0.95}O_5$$

in which the concentration of vanadium(IV) and vanadium(V) ions is practically the same. Double channels run parallel to the crystallographic b-axis with iron-iron distance of 0.368nm. The shortest possible iron-iron distance is 0.362nm. All other iron-iron distances are longer than 0.611nm [8].

MAGNETIC SUSCEPTIBILITY

Because of the close structural and electrical similarity of θ-$Fe_{0.35}V_2O_5$ and θ-$Al_{0.35}V_2O_5$, it was assumed that the two magnetic subsystems, iron(III) and vanadium(IV) are practically virtually decoupled and contribute additively to the total magnetic susceptibility. The iron component of the susceptibility,

$$X_{Fe} = X(Fe_{0.35}VOB) - X(Al_{0.35}VOB),$$

only shows a weak maximum near 6K in a low external magnetic field of 0.15T as shown in Figure 2). With increasing applied field, the maximum becomes more diffuse. This behavior is believed to be one of the characteristic features of spin glasses. The corresponding Wohlfarth distribution function [10] for blocking temperatures, is shown in Figure 3. The blocking

$$W(T) = (1/c)\bullet d/dT(T\bullet X_{Fe}(T)), \tag{1}$$

temperature, T_B, of small magnetic clusters is defined as the temperature below which the cluster behaves like a magnetically ordered crystal. T_B is not uniquely defined but is related to the time scale of the experimental technique used. The distribution function F(f(T)) given in Figure 3 was calculated on the basis of the Mössbauer spectral results presented below.

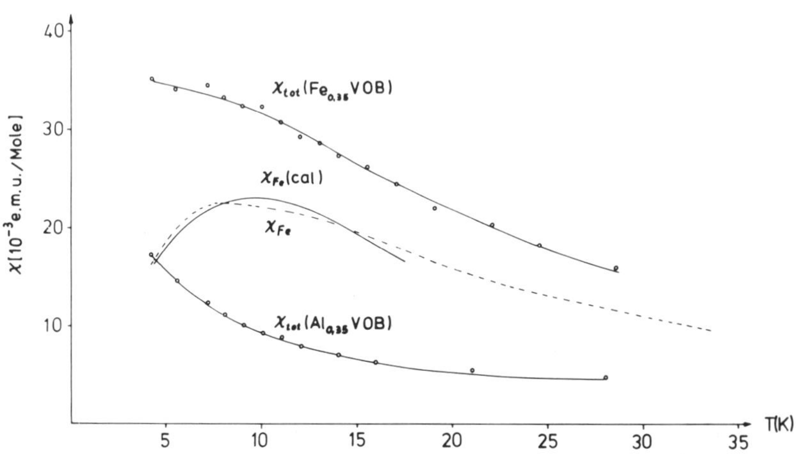

Figure 2. The magnetic susceptibility of $Fe_{0.35}V_2O_5$, $Al_{0.35}V_2O_5$, the experimental X_{Fe} and the X_{Fe} values calculated on the basis of the distribution function (see equation 5) obtained from Mössbauer measurements.

Table I. Mössbauer Spectral Parameters and relaxation Rate for θ-$Fe_{0.35}V_2O_5$ at 4.2K

Component	δ	H_{int}	ω
1	0.160(8)mm/s	45.0(3)T	4.0MHz
2	0.158(8)mm/s	38.5(5)T	0.9MHz

MÖSSBAUER EFFECT

We have studied samples of θ-$Fe_xAl_{0.35-x}V_2O_5$ with x values of 0.35, 0.175, 0.09, and 0.009 by Mössbauer spectroscopy between 4.2 and 300K. In the temperature range from 20 to 300K, the Mössbauer spectra of θ-$Fe_{0.35}V_2O_5$ show two quadrupole doublets with splittings of 1.10 and 0.69mm/s. Both spectra reveal slight deviations from a one to one line intensity ratio, probably, as a result of a more extended iron site distribution [11]. The corresponding isomer shifts are 0.23 and 0.24mm/s at 293K relative to metallic iron. The integral intensity ratio of the doublets A_1A_2 is temperature independent and approximately equal to 0.75. The spectra obtained at 293K for x=0.175, 0.09, and 0.009 show approximately the same quadrupole splitting and isomer shift values as for θ-$Fe_{0.35}V_2O_5$. Thus the oxide octahedra which surround the iron(III) are not disturbed by the aluminum(III) admixture.

The onset of magnetic order below 18K for the compositions with x=0.35, 0.175, and 0.09, shows up as broadened magnetic hyperfine patterns. The magnetic hyperfine spectra are fully symmetric around their center of gravity below spin glass temperature of 18K, in apparent contradiction with the high temperature quadrupole interaction. However, the linewidths of the spectral components are unequal. The sum of Lorentzian fits provides linewidths of 1.50, 1.08, and 0.75mm/s for the outer, middle and inner components, respectively at 4.2K.

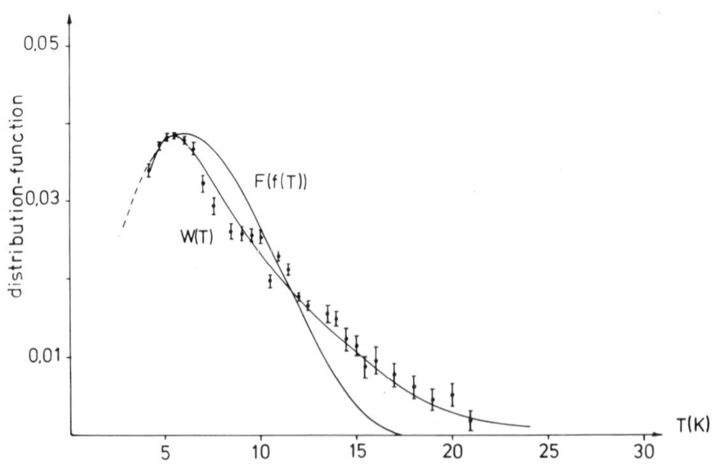

Figure 3. Blocking distribution function W(T) and F(f(T)) deduced from magnetic susceptibility measurements and Mössbauer spectroscopy, respectively.

CALCULATION OF THE MAGNETIC SPECTRA

The Mössbauer patterns could be successfully fit by assuming superparamagnetic relaxation of a cluster magnetization. We used a relaxation model [12] which is appropriate for our systems [13]. It allows for a flipping of the hyperfine field into the reverse direction with a rate w. From the resolved spectra of θ-$Fe_{0.35}V_2O_5$ it is apparent that clusters with different relaxation frequencies must be present. At 4.2K this shows up as asymmetric line shapes as seen in Figures 4 and 5. Between 6 and 8K, however, there appears a central resonance line near zero relative velocity due to faster relaxation clusters. The fit parameters derived from the spectrum of θ-$Fe_{0.35}V_2O_5$ at 4.2K are given in Table I. Although a distribution function for the cluster magnetization has to take into consideration (see Figure 3) for optimum fits it was sufficient to assume that two types of clusters with different relaxation frequencies and different hf

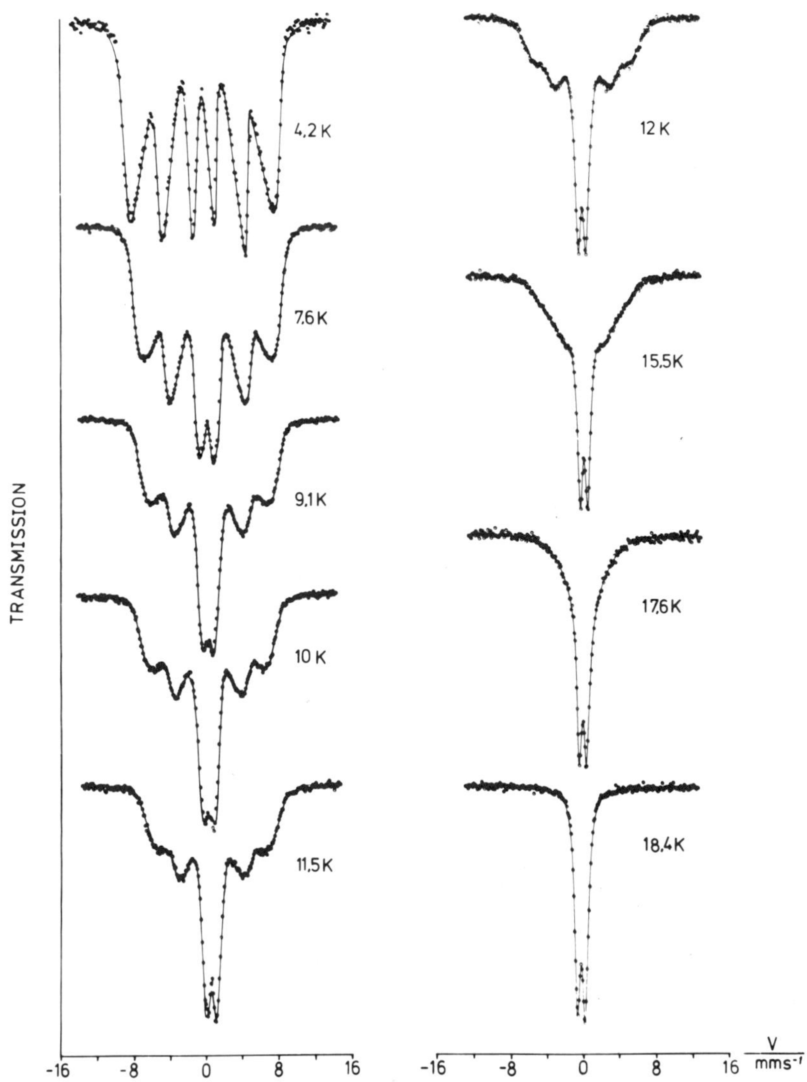

Figure 4. Mössbauer spectra of iron-57 in $Fe_{0.35}V_2O_5$ within the spin glass phase.

fields are present. The calculated values of isomer shift IS, magnetic hf field H and relaxation frequency w are given in Table I.

The integral intensity ratio S_2/S_1 of the two magnetic hyperfine spectra is equal to ca. 0.8. The evaluated magnetic fields H_1 and H_2 obtained for iron(III) in θ-$Fe_{0.35}V_2O_5$ are about 18 and 30 percent smaller than the values established for paramagnetic iron(III) [14]. This difference may be a result of dipole-dipole and exchange interaction between the 5/2 state of iron(III) and the spins of the vanadium(IV) ions, as has been observed previously [15].

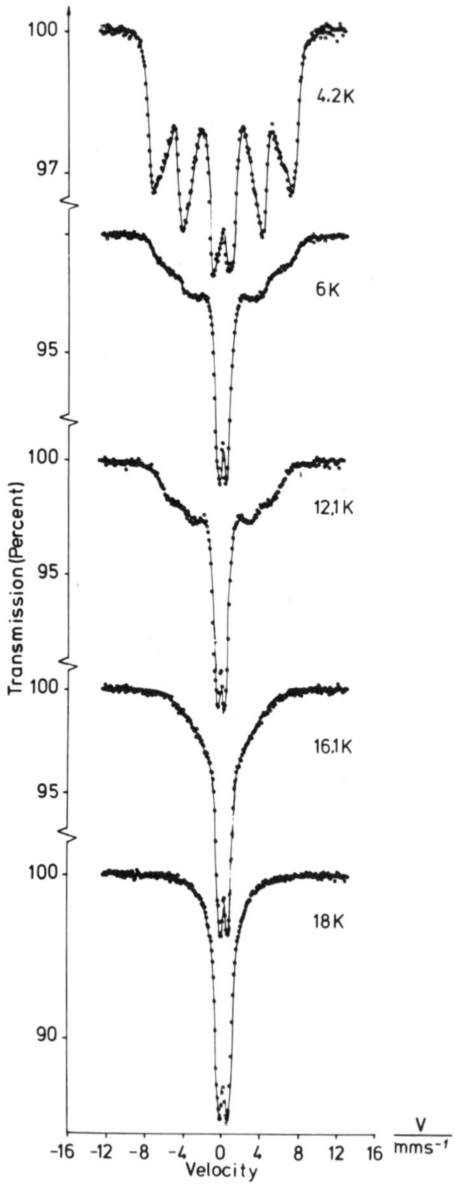

Figure 5. Mössbauer spectra of iron-57 in $Fe_{0.09}Al_{0.26}V_2O_5$ within the spin glass phase.

CLUSTER SIZE DISTRIBUTION

Our data support the presence of a cluster-type magnetic order in these vanadium oxide bronzes. Although their crystallographic structure is a single phase, the break-up of the material into magnetic clusters with different relaxation frequencies may be induced by fluctuations in the distribution iron over the octahedral lattice sites. The collective spin dynamics of the clusters in the partially frozen magnetic state can be quantitatively evaluated from our Mössbauer data. If there are only some tens or hundreds of magnetic ions in the cluster, the thermal excitation energy may be enough to invert all of the spins simultaneously, thereby reversing the magnetization of the sublattices. In order to estimate the size of the clusters, we introduce a critical flip frequency, w_c, with the understanding that the Mössbauer spectrum is a magnetically split sextet for $w < w_c$ and a paramagnetic doublet for $w > w_c$ [16]. In the uniaxial case, the magnetization flips with a frequency,

$$w = w_0 \exp(-KV/k_B T), \tag{2}$$

along the easy axis, where w_0 is ca. $10^{11}\,\mathrm{sec}^{-1}$. In the absence of any detailed information about K, the crystalline anisotropy constant, except that K is zero at T_g, the spin glass transition temperature, we assume [17] a temperature dependence for K of

$$K = K_0(1-T/T_g)$$

where K_0 is ca. $1.4*10^5\,\mathrm{J/cm^3}$. Then, with k_BT approximately equal to kT_g and $w=w_c$, the size diameter can be critical dimensionless cluster expressed as

$$f_c = (1-T/T_g)^{-1/3} = (K_0/(k_B T_g \ln(w_0/w_c)))^{1/3} V^{1/3}. \tag{3}$$

The relative intensity D of the doublet is given by

$$D(T) = \int_0^{fc(T)} (f^3 F(f)df) / \int_0^{\infty} (f^3 F(F)df). \tag{4}$$

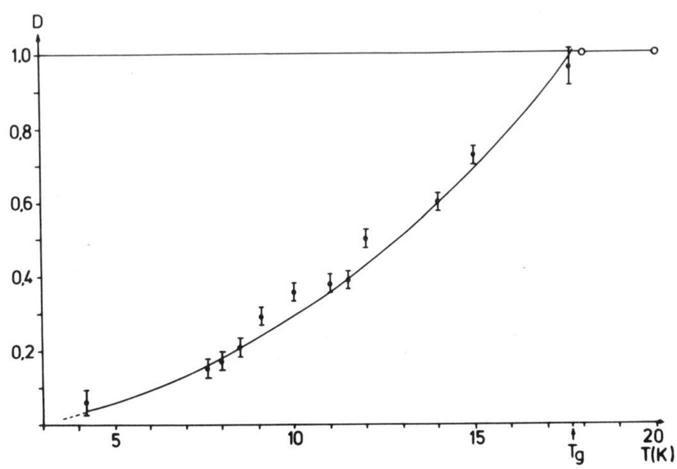

Figure 6. The relative intensity D of the doublets as a function of temperature.

The temperature dependence of D, which can be deduced from the Mössbauer spectra (Figure 4), is shown in Figure 6. The results can be approximated by the function $D(T) = aT^b$ with the fit-parameters of $0.18 \cdot 10^{-2}$ and 2.2 for a and b, respectively. From this fit the value of the extrapolated spin glass transition temperature, T_g, is 17.7K. By taking the derivation of both sides of equation 4 with respect to T, we obtain

$$F(f_c(T) = 3abT_g^{b-1}((1-f_c(T)^{-3})^b)/(f_c(T)^7-f_c(T)^4). \tag{5}$$

From the evaluation of our data we obtain

$$V_c^{1/3} = 2.30 \cdot f_c(nm). \tag{6}$$

From reasonable arguments we put

$$F(f_c(T)) \approx F(V^{1/3}) \tag{7}$$

and obtain the cluster distribution function,

$$F(V^{1/3}) = 3abT_g^{b-1}((1-(V^{1/3}/23)^{-3})^b/((V^{1/3}/23)^7-(V^{1/3}/23)^4), \tag{8}$$

shown in Figure 7, with a mean value for the cluster size of 2.95nm. If we replace W(T) in equation 1 by $F(f_c(T))$ from equation 5, we can derive the temperature dependence of the susceptibility, X_{Fe}, below T_g (see Figure 3). By assuming a close packed array of iron(III) ions occupying the octahedral sites of the double channels shown in Figure 1, we can estimate from our cluster size distribution an average cluster volume of $25.7nm^3$, corresponding to an average of 57 unit cells are 227 iron(III) ions for $Fe_{0.35}V_2O_5$. The determination of the cluster size distribution from Mössbauer spectroscopy results in a range between 27 and 420 unit cells per cluster corresponding to an iron(III) distribution between 107 and 1700 ions per cluster.

ACKNOWLEDGMENTS

The authors are grateful for the financial support from the Deutschen Forschungsgemeinschaft and the Fonds der Chemischen Industrie.

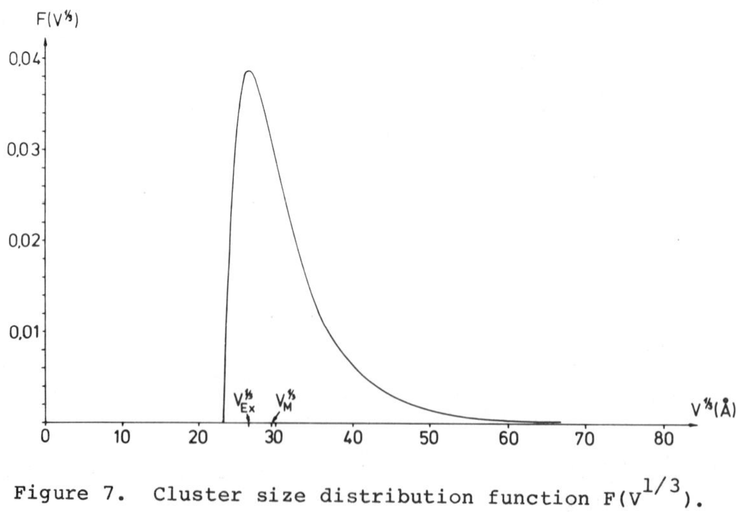

Figure 7. Cluster size distribution function $F(V^{1/3})$.

REFERENCES

1. S. F. Edwards and P. W. Anderson, J. Phys F 5, 965 (1975).
2. J. L. Tholance and R. Tournier, J. de Phys. (Paris) 35, C4-229 (1974).
3. L. Neel, Ann. Geophys. 5, 99 (1949).
4. P. A. Beck, Metal. Trans. 2, 2015 (1971).
5. E. P. Wohlfarth, Physica 86, 852 (1977).
6. G. Herber, Appl. Phys. 10, 101 (1976).
7. C. N. Guy, J. Phys. F7, 1505 (1977).
8. A. Polaczek and M. Pekala, Phys. Stat. Solidi (a) 56, 457 (1979); M. Pekala and A. Polaczek, Phys. Stat. Solidi (a) 58, 533 (1980); M. Pekala and A. Polaczek, Solid State Comm. 34, 275 (1980); J. Korecki, A. Polaczek, and M. Pekala, Nuclear Instruments and Methods 199, 209 (1982).
9. J. Galy, M. Pouchard, A. Casalot, and P. Hagenmuller, Bull. Soc. Fr. Min. Cristallogr. 544 (1967).
10. E. P. Wohlfarth, Phys. Lett. 70A, 489 (1979).
11. J. J. Bara, B. F. Bogacz, M.Pekala, and A. Polaczek, to be published.
12. A. M. Afanasev and E. V. Onishchenko, Sov. Phys. JETP 43, 322 (1976)
13. J. Pebler, Phys. Stat. Solidi (a) 78, 589 (1983).
14. H. H. Wickman, M. P. Klein, and D. A. Shirley, Phys. Rev. 152, 345 (1966).
15. J. Pebler, K. Schmidt and H. Wasinski, Z. Naturforsch. 32b, 507 (1977); J. Pebler and K. Schmidt, Phys. Stat. Sol. (a) 56, 75 (1979); J. Pebler, K. Schmidt, and G. Weiser, Phys. Stat. Sol. (a) 56, 457 (1979).
16. I. P. Suzdalev, Proceedings of the Conference on the Application of the Mössbauer Effect, Tihany (1960).
17. H. L. Wehner, G. Ritter, and H. H. F. Wegener, Phys. Let. 46A, 333 (1974).

APPLICATION OF MÖSSBAUER EFFECT ON THE CRYSTALLIZATION OF SEVERAL GLASSES

Tetsuaki Nishida and Yoshimasa Takashima

Department of Chemistry
Faculty of Science
Kyushu University, 33
Hakozaki, Higashiku
Fukuoka 812, Japan

INTRODUCTION

The structural change of glass from a non-crystalline state to a polycrystalline state caused by controlled crystallization is a very important and interesting phenomenon both from the scientific and industrial point of view. Crystallized glasses are extensively utilized in many industrial fields such as mechanical, electrical, and nuclear engineering because of their ultra low thermal expansion and high mechanical strength [1,2]. Crystallization of glasses is generally known to proceed by two steps, nucleation and crystal growth. The crystallized glasses are known to consist of many crystalline particles mixed into the remaining glass phase. Because of this, crystallized glasses are still regarded as non-crystalline, and are called glass-ceramics [1,2]. As for the crystallization of glasses, several techniques [3-8] have so far been utilized to study glass ceramics. Differential thermal analysis [4-6] and differential scanning calorimetry [7] are known to be very useful. Mössbauer studies on the crystallization of glasses [9-11] have also been performed, and the method proved to be extremely useful for the structural study of borate [12-15], borosilicate [16,17], phosphate [18,19], borophosphate [20], germanate [21,22], zinc chloride [23], and zirconium tetrafluoride [24] glasses. Structural studies of glasses by Mössbauer spectroscopy and NMR have been reviewed by Müller-Warmuth and Eckert [25]. Several kinds of glasses are generally known to consist of three-dimensional networks lacking periodicity, and therefore the structural features of glass are often described as short-range order and long-range disorder. The three-dimensional networks of the oxide glasses are usually composed of tetrahedral network-forming units such as BO_4, SiO_4, PO_4, and GeO_4 units. In case of borate and germanate glasses, triangular BO_3 and octahedral GeO_6 units are also known to be present [2]. NMR studies performed by Bray and O'Keefe [26] revealed that the fraction of BO_4 units in borate glasses reaches a maximum when the alkali oxide concentration is about 33 mole percent, and the decrease in the BO_4 fraction in the alkali oxide content region higher than 33 mole percent was ascribed to the formation of nonbridging oxygen in BO_3 units. In contrast, Mössbauer studies by Nishida et al. [12-14] revealed that the nonbridging oxygen is also present in BO_4 and FeO_4 tetrahedra when the alkali oxide content of borate glasses exceeds about 20 mole percent. Mössbauer studies of borosilicate [16,17] and germanate [21,22] glasses also revealed that the nonbridging oxygen is present in BO_4 and GeO_4 tetrahedra when the alkali oxide contents of borosilicate and germanate glasses exceed about 8 and 16 mole percent, respectively. These conclusions have been obtained on the basis of the distinct composition dependences of the Mössbauer parameters and the absorption area for iron(II), iron(III), and tin(IV). It is very important to elucidate the presence and also the fraction of nonbridging oxygen in glasses because the abnormal composition dependences of some physical properties observed in those glasses, e.g., the borate anomaly [27-29] and the germanate anomaly [21,22,30-34] are considered to be closely correlated with the presence of nonbridging oxygen.

The present study was carried out to elucidate the structural change of borate and borosilicate glasses from glass to glass-ceramic, caused by the heat-treatments at different temperatures.

EXPERIMENTAL METHODS

A series of borate glasses, denoted by the formula $xR_2O \bullet (100-x)B_2O_3 \bullet 7Fe_2O_3$, were prepared by melting the individual 1.0 g batches of R_2CO_3 where R is Na, K, and Rb, H_3BO_3, and Fe_2O_3 of guaranteed reagent grade at 1000°C for 3h in an electric muffle furnace. The x value was chosen to be 15, 20, 23, and 30. Transparent and brown glasses were then prepared by quenching the melts in platinum crucibles with ice-cold water.

Borosilicate glasses, denoted by $xK_2O \bullet (100-x)(B_2O_3,SiO_2) \bullet 7Fe_2O_3$, were prepared by quenching the individual melts fused at 1300°C. The molar ratio between SiO_2 and B_2O_3 was kept at 2.0 in each borosilicate glass. All the glass samples prepared in this way have been kept in a desiccator to protect them from atmospheric moisture.

Annealing experiments were performed in a closed-type electric furnace made for the purpose. The electric furnace is made of a transparent quartz tube and an electrothermal wire. Each end of the electric furnace is stoppered with silicone rubber. Each glass sample placed in a boat-type alumina crucible was annealed under a dry nitrogen atmosphere. As soon as each heat-treatment was completed, the annealed glass sample (powder) was quenched by shifting the position of the glass sample from the center to the end of the electric furnace. The surroundings of the end of the electric furnace are chilled with ice-cold water, instead of being wound up with an electrothermal wire. At the same time, chilled nitrogen gas was also blown through the electric furnace to quench the glass sample. These procedures are necessary to stop the crystallization reaction as quickly as possible at the end of each annealing experiment.

Mössbauer measurements were performed by the constant acceleration method at room temperature by using a 10mCi source of cobalt-57 diffused into a palladium foil. Metallic iron foil enriched with iron-57 was used as a reference for isomer shift and for calibrating the velocity of the spectrometer. All the Mössbauer spectra were fitted to Lorentzian lineshapes by using a least-squares method.

RESULTS AND DISCUSSION

The Mössbauer spectrum for the borate and borosilicate glasses studied consists of a quadrupole doublet as is shown in Figure 1, which illustrates the change in the Mössbauer spectrum for $15K_2O \bullet 85B_2O_3 \bullet 7Fe_2O_3$ glass caused by heat-treatments at different temperatures. It is obvious from Figure 1 that the quadrupole splitting shows a continuous decrease with increasing annealing temperature. This result suggests that the annealed glasses are still regarded as homogeneous from the Mössbauer spectroscopic point of view. It is noted that the annealing experiments at higher temperatures, e.g., at 480°C in the case of $15K_2O \bullet 85B_2O_3 \bullet 7Fe_2O_3$ glass, result in the formation of a magnetic phase, which can be confirmed to be α-Fe_2O_3 from the magnitude of the internal magnetic field. The experimental results shown in Figure 1 therefore suggest that a kind of structural change occurs within the paramagnetic glass phase. Observation of the annealed glasses under a polarized microscope was performed to obtain the information on the structural change of the glasses. Figure 2 shows a typical polarized micrograph of $20K_2O \bullet 80B_2O_3 \bullet 7Fe_2O_3$ glass annealed at 420°C for 60min. Transparent areas in the glass particles shown in Figure 2 indicate the crystallized glass-ceramics, and the less transparent, dark area indicates the original glass phase. These results lead to the conclusion that the structural change from glass to glass-ceramic is reflected in the Mössbauer spectrum as a continuous decrease in the quadrupole splitting. All the isomer shifts for the quadrupole doublets in borate and borosilicate glasses lie in the range of 0.23 to 0.36mm/s with respect to metallic iron.

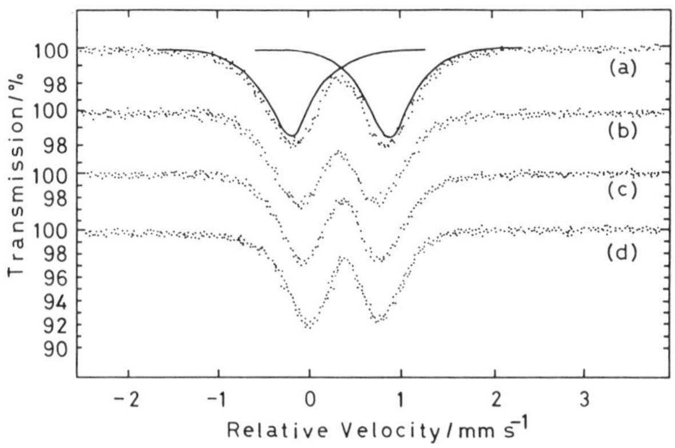

Figure 1. Mössbauer spectra for $15K_2O \bullet 85B_2O_3 \bullet 7Fe_2O_3$ glass annealed at different temperatures. (a) unannealed, (b) at 400°C for 30min, (c) at 450°C for 30min, and (d) at 470°C for 10h.

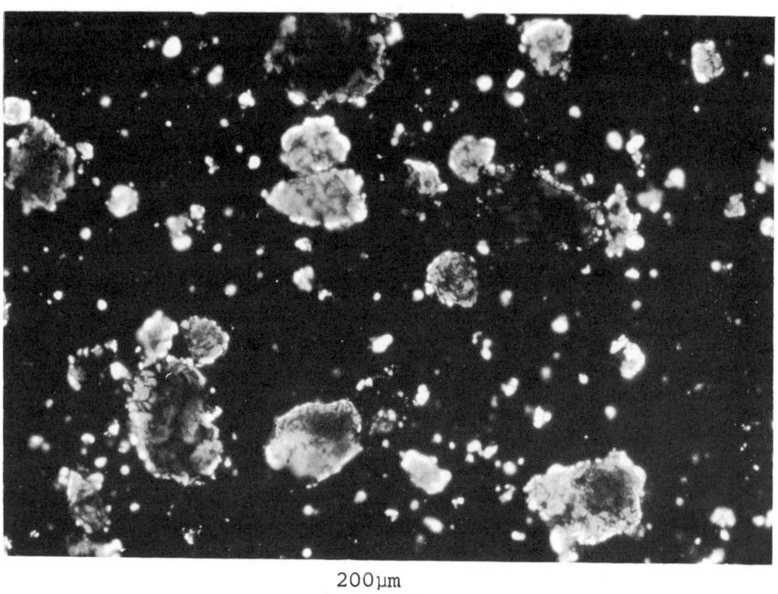

Figure 2. Polarized micrograph of the $20K_2O \bullet 80B_2O_3 \bullet 7Fe_2O_3$ glass annealed at 420°C for 60 min.

The error is estimated to be ±0.01mm/s. All the iron atoms are therefore considered to be present as iron(III) at the substitutional sites of tetrahedral silicon or boron, as a network former, because the isomer shifts are smaller than the ca. 0.4mm/s which can be a criterion for the tetrahedral iron(III) species [12]. As is also seen from Figure 1, the heat-treatment of these glasses results in a small increase in the isomer shift, but no systematic change has been observed.

Figures 3 to 5 demonstrate the continuous decreases in the quadrupole splitting of iron(III) in several borosilicate glasses. The experimental error for the quadrupole splitting is estimated to be ±0.02mm/s. It is

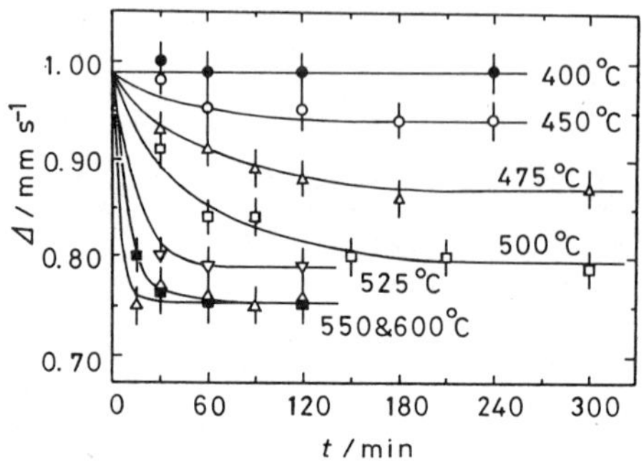

Figure 3. Change in the quadrupole splitting of iron(III) caused by isothermal annealing of the $15K_2O \cdot 85(B_2O_3,SiO_2) \cdot 7Fe_2O_3$ glass.

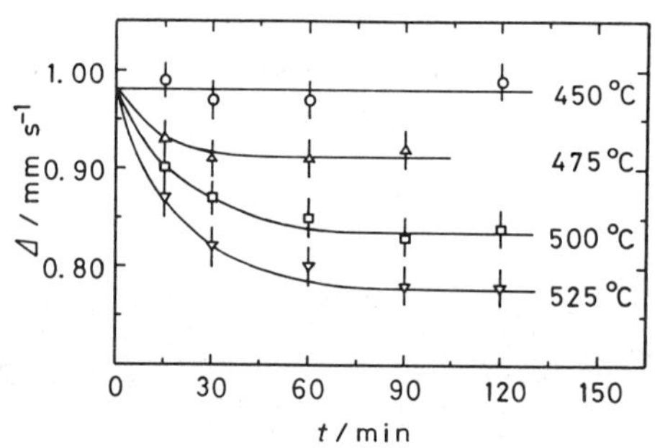

Figure 4. Change in the quadrupole splitting of iron(III) caused by isothermal annealing of the $15Rb_2O \cdot 85(B_2O_3,SiO_2) \cdot 7Fe_2O_3$ glass.

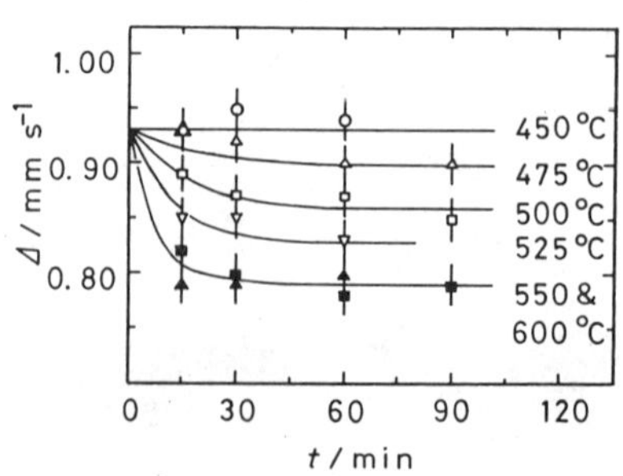

Figure 5. Change in the quadrupole splitting of iron(III) caused by isothermal annealing of the $20K_2O \cdot 80(B_2O_3,SiO_2) \cdot 7Fe_2O_3$ glass.

obvious from these figures that the quadrupole splitting shows a remarkable decrease when the glasses are annealed at higher temperatures and also for the prolonged times. It is noted that the decrease of the quadrupole splitting is not unlimited, but becomes saturated depending on the annealing temperature. This is explained by the saturation effect in the nucleation [1], i.e., the number of crystal nuclei becomes constant depending on the annealing temperature. The decrease in the quadrupole splitting of iron(III) suggests that the symmetry around iron nucleus is increased by the heat-treatment. Because high spin iron(III) has a symmetric $3d^5$ electron configuration, only the electric field gradient caused by the neighboring atoms or ions affects the quadrupole splitting. Therefore, the decrease in the quadrupole splitting of iron(III) caused by the heat-treatment indicates a rearrangement of oxygen atoms around the iron(III) ion constituting an FeO_4 pseudo tetrahedron.

A kinetic study was then performed to get quantitative information on the crystallization from the decrease in the quadrupole splitting. If we define the energy levels of glass and crystalline particles (Figure 6) as $\varepsilon_i(g)$ and $\varepsilon_i(c)$, respectively, individual free energies can be expressed by using the partition functions:

$$Z(g) = \sum_i \exp(-\varepsilon_i(g)/kT) \tag{1}$$

and

$$Z(c) = \sum_i \exp(-\varepsilon_i(c)/kT). \tag{2}$$

When equilibrium is established in the transition reaction from glass to glass-ceramic, the probability that we can find one crystalline particle in the entire glass-ceramic matrix can be expressed by using the corresponding ratio:

$$\frac{N_i(c)}{N(c)+N(g)} = \frac{\exp(-\varepsilon_i(c)/kT)}{Z(c)+Z(g)}. \tag{3}$$

In such a condition, the number of crystalline particles is considered to be constant because the crystallization reaction is artifically quenched at the end of each annealing experiment as described above. In equation (3), N(c) and N(g) indicate the fractions of crystalline particles and glass phase, respectively. The probability that we can find all the crystalline particles in the entire glass-ceramic matrix is likewise expressed as

$$\frac{N(c)}{N(c)+N(g)} = \frac{Z(c)}{Z(c)+Z(g)}. \tag{4}$$

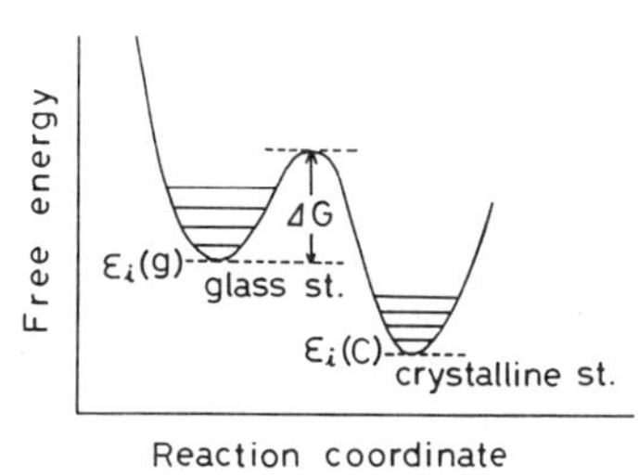

Figure 6. Energy diagram for the glass and crystalline states.

The probability that we can find the fraction of the residual glass phase in the entire glass-ceramic is also expressed as

$$\frac{N(g)}{N(c)+N(g)} = \frac{Z(g)}{Z(c)+Z(g)}. \tag{5}$$

Dividing equation (4) by equation (5) results in the fractional ratio between the crystalline phase and the glass phase, i.e.,

$$N(c)/N(g) = z(c)/Z(g). \tag{6}$$

This equation suggests that the fractionation ratio of the crystalline particle to the remaining glass phase is equivalent to the corresponding energy ratio, only if an equilibrium is established. These considerations lead to a conclusion that the relative fraction of the crystalline particles in each glass-ceramic can be approximated by the energy change, i.e., by the continuous decrease in the quadrupole splitting shown in Figures 3 to 5. Kinetic study of the crystallization of glasses can therefore be performed by utilizing a change in the quadrupole splitting instead of the fractional ratio of the glass to the crystalline particles.

The rate constant, k, for the crystallization of glasses can be obtained from the first order equation,

$$kt = \ln(\Delta_0/\Delta), \tag{7}$$

in which t is the annealing time, and Δ_0 and Δ represent the quadrupole splitting of the iron(III) observed before and after the isothermal annealing experiment, respectively. The application of the first order equation to the crystallization of glasses is considered to be valid because the crystallization is generally known to be controlled by the diffusion or migration of alkali metal ions. The rate constant can be obtained if $\ln(\Delta_0/\Delta)$ is plotted against the annealing time. This is shown in Figure 7, in which rate constant is calculated from the slope of the tangent to each curve. It may seem strange that each plot of $\ln(\Delta_0/\Delta)$ shown in Figure 7 shows a non-linear time dependence, being contrary to the prediction based on equation (7). This is due to the temperature-dependent saturation effect in the crystallization reaction [1] described above, and therefore the rate constant obtained from the first order plot at the unsaturated crystallization stage will really reflect the reaction rate for the crystallization. The correlation between the rate constant and the temperature is expressed by the Arrhenius equation. Figure 8 shows the Arrhenius plot of the rate constants obtained from the first order plot shown in Figure 7. It is obvious from Figure 8 that the Arrhenius plot consists of two straight lines with different slopes, and that the slope is steeper in lower temperature region, below 525°C, than in higher region.

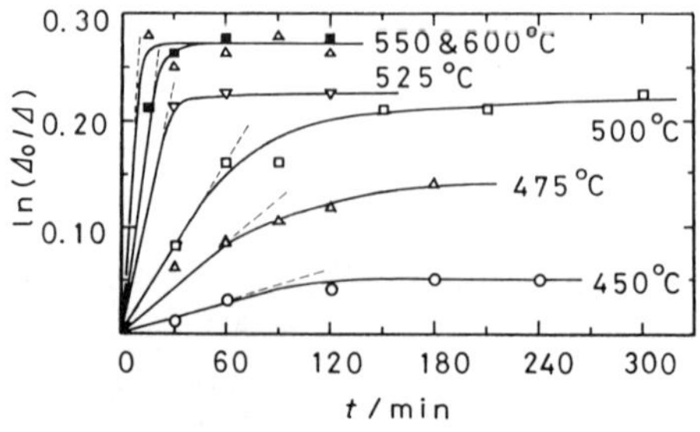

Figure 7. First order plot of the change in the quadrupole splitting of iron(III) for the $15K_2O\cdot85(B_2O_3,SiO_2)\cdot7Fe_2O_3$ glass.

Table I. Activation Energies for the crystallization of borosilicate glasses

Composition	E^a
$15Na_2O \cdot 85(B_2O_3, SiO_2) \cdot 7Fe_2O_3$	∞
$15K_2O \cdot 85(B_2O_3, SiO_2) \cdot 7Fe_2O_3$	41 & 28
$15Rb_2O \cdot 85(B_2O_3, SiO_2) \cdot 7Fe_2O_3$	17
$15Cs_2O \cdot 85(B_2O_3, SiO_2) \cdot 7Fe_2O_3$	15
$20Na_2O \cdot 80(B_2O_3, SiO_2) \cdot 7Fe_2O_3$	∞
$20K_2O \cdot 80(B_2O_3, SiO_2) \cdot 7Fe_2O_3$	46 & 30
$30K_2O \cdot 70(B_2O_3, SiO_2) \cdot 7Fe_2O_3$	∞

[a] Activation energy in kcal/mole obtained from the first order plot.

These results lead to the conclusion, as is generally known, that the crystallization of the borosilicate glasses studied herein also proceeds in two stages; the first stage being nucleation and the second crystal growth. Activation energies for the crystallization of borosilicate glasses obtained in this way are summarized in Table I, from which we can see that the activation energy increases with decreasing ionic radius of the alkali metal ion and also with increasing alkali oxide content of the glasses. As is seen in Table I and Figure 8, the larger values of the activation energy, e.g., 41 and 46kcal/mole, correspond to the nucleation and the smaller values of 28 and 30kcal/mole correspond to the crystal growth. In the cases of rubidium and cesium borosilicate glasses, the Arrhenius plot gives only one straight line corresponding to the activation energy for the nucleation, probably because of the small number of experimental points obtained at higher temperatures.

The crystallization of several kinds of borate glasses has also been studied. The change in the quadrupole splitting of iron(III) caused by the isothermal annealing of $20K_2O \cdot 80B_2O_3 \cdot 7Fe_2O_3$ glass is shown in Fig. 9. It is obvious from Fig. 9 that the decrease in the quadrupole splitting of iron(III) becomes

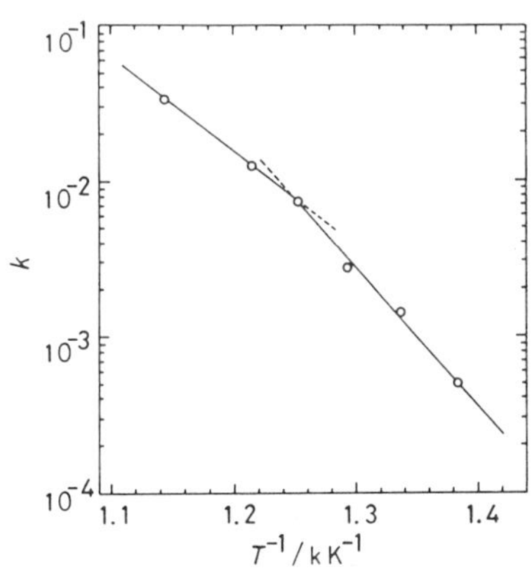

Figure 8. Arrhenius plot of the rate constants obtained from the first order plot of the decrease in the quadrupole splitting (Figure 7) for the $15K_2O \cdot 85(B_2O_3, SiO_2) \cdot 7Fe_2O_3$ glass.

pronounced with increasing annealing temperature and time, and that the decrease is not infinite, but becomes saturated depending on the annealing temperature as was found above. It is also obvious from Figure 9 that the change in the quadrupole splitting of iron(III) in the potassium borate glass is more distinct than that in the several borosilicate glasses shown in Figures 3 to 5. This is also the case for the other kinds of borate glasses containing different types of alkali metal ions. The first order plot of the change in the quadrupole splitting of iron(III) in $20K_2O \cdot 80B_2O_3 \cdot 7Fe_2O_3$ glass is shown in Figure 10, and each rate constant can be obtained from the slope of the tangent to each curve. Arrhenius plot of the rate obtained from the first order plot of the decrease in the quadrupole splitting (Figure 10) is shown in Figure 11. Similarly to the results obtained for the borosilicate glasses, Figure 11 also consists of two straight lines with different slopes. In this figure the open and solid circles refer to the results obtained from the decreases in the quadrupole splitting (QS) and linewidth (LW), respectively. All the activation energies obtained in this way for the crystallization of borate glasses are summarized in Table II, in which the larger values and the smaller values correspond to the activation energies for the nucleation and crystal growth, respectively.

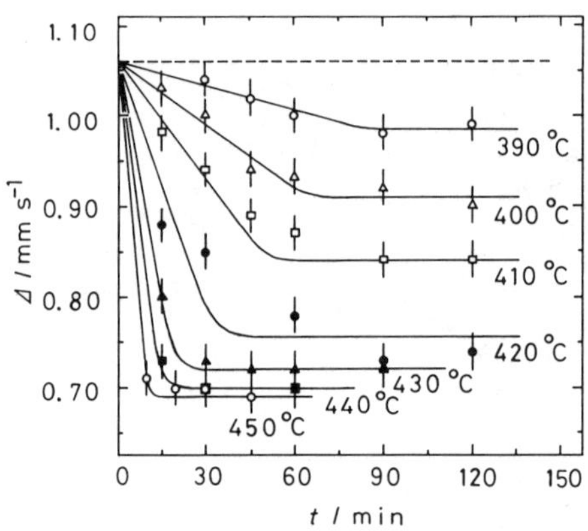

Figure 9. Change in the quadrupole splitting of iron(III) caused by isothermal annealing of the $20K_2O \cdot 80B_2O_3 \cdot 7Fe_2O_3$ glass.

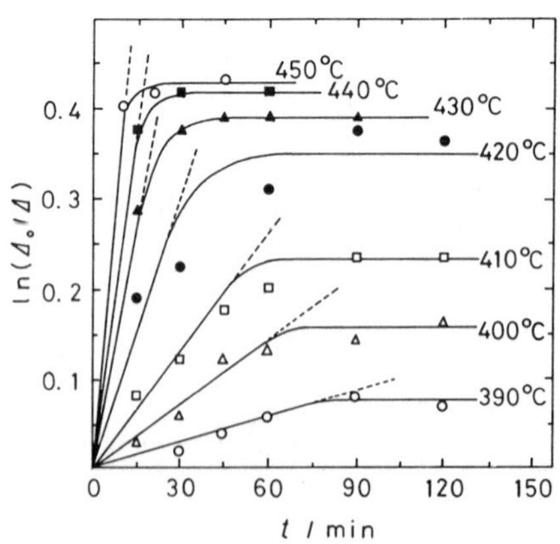

Figure 10. First order plot of the change in the quadrupole splitting of iron(III) for $20K_2O \cdot 80B_2O_3 \cdot 7Fe_2O_3$ glass.

Table II. Activation Energies for the Crystallization of Borate Glasses

Composition	E^a		
$15Na_2O \cdot 85B_2O_3 \cdot 7Fe_2O_3$	∞	&	121
$15K_2O \cdot 85B_2O_3 \cdot 7Fe_2O_3$	64	&	20
$15Rb_2O \cdot 85B_2O_3 \cdot 7Fe_2O_3$	59	&	20
$20K_2O \cdot 80B_2O_3 \cdot 7Fe_2O_3$	72	&	26
$23K_2O \cdot 77B_2O_3 \cdot 7Fe_2O_3$	105	&	66
$30K_2O \cdot 70B_2O_3 \cdot 7Fe_2O_3$	∞	&	∞

[a]Activation energy in kcal/mole obtained from the first order plot.

Activation energies shown in Table II suggest that the crystallization of borate glasses occurs less easily when the ionic radius of alkali metal ion is small and also when the alkali oxide content is high. This conclusion is the same as that obtained for the borosilicate glasses described in Table I. This is explained by the increased electrostatic force between the alkali metal ion and the neighboring oxygen atoms as a result of the decreased interatomic distance between the alkali metal ion and the oxygen atom when the ionic radius of the alkali metal ion is decreased. The increase in the electrostatic force is also ascribed to the presence of nonbridging oxygen in the BO_4, SiO_4, and FeO_4 units when the alkali oxide contents of borate and borosilicate glasses are higher than 20 [11-13] and 8 [15,16] mole percent, respectively.

In the crystallization studies of glasses, the Johnson-Mehl-Avrami, JMA, equation [35-38] is often used to follow the reaction kinetics, and it is expressed as

$$x = 1 - \exp[-(kt)^n] \tag{8}$$

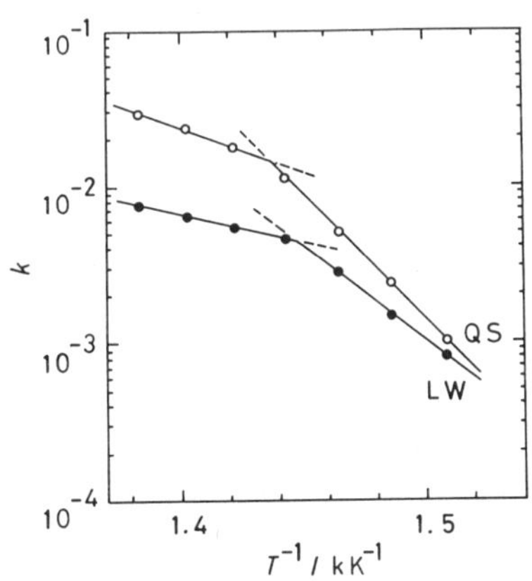

Figure 11. Arrhenius plot of the rate constants obtained from the first order plot of the decrease in the quadrupole splitting (Figure 10) for the $20K_2O \cdot 80B_2O_3 \cdot 7Fe_2O_3$ glass.

or

$$\ln[-\ln(1-x)] = n\ln t + n\ln k, \tag{9}$$

where x and n are the crystallized fraction and the Avrami exponent, respectively. The latter is known to give some information on the crystallization mechanism, such as whether the number of crystal nuclei is constant or not during the crystallization, and whether the crystallization begins from the surface or from the bulk. If the JMA equation is applicable to the change in the quadrupole splitting, a plot of ln[-ln(1-x)] against the annealing time, t, on a log scale will result in a straight line having a slope n. In the present study, the crystallized fraction x can be replaced by

$$x = \frac{\Delta_0 - \Delta}{\Delta_0 - \Delta_m}, \tag{10}$$

where Δ_m indicates the minimum quadrupole splitting value, which depends only on the composition of the glasses, and is therefore inherent in each glass. Therefore, the Δ_m value must be individually obtained for each glass.

The JMA plot for the change in the quadrupole splitting of iron(III) observed in the potassium borate glass is shown in Figure 12, in which each solid line is drawn so that the slope should be parallel to each other, because the reaction mechanism for the crystallization should be invariable with respect to the annealing temperature. The rate constant obtained from the intercept of Figure 12 is plotted against the annealing temperature in Figure 13, which also consists of two straight lines similar to the Arrhenius plot (see Figure 11) of the rate constants obtained from the first order equation. Activation energies obtained from the JMA plot are summarized in Table III. They are relatively well consistent with the activation energies obtained from the first order plot shown in Table II, except for the result for $15Na_2O \cdot 85B_2O_3 \cdot 7Fe_2O_3$ in which no significant change has been observed in the quadrupole splitting after the annealing experiments. It is seen from the Avrami exponent n, that the crystallization of the borate glasses proceeds two or three dimensionally, and that the number of crystal nuclei is constant during the crystallization [35].

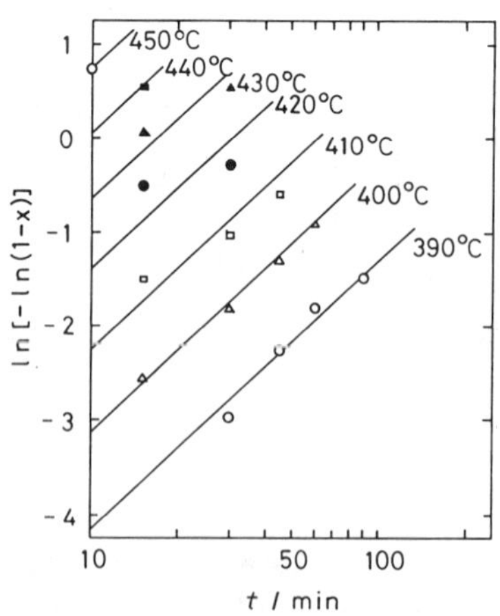

Figure 12. JMA plot of the change in the quadrupole splitting of iron(III) for the $20K_2O \cdot 80B_2O_3 \cdot 7Fe_2O_3$ glass.

Table III. Activation Energies for the Crystallization of Borate Glasses

Composition	E^a	n^b
$15Na_2O \bullet 85B_2O_3 \bullet 7Fe_2O_3$	∞ & 66	1.4
$15K_2O \bullet 85B_2O_3 \bullet 7FE_2O_3$[6]	63 & 12	1.1
$15Rb_2O \bullet 85B_2O_3 \bullet 7Fe_2O_3$	59 & 23	1.0
$20K_2O \bullet 80B_2O_3 \bullet 7Fe_2O_3$	69 & 38	1.3
$23K_2O \bullet 77B_2O_3 \bullet 7Fe_2O_3$	81 & 55	1.7
$30K_2O \bullet 70B_2O_3 \bullet 7Fe_2O_3$	∞ & ∞	——

[a] Activation energy in kcal/mole obtained from the JMA plot.
[b] Avrami exponent.

Observation of the crystallized glass-ceramics under a polarized microscope also supports this conclusion, and the crystallization of these glasses seems to be initiated around iron(III), because the color of the crystallized glasses observed under a polarized microscope is reddish brown. It is also obvious from Table III that the activation energy increases with increasing Avrami exponent. This means that the two-dimensional crystallization, which is assumed to be closely correlated with the structural change around planar BO_3 units, occurs more easily than the three-dimensional crystallization correlated with the structural change around BO_4 units.

A systematic change in the Mössbauer linewidth of iron(III) has also been observed in the borate glasses by the isothermal annealing experiments. Some results obtained from the change in the linewidth (LW) are shown in Figures 11 and 13. The decrease in the linewidth with the crystallization of the glass can be explained by the increased regularity in the configuration of the FeO_4, BO_4, and SiO_4 tetrahedra. The bond length between iron(III) and oxygen, and also the bond angle around iron in the FeO_4 unit will become uniform as a result of the crystallization. This is also the case for the tetrahedral BPO_4 and SiO_4 units. The crystallization of borate glasses, however, resulted in a smaller decrease in the linewidth than that in the

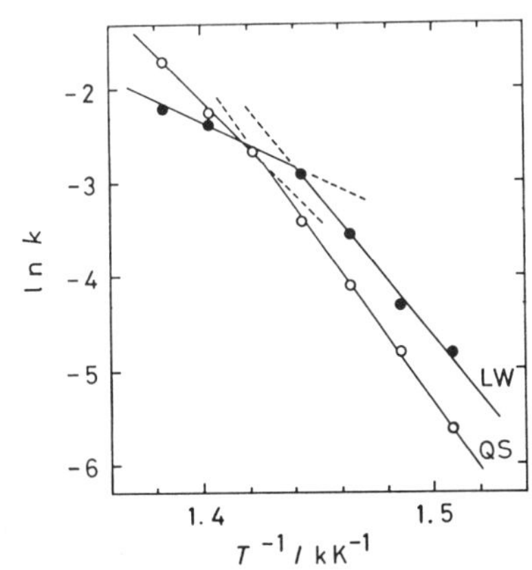

Figure 13. Arrhenius plot of the rate constants obtained from the JMA plot (Figure 12) for the $20K_2O \bullet 80B_2O_3 \bullet 7Fe_2O_3$ glass.

quadrupole splitting. Therefore, the experimental results obtained from the change in the linewidth are not discussed in the present paper.

In conclusion, Mössbauer spectroscopy proves to be a very effective method to study the crystallization of glasses. Diffusion of alkali metal ions followed by a rearrangement of oxygen atoms around a crystal nucleus such as iron(III) is concluded to be a rate-determining step. Therefore the activation energies for the nucleation and the crystal growth are concluded to be closely correlated with the electrostatic force between the alkali metal ion and the neighboring oxygen atoms. It is also concluded that the crystallization of borate glasses, and probably of borosilicate glasses, occurs more easily when the fraction of the planar BO_4 units.

ACKNOWLEDGMENTS

The authors are indebted to Toshiharu Hirai and Toshiaki Nonaka for their kind help in the experiment. One of the authors (TN) would like to express his thanks to Nippon Sheet Glass Foundation for Materials Science for the financial support.

REFERENCES

1. P. W. McMillan, in "Glass Ceramics", Academic Press, Chaps. 1 and 2 (1979).

2. N. J. Kreidl, in "Glass: Science and Technology", D. R. Uhlmann and N. J. Kreidl, eds., Vol. 1, Chaps. 3 and 7 (1983).

3. T. Komatsu and N. Soga, J. Chem. Phys. 72, 1781 (1980).

4. J. Gotz, Phys. Chem. Glasses 18, 32 (1977).

5. J. Briggs and T. G. Carruthers, Phys. Chem. Glasses 17, 30 (1976).

6. J. G. Morley, Glass Technology 6, 77 (1965).

7. K. Matusita and S. Sakka, Phys. Chem. Glasses 20, 81 (1979).

8. T. Furukawa and W. B. White, J. Am. Ceram. Soc. 64, 443 (1981).

9. A. Prasad, D. Bahadur, R. M. Singru, and D. Chakravorty, J. Mater. Sci. 17, 2687 (1982).

10. T. Nishida, T. Hirai, and Y. Takashima, Phys. Chem. Glasses 24, 113 (1983).

11. T. Nishida, T. Hirai, and Y. Takashima, Mem. Fac. Sci., Kyushu Univ., Ser. C 14, 91 (1983).

12. T. Nishida and Y. Takashima, J. Non-Cryst. Solids 37, 37 (1980).

13. T. Nishida, T. Shiotsuki, and Y. Takashima, J. Non-Cryst. Solids 41, 161 (1980).

14. T. Nishida, T. Hirai, and Y. Takashima, J. Non-Cryst Solids 43, 221 (1981).

15. T. Nishida, T. Nonaka, T. Isobe, and Y. Takashima, Phys. Chem. Glasses 24, 88 (1983).

16. T. Nishida, T. Hirai, and Y. Takashima, Phys. Chem. Glasses 22, 94 (1981).

17. T. Nishida, T. Hirai, and Y. Takashima, Bull. Chem. Soc. Jpn. 54, 3735 (1981).

18. T. Nishida, T. Shiotsuki, and Y. Takashima, J. Non-Cryst. Solids 43, 115 (1981).

19. T. Nishida, T. Shiotsuki, and Y. Takashima, J. Non-Cryst. Solids 43, 123 (1981).

20. T. Nishida, Y. Miyamoto, and Y. Takashima, Bull. Chem. Soc. Jpn. 56, 439 (1983).

21. T. Nishida, M. Katada, and Y. Takashima, Bull. Chem. Soc. Jpn. 57, 3566 (1984).

22. T. Nishida, M. Katada, and Y. Takashima, Bull. Chem. Soc. Jpn. 58, 1745 (1985).

23. T. Nishida, A. Nakamura, and Y. Takashima, Chem. Lett., 1683 (1984).

24. T. Nishida, T. Nonaka, and Y. Takashima, Bull. Chem. Soc. Jpn. 58, 2255 (1985).

25. W. Müller-Warmuth and H. Eckert, Phys. Rep. 88, 91 (1982).

26. P. J. Bray and J. O'Keefe, Phys. Chem. Glasses 4, 37 (1963).

27. E. J. Gooding and W. E. S. Turner, J. Soc. Glass Technol. 18, 32 (1934).

28. L. Shartsis, W. Capps, and S. Spinner, J. Am. Ceram. Soc. 36, 35 (1953).

29. L. Shartsis, W. Capps, and S. Spinner, J. Am. Ceram. Soc. 36, 319 (1953).

30. A. O. Ivanov and K. S. Yestropovb, Dokl. Akad. Nauk SSSR 145, 797 (1962).

31. M. K. Murthy and J. Ip, Nature 201, 285 (1964).

32. M. K. Murthy and B. Scroggie, Phys. Chem. Glasses 6, 162 (1965).

33. J. E. Shelby, J. Am. Ceram. Soc. 57, 436 (1974).

34. J. E. Shelby, J. Appl. Phys. 46, 193 (1975).

35. W. A. Johnson and K. F. Mehl, Trans. Am. Inst. Mining Met. Engns. 135, 315 (1981).

36. M. Avrami, J. Chem. Phys. 7, 1103 (1939).

37. M. Avrami, J. Chem. Phys. 8, 212 (1940).

38. M. Avrami, J. Chem. Phys. 9, 177 (1941).

APPLICATION OF IRON-57 MÖSSBAUER SPECTROSCOPY TO THE CHARACTERIZATION OF SILICATE GLASSES USED FOR HIGH TEMPERATURE INCINERATED LOW-LEVEL RADIOACTIVE WASTE PRODUCTS

E. De Grave*

Laboratory of Magnetism
Gent State University
B-9000 Gent, Belgium

P. Van Iseghem

Materials Science Department
Nuclear Energy Research Establishment
B-2400 Mol, Belgium

ABSTRACT

Above seventy glass samples with the general compositions $(84\text{-}y\text{-}x)SiO_2 \bullet xFe_2O_3 \bullet yAl_2O_3 8M''O \bullet 8M'_2O$ and $(84\text{-}y\text{-}x)SiO_2 \bullet xFe_2O_2 \bullet yAl_2O_3 \bullet 8M''O \bullet 8M'_2O$, where M" is calcium, magnesium, or barium and M' is lithium or potassium and with y either zero or equal to x and with x + y up to 30, have been investigated by means of iron-57 Mössbauer spectroscopy at room temperature. The spectra of the iron(III) glasses are composed of a weak broad iron(II) doublet, arising from a wide spread in local environment, and two iron(III) doublets attributed to ions in essentially octahedral and tetrahedral positions. The variations of the iron(II) to iron(III) ratio, the iron(III) site occupancy and the hyperfine interactions are determined as a function of the composition x and y, and for a particular composition, as a function of the preparation conditions. The results are discussed in terms of the structural and chemical properties of the glass network. The crystalline phase, which segregates when the total amount of iron and aluminum exceeds a certain limiting value, is characterized by both its hyperfine parameters and X-ray diffraction patterns. The influence of the composition and quenching mode upon the crystallization process is determined. The iron(II) glasses were prepared in an argon atomsphere which results in high iron(II) concentrations. The structure of these glasses is different in that much more iron and aluminum may be incorporated and that iron(III) ions seem to be situated in the larger holes with a coordination number of six or higher. The iron(II) hyperfine parameters exhibit a broad distribution which is accounted for by fitting the absorption envelope with a sum of five quadrupole doublets. The coordination of the iron(II) cations is discussed and the variation of the Mössbauer parameters with the composition is investigated.

---

*Research Associate of the National Fund for Scientific Research, Belgium.

INTRODUCTION

In the past decade, amorphous systems have become a very popular research subject among scientists from many different fields in natural science. Not only have the fundamental physical and chemical properties of these materials received a lot of attention, but their technical applications have also been extensively studied. This has occurred because some of those systems are found to exhibit remarkable mechanical, electrical, and/or magnetic properties. Well known examples are the so-called met-glasses. Silicate and borosilicate glasses are two other amorphous systems which have attracted the interests of many research scientists. These glasses have proved to be suitable immobilizing media for especially long-living radioactive waste [1].

Nuclear waste is any by-product of an industrial process which is significantly radioactive and which is of no further use for any subsequent process. Such waste is currently being generated by nuclear energy production, defense production, medical diagnostics, and scientific research. Each of these specific waste products requires a proper management as far as control and disposal are concerned. There is at present general agreement among specialists that potentially hazardous waste must be isolated from the biosphere, so that the release of radioactivity to the biosphere does not exceed the limits proposed by the national and/or local governmental authorities.

The anticipated procedure consists of building three main protective barriers around the radioactive elements. The first one is the incorporation of these elements into a solid state phase, thus making them immobile, at least to a large extent. The next step is to seal the conditioned waste in a container which itself may consist of different successive steel and concrete shields. After interim storage for periods long enough for the temperature to reach reasonably low values, the canisters and their contents are ready for final disposal in a suitable geological formation.

Many different options within each of these three principal steps have been proposed and thoroughly investigated in the past decade. For the solidification, however, only two different processes now remain under serious consideration based on (i) the ease and the flexibility of the production and, (ii) the mechanical and chemical stability of the end-product. The first process produces the so-called SYNROC material (the abbreviation is for synthetic rocks) which is a mixture of synthetic minerals such as perovskite, hollandite, zirconolite, etc. Each of the constituent minerals is used to isolate a very limited number of different radionuclides. The production process is rather inflexible because the composition of the starting oxide mixture must be adjusted whenever the composition of the waste changes. The advantage of synroc is that it can withstand high temperatures and pressures so that it can be buried at great depths where it is not continuously attacked by groundwaters. A significant point of debate, however, still remains concerning the stability of the minerals after long burial times when a great part of the radioactive nuclei have decayed to other elements which are normally more difficult to dissolve in the structure and thus might affect the initially high resistivity of SYNROC to corrosion.

The second solidification process under consideration is the vitrification of the waste products. This method is currently being used in France where, after mixing the waste with additional oxides, the radioactive elements are isolated in a borosilicate glass. Another vitrification process has been developed over the last eight years at the Nuclear Energy Research Establishment, Mol, Belgium, for the processing of predominantly low-level radioactive wastes. The mixture, containing both combustible and noncombustible contaminated materials, is melted at high temperatures (1500

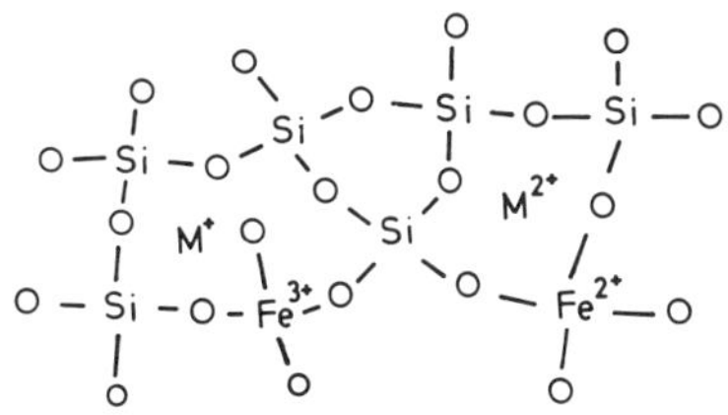

Figure 1. Schematic representation of a silicate glass network with incorporated cations

to 1600°C) in a FLK incinerator [2] and subsequently quenched in water. The resulting solid state phase consists of basalt-like granules with silicon as a major constituent, and with considerable amounts of iron and aluminum and, to a lesser extent, alkali and alkaline-earth elements. More details about the industrial process involved may be found elsewhere [3].

An inevitable consequence of the vitrification method is the large composition range of the finally conditioned waste. It is therefore important to determine the influence of the various compositional parameters on the final chemical and structural properties of the glass. For that purpose a large number of simple base or more complex reference glasses have been prepared and characterized by various techniques [4,5,6]. This paper deals with an iron-57 Mössbauer spectroscopy analyses of a large number of relatively simple base silicate glasses.

Structure and Composition of Silicate Glasses

A silicate glass consists of an irregular network of highly deformed SiO_4 tetrahedra with random orientations, as schematically represented in Figure 1. The tetrahedra are linked to each other by sharing a common oxygen ion. Holes with widely varying size and coordination symmetry are present and it is this structural property which makes these systems suitable for the isolation of radionuclei.

It is generally assumed [7] that a foreign ion may be incorporated in the glass structure either substitutionally or interstitially. In the first case, the ion involved is called a network former, in the second, a network modifier. The presence of modifying species is, however, interrelated with the presence of substitutional species. Indeed, if a silicon(IV) ion is replaced by an iron(III) ion, another monovalent cation must be located in the vicinity in order to provide charge neutrality. If the substitutional element were divalent, then either another divalent or two monovalent interstitial cations are required to compensate for the excess negative charge [8] as illustrated in Figure 1.

Not every kind of cation can function as a network former. A number of basic requirements must be fulfilled, the two most important being the electronegativity of the cation which is preferentially between 1.5 and 2.1, and its ionic radius, which should be between 0.3 and 0.6Å [9,10]. The values of these parameters for a number of cations involved in this study are listed in Table I. Silicon(IV) is an ideal network former. The species from lithium(I) to uranium(IV) are modifying cations because both the electronegativity and the ionic radius are unfavorable for glass formation. Iron might be an intermediate case, both in divalent and trivalent state. Aluminum is a network former, provided the $AlO^{-}_{4/2}$-tetrahedron is charge compensated through a positive charge [11].

Table I. The Electronegativity and Ionic Radius for a Number of Cations and the Empirical Range Expected for Glass Network Formation [7,10]

Cation	Electronegativity	Radius(Å)
Network former	1.5-2.1	0.30-0.60
Si^{4+}	1.8	0.42
Fe^{2+}	1.8	0.77
Fe^{3+}	1.9	0.65
Al^{3+}	1.5	0.53
Li^{+}	1.0	0.68
Na^{+}	0.9	0.95
K^{+}	0.8	1.33
Ca^{2+}	1.0	1.02
Mg^{2+}	1.2	0.64
U^{4+}	1.7	0.97

Table II. Composition of the FLK Slags

Component	Weight Percent
SiO_2	50-70
$FeO+Fe_2O_3+Al_2O_3$	10-45
Alkali oxides	5-10
Alkaline-earth oxides	5-10
Others (mainly oxides of Cu, Mo, Cr, Ni, Ti)	0-5

By taking into account these general rules, it is possible to distinguish the main compositional groups for the FLK slags which are described in Table II. A typical difference with the commercial glasses is the large concentration of intermediates (Al^{3+}, Fe^{2}, Fe^{3+}) in the slags. The simultaneous presence of these intermediates moreover must induce competition between them. In the literature there is a general agreement that aluminum(III), iron(II), and iron(III) may be glass formers provided the ($M^{3+}O_{4/2}$)$^-$ tetrahedra are charge compensated by e.g. one cation [12]. The structural behavior of iron(II) in silicate glasses is far less conclusive. However, there are indications that iron(II) is only incorporated into the octahedral sites [13-21] while other authors conclude that iron(II) is at least partly found on tetrahedral sites [8,22-27]. Therefore, it was decided to focus the basic structural studies on the incorporation of aluminum(III), iron (II) and iron(III) into silicate glasses.

Three parameters were considered. The first was the amount of Fe_2O_x (between 2.5 and 30 mole percent with x of 2 or 3), in balance with SiO_2.

Table III. Nominal Mole Percent Composition and Preparation Conditions of the Investigated Base Glasses

Sample	SiO_2	Al_2O_3	Fe_2O_3	Fe_2O_2	Li_2O	Na_2O	K_2O	MgO	CaO	BaO	Q[a]
VIIIa											
115	81.5	-	2.5	-	4.0	-	4.0	4.0	4.0	-	a,w
67	79.0	-	5.0	-	4.0	-	4.0	4.0	4.0	-	a,w
116	76.5	-	7.5	-	4.0	-	4.0	4.0	4.0	-	a,w[b]
52	74.0	-	10.0	-	4.0	-	4.0	4.0	4.0	-	a,w
86	74.0	-	10.0	-	4.0	-	4.0	---	4.0	4.0	a
87	74.0	-	10.0	-	4.0	-	4.0	4.0	---	4.0	a
117	71.5	-	12.5	-	4.0	-	4.0	4.0	4.0	-	a,w
69	69.0	-	15.0	-	4.0	-	4.0	4.0	4.0	-	a,w
53	64.0	-	20.0	-	4.0	-	4.0	4.0	4.0	-	a,w
54	54.0	-	30.0	-	4.0	-	4.0	4.0	4.0	-	a,w
VIIIb											
69	79.0	2.5	2.5	-	4.0	-	4.0	4.0	4.0	-	a,w
55b	74.0	5.0	5.0	-	4.0	-	4.0	4.0	4.0	-	a,w
109	74.0	5.0	5.0	-	4.0	-	4.0	4.0	---	4.0	a
70	69.0	7.5	7.5	-	4.0	-	4.0	4.0	4.0	-	a,w
56b	64.0	10.0	10.0	-	4.0	-	4.0	4.0	4.0	-	a,w
57b	54.0	15.0	15.0	-	4.0	-	4.0	4.0	4.0	-	a,w
VIIIc											
55	72.5	4.7	4.7	-	3.7	3.2	3.7	3.7	3.7	-	a,w
56	60.0	9.3	9.3	-	3.7	6.5	3.7	3.7	3.7	-	a,w
57	61.2	10.3	10.3	-	2.7	7.1	2.7	2.7	2.7	-	a,w
IXa											
135	81.5	-	-	2.5	4.0	-	4.0	4.0	4.0	-	A,w
136	79.0	-	-	5.0	4.0	-	4.0	4.0	4.0	-	A,w
137	76.5	-	-	7.5	4.0	-	4.0	4.0	4.0	-	A,w
138	74.0	-	-	10.0	4.0	-	4.0	4.0	4.0	-	A,w
139	71.5	-	-	12.5	4.0	-	4.0	4.0	4.0	-	A,w
140	69.0	-	-	15.0	4.0	-	4.0	4.0	4.0	-	A,w
141	66.5	-	-	17.5	4.0	-	4.0	4.0	4.0	-	A,w
142	64.0	-	-	20.0	4.0	-	4.0	4.0	4.0	-	A,w
143	54.0	-	-	30.0	4.0	-	4.0	4.0	4.0	-	A,w
IXb											
144	79.0	2.5	-	2.5	4.0	-	4.0	4.0	4.0	-	A,w
145	74.0	5.0	-	5.0	4.0	-	4.0	4.0	4.0	-	A,w
146	69.0	7.5	-	7.5	4.0	-	4.0	4.0	4.0	-	A,w
147	64.0	10.0	-	10.0	4.0	-	4.0	4.0	4.0	-	A,w
148	54.0	15.0	-	15.0	4.0	-	4.0	4.0	4.0	-	A,w

[a]All glasses were melted at 1550°C for three hours and quenched in an air, a; water, w; or an argon, A atmosphere. [b]This glass was also melted at 1400°C for one hour.

For each Fe_2O_x concentration the glass was prepared either under oxidizing conditions as currently done in practical glass preparation, or under reducing conditions, corresponding to the actual FLK incinerator operation as is indicated by iron(II) to total iron ratio of 0.85 in the FLK slags [4].

The second parameter was the equimolar replacement of Fe_2O_x (x being 2 or 3) by Fe_2O_x and Al_2O_3. The third was the cooling rate. All glasses were cooled either in the ambient atmosphere, as is currently the practice, or in water, as occurs in the FLK incinerator. The various glass compositions are listed in Table III.

The amount and distribution of modifiers (alkali and alkaline earth oxides) in the base glass were generally kept constant. Equimolar amounts of Li_2O, K_2O, MgO, and CaO were chosen, with a total concentration of 16 mole percent. This is somewhat larger than the M_2O + MO content currently measured in the FLK slags [5], but was the minimum amount necessary to be able to melt the low Fe_2O_x and Al_2O_3 content glasses at 1550°C. This corresponds to the higher FLK incinerator operating temperature. Li_2O replaced Na_2O, in some cases, to include a larger range in the dimensions of the alkali ions.

For two particular glass compositions (52 and 55b) the influence of the type of the alkaline earth oxides was considered. The influence of the melting time (1-3 hours) and the melting temperature (1400-1550°C) was investigated for glass 52. Glasses 55-57 (series VIIIc), with an equimolar concentration of Fe_2O_3 and Al_2O_3 were prepared with feldspar as the Al_2O_3 generating product. As feldspar also contains other components such as Na_2O, K_2O, etc, these glasses will be considered only with respect to the characteristics of the crystals formed upon melting. These glasses are not discussed with respect to possible compositional effects.

All glasses were prepared by starting with a pelletized mixture of the appropriate fine grained powders. Iron(III) was added as Fe_2O_3, iron(II) as $Fe(C_2O_4)\cdot 2H_2O$. The Fe_2O_3 containing glasses were melted in Pt-10 percent Rh crucibles in an air atmosphere. The Fe_2O_2 containing glasses were melted in zirconia grain stabilized (zgs) Pt-10 percent Rh crucibles under an argon atmosphere; the amount of oxygen in this atmosphere was below 1000ppm.

Preliminary Mössbauer measurements on mainly iron(III) containing base glasses, with M_2O + MO content of24 mole percent have been presented in two earlier papers [4,29]. One of the earlier conclusions was the dependence of at least some of the relevant Mössbauer parameters on the composition of the glass and on the melting temperature. Besides the discussion of the Mössbauer spectra for the new base glasses, attention will be devoted to the characterization of the iron-containing crystalline phase obtained when the conditions for glass formation are no longer fully met. A later paper, will attempt to correlate the results from the structural studies with the phycochemical properties, mainly the corrosion resistance, of the glasses.

EXPERIMENTAL

Mössbauer spectra were obtained at room temperature on a conventional time-mode spectrometer with cobalt-57 on either palladium or rhodium as a source. The spectrometer was calibrated with metallic iron foil and all isomer shifts are referred to iron. The scale was approximately 0.016mm/s/channel except for those cases where a magnetically split contribution was obvious, in which case, a scale of approximately 0.45mm/s/channel was selected. The absorbers consisted of the finely ground glass compounds spread out to a uniform thickness of 5 to 6mg/cm^2 of iron.

Approximately one million counts per channel were measured in the unfolded spectrum. The spectra were numerically analyzed by fitting to a sum of Lorentzian lines. Restrictions were imposed on the parameters in order to make the quadrupole doublets symmetric.

X-ray diffraction (XRD) patterns were recorded on a Philips powder diffractometer with FeK_α radiation.

RESULTS AND DISCUSSION

Series VIIa Iron(III) Glasses

Typical spectra for this series of glasses are shown in Figure 2 which illustrates samples 116E and 116F. They were both melted for three hours at 1550°C and quenched in air (E) and water (F), respectively. The spectra were adequately fit with two iron(III) doublets and one broad iron(II) doublet. The fitting function did improve slightly when, as seen in Figure 2, a second iron(II) doublet was introduced. However, because of the weak intensities and the strong overlap with the iron(III) doublets, the iron(II) spectral parameters, resulting from the four doublet fit, are found to be widely scattered and therefore are not very reliable. We therefore prefer to consider only a single iron(II) quadrupole doublet. Its hyperfine parameters should then be regarded as being characteristic of an average iron(II) state within the glass structure. As will be clear from the results obtained for iron(II) glasses, this picture is indeed justified.

The effect of the preparative conditions on the Mössbauer parameters is presented in Table IV. The data are for samples containing 10 mole percent Fe_2O_3. The temperature of the melt was either 1400°C for samples A to D or 1550°C for samples E to H. The melting time was three hours for samples A, B, and E, F or one hour for samples C, D and G, H. After firing, the melts were quenched in air for samples A, C, E and G or in water at room

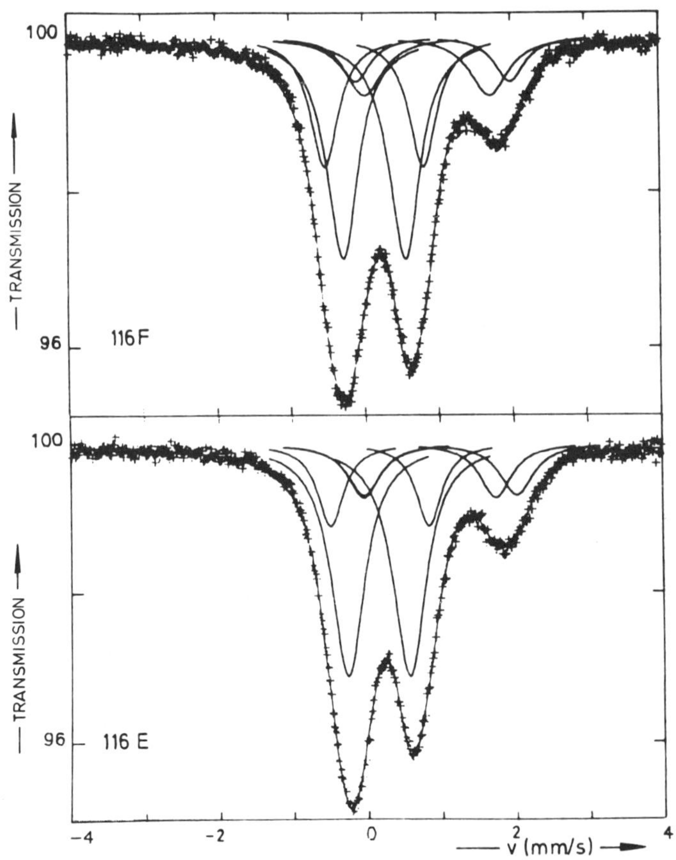

Figure 2. Mössbauer spectra obtained at room temperature for water-quenched (116F) and air-quenched (116E) iron(III) glasses. Full lines represent fitted Lorentzian lines and their sum.

Table IV. Mössbauer Spectral Parameters for a Glass Sample Containing 10 Mole Percent Fe_2O_3 and 16 Mole Percent M"O + M'_2O Melted and Cooled Under Different Conditions.[a]

Sample	Doublet	δ	ΔE_Q	Γ	Area	R^b
52A	D1	0.30(1)	1.33(2)	0.46(3)	0.25(4)	
	D2	0.29(1)	0.82(2)	0.53(2)	0.61(4)	0.29(5)
	D3	1.05(2)	1.95(3)	0.60(3)	0.14(1)	
52B	D1	0.29(1)	1.37(2)	0.43(2)	0.29(4)	
	D2	0.28(1)	0.85(2)	0.51(2)	0.57(4)	0.34(4)
	D3	1.01(2)	1.99(3)	0.60(3)	0.14(1)	
52C	D1	0.30(1)	1.33(2)	0.43(2)	0.29(4)	
	D2	0.29(1)	0.82(2)	0.54(2)	0.62(4)	0.28(4)
	D3	1.03(2)	2.00(3)	0.64(3)	0.14(1)	
52D	D1	0.27(1)	1.39(2)	0.44(2)	0.31(3)	
	D2	0.27(1)	0.88(2)	0.51(1)	0.58(3)	0.34(4)
	D3	1.02(2)	2.05(3)	0.61(3)	0.13(1)	
52E	D1	0.33(2)	1.28(3)	0.46(3)	0.26(5)	
	D2	0.29(1)	0.78(3)	0.51(2)	0.46(5)	0.36(7)
	D3	1.03(2)	2.03(3)	0.68(2)	0.28(1)	
52F	D1	0.30(1)	1.38(2)	0.45(3)	0.25(4)	
	D2	0.29(1)	0.87(2)	0.52(2)	0.48(4)	0.35(6)
	D3	1.03(1)	1.93(2)	0.65(2)	0.27(1)	
52G	D1	0.30(1)	1.30(2)	0.43(3)	0.24(4)	
	D2	0.30(1)	0.71(2)	0.54(1)	0.51(4)	0.32(5)
	D3	1.01(2)	2.09(4)	0.65(2)	0.25(1)	
52H	D1	0.28(1)	1.35(2)	0.43(3)	0.26(4)	
	D2	0.27(1)	0.85(2)	0.51(2)	0.48(4)	0.35(5)
	D3	1.00(1)	1.95(2)	0.61(2)	0.26(1)	
86E[c]	D1	0.28(1)	1.21(3)	0.40(2)	0.25(3)	
	D2	0.27(1)	0.75(2)	0.47(2)	0.55(4)	0.31(5)
	D3	0.98(1)	1.83(3)	0.55(2)	0.20(1)	
87[d]	D1	0.28(1)	1.26(2)	0.38(2)	0.18(3)	
	D2	0.26(1)	0.79(2)	0.49(2)	0.59(3)	0.25(5)
	D3	0.96(1)	1.81(2)	0.60(2)	0.23(1)	

[a]All data in mm/s relative to iron foil. The numbers within parenthesis are three times the computed standard deviations. [b]The tetrahedral fraction. [c]Contains 4 mole percent BaO instead of MgO in samples 52. [d]Contains 4 mole percent BaO instead of CaO in samples 52.

temperature for samples B, D, F and H. The resulting hyperfine parameters are very similar to those previously observed [29] for a similar series of iron(III) glasses which contained a significantly higher amount of glass modifying M"O and M'O oxides, i.e., 24 mole percent instead of 16 mole percent for the current series. It appears that the relative amount of glass-modifying components in the initial oxide mixture does not significantly affect the local structural and chemical environment of the iron species, at least in the range 16 to 24 mole percent for M_2^IO + M"O and for Fe_2O_3 equal to 10 mole percent.

Doublets D1 and D2 clearly arise from iron(III) ions and on the basis of literature data [17,30,31], the doublet with the larger quadrupole splitting (D1), may be assigned to a tetrahedral coordination site and the second doublet (D2) to an octahedral site. Except for glass 81E, the fraction R of four-fold coordinated iron(III), relative to the total iron(III) content, is not affected by the preparative conditions and is on the average 0.33 ± 0.02. This means that about one-third of all iron(III) ions take glass-forming positions, whenever preparation method is used. This finding is rather surprising in view of the observation, to be presented in a later section of this paper, that, on quenching in water, the glasses can incorporate a larger amount of Fe_2O_3 so that one would expect a significant influence of the quenching mode on R.

The only parameter which is clearly related to the melting conditions is the iron(II) concentration which increases from 14 percent when melted at 1400°C to 25 to 28 percent for melting at 1550°C. This feature is generally observed for silicon based glasses [17] and must be ascribed to the redox equilibrium of an iron ion in a silicate melt [32-34]:

$$4Fe^{2+} + O_2 \rightleftharpoons 4Fe^{3+} + 2O^{2-}$$

which is shifted to the left with increasing temperatures.

The isomer shift, quadrupole interaction and line width are not affected by the preparative method. Seemingly significant changes are observed occasionally but are, in our opinion, irrelevant. Both iron(III) doublets have the same isomer shift and, although this feature has been observed by other workers [17] on similar glass systems, the results seem somewhat puzzling in view of the larger covalency, and smaller isomer shift, one would expect for the iron-oxygen bond in a tetrahedral coordination geometry.

The iron(II) quadrupole interaction is not typical for either a four-fold or a six-fold oxide environment. Its value may suggest a superposition of quadrupole doublets arising from iron(II) in both types of network sites. This would also account for the large line width of the iron(II) doublet as compared to the iron(III) doublets. The latter ones have an average line width of 0.439(11) and 0.524(15)mm/s for tetrahedral and octahedral species, respectively. These values are high compared to the line width of 0.25mm/s found for the inner lines of the iron foil calibration. This line broadening is known to result from the wide variations in magnitude and type of distortions of the local symmetry at the absorbing nuclei and is typical of amorphous substances. Our results unequivocally indicate that these symmetry variations are somewhat broader for six-fold coordination and that they are unaffected by any of the relevant preparation parameters.

The influence of introducing the much larger barium(II) ions in place of calcium(II) or magnesium(II) is very pronounced in the observed iron(II) concentration values, which are significantly smaller for glasses 86E and 87E as compared to glass 52E. This indicates that a geometric factor, such as the average ionic radius of the glass-modifying constituents, affects the above mentioned iron redox equilibrium as was predicted, although not in a straightforward fashion, by the semiempirical work of Lauer [33,34]. Qualitatively one might explain this behavior on the basis of the large barium(II) ions, which compress slightly the glass network and favoring the formation of the smaller iron(III) ions over the larger iron(II) ions. This mechanism might affect the distribution of iron(III) between the network-forming and network-modifying sites. Obviously additional selected samples should be measured in order to further support this conclusion.

A second effect caused by the barium(II) ion substitution is the reduction of the quadrupole splitting, especially for iron(II). This most probably is not due to an increase in the average site symmetry, because it has resulted from the incorporation into the glass network of a larger, and hence more distorting, cation. An acceptable explanation is not available and will probably require a detailed study of the temperature dependence of the iron(II) quadrupole splitting which is beyond the scope of the present paper.

From the other compositions of the VIIIa system, only those glasses fired at 1550°C for three hours, and denoted E and F, have been considered in studying the correlation, if any, between the room temperature Mössbauer spectra and the Fe_2O_3 content. The first important finding is that all the samples containing 15 mole percent or more of Fe_2O_3 in the starting oxide mixture, more or less clearly show a six lines in their Mössbauer spectra. These samples are partially crystallized and will be discussed below. Nearly all the remaining spectra, for the compositions involved, could be fitted in the same way as described above for the 10 percent samples. An exception to this was sample 115, containing 2.5 mole percent of Fe_2O_3, for which the computer fitting program did not converge with a three doublet fit. The total iron(III) absorption band is unusually broad at more than one mm/s when described as a single doublet, indicating the existence of a wide quadrupole distribution with an average quadrupole splitting of ca. 0.8mm/s, a value which is close to the quadrupole splitting values observed for iron(III) in the six-fold positions in samples with 5, 7.5, 10 (see Table IV), and 12.5 mole percent of Fe_2O_3. All this might suggest that the "structure" of the glasses involved changes, more or less drastically, when the amount of incorporated Fe_2O_3 becomes smaller than a certain, not yet defined limiting value. It should be noted that the Mössbauer parameters of the iron(III) glasses reported on in a previous paper[29] also exhibited the largest variation in the Fe_2O_3 concentration range below ca. four mole percent. It would certainly be worthwhile to explore this field in somewhat more detail in a future investigation.

In the concentration range of 5 to 12.5 mole percent Fe_2O_3, little variation was observe in the room temperature Mössbauer parameters and there is no indication of a correlation between the Fe_2O_3 content and any of the hyperfine parameters. Hence the local structure around the iron species is quite insensitive to the glass composition.

The iron(II) concentration is not affected by the quenching method and is within the range 0.26±0.03mm/s for all compositions involved. The distribution of iron(III) between the octahedral and tetrahedral sites is not influenced by the quenching method and the average R value was found to be 0.3, which is in good agreement with previous results.

Series VIIIb. Iron(III) Glasses

The aim of preparing this series of glasses was to check whether the incorporation of aluminum into the glass network alters the structural properties as observed by Mössbauer spectroscopy. Samples fired at 1550°C, quenched in air or in water, and with an Al_2O_3 and an Fe_2O_3 content of 2.5, 5, 7.5, 10, and 15 mole percent have been considered. Their room temperature Mössbauer spectra are found to be very similar to those of the corresponding compositions of series VIIIa. The final results may be summarized as follows. First, the compositions 70, 56b, and 57 show a six spectrum at room temperature. Second, there is no correlation between the tetrahedral fraction R and the composition or quenching mode. R was found to be 0.36(3) for glasses 55b and 109, 0.24 for glass 69, and 0.41 for the E and F samples. Third, the results obtained for glass 109E are in agreement with the above mentioned influence of the large barium(II) cations, which lower the iron(II) quadrupole splitting (1.77mm/s), and the iron(II) concentration (to 23 percent as compared with 26 percent for 55E). Fourth, the iron(II) content is once again unaffected by the quenching mode, but differs drastically for glasses 69 and 55b. A similar observation was made earlier in a comparable glass series [29] in which samples with less than 4 percent of Fe_2O_3 and Al_2O_3 exhibit a significantly higher iron(II) concentration.

In conclusion, the incorporation of aluminum(III) ions in the glass network has, in general, no significant effect on the chemical and structural environment of the different iron species nor on the valence state of the iron. Only when the oxide mixture contains very small amounts of Al_2O_3 and Fe_2O_3, is a drastic enhancement of the iron(II) concentration in the resulting glass observed.

Iron(II) Glasses

The similarity of the Mössbauer spectra of the iron(II) glasses prepared under the required conditions to those of the real FLK granules (sample FLK 77, which was taken from a test-run of the industrial FLK incinerator) is shown in Figure 3a. The spectrum was acceptably fitted with a superposition of two iron(II) and one iron(III) doublets, as is shown in Figure 3. The physical meaning of this fit is questionable, however, and because the fit is of no relevance to the present discussion, no further analysis of the line shape was attempted. Also, the weak magnetically split contribution which is observed can be neglected for the time being. The important observation is that the iron(II) content, which is 85 percent of the total iron, is much higher in this industrial compound as compared with the glass studied in the previous section of this paper.

In order to determine the proper melting conditions for obtaining glasses with similar iron(II) concentration, a FLK base glass (WG 122 in Ref. 35) has been prepared starting from an identical feed, with iron added as iron(II) oxalate, but melted at 1550°C under ambient air (see Figure 3b),

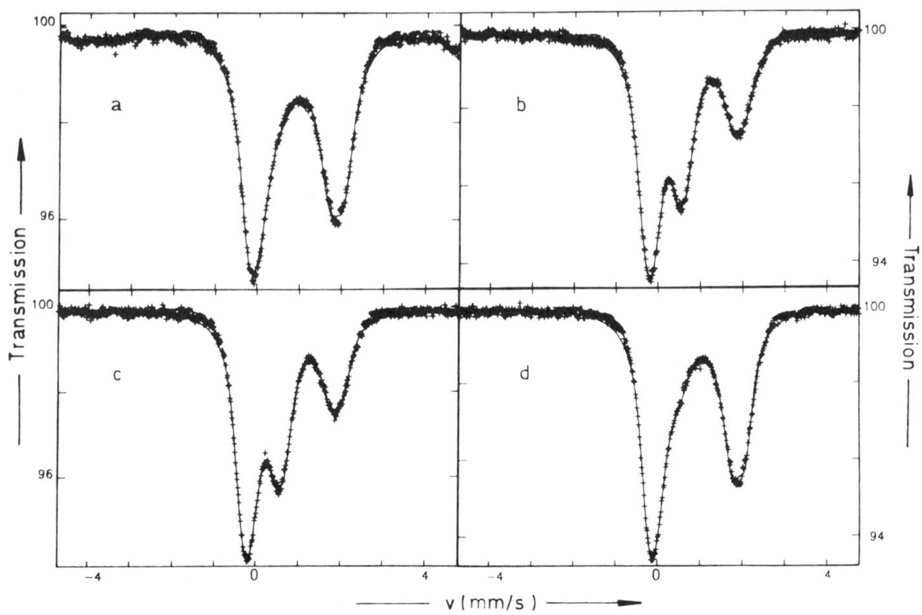

Figure 3. The Mössbauer spectrum for a sample taken from the amorphous slags produced by the FLK incinerator in Mol, Belgium (a) and of reference glasses with the same composition and prepared in an ambient (b), nitrogen (c), and argon (d) atmosphere.

under nitrogen (Figure 3c), or under argon (Figure 3d). The spectra clearly show why the iron(II) reference glasses used for further, more detailed Mössbauer analyses, have all been prepared under an argon atmosphere.

Series IXa

This series contains nine different specimens with an initial Fe_2O_2 content ranging from 2.5 to 30 mole percent. A typical room temperature Mössbauer spectrum in shown in Figure 4 and refers to the water-cooled sample 143, which contains 30 mole percent Fe_2O_2. All other spectra exhibit very similar line shapes, characteristic of a high iron(II) to iron(III) ratio and

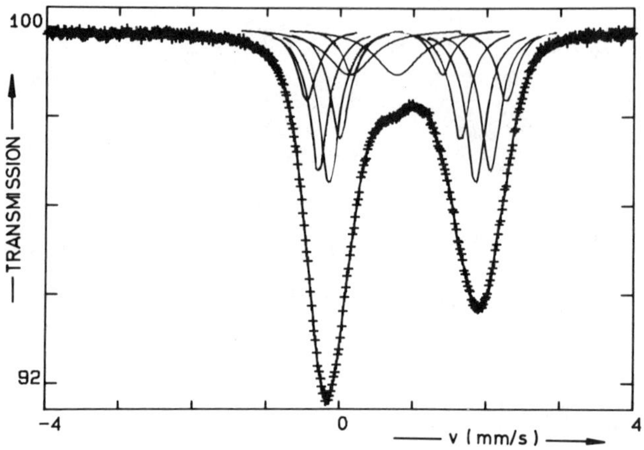

Figure 4. A typical Mössbauer spectrum for an iron(II) reference glass.

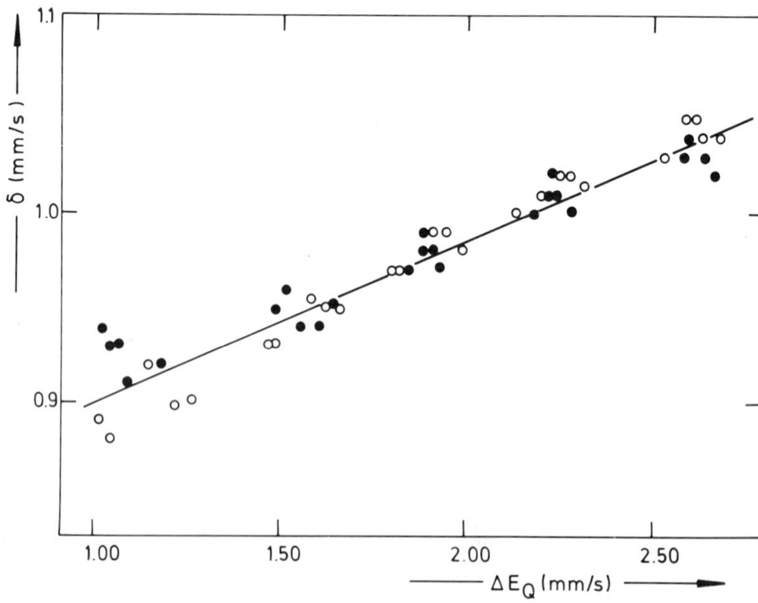

Figure 5. Isomer shift versus quadrupole splitting for the five distinct iron(II) quadrupole doublets found in the experimental spectra of some iron(II) glasses. The different symbols indicate different quenching modes.

a distribution of hyperfine interaction parameters. One exception to this behavior is the argon cooled 30 mole percent Fe_2O_2 sample, 143, which displays the presence of an additional phase indicated by a very broad six line absorption profile. It is clear that the iron(II) glasses allow the incorporation of a considerably larger amount of iron, up to 30 mole percent, as compared to the iron(III) glasses. This conclusion is even more puzzling, when one realizes that iron is predominantly in the divalent state whose electronic and geometric properties are presumably less favorable for glass formation than for iron(III). The observation may, however, indicate that the structure and bonding properties of these iron(II) glasses are basically distinct from those of the iron(III) glasses discussed earlier.

As shown in Figure 4, the purely amorphous spectra are reasonably well described by a sum of five symmetric iron(II) quadrupole doublets, with an average line width of 0.35mm/s, and a weak, broad iron(III) doublet, which could not be further resolved. It is reasonable to assume that the five iron(II) doublets are actually an approximation of a continuous distribution of quadrupole splittings.

Figure 5 is a plot of the isomer shift, δ, versus the quadrupole splitting, ΔE_Q, for the five distinct doublets fitted to the iron(II) absorptions of both the water quenched and argon quenched reference samples. The glasses 136 and 138 were left out for reasons discussed below. The plot clearly displays a linear correlation between the two hyperfine parameters. The lowest (ΔE_Q,δ) values of δ and ΔE_Q are more scattered because of the larger inaccuracy of their computed parameters.

The observed correlation between the isomer shift and quadrupole splitting is possibly the result of a non-uniform covalent character in the iron(II)-oxide bond because both δ and ΔE_Q are known to decrease with increasing covalency [36,37]. It further suggests that the majority of the iron(II) ions have basically the same coordination number. The calculated hyperfine parameters, however, do not allow an unequivocal assignment. The observed quadrupole interaction values can be explained by either a tetrahedral or an octahedral oxide coordination geometry [36]. Also, the isomer shifts are not typical of any particular coordination number, although it must be mentioned that the observed values of less than 1.05mm/s are somewhat smaller than the 1.10 to 1.30mm/s values found for octahedrally coordinated iron(II) in many different crystalline solids. This could be an indication that the coordination number of iron(II) in these glasses is smaller than six but this argument is, in our opinion, far from conclusive. More convincing indications with respect to the coordination geometry may perhaps be obtained from Mössbauer measurements as a function of temperature and in high external magnetic fields. Such measurements are planned for the future.

Some further quantitative data concerning the iron(II) absorption are presented in Table V for both water and argon cooled samples. In this table, $\langle\Delta E_Q\rangle$ is the weighed average value of the quadrupole splitting and σ is the second order moment of the discrete distribution, the latter being a measure for the spread of the distribution. At this point it should be mentioned that, in the initial stages of operating the laboratory melting equipment, the argon atmosphere was not well regulated, resulting in higher partial oxygen pressures than anticipated. Samples 136 and 138 were prepared under these conditions, and the results (Table V) clearly show the significant influence of the atmosphere conditions on both the iron(II) content and the quadrupole distribution profile.

With the exception of glass 135, none of the listed iron(II) hyperfine parameters seems to be sensitive to the cooling mode and the composition itself does not significantly affect the percent of iron(II). The spread of the distribution remains essentially constant whereas the average quadrupole

Table V. Mössbauer Spectral Parameters for the Iron(II) References Glasses[a]

Sample	Mole Percent Fe_2O_3	Mole Percent Al_2O_3	Percent Iron(II)[b]	$\langle\Delta E_Q\rangle$	σ[c]
			Argon Cooled		
135	2.5	-	87	1.84	0.39
136	5[d]	-	70	1.94	0.33
137	7.5	-	85	1.92	0.41
138	10[d]	-	79	1.96	0.39
139	12.5	-	84	1.96	0.40
140	15	-	79	2.02	0.40
141	17.5	-	86	2.00	0.40
142	20	-	83	2.02	0.40
144	2.5	2.5	85	1.87	0.40
145	5	5.0	73	2.00	0.42
146	7.5	7.5	78	2.06	0.43
147	10	10.0	77	2.03	0.42
148	15	15.0	82	1.98	0.45
			Water Cooled		
135	2.5	-	82	1.84	0.36
136	5[d]	-	73	1.88	0.41
137	7.5	-	83	1.89	0.42
138	10[d]	-	78	1.99	0.43
139	12.5	-	85	1.93	0.42
140	15	-	80	1.97	0.41
141	17.5	-	87	1.96	0.43
142	20	-	83	1.99	0.42
143	30	-	82	2.05	0.42
144	2.5	2.5	81	1.85	0.40
145	5	5.0	62	2.05	0.42
146	7.5	7.5	75	2.00	0.44
147	10	10.0	79	1.99	0.43
148	15	15.0	84	1.99	0.44

[a]The exact compositions are given in Table III. [b]Error $\pm$ 3 percent. [c]Data in mm/s with an error of $\pm$0.05mm/s. [d]Oxygen partial pressure in argon too high.

splitting increases slightly with increasing mole percent Fe_2O_2. This is probably due to small structural changes in the glass network resulting from the incorporation of larger amounts of iron(II).

The iron(III) Mössbauer parameters show no obvious relationship with the composition. For all samples the line width is scattered between 0.7 and 0.9mm/s, and this considerable line broadening indicates a wide distribution in the iron(III) hyperfine interactions. The average quadrupole splitting is found in the range 0.52(3)mm/s for the argon cooled series and 0.66(5)mm/s for the water cooled series. All isomer shift values are between 0.45 and 0.58mm/s. Again, the 2.5 mole percent Fe_2O_2 is unusual, because its quadrupole splitting was found to be around one mm/s. However, an unusually large line broadening, as in the case for the comparable iron(III) glasses, was not observed.

Although the iron(III) hyperfine parameters in the iron(II) glasses are, for obvious reasons, rather inaccurate, they do show significant differences from those obtained for the corresponding iron(III) glasses. The much higher isomer shifts indicate that the iron(III) oxide bonds are less covalent in the iron(II) glasses. A plausible explanation is that the iron(III) ions in these glasses are situated in larger interstitial sites with a coordination number of six or, possibly, eight. The smaller quadrupole splitting, which is considerably smaller than for the iron(III) doublets assigned to six-fold positions in the iron(III) glasses, would then be accounted for by the greater iron-oxygen distance and the smaller overlap contribution to the electric field gradient [38]. The apparent distinction between the iron(III) and iron(II) glasses, as suggested by the iron(III) Mössbauer parameters, is in agreement with an earlier conclusion based on the observation that the iron(II) glasses can accommodate a substantially larger amount of iron than the iron(III) glasses. However, it is not yet understood why the iron(II) hyperfine parameters are very similar for both iron(II) and iron(III) glasses because, as one would expect, the differences in chemical and structural properties of the glass systems involved should affect iron(II) to the same or greater extent than iron(III).

Series IXb

As for the iron(III) glasses, the effect of aluminum was studied by using glass samples prepared by exactly the same method as those of the previous series, except that half of the Fe_2O_2 in the oxide mixture was replaced with Al_2O_3. Compositions of 2.5, 5, 7.5, 10, and 15 mole percent each of Fe_2O_2 and Al_2O_3 have been considered. Again it is observed (see Table V) that the aluminum ions do not drastically affect those features which are accessible to the Mössbauer effect technique. The results for the iron(III) concentration are even more scattered, the average value of 22 percent (with a variance of 6 percent) being higher than for the aluminum free samples. However, in view of the high sensitivity to the partial oxygen pressure, it is difficult to assess the validity of this result.

For the iron(III) doublet, the average quadrupole splitting is 0.72(9)mm/s for the argon cooled samples and 0.79(2)mm/s for the water cooled samples. These results are more in line with those obtained for six fold coordinated iron(III) in the corresponding iron(III) glasses. The average isomer shift remains, however, relatively high at 0.55(4)mm/s.

Finally, it should be noted that, in contrast to the corresponding composition 143, the argon cooled sample 148, with 15 mole percent Al_2O_3 and 15 mole percent Fe_2O_2, does not have an additional crystalline phase in its Mössbauer spectrum. This suggests that aluminum substitution more or less stabilizes the glass structure. This effect will become even more obvious from a detailed compilation of Mössbauer results obtained for the partially crystallized glass samples.

Partially Crystallized Glasses

As mentioned in the above section dealing with iron(III) glasses, spectra obtained on reference glasses with a high Fe_2O_3 and Al_2O_3 content show, in addition to the central absorption, which is characteristic for the vitreous phase, a six line contribution, the shape of which strongly depends on the composition and preparative conditions. Here the crystalline phase giving rise to this six line subspectrum will be studied in more detail with regard to its characterization and its dependence on the external parameters. The composition of the samples used for this purpose is listed in Table VII and as seen, a few extra samples were prepared in addition to those mentioned above. For each composition an air cooled, E, and water cooled, F, specimen was obtained.

Table VI. Room Temperature Mössbauer Spectral Parameters for the Magnetic Phase in Some Reference Glasses as Compared to Fe_3O_4[a]

Site	Parameter	Fe_3O_4[b]	57E	54E	53E
A	H_{hf}	493(3)	490(4)	490(4)	488(4)
	δ	0.25(2)	0.28(2)	0.25(2)	0.28(2)
	2ε[c]	0.00(2)	-0.03(2)	0.00(2)	-0.03(2)
B	H_{hf}	460(2)	464(4)	462(4)	460(4)
	δ	0.65(2)	0.67(2)	0.68(2)	0.66(2)
	2ε[c]	0.00(2)	-0.01(2)	0.04(2)	0.04(2)

[a] Data are in kOe and mm/s and the numbers within parentheses are three times the computed standard deviation. [b] Data taken from reference 15. [c] $e^2qQ(3\cos^2\theta-1)/4$.

Characterization of the Crystalline Phase

A few typical room temperature Mössbauer spectra are shown in Figure 6. Figure 7 shows the x-ray powder diffraction patterns recorded on the same samples. These show that a crystalline phase is formed and that its

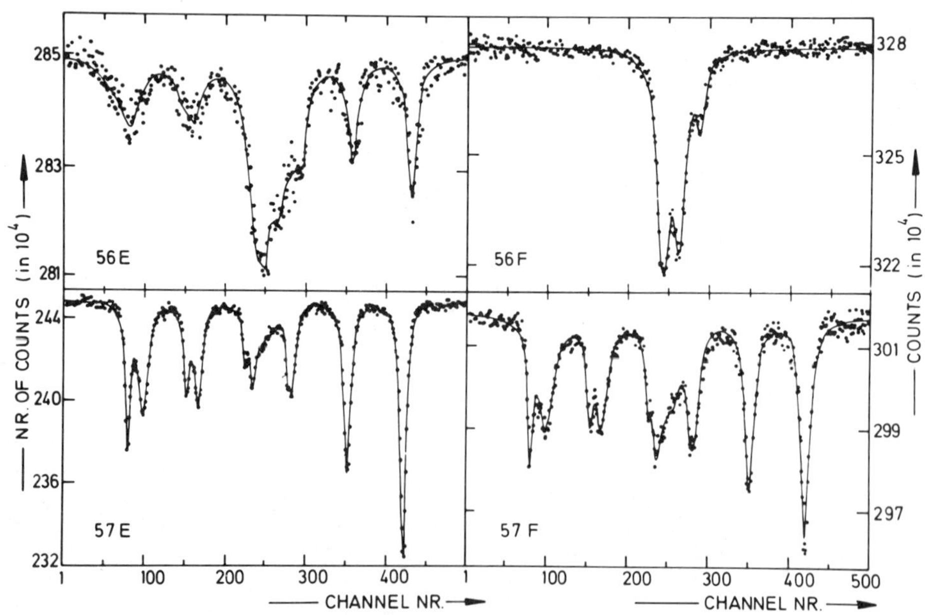

Figure 6. Mössbauer spectra obtained at room temperature for a number of partially crystallized iron(III) glasses.

Table VII. The Composition, Relative Area of the Glass-Subspectrum and Partition Factor, N (See Text), for a Number of Reference Samples Melted at 1550°C and Quenched in Water, F, or in Air, E

Sample	Weight Percent Composition				Relative Area of Glass		N	
	Al_2O_3	Fe_2O_3	M'_2O [a]	$M''O$ [b]	E	F	E	F
52	-	10	8	8	1.00	1.00	0.27	0.27
68	-	15	8	8	0.45	1.00	0.20	0.43
53	-	20	8	8	0.25	-	0.16	-
54	-	30	8	8	0.16	0.25	0.18	0.28
55[b]	5	5	8	8	1.00	1.00	0.14	0.14
70	7.5	7.5	8	8	0.23	0.71	0.05	0.15
56[b]	10	10	8	8	0.16	0.38	0.05	0.12
37[b]	15	15	8	8	0.11	0.16	0.06	0.09
55	4.7	4.7	10.6	7.4	1.00	1.00	0.13	0.13
56	9.3	9.3	13.9	7.4	0.45	1.00	0.14	0.31

[a] Here M' is sodium, lithium, or potassium. [b] Here M" is calcium or magnesium.

evolution is related to the changes observed for the overall Mössbauer spectral shape.

For glass 56F, both the Mössbauer and the powder diffraction pattern are characteristic of an amorphous material. The broad reflection band centered at a 2Θ value of ca. 30° arises from short range atomic ordering with widely

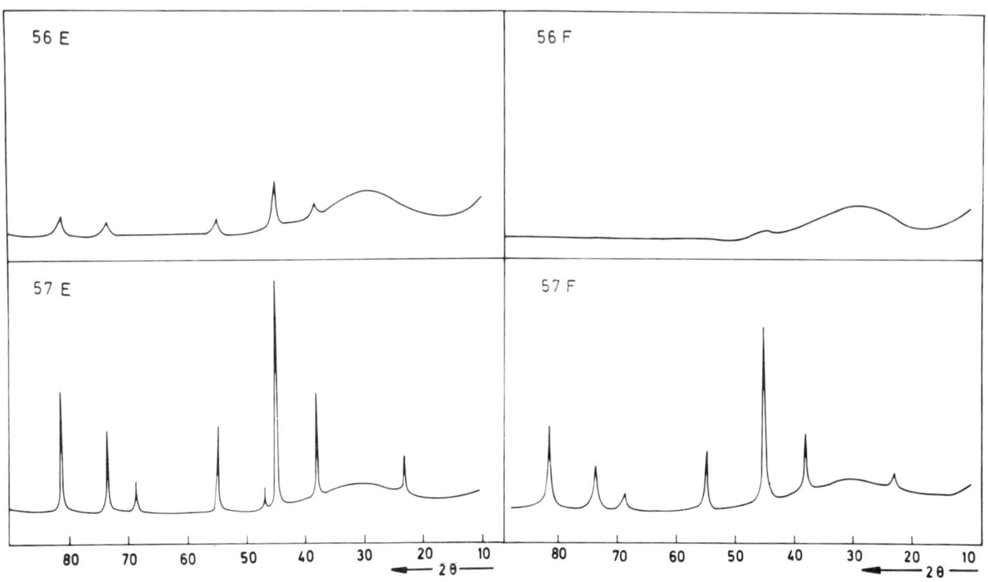

Figure 7. X-ray powder diffraction patterns for the samples referred to in Figure 6.

varying interatomic distances. The same composition, when air cooled at a slower cooling rate, exhibits a few low-intensity broad diffraction lines superimposed on the broad amorphous peak. This indicates that part of the sample has crystallized. However, as indicated by the powder diffraction line broadening, and by the shape of the six-line Mössbauer spectrum, the degree of crystallinity is still low. It is clear from these observations that composition 56 (see Table VII) is on the edge of the composition range for which a completely amorphous substance can be formed. In such cases the cooling rate of the melt has a substantial influence on the crystallinity of the end-product. It should be noted here that the Mössbauer spectrum, recorded on glass 56F at 78K shows the typical spectrum of the vitreous state with no sign, of a crystalline contribution. Based on the nature of the crystalline phase involved (see below), this indicates that nucleates as small as 50Å have not even been formed in the sample.

The influence of the composition on the crystallization process is nicely illustrated by the spectra for samples 57E and 57F. As compared to 56, these samples have approximately the same SiO_2 content, but one mole percent more of both Fe_2O_3 and Al_2O_3. The crystalline phase is clearly much better defined and its contribution has increased markedly. Both features are even more apparent in the air-cooled sample.

A close look at the Mössbauer spectra of the partially-crystallized reference glasses reveal a rather complex spectral structure. Some of the spectra are acceptably fitted by a superposition of two magnetic sextets and two quadrupole doublets, the latter arising from iron(II) and iron(III) ions in the vitreous phase. As an example, the spectrum of glass 54E, is presented in Figure 8.

For many spectra, however, it was found impossible to describe the central part of the absorption in the detail shown in Figure 8. In these cases, the hyperfine parameters of the iron species in the glass phase cannot be determined. The outer lines of both magnetic subspectra are usually well fit with Lorentzian lines which allows the determination of the hyperfine

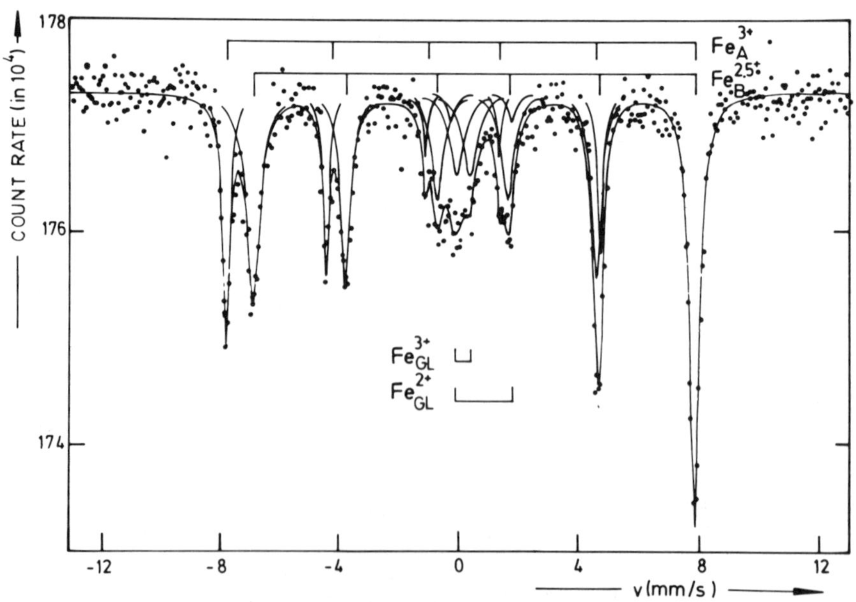

Figure 8. Detailed fit of the room temperature Mössbauer spectrum for sample 54E. The assignment of doublets and sextets is indicated by the horizontal bars.

parameters of the iron species in the crystalline phase. Because the absorption area of the outer lines is then well known, it is possible to evaluate the total area, S_{GL}, of the amorphous contribution from S_{TOT}, the total area of the spectrum. RA_{GL}, which is the ratio of S_{GL} to S_{TOT}, is to a good approximation proportional to the fractional iron concentration in the glass phase and is of great importance for later discussion.

The overall shape of the magnetic subspectrum is very similar to that of the well-known and characteristic room temperature Mössbauer spectrum of magnetite [39] or substituted magnetite [40]. The observed powder diffraction patterns support this similarity because all the reflection lines and corresponding d-spacings are consistent with the spinel structure of magnetite. In pure magnetite, Fe_3O_4, the iron ions are distributed between the tetrahedral A sites, and the octahedral B sites. There are twice as many B sites as A sites. The A sites contain only iron(III), while the B sites contain equal amounts of iron(III) and iron(II). At room temperature, a fast electron hopping between the B site iron(II) and iron(III) takes place and the Mössbauer effect cannot distinguish between the two valence states. The resulting hyperfine parameters are characteristic of an intermediate valence state.

This characterization of the crystalline phase is confirmed by the hyperfine interactions observed for some samples in which the magnetic subspectra are most sharply resolved. As seen in Table VI, all parameters for the three reference glasses involved are indeed very close to the values observed for Fe_3O_4.

For several compositions both the absorption and reflection lines in the Mössbauer spectra and the x-ray powder diffraction patterns exhibit considerable line broadening. This is especially true for compositions slightly above the solubility range and quenched in water. This may be attributed to the occurrence of lattice defects and/or to the small dimensions of the segregates.

By using the Scherrer approximation [41] which relates the x-ray diffraction line broadening to the average particle size, a lower size limit of ca. 200Å is obtained. A distribution of such small particles results in

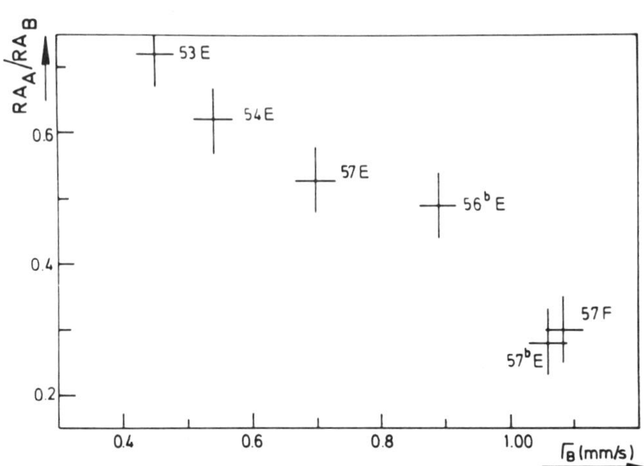

Figure 9. Area ratio of the A and B site magnetic sextets in the magnetite-like phase, segregated from the glass network, versus the outer B site line width for a number of reference samples.

fluctuating magnetic fields due to surface effects and collective excitations of the spins [42], both effects being dependent on the particle volume. However, lower limit of 200Å excludes the possibility of significant super-paramagnetic relaxation effects at 300K [43]. This was indeed confirmed by the Mössbauer spectrum obtained at 80K, on a sample with a poorly defined room temperature Mössbauer spectrum and diffraction pattern, which yielded an unchanged S_{GL}/S_{TOT} ratio. These results justify the assignment of the central part of the absorption to iron species in the vitreous phase.

Samples 53E, 54E, 57E, 56E, 57E, and 57F (see Table VII) all contain a spinel phase with a relatively high degree of crystallinity. Two observations, apparently related, let us conclude, however, that this phase is not pure Fe_3O_4. They concern the ratio of the B site absorption area to the A site absorption area, which should be close to two [39,40], and the B site line width Γ_B. Both parameters are plotted in Figure 9. For the A site, the line width was found to be constant at 0.32(2)mm/s. The behavior of the area ratio and Γ_B can be ascribed to the significant amount of non-iron cations which go into the Fe_3O_4 phase during the crystallization process. Such substitution, if present on the A site of a spinel material, even in small amounts, causes an asymmetric distribution of the magnetic hyperfine fields on the B site, thus resulting in a considerable B site line broadening. Due to structural features,the A site hyperfine parameters are much less sensitive to B site substitutions [44].

From those cations present in the melt and possibly substituting into the magnetite phase, lithium(I), magnesium(II) and aluminum(III) prefer octahedral positions [45,46,47] whereas calcium(II) will mainly be confined to tetrahedral positions [48]. It is however not possible to draw any conclusions regarding the type and the amount of cations present in the Fe_3O_4 phase. However, a distinction between the aluminum containing and aluminum free samples seems to be apparent. Samples 53E and 54E have no Al_2O_3 in the initial oxide mixture. Their magnetic subspectra area ratio is higher than the ideal value, indicating that the B sites are more substituted, with magnesium(II) and/or lithium(I) than is the A site. For similar aluminum containing samples (56b, 57b), the area ratio is smaller and Γ_B higher, both indicating a higher A site substitution. Because calcium(II) is the only cation likely to be found on the A sites,it must be concluded that the presence of aluminum(III) causes more calcium(II) ions to precipitate into the magnetite phase while aluminum(III) prefers to remain in the vitreous phase.

Influence of Composition and Quenching Mode on Crystallinity

In order to gather some additional information regarding the influence of the composition and the quenching mode upon the crystallization process, we define a distribution parameter N,

$$N = RA_{GL} \cdot \frac{n_{Fe}}{n_{Si}},$$

where n_{Fe} and n_{Si} are the number of iron and silicon atoms in the melt. As such, N is proportional to the number of iron atoms per silicon atom in the glass phase. The value of the proportionate factor is determined by the Mössbauer fractions of the phases involved but its exact value is of no importance for the following discussions.

For a number of reference samples quenched in water and air, the calculated RA_{GL} and N values are listed in Table VII, together with the composition of the compounds. It is immediately clear that the amount of iron remaining in the glass phase increases with increasing cooling rate. This results because, at lower cooling rates, there is more time available for the atoms to rearrange into a crystalline lattice. N is usually considerably smaller for the partially crystallized samples than for the samples in the same series which are completely amorphous. This is consistent with the well-known fact that the initial formation of crystallites triggers the crystallization of more material.

The incorporation of aluminum has a substantial effect on the partition of iron among the crystalline and vitreous phases. If the N values of those samples for which all M_2O_3 is Fe_2O_3 are compared with those of the corresponding aluminum containing samples, it is observed that for the latter specimens the decrease significantly exceeds the expected factor of two and is, in fact, much closer to a factor three. This means that, in the presence of aluminum(III) ions, a relatively large amount of iron crystallizes into the magnetite-like phase. This further implicates that the aluminum(III) is more likely to be incorporated into the glass network than iron(III), which is probably explained by the larger ionic radius of iron(III).

A final note concerns the well-established importance of alkaline and alkaline-earth elements with respect to the stability of the glass network. This effect explains why composition 56, with more than 18 mole percent M_2O_3, still is completely amorphous (water-quenched) whereas sample 70F, with only 15 mole percent M_2O_3, but with a lower $M_2'O$ and M"O concentration, already shows a significant degree of crystallization. Table VII further suggests that the mole percent $M_2'O$ and M"O also affect the distribution of iron among the two phases, which is of course related to the degree of crystallization. At present, however, too few systematic data are available in order to assess the role of the glass-modifying constituents upon the crystallization process as seen from the Mössbauer spectra.

CONCLUSIONS

A detailed Mössbauer study of the glasses $(84-x)SiO_2$-xFe_2O_3-8M"O-$8M_2'O$ where M" is Mg, Ca; and M' is K, Li, with x between 2.5 and 30, has shown that, for compositions below a certain threshold, the hyperfine interactions of both iron(II) and iron(III) are not affected by the melting time or quenching mode. This indicates that the chemical and structural environment of the iron species in the glass network remains essentially unchanged upon the variation of these preparative parameters.

The iron(II) concentration relative to the total iron content is affected only by the melting temperature, and increases from 14 percent at 1400°C to 27 percent at 1550°C. The coordination of the iron(II) ions could not be determined, but the data suggest that both fourfold and sixfold coordination is likely.

Iron(III) is situated in essentially two different network positions. The distinct hyperfine parameters are consistent with octahedral and tetrahedral oxide coordination. Approximately one third of the iron(III) ions are found on glass-forming positions and the number is not affected by any of the external parameters, not even by the cooling rate, although the latter does affect the total amount of M_2O_3 which can be incorporated into the glass phase.

Substituting barium for magnesium or calcium causes the iron(II) contribution and its quadrupole splitting to decrease. These effects are ascribed to the large ionic radius of barium(II). Replacement of one half of the Fe_2O_3 content by Al_2O_3 does not yield significant changes in the Mössbauer parameters.

Glasses with a low mole percent Fe_2O_3 in the oxide mixture usually behave differently. It is believed that these glasses have a somewhat modified structure, at least as far as the iron coordination is concerned. The reason for this behavior is not yet understood.

Compositions with a high mole percent M_2O_3 and melted at 1550°C are partly crystalline. The limiting concentration of M_2O_3 for obtaining a completely amorphous substance depends upon the amount of glass modifying elements present and shifts to higher values when quenching occurs in water instead of in air.

The crystalline phase yields a magnetic contribution to the Mössbauer spectrum. From its hyperfine parameters it is concluded that it is a magnetite-like phase, in which a substantial part of the iron ions have been substituted. The relative amount of iron remaining in the vitreous state is higher upon quenching in water. It is significantly lower for the aluminum containing samples. Several observations suggest that aluminum(III) ions are more easily incorporated in the glass network than are iron(III) ions.

The system $84SiO_2$-xFe_2O_2-$8M''O$-$8M'_2O$, prepared in an argon atmosphere can contain up to 30 mole percent Fe_2O_2. Their Mössbauer spectra closely resemble the spectrum obtained for a sample of the actual end-product of industrially conditioned low-level α-waste. The iron(III) concentration is low and strongly depends on the purity of the argon atmosphere. The structure of these iron(II) glasses seems to differ from the iron(III) glasses. Iron(III) ions exhibit a high isomer shift and low quadrupole splitting indicating large interstitial positions with coordination numbers of at least six and possibly higher. The iron(II) quadrupole splitting has values distributed between 1.1 and 2.6mm/s, indicating a distribution of local site symmetries around the iron(II) ions. The width of the distribution is not sensitive to the composition, whereas the average value increases slightly with increasing Fe_2O_2 content. The results indicate that most of the iron(II) cations are located at one particular type of site, but it was found impossible to determine its coordination number.

The substitution of one half of the Fe_2O_2 content with Al_2O_3, does not affect the observed Mössbauer parameters, except that the average iron(III) quadrupole splitting increases by 0.1 to 0.2mm/s.

Finally, none of the observed Mössbauer spectra of the iron(II) glasses seems to be sensitive to the quenching mode.

ACKNOWLEDGEMENTS

The authors wish to thank Dr. R. De Batist for stimulating this research, Ing. W. Timmermans and Mr. L. Peters for preparing the glasses, and Drs. R. E. Vanderberghe, D. Chambaere, and J. Vanderbruwaene, who recorded and analyzed some of the Mössbauer spectra. The technical assistance of Ms. C. Van de Veire is acknowledged. This work was supported in part by the Fund for Joint Fundamental Research of Belgium.

REFERENCES

1. The evaluation and review of alternative waste forms for immobilization of high-level radioative wastes, DOE/TIC-11472 (1981).

2. FLK LACORAY S. A., Environmental Control Department, CH-1211 Geneva 11.

3. N. Van de Voorde, G. Dumont, J. Claes, A. Taeymans, J. Vangeel, and P. Dejonghe, Proc. IAEA-NEA Intern. Symp. on the Manag. of Rad. wastes from the nucl. fuel cycle, Vienna, Austria, 141 (1976).

4. R. De Batist, E. De Grave, W. Timmermans, J. Vangeel and P. Van Iseghem, in Scientific basis for nulcear waste management, vol. 2, ed. C. J. M. Northrup, Plenum Press, New York, 351 (1980).

5. P. Van Iseghem, W. Timmermans, and R. De Batist, EUR 8979, 27 (1984).

6. P. Van Iseghem, W. Timmermans, and R. De Batist, Mat. Res. Soc. Symp. Proc. 26, Elsevier Sci. Publ., 527 (1984).

7. W. D. Kingery, H. K. Bowen, and D. R. Uhlmann, Introduction to ceramics, J. Wiley & Sons, NY (1976).

8. A. A. Belyustin, Yu. M. Ostanevich, A. M. Pisarevskii, S. B. Tomilov, U. Bai-shi, and L. Cher, Soviet Phys. - Solid State 7, 1163 (1965).

9. J. E. Stanworth, J. Am. Cer. Soc. 54, 61 (1971).

10. F. V. Tooley, The handbook of glass manufacture, vol. II, Books for Industry, NY (1974).

11. D. E. Day and G. E. Rindone, J. Am. Cer. Soc. 45, 579 (1962).

12. C. R. Kurkjian and E. A. Sigety, Phys. & Chem. of Glasses 9, 73 (1968).

13. W. A. Weyl, Colored glasses, Society of Glass Technology, Sheffield (1951).

14. F. N. Steele and R. W. Douglas, Phys. & Chem. of Glasses 6, 246 (1965).

15. J. P. Gosselin, U. Shimony, L. Grodzins, and A. R. Cooper, Phys. & Chem. of Glasses 8, 56 (1967).

16. J. A. Boon and W. S. Fyfe, Chem. Geol. 10, (1972).

17. R. Al Levy, C. H. P. Lupis, and P. A. Flinn, Phys. & Chem. of Glasses 17, 94 (1976).

18. C. R. Kurkjian and D. N. E. Buchanan, Phys. & Chem. of Glasses 5, 63 (1964).

19. L. Pargamin, C. H. P. Lupis, and P. A. Flinn, Met. Trans. 3, 2093 (1972).

20. B. O. Mysen, F. Seifert, and D. Virgo, Am. Mineral 65, 867 (1980).

21. K. E. Fox, T. Furukawa and W. B. White, Phys. & Chem. of Glasses 23, 169 (1982).

22. M. M. Nasrallah, S. K. Arif, and A. M. Bishay, Mat. Res. Bull. 19, 269 (1984).

23. V. M. Firsov, N. V. Petrovykh, and P. I. Litvinov, Proc. Xth Int. Conf. on Glass, Kyoto, 1974, 7, 73.

24. A. B. Klyukin, T. A. Filatova, A. D. Mokrushin, V. V. Gorbachev, I. V. Razamovskaya, L.I. Shvorneva, and V. F. Blokhin, Soviet J. Glass Phys. Chem. 7, 269 (1982).

25. J. M. F. Navarro and R. Brückner, Glastechn. Ber. 49, 82 (1976).

26. C. Labar and P. Gielen, J. Non-Cryst. Solids 13, 107 (1973/74).

27. A. K. Bandyopadhyay, P. Auric, J. Phalippou, and J. Zarzycki, J. Mat. Sci. 15, 2081 (1980).

28. G. Calas and J. Petiau, Solid State Comm. 48, 625 (1983).

29. E. De Grave, P. Van Iseghem, R. De Batist, and D. Chambaere, J. Phys. Colloq. 41, C1-269 (1980).

30. C. R. Kurkjian, J. Non-Crystalline Solids 3, 157 (1969).

31. T. Raman, G. N. Rao, and D. Chakravorty, J. Non-Crystalline Solids 29, 85 (1978).

32. W. D. Johnston, J. Am. Cer. Soc. 47, 198 (1964).

33. H. V. Lauer, Jr., Phys. Chem. Glasses 18, 49 (1977).

34. H. V. Lauer, Jr. and R. V. Morris, J. Am. Cer. Soc. 60, 443 (1977).

35. R. De Batist, N. Van de Voorde, and P. Van Iseghem, Nucl. waste Manag. 1, 171 (1980).

36. R. Ingalls, Phys. Rev. 133A, 787 (1964).

37. R. Vanleerberghe and R. E. Vandenberghe, Hyperfine Interactions 23, 75 (1985).

38. G. A. Sawatzky, F. van der Woude, and J. Hupkes, J. Phys. Colloq. 32, C1-276 (1971).

39. B. J. Evans and S. S. Hafner, J. Appl. Phys. 46, 1411 (1969).

40. E. De Grave, R. Leyman, and R. Vanleerberghe, Phys. Lett. 97A, 354 (1983).

41. H. P. Klug and L. E. Alexander, X-ray diffraction procedures for poly-crystalline and amorphous materials, 2nd edition, Wiley & Sons, New York (1974).

42. S. Morup and H. Topsoe, Appl. Phys. 11, 63 (1976).

43. T. K. McNab, R. A. Fox, and J. F. Boyle, J. Appl. Phys. 39, 5703 (1968).

44. E. De Grave, A. Govaert, D. Chambaere, and G. Robbrecht, Physica 96B, 103 (1979).

45. C. Boekema, F. van der Woude, and G. A. Sawatzky, J. Phys. C: Solid State Phys. 9, 2439 (1976).

46. E. De Grave, J. De Sitter, and R. E. Vandenberghe, Appl. Phys. 7, 77 (1975).

47. J. A. Schulkes and G. Blasse, J. Phys. Chem. Solids 24, 1651 (1963).

48. J. De Sitter, A. Govaert, E. De Grave, D. Chambaere, and G. Robbrecht, Phys. Status Sol. (a) 43, 619 (1977).

MÖSSBAUER ANALYSIS OF Q-BASIC OXYGEN PROCESS SLAGS

F. E. Huggins and G. P. Huffman

U.S. Steel Technical Center
Monroeville, PA 15146

INTRODUCTION

The Q-Basic Oxygen Process (Q-BOP) is one of a number of alternate oxygen steelmaking processes that can be utilized to refine the so-called hot metal or carbon-saturated liquid iron produced in the conventional blast furnace. In the steelmaking process, the carbon and other dissolved impurities in the hot metal are oxidized by blowing oxygen into the bath. The oxidation products separate from the liquid steel as gases (e.g., carbon monoxide) or by partitioning as oxides into the slag layer on the top of the liquid steel. In the Q-BOP, the oxygen, as well as burnt lime that forms the slag, is blown through the hot metal from underneath in contrast to other major steelmaking processes such as the open hearth, electric-arc furnace, and the BOF (also known as BOP or LD), in which oxygen is blown into the hot metal from above.

Although the final composition of the steel is usually adjusted in the subsequent step of ladle refining, the steelmaking process is also important for determining the steel composition. Many of the elements partition between the slag and liquid steel according to the following generalized reaction:

$$x(E)_m + y(FeO)_s \rightleftarrows (E_xO_y)_s + y(Fe)_m \qquad (1)$$

where E is an element that can be oxidized during steelmaking and the subscripts m and s refer to the liquid metal and the slag phase, respectively. At a given temperature, the equilibrium relation, k_E, can be written in terms of weight percentages as, after Turkdogan [1]:

$$k_E = \frac{(\%E_xO_y)_s}{[\%E]_m^x(\%FeO)_s^y} \qquad (2a)$$

Alternatively, the distribution of element, E, between slag and metal can be written as:

$$\frac{(\%E_xO_y)_s}{[\%E]_m^x} = k_E(\%FeO)_s^y \qquad (2b)$$

It should be noted that K_E is defined in terms of concentrations and not activities and because the activity coefficients have been shown to vary significantly with such compositional factors as slag basicity [1], k_E is also a function of slag composition as well as temperature.

The form of equation 2b emphasizes the importance of the FeO content of the slag in determining element partitioning during steelmaking. For an element such as phosphorus, for which the following partitioning reaction may be written:

$$2(P)_m + 5(FeO)_s \quad (P_2O_5)_s + 5(Fe)_m, \tag{3}$$

equation 2b becomes:

$$\frac{(\%P_2O_5)_s}{[\%P]_m^2} = k_P(\%FeO)_s^5 \tag{4}$$

Thus, for phosphorus partitioning, there is a dependency on the fifth power of the FeO concentration in the slag.

In current practice, the operation of Q-BOP steelmaking furnaces is increasingly being controlled by computer. In order to optimize steel composition and minimize usage of consumables such as oxygen, lime, and energy, elaborate computer models of Q-BOP operation are being implemented. As part of the input to such models, slag chemistries are routinely obtained. As indicated by equations 2b and 4 above, one of the more important parameters to be included in such models is the FeO concentration in the slag. However, because of imperfect sampling procedures and the indeterminate nature of the slag/metal interface due to the agitation of the bath, some metallic iron is often included in the slag sample. Hence, the assumption of total iron in the slag sample determined by X-ray fluorescence (XRF) as equal to iron in oxide form could cause significant errors in the computer model, unless correction for the metallic iron content of the slag sample is made.

As will be developed later in this paper, Mössbauer analyses of suites of Q-BOP slags were instrumental in deciding upon a method for rapidly determining metallic iron in Q-BOP slags, so that a more accurate value of $[FeO]_s$ could be used in the computer-control models. However, in the next sections, details of the Mössbauer analysis of Q-BOP slags will be discussed.

SAMPLES

In this work, Mössbauer data are presented for three series of slags from Q-BOP furnaces in U.S. Steel plants at Gary, Indiana, and Fairfield, Alabama. In addition, data from a fourth series of slags from a BOF in U.S. Steel's plant at Lorain, Ohio, are also included for comparison purposes. The Mössbauer effect spectral parameters are given in mm/s and the isomer shifts are reported relative to room temperature α-iron foil.

Different methods of sampling were used at the different furnaces. At the Gary Q-BOP, samples were taken by a method known as 'sculling' or 'poling'. In this method, a steel pole is inserted into the slag and removed. Liquid adhering to the pole is permitted to solidify and is then knocked off when cool. This method provides a much more homogeneous sample than the more usual 'dip' method, which often includes significant quantities of solid, unreacted lime. The Fairfield Q-BOP samples were obtained by taking a 'spoon' sample, which is poured on a cold steel plate to solidify. This method cools the sample more rapidly than sculling. Thus, in addition to coming from

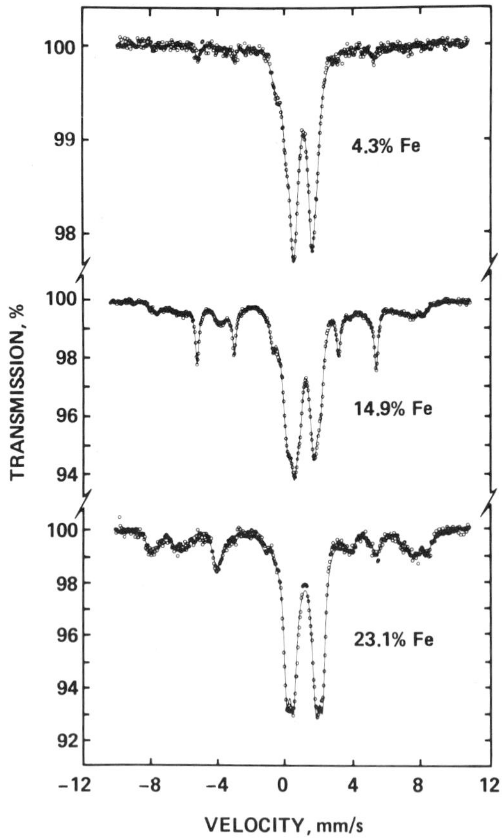

Figure 1. Mössbauer spectra of three Q-BOP slags with low (4.3%), intermediate (14.9%), and high (23.1%) iron contents.

different furnaces, the series of Q-BOP slag samples have also experienced significantly different cooling rates.

As the metallic iron content of these slags was the principal item of investigation, full chemical compositional data are only available for the Fairfield and Lorain series of slags. However, chemical data on the forms of iron are available for one of the other series of slags and chemical determinations of the metallic iron content are available for all samples. Comparisons between Mössbauer an chemical data for the forms of iron will be discussed in a subsequent section.

FORMS OF IRON IN Q-BOP SLAGS

The Mössbauer spectra of iron in Q-BOP slags appear similar to those of slags from other steelmaking furnaces. Previous investigations of slags from an LD (BOF) furnace [2] and from an electric furnace [3] have shown that the Mössbauer spectra are complex and variable and consist of both magnetic and nonmagnetic components. Generally, dicalcium ferrite (DCF), $Ca_2Fe_2O_5$, and metallic iron constitute the magnetic components, but the nonmagnetic component in the spectrum is not as readily identified and is usually fit with two ferrous absorptions and one ferric absorption. There is agreement that iron(II) in a complex oxide phase of composition (Fe,Ca,Mg,Mn)O must be present. However, whereas Gohy et al. [2] attribute all the nonmagnetic component to this phase, Ramous et al. [3] attribute the outer ferrous absorption to iron(II) in a 'silicate glass' phase and were not certain about the origin of the ferric absorption. This apparent difference in interpretation of the spectrum derives largely from the fact that Gohy et al. examined only one low-silica (11.2 weight percent) slag, whereas Ramous et al examined 13 slags, which varied in composition from 9.7 to 29.0 weight percent SiO_2.

In our examination of Q-BOP slags by Mössbauer spectroscopy, we have been as consistent as possible in fitting the spectra. The nonmagnetic region of the spectrum has been uniformly fit with three quadrupole doublets of Lorentzian line-shape. Although this procedure is an imperfect approximation to the actual spectrum, all spectra were consistently fit. When present, the magnetic absorptions were fit as six-line subspectra and were

Table I. Distribution of Iron Among Mössbauer Spectral Components for Four Slag Series

Sample		Inner Doublet	Outer Doublet	Fe(III) Doublet	Magn.[a] Oxide	Fe Metal	X_{eff}[b]
Fairfield	1	49	11	13	<3	27	14.6
	2	51	18	10	12	10	25.0
	3	36	13	11	25	16	22.3
	4	42	19	13	18	9	20.4
	5	41	15	12	22	9	24.6
	6	33	11	19	17	20	17.3
	7	44	18	11	21	6	21.8
	8	42	18	17	14	9	20.5
	9	47	13	11	20	9	25.9
	10	38	11	14	17	21	18.4
	11	44	16	9	30	<3	20.4
	12	34	18	15	9	25	14.3
Gary 1981	1	67	13	9	6	6	15.5
	2	56	16	5	14	9	16.9
	3	53	15	6	21	5	25.6
	4	70	15	9	<3	6	6.4
	5	67	14	13	<3	6	6.3
	6	46	26	3	26	<3	37.4
Gary 1982	1	53	14	15	18[a]	<3	12.8
	2	49	21	<3	30	<3	16.1
	3	52	17	21	10	<3	11.2
	4	48	27	<3	25	<3	33.8
	5	44	22	<3	34	<3	31.7
Lorain BOF	1	68	8	7	17	<3	13.4
	2	38	27	4	31	<3	26.6
	3	45	26	3	25	<3	23.8
	4	44	25	3	28	<3	27.4
	5	43	20	3	34	<3	24.9

[a]Except for the Gary 1982-1 sample, in which the magnetic oxide was magnetite, substituted dicalcium ferrite is the magnetic oxide phase present in these samples. [b]X_{eff}--total effective Mössbauer thickness per gram.

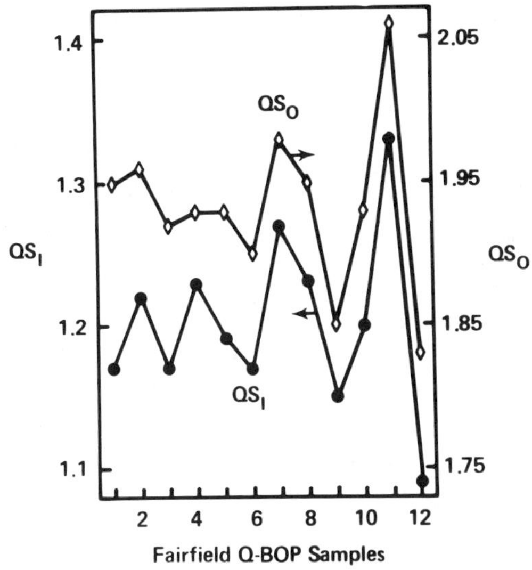

Figure 2. Variation of quadrupole splittings of inner and outer iron(II) doublets for the 12 Fairfield Q-BOP slags.

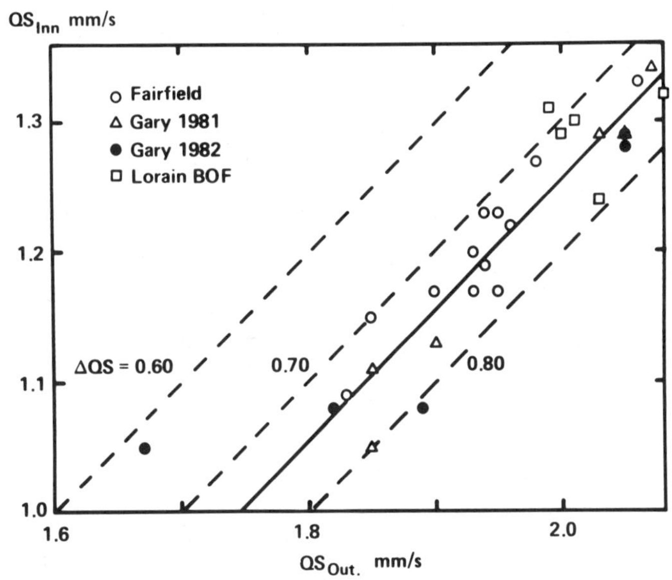

Figure 3. Plot of inner-doublet quadrupole splitting against outer-doublet quadrupole splitting. Lines indicate constant values for the difference of the two parameters.

generally adequately fit except when the peaks of the dicalcium ferrite phase were greatly broadened.

Mössbauer spectra of three Q-BOP slags of low-, intermediate-, and high-iron content are shown in Figure 1. It is clear from these spectra that there is considerable variation among the slags; however, as summarized in Table 1, the variation reflects the relative distribution of iron among the phases that give rise to the five spectral components. These components will now be examined individually.

(a) Ferrous quadrupole doublets

Mössbauer data for the two ferrous quadrupole doublets are summarized in Table II for each series of slags. These two doublets normally account for between 45 percent and 85 percent of the spectral absorption. The isomer shift values for the two doublets are usually very close to each other (the values differed by more than 0.03mm/s in only 3 of 28 cases), and are similar to those measured for wustite (isomer shift: 0.90 to 1.07mm/s for $Fe_{1-x}O$ depending on the value of x) [4], but less than that expected for iron(II) in octahedral sites in most crystalline silicates (isomer shifts: 1.10 to 1.20mm/s). The quadrupole splittings of the two doublets show more variation than the isomer shift, but the variation in the two values is similar (Figure 2) and highly correlated (Figure 3). The width of the inner doublet (smaller quadrupole splitting) is usually in the range 0.50 to 0.65mm/s, whereas the width of the outer doublet is narrower and usually varies between 0.35 and 0.45mm/s.

The area ratio of the two ferrous doublets varies between 1.5 and 5.0, with the area of the inner doublet always greater (Table I). As shown in Figure 4, there appears to be a weak negative correlation of this area ratio with the total iron content of the Q-BOP slags.

Our interpretation of these systematics in the Mössbauer parameters of the two ferrous doublets is that both doublets derive from iron(II) in the complex MO oxide phase. Unless the absorptions come from the same phase, it appears improbable that the isomer shifts and quadrupole splittings of the two doublets would be so highly correlated with each other. The variation in the parameters of the doublets is attributed to compositional variation of the complex oxide phase; however, attempts to quantify the variation in terms of the slag composition have not been successful. Further evidence that the doublets arise from the complex oxide phase is forthcoming from low-temperature spectra, which show that both doublets broaden and become magnetic absorptions at temperatures above 80K. This behavior is not expected for

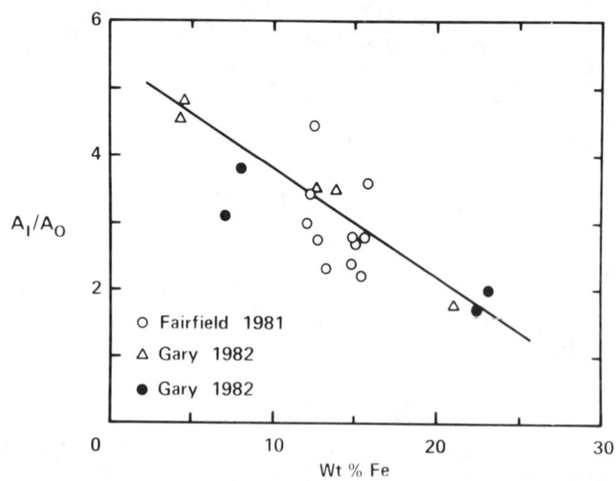

Figure 4. Plot of the area ratio of the two ferrous doublets (Inner/Outer) against weight percent total iron for Q-BOP samples. N.B. Three samples rejected for exceptionally broad outer-doublet widths (>0.45mm/s).

Table II. Mössbauer Data for Iron(II) Quadrupole Doublets in Slag Spectra

Sample[a]		Isomer Shift		ΔIS	Quadrupole Splitting		ΔQS	%Fe	
		Inner	Outer	(Out-Inn)	Inner	Outer	(Out-Inn)	Inner	Outer
Fairfield	R:[b]	0.95-1.04	0.95-1.04	-0.01-0.03	1.09-1.33	1.82-2.06	0.70-0.78	33-51	11-19
(12)	M:	1.01±0.03	1.02±0.02	0.01±0.01	1.20±0.06	1.94±0.06	0.72±0.02		
Gary 1981	R:	0.99-1.05	0.95-1.03	-0.06-0.02	1.05-1.34	1.85-2.07	0.73-0.80	46-70	14-26
(6)	M:	1.02±0.02	1.00±0.03	-0.02±0.03	1.20±0.12	1.96±0.10	0.76±0.03		
Gary 1982	R:	1.03-1.05	1.00-1.04	-0.04-0.01	1.05-1.29	1.67-2.05	0.62-0.81	44-53	17-27
(5)	M:	1.04±0.01	1.02±0.02	-0.01±0.02	1.16±0.12	1.90±0.16	0.74±0.07		
Lorain BOF	R:	1.02-1.04	0.99-1.03	-0.03-0.01	1.24-1.32	1.99-2.08	0.68-0.79	38-68	8-27
(5)	M:	1.03±0.01	1.02±0.02	-0.01±0.02	1.29±0.03	2.02±0.04	0.73±0.04		

[a]Values in parentheses indicate number of samples included in mean and range calculations. [b]R: Range in values. M: Mean ± standard error. $\Delta IS = (IS_{outer} - IS_{inner})$. $\Delta QS = (QS_{outer} - QS_{inner})$.

Table III. Mössbauer Data for Ferric Quadrupole Doublet in Slag Spectra

Sample		Isomer Shift	Quadrupole Splitting	%Fe
Fairfield	R:	0.16-0.31	0.77-0.96	9-19
(12)	M:	0.27±0.04	0.86±0.06	
Gary 1981	R:	0.05-0.29	0.78-1.18	<3-13
(5)	M:	0.15±0.11	1.01±0.18	
Gary 1982	R:	0.21-0.30	0.99-1.00	<3-21
(2)	M:	0.26	1.00	
Lorain BOF	R:	0.12-0.26	0.75-0.86	<3-7
(3)	M:	0.21±0.08	0.79±0.06	

iron(II) in silicate glass or in silicates, but is compatible with substituted wustite.

(b) Ferric quadrupole doublet

Mössbauer data for the ferric quadrupole doublets are summarized in Table III. This absorption, although present in most spectra, does not exceed 20 percent of the total absorption in any spectrum. It appears more abundant in the Fairfield Q-BOP samples than in any other series, and is particularly low in the Lorain BOF series and certain Gary 1982 samples. Samples very rich in iron ($X_{eff}/g>30$; weight percent iron>20) tend to have the lowest amount of this absorption.

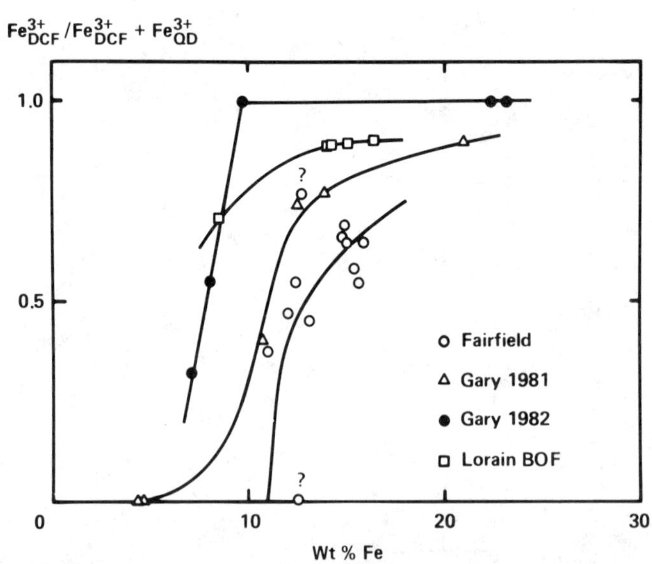

Figure 5. Plot of the ferric-iron ratio (iron(III) in DCF/total iron(III)) against weight percent iron. See text for discussion of separate trends for different slag series and the two Fairfield samples marked with a ?.

There is considerable variation in the Mössbauer parameters of the ferric quadrupole doublet. Some of this variation can be attributed to inadequacy of the least-squares fitting procedure to define properly a weak doublet under the more intense ferrous doublets; however, there may be more than one interpretation for this absorption, which would also result in variability in the Mössbauer parameters. Gohy et al. [2] have attributed this absorption to iron(III) in the complex oxide phase. However, as noted above, there is a tendency for the ferric doublet to be absent in high iron samples. Thus, as the ferrous content of the oxide phase increases, the ferric content decreases. This trend is contrary to expectation because one would anticipate increasing ferrous iron would facilitate incorporation of ferric iron in the oxide phase. Moreover, as shown in Figure 5, the ferric iron in the slags is increasingly incorporated into the magnetic dicalcium ferrite phase as the iron in the slags increases. A possible explanation for these relationships is that the iron(III) quadrupole doublet largely arises from iron(III) in superparamagnetic dicalcium ferrite and that the total iron content of the slag and the rate of cooling determines how much of the substituted dicalcium ferrite appears magnetic. As noted above, the method of sampling of the slags was different at each furnace and the resulting different cooling rates for the series slag samples is reflected in separate trends displayed in Figure 5 for each slag series. The large value for the quadrupole splitting would also appear to be consistent with this interpretation as the quadrupole splittings of synthetic $Ca_2Fe_2O_5$ and $Ca_2(Fe,Al)_2O_5$ phases measured at temperatures above their Neel temperatures are very large (1.1 to 1.6mm/s) [5,6]. In contrast iron(III) in wustite has a much smaller quadrupole splitting, and often is fit as a single peak (i.e., quadrupole splitting <0.5mm/s) [4]. However, we cannot eliminate the possibility that the iron(III) absorption does include some contribution from iron(III) in the complex oxide phase.

(c) Substituted dicalcium ferrite

Mössbauer parameters for the magnetic substituted dicalcium ferrite (DCF) are summarized in Table IV. These data are for absorptions in which both a tetrahedral and an octahedral component were clearly resolvable in the spectra. Each component was usually fit as one six-line subspectrum, but occasionally it was sufficiently intense and well defined to warrant fitting with two subspectra. In such situations, the values of the Mössbauer parameters for the two subspectra have been averaged for inclusion in Table IV. In other spectra, the magnetic absorption due to DCF was very weak and broad and a single, broad six-line subspectrum was used to approximate it. Parameters for this type of fit have not been included in Table IV.

The magnetic hyperfine splittings for DCF in slags are considerably smaller than those found for the octahedral and tetrahedral components in synthetic $Ca_2Fe_2O_5$ for which the room-temperature values are 510 and 435kG, respectively [6,7]. Similar differences were also reported by Gohy, et al. for DCF in a LD slag [2]. These differences are due to combined superparamagnetic and substitutional effects. The principal substituting element for iron(III) in DCF in the slag is most probably aluminum(III) as extensive solid solution between $Ca_2Fe_2O_5$ and the hypothetical phase $Ca_2Al_2O_5$ is known to exist [5] and other trivalent ions are not present in significant quantities in the slag. Relative to iron(III), aluminum(III) has a preference for the tetrahedral site in the $Ca_2Fe_2O_5$ structure and the Mössbauer spectra of DCF in slags do exhibit somewhat more variability for this site. Evidence that aluminum(III) plays a significant role in the formation of DCF comes from the plot shown in Figure 6. In this figure, except for one sample, there is a highly correlated relationship between the fraction of iron(III) in DCF and the Al_2O_3 to Fe_2O_3 ratio for the Fairfield Q-BOP and Lorain BOF slags. As the ratio of Al_2O_3 to Fe_2O_3 increases, the fraction of ferric iron in magnetic DCF decreases. It should be noted that the points marked with a '?' in Figure 5, which are well removed from the

Table IV. Mössbauer Data for Substituted Dicalcium Ferrite in Slag Spectra

		Octahedral Site			Tetrahedral Site			
		IS	QS	Ho[a]	IS	QS	Ho[a]	%Fe
Fairfield (6)	R:	0.27-0.39	-0.30--0.10	452-486	0.27-0.40	-0.09-0.18	369-418	<3-30
	M:	0.33 ± 0.04	-0.22 ± 0.07	472 ± 15	0.32 ± 0.05	0.06 ± 0.09	403 ± 18	
Gary 1981 (2)	R:	0.33-0.34	-0.30--0.16	472-474	0.24-0.25	0.04-0.20	397-403	<3-26
	M:	0.34	-0.23	473	0.25	0.12	400	
Gary 1982 (2)	R:	0.39-0.41	-0.32--0.30	478-479	0.11-0.14	0.34-0.36	401-403	<3-30
	M:	0.40	-0.31	479	0.13	0.35	402	
Lorain BOF (4)	R:	0.32-0.37	-0.30--0.26	465-475	0.17-0.22	0.15-0.31	391-401	17-34
	M:	0.34 ± 0.02	-0.28 ± 0.02	470 ± 4	0.20 ± 0.02	0.24 ± 0.07	397 ± 5	

[a]Magnetic hyperfine splitting in kGauss.

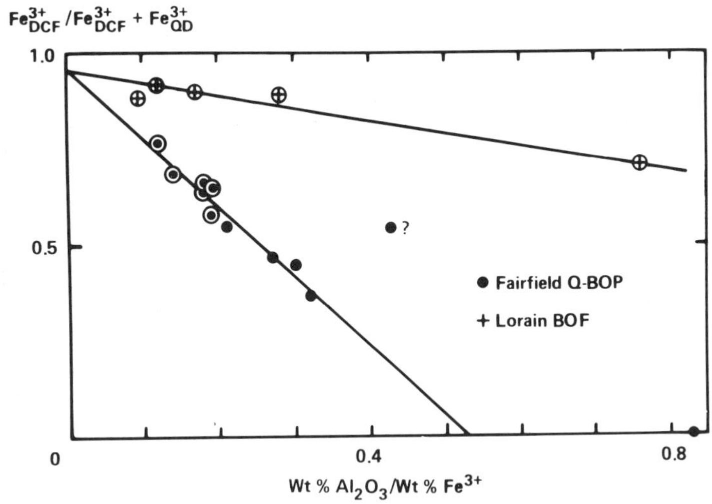

Figure 6. Plot of the ferric-iron ratio (Iron(III) in DCF/total iron(III)) against the Al_2O_3/total iron(III) ratio for two series of slags. Circled data points indicate well resolved DCF absorptions fit with two magnetic subspectra.

trend of increasing iron(III) with increasing total iron for the Fairfield Q-BOP slags, have the highest and lowest values for the Al_2O_3 to Fe_2O_3 ratio. The difference in trend between the two slag series can be attributed to differences in quenching technique. The trends in Figure 6 are additional evidence that the iron(III) quadrupole doublet and the magnetic DCF absorption are superparamagnetic and antiferromagnetic manifestations of the same phase. The relative abundance of these different manifestations is determined both by composition and by the quenching rate of the sampling procedure.

Table V. Mössbauer Data for Metallic Iron in Slag Spectra

		IS	QS	Ho[a]	%Fe
Fairfield	R:	0.00-0.04	--	328-331	<3-27
(11)	M:	0.01±0.02	--	329±1	
Gary 1981	R:	0.00-0.04	--	329-331	<3-9
(5)	M:	0.02±0.02	--	330±1	
Gary 1982 (0)		---------	--	-------	<3
Lorain BOF (0)		---------	--	-------	<3

[a]Magnetic hyperfine splitting in KGauss.

(d) Metallic iron

The absorption arising from metallic iron is the most consistent of any in Mössbauer spectra of slags. As summarized in Table 5, the hyperfine parameters show relatively little variation from slag to slag or from a metallic iron standard (isomer shift = 0.00mm/s, magnetic hyperfine splitting = 331kG). The phase is consistently present in two series of slags, but absent in the other two series. It would therefore appear that the presence or absence of metallic iron may be largely related to sampling practice at the different furnaces. Whether the metallic iron content merely reflects inclusion of liquid steel in the slag sample or, more interestingly, differences in sample cooling rate in the various sampling practices remains to be established.

DISCUSSION

Phase Equilibrium Considerations

From the above interpretation of the Mössbauer data, the forms of iron in quenched Q-BOP slags consist only of a complex metal(II) oxide phase, substituted dicalcium ferrite and metallic iron. Moreover, the last phase may be largely carried over the liquid metal into the slag sample due to imperfect sampling procedures, so that the actual iron-bearing phases in the slag itself number only two. This result appears consistent with the path of solidification of Q-BOP slags predicted from phase equilibrium diagrams.

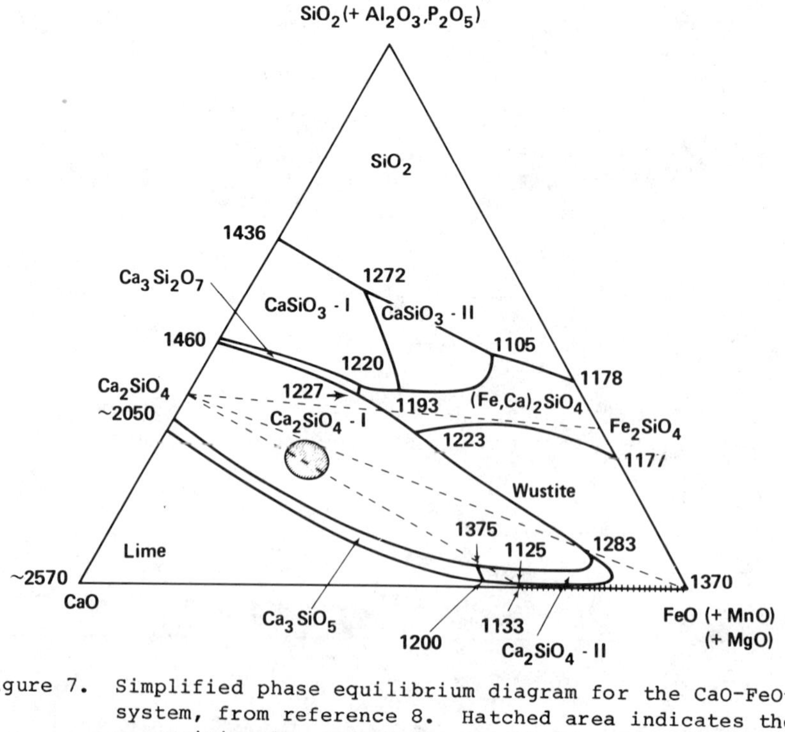

Figure 7. Simplified phase equilibrium diagram for the CaO-FeO-SiO_2 system, from reference 8. Hatched area indicates the compositional range of the Fairfield Q-BOP slags with minor oxides apportioned to either FeO or SiO_2.

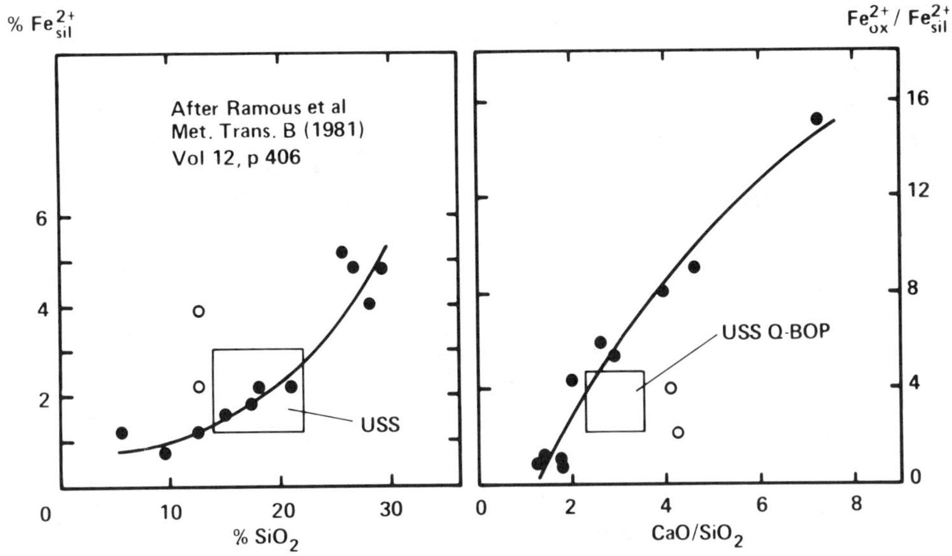

Figure 8. Plots taken from reference 3 with superimposed rectangles indicating the range shown by the Fairfield Q-BOP samples. Fe^{2+}_{ox}, Fe^{2+}_{sil} are the weight percentages of iron in the slag that give rise to the inner and outer quadrupole doublets.

Shown in Figure 7, is the equilibrium diagram for the CaO-FeO-SiO_2 system [8] and plotted on this diagram, with minor oxides (Al_2O_3, P_2O_5, MnO, MgO) apportioned to either SiO_2 or FeO, is the field in which the compositions of the Fairfield Q-BOP slags occur. This compositional field lies completely within the liquidus area for dicalcium silicate, Ca_2SiO_4. Therefore, as molten slag of such composition cools, Ca_2SiO_4 is first to solidify and the composition of the remaining liquid shifts towards the FeO corner. As the temperature lowers, the liquid becomes increasingly iron-rich and eventually the composition reaches the boundary between the Ca_2SiO_4 and wustite liquidus regions. At this point, wustite solidifies from the liquid along with Ca_2SiO_4 as the temperature decreases. Because the solid solution of CaO in wustite is so extensive at these temperatures [1100-1200°C] (as indicated by the crossed length on the CaO-FeO join in Figure 7), most compositions that fall in the hatched area will crystallize completely to a mixture of calciowustite and dicalcium silicate. Compositions low in SiO_2 will crystallize a minor amount of Ca_3SiO_5 or CaO in addition to dicalcium silicate and calciowustite. Therefore, calciowustite should be the only major iron-bearing phase to crystallize from Q-BOP slag compositions that plot in or near the hatched area in Figure 7.

However, dicalcium ferrite is also prominant in the Mössbauer spectra of quenched Q-BOP slags in addition to wustite. It forms because upon further cooling calciowustite undergoes a subsolidus decomposition to dicalcium ferrite and calcium-poor wustite [8,9]. Aluminum ions, which are largely incompatible with the wustite structure, would also be incorporated and formation of substituted dicalcium ferrite, $Ca_2(Fe,Al)_2O_5$, results. As the formation of this phase occurs by a subsolidus reaction, the resulting particle size of the dicalcium ferrite will be greatly influenced by the rate of cooling during the quenching process. This variation in particle size is reflected in the magnetic properties of the dicalcium ferrite, as documented above.

The calciowustite decomposition can also be regarded as a disproportionation reaction. By ignoring ferric iron in either calciowustite or wustite and any calcium in the product wustite, the disproportionation can be simplified to the following reaction:

$$(Ca_{2x}Fe_{1-2x})O \longrightarrow xCa_2Fe_2O_5 + xFe + (1-5x)FeO \qquad (5)$$

This reaction would give rise to all three major forms of iron in Q-BOP slags and leads to the speculation that the metallic iron found in slag samples is not included liquid steel but is formed during quenching. However, on the basis of the written reaction, a positive correlation between iron in the form of dicalcium ferrite and metallic iron would be expected. A cursory inspection of Table I establishes that no such correlation is found and therefore the importance of the disproportionation reaction in forming metallic iron appears minor. It should be noted, though, that inclusion of ferric iron and calcium in the product wustite would weaken the correlation anticipated between metallic iron and dicalcium ferrite, possibly to the point of nonexistence.

Comparison with Other Steelmaking Slags

The electric furnace slags investigated by Ramous et al. [3] showed a much wider variation in composition than the Q-BOP slags of the current study. This difference is indicated in Figure 8, which shows two plots from their investigation. Superimposed upon these plots are rectangles indicating the range exhibited by slags from U.S. Steel Q-BOP furnaces. Although the range in composition is much less, the data from the Q-BOP slags appear compatible with electric-furnace slag data of intermediate composition.

Our interpretation of the Mössbauer spectrum of slags differs from that of Ramous et al. [3] in that they attribute the outer ferrous quadrupole doublet as arising from iron(II) in silicate glass. From phase equilibrium relationships, it would be predicted that such phases would be observed for slags with high silica contents (SiO_2 > 25 weight percent) that plot above the Ca_2SiO_4-FeO line in Figure 7. However, slags that plot below that line should form the complex (Fe,Mn,Mg,Ca)O phase and not ferrous silicates. For this reason, as well as the reasons given above, we interpret the outer quadrupole doublet in all slags of lower silica contents (<25 weight percent) as being derived from the oxide phase and not from any ferrous silicate phase.

Our interpretation of the Mössbauer spectra of Q-BOP slags differs also from that of an LD (BOF) slag made by Gohy et al. [2]. The principal differences involve the ferric quadrupole doublet, which Gohy et al. attribute to iron(III) in wustite and which we attribute principally to superparamagnetic dicalcium ferrite. Our data on suites of samples provide convincing evidence (notably Figure 6) that the paramagnetic and magnetic ferric absorptions are different manifestations of the same phase.

Despite these differences in interpretation, however, the Mössbauer spectra of slags from the different steelmaking processes appear very similar. This similarity reflects not only the similar slag compositions and sampling practices used in the different steelmaking processes, but also the function of the slag that is common to these processes.

Comparison of Metallic Iron Determinations

The principal reason for investigating Q-BOP slags was to develop a method for determining metallic iron contents of slag samples. These data would then be used with the total iron content to give the total iron-oxide content for application in computer-control models of Q-BOP operation. In full operation, as many as 50 such analyses might be required daily in a Q-BOP shop. Hence, any method would have to be rapid and easy to perform.

Table VI. Comparison of Chemical and Mössbauer Determinations of Forms of Iron in Basic Slags

		Wt.% in Slag (Chemical)				Wt.%Fe in Slag (Mössbauer)		
		Total Fe	Fe Metal	Ferrous	Ferric	Fe Metal	Ferrous	Ferric
Fairfield	1	12.51	5.69	4.01	3.61	3.37	7.38	1.75
	2	12.60	3.56	7.40	4.64	1.56	10.61	3.43
	3	14.90	5.24	4.84	4.36	2.38	7.15	5.36
	4	15.36	4.30	6.54	4.53	1.38	9.22	4.76
	5	14.98	4.08	5.85	5.05	1.35	8.54	5.09
	6	11.99	4.87	3.22	3.90	2.40	5.28	4.32
	7	14.75	2.63	6.69	5.43	0.89	9.15	4.72
	8	13.07	2.73	5.35	4.99	1.18	7.84	4.05
	9	15.87	2.34	7.19	6.34	1.43	9.52	4.92
	10	12.39	4.69	3.58	4.12	2.60	5.95	3.84
	11	12.65	1.90	5.77	4.98	<0.4	7.72	4.93
	12	10.99	4.11	3.73	3.15	2.75	5.61	2.64
Gary 1981	1	10.71	2.21	5.78	2.71	0.64	8.57	1.50
	2	12.46	3.62	5.54	3.30	1.12	8.97	2.37
	3	13.87	1.95	7.25	4.67	0.70	9.43	3.74
	4	4.30	0.81	2.18	1.31	0.25	3.66	0.30
	5	4.17	1.02	2.08	1.37	0.27	3.62	0.58
	6	21.05	1.57	12.96	6.52	<0.6	15.12	5.88
Gary 1982	1	7.98	0.26	a	a	<0.3	5.35	2.63
	2	9.82	0.77	a	a	<0.3	6.87	2.95
	3	6.93	0.40	a	a	<0.3	4.78	2.15
	4	22.40	2.16	a	a	<0.8	16.80	5.60
	5	23.10	1.85	a	a	<0.8	15.25	7.85
Lorain BOF	1	8.50	1.35	4.19	2.96	<0.3	6.46	2.04
	2	15.10	0.99	9.08	5.01	<0.5	9.81	5.29
	3	14.32	1.01	8.48	4.83	<0.5	10.31	4.01
	4	16.36	1.36	10.19	4.81	<0.5	11.29	5.07
	5	14.07	1.36	7.63	5.08	<0.5	8.86	5.21

[a]not determined

Before any method could be selected, however, the precision and accuracy of various possible methods had to be assessed and a round-robin study of the Fairfield Q-BOP slags was performed by using Mössbauer, X-ray diffraction, wet chemical, and magnetic methods for determining metallic iron. Traditionally, the wet chemical method has been used to determine metallic iron contents of basic furnace slags. This method employs a bromine-methanol (Br-MeOH) reagent to extract the metallic iron from the slag sample prior to standard chemical determinations of the ferrous iron and total iron-oxide contents by dichromate titration before and after reduction of ferric iron with stannous chloride.

A comparison of forms of iron by Mössbauer and wet chemical methods was made by dividing the total iron content determined chemically into metallic, ferrous, and ferric iron components as determined by Mössbauer spectroscopy. This comparison is shown in Table VI for all 28 slag samples. It is clear upon examining Table VI that the metallic iron contents by the Br-MeOH

Table VII. Correlation Matrix (r) for Methods of Determining Metallic Iron Contents of Q-BOP Slags

Method	Mössbauer	Chemical	Magnetic	X-ray Diffraction
Mössbauer	1.0			
Chemical	0.86	1.0		
Magnetic	0.98	0.89	1.0	
X-ray Diffraction	0.93	0.71	0.94	1.0

extraction method are consistently 1-3 weight percent higher than those estimated from Mössbauer spectroscopy. Even those samples for which the Mössbauer technique did not detect any metallic iron gave positive indications of metallic iron by the Br-MeOH extraction. The relationship between the two determinations of metallic iron is shown graphically in Figure 9. This figure also shows that the bias in the determinations of this component is compensated by a similar, but opposite, bias in the ferrous iron determination.

To resolve this difference between the techniques, residues after the Br-MeOH extraction were examined chemically and by Mössbauer spectroscopy. The chemical analysis showed that the Br-MeOH reagent also extracted significant amounts of the calcium-bearing phases in addition to metallic iron. As both dicalcium ferrite and the complex oxide phase are calcium-bearing phases that also contain iron oxides, it is likely that the excess 'metallic iron' resulted from extraction of one or both of these phases. Mass balance calculations employing Mössbauer effective thickness data on the slags and on the residues after Br-MeOH extraction indicated that significant nonmetallic iron was extracted by the Br-MeOH reagent. However, the Mössbauer spectra of the residues after extraction were very similar to the original spectra of the slags and no particular ferrous or ferric phase could be identified as

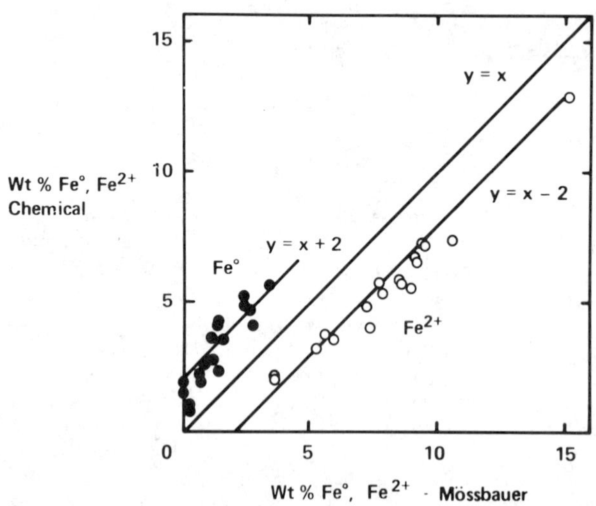

Figure 9. Comparison of wet chemical and Mössbauer determinations of metallic and ferrous forms of iron for Q-BOP slag series with measurable amounts of metallic iron determined by Mössbauer spectroscopy.

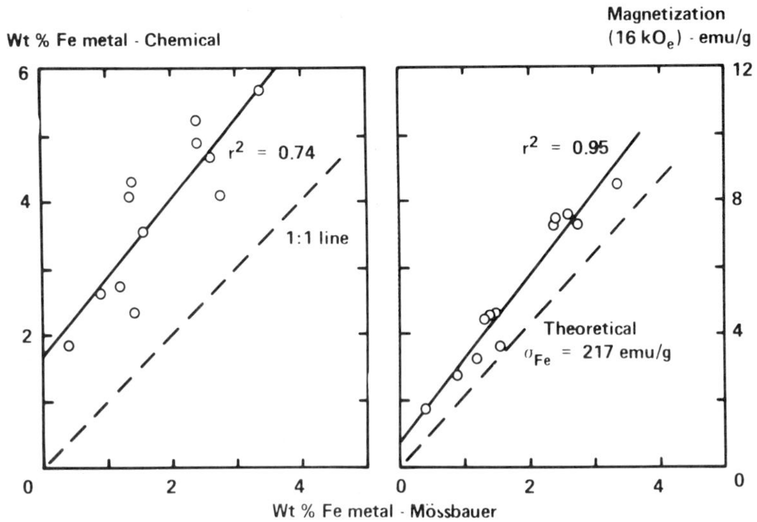

Figure 10. Comparison of correlations between magnetic and Mössbauer determinations and between chemical and Mössbauer determinations of metallic iron in Q-BOP slags.

having been preferentially removed during the extraction. Hence, the reagent must extract the calcium-rich phases at approximately the same rate. This observation may indicate that the extraction of nonmetallic phases by Br-MeOH occurs by a physical rather than a chemical process.

Other evidence that the Mössbauer technique is a more accurate measure of the metallic iron content comes from the other two independent determinations of the metallic iron content. As shown in Table VII, the correlation coefficients between the X-ray diffraction and magnetization determinations and the Mössbauer determination are uniformly higher than with the chemically determined values. This is also illustrated in Figure 10, which compares the magnetization-Mössbauer and chemical-Mössbauer correlations.

As a result of the comparative study, the choice for the method for use in the Q-BOP shop narrowed to the saturation magnetization measurement or a Mössbauer-based method. Although rapid variations of Mössbauer spectroscopy, such as constant-velocity scans and the rapid on/off method described elsewhere [10], were considered, it was clear that the magnetic method was quicker and better suited for inhomogeneous samples (i.e., required less sample preparation) and for in-plant use. In addition, it was available as a commercial instrument for about the same price as the rapid variations of the Mössbauer method would cost to assemble.

CONCLUSIONS

Mössbauer analysis of forms of iron in Q-BOP slags show that the iron is predominantly distributed among three phases: a complex oxide of composition (Fe(II),Ca,Mg,Mn)O; substituted dicalcium ferrite, $Ca_2(Fe,Al)_2O_5$, which, depending on aluminum content and the rate at which the sample is cooled, may be largely superparamagnetic; and metallic iron. There is no compelling evidence that iron-bearing silicates or glasses are present in measurable amounts in the slags investigated in this study. Although this conclusion appears to be at variance with the results of Ramous et al. [3], it must be remembered that the slags of their work covered a considerably wider range of composition than those of this study. With silica-rich slags, phases such as $CaFeSiO_4$ are expected to be formed. Indeed, this phase has been observed in a slag not included in this study [7].

As long as the compositions are similar, and despite differences in interpretation, there do not appear to be any really significant differences among the Mössbauer spectra of slags from different steelmaking furnaces. The spectra of the BOF slags investigated in this study are closely similar to the Q-BOP slags and, as far as we can tell, to the LD (BOF) slag investigated by Gohy et al. [2] and the electric furnace slags of intermediate silica contents investigated by Ramous et al. [3]. This is not too surprising as the composition and function of the slags are generally similar for each kind of steelmaking furnace.

This investigation of the forms of iron in Q-BOP slags had its origin in the need to develop a rapid method for the determination of the metallic iron contents of slags. As we have shown, the Mössbauer technique was instrumental in establishing that a bias was present in the traditional chemical method for this determination. Consequently, the Mössbauer determination became the benchmark to which other methods were compared. From this comparison, it was concluded that a saturation magnetization method was best suited in terms of speed, convenience, and cost for metallic iron determinations for use in computer-control models of Q-BOP operation at steelmaking facilities.

Although much has been learned from this and other Mössbauer studies of steelmaking slags, there is still considerable work that could be done in relating the forms-of-iron data to thermodynamic analysis of element partitioning between slag and liquid steel. In particular, given the importance of the FeO concentration on the partitioning of certain elements, such relationships should be reexamined with data from Mössbauer analysis.

ACKNOWLEDGMENTS

We would like to acknowledge the contributions of a number of our former U.S. Steel Colleagues, who have supplied data and information relating to this study. These colleagues include J. D. Selvaggio and H. A. Barnett (chemical data), R. P. Bacco (X-ray diffraction), C. E. Spangler and F. C. Schwerer (magnetic data), and also E. T. Turkdogan and the late M. J. Papinchak, who provided much theoretical and practical insight into Q-BOP operation.

REFERENCES

1. E. T. Turkdogan, Physical Chemistry of High Temperature Technology, Chapter 9, 342, Academic Press, New York, 1980; Physicochemical Properties of Molten Slags and Glasses, Chapter 8, 390, The Metals Society, London, 1983.

2. C. Gohy, A. Gerard, and F. Grandjean, Mater. Res. Bull. 18, 285 (1983).

3. E. Ramous, G. Principi, M. Magrini, and A. Taziani, Met. Trans. 12B, 403 (1981).

4. C. Gohy, A. Gerard, and F. Grandjean, Phys. Stat. Solidi 74A, 583 (1982).

5. R. W. Grant, J. Chem. Phys. 51, 1156 (1969).

6. M. Eibschutz, U. Ganiel, and S. Shtrikman, J. Mater. Sci. 4, 574 (1969).

7. F. E. Huggins, unpublished data on slags and standards.

8. E. M. Levin, C. R. Robbins, and H. F. McMurdie. Phase Diagrams for Ceramists (3rd edition), The American Ceramic Society, Columbus, Ohio (1974).

9. W. C. Allen and R. B. Snow, J. Am. Ceramic Soc. 38, 268 (1955).

10. F. E. Huggins and G. P. Huffman, J. Appl. Phys. 55, 1404 (1984).

SLAG ATTACK OF FIREBRICK IN A TIN SMELTING FURNACE

Pamela McNamara*, John D. Cashion and Mary S. J. Gani

Department of Physics, and the Department of Materials Engineering
Monash University, Clayton
Victoria, 3168, Australia

ABSTRACT

A combined tin-119 and iron-57 Mössbauer study has been made of the slag and an aluminosilicate hearthbrick taken from a reverberatory tin furnace. Three distinct tin(II) glassy phases were observed, one of which also contains iron(II) and appears to be the phase responsible for the attack on the furnace hearth. Some SnO_2 was also found to have penetrated 4cm past the reaction zone in the brick, presumably by vapor transport.

INTRODUCTION

The extraction of tin from its ores is accomplished by a variety of methods including reduction of the ores in reverberatory furnaces. Reviews of modern tin smelting techniques have been given by Wright [1] and Floyd [2]. Tin smelting is complicated by the presence of iron-containing minerals in the ore. The iron, depending on the furnace conditions and slag composition, may either enter the slag, or be reduced to metallic iron which alloys with the tin. A two stage smelting process is normally employed. The first is a 'concentrate smelt' at ca. 1200°C to produce metallic tin. The slag composition is adjusted so that it contains the iron oxide impurities, but this results in an appreciable solubility of the tin oxides in the slag. The second is a "slag smelt" at ca. 1300°C, in which the tin is extracted from the slag produced in the concentrate smelt by increasing the basic content of the slag. The metal produced from the slag smelt (hardhead) contains roughly equal amounts of iron and tin. Other impurities present in the tin ores, such as SiO_2, Al_2O_3, TiO_2, and MnO_2, are not reduced during the smelting and become incorporated into the slag.

During the tin smelting process, the refractory linings are subject to attack by the metal and slag liquid phases and also by gaseous phases such as SnO and ZnO. This attack not only leads to the degradation of the refractories, but also results in a considerable amount of the tin bearing phases being absorbed by the refractories. The used refractories contain approximately 16 weight percent of tin, making recovery economically viable when the

*Present address: Telecom Research Laboratories, 770 Blackburn Road, Clayton, Victoria 3168, Australia.

furnace hearth is ultimately replaced. Thus the selection of refractories for use in tin smelting reverberatory furnaces is governed not only by the cost of the refractories and their susceptibility to attack, but also by the ease of extraction of the tin contained in the used linings.

It is because of these considerations that aluminosilicate linings, containing ca, 42 weight percent Al_2O_3, are normally used in the hearth, despite the observations reported by Ignatova et al [3] that the more expensive sillimanite, $Al_2O_3 \cdot SiO_2$, or corundum-mullite, Al_2O_3-$3Al_2O_3 \cdot 2SiO_2$, refractories show increased resistance to attack by tin and its oxides. Basic refractories, such as chrome-magnesite are highly resistant to slag attack, and are thus used at the slag line, but these present problems in the extraction of tin from the bricks after use.

The primary aim of our investigation was to identify the tin bearing phase which had penetrated the aluminosilicate hearthbrick, because this would facilitate the subsequent extraction of tin from the used refractories. Because tin and iron are the two most common Mössbauer isotopes, the problem seemed ideally suited for Mössbauer effect analysis. Other techniques had been used to investigate this system, including x-ray diffraction, optical microscopy on thin sections, and scanning electron microscopy with energy dispersive analysis; but in order to interpret the results from these techniques, the information obtained from Mössbauer studies was needed. For example, the large number of phases present and the poor crystallinity of at least some of the phases, made analysis of the x-ray diffraction pattern difficult and the results ambiguous. Scanning electron microscopy, with back-scattered electrons to give atomic number contrast, combined with x-ray energy dispersive analysis, gave the elements present in each phase, but gave no information about the crystallinity of the phase or the oxidation state of the elements present in each phase. In order to identify the tin and iron bearing phases within the attacked portion of the brick, it was also necessary to identify the phases present in the unattacked portion of the brick and in the slag which was in contact with the brick.

The hearthbrick investigated had been removed from a furnace in which the lining had suffered premature failure. Thus analysis of the attack on the brick would not only aid in the subsequent tin extraction, but also provide information on the probable cause of the expensive premature failure.

SAMPLE DESCRIPTION

Because the furnace hearth had failed prematurely, the average penetration depth was not large. The unpenetrated brick had a light creamy brown matrix mixed with small particle hard chamotte grains about 1-2mm in diameter. The particles were variously cream, white, lilac with a white rim (in cross-section), dark grey and red-brown in color.

X-ray diffraction revealed that the crystalline phases present were mullite, $3Al_2O_3 \cdot 2SiO_2$, and cristobalite, SiO_2. The brick surface which had been on the interior of the furnace bore a strongly adhering crust of black, sintered slag about 5mm thick containing β-tin either as many bright, silvery, metallic lumps with sizes from 0.5 to 20mm or sometimes as a thin ca. one mm layer on the surface furthest from the brick. The slag was generally brittle and porous, with a few quite large cavities, some of which contained dark red crystallites. Red iron oxide could be observed in some places on the surface of the slag, overlying the tin layer where present.

The brick immediately under the slag crust was darkened to varying depths up to 25mm due to slag penetration. It appeared that the matrix had become black, while the aggregate particles were unaffected and stood out

Table I. Chemical Analyses[a] of an Unattacked Cold Face of the Hearthbrick and of the Slag

Component	Brick	Slag
SiO_2	67.0	66.8
Al_2O_3	33.0	33.2
Fe_2O_3[b]	0.7	29.6
TiO_2	1.2	16.3
K_2O	0.4	1.1
Na_2O	0.3	0.5
MgO	0.4	3.3
MnO_2	0.1	0.7
SnO_2[B]	-	54.0
CaO	-	23.3

[a]The analyses are given in mole percent as Niggli values relative to the total weight of SiO_2 and Al_2O_3. [b]The iron and tin concentrations in the slag are the total iron and tin present as metal and oxides.

against the darkened background. The black region had a sharp edge and did not fade gradually into the adjacent brick.

The chemical analyses for the slag and the unattacked brick are given in Table I. The results are presented as Niggli values relative to the sum of Al_2O_3 and SiO_2 because this enables easier comparison of the differences between the two samples.

Absorbers for the Mössbauer studies were prepared from material cut from the three distinct regions, the black slag crust, the discolored penetrated brick, and the apparently unpenetrated brick about 4cm from the penetrated zone. Each piece was initially crushed and then powdered with an agate mortar and pestle. Crushing the slag sample left large pieces of metallic tin up to 5mm in diameter and these were picked out before putting the sample into the mortar. The approximate optimum thickness of each sample was determined for both iron-57 and tin-119 gamma rays and these quantities placed in perspex holders.

OPTICAL MICROSCOPE EXAMINATION

Examination of a cross-section of the brick both before and after polishing revealed, as might be expected, an enormous variety of structures, both glassy and crystalline. A distinct reaction zone, ca. 50μm wide, separated the slag and aluminosilicate grains. The penetrated brick matrix was obviously harder than either the slag or the unpenetrated matrix as evidenced by its polishing characteristics and seemed to be of similar hardness to the aggregate particles.

Small crystallites, about 0.1 mm in diameter, yellow in color and similar to pyrite in appearance could be clearly seen. The black region of the penetrated brick contained rectangular grey crystals interspersed with white needle-like crystals of mullite.

TIN-119 MÖSSBAUER SPECTRA

Because the tin-119 Mössbauer spectra are considerably easier to understand we shall describe them first. At room temperature, the spectra of the three samples are shown in Figure 1. The slag sample (Figure 1c) shows a broadened line at zero velocity, with what appears to be a doublet with peaks at 2.0 and 4.0mm/s. The relative proportions of these components are changed in the spectrum of the penetrated brick (Figure 1b) and a more careful examination shows that the peaks of the doublet are shifted to higher velocity by approximately 0.2mm/s. Surprisingly, some tin had penetrated into the brick sample cut from well past the reaction zone (Figure 1a) and this showed the same broadened line at zero velocity as the other two samples, with possibly a very weak contribution from the other doublet. The parameters for the computer fits to these spectra are given in Table II and an interpretations will be given after we have discussed all the spectra.

In an attempt to check for association of tin with the iron in the sample, spectra of the slag and penetrated brick samples were run at room temperature in a magnetic field of 1.7T. The spectra are shown in Figure 2 and the fitted parameters are also given in Table II. The parameters for the fits are in good agreement with the value for the zero field spectra showing that the essentials of the system have been taken into account. However the interesting feature is the enhancement of the applied field shown by the phase producing doublet D3.

As a further check, spectra of both the penetrated brick and the slag were run at 4.2K at which temperature the iron is magnetically ordered as we shall see in the next section. An appreciable line broadening was observed in the spectrum from the penetrated brick, again showing the association of doublet D3 with the iron. A magnetic separation using a bar magnet was performed on the slag samples and the room temperature spectrum of this fraction showed an enhanced peak near zero velocity compared to the slag spectrum. The peak was now asymmetric due the existence of an additional line at +0.2mm/s.

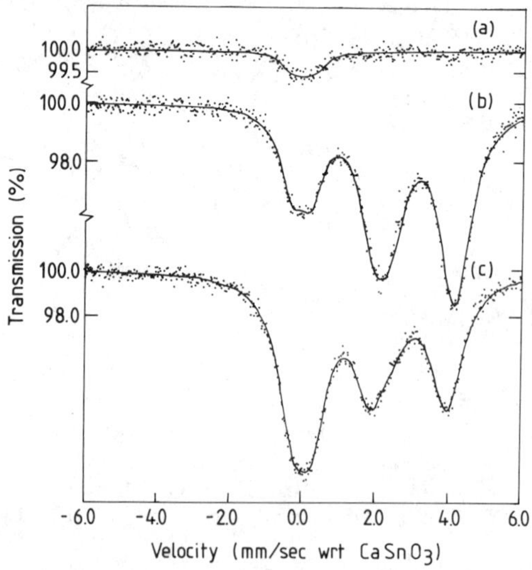

Figure 1. Room temperature tin-119 spectra of samples of (a) hearthbrick, (b) penetrated brick, and (c) slag.

Table II. Parameters for the Tin-119 Mössbauer Spectra of the Samples Under the Conditions Stated[a]

Sample	T,K	H_{app},T	Doublet D1					Doublet D2					Doublet D3				
			δ	ΔE_Q	Γ	B	%A	δ	ΔE_Q	Γ	B	%A	δ	ΔE_Q	Γ	B	%A
Brick	RT	0	0.00	0.52	0.45	-	100										
Penetrated	RT	0	-0.03	0.55	0.41	-	24	2.99	2.10	0.43	-	33	3.33	1.88	0.49	-	43
Brick	RT	1.7	0.00	0.34	0.60	1.91	26	2.64	2.25	0.64	1.50	40	3.22	1.76	0.43	2.12	34
	4.2	0	0.02	0.47	0.52	-	12	3.07	2.05	0.60	-	61	3.54	1.96	0.51	-	27
Slag	RT	0	0.00	0.55	0.51	-	49	2.89	2.02	0.53	-	41	3.20	1.56	0.46	-	10
	RT	1.7	-0.01	0.29	0.70	1.79	56	2.74	1.99	0.66	1.72	38	3.31	1.61	0.37	2.02	6
	4.2	0	0.06	0.31	0.72	-	27	2.90	1.97	0.60	-	60	3.39	1.86	0.41	-	13

[a] All data in mm/s with the isomer shift relative to $CaSnO_3$ and the fields in tesla.

IRON-57 MÖSSBAUER SPECTRA

The room temperature spectrum of the brick sample, given in Figure 3, shows principally a doublet flanked by a weak magnetic component which seems to be partly relaxed. These constituents are also expected to be weakly present in the spectrum of the penetrated brick, the central part of which is shown in Figure 4a. Although basically a doublet, the two lines clearly have very different line shapes and intensities and were extremely difficult to fit with a sum of Lorentzian lineshapes. The spectrum of the slag (Figure 4b) is very similar although the high velocity component is more rounded.

At 77 K the spectra had sharpened considerably (Figure 5) and the splitting had also widened leaving a line at 1mm/s clearly visible in Figure 5b. The transformation in the spectrum was more complete for the slag sample which also showed the inner peaks of a magnetic sextet. The parameters for these spectra are given in Table III.

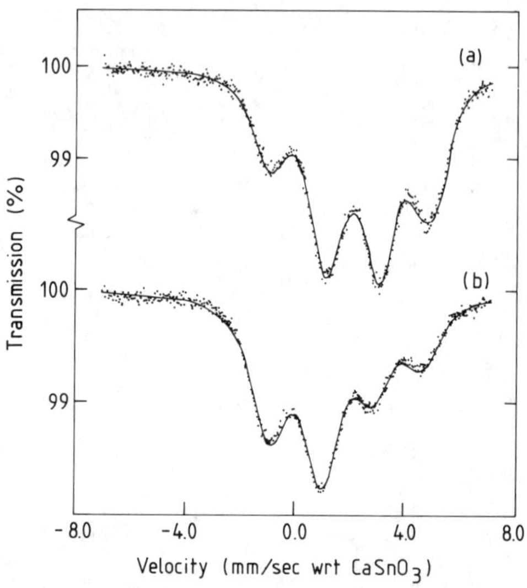

Figure 2. Room temperature tin-119 spectra taken in a transverse magnetic field of 1.7T for (a) penetrated brick and (b) slag samples.

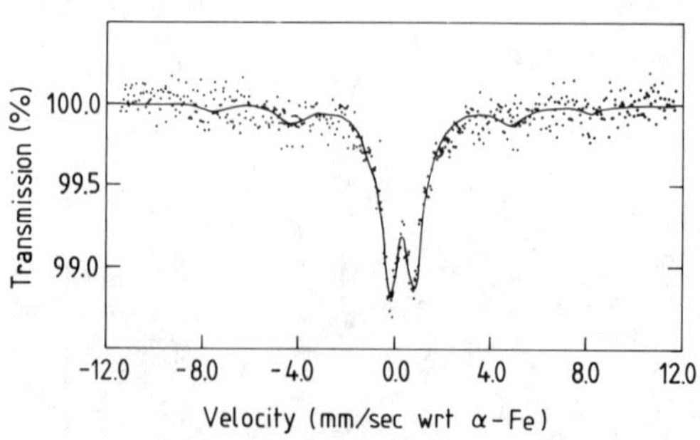

Figure 3. Room temperature iron-57 spectrum of the hearthbrick sample.

Table III. Parameters for the fits to the iron-57 Mössbauer Spectra[a]

Sample	T,K	Doublets			Sextets			
		δ	ΔE_Q	%A	δ	ε	B	%A
Brick	Room	0.31	0.98	51	0.25	0.00	49.7	41
		0.95	2.28	8				
Penetrated Brick	77	0.26	0.80	15				
		1.15	1.81	36				
		1.21	2.45	39				
		1.22	3.04	10				
	4.2	1.47	1.76	6	0.93	0.11	21.0	52
					0.33	-0.06	47.9	42
Slag	77	0.34	1.12	9	0.48	0	48.3	14
		1.13	1.80	17				
		1.18	2.46	40				
		1.07	2.98	21				
	4.2	1.62	1.90	7	0.93	0.18	20.9	54
					0.51	-0.10	49.2	39

[a]All data in mm/s with the isomer shift relative to α-iron and the field in tesla. These parameters should be regarded as approximations for the reasons explained in the text.

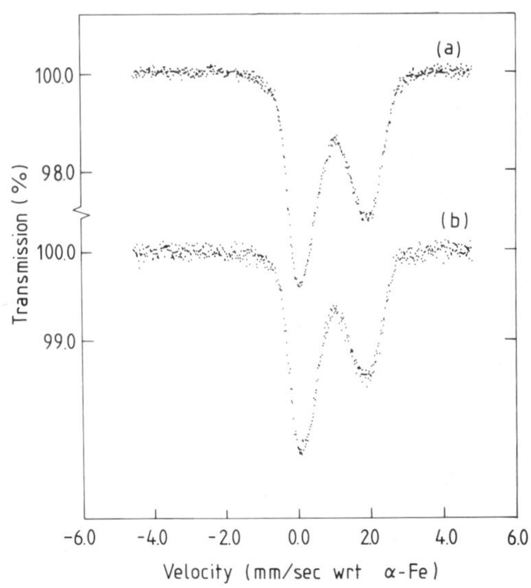

Figure 4. Room temperature iron-57 spectra of the (a) penetrated brick and (b) slag samples.

To complete the picture of changing temperature, 4.2 K spectra of the two samples (Figure 6) showed a much more drastic change indicating that the samples had passed through a magnetic ordering transition. Spectra of both samples were again very similar to each other, and were well approximated by the sum of a trivalent sextet, a divalent sextet, and a doublet. The application of a magnetic field of 1.7 T at room temperature resulted in a sharpening of the right hand line of the doublet in Figure 4a and its intensity for the penetrated brick sample increased from 68 to 80 percent of the height of the left hand peak.

The magnetically separated fraction of the slag described in the previous sections was also run at room temperature. One aim of this was to detect any small concentrations of pure iron phases such as magnetite, but,

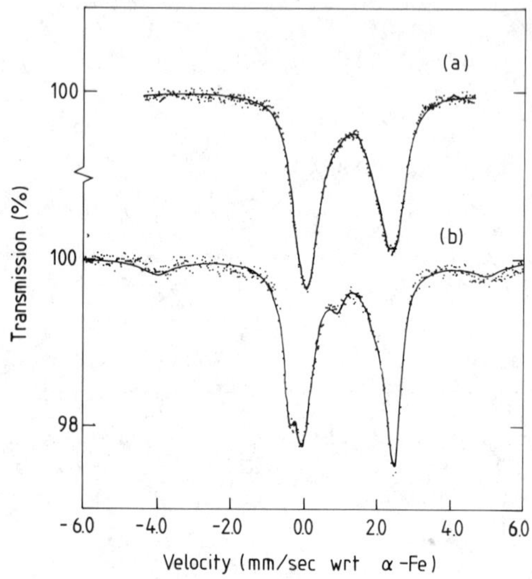

Figure 5. Iron-57 spectra obtained at 77K of the (a) penetrated brick and (b) slag samples

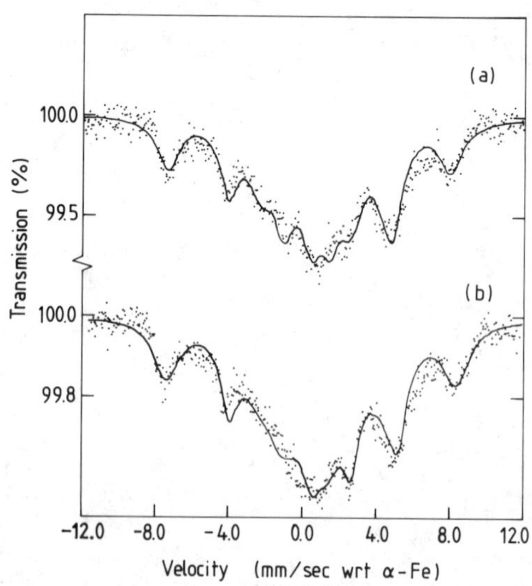

Figure 6. Iron-57 spectra obtained at 4.2K of the (a) penetrated brick and (b) slag samples.

although an enhanced magnetic sextet was observed, its parameters (Table III) did not correspond to those of magnetite and in fact most of the intensity was still contained in a doublet identical to that in Figure 4b.

INTERPRETATION

The broadened line at zero velocity present in all three samples in Figure 1 can be ascribed to SnO_2 and the parameters agree well with the literature values of 0.00 and 0.54mm/s for the isomer shift and quadrupole splitting respectively. The spectrum of β-tin is a single line at 2.56mm/s and its absence in the slag spectrum was a little surprising. This shows that the metallic tin coagulates very rapidly so that only a small fraction of the tin can be present as particles smaller than could be picked out with tweezers and the naked eye.

The other easily identifiable subspectrum is the iron doublet in the brick (Figure 5) which is due to mullite whose typical isomer shift is 0.3mm/s and quadrupole splitting is 0.8mm/s. Mullite sometimes contains divalent iron as well and it may be weakly present in this sample.

The remaining components of the spectra we believe are entirely or almost entirely due to glassy phases present in the slag. The doublet D2 observed in the tin spectra is very similar to that observed by Mitrofanov and Sidorov [4] in a pure SiO_2-SnO_2 glass which had an isomer shift of 2.8mm/s and a quadrupole splitting of 2.0mm/s.

The other tin doublet, D3, we ascribe also to a glassy phase, but composed of SiO_2-SnO_2-FeO. The intensity of this phase is much larger in the penetrated brick than in the slag and consequently this is probably the phase which is responsible for the initial destructive penetration of the brick.

Mitrofanov and Sidorov [4] observed a broadened line centered at 0.02mm/s from their SiO_2-SnO_2 glass in which the tin(IV) atom must be in a similar configuration to that in SnO_2. This phase may also be present in our samples as well, but would be inseparable from the SnO_2 contribution. This likelihood is enhanced by the observation of the asymmetric line in the magnetically separated sample. This asymmetry is due to an extra line at 0.2mm/s, with its isomer shift being due to its association with iron as described for doublet D3. Because this line is attributed to a modification of the phase which produces the zero velocity line, we regard this as strong evidence for the existence of the SnO_2-like phase.

The dominant iron spectrum is also due to this SiO_2-SnO_2-FeO glassy phase. Hinckley et al. [5] published a spectrum almost identical to our Figure 4 in their study of a glassy phase from a power station furnace. Iron in a glassy rad-waste matrix also gives this characteristic asymmetric doublet. The parameters of the two doublets fitted to our 77K spectra are similar to those of iron (II) in octahedrally coordinated silicate phases. Such configurations have been observed by Ramous et al [6] in a basic steel-making slag with isomer shifts of 0.7 to 1.1mm/s, and quadrupole splitting of 2.1 to 2.6mm/s, and two similar pairs of doublets were found recently by Hollitt et al [7] in a PbO-FeO-SiO_2 synthetic slag with isomer shifts of 1.17 and 1.1mm/s and quadrupole splitting of 2.9 and 2.7mm/s respectively.

The unusual behavior of the iron(II) spectrum with temperature is due to a partial population of an excited crystal field level at room temperature. Because the electric field gradient is proportional to $\langle 3J_z^2 - J(J+1) \rangle$, this gives rise to a temperature dependence of the splitting, with additional broadening being induced because there will be small variations in the crystal field levels of different iron atoms due to local differences in the glass structure. At 77K this upper level has become almost completely depopulated, resulting in a sharpening of the spectrum. However some vestige of it still remains in the curved inner side of the high velocity line. Application of the magnetic field also reduced the spread of values of $\langle J_z^2 \rangle$., resulting in a sharpening of the spectrum.

Based on the work of Mitrofanov and Sidorov, [4] the most likely configuration for tin atoms in the glassy phase represented by D2 is a fourfold coordination through oxygen atoms to silicon atoms. The large quadrupole splitting suggests that the coordination is in a planar arrangement similar to SnO rather than a tetrahedral one.

The doublet D3 is a modification of this configuration to incorporate the iron and thus has the arrangement $(Si\text{-}O\text{-})_n\text{-}Sn\text{-}(\text{-}O\text{-}Fe)_{4-n}$ with the increased isomer shift and hence increased s-electron density being caused by the electrons donated by the iron(II) ions. These would be expected to make a larger contribution to the electron density at the tin atom than the covalent silicon atoms. Mitrofanov and Sidorov [4] did not see any change in either the isomer shift or quadrupole splitting with changing relative proportions of the constituents, so there is no information which can be obtained about this. However they did observe changes on adding different alkali metal ions, which is consistent with interpreting the change in our spectrum as being due to the incorporation of iron. Computer fitting of the doublets in Figure 1b,c indicated that only two doublets were present rather than the five possibilities suggested by the above co-ordination arrangement, which would be expected to give a smooth broadening of the lines. This means that the parameter n, giving the relative number of silicate and ferrate ligands, is probably fixed but we cannot at present determine its value. At 4.2K the iron spectra (Figure 6) consist of the sum of broadened iron(III) sextet and at least one, but more probably a distribution of iron(II) magnetic sextets. The doublet remaining in the centre of the spectrum probably originates from part of the sample which is more dilute in iron so that the ordering temperature has dropped below 4.2K.

The iron(III) magnetic splittings are harder to interpret but the very large hyperfine fields make it likely that at least some of these are also associated with a glassy phase. Because this component must be completely contained in the left hand peak in the room temperature and 77K spectra, these two constraints eliminate most possible compounds which can be produced from the elements in the chemical analysis (Table 1). The only likely contenders would seem to be one of the Fe-Ti-O compounds, either ulvöspinel ferrian, $3FeO{\bullet}Fe_2O_3{\bullet}TiO_2$, or pseudobrookite, $Fe_2O_3{\bullet}TiO_2$. The former compound does not seem to have been studied by Mössbauer spectroscopy, but its x-ray diffraction pattern is in agreement with a spinel type phase observed in this analysis. At room temperature $Fe_2O_3{\bullet}TiO_2$ has an isomer shift of 0.3mm/s and a quadrupole splitting of 0.7mm/s [8,9] with a hyperfine field at 4.2K of 45T. The linewidths in the ordered phase suggest that more than one trivalent species is present.

Although the broadened line at zero velocity in the tin spectra of the slag and penetrated brick may be due to a glassy phase as well as to SnO_2, that observed in the unpenetrated brick, well past the reaction zone of the slag penetration should be solely due to SnO_2. This is created by vapor phase penetration by SnO which would normally disproportionate into SnO_2 and tin on cooling. However since no β-tin was observed in this sample, we assume that sufficient oxygen must be available to oxidize the metal in this region.

CONCLUSIONS

The iron and tin Mössbauer spectra combine to form a consistent picture of the constituents of the slag, penetrated brick, and unattacked brick samples. The major tin phases are two tin(II) silicate glasses, tin metal, and SnO_2 with a probable tin(IV) silicate glass which has very similar parameters to SnO_2. The principal iron phase is an iron(II) silicate glass which appears identical in both the slag and the penetrated brick.

We conclude that the attack on the firebrick was initiated by a mixed tin(II) - iron(II) silicate glassy phase whose concentration as measured by the tin spectra was much larger in the penetrated brick than in the slag. Some SnO_2 was observed in the brick beyond the reaction zone but this vapor phase penetration is not believed to have contributed to the failure.

ACKNOWLEDGMENTS

We are grateful to L. J. Brown, J. R. Fraser and A. C. McGrath for technical assistance during this work. The project was supported by the Australian Research Grants Scheme.

REFERENCES

1. P. A. Wright,, "Extractive Metallurgy of Tin", Elsevier, New York, Chapter 7, (1982).

2. J. M. Floyd, 1980, in "Lead-Zinc-Tin" '80, Proc. World Symp. Metall. Environ. Control, Las Vegas, ed. Cigan, J. M., Mackay, T. S. and O'Keefe, T. J., Metall. Soc. AIME, Warrendale, Pa, P. 508, (1980).

3. T. S. Ignatova, T. N. Kudryavtseva, T. I. Nazarova, L. P. Sudakova, A. S. Makhnavetskii, M. I. Novomlinova, B. V. Tarasov, and N. A. Yavtukhovich, Ogneupory 9, 23 (1968), Eng. Transl., Refractories 9, 584, (1968).

4. K. P. Mitrofanov, and T. A. Sidorov, Sov. Phys. Solid State, 9, 693, (1967).

5. C. C. Hinckley, G. V. Smith, H. Twardowska, M. Saporoschenko, R. H. Shiley, and R. A. Griffin, Fuel, 59, 161 ((1980).

6. E. Ramous, G. Principi, M. Magrini,and A. Tiziani, Metall. Trans. B 12B, 403 (1981).

7. M. J. Hollitt, J. D. Cashion, and L. J. Brown, Proc. Int. Symp. on Metall. Slags and Fluxes, Reno 1984, Metall. Soc. AIME, Warrendale, Pa, to be published (1985).

8. T. C. Gibb, N. N. Greenwood, and W. Twist, J. Inorg. Nucl. Chem. 31, 947 (1969).

9. T. Birchall, N. N. Greenwood, and A. F. Reid, J. Chem. Soc. A, 2382 (1969).

MÖSSBAUER EFFECT STUDY OF DECREPTATION PHENOMENA IN IRON ORES

Joice Terra, Elisa B. Saitovitch, Jacques Danon
and Nilce G. Souza

Centro Brasileiro de Pesquisas Fisicas
150 rua Xavier Sigaud
22290 Rio de Janeiro, Brasil

INTRODUCTION

Hematite, α-Fe_2O_3, was one of the first iron compounds studied by Mössbauer spectroscopy and the study of its properties as well as its participation in the formation of the other minerals have been object of several such investigations. Brazil has one of the world's largest mineral deposits of hematitic iron among which, the compact hematite, is the subject of the present investigation.

Compact hematite corresponds to the ore whose iron content is greater than 66 percent. In general, it forms the summits of mountains because of its resistance to erosion. From the economic point of view, it presents characteristics which makes its direct utilization troublesome in blast furnaces. This occurs because this material suffers considerable granulometric degradation during the heating required for reduction. Decreptation is a phenomenon that occurs in the temperature range of 573 to 773K and it is found to be practically independent of the heating rate and is characterized by its non-repetitive nature [1-2].

The problem occurring from decreptation has been resolved by using synterized and pelletized ore, a procedure which obviously raises the price of the final product. Decreptation phenomenon has been studied but not to the extent that the problem requires.

Two mechanisms have been proposed to explain the decreptation phenomena [3]. The first of relates the decreptation mechanism to the internal pressure of water vapor released by the decomposition of hydrated phases, mainly goethite, α-FeOOH. The other mechanism is of a physical nature and assumes that the decreptation is a consequence, in the case of nonhydrated ore of the dilatation anisotropy in hematites. Such a characteristic would produce stress and consequently fracture, because hematite has a large value for Young's modulus. Although neither mechanism has support which can confirm it, the first is currently the accepted mechanism of decreptation. This is surprising because Brazilian iron ore is normally considered to be nonhydrated [4].

By using mainly Mössbauer spectroscopy, we have tried to find some characteristics which may be related to the decreptation process. In this work we have studied and compared samples of compact decreptate hematite, HD, non-decreptable hematite, HND, as well as the compact decreptate after undergoing decreptation, HAD. Mössbauer spectra of these three types of samples were recorded as a function of the temperature, from 295 to 673K and the results have been analyzed in terms of the temperature dependence of the isomer shift, δ, quadrupole interaction, ε, effective magnetic field, H_{eff}, and the linewidth, Γ. Further, we have studied the Morin transition [5] in the temperature range between 177 and 300K to observe the possible influence of impurities or defects in the crystals of hematite [6,7].

The results by Mössbauer spectroscopy have been compared with those obtained by other techniques such as differential thermal analysis, DTA, gravimetric thermal analysis, GTA, differential scanning calorimetry, DSC, x-ray powder diffraction, and infrared spectroscopy.

EXPERIMENTAL

The hematite samples were obtained from the National Steel Company, and are from the Congonhas site, in the south of the iron belt, where the iron content varies from 67.3 to 68.0 percent with 2 percent of Al_2O_3. Their classification in relation to the decreptation phenomenon, that is decreptate and non-decreptate hematite, was done by the Research Center of the National Steel Company.

When one compares the two kinds of samples one observes obvious differences. While the decreptate hematite is hard, dark gray, and has metallic brightness, whereas non-decreptate hematite shows porosity and yellow incrustations due to goethite.

In general, we used 30mg of sample with an average particle size of 0.149mm. This gave Mössbauer effect absorbers containing $12mg/cm^2$ of natural iron. Spectra were obtained on a conventional Mössbauer spectrometer which utilized at room temperature a 15mCi cobalt-57 on rhodium source. The isomer shift data presented in this paper are referred to room temperature metallic iron. The Mössbauer spectra at high temperature were obtained with a Ricor furnace and at low temperatures with a cold finger cryostat.

For infrared measurements the samples was prepared by mixing with KBr, in a proportion of 1mg of compact hematite to 100mg of KBr. The DTA and GTA experiments were performed under nitrogen with Al_2O_3 as a thermal reference standard which does not show any thermal variation from 300 to 1223K. The heating velocity was 10K/min. The granularity of the sample was of 0.149mm. The DSC measurements covered the 323 to 973K temperature range with a heating rate of ca. 10K/min.

RESULTS

Some differences were observed between compact decreptate hematite and non-decreptate hematite by spectrographic semi-quantitative analysis. Compact decreptate hematite contains small amounts of titanium and manganese. It has been found that the different concentration of titanium and manganese can be responsible for the difference observed in the Néel temperature for two types of hematite [8]. X-ray powder diffraction shows that the samples contain essentially pure Fe_2O_3, SiO_2, Al_2O_3, and goethite are present only in non-decreptate hematite to the extent of about two percent. The infrared spectra of compact decreptate hematite showed only the absorption bands corresponding to pure Fe_2O_3 [9]. Non-decreptate hematite showed, in addition, bands corresponding to goethite.

Differential thermo analysis shows the endothermic peak in the non-decreptate hematite sample at ca. 573K as is illustrated in Figure 1a. This peak is associated with the removal of the hydroxyl groups found in goethite and corresponds, in the GTA, to a weight loss of 2mg (Figure 1b). The compact decreptate hematite sample does not show any significant variation in the DTA or GTA. The endothermic peak associated with the magnetic phase transition at the Néel temperature is the only peak observed in the DTA of α-Fe_2O_3 [8]. The Néel temperatures of 933K for compact decreptate hematite and 988K for non-decreptate hematite revealed differences between the two types of hematites. This can be explained by the different concentrations of titanium and manganese in the two samples [8]. The differential scanning calorimetric study has revealed one endothermic peak related to the decomposition (Figure 1c) of non-decreptate hematite.

MÖSSBAUER SPECTROSCOPY AT HIGH TEMPERATURE

The Mössbauer hyperfine parameters obtained at room temperature revealed no differences between the two types of hematites and are in good agreement with those reported in the literature. Mössbauer spectra of compact decreptate hematite, non-decreptate hematite, and compact decreptate hematite after undergoing decreptation have been recorded in the temperature between 300 to 673K. The variation of the quadrupole splitting and magnetic hyperfine interaction with temperature was found to be the same for all three types of hematites.

The absorption area of a Mössbauer spectrum can be written as

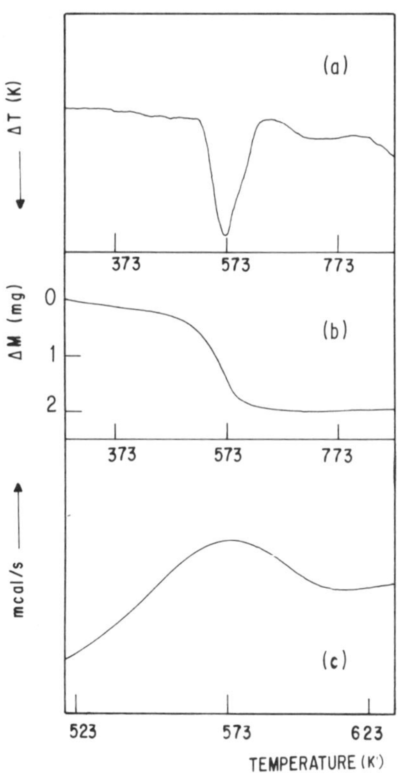

Figure 1. Thermal analysis of the non-decreptate hematite: (a) DTA, (b) GTA, (c) DSC

$$S = \sum_i A(i)\Gamma(i), \tag{1}$$

where A(i) is the intensity and Γ(i) is the linewidth of a Mössbauer line. The area is important because it is proportional to the f factor, the proportionality constant being dependent upon such external variables as the absorber thickness, background effects, etc. Because the temperature was the only parameter varied during the experiments, we can write

$$\frac{d\ell nS}{dT} = \frac{d\ell nf}{dT} \tag{2}$$

where T is the temperature. Within the Debye model of a solid the temperature dependence of ℓnS is linear for temperatures greater than $0.5\theta_D$. It is rather easy, therefore, in experiments which satisfy this condition, to identify any deviation from linearity. After fitting the spectra, we observed a discontinuity in ℓnS as a function of temperature in the compact decreptate hematite sample. The deviation occurred at ca. 582K with a width of ca. 62K, or more precisely between 561 and 662K as is illustrated in Figure 2a. In the non-decreptate hematite and compact decreptate hematite after undergoing decreptation samples no meaningful drift from the linear behavior of greater than the experimental error has been observed (see Figures 2b and 2c).

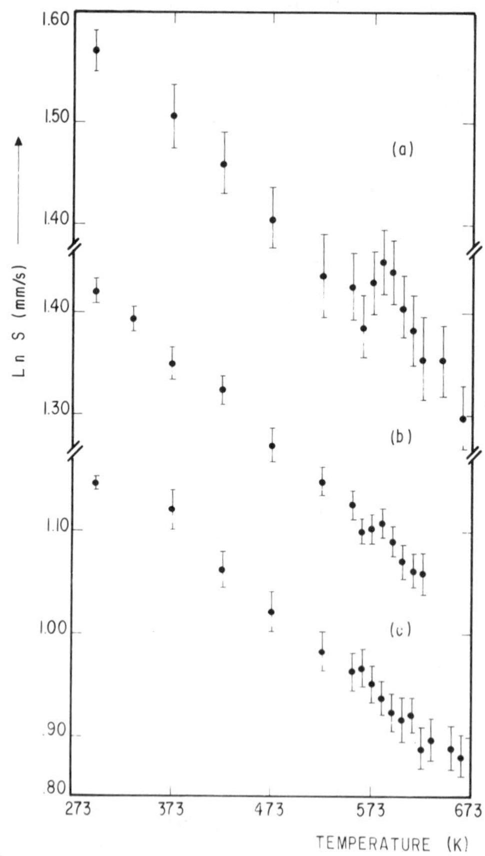

Figure 2. Logarithm dependence of the absorption area as a function of the temperature for (a) compact decreptate hematite, (b)compact decreptate hematite after undergoing decreption, and (c) non-decreptate hematite.

The linewidth also showed the same behavior described above, and for compact decreptate hematite, we have observed a discontinuity in the linewidth over the same temperature range as the deviation in ℓnS. Again, within experimental accuracy, no deviation was observed for the non-decreptate hematite and compact decreptate hematite after undergoing decreptation samples as shown in Figure 3.

MÖSSBAUER SPECTROSCOPY AT LOW TEMPERATURE. THE MORIN TRANSITION

The magnetic properties of hematite have been studied intensively by Mössbauer spectroscopy. Hematite has a rhombohedral structure and is isomorphous with corundum. Its unit cell contains four iron(III) ions located along the diagonal of the rhombohedral [111] axis. Between the Néel and Morin temperatures, hematite is a non-collinear antiferromagnet and

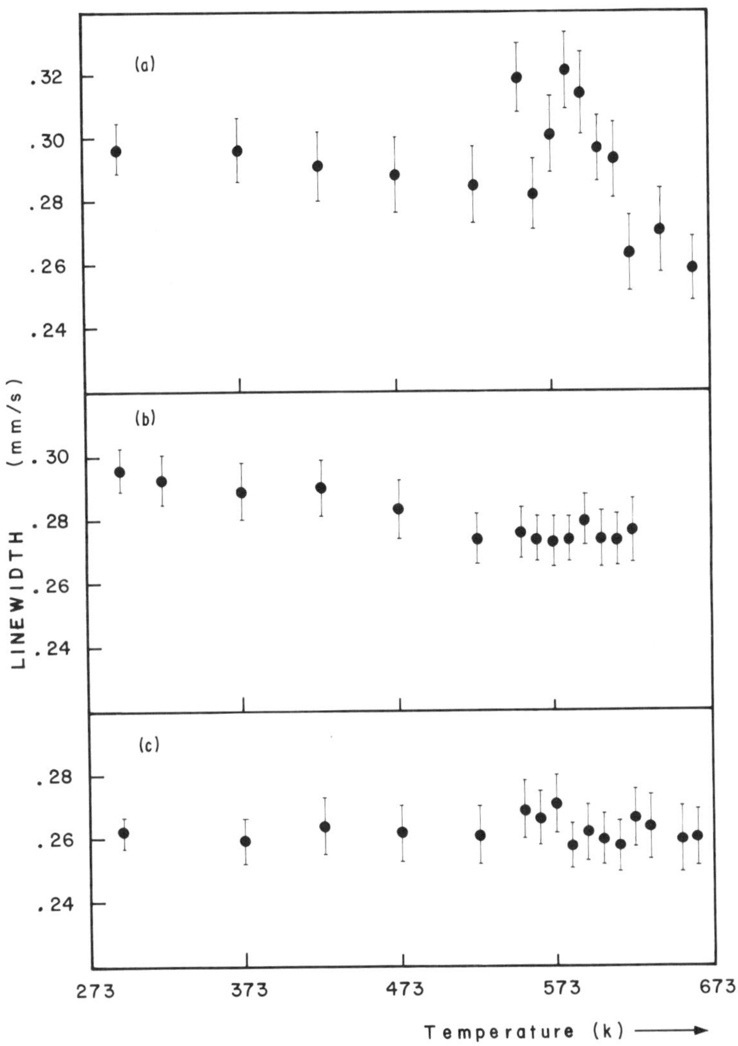

Figure 3. Lindwidth dependence as a function of the temperature for (a) compact decreptate hematite, (b) compact decreptate hematite after undergoing decreptation, and (c) nondecreptate hematite.

exhibits weak ferromagnetism. In this state the spins of the iron(III) ions lie close to the basal (111) plane, with a small magnetic moment perpendicular to the trigonal [111] axis. Below the Morin temperature the weak ferromagnetism vanishes and hematite becomes a pure antiferromagnetic with the local magnetization directed along the [111] axis. The phase transition at the Morin temperature, characterized by a rotation of the magnetic moments of the iron(III) ions by 90°, takes place spontaneously. Although several theoretical and experimental investigations have been carried out on the Morin transition, there are still some open questions.

The Mössbauer effect hyperfine parameters are strongly affected by the Morin transition. The Morin temperature, T_M, is highly sensitive to small variations of the crystalline lattice. A considerable increase, of several degrees, is observed in T_M as a result of imperfections in the crystal lattice, which may be produced by shock waves [7], impurities, or other defects. The Morin transition is also broadened by grain size, because T_M decreases with the reduction of the particle dimensions [6].

The behavior of the hyperfine parameters with temperature is illustrated in Figure 4, for compact decreptate hematite. The experimental error in the isomer shift, quadrupole interaction, and linewidth outside the Morin

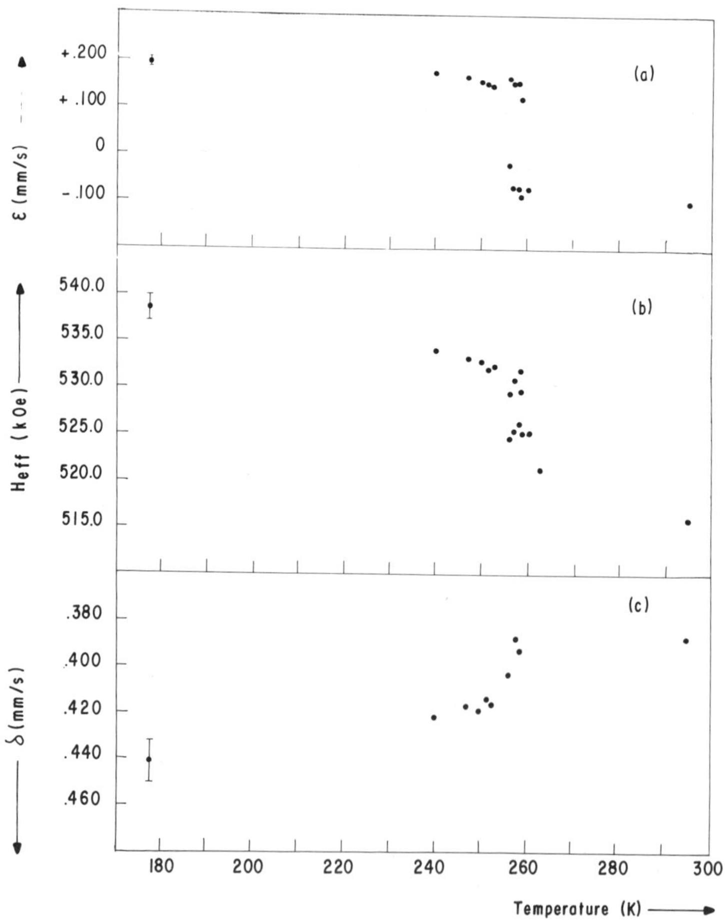

Figure 4. Mössbauer parameters at the Morin transition for compact decreptate hematite.

Table I. Discontinuities found in the Mössbauer Spectral Parameters at the Morin Transition.

Sample	Quadrupole Interaction mm/s	Magnetic Interaction kOe	Isomer Shift
Compact decreptate hematite	0.198 and -0.098mm/s	5.1kOe	-0.012mm/s
Compact decreptate hematite after undergoing decrep-tation	0.204 and -0.111mm/s	5.7kOe	-
Non-decreptate hematite	0.172 and -0.098	10.5	-

transition range is ca. 2.5 percent and within the transition range it is ca. 4 percent. The error in effective field is ca. ±3kOe.

The quadrupole interaction, as shown in Figure 4, is the Mössbauer parameter which is most sensitive to the Morin transition. This results because, when the quadrupole interaction is combined with the magnetic interaction, it has an angular dependence. This angular dependence will result in a discontinuity which is well defined at the transition. The quadrupole interaction is positive in the antiferromagnetic phase and negative in the weak ferromagnetic phase. The ratio of the magnitude of the quadrupole interaction above and below T_M must be two. This is illustrated in Figure 4a. Figure 4b shows the abrupt reduction in the effective magnetic field within the Morin transition region. The average effective hyperfine field on both magnetic phases shows a decrease of 5.1, 5.7, and 10.5kOe for the compact decreptate hematite, compact decreptate hematite after undergoing decreption, and non-decreptate hematite, respectively. These values can be compared with those reported earlier [5] which indicated a decrease of ca. 8.0kOe. Nininger et al. [6] found a decrease of 7.0kOe in the effective field.

The isomer shift is less sensitive to the spin reorientation process. van der Woude [5] has found a small discontinuity in T_M of ca. 0.007mm/s. Nininger et al. [6] obtained 0.14mm/s. Table I summarizes the results obtained for the various hematite samples.

The transition range was determined by a procedure that we call composed fitting. The two sets of magnetic hyperfine structure lines corresponding to the antiferromagnetic and weak ferromagnetic hematite phases are considered. We calculated the area, S, of each magnetic phase and normalized to the total spectral area. The results are shown in Figure 5. The points represented by triangles, correspond to the temperatures at which the non-predominant phase was so small that it was not possible to use the composed fitting technique. In these cases, we obtained a simple fit by using the continuous broadening of the sixth line during the reorientation process. This line will reach its highest width at the Morin temperature. Figure 6 shows the behavior of this line as a function of the temperature for the three samples. The transition region was defined for each fitting method as follows. In the composed fit we define this region as that temperature range that begins with 90 percent and finishes with 20 percent of the antiferromagnetic phase. In the simple fit we defined the Morin transition as the temperature range at half height of the curves shown in Figure 6. Morin temperature corresponds to the point where we have 50 percent of each phase present in the sample. Table II reproduces the results obtained for each hematite. It is important to observe that a narrow transition range is associated with compact decreptate hematite as compared to the non-decreptate hematite.

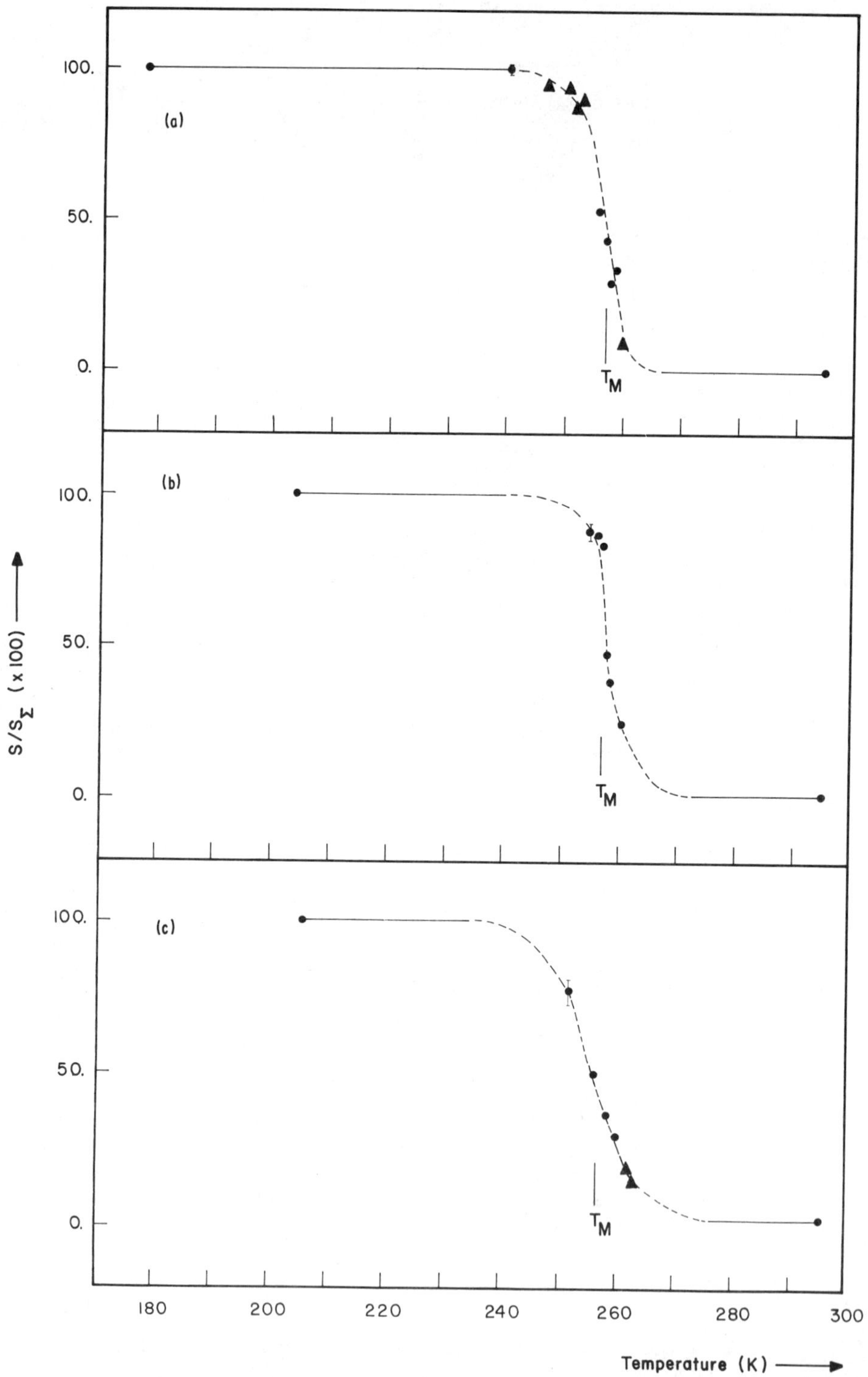

Figure 5. Spectral area S/S_{Σ}(x100) variation associated with the antiferromagnetic phase of (a) compact decreptate hematite, (b) compact decreptate hematite after undergoing decreptation, and (c) nondecreptate hematite.

Table II. The Transition Region, ΔT, and the Morin Temperature, T_M, as Determined by the Composed and Simple Fit

Sample	Composed Fitting		Simple Fitting	
	ΔT(K)	T_M(K)	ΔT(K)	T_M(K)
Compact decreptate hematite	7.5	257.0	2.4	257.0
Compact decreptate hematite after undergoing decrep-tation	7.5	258.0	4.0	258.5
Non-decreptate hematite	15.5	256.0	10.0	256.0

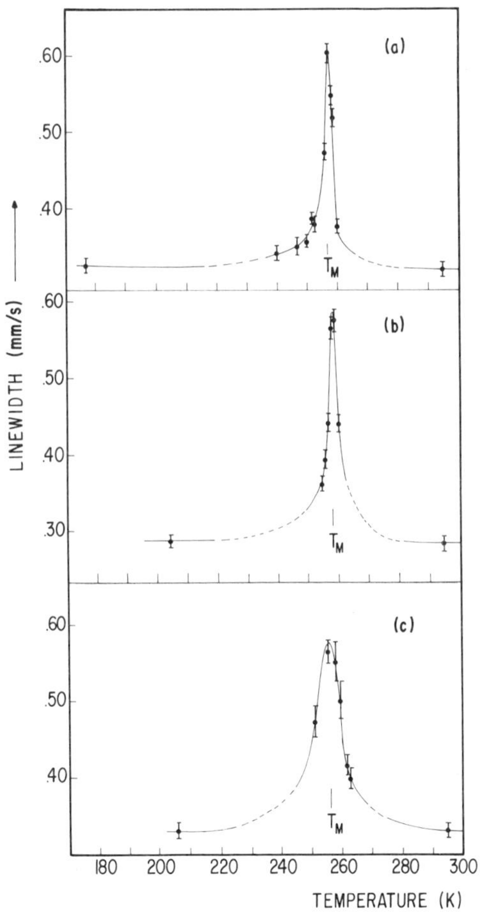

Figure 6. The sixth line variation of the Mössbauer spectrum as a function of the temperature for (a) compact decreptate hematite, (b) compact decreptate hematite after undergoing decreptation, and (c) nondecreptate hematite.

DISCUSSION

Compact hematite was chosen as subject of this investigation in order to determine whether Mössbauer spectroscopy could be useful in understanding the decreptation phenomenon. The simple comparison of the Mössbauer spectra for compact decreptate hematite and non-decreptate hematite does not reveal any significant difference because the hyperfine parameters are about the same for both types of sample. More relevant information has been obtained from the analysis of the Mössbauer spectra as a function of the temperature.

The analysis of the variation of the Mössbauer absorption area by heating was studied for each of the hematite samples. Only with compact decreptate hematite during the first heating we have observed a deviation in the linear behavior of the logarithm of resonance area as a function temperature. This deviation was observed for all samples of compact decreptate hematite, and it occurs at the decreptation region. Thus the decreptation manifests itself as a discontinuity in the Mössbauer effect absorption area.

With non-decreptate hematite calorimetric measurements have shown a peak corresponding to the goethite decomposition,

$$2FeOOH \rightarrow Fe_2O_3 + H_2O$$

This peak occurs within the same temperature range that decreptation is observed, However, no such peak is observed in calorimetric measurements with

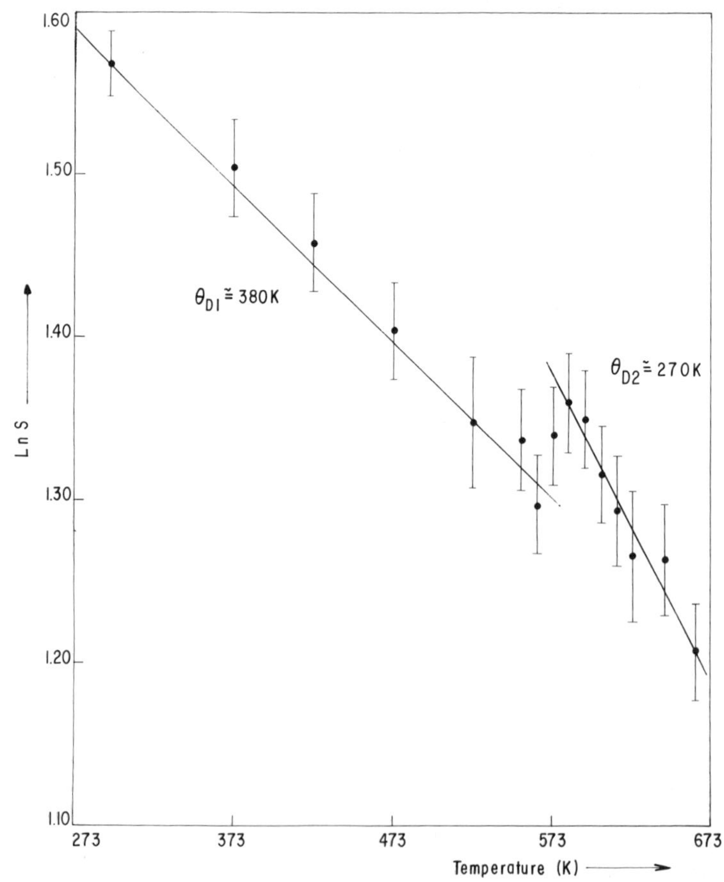

Figure 7. The temperature dependence of ℓnS for compact decreptate hematite.

decreptate hematite. If a mechanism such as water release is responsible for decreptation it must be present at a level below that of the calorimetric measurements.

The Morin transition which shows narrow lines, indicates that compact decreptate hematite has a high degree of crystallinity, in agreement with the x-ray diffraction results. The deviation observed in ℓnS with temperature can be explained by the presence of impurities or defects.

It is possible to analyze the variation of the natural logarithm of the resonance area in two regions, with linear behavior in each, as shown in Figure 7. Thus two Debye temperatures would be associated with compact decreptate hematite, one resulting before decreptation (θ_D ca. 380K) and the other after decreptation (θ_D ca. 270K). It is interesting to observe that this result is similar to the effect of shock waves in the hematite [7] which lead to a decrease in the Debye temperature. The effect of decreptation in hematite is a violent fracture that occurs as the ore is reduced at the level of the crystalline lattice; this violent fracture is the same as a shock wave. Micro-inclusions of goethite, decomposing to give water, even in amounts not observable by the calorimetric measurements, will suddenly increase the internal pressure in the hematite. This is equivalent to a shock wave which will disrupt the crystal lattice of the mineral.

REFERENCES

1. I. Masuda, Metalurgia - ABM 29, 783 (1973).

2. I. Masuda, R. P. Marroquim, and M. Franco, Metalurgia - ABM 31, 159 (1975).

3. S. Wanatabe and M. Yoshinaga, Trans. Soc. Mining Engrs. Aime 241, 1 (1968).

4. K. Kunii et al., Tetsu-to-Haganè 52, 228 (1966).

5. F. van der Woude, Phys. Stat. Sol. 17, 417 (1966).

6. R. C. Nininger, Jr. and D. Schroeer, J. Phys. Chem. Sol. 39, 137 (1978).

7. S. V. Koptsik, V. A. Povitskii, M. Yu. Messinev, Yu. V. Baldokhin, and N. F. Shchukin, Phys. Stat. Sol. 47, 39 (1978).

8. W. Smykatz-Kloss, in "Differential Thermal Analysis, ed. Springer-Verlag, Berlin, p. 36 (1974).

9. J. A. Gadsden, in "Infrared Spectra of Minerals and Related Inorganic Compounds, ed. Butterworth Group, London, p. 45 (1975).

COAL TREATMENT AND DESULFURIZATION

H. Pollak

Physics Department
University of Kinshasa
Kinshasa XI, Zaire

J. G. Stevens

Mössbauer Effect Data Center
University of North Carolina
Asheville, NC 28804, USA

ABSTRACT

Pyrite, an impurity commonly found in coal, will react with air to form sulfur oxides, which are poisonous to humans and can cause severe damage to industrial equipment. Because of the relatively high percentage of pyrite found in many coals, it is necessary to treat coal to satisfy environmental regulations. Mössbauer spectroscopy can be effectively used to monitor the pyrite and other iron sulfides through various coal treatment processes.

Mechanical treatment of coal, such as froth flotation and float-sink separation, is today the most commonly used procedure. For pulverized coal, it is possible to use magnetic desulfurization in which pyrite is transformed into pyrrhotite. This enhances the magnetic susceptibility of the inorganic sulfur containing material.

Other treatment processes include several chemical methods and biological leaching, which still needs development, but is very promising. One of the chemical methods is the Meyer's process, TWR, and is perhaps the best because it recycles the chemical, it does not require fine crushing of the coal, and sulfur removal is very efficient.

INTRODUCTION

Our society is conditioned to consume vast quantities of energy. As our consumption has increased, so has the amount of pollution. Because coal is one of our major sources of energy, it is also the primary cause of atmospheric pollution.

Man has used coal for the last 100,000 years, but it is only in the last decade that it has been necessary to focus on the drawbacks of coal utilization, i.e., the pollution problems, mostly due to the presence of sulfur compounds in coals. We must question whether the industrial growth and increased coal consumption of this century, to a level never before attained, is worthwhile in view of the corresponding increase in atmospheric pollution.

To preserve our quality of life, governments have regulated the sulfur content of coal used by power plants. A limit of one percent by weight is the limit established in some German states. This corresponds to an emission limit of 2.75kg of SO_2 per megawatt hours of power generated. It was not until 1960 that the federal government of the United States became involved with the problem of pollution and established air quality regulations under the control of the U.S. Environmental Protection Agency, EPA. The current federal standards require that the emissions from new steam boilers of larger than 25MWh capacity, be limited to 1.84kg of SO_2/MWh. There are other regulations which apply to the nitrogen dioxide and particulate matter level. It is expected that there will be further reductions in the future. The EPA, through the 1970 Clean Air Act, sets a maximum limit of sulfur content in coal of 0.6 to 0.8 percent for use by power plants.

To achieve these limits two approaches are possible. They consist of cleaning the output gas [1] by scrubbers (see Figure 1) or purifying the coal to a very low sulfur content. In this paper, the later procedure will be discussed.

HISTORICAL SURVEY OF COAL TREATMENT STUDIES

The necessity of coal cleaning goes back several centuries. The idea of an epuration using high density liquids dates from the middle of the 19th century. The first processes to remove sulfur used iron(III) chloride and sulfuric acid. It was done during the time when the coal was separated from the stones and rocks. Industrial plants to specifically do this were built in Great Britain. Berard tried at the end of the last century to use calcium chloride to remove the sulfur [6]. In 1902, the DuPont Company proposed the

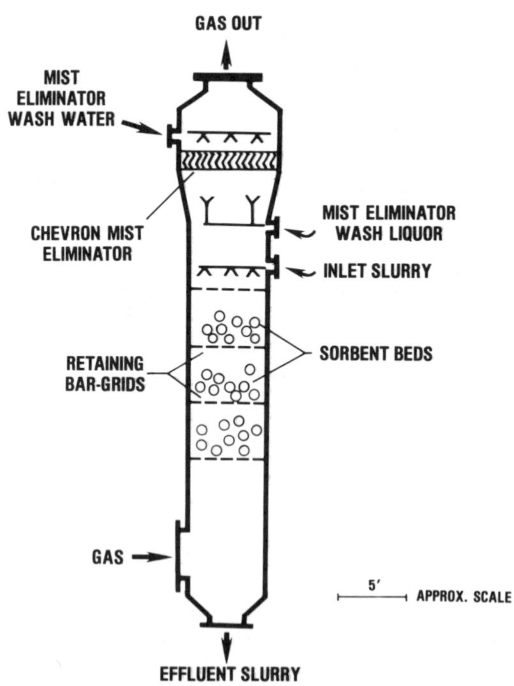

Figure 1. Lime/limestone scrubber. This three-bed scrubber has been tested by the EPA on flue gas from a 10MWh boiler at the TVA Shawnee station, Paducah, KY [1].

use of organic fluids. It was not until this time that the first real operating plants were built. Froth flotation techniques appeared between 1910 and 1920.

After World War II, the problem of sulfur became important and people started to work intensively on coal desulfurization. In order to compare results from different laboratories, common methods of data analysis were established.

Demonstrations of coal epuration or cleaning were first shown by Henry (1903) and Blanc (1909) in France and by Rheinard (1911) in Germany. Scientific research appears in the literature [2] as early as 1928.

To represent the stages of an epuration, Wolf [3] drew leaching curves, some examples of which are given in Figure 2. The percentage of the float and the sink part of coal is plotted as a function of the density. In the same graph, the ash yield is often also plotted versus the density of the coal after gravity separation.

The initial use of these separation curves and separation densities was demonstrated by Tromp [4]. Their importance was demonstrated by Terra [5] who introduced statistical methods in examining the leaching process. The theoretical "separation point" is defined as the density at which 50 percent of the product is equally floating and sinking. The use of these curves is not easy in practice. At the CERCHAR company in France, Belougou and Ulmo [6] in 1950 proposed standardization of the methods to enable comparison of results from different sources. Leaching curves are widely used to characterize the cleaning of coal as the result of its separation from impurities such as rocks and ashes. These curves do not give any information on the intrinsic quality of the coal nor on its sulfur content.

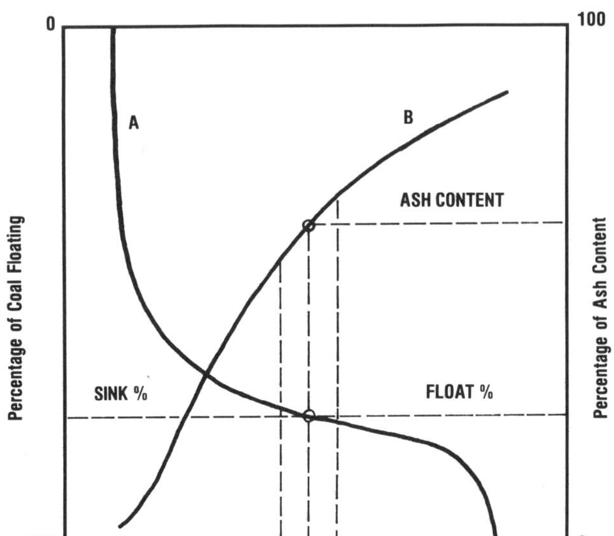

Figure 2. Leaching curves. The leaching curves give the percentage of float and sink part of coal, as well as the ash content as a function of the density of the washing media [6]. Line A gives the percentage of coal floating (left scale), while line B indicates the ash content.

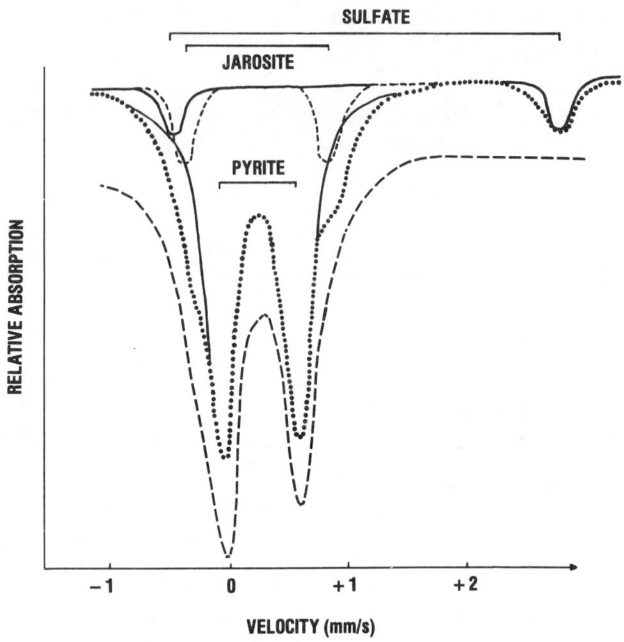

Figure 3. Natural coal oxidation. Mössbauer spectra of a fresh coal sample from the mine of Luena, Zaire, (lower line). The spectrum only shows the pyrite quadrupole doublet, while the same sample after a few months in open air, it displays new lines attributed to various sulfates and jarosite (upper line).

The methods commonly used to clean coal are jigs and shaking tables, float-sink separation in heavy media, hydrocycloning, froth flotation, and chemical cleaning. The mechanical methods of cleaning work best with coals that have a high inorganic or pyritic sulfur content and a low organic or chemically bonded surfur content. Otherwise, more elaborate techniques, such as fluidized bed combustion, or solvent refining, are more efficient.

SULFUR IN COAL

The structure of coal can be thought of as being composed of hydroaromatic rings in which the aromaticity increases with the coal rank. Coal, mainly composed of carbon, also contains sulfur, oxygen, and nitrogen as impurities. More generally coal is a heterogeneous, hydrogen-deficient, organic rock.

Sulfur appears in coal in the form of organic and inorganic matter, both of which come from chalcopyrite at the time of the formation of the coal. Sulfur usually is not directly bonded to the coal, but it is found in the various sulfur compounds that exist in the coal. The inorganic forms of sulfur are usually pyrite, marcasite, and sulfates. These sulfates are, in order of increasing water content, szomolnokite, rozenite, and melanterite. Jarosite can also be present. All sulfates are weathered products of pyrite and/or marcasite. Figure 3 shows the Mössbauer spectrum of a coal sample that has been exposed to air in the laboratory for a few months during which sulfate has been formed. Depending on the type of sample, and the degree of weathering, the following sulfates can be found:

szomolnokite	$FeSO_4 \cdot H_2O$
rozenite	$FeSO_4 \cdot 4H_2O$
melanterite	$FeSO_4 \cdot 7H_2O$
coquimbite	$Fe(SO_4)_3 \cdot 9H_2O$
roemerite	$FeSO_4 \cdot Fe(SO_4)_3 \cdot 12H_2O$.

Generally these sulfates, because of their small particle size, are not detected by x-ray diffraction. If rozenite is an altered product of pyrite, it disappears by low temperature ash treatment at 90°C and is transformed into an iron(III) compound [7]. When FeS_2 is exposed to air, it oxidizes readily to various sulfates, which are dissolved by water, and given rise to an acid mine drainage. Figure 4 shows the reactive and unreactive portions from the Dipple surface mine in West Virginia [8]. Note in the Mössbauer spectrum the absence of szomolnokite in the unreactive portion.

Pyrite particles range in size from a few centimeters to submicron particles. Pyrite can also exist in the form of clusters of small framboidal crystals of about 15μm in diameter. In this article the word pyrite is also used for FeS_2, even though it actually may be both pyrite and marcasite. In practice, different FeS_2 forms are very difficult to distinguish. Figure 5 shows the Mössbauer spectra of pure pyrite, pure marcasite, and a 50-50 mixture of pyrite and marcasite [9]. The slight asymmetry of the Mössbauer lines suggests the presence of marcasite.

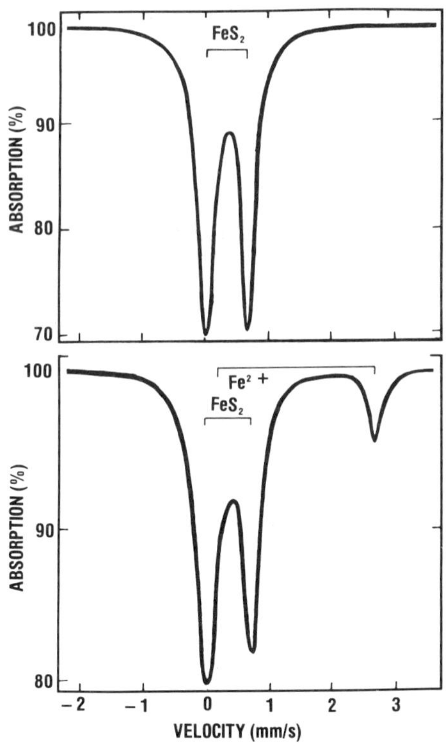

Figure 4. In situ coal oxidation. The two Mössbauer spectra are taken of a fresh coal sample and a reacted coal sample, surface part of the coal from the Dipple mine in West Virginia [8].

For the organic sulfur, the two most important compounds are thioether and thiophene. Depending on the origin of the sulfur in coal, siderite may also be found [10]. In some coals besides the pyrite, which is the most common sulfur compound, jarosite can also be found in relatively large amounts. Mössbauer spectra have also shown the presence of various clays in some coals. Additional information on the minerals in coal is given by Montano [7]. Depending on the type of coal and its geological origin, the organic sulfur content varies from 20 to 50 percent of the total sulfur content. Organic sulfur occurs in coal generally as very fine powered particles, not easily separated from the pure coal itself.

Selective removal of organic sulfur is not possible without affecting the coal structure. However, by hydrogenation organic and pyritic sulfur can be removed. In practice pyrite can only be separated from particles of dimension larger than 0.10mm by physical and physico-chemical methods of separation. Coal desulfurization can only be possible by reducing the content of epigenetic coarsened-crystalline pyrite and syngenetic fine crystalline pyrite.

All methods of pyrite separation are based on the physical properties of coal listed in Table I. It is the difference in the properties of coal, pyrite, and other minerals that enables separations. The major properties to consider are the difference in density between coal, pyrite, and other minerals used in float-sink (gravi-float separation) methods, the difference of the Mohr index, which is related to the ability of the coal to be broken into smaller pieces, and the difference in magnetic susceptibility. The degree of pyrite removal is determined mostly by the grain size of the coal, e.g., 50 percent removal of the sulfur for 10mm particles. As the particles become smaller, the percent of sulfur removal increases.

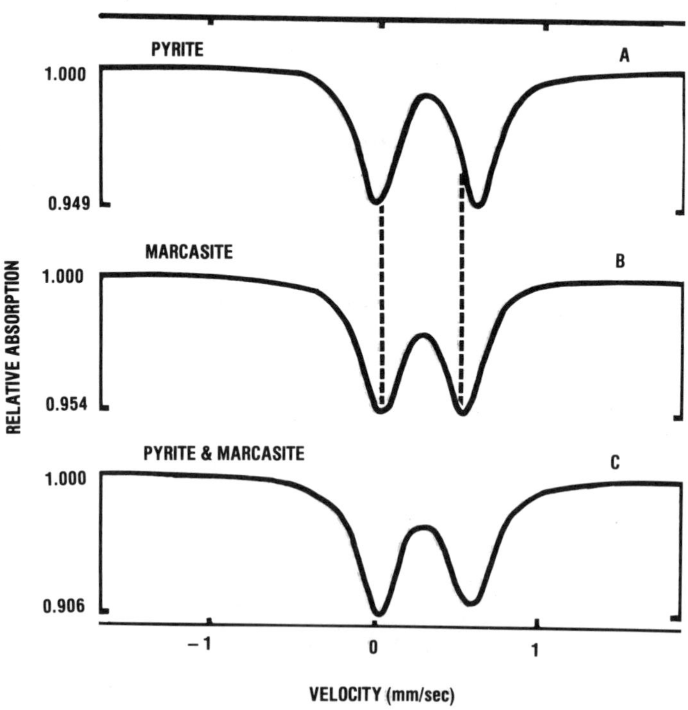

Figure 5. Mössbauer spectra of pyrite and marcasite. The spectra are respectively pure pyrite (A), pure marcasite (B) and a 50-50 mixture of pyrite and marcasite (C) [9].

Table I. Characteristics of Coal

Characteristics	Coal	Other Minerals	Pyrite
Strength or Mohr's scale hardness	2.0-2.5	2.0-4.5	6.0-6.5
Cleavability	good	good	bad
Color	Black	grey	gold
Density, g/cm^3	1.1-1.5	2.4-3.9	4.8-5.0
Dielectric constant	2.0-2.5	4.7-7.8	5.2-8.6
Susceptibility ($x10^6$ cqsu)	0.0-0.6	20-400	1.5-127
Thermal disintegration, $^\circ C$	250-450	450-700	250-450

Table II. ASTM and ISO Sulfur Dosage in Coal

	ASTM	ISO
HCl density	$1.19 g/cm^3$	$1.18 g/cm^3$
NH_4OH density	$0.90 g/cm^3$	$0.88 g/cm^3$
Quantity of Coal	5g	5g
Size	<0.25mm	<0.20mm
Bromide water add	10ml	10ml
Oven temperature	$925^\circ C$	$80^\circ C$
H_2SO_4-H_3PO_3 mixture	20ml	15ml

SULFUR DETERMINATION IN COAL

The determination of the amount of sulfur in coal is not easy. Many methods have been proposed, and there are standard ones used for comparing coals from different industrial plants.

Chemical Methods

The American Standard Testing Materials Society has established procedures for the determination of total sulfur and pyritic sulfur. The Eshka method is used to determine total sulfur following the ASTM procedure D-271-70. The ASTM procedure D-2492-68 is used to determine the total amount of sulfates. The permissible difference in the results of sulfur determination in coals containing less than 2 percent of sulfur is 0.05 percent for samples analyzed in the same laboratory and 0.10 percent for interlaboratory comparison. In these standard methods the coal is ground to 60 mesh or 250μm. If a sulfur particle is much smaller than the average coal particle, the sulfur can be totally encapsulated in the coal matrix and therefore not detected.

The International Standardization Organization, ISO, published recommendation on sulfur determination when it exists in different forms; ISO/R157 is for pyrite and sulfates. These were further detailed by the Technical Committee Report, ISO/TC27, of the British Standards Institution. After numerous amendments, on June 29, 1957, recommendations were approved by 24 countries including the United States. These recommendations are outlined in Table II.

This method is illustrated in Figure 6. Initially the coal is caused to react with hydrochloric acid. At this stage the sulfates are dissolved. The amount of pyritic sulfur, which is insoluble in HCl, but soluble in nitric acid, is determined indirectly by the determination of the amount of iron. The organic sulfur is evaluated by the difference between the amount of total sulfur and pyritic sulfur. This method assumes that all the sulfates are dissolved by HCl, and that all pyritic sulfur is dissolved in HNO_3. The Mössbauer experiments of Levinson and Jacobs [11] show that the HCl treatment of coal indeed carries away some pyrite as shown in Figure 7. They also observed that after HNO_3 treatment, some pyrite remains in coal. This experiment was also reported in 1981 by Williamson et al. [9] with the same conclusions, and especially that a 25 percent difference occurred by using Mössbauer spectroscopy. It appears that the ASTM and ISO methods underestimates the amount of pyritic sulfur. This has been further confirmed by neutron activation analysis [12] and also by scanning electron microprobe analysis [13].

Physical Methods

Other methods of analysis have been proposed each having various advantages and drawbacks. These include neutron activation analysis, [12] x-ray florescence [14], x-ray diffraction [15], and scanning electron microscopy, SEM [16]. SEM analysis give results higher than those of the ASTM, and higher than the results obtained by using Mössbauer spectroscopy. The Mössbauer data of Huffman and Huggins [17] was interpreted by using a low value for the recoil free fraction of 0.48. This value was arrived by using a similar matrix of charcoal [16]. The result is that the pyrite content is depressed. Montano [18] determined the Mössbauer f value in coal to be 0.60. Williamson's determination [9] of f yields a value of 0.53. Also it has been shown that f varies 10 to 40% depending on the sample [18]. This implies that a different correction for each specific coal must be made.

Montano [18] has shown that, when grinding coal to a smaller size than that described in the ASTM D-2492, there is agreement between the chemical (ASTM) and Mössbauer methods.

MECHANICAL COAL TREATMENTS

Mechanical treatments of coal for desulfurization are based on the fact that the density of pyrite is much higher than that of coal, and also that a certain amount of the sulfur agglomerates. The sulfur that is finely divided in the coal cannot be removed by any mechanical methods unless the grain size of the sample is reduced to the submicron range. This is unrealistic for the final state of coal before it is to be used. Also the cost involved in such a process would be prohibitive [19].

Mechanical Tables and Jigs

The crushing of coal is effective due to the stratification structure of coal. The results will be the elimination of fines, shales, refuses, and also the reduction of sulfur content due to agglomeration. A typical method consists of vibrating or shaking concentration tables. But, only a 15 percent sulfur reduction in Donetz, USSR, coal [19] was obtained. This coal normally contains 6 percent sulfur by weight. This method of coal concentration is widely used in the United States. German studies have revealed the possibility of sulfur elimination by this method of up to 19 percent.

Heavy Media

The "float and sink" method requires heavy density liquids to separate the lighter density of coal, around $1.2g/xm^3$ and that of pyrite having a density larger than $4g/cm^3$. In the laboratory organic liquids such as bromoform or carbon tetrachloride can be used. These are expensive. On an

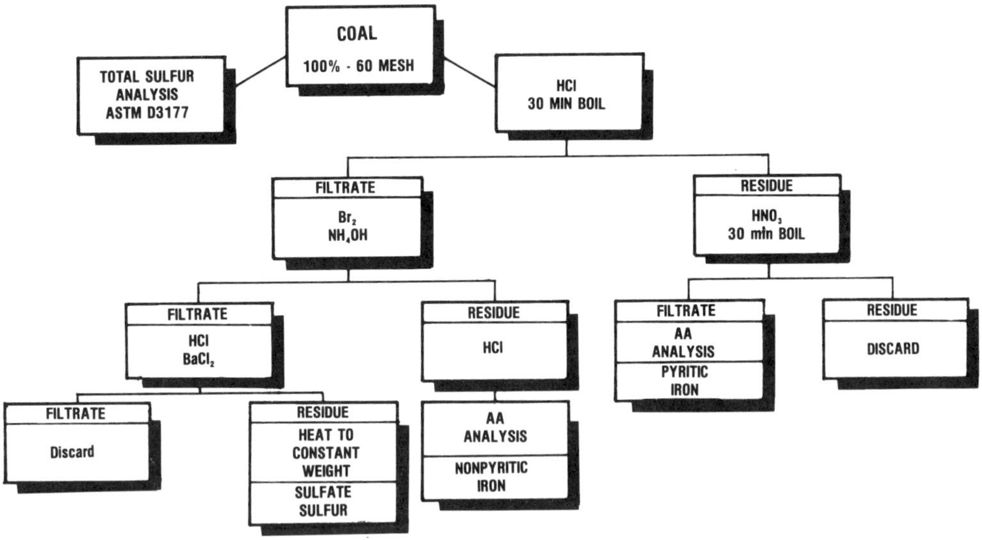

Figure 6. ISO/ASTM sulfur chemical analysis [9].

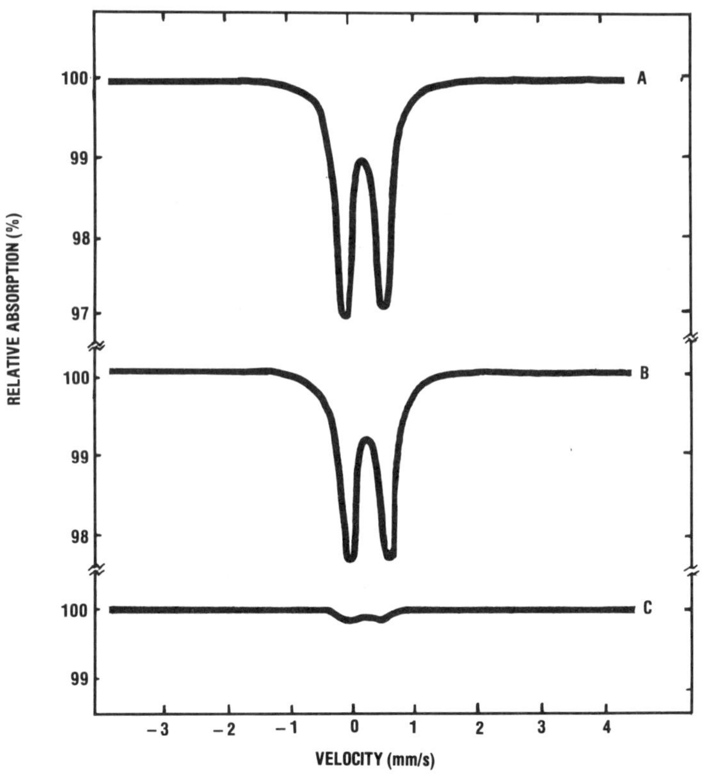

Figure 7. Mössbauer spectra of coal taken during the ISO/ASTM treatment, (A) untreated, (B) after HCl treatment, and (C) after HNO_3 treatment. The relative transmission is proportional to the sulfur content [11].

Table III. The Effect of Coal Size on Washing Characteristics at 1.6g/cm^3 of Three West Virginia Coals [21].

Coal	Size,	Pyrite	Sulfur Weight Percent Organic	Total	Removed
1	5.0	1.42	1.70	3.12	23.9
	0.8	1.27	1.70	2.97	27.6
	0.15	0.82	1.70	2.52	38.5
2	5.0	1.59	2.25	3.84	28.4
	0.8	1.59	2.25	3.84	28.4
	0.15	0.81	2.25	3.11	42.0
3	5.0	1.07	0.56	1.63	58.8
	0.8	0.97	0.56	1.63	61.4
	0.15	0.36	0.56	0.92	76.8

industrial scale, zinc chloride in water solution is used, which has an approximate 1.9g/cm^3 density. But his chemical treatment changes the coking properties of the coal.

An epuration is completely defined when the washing curve is known. Figure 8 is a Mössbauer spectrum of the float part of Luena coal after separation at a density of 1.4g/cm^3. Note the appearance of the jarosite lines which were completely masked by the pyrite line in the original sample. This shows the specificity of the separation of pyrite from jarosite. In Table III, the results of an epuration of a density of 1.6g/cm^3 as a function of the coal size is given. It was concluded [20] that the degree of pyrite liberation is mainly determined by the coal size. Further, the organic

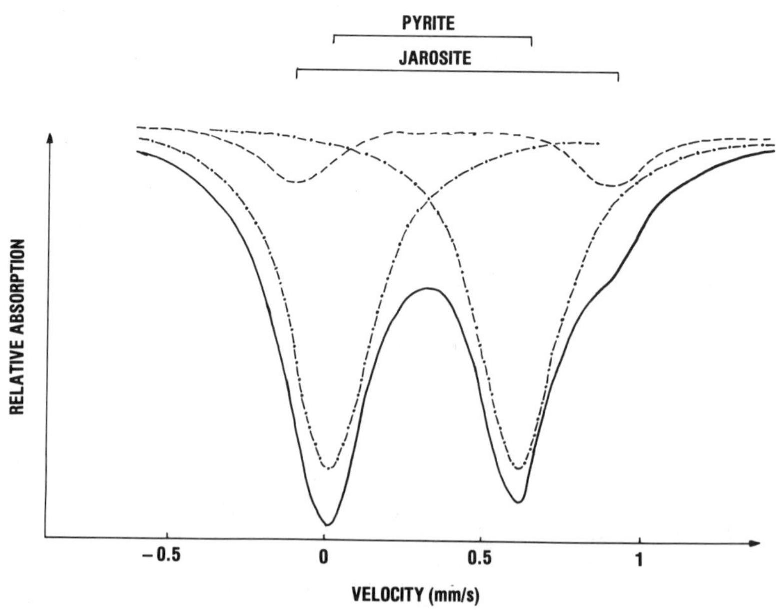

Figure 8. Float part of Luena coal at 1.45g/cm^3. Notice the appearance of the jarosite lines, which were masked in fresh coal, the lower relative intensity of the pyrite.

desulfurization is independent of the grain size because of the tight binding of the sulfur compounds to the coal. Finally, most of the pyrite removal is the result of the small size of the particles < 100 microns. To get the removal of pyrite up to about 80 percent, takes a reduction in coal size which is too expensive for known mechanical processes.

Hydrocycloning

Froth flotation techniques are commonly used in coal desulfurization with hydorcyclones using heavy liquids. Pyritic sulfur can be reduced substantially, and the sulfates reduced dramatically. At the same time, the ash content is reduced in half, but the moisture content of the coal is increased by about 10 percent. Nevertheless the caloric value increases by more than 10 percent [21]. The hydrocyclones are more widely used around the world, even though sulfur reduction yield is a maximum of 35 percent, and diminishes to 17 percent for some coals where the sulfur is present in the organic state.

In centrifugal separation with a heavy liquid media, a solution of calcium chloride removes up to 45 percent of the sulfur, but this method is relatively expensive because of the necessity of regenerating the solution. Using water solution reduces the cost, but the coal needs to be ground to fine particles of less than 3mm. In this case the removal of sulfur can be as high as 40 percent [21]. To increase the desulfurization further requires a lower separation density, but then the yield of the concentrated coal decreases. For example, in the case of an Illinois coal, the lowering of the separation density from 1.45 to 2.35g/cm^3 decreases the sulfur from 2.32 percent to 2.20 percent, but the yield falls from 74 percent to 60 percent.

A multistream coal cleaning system, which uses two hydrocyclones in cascade, is in operation in several places. The two stage process first separates medium-sulfur coal from low-sulfur by cycloning with a medium at a density of 1.3g/cm^3 that separates the ash and pyrite [22] (Figure 9). For an input coal initially containing 2.7 percent sulfur, after the first stage, the coal contains 1.7 percent sulfur, and 0.8 percent sulfur is found in the coal at the output of the second stage. The heavy density media is obtained by adding magnetite water. This allows the magnetite to be recovered by large magnets. Typically this multistream process will remove 70 percent of

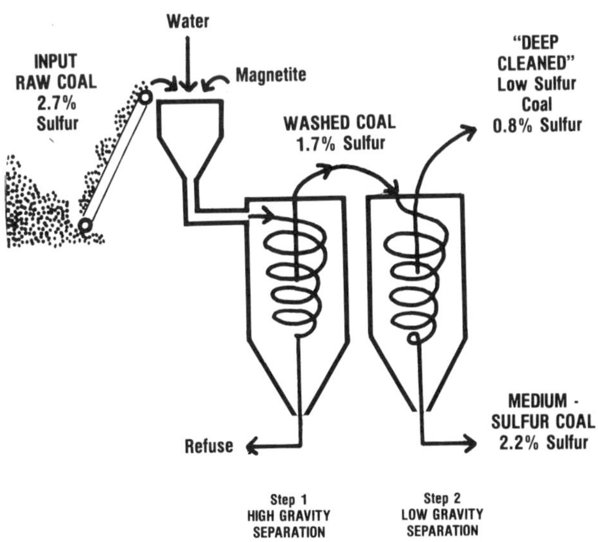

Figure 9. Multistream coal cleaning operating plant [22].

the sulfur. In practice this is the best that can be obtained from any mechanical system.

In conclusion mechanical treatment generally works very well for the removal of 35 percent of pyritic coal. Hydrocycloning with a heavy liquid gives the best results. The value of 35 percent removal can be raised to 50 percent without too much cost by using more recent techniques [23]. For better sulfur removal more elaborate methods other than the mechanical ones need to be used. These include bacterial and magnetic techniques. Before any of these methods can be used, it is necessary to crush and/or grind the coal to small particles by mechanical means such as shaking tables or ball grinders. In order to reduce the cost of crushing and grinding, chemicals methods of particle size reduction are possible and will be discussed in a section below. Crushing of the coal releases much of the ash and other impurities for later separation, but it produces a large amount of fine coal which is expensive to recover.

CHEMICAL COAL PREPARATION AND DESULFURIZATION

Pyrite is a very stable compound and its solubility in water at room temperature is only of $5.9*10^{-3}$ g/l. Its solubility increases with increasing temperature, but does not double until 100°C. Pyrite is soluble in an alkaline sulfide solution, to the extent of 0.55g/l at room temperature in a 1.5 molar solution of sulfide. Organic solvents appear to be inefficient for the removal of pyrite in coal. At room temperature, pyrite is attacked rapidly only by strong oxiding agents such as nitric acid, and not by hydrochloric acid. A summary of pyrite chemical decomposition are given in Table IV and the methods are compared in Table V.

Chemical methods are basically ineffective for the removal of pyrite in coal on an industrial scale. The Meyers process [27] is one exception. Nevertheless, some of the proposed methods will be discussed briefly. All chemical procedures require the complete pulverization of the raw coal.

CHEMICAL PARTICLE SIZE REDUCTION OF COAL

Methods of chemical particle size reduction of coal are being used in industry [24]. They are more efficient for lowering the pyritic sulfur than the mechanical methods, and prevent reducing the coal to a size below the desirable value. Up to 80 percent of the sulfur can be eliminated, even for particles that are of the order of 12 (1.4mm) to 50 mesh (2.7mm). It takes about 50kg of the proper chemicals for the sulfur removal per ton of coal. Up to 99 percent of these chemicals can be recycled.

One method uses ammonia gas and another ammonia solution. It is understood that the hydrogen from the ammonia breaks the weak covalent bond of pyrite. Because of the need to work in air tight tanks, the cost of this procedure does not make it practical. Nevertheless these approaches should not be ruled out for the future.

GAS STREAM DESULFURIZATION

Different gases have been evaluated to their effectiveness in stream desulfurization of coal. The following conclusions have been reached [25]. The desulfurization increases as the duration of degasing decreases. The degree of desulfurization increases as the grain size becomes smaller, Steam is needed in the desulfurization process. The reaction time affects the efficiency of desulfurization.

Table IV. Various Methods of Chemical Pyrite Decomposition

T°C	Agents	Decomposition Products or Observations
100-200	Copper sulfate, copper chloride, or silver nitrate	In sealed tube decomposition
140	S_2Cl_2	Iron(III) chloride
185	Sodium bicarbonate, thiosulfate	Sodium sulfide
250	Carbon tetrachloride	Decompose in vapor
300-400	Steam, hydrogen, sulfide	Iron(II) oxide
335	Ammonium chloride[a]	Ammonium sulfide
400	Oxygen	S removed in 2 to 3 hours
500	Carbon + carbon monoxide	Hydrogen sulfide
500	Carbon + carbon monoxide + steam	Hydrogen sulfide + carbon oxysulfide
600	Sulfur dioxide	Sulfur iron(III) oxide

[a]Seven percent of the sulfur is lost in 25 min.

Table V. Comparison of Some Chemical Methods of Coal Desulfurization

Method	Agent	T,°C	Pressure, Pa	Size	Percent S Recovered	Time, min.
Meyers	$Fe_2(SO_4)_3$	60	N	~mm	80	40
Wet oxidation	NO_x	160	250	75µm	90	120
Ammonia/O_2	NH_3+O_2	130	40	150µm	90	
Hydrothermal	NaOH+CaO	300	500	275µm	90	
Chlorinolosis	methyl-chloroform	450	N	pulver.		90
KVB	O_2+NaOH					
Hydrogenation	H_2	750				
Oxygenation	O_2	150	10^6	95	60	

At temperatures of the order of 650°C, stream acts as an active desulfurizing agent by producing iron oxide, sulfur, and hydrogen sulfide by the reaction

$$2FeS_2 + 3H_2O \rightarrow Fe_2O_3 + S + 3H_2S.$$

In a stream of oxygen and hydrogen, two following reactions occur,

$$4FeS_2 + 11O_2 \rightarrow 2Fe_2O_3 + 8SO_2,$$

$$FeS_2 + xS(organic) + (x+2)H_2 \rightarrow Fe + (x+2)H_2S,$$

where the first takes place in the presence of water. By operating at pressures of 330kg/cm^2 at 150°C, this method shows promise. However, because this method requires high pressure vessels, it is too expensive to be of industrial use.

The ammonia/oxygen system leads to the following chemical reaction,

$$4FeS_2 + 16NH_3 + 14H_2O + 15O_2 \rightarrow 16NH_4 + 8SO_4 + 4Fe(OH)_3.$$

The oxygen also reacts with the coal, which leads to some loss in the total caloric content of the coal. The reaction takes place at 130°C at an oxygen partial pressure of 1MPa or 100kg/cm^2. After two hours the reaction is complete and the slurry goes through a solid/liquid separator. The mixed carbonate/sulfate stream must be recycled to recover the ammonia. Almost all pyritic sulfur is removed along with 25 percent of the organic sulfur.

Solvent Refining Process

In the solvent refining process [26] for the desulfurization of coal, the pulverized coal is mixed with a coal solvent to make a slurry that is pumped with hydrogen through a preheater stage at a pressure of 12.5MPa, 125 bars, and then it is heated up to 440°C.

The TWR or Meyers Process

The Meyers desulfurization process [27] is one of the most recent methods proposed for coal desulfurization. It appears to be the most efficient and most economical. This method removes only the inorganic sulfur by using an aqueous solution of iron(III) sulfate and oxidizing the pyritic sulfur to elementary sulfur. The chemistry of this process can be summarized by the reactions first for leaching

$$2Fe^{+3} + FeS_2\text{-Coal} \rightarrow 3Fe^{+2} + 2S\text{-Coal},$$

and for sulfur recovery

$$S\text{-Coal} \rightarrow S(elemental) + Coal.$$

This is followed by regeneration of iron(III) by the reaction

$$12Fe^{+2} + 3O_2 \rightarrow 12Fe^{+3} + 6O^{-2}.$$

The pyrite will reduce the iron(III) formed in the last step by the reaction

$$FeS_2 + 14Fe^{+3} + 8H_2O \rightarrow 15Fe^{+2} + 2SO_4^{-2} + 16H^+.$$

The leaching time of coal is only about 7 hours when the operating temperature is 70 to 10°C. The coal does not need to be crushed in small size particles as the Meyers process works with particles up to 0.8cm. Mössbauer spectroscopic results are given in Table VI for this method on Luena coal. Figure 10 shows the difference between the original and the coal treated for 40 minutes. Figure 11 shows the configuration of a TWR installation for coal desulfurization. The regeneration of the leaching solution is readily accomplished by air oxidation. The sulfur dioxide emission of the treated Meyers coal is greatly reduced in comparison with the raw coal.

Table VI. Luena Coal Treated by the TWR Process

Duration of Treatment	Weight Percent Pyrite Remaining
	2.8
40 minutes	1.7
24 hours	0.6

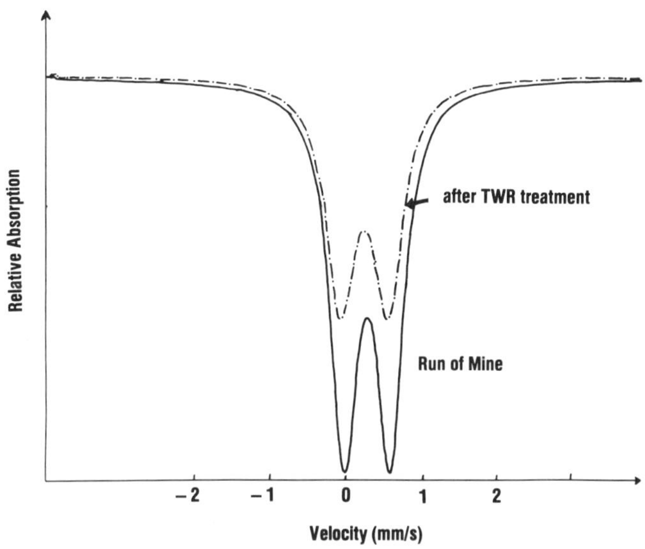

Figure 10. TWR results obtained on Luena coal. Notice that the two spectra have almost equal counting statistics. The reduction of the pyrite lines is evident after the TWR treatment

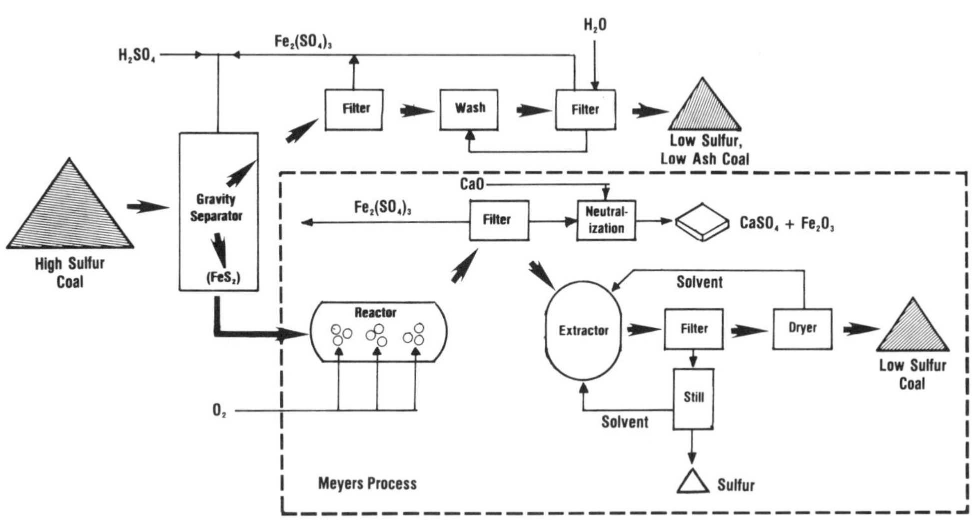

Figure 11. TWR coal desulfurization plant [28].

The Gravi-Melt Process

This process is particularly efficient for coal with a high concentration of organic sulfur, but this method requires high temperatures. The sulfur removal occurs through dry caustic impregnation of the coal. This method has not been applied on an industrial scale [28].

The Oxygen Leaching Process

The oxygen leaching process [28], also called the Ledgemont process, is in principle very simple, but the technology involved is rather sophisticated. Pyrite under normal atmospheric conditions is oxidized into various sulfates, although these reactions are very slow. By choosing the appropriate conditions of temperature and pressure, it is possible to increase the rate of these reactions. At this stage the iron(III) sulfates, which are soluble in water, are leached from the coal. Finally the remaining water is neutralized with limestone and the water removed from the coal.

This process for certain coals can compete with the TWR process. The technology involved is more complicated because of the use of oxygen at moderated temperature (130°C) and the relatively high pressures of about 900kg/cm^2. The recovery of elementary sulfur, which is a real problem in the TWR method, is avoided in he Ledgemont process by the formation of soluble sulfates.

MAGNETIC DESULFURIZATION

Table VII lists the magnetic susceptibility for coal and some of the mineral components of coal. Although the susceptibility of pyrite is higher than that of coal, it is still too low for magnetic separation. But pyrite can be readily transformed into pyrrhotite which has a larger susceptibility, enabling magnetic separation. The pyrite can be almost 100 percent converted into pyrrhotite in a reactor vessel held at 440°C for about 15 minutes under

Table VII. Magnetic Susceptibilities of Some Minerals of Coal

Mineral	Susceptibility[a]	Type of Magnetism
Coal	0.25 to -0.6	
Quartz	-20 to 12	
Calcite	-12 to 1	Moderate para
Pyrite	1.5 to 127	Weakly para
Dolomite	1.8 to 2.5	
Siderite	200 to 7000	Moderate para
Hematite	70 to 13000	Moderate para
Magnetite	3200 to 80000	Strongly ferri
Pyrrhotite	7000 to 20000	Strongly ferri
Sulfate	10000 to 12000	Moderate para
Clay	-50 to 10	Moderate para
Troilite	1074	Moderate para
Siderlite	200	
Calcite	-12 to 1	Moderate para
Pyrite	1.5 to 127	Weakly para
Dolomite	1.8 to 2.5	

[a]Values of 10^{-6}cqs units.

a nitrogen atmosphere [29]. Figure 12 is the Mössbauer spectrum of pyrrhotite with its typical magnetic sextet. The major effect on the iron to sulfur ratio is the total amount of sulfur available for H_2S formation. The partial pressure of this gas is the crucial quantity [2] controlling the stoichiometry of the reaction.

Magnex is an industrial method of magnetic separation that has been proposed [30]. The process consists of the four steps outlined in Figure 13. The coal is crushed to less than 14 mesh (1mm), heated to 170°C and exposed to iron pentacarbonyl, which transforms pyrite into pyrrhotite by the reaction

$$FeS_2 + Fe(CO)_5 \rightarrow 2Fe_{(1-x)}S + 5CO + (x+1)S.$$

Dry magnetic separation at 10,000 gauss produces 85 percent clean coal from the raw coal.

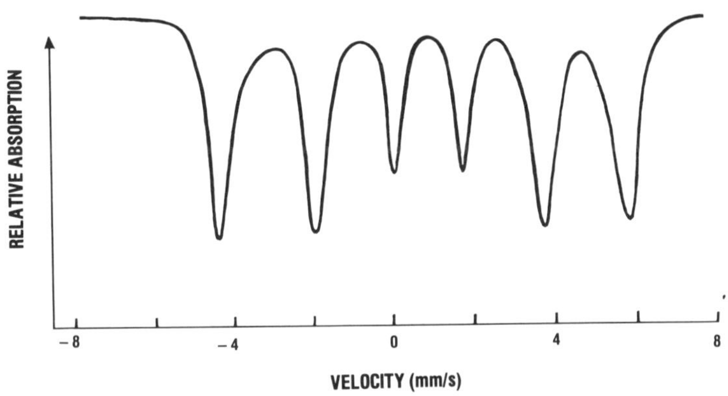

Figure 12. Mössbauer spectrum of pyrrhotite.

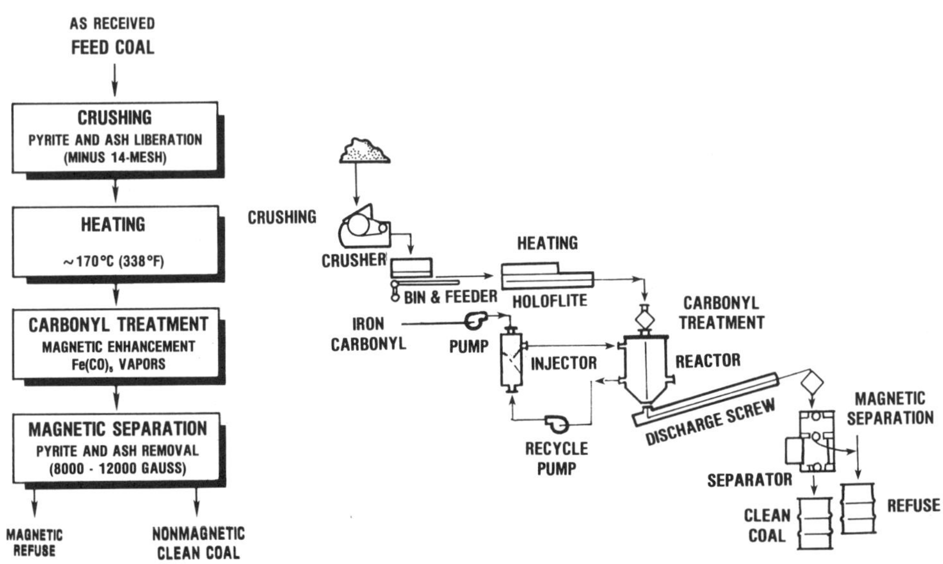

Figure 13. Magnex coal desulfurization [30].

BACTERIOLOGICAL COAL UPGRADING

It has been determined by Ashmead [31] that in mines where acidic waters is found the relative percentage of sulfate versus pyrite is higher than in other mines. This is due to the natural microbial flora of these mines. Table VIII lists some of the acidophilic chemoautotrophic bacteria known to be active in pyrite conversion [32]. As early as 1959, a group of Soviet scientists discovered that for Donetz coal, from one of the main coal basins of the USSR, the bacterium Thiobacillus ferrooxidans oxidizes the pyrite and transforms it into sulfates. In 30 days the pyrite removal is about 25 percent complete. Silverman et al. [33] found, in agreement with other authors, that the bacterium Thiobacillus thiooxidans is unable to oxidize pyrite though it is an iron oxidizing bacterium. They also confirmed the importance of the Ferrobacillus ferrooxidans, whose natural habitat is acid mine waters, for converting pyrite to iron sulfate. The chemoautotrophic bacteria are active in the 2 to 4.5 pH range. Coal is used as the carbon source. But their role is not exactly known. Organic sulfur, especially heterocyclics, is quite resistant to microbial attack and therefore cannot be removed from coal by bacterial action [34].

The action of the acidiphilic bacterium, Ferrobacillus ferrooxidans, is rapid in the pH range of 2.5 to 4.2, and all conversion occurs within 4 days. Longer action does not increase significantly the removal of sulfur. Approximately 50 percent removal seems to be a practical limit. The removal of pyritic sulfur is only efficient if the coal particle size is less than 80μm. The value is somewhat dependent on the type of coal.

Another bacterium, Thiobacillus ferrooxidans, works by a similar method at a pH of 2.6 with pyrite conversion of 85 percent. The reaction time is a function of the particle sizes, but in 1 to 3 days 50 percent of the sulfur can be removed.

Iron(III) sulfate, added in acid coal treatment, has the general effect of increasing the desulfurization process, even though the exact mechanism is not yet well understood. The bacteria can act to catalyze the aerobic pyritic oxidation of iron(II) ions to iron(III) ions, and then regenerate the primary oxidant. A process similar to the Meyers methods apparently takes place.

This type of desulfurization seems to be very promising, even though one of the major drawbacks is the relatively small size of the coal particle required. Table VIII gives a summary of bacteriological coal upgrading.

Table VIII. Biological Desulfurization (Pyrite)

Chemoautotrophic Bacteria	Percent Desulfurization	Time, Hours
Thiobacillus thiooxidans	0	24
Ferrobacillus ferrooxidans	50	72
Thiobacillus ferrooxidans	50[b]	24-72
Ferrobacillus sulfooxidans	?	?

[a]For pulverized coal at basic or neutral pH. [b]With a limit of 85 percent.

CONCLUSION

Mössbauer spectroscopy has proved to be of great help in the characterization and the treatment of coal. It has been shown that the well known ASTM pyritic sulfur determination usually underestimates the pyritic content of coal. To find agreement between Mössbauer spectroscopy and chemical analysis it is suggested that the coal should be ground to a small grain size.

Clean burning coal does not have too be a dream. Ecologists and the public have put pressure upon governments to write and enforce laws to minimize or eliminate atmospheric pollution. There are several methods that exists today for coal desulfurization. Some of these are economically feasible in that only a slight increase in the price of coal would result. This increase in the long term will be more than economically compensated by the reduced maintenance of the power plants.

The removal of the pyrite content of coal by up to 60 percent can be easily achieved [35] by conventional method such as jigs, concentrating tables, froth flotation, gravity separation, and hydrocyclones. In almost all these conventional methods, 10 percent of coal is lost during the treatment. To obtain even higher removal of the sulfur at reasonable costs, more elaborate methods are needed. These other approaches are magnetic and chemical.

Magnetic separation is very effective on pulverized coal provided the pyrite has been first converted into pyrrhotite. At this time magnetic separation methods are not economical.

Table VI lists some of characteristics of chemical desulfurization methods. Among the chemically proposed methods,oxygen leaching and the TWR, are the best and the easiest to develop and use.

The bacteriological methods show promise for the future and require more research.

In coal studies Mössbauer spectroscopy has been used in many laboratories to study sulfur reactions in coal and coal liquefaction, pyrite dosage, and coal cleaning. It has been shown that the Mössbauer effect is suitable for determining the pyrite content and that it can monitor coal desulfurization. Engineers are advised to introduce this technique in field treatment and to apply it also to mechanical coal treatment. It is expected that, on a routine basis, pyrite content through the Mössbauer measurement will give a faster answer than the ASTM procedure for the same accuracy.

ACKNOWLEDGMENT

The authors thank the assistance of Sindano for carefully reading this manuscript.

REFERENCES

1. Unleashing A Black Beauty, Air Magazine, April, pp. 5-11 (1978).

2. H. Heidenreich, Wissenschaftliche und wirtschaftliche Untersuchungen in der Steinkohleaufbereitung, Glueckauf 65(28), 13 (1929); 65(29), 20 (1929).

3. C. Wolf, Rev. Ind. Min., 326 (1934).

4. K. F. Tromp, Neue Wege für die Beurteilung der Aufbereitung von Steinkohlen, Glueckauf 33, 60 (1937).

5. A. Terra, Rev. Ind. Min., 411 (1937).

6. P. Belugou and J. Ulmo, in First International Conference on Coal Preparation, Paris (1950).

7. P. A. Montano, in Coal Structure, M. L. Gorbaty and K. Ouchi, eds., Advances in Chemistry Series, No. 192, 338 (1979).

8. P. A. Montano, in Nuclear and Electron Resonance Spectroscopies Applied to Materials Science, E. Kaufman and G. K. Shenoy, eds., pp. 171-183 (1981).

9. D. L. Williamson, T. W. Guettinger, and D. W. Dickerhoof, in Mössbauer Spectroscopy and Its Chemical Applications, J. G. Stevens and G. K. Shenoy, eds., Advances in Chemistry, No. 194, pp. 179-208 (1981).

10. L. Wuckowska,, in Solid Fuels, Prace Glownego Insyytutu Gorniectwa, Wydawnictwo Slak, Katowice, pp 3-12 (1984).

11. L. M. Levinson and I. S. Jacobs, Fuel 56, 453 (1977).

12. C. G. Clayton, A. M. Hassan, and M. R. Wormald, Int. J. Appl. Radiat. Isot. 34(1), 83 (1893).

13. P. R. Solomon and A. V. Manzione, Coal and Char. Fuel 56, 393 (1977).

14. M. Berman and S. Ergun, Analysis of Sulphur in Coals by X-Ray Fluorescence (1967).

15. S. S. Pollack, Fuel 50, 453 (1971).

16. F. E. Huggins, D. A. Kosmack, G. P. Huffman, and R. J. Lee, 1, 531 (1980).

17. G. P. Huffman and F. E. Huggins, Fuel 57, 592 (1978).

18. P. A. Montano, Application of Mössbauer Spectroscopy to Goal Characterization and Utilization, in Mössbauer Spectroscopy and Its Chemical Applications, J. G. Stevens and G. K. Shenoy, ed., Advances in Chemistry, No. 194, pp. 135-175 (1981).

19. Methodes de Reduction de la Teneur en Soufre des Combustibles Solides, Commission Economique pour l'Energie, Comite du Charbon, Rapport ST/ECE/Coal/63, pp. 29, Released on August 17, 1973.

20. J. W. Tieman, The Sulphur Problem in Coal: What's being done to get rid of it, Industrial Coal Conference, University of Kentucky, April 12-13 (1967).

21. F. M. Monstry, Glueckauf 115(18), 420 (1979).

22. Multi-Stream Coal Cleaning System promises Help with Sulfur Problem, Coal Age (January), pp. 86-88 (1976).

23. T. D. Wheelock, Coal Desulfurization, Chemical and Physical Methods, in ACS Symposium Series 64, R. F. Gould ed., American Chemical Society, Washington D. C., p. 332 (1977).

24. Will Chemicals Replace Crushing?, Coal Age (Febuary), pp. 176-179 (1978).

25. Solvent Refining gets the Sulphur out of Coal, World Mining (June), 32-33 (1973).

26. L. Lorenzie Jr., J. S. Land, L. J. Van Nice, E. P. Koustoudoand, R. A. Meyers, Coal Age, 76 (1972).

27. R. A. Meyers, Hydrocarbon Processes, 123 (1979).

28. J. C. Agarwal, R. A. Giberti, P. F. Irminger, L. F. Petrovic, and S. S. Sareen, Mining Congress J. 61, 40 (1975).

29. A. R. Bommannavar and P. A. Montano, Fuel 61, 523 (1982).

30. C. R. Porter and D. N. Goens, Mining Engineering (Febuary), 175 (1979).

31. D. Ashmead, Colliery Guardian J. Coal Iron Trades 190, 694 (1955).

32. C. E. Capes, A. E. Mclliney, A. F. Puddington, and I. E. Puddington, The Canadian Mining and Metallurgical Bulletin (November), 88 (1973).

33. M. P. Silverman, M. H. Rogoff, and I. Wender, Fuel 41, 113, (1963).

34. R. B. Baker, Water Research 6, 9 (1972).

35. R. von Der Gathem, Glueckauf 115(3), 52 (1979).

IRON PHASES IN BROWN COAL ASHES

J. D. Cashion and L. Joan Brown

Department of Physics
Monash University
Clayton, Vic. 3168, Australia

A. L. Ottrey

Herman Research Laboratory
State Electricity Commission of Victoria
Richmond, Vic. 3121, Australia

ABSTRACT

Iron-57 Mössbauer spectroscopy has been used in three areas of analysis of precipitator ash and deposits resulting from the combustion of Victorian brown coals. The principal iron-containing constituents identified in the samples were two mixed oxides (calcium aluminoferrite and magnesioaluminum ferrite). The changes occurring in some of the phases by altering the combustion conditions and during wet agglomeration are described. This work is aimed at developing a systematic description of relative phase stabilities for different flyash compositions, based on chemical analysis of the coal.

INTRODUCTION

Investigations are being carried out on three different aspects important in the utilization of brown coal (lignite), which is Victoria's principal energy resource. The first area covers the analysis of the iron-containing phases present in bed-moist brown coal [1,2] and their changes during drying and pyrolysis. This information is essential background for the second area involving hydroliquefaction of the coals using iron and tin promoters [3,4]. The third investigation is the analysis of the ash and deposits formed during utilization of the coal for power generation. This paper presents an overview of some of the conclusions reached from a study of these ashes, which were produced at different plants and under different conditions, and emphasizes the transition of the use of Mössbauer spectroscopy in this area from a research technique to a more routine analytical tool.

The production of ash is an unavoidable result of the burning of any variety of coal, but its constituents vary considerably with the rank of the coal (lignite, sub-bituminous, bituminous), from seam to seam and within the seam for the one type of coal. The weight of ash produced by black coal is usually larger than currently used brown coals and because black coal burns at a higher temperature, its ash produces a higher proportion of glassy phases - mainly from the clay material in the coal. Victorian brown coals generally contain little clay or pyrite, FeS_2, and the ash is principally formed from the acid-extractable inorganic cations derived from iron, calcium, magnesium, aluminum, and sodium and bound to the organic coal matrix, rather than from discrete minerals as in black coals. This results in the ash having a predominantly skeletal structure of unfused particles of oxides, with some fused spheres and char.

Several reports have been published of the application of the Mössbauer effect to analysis of flyashes and bottom ashes resulting from the combustion of North American coals [5-9]. Only one of these [9] has mentioned brown coal in a study of a bottom ash from a lignite/natural gas fired boiler. There is not a strong correspondence between the products which we observe [10] and the North American work. The main similarity is in the difficulty of fitting some of the spectra; imperfect and ambiguous fits frequently occur, particularly with room temperature magnetically split spectra.

AIMS AND PROCEDURE

Three main aims are covered in this work on the ash deposits. The first is the identification of the iron bearing phases in the deposits, several of which are of such small particle size, less than 10nm, as to make them amorphous to x-ray diffraction. The large number of phases present and the similar structures of several, e.g., spinel phases, all combine to make a complete analysis by x-ray diffraction almost impossible. Although iron, which is generally 5-20 weight percent, but sometimes as high as 45 weight percent expressed as Fe_2O_3, is not usually the dominant metal atom present, its association with calcium, and aluminum, magnesium as mixed oxides usually ensures that its compounds account for more than half the total mass. In such complicated systems, a high resolution technique which is selective to only one element can have a considerable advantage in unravelling the overall picture.

The second aim is to observe the modifications in the identified phases with changes in burning conditions. Ash fouling is a problem in all coal burning systems and the ash constituents determine the dominant mechanism of the ash fouling, and consequently the physical properties, such as strength and thermal conductivity, of the resultant deposit. We are measuring the changes of the nature of the ash against two of the burning conditions, flame temperature and flue gas oxygen content. The samples for this part of the work were produced in an experimental furnace at the S.E.C.V. Herman Research Laboratory, whereas the remainder are from Latrobe Valley power stations which form a major part of the S.E.C.V. generating capacity.

The final aim concerns the development of wet agglomeration as an alternative technique for the disposal of electrostatic precipitator flyash. These samples were prepared at the C.S.I.R.O. Division of Mineral Engineering by placing pre-moistened flyash into a 600mm diameter rotating drum under conditions of constant temperature and humidity. No added binders were used and pellets with an equilibrium diameter of ca. 10mm were usually obtained, as described in detail in Hall et al. [11]. Some pellets were also autoclaved to study the hydration process further. The Mössbauer spectra provided information on the chemical and particle size changes which occurred.

The study of the effects of different burning conditions uses Mössbauer spectroscopy in the uncommon role of a routine analysis tool, operating on a large number of samples. This necessitated the development of a procedure which would produce the desired information most economically. Unfortunately, superparamagnetic effects, due to small particle size, and the distribution of hyperfine fields, due to aluminum substitution, combined to prevent the spectra obtained at room temperature from being easily fitted to produce

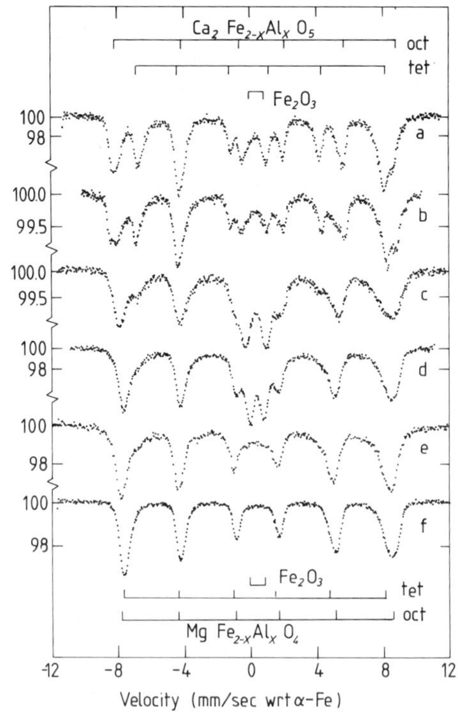

Figure 1. Mössbauer spectra taken at 4.2K of precipitator ash from (a) the Morwell coal/Hazelwood power station (at 77K), (b) the Morwell coal/Morwell power station, (c) Morwell coal/experimental furnace, (d) Yallourn number 4 cut coal/Yallourn E power station, (e) Yallourn coal/experimental furnace, and (f) the superheater deposit from the Hazelwood power station.

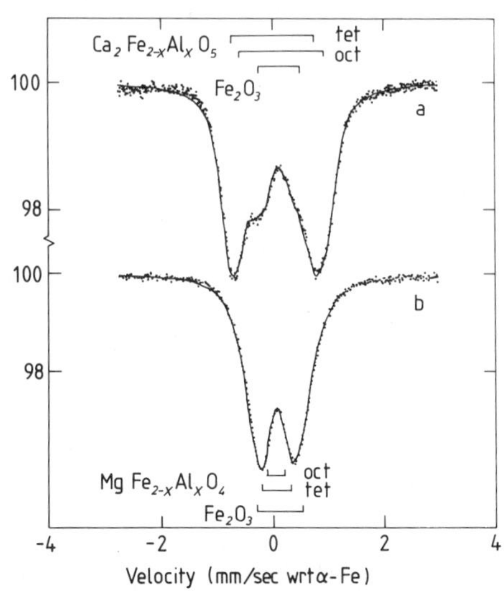

Figure 2. Mössbauer spectra obtained at 576K for precipitator ash from Yallourn E power station burning (a) Yallourn number 2 cut coal and (b) Yallourn number 4 cut coal.

the necessary analytical data. The analysis is made more difficult by the fact that the most common species each have two iron sites. We found that a combination of runs taken at 4.2K (Figure 1) and at a temperature above the ordering temperature or superparamagnetic blocking temperature (usually ca. 600K), (Figure 2) were necessary to determine the iron-containing phases.

Fortunately, for each series of samples from a particular source, details of the changes in the relative proportions of the phases as well as changes in mean particle size could be determined from room temperature spectra, once it had been ascertained to which parts of the spectra they contributed. This technique meant that the spectra had to be approximated by sums of doublets, sextets and sometimes a Gaussian baseline for a magnetically relaxing contribution and then these area components distributed amongst the various phases. The only information not obtained by the technique is the distribution of particle sizes which would have required many more experiments than the value of the information could justify.

RESULTS

Power Station Ashes

Precipitator ash from four different power station and coal seam combinations have been analyzed. These are Hazelwood power station/Morwell coal, Morwell/Morwell coal, Yallourn E/Yallourn number 2 cut coal and Yallourn E/Yallourn number 4 cut coal. In addition, deposits from the gas off-take and the convection pass heat exchangers at Hazelwood power station were examined and have been described by Aikin et al. [10].

The dominant iron-containing phase was always an aluminum substituted mixed oxide, either the brownmillerite structure calcium aluminoferrite, $Ca_2Fe_{2-x}Al_xO_5$, or the spinel structure magnesioaluminum ferrite, $(Mg,Fe)[Fe,Al]_2O_4$, where the parenthesis indicate the tetrahedral sites, and the brackets the octahedral sites. It is clear that the calcium aluminoferrite is the preferred phase, with the controlling parameter being the availability of calcium. The most reactive anion is sulfate which combines with sodium, calcium and to some extent magnesium. The amount of calcium remaining after these reactions determines the nature of the iron-containing phase.

An indication of the strength of the preference and the effects of the speed of formation can be obtained from the results on the Hazelwood samples. The tetrahedral to octahedral cation distribution in the gas off-take slag deposit of calcium aluminoferrite was significantly closer to equilibrium than that in the precipitator ash, because the former was maintained at a temperature of ca. 1000°C for sufficient time to allow solid state diffusion to occur. The phases in this deposit represent the ultimate, thermodynamically preferred phases to be formed in the flyash at the flame temperature. However, these reactions do not necessarily proceed to completion in the flyash, as its residence time of ca. one second in this region is too brief.

The material deposited on the superheater and maintained at a continuous temperature of ca. 500-800°C, can undergo post-depositional reaction with the sulfurous flue gas and alter the calcium into a calcium sulfate phase. This leaves the iron and aluminium components to combine with periclase, MgO, in a solid-state reaction to form magnesioaluminium ferrite.

Figure 1 shows the low temperature spectra of six of the samples with the top three being dominated by calcium aluminoferrite and the lower three by magnesioaluminium ferrite. The calcium aluminoferrite gave an isomer shift of 0.27(2)mm/s, a quadrupole shift of +0.35(2)mm/s and a hyperfine field of -46.4(13)T for the tetrahedral site and an isomer shift of

0.47(3)mm/s, a quadrupole shift of -0.24(2)mm/s and a hyperfine field of -52.3(7)T for the octahedral site. Here the quadrupole shift is defined as $e^2qQ/4$ and the numbers in parenthesis include both the error in the last digit and the range of observed values. The isomer shifts are measured with reference to iron. At 576K the magnetic splitting had disappeared in all cases, as illustrated in Figure 2a, and the spectrum gave an isomer shift of -0.02(2)mm/s, and a quadrupole splitting of 1.52(4)mm/s for the tetrahedral site and an isomer shift of 0.19(4)mm/s, and splitting of 1.55(5)mm/s for the octahedral site. Superparamagnetic effects were evident in all samples (Figure 3) with an appreciable fraction of the intensity being in the doublet at room temperature. The parameters agree well with those observed by Geller et al. [12] for $Ca_2Fe_{2-x}Al_xO_5$ and suggest that x is ca. 0.6-0.7 for all of our samples.

The magnesioaluminium ferrite was characterized by the 4.2K spectra with an isomer shift of 0.34(3)mm/s, a quadrupole shift of 0.0mm/s, and a hyperfine of 49.0(13)T for the tetrahedral site and an isomer shift of 0.37(7)mm/s, a quadrupole shift of 0.0mm/s, and a hyperfine field of 51.0(5)T for the octahedral site. Superparamagnetic effects were usually even more evident than for the calcium aluminoferrite and at 576K the spectra were fitted to two unresolved doublets (Figure 2b) with an isomer shift of 0.08(2)mm/s and quadrupole splitting of 0.52(3)mm/s for the tetrahedral site and an isomer shift of 0.08(4)mm/s and quadrupole splitting of 0.28(5)mm/s for the octahedral site. These values are in reasonable agreement with those obtained by de Grave et al. [13].

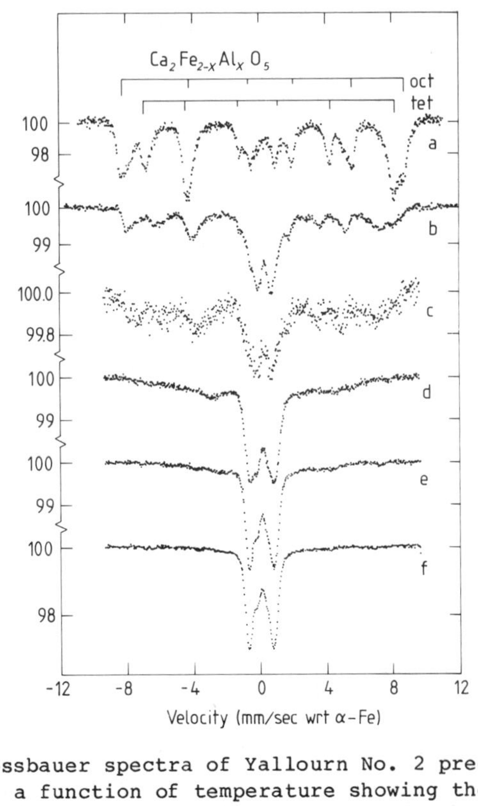

Figure 3. Mössbauer spectra of Yallourn No. 2 precipitator ash as a function of temperature showing the supermagnetic behavior of the calcium aluminoferrite. (a) 4.2K, (b) 290K, (c) 350K, (d) 500K, (e) 576K, (f) 590K.

Victorian brown coals contain small concentrations of pyrite, FeS_2, compared to most other coals and it is known that this phase produces hematite, α-Fe_2O_3 or magnetite Fe_3O_4 on combustion, as already observed in the Mössbauer spectra of United States coals [5]. We identified large particle size hematite and magnetite in one Hazelwood power station ash sample, in which the two phases could be almost completely separated from the remainder of the ash with a bar magnet. This provided further evidence that their origin was different from the inorganic iron which makes up the other iron phases described so far.

However, the spectra of many of the precipitator ashes examined also contained a doublet, with an isomer shift of 0.29(3)mm/s and a quadrupole splitting of 0.80(5)mm/s, which can persist down to 4.2K as shown in Figures 1d and 2a. It is possible that this absorption is due to aluminum substituted superparamagnetic hematite, α-Fe_2O_3, produced from the inorganic iron, or to iron substituted into one of the paramagnetic phases such as spinel, $MgAl_2O_4$ or periclase, MgO. Superparamagnetic hematite has been observed with a variety of Mössbauer parameters; those reported by Kündig et al. [14] with an isomer shift of 0.32mm/s, and a quadrupole splitting of 0.6 to 0.8mm/s are in closest agreement with our values. Even at low temperature the magnetically split spectrum of aluminum substituted hematite would be difficult to detect in our spectra, due to the presence of large contributions from either of the two double oxides. However, coals currently being examined in the experimental furnace are producing iron concentrations in the precipitator ash of up to 45 weight percent (expressed as Fe_2O_3) so this phase should become more evident if it is due to hematite.

Interesting magnetic effects were observed with two of the samples in the presence of magnetic fields. The first, as reported by Aikin et al. [10] was the effect of surface domains on the fine particle magnesioaluminium ferrite which resulted in an appreciable proportion of Mössbauer spectral lines 2 and 5 remaining even in the presence of a 4T applied field. The second effect was obtained with samples containing calcium aluminoferrite, when strong magnetic fields were applied to try and unravel the various fields evident in the 40 to 52T range. We found that the line positions for the calcium aluminoferrite subspectra remained surprisingly unaffected by fields up to 6T, whereas other antiferromagnetic phases, e.g., magnesioferrite, showed the characteristic increase and decrease of the observed fields on the two sublattices. This behavior can be understood in terms of the magnetic behavior of $Ca_2Fe_{2-x}Al_xO_5$ as a function of x, as observed by Geller et al. [12]. Our samples typically have x in the region of 0.7 which is the critical region in which the magnetization changes from being parallel to the a-axis to being parallel to the c-axis; correspondingly the anisotropy energy difference between these two configurations is very small. The EFG axis remains parallel to the b-axis independent of the value of x.

With the application of a magnetic field along the a-axis it would be energetically favorable for the sublattice magnetizations to flop to the c-axis. Similarly if the field is applied along the c-axis then they will flop to the a-axis. For the field parallel to the b-axis the magnetizations can stay along either the a or c-axis. In all cases the magnetization is perpendicular to the applied field, which accounts for the failure to observe any change in the measured hyperfine field. The magnetization also remains perpendicular to the EFG axis so that the spin flop still does not affect the line positions.

CHANGES WITH BURNING CONDITIONS

As part of a program to investigate the effects of changing the burning conditions, both for presently utilized coals and coals for possible future development, tests were carried out at the S.E.C.V. Herman Research Laboratory using an experimental furnace burning coal at a rate of 35kg/h. Two of the coals in the experimental furnace combustion program were run-of-mine Yallourn coal and Morwell coal, although they were mined at least a year after those burnt to produce the power station ash samples described above.

The changes in the ash nature for these two coals have been studied for flame temperatures of 1100, 1200, and 1300°C with overall flue gas oxygen contents of 3 and 6 percent on a wet volume basis. Changes in the flyash were immediately obvious on visual inspection, with the Yallourn sample becoming darker with increasing flame temperature and increasing oxygen, while the Morwell ashes become darker with increasing temperature but lighter with increasing oxygen.

A typical room temperature spectrum for the Yallourn sample is shown in Figure 4 and the approximation fitting needed, as described above can be seen

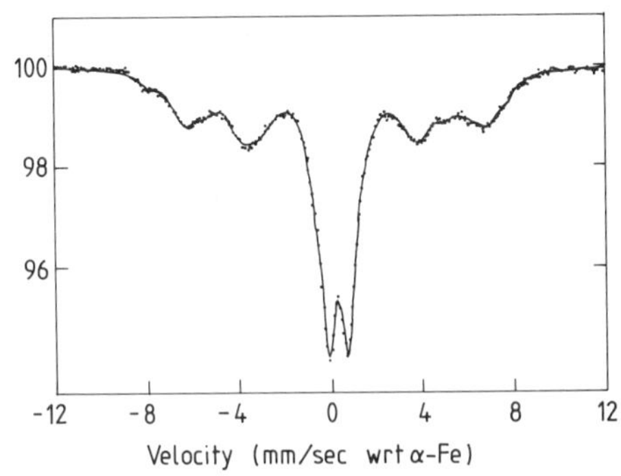

Figure 4. Room temperature spectrum of Yallourn coal precipitator ash from the experimental furnace operating at 1100°C and 3 percent excess oxygen.

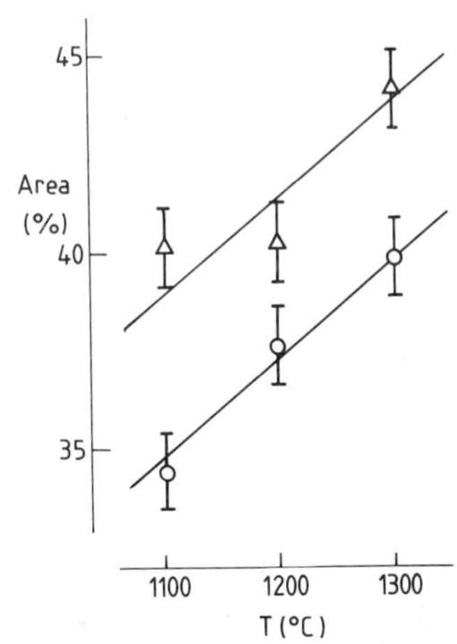

Figure 5. Percentage of the spectral area at room temperature found in the superparamagnetic doublet for Yallourn coal precipitator ash as a function of flame temperature and 3 percent, upper line, and 6 percent, lower line, flue gas oxygen content.

from the figure. The fraction of the total spectral areas appearing under the superparamagnetic doublet for the six samples is shown in Figure 5 and it can be seen that it increases smoothly with both increasing flame temperature and increasing oxygen. Such a change could be caused by either decreasing particle size or increased aluminum substitution. Because the bulk elemental analysis shows that the aluminum to iron decreases for these samples, it is assumed that there has been a reduction in particle size sufficient to compensate for this as well as to produce the change seen in Figure 5.

For the Morwell coal, which produces calcium aluminoferrite as the principal iron phase, it was found, in contrast to the Yallourn sample, that the doublet area decreased with increasing temperature. The change in aluminum to iron concentrations are less well defined, and we believe that the difference in behavior is due to changes in particle size and degree of crystallization.

CHANGES DURING WET AGGLOMERATION

The three samples produced from the Morwell and Yallourn E power stations were used for wet agglomeration tests. Mössbauer spectra of the initial precipitator ash have already been dealt with above and this section describes the changes which occurred following hydration, pelletizing, and autoclaving.

Hydration was found to increase the fraction of the spectral area in the superparamagnetic doublet of the Morwell sample, presumably due to either increased particle-particle separation or partial dissolution or hydration of the outside of the particle which reduces its volume dependent magnetic anisotropy. Further changes were observed on pelletizing and also after autoclaving, by which stage the calcium aluminoferrite spectrum had been completely replaced by a spectrum of ferrihydrite. A similar change has been observed recently by Fortune and Coey [15] in a study of changes in Portland cement. The calcium aluminoferrite in the precipitator ash is very close in composition to the C_4AF constituent of Portland cements. Fortune and Coey also observed that the iron content of the other product, hydrogarnet, $Ca_3Fe_xAl_{2-x}(OH)_{12}$, decreased to zero in the presence of 5 percent excess gypsum; our sample did not contain hydrogarnet.

The ash sample from Yallourn number 2 coal produced similar reactions, but they were not as complete, and this can be rationalized in terms of the specific surface areas and bulk densities of the two initial ashes.

The magnesioaluminum ferrite phase in the ash sample from Yallourn number 4 coal did not undergo such changes and the pellet strength was markedly less than that of the other samples. Our results suggest that, although crystal growth occurs for other phases such as gypsum, $CaSO_4 \cdot 2H_2O$, during pelletizing and autoclaving, it is likely that successful pelletizing may only be achieved for those ashes which contain calcium aluminoferrite. Complete details of these experiments will be published elsewhere.

CONCLUSIONS

We have demonstrated that Mössbauer spectroscopy can provide useful complementary information to that obtainable by other techniques. This information greatly assists in understanding the processes of the production and disposal of ash products formed during coal combustion. With six more coal types currently under investigation, it is hoped that this work will result in an understanding of the systematics of ash formation, allowing prediction of the phases from chemical analysis of the coal.

ACKNOWLEDGEMENTS

We are greatful to J. S. Hall and C. S. Hannaford for providing all the pelletizing samples and for many helpful discussions. We would also like to thank P. McNamara and J. R. Fraser for technical assistance with part of the work. The research was supported by the S. E. C. V. Herman Research Laboratory through a grant from the National Energy Research Demonstration and Development Council and by a grant from the Australian Research Grants Scheme.

REFERENCES

1. J. D. Cashion, B. Maguire, and L. T. Kiss, in "Mössbauer Spectroscopy and Its Chemical Applications", J. G. Stevens and G. K. Shenoy, eds., Am. Chem. Soc. Adv. Chem. Series, Vol. 194, p. 209 (1981).

2. P. S. Cook and J. D. Cashion, submitted for publication (1985).

3. J. D. Cashion, P. E. Clark, P. S. Cook, F. P. Larkins, and M. Marshall, in "Nuclear and Electron Resonance Spectroscopies Applied to Materials Science", E. N. Kaufmann and G. K. Shenoy, eds., North Holland, New York, p. 277 (1981).

4. P. S. Cook, J. D. Cashion, and P. J. Cassidy, Fuel 64, 1121 (1985).

5. F. E. Huggins and G. P. Huffman, in "Analytical Methods for Coal and Coal Products", C. J. Karr, ed., vol. 3, p. 371, Academic Press, New York (1979).

6. C. C. Hinckley, G. V. Smith, H. Twardowska, M. Saporoschenko, R. H. Shiley, and R. A. Griffin, Fuel 59, 161 (1980).

7. G. P. Huffman, F. E. Huggins, and G. R. Dunmyre, Fuel 60, 585 (1981); F. E. Huggins, D. A. Kosmack, and G. P. Huffman, Fuel 60, 577 (1981).

8. S. P. Taneja and C. H. W. Jones, Fuel 63, 695 (1984).

9. M. D. Patil, H. C. Eaton, and M. E. Tittlebaum, Fuel 63, 788 (1984).

10. T. L. H. Aikin, J. D. Cashion, and A. L. Ottrey, Fuel 63, 1269 (1984).

11. J. S. Hall, C. S. Hannaford, J. D. Cashion, L. J. Brown, A. L. Ottrey, and D. Tang, 1985 Proc. 4th Int. Conf. on Agglomeration, Toronto, Nov. 1984.

12. S. Geller, R. W. Grant, and L. D. Fullmer, J. Phys. Chem. Solids 31, 793 (1970).

13. E. de Grave, A. Govaert, D. Chambaere, and G. Robbrecht, Physica 96B, 103 (1979).

14. W. Kündig, H. Bömmel, G. Constabaris, and R. H. Lindquist, Phys. Rev. 142, 327 (1966).

15. J. M. Fortune and J. M. D. Coey, Cem. Concr. Res. 13, 696 (1983).

MÖSSBAUER SPECTROSCOPY AS A TOOL FOR STUDYING HYDROMETALLURGICAL TREATMENTS OF COPPER-IRON-SULFIDES

Stuart A. Fysh

Broken Hill Proprietary Co. Ltd.
Central Research Laboratories
P.O. Box 188, Wallsend
New South Wales, 2287, Australia

INTRODUCTION

Chalcopyrite, $CuFeS_2$, is by far the most abundant copper-bearing mineral, and forms the basis of the major copper deposits currently being exploited. In flotation concentrates chalcopyrite is often accompanied by one or more of the following minerals; bornite, Cu_5FeS_4, pyrite, FeS_2, pyrrhotite, $Fe_{1-x}S$, or hematite, α-Fe_2O_3. Present copper extraction technology relies on smelting of concentrates, which is polluting and energy inefficient, and substantial effort has been directed towards hydrometallurgical treatment of copper concentrates [1,2].

Most reported studies of hydrometallurgical copper extraction are based on naturally occurring mineral specimens. Quantitative mineralogical analysis of such materials by automated image analysis has been demonstrated [3], but the majority of investigations have relied on x-ray diffraction and electron microprobe analyses for characterization of samples and leach residues [4-10].

Mössbauer spectroscopy is an attractive tool for studying hydrometallurgical treatment of copper concentrates for two reasons. First, the technique is able to distinguish the iron-containing species present, giving a quantitative estimate of their relative abundance. Because the majority of copper is usually associated with iron-containing minerals, Mössbauer spectroscopy is an effective means of monitoring copper behavior in many situations. Second, the iron contained in copper-bearing minerals must be liberated at some stage of leaching, and its cost-effective removal from solution has proven a difficulty with a number of proposed leach routes [2]. Mössbauer spectroscopy has been shown to be of value for characterizing the iron species produced during acid leaching of bauxite [11], and may be of similar use in studying the role of iron in various hydrometallurgical approaches to copper extraction.

McDonald et al. [12] recently reported the use of Mössbauer spectroscopy to study the reaction product of copper(II) chloride leaching of chalcopyrite. Pyrite was the only iron-containing phase found to be present. Dimitrov et al. [13] have used Mössbauer analysis to show that the mineralogy of calcined copper concentrates is not affected by treating in a magnetic field prior to calcination. Calcination and heating in the presence of sulfur are "activation" techniques employed to increase the reactivity of copper concentrates during leaching [14], and Mössbauer analysis of activated concentrates may prove useful in explaining the process.

The use of Mössbauer spectroscopy for quantitative analysis of multiphase copper concentrates and their leach products does not appear to have been reported. Quantitative Mössbauer analysis requires reliable standard data for the species involved. In particular, it is important that the different recoil free fractions, f (the efficiency of iron nuclei as Mössbauer absorbers), of minerals concerned be taken into account. Although Mössbauer studies of the individual iron-containing minerals found in copper concentrates have appeared, recoil free fraction data is only available for hematite [15] and pyrite [16], and that for the latter is the subject of some debate. The present paper provides this data for all of the iron-containing minerals commonly encountered in copper concentrates. Quantitative analysis of a number of Australian and of two acid-leached copper concentrates is illustrated. Quantitative Mössbauer analysis of activated copper concentrates has also proven useful and is reported herein.

EXPERIMENTAL

The measurement and fitting procedures employed were similar to those previously described [17], with the only significant difference being the use of a laser interferometer to determine the exact folding points of the present spectra. All isomer shifts are referred to metallic iron at 300K. Recoil free fractions have been determined by determining the ratio of spectral areas for a two-phase absorber of known composition [18]. The unknown f is calculated from that of the reference phase and the absorber composition. A well-characterized hematite specimen, with f of 0.64(2) [15] was used as the reference material. The area, A_a, under an unsplit Mössbauer absorption line is given by Dehet et al. [19] as

$$A_a = 1/2\, \Gamma_{nat}\, \pi\, f_s t_a [I_o(1/2 t_a) + I_1(1/2\, t_a)] \times e^{-1/2 t_a}, \tag{1}$$

where Γ_{nat} is the natural Lorentzian linewidth of the transition, $t_a = f_a n_a \sigma_o$ is the effective absorber thickness, I_m is the mth-order modified Bessel function of imaginary argument, n is the number of resonant nuclei per unit area, σ_o is the absorption cross section at resonance, and s and a are source and absorber subscripts, respectively. Previously [15] the known value of f_1 was calculated from the area ratio by using the approximation

$$A_1/A_2 \simeq n_2 f_1/n_2 f_2. \tag{2}$$

However, Meisel [20] has used a series expansion of the Bessel functions in (1) to derive expressions for A_1/A_2 for various combinations of subspectra having well resolved spectral lines. These expressions are easily inverted to give a more accurate estimate of the unknown recoil free fraction. Because of the uncertainty in associating the f value of a reference mineral with that of a mineral in a copper concentrate, and the requirement of a known iron content, the use of the Meisel approximations has been restricted to deriving f data. Equation (2) has been used for subsequent analyses. Meisel gives no approximation for the case of a singlet superimposed on a sextet spectrum. The approximate expression (required here for bornite) is readily shown to be

$$A_1/A_2 = f_1/f_2\, [1 + 0.201\, (t_2 + 0.241(t_2 - t_1)) + \cdots]. \tag{3}$$

Table I. Fitted Mössbauer Spectral Parameters at 300K for the Minerals Studied.

Mineral	Origin	H kOe	Δ mm/s	δ mm/s	f
Bornite, Cu_5FeS_4	Mt. Lyell, Tasmania	-	-	0.385(2)	0.42(3)
Chalcopyrite, $CuFeS_2$	Karangi, NSW	353(1)	0.000(3)	0.247(4)	0.48(4)
Pyrite, FeS_2	Liena, Tasmania	-	0.614(3)	0.311(4)	0.58(3)
Pyrrhotite, $Fe_{1-x}S$	Renison, Tasmania	301(1), 40% 253(1), 30% 228(1), 30%	+0.06(1) 0.00(1) +0.15(1)	0.69(1) 0.68(1) 0.66(1)	0.45(5)[a]
Hematite	-	518(1)	-0.22(1)	0.36(1)	0.64(2)

[a]This f value is an average for the three different iron sites.

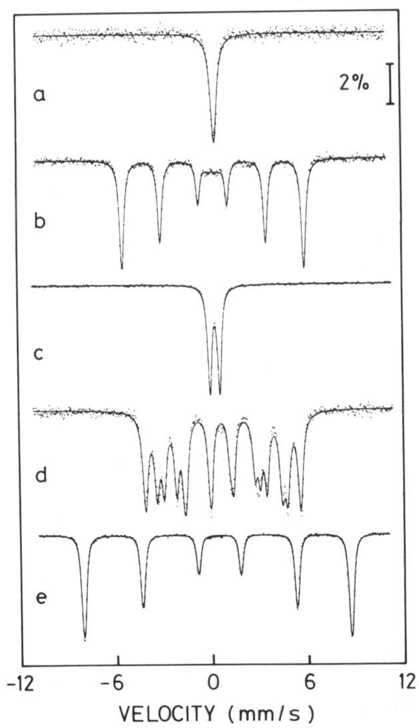

Figure 1. Mössbauer spectra obtained at 300K of the iron-containing minerals commonly found in copper concentrates, (a) bornite, (b) chalcopyrite, (c) pyrite, (d) pyrrhotite, and (e) hematite.

With the exception of hematite, all of the individual minerals studied are natural specimens (see Table I). Polished blocks of each were examined by electron microprobe analysis, and elemental concentrations for all of the minerals were found to be within one atomic percent of the stoichiometric values. The iron content of these specimens were determined by standard atomic absorption spectroscopy techniques. Polished blocks of the copper concentrates studied were also examined, again indicating mineral compositions within one atomic percent of the expected values.

RESULTS AND DISCUSSION

INDIVIDUAL MINERAL MÖSSBAUER SPECTRA

Shown in Figure 1 are the 300K spectra of the iron-containing minerals found in copper concentrations. It may be seen that contributions of the various minerals will be sufficiently well resolved in the spectrum of a mixture to allow unambiguous computer fitting. Spectra of all of the minerals except hematite have been rerun at lower velocities to establish accurate spectral parameters presented in Table I.

The Mössbauer spectra of sulfides have been well studied, and recent papers have reviewed previously published results for chalcopyrite [21], bornite [22], pyrite [23], pyrrhotite [24], and hematite [15]. Variations between natural samples are often observed to affect their Mössbauer parameters, and Goodenough and Fatseas [25] have discussed the electronic interactions important to the Mössbauer spectra of sulfides, giving some explanation for the spectral effects of stoichiometric changes. Measured parameters for all of the minerals are well within the range of values given in the studies previously cited. The only mineral considered here which commonly exhibits large spectral variations is pyrrhotite. Pyrrhotite exists in several structural forms, and its composition and Mössbauer spectrum can vary over a wide range. Three magnetically split subspectra can be resolved in the spectrum of the present sample, and the parameters of these are in reasonable agreement with values given by Huggins and Huffman [24].

RECOIL FREE FRACTION MEASUREMENTS

Compositional variation may have a substantial effect on the recoil free fraction values which should be used for quantitative analysis. The only reported studies of the effects of mineral composition on recoil free fraction [15,17] showed that the addition of aluminum to the structures of hematite and goethite, α-FeOOH, results in changes of up to 30 percent in f. This sort of variation can be studied with synthetically prepared samples, and such a study of the present suite of minerals would be desirable. However, a wide range of Australian copper concentrates have been examined as part of the present study, and both electron microprobe and Mössbauer analyses indicate that the present specimens are reasonably representative of the phases likely to be encountered in such materials.

The measured recoil free fractions are given in Table I. The value for pyrrhotite has not been corrected by Meisel's method because of strong overlap of the spectral lines. Further, the value given for this mineral is an average for each of the three distinct iron sites. The true f of each is likely to differ from this due to variations in bond strength. Further study of this complex mineral did not seem warranted because of the efficiency with which it is removed from copper concentrates by magnetic separation.

Williamson et al. [16] summarize previous measurements of the f factor for pyrite and obtained values of 0.53(4) and 0.80(8) depending on the technique used. They suggest the latter value to be more reliable while the value of 0.58(4) obtained here favors the former value.

Contaminant iron present in the reference sample of chalcopyrite is evident from a small central doublet of less than 4 percent spectral area and another very low intensity doublet of much less than one percent spectral area and having its high velocity peak centered near +2mm/s in the chalcopyrite spectrum of Figure 1. These contributions are probably due to pyrite and some form of iron(II) sulfate respectively, and have been allowed for by assuming they have the same f as chalcopyrite. Even if this assumption is in error by 10 percent, the effect on the calculated f factor for chalcopyrite is less than one percent, i.e., less significant than the error in determining the absolute iron content of the absorber.

ANALYSIS OF AUSTRALIAN COPPER CONCENTRATES

The 300K Mössbauer spectra of a number of Australian copper concentrates are shown in Figure 2. The first three are from chalcopyrite-rich deposits, the only other major iron-containing mineral present being pyrite. In the spectrum of the Cobar concentrate (Figure 2a) a minor contribution due to pyrrhotite (not included in the fit) is evident as a departure of the fit from the data near +5mm/s. This is the only pyrrhotite seen in any of the concentrates, and represents less than one percent of the total iron. A second line near +2.4mm/s in this spectrum is due to a doublet having its low velocity line almost coincident with that of the central pyrite doublet as can be noted by the greater intensity of the low velocity pyrite line as

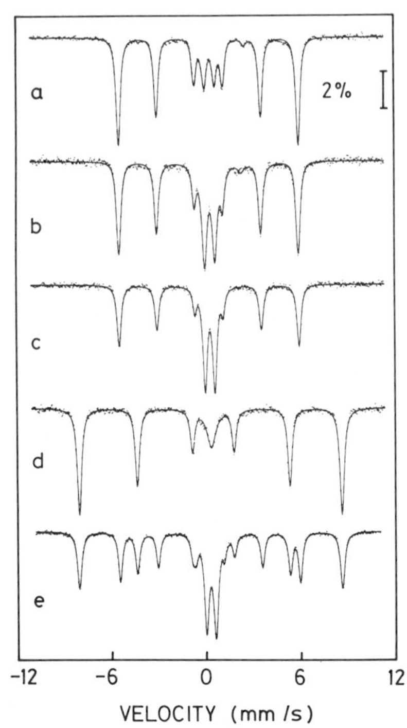

Figure 2. Mössbauer spectra obtained at 300K of five Australian copper concentrates, (a) Cobar, (b) Cleveland, (c) Woodlawn, (d) Olympic Dam OD1, and (e) Olympic Dam, OD2.

Table II. Calculated Iron Distributions for the Various Australian Concentrates and Leached Concentrates Studied[a]

Concentrate	Cu_2FeS_2	Cu_5FeS_4	FeS_2	$FeSO_3$	$FeSO_4 \cdot H_2O$	$Fe(OH)_3$
Cobar	82	-	15	-	2	-
Cleveland	66	-	31	-	3	-
Woodlawn	59	-	41	-	-	-
Olympic Dam #1	-	21	-	79	-	4
leach product		17 (-19%)[b]				
Olympic Dam #2	35	3	29	33	-	-
leach product	11 (-65%)[b]	-	59[c]	30	-	-

[a] The error associated with these values is discussed in the text. [b] The figures in brackets are calculated reductions in the content of the major copper-bearing minerals during leaching. [c] The total iron content of the pyrite, bornite, and ferric hydroxide present.

compared with the high velocity line, despite the greater displacement of the former from the neighboring inner chalcopyrite line. This doublet, of less than three percent of the spectral area, is also present to a lesser degree in the Cleveland and Woodland spectra. Its quadrupole splitting of 2.6(1)mm/s and isomer shift of +1.10(5)mm/s, indicate that it is due to iron(II). The only other iron-containing mineral indicated by x-ray diffraction analysis is a trace quantity of beaverite $Pb(Cu,Fe,Al)_3(SO_4)_2(OH)_6$, in the Cobar sample. Because the iron in this mineral is iron(III) [26], it cannot account for the observed doublet, which is most likely due to szomolnokite, $FeSO_4 \cdot H_2O$. The 300K spectral parameters of szomolnokite given by Montano [27] are 2.69(3)mm/s for the quadrupole splitting and +1.18(2)mm/s for the isomer shift, in reasonable agreement with those observed here, and it is likely that this mineral has been formed by decomposition and hydration of finely ground pyrite.

Copper in the first Olympic Dam concentrate, OD1, is mainly present as chalcocite, Cu_2S, with a lesser quantity of bornite. The other iron-

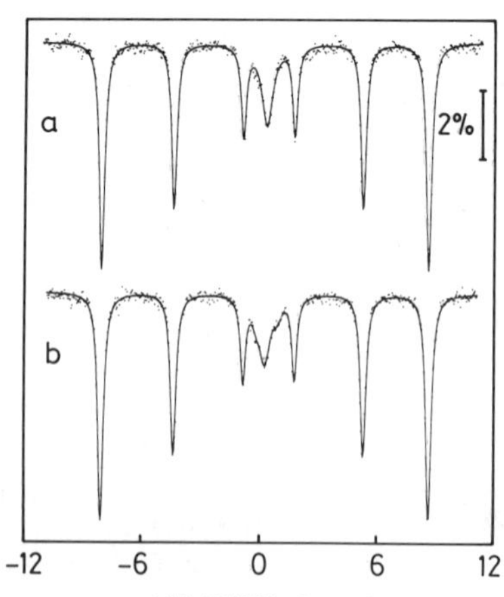

Figure 3. Mössbauer spectra of OD1 obtained at 300K (a) before and (b) after acid leaching. Reprecipitated iron from leached bornite is evident as a central doublet.

containing mineral present is hematite. Hematite is also the major iron-containing mineral in the second Olympic Dam concentrate, OD2, together with chalcopyrite, pyrite, and a small amount of bornite. The latter is not readily apparent from the spectrum (Figure 2e), but failure to account for it during the fitting results in a poor fit to the low velocity pyrite line.

Quantitative analyses of the five concentrates are given in Table II. The value of f for szomolnokite of 0.59(9) was taken from Montano [27]. As pointed out previously, because of the uncertainty in identifying concentrate constituents with the reference minerals, minor overlap corrections are not justified for these spectra. In any case the absorber thicknesses used were such that t_a is less than one for all lines, so that thickness effects will be negligible [28].

ANALYSIS OF LEACHED CONCENTRATES

The two Olympic Dam concentrates were subjected to an oxidizing acid leach, and the products analyzed by Mössbauer spectroscopy (see Table II). The spectra of OD1 before and after leaching are shown in Figure 3. The leaching conditions would not be expected to dissolve hematite, and this is confirmed by the analysis. Careful examination of the leached product spectrum reveals a central doublet with a quadrupole splitting of Δ = 1.10(2)mm/s, and an isomer shift of 0.37(1)mm/s, almost certainly due to amorphous ferric hydroxide similar to that precipitated during acid leaching of bauxite [11]. The recoil free fraction assumed for this phase is 0.65(2) [17,18]. The broadening of the central doublet in the spectrum of the OD2 leach product (Figure 4) also indicated reprecipitation of iron as an amorphous gel phase. Bornite appears to represent less than one percent of the iron present, but it is not possible to resolve the separate bornite, pyrite, and gel contributions at this temperature. As it had already been shown that hematite was not dissolved from OD1 under identical leach conditions, the reduction in chalcopyrite content was calculated relative to this mineral. The gel phase would be expected to order magnetically, producing a characteristic six line spectrum at reduced temperatures, thus permitting its quantification. However, this analysis would require remeasurement of low temperature f values for the minerals concerned and was not pursued.

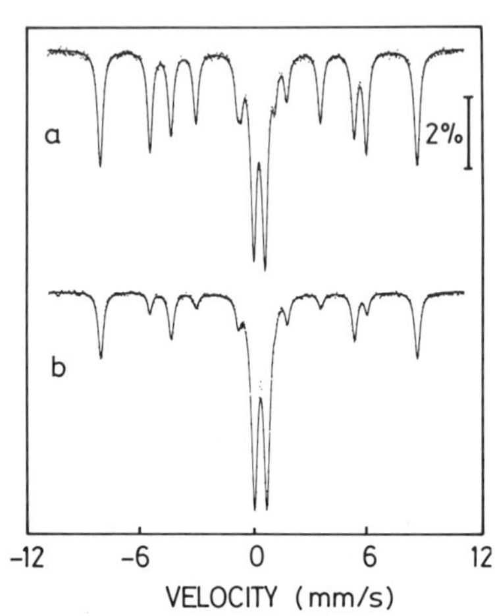

Figure 4. Mössbauer spectra of OD2 obtained at 300K (a) before and (b) after acid leaching. The significant reduction in chalcopyrite content is obvious.

Table III. Calculated Iron Distributions for the Activated Copper Concentrates Studied

Weight Percent Sulfur	Relative Percentage of Total Iron Present as	
	Chalcopyrite	Pyrite
Original concentrate	80	20
12	57	43
15	9	91

ANALYSIS OF ACTIVATED COBAR CONCENTRATE

The activation of copper concentrates aims to increase both the copper leaching rate and the selectivity for copper extraction over iron by reducing the iron content of the copper-bearing phases present. As noted by Demopoulos and Distin [14], it has been shown that transformation of chalcopyrite to mixtures of pyrite and covellite and/or nukundamite, $Cu_{5.5}FeS_{6.5}$, and/or bornite can be achieved by heating in the presence of sulfur. These investigators found that small variations in activation conditions may substantially affect the product mineralogy and hence the leaching behavior.

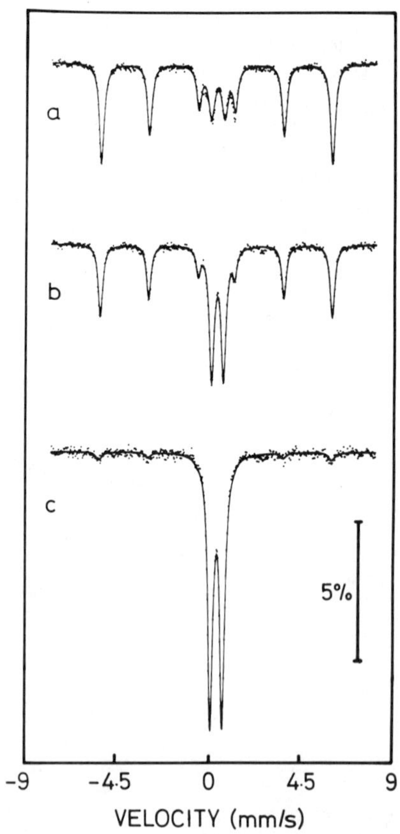

Figure 5. Mössbauer spectra of Cobar copper concentrate obtained at 300K (a) before and after activation with (b) 12 and (c) 15 weight percent sulfur.

The spectra of a Cobar copper concentrate are shown in Figure 5 before and after activation for about one hour at 340°C, in a nitrogen atmosphere. Spectrum (a) is the initial sample, while (b) and (c) represent 12 and 15 weight percent sulfur mixtures, respectively. Pyrite is the only iron-containing product formed, with relative pyrite content increasing with sulfur content in the mixture (Table III). Mössbauer analysis is an easy way in which to quantitatively determine the extent of mineral transformation during activation, and the simple form of the spectra suggest that rapid Mössbauer analysis techniques [29] may be applicable in this situation. It is difficult to estimate the error associated with all of the above quantitative analyses. Considering the errors in the measured recoil free fractions, and the potential error in associating these values with supposedly similar minerals in a particular specimen, an absolute error of ±10 to 20 percent is not unreasonable. However, this is at least as good as can be routinely achieved by other instrumental techniques such as x-ray diffraction analysis. Of more interest is the relative error between, for example, determinations made on similar samples subjected to different treatments. In this case relative errors of ±2 to 5 percent are easily achieved, the major uncertainty being introduced by the sampling procedure.

CONCLUSIONS

Quantitative mineralogical analysis of copper concentrates has been demonstrated to be readily performed by Mössbauer spectroscopy. The recoil free fraction data reported here can be used in conjunction with the spectra of thin copper concentrate absorbers to give quantitative mineralogical analyses of use in leaching and activation studies. Greatest accuracy is obtained when Mössbauer analysis is used as a relative technique, applied to a suite of similar samples. This technique is likely to prove of great use in copper extraction research, where accurate quantification of iron phase redistribution is important. Further, the well-resolved nature of the various mineral subspectra contributing to copper concentrate Mössbauer spectra indicates that mineralogical analysis using rapid Mössbauer techniques may be feasible. This possibility is currently under investigation.

ACKNOWLEDGMENTS

The author wishes to express his appreciation to Drs. K. A. Chouzadjian and N. V. Quy for helpful discussions and to the Broken Hill Proprietary Co. Ltd. for permission to publish this work.

REFERENCES

1. F. Habashi, in "Chalocopyrite, Its Chemistry and Metallurgy", pp. 45-135, McGraw-Hill, New York (1978).

2. G. Thorsen, in "Topics in Nonferrous Extractive Metallurgy", R. Barkin, ed., Critical Reports on Applied Chemistry, vol. 1, pp.1-41, Blackwell, Oxford (1980).

3. W. Petruk, CIM Bull., 146 (1976).

4. D. C. Price and J. P. Chilton, Hydrometall. 5, 381 (1980).

5. M. Borran, J. M. Demarthe, H. Renon, and F. Boratin, Metall. Trans. B. 12B, 269 (1981).

6. J. E. Dutrizac, Metall. Trans. B. 12B, 371 (1981).

7. J. E. Dutrizac, Metall. Trans. B. 13B, 303 (1982).

8. G. W. Warren, M. E. Wadsworth, and S. M. El-Raghy, Metall. Trans. B. 13B, 571 (1982).

9. B. Pesic and F. A. Olson, Metall. Trans. B. 14B, 577 (1983).

10. B. Pesic and F. A. Olson, Hydrometall., 12, 195 (1984).

11. S. A. Fysh and P. E. Clark, Hydrometall. 10, 285 (1983).

12. G. W. McDonald, T. J. Volovic, J. A. Dumesic, and S. H. Langer, Hydrometall. 13, 125 (1984).

13. R. Dimitrov, A. Hekimova, S. Asenov, T. Ruskov, and T. Tomov, Thermochim. Acta 40, 349 (1980).

14. G. P. Demopoulos and P. A. Distin, Hydrometall. 1, 111 (1983).

15. S. A. Fysh and P. E. Clark, Phys. Chem. Minerals 8, 257 (1982).

16. D. L. Williamson, T. W. Guettinger, and D. W. Dickerhoof, in "Mössbauer Spectroscopy and Its Chemical Applications", Adv. in Chemistry Series, 194, pp. 177-208, Amer. Chem. Soc., Washington (1981).

17. S. A. Fysh and P. E. Clark, Hydrometall. 10, 285 (1983).

18. S. A. Fysh, J. D. Cashion, and P. E. Clark, Clays Clay Min. 31, 293 (1983).

19. G. Dehe, B. Siedel, and W. Meisel, Nucl. Instr. and Meth. 133, 381 (1976).

20. W. Meisel, in "Proc. 5th Int. Conf. on Mössbauer Spectroscopy Bratislava", M. Hucl and T. Zencik, eds., Nuclear Information Centre, Prague (1973).

21. H. N. Ok and C. S. Kim, Nuovo Cimento 28B, 138 (1975).

22. M. S. Jaghadeesh, H. M. Nagarathna, P. A. Montano, and M. S. Seehra, Phys. Rev. B, 23, 2350 (1981).

23. B. J. Evans, R. G. Johnson, F. E. Senftle, C. Blaine-Cecil, and F. Dulong, Geochim. Cosmochim. Acta 46, 761 (1982).

24. F. E. Huggins and G. P. Huffman, in "Analytical Methods for Coal and Coal Products", pp. 371-423, vol. III, C. Karr, Jr., ed., Academic Press, New York (1979).

25. J. B. Goodenough and G. A. Fatseas, J. Solid State Chem. 41, 1 (1982).

26. J. L. Jambor and J. E. Dutrizac, Can. Mineral., 21, 101 (1983).

27. P. A. Montano, Fuel 56, 397 (1977).

28. G. K. Shenoy, J. M. Friedt, M. Maletta, and S. L. Ruby, Mössbauer Effect Methodology 9, 277 (1974).

29. S. A. Fysh and J. B. Lee, Proc. 5th Process Technology Conference, Iron and Steel Society, AIME, 187 (1985).

MÖSSBAUER EFFECT STUDY OF THE THERMAL REACTIONS OF PHOSPHATE MINERALS FROM CHRISTMAS ISLAND FERTILIZER ROCKS

I. W. M. Brown and J. E. Davies

Chemistry Division, Department of
Scientific and Industrial Research
Lower Hutt, New Zealand

ABSTRACT

Phosphate rock from Christmas Island is processed for agricultural purposes by sulfuric acid treatment, which converts it into soluble superphosphate. The presence of iron in the ore can interfere with this reaction. A Mössbauer effect study undertaken to identify the crystallographic environment of the iron in these rocks has focussed on crandallite, millisite, and goethite, the principal impurity components. A combination of Mössbauer spectroscopy, thermal analysis, and X-ray diffraction has been used to follow the structural course of the iron throughout the thermal reactions of these phosphate rocks, providing information about the initial state of iron in these minerals and the final environment of the iron in the calcined phosphate ores.

INTRODUCTION

Until the late 1960's, New Zealand superphosphate was made mainly from very pure Nauru and Ocean Island rocks. From 1968, with the gradual depletion of the Ocean Island deposit, New Zealand began using increasing quantities of phosphate rock from Christmas Island in the Indian Ocean. Since the mid 1970's, most New Zealand superphosphate has been prepared from a 50:50 blend of pure Nauru and relatively impure Christmas Island rock. Depletion of the purest grade of Christmas Island rock is now imminent and the remaining reserves of "A" grade quality phosphate rock are becoming difficult and expensive to recover. The overall reduction in quality of the Christmas Island rock can barely be accommodated by the pure Nauru rock, particularly with increased prices being applied to this latter resource over recent years. These factors are causing New Zealand fertilizer manufacturers to search elsewhere for raw phosphate rock of high quality and the recent trend to import phosphate rock from USA and Jordan is sure to continue. However, vast tonnages of lower grade phosphate ore remain on Christmas Island and the challenge of utilizing this Australian operated resource, with its inherently high iron and aluminum impurity content, has confronted manufacturers and scientists for several years.

Fertilizer manufacture in New Zealand is almost entirely designed to produce "single" superphosphate by the addition of sulfuric acid to raw phosphate rock. A simplified scheme, assuming the addition of pure sulfuric acid to pure fluoroapatite, can be represented by a two step reaction as

$$2Ca_5(PO_4)_3F + 9H_2SO_4 \rightarrow 6H_3PO_4 + 9CaSO_4 + CaF_2 \quad (1)$$

$$2Ca_5(PO_4)_3F + 12H_3PO_4 \rightarrow 9Ca(H_2PO_4)_2 + CaF_2 \quad (2)$$

to give a net reaction of

$$2Ca_5(PO_4)_3F + 6H_2SO_4 \rightarrow 3Ca(H_2PO_4)_2 + 6CaSO_4 + CaF_2.$$

In reality, the monocalcium phosphate (MCP) crystallizes as the monohydrate, but this crystallization reaction and much other reaction chemistry often proceeds long after the initial manufacturing process while the material is in storage. Further, carbonate and hydroxy substitution in the apatite causes a diversity in the product composition that varies with the nature of the source rock. However, some of the most profound effects on the reaction chemistry are the result of iron and aluminum contamination of the phosphate rock. The main contaminants of Christmas Island rock, crandallite, millisite, and goethite, react with sulphuric acid to form hydrous, gelatinous iron aluminum phosphates. The formation of these compounds has three major effects on the reaction chemistry. First, they compete with the apatite for the free acid. Second, they reduce the reaction rate between the rock and the acid by physically coating the apatite particles. Third, they increase the liquid phase in the product, preventing the short term crystallization of MCP hydrate thus causing handling problems. On storage, often after weeks or months, various iron aluminum phosphates crystallize out. These are water-insoluble compounds whose presence reduces the amount of plant-available phosphorus in the superphosphate. The synthesis and identification of some of these compounds was originally investigated by White [1], who identified several compounds of particular importance, but the characterization of these phases is outside the scope of this investigation.

Solutions to circumvent problems associated with the effects of iron and aluminum contamination have included calcination [2] of the phosphate rock at temperatures as high as 1100°C. The calcined material is then suitable for conventional acid processing into superphosphate as the solubility of both iron and aluminum is significantly reduced. Further, materials calcined at rather low temperatures have been marketed as direct application fertilizers, i.e., no sulphuric acid treatment has been undertaken. For example, "Calciphos" and "Citraphos" are retailed versions of Christmas Island "C" rock calcined at 550°C.

A detailed analysis of these thermal reactions and assignments of the differential thermal analysis, DTA, and thermogravimetric, TG, curves has not previously been undertaken and is seen as a vital step in understanding the role of iron-containing species in fertilizer chemistry. Mössbauer spectroscopy and X-ray powder diffraction, XRD, have been used to identify these iron species so as to help understand where and why processing problems occur.

EXPERIMENTAL

Up to four grades of Christmas Island phosphate rock may be arbitrarily assigned on the basis of iron, aluminum and phosphorous content [3] and typical analyses for these rock grades together with the principal mineral phases found in the rocks are summarized in Table I.

Table I shows that although "A" grade rock is predominantly apatite, the lower grades are heavily contaminated with three main mineral phases,

Table I. Typical Chemical and Mineral Analyses of Christmas Island Rocks

Grade	%Fe	%Al	%P	Principal Mineral Phases
A	1.0	1.9	15.7	Apatite >> Crandallite ~ Millisite > Goethite
B	3.3	6.1	14.9	Apatite > Crandallite > Millisite > Goethite
C	7.7	12.7	12.7	Crandallite > Millisite > Apatite ~ Goethite
D	9.1	14.9	12.2	Crandallite > Millisite > Apatite ~ Goethite

$Ca(Al,Fe)_3(PO_4)_2(OH)_5 \cdot H_2O$, millisite, $(Na,K)Ca(Al,Fe(_6(PO_4)_4(OH)_9 \cdot 3H_2O$, crandallite, and goethite, α-FeOOH. Other minor mineral phases present include barrandite, montgomeryite, and wavellite. More detailed mineralogical and geological information has been presented by Trueman [4]. Representative samples of all four rock grades of Christmas Island rock were investigated by thermal analysis and Mössbauer spectroscopy.

Mössbauer spectra were recorded with an Elscint AME-50 spectrometer with a 10mCi cobalt-57 in rhodium source, calibrated with respect to natural iron foil, and computer fit to Lorenzian line shapes. In all cases, the spectra were fit to symmetric doublets and sextets.

Combined thermogravimetry-mass spectroscopy (TGMS) experiments were undertaken with a Stanton TG-770 thermobalance coupled to an Extranuclear quadrupole mass spectrometer. The samples were heated at $10^{\circ}C$/min in 0.04 liters/min of helium carrier gas. X-ray diffraction of the product phases was undertaken with a Philips automatic powder diffractometer system which used a graphite monochromator and cobalt Kα radiation.

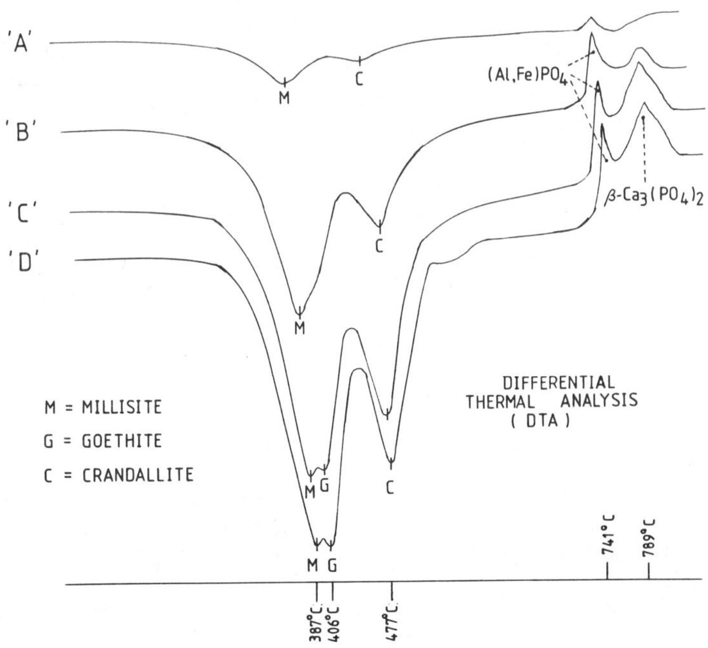

Figure 1. DTA curves for the four basic grades of Christmas Island rock.

RESULTS AND DISCUSSION

Thermal Analysis Experiments

Differential thermal analysis traces for the four basic rock grades are shown in Figure 1. The DTA curves are characterized by a number of endothermic events clustered in the region of 400 to 500°C and several exothermic events in the region of 700 to 800°C. The materials were heated in air to selected temperatures chosen from the DTA curve and the fired products were analyzed by XRD and Mössbauer spectroscopy.

The lowest temperature feature, seen as a broad weak endothermic peak in "A" grade, develops into a sharp resolved doublet in lower grade rock with peaks at 387 and 406°C. XRD analysis shows that these peaks are due to the decomposition of millisite and goethite, respectively. The second prominent endothermic event, at 477°C, is due to the decomposition of crandallite. Although the goethite decomposition leads to the formation of an initially poorly crystalline hematite, neither millisite nor crandallite decompose to form crystalline phases. Instead, they form an X-ray amorphous calcium-iron-aluminum phosphate phase.

Although no previous thermal analysis data for millisite have been reported, DTA data for a Florida crandallite has been described by Blanchard [5], who measured a decomposition endotherm at 525°C followed by an amorphous phase which recrystallized at 845°C to $AlPO_4$. For the present data, the exothermic events, seen more prominantly in the lower quality rock, represent recrystallization reactions from this amorphous phase. The peak at 741°C represents the crystallization of an $(Al,Fe)PO_4$ phase having the cristobalite structure. The peak at 789°C is due to the crystallization of whitlockite, β-$Ca_3(PO_4)_2$. The fact that the observed temperatures are very different to those reported for the Florida crandallite reflects the very great chemical differences between these two materials. In particular, the Florida crandallite had a high level of SrO substitution and was contaminated by various clay minerals.

Further heating of the Christmas Island rock leads to the formation of corundum, α-Al_2O_3. By 1100°C the stable phases present are hematite, $(Al,Fe)PO_4$, whitlockite, corundum, and the original apatite.

TGMS experiments were undertaken on samples of the four rock grades. By way of example, the curves for grade "C" rock are reproduced in Figure 2. It may be seen that the DTG curve closely mimics the water loss curve (m/e =17), suggesting that the dominant reactions observed in both TG and DTA experiments in the region up to 600°C are due to dehydration and dehydroxylation reactions. Note however that there is a small but significant CO_2 loss (m/e =44) coincident with the major water loss reactions. X-ray diffraction evidence suggests that the major decompositions at ca. 350 and 475°C are due to the decomposition of millisite and crandallite, respectively, and the TGMS data indicates that there is a small amount of carbonate substitution in both of these phases. The higher temperature CO_2 evolution peak at ca 780°C is due to either free $CaCO_3$, associated with the apatite phase, or to carbonate substitution in the apatite phase itself. Although the work of Silverman et al. [6] would tend to suggest the former option, albeit at a very dilute level, more recent evidence suggests that lattice substitution of carbonate is possible in at least two discrete types of site [7].

Mössbauer Spectroscopy

Mössbauer spectra were recorded for each of the four rock grades at various temperatures as determined from the DTA and TGMS curves. Figure 3 shows spectra for grades "B" to "D" recorded at room temperature. The relatively low iron level in grade "A" rock results in a poor signal to noise ratio for this material which precludes accurate measurement of the spectral parameters. The spectrum is not reproduced here. All spectra may be fit to three overlapping iron(III) quadrupole doublets with the proportion of the innermost doublet increasing as the rock grade becomes less pure. Magnetically split spectra are virtually absent although there is a hint of a broad six-line absorption in the "D" spectrum.

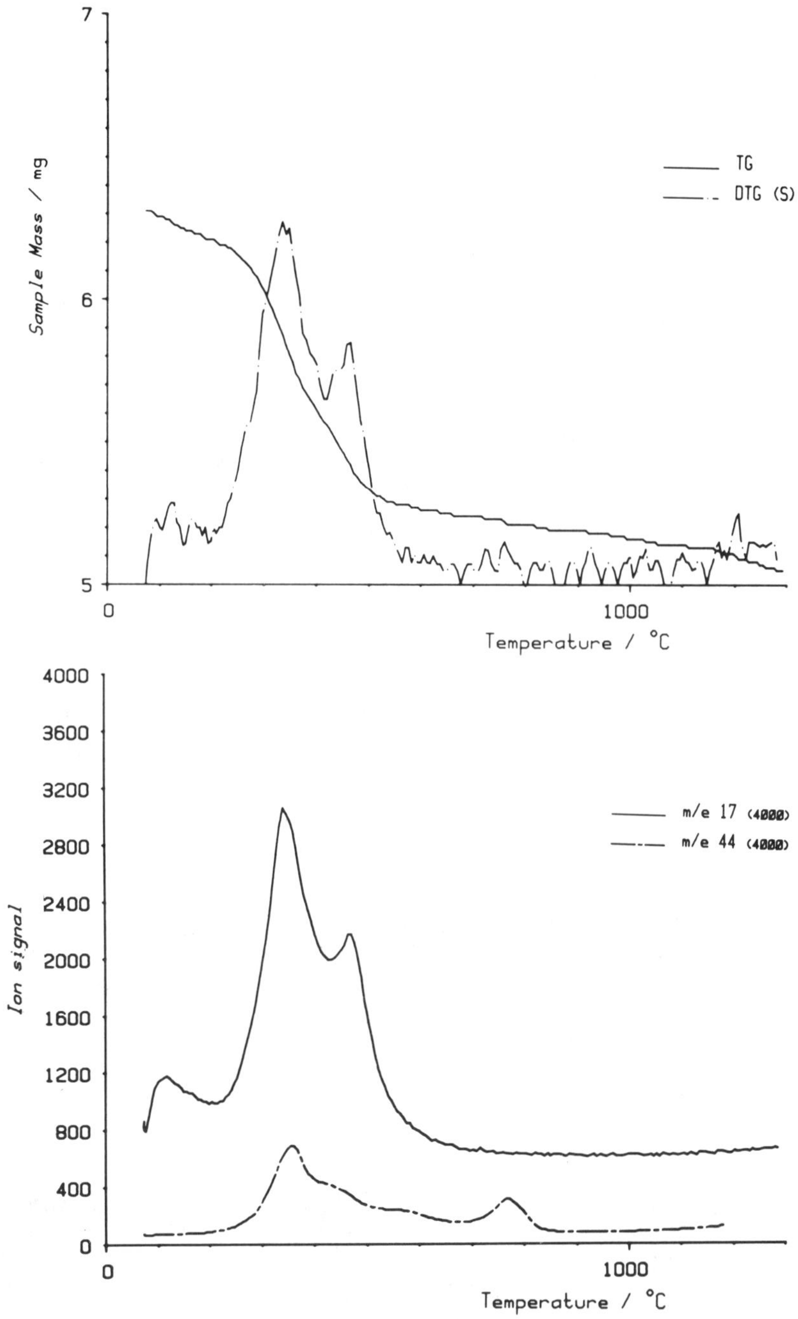

Figure 2. TGMS curves for C grade Christmas Island rock. TG and DTG curves, above, and H_2O (m/e =17) and CO_2 (m/e = 44) ion signal curves.

XRD of the lower grade rocks shows clear evidence of goethite, α-FeOOH, but this is surprisingly absent as a magnetically split species in any of the Mössbauer spectra. Conventional interpretation would suggest that if the

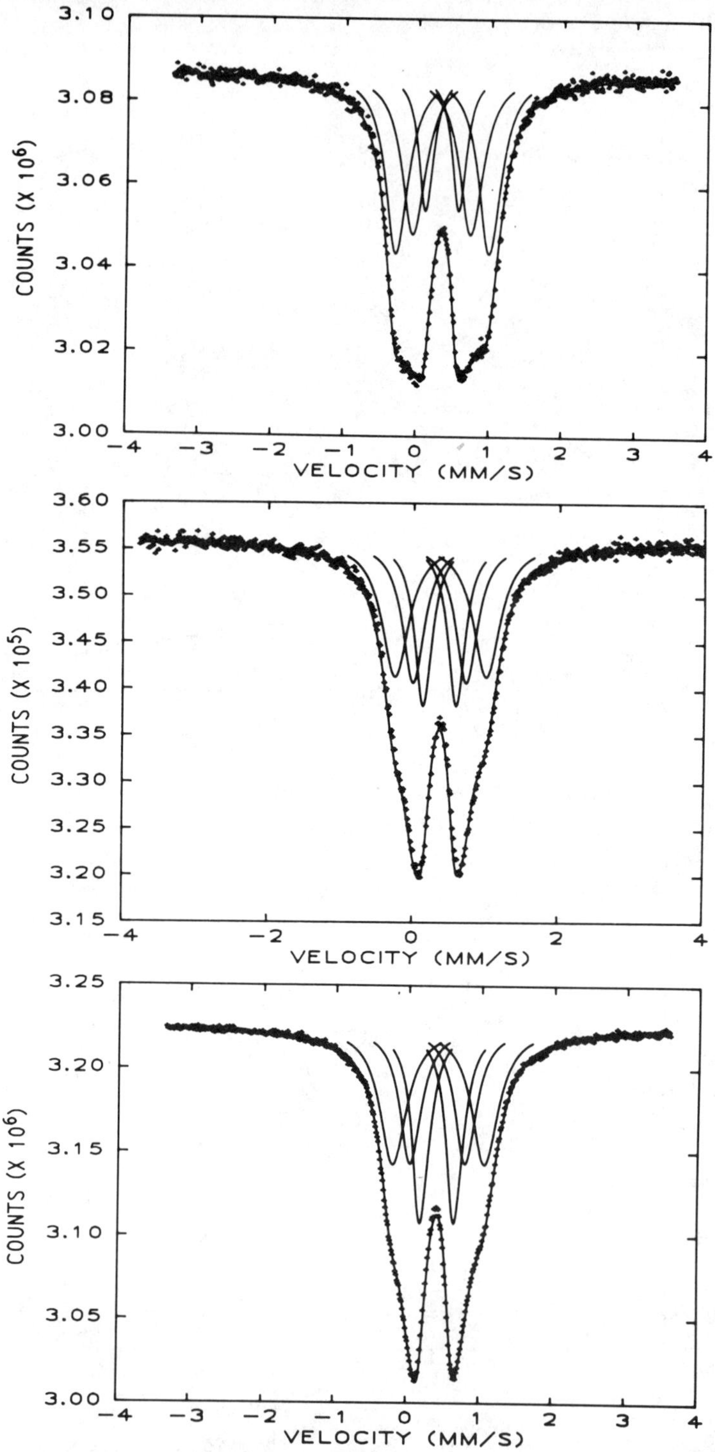

Figure 3. Mössbauer spectra of Christmas Island rock, grades B, top, C, center, and D, bottom.

Table II. Parameters and Assignments for Christmas Island Phosphate Rocks

Grade	ΔE_Q mm/s	δ mm/s	Percent Area	Assignment
B	1.28	0.35	41	Crandallite
	0.79	0.34	39	Millisite
	0.45	0.34	20	Goethite
C	1.25	0.36	38	Crandallite
	0.72	0.35	33	Millisite
	0.45	0.35	29	Goethite
D	1.24	0.34	37	Crandallite
	0.74	0.33	30	Millisite
	0.45	0.33	33	Goethite

goethite particles were large enough to be seen by XRD then a six-line Mössbauer spectrum should be observed also. The solution to this paradox maybe found by careful measurement of the XRD parameters of the goethite, which show a clear shift from the expected positions for pure α-FeOOH. Based on recent XRD work by Schulze [8] on the system FeOOH-AlOOH we may predict that the Christmas Island goethite is aluminum-substituted to a level of ca. 23 percent. This estimate is based on observation of the 110 and 221 reflections as the preferred 111 reflection is obscured by a crandallite peak. These observations are consistent with those of Murad and Schwertmann [9] who have recorded two-line goethite spectra at room temperature only in cases where the level of aluminum substitution is of the order of 12 percent or higher. It must be acknowledged that aluminum substitution versus crystallinity remains an area of some debate for goethite and even in the present case, where the material has been characterized by two independent techniques, there remain some minor doubts that the material is of natural origin and not synthetic, i.e. the level of impurity substitution is unknown. However, we have been able to further clarify the peak assignments in the spectra of the unheated phosphate rock by two additional experiments.

First, we have applied the well-known sodium citrate-sodium dithionite extraction technique to samples of "B", "C" and "D" grade rock with the expectation that this would leach out any free iron oxide or hydroxide species present. Figure 4 shows the spectra for unleached and leached samples of Christmas Island "B" rock. Three iron(III) quadrupole doublets can be readily fit and assigned under the envelope for the unleached rock. While the innermost doublet with a quadrupole splitting of 0.45mm/s and an isomer shift of 0.34mm/s has clearly been leached out and may therefore be assigned to the aluminum-substituted goethite, the assignment of the two remaining doublets is less direct and relies on the observation of the partially thermally decomposed rock spectra presented in the next section. The Mössbauer parameters and assignments for rock grades "B" to "D" are shown in Table II. For the percentage area under the fitted peaks to be equated directly with the percentage of iron in the phases present, the assumption must be made that the recoil-free fraction is equal for each of the phases present.

Second, we have measured the spectrum of Christmas "D" rock at both 78 and 4.2°K. Figure 5 shows these low temperature spectra together with that for the room temperature material obtained at the same velocity scale for comparison. In both low temperature spectra the inner goethite doublet is absent and is replaced by magnetic sextet. Note that the residual crandallite/millisite spectrum is still present. The measured magnetic hyperfine field for this goethite is 467kOe at 78°K and 490kOe at 4.2°K. These values may be compared with typical values of 498kOe at 78°K and 506kOe at 4.2°K for a well crystallized goethite free from impurity substituents [9]. On the basis of numerous experimental data, Murad and Schwertmann [9] have evolved a relationship (3) between the magnetic hyperfine field, H_i, the mean crystallite diameter, MCD_{111}, calculated from the half-width of the 111 XRD reflection, and the degree of aluminum substitution, Al_s.

$$H_i = 506.5 - 42Al_s - 87/MCD_{111} \qquad (3)$$

As noted earlier, we have been unable to measure the 111 goethite reflection accurately due to interference from an adjacent crandallite peak and therefore have used the alternative (but less preferred) 110 reflection. Entering our observed data of MCD_{110} of 19nm into equation (3) gives an aluminum substitution of ca. 28 percent, somewhat higher than the 23 percent calculated from the XRD peak shift measurement, but within acceptable error bearing in mind the problems imposed by the obscured 111 reflection.

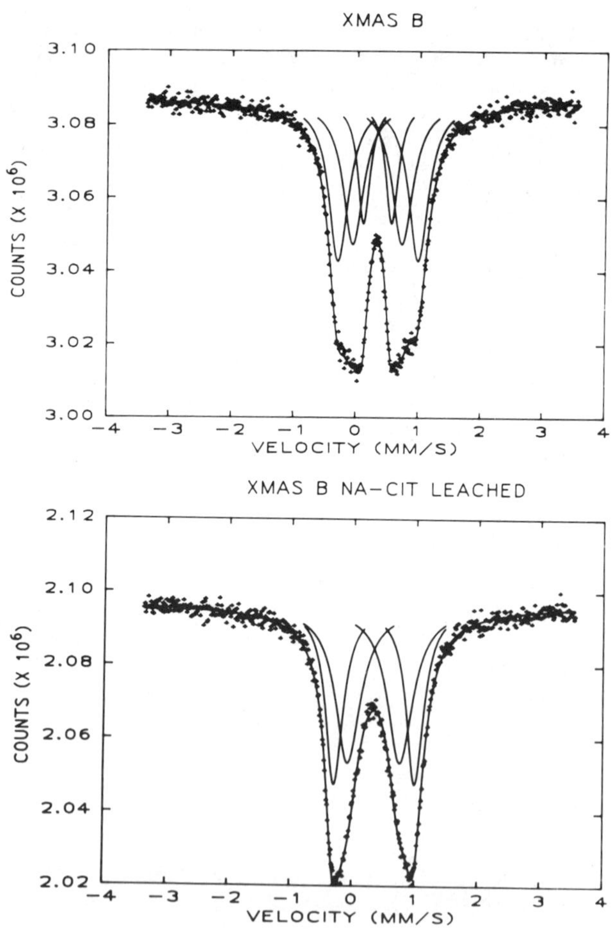

Figure 4. Mössbauer spectra of B grade Christmas Island rock, above, and of sodium citrate/sodium dithionite leached rock, below.

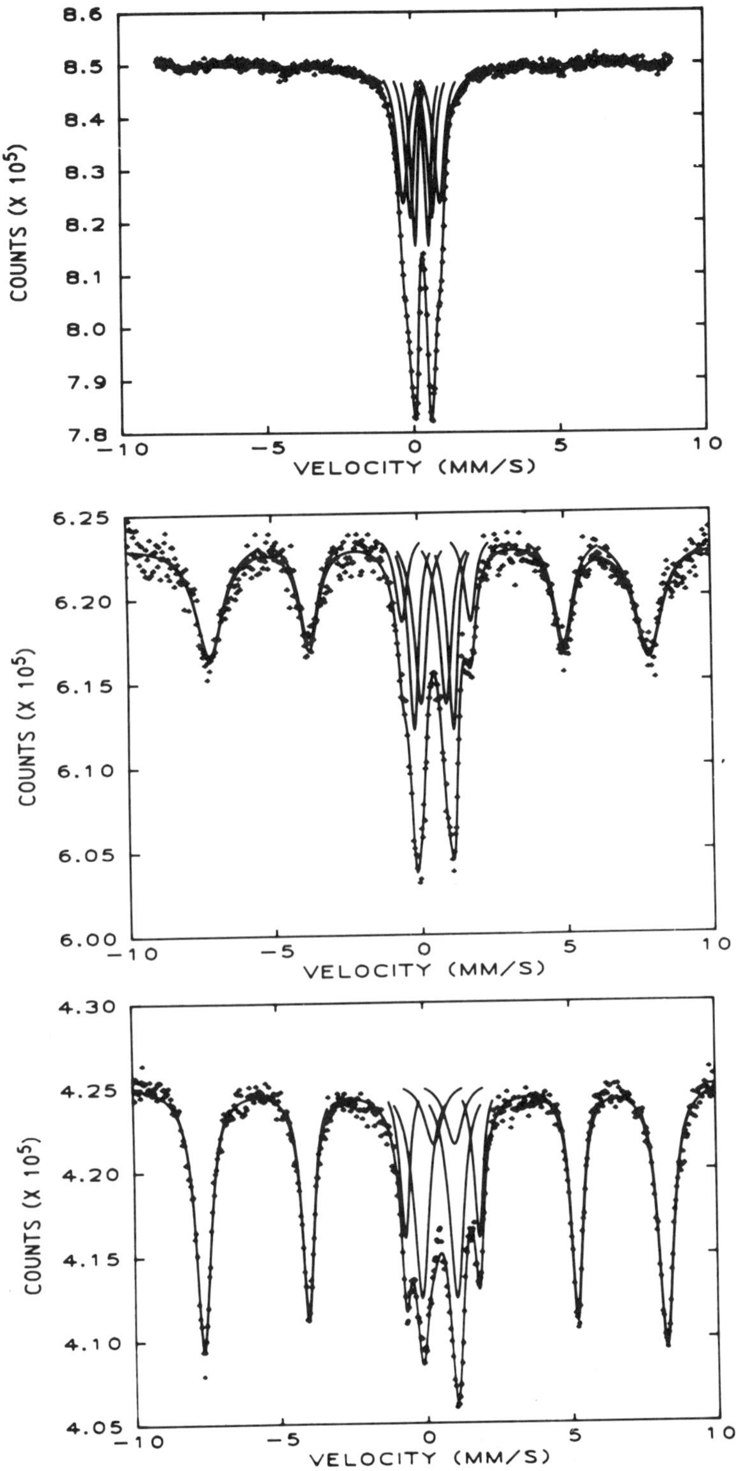

Figure 5. Mössbauer spectra of D grade Christmas Island rock, at room temperature, top, 78K center and 4.2K, bottom.

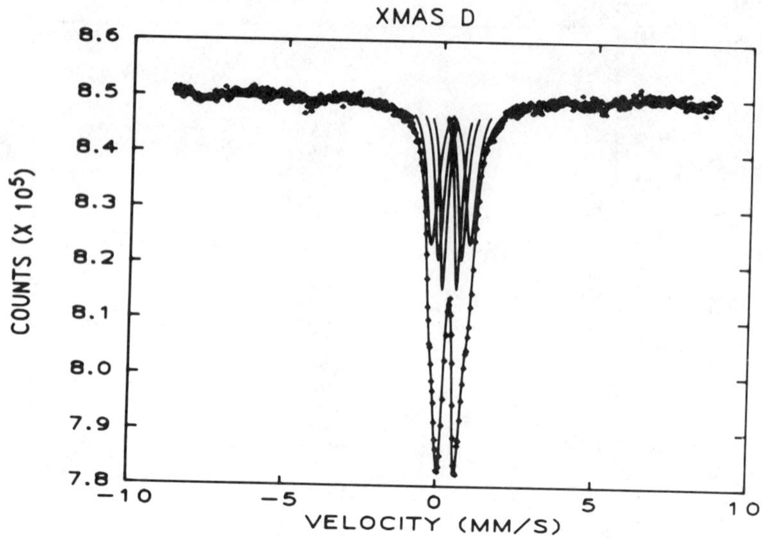

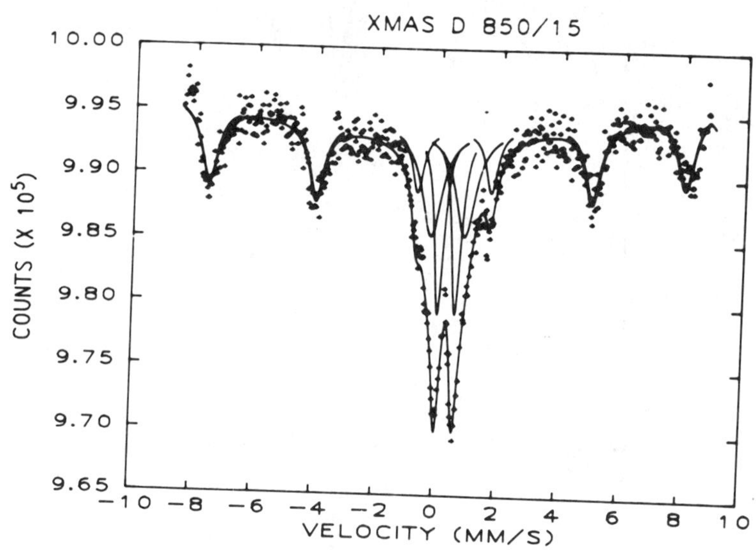

Figure 6. The effect of calcination on the Mössbauer spectra of D grade Christmas Island rock; lower Mössbauer spectrum, "Calciphos" fertilizer.

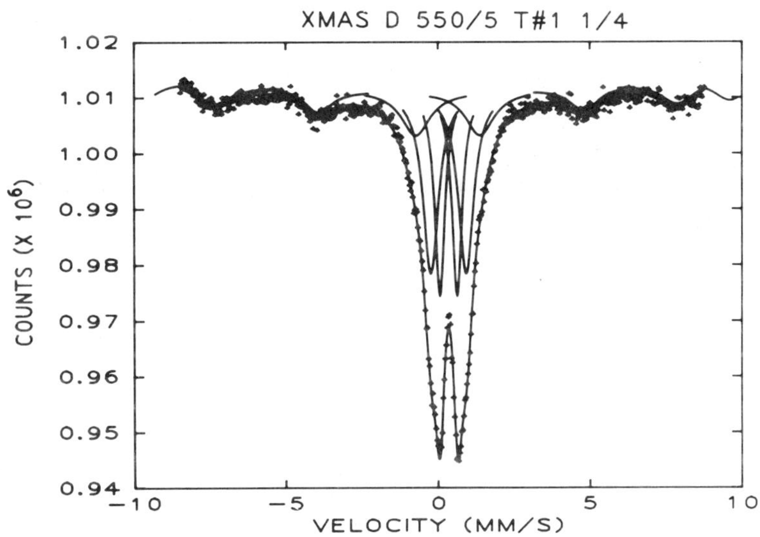
XMAS D 550/5 T#1 1/4
COUNTS (X 10^6)
1.02
1.01
1.00
0.99
0.98
0.97
0.96
0.95
0.94
−10
−5
0
5
10
VELOCITY (MM/S)

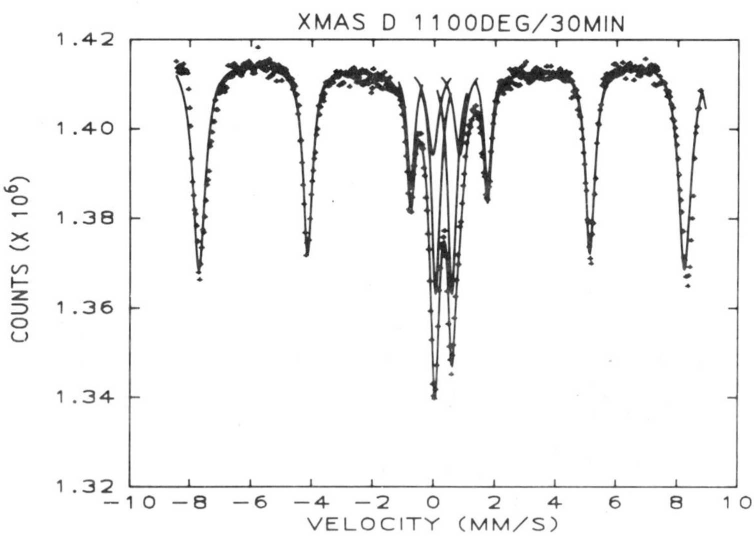
XMAS D 1100DEG/30MIN
COUNTS (X 10^6)
1.42
1.40
1.38
1.36
1.34
1.32
−10
−8
−6
−4
−2
0
2
4
6
8
10
VELOCITY (MM/S)

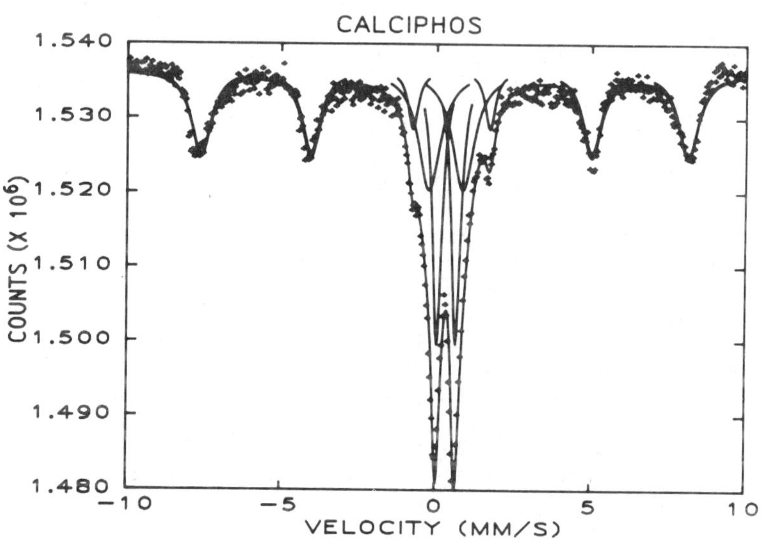
CALCIPHOS
COUNTS (X 10^6)
1.540
1.530
1.520
1.510
1.500
1.490
1.480
−10
−5
0
5
10
VELOCITY (MM/S)

Table III. Mössbauer Parameters and Assignments for Heated Christmas Island "D" Rock

Temp. °C	Time Min.	ΔE_Q mm/s	δ mm/s	H kOe	Percent Area	Assignment
Unheated	-	1.24	0.34		37	Crandallite
		0.75	0.33		30	Millisite
		0.45	0.33		33	Goethite
550	5	1.15	0.36		37	Crandallite residue
		0.57	0.36		27	Amorphous Fe,Al phosphate
		-	-	471	36	Hematite
850	15	1.10	0.32		30	Poorly cryst. Fe,Al phosphate
		0.57	0.33		25	Poorly cryst. Fe,Al phosphate
		-	-	484	45	Hematite
1100	30	0.88	0.40		12	$(Fe,Al)PO_4$
		0.53	0.33		23	$(Fe,Al)PO_4$
		-	-	496	65	Hematite
Calciphos		1.08	0.31		29	Poorly cryst. Fe,Al phosphate
		0.60	0.34		28	Poorly cryst. Fe,Al phosphate
		-	-	487	43	Hematite

Mössbauer Spectroscopy of High Temperature Phases

Mössbauer spectra were recorded for each rock grade after heating to temperatures representing each major thermal event on the DTA and TGMS curves. Figure 6 shows a typical set of spectra for Christmas "D" rock for selected temperatures and heating durations. By 550°C the dehydration/dehydroxylation phase of the reaction is complete. The Mössbauer spectrum shows that the goethite has been converted to a primitive hematite and the absence of both goethite and millisite is confirmed by XRD. A small amount of undecomposed crandallite remains after 5 minutes heating at this temperature but is totally absent after 10 minutes at 550°C, by which stage the material has become X-ray amorphous with the exception of poorly crystalline hematite and a trace of TiO_2, anatase. This latter phase may have been present in the raw material, being swamped by more intense diffracting species. This stage of the heating process is characterized by a rather broad iron(III) Mössbauer resonance due to the amorphous Ca-Al-Fe-PO_4 phase present.

The Mössbauer spectrum of Christmas Island "D" heated to 850°C for 15 minutes shows a further development of the crystalline hematite phase and at least two other iron(III) doublet resonances whose parameters reflect an increase in crystallinity from the previous heating, that is the quadrupole splitting values decrease with increasing firing temperature. The implication of this finding is that the iron ions are in sites which are becoming less distorted from regular symmetry as the crystallization reactions proceed. This trend continues up to 1100°C, where the hematite crystallinity is fully developed and the iron(III) quadrupole doublets have their narrowest linewidths and splittings. By this stage, not only have the iron sites become more regular, but the range of site distortions has been reduced considerably. Peak positions and assignments for the Mössbauer spectra of heated "D" grade rock are summarized in Table III. Also included in Table III is data for a commercial "Calciphos" sample, whose Mössbauer spectrum is also shown in Figure 6.

Table III also shows the observed hyperfine fields for the hematite phase. It can be seen that these values increase as the crystallization proceeds. Although the hematite generated at 1100°C is very well crystallized as judged by both its Mössbauer spectrum and X-ray powder diffraction pattern, the hyperfine field is below the room temperature value for pure hematite which is typically 515kOe at 298°K. The implication is that the hematite, like its parent goethite, is significantly aluminum-substituted. After heat treatment at 1100°C a large degree of particle sintering is certain to have occurred, thereby eliminating particle size effects as the cause of the reduced hyperfine field. Note that the relative peak area of the hematite doubles during the course of the crystallization reactions. this suggests that either the recoil-free fraction of the various phases is changing during the crystallization, and therefore that the peak areas cannot be related directly to the iron distribution between these phases, or that fresh hematite is being crystallized from the amorphous iron and aluminum phosphate phase. Although this latter option seems reasonable it would imply that the hematite produced had an iron to aluminum ratio very similar to that produced from the decomposing goethite. It would therefore appear that this latter option is unlikely because there is no serious broadening or splitting of the 1100°C hematite peaks.

The "Calciphos" parameters show a marked similarity to the mid-range heatings of Christmas Island "D" rock. As described earlier in this paper, "Calciphos" is "C" grade rock calcined at 550°C. Clearly, this material contains a mixture of poorly crystalline hematite, and amorphous iron and aluminum phosphates, a fact readily confirmed by XRD analysis. Both the observed hyperfine field and the percentage hematite tend to suggest that this material has been calcined more severely than the usual 550°C processing conditions.

Several unsatisfactory attempts were made to generate an $(Fe,Al)PO_4$ phase with the cristobalite structure for direct comparison with the high temperature phases observed above. A sample of metavariscite, $(Fe,Al)PO_4 \cdot 2H_2O$, from Christmas Island gave an amorphous phase when heated

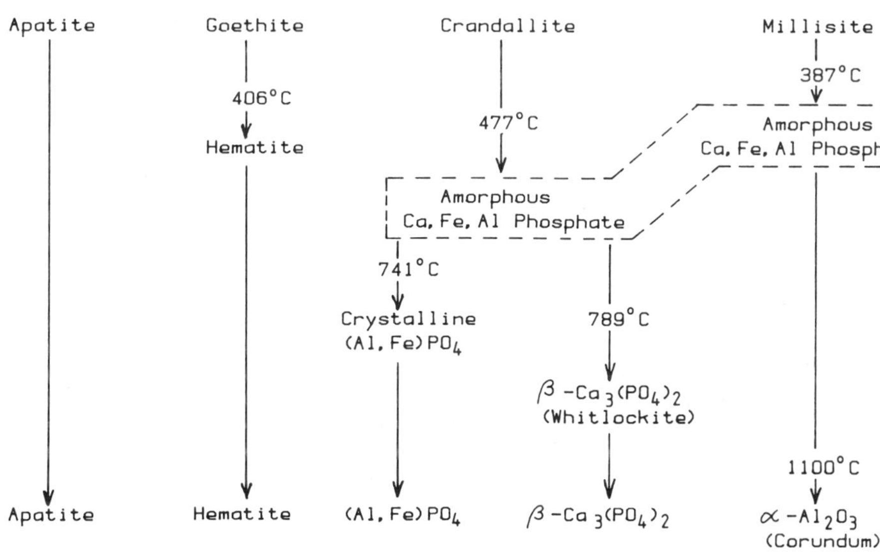

Figure 7. Summary of the thermal reaction sequence of Christmas Island phosphate rocks.

above its dehydroxylation temperature, which was encouraging, but formed the quartz analogue only when recrystallized at higher temperatures. Data for these phases are not included here.

SUMMARY AND CONCLUSIONS

A summary of the thermal reactions of Christmas Island phosphate rocks heated to 1100°C is shown in Figure 7. Although the thermally inert apatite is still present after high temperature calcination all other mineral phases initially present in the raw rock have decomposed. In terms of current superphosphate processing technology calcining Christmas Island rock at 1100°C has been seriously advanced as a means of reducing the acid solubility of iron and aluminum containing phases [2]. Comparing the iron and aluminum containing phases in the raw material, which, in the case of goethite, has a relatively small particle size, with the reaction products in Figure 7 it is clear why this should be the case. Hematite, (Al,Fe)PO_4, and corundum have relatively low acid solubilities when compared with apatite and whitlockite and the natural sintering processes occurring during calcination are certain to enhance this solubility difference.

The other calcined rock phosphate product of commercial interest, "Calciphos", which is used as a direct application fertilizer, has been shown to consist of a mixture of apatite, poorly crystallized hematite and amorphous calcium-iron-aluminum phosphates. The apatite and amorphous phosphate components are relatively insoluble in water but under acid soil conditions there is sufficient phosphate solubility to enable this material to be a very effective slow release fertilizer.

The salient features of the thermal reaction chemistry of Christmas Island phosphate rock are as follows. The impure rock grades decompose in three discrete steps; dehydration/dehydroxylation in the region 350 to 500°C, formation of an X-ray amorphous Ca-Fe-Al-PO_4 phase through to 740°C; and crystallization of phosphate and oxide components above 740°C. The goethite impurity component is aluminum substituted to the extent of 23 to 28 percent as judged by independent methods involving XRD and Mössbauer spectroscopy. Mössbauer spectroscopy can be used to advantage to monitor the progress of the thermal reactions throughout their entire duration, including through the X-ray amorphous region, and when coupled with XRD analysis, provides a very powerful structural tool for the identification and characterization of solid state reactions in iron-substituted phosphate mineral systems.

REFERENCES

1. M. S. White, N.Z.J. Science 19, 421 (1976).

2. M. S. White N.Z.J. Science 14, 971 (1971).

3. H. P. Rothbaum, N.Z.J. Science 12, 828 (1969).

4. N. A. Trueman, J. Geol. Soc. Aust. 12, 261 (1965).

5. F. N. Blanchard, Am. Min. 57, 473 (1972).

6. S. R. Silverman, R. K. Fuyat, and J. D. Weiser, Am. Min. 37, 211 (1952).

7. Y. Nathan, in "Phosphate Minerals", J. O. Nriagu and P. B. Moore, eds., Springer-Verlag, Heidelberg (1984).

8. D. G. Schulze, Clays Clay Miner. 32, 36 (1984).

9. E. Murad and U. Schwertmann, Clay. Min. 18, 301 (1983).

MÖSSBAUER STUDIES OF THERMAL REACTION MECHANISMS IN HYDROUS IRON LAYER SILICATES

K. J. D. MacKenzie, M. E. Bowden, and R. M. Berezowski

Chemistry Division
DSIR,
Private Bag, Petone, New Zealand

ABSTRACT

To shed light on mechanistic aspects of the thermal reactions of industrially important clay minerals such as kaolinite and serpentine, Mössbauer studies have been made of analogous iron-containing minerals. The relatively rare iron layer-lattice silicates amesite, cronstedtite and berthierine form a series structurally related to serpentine, and containing various substitutions of iron(II) and iron(III) in both octahedral and tetrahedral sites. Comparative Mössbauer studies of the thermal reactions of these minerals in both oxidizing and reducing conditions have been used to identify trends in the thermal stability of the various aspects of the thermal decomposition mechanisms of the related economic minerals.

INTRODUCTION

The layer-lattice hydrous aluminosilicates form a group of minerals of considerable significance in materials technology, because they include kaolinite, serpentine, talc, and pyrophyllite, all of which are important raw materials for ceramics manufacture. Studies have been made of the thermal reactions undergone by many of these materials when heated [1], but details of the reaction mechanisms are still debated, even in the case of the much-studied kaolinite reaction. Although most of the important clay minerals are not iron minerals per se, because they are natural materials they often contain appreciable substitutional iron impurities which influence their firing behavior and fired properties. Studies of the iron behavior in these minerals are therefore interesting and relevant.

Mössbauer studies have been made of the thermal behavior of adventitious substitutional iron in layer-lattice aluminosilicates such as the smectites (montmorillonite-type materials) [2] and the kandites (kaolinite-type minerals) [3]; the aim of the more recent work has been to provide information about the composition and firing conditions of ancient ceramic artifacts.

The extraction of more fundamental mechanistic information by studying adventitious iron impurities in these minerals can be hampered by the typically low iron contents encountered and the need to assume that the behavior of the iron parallels the behavior of the ion for which it is substituted. The true situation is often more complicated than this, because in impure kaolinites, a proportion of the iron can occur as a fine oxyhydroxide coating in the kaolinite particles which is resistant to chemical removal and has a room temperature Mössbauer spectrum which is very similar to that arising from iron in structural sites [4].

An alternative approach to derive mechanistic information is to study the thermal behavior of structurally-related layer-lattice iron hydrous silicates. Kaolinite (Figure 1A) has a dioctahedral structure in which all the octahedral ions are aluminum(III). A true iron(III) analogue of this mineral is unknown. The related trioctahedral mineral serpentine has an octahedral layer containing magnesium(II) (Figure 1B), giving more scope for partial or full replacement by iron(II), iron(III), or other ions such as aluminum(III). Figure 2 shows the range of minerals which result from such replacements.

Partial replacement by aluminum of some of the magnesium and silicon in serpentine gives amesite, which can then undergo a further series of replacements of both octahedral and tetrahedral magnesium(II) and aluminum(III) by iron(II) and iron(III), resulting in a series of minerals of varying octahedral and tetrahedral iron(II) and iron(III) contents. These minerals all have the 7Å repeating structural unit of kaolinite and serpentine but their chemical compositions are similar to those of the chlorite minerals, with which they were sometimes confused in the earlier literature. Attempts to clarify their nomenclature by referring to the group as the 7Å chlorites or septechlorites have not found favor with clay mineralogists. Although most of these minerals are relatively uncommon, the series presents an opportunity of studying thermally-induced changes in a group of related materials containing a range of iron(II) and iron(III) environments related to those occurring in commercially-important clay minerals.

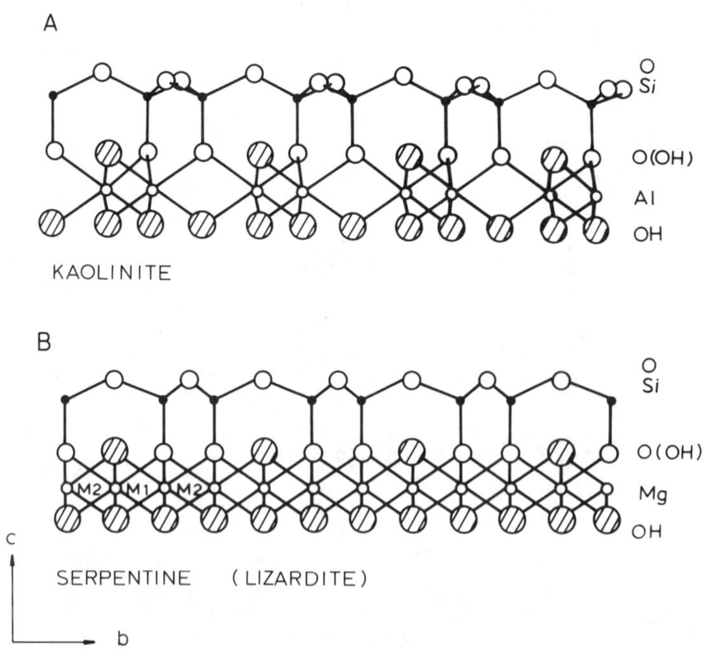

Figure 1. Computer-generated structures of representative representative examples of (A) dioctahedral and (B) trioctahedral 1:1 hydrous layer silicates, viewed down the a-axis. Lizardite structural data are taken from reference 13.

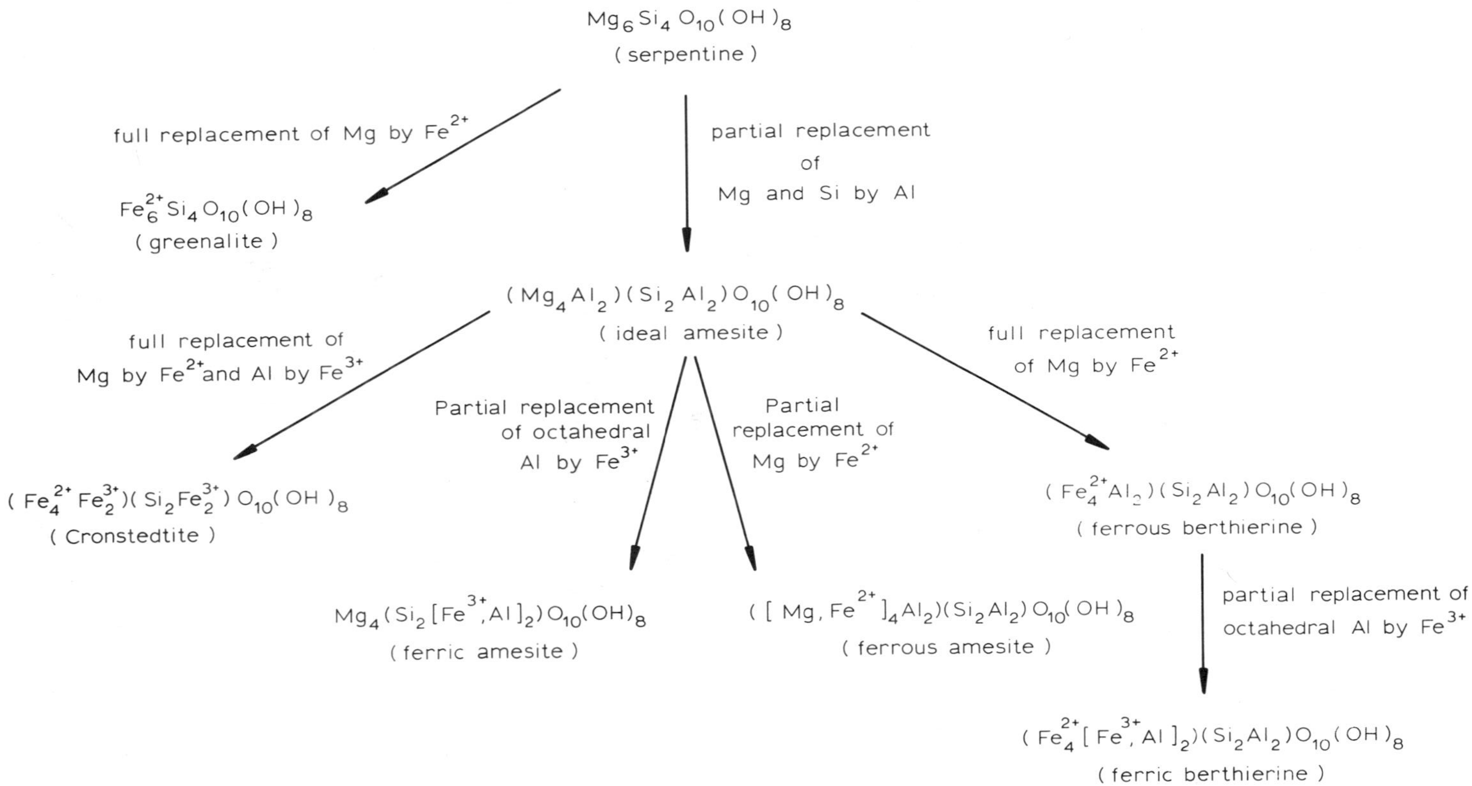

Figure 2. Typical composition relationships between the 7Å chlorite-type minerals and serpentine.

This paper brings together the results of Mössbauer studies which have been undertaken over a period of several years, and correlates these results with structural and mechanistic considerations for the thermal reactions.

EXPERIMENTAL

Included in these studies are two amesites, two berthierines and one cronstedtite sample. All specimens are reputable minerals, mostly from museum collections, and were characterized by x-ray diffraction and chemical analysis. One of each of the amesites and berthierines was of predominantly iron(II) content, whereas the other of each pair contained a significant proportion of iron(III). For convenience, these are described here as ferrous and ferric amesite and berthierine respectively, but these terms are not intended to indicate the occurrence of iron solely in that valence state. The salient details of these samples are shown in Table I, from which the unit cell compositions are calculated to be

Ferrous amesite, $[Mg_{3.28}Fe^{2+}_{0.66}Al_{2.0}](Si_{2.01}Al_{1.99})O_{10}(OH)_8$

Ferric amesite, $[Mg_{4.08}Fe^{2+}_{0.08}Fe^{3+}_{1.14}Al_{1.82}](Si_{2.02}Al_{1.98})O_{10}(OH)_8$

Ferrous berthierine, $[Mg_{0.29}Fe^{2+}_{3.81}Al_{1.72}](Si_{2.60}Al_{1.40})O_{10}(OH)_8$,

Ferric berthierine,
$[Mg_{0.40}Ca_{0.14}Fe^{2+}_{1.96}Fe^{3+}_{2.30}Al_{0.48}](Si_{2.98}Al_{1.02})O_{10}(OH)_{7.66}$,

and cronstedtite,

$[Mg_{0.13}Fe^{2+}_{4.29}Fe^{3+}_{1.34}](Si_{1.87}Al_{0.16}Fe^{3+}_{1.97})O_{10}(OH)_{9.05}$,

where the octahedral ions are contained in square brackets. It should be noted that the unit cell composition of the ferrous berthierine is based on a published electron microprobe analysis, in which all the iron was assumed to be iron(II). On the basis of 3.1 percent Fe_2O_3 estimated from the Mössbauer spectrum, a more realistic unit cell composition for this phase is

$[Mg_{0.29}Fe^{2+}_{3.54}Fe^{3+}_{0.27}Al_{1.72}](Si_{2.60}Al_{1.40})O_{10}(OH)_8$.

Details of the heating experiments are given in references 5-7. After thermal treatment in either oxidizing or reducing atmospheres, the room temperature Mössbauer spectra were obtained and fitted to Lorentzian lineshapes by using a program by Aldridge [8] based on that of Stone. The fitting strategy was progressively to remove all constraints except the pairing of widths and areas of quadrupole doublets. In each case, a minimum number of doublets was fitted, consistent with an acceptable χ^2 value. If the inclusion of further doublets gave no significant improvement in χ^2, the peaks were only added if there was sound crystallochemical reasons for including them. All the quoted isomer shift values are relative to soft iron.

Complementary information on the course of the thermal reactions, obtained by thermal analysis, x-ray diffraction, and infrared spectroscopy is reported elsewhere [5-7].

RESULTS AND DISCUSSION

Mössbauer Spectra and Site Assignments in the Unheated Minerals

The spectra of the unheated minerals are all very similar, and can be fitted by a single iron(II) doublet, with one or more iron(III) doublets also

Table I. Mineral Samples Investigated

Mineral	Locality	Museum No.	Color	% FeO	% Fe_2O_3	Principal Impurities	Ref.
Iron(II) amesite	Chester, USA	Smithsonian, 80715	Pale green	8.28		Chlorite[a]	5
Iron(III) amesite	Saranovskoye, USSR	Smithsonian, 103312	Pale violet	0.95	0.63	Chlorite[a]	5
Iron(II) berthierine	Utatsu, Japan		Dark grey	37.05	3.1[b]	Quartz[c]	6
Iron(III) berthierine	Kongsberg, Norway	British Museum BM34121	Greenish-brown	20.20	27.25	None	6
Cronstedtite	Kisbanya, Rumania	Smithsonian, B13823	Black crystals	39.22	33.70	None	7

[a]Trace. [b]Estimated by Mössbauer spectroscopy. [c]28-37 percent.

occurring in all but the ferrous amesite. Because it has previously been suggested [9] that two iron(II) sites can be present in related serpentine minerals, considerable efforts were made to computer-resolve such iron(II) sites. The lack of a significant improvement in the spectra fit with two iron(II) doublets led to the conclusion that only one doublet is present. The Mössbauer parameters of all the unheated minerals are shown in Table II.

Differences between the parameters for unheated cronstedtite shown in Table II and those given in reference 7 arose during a refitting of the original data, as a result of relaxing the severe width constraints which were originally applied [7]. The resulting freer fit confirms the previous finding that a significantly improved χ^2 value can be obtained by including a second iron(III) doublet, but the inclusion of a second iron(II) doublet does not improve the fit. The isomer shift of the free-fitted second doublet is 0.53mm/s, and although large, is within the range previously reported for other serpentine minerals [9], but its width of 1.06mm/s suggests a relaxation lineshape arising from an electron-hopping mechanism such as previously reported in cronstedtite [10].

The parameters given in Table II are all very similar to those previously reported for a number of serpentine minerals, such as chrysotile, antigorite, and lizardite [9,11] as well as for berthierine [9,12] and cronstedtite [12]. These literature values are also included in Table II. Rozenson, et al. [9] note that two different iron(II) absorptions can occur in the serpentines; one has a quadrupole splitting value typically of ca. 2.7mm/s, the other being smaller and ca. 2.0 to 2.2mm/s. Some lizardites exhibit both types of absorption in the same sample, whereas in other serpentines only the higher quadrupole splitting is found [9]. The iron(II) spectra of the present minerals fall into two similar groups, the higher quadrupole splitting values being found for all except ferric berthierine and cronstedtite. Although most of the present work was done at room temperature, the few unheated mineral spectra which were run to 77°K were consistent with iron(II) in distorted octahedral lattice sites.

The present minerals also show a similar range of iron(III) parameters to those reported for the serpentine minerals. Both the iron(III) parameters measured herein and the literature values show considerably more scatter than do the iron(II) results.

To make a sensible structural interpretation of the changes in the Mössbauer parameters caused by thermal reaction of these minerals, the iron sites in the unheated material must be identified with known crystallographic positions. In attempting to assign the absorption in their serpentine spectra, Rozenson et al. [9] noted that the serpentine structure contains two distinguishable octahedral sites, designated designed M1 and M2 (Figure 1B). They suggested that the M2 site is more distorted than the M1, and associated this with the iron(II) doublet of larger splitting in contrast with the more usual assumption that iron(II) sites of lowest distortion will exhibit the largest quadrupole splitting values [9]. In many serpentines, the distortions of M1 and M2 are probably too similar to allow the two iron(II) doublets to be resolved; this is certainly the case for the idealized lizardite structure of Rucklidge and Zussman [13] illustrated in Figure 1B. However, more recently it has been suggested that the M1 and M2 sites of serpentines may become nonequivalent due to buckling of the octahedral plane resulting from mismatch between the octahedral and tetrahedral layers [14], thus raising the possibility that in both fibrous and platy serpentines, two different octahedral sites may be distinguished.

Based on their initial assumption that the M2 site is more distorted than the M1, Rozenson et al. [9] assign serpentine iron(III) doublets with splittings of about 0.7 to 1.08mm/s to the M1 sites and doublets with splittings of about 0.4 to 0.5mm/s to M2. It is not clear, however, why it

Table II. Mössbauer Parameters for Unheated 1:1 Iron Hydrous Layer Silicates and Related Minerals[a]

Mineral	Iron(II)		Iron(III)		Iron(III)		Ref.
	ΔE_Q	δ	ΔE_Q	δ	ΔE_Q	δ	
Iron(II) amesite	2.59	1.13					5
Iron(III) amesite	2.61	1.11			0.57	0.35	5
Iron(II) berthierine	2.56	1.12	1.08	0.24			6
Iron(III) berthierine	2.26	1.24	1.11	0.35	0.76	0.29	6
Cronstedtite	2.11	1.18	0.84	0.53	0.47	0.19	7
Antigorite[b]	2.70-2.75	1.12-1.15	0.70-0.80	0.34-0.43			9
Chrysotile[c]	2.74-2.76	1.14	0.82-1.08	0.37-0.38	0.30	0.27	9
Chrysotile	2.65	1.12	0.75	0.34	0.34	0.20	11
Lizardite[d] (M2)	2.70-2.79	1.13-1.15	0.65	0.37	0.53	0.39	9
(M1)	1.98-2.21	1.07-1.16					
Berthierine (M2)	2.68	1.13	0.65	0.42			9
(M1)	2.28	1.13					
Berthierine	2.57	1.13	0.78	0.38			12
Iron(III) cronstedtite			0.67	0.38	0.69	0.17	12

[a]All data obtained at room temperature in mm/s with the isomer shift reported relative to natural iron. [b]Five samples. [c]Two samples. [d]Three samples.

should be possible to resolve iron(III) in the M1 and M2 sites in samples containing only one iron(II) doublet. One possible explanation is that iron(III) may be more sensitive to site distortion than iron(II), or alternatively the M2 sites may be preferred by iron(II) for steric reasons.

In the case of amesite and the related minerals, which are the subject of this study, site assignments cannot be made by direct analogy with the serpentines, because the mismatch between the octahedral and tetrahedral layers can be relieved by substitutions of aluminum for magnesium and silicon in the amesite structure [14]. Further substitution of iron, as in the present minerals, has been suggested to affect the mismatch of the layers in various ways, depending on the nature of the substituted site [14]. Thus, substitution of iron(III) in the tetrahedral layers is predicted to very efficiently relieve mismatch, whereas substitution of iron(III) in octahedral sites is expected to be less effective [12]. Substitution of iron(II) is predicted to increase the degree of mismatch [14]. A further complication in amesite is the effect of cation ordering which has been reported in a recently published crystal structure of an Antarctic amesite [15] and which has also been demonstrated by infrared spectroscopy in the present amesites [5].

Figure 3 shows a computer-generated model of the Antarctic amesite structure, the formula of which was found by electron microprobe analysis [18] to be

$$[Mg_{3.404}Al_{1.902}Fe^{2+}_{0.63}Mn_{0.028}\square_{0.036}](Si_{2.17}Al_{1.83})O_{10}(OH)_8.$$

The computer simulation of this structure shows that some of the octahedral M1 and M2 sites have bond lengths of about 1.95Å, whereas the remaining M1 and M2 bond lengths are about 2.09Å. Furthermore, if the deviation of the octahedral bond angles from 90° is taken as a measure of distortion from octahedral symmetry, the sites of shorter bond length are substantially less distorted than those of longer bond length. The former have been suggested [15] as corresponding predominantly to aluminum containing sites and the

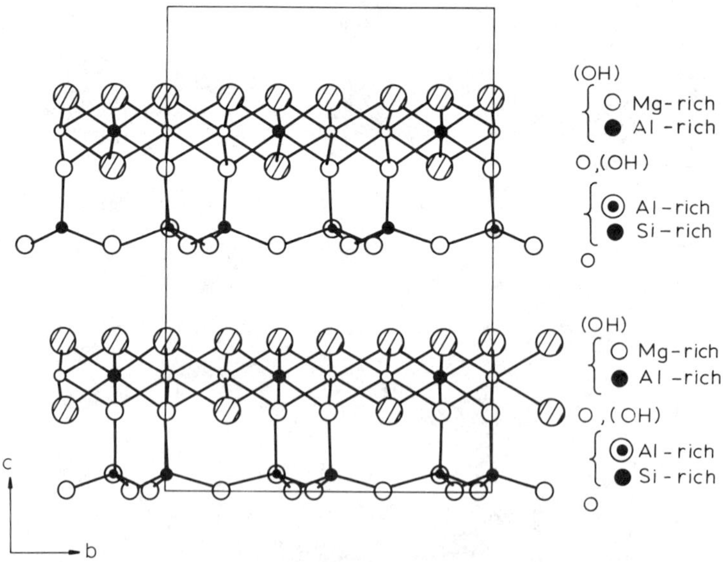

Figure 3. Computer-generated diagram of the structure of Antarctic amesite viewed along the a-axis from the structural data of Hall and Bailey [15].

latter to magnesium-rich sites (Figure 3). Ordering of the cation substitution pattern has also been demonstrated in the tetrahedral sites of this amesite [15]. Thus, differences in the distortions of the cation sites in amesite, and probably in the similarly substituted minerals, reflect the cation substitution pattern, in contrast with the serpentines, in which interlayer mismatch appears to be a more important factor. Analysis of the x-ray scattering powers of the octahedral sites of Antarctic amesite suggests that the heavy ion substituents such as iron are randomly distributed over all the available octahedral sites and exhibit no apparent site preferences [15]. Thus, in amesite and the related minerals, the following possibilities exist exist for iron substitution. (1) Ferrous is substituted on the magnesium-rich M1 or M2 octahedral sites. Based on the electric field gradient analysis of Rozenson, et al [9], these more highly-distorted sites should have splittings of ca. 2.6 to 3.0mm/s. (2) Ferrous is substituted on the less distorted aluminum rich M1 or M2 sites. These should have splittings of 2.0 to 2.5mm/s. (3) Ferric is substituted in magnesium rich M1 or M2 octahedral sites, with splittings of ca. 0.6 to 1.1mm/s. (4) Ferric is substituted on the aluminum rich M1 or M2 octahedral sites, with splittings of ca. 0.3 to 0.6mm/s. (5) Ferric is substituted in tetrahedral sites. On the basis of data for the serpentines, the quadrupole splittings of these species are probably similar to (4) above, but the isomer shifts of ca. 0.15 to 0.3mm/s are also rather small. By using the relationship deduced by Annersten and Olesch [16] between tetrahedral iron(III) isomer shift values and the tetrahedral M^{3+}-O bond length, the isomer shift values for tetrahedral iron(III) in the present iron(III) berthierine and cronstedtite suggest bond lengths of 1.76 and 1.68Å, respectively. These bond lengths correlate well with the two types of tetrahedral sites which have been distinguished in amesite [15], in which tetrahedral bond lengths of ca. 1.64 and 1.73Å have been identified with the silicon and aluminum rich sites, respectively. On this basis, the tetrahedral iron(III) in ferric berthierine and cronstedtite appears to substitute for tetrahedral aluminum and silicon, respectively.

The assignments of the Mössbauer lines of the unheated minerals studied here, made on the basis of the above, are shown in Table III. In this table the degree of distortion of the amesite sites is equated with the cation ordering known to exist in these samples. Even though significant cation ordering has not been demonstrated in the present berthierines and cronstedtite, it appears reasonable that, by analogy with amesite, the sites described in Table III as distorted and less-distorted represent iron-substituted magnesium and aluminum sites, respectively.

CHANGES IN THE MÖSSBAUER SPECTRA ON HEATING UNDER OXIDIZING CONDITIONS

Changes in the relative occupancy of the various sites and in their quadrupole splitting values are shown as a function of heating temperature in Figures 4 and 5, respectively. the general points of these figures can be summarized as follows.

The octahedral iron(II) initially present is progressively oxidized, being replaced by iron(III) with splitting values of 1.26 to 1.78mm/s which are indicative of severe distortions. The thermal stability of the iron(II) varies considerably from mineral to mineral, being greatest in the low-iron amesites. In ferric amesite and ferric berthierine the resulting iron(III) sites increase in distortion until dehydroxylation has occurred, after which the distortion is slightly relieved by the formation of new crystalline phases. In ferrous amesite and ferrous berthierine, the highly distorted iron(III) initially formed is joined at higher temperatures by a second, less-distorted iron(III) doublet. The iron(III) doublet resulting

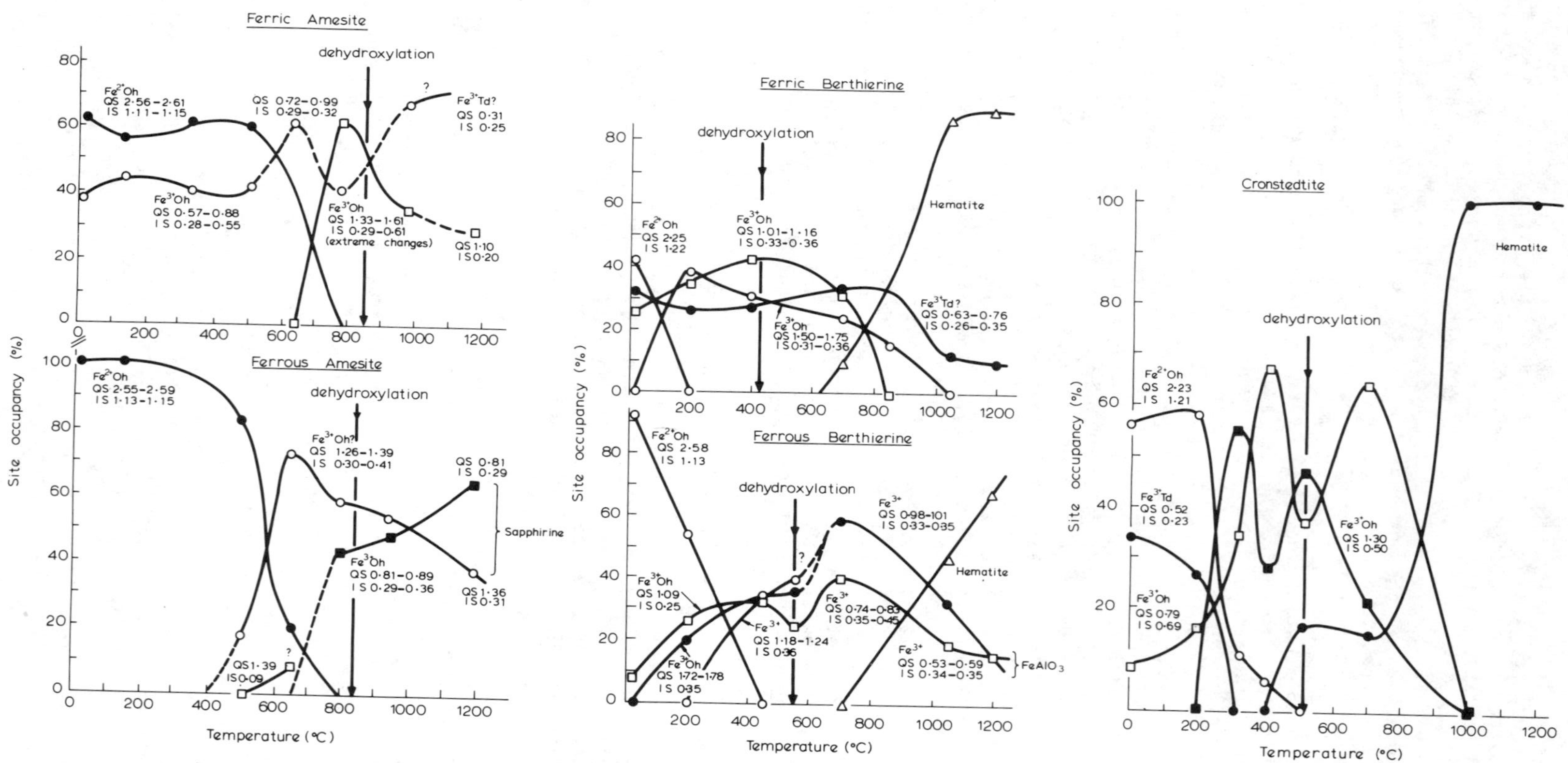

Figure 4. (A) Changes in the iron site occupancy of iron(III) and iron(II) amesite as a function of heating temperature in air. Deduced from Mössbauer peak areas, assuming recoil-free fractions of iron(II) and iron(III) to be approximately equal. (B) Changes in the iron site occupancy of iron(III) and iron(II) berthierine as a function of heating temperature in air. (C) Temperature dependence of the site occupancy in cronstedtite under oxidizing conditions.

Table III. Mössbauer Spectral Assignments for the Unheated 1:1 Hydrous Layer Silicates

Mineral	Substituent	ΔE_Q, mm/s	Site	% of Total Fe on Site[a]
Ferrous amesite	Fe^{2+}	2.59	Mg-rich octahedral (distorted)	100
Ferric amesite	Fe^{2+}	2.61	Mg-rich octahedral (distorted)	62
	Fe^{3+}	0.57	Al-rich octahedral (?) (less distorted)	38
Ferrous berthierine	Fe^{2+}	2.56	Distorted octahedral	92
	Fe^{3+}	1.08	Distorted octahedral	8
Ferric berthierine	Fe^{2+}	2.26	Less-distorted octahedral	42
	Fe^{3+}	1.11	Distorted octahedral	26
	Fe^{3+}	0.76	Tetrahedral (in Al sites?)	32
Cronstedtite	Fe^{2+}	2.11	Less-distorted octahedral	51
	Fe^{3+}	0.84	Distorted octahedral	22
	Fe^{3+}	0.47	Tetrahedral (in Si sites?)	27

[a]Estimated from Mössbauer peak areas, by assuming similar recoil-free fractions for iron(II) and iron(III).

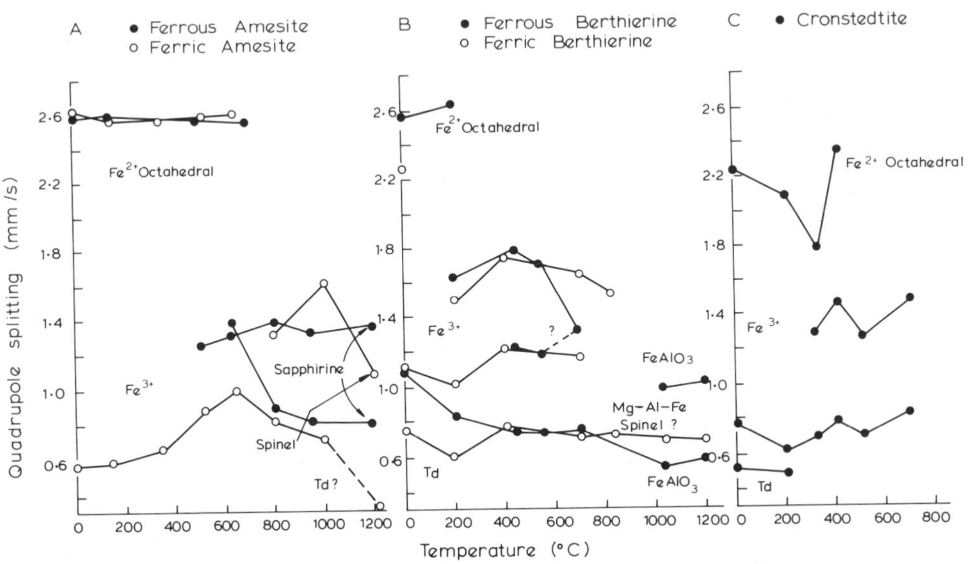

Figure 5. The temperature dependence of the quadrupole splitting for the minerals under oxidizing conditions.

from the iron(II) oxidation in cronstedtite is comparatively short-lived because the structure completely breaks down to hematite and silica at ca. 700°C [7].

The octahedral iron(III) initially present persists to high temperatures in all except cronstedtite and ferric berthierine, in which it is oxidized to hematite. Although progressive changes in the distortion of these sites occur throughout the reaction, as reflected in the quadrupole splitting, no dramatic changes in splitting are found to accompany hydroxylation.

The tetrahedral iron(III) initially present in iron(III) berthierine and cronstedtite behaves differently in these two minerals; in the former it persists to high temperatures, eventually ending up in one of the crystalline products, probably an iron-substituted spinel [6]. The tetrahedral iron(III) resonance in cronstedtite is very short lived, having disappeared by 300°C. Although the fate of this iron is unclear, it probably moves into octahedral magnesium sites, the iron population of which increases at about this temperature (Figure 4).

CHANGES IN THE MÖSSBAUER SPECTRA ON HEATING UNDER REDUCING CONDITIONS

Changes in the site occupancy and quadrupole splitting values are shown as a function of temperature in Figures 6 and 7 and are summarized as follows.

The octahedral iron(II) initially present is progressively lost prior to dehydroxylation, probably by a lattice oxidation-reduction involving hydroxyl groups. The appearance of a transitory iron(III) doublet in ferrous amesite, ferric berthierine, and cronstedtite at about the dehydroxylation temperature is consistent with such a process. The loss of the original iron(II) doublet is accompanied by the appearance of one or more new iron(II) doublets of smaller splitting. On the basis of the analysis by Rozenson et al. [9], this would indicate the migration of iron(II) into less-distorted octahedral aluminum sites. In iron(III) berthierine and cronstedtite, in which the original iron(II) is already in the less-distorted sites, rereduction of the original sites appears to occur at about the dehydroxylation temperature. Additional iron(II) doublets which appear in the berthierines and cronstedtite at about dehydroxylation eventually develop into the absorption expected of olivine, one of the high-temperature crystalline products formed under reducing conditions [6,7]. At increasing temperatures, further reduction of the non-olivine iron(II) to iron metal occurs, as reflected by the appearance of the sextet magnetic spectrum, and in the case of the amesites, by an additional unusual singlet of almost zero isomer shift, which has been identified with fine iron metal particles embedded in the insulating silicate matrix [17].

The octahedral iron(III) initially present persists to well above the dehydroxylation temperature in the two berthierines, in which it occurs in the more distorted magnesium sites. In ferric amesite and possibly in cronstedtite, the iron(III) initially occurs in the less-distorted octahedral aluminum sites, which are much less stable to thermal reduction, being lost at about the dehydroxylation temperature. The initial reduction products are iron(II), which either undergoes further reduction to iron metal, as in metal, as in ferric amesite may become one of the ferrous sites in olivine, as in ferric berthierine.

The tetrahedral iron(III) initially present in ferric berthierine and cronstedtite are relatively short-lived species, scarcely surviving dehydroxylation, at which temperature they are reduced. Since no tetrahedral iron(II) species can be identified in any of the spectra, this reduction is

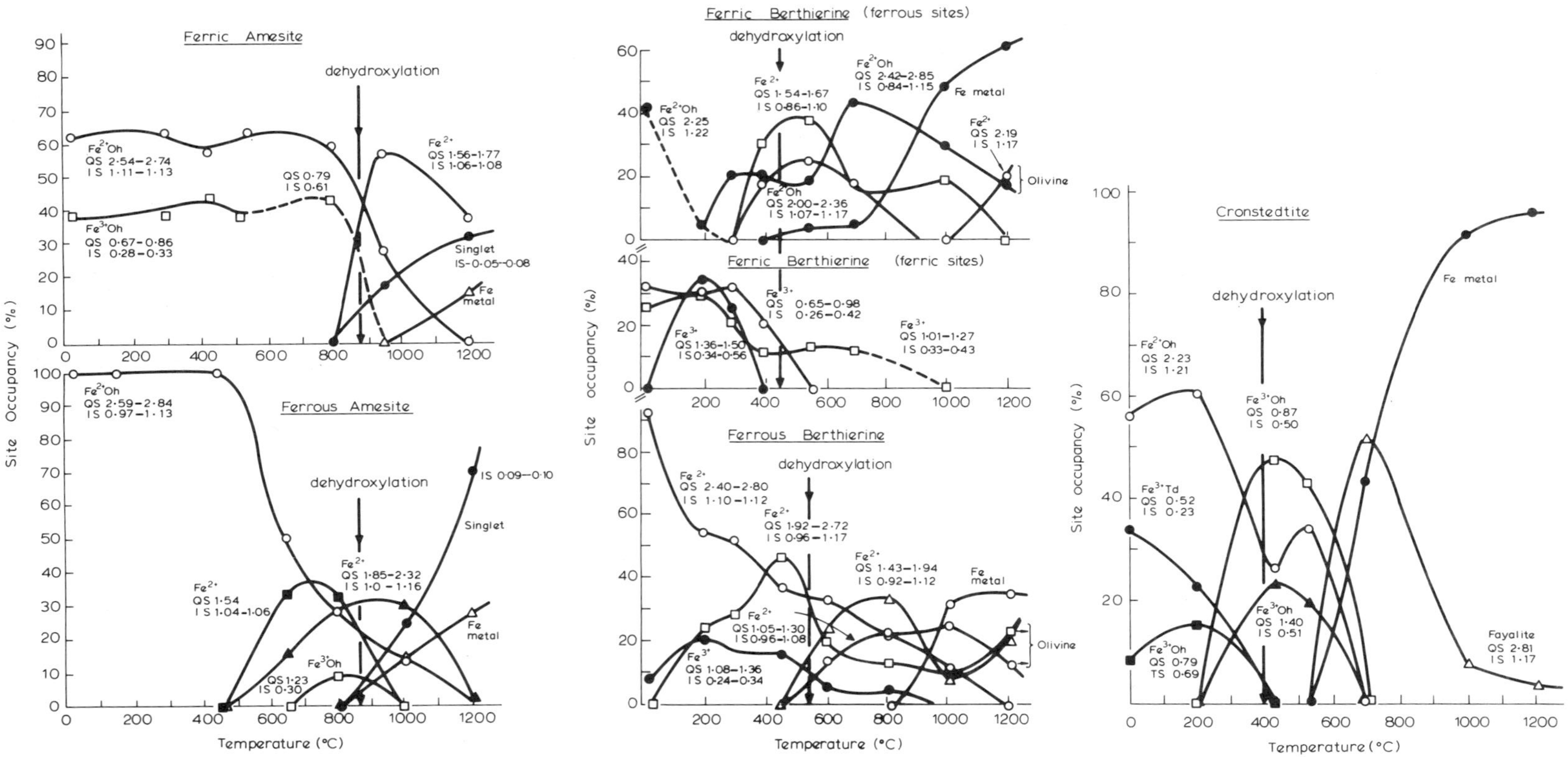

Figure 6. (A) Changes in the iron site occupancy of iron(III) and iron(II) amesite as a function of heating temperature in reducing conditions. (B) Changes in the iron site occupancy of iron(III) and iron(II) berthierine as a function of heating temperature in reducing conditions. (C) Temperature dependence of the site occupancy in cronstedtite under reducing conditions.

apparently accompanied by migration into the octahedral sites, consistent with the known octahedral site preference of iron(II) [18].

IMPLICATIONS FOR THE THERMAL REACTIONS OF THESE MINERALS

The Dehydroxylation Process

When hydrous layer silicates of magnesium or aluminum are heated, they lose hydroxyl water by the reaction

$$2OH^{-} \rightarrow O^{2-} + H_2O \tag{1}$$

The precise mechanism of this reaction, i.e., whether by counter-migration of protons and cations or by bulk diffusion of water, varies from mineral to mineral, although in general the magnesium silicates appear to dehydroxylate by a counter-migration process, the so-called "inhomogeneous" mechanism [1].

In materials containing variable-valence lattice substituents such as iron, oxidation reduction reactions with hydroxyl groups can occur in addition to, and usually prior to, reaction (1), leading also to the discharge of hydroxyl water. In oxidizing conditions the reaction can be written as

$$4Fe^{2+} + 4OH^{-} + O_2 \rightarrow 4Fe^{3+} + 4O^{2-} + 2H_2O \tag{2}$$

The presence of oxygen is not necessary for intralattice oxidation of iron(II) to occur, with the elimination of hydroxyl groups. Under inert or reducing conditions, a so-called "dry" reaction, can be written as

$$Fe^{2+} + OH^{-} \rightarrow Fe^{3+} + O^{2-} + H \tag{3}$$

where the hydrogen atoms thus produced may either react with any surface oxygen to form water, or under vacuum conditions, combine to form H_2.

Reactions (2) and (3) can occur prior to complete dehydroxylation, and have been observed by Mössbauer spectroscopy in a number of other iron-

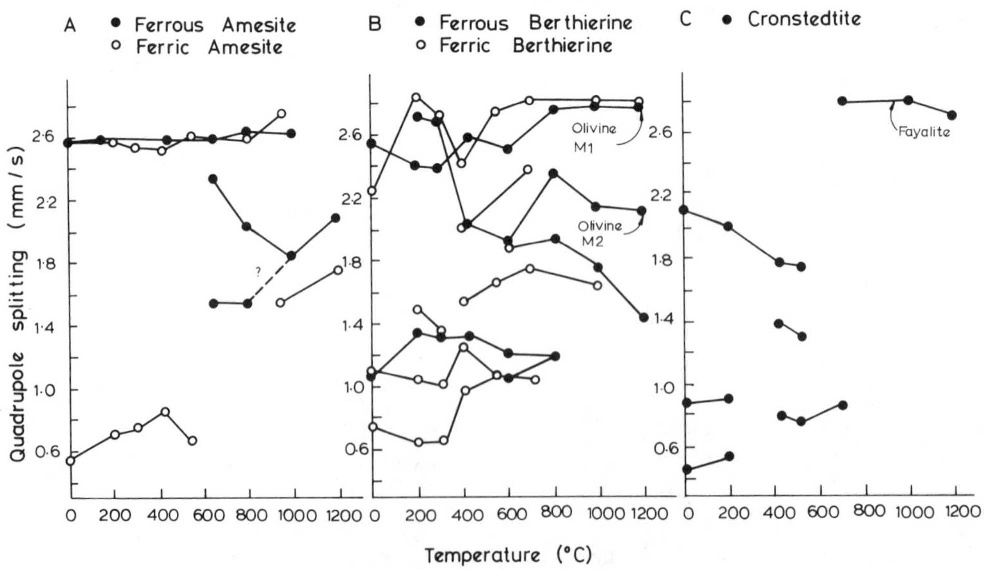

Figure 7. The temperature dependence of the quadrupole splitting for the minerals under reducing conditions.

containing minerals, including biotite [19,20], various chlorites [21,22] and several Russian berthierines [23]. A general feature in all these previously reported studies is the development of highly-disordered iron(III) sites with large quadrupole splitting values on oxidation of the iron(II) sites. The distortions giving rise to these large splittings have been suggested by Gendler, et al [20] to arise from a shortening of the M-O bond to the oxygen which was originally a hydroxyl group, the bonds to the other oxygens being simultaneously lengthened. An alternative suggestion by Hogg and Meads [19] is that partial dehydroxylation (of biotite) can result in distorted $Fe^{3+}(O_5OH)$ configurations. In the present 1:1 minerals the corresponding configuration would be $Fe^{3+}O_3(OH)_3$. In biotite, further dehydroxylation to $Fe^{3+}(O_5OH)$ can occur where two iron(II) ions are located in adjacent sites, and this has been advanced as an explanation for the later appearance of a second iron(II) doublet of smaller splitting [19].

Applying similar arguments to the present minerals, and taking into account preferences of iron(II) for either aluminum or magnesium sites in the various samples, leads to the following conclusions. (1) Even if full octahedral substitution by iron(II) were to occur, complete dehydroxylation will never occur by reactions (2) or (3) alone, because there are always more hydroxyl groups than iron(II) (Figure 3) present. (2) Oxidation of isolated iron(II) ions located in magnesium or aluminum sites should lead to distorted $Fe^{3+}(O_3(OH)_3)$ configurations, whereas $Fe^{+3}(O_4(OH)_2)$ configurations will result when iron(II) ions are located in adjacent sites. In cation-ordered minerals such as amesite,this will only occur where the ions occupy magnesium sites, because iron(II) ions in the aluminum sites are always isolated (Figure 3). (3) Subsequent thermal dehydroxylation by reaction (1) will not necessarily reduce the distortion of the octahedral ion further, because it introduces vacancies into the octahedral coordination sphere, effectively changing the coordination number.

The practical consequence of (2) above is that the oxidation of iron(II) in the magnesium sites can lead to two types of iron(III) environment, one more distorted than the other, whereas the oxidation of iron(II)in aluminum sites should produce only the more distorted site. This prediction is borne out in the present experiments. When heated in air, ferrous amesite and ferric berthierine, with iron(II) in magnesium sites only, initially oxidize to very distorted iron(III) with a splitting of ca. 1.4 to 1.8mm/s, with a second, less-distorted site with a splitting of ca. 0.8 to 1.2mm/s, appearing about 200°C later (Figure 4). By contrast, ferric berthierine and cronstedtite, with iron(II) in only the aluminum sites produce only one iron(III)site on air oxidation with a splitting of ca. 1.3 to 1.7mm/s. The only apparent exception to this rule is ferric amesite with iron(II) only in the magnesium sites, which produces only one distorted iron(III) site on air oxidation. The extremely low iron content of this sample may argue against the statistical chances of iron(II) occurring in adjacent sites, even though such an occurrence is allowable in terms of the cation ordering sequence.

The Higher-Temperature Reactions

The complete dehydroxylation of the 1:1 layer silicates results in the octahedral cations becoming tetrahedrally coordinated. Because tetrahedral coordination is not favored by magnesium, a stable dehydroxylated phase is not formed by the serpentine minerals, which eject silica and form crystalline magnesium silicate almost immediately on dehydroxylation [1]. By contrast, tetrahedral coordination is tolerated by aluminum as is illustrated by the formation in the 1:1 aluminum mineral, kaolinite, of a dehydroxylated phase which, although possessing almost no crystallinity, is stable for several hundred degrees. Eventually this x-ray amorphous phase also decomposes to form crystalline products.

Because the minerals of the present study constitute a series in which varying degrees of iron and aluminum substitution for magnesium occur in the serpentine structure, they provide an opportunity to observe the effects of such substitutions on the formation of the higher-temperature phases. The least heavily-substituted minerals of the series are the amesites, in which approximately two to three out of six octahedral sites are substituted by aluminum or iron(III). The behavior of the amesites is similar to that of the serpentines as dehydroxylation occurs at relatively high temperatures, with the immediate formation of crystalline products such as a spinel, which in air transforms to sapphirine, $Mg_{3.5}Al_9Si_{1.5}O_{20}$ [5]. Under oxidizing conditions, the iron in the structure remains in the spinel and sapphirine phases but in reducing atmospheres it is reduced to iron metal.

By contrast, the two berthierines are comparatively highly substituted; approximately two out of six octahedral sites in the ferrous berthierine contain aluminum with the bulk of the more distorted sites containing iron(II) which, after dehydroxylation in air, becomes iron(III). This mineral behaves similarly to kaolinite, forming an x-ray amorphous phase which is stable for several hundred degrees before transforming to mullite, $Al_6Si_2O_{13}$, and $FeAlO_3$, in air, or cordierite $(Mg,Fe)_2Al_4Si_5O_{18}$ and olivine $(Fe,Mg)_2SiO_4$ under reducing conditions [6]. The octahedral aluminum content of the ferric berthierine is only about 0.5 atoms in six, the remainder being largely iron(III) and iron(II). In addition, the Mössbauer spectrum indicates significant tetrahedral substitution by iron(III). The fully dehydroxylated structure is thus rather unstable, and like serpentine, immediately decomposes, forming hematite, Fe_2O_3 in air or a cubic silicate spinel under oxidizing conditions [6].

Cronstedtite, which contains no octahedral aluminum and very little octahedral magnesium behaves similarly to the ferric berthierine, decomposing immediately on dehydroxylation. However, unlike the ferric berthierine, in which the tetrahedral iron(III) persists in air to high temperatures, probably ending up in the tetrahedral sites of a cubic magnesium-aluminum-iron spinel, the tetrahedral iron(III) in cronstedtite is a short-lived species, migrating into the octahedral sites at an early stage. This behavior is similar to that reported for a Russian serpentine [22], in which an increase in the concentration of octahedral iron(III) at the expense of tetrahedral iron(III) was noted on heating.

Although the nature of the high-temperature reactions and products clearly depends on a number of compositional features in addition to the distribution of the iron atoms, two trends can be discerned in the Mössbauer results. First, in air, the minerals in which the iron(II) sites have larger splitting values appear to be more thermally stable. When eventually formed, the resulting iron(III) becomes part of the crystalline aluminate or silicate product. Minerals in which the iron(II) occurs in sites of smaller splitting generally oxidize at lower temperatures, the product being hematite. Second, in reducing atmospheres, the minerals in which the iron(II) is in sites with larger quadrupole splitting are also more stable to internal oxidation. The eventual products of rereduction of these sites, iron metal or olivine are however the same as those formed from minerals with smaller splitting sites and are independent of the initial degree of distortion present in these sites.

REFERENCES

1. N. H. Brett, K.J. D. MacKenzie, and J. H. Sharp, Quart. Revs. Chem. Soc. 24, 185 (1970).

2. L. Heller-Kallai and I. Rozenson, Clays Clay Mins. 28, 355 (1980); I. Rozenson and L.Heller-Kallai, Clays Clay Mins. 28, 391 (1980); M. Takeda, O. Kawakami, and T. Tominaga, J. Physique, Colloque C2 40, 472 (1979).

3. K. J. D. MacKenzie, Clays Clay Mins. 8, 151 (1969); A. Simopoulos, A. Kostikas, I.Sigalas, N. A. Gangas, and A. Moukarika, Clays Clay Mins. 23, 393 (1975); Y. Maeda, H. Sakai, S. Onoyama, and E. Yoshida, J. Physique, Colloque C2 40, 485 (1979); M. Takeda, O. Kawakami, H. Kobayashi, and T. Tominaga, J. Physique, Colloque C2 40, 483 (1979).

4. S. A. Fysh, J. D. Cashion, and P. E. Clark, Clays Clay Mins. 31, 285 (1983).

5. K. J. D. MacKenzie and M. E. Bowden, Thermochim. Acta 64, 83 (1983).

6. K. J. D. MacKenzie and R. M. Berezowski, Thermochim. Acta 74, 291 (1984).

7. K. J. D. MacKenzie and R. M. Berezowski, Thermochim. Acta 44, 171 (1981).

8. L. P. Aldridge, AMOSS - A Fortran Program for Fitting Mössbauer Spectra, Rept. CD 2336, Dept. of Sci. and Ind. Res., New Zealand (1984).

9. I. Rozenson, E. R. Bauminger, and L. Heller-Kallai, Amer. Mineral. 64, 893 (1979).

10. J. M. D. Coey, A. Moukarika, and C. M. McDonagh, Solid State Comm. 41, 797 (1982).

11. C. Blaauw, G. Stroink, W. Leiper, and M. Zentilli, Can. Mineral 17, 713 (1979).

12. J. M. D. Coey, O. Ballet, A. Moukarika, and J. L. Soubeyroux, Phys. Chem. Minerals 7, 141 (1981).

13. J. C. Rucklidge and J. Zussman, Acta Cryst. 19, 381 (1965).

14. F. J. Wicks and E. J. W. Whittaker, Can. Mineral. 13, 227 (1975).

15. S. H. Hall and S. W. Baily, Clays Clay Mins. 27, 241 (1979).

16. H. Annersten and M. Olesch, Can. Mineral. 16, 199 (1978).

17. K. J. D. MacKenzie and M. E. Bowden, J. Mater. Sci. Lett. 2, 317 (1983).

18. A. Navrotsky and O. J. Kleppa, J. Inorg. Nucl. Chem. 29, 2701 (1967).

19. C. S. Hogg and R. E. Meads, Mineral. Mag. 40, 79 (1975).

20. T. S. Gendler, L. G. Daynyak, and R. N. Kuzmin, Geochem. Int. 15, 17 (1978).

21. O. K. Borggard, H. B. Lindgreen, and S. Mørup, Clays Clay Mins. 30, 353 (1982).

22. T. V. Malysheva, L. M. Satarova, and N. P. Polyakova, Geochem. Int. 14, 117 (1977).

23. Z. P. Yershova, A. P. Nikitina, Yu. D. Perfil'ev, and A. M. Babeshkin, Proc. Int. Clay. Conf., p. 211, Mexico City, (1975).

A STUDY OF THE INITIALLY-FORMED HYDROLYSIS SPECIES AND INTERMEDIATE POLYMERS AND THEIR ROLE IN DETERMINING THE PRODUCT IRON OXIDES FORMED IN THE WEATHERING OF IRON

James H. Johnston

Department of Chemistry
Victoria University of Wellington
Wellington
New Zealand

David G. Lewis

Department of Soil Science
Waite Agricultural Research Institute
Glen Osmond
South Australia 5064

INTRODUCTION

Iron oxides and hydrous oxides (oxyhydroxides and hydroxides) are ubiquitous in soils and weathered iron-containing material. In the weathering process iron is initially released to form aquated iron(II) and/or iron(III) ions in solution. Under appropriate pH and redox conditions aquated iron(II) is readily oxidized to form iron(III) ions which may then lose proton(s) to form $FeOH^{2+}$ and/or $Fe(OH)_2^{+}$ species as well as the dimer, $Fe_2(OH)_2^{4+}$. The true nature of these species are probably $[Fe(H_2O)_6]^{3+}$, $[Fe(H_2O)_5OH]^{2+}$, $[Fe(H_2O)_4(OH)_2]^{+}$, and $[(H_2O)_4Fe(OH)_2Fe(H_2O)_4]^{4+}$, respectively, but they are usually abbreviated to the former. Such species have been identified in solution by spectrophotometric and potentiometric methods [1-9], and by magnetic methods [10]. However, in other than very acidic solution, hydrolysis continues beyond these simple products and based upon ultracentrifuge analyses [11] large polymers form. Pathways for polymer formation have been suggested by Dousma and deBruyn [7]. With time, and a continued supply of monomeric species such polymers continue to grow and ultimately yield precipitates of iron oxides or hydrous oxides, [12,13] the nature of which can vary depending on the conditions in the hydrolyzing medium. Factors influencing the product distribution include pH, temperature, iron concentration, rate of hydrolysis and the nature and concentration of other solutes [12]. Possible pathways for the formation of the product iron oxides and hydrous oxides are shown in Figure 1 and the general characteristics of these oxides [14] are presented in Table I. In some cases previous workers have used the term iron oxides to include collectively both the oxides and hydrous oxides [14]. For simplicity this approach will be adopted here.

Table I. General Characteristics of the Iron Oxide Minerals[a]

Mineral	Formula	Crystal System	Crystal Morphology
Goethite	α-FeOOH	Orthorhombic	Acicular
Lepidocrocite	γ-FeOOH	Orthorhombic	Laths, serrated, elongated plates
Ferrihydrite	$Fe_5HO_8 \bullet 4H_2O$	Rhombohedral (Hexagonal)	Sphere
Hematite	α-Fe_2O_3	Rhombohedral (Hexagonal)	Hexagonal plates
Maghemite	γ-Fe_2O_3	Cubic or Tetragonal	Cubes
Magnetite	Fe_3O_4	Cubic	Cubes

[a]Adapted from Schwertmann and Taylor [14]

In soils these iron oxides (Table I) generally lack any significant degree of crystallinity and range in particle size from what may be termed microcrystalline (ca. < 1000nm) to poorly ordered x-ray amorphous material (ca. 3-10nm). However, electron microscope studies show that each oxide has a distinctive morphology (Table I) which is related to its basic crystal structure.

In rapid hydrolysis the initial precipitate is usually amorphous to x-ray diffraction and is therefore considered to be a polymeric type material. Protoferrihydrite which exhibits only two very broad x-ray diffraction (XRD) maxima centered at 2.6A and 1.5A displays some slight structural order [15,16]. Ferrihydrite, $Fe_5HO_8 \bullet 4H_2O$, which gives six slightly sharper XRD peaks, is more ordered and is considered to have a defect hematite, α-Fe_2O_3, structure [17]. Johnston and Lewis [18] showed that ferrihydrite particles

$$\text{Metallic Iron, Iron-containing minerals} \xrightarrow[\text{oxidation}]{\text{weathering}} Fe^{3+}(aq)$$

$$\underset{\text{monomer}}{Fe^{3+}(aq)} \xrightarrow{\text{ages}} [Fe,\ O,\ OH,\ H_2O]_n^{m+}$$

n = small - polycations
n = large - polymers

polycations, polymers may age in solution to any one or more of the following depending upon prevailing conditions:

Species

1. Ferrihydrite $Fe_5HO_8.4H_2O$
2. Goethite α-FeOOH
3. Lepidocrocite γ-FeOOH
4. Akaganéite β-FeOOH
5. Hematite α-Fe_2O_3

Figure 1. Possible Pathways in the weathering of iron

could coalesce and transform directly to hematite at 92°C in aqueous medium, thus verifying the existence of a structural relationship between these two oxides. Suspensions of ferrihydrite aged at ambient temperature generally produce goethite and/or hematite [19], with the proportion of goethite to hematite, as well as the size of the crystals being controlled, to some extent, by the hydrolysis rate and pH. Goethite may readily incorporate up to 30 mole percent AlOOH in its structure [20-22]. It is interesting to note that aluminum-substituted goethite cannot as yet be synthesized in the laboratory at near common soil conditions from an aluminum(III)-iron(III) hydroxy system. In contrast aluminum promotes the formation of aluminum-substituted goethite during the oxidation of an iron(II)-aluminum(III) system under such soil conditions, which in the absence of aluminum, would normally lead to other iron oxide phases [23]. The formation of lepidocrocite, γ-FeOOH, also appears to proceed in nature via an iron(II)-iron(III) hydroxy species and usually requires mildly reducing conditions. Maghemite, γ-Fe_2O_3, can form by the dehydration of either this iron(II)-iron(III) hydroxy species or lepidocrocite, and also by the thermal transformation of other oxides in the presence of organic matter [14]. Magnetite, Fe_3O_4, is not a common weathering product and usually requires conditions of low moisture content, alkaline pH, and a high soluble salt content [24]. It is also formed from the continuous slow oxidation of an initially pure iron(II) system [22]. It is, however, a common product of volcanic activity. Akaganèite, β-FeOOH, is a much less common iron oxide in nature and because it requires the presence of chloride ions to stabilize its channel-type structure, its occurence is generally restricted to high chloride environments, e.g. HCl dominated volcanic areas [25].

In nature the amorphous polymeric materials, ferrihydrite and goethite, are the most common forms occurring under temperate climatic conditions, whereas in warmer areas hematite is common. Because of their intense color and fine particle size the iron oxides have a high pigmenting power and even at low concentrations effectively determine the color of many soils. This provides an important criteria for naming and classifying soils [26]. Also, the chemical nature and generally high specific surface area of these oxide particles readily enable them to adsorb anions such as phosphate, molybdate, etc.

In the corrosion of ferrous metals the product oxide, which is often amorphous to XRD, forms a spongy coating that readily retains moisture and consequently promotes further corrosion or rusting. The presence of other ions such as those in sea water increases the conductivity of the aqueous medium and enhances the corrosion process.

If the particle size of the naturally occurring iron oxides is large enough, their respective forms can be distinguished by XRD. The identification of certain characteristic absorption peaks in their infrared spectra are also useful to some extent, for this purpose [27]. Iron-57 Mössbauer spectroscopy has been extensively used in recent years to identify and characterize the iron oxides [28]. At room temperature protoferrihydrite, ferrihydrite, poorly-crystalline (superparamagnetic) goethite, and lepidocrocite all give a broadened doublet resonance characteristic of iron(III) in octahedral coordination with O, OH, or H_2O. When the sample is cooled to 78K goethite and the more ordered ferrihydrite show magnetic relaxation and give a hyperfine sextet resonance. Protoferrihydrite and poorly ordered ferrihydrite lack sufficient order to enable such magnetic relaxation at 78K, whereas lepidocrocite, although structurally more ordered, also exhibits an iron(III) doublet at 78K. Goethite with a larger particle size, and also hematite, give a magnetic hyperfine sextet at room temperature [18,22,28-30]. The asymmetric lineshapes in the hyperfine spectra, particularly for goethite, arise from a range of particle sizes within the material [31].

Most of the research to date in the field of iron oxides has been directed towards characterizing the early-formed hydrolysis species in solution, determining the associated reaction kinetics and identifying the ultimate solid phase product precipitates. The intermediate sized polymers have been largely overlooked. The nature of these are, however, of interest since they frequently exist as stable colloidal solutions and so are capable of mobilizing iron from one point in a weathering system to another. Spatial variation of some of the factors controlling the growth can thus lead to the formation of different products. Little work has been carried out on the linking of the early-formed monomeric and dimeric species, the structural relationships of the intermediate polymers and the way in which these determine the solid phase products, i.e. ferrihydrite, goethite, lepidocrocite, hematite, etc.

In this paper we present the results of a study of the hydrolysis of iron(III) nitrate solution which is used as a model system for the hydrolysis of iron. Various low molecular weight fractions have been separated by passing the solution through appropriately sized Amicon Diafilters and freeze drying the filtrate to give the respective solid intermediate fractions. These were then characterized directly by iron-57 Mössbauer spectroscopy. The ways in which the early-formed dimeric species may link to eventually determine the product iron oxide are discussed and a hydrolysis mechanism suggested.

EXPERIMENTAL

In this study essentially two groups of samples were prepared from the hydrolysis of iron(III) nitrate solution. The first group was a range of hydrolysis fractions with molecular weights (MW) nominally of: 1k-10k, 10k-20k, 20k-50k, 50k-100k, 100k-300k and greater than 300k, with NaOH being used in the hydrolysis. These were prepared as follows: A freshly prepared solution of iron(III) nitrate (0.1M, 500 ml, initial pH=1.43) was stirred vigorously and partially hydrolyzed to an OH/Fe ratio of 2 by slow addition of 100 ml of 1M NaOH. Immediately after NaOH addition, the pH of the deep red clear solution reached 2.03, but with continuous stirring, polymerization occurred and the pH dropped to 1.80 within one hour and subsequently more slowly, reaching a stable value of 1.70 ± 0.05 within 48 hours. After four days aging at 20 degree C, 500 ml of the still clear solution was passed through an XM 300 Amicon Diafilter and the material retained by the filter washed with distilled water until the filtrate was colorless. The dark red sediment of nominal MW greater than 300k was washed off the filter and resuspended in water for analysis of the iron content. The filtrate was further fractionated in a similar manner by passing successively through a series of Amicon Diafilters of decreasing pore size (Table II). Subsamples taken from the various resuspended fractions were digested with 6M HCl to ensure complete reversion to monomeric aqueous iron(III) before determination of total iron. Finally each fraction was freeze dried and stored in polystyrene vials with closely-fitting polythene lids.

The second group consisted essentially of the MW less than 500 fractions separated from 0.1M iron(III) nitrate solutions which had been subjected to different periods of aging and extents of neutralization with NaOH or Na_2CO_3 to give appropriate hydroxide/iron or carbonate/iron ratios, before being passed through UM 05 Amicon Diafilters. No neutralization was carried out on the first two samples (Table II). The filtrates (of nominal MW less than 500) were analysed for total iron content before freeze drying and storage in tightly sealed vials (Table II). A further fraction of nominal MW range 500-1k was obtained by re-suspending the UM 05 sediment in water and collecting the material passing through a UM 2 Diafilter. For one sample, which had

Table II. Experimental Data for the Hydrolysis of 0.1M Ferric Nitrate Solution

A. Range of MW Fractions

Amicon Diafilter	Nominal MW Range of Fraction	% of Total
XM 300	>300k	31
XM 100	100k-300k	5
XM 50	50k-100k	41
UM 20	20k-50k	3
PM 10	10k-20k	0.3
UM 2	1K-10k	2
	<1k	18

B. Ageing of MW<500 Fractions

Neutralization Conditions	Aging Period Hours	pH when[a] Filtered	Fe in MW<500 % Fraction
	0[b]	1.75	
	48	1.65	88
OH/Fe = 0.2	52	1.83	88
OH/Fe = 0.5	76	1.90	62
$CO_3^{=}$/Fe = 0.1	143	1.89	84
$CO_3^{=}$/Fe = 0.2	119	2.06	78
$CO_3^{=}$/Fe = 0.5	41	1.97	56

[a] initial pH = 1.75. [b] solution not fractionated - freeze dried immediately

been neutralized by Na_2CO_3 to a carbonate/iron ratio of 0.5 before aging for two days, the MW greater than 1k fraction was also recovered and freeze dried. The changes in pH during hydrolysis and aging are given in Table II.

Iron-57 Mössbauer spectra were recorded at room temperature (approximately 293K) and 78K by using a rhodium matrix cobalt-57 source. Zero velocity is defined by the midpoint of the magnetic hyperfine spectrum for natural iron recorded at room temperature. Care was taken to protect the sample against hydrolysis during data collection.

An appropriate number of Lorentzian peak lineshapes were computer fitted to the experimental spectra by using a non-linear least squares minimization procedure. For all the computer-fits the resulting X^2 value was satisfactorily between the 1% and 99% confidence values. The resulting calculated values for the isomer shift, δ, quadrupole interaction, Δ, magnetic hyperfine splitting, H, and linewidth, Γ, are given in Tables III and IV.

The sample with a MW less than 500 and aged for 48 hours before freeze drying (Table IIB) was also allowed to further hydrolyze in the Mössbauer sample holder. The respective spectra recorded continuously during this

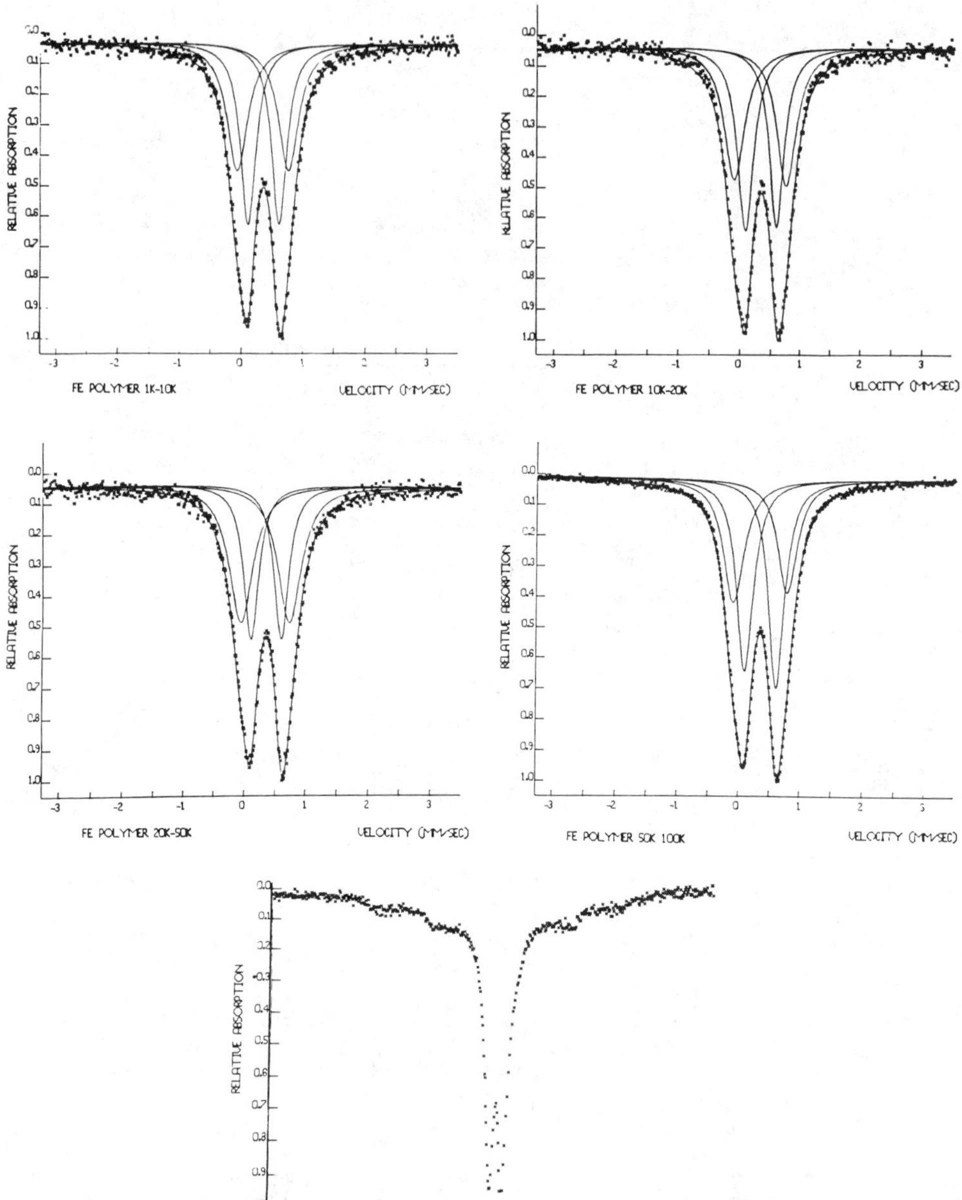

Figure 2. Mössbauer spectra recorded at room temperature for the polymeric fractions in the 1k-300k molecular weight range isolated and freeze dried from a hydrolyzed 0.1M iron(III) nitrate solution

period were output after recording intervals of 0-2 hours, 0-45, hours, 0-66 hours and 93-96 hours.

RESULTS AND DISCUSSION

The Mössbauer spectra recorded at room temperature for the fractionated polymers isolated from the bulk aged iron(III) nitrate solution are shown in Figure 2. The experimental spectra for the 1k-10k, 10k-20k, 20k-50k, and 50k-100k molecular weight fractions are all essentially a broadened iron(III) doublet similar to that of ferrihydrite. These have been computer-fitted with two closely overlapping iron(III)-doublet resonances which are considered to represent the mean extremes of a continuum of doublets arising from the slightly differing site geometries in this poorly ordered polymeric material, rather than two well-defined crystallographic sites [18,30]. As with ferrihydrite the linewidth for the outer doublet is larger than that for the inner doublet. However the Δ value for the outer doublet of these fractionated polymers is slightly larger than the corresponding value for ferrihydrite (Table III) which suggests these fractionated polymers are not ferrihydrite-like in structure. The 100k-300k (Figure 2) and greater than 300k fractions show a partially relaxed magnetic hyperfine goethite resonance as well as the central broadened iron(III) doublet. Such a goethite resonance was not detected at room temperature in the above four smaller fractions when the Mössbauer spectra were similarly recorded over a wider velocity range. However, x-ray diffraction (XRD) showed the presence of very poorly ordered goethite in the 50k-100k fraction with a progressive increase in order for the 100k-300k and greater than 300k fractions. Goethitic absorption peaks could also be identified in these latter three fractions. The typical broad six line XRD pattern of ferrihydrite was not observed in any of these fractions.

The Mössbauer spectra recorded at 78K for the 10k-20k, 50k-100k and 100k-300k fractions are shown in Figure 3. The 10k-20k spectrum shows only the central ferric doublet resonance which for simplicity has only been computer-fitted with a single doublet. The 50k-100k spectrum shows a well-resolved magnetic hyperfine component as well as the prominent central doublet. For the 100k-300k spectrum the magnetic hyperfine resonance predominates (Figure 3), as it does for the greater than 300k fraction (not shown here). At 78K the magnetic field strengths for all these polymer fractions are the same (ca.487kOe) and are significantly greater than that for ferrihydrite but slightly less than microcrystalline goethite (Table III). The δ values for the 78K spectra of ferrihydrite, microcrystalline goethite and the polymer fractions are all identical (0.48 mm/sec), as are the quadrupole splitting values (-0.24 mm/sec) again with the exception of ferrihydrite which has a larger value of -0.10 mm/sec (Table III). The above results therefore suggest these polymer fractions are goethitic in their structural character and with increasing aging or hydrolysis they simply increase in size and perhaps order, to actually form goethite.

The room temperature Mössbauer spectra for the less than 500 MW fraction isolated and freeze dried from 0.1M iron(III) nitrate solution immediately after preparation, and after hydrolysis in solution for 48 and 52 hours are shown in Figures 4b-d. The corresponding computer-fitted data are presented in Table IV. The spectrum for the fraction isolated immediately (Figure 4b) has been computer-fitted with a broad singlet resonance and a doublet with a narrow linewidth (ca. 0.21mm/sec). After ageing for 48 hours this singlet peak is reduced considerably in intensity (Figure 4c) and after 52 hours with partial neutralization to OH/Fe=0.2 only the doublet is observed (Figure 4d). Hence the addition of even a relatively small amount of base increases the rate of hydrolysis notably. No magnetic hyperfine resonance is observed at 78K for the 48 hour sample (Figure 4e).

The singlet resonance which is characteristic of iron(III) in a symmetric octahedral oxygen environment has a room temperature isomer shift value (0.38mm/sec) identical with that of $Fe(NO_3)_3 \cdot 9H_2O$ (Table IV, Figure 4a), where the iron(III) is coordinated to six surrounding water molecules. It therefore seems that the singlet resonance in the less than 500 MW fractions arise from the initial monomeric $[Fe(H_2O)_6]^{3+}$, i.e. aqueous iron(III) species, rather than $Fe(OH)^{2+}$ or $Fe(OH)_2^{+}$ as the latter two would be expected to yield a spectrum with a small quadrupole splitting.

The doublet also has a δ value characteristic of iron(III) in octahedral coordination with oxygen (Table III). However the Δ value (1.20mm/sec) is

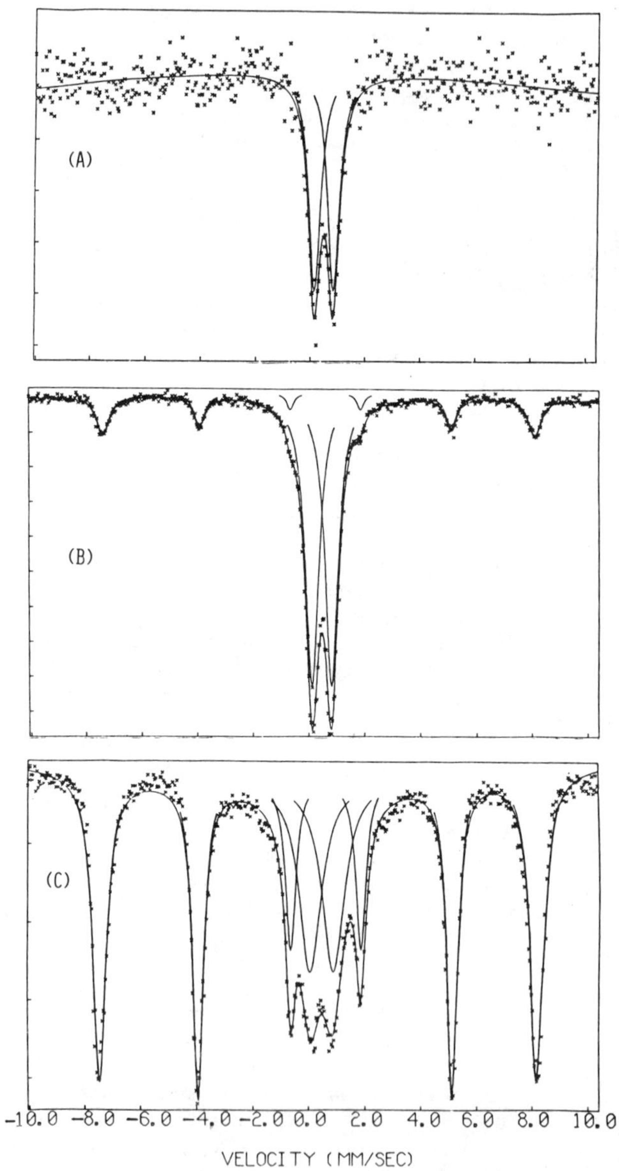

Figure 3. Mössbauer spectra recorded at 78K for selected polymeric fractions isolated and freeze dried from a hydrolyzed 0.1M iron(III) nitrate solution: (a) 10k-20k MW, 50k-100k MW, (c) 100k-300k MW.

Table III. Mössbauer data for the isolated polymeric/goethite fractions with a molecular weight greater than 1000

Molecular Wt. Range	Spectrum Figure	Species[a]	Room Temperature-Doublets			78K Magnetic Sextets		
			δ (mm/sec)	Δ (mm/sec)	Γ (mm/sec)	δ (mm/sec)	Δ (mm/sec)	H(kOe)
1k-10k	2	P	0.37(1) 0.36(2)	0.51(1) 0.84(1)	0.30(2) 0.41(2)		Doublet	
1k-10k[b]	2	P	0.364(2)	0.632(2)	0.42(1)			
10k-20k	2,3	P	0.37(2) 0.35(2)	0.51(2) 0.87(2)	0.31(2) 0.41(2)		Doublet	
20k-50k	2	P+G	0.37(2) 0.35(3)	0.51(2) 0.81(3)	0.30(2) 0.47(2)	0.48(2)	-0.26(2)	488
50k-100k	2,3	P+G	0.37(1) 0.36(1)	0.51(1) 0.88(1)	0.32(1) 0.40(1)	0.49(2)	-0.24(2)	486
100k-300k	2,3	P+G	0.37(1) 0.34(2)	0.52(1) 0.91(2)	0.37(1) 0.63(3)	0.48(2)	-0.24(2)	486
>300k[b]		P+G	0.35(1)	0.63(1)	0.61(1)	0.48(2)	-0.24(2)	485
Ferrihydrite[c]			0.36(1) 0.35(1)	0.50(1) 0.83(1)	0.38(1) 0.40(1)	0.48(2)	-0.10(2)	451
Microcrystalline Goethite[b,d]			0.37	0.68	0.44	0.48(1)	-0.24(1)	495

[a]P = Polymer, G = Goethite. [b]Two peaks only fitted to the central doublet. [c]Reference 18. [d]Reference 31.

Table IV. Mössbauer data[a] for the hydrolysis product fractions with a molecular weight less than 500

Spectrum Figure	Aging Period Hours	Neutralization Conditions	Species[b]	δ (mm/sec)	Δ (mm/sec)	Γ (mm/sec)	Relative % Area
4a	0		$Fe(NO_3)_3 \cdot 9H_2O$	0.38(1)		1.5(1)	100
4b	0		M	0.390(4)		0.94(2)	73(1)
			D	0.451(3)	1.205(3)	0.24(1)	27(1)
4c	48		M	0.384(11)		0.84(6)	22(1)
			D	0.448(2)	1.202(2)	0.27(1)	78(1)
4d	52	OH/Fe = 0.2	D	0.449(2)	1.201(2)	0.29(1)	100
5a	0-2[c]		M	0.41(2)		0.40(8)	12(4)
			D	0.450(4)	1.195(4)	0.26(1)	88(4)
5b	0.45[c]		D	0.449(3)	1.201(3)	0.23(1)	57(3)
			P	0.35(1)	0.62(1)	0.57(2)	43(3)
5c	0-66[c]		D	0.449(3)	1.199(3)	0.24(1)	36(3)
			P	0.36(1)	0.64(1)	0.50(1)	54(3)
5d	93-96[c]		P	0.369(6)	0.687(6)	0.44(1)	100
6a	119	$CO_3^=/Fe = 0.2$	M	0.388(5)		0.87(2)	41(1)
			D	0.447(2)	1.204(2)	0.24(1)	59(1)
6b	119[d]	$CO_3^=/Fe = 0.2$	D	0.447(2)	1.207(4)	0.25(1)	59(2)
			P	0.35(2)	0.61(2)	0.56(4)	41(2)
6c	40	$CO_3^=/Fe = 0.5$	D	0.447(2)	1.203(2)	0.25(1)	90(1)
			P	0.35(1)	0.64(1)	0.36(1)	10(1)
6d	40[e]	$CO_3^=/Fe = 0.5$	P	0.37(1)	0.52(1)	0.33(1)	29(4)
			P	0.36(2)	0.79(2)	0.57(1)	71(7)

[a]All data are from spectra recorded at room temperature. [b]M = monomer, D = dimer, P = polymer. [c]Sample hydrolyzed further in Mössbauer holder (see text). [d]500-1k MW fraction. [e]>1k MW fraction.

rather large for an iron(III) entity and is indicative of the iron(III) ion being subjected to an equally large electric field gradient arising from some considerable asymmetry in the surrounding coordination sphere. The narrow linewidths of these doublet peaks suggest the species is well-ordered structurally. Table V lists the molecular weights of likely monomeric, dimeric and trimeric hydrolysis species with the associated nitrate anion.

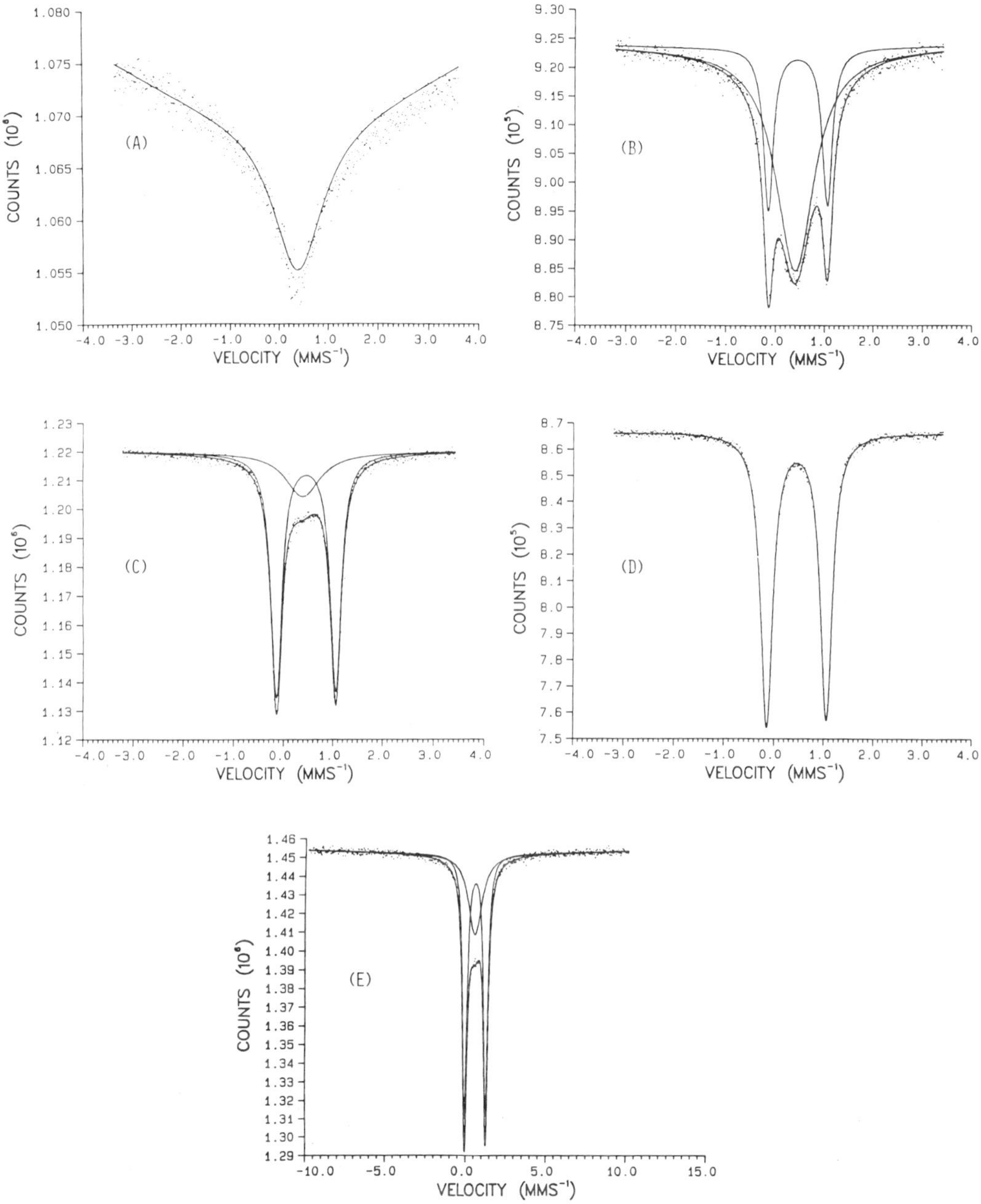

Figure 4. Mössbauer spectra recorded at room temperature for (a) $Fe(NO_3)_3.9H_2O$, and the <500 MW fractions and freeze dried from a hydrolyzed 0.1M iron(III) nitrate solution aged as follows: (b) isolated immediately, (c) 48 hours, (d) 52 hours, OH/Fe = 0.2, (e) sample c at 78K.

Table V. Molecular weights of possible initial hydrolysis species

Species	Molecular Weight[a]
2+ Monomer	287
3+ Monomer	350
3+ Dimer	475
4+ Dimer	538
2+ Trimer	519
4+ Trimer	645

[a]Includes the appropriate number of nitrate anions.

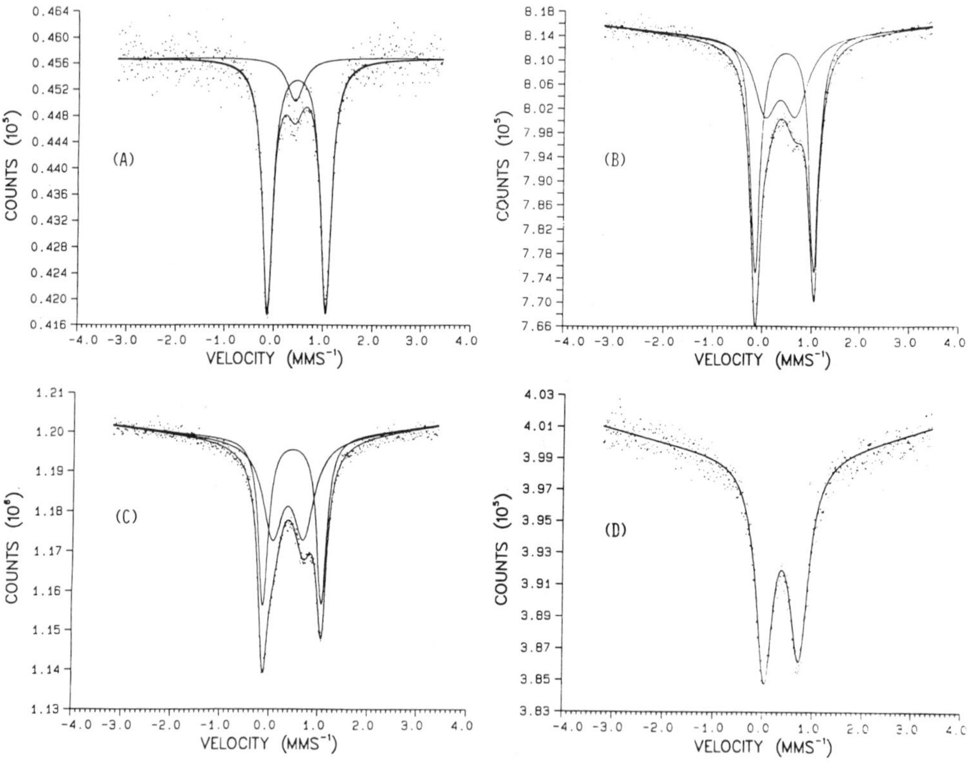

Figure 5. Mössbauer spectra recorded at room temperature for the <500 MW fraction isolated and freeze dried after 48 hours hydrolysis of a 0.1M iron(III) nitrate solution. The sample was then further hydrolyzed in the Mössbauer holder and the spectra output at recording intervals of (a) 0-2 hours, (b) 0-45 hours, (c) 0-66 hours, (d) 93-96 hours.

Possible structures for these species are given in Figure 7. From molecular weight considerations the iron(III) dimer appears to be the most likely structure giving rise to this relatively wide quadrupole-split doublet. In this dimer each iron(III) has an identical coordination sphere of four water molecules and bridging OH and O groups. These groups could certainly impose a large electric field gradient on the iron(III) ions and hence produce a large quadrupole splitting in the Mössbauer spectrum. The bridging OH and O groups provide points for addition of further monomers or dimers in the polymerization process. It is possible that this doublet could arise from a 2+ trimer species rather than a 3+ dimer. However the three iron(III) ions in the 2+ trimer do not have identical environments and in view of the narrow experimental doublet peak linewidths the trimer is considered unlikely. It is therefore concluded that the initial hydrolysis species in this less than 500 MW fraction are the $[Fe(H_2O)_6]^{3+}$ monomers and discrete well-ordered $[(H_2O)_4FeO(OH)Fe(H_2O)_4]^{3+}$ dimers. With increasing aging in solution the monomeric species readily link up to form dimers which further link to form small polymers (polycations). When the dimeric material is all used up, these small polymers may link in a similar manner to form large polymers and eventually goethite. The assignment of these Mössbauer resonances to monomeric and dimeric species is entirely consistent with the conclusions from earlier potentiometric and spectrophotometric measurements [7-9].

In order to investigate subsequent aging of the less than 500 MW fraction the above 48 hour sample was permitted to hydrolyze in the Mössbauer sample holder by contact with atmospheric moisture. The spectrum obtained after 2 hours recording time (Figure 5a) during which the sample was hydrolyzing continuously, is essentially the same as that of the initial sample (Figure 4c). After 45 hours the monomeric material is completely depleted, the dimeric species predominates, and there is a contribution of a further iron(III) doublet with a broader linewidth and a narrower quadrupole splitting (Figure 5b). This latter doublet corresponds exactly with the similarly broadened experimental iron(III) doublet in the 1k-100k range of polymer fractions discussed above (Figure 2 and Table III). Because of the complex overlap it was not possible to further resolve this broadened doublet for the hydrolyzed less than 500 MW fraction into the two component closely overlapping doublets, as was the case for the polymeric material discussed earlier. After 66 hours recording time and continued hydrolysis (Figure 5c) the relative amount of polymeric material increases and the spectrum recorded from 93-96 hours (Figure 5d) shows only polymeric material.

Figure 6 shows the effect of neutralization with Na_2CO_3 and different aging periods prior to fractionation and freeze drying. After 119 hours with an added carbonate to iron ratio of 0.2 the less than 500 MW fraction still contains a significant proportion of monomer (Figure 6a) but the 500-1k MW fraction contains the dimer plus a smaller proportion of polymer material (Figure 6b). With increased addition of Na_2CO_3 to give an added carbonate to iron ratio of 0.5, and after only 40 hours ageing, the less than 500 MW fraction shows essentially dimeric material with only a very small amount of polymer (Figure 6c). The monomeric material has reacted completely, which further demonstrates that added base enhances the hydrolysis process. The greater than 1k MW fraction of this latter sample shows only polymeric material and the Mössbauer spectrum has been computer-fitted with two closely overlapping doublets (Figure 6d) consistent with the spectra shown in Figure 2. The iron analysis figures given in Table II confirm that with increased hydrolysis the proportion of the total iron in the less than 500 MW fractions decreases with increasing ageing as more of the larger polymeric material is formed.

Hence it is concluded that the initial monomers combine to form dimers which can be isolated as stable species. These dimers may exist as hydroxyl bridged entities or more likely may lose a proton to form the O, OH bridged species (Figure 7). It is possible the ageing sequence continues with the dimers linking longitudinally to form single linear chains. Alternatively the dimers can link both laterally and longitudinally via bridging OH and O groups to form the double chain which is the basis of the goethite structure (Figure 7). This imparts the goethitic character to the ensuing polymeric species, which continue to grow and ultimately, in this case, form goethite.

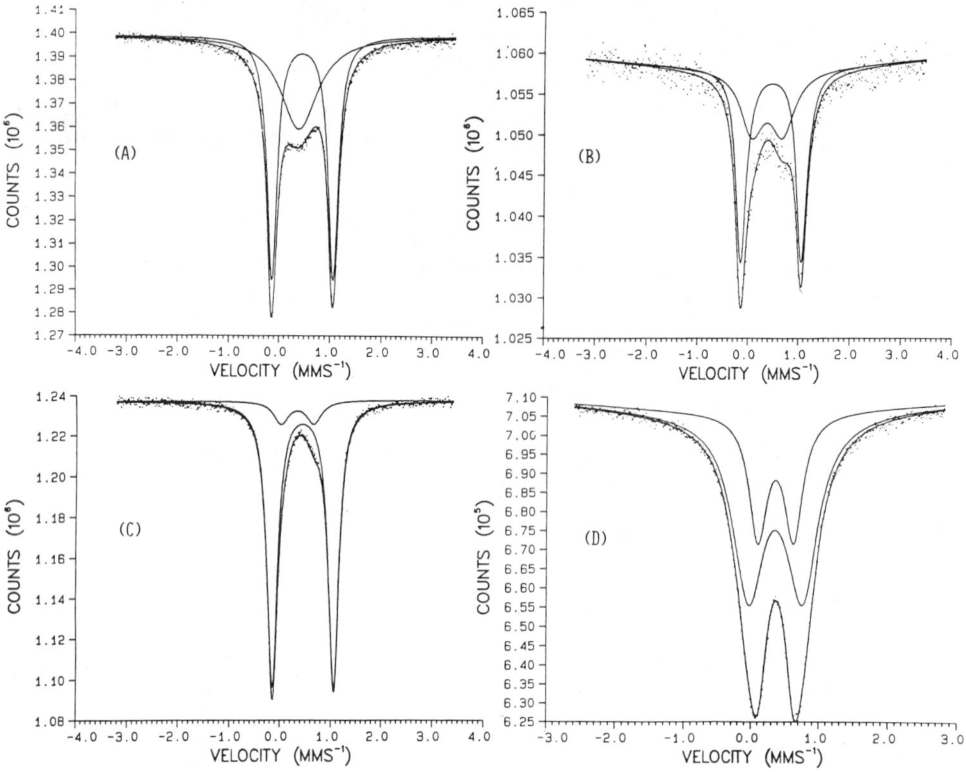

Figure 6. Mössbauer spectra recorded at room temperature for the following fractions isolated and freeze dried from a hydrolyzed 0.1M iron(III) nitrate solution: (a) <500 MW, 119 hours, added $CO_3^{=}/Fe = 0.2$, (b)500-1k MW, 119 hours, added $CO_3^{=}/Fe = 0.2$, (c) <500 MW, 40 hours, added $CO_3^{=}/Fe = 0.5$, (d) >1k MW, 40 hours, added $CO_3^{=}/Fe = 0.5$.

It could also be possible for a monomer to add onto the dimer and form a trimer intermediate (Figure 7), onto which further monomers, dimers or even trimers themselves could add, again forming double chains. This latter mechanism seems unlikely as the only intermediates which could be isolated and characterized both in this Mössbauer study and earlier spectrophotometric and potentiometric studies are monomers and dimers [6].

It is now of interest to consider the significance of the dimeric intermediate in relation to the structures of the more important iron oxides existing in weathered systems, i.e. goethite, lepidocrocite, ferrihydrite, and hematite. The goethite structure consists of double chains of (Fe,O,OH) octahedra which are linked to neighboring double chains by sharing octahedral apices and also by H-bonding (Figure 8a). Each double chain may be considered as being formed by individual octahedra (monomers) linked by sharing five out of the six O or OH groups along the edges of the octahedra (Figure 8c) or alternatively by dimers linked similarly. Hence it is easy to see how a polymeric structure formed initially from dimers linked to create goethite-like double chains, can grow and order progressively to goethite.

The lepidocrocite structure which is similar, is formed from these same double chains linking together by sharing edges of the octahedra to form sheets. The sheets are then H-bonded together (Figure 8b). Lepidocrocite is usually formed under reducing conditions [14] but the role which the iron(II)

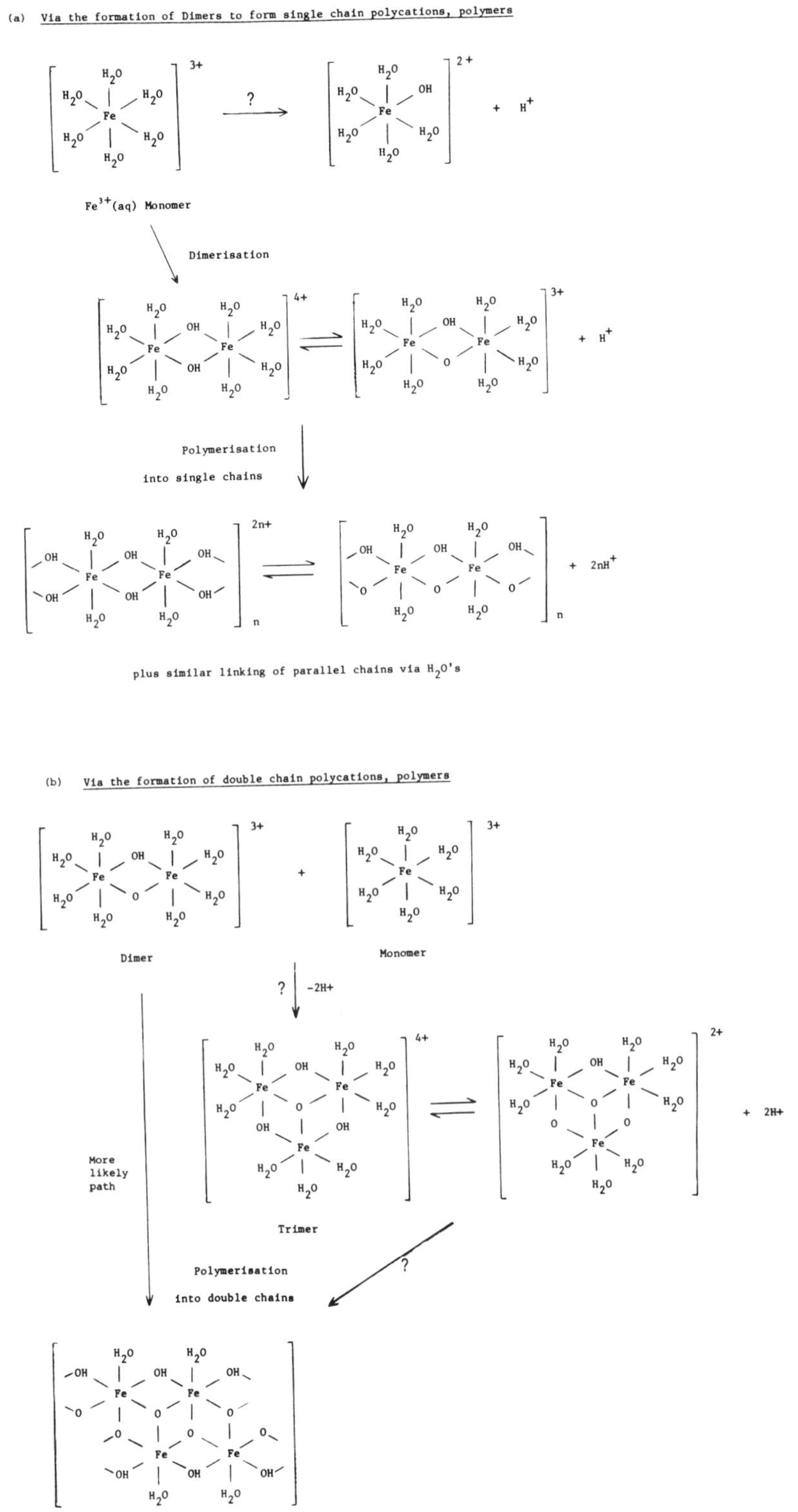

Figure 7. Possible mechanisms for the linking of monomeric and dimeric species to form polymer chains.

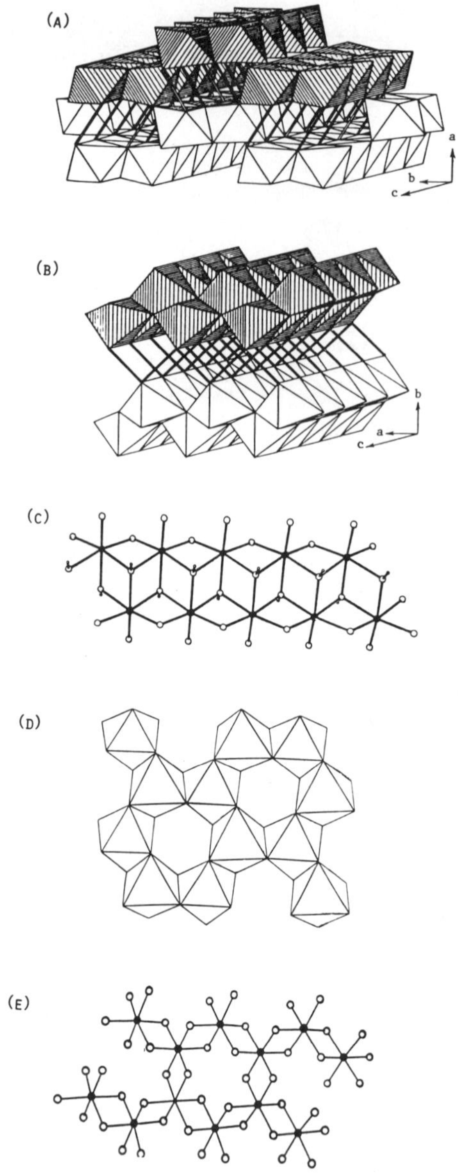

Figure 8. Diagrammatic representations of (a) the goethite structure, (b) the lepidocrocite structure, (c) the double chain unit of the goethite and lepidocrocite structures, (d) the hematite structure, (e) the arrangement of linked dimers in the hematite and ferrihydrite structures. In (a) and (b) the H-bonding is depicted by the double lines linking appropriate octahedra. In (c) and (e) iron - large solid circles, oxygen - large open circles, hydrogen - small open circles.

ions actually play in the mechanism of lepidocrocite formation is not understood. Presumably they are important in directing the double chains to link by sharing octahedral edges to generate the lepidocrocite structure rather than by sharing octahedral apices to form the goethite structure. However, lepidocrocite has been observed to form from dilue iron(III) solutions under certain conditions of iron(III) concentration and total ionic strength [33].

The hematite structure consists of a hexagonal network of octahedra (Figure 8d) and it is easy to see how the dimers may be linked to form such an arrangement (Figure 8e). This iron oxide structure does not have any protons and is formed generally at temperatures above that of goethite and lepidocrocite. At these slightly higher temperatures the hematite structure is favored energetically. Ferrihydrite which has both OH groups and water molecules is considered to have a much less ordered defect hematite structure [34]. Ferrihydrite readily forms at ambient temperature, particularly under conditions of rapid hydrolysis [14] and at higher temperatures this transforms to hematite [18].

Presumably the protons and water molecules stabilize the ferrihydrite structure at ambient temperatures. This structure is favored kinetically over that of goethite and is therefore formed under conditions of rapid hydrolysis. With slower hydrolysis at ambient temperature, when the ferrihydrite solubility product is not exceeded the goethite linkages are energetically favored and the dimers join to form goethite double chains.

These results therefore suggest that the particular iron oxide formed as the ultimate hydrolysis product under a certain set of conditions is determined at the onset from the way in which the dimers link together initially. The nature of these linkages then propagate through the intermediate polymeric material which has the essential characteristic structure of final product iron oxide. The goethitic-like polymers characterized in this study do not resemble ferrihydrite, and goethite is the product formed. It follows that ferrihydrite is not a precursor to or a poorly ordered form of goethtite, and that ferrihydrite may only form goethite by dissolution back to monomeric and dimeric species followed by subsequent slower hydrolysis. Ferrihydrite is now well recognised as a stable hydrolysis product at ambient temperature but it may transform to hematite at slightly elevated temperature with minimal structural rearrangement. Hematite can also form by dehydration of goethite but this requires a substantial structural rearrangement and takes place at higher temperatures of ca. 200-300 degrees C.

CONCLUSION

This study has indeed enabled the early-formed monomeric and dimeric species in the hydrolysis of iron to be isolated and characterized. It has also provided an insight into how these dimeric species link and thus determine the structural characteristics of the intermediate polymers and hence the particular product iron oxide formed. From the results the following conclusions can be drawn:

1. Two initial monomeric aqueous iron(III) species combine rapidly to form a dimer which may lose a proton. The dimers appear to be a relatively stable intermediate under certain conditions.

$$\underset{\text{monomers}}{[Fe(H_2O)_6]^{3+}} \rightarrow \underset{\text{dimers}}{[(H_2O)_4Fe\overset{O}{\underset{OH}{<>}}Fe\,(H_2O)_4]^{3+}}$$

2. The dimers can then link both laterally and longitudinally in different ways to form the double chains with either the goethite or

lepidocrocite structures, or the hexagonal ferrihydrite or hematite networks. These initial structural entitites progressively increase in size by the addition of further dimers, while they persist, to form small polymers (polycations) which may then link similarly to form larger size polymers and ultimately form the product iron oxide.

3. The addition of base promotes hydrolysis.

4. The particular product iron oxide formed is determined by the way in which the dimers initially link at the onset of hydrolysis.

5. Ferrihydrite is not a precursor to goethite but may be to hematite.

REFERENCES

1. A. B. Lamb and A. G. Jacques, J. Am. Chem. Soc. 60, 967 (1938).

2. T. H. Siddall and W. C. Vosburgh, J. Am. Chem. Soc. 73 4270 (1951).

3. R. M. Milburn and W. C. Vosburgh, J. Am. Chem. Soc. 77 1352 (1955).

4. R. M. Milburn, J. Am. Chem. Soc. 79, 537 (1957).

5. R. C. Turner and K. E. Miles, Canad. J. Chem., 35 1002 (1957).

6. C. F. Baes and R. E. Mesmer, 'The hydrolysis of cations'. Wiley Interscience, New York, 1976.

7. J. Dousma and P. L. de Bruyn, J. Colloid. Interface Sci. 56, 527 (1976).

8. J. Dousma and P. L. de Bruyn, J. Colloid. Interface Sci. 64, 154 (1978).

9. J. Dousma and P. L. de Bruyn, J. Colloid. Interface Sci. 72, 314 (1979).

10. L. N. Mulay and P. W. Selwood, J. Am. Chem. Soc. 77, 2693 (1955).

11. T. G. Spiro, S. E. Allerton, J. Renner, A. Terzis, R. Bils, and P. Saltman, J. Am. Chem. Soc. 88, 2721 (1966).

12. R. N. Sylva, Rev. Pure Appl. Chem. 22, 115 (1972).

13. R. J. Knight and R. N. Sylva, J. Inorg. Nucl. Chem. 36, 591 (1974).

14. U. Schwertmann and R. M. Taylor, Iron oxides, Chapter 5 in 'Minerals in Soil Environments', ed., J. B. Dixon and S. B. Weed., Soil Sci. Soc. Amer., 145 (1977).

15. U. Schwertmann and W. R. Fischer, Geoderma. 10, 237 (1973).

16. L. Carlson and U. Schwertmann, Geochim. Cosmochim. Acta 45, 421 (1981).

17. F. V. Chukhrov, B. B. Zvyagin, L. P. Ermilova, and A. I. Gorshkov, Proc. Int. Clay. Conf., Madrid, 397 (1972).

18. J. H. Johnston and D. G. Lewis, Geochim. Cosmochim. Acta 47, 1823 (1983).

19. U. Schwertmann and E. Murad, Clays Clay Miner. 31, 277 (1983).

20. R. Thiel, Zeit. anorg. allg. Chemie 326, 70 (1963).

21. M. V. Fey and J. B. Dixon, Clays Clay Miner. 29, 91 (1981).

22. B. A. Goodman and D. G. Lewis, J. Soil Sci. 32, 351 (1981).

23. R. M. Taylor, Proceedings of the NATO workshop on soil colloids and their association in soil aggregates (1984).

24. T. Misawa, K. Hashimoto, and S. Shimodaira, Corrosion Sci. 14, 131 (1974).

25. J. H. Johnston, Geochim. Cosmochim. Acta 41, 539 (1977).

26. U. Schwertmann and W. Lentze, Z. Pflanzenernähr. Dueng. Bodenkd. 115, 209 (1966).

27. V. C. Farmer (ed), Infrared Spectra of Minerals. Mineralogical Society Monograph 4. Mineral Soc., London, 580 pp. (1974).

28. L. H. Bowen, Mössbauer Effect Ref. Data J. 2, 76 (1979).

29. C. W. Childs and J. H. Johnston, Aust. J. Soil Res. 18, 245 (1980).

30. E. Murad and U. Schwertmann, Am. Mineral. 65, 1044 (1980).

31. S. Mørup, 1981. Paramagnetic and superparamagnetic relaxation phenomena studied by Mössbauer spectroscopy, PhD thesis, Technical University of Denmark, Lyngby, Denmark 260 pp.

32. C. W. Childs and J. S. Baker-Sherman, 'Mössbauer spectra and parameters of standard samples 1. Results of standards examined at N. Z. Soil Bureau to 31 July 1983.' New Zealand Soil Bureau Report 66, Department of Scientific and Industrial Research, Lower Hutt, New Zealand, 1984, 50 pp.

33. D. G. Lewis, unpublished results.

34. U. Schwertmann and W. R. Fischer, Zeit. anorg. allg. Chemie 346, 137 (1966).

THE OXIDATION OF HYDROXYCARBONATE GREEN RUSTS

Enver Murad

Lehrstuhl für Bodenkunde, Technische Universität München,
D-8050 Freising-Weihenstephan, Federal Republic of Germany

Reginald M. Taylor

CSIRO Division of Soils, Private Bag No. 2, Glen Osmond,
South Australia 5064

INTRODUCTION

Green rusts are elusive compounds. These hydroxy salts are members of the pyroaurite group. Their structure consists of brucitic layers containing the cations that carry a net positive charge in alternation with anion plus water containing interlayers. The brucitic layers can contain both iron(II) and iron(III). Their structure and the presence of substantial amounts of iron(II) in them renders the green rusts highly susceptible to rapid oxidation and breakdown. This is probably one of the main factors that has so far prevented their identification in natural systems, should they exist there (e.g. in soils, where their presence has been long suspected). The formation of green rusts as intermediate phases during the corrosion of iron has also been indicated [1,2]. Positive identification of such a phase as a corrosion product has, however, been effected in only one case [3]. It is nevertheless conceivable that green rusts are much more frequently formed in the course of corrosion processes than is hitherto known.

In the laboratory green rusts can be synthesized by a variety of methods: by exposing metallic iron to different corrosive reagents, thus simulating natural corrosion conditions [1], by the controlled oxidation of iron(II) compounds [4], or by instigating a reaction between iron(II) and iron(III) compounds [5]. The last-named method has the advantage of allowing the initial iron(II) to iron(III) ratio to be varied systematically, thus giving more controlled synthetic conditions.

Because of their rapid oxidation and structural breakdown when exposed to air, green rusts are difficult to characterize by analytical methods. Storage under vacuum causes decomposition by allowing the interlayer water and anions to diffuse out of the lattice, in some cases resulting in a topotactic solid-state transformation to magnetite [6]. Storage in an aqueous suspension containing only a few ppm of dissolved oxygen will also cause a rapid transformation to magnetite [7], while a more rapid oxidation or higher oxygen potential results in the formation of other common iron oxide phases. Thus, if completely anoxic conditions cannot be guaranteed during the storage of such samples, it may be more efficacious to reduce the rate of oxidation to such a degree (e.g. by immersion of the samples in liquid nitrogen), that these compounds remain unchanged for an adequately long period of time. This may be more successful than attempting to prevent their oxidation altogether.

EXPERIMENTAL DETAILS

Taylor [6] synthesized iron(II)-iron(III) hydroxycarbonate green rusts of different crystallinities and stabilities towards oxidation by adding iron(III) as the soluble nitrate or ferrihydrite to iron(II) carbonate solutions. X-ray diffraction results showed that the hydroxycarbonates formed corresponded to the natural and synthetic corrosion products described by Stampfl [3] and by McGill et al. [2], respectively.

The green rusts studied here were prepared by the procedure of Taylor [6]. The precipitates were separated from the reactant solutions as described previously [8]. An aliquot of each sample was, without being dried, placed into a tightly stoppered Plexiglas container which was

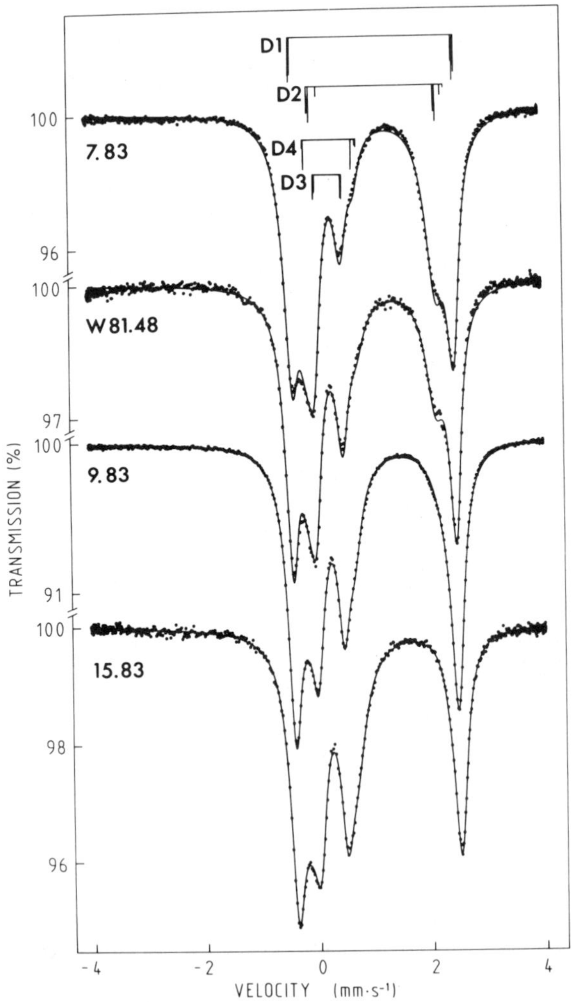

Figure 1. Mössbauer spectra obtained at 120K of different iron(II)-iron(III) hydroxycarbonate green rusts.

immediately immersed in liquid nitrogen, and subsequently used for Mössbauer spectroscopy.

Mössbauer spectra were taken by using cobalt-57 on rhodium foil, moved in a sinusoidal mode, as a radiation source. Both source and absorber were cooled in a bath cryostat to temperatures between 114 and 121K by using helium gas as the thermal exchange medium. Oxidation was effected in stages by removing the drive system and sample from the cryostat and exposing the sample holder to the atmosphere for a certain length of time without opening. Thus, contact of the sample to the atmosphere was restricted to air diffusing into the stoppered sample holder. The data was accumulated in a 1024 channel analyzer. After folding the spectra, computer fits were made by using pairs of Lorentzians lines. The component lines of each doublet were constrained to have equal widths and intensities. All isomer shifts are given relative to metallic iron.

X-ray diffraction was carried out on a Philips PW 1050/70 instrument using graphite-monochromatized CoKα radiation. Diffraction patterns for the oxidized samples were run by step-scanning in 0.02° 2θ-intervals; peak positions and half-widths were determined by computer-fitting the data with combinations of Gaussian and Lorentzian functions.

RESULTS

X-ray diffraction results show the iron(II)-iron(III) hydroxycarbonates studied here to have a (003) d-spacing of about 7.8Å giving a $\underline{c}$ unit-cell dimension of approximately 23.4Å. These samples therefore belong to the green rust I family of Bernal et al. [9] and correspond to the 3R polytype double hydroxide, pyroaurite [10] having the general composition,

TABLE I. Mössbauer spectral parameters for unoxidized iron(II)-iron(III)hydroxycarbonate green rusts obtained at 120K.

Sample Number	Spectral Component	δFe mm/s	ΔE_Q mm/s	τ mm/s	I_{rel}
7.83	D1	1.15(1)	2.85(1)	0.29(1)	0.38(1)
	D2	1.15(1)	2.23(1)	0.40(1)	0.37(1)
	D3	0.37(1)	0.47(1)	0.27(1)	0.18(1)
	D4	0.39(1)	0.92(2)	0.26(2)	0.07(1)
9.83	D1	1.15(1)	2.88(1)	0.28(1)	0.46(1)
	D2	1.27(1)	2.17(2)	0.47(2)	0.10(1)
	D3	0.35(1)	0.47(1)	0.26(1)	0.24(1)
	D4	0.36(1)	0.85(1)	0.36(2)	0.19(1)
W81/48	D1	1.14(1)	2.89(1)	0.26(1)	0.40(1)
	D2	1.13(1)	2.23(1)	0.40(1)	0.28(1)
	D3	0.36(1)	0.49(1)	0.28(1)	0.24(1)
	D4	0.40(1)	0.95(3)	0.35(4)	0.09(1)
15.83	D1	1.16(1)	2.90(1)	0.28(1)	0.41(1)
	D2	1.23(2)	2.41(3)	0.28(5)	0.03(1)
	D3	0.36(1)	0.48(1)	0.27(1)	0.26(2)
	D4	0.35(1)	0.86(2)	0.45(1)	0.31(2)

[a] Numbers in parentheses give standard deviation on last digit.

$$Fe_x^{2+}Fe_2^{3+}(OH)_{2x+4}\ CO_3(x-2)H_2O$$

where x is in the range of four to six. The spectra of four different unoxidized iron(II)-iron(III) hydroxycarbonates are shown in Figure 1. In accordance with observations of Murad and Taylor [11] on one iron(II)-iron(III) green rust (sample W81/48), four doublets, two each for iron(II) and iron(III) could be observed in each spectrum. The positions and relative areas of the constituent lines of these doublets, designated D1 to D4, are shown at the top of Figure 1, and the resulting parameters for the fits are given in Table I.

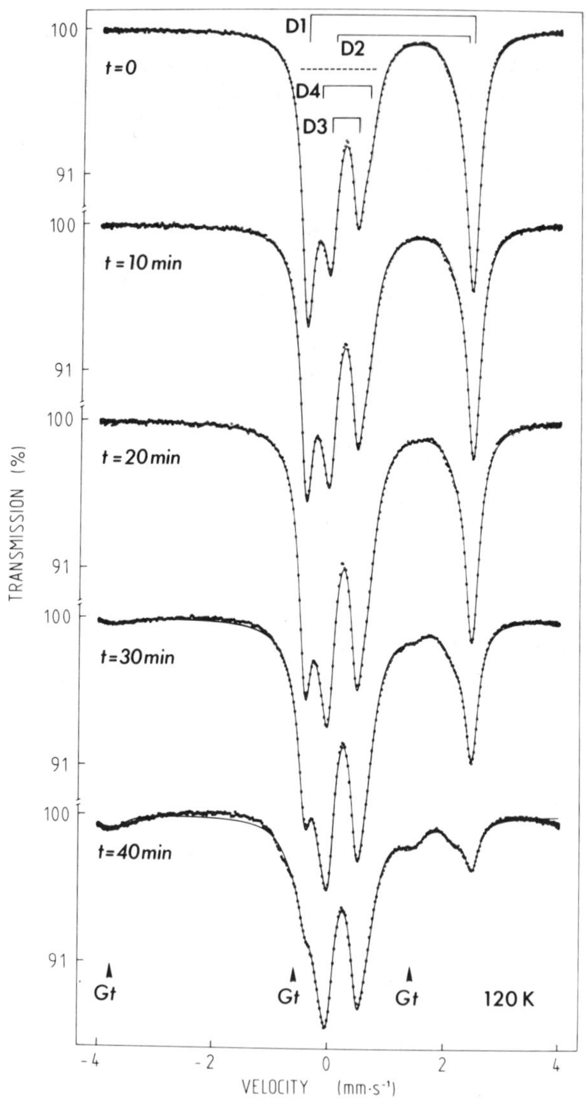

Figure 2. Mössbauer spectra obtained at 120K of hydroxycarbonate 9.83 initially and after four 10- minutes steps of exposure to the atmosphere. Goethite components are marked by Gt.

Table II. Mössbauer spectral parameters obtained at 120K for iron(II)-iron(III)hydroxycarbonate green rust 9.83 after stepwise oxidation.

Oxidation Min.	Spectral Component	δFe mm/s	ΔE_Q mm/s	τ mm/s	I_{rel}
0	D1	1.15(1)	2.88(1)	0.28(1)	0.46(1)
	D2	1.27(1)	2.17(1)	0.47(1)	0.10(1)
	D3	0.35(1)	0.47(1)	0.26(1)	0.24(1)
	D4	0.36(1)	0.85(2)	0.36(2)	0.19(1)
10	D1	1.15(1)	2.90(1)	0.28(1)	0.43(1)
	D2	1.27(1)	2.12(2)	0.46(2)	0.07(1)
	D3	0.35(1)	0.47(1)	0.25(1)	0.24(1)
	D4	0.37(1)	0.79(1)	0.38(1)	0.25(1)
20	D1	1.15(1)	2.89(1)	0.28(1)	0.35(1)
	D2	1.29(1)	2.12(3)	0.52(2)	0.08(1)
	D3	0.36(1)	0.49(1)	0.26(1)	0.26(2)
	D4	0.37(1)	0.81(1)	0.40(1)	0.30(2)
30	D1	1.15(1)	2.89(1)	0.28(1)	0.28(1)[a]
	D2		2.30(2)	0.37(3)	0.07(1)[a]
	D3	0.37(1)	0.49(1)	0.28(1)	0.31(4)[a]
	D4		0.80(3)	0.44(1)	0.35(4)[a]
40	D1	1.16(1)	2.89(1)	0.28(1)	0.16(1)[a]
	D2		2.30(3)	0.27(5)	0.05(1)[a]
	D3	0.37(1)	0.52(1)	0.29(1)	0.43(6)[a]
	D4		0.87(4)	0.42(3)	0.36(7)[a]

[a]paramagnetic constituents only

The quadrupole splittings of the iron(III) doublets averaged 0.48(1) and 0.90(4)mm/s, varying less for the inner than the outer of these doublets. The inner iron(III) doublets also had narrower line than the outer doublets (0.27(1) vs. 0.35(7)mm/s). In contrast, the isomer shifts, 0.36(1) and 0.37(2)mm/s, respectively, did not vary significantly. These two iron(III) doublets are not diagnostic for this compound. They merely indicate high-spin iron(III) in either one highly distorted, or in several discrete coordinations and, for example, resemble spectra observed for the poorly ordered iron oxide, ferrihydrite, at such temperatures.

The iron(II) doublets had quadrupole splittings of 2.88(2) and 2.26(9)mm/s and isomer shifts of 1.15(1) and 1.19(6)mm/s, respectively. The outer iron(II) doublet obviously has considerably better defined spectral parameters than the inner one. The outer doublet also has significantly narrower lines (average 0.28(1) vs. 0.39(7)mm/s) than the inner iron(II) doublet. There is almost no variation in the intensity of this doublet (relative area 0.41(3)).

Sample 9.83 was oxidized in four successive 10-minute steps. In the course of this oxidation the color changed from blue-green, to olive-green, to olive-green with slightly brown stains, to greenish-brown, to brown with greenish mottles. Mössbauer spectra taken after each such oxidation are shown in Figure 2 and the resulting parameters are given in Table II. The first two oxidation steps resulted in a distinct increase of the iron(III) doublet, D4, at the cost of the iron(II) doublet, D1 (see Table II). In

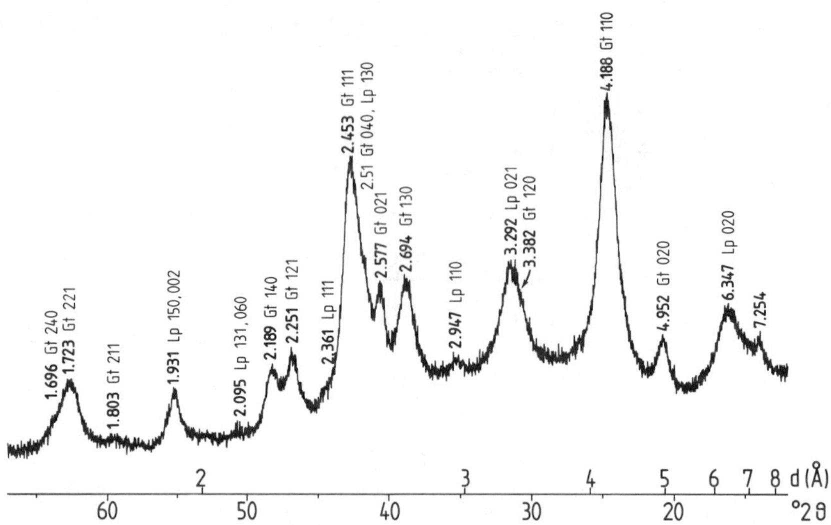

Figure 3. X-ray powder diffraction pattern of sample 9.83 after one year of exposure to the atmosphere. Goethite peaks marked Gt, Lepidocrocite peaks marked Lp.

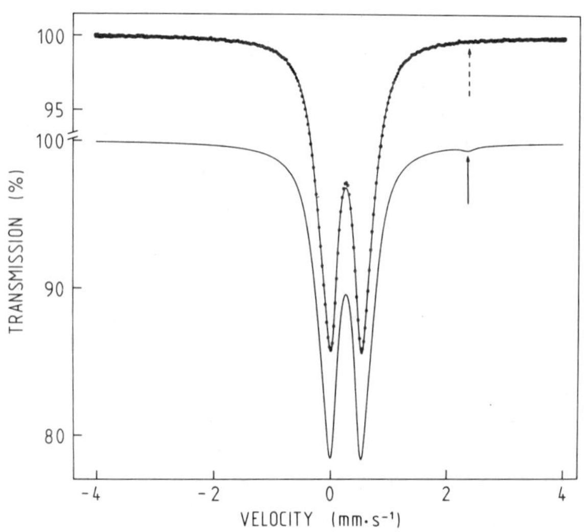

Figure 4. Room temperature Mössbauer spectrum of 9.83 after one year of exposure to the atmosphere (top), and simulated spectrum with the same parameters plus one percent of an iron(II) component (bottom).

contrast, the area of the iron(II) doublet, D2, did not decrease significantly, nor did that of the iron(III) doublet, D3, increase. After 30 minutes of oxidation, the presence of magnetically ordered goethite (α-FeOOH) became evident. The relative areas of both D3 and D4 increased to 0.31 and 0.35 at the cost of D1, which was 0.28 when only paramagnetic components are considered. Finally, after 40 minutes of oxidation, the spectrum showed a major contribution of magnetically ordered goethite, and the relative area of D1 was down to 0.16 of that of the total paramagnetic components. The significant increase in the relative area of D3 can be attributed to paramagnetic lepidocrocite (γ-FeOOH, see below). Possibly the slight increase in D3 observed after 30 minutes of oxidation is also due to minor amounts of lepidocrocite. In contrast to the variation in the intensities of the spectral components, neither the isomer shifts nor the quadrupole splittings changed significantly during oxidation.

The susceptibility of this sample to oxidation limits the application of X-ray diffraction to rapid scans at selected angles using precautionary measures described by Taylor [8]. After complete oxidation by a one year exposure to the atmosphere, X-ray diffraction showed the sample to consist of poorly crystallized goethite and lepidocrocite (Figure 3). One peak with a spacing of 7.254Å could not be attributed to either of these phases; this is tentatively identified as a relic of the original green rust. Although a room-temperature Mössbauer spectrum (Figure 4) showed all the iron(II) to have been oxidized to iron(III), a small percentage of the original layers structure may have been preserved in spite of the oxidation.

Another sample, W81/48, was studied before and after 2 hours of oxidation (Figure 5). The effects of this oxidation on the Mössbauer spectrum have been described by Murad and Taylor [11]. The results are similar to those observed during the stepwise oxidation of sample 9.83, the areas of the iron(III) doublets, D3 and D4, increasing at the cost of the iron(II) doublet, D1.

These observations are in agreement with an ordered arrangement of iron(II) and iron(III) in the brucite layer producing the outer iron(II) and

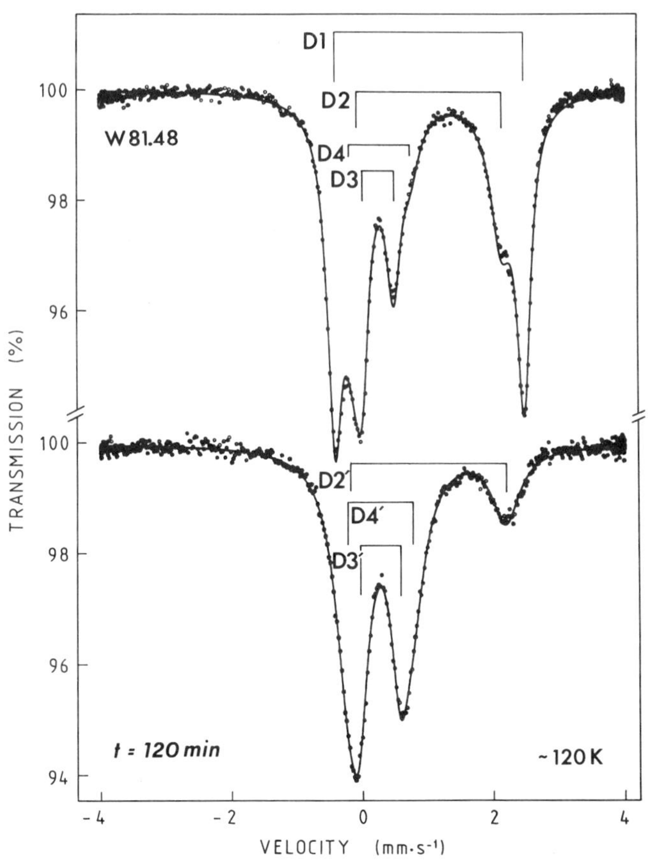

Figure 5. Mössbauer spectra obtained at 120 K of hydroxycarbonate 81.48 and after two hours of exposure to the atmosphere.

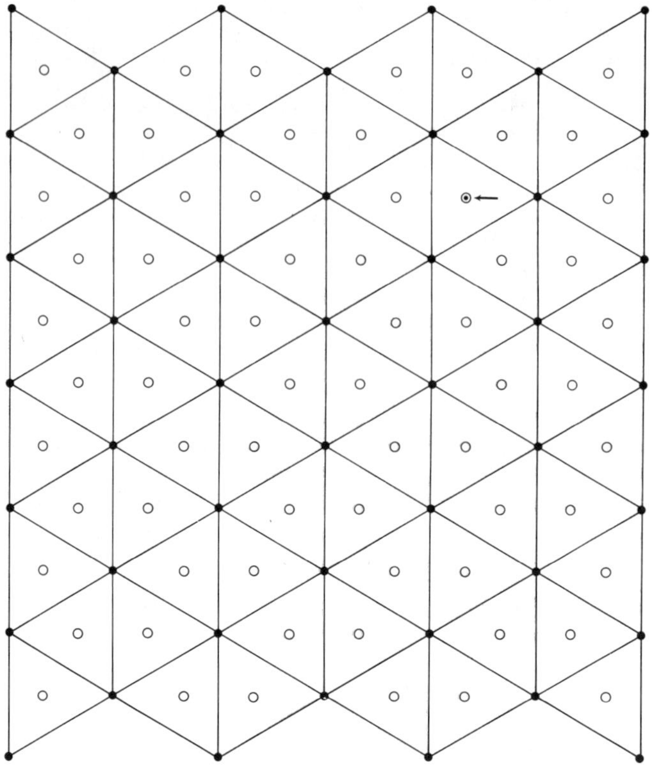

Figure 6. Idealized model of the brucite layer in a 2:1 iron(II)-iron(III) green rust assuming trioctahedral structure and cation ordering. Open circles: iron(III), closed circles: iron(II). Arrow points to an iron(III) on an iron(II) site.

inner iron(III) doublets [11]. If a 2:1 stoichiometry, as has been observed by Allmann and Jepsen [13] for hydrotalcite is assumed, such an ordered arrangement in the trioctahedral brucite layer is readily conceivable (Figure 6). Replacement of an iron(II) ion by iron(III) (e.g. atom marked with arrow in Figure 6) would change the environments of both the neighboring iron(II) and iron(III) ions, giving rise to new spectral components. The doublets D2 and D4 are considered to be due to such deviations from ideal structural order. The failure of the relative intensity of D2 to increase during the course of oxidation indicates that D2 is due to excess ions of the "wrong" valence initially incorporated in the structure. Subsequent oxidation probably proceeds in a broad front rather than occurring at iron(II) ions at random.

CONCLUSIONS

Mössbauer spectra of iron(II)-iron(III) green rusts taken in the paramagnetic state can be adequately fitted with four Lorentzian doublets. Of these, the outer of the two iron(II) doublets, D1, which has a quadrupole splitting of 2.88mm/s and narrow lines of 0.28mm/s, can be considered characteristic. The degree of oxidation of iron(II)-iron(III) green rusts can be observed and quantified by Mössbauer spectroscopy. As such samples oxidize, changing color from green to brown, the iron(II) double D1 gradually disappears, giving way initially to one, and then, after further oxidation, to two iron(III) doublets. In spite of the changes in relative intensity, the quadrupole splitting, isomer shift, and line width of the D1 doublet do not change. Thus Mössbauer spectroscopy is suitable for the identification of such metastable compounds formed in nature and in the course of corrosion processes.

ACKNOWLEDGMENTS

We are indebted to Prof. F. E. Wagner for making the 120K Mössbauer measurements at the Physik Department of T. U. München possible. Thanks are also extended to the Deutsche Forschungsgemeinschaft and the Deutscher Akademischer Austauschdienst for financial support.

REFERENCES

1. C. Carius, Korrosion und Metallschutz 7, 181 (1931).
2. I. R. McGill, B. McEnaney, and D. C. Smith, Nature 259, 200 (1976).
3. P. P. Stampfl, Corrosion Sci. 9, 185 (1969).
4. W. Feitknecht and G. Keller, Zeitschr. anorg. Chemie 262, 61 (1950).
5. E. Deiss and G. Schikorr, Zeitschr. anorg. allg. Chemie 172, 32 (1928).
6. R. M. Taylor, Clay Miner. 15, 369 (1980).
7. R. M. Taylor and Willett, unpublished results.
8. R. M. Taylor, Clay Miner, 17, 369 (1982)
9. J. D. Bernal, D. R. Dasgupta, and A. L. Mackay, Clay Miner. Bull. 4, 15 (1959).
10. H. F. W. Taylor, Miner. Mag. 39, 377 (1973).
11. E. Murad and R. M. Taylor, Clay Miner 19, 77 (1984).
12. E. Murad and U. Schwertmann, Amer. Miner. 65, 1044 (1980).
13. R. Allmann and H. P. Jepsen, N. Jahrb. Miner. Mh., 544 (1969).

METAL ADSORPTION BY ACTIVATED CARBON

J. D. Cashion, D. J. Cookson, L. J. Brown and D. G. Howard*

Department of Physics
Monash University
Clayton, Victoria, 3168, Australia

ABSTRACT

Iron-57 and gold-197 Mössbauer spectra have been taken of activated carbon loaded from aqueous $Fe_2(SO_4)_3$ and $KAu(CN)_2$ solutions, respectively. The iron-57 spectra for six different samples of activated carbons exhibited considerable variations, showing more than one active site as well as different sites for the inherent iron. The spectra did not resemble those known previously for either $Fe_2(SO_4)_3$ or intercalated species. In contrast, the gold-197 parameters agreed very closely with those for $KAu(CN)_2$, consistent with the gold being adsorbed as either $Au(CN)_2^-$ or $KAu(CN)_2$.

INTRODUCTION

Activated carbons made from various starting materials, have been known for nearly two hundred years. They exhibit an enormous capacity to adsorb many different kinds of materials and this has resulted in a variety of uses. For example, its gas adsorbent properties have been used in the manufacture of gas masks, while its ability to absorb many organic molecules have resulted in use as a decolorizing agent in laboratories and sugar and oil refineries and for water purification.

However, over the past decade its ability to accomodate metal atoms or ions has led to extensive use in the extraction of metals from dilute ores, for which the concentrations of metal are in the region of a few parts per million, or a few grams per tonne in mining terminology.

The most obvious useful characteristic of activated carbon is its very large surface area which is usually in excess of $1000m^2/g$. However, its adsorption mechanism or type of bonding is poorly understood on an atomic scale and is not necessarily the same for all adsorbents. Activated carbon is typically made from charcoal obtained from wood, bone, or other carbonaceous material, by heating it at 800 to 1000°C in an atmosphere of steam, carbon dioxide, or a limited air supply. An alternative to charcoal as the starting material is to carbonize an organic material, such as coal, by heating it in zinc chloride solution or phosphoric acid before using the steam activation technique.

---

*Permanent address: Department of Physics, Portland State University, Portland, Oregon, USA.

The structure of activated carbon is basically a highly defective graphite interspersed with a high density of dislocations and pores. Some degree of graphitic order remains in x-ray diffraction patterns, and the pores exist in a continuous range from above 100nm to below 1nm. For metal adsorption, the best types have a very high proportion of micropores in the region 1 to 2nm.

It has been found that the best type of activated carbon for gold adsorption is that made from either coconut shells or peach pits which contain ca. 95% of their cavities in the micropore size in a porous skeletal structure. Unfortunately each manufacturer of activated carbon has his own starting materials and partly proprietary methods of manufacture. Comparison between carbons is done on the basis of many different tests which typically determine an effective surface area as measured by the quantity of a particular gas or chemical entity, e.g., iodine or carbon tetrachloride, which it will absorb. Unfortunately none of these tests give a good indication of the ability of an activated carbon to adsorb gold (except by using gold!) and so clearly are all measuring different subsets of the properties of the carbon.

Our first aim in this investigation was to determine what were the essential characteristics of an "active site" in the metals context. It is well known that the pore surfaces of activated carbons are covered by oxygen adsorbed from the air and this not only assists in the bonding of subsequent additives, but in the case of organic molecules, frequently oxidizes them as well.

It is hoped that a better understanding of the bonding of metals onto the activated carbon could lead to improvements in our understanding of the characteristics of activated carbons which are important in gold adsorption and may lead to a better testing scheme or enhanced performance; to improved techniques for removing metals from activated carbon without destroying or degrading the carbon; and to the separate removal of metals such as cobalt and nickel.

Mössbauer spectroscopy is well suited to this problem because of its sensitivity to nearest neighbor configurations and its lack of sensitivity to characteristics important for other techniques, such as poor crystallinity, optical opacity in the visible and infrared, etc.

PREVIOUS WORK

In a separate part of this project [1], we have used iron-57 Mössbauer spectroscopy to study the effects of different iron ligands on the absorption spectra. Nine different iron compounds were used as aqueous solution starting materials, including both organic and inorganic anions and one, potassium ferricyanide, in which iron was part of the anion. Three samples were also adsorbed from non-aqueous solvents. Room temperature spectra of the samples from aqueous solution comprised several doublets and a singlet with the parameters showing little dependence on the anion. The three most frequent and dominant contributions were two doublets, with isomer shifts of 0.32 to 0.37mm/s and a quadrupole splitting of 0.61 to 0.74mm/s or 0.94 to 1.20mm/s, and a singlet with an isomer shift close to zero with respect to α-iron.

At 4.2K, the spectra showed varying degrees of magnetic ordering whose intensity and hyperfine field depended strongly on the original anion. The f-value also showed a strong dependence on the anion.

On the basis of this work we chose an aqueous solution of iron(III) sulfate, $Fe_2(SO_4)_3$, for the current tests involving differing concentrations and different activated carbons. Our principal aim was to use a system as similar as possible to that used for cobalt and nickel mining for which we hoped that the chemistry of iron would be sufficiently similar to provide useful results. The best compound under this criterion would be $FeCl_3$, however, its f-value was so low that we were deterred from using it for an exploratory investigation. Iron(III) sulfates also occur commonly in ores.

Several workers [2-10] have studied the Mössbauer spectra of $FeCl_3$ intercalated in graphite. Typically their spectra consist of a single line centered at 0.55 to 0.60mm/s for the intercalated $FeCl_3$, and an extra two doublets for intercalated $FeCl_2$ which can be obtained by reduction of $FeCl_3$. These bear no resemblance to any of the spectra which we have observed and hence will not be discussed further here.

SAMPLES AND PREPARATION

Five different activated carbons were selected for these experiments. Two were called "decolorizing charcoal powder" and manufactured by AJAX Chemicals Ltd and May and Baker Pty Ltd, M & B, and a similar "activated charcoal powder" from British Drug Houses, Ltd, BDH. All were in the form of very fine powders. A similar AJAX material was used for the earlier work described by Howard et al. [1], however, they were purchased separately and turned out to have completely different initial iron concentrations and iron-57 Mössbauer spectra. These powders will be referred to by their manufacturers names throughout this paper with the two AJAX samples being denoted AJAX(1) and AJAX(2) in chronological order.

The fourth sample was a commercial coconut char produced by Kintal in Japan for metals processing. It was originally in the form of granules to enable filtration of the carbon from the crushed ore during processing. For our experiments, the material was crushed before loading with metal both for ease of making the Mössbauer absorbers and, in the case of the gold, to increase the amount of metal adsorbed. It is not expected that this will affect the nature of the results. The final activated carbon was an experimental sample produced at the Gippsland Institute of Advanced Education, Churchill, Victoria from Morwell, Victoria, brown coal. It was already in the form of a fine powder and needed no further crushing.

IRON LOADING

Approximately two grams of each activated carbon was immersed in an aqueous solution of iron(III) sulfate for 30 minutes. The carbon was then separated by filtering, washed with distilled water to remove any iron(III) sulfate on the surface and dried in air at 100°C for 24 hours.

The amount of iron(III) sulfate adsorbed was determined by measuring the iron concentration in the solution with a spectrophotometer before and after loading the carbon. Thiocyanate was used to color the solution and the absorption of the $[Fe(SCN)_6]^{3-}$ ion measured at 470nm. The accuracy of this method we estimate to be ±5 percent and good reproducibility was obtained. Iron loadings of up to 10mg of iron per gram of carbon were obtained and absorbers typically contained about one gram of carbon.

GOLD LOADING

The activated carbon was placed in an aqueous solution containing two to three grams of $KAu(CN)_2$ and maintained at a pH of 11. After standing for 24 hours it was filtered, washed, and dried as for the iron loaded samples. Due to the greater difficulty of obtaining gold spectra, because of the 18h half life of platinum-197, only two samples were studied. These were the AJAX(2) decolorizing powder and the coconut char. The amount of gold loaded was determined by electron microprobe analysis to be 49 and 60mg of gold per gram of carbon, respectively. Thus the more heavily loaded sample contained over 0.3g of gold in a total mass of nearly 6g, compared to a gold mass of approximately 1g which would be required to optimize Mössbauer absorption percent effect. This loading is higher than that normally achieved commercially although comparable with laboratory scale tests. We do not expect that these differences in concentration will affect the results.

RESULTS FOR IRON LOADED SAMPLES

All the iron-57 Mössbauer spectra were taken with a cobalt-57 in rhodium source, but the isomer shift values are referred to α-iron at room temperature. Room temperature spectra were taken on the AJAX(2) carbon at loadings of 0, 0.8, 2.1, and 5.0mg of iron per gram of carbon. All spectra showed a broadened doublet whose mean isomer shift decreased from 0.36 to 0.33mm/s and quadrupole splitting increased from 0.81 to 0.86mm/s with increasing iron concentration. Extrapolation of the graph of spectral area against amount of added iron gives the amount of iron in the untreated sample as 3.6(5)mg/g of carbon. This is much higher than in our earlier study on AJAX(1) in which the initial iron concentration as negligible.

The spectra could all be well fit by two doublets with the intensity of the doublet with the larger quadrupole splitting increasing with added iron. Because of ambiguity in fitting the overlapping spectra, it was not clear

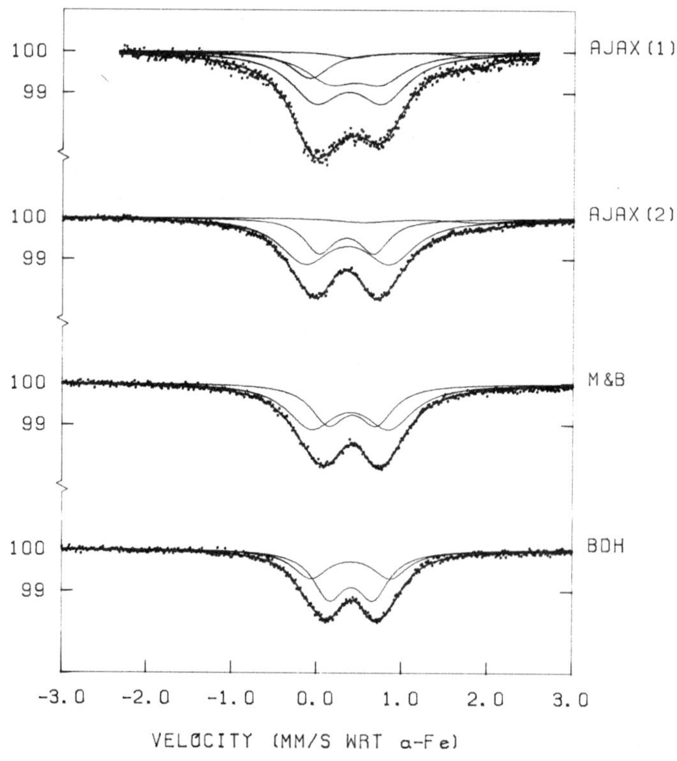

Figure 1. The room temperature iron-57 Mössbauer spectra of the four decolorizing carbons.

whether the other component was also increasing, but at a slower rate. Two doublets are seen in the spectra of all decolorizing carbons, as shown in Figures 1b to 1d for samples of similar iron content. The spectra show strong similarities, and the parameters for the fits are given in Table I with the smaller quadrupole splitting value of the M & B inner doublet being the major difference. The intensities vary also with the outer doublet dominating in the AJAX(2) sample and the inner dominating in the BDH sample.

Also shown in Figure 1a is the spectrum of the original AJAX(1) material loaded with iron(III) sulfate [1]. This spectrum is asymmetrical due to a single line near zero velocity and the parameters of which are also given in Table I.

At 4.2K the spectra had changed drastically with evidence of magnetic splitting occurring in all samples (Figure 2). The AJAX(2) sample, Figure 2b, is strongly magnetic and has been fitted to the sum of one doublet, three sextets, and a gaussian baseline to account for the relaxation which is occurring (Table II). The three sextets and the gaussian were the minimum number of components for which we could obtain a reasonable fit, but they are merely an approximation to the true magnetically relaxing state and enabled us to obtain the relative proportions of the magnetic and non-magnetic contributions. The doublet parameters agree well with the mean of the values for the room temperature spectrum and its linewidth is correspondingly large.

The magnetic contribution to the spectrum of the M & B sample, Figure 2c, is completely structureless due to relaxation effects. It has been

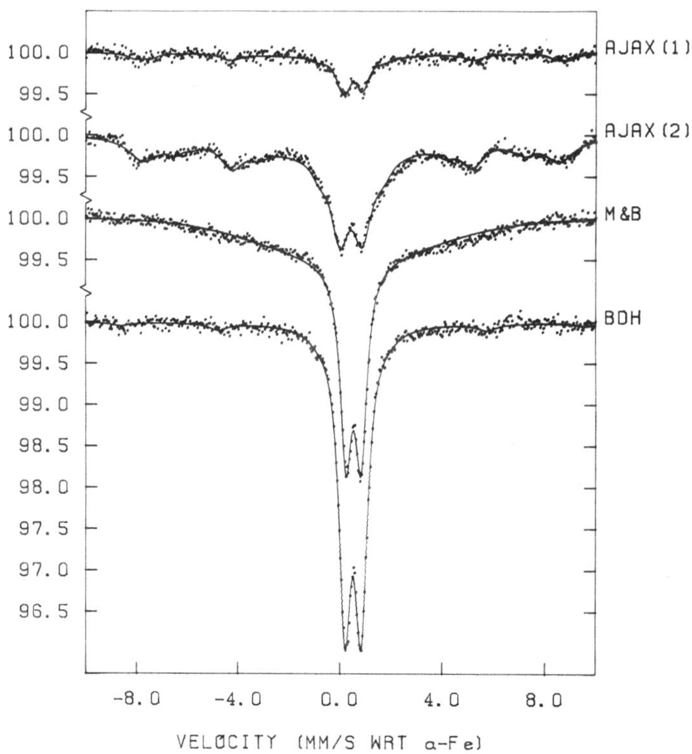

Figure 2. Iron-57 Mössbauer spectra obtained at 4.2K for the four decolorizing carbons. The fitted curves are explained in the test.

Table I. Iron-57 Mössbauer Spectra for the Iron Loaded Activated Carbons[a] Taken at Room Temperature.[a]

Sample	Iron Loading mg/g	Doublet 1				Doublet 2				Doublet 3			
		δ mm/s	ΔE_Q mm/s	Γ mm/s	Area %	δ mm/s	ΔE_Q mm/s	Γ mm/s	Area %	δ mm/s	ΔE_Q mm/s	Γ mm/s	Area %
AJAX(1)[b]	19	0.43	0.60	0.71	32	0.36	0.79	0.71	53	1.13	1.46	0.52	4
AJAX(2)	0.8	0.34	0.65	0.44	30	0.35	1.00	0.72	64	1.22	1.38	0.76	6
M & B	3.6	0.41	0.57	0.50	40	0.39	0.92	0.68	60				
BDH	3.5	0.41	0.52	0.44	60	0.40	0.94	0.50	40				

[a]Isomer shift given relative to α-iron. Fitting errors are typically ±2 in the last digit. [b]This sample also has a singlet with δ of −0.10mm/s, Γ of 0.54mm/s, and a percent area of 11.

Table II. Iron-57 Mössbauer Spectral Parameters for the Iron Loaded Activated Carbons Obtained at 4.2K.[a]

Sample	Iron Loading mg/g	Doublet δ mm/s	Doublet ΔE_Q mm/s	Doublet Γ mm/s	Doublet Area %	Magnetic Components δ mm/s	Magnetic Components Field T	Magnetic Components Γ mm/s	Magnetic Components Area %
AJAX(I)	19	0.48	0.79	0.60	43	0.44	52.9	0.68	17
						0.49	44.5	0.90	18
						-0.08	c	8.76	19
AJAX(2)	0.8	0.44	0.85	0.80	30	0.47	52.3	0.78	19
						0.55	46.4	1.00	18
						0.58	40.1	1.00	11
						0.85	c	8.10	22
M & B	3.6	0.54	0.59	0.54	49	0.36	c	9.28	51
BDH	3.5	0.52	0.64	0.58	93	0.50	56.7	0.86	7

[a] Isomer shifts reported relative to α-iron at room temperature. The fitting errors are typically ±3 in the last digit. [b] The electric field gradient was set to zero for the magnetically split spectra. [c] The relaxing part of the magnetic component has been approximated by a gaussian.

approximated by a gaussian for the purposes of determining its area. Approximately half the total intensity remains in the doublet which has the same parameters as the original inner doublet.

The spectrum of the BDH sample, Figure 2d, is dominated by the doublet with a weak but well resolved magnetic sextet with a very large hyperfine field of 56.7T making up the remainder. This field is larger than any other which we have observed in this study. The quadrupole splitting of the doublet corresponds to the value for the inner doublet in the room temperature spectrum, leading one to the conclusion that it is the outer doublet which constitutes the field subspectrum. However for this explanation to be true, there must be a very large change in the relative f-values of the two sites between 4.2K and room temperature and this has not yet been confirmed.

The electric field gradient was constrained to be zero for the magnetic parts of all spectra because we have found [1] that fitting always returned a value not significantly different from zero, presumably because of averaging effects caused by random angles between the hyperfine field and electric field gradient directions.

The two remaining samples made from coconut char and coal char provided a surprise by having a considerable proportion of their spectral intensity subject to magnetic interactions at room temperature. A further unexpected feature was a large singlet centered at ca. zero velocity. Both samples contained an appreciable concentration of iron in the as-received state and representative spectra are shown in Figures 3a, 4a, and 4b. The parameters for all the spectra are given in Table III. Loading the coconut sample with 4.5mg of iron per gram of carbon did not produce any major change in its profile, although some enhancement of the peak at 0.8mm/s (Figure 3b) could be seen relative to the remainder of the spectrum. At 80°C (Figure 3a), two of

Table III. Iron-57 Mössbauer Spectral of the Activated Coconut Char and Activated Brown Coal Char.[a]

Sample	Iron Leading mg/g	T, K	Singlet δ mm/s	Singlet Γ mm/s	Singlet Area %	Doublet δ mm/s	Doublet ΔE_Q mm/s	Doublet Γ mm/s	Doublet Area %	Magnetic Components[b] δ mm/s	Magnetic Components[b] Field T	Magnetic Components[b] Γ mm/s	Magnetic Components[b] Area %
Coconut	Nil	353	-0.16	0.76	48	1.50	1.78	1.06	40	0.00	32.5	0.50	12
Coconut	Nil	RT	-0.08	0.42	19	1.09	1.69	1.12	32	0.40	46.8	1.76	29
										0.73	44.8	0.92	3
										-0.01	33.1	0.34	19
Coconut	4.5	RT	-0.05	0.50	31	1.28	1.36	1.18	35	0.22	49.2	0.54	8
										0.78	45.7	0.74	12
										-0.02	33.1	0.30	15
Coconut	4.5	4.2	0.05	0.64	27	1.71	1.53	0.84	13	0.45	51.7	0.88	17
										0.41	47.1	1.22	10
										0.12	33.9	0.34	17
										0.54	c	9.12	17
Coal	Nil	RT	-0.26	0.34	11	0.25	0.66	0.66	47	-0.06	48.1	1.16	8
						0.79	1.56	0.90	6	0.06	20.5	0.54	29
Coal	Nil	4.2	-0.03	0.44	20	0.77	0.17	0.94	7	0.45	50.3	0.84	21
						1.51	2.14	0.44	4	0.51	45.3	0.98	19
										0.11	34.3	0.32	6
										0.31	25.3	0.42	23

[a] Isomer shifts given relative to α-iron at room temperature. The fitting errors are typically ±3 in the last digit. [b] Zero electric field gradient for magnetic subspectra as for Table II. [c] Gaussian approximation as for Table II.

the magnetic subspectra visible in the room temperature spectrum, which is almost identical to Figure 3b, have collapsed down to the central region. Cooling to 4.2K (Figure 3c) produced the expected increases in the hyperfine field values but made little difference to the central singlet. Applying a magnetic field of 1.7T at room temperature also produced only a slight broadening of this component suggesting that it is at most, weakly magnetic.

The spectra of the unloaded coal char at room temperature and 4.2K (Figure 4) are very similar to those of the coconut char. As can be seen from Table III, all samples exhibited at least one divalent iron doublet in contrast to the decolorizing powders in which only trivalent iron was observed.

RESULTS FOR GOLD LOADED SAMPLES

The source used for these experiments was 240mg of natural platinum irradiated for 12h in a flux of $5*10^{13}$ n/cm^{2}s at the Australian Atomic Energy Commission HIFAR reactor. The two samples gave virtually identical spectra at 4.2K showing one quadrupole split doublet with isomer shift of 4.6mm/s with respect to gold metal, and a quadrupole splitting of 10.3mm/s for the AJAX(2) sample and 10.4mm/s for the coconut char. The spectrum of the coconut char is shown in Figure 5. Although this has been fitted with a symmetric doublet there is a possible asymmetry of up to 10 percent in the peak intensities which is not apparent in the AJAX(2) spectrum. The linewidths of both spectra were 1.06mm/s compared to the natural linewidth for gold-197 of 0.924mm/s.

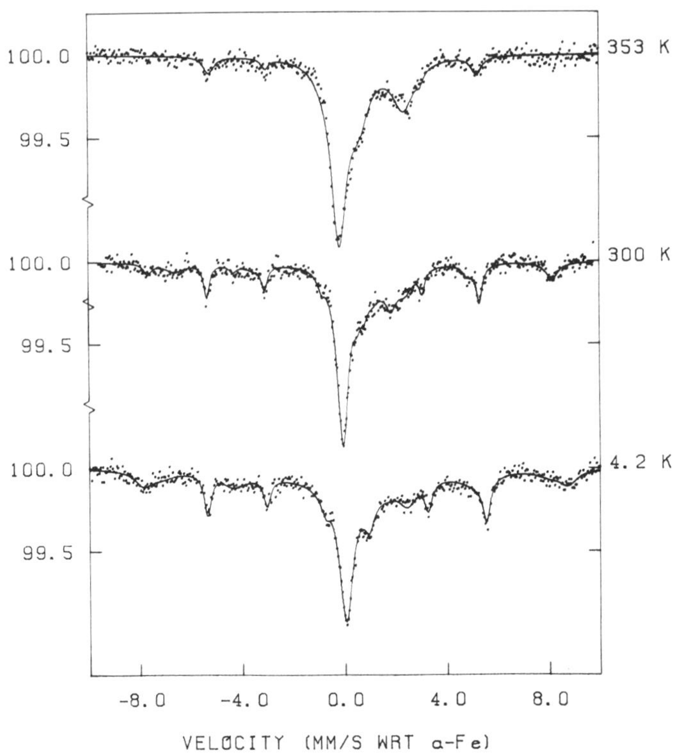

Figure 3. Iron-57 Mössbauer spectra of the coconut char activated carbon taken (a) as received at 80°C, (b) iron loaded at room temperature, and (c) iron loaded at 4.2K.

INTERPRETATIONS OF IRON SPECTRA

The two doublets observed in the room temperature spectra of the three commercial decolorizing powders must correspond to two different configurations of iron bonded to the carbon surface. Because the parameters agree with those for the two dominant doublets observed by Howard et al. [1], which were almost independent of the anion used, we propose that these sites correspond to iron atoms bonded on one side to the carbon, probably through intervening oxygen atoms, and on the other side to the ligand. The geometric arrangement of ligands cannot change very much for different anions in view of the similarities of the spectra and the most likely configuration is probably a distorted octahedron. In view of the large f-value it is possible that the sulfate ions form bridges across the back to nearby iron atoms. The spectral parameters for bulk $Fe_2(SO_4)_3$ are an isomer shift 0.33mm/s and a quadrupole splitting of 0.44 mm/s, which is sufficiently different from the parameters in Table I to indicate that no iron(III) sulfate remained after the washing procedure. Infrared spectra taken of one of the samples, showed no evidence of any OH bonds indicating that all the water had been removed by the drying procedure.

All the Mössbauer absorption lines are appreciably broadened, indicating that there must be small differences between the members of each of those two basic sites. All the iron is in the trivalent form, and the results of Howard et al. [1], showed that divalent iron was not adsorbed by activated carbon. The low temperature spectra, however, show considerable differences between the carbons, apparently with the outer doublet splitting magnetically in the M & B and BDH samples while both doublets split in the AJAX(2) sample.

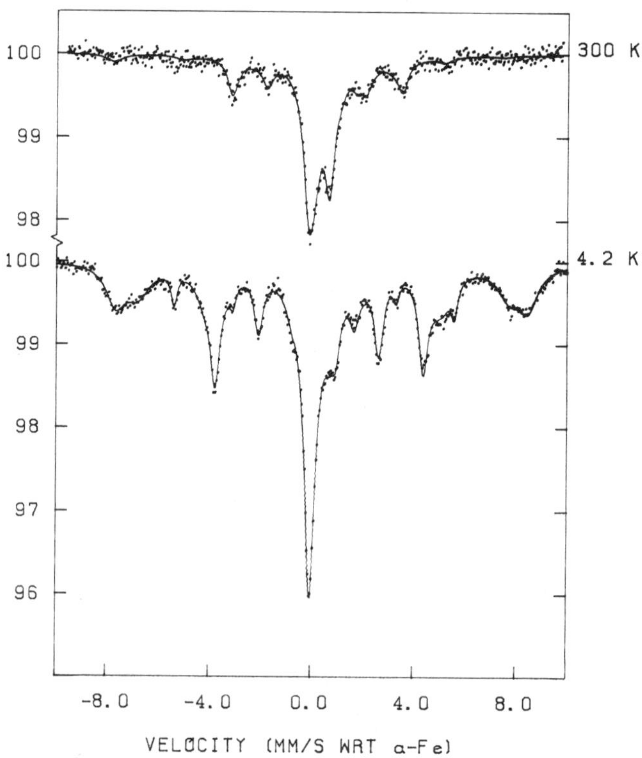

Figure 4. Iron-57 Mössbauer spectra of the as received coal char activated carbon taken at (a) room temperature, and (b) 4.2K.

The magnetic splitting of the outer doublet is consistent with the observation that this was a site which became more populated with iron loading, because this would reduce the mean iron-iron distances and hence enhance the magnetic interaction. The apparent large change in the relative intensities of the two spectral components in the BDH sample suggest that the inner doublet must be due to a very loosely bound iron species of a type not found in the other samples.

Hyperfine fields in excess of 56T are uncommon in iron-57 Mössbauer spectra and much too large for any of the conventional bulk oxides, although they have been observed, for example, in frozen solutions containing the $[Fe(H_2O)_6]^{3+}$ ion [11] and even larger fields are found for iron isolated in argon or xenon matrices [12,13]. Because infrared spectra of an earlier sample showed no evidence for OH species after drying the sample, we propose that the magnetic part of the BDH 4.2K spectrum is due to iron atoms hexacoordinated to oxygen atoms and probably in a micropore so narrow that this arrangement completely fills the pore.

The relaxation in the M & B sample and the probably continuous range of fields in the AJAX(2) sample are both due to differences in the configurations caused by modifications in the internal structure of the carbons. There is much information to be gained about pore size distributions from further experiments of this kind when we understand more about how to interpret the results.

The unexpected observation of divalent iron already present in the coconut and coal chars can be explained by consideration of the activation process. It is known that iron(III) in graphite can be reduced to iron(II) by heating at 400°C in hydrogen. The activation process of heating to 800°C in

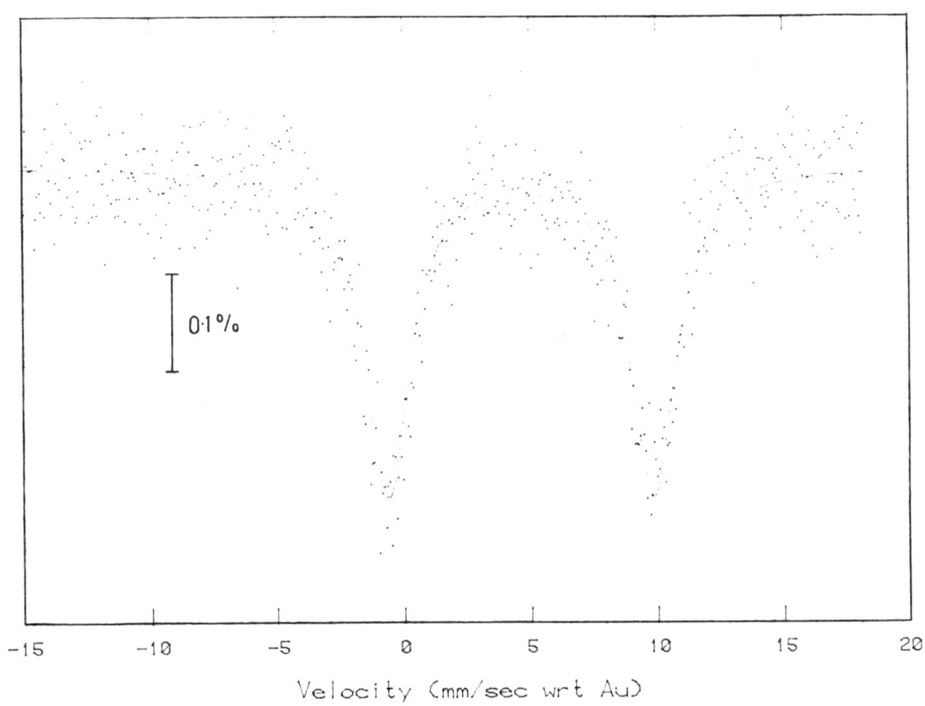

Figure 5. Gold-197 Mössbauer spectrum of the gold-loaded coconut char activated carbon.

the presence of steam will produce water gas ($CO + H_2$) which will achieve the same result.

The subspectrum in these samples with a field of 33T at room temperature and an isomer shift of zero is clearly due to iron metal, although we cannot tell if it is on the surface of the carbon or intercalated between the graphite planes. Intercalated iron has been produced by Tricker et al. [9] and Stukan et al. [5] by reduction of intercalated $FeCl_3$ with both groups obtaining parameters identical to that of α-iron.

We initially considered that the singlet with an isomer shift of zero or a small negative value may have been due to superparamagnetic iron, but because it appears to be non-magnetic, this cannot be the case. Stukan, et al [5] published one spectrum displaying a similar peak (see their Figure 2b), but did not give its parameters. However, the isomer shift can be easily estimated as -0.1mm/s by comparison with the α-iron component in the same figure.

A singlet with a small negative isomer shift has been observed in iron-57 Mössbauer spectra of both black coal [14] and brown coal [15]. Huffman and Huggins suggested [14] that the line may be due to superparamagnetic iron, an iron sulfate or an austenite iron-carbon phase. We believe that this last explanation is essentially the correct one for all four studies, with the γ-iron structure in our case being stabilized by carbon passing into solution in the iron during the activation process at 800 to 1000°C. This assignment is consistent with the unusual negative isomer shift, zero quadrupole splitting, and the fact that the phase is almost non-magnetic, an observation which was not checked by other workers. Workers in our laboratory have also shown recently that the γ-iron phase can be stabilized in brown coal by heating at 900°C in a steam atmosphere [16].

The species producing the large fields at room temperature but which disappeared by 80°C are most likely due to surface iron oxides which have become superparamagnetic at this temperature. With hyperfine fields of well over 40T at room temperature, their true ordering temperature must be above 600K. The sextet observed with hyperfine fields of 20.5 and 25.3T shown in Figures 4a and b, respectively, is due to cementite, Fe_3C.

INTERPRETATION OF GOLD SPECTRA

In contrast to the confusing but probably eventually rewarding iron spectra, the gold spectra are more straightforward to interpret. The parameters for bulk $KAu(CN)_2$ are an isomer shift of 4.2mm/s, and a quadrupole splitting of 10.1mm/s and these are very close to our values of 4.6 and 10.3mm/s for gold on the activated carbon. The structure of $KAu(CN)_2$ consists of layers of potassium ions bridged by linear $(N\text{-}C\text{-}Au\text{-}C\text{-}N)^-$ molecules [17]. Thus we could envisage that the $Au(CN)_2^-$ molecules bond onto the carbon structure at the nitrogen atoms, possibly through an intervening oxygen atom, without appreciably affecting the nearest neighbor configuration of the gold atoms and hence the gold-197 Mössbauer spectrum. It is interesting to note that the nitrogen-nitrogen distance in the $Au(CN)_2^-$ molecule is 0.66nm, almost exactly equal to the c-axis spacing of 0.669nm in graphite. The Mössbauer parameters of $KAu(CN)_2$ place it in a very underpopulated region of the isomer shift-quadrupole splitting diagram [18-19], however more measurements, for example with $NaAu(CN)_2$, are necessary to determine whether and where the potassium ions are incorporated in the adsorption.

There have been many different postulates about the way in which gold is absorbed onto carbon from aurocyanate solutions and recent summaries may be found by McDougall et al. [20] and Tsuchida et al. [21]. The three theories still current are that the gold is present as (a) the aurocyanide ion,

(b) a compound or degradation product other than the aurocyanide ion, such as precipitated AuCN, or (c) as metallic gold. Our results definitely rule out the last two methods as the major mechanism. The small asymmetry in Figure 5 mentioned earlier could be due to a small amount of either AuCN or gold metal although the shifts of the peak positions from those for bulk $KAu(CN)_2$ do not favor these explanations. Other possible explanations are the presence of different cations, such as potassium or hydrogen ions, because it is known that the potassium concentration is usually less than the gold concentration [21] or Goldanskii-Karyagin vibrational anisotropy in the linear $Au(CN)_2$ molecules.

Our results show that the majority of the gold is present as either $Au(CN)_2^-$ ions or as $KAu(CN)_2$ and further experiments are required to distinguish between these possibilities. Our results do not support the conclusion of McDougall et al [20] that the gold atoms have an oxidation number of 0.3.

CONCLUSIONS

We have shown that iron atoms can be adsorbed onto activated carbon in several different configurations. These configurations seem to be relatively independent of the anion used but there are marked differences in the spectra of the six carbons studied, both for the inherent iron and for the adsorbed iron contributions. The variety in the latter contributions are due to either differences in the nature of the active sites or in the micropore nature of the activated carbons. In contrast, the gold spectra showed that gold is absorbed onto carbon as either the $Au(CN)_2^-$ molecule or as $KAu(CN)_2$ in agreement with some current theories. It is expected that further gold Mössbuer spectra will elucidate more details of the gold adsorption mechanism, while the iron spectra show promise of being a sensitive probe of the details of either bonding or structure in the activated carbons.

ACKNOWLEDGMENTS

We would like to thank Neil Thomas and Colin McAllan for helpful discussions and also for providing the coconut char and coal char samples, respectively. We are grateful to Roman Liebach for the electron microprobe analysis and to Jim Fraser for technical assistance. The financial support of the Australian Research Grants Scheme and the Australian Institute for Nuclear Science and Engineering is also gratefully acknowledged.

REFERENCES

1. D. G. Howard, J. D. Cashion, and L.J. Brown, to be published (1986).

2. B. V. Liengme, M. W. Bartlett, J. G. Hooley, and J. R. Sams, Phys. Lett. 25A, 127 (1967).

3. J. G. Hooley, M. W. Bartlett, B. V. Liengme, and J. W. Sams, Carbon 6, 681 (1968).

4. A. G. Freeman, Chem. Comm. 193 (1968).

5. R. A. Stukan, V. A. Prusakov, Yu. N. Novikov., M. E. Vol'pin, and V. I. Goldanskii, J. Struct. Chem. (USSR), 12, 567 (1971).

6. D. Hohlwein, P. W. Readman, A.Chamberod, and J. M. D. Coey, Phys. Stat. Sol. (b) 64 305 (1974).

7. V. G. Jadhao, R. M. Singhu, G. M. Joshi, K. P. R. Pisharody, and C. N. R. Rao, Z. Phys. Chem. 92, 139 (1974).

8. K. Ohhashi and I. Tsujikawa, J. Phys. Soc. Jap. 36, 422 and 37, 63 (1974).

9. M. J. Tricker, E. L. Evans, P. Cadman, N. C. Davies, and B. Bach, Carbon 12, 499 (1974).

10. R. H. Herber and M. Katada, J. Inorg. Nucl. Chem. 41, 1097 (1979).

11. A. Vertes, L. Korecz, and K. Burger, in "Mössbauer Spectroscopy", Elsevier, Amsterdam, p. 308 (1979).

12. P. A. Montano, P. H. Barrett, and Z. Shanfield, Solid State Commun. 15, 1675 (1974).

13. P. A. Montano, P. H. Barrett, and Z. Shanfield, J. Chem. Phys. 64, 2896 (1976).

14. G. P. Huffman and F. E. Huggins, Fuel 57, 592 (1978).

15. P. S. Cook and J. D. Cashion, to be published (1986).

16. G. S. Hang, private communication.

17. A. Rosenzweig and D. T. Cromer, Acta Cryst. 12, 709 (1959).

18. H. D. Bartunik, W. Potzel, R. L. Mössbauer, and G. Kaindl, Z. Phys. 240, 1 (1970).

19. R. V. Parish, in "Mössbauer Spectroscopy Applied to Inorganic Chemistry", G. J. Long, ed., Plenum Press, New York, Vol. 1, p. 527 (1984).

20. G. J. McDougall, R. D. Hancock, M. J. Nicol, O. L. Wellington, and R. G. Copperthwaite, J. S. Af. Inst. Mining. Metall., 344 (1980).

21. N. Tsuchida, M. Ruane, and D. M. Muir, Proc. Int. Conf. Recent Adv. in Mineral Science and Technology, 1984 MINTEK, Johannesburg, in press (1985).

MÖSSBAUER SPECTROSCOPY OF IRON AND IRON ALLOY FISCHER-TROPSCH CATALYSTS

A. M. van der Kraan

Interuniversitair Reactor Instituut
2629 JB Delft, The Netherlands

J. W. Niemantsverdriet

Laboratory for Inorganic Chemistry and Catalysis
Eindhoven University of Technology
5600 MB Eindhoven, The Netherlands

I. INTRODUCTION

Mössbauer spectroscopy is an excellent technique for in situ investigations of iron-containing catalysts, because of the high penetrating power of the γ-radiation and the high sensitivity of the spectral parameters to the chemical state and the local environment of the Mössbauer atom. Applications of Mössbauer spectroscopy in the field of catalysis up to 1980 have extensively been reviewed by Dumesic and Topsøe [1], and by Topsøe, Dumesic, and Mørup [2]. Here we will review Mössbauer investigations of iron and iron alloy Fischer-Tropsch catalysts, carried out at the Interuniversitair Reactor Instituut at Delft.

The Fischer-Tropsch process provides a means to produce fuels, chemical feedstocks and special chemicals from synthesis gas over iron and other group VIII metal catalysts. The reactants, a mixture of carbon monoxide and hydrogen called synthesis gas, can be obtained from coal gasification or from steam-reforming of natural gas. Fischer-Tropsch synthesis represents proven technology. Germany used this process during the Second World War to produce a part of its gasoline, diesel oil, lubricants and waxes [3], and today South Africa applies Fischer-Tropsch synthesis in the Sasol plants [4], and exploits in this way its large quantities of easily mineable coal to become independent of external oil suppliers.

The oil crises of the seventies and the growing realization that oil supplies are not inexhaustible, are the main reasons that Fischer-Tropsch technology has been considered for use in the Western world also. However, the relatively high price of synthesis, which requires costly separations of the various hydrocarbon products, make production of liquid fuels from synthesis gas economically unattractive under the present conditions. At present, the only viable possibilities for the Fischer-Tropsch synthesis would be the production of light olefins, ethylene and propylene [5], and of oxygen-containing hydrocarbons, such as ethanol, ethylene glycol, and acetic acid [6]. In this respect, catalysts based on iron offer the best prospects for conversion of carbon monoxide and hydrogen into light olefins [7], whereas rhodium-based catalysts, among which bimetallic combinations of rhodium with iron, offer good potential for production of oxygenated hydrocarbons [6,8].

Much of the academic research on Fischer-Tropsch synthesis has focussed on the understanding of the mechanism of the reaction between carbon monoxide and hydrogen on the different group VIII metal catalysts [9,10]. Iron behaves differently from the other metals,as it is the only metal which forms bulk carbides during the synthesis. The relation between composition and catalytic properties of iron and iron alloy catalysts has attracted a great deal of attention. As we will show in this paper, Mössbauer spectroscopy has provided indispensable information on this subject.

'IN SITU' MÖSSBAUER SPECTROSCOPY BETWEEN 2 AND 900K

Fischer-Tropsch reactions are generally carried out at temperatures between 500 and 650K. In principle, it is possible to measure Mössbauer spectra during the reaction, as has been elegantly demonstrated by Raupp and Delgass [11], but the common procedure is to measure a spectrum of a catalyst after quenching the reaction by rapidly cooling down to room temperature, or even to cryogenic temperatures as low as 4K. It is essential, however, to cool the sample and measure spectra in situ or at least under controlled atmosphere so as to preserve the chemical state of the highly reactive catalyst surface.

In investigating highly dispersed iron catalysts with their characteristically low Debye temperatures, application of in situ Mössbauer spectroscopy at cryogenic temperatures has proven to be very valuable. At liquid nitrogen or liquid helium temperatures the recoilless fraction increases significantly (Figure 1) and hence the quality of the spectra may improve considerably. Analysis of spectral areas at different temperatures enables one to estimate the Debye temperatures of the iron phases present. Somorjai [12] has noted that in single crystals the Debye temperature of a surface is roughly equal to half that of the bulk. For small particles with a less perfect surface structure, lower values may be expected for the surface Debye temperature. As bulk iron and iron oxides have Debye temperatures of about 450 to 500K [13,14], application of the Somorjai rule indicates that Debye temperatures on the order of 200K are expected for the surfaces of these substances. Figure 1 shows that the corresponding recoilless fractions depend strongly on temperature. Hence, analyzing resonant absorption areas of Mössbauer spectra as a function of temperature may be a way to recognize iron compounds that are located at the surface. A more elaborate discussion on the recoilless fraction of surface atoms has been presented in [15]. Another advantage of Mössbauer spectroscopy at cryogenic temperatures is that when the spectra of components which are paramagnetic or superparamagnetic at room temperature become magnetically split at low temperature, identification or iron phases in the spectra can be greatly facilitated.

In order to be able to measure Mössbauer spectra of catalysts in situ at cryogenic temperatures, we use an absorber holder that can be closed vacuum-tight inside a chemical reactor, as shown in Figure 2. the absorber holder is equipped with beryllium windows, vacuum-brazed on stainless steel. the sealed absorber holder can be mounted in a cryostat, after which transmission Mössbauer experiments can be performed at any desired temperature. A detailed description of the apparatus has been given elsewhere [16].

A note of caution concerning the use of in situ cells at cryogenic temperatures may be appropriate. It is our experience that in situ cells which are vacuum-tight at room temperature, may leak at cryogenic temperature,

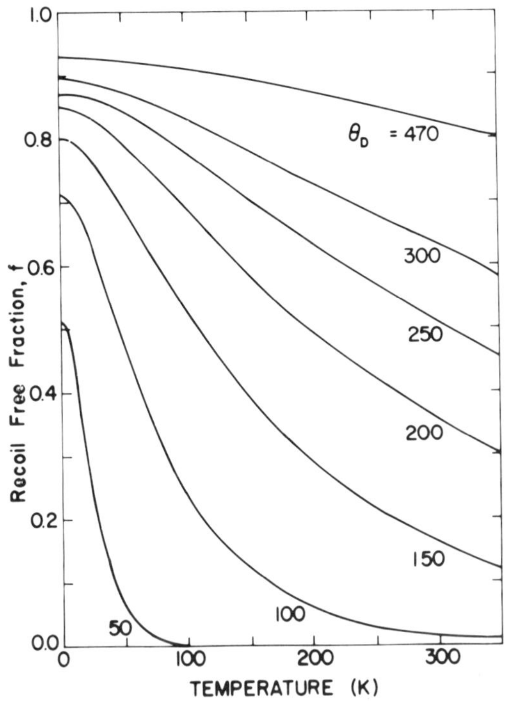

Figure 1. The recoil free fraction as a function of temperature for different Debye temperatures. Reproduced from [15] with permission.

implying that the catalyst is exposed to the residual gas of the cryostat vacuum. An additional spectrum at room temperature after completion of the low temperature experiments provides an easy check on whether or not cryogenic leaks have influenced the results [15].

For measuring Mössbauer spectra in situ at or above room temperature we use the simple and inexpensive reactor shown in Figure 3. The reactor is cooled on the outside with water to permit the use of mylar windows, which are protected on the inside by thin aluminum foils. the heating section around the catalyst wafer is made of copper, which has the advantage that it does not react with the carbon monoxide in the synthesis gas. Heat losses from the heating block to the cooled reactor wall are minimized by using a thin stainless steel connection between them [17,18].

II. IRON CATALYSTS IN THE FISCHER-TROPSCH SYNTHESIS

Iron metal is not stable under the Fischer-Tropsch reaction conditions, but is converted into iron carbides. Mössbauer spectroscopy is an ideal technique to analyze the carbide products which form during the Fischer-Tropsch process, and we have used this method in combination with X-ray diffraction, carbon content analysis, and reaction kinetic measurements to study the behavior of a model system consisting of unpromoted and unsupported iron in atmospheric Fischer-Tropsch synthesis [19]. Carburization studies of supported iron catalysts by means of Mössbauer spectroscopy have been reported by Amelse et al. [20], Raupp and Delgass [11,21], Nahon et al. [22], Jung et al. [23], Bianchi et al. [24], and Tau et al. [25], whereas Mössbauer investigations of various promoted iron catalysts have been presented by Maksimov et al. [26], van Dijk et al. [27], Vogler et al. [28], and Jaggi et al. [29].

Figure 4 contains Mössbauer spectra of an initially 100 percent α-iron catalyst after different periods of Fischer-Tropsch synthesis at 513K. Visual inspection shows that 2.5 hours of synthesis is sufficient to convert all metallic iron into a mixture of iron carbides. Figure 5 shows the composition of the Mössbauer spectra, along with the activity of he iron catalyst for hydrocarbon formation, as a function of time on stream. Before we discuss the mechanistic implications of these results, we will note the way in which the complex spectra of Figure 5 have been analyzed and the type of iron carbides that appear during Fischer-Tropsch synthesis.

Lorentzian fitting of Mössbauer spectra of partially carburized iron catalysts, which, as seen in Figure 5b,consist of more than 40 lines, is a cumbersome and time-consuming procedure. Our approach has been to prepare the different iron carbides in a single phase and to use the Mössbauer

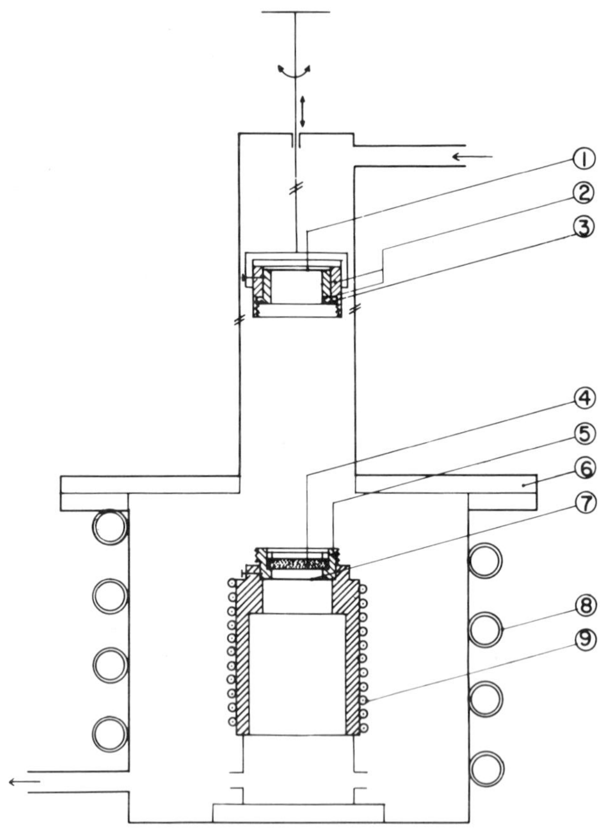

Figure 2. Chemical reactor and absorber holder for in situ Mössbauer studies at cryogenic temperatures: 1, beryllium window; 2, upper part absorber holder; 3, indium O-ring; 4, catalyst wafer; 5, lower part absorber holder; 6, flange with O-ring; 7, beryllium window; 8, water cooling; 9, heating wire. The outside diameter of the reactor is 100mm. Reproduced from [16] with permission.

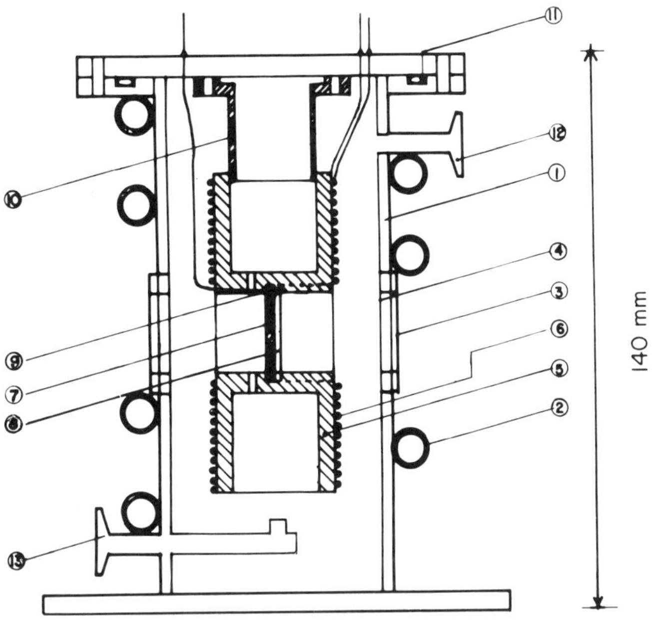

Figure 3. Mössbauer in situ reactor for experiments between 300 and 900K: 1, brass container; 2, water cooling; 3, mylar window; 4, aluminum foil; 5, copper heating block, 6, electrical heating wires; 7, catalyst wafer; 8, copper ring; 9, thermocouple; 10, stainless steel connection; 11, viton O-ring; 12, gas inlet; 13, gas outlet.

spectra of these compounds as base spectra. The complex spectra of mixtures of iron carbides are then considered as a linear combination of these base spectra, and solving for the coefficients of the base spectra is only a simple least squares problem, which can be handled by almost any computer. It is our experience that this procedure, which has been described in detail in [19], is not only more convenient but also more reliable than the typical Lorentzian fitting routines, when dealing with complex spectra such as those seen in Figure 4.

Three different iron carbides appeared in our Fischer-Tropsch experiments, the well-known χ-Fe_5C_2 [30], the less-known ε'-$Fe_{2.2}C$ [20], and the until then unknown Fe_xC [19]. The carbide, θ-Fe_3C [30], is only formed in Fischer-Tropsch synthesis at temperatures above 600K [19]. Comparison with results of other investigators shows that in unsupported iron catalysts, and in supported catalysts with rather large iron particles, χ-Fe_5C_2 is the dominant carbide, whereas ε'-$Fe_{2.2}C$ is mainly formed in well-dispersed iron [19,21,23-25].

Fe_xC represents a carbide in which a broad distribution of hyperfine fields occurs with values up to 270 to 280kOe. This carbide is observed during carburization at relatively low temperatures of 430 to 460K, and at higher temperatures in an early stage of the Fischer-Tropsch synthesis. The X-ray diffraction pattern of this phase shows broad lines, with a weaker intensity than would have been expected from the corresponding Mössbauer spectra. Another noteworthy results about the Fe_xC phase is that it is always observed together with metallic iron, or shortly after the moment that α-iron

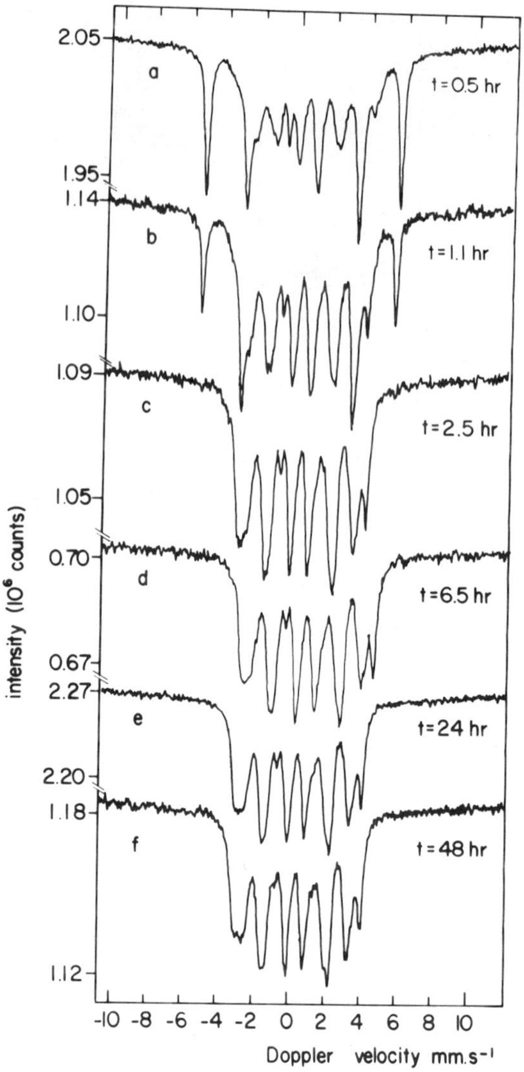

Figure 4. Mössbauer spectra measured at 295K of a metallic iron catalyst after different periods of Fischer-Tropsch synthesis at 513K. Reproduced from [19] with permission.

disappears from the spectra [19]. These results suggest that the Fe_xC carbide represents poorly defined structures intermediate between α-iron and a real crystallographic carbide structure. The time-dependent appearance of the Fe_xC carbide in relation to α-iron and the other carbides, as shown in Figure 5, is also characteristic for a transient phase. Fe_xC has also been observed by other authors [23,25,27,31]. In fact, the subspectrum which Raupp and Delgass [21] originally attributed to a carbide ε-Fe_2C, has the same parameters as the Fe_xC phase, according to Delgass [32].

Doubt exists about the correct stoichiometry of the hexagonal ε'-$Fe_{2.2}C$ carbide. The Mössbauer spectrum of this carbide consists of a single sextet with a magnetic field of 173kOe [20], indicating that all iron atoms in this carbide have the same local environment. Le Caer, et al. [33] have raised the question whether the existence of an unique crystallographic iron site is consistent with the stoichiometry $Fe_{2.2}C$. This question certainly needs further investigation.

CATALYTICALLY ACTIVE PHASE: IRON OR IRON CARBIDE?

Figure 5 shows that as soon as an iron catalyst is brought into contact with a mixture of carbon monoxide and hydrogen, metallic iron is rapidly converted into the carbides Fe_xC, χ-Fe_5C_2, and ε'-$Fe_{2.2}C$. Apparently, carbon atoms deposited at the iron surface by dissociative adsorption of carbon monoxide migrate into the bulk of the catalyst at a high rate. the rate of hydrocarbon formation, however, starts at zero and increases slowly towards a maximum , which in our case, is reached after about one hour [19].

The now important question is whether the Fischer-Tropsch activity of iron is initially low because iron atoms have no intrinsic catalytic activity for the reaction between carbon monoxide and hydrogen, or is it initially low because most of the carbon is consumed in the carburization process? For comparison we note that cobalt and nickel metal are active for Fischer-Tropsch synthesis from the beginning of the process, whereas bulk carbides are not formed in these metals under carbon monoxide and hydrogen [34-37]. Hence, a second question may be asked. Why does iron behave so much differently from cobalt and nickel in the Fischer-Tropsch synthesis?

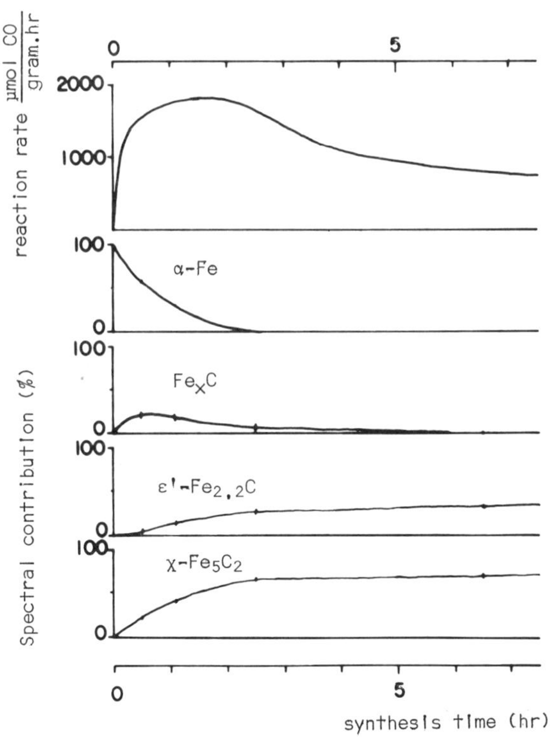

Figure 5. Fischer-Tropsch reaction rate (upper curve) and relative contributions of α-iron and the various iron carbides to the Mössbauer spectra of iron catalysts during Fischer-Tropsch synthesis at 513K. Reproduced from [19] with permission.

Several authors favor the "carbide model" to explain the behavior of iron in the Fischer-Tropsch synthesis [11,20,39,40]. In this model iron itself is not active for Fischer-Tropsch reactions. The active catalyst has an iron carbide bulk structure and it is believed that the number of catalytically active sites at the surface of the catalyst is somehow determined by the state of the bulk material. Strong support for the carbide model was provided by an elegant single-velocity Mössbauer experiment carried out by Raupp and Delgass [11], who observed a linear relation between the extent of carburization and the Fischer-Tropsch activity of a supported iron catalyst.

Another explanation in which iron itself is considered inactive, is the so-called "slow activation model". Here the catalytic activity is attributed to surface complexes consisting of iron, carbon, and hydrogen atoms in a certain configuration, which form at a slow rate. Carburization and Fischer-Tropsch synthesis occur independently in this model [19,38,41].

Several experiments reported in the literature make the carbide and slow activation models unlikely. For example, it has been found that iron atoms in a carbon free surface on a carburized bulk structure [11,42], an oxidized iron film [43], a supported iron oxide [44], or iron ruthenium alloy [45,46] are immediately active for Fischer-Tropsch synthesis when they are exposed to carbon monoxide and hydrogen. This shows that the iron-carbon-hydrogen complexes which are necessary for hydrocarbon formation can be formed on a much shorter time scale than predicted by the slow activation model. The results referred to demonstrate also that iron atoms located at the surface of the catalyst can be very active for Fischer-Tropsch synthesis without the presence of a bulk carbide structure underneath. The general trend is that when conversion of carbon into iron carbides can be slowed down, as in iron nitrides [47,48] or iron oxides [43,44], or when carbide formation is prohibited, as in FeRu alloys [45,46] or in precarbided iron [11,42], the Fischer-Tropsch activity is high from the moment that synthesis gas is passed over the catalyst.

All results are in agreement with the "competition model" [19,38]. According to this model, the surface of iron itself is active for Fischer-Tropsch synthesis. After adsorption and dissociation of carbon monoxide and hydrogen three reactions compete for carbon atoms at the surface (denoted by C*):

$$C^* + \text{metal} \rightarrow \text{bulk carbides} \qquad (i)$$

$$C^* + x\,H^* \rightarrow CH_x^* \qquad (ii)$$

$$C^* + C^* \rightarrow \text{inactive carbon} \qquad (iii)$$

Reaction (i), the carburization reaction, is the key to understanding why iron exhibits different Fischer-Tropsch behavior than cobalt and nickel. In case of iron, reaction (i) is extremely fast, due to the low activation energy of 10 to 16 kcal/mole for carbon diffusion in α-iron. Carbon diffusion in nickel and cobalt on the other hand, occurs much slowly with an activation energy of 33 to 35 kcal/mole, and moreover, bulk carbides of these metals are unstable in hydrogen [38]. With iron catlaysts, reaction (i) is faster than the Fischer-Tropsch reaction (ii), and in the early stage of the Fischer-Tropsch synthesis most of the carbon is consumed in the carburization process, as Figure 6 clearly illustrates. When the interior of the iron catalyst becomes more and more carbided, progressively more C* remains at the surface and becomes available for the slower reactions (ii) and (iii). In the case of cobalt and nickel bulk carburization of the metals is not expected, as reaction (i) is a factor of about 10^5 slower than in iron catalysts. Hence, all carbon deposited at the surface of cobalt and nickel is available for reaction (ii) and (iii) from the beginning of the Fischer-Tropsch synthesis and the metals show a high initial catalytic activity. Full details has been given in [38].

Deactivation of the Fischer-Tropsch synthesis has generally been attributed to the blocking of catalytic sites by inactive carbon and by hydrocarbons with high boiling points [10,11,19,38,42,43]. Reaction (iii) accounts for the deactivation process in a simplified way.

THE ROLE OF OXYGEN AND HYDROGEN IN THE CARBURIZATION OF IRON

So far we have only considered the interaction of the iron catalyst with the carbon from the carbon monoxide in the synthesis gas. However, dissociation of carbon monoxide also yields oxygen. In Fischer-Tropsch reaction mechanisms it is usually assumed that the oxygen from carbon monoxide is removed from the catalyst in the form of H_2O or CO_2 [9,10]. For a complete understanding of the Fischer-Tropsch reaction over iron it is important to know if the oxygen from dissociated carbon monoxide can also react with the iron catalyst. We note that oxidation or iron catalysts in industrial Fischer-Tropsch synthesis is very common [3]. At high carbon monoxide conversion levels the gas environment of the catalyst may become oxidizing in nature due to the formation of relatively large amount of H_2O and CO_2. At differential carbon monoxide conversions, as in most fundamental studies, the net atmosphere around the catalyst is reducing and oxidation of the iron catalyst cannot generally be predicted. Yet another good reason to investigate the influence of oxygen from carbon monoxide on iron is the observation that deliberate oxidation of an iron catalyst may substantially enhance its Fischer-Tropsch activity [43,44]. The interaction between oxygen and iron under Fischer-Tropsch conditions has been studied by in situ Mössbauer spectroscopy at liquid helium temperature, in the reactor of Figure 2 [16].

The need to study surface oxidation of iron by Mössbauer spectroscopy at 4K was suggested by the results obtained with a spent catalyst, which had

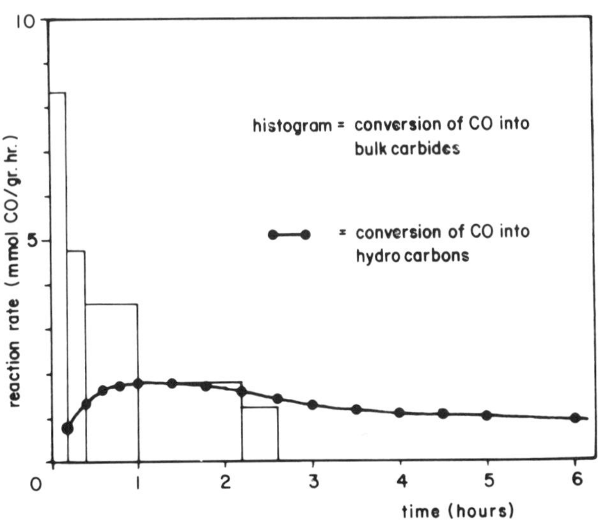

Figure 6. Conversion of carbon monoxide into iron carbides and into hydrocarbons during Fischer-Tropsch synthesis over iron catalysts at 513K. Reproduced from [19] with permission.

been exposed to the air after use in the Fischer-Tropsch synthesis. Spectra of this sample at 295 and 77K showed exclusively contributions from an iron carbide, χ- e_5C_2. At 4K, however, small magnetic contributions due to an iron(III) oxide appeared, corresponding to 10 percent of the total spectral area [19]. Additional spectra between 4 and 77K confirmed that the recoil-free fraction of this iron oxide decreased rapidly with increasing temperature and indicated that the Debye temperature of the oxide was only about 50K [16]. Several considerations supported the conclusion that the iron(III) oxide is located at the surface of the catalyst. Iron oxides with low Debye temperatures in iron catalysts have also been observed by Amelse, et al. [49] and by Shen and Dumesic [50].

Experiments to investigate surface oxidation of iron under Fischer-Tropsch conditions, carried out in the in situ reactor of Figure 2, are shown in Figure 7. The Mössbauer spectrum of the reduced iron catalyst, measured at 4K, confirms that all iron is present as α-iron at the start of the Fischer-Tropsch synthesis. Exposure of the catalyst to oxygen-free carbon monoxide at 500K results in significant oxidation of the iron, as the 4K spectrum in Figure 7b shows. This oxide contribution is absent in the spectra at 77 and 295K, indicating that the Debye temperature of the iron oxide is low and that the oxide is most probably located at the surface of the iron catalyst [16]. This experiment demonstrates that the surface of metallic iron can be oxidized by the oxygen from dissociated carbon monoxide.

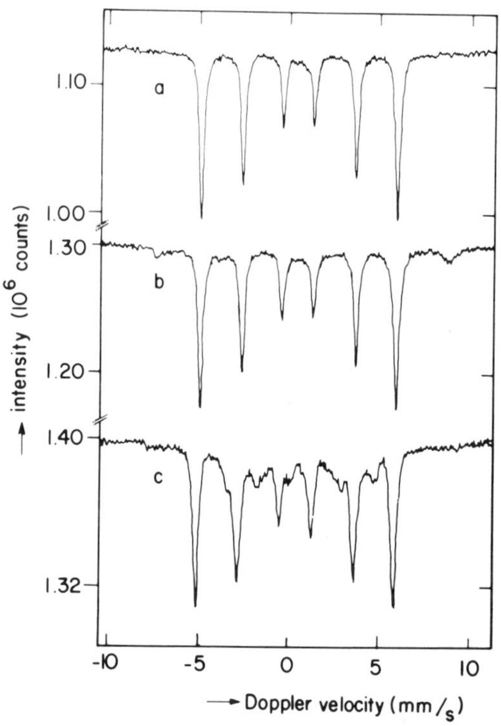

Figure 7. In situ Mössbauer spectra measured at 4K of (a) a reduced iron catalyst, (b) a reduced catalyst after 3h carbon monoxide treatment at 500K, and (c) a reduced iron catalyst after 3h Fischer-Tropsch synthesis in one part carbon monoxide and 3 parts hydrogen at 500K. Reproduced from [16] with permission.

Oxidation of the iron surface is much less pronounced in a mixture of three parts of hydrogen and one of carbon monoxide, as Figure 7c shows. Nevertheless, the spectrum contains a small contribution of iron oxide, corresponding to about 3 percent of the total area, which is just above the detection limit [16]. These results indicate that oxidation of iron can occur under Fischer-Tropsch conditions, and that reactions between iron and oxygen should be taken into account in reaction mechanisms for Fischer-Tropsch synthesis over iron catalysts. However, as Figure 7 also shows, the main difference between treatment of iron in carbon monoxide alone or in a mixture of carbon monoxide and hydrogen is that in the latter case significantly more iron carbide is formed than in carbon monoxide alone. This suggests that hydrogen also plays a role in the carburization process.

The influence of hydrogen on the carburization of iron by synthesis gas was studied by exposing fully reduced iron catalysts to mixtures of carbon monoxide and hydrogen with different hydrogen to carbon monoxide ratios, at 515 and 575K [17,18]. the experiments showed that the rate of carburization increases with increasing hydrogen content of the synthesis gas, and also with increasing partial pressure of carbon monoxide in synthesis gas with a fixed hydrogen to carbon monoxide ratio. Figure 8 illustrates these results and shows in particular that carburization in carbon monoxide alone proceeds at a much slower rate than in mixtures of carbon monoxide and hydrogen, in agreement with the results shown in Figure 7. As Figure 8 shows, relatively small amounts of hydrogen, corresponding to hydrogen to carbon monoxide ratios as low as 0.05, accelerate the carburization process considerably. Correlations between the synthesis gas composition and the kind of carbides formed were not observed. In all cases χ-Fe_5C_2 was the dominant carbide phase [17,18], as expected in unsupported iron catalyst [19,21].

The slow carburization of iron in pure carbon monoxide has been related to the formation of an iron oxide layer on the surface of the iron catalyst [16]. The oxide skin prevents diffusion of carbon into the bulk of the catalyst and hence, a substantial fraction of the deposited carbon remains at the surface of the catalyst. This would also explain why carbon monoxide pretreated iron catalysts show a high initial Fischer-Tropsch activity, as compared to freshly reduced iron [19]. However, when iron is exposed to

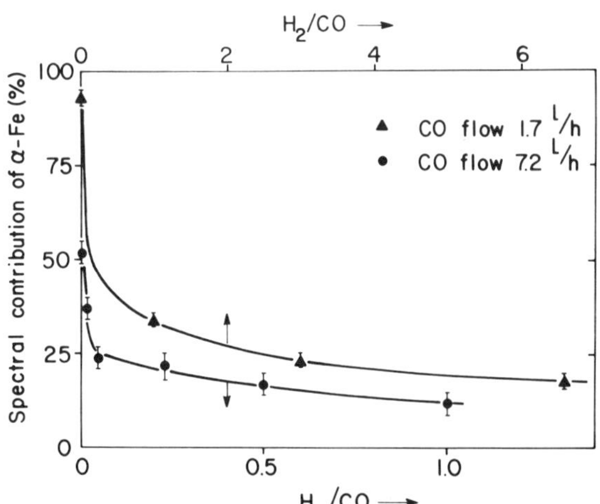

Figure 8. Fraction of uncarbided iron in iron catalysts after 2.5h of Fischer-Tropsch synthesis at 575K, as a function of hydrogen to carbon monoxide ratio in the synthesis gas.

carbon monoxide in the presence of hydrogen, most of the oxygen originating from dissociated carbon monoxide will effectively be removed from the surface by hydrogen in the form of water, and only a small amount of iron oxide can be formed, as indicated in Figure 7c. In this case carbon diffusion from the surface into the bulk of the catalyst is no longer hindered by an iron oxide skin, and substantial carburization is observed. A more detailed description of this mechanism has been given elsewhere [16].

The results discussed above demonstrate how useful Mössbauer spectroscopy has been elucidating the mechanism of Fischer-Tropsch synthesis and carburization with iron catalysts. The technique allows for a straightforward characterization of the iron carbides which are formed during Fischer-Tropsch synthesis. The combination of Mössbauer spectroscopy with reaction kinetic measurements has led to a detailed understanding of the reaction mechanism in a simplified model system for the industrially important reaction between carbon monoxide and hydrogen over iron catalysts.

III. SUPPORTED BIMETALLIC CATALYSTS

In catalysis it has long been known that alloying of metals provides a powerful means to influence and optimize the performance of catalysts [51-53]. As catalysts should preferably have high surface areas, small particle alloys supported on high surface area materials such as SiO_2 and Al_2O_3, usually called supported bimetallic catalysts, are of great practical interest. Examples of industrially successful bimetallic systems are Al_2O_3-supported PtRe and PtIr catalysts which are used in the reforming of naphta [53].

In the field of synthesis gas reactions, some iron-containing bimetallic catalysts have shown attractive selectivity improvements. Bhasin et al. [8] reported that FeRh supported on SiO_2 catalyzed the formation of methane, methanol, and ethanol from carbon monoxide and hydrogen in Fischer-Tropsch synthesis at 575K and 68atm, whereas the monometallic iron and rhodium catalysts yielded much broader product distributions. The formation of substantial amounts of ethanol is particularly interesting, as a direct route from synthesis gas to an added-value chemical, such as ethanol, might be attractive commercially [6]. However, high pressure appears to be a necessary condition for alcohol formation, as in atmospheric Fischer-Tropsch synthesis over FeRh supported on SiO_2 methane was by far the dominant product [54,55].

Another interesting bimetallic catalyst for Fischer-Tropsch synthesis is FeRu supported on SiO_2. Vannice et al. [56] observed a considerably enhanced selectivity for ethylene and propylene over this bimetallic FeRu catalyst as compared to the monometallic iron and ruthenium SiO_2 supported catalysts. Recent work by Stoop [57] comfirmed these results and revealed that TiO_2-supported FeRu catalysts are even more selective for formation of low olefins than the SiO_2-supported catalysts.

Investigations of the structure of supported bimetallic catalysts are seriously hindered by their small dimensions and by the presence of a nonconducting, porous support. Therefore, these catalysts are not amenable to study by means of techniques such as X-ray diffraction or Auger electron spectroscopy. In this respect, iron-containing catalysts have the advantage that Mössbauer spectroscopy can be applied. Several Mössabauer investigations of iron-containing bimetallic catalysts have been presented in the literature, and have been reviewed by Garten [58], Guczi [59], and Topsøe et al. [2]. In this paper we will limit ourselves to supported bimetallic catalysts consisting of iron and one of the more noble group VIII metals ruthenium, rhodium, palladium, iridium, and platinum, and in cases where generalization is permitted we will denote the noble group VIII metal by "M".

We mention that Mössbauer investigations of supported FeNi and FeCo have been carried out in the laboratories of Butt and Schwartz [37,60,61], Delgass [21,31], and Mørup and Topsøe [62].

REDUCED FeM SUPPORTED ON SiO_2 CATALYSTS

Figure 9 shows Mössbauer spectra of SiO_2 supported FeRu, FeRh, FePd, FeIr, and FePt catalysts, which contain equal molar amounts of iron and the noble group VIII metal [63]. The catalysts have been reduced in hydrogen at temperatures between 725 and 800K, in the reactor shown in Figure 3. The

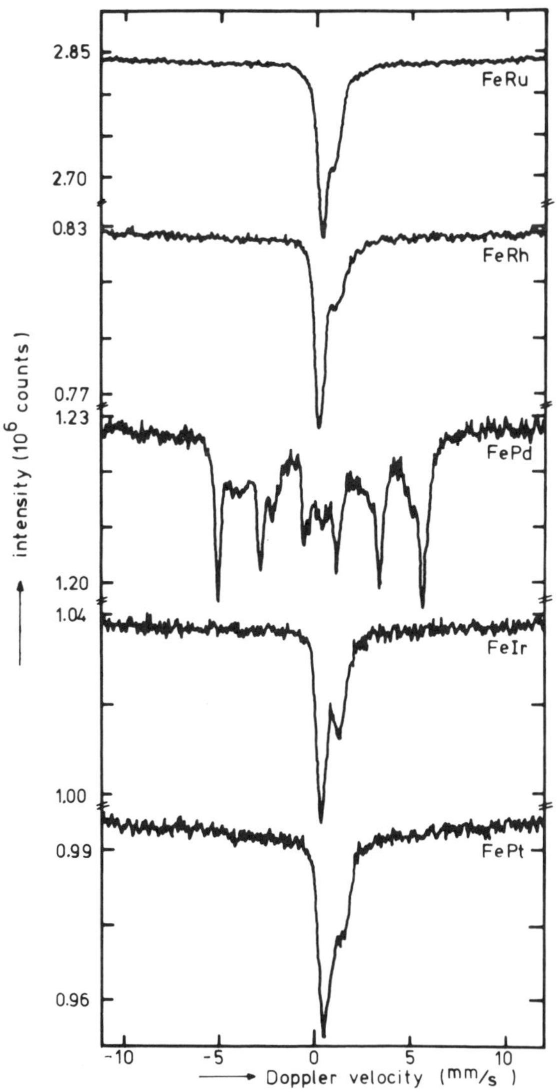

Figure 9. Mössbauer spectra of silica-supported bimetallic FeM catalysts after reduction, measured in situ under H_2 at room temperature. Reproduced from [63] with permission.

spectra of FeRu, FeRh, FeIr, and FePt on SiO_2 look very similar. Comparison with spectra of similar bimetallic catalysts with iron to M ratios smaller than one indicates that, although the shape of the spectra can vary somewhat with composition, particle size and pretreatment, the spectra generally consist of a single peak or unresolved doublet characteristic of iron in an alloy and a shoulder or resolved peak at higher velocity, as shown in Figure 10. The spectrum of FePd supported on SiO_2 (Figure 9) is obviously an exception, but is consistent with the FePd phase diagram which predicts the bcc structure for alloys with a iron to palladium ratio of one [64]. As a consequence, the spectrum of 1 to 1 FePd supported on SiO_2 shows magnetic splitting. FePd alloys with higher palladium to iron ratios, as used by Garten [65,66], have the fcc structure and the Mössbauer spectra of such FePd catalysts show a single peak for the FePd alloy.

The interpretation of the Mössbauer spectra of reduced supported FeM catalysts has been a matter of controversy in the literature. The assignment of the single peak near zero velocity to the FeM alloy is straightforward and has been made by all authors. For the high-velocity peak several interpretations can be considered. Most authors agree that the high velocity peak is the right half of a doublet, the left of which coincides with the alloy peak, as indicated by the bar diagrams in Figure 10. Isomer shift and quadrupole splitting of the latter correspond with those of either high-spin iron(III) or low-spin(II) ions, but Lam and Garten [67] have suggested that the doublet could also correspond to zero-valent iron atoms at sites of low symmetry in the surface of the alloy particles. Another interpretation for the high velocity peak is that it represents an unresolved doublet of high-spin iron(II) ions, in which electron and lattice contributions to the electric field gradient are equal but of opposite sign, as encountered recently in carbon-supported iron catalysts [68].

Although, from the point of view of Mössbauer spectroscopy, assignment of the high-velocity peak to a doublet of high-spin iron(III) is the most straightforward interpretation, and the consideration of possibilities such as low-spin iron(II), surface iron(0), or an iron(II) singlet may seem far-fetched, one should realize that from the viewpoint of chemistry the presence of iron(III) in reduced FeM supported on SiO_2, or Al_2O_3 is unexpected. The common experience with iron supported on SiO_2 or Al_2O_3 catalysts is that iron

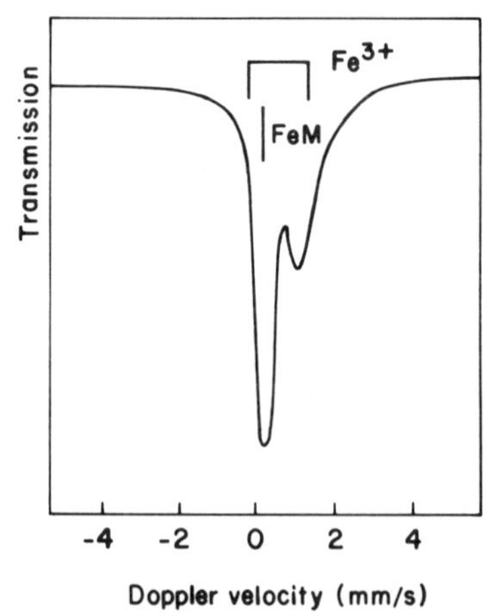

Figure 10. Typical Mössbauer spectrum of a reduced supported bimetallic catalyst consisting of iron and a more noble group VIII metal.

can always be reduced at least to the iron(II) state [2], and it is not readily apparent why in bimetallic catalysts an iron(III) phase would be stabilized under the influence of a noble metal, which is expected to facilitate the reduction of iron [69].

Mössbauer spectra of the reduced FeM on SiO_2 catalysts measured in situ at 77 and 4K in the cell shown in Figure 2, help to distinguish between the possible interpretations for the high velocity peak in the room temperature spectra. As an example we discuss the spectra of 1 to 1 FeRh supported on SiO_2 in Figure 11 [15]. The spectrum of reduced FeRh on SiO_2 obtained at 77K is very similar to the room temperature spectrum and shows in particular that the high velocity peak has not split at 77K. This excludes the assignment of this peak to an unresolved doublet of high-spin iron(II) caused by cancellation of electronic and lattice contributions to the electric field gradient. As the former contribution depends on temperature and the latter does not, an unresolved doublet at 295K should split at 77K. Hence, the high velocity peak is indeed the right half of a doublet. The fact that the quadrupole splitting at most depends slightly on temperature, would be consistent with its assignment to either high-spin iron(III) or low-spin iron(II) [70].

The spectrum of the FeRh on SiO_2 catalyst at 4K (Figure 11) shows that both the singlet of fcc FeRh and the doublet present at 77 and 295K have

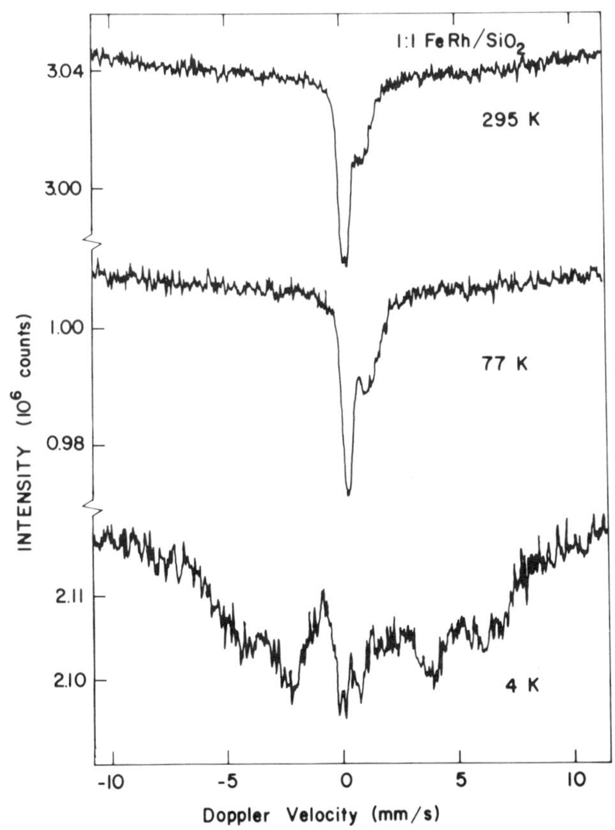

Figure 11. In situ Mössbauer spectra at 295, 77, and 4K of an FeRh on SiO_2 catalyst after reduction in hydrogen. Reproduced from [15] with permission.

magnetically split, with a broad distribution of hyperfine fields. As a low-spin iron(II) ion has no magnetic moment, it cannot exhibit magnetic splitting and hence the possibility of low-spin iron(II) can be discarded. The values of the magnetic splitting of up to 450kOe, are much higher than the magnetic fields of zero-valent iron in either fcc FeRh alloys or metallic iron. Therefore, it is believed that the 4K spectrum contradicts the suggestions made by Lam and Garten [67] and by Garten and Sinfelt [71] that the doublets observed in the room temperature spectra of supported FeRu, FeIr, and FePt catalysts are due to zero-valent iron in the surface of the alloys. However, the presence of a broad distribution of hyperfine fields with values up to 450kOe is consistent with the assignment of the corresponding phase to iron(III) in a highly dispersed iron(III) oxide [15].

So we conclude that the Mössbauer spectrum of reduced FeM supported on SiO_2 catalyst as given in Figure 10 consists of a singlet, or an unresolved doublet in the case of FeRu, due to the FeM alloy, and a doublet of iron(III) oxide.

NATURE OF THE IRON(III) IN THE FeM SUPPORTED ON SiO_2 CATALYSTS

We will now discuss three series of experiments which provide evidence that the iron(III) oxide in FeM on SiO_2 catalysts is present in a highly dispersed phase, whereas the alloy occurs in particles of lower dispersion.

The first experiments studied the particle size dependence. Aschenbeck [55,72] prepared a series of FeRu on SiO_2 catalysts with different

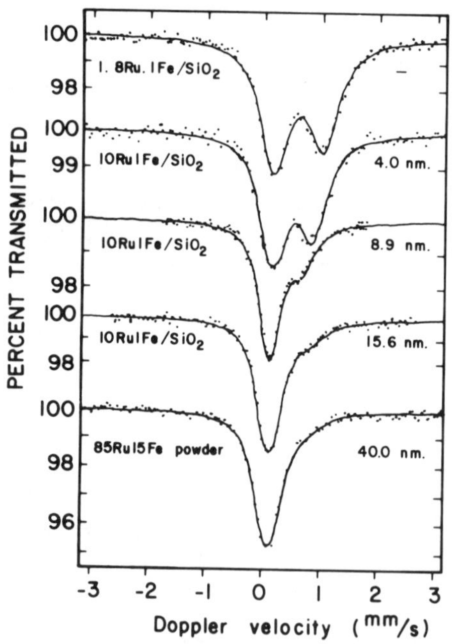

Figure 12. In situ Mössbauer spectra of a series of reduced FeRu on SiO_2 catalysts of different dispersion. The average diameters indicated in the figure were determined from X-ray diffraction line broadening. Reproduced from [72] with permission.

dispersions and used X-ray line-broadening to estimate the diameter of the alloy particles formed after reduction. Figure 12 clearly shows, in agreement with Vannice, et al. [56], that the FeRu on SiO_2 catalysts with the largest FeRu particles show mainly the unresolved doublet of hcp FeRu alloys, whereas the iron(III) doublet becomes more pronounced when the particle size decreases. Thus, the doublet is associated with an interface or surface phase.

The second experiments studied the influence of gas adsorption. Further evidence for the surface character of the iron(III) oxide is provided by Mössbauer spectra of FeM on SiO_2 catalysts after exposure to strongly adsorbing gases such as NH_3 and carbon monoxide [72]. Figure 13 and 14 show that chemisorption of these gases at room temperature results in the conversion of a substantial fraction of the iron(III) into iron(II), as is observed in the spectra by the peak at about 2.5mm/s which belongs to a doublet of high-spin iron(II). Similar conversions of iron(III) to iron(II) by carbon monoxide at 295K were observed with reduced FeRu, FeIr, and FePt on SiO_2 [63]. The fact that a part of the iron(III) oxide is affected by exposure of the bimetallic catalysts to a gas at room temperature is another indication that the oxide is present in a phase of high dispersion.

The third experiment studied the Debye temperatures. Analysis of the intensities in the spectra of FeRh on SiO_2 (Figure 11) as a function of temperature yields Debye temperatures for the FeRh alloy and the iron(III) oxide of 450(50)K and 175(25)K, respectively [15]. The former value is about equal to the Debye temperature of bulk metallic iron and iron oxides [14]. This means that the FeRh alloy behaves like bulk material with respect to lattice vibrations, implying that the alloy particles must be sufficiently large to

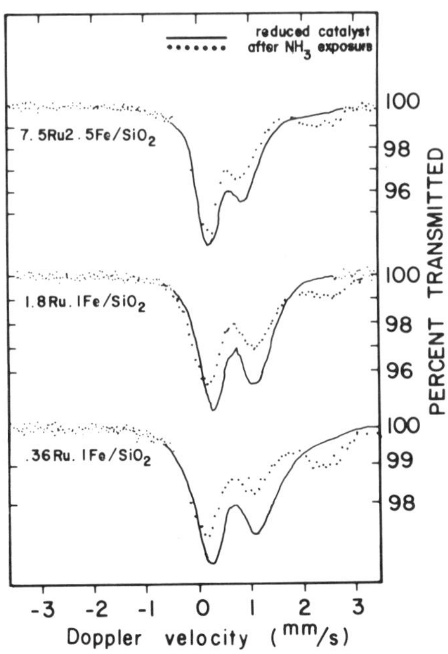

Figure 13. In situ Mössbauer spectra of FeRu on SiO_2 catalysts. Full line spectra, after reduction in hydrogen; dotted spectra, after exposure to NH_3 at room temperature. Reproduced from [72] with permission.

accommodate the majority of the iron atoms in the interior of the particle. The iron oxide in FeRh on SiO_2, on the other hand, has a Debye temperature which is about one third than that of bulk iron(III) oxides. This suggests that the iron(III) oxide in reduced FeM on SiO_2 catalysts has very small dimensions, and that most of the iron(III) ions are located at an interface or at the surface.

REDUCTION OF FeM SUPPORTED ON SiO_2 CATALYSTS

The degree of reduction of the iron in reduced FeM on SiO_2 catalysts is often surprisingly small, particularly in view of the fact that noble metals usually facilitate he reduction of less noble metals. For example, the

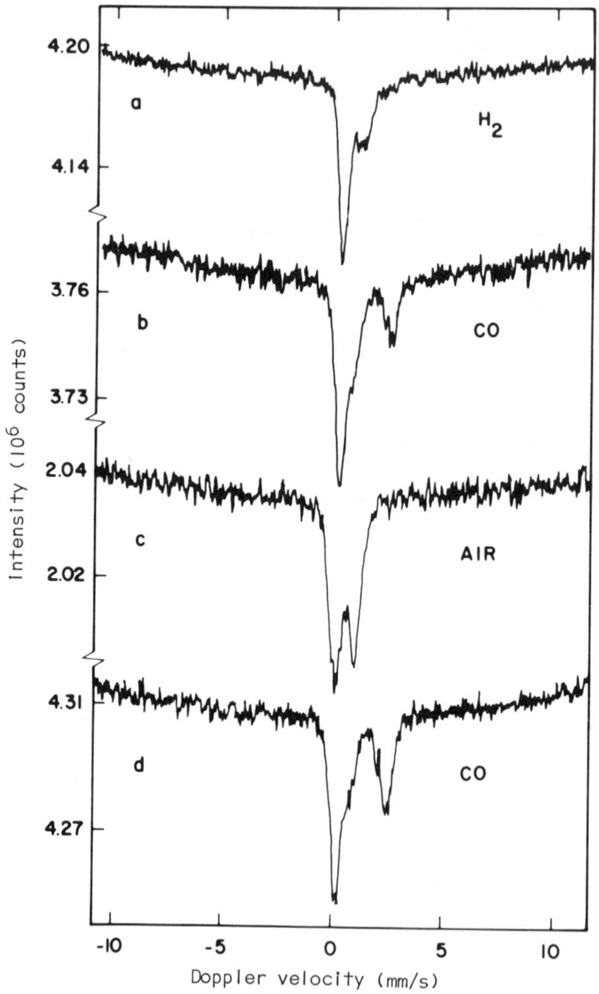

Figure 14. In situ Mössbauer spectra of FeRh on SiO_2 after a series of consecutive treatments, (a) reduction in hydrogen at 725K (b) carbon monoxide chemisorption at 295K, (c) exposure to air at 295K, and (d) carbon monoxide chemisorption at 295K. Reproduced from [72] with permission.

contribution of zero-valent iron to the Mössbauer spectra of SiO_2 supported FePt, FeIr, and FeRu as shown in Figure 9, are only 45, 26, and 18 percent, respectively [63]. The actual degree of iron reduction will be even lower, as the recoilfree fractions of the unreduced iron are substantially lower than those of the reduced iron [15].

Particle size appears to be an important parameter in the reduction of the FeM on SiO_2 catalysts. As Figure 12 shows, FeRu on SiO_2 catalysts with a low dispersion contain most of the iron in the zero-valent state, whereas iron(III) prevails in catalysts with higher dispersions [72]. Similar results were obtained with other supported iron and M combinations. The Mössbauer spectrum of reduced FeRh on SiO_2 with 8-10nm alloy particles consisted for 75 percent of contributions due to iron(0) [73,74], that of FeRh on SiO_2 with 3-4nm alloy particles only 45 percent were due to iron(0) [15]. Furthermore, as discussed above, the iron which cannot be reduced by hydrogen is indeed present in a highly dispersed phase. Thus in FeM on SiO_2 catalysts, dispersion and degree of iron reduction are inversely related.

Nevertheless, promotion of the reduction of iron by the noble metal does take place. Several Mössbauer investigations have shown that in supported FeM catalysts the reducible iron, i.e., the iron which forms an alloy with the noble metal upon reduction, reduces at temperatures which are significantly lower than needed for reduction of monometallic iron catalysts [2,58,59]. Perhaps the best illustration of this phenomenon comes from a combination of Mössbauer spectroscopy and temperature-programmed reduction, TPR, a technique in which the hydrogen consumption of a catalyst is measured as a function of temperature [75].

Van't Blik and Niemantsverdriet [74] applied TPR an FeRh on SiO_2 catalyst of rather low dispersion (16 percent based on hydrogen adsorption) in order to study more specifically the reduction behavior of the FeRh alloy particles. It appeared that the bimetallic catalyst reached its maximum degree of reduction at temperatures below 500K, whereas temperatures well above 600K were necessary to observe appreciable reduction in the monometallic iron on SiO_2 catalysts. TPR studies of SiO_2 supported FeRu [76], FePd, and FeIr [77] yielded similar results as were obtained with FeRh on SiO_2. Bimetallic catalysts reduce at significantly lower temperatures than does monometallic iron on SiO_2.

The catalytic effect of the noble group VIII metal, M, in the reduction of the bimetallic FeM on SiO_2 catalysts can be explained by a mechanism called "intra-particle hydrogen spillover" [75], which we will briefly describe below.

Reduction of highly dispersed iron oxide on supports such as SiO_2, Al_2O_3, or MgO is difficult due to inhibition of the nucleation, i.e., the reaction between molecular hydrogen and iron oxide to form zero-valent iron and water [69,75]. Once iron metal nuclei are present in the surface of an iron oxide particle, molecular hydrogen dissociates on these nuclei and the reduction proceeds by hydrogen atoms, which are much more reactive. Relatively high temperatures of 500 to 600K are required to start nucleation in iron oxide. When the iron oxide is present in a phase of high dispersion, this nucleation process can be inhibited [69]. Supported noble metals, on the other hand, do not suffer from nucleation inhibition. Often, as for example in ruthenium and palladium supported on SiO_2, the reduction starts even at ambient temperatures and proceeds to virtually 100 percent of reduction at temperatures which do not exceed 500K [74,75,77]. Also in bimetallic FeM on SiO_2 catalysts, noble metal atoms at the surface can be readily reduced by molecular H_2. These atoms are thought to act as nucleation centers where H_2 dissociates to yield hydrogen atoms which are spilled-over directly on to iron oxide in contact with the noble metal.

These hydrogen atoms are sufficiently reactive to reduce oxidic iron at relatively low temperatures, provided that the dispersion of the oxidic iron is not too high.

Although this intra-particle hydrogen spillover mechanism, combined with the concept of nucleation inhibition in highly dispersed supported iron oxide, can easily account for the presence of both alloy particles and highly dispersed iron oxide, the question remains why the latter is stabilized in the iron(III) state and not as iron(II), as in iron on SiO_2 catalysts. In this respect, Guczi and coworkers have presented relevant information. Guczi et al. [78] have found that in FeRu on SiO_2 catalysts with metal loadings below one percent, where alloy formation is not observed, reduction of iron(III) to iron(II) is suppressed, in contrast to iron on SiO_2 catalysts where this partial reduction occurs even at loadings as low as 0.1 percent. Guczi [59] suggests in bimetallic catalysts with low metal loadings, in which the bulk alloy formation does not occur, iron oxide and ruthenium mutually impede surface mobility, thus stabilizing irreducible iron oxide particles, separated from each other by ruthenium.

The picture proposed by Guczi has two interesting implications. First, it suggests that particle growth is involved in the iron(III) to iron(II) reduction step in monometallic iron on SiO_2 catalysts. As temperatures of the order of 600 to 700K are required for this reduction, the notion that the reduction process is acccompanied by diffusion of iron over the support seems likely. Second, noble metal atoms capable of preventing migration of iron over the support in bimetallic catalysts, are probably in contact with both the support, and the iron oxide. Close promimity of the noble metal and the iron(III) in reduced FeM on SiO_2 catalysts would be in agreement with the fact that chemisorption of carbon monoxide can convert iron(III) to iron(II) at room temperature, a transition which is absent in monometallic iron catalysts [63].

Finally, the influence of hydroxyl groups, an oxidizing agent which is always present on SiO_2 supports, must be considered. It seems likely that these oxidizing species will have a more profound influence on smaller than on larger particles. This might explain why iron in the highly dispersed iron oxide in bimetallic catalysts is stabilized in the iron(III) state, instead of the expected iron(II) state. This point requires further investigation.

PASSIVATION AND REREDUCTION AT ROOM TEMPERATURE

When reduced bimetallic FeM on SiO_2 catalysts, with Mössbauer spectra as shown in Figure 9, are passivated by exposing them to air at room temperature, almost all alloyed iron is oxidized to iron(III) oxide. An example is shown in Figure 14c. The noble metal, on the other hand, remains largely reduced, as was concluded from X-ray photoelectron experiments [63]. A combination of TPR, TPO, and Mössbauer spectroscopy on FeRh on SiO_2 catalysts indicated that the oxidation process induces segregation of the iron in the FeRh alloy to the surface, leading to particles with an outer shell of mainly iron oxide, and a core of rhodium rich FeRh alloy [74]. Oxidation of the rhodium in the FeRh on SiO_2 catalyst required temperatures as high as needed for complete oxidation of monometallic rhodium on SiO_2 catalysts. This shows that, although the noble metal catalyzes the reduction of the less noble iron, the reverse is not true and that iron does not promote the oxidation of the noble metal.

Many authors have shown that passivated supported FeM catalysts can be reduced to a variable extent by H_2 at room temperature [2,58,59,63,73,79]. In systems which as FeRh and FeRu on SiO_2 the reduction proceeds no further than to iron(II). In FeI_2 and FePt on SiO_2, however, some rereduction occurs even to iron(0). Platinum appeared to be the best promoter for reduction of iron(III) at room temperature. Exposure of passivated FePt on SiO_2 to H_2 at 295K was found to be as effective as reduction in H_2 at 725K [63]. The extent of rereduction of passivated FeM on SiO_2 increased in the order ruthenium < rhodium = palladium < iridium < platinum, which correlates with the position of these metals in the periodic system. The remarkable reducibility of iron(III) in passivated bimetallic catalysts in H_2 at room temperature forms probably the most direct evidence for the proximity and influence of the noble metal.

Obviously, the noble metal atoms in the passivated catalysts provide sites where H_2 can be dissociated, after which reduction of iron occurs according to the intra-particle hydrogen spillover mechanism discussed above. As this mechanism does not explain why, for instance, platinum is a better promoter for iron oxide reduction than ruthenium, it is likely that the noble metals also play a role in weakening the iron to oxygen bonds in the partially oxidized bimetallic catalysts, as has been suggested by Garten and Ollis [65].

As discussed earlier in this paper, chemisorption of CO on reduced bimetallic catalysts converts a fraction of the iron(III) to iron(II) at 295K, indicating that the iron(III) ions involved are accessible to the gases. Passivated bimetallic catalysts show the same iron(III) to iron(II) transition under CO at 295K, but this time significantly more iron(II) is formed than in the reduced catalyst. Figure 14 shows this for FeRh on SiO_2 catalysts. Similar behavior was found with FeIr and FePt on SiO_2 and FeRu on TiO_2 catalysts [63,80]. These results show that the iron(III) to iron(II) reduction by CO at room temperature does not only occur in the unreducible iron oxide present in the reduced catalysts, but on the iron oxide in the passivation layer on the alloy particles as well [72].

BIMETALLIC CATALYSTS UNDER SYNTHESIS GAS

Mössbauer spectroscopy has also been used to investigate the state of iron in FeM on SiO_2 catalysts during Fischer-Tropsch reactions. The effect of exposing a reduced FeM on SiO_2 catalyst to synthesis gas at elevated temperatures of 500-700K appears to be similar to the effect of CO at 295K. Iron(III) oxide is partially converted into a iron(III) species and the zero-valent or alloyed iron remains unaffected [15,17,80,81]. Carburization as in monometallic iron catalysts is absent, except in those cases as 1 to 1 FePd on SiO_2 or 10 to 1 FeRu on TiO_2, in which the reduced catalysts contain metallic iron [55,80].

Mössbauer spectra of FeRh on SiO_2 after Fischer-Tropsch synthesis, measured at 295, 77, and 4K, indicate that the iron(II) compound has a low Debye temperature, confirming that it is mainly a surface phase [15]. The Mössbauer parameters indicate that the coordination of the iron(II) ion is octahedral. As surface atoms are not expected to possess coordinatively saturated environments unless species have been adsorbed from the gas phase, it was proposed that the octahedral coordination of the iron(II) ions is due to adsorption of CO or CO-derived species [15]. The question whether the iron(II)-CO species on the support play a role in the Fischer-Tropsch reactions over the FeM on SiO_2 catalysts is an intriguing one, particularly in the light of the suspected role of the ions in the formation of alcohols from synthesis gas [82].

Minai et al. [81] reported that the iron(II) species formed in FeRh on SiO_2 under synthesis gas are not stable towards reduction in H_2 at 675K, but are converted back to the iron(III) state. This remarkable result, in combination with the interpretation given above, would imply that the iron(II)-CO complex on the support is destroyed by H_2, as a result of which the iron(II) is no longer stabilized and oxidizes back to the iron(III) state, possibly by hydroxyl groups on the SiO_2 support.

STRUCTURE OF FeM SUPPORTED ON SiO_2 CATALYSTS

The combination of Mössbauer spectroscopy, X-ray photoelectron spectroscopy, and TPR/TPO yields information on the type of species present in the bimetallic catalysts, but also has implications for the location of the species with respect to each other. An example is the result that iron(III) in the reduced catalysts is accessible for CO and NH_3 at room temperature. Furthermore, the fact that passivated catalysts contain more iron(III), which is accessible to gases than reduced catalysts, indicates that the unreducible iron(III) in the reduced catalysts does not cover the alloy particles, but forms a separate phase on the support [72].

Three possibilities for the structure of bimetallic FeM on SiO_2 catalyst are visualized in Figure 15. All three models account for the presence of alloy particles together with highly dispersed, and therefore unreducible iron oxide, part of which is accessible to adsorbing gases. Note that the models are also consistent with the results discussed above that air-passivated catalysts have more iron(III) oxide exposed to the gas phase than reduced catalysts.

In Figure 15a the alloy particles rest on a thin layer of iron(III) oxide. CO-induced iron(II) formation is thought to occur on the iron(III) in the periphery of the alloy particle. Figure 15b is a generalization of the model proposed by Yermakov and Kuznetsov [83], in which iron(III) ions form anchors between the bimetallic particle and the support. In order to account for the iron(III) to iron(II) transition by chemisorbed carbon monoxide and NH_3, however, the presence of uncovered iron(III) anchors has to be invoked. Figure 15c is based upon the idea of a bimodal particle size distribution in the bimetallic catalyst. The larger particles can be reduced to an alloy, the small particles cannot be reduced due to the inhibition of nucleation. A more detailed discussion of this model and the changes which take place when the catalysts are exposed to oxygen and carbon monoxide has been given elsewhere [72].

IV. FINAL REMARKS

Mössbauer spectroscopy is an almost ideal technique for in situ studies of iron-containing Fischer-Tropsch catalysts. As shown in this paper, the combination of Mössbauer spectroscopy with reaction kinetic measurements has revealed new and highly relevant information on the mechanisms of Fischer-Tropsch synthesis over and carburization in iron catalysts. Combination of Mössbauer spectroscopy with X-ray photoelectron spectroscopy and temperature-programmed reduction and oxidation has provided detailed insight into the composition and the structure of iron-containing bimetallic catalysts under various gaseous environments. We would like to stress that application of Mössbauer spectroscopy at temperatures as low as 4K is a most valuable extension of the more common room temperature work. Many examples discussed in this paper illustrate that equipment as shown in Figure 2, which permits the measurement of spectra in situ at cryogenic temperatures is an absolute necessity for complete Mössbauer studies of highly dispersed catalysts.

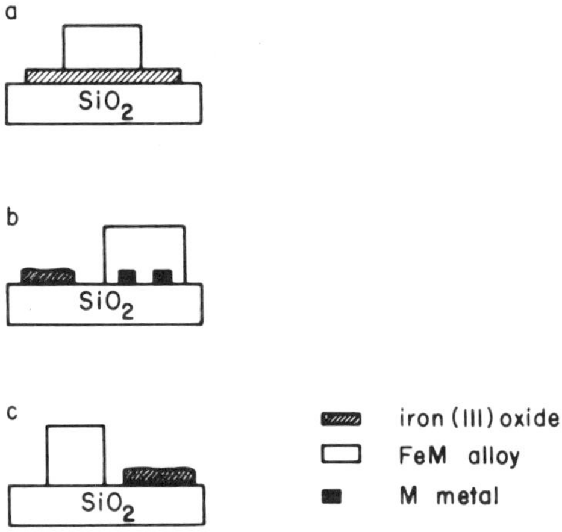

Figure 15. Models for the structure of a reduced bimetallic FeM on SiO_2 catalyst, see text for explanation.

ACKNOWLEDGEMENT

One of the authors (J. W. N.) is supported by a "C en C Huygens-Fellowship" granted by the Netherlands Organization for the Advancement of Pure Research.

REFERENCES

1. J. A. Dumesic and H. Topsøe, Adv. Catal. 26, 121 (1977).

2. H. Topsøe, J. A. Dumesic, and S. Mørup, in "Applications of Mössbauer Spectroscopy", R. L. Cohen, ed., vol. II, p. 55, Academic Press, New York (1980).

3. H. H. Storch, N. Golumbic, and R. B. Anderson, "The Fischer-Tropsch and Related Synthesis", Wiley, New York (1951).

4. M. E. Dry and J. C. Hoogendoorn, Catal. Rev. - Sci. Eng. 23, 265 (1981).

5. B. Büssemeier, C. D. Frohning, and B. Cornils, Hydrocarbon Processing 55, 105 (1976).

6. D. L. King, J. A. Cusumano, and R. L. Garten, Catal. Rev. - Sci. Eng. 23, 233 (1981).

7. B. Büssemeier, C. D. Frohning, G. Horn, and W. Kluy, Ger. Offen. 2536488 (1976).

8. M. M. Bhasin, W. J. Bartley, P. C. Ellgen, and T. P. Wilson, J. Catal. 54, 120 (1978).

9. A. T. Bell, Catal. Rev. - Sci. Eng. 23, 203 (1981).

10. P. Biloen and W. M. H. Sachtler, Adv. Catal. 30, 165 (1981).

11. G. B. Raupp and W. N. Delgass, J. Catal. 58, 361 (1979).

12. C. A. Somorjai, "Chemistry in Two Dimensions: Surfaces", Cornell Univ. Press, Ithaca, New York (1981).

13. C. Kittel, "Elementary Solid State Physics: A Short Course", Wiley, New York (1961).

14. J. W. Niemantsverdriet, C. F. J. Flipse, B. Selman, J. J. van Loef, and A. M. van der Kraan, Phys. Lett. A100, 445 (1984).

15. J. W. Niemantsverdriet, A. M. van der Kraan, and W. N. Delgass, J. Catal.89, 138 (1984).

16. J. W. Niemantsverdriet, C. F. J. Flipse, A. M. van der Kraan, and J. J. van Loef, Appl. Surface Sci. 10, 302 (1982).

17. J. W. Niemantsverdriet, thesis, Delft (1983).

18. A. M. van der Kraan and J. W. Niemantsverdriet, "Appl. of the Mössbauer Effect", Yu. M. Kagan and I. S. Lynbatin, ed., Gordon and Breach Science Publishers LTD, New York, p. 1795 (1985).

19. J. W. Niemantsverdriet, A. M. van der Kraan, W. L. van Dijk, and H. S. van der Baan, J. Phys. Chem. 84, 3363 (1980).

20. J. A. Amelse, J. B. Butt, and L. H. Schwartz, J. Phys. Chem. 82, 558 (1978).

21. G. B. Raupp and W. N. Delgass, J. Catal. 58, 348 (1979).

22. N. Nahon, V. Perrichon, P. Turlier, and P. Bussiere, J. Phys:, (Orsay, Fr.), 41, C1-339 (1980).

23. H.-J. Jung, M. A. Vannice, L. N. Mulay, R. M. Stanfield, and W. N. Delgass, J. Catal. 76, 208 (1982).

24. D. Bianchi, S. Borcar, F. Teule-Gay, and C. O. Bennett, J. Catal. 82, 442 (1983).

25. L. M. Tau,S. Borcar, D. Bianchi, and C. O. Bennett, J. Catal. 87, 36 (1984).

26. Y. V. Maksimov, I. P. Suzdalev, R. A. Arents, and S. M. Loktev, Kinet. Katal. 15, 1293 (1974).

27. W. L. van Dijk, J. W. Niemantsverdriet, A. M. van der Kraan, and H. S. van der Baan, Appl. Catal. 2, 273 (1982).

28. G. L. Vogler, X.-Z. Jiang, J. A. Dumesic, and R. J. Madon, J. Catal. 89 116 (1984).

29. N. K. Jaggi, L. H. Schwartz, J. B. Butt, H. Papp, and M. Baerns, Appl. Catal. 13, 347 (1985).

30. G. Le Caer, J. M. Dubois, and J. P. Senateur, J. Solid State Chem. 19, 19 (1976).

31. R. M. Stanfield and W. N. Delgass, J. Catal. 72, 37 (1981).

32. W. N. Delgass, personal communication.

33. G. Le Caer, J. M. Dubois, M. Pijolat, V. Perrichon, and P. Bussiere, J. Phys. Chem. 86, 4799 (1982).

34. A. O. I. Rautavuoma, thesis, Eindhoven (1979).

35. W. M. Shen, J. A. Dumesic, and C. G. Hill, J. Catal. 84, 119 (1983).

36. M. Araki and V. Ponec, J. Catal. 44, 439 (1976).

37. E. E.Unmuth, L. H.Schwartz, and J. B. Butt, J. Catal. 63, 404 (1980).

38. J. W. Niemantsverdriet and A. M. van der Kraan, J. Catal. 72, 385 (1981).

39. H. Schäfer-Stahl, Angew. Chem. Int. Ed. (English) 19, 729 (1980).

40. H. Matsumoto, J. Catal. 86, 201 (1984).

41. W. L.van Dijk, thesis, Eindhoven (1981).

42. H.Matsumoto and C. O. Behnett, J. Catal. 53, 331 (1978).

43. D. J. Dwyer and G. A. Somorjai, J. Catal. 52, 291 (1978).

44. J. P. Reymond, P. Meriaudeau, and S. J. Teichner, J. Catal. 75, 39 (1982).

45. G. L. Ott, T. Fleisch, and W. N. Delgass, J. Catal. 60, 394 (1979).

46. G. L. Ott, T. Fleisch, and W. N. Delgass, J. Catal. 65, 253 (1980).

47. W. L. van Dijk and J. W. Niemantsverdriet, unpublished results.

48. D. J. Knoechel and J. W. Niemantsverdriet, unpublished results.

49. J. A. Amelse, G. Grynkewich, J. B. Butt, and L. H. Schwartz, J. Phys. Chem. 85, 2484 (1981).

50. W. M. Shen and J. A. Dumesic, personal communication.

51. V. Ponec, Catal. Rev. - Sci. Eng. 11, 41 (1975); idem, Surf. Sci. 80, 352 (1979).

52. W. M. H. Sachtler, Catal. Rev. - Sci. Eng. 14, 193 (1976).

53. J. H. Sinfelt, in "Catalysis, Science and Technology", J. R. Anderson and M. Boudart, eds., vol. 1, Springler, Berlin, Heidelberg, p. 257 (1981).

54. H. F. J. van 't Blik and J. W. Niemantsverdriet, unpublished results.

55. D. P. Aschenbeck, Ph.D. thesis, Purdue University (1979).

56. M. A. Vannice, Y. L. Lam, and R. L. Garten, Adv. Chem. 178, 25 (1979).

57. F. Stoop, to be published.

58. R. L. Garten, in "Mössbauer Effect Methodology", I. J. Gruverman, ed., vol. 10, Plenum Press, New York, p. 69 (1976).

59. L. Guczi, Catal. Rev. - Sci. Eng. 23, 329 (1981).

60. E. E. Unmuth, L. H. Schwartz, and J. B. Butt, J. Catal. 61, 242 (1980).

61. J. A. Amelse, L. H. Schwartz, and J. B. Butt, J. Catal. 72, 95 (1981).

62. P. H. Christensen, S. Mørup, B. S. Clausen, and H. Topsøe, Proc. Int. Conf. on the Appl. of the Mössbauer Effect (Alma Ata, 1983); idem, Proc. 8th Int. Congr. on Catal., Berlin (1984).

63. J. W. Niemantsverdriet, J. A. C. van Kaam, C. F. J. Flipse, and A. M. van der Kraan, J. Catal., 96, 58 (1985).

64. O. Kubaschevski, "Iron - Binary Phase Diagrams", Springer-Verlag, Berlin (1982).

65. R. L. Garten and D. F. Ollis, J. Catal. 35, 232 (1974).

66. R. L. Garten, J. Catal. 43, 18 (1976).

67. Y. L. Lam and R. L. Garten, Proc. of the 6th Ibero-American Symposium on Catalysis", Rio de Janeiro (1978).

68. J. W. Niemantsverdriet, A. M. van der Kraan, W. N. Delgass, and M. A. Vannice, J. Phys. Chem. 89, 67 (1985).

69. J. R. Anderson, "Structure of Metallic Catalysts", Academic Press, New York (1975).

70. J. C. Travis, in "An Introduction to Mössbauer Spectroscopy", L. May, ed., Plenum, New York (1971).

71. R. L. Garten and J. H. Sinfelt, J. Catal. 62, 127 (1980).

72. J. W. Niemantsverdriet, D. P. Aschenbeck, F. A. Fortunato, and W. N. Delgass, J. Mol. Catal. 25, 285 (1984).

73. J. W. Niemantsverdriet, A. M. van der Kraan, J. J. van Loef, and W. N. Delgass, J. Phys. Chem. 87, 1292 (1983).

74. H. F. J. van't Blik, and J. W. Niemantsverdriet, Appl. Catal. 10, 155 (1984).

75. N. W. Hurst, S. J. Gentry, A. Jones, and B. D. McNicol, Catal. Rev. - Sci. Eng. 24, 233 (1982).

76. F. Stoop, unpublished results.

77. J. W. Niemantsverdriet, J. Van Grondelle, and A. M. van der Kraan, Hyperfine Interactions, in press (1986).

78. L. Guczi, K.Matusek, I. Manninger, J. Kiraly, and M. Eszterle, in "Preparation of Catalysts", vol. II, Elsevier, Amsterdam, p. 391 (1979).

79. C. H. Bartholomew and M. Boudart, J. Catal. 29, 278 (1973).

80. A. M. van der Kraan, R. C. H. Nonnekens, F. Stoop, and J. W. Niemantsverdriet, Hyperfine Interactions, in press (1986).

81. Y. Minai, T. Fukushima, M. Ichikawa, and T. Tominaga, J. Radioanal. Nucl. Chem. Lett. 87, 189 (1984).

82. V. Ponec, in "Metal Support and Metal-Additive Effects in Catalysis", B. Imelik, ed., Elsevier, Amserdam, p. 233 (1982).

83. Y. I. Yermakov and B. N. Kuznetsov, J. Mol. Catal. 9, 13 (1980).

MÖSSBAUER EFFECT CHARACTERIZATION OF SILICA SUPPORTED RHODIUM-IRON CATALYSTS

Yoshitaka Minai and Takeshi Tominaga

Department of Chemistry
Faculty of Science
The University of Tokyo
Hongo, Tokyo, Japan

Takakazu Fukushima and Masaru Ichikawa

Sagami Chemical Research Center
Nishi-Ohnuma, Sagamihara
Kanagawa, Japan

ABSTRACT

Silica supported rhodium-iron catalysts were characterized with iron-57 Mössbauer spectra indicated that iron in the catalysts existed as iron(0) in the rhodium-iron alloy and iron(III) stabilized through the metal-support interaction. A part of iron(III) combined the rhodium-iron alloy and the silica as an anchor. The iron(III) to iron(0) ratio and Mössbauer effect parameters were found to depend on the iron to rhodium ratio in the catalysts. Their dependences correspond to the change of the catalytic properties of rhodium-iron catalysts with the iron content. A schematic model for the surface structure of the rhodium-iron catalysts is proposed. The Mössbauer spectra of rhodium-iron catalysts with various metal loading supported this model.

INTRODUCTION

Silica supported noble metals, such as rhodium have been widely used as catalysts for synthesis of hydrocarbons, methanol, ethanol, and other oxygenated compounds from carbon monoxide with hydrogen reduction [1]. The yields and selectivity for these products depend on the particle size of metallic clusters on the support and the kinetic parameters such as pressure, temperature, and carbon monoxide/hydrogen ratio, for the carbon monoxide/hydrogen conversion reaction. By controlling these factors, we can selectively obtain desired organic compounds with good yields. Recently another factor has been found to govern properties of noble metal heterogeneous catalyst systems. The addition of iron, titanium, zirconium or other lithophile elements to the noble metal cluster on the support has modifying effects on the catalytic properties [2,3]. With such effects of additives we may obtain more effective and selective catalysts for carbon monoxide-hydrogen conversion than noble metal catalysts themselves. Many works are in progress to prepare new excellent bimetallic or trimetallic catalysts.

While the properties of bimetallic catalysts are benefitable to convert carbon monoxide and hydrogen synthesis gas into industrially useful organic compounds, the active-sites of the bimetallic catalysts and the mechanisms of the catalytic reactions on the surface of the catalysts are not well understood. In order to elucidate the modifying role of the additives, we need to characterize such bimetallic systems on the support and acquire information regarding chemical states of supported bimetallic clusters.

Mössbauer spectroscopy is a useful technique for characterization of catalyst systems in spite of its limitation in the number of applicable elements [4-7]. In transmission geometry, we can characterize chemical states of Mössbauer elements in bulk catalysts under various gaseous atmospheres because of the high penetrability of Mössbauer gamma ray radiation. In contrast, electron spectroscopic techniques for surface characterization, such as x-ray photoelectron spectroscopy require degassing before analysis of catalysts samples. However, the phenomenon occurring on the catalyst surface during the catalytic reactions should be examined under in situ conditions if possible. While relatively few techniques are available which provide useful information on the characteristics of catalysts, Mössbauer spectroscopy along with infrared spectroscopy and extended x-ray absorption fine structure are useful means for chemical state analysis. Unique information regarding the oxidation state, spin state, coordination state and magnetic properties as derived from Mössbauer spectra may afford clues to understanding the surface structure of catalysts and the mechanisms of catalytic reactions on supported metal cluster surface.

We have recently started iron-57 Mössbauer effect characterization of noble metal-iron bimetallic catalyst systems so as to clarify the role of iron in modifying catalytic properties of noble metals and the structural change of metal clusters on the support with addition of iron. We have analyzed bimetallic and trimetallic catalyst systems such as rhodium-iron [8], rhodium-iron-titanium, palladium-iron, iridium-iron, platinum-iron, and ruthenium-iron at various iron to noble metal atomic ratios below unity. In this article, we report the results of Mössbauer characterization of the rhodium-iron catalysts and discuss the surface structure of such catalysts on the support. Mössbauer studies on the other catalysts will be published elsewhere.

EXPERIMENTAL

The rhodium-iron bimetallic catalysts were prepared by the impregnation method. Silica-gel powder (AEROSIL-300) was added to a methanol solution of $RhCl_3 \cdot 3H_2O$ and $FeCl_3 \cdot 6H_2O$. The iron to rhodium atomic ratio in the sample ranged from 0.03 to 1. Samples with the rhodium to iron ratio below 0.3 were prepared from the iron-57 enriched salt. After removal of the solvent in a rotary evaporator, the silica which absorbed rhodium(III) and iron(III) ions were pressed to give a 350mg 20mm diameter sample wafer for Mössbauer measurement. The sample was placed in a reaction vessel with two windows for transmission of Mössbauer gamma-rays (Figure 1c) for 2 hours at 300°C and then for 6 to 15 hours at 400°C in a hydrogen gas flow. For Mössbauer measurement under the reactant gaseous atmosphere, the sample wafer was transferred to the window part (Figure 1g) without being exposed to air.

Mössabuer spectra were recorded at 293K with a Shimadzu MEG-2 or Austin Science S-600 spectrometer with a 10mCi cobalt-57 on rhodium Mössbauer source. The spectra were fit with linear combinations of Lorentzian lines with the least squares method to determine Mössbauer parameters and relative absorption area [9].

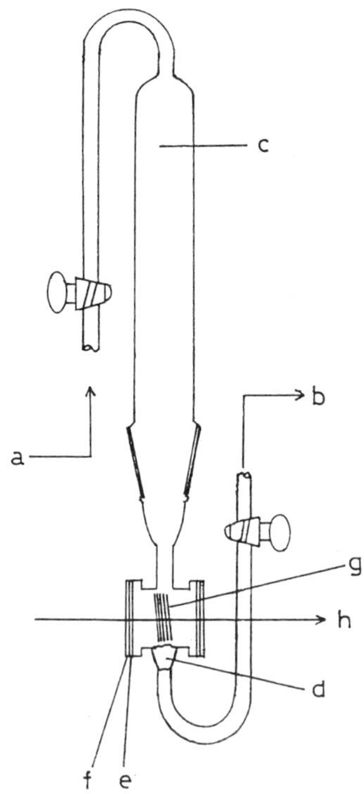

Figure 1. Apparatus for in situ Mössbauer measurement of catalysts under reactant gas atmosphere. (a) gas inlet, (b) gas outlet, (c) reactor, (d) glass wool, (e) aluminum film (f) Mylar film, (g) sample wafer, (h) gamma-rays.

After Mössbauer measurements, samples were allowed to react with carbon monoxide-hydrogen at 210°C, as a conversion catalyst. Then Mössbauer spectra were taken so as to examine the change in chemical states of iron in the catalysts after conversion. Mössbauer spectra of the catalysts after exposure to air at room temperature, in contact with carbon monoxide at 210°C and reduced in hydrogen gas flow at 400°C for regeneration, were also measured to obtain information regarding characteristics of the iron species in various gas atmospheres and surface structure of the supported bimetallic catalysts.

RESULTS AND DISCUSSION

Chemical States of Iron in Rhodium-Iron Catalysts

Typical Mössbauer spectra of the silica supported rhodium-iron catalysts are illustrated in Figure 2. The amount of rhodium on the silica support was kept at four weight percent except for the runs whose metal loading was described otherwise. The isomer shifts are reported relative to the centroid of α-iron. While no magnetic hyperfine structure was observed in the spectra, two non-magnetic envelopes were observed near zero velocity. These absorption peaks could be deconvoluted into a singlet and a quadrupole doublet. Mössbauer parameters determined from the spectra are listed is Table I. Based on the isomer shifts and quadrupole splittings, the singlet and the quadrupole doublet may be ascribed to iron(0) and iron(III) states, respectively, but no iron(II) quadrupole doublet was observed in the spectra. The rhodium-iron bimetallic catalyst system was also studied by the Mössbauer technique in another laboratory, where the silica supported rhodium-iron catalyst was measured only for the iron to rhodium ratio of one [10-13]. Their spectra were similar to those obtained in the present work and the peaks were assigned to an iron(0) singlet and an iron(III) quadrupole doublet.

Thus the Mössbauer absorption spectra suggest the presence of at least two kinds of iron-bearing species on the silica support. The isomer shifts for iron(0) agree with those reported for rhodium-iron alloys [14] and the agreement of Mössbauer parameters implies that iron(0) in our samples was contained as rhodium-iron alloys in the bimetallic catalysts. The isomer shift and quadrupole splitting of the iron(III) doublet were typical of those for a high spin species. Although the presence of the iron(III) species under highly reductive condition is suprising, we find both iron(0) and iron(III) states can be stabilized with rhodium on the silica support as will be explained later in terms of a proposed model. We also presume that at least two kinds of iron(III) species are present in the rhodium-iron catalysts, one of which can be stabilized as an anchor on the silica surface

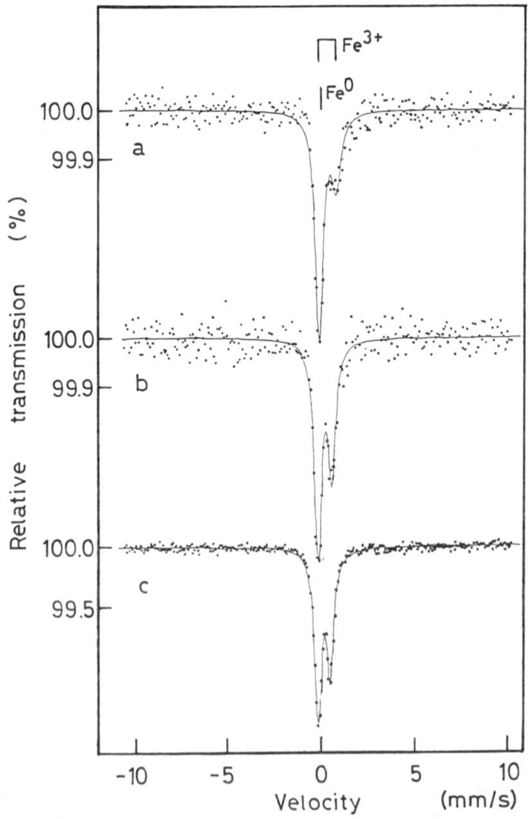

Figure 2. Typical Mössbauer spectra obtained at 293K for the rhodium-iron catalysts under H_2 atmosphere with an iron to rhodium ratio of 1 (a), 0.3 (b), and 0.03 (c).

Table I. Mössbauer Parameters at 293K of the Rhodium-Iron Catalysts Under Hydrogen[a]

Iron/Rhodium Ratio	Species	δ	ΔE_Q	Γ	Percent Area
1	iron(III)	0.44(3)	1.00(6)	0.54(7)	52(8)
	iron(0)	0.08(2)	0.0	0.45(6)	49(7)
0.5	iron(III)	0.41(2)	0.89(4)	0.53(5)	68(8)
	iron(0)	0.09(3)	0.0	0.44(9)	32(7)
0.3	iron(III)	0.37(2)	0.82(3)	0.47(4)	77(8)
	iron(0)	0.12(2)	0.0	0.33(7)	23(5)
0.1	iron(III)	0.38(2)	0.72(5)	0.48(2)	83(5)
	iron(0)	0.22(2)	0.0	0.4(1)	17(5)
0.05	iron(III)	0.37(1)	0.71(1)	0.44(1)	84(2)
	iron(0)	0.17(1)	0.0	0.30(2)	16(2)
0.03	iron(III)	0.38(1)	0.70(1)	0.44(1)	86(2)
	iron(0)	0.16(1)	0.0	0.29(3)	14(2)

[a]All data obtained at 293K and reported in mm/s with the isomer shift relative to α-iron. Figures in parentheses denote standard deviations.

through the interaction between the silica support and the metal cluster in the rhodium-iron alloy.

DEPENDENCE OF MÖSSBAUER PARAMETERS AND IRON(III) TO IRON(0) RATIO ON THE COMPOSITION OF IRON-RHODIUM RATIO IN THE CATALYSTS

As shown in Table I, the Mössbauer parameters, based on the fit assuming the components of iron(0) and iron(III), depend on the iron to rhodium in the catalysts. The isomer shift for iron(0) was about 0.1mm/s when the iron to rhodium ratio was greater then 0.3, while it was about 0.2mm/s for less than 0.3. The Mössbauer parameters of the iron(III) quadrupole doublet remained constant over the iron to rhodium below 0.1, but increased with the ratio above 0.3. Such slight but systematic changes in Mössbauer parameters with the iron to rhodium ratio may be explained in terms of the change in iron to rhodium ratio of the rhodium-iron alloy in the catalysts. The isomer shift in the alloy is reported [14] to increase with the decrease in the iron content and is consistent with our results.

The Mössbauer parameters for the iron(III) doublet also varied with the iron to rhodium ratio. The line width of each iron(III) doublet peak was larger than that of the iron(0) singlet for all samples, suggesting that the iron(III) doublet was composed of at least two components. If the observed iron(III) doublet indicates overlapping of two iron(III) components, denoted as iron(III)A and iron(III)B hereafter, whose peak positions are too close to be resolved instrumentally, Mössbauer parameters of the composite iron(III) doublet indicate mean values of those of two unresolved iron(III) doublets. Thus the variation in the Mössbauer parameters with the iron to rhodium ratio may be explained in terms of the change in the relative intensity for each iron(III) component. However, the major component of the iron(III) species does not remain the same for low iron content and high iron content catalysts. It is obvious that the model proposed previously for the structure of the catalysts at an iron to rhodium ratio of one [13] cannot be adequately applied to the low iron content catalysts.

Besides such changes in Mössbauer parameters, a marked change was observed in the iron(III) to iron(0) ratio on the assumption that the recoilless fraction remains nearly equal for all components. The iron(III) to iron(0) ratio decreases with the increase in the iron to rhodium ratio from 86/14 to iron to rhodium ratio 0.03 to 51/49 at iron to rhodium = 1.0 (see Table I). The amounts of iron(III) and iron(0) on the silica support estimated from the iron(III) to iron(0) ratio, and the total amount of iron as additive, are shown plotted against the iron to rhodium ratio in Figure 3. The specific rate of methanol formation from carbon monoxide-hydrogen gas was reported to increase with the iron to rhodium ratio in the range above 0.1 [2], and this trend was in parallel to the increment of the iron(0) in the catalysts. The similar behavior of the methanol formation rate and the amount of the iron(0) may suggest that the iron(0) as rhodium-iron alloy in the bimetallic catalysts are probably associated with the active sites for catalyzing the methanol synthesis from carbon monoxide and hydrogen. This assumption implies that the rhodium-iron alloy may exist on the surface of the supported catalysts easily accessible to gaseous reactants. In the next section, we discuss the change in the Mössbaeur spectra of the catalyst after exposure to air, an experiment which also confirms the presence of the rhodium-iron alloy on the surface.

CHANGE OF THE CHEMICAL STATES OF IRON UNDER VARIOUS GASEOUS ATMOSPHERES

Typical Mössbauer spectra of the rhodium-iron catalysts after the reaction with carbon monoxide-hydrogen gas at 210°C are presented in Figure 4. The spectra are composed of the iron(0) and iron(III) envelopes and a newly developed quadrupole doublet after the reaction. The Mössbauer parameter were listed in Table II. The results for the catalyst at a iron to rhodium

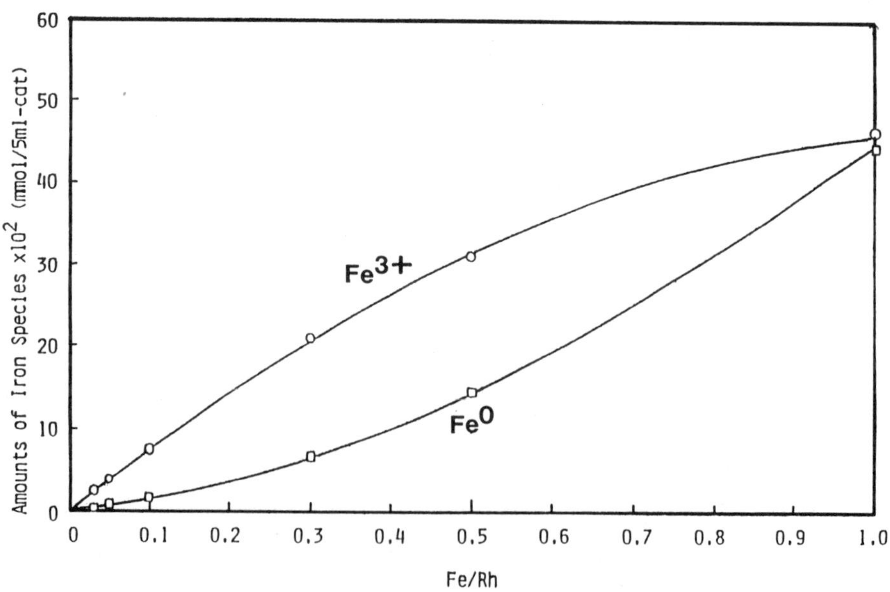

Figure 3. The relative amounts of iron species estimated from Mössbauer spectra of rhodium-iron catalysts at various iron to rhodium ratios.

Table II. Mössbauer Parameters of the Rhodium-Iron Catalysts Under a Hydrogen-Carbon Monoxide Atmosphere[a]

Iron/Rhodium Ratio	Species	δ	ΔE_Q	Γ	Percent Area
1	iron(III)	0.32(4)	0.91(7)	0.50(9)	26(5)
	iron(II)	1.23(5)	1.78(9)	0.64(8)	29(4)
	iron(0)	0.03(1)	-	0.48(5)	45(8)
0.5	iron(III)	0.30(6)	0.8(1)	0.49(6)	39(6)
	iron(II)	1.1(1)	2.3(2)	0.57(6)	33(4)
	iron(0)	0.13(3)	-	0.5(1)	28(7)
0.05	iron(III)	0.32(1)	0.70(1)	0.49(1)	78(2)
	iron(II)	0.92(2)	2.47(3)	0.28(4)	9(1)
	iron(0).	0.14(1)	-	0.27(2)	13(1)

[a]All data obtained at 293K and reported in mm/s with the isomer shift relative to α-iron. Figures in parentheses denote standard deviations.

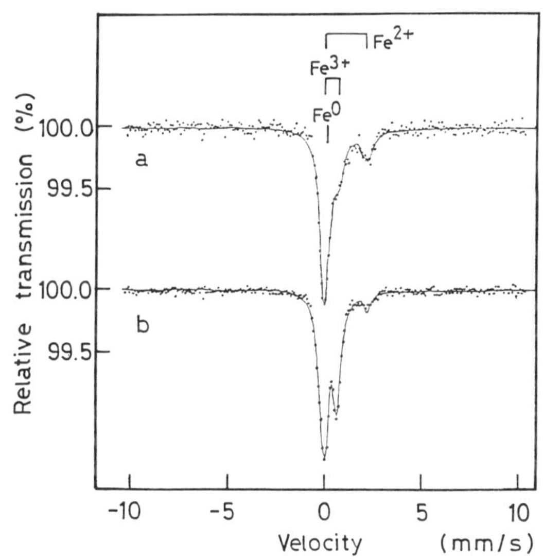

Figure 4. Typical Mössbauer spectra obtained at 293K of rhodium-iron catalysts under hydrogen-carbon monoxide atmosphere for a ratio of 1 (a) and 0.05 (b).

ratio of one is consistent with those reported previously [10,13]. The new absorption peak was assigned to a high spin iron(II) component based on its Mössabauer parameters. The comparison of relative peak areas of the iron(0), iron(III), and iron(II) absorptions before and after the reaction with carbon monoxide and hydrogen gas (see Table III) indicates that the amount of iron(0) was scarcely affected by the reaction whereas the iron(II) species

Table III. Comparison of Relative Peak Areas in the Rhodium-Iron Catalysts Before and After the Hydrogen-Carbon Monoxide Treatment

Iron/Rhodium	Species	Percent Area Before	Percent Area After
1	iron(III)	51(8)	26(5)
	iron(II)	-	29(4)
	iron(0)	49(7)	45(8)
0.5	iron(III)	68(8)	39(6)
	iron(II)	-	33(4)
	iron(0)	32(7)	28(7)
0.05	iron(III)	84(2)	78(2)
	iron(II)	-	9(1)
	iron(0)	16(2)	13(1)

was produced at the expense of iron(III) species. Such a change in the relative peak area implies that the iron(II) species formed after the carbon monoxide-hydrogen conversion reaction were mainly derived from the iron(III) species. After the reaction with carbon monoxide, the formation of the iron(II) species in the catalyst was enhanced as compared with the iron(II) formation after carbon monoxide-hydrogen conversion. The iron(II) component appears to be stabilized through bonding with carbon monoxide and electron donation from carbon monoxide to the iron(III) species.

The amount of iron(II) converted from iron(III) depends on the iron to rhodium ratio in the catalysts. The iron(II) formation was obviously suppressed in the low iron content catalysts where the iron(III) species was predominant. If the major iron(III) component is the same species throughout the whole range of the iron to rhodium ratio, more formation of iron(II) will be required at low iron to rhodium ratio. As was mentioned before, however, the change in Mössbauer parameters suggests that the amount of each component of the iron(III) species varied with the iron to rhodium ratio. It may be possible that iron(III)A, the major iron(III) species in low iron catalysts, was inert in the reaction with carbon monoxide-hydrogen or carbon monoxide gas and was not reduced with hydrogen, whereas iron(III)B, the iron(III) species enhanced with the increase in the iron to rhodium ratio, was reduced to iron(II) in the presence of carbon monoxide containing gas. We may assume that the iron(III)B exists on the surface of the supported catalyst because of its high reactivity toward the gas. Unlike iron(III)B, iron(III)A appears to be covered with other species which prevents direct contact with gas.

The change of chemical states of iron in the catalysts after air-oxidation and re-reduction with hydrogen at room temperature was also investigated by means of the Mössbauer technique as shown in Figure 5. The oxidation of iron(0) to iron(III) was clearly observed in the spectrum after only a brief ten second exposure to air, implying that the rhodium-iron alloy may exist on the surface of the catalysts. As mentioned above, the parallel between the specific methanol formation rate and the amount of iron(0) confirms the same conclusion. After contact with hydrogen, the iron(III) component was converted to iron(0) through the iron(II) state. This iron(III) species formed by air-oxidation may exist in an environment which is different from both iron(III)+A and iron(III)+B because of its reducibility with hydrogen.

THE SURFACE STRUCTURE OF THE SUPPORTED RHODIUM-IRON CATALYSTS

The Mössbauer spectra of silica supported rhodium-iron catalysts indicate that the iron in the rhodium-iron alloy exists as iron(0) and iron(III)A which is nonreducible with hydrogen, carbon monoxide and carbon monoxide-hydrogen, and the iron(III)B species, which is reducible to iron(II) with carbon monoxide and carbon monoxide-hydrogen gas. Gas exposure experiments showed that iron(0) and iron(III)B were present on the surface of the catalyst while iron(III)A was prevented from contacting with gas. The distribution of iron among these components was strongly affected by the iron to rhodium ratio in the catalysts.

Based on these experimental results, we propose a schematic model, as illustrated in Figure 6, for the surface structure of the rhodium-iron bimetallic catalysts on the silica support. The rhodium-iron alloy formed by hydrogen reduction is located on the iron(III)A anchor which combines the alloy and the silica support. The iron(III)A state was stabilized through the interaction with the silica surface and was prevented by the rhodium-iron

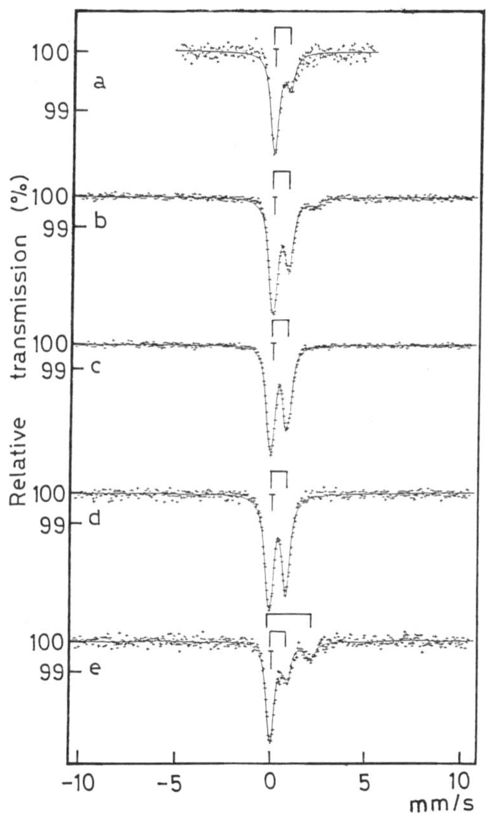

Figure 5. Mössbauer spectra of the rhodium-iron catalysts before and after exposure to air and after regeneration with hydrogen contact. (a) before air-oxidation (b) Passivated to air for ca. 10 seconds. (c) exposed to air for 1 minute, (d) exposed to air for 1 hour, and(e) exposed to hydrogen at room temperature after exposure to air for 1 hour.

alloy coating from contacting the reactant gas. Because iron(III)B sites on the surface of the catalyst and is inert to hydrogen reduction, we presume that iron(III)B is located along the rim of the rhodium-iron catalyst particles easily accessible to the reactant gas and that the unusual stability of such oxidative state in the hydrogen reductive environment is attributed to its bonding to the silica support through oxygen. Such iron-oxygen as well as iron-rhodium bonding have been observed in a recent EXAFS work [15]. With the addition of iron, the distribution of iron among these species is changed according to the affinity of iron to each component.

The addition of small amounts (an iron to rhodium ratio of less than 0.1) of iron to rhodium ratio which is itself a useful catalyst for ethanol formation, enhances the specific rate of ethanol formation and most of iron was in the form of iron(III)A. Due to the difference in surface energies between iron and rhodium, iron tends to occupy the surface of the rhodium-iron cluster and combine with oxygen on the silica surface because of its affinity to oxygen. Accordingly, the iron added to the rhodium cluster preferably exists in the surface of the cluster close to the silica support. The iron in contact with the support is oxidized by silica. This explanation is in accordance with the iron distribution in the low iron content catalysts estimated from Mössbauer spectra. The presence of iron(III)A may increase the affinity of the rhodium-iron cluster to silica. However, the impregnation method employed in the present study could only produce spherical rhodium cluster particles stabilized on the support in the absence of iron. Because the shape of the metal cluster on the support is controlled by the surface tension of metal and the affinity to the support, the addition of iron increases the affinity of the metal cluster, which eventually overcomes the surface tension so that the cluster becomes flat rather than spherical. Such a change of the shape of the cluster may explain the increase in the specific rate of ethanol formation because of the increased surface area of metal cluster with the iron addition. The surface area of the cluster is obviously increased as its shape becomes flat.

As the iron content is increased in the catalyst, the excess iron does not form the iron(III)A anchor but is dissolved in the rhodium-iron alloy. The growth of the iron(0) fraction and the change in the iron(0) isomer shift with the increase in the iron/rhodium ratio can be accounted for by this assumption. At the same time, the other iron(III)B species develops on the silica support. The parallel increase in the iron(0) and iron(III) amounts may suggest certain correlation between the behavior of iron(0) and iron(III)B. The iron(III)B component can be located along the rim of the

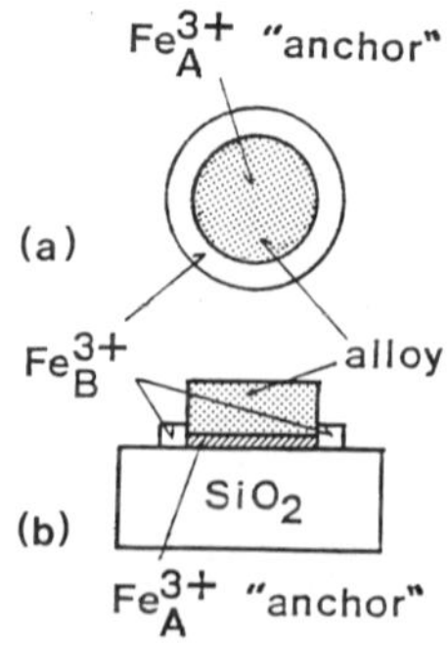

Figure 6. Schematic model for the surface structure of the rhodium-iron bimetallic catalysts from above (a) and in cross-section (b).

cluster on the support. As mentioned above, the iron in the alloy is enriched in the surface of the alloy. If iron(III)B is formed by the oxidation of the rhodium-iron alloy overflowing from the anchor onto the silica support surface along the rim of the catalyst particle, the correlation between iron(0) and iron(III)B formation can be accounted for. On addition of iron, the excess iron is enriched in the surface of the alloy. Because the iron(III)B component is also located on the surface of the rhodium-iron alloy, the increment of iron(0) may also increase iron(III)B. The Mössbauer spectra of the rhodium-iron catalysts obtained for various iron to rhodium ratios were remeasured recently with better resolution and clearly indicated the presence of at least two kinds of iron(III) species and a variation of their relative peak areas with iron to rhodium ratio.

Assuming that the above model for the structure of the rhodium-iron catalyst is applicable as the total metal loading is varied, while keeping the iron to rhodium ratio constant, the size of the supported cluster, or the distribution of iron in each component should be changed. As the total loading decreases, the size of the cluster also decreases. If the width of the rim remains narrow and constant irrespective of the cluster size, the surface area for iron(III)A anchor is proportional to the square of the cluster radius while the surface area for rim iron(III) is proportional to the cluster radius. It suggests that the amount of iron(III)A is decreased more rapidly than iron(III)B if the cluster size is decreased. The decrement of iron(0) may correlate with that of iron(III) because iron(0) is present in the iron(III) anchor zone. On this assumption, as the metal loading increases, increments of iron(III)A and iron(0) may exceed that of iron(III)B. Because horizontal growth of the cluster is limited by the surface tension, the cluster may grow upward to increase the iron(0) component.

The Mössbauer spectra of the rhodium-iron catalysts at an iron to rhodium ratio of one with various metal loadings are presented in Figure 7.

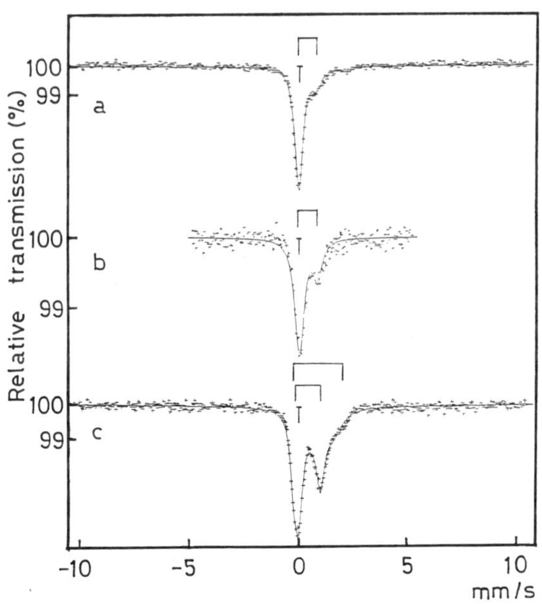

Figure 7. Mössbauer spectra of the rhodium-iron catalysts with the iron to rhodium ratio of one at various total metal loading, (a) 8 weight percent, (b) 4 weight percent, and (c) 0.125 weight percent.

Iron(0) increases with the increase in the total metal loading on silica and the Mössbauer parameters of iron(III) in the low metal content catalysts differ from those of iron(III) in the rhodium-iron catalysts with higher metal loading. The parameters were characterized by high isomer shift and quadrupole splitting. This iron(III) component might be the iron(III)B component in the spectra for other rhodium-iron catalysts.

CONCLUSIONS

Iron in the rhodium-iron catalysts with the iron to rhodium ratio ranging from 0.03 to 1 was either in the form of iron(0) in the rhodium-iron alloy or iron(III) and the latter was stabilized through interaction with the silica support. A part of the iron(III) combines the metal cluster and the support as an anchor.

The iron(III) to iron(0) ratio depends on the iron to rhodium ratio and the total metal loading in the catalysts. Mössbauer parameters also change slightly with the iron to rhodium ratio. This suggests changes in the chemical composition regarding iron(0) in the iron-rhodium alloy and in the distribution of iron between two or more iron(III) species.

The dependence of the methanol formation rate on the iron to rhodium ratio correlates with the dependence of the iron(0) amount of the iron to rhodium ratio.

The iron(II) species which is unstable in air or a hydrogen atmosphere was produced through the reaction between iron(III) on the catalyst surface and carbon monoxide-hydrogen or carbon gas.

A model for the surface structure of rhodium-iron catalysts was proposed on the basis of Mössbauer spectra of various rhodium-iron catalysts and gas exposure experiments.

REFERENCES

1. M. Ichikawa, Chemtech., 680 (1982).

2. M. M. Bhasin, W. J. Bartley, P. C. Ellgen, and T. P. Wilson, J. Catal. 54, 120 (1978); T. Fukushima, Y. Ishii, Y. Onda, and M. Ichikawa, J. Chem. Soc., Chem. Commun, 1752 (1985).

3. M. Ichikawa, K. Sekizawa, K. Shikakura, and M. Kawai, J. Mol. Catal. 11, 167 (1982).

4. J. A. Dumesic, J. Phys. (Paris), Colloq., C6, 233 (1976).

5. R. L. Garten, in "Mossbauer Effect Methodology", Vol. 10, I. J. Gruverman and C. W. Seidel, eds., Plenum Press, New York (1976).

6. L. Guczi, Catal. Rev. Sci. Eng. 23, 329 (1981).

7. M. L. Good, M. D. Patil, J. T. Donner, and C. P. Madhusudahan, in "Mössbauer Spectroscopy and Its Chemical Applications", J. G. Stevens and G. K. Shenoy, eds., Am. Chem. Soc., Washington, D.C. (1981).

8. Y. Minai, T. Fukushima, M. Ichikawa, and T. Tominaga, J. Radioanal. Nucl. Chem. Letters 87, 189 (1984).

9. The program had been written originally in ALGOL by B. J. Duke and T. C. Gibb, J. Chem. Soc. (A), 1478 (1967), and it was rewritten by M. Takeda in Fortran.

10. J. W. Niemantsverdriet, A. M. van der Kraan, J. J. van Loef, and W. N. Delgass, J. Phys. Chem. 87, 1292 (1983).

11. J. W. Niemantsverdriet, A. M. van der Kraan, and W. N. Delgass, J. Catal. 89, 138 (1984).

12. H. F. J. van't Blik and J. W. Niemantsverdriet, Appl. Catal. 10, 155 (1984).

13. J. W. Niemantsverdriet, D. P. Aschenbeck, F. A. Fortunato, and W. N. Delgass, J. Mol. Catal. 25, 285 (1984).

14. C. C. Chao, P. Duwez, and c. c. Tsuei, J. Appl. Phys. 42, 4282 (1971).

15. M. Ichikawa, T. Fukushima, T. Yokoyama, N. Kosugi, and H. Kuroda, J. Chem. Soc., Chem. Commun., submitted for publication.

THE APPLICATION OF MÖSSBAUER SPECTROSCOPY TO METALLIC, BIMETALLIC, AND MIXED METAL OXIDE CATALYSTS

Frank J. Berry

Department of Chemistry
University of Birmingham
Birmingham B15 2TT, England

INTRODUCTION

The ability of Mössbauer spectroscopy to examine authentic solid catalysts in their working gaseous environments has conveyed a special power to the technique for the characterization of heterogeneous catalysts. The application of Mössbauer spectroscopy in this particular field has been comprehensively reviewed on two occasions in the recent past [1,2] and in this article of restricted length only a few systems have been selected, mainly from work reported since the last reviews, to demonstrate the suitability of Mössbauer spectroscopy for the study of solid catalysts and to illustrate its potential power in the examination of metallic, bimetallic, and mixed metal oxide catalysts.

Although recent developments in techniques such as x-ray diffraction, electron microscopy, electron spin resonance, photoelectron- and Auger-spectroscopy have lead to improved understandings of both the bulk and surface properties of catalytic solids, the frequent requirement that the samples be in a specially prepared form, for example as single crystals or as thin films, has limited the usefulness of some of these techniques because authentic catalysts tend to be poorly crystalline and sometimes multiphasic solids. It is also pertinent to note that several of these techniques only operate under conditions, such as ultrahigh vacuum, which are far removed from the gaseous atmospheres at elevated temperatures which frequently characterize the working environment of the real catalyst. Hence the ability of the high energy gamma-radiation used in Mössbauer spectroscopy to penetrate a gaseous environment and the suitability of the technique for the study of crystalline and non-crystalline solids, including high dispersions of supported small particles, has conveyed a special power to Mössbauer spectroscopy for the in situ study of catalytic solids. The capacity of Mössbauer spectroscopy for in situ characterization is important because solid heterogeneous catalysts have surfaces which are highly sensitive to their gaseous surroundings and information pertaining to the solid state properties, which may be related to catalytic performance, are best recorded under conditions in which the gaseous environment, temperature, and pressure can be controlled.

A number of cells which enable the acquisition of Mössbauer spectra, sometimes at low temperature, from materials in gaseous atmospheres at pressures of ca. 1.5atm following treatment at high temperatures have been designed. The cell depicted in Figure 1a is illustrative of the equipment which may be used for studies at room temperature [3]. The cell is made from either quartz or pyrex glass with windows made thin enough to transmit γ-rays. The catalyst is supported between the windows (a), a heating coil (f) heats the gas which flows (d and e) over the catalyst (1) at a temperature which can be monitored (c) and regulated.

Cells in which a catalyst can be treated at high temperature in a gaseous environment, cooled and then examined in situ by Mössbauer spectroscopy at 77K in the presence of applied magnetic fields have also been developed [4]. Such a cell is depicted in Figure 1b which shows the side and front view of the device which is connected directly to the bottom of a vacuum insulated nitrogen cryostat.

Reactors which allow the catalyst to be treated in a gaseous atmosphere at elevated temperature and then permit the catalyst to be removed and studied in situ by Mössbauer spectroscopy at low temperature have also been developed [5] (Figure 1c). After treatment in the chemical reactor the upper part of the absorber holder (1), which is mounted on a stainless steel rod extending to the exterior of the reactor, is lowered and screwed vacuum tight onto the lower fixed part of the absorber holder (2). The absorber in its holder is then removed and the Mössbauer spectra may then be recorded from the absorber at temperatures down to 4K.

It is relevant to note that good quality Mössbauer spectra may be recorded from a wide range of elements which are common components of catalytic solids, for example iron, tin, antimony, tellurium, iridium, ruthenium, tungsten, gold, platinum, and europium. It should also be noted that some materials not containing Mössbauer nuclei and which are difficult to study by other techniques may be advantageously examined by incorporating Mössabuer active atoms into the structure to enable the acquisition of Mössbauer spectra which can be interpreted in terms of the chemistry and structure of the whole solid. It is also pertinent to record that catalysts may be studied by performing the Mössbauer experiments in both traditional and unusual modes of operation. For example, supported bimetallic catalysts for Fischer-Tropsch reactions have been examined [6] in situ by conventional transmission mode Mössbauer absorption spectroscopy, where gamma-radiation from the radioactive source is resonantly absorbed by the Mössbauer isotope in the catalyst, whilst cobalt molybdate hydrodesulfurization catalysts have been investigated [7] by Mössbauer emission spectroscopy in which the catalyst contains the radioactive isotope and is used as the source. In principle the surface layers of catalytic solids may also be examined by conversion electron Mössbauer spectroscopy [8].

The Mössbauer spectra recorded during the characterization of catalytic solids are normally interpreted in terms of the three principal Mössbauer parameters. The chemical isomer shift is an informative parameter by which the nature and proportion of cationic oxidation states and the character of chemical bonding can be examined. Quadrupole splitting data may be interpreted in terms of the structural properties of a catalyst, the distribution of electronic charge, and the effects of doping the catalyst with foreign atoms to improve catalytic performance. Magnetically split Mössbauer spectra may be used for phase identification, the investigation of magnetic ordering and the characterization of relationships between phases in a multiphasic catalyst. Changes in the catalyst following the adsorption of gases are easily detected by Mössbauer spectroscopy and the data may be related to changes in cationic oxidation states, the formation of surface phases, and the modification of catalyst structure.

Despite the numerous advantages of Mössbauer spectroscopy for the study of catalytic solids it is necessary to note that the low natural abundance of some Mössbauer isotopes necessitates the isotopic enrichment of catalysts

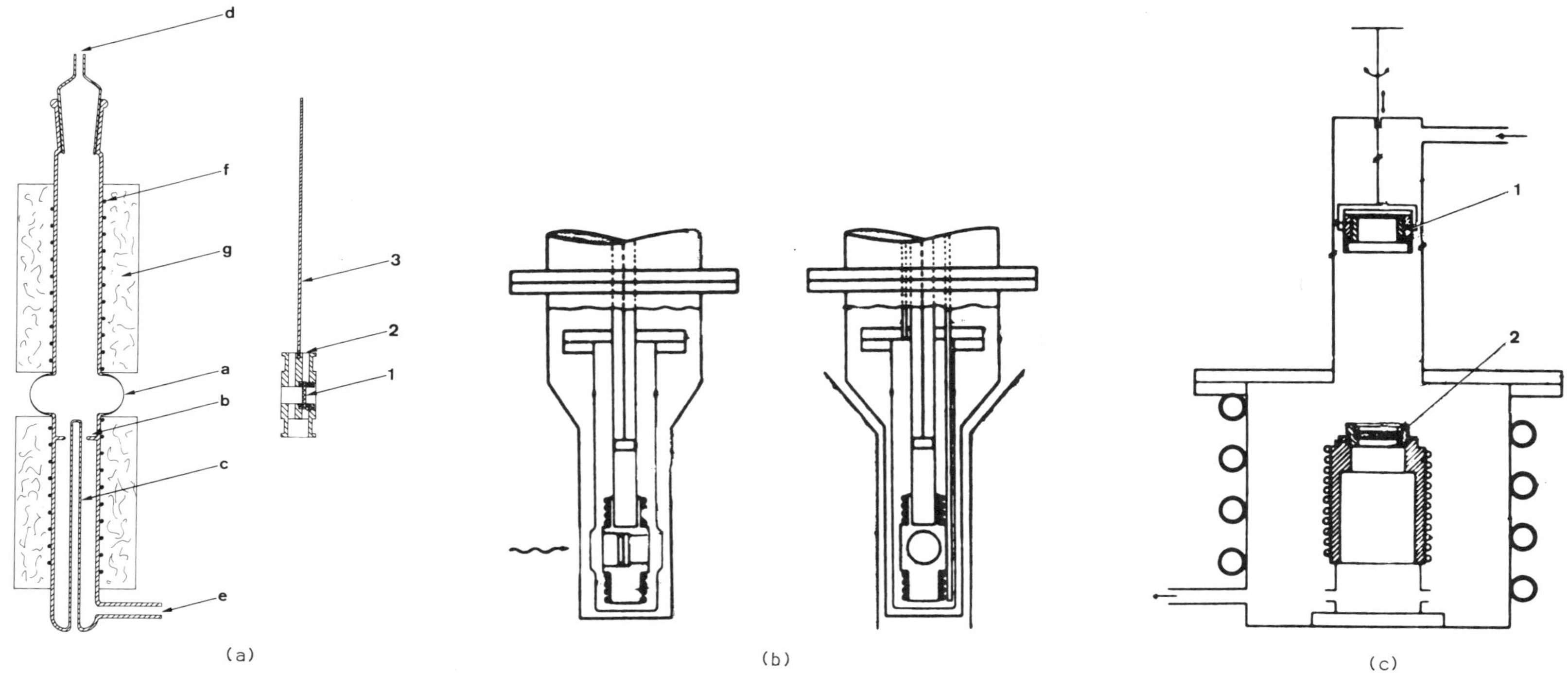

Figure 1. Cells for recording Mössbauer spectra from catalysts in situ following their treatment at elevated temperatures in gaseous environments. Reproduced with permission from [3], [4], and [5].

which contain these elements before useful spectra can be recorded and that the short half lives of some sources can result in expensive investigations when detailed and sophisticated studies are required.

It must also be acknowledged that the conditions under which catalysts are prepared and treated can give rise to lower quality spectra than are normally recorded in Mössbauer spectroscopy. This is, in some cases, a reflection of the need to support the small catalytic particles containing the Mössbauer isotope on materials such as alumina, silica, or titania. It must also be appreciated that the in situ conditions impose restrictions on, for example, the adjustment of the catalyst once it has been introduced into the cell. Such difficulties are reflected in some of the spectra illustrated later.

The remainder of this article is devoted to a consideration of the application of Mössbauer spectroscopy to a number of different catalyst systems to illustrate the type of information which can be obtained and its interpretation in terms of solid state properties which are associated with catalytic performance.

METALLIC CATALYSTS

IRON CATALYSTS

Metallic iron has been used as a catalyst for gaseous reactions for many years, for example it has been used commercially in the reaction between nitrogen and hydrogen to produce ammonia. However, recent problems associated with the supply of petroleum have focused attention on the need to obtain fuels and chemicals from routes derived from coal. One of these involves the thermally induced degradation of coal into carbon monoxide and hydrogen which then react over an iron catalyst to produce hydrocarbons in a process known as Fischer-Tropsch synthesis.

Given the amenability of iron to examination by Mössbauer spectroscopy, the technique is clearly well suited to the examination of iron-containing Fischer-Tropsch catalysts. It is therefore pertinent to cite a carefully performed iron-57 Mössbauer study [5] of an iron catalyst which had been removed from the catalytic reactor after 24 hours treatment under Fischer-Tropsch conditions and which showed, at 77K, a pattern (Figure 2a) which was attributed to the presence of the iron carbide X-Fe_5C_2. However, the spectra recorded from the sample at temperatures of 40K and below showed additional broad lines at positions which roughly coincided with the outermost lines of α-Fe_2O_3. The broadness of the lines was associated with a distribution of hyperfine fields which were slightly lower than that of bulk α-Fe_2O_3 and similar to those which characterize small particles or surface layers of α-Fe_2O_3. Hence the additional component of the spectra was associated with an iron oxide located at the surface of the catalyst. Further evidence for the presence of a surface oxide was obtained when the resonant absorption area of the oxide subspectrum was found to decrease more rapidly with increasing temperature than the total spectrum. The result was interpreted in terms of the low recoil-free fraction of iron atoms in the iron oxide resulting from their location in the superficial regions of the catalyst.

However, this ex situ study of the catalyst was unable to show whether the surface iron oxide was formed during the Fischer-Tropsch process or immediately after the sample was removed from the reactor. Hence an in situ Mössbauer study was undertaken (Figure 2b) on an iron catalyst produced by the reduction of α-Fe_2O_3 in hydrogen. The Mösssbauer spectrum, recorded at 4.2K in situ, was characteristic of pure metallic iron and showed no evidence for the presence of an iron oxide. Hence the reduction conditions were shown to be sufficient to reduce the whole catalyst, including the surface, to metallic iron. When the catalyst was treated in a flow of carbon monoxide gas at 500K for 3 hours a used catalyst was obtained which at 4.2K, gave a spectrum showing the presence of the additional broad line iron oxide phase and a small amount of iron carbide. This iron oxide subspectrum, which constituted about 11 percent of the total spectrum, disappeared at a temperature between 4.2 and 77K and was associated with a surface layer of iron oxide formed when the carbon monoxide was brought into contact with the iron catalyst at 500K. When the pure iron catalyst was treated with a mixture of carbon monoxide and hydrogen at a similar temperature and for an identical period of time, a spectrum was recorded at 4.2K showing the development of more carbide but hardly any iron oxide. The results of the in situ Mössbaur experiments therefore showed that the carbon monoxide in the synthesis gas mixture used in the Fischer-Tropsch reaction undergoes dissociation and that the iron catalyst is converted, at least partially, into iron carbides containing iron oxides on the surface. The results also showed that hydrogen in the synthesis gas mixture plays an important role because, in its absence, a considerable quantity of surface oxide is formed whilst in the carbon monoxide and hydrogen mixture only small amounts of the superficial iron oxide are produced.

Finally it is interesting to note that the treatment of the used catalyst containing carbide and iron oxide in hydrogen gave a material which, when examined in situ by Mössbauer spectroscopy, showed the reduction of the iron carbide to iron but little change in the iron oxide subspectrum. The result suggests that iron oxides at the surface are more stable with respect to hydrogen than iron carbides in the bulk.

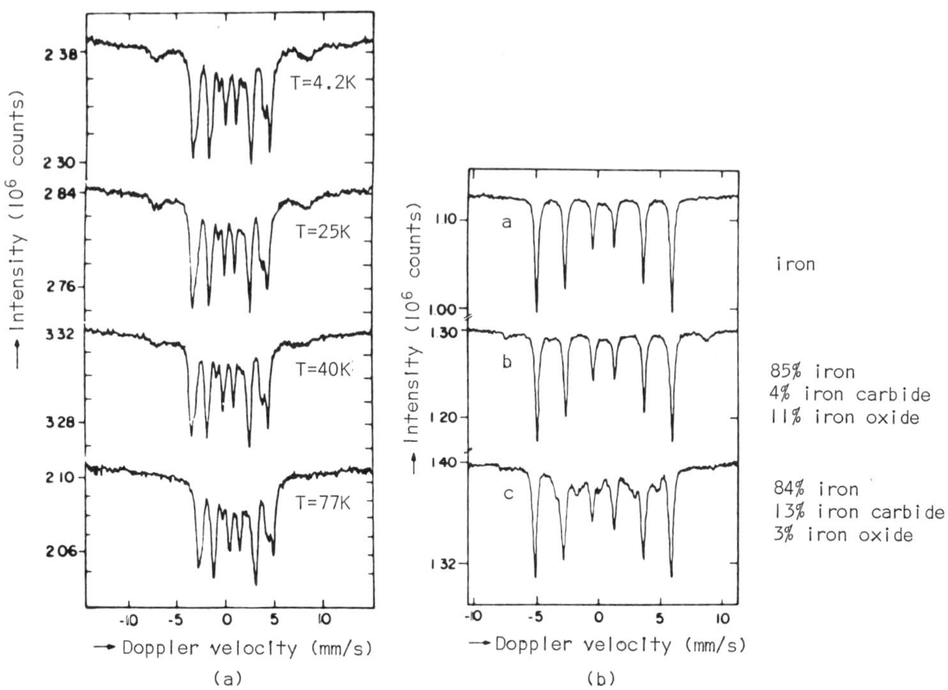

Figure 2. Mössbauer spectra recorded (a) ex situ, and (b) in situ from an iron catalyst following thermal treatment in carbon monoxide and hydrogen. Reproduced with permission from [5].

GOLD CATALYSTS

The impression must not be given that iron is the only metal suitable for investigation by Mössbauer spectroscopy. For example, gold-197 Mössbauer studies [9] of solids formed by the impregnation of magnesia and alumina with solutions of gold cyanide and gold chloride showed the formation of materials which were partially decomposed into metallic gold when calcined at temperatures exceeding 140°C. Given the inert nature of both gold and alumina it is interesting to note than an uncharacterized electron deficient gold species, which was attributed to the interaction of gold with alumina, was identified when the catalyst was heated to 370°C.

BIMETALLIC CATALYSTS

Although iron has been found to be a very active Fischer-Tropsch catalyst it has lacked selectivity in that it converts the carbon monoxide and hydrogen into a variety of products ranging from short chain to long chain hydrocarbons, including alcohols, aldehydes, ketones, and other organic compounds. Recent work has therefore been directed towards the development of catalysts which are not only active but also selective and give only a restricted range of products. Some success has been achieved by modifying the active iron catalyst by incorporating another element.

IRON-RUTHENIUM CATALYSTS

The value of Mössbauer spectroscopy in such investigations is well illustrated by recent study [6] of catalysts composed of iron and ruthenium supported on alumina and silica in which the changes in catalytic activity and selectivity were associated with fundamental solid state properties of the bimetallic catalyst which were identified in situ by iron-57 Mössbauer spectroscopy. The supports, which in the past have often been considered to be inert, maintain high dispersions of the small particles of the catalytically active material so that the high surface areas of the catalyst particles offer a large number of surface sites on which gaseous molecules can adsorb and react. The initial Mössbauer experiments, on catalysts prepared by the impregnation of alumina and silica with aqueous solutions of iron(III) nitrate and ruthenium(III) chloride, followed by calcination and in situ pretreatment in flowing hydrogen, were performed to investigate the nature of the prereduced catalyst and changes following treatment in carbon monoxide and hydrogen.

The iron-57 Mössbauer spectra recorded from catalysts composed of pure iron on alumina showed the materials formed by calcination at 480°C in air to contain iron(III) and were similar to other materials which have been described as small particle, superparamagnetic, α-Fe_2O_3. These were reduced in hydrogen at 480°C to materials containing both iron(III) and iron(II) with the concentration of iron(II) species increasing as the temperature was increased to 600°C. The observations differed considerably from those recorded from silica supported iron where the catalyst calcined in air gave an eight line Mössbauer spectrum in which the doublet characteristic of small particle superparamagnetic α-Fe_2O_3 was superimposed on a sextet characteristic of large particle magnetically ordered α-Fe_2O_3. Treatment of the silica supported iron catalyst in hydrogen lead to a reduction which, at 600°C, indicated the formation of an iron(II) species and metallic iron. The result contrasts with the alumina supported catalyst in which reduction ceased with the formation of iron(II) and clearly indicates the weaker interactions between the iron and the silica support which enables the sintering of small Fe_2O_3 particles to form large particle magnetically ordered α-Fe_2O_3 which is amenable to reduction to iron(0). The results demonstrated that, contrary to

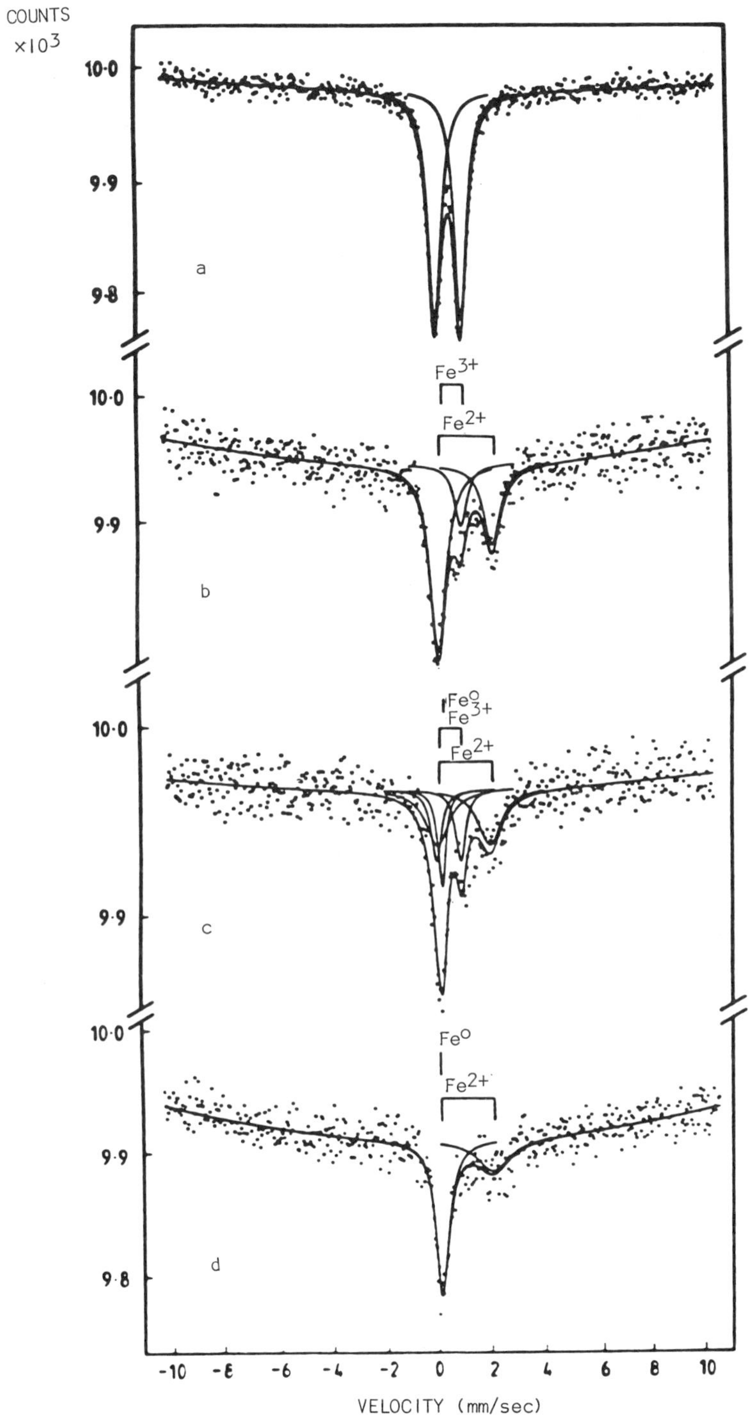

Figure 3. Iron-57 Mössbauer spectra, recorded at 298K, from 5 percent iron-5 percent ruthenium supported on alumina and heated at (a) 480°C for 4h in air; (b) 235°C for 4h in hydrogen; (c) 480°C for 2h in hydrogen; and (d) 600°C for 2h in hydrogen. Reproduced with permission from [6].

earlier beliefs, alumina is not an inert support and that the nature of the prereduced catalyst depends on the nature of the support.

The effect of the inclusion of ruthenium into the iron catalyst is reflected in the Mössbauer spectra recorded from catalysts containing equimolar ratios of iron and ruthenium supported on alumina (Figure 3). Although the material formed in air gave a Mössbauer spectrum (Figure 3a) similar to that recorded from pure iron on alumina, which was partially reduced at 480°C to iron(II), the catalyst containing both iron and ruthenium was converted to iron(II) at only 235°C (Figure 3b) and at 480°C to iron(0) (Figure 3c). The iron(0) component resembled the single peak Mössbauer spectrum recorded from an iron-ruthenium alloy which was prepared by the fusion of a milled mixture of the metals in an argon arc furnace. Further reduction at 600°C gave a material which was shown by Mössbauer spectroscopy (Figure 3d) to be predominantly reduced to the iron-ruthenium alloy with the non-alloyed iron being present as iron(II). When a ruthenium-rich sample was examined the Mössbauer spectra showed that reduction of iron(III) occurred at even lower temperatures and the results demonstrated that the ease of hydrogen reduction is highly dependent on the ruthenium content of the catalyst. However, it is pertinent to record that there is some recent evidence [10] that the behavior of iron-ruthenium catalysts in reducing atmospheres is far from simple and

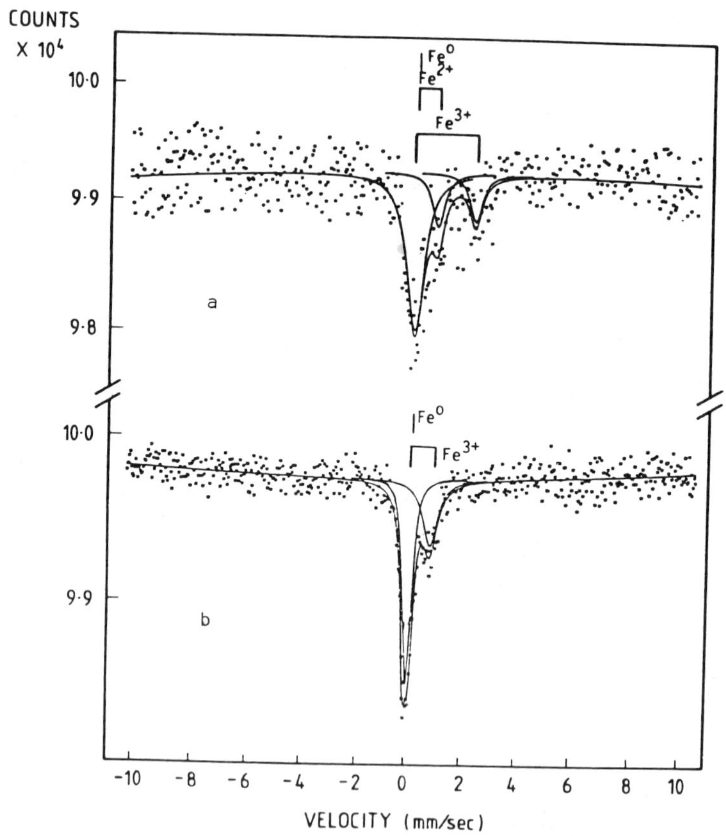

Figure 4. Iron-57 Mössbauer spectra, recorded at 298K, from 5 percent iron-5 percent ruthenium supported on (a) alu mina and (b) silica following treatment at 452°C for 6h in hydrogen. Reproduced with permission from [6].

that under some circumstances partial reoxidation may occur. No evidence of such phenomena were observed in the initial studies [6] of iron-ruthenium catalysts under the conditions described here, indeed the Mössbauer spectra recorded from the silica supported iron-ruthenium particles showed them (Figure 4a) to be more extensively reduced than their alumina-supported counterparts (Figure 4b). Hence the results from the in situ Mössbauer studies of supported iron-ruthenium suggested [6] that the amenability of iron to reduction is dependent on both the ruthenium concentration and the nature of the support with more facile reduction being achieved on silica where the metal support interaction is weaker.

The influence of these solid state properties on catalytic performance was evaluated by the examination of a high metal loaded series containing 5 weight percent ruthenium with varied iron contents, and a low metal loaded series which contained 1 weight percent ruthenium with similar iron to ruthenium ratios. The mean particle sizes of the bimetallic components of the alumina-supported catalysts were determined by transmission electron microscopy and found to be smaller than the silica-supported catalysts, an observation which is also consistent with strong metal support interactions in alumina supported catalysts. The catalytic activities of the iron-ruthenium catalysts prepared for Fischer-Tropsch synthesis are depicted in Figure 5 as a function of the iron content and showed that the alumina-supported materials have higher catalytic activities than those supported on silica. The higher activity may be associated with the Mössbauer results which show that the interaction of the metallic phase with alumina is stronger than that with silica and gives rise to the smaller particles in which the reduction of iron is inhibited.

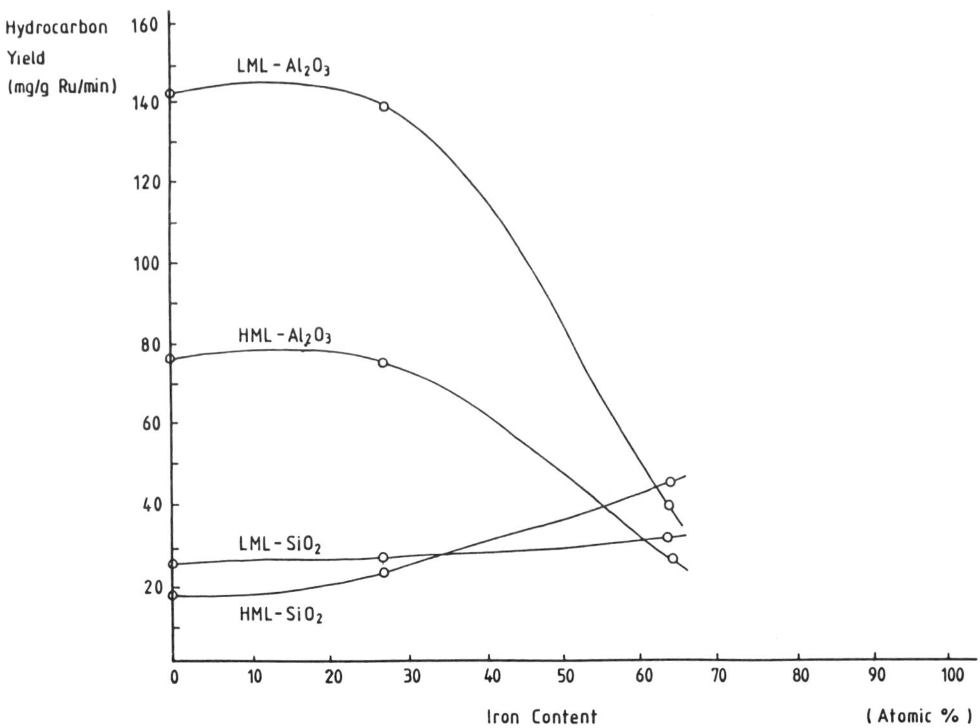

Figure 5. Catalytic activity of alumina- and silica-supported iron-ruthenium catalysts as a function of the iron content. Reproduced with permission from [6].

The decrease in catalytic activity which accompanies an increase in the iron content of the alumina-supported series (Figure 5) is not reflected in the silica supported catalysts and the phenomenon is best considered in conjunction with the effect of the iron on the selectivity of these catalysts. The hydrocarbon product distribution in the low metal loaded silica supported catalysts (Figure 6a) in which the metal support interaction is weak were concentrated towards the lower molecular weight hydrocarbons and their selectivities were fairly insensitive to changes in the iron concentration. However, the selectivities of the alumina-supported catalysts (Figure 6b) were particularly sensitive to the iron content of the catalysts with increases in the iron concentration producing higher yields of the shorter chain hydrocarbons. Indeed, a most striking feature of the changes in activity and selectivity to C_3 hydrocarbons was the performance of the alumina-supported one percent ruthenium and one percent iron catalyst which approached that of pure ruthenium supported on silica. Some indication of the role of iron in the prereduced alumina-supported catalyst, which alters the activity and selectivity can be obtained from a consideration of the Mössbauer spectra recorded in situ from the prereduced catalyst which showed that the iron had been partially reduced to iron(II). Of special importance

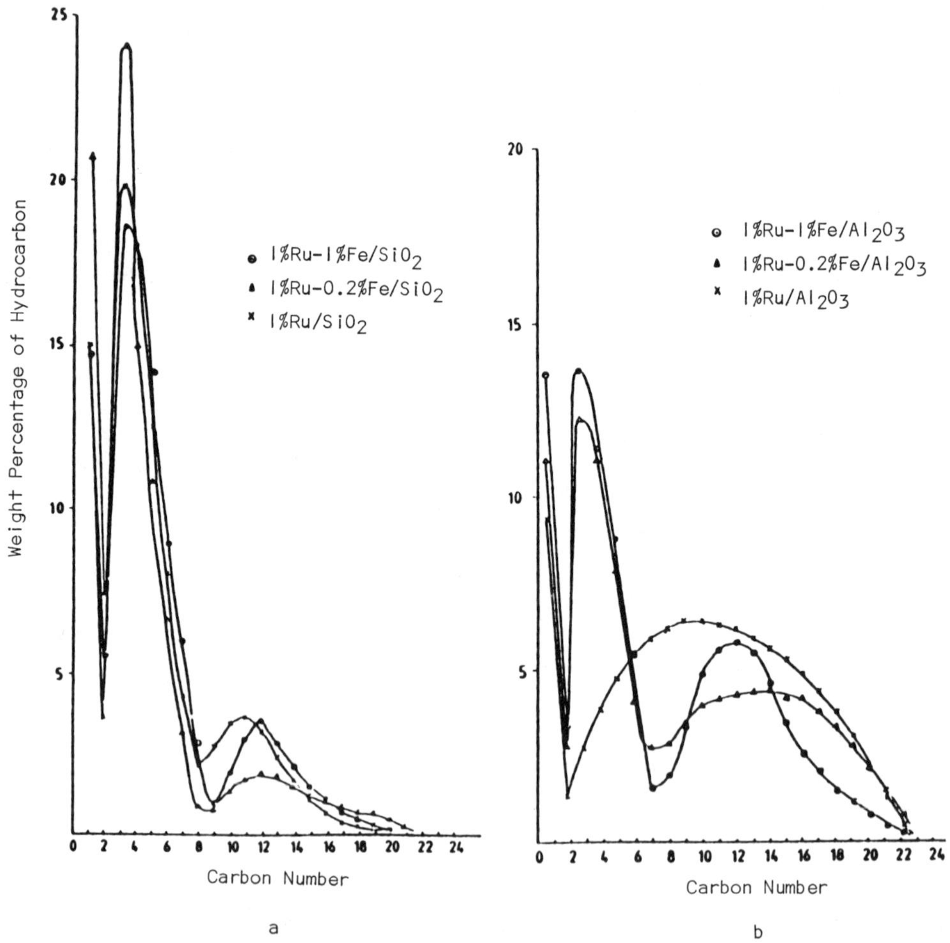

Figure 6. Hydrocarbon product distribution from low metal loaded iron-ruthenium catalysts supported on (a) silica and (b) alumina. Reproduced with permission from [6].

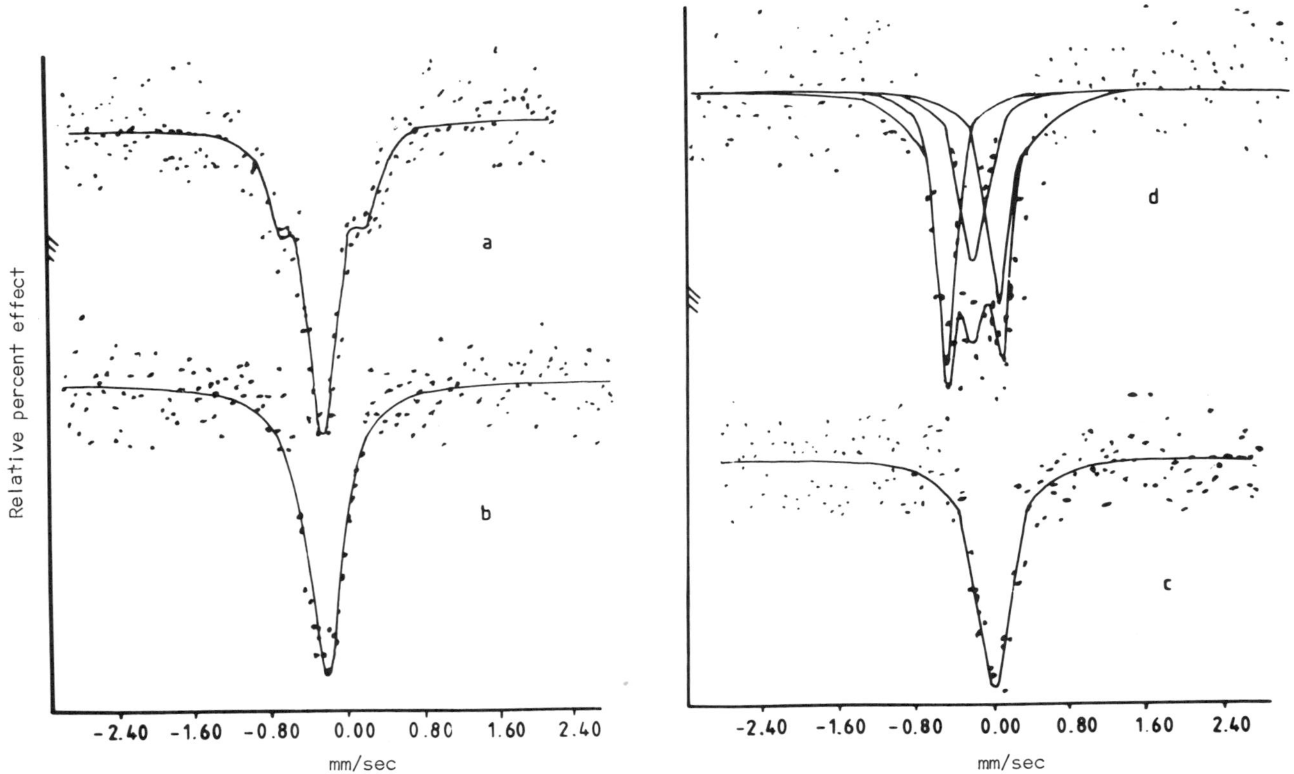

Figure 7. Ruthenium-99 Mössbauer spectra recorded from (a) barium ruthente catalysts, (b) barium ruthenate, (c) catalysts following use in simulated automobile exhaust gas, (d) subsequent treatment in oxygen. Reproduced with permission from [11].

was the iron(II) chemical isomer shift data which decreased in magnitude as the ruthenium concentration increased, a trend which corresponds to an increase in the s- electron density at the iron nucleus as a result of a decrease in the d-electron population. Hence the interaction between iron and ruthenium in the prereduced alumina-supported catalyst appears to involve the transfer of d-electrons from the iron(II) to the ruthenium. If the catalytic performance of alumina-supported ruthenium is tentatively associated with the acidity of alumina which gives rise to electron transfer from ruthenium to the acid sites, then the resulting strong interaction between ruthenium and the support can be envisaged in terms of the formation of electron deficient ruthenium species, ruthenium(δ+). Hence the transfer of electrons from iron to ruthenium as indicated by the Mössbauer chemical isomer shift data would mitigate against the formation of the electron deficient ruthenium(δ+) species and thereby give rise to a weakening of the ruthenium-alumina interaction and a corresponding decrease in activity and a shift in selectivity towards the formation of lower hydrocarbons. Such a scheme is supported by a consideration of the lower degree of electron transfer between ruthenium and the less acidic silica which results in the activity and selectivity being less susceptible to the influence of the incorporation of iron. Hence the catalytic performance of the alumina-supported iron-ruthenium catalysts is emulated by the silica-supported ruthenium catalysts when the former are iron-rich. These investigations [6] of the iron-ruthenium system illustrate how the catalytic properties of supported bimetallic catalysts reflect metal-metal and metal-support interactions which are amenable to examination by Mössbauer spectroscopy.

BARIUM RUTHENATE CATALYSTS

A very interesting study, involving the use of ruthenium-99 Mössbauer spectroscopy, has been reported [11] of a barium ruthenate catalyst which can be used for automobile exhaust emission control. The catalyst which was supported on alumina and rapidly calcined at 900°C gave a spectrum (Figure 7a) showing a broad singlet, characteristic of barium ruthenate (Figure 7b), with shoulders characteristic of ruthenium dioxide. Treatment of the catalyst in a simulated automobile exhaust gas at 973K for 10 hours gave a spectrum (Figure 7c) consistent with reduction to ruthenium metal. However, experiments involving the cycling of the catalysts between this reducing gaseous environment and an oxidizing one in which carbon monoxide was replaced by oxygen produced a catalyst which gave a Mössbauer spectrum (Figure 7d) showing the presence of barium ruthenate and ruthenium dioxide.

MIXED METAL OXIDE CATALYSTS

The catalytic properties of mixed metal oxides, consisting of two metals and oxygen, for the oxidation of hydrocarbons has been known for many years.

TIN-ANTIMONY OXIDE CATALYSTS

Some recent Mössbauer investigations [12] of tin-antimony oxides are a suitable illustration of the power of the technique for the study of mixed metal oxide catalysts because it is a system which contains two Mössbauer active elements and the solids have, in the past, been commercially developed as catalysts for the oxidation and ammoxidation of propene and for the oxidative dehydrogenation of butene. Typical antimony-121 Mössbauer spectra recorded [13] from these catalysts are depicted in Figure 8. The system also provides a good example of the power of Mössbauer spectroscopy, when used as a complementary technique to accompany investigations by x-ray diffraction and electron microscopy which examine crystallinity, phase composition, and structure, and by x-ray photoelectron spectroscopy which examines the composition and structure of the surface regions. The results emanating from the combined use of these techniques in the examination of tin-antimony oxides prepared by co-precipitation techniques involving the simultaneous addition of tin tetrachloride and antimony pentachloride to ammonium hydroxide are summarized in Figure 9.

The white precipitates dried at 25°C were amorphous to x-rays and were shown by tin-119 and antimony-121 Mössbauer spectroscopy to contain tin(IV) and antimony(V) in hydroxyl environments. It would seem therefore that these materials are homogeneous amorphous hydrated precipitates containing random

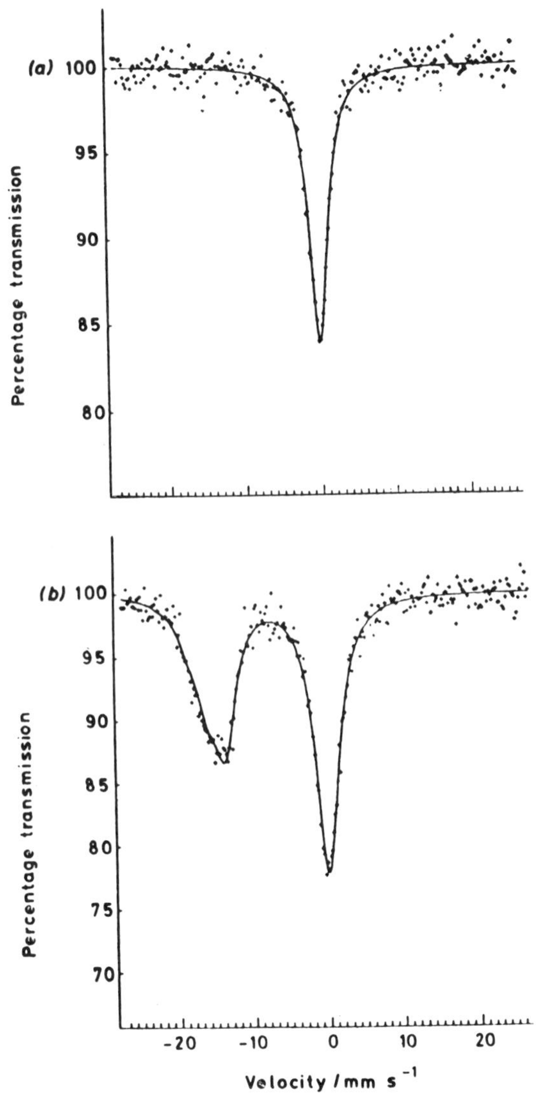

Figure 8. Antimony-121 Mössbauer spectra (relative to barium stannate) of tin-antimony oxide containing: (a) 4 percent antimony and calcined at 300°C for 16h; (b) 20 percent antimony and calcined at 1000°C for 14d. Reproduced with permission from [13].

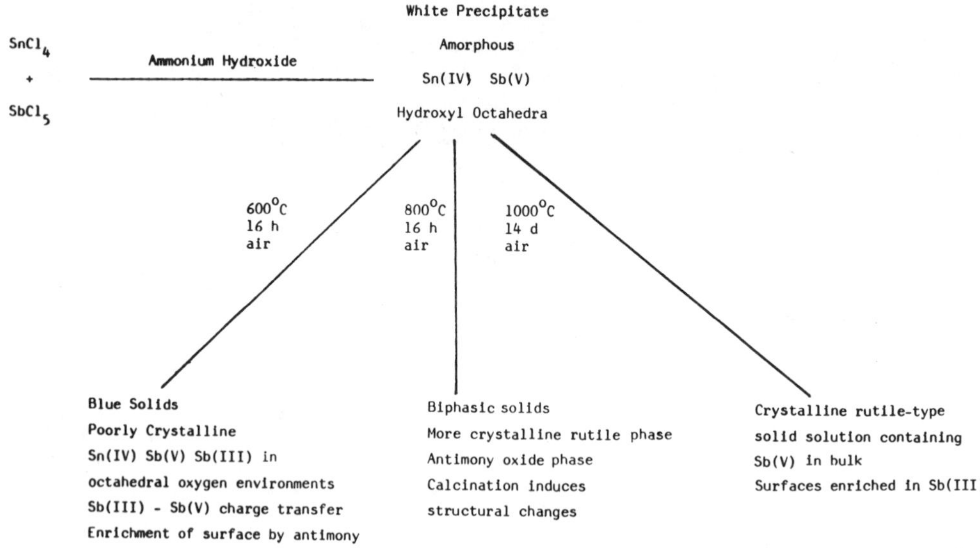

Figure 9. The preparation of tin-antimony oxide catalysts.

assays of tin(IV) and antimony(V) hydroxyl units which are probably hydrogen bonded together. The first significant change in these materials is observed when they are heated at temperatures exceeding ca. 400°C when they turn from white to blue. X-ray diffraction showed such materials heated at ca. 600°C to be poorly crystalline solids with rutile-related structures, similar to that adopted by pure tin(IV) oxide. The Mössbauer spectra were characterized by broad peaks indicative of the presence of tin(IV), antimony(V), and antimony(III) in roughly octahedral oxygen environments. The antimony(V) and antimony(III) species may therefore be envisaged as occupying similar sites in close proximity in the highly defective, disordered, poorly crystalline rutile-type solids such that charge transfer between antimony(V) and antimony(III) can readily occur and give rise to the blue coloration. It was also found from x-ray photoelectron spectroscopy that, at temperatures exceeding ca. 400°C, the surfaces of these materials became enriched in antimony. This phenomenon subsequently proved to be an important feature of the catalytic properties of these mixed oxides.

Calcination of the tin-antimony oxides at 800°C gave segregation into two-phases, a rutile-type phase which gave narrow line quadrupole split tin-119 Mössbauer spectra consistent with a very crystalline rutile-related solid, and an antimony oxide phase which was identified by antimony-121 Mössbauer spectroscopy as antimony tetroxide. It seems therefore that calcination induces structural changes in the initially homogeneous precipitate by which the tin dioxide octahedra aggregate to form a crystalline rutile-type phase and a simultaneous migration of antimony to the surface where it nucleates to form a macroscopically distinguishable antimony oxide phase. Following calcination at 1000°C for prolonged periods, such as 14 days, all the tin-antimony oxides were shown by x-ray diffraction to be crystalline monophasic rutile-type solids. It would seem that at these temperatures any antimony which has segregated to form antimony tetroxide is volatilized and the materials which are left are the equilibrated rutile-type solid solutions of antimony within the tin(IV) oxide lattice. Indeed, the x-ray, tin-119 and antimony-121 Mössbauer, and electron microscopy data were all indicative of a low tolerance of the tin(IV) oxide lattice for antimony and were all consistent with about 4 percent antimony being accommodated within the lattice as antimony(V). However, x-ray photoelectron spectroscopy showed the surfaces of these solid solutions containing 4 percent antimony within the tin(IV) oxide lattice to be significantly different in that about 25 percent of the surface was composed of antimony. However x-ray photoelectron spectroscopy, although very sensitive to the concentrations of antimony in the surface sites, is notoriously insensitive to the differentiation between antimony(V) and antimony(III) and in this respect it is pertinent to note how these catalysts, which had been heated to 1000°C for long periods of time so that all the excess antimony oxide had volatilized and which had surfaces enriched in antimony, gave antimony-121 Mössbauer spectra exhibiting high antimony(III) to antimony(V) ratios. Hence the increasing concentration of antimony(III) with increasing temperature, which was detected by Mössbauer spectroscopy was associated with the x-ray photoelectron spectroscopy data and interpreted in terms of a surface layer in which each antimony(III) species was surrounded by tin(IV).

An evaluation of the specific activity of the tin-antimony oxides for the formation of butadiene and its dependence on catalyst calcination temperature showed that when the catalyst contained a rutile-type phase the specific activity for butadiene formation increased with higher calcination temperature as increasing amounts of antimony(III) segregated to the surface. It was also found that the pure antimony oxides were relatively inactive and that their presence in catalysts, as could be identified by Mössbauer spectroscopy, generally produced significant changes in catalytic activity.

COBALT MOLYBDATE CATALYSTS

The power of Mössbauer emission spectroscopy has been excellently demonstrated in the recent review [7] of cobalt molybdate hydrodesulfurization catalysts. Hydrodesulfurization involves the removal of sulfur from petroleum compounds and is important because it enables the removal of sulfur from fuel oils and feedstocks for the chemical industry which would otherwise

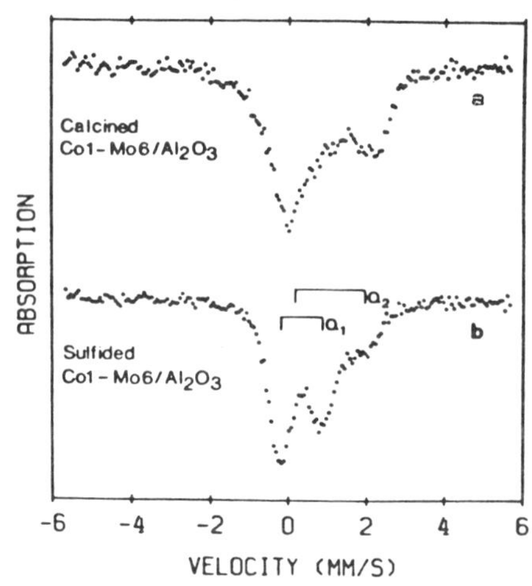

Figure 10. Mössbauer emission spectra of an alumina-supported cobalt molybdate catalyst (a) after calcination, (b) after treatment in a gaseous mixture of 2 percent H_2S/98 percent H_2. Reproduced with permission from [3].

lead to the production of atmospheric pollutants such as sulfur dioxide or to the formation of sulfides which would poison catalysts in the chemical industry. Various models of cobalt molybdate catalysts have been proposed and in an elegant study [3] which because of the restricted length of this article is only briefly discussed, the Mössbauer emission spectra were recorded from materials representing the different models. A sulfided alumina-supported cobalt molybdate catalyst was also examined in situ by Mössbauer emission spectroscopy and the spectrum (Figure 10) showed the cobalt to be present in two configurations which gave rise to the two doublets designated as Q_1 and Q_2. The broad line absorption, Q_2, which represented about 30 percent of the cobalt was assigned to cobalt located in sites of slightly differing symmetry inside the alumina support and showed that a significant amount of cobalt is unaffected by the sulfiding treatment. The other spectral component, Q_1, was found to correspond to a model compound in which the cobalt had been intercalated in the molybdenum sulfide layers. The results therefore gave for the first time, direct evidence for the presence of a cobalt-molybdenum-sulfur phase in real cobalt molybdate catalyst. Subsequent investigations using other techniques have sought to elucidate more fully the location of cobalt in this phase and the working mechanism of these cobalt molybdate catalysts.

CONCLUSION

Mössbauer spectroscopy is a very powerful technique for the characterization of real catalysts and for the in situ examination of the changes which are induced by treatment under realistic industrial conditions. It should be mentioned however that there are many other catalytic systems, for example zeolites, and numerous problems, such as the role of alumina as a textual promoter of iron catalysts, which are equally amenable to examination by Mössbauer spectroscopy but to which it has not been possible to give attention in this article. Much scope remains for the application of this technique in catalyst research and it is quite reasonable to expect significant developments in the application of Mössbauer spectroscopy to various catalytic systems in forthcoming years.

ACKNOWLEDGEMENT

The Award of a Sir Eric Rideal Bursary is gratefully acknowledged. The author also records his appreciation to Messrs. B. Emson and S. Vernals for assistance with the reproduction of the figures used in this chapter.

REFERENCES

1. F. J. Berry, in "Mössbauer Spectroscopy Applied to Inorganic Chemistry", Vol. 1, G. J. Long, ed., Plenum, New York, p. 391 (1984).

2. J. A. Dumesic and H. Tøpsoe, in "Advances in Catalysis", Vol. 26, D. D. Eley, H. Pines and P. B. Weisz, eds., Academic Press, New York, p. 121 (1976).

3. H. Tøpsoe, B. S. Clausen, R. Candia, C. Wirel, and S. Mørup, J. Catalysis 68, 433 (1981).

4. B. S. Clausen, S. Mørup, P. Nielson, N. Thrane, and H. Tøpsoe, J. Phys. E: Sci. Instrum. 12, 439 (1979).

5. J. W. Niemantsverdriet, C. F. J. Flipse, A. M. van der Kraan, and J. J. van Loef, Appl. Surf. Sci. 10, 302 (1982).

6. F. J. Berry, L. Liwu, W. Chengyu, T. Renyuan, Z. Su, and L. Dongbai, J. Chem. Soc. Faraday Trans. 1, 81, 2293 (1986).

7. H. Tøpsoe, B. S. Clausen, R. Candia, C. Wivel, and S. Mørup, Bull. Soc. Chim. Belg. 90, 1189 (1981).

8. F. J. Berry, Transition Met. Chem. 4, 209 (1979).

9. W. N. Degass, M. Doudart, and G. Parravano, J. Phys. Chem. 72, 3563 (1968).

10. A. M. Van der Kraan, personal communication.

11. C. A. Clausen and M. L. Good, J. Catalysis 46, 58 (1977).

12. F. J. Berry, J. Catalysis 73, 349 (1982).

13. F. J. Berry and B. J. Laundy, J. Chem. Soc. Dalton, 1442 (1981).

INFLUENCE OF PROMOTERS ON THE PHYSICAL PROPERTIES OF IRON ZEOLITE CATALYSTS

Friedrich Schmidt

Institute of Physical Chemistry
University of Hamburg
Laufgraben 24
D-2000 Hamburg 13
Federal Republic of Germany

INTRODUCTION

In addition to classical Fischer-Tropsch, F-T, gasoline production, many new efforts in technology are being directed towards production of basic hydrocarbons, because of the need of an alternative source for oil and gas. The main components of F-T catalysts usually are iron or cobalt. Promoters present, in general, simultaneously influence the chemisorption of the reactants, promote the formation of the primary complex, and improve rates of all carbon monoxide reactions.

The types of promoters are used in the classical F-T synthesis. Structural promoters such as alumina, thoria, and magnesia are present in order to produce large metal surface areas and to reduce recrystallization. Using molecular sieves, such as ZSM-5, as porous supports, the zeolite promoter appears to exhibit an additional function by controlling the chain growth of the hydrocarbons by means of the porous structure of the zeolite [1,2]. Energetic promoters are the second class of promoters. They influence the molecular electronic structures. Alkali cations are partially good electronic promoters for iron catalysts. In the bifunctional zeolite-based iron catalysts, which combine F-T activity of iron and the shape selectivity of the zeolite, the electronic promoter properties of the zeolite ions may also play an important role.

In the present study two types of experiments will be discussed. They are the promoter effect of the zeolite on the rate determining step of the F-T reaction, and the promoter effect of alkali metal on the physical properties of iron clusters.

Usually, two steps are necessary in order to prepare a F-T catalyst. These are formation of iron- or iron-oxide containing material and the carburization of the catalyst. Recent investigations [3,7] have shown that reduced iron is not a very active synthesis catalyst and that an activation period is accompanied by carburization of iron, the bulk reaction leading to the formation of different iron carbides. The presence of χ=Fe_5C_2 (Hägg) and Fe_3O_4 in the iron-FAU, iron-MOR and iron-ZSM systems, prepared by multinuclear iron carbonyls in conjunction with an "extraction method", was shown by Melson et al. [8]. A mnemonic code was used to identify the zeolite structure type [9]. In iron ZSM-5, after the carburization period, iron is mainly in the metallic form. However, in catalysts, prepared by means of the $Fe_3(CO)_{12}$ "extract method", most metal/metal-oxide [10-14] is found to be outside the cavities. The same holds true for the classical impregnation method [15].

An alternative route for preparing iron containing zeolites is the adsorption and decomposition of zero valent mononuclear iron complexes [16-22]. Another novel method of introducing iron into the zeolite has been published by Scherzer, et al. [23]. In all the above mentioned cases, at least a bidisperse iron phase is formed. In some cases, additional iron oxides are formed, by oxidation of the zero valent iron complexes by the acidic properties of the zeolites such as H-ZSM, H-FAU. Furthermore, during decomposition of iron carbonyls, some carbides may be produced. The chemical composition of these carbides depends on the conditions of decomposition, either in vacuum or in hydrogen [24].

In the first part of the present study, the influence of the zeolite promoter on the rate determining step in F-T catalysis will be investigated.

Recently it was found, that the rate determining, step is the dissociative adsorption of carbon monoxide and hydrogen on the iron surface. This step is based on the electron donation from the iron into the antibonding orbitals of the carbon monoxide [25,26]. Electropositive promoters present on the surface, which increase this electron transfer, such as potassium or sodium ions, K^+, Na^+ favor this step, whereas electronegative ions, which decrease this transfer, such as oxide ion make the electron transfer more difficult. In all commercial F-T catalysts both components are present. Zeolite-based F-T catalysts also contain alkali cations and oxygen anions. By using zeolites as model systems, the location of the respective ions are known. However, the location of the atoms of the iron cluster with respect to those of the matrix is not known. Therefore it is useful to apply a macroscopic approach, namely the use of the overall Sanderson electronegativity SEN [27] of the zeolite. In the present study, the problem of the promoter effect of the zeolite on the rate determining step of the F-T reaction was turned around. Instead of investigating the adsorption of carbon monoxide on iron, we studied the decomposition of iron carbonyl, especially the stability of the last intermediates of the stepwise process, as a function of the overall SEN of the zeolite.

Recent key experiments concerning the influence of the alkali on the dissociative adsorption of carbon monoxide on iron [26] have been performed in UHV systems by means of promoting iron single crystals by adsorption of alkali metal vapor. In order to transfer this information to real catalyst systems such as zeolite supported iron, the second part of this study deals with the influence of alkali metal on the physical properties of iron clusters.

The iron clusters have been prepared by the classical method of ion exchange, dehydration, and reduction. Iron(II) ion exchange of zeolites leads to definite systems in unique states [28,29]. However, an appropriate reducing agent has to be found. The reduction of iron(II) ion containing zeolites by means of hydrogen is possible only at and above 1023K [30], but these conditions lead to lattice destruction. However, photoreduction of iron(II) FAU, and X-type zeolite, is possible at 700K [31]. The Mössbauer effect spectrum of the degassed iron(II)-NaX sample is in agreement with that observed by Rees et al. [28].

After five hours of reduction the doublet, which was attributed to iron ions located in S_I' sites, disappeared. The beginning of the reduction is indicated by a small doublet with close to the isomer shift of α-iron [29].

The area of this doublet is much smaller than the area of the doublet which disappeared after five hours of reduction. Hence, a rearrangement of the iron(II) ions can be assumed. After a reduction period of 52 hours the degree of reduction is found to be larger than 50 percent. X-ray diffraction measurements on this sample indicate extended lattice destruction.

By using hydrogen as a reducing agent, the acidic properties of the support are changed and the solid acid hydrogen-zeolite tends to reoxidize the metal phase and to destroy the structure in the case of aluminum-rich zeolites. In the case of Na/NH_3 as a reducing agent [32], the catalyst is contaminated with NH_3 if high temperature degassing is not applied, which usually produces bi-disperse systems. On the other hand, the acidity of the support is not changed. However, in order to prepare a system as unique as possible in the present study, sodium or potassium vapor reduction has been chosen. The location of the alkali cations, which are still present or formed upon reduction, inside the matrix is known.

It is a prerequisite for the understanding of the state of a working catalyst to obtain data on the state of the system before reaction. Therefore, in the second part of this study the metal phase formation is followed by means of Mössbauer effect spectroscopy in iron(II) ion exchanged LTA, FAU, and MOR. Subsequently, the influence of sodium on the metal phase is studied. Work is still in progress to investigate the carburization of the alkali-doped iron zeolite catalysts.

EXPERIMENTAL

In order to prepare iron clusters via adsorption and decomposition of $Fe(CO)_5$, synthetic NaY (Na-FAU; $Na_{55.5}$-$Al_{55.5}Si_{136.5}O_{384}$) and NaX (Na-FAU; $Na_{88}Al_{88}Si_{104}O_{384}$) were treated with 0.1M NaCl solution, washed, air dried, and stored over saturated NH_4Cl solution. Iron pentacarbonyl from Ventro (99.5 percent) was coldly distilled in the dark and stored in a stainless steel tube over a 5A molecular sieve. A stainless steel gas-dosing system [33], connected to a quartz reactor, allowed degassing and loading of the zeolite samples with vapor of $Fe(CO)_5$. The saturated $Fe(CO)_5$/zeolite adducts were thermolyzed. Fluidized bed conditions were achieved by a horizontal vibration of the reactor. Decomposition was carried out at a heating rate of 0.00333K/s under a vacuum of 10^{-3} Pa. Subsequently, the adducts were degassed for 50ks at the respective final temperatures 373K and 453K.

The samples which were prepared via ion exchange, dehydration and reduction were obtained according to the following procedure. Sodium FAU zeolites were supplied by LINDE Division of Union Carbide. MOR was from Norton. The zeolites were treated with 0.1M NaCl solution to remove possible cation deficiencies, washed, air-dried and stored over saturated NH_4Cl solution. All steps of ion exchange and reduction were made in an argon atmosphere. The iron exchanged FAU zeolites and MOR were prepared by treating the sodium forms with an aqueous solution of iron(II) sulfate at a pH of 5 for three hours [29]. As an example, the LTA ion-exchange is described. For preparing the iron(II) exchanged form of LTA-type zeolite, the iron(II) salt solution was made by dissolving between 5 and 200mg iron in a calculated volume of 2N sulfuric acid, diluted in approximately 6cm^3 of water. Thus, a pH of 5 was reached. A suspension of 0.5g oxygen free LTA zeolite in about 2cm^3 water was initially adjusted to pH 5 with about 6cm^3 of standard acetic acid buffer solution. Then the zeolite was ion-exchanged.

The ion-exchanged zeolites were dehydrated at 673K for 24 hours at 0.1mPa. After this degassing process all samples were reduced by means of a small excess with respect to the calculated amount of sodium or potassium necessary for complete reduction. The alkali metal was cut under n-hexane,

Table I. Nomenclature and Zeolite Sample Composition

Sample	Composition
Na-X	$Na_{88}Al_{88}Si_{104}O_{384}$
Na-Y	$Na_{55.5}Al_{55.5}Si_{136.5}O_{384}$
Fe-LTA 1	$Na_{11.94}Fe(II)_{0.78}Al_{12}Si_{12}O_{48}$
Fe-LTA 36	$Na_{7.67}Fe(II)_{2.16}Al_{12}Si_{12}O_{48}$

degassed, and placed at the bottom of a glass tube. The dehydrated zeolite was supported 3cm above it on a small glass frit. The top of the tube was formed as a Mössbauer cell. The full length heating mantle of the glass tube was brought to the reduction temperature. After the respective heating period the oven was removed and the tube was allowed to cool. Then the zeolite was moved into the Mössbauer cell. After taking the spectrum, the sample was transferred back onto the frit and the reduction was continued.

All completely reduced samples were analyzed by means of x-ray fluorescent spectroscopy, neutron activation analysis, and chemical analysis. The crystallinity of the iron zeolites after ion exchange, after dehydration, and after complete reduction was examined by means of x-ray powder diffraction. No change of the linewidth of the Bragg peaks of Fe-FAU and Fe-MOR and small changes in the case of Fe-LTA indicated that no and very small structural breakdown had occurred, respectively. The composition of the samples is given in Table I.

A reference sample was prepared by placing α-iron powder on the glass frit and alkali metal at the bottom of the glass tube below the frit. After the respective heating time and temperature, the iron sample was transferred into the Mössbauer cell at the top of the tube and sealed off.

All Mössbauer spectra were obtained by using a constant acceleration Mössbauer spectrometer with a 25mCi source of cobalt-57 in copper. The spectrometer was calibrated by use of a 10μm foil of α-iron at room temperature. Isomer shifts are reported to the source at 300K. The spectra were analyzed by least squares fitting of the appropriate line shapes.

RESULTS AND DISCUSSION

Adsorption and Decomposition of $Fe(CO)_5$

The surface of iron clusters, which are encapsulated in zeolites is covered by the walls of the zeolite cages. Therefore, the neighboring ions of the zeolite, in particular Na^+, K^+, Cs^+, and oxide ion are expected to influence the cluster. Hence, the support acts no longer only as a structural promoter.

Adsorption of carbon monoxide on this type of clusters should be described by means of a modified Blyholder model [25,26]. By going for instance from Na^+ to Cs^+, the Fe-C bond strength should be enhanced for electronic [34] and electrostatic [35] reasons. We turned the problem around and

Table II. Thermogravimetric Results

Sample	Si/Al	SEN[a]	T_d,K	E_d,kJ	Heating Rate, K/s	Ref.
Na-Y	2.48	3.66	364	112	0.00833	21
Na-Y	2.46	3.55	390	116	0.03333	19
Na-Y	2.46	3.55	368	116	0.00333	This work
Cs-Y	2.46	3.36	400	119	0.03333	19
Na-X	1.22	3.24	373	120	0.00333	This work
Na-X	1.17	3.35	456	141	0.00833	21

[a]Sanderson electronegativity [27].

investigated the stability of the last intermediates of the stepwise decomposition of $Fe(CO)_5$/zeolite adducts [36] as a function of the chemical composition of the support.

Recently we have shown, that changing the electronegativity of the support by changing the cations from sodium to cesium while preserving the structure (in both cases FAU was used with the same Si/Al ratio), the stability of the iron-carbonyl bond was shifted towards higher temperatures of decomposition [19]. In addition, oxidation of the iron, most probably by means of the oxygen, which was generated by carbon-oxygen bond dissociation is more pronounced is Cs-FAU than in Na-FAU.

In the present study we investigated the influence of the silicon to alumina ratio on the stability of the last intermediates of the stepwise $Fe(CO)_5$/zeolite decomposition. Thermogravimetric measurements, published by Trautwein et al. [21] have given desorption temperatures, T_d, of 364K for Y-type zeolite and 456K for X-type zeolite, as can be taken from the derivative of their thermogram. According to our model, we correlate the SEN of their samples, which can be calculated from the given compositions, with the activation energies, E_d, of the carbon monoxide desorption, which can be calculated from T_d, by means of the formula given by Redhead [37], if the heating rate is known. In Table II the results are summarized and compared with those data, which can be calculated from the measurements of Bein, et al. [19] for Na-FAU and Cs-FAU.

In order to prove our model, we decomposed $Fe(CO)_5$/zeolite adducts with smaller differences of the SEN than the samples of Trautwein et al. [21]. Two decomposition temperatures were chosen, 373K, which is just close to the desorption temperature, T_d, and 453K, which is sufficiently above T_d. T_d was determined from the thermogram, which was simultaneously monitored during the decomposition.

The top of Figure 1 shows the room temperature spectrum of $Fe(CO)_5$/Na-FAU, X-type, which was measured after complete decomposition at 453K. The spectrum is identical with that of the respective Y-type samples and that of α-iron. The spectra of those samples, which were decomposed at 453K exhibit two systems: α-iron and an additional central component, which we attribute to subcarbonyl or to very small clusters of iron, which are covered by carbon monoxide. Because of the small differences in the SEN values between the X-type and the Y-type zeolites of the samples we used, the differences in the area fractions of the central component are much less pronounced than in the case of the samples Trautwein used, reference 21, Figures 3b and 6b, as seen also in Table II.

Furthermore, samples with a strong iron-carbon bond strength, which therefore decompose at higher temperatures, such as Na-FAU which is X-type with a silicon/aluminum of 1.17 and a SEN of 3.35, and Cs-FAU, which is Y-type, with a silicon/aluminum of 2.46 and a SEN of 3.36, undergo internal oxidation as can be seen from the Mössbauer-spectra (reference 19, Figures 3e and 1b). This behavior can be explained in terms of the modified Blyholder model. Due to electronic charge transfer, and/or electrostatic reasons from the promoter to the iron and from iron to antibonding orbitals of the adsorbed carbonyl, the iron-carbon bond strength is enhanced and the carbon-oxygen bond strength is weakened. Therefore, at higher temperatures, carbon-oxygen bond dissociation can occur and the oxygen generated can oxidize the metal atoms.

In order to study the charge transfer between promoter and metal in more detail, the promoter effect of alkali metal on the physical properties of iron clusters was investigated. In contrast to the situation in ultra-high vacuum clean surface studies [26,34], where the transferred charge is

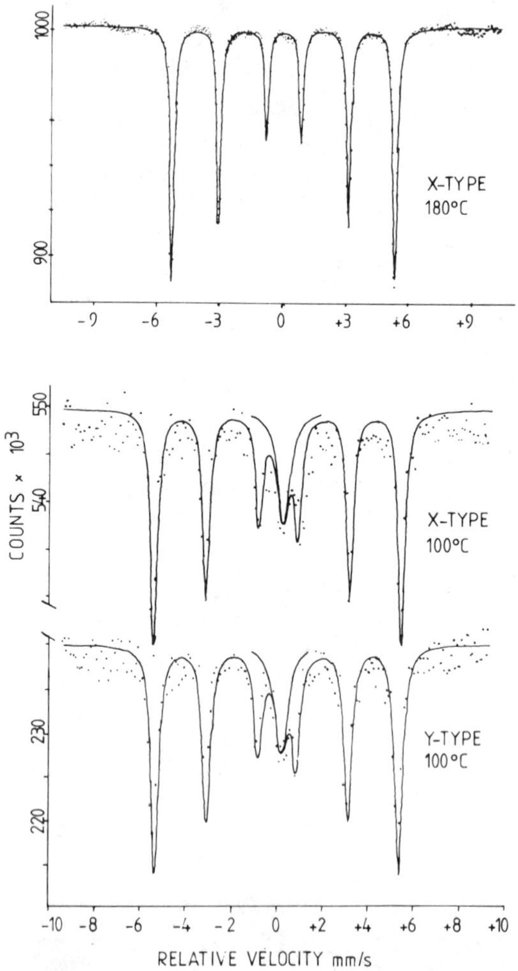

Figure 1. Mössbauer spectra of $Fe(CO)_5$/zeolite adducts after decomposition at 453K (X-type) and at 373K (X- and Y-type). The spectra are measured at room temperature. The isomer shift is calibrated vs. α-iron foil.

distributed over a very high number of atoms near the sorption site of the alkali metal, in case of carbon monoxide adsorption on very small clusters, the transferred charge is expected to change the electron density of the whole cluster remarkably. Hence, a change of the average atomic magnetic moment of the atoms of ferromagnetic metals should be expected.

In the following section the preparation of very small iron clusters in zeolites is shown and subsequently the influence of alkali metal on the iron clusters is studied.

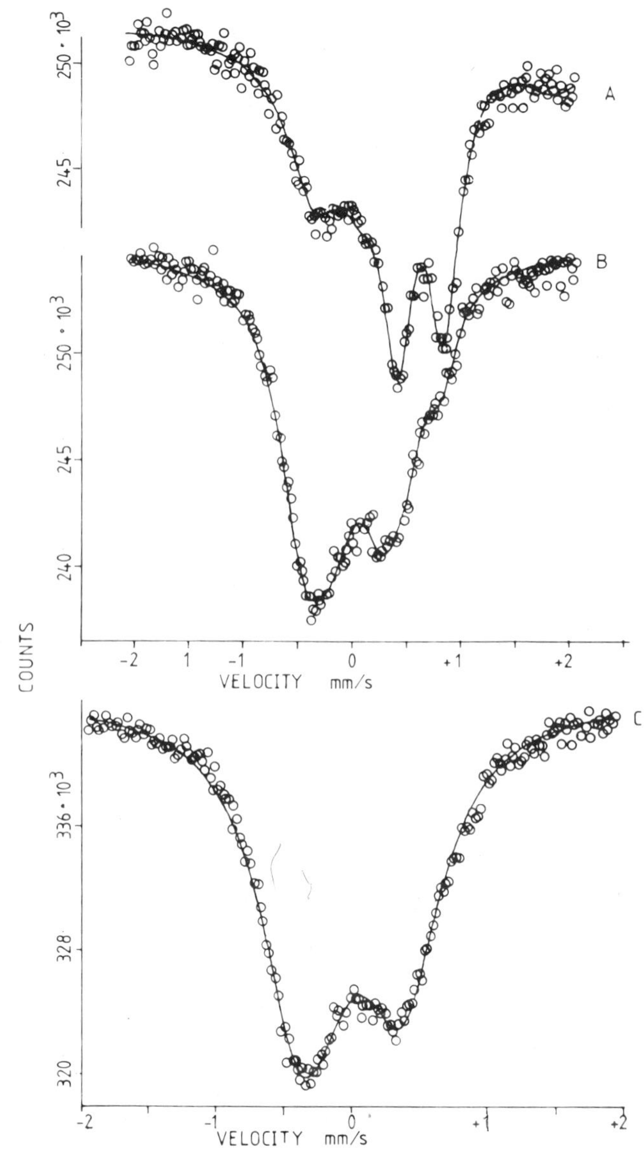

Figure 2. Mössbauer spectra of Fe-LTA 1, taken at various stages of sodium vapor reduction at 573K. The spectra are measured at room temperature. The isomer shift is calibrated vs. α-iron. A, after a time of reduction of 9.15 h, B, after 46 h; and C: after 85 h.

Metal Formation in Fe-LTA

Representative room temperature Mössbauer effect spectra of iron(II)-57 exchanged LTA, with a degree of ion exchange of 1.3 percent, during sodium vapor reduction, are given in Figure 2. The Mössbauer effect parameters of this series are presented in Table III. A small amount of nonstoichiometric exchanged ions is present, in system 2, (see also Figure 4). The parameters in Table III have been derived from a least squares fit by using Lorentzian line shapes for all lines. The doublet with the isomer shift close to the value of α-iron is attributed to zero valent iron clusters. The fact that a six-line pattern is not formed at room temperature is due to the small size of the clusters. The quadrupole splitting of the superparamagnetic component is a consequence of the location of the clusters inside the zeolite matrix, which possesses large electric field gradients. The formation of the metal phase has been followed by plotting the area fraction of the zero valent component vs. time of reduction as depicted in Figure 3. From this figure, it can be concluded, that after a period of 90 hours the sodium vapor reduction at 573K is complete.

Figure 3 also shows that the kinetics of metal formation is complicated. The data only permit the general conclusion that the degree of reduction linearly increases between 10 and 45 hours and between 45 and 90 hours, but in the latter case with a rate constant roughly half the value of the first period.

The behavior reported so far is also found in iron(II) exchanged LTA zeolite, with a degree of exchange of 36.1 percent - hereafter referred to as Fe-LTA 36, Figure 4A. This figure shows the room temperature Mössbauer effect spectrum of this sample after a reduction time of 85 hours at a temperature of reduction of 573K, (Figure 4B). The parameters for this spectrum are also given in Table III. The parameters of all reduced phases are identical within the limit of the experimental error.

State of the Iron in Reduced Fe-LTA

Measurements of the magnetization of the sample Fe-LTA 36 already have been published [17]. In the first paper the total magnetization has been attributed to iron. The total magnetization of those samples, which were measured in the field and temperature range, which were not sufficient for saturation, has been interpreted as being due to iron clusters exhibiting the same spontaneous magnetization as bulk iron. However, later measurements on this sample in fields up to 7T and temperatures down to 1.7K have shown that the magnetization was higher than in bulk material. After subtracting the magnetization of the fraction with a spin S of 0.5, the saturation magnetization of the iron was found to be 10 percent lower than in bulk iron. This has been discussed in terms of a size effect [38] of the smallest particles. The size distribution of the sample Fe-LTA 36 was found to be bidisperse, with ca. 75 atomic percent smaller than 0.7 nm, and the remainder with an average value of ca. 2.5 nm. A remarkable number of the clusters were found to be built up by only a few atoms.

In order to come to a better understanding of the state of the iron, the Mössbauer effect spectra have been analyzed in more detail. For this purpose, the spectra of the iron exchanged sample, Fe-LTA 36, have been used. The spectrum measured at 300K (39, Figure 2) can be fit by means of an asymmetric doublet or, alternatively, by means of a symmetric doublet and a singlet (asymmetric doublet: is = -0.073mm/s ± 0.006mm/s; singlet: is = -0.65 ± 0.01mm/s; area fraction approx. 10 percent, symmetric doublet: is = -0.050 ± 0.001mm/s; area fraction approx. 90 percent, see Figure 4B).

Table III. Mössbauer Data of Iron Clusters Reduced by Means of Sodium Vapor in Zeolites[a]

Sample	Si/Al Ratio	Weight % Iron	Reduction T,K	Time, Hr.	System 1		System 2		System 3		System 4[b]
					δ mm/s	ΔE_Q	δ mm/s	ΔE_Q	δ mm/s	ΔE_Q	δ mm/s
LTA	1	0.26	573	0	0.619	0.494	0.859	2.125	--	--	--
LTA	1	0.26	573	9.2	0.617	0.424	0.703	2.218	-0.147	0.668	-0.177
LTA	1	0.26	573	14.2	0.618	0.422	0.940	1.855	-0.177	0.679	-0.300
LTA	1	0.26	573	20	0.632	0.409	0.965	1.945	-0.148	0.710	-0.276
LTA	1	0.26	573	24.2	0.631	0.387	0.822	2.003	-0.116	0.731	-0.229
LTA	1	0.26	573	29	0.629	0.356	0.904	1.863	-0.111	0.761	-0.238
LTA	1	0.26	573	40	0.642	0.361	0.691	2.232	-0.094	0.759	-0.201
LTA	1	0.26	573	46	0.654	0.318	0.855	1.818	-0.084	0.740	-0.221
LTA	1	0.26	573	52.2	0.673	0.284	0.954	1.805	-0.072	0.768	-0.213
LTA	1	0.26	573	59	0.654	0.278	0.728	2.409	-0.060	0.763	-0.195
LTA	1	0.26	573	78	--	--	--	--	-0.026	0.793	-0.152
LTA	1	0.26	573	85	--	--	--	--	-0.026	0.784	-0.174
LTA	1	36.7	573	85	--	--	--	--	-0.05	0.77	-0.65[c]
FAU	2.4	13.4	673	10	--	--	--	--	-0.01	0.77	--
FAU	2.4	13.4	473[d]	108	--	--	--	--	-0.04	0.78	--
MOR	5	1.54	573	50	--	--	--	--	-0.05	0.76	--
MOR	5	1.49	473[d]	108	--	--	--	--	-0.05	0.77	--

[a]Data give in mm/s obtained at 295K relative to cobalt-57 in copper. [b]Alternatively an asymmetric doublet can be fit with a δ of -0.015(±) and a ΔE_Q of 0.7065(±49)mm/s. [c]Alternatively an asymmetric doublet can be fit with a δ of -0.073(±6) and a ΔE_Q of 0.83(±1)mm/s. [d]The reducing agent is potassium vapor.

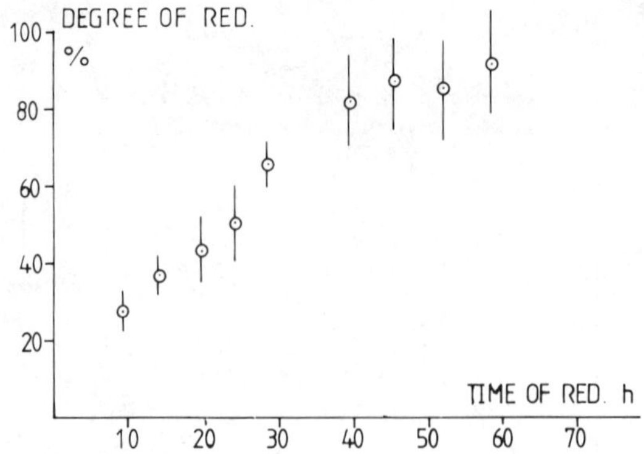

Figure 3. Area fraction of the zero valent component of the room temperature Mössbauer spectra of Fe-LTA 1 vs. time of reduction.

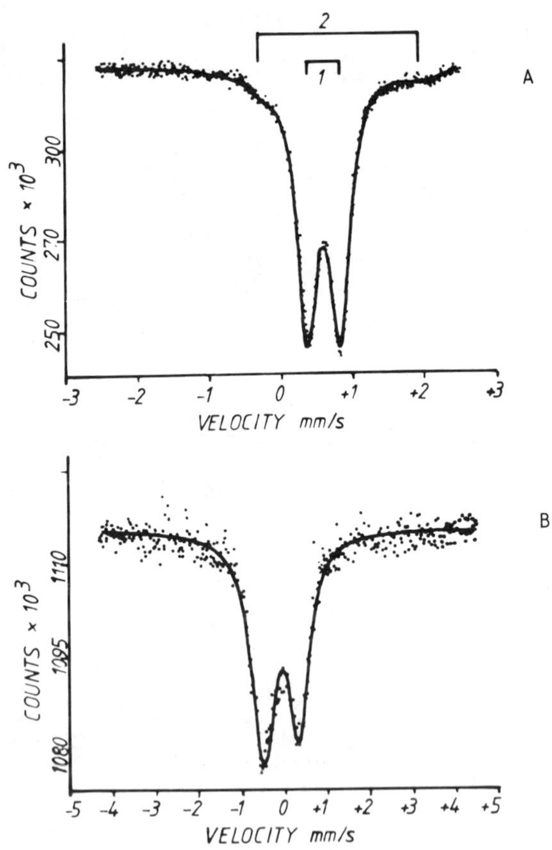

Figure 4. The room temperature spectra of Fe-LTA 36, taken before (A) and after (B) 85 hours of sodium vapor reduction at 573K. Component 2 is due to a nonstoichiometric exchanged iron phase. The isomer shift is calibrated versus α-iron.

These two possibilities correspond to two different models. In the first, the asymmetry of the doublet is due to an anisotropy of the interface atom vibrations. This is caused by the different lattice vibrations of the cluster and of the matrix, which on the other hand are strongly coupled. In the second model, the singlet is attributed to the central atoms of the clusters, whereas the doublet may be due to the surface atoms. The second model has been excluded [39]. In the first model the small, as compared to bulk iron, isomer shift was assigned to the second order Doppler shift. The resulting chemical shifts of iron clusters in the zeolite matrix and in the bulk materials appeared to be identical. This was in line with the first results of the magnetic measurements on this sample [17]. However, the new magnetic and Mössbauer effect spectra results demand a correction of the picture developed so far.

Figure 5 shows the Mössbauer effect spectrum of sample Fe-LTA 1 which was reduced for 85 hours at 573K by means of a clear excess of sodium. The spectrum was taken at 4.2K in an external field of 1T. The fit was obtained by using a distribution of hyperfine fields, which is also given in Figure 5. Two interesting observations can be made. A small number of particles exhibits an increased hyperfine field as compared to bulk iron whereas the mean value was found to be 21T. After making corrections for the external field, the average hyperfine field amounts to 20T. This value is much smaller than the value of bulk iron at 4.2K. By increasing the external field up to 3T, the same value of the hyperfine field was found. Therefore, the magnetization of the sample seems to be saturated at 1T.

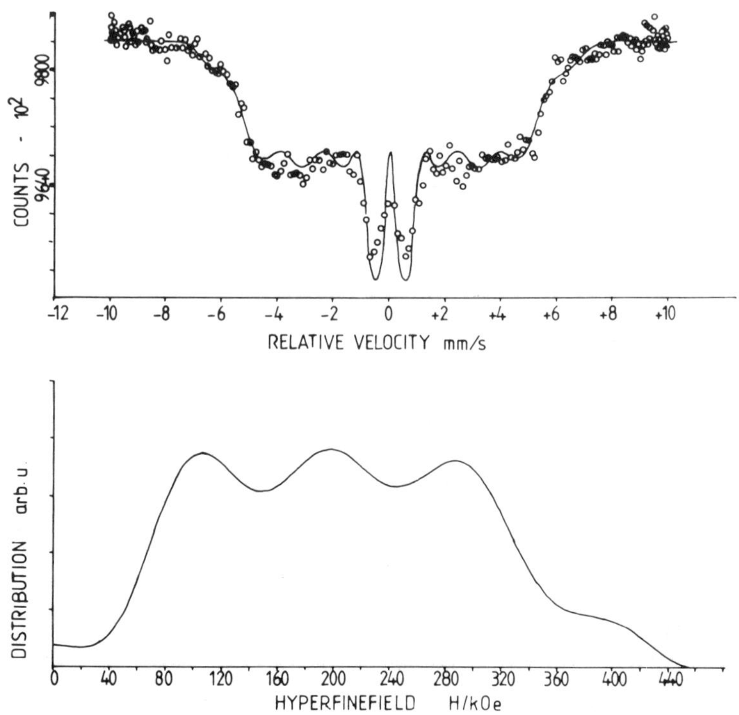

Figure 5. Mössbauer spectrum and hyperfine field distribution obtained at 4.2K in a 1T external field of sample Fe-LTA 1, which was reduced by means of a clear excess of sodium vapor for 85h at 573K.

Effect of the Sodium Promoter

In the following section, the decrease of the average hyperfine field and hence, the decrease of the average magnetic moment of the iron will be discussed. Two sources may cause the effect, the matrix and the sodium. Medium metal support interaction as was observed in other electron donor acceptor systems [40] should cause clusters with slightly different electron concentrations. This may result in a change of the isomer shift, but this is not necessarily the case. Alumina-doped iron as well as iron on magnesia exhibit no change in the isomer shift [41,42]. In the iron zeolite system, the isomer shift is very close to the bulk value after corrections for the second order Doppler effect have been made (see above). On the other hand, phonon coupling of the clusters to the matrix was found (see above). The hyperfine splitting of iron clusters supported on zeolites which have been prepared by means of various other techniques [19,21,29] is the same as in bulk iron. Therefore, the decrease of the hyperfine field, which was observed in the present study, may be due to the size of the clusters or to the different preparation techniques. The size effect can be excluded by the simple argument that with decreasing size the moment of iron increases [43] as seen also in Figure 5 at the high field of ca. 40T. For very small isolated and unsupported clusters, the saturation hyperfine field should be reduced according to calculations by Levin et al. [44]. On the other hand, according to calculations by Johnson [45], which have been established by measurements by Montano et al. [46] and recently by Whetten et al. [47], the magnetic moment of small iron clusters should be enhanced.

The decrease of the average hyperfine field, shown in Figure 5 can easily be caused by the sodium promotor present in the sample. It is well-known from recent model experiments, that the adsorption of zero valent alkali metals on ultra high vacuum clean iron surfaces results in local electron transfer from the alkali metal to the surface [26]. In case of potassium, 0.6 electrons per alkali atom are transferred to the surface of the iron [35]. The iron clusters in LTA which are located inside the cavities grow up to diameters of typically 0.7nm. These clusters contain 13 to 15 atoms on the average. Assuming those clusters are covered by the same number of sodium atoms, every iron atom takes 0.7 electrons from the sodium. If these electrons fill the d-band-like structure of the metal cluster, the decrease of the average magnetic moment and, hence, the decrease of the average hyperfine field can be understood.

Further investigations are necessary for more detailed quantitative data. The electron configuration of the iron can be calculated from the measured isomer shift, which is close to the value of bulk iron, if the number of electrons per iron atom after the charge transfer is known.

CONCLUSION

In the present study it is shown, that a correlation exists between the overall Sanderson electronegativity of a structural promotor and the stability of the bond strength between iron atoms in small clusters and carbon monoxide. The Fe-CO-zeolite system is a suitable F-T model catalyst for studying the electronic and structural promotor effects. The experiments allow one to investigate the interaction of alkali metal with iron clusters. Evidence was found for electron transfer from sodium to iron.

This work demonstrates, that Mössbauer spectroscopy is a suitable method for studying promotor effects on the physical properties of iron catalysts. The big advantage of Mössbauer spectroscopy as compared to the methods such as AES, PES, etc. lies in the possibility of performing high pressure in situ measurements under technical relevant reaction conditions in the steady state of a working catalyst.

ACKNOWLEDGEMENTS

Financial support of this work by the Deutsche Forschungsgemeinschaft is gratefully acknowledged. We thank the University of Bochum and Prof. Dr. Micklitz for some of the Mössbauer spectra.

REFERENCES

1. P. A. Jacobs and D. van Wouve, J. Mol. Catl. 17, 145 (1982) and literature cited therein.

2 W. D. Deckwer and M. Ralek in: "Metal Microstructures in Zeolites", eds. P. A. Jacobs, N. I. Jaeger, P. Jiru and G. Schulz-Ekloff, Elsevier p. 267 (1982).

3. G. B. Raupp and W. N. Delgass, J. Catal. 58, 337, 348, 361 (1979).

4. E. E. Unmuth, L. H. Schwartz, and J. B. Butt, J. Catal. 61, 242 (1980).

5. J. A. Amelse, J. B. Butt, and L. H. Schwartz, Z. Phys. Chem. 82, 558 (1978).

6. J. W. Niemantsverdriet, A. M. van der Kraan, W. L. van Dijk, and H. S. van der Baan, J. Phys. Chem. 84, 3363 (1980).

7. J. W. Niemantsverdriet, C. F. J. Flipse, A. M. van der Kraan and J. J. van Loef in "Applications of Surface Science", 10, 302 (1982).

8. G. M. Melson, J. E. Crawford, J. W. Crites, K. J. Mbadcam, J. M. Stencel, and V. S. Rao in "Intrazeolite Chemistry", ACS Symposium Series 218, 398 (1983).

9. D. H. Olson and W. M. Meier, "Atlas of Zeolite Structure Types", (IZA, Polycrystal Book Service, Pittsburgh, 1978).

10. J. M. Stencel, J. R. Diehl, L. J. Douglas, C. A. Spitler, J. E. Crawford, and G. A. Melson in "Colloids and Surfaces", 4, 331 (1983).

11. R. T. Obermyer, L. N. Mulay, C. Lo, M. Oskovie-Tabrizki, V. S.Rao, J. Appl. Phys. 53, 2683 (1982).

12. D. Ballivet-Tkatchenko and G. Coudurier, Inorg. Chem. 18, 558 (1979).

13. D. Ballivet-Tkatchenko, G. Coudurier, and Nguyen Duc Chan, in "Metal Microstructure in Zeolites", eds. P. A. Jacobs, N. I. Jaeger, P. Jiru and G.Schulz-Ekloff, Elsevier, p. 123 (1982).

14. D. Ballivet-Tkatchenko, G. Coudurier, H. Mozzanega, and I. Tkatchenko in "Fundamental Research in Homogeneous Catalysis", ed. M. Tsutsui, Plenum, p. 257, (1979).

15. S. A. Butter, W. W. Kaeding, U. S. Patent 1979, 3, 972, 832.

16. Th. Bein and P. A. Jacobs, J. Chem.Soc. Faraday Trans. I, 80, 1391 (1984).

17. F. Schmidt in "Metal Microstructures in Zeolites", eds. P. A. Jacobs, N. I. Jaeger, P. Jiru, and G. Schulz-Ekloff, Elsevier, p. 191 (1982).

18. S. L. Suib, K. C. McMahon, and D. Psaras in "Intrazeolite Chemistry", ACS Symposium Series 218, p. 301 (1983).

19. Th. Bein, F. Schmidt, W. Gunsser, and G. Schmiester, Surface Sci. 156, 57 (1985).

20. J. B. Nagy, M. van Eenoo, and E. G. Derouane, J. Catal. 58, 230 (1979).

21. A. X. Trautwein, E. Bill, R. Bläs, G. Doppler, F. Seel, B. Wolf, R. Klein, and U. Gonser, Surface Sci. 156, 140(1985).

22. Th. Bein, F. Schmidt, and P. A. Jacobs, Zeolites 5, 240 (1985).

23. J. Scherzer and D. Fort, J. Catal. 71, 111 (1981).

24. D. Rojas, P.Bussiere, J. A. Dalmon, A. Choplin, J. M. Basset, and D. Olivier, Surface Sci. 156, 516 (1985).

25. G. Blyholder, J. Phys. Chem. 68, 2772 (1964).

26. H. P. Bonzel, J. Vac. Sci. Technol. A2(2), 866 (1984) and literature cited therein.

27. R. T. Sanderson, Chemical Bonds and Bond Energy, Academic Press, New York, 2nd ed.,(1976).

28. L. V. C. Rees in "Structure and Reactivity of Modified Zeolites", eds. P. A. Jacobs et al., Elsevier, p. 1, (1984) and literature cited therein.

29. F. Schmidt, W. Gunsser and J. Adolph in "Molecular Sieves", eds. J. R. Katzer et al., ACS Symposium Series 40, 291 (1977).

30. B. Küspert, Dissertation, Berlin, FRG (1970).

31. F. Schmidt, W. Gunsser, and A. Knappwost, Z. Naturforsch. 30a, 1627 (1975).

32. F. Schmidt, Dissertation, Universität Hamburg, FRG (1973); F. Schmidt, W. Gunsser, and A. Knappwost, Ber. Bunsenges. Phys. Chem. 77, 1022 (1973).

33. Th. Bein, F. Schmidt, and P.A. Jacobs, Proc. of the Intern. Symposium on Zeolites, Portorose, Elsevier, accepted (1984).

34. G. Broden, G. Gafner, and H. P. Bonzel, Surface Sci. 84, 295 (1979).

35. J. K. Norskov, S. Holloway, and N. D. Lang, Surface Sci. 137, 65 (1984).

36. F. Schmidt, Th. Bein, U. Ohlerich, and P. A. Jacobs in Proc. of the Sixth Intern. Zeolite Conf., eds. D. Olson et al., Butterworths, p. 151 (1984).

37. P. A.Redhead, Trans. Faraday Soc. 57, 641 (1961).

38. F. Schmidt, U. Stapel, and H. Walther, Ber. Bunsenges. Phys.Chem. 88, 310 (1984).

39. F. Schmidt and J. Adolph, Proc. Int. Conf. Mössbauer Spectroscopy, Jaipur, India, p. 454 (1981).

40. P. P. Vasihnava and P. A. Montano, J. Phys. Chem. Solids 43, No. 9, 808 (1982).

41. M. Boudart, A. Delbouille, J. A. Dumesic, S. Khammouma, and H. Topsoe, J. Catal. 37, 486 (1975).

42. T. Topsoe, J. A. Dumesic, and M. Boudart, J. Catal. 28, 477 (1973).

43. P. A. Montano, P. H. Barrett, and Z. Shanfield, Solid State Comm. 15, 1675 (1974).

44. K. Levin, A. Liebsch, K. H. Bennemann, Phys. Rev. B7, 3066 (1973).

45. K. H. Johnson, CRC Crit. Rev. in Solid State and Mat. Sci. (1978).

46. P. A. Montano, P. H. Barrett, and Z. Shanfield, J. Chem. Phys. 64, No. 7, 2896 (1976).

47. R. L. Whetten, D. M. Cox, D. J. Trevor, A. Kaldor, Surface Sci. 156, 8 (1985).

APPLICATION OF MÖSSBAUER EMISSION SPECTROSCOPY TO THE STUDY OF COBALT-MOLYBDENUM HYDRODESULFURIZATION CATALYSTS

Steen Mørup

Laboratory of Applied Physics II
Technical University of Denmark
DK-2800 Lyngby, Denmark

Bjerne S. Clausen and Henrik Topsøe

Haldor Topsøe Research Laboratories
DK-2800 Lyngby, Denmark

INTRODUCTION

The need for reducing sulfur pollution has made hydrodesulfurization an important catalytic process for removing sulfur from oil fractions. Hydrodesulfurization, HDS, is commonly carried out by passing oil, together with hydrogen, over a sulfided catalyst consisting of cobalt and molybdenum supported on a high surface area alumina. In this way the sulfur is removed as H_2S. For many years the structure of HDS catalysts has been the subject of extensive research efforts. Nevertheless, until recently, the nature of the catalytically active phase has been a subject of great controversy [1-5]. The first investigation of HDS catalysts using Mössbauer spectroscopy was published in 1975 [6]. In this study part of the cobalt atoms in the catalysts were replaced by iron-57 atoms in order to allow absorption spectra to be obtained. In subsequent investigations the catalysts have been doped with cobalt-57 and used as sources in Mössbauer emission studies [7-14]. Catalysts in which all the cobalt has been replaced by iron have also been studied [9,12].

In this paper we give a short review of the most important results that have been obtained by Mössbauer emission spectroscopy studies on cobalt-57 doped HDS catalysts.

PREPARATION OF HDS CATALYSTS

Hydrodesulfurization catalysts are usually prepared on a support of an alumina such as γ-Al_2O_3 or η-Al_2O_3 with a high surface area. The alumina is impregnated with cobalt and molybdenum, for example, in an ammoniacal solution of cobalt nitrate and ammonium heptamolybdate. After impregnation, the catalysts are dried and subsequently calcined in air at about 700 to 800K. The catalysts are then activated by sulfiding at about 600 to 700K in a gas mixture of hydrogen and about two percent H_2S. Typical catalysts contain about 2 weight percent cobalt and about 8 weight percent molybdenum.

MÖSSBAUER SPECTROSCOPY

Because the catalytically active metal atoms in cobalt-molybdenum/Al_2O_3 HDS catalysts contain no Mössbauer isotopes, it is not possible to obtain normal Mössbauer absorption spectra of the samples. Therefore, in most of the investigations, the catalysts have been doped with cobalt-57 during the preparation and used as sources in the Mössbauer spectroscopy measurements. It must be realized that the gamma radiation emitted from the source only gives direct information about iron atoms produced by decay of the cobalt-57, whereas the valence and spin states of the parent cobalt atoms cannot in general be elucidated. However, the Mössbauer spectra give information about the environments of the iron atoms which are expected to be identical to those of the parent cobalt atoms [9]. Therefore, Mössbauer emission spectroscopy gives useful information about the structure of the cobalt containing phases in the catalysts.

In studies of catalysts it is of great importance that the measurements are carried out with the catalysts exposed to a desired gas or gas mixture, i.e., the measurements are carried out in situ. For the investigations of cobalt-molybdenum/Al_2O_3 catalysts, a simple in situ cell like the one shown in Figure 1 has proven very convenient for studies at atmospheric pressure at room and elevated temperatures [10,15]. The cell can be made of either

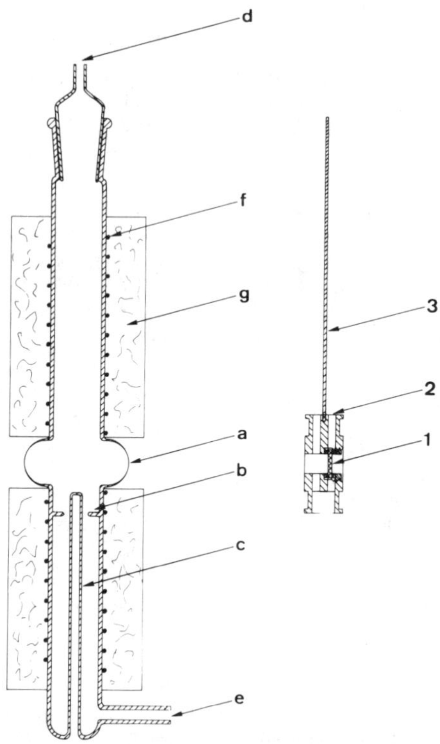

Figure 1. In situ cell for Mössbauer emission spectroscopy studies. (a) Window for transmission of γ-rays; (b) basis for sample holder; (c) tube for temperature measurement; (d) inlet of gas; (e) outlet of gas; (f) heating coil; (g) insulating material; (1) catalyst; (2) channels for gas flow; (3) rod for removal of the absorber holder. Reproduced from [10] with permission.

quartz or Pyrex glass with windows thin enough to allow transmission of the γ-rays.

During the Mössbauer studies, the catalysts were fixed in the in situ cell and the spectra were recorded on a constant acceleration Mössbauer spectrometer with a moving absorber of $K_4(Fe(CN)_6)\cdot 3H_2O$ enriched with iron-57. Zero velocity is defined as the centroid of the room temperature spectrum of a source of cobalt-57 in metallic iron. Positive velocity corresponds to the absorber moving away from the source.

COBALT-CONTAINING PHASES IN HDS CATALYSTS

A large number of different proposals have been made for the structure of HDS catalysts [1-5]. The reason for the controversies was mainly the lack of suitable experimental techniques for structural studies of highly dispersed catalysts. For example, x-ray diffraction, XRD, studies of typical HDS catalysts only reveal the presence of the alumina support. Normally the presence of cobalt and molybdenum containing phases cannot be detected by this technique. This shows that the catalytically active atoms are present in amorphous structures, in extremely small crystalline particles, or possibly dissolved in the alumina.

Mössbauer spectroscopy is, however, a very suitable technique for studies of both amorphous and microcrystalline materials and of impurity atoms in crystalline materials [15,16]. Therefore, it is an ideal technique

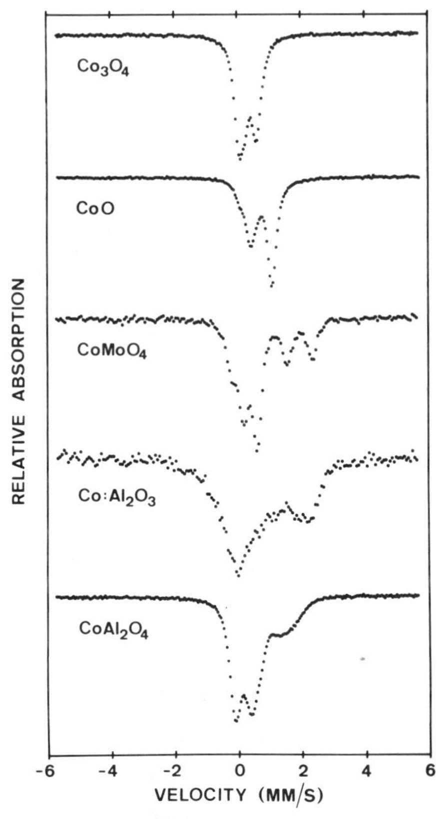

Figure 2. Mössbauer emission spectra of oxidic model compounds obtained at room temperature. Reproduced from [10] with permission.

for studies of highly dispersed catalysts. Because Mössbauer spectroscopy is mainly sensitive to the nearest neighbor atoms of the Mössbauer nucleus, the spectrum of atoms in extremely small microcrystals may be quite similar to those atoms in large crystals.

In our studies of HDS catalysts, we have therefore, as a starting point, used Mössbauer spectroscopy as a fingerprint technique in order to identify the cobalt containing phases present [8,10,13]. Thus, we have obtained spectra of a large number of compounds, which might be present in the catalyst, and we have compared the spectra of these model compounds with the spectra of the catalyst in the calcined and the sulfided states. Some of the spectra of the model compounds are shown in Figures 2 and 3. It is quite obvious that the spectra of both the oxide and the sulfide like compounds are significantly different, and therefore the fingerprint technique was found to be a very useful method.

Typical spectra of a catalyst in the calcined and in the sulfided states are shown in Figure 4. In both states of the catalyst, the spectra contain more than one principal component. This suggests that the cobalt atoms are present in more than one type of environment. Furthermore, the rather broad lines in the spectra indicate, in accordance with the XRD results, that the

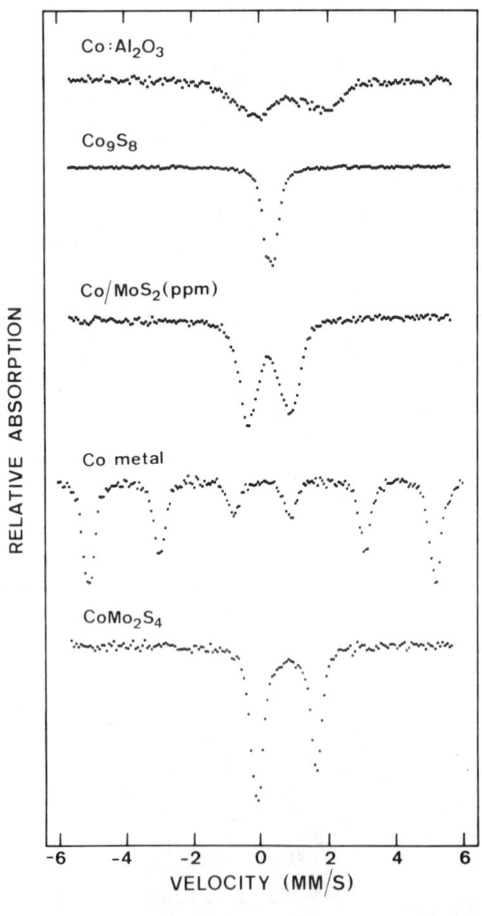

Figure 3. Mössbauer spectra of sulfided and reduced model compound obtained at room temperature. Adapted from [12].

cobalt atoms are not present in well-crystallized phases. The spectrum of the catalyst in the calcined state (Figure 4a) is quite different from the spectra of the crystalline cobalt oxides, $CoAl_2O_4$ and $CoMoO_4$. However, it has many similarities to the spectrum of cobalt diffused into the alumina, $Co:Al_2O_3$. A detailed analysis of such spectra shows that cobalt atoms in alumina may decay to both iron(II) and iron(III) in different tetrahedral or octahedral sites, and the very broad lines presumably indicate that cobalt may be associated with different types of defects in the alumina lattice or may be located on different surface sites [13].

In the sulfided state of the catalyst there seems to be two major components in the spectra, Q_1 and Q_2, with different isomer shifts and quadrupole splittings, as shown in Figure 4b. The component, Q_2, does not resemble any of the sulfide like model compounds, but is quite similar to the iron(II) component in the spectrum of cobalt in alumina. Therefore, this component seems to be due to atoms which have not been sulfided [10]. The second component, Q_1, has an isomer shift at 300K of about 0.33mm/s and a quadrupole splitting of about 1.1mm/s. These parameters suggest that the iron atoms are present in the trivalent state. Studies of the temperature dependence of the spectrum shows that the quadrupole splitting decreases with increasing temperature and thus the results can only be explained by iron(III) in the low-spin state [7,10]. Iron atoms in oxides are generally in the high spin state, but in sulfides, iron in the low-spin state is often observed.

A comparison with the spectra of the sulfide model compounds, shown in Figure 3, illustrates that the Q_1 component is similar to the spectrum of cobalt in MoS_2, but is significantly different from those of the other sulfides [7,10]. A detailed investigation has shown that the temperature dependencies of the quadrupole splitting and the isomer shift of cobalt in MoS_2 are essentially identical to those of the Q_1 component in the spectrum of the catalyst. The absolute value of the quadrupole splitting, however, seems to depend on the cobalt concentration [10,12]. The Debye temperatures are also found to be quite similar, and significantly lower than those of the other sulfides investigated [10]. It has therefore been concluded that the Q_1 component is due to cobalt atoms associated with a phase with a structure similar to that of MoS_2, the so-called Co-Mo-S phase [10]. This has recently been confirmed by extended x-ray absorption fine structure EXAFS, measurements [17,18].

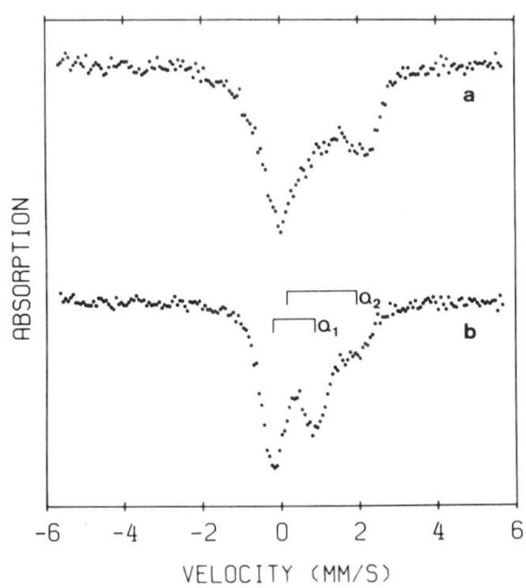

Figure 4. Mössbauer spectra of a HDS catalyst obtained at room temperature after calcination (a) and after sulfidation (b). Adapted from Reference [12].

Because cobalt in Al_2O_3 is not catalytically active [11], it seems likely that this Co-Mo-S phase is responsible for the catalytic activity. This is further elucidated by the studies of catalysts of various cobalt to molybdenum ratios, discussed in the next section.

INFLUENCE OF THE COBALT LOADING

The activity of cobalt-molybdenum/Al_2O_3 HDS catalysts depends strongly on the cobalt/molybdenum ratio. Catalysts without cobalt exhibit a very low HDS activity, but the activity increases drastically when small amounts of cobalt are added. The maximum activity is usually achieved at cobalt/-molybdenum atomic ratios of the order of 0.5 to 1.0. At higher cobalt loadings the activity decreases [1-5]. In order to determine whether this independence of the activity on the cobalt loading is related to structural properties in the catalysts, we have obtained Mössbauer spectra of a number of catalysts with a fixed molybdenum loading but different cobalt loadings [11]. Room temperature Mössbauer spectra of the resulting catalysts in the calcined and the sulfided states are shown in Figures 5 and 6, respectively.

In the sulfided state as illustrated in Figure 6, the spectra of the catalysts reveal, in addition to the Co-Mo-S phase and Co:Al_2O_3, the presence of Co_9S_8 when the cobalt to molybdenum atomic ratio is higher than about 0.4. The amount of cobalt in Co-Mo-S, Co:Al_2O_3, and Co_9S_8, estimated from the Mössbauer spectra are shown in Figure 7 as a function of the cobalt to molybdenum ratio. For a comparison, the catalytic activity of the samples is shown in Figure 8. From this figure and in particular from Figure 9, it is seen that the activity is related to the amount of cobalt present in the Co-Mo-S phase. The linear relation between the amount of cobalt in Co-Mo-S and the catalytic activity, illustrated in Figure 9, strongly suggests that the

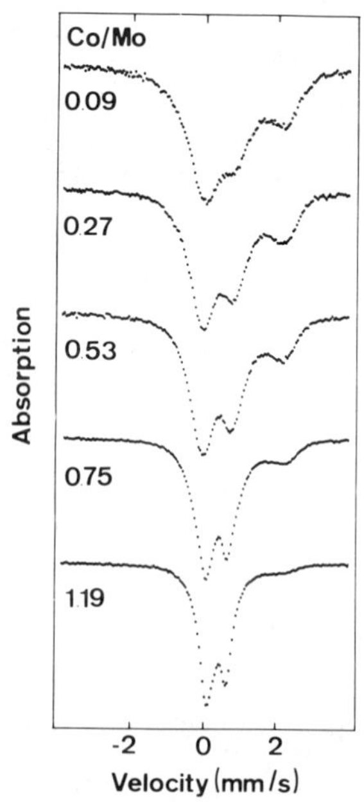

Figure 5. Room temperature Mössbauer emission spectra of calcined cobalt-molybdenum/Al_2O_3 catalysts with the different cobalt to molybdenum ratios indicated in the figure. Adapted from [13].

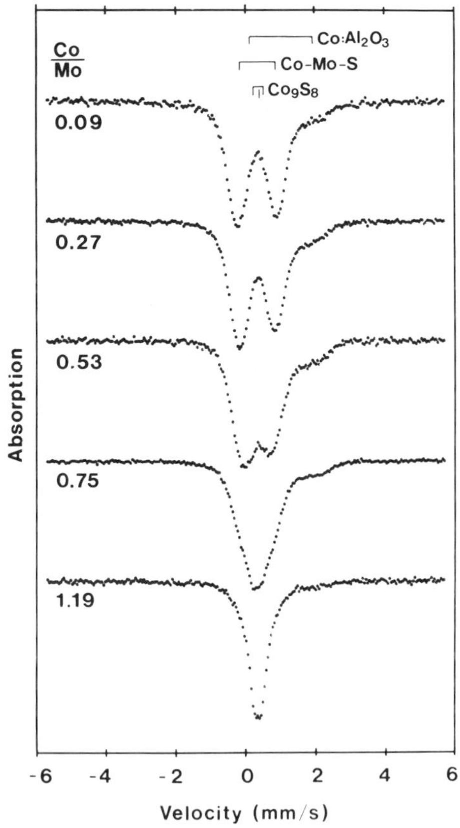

Figure 6. In situ Mössbauer spectra obtained at 300K of sulfided catalysts with different cobalt to molybdenum ratios. The bar diagrams in dicate the line positions of the components of the three cobalt-containing phases. Re produced with permission from reference [11].

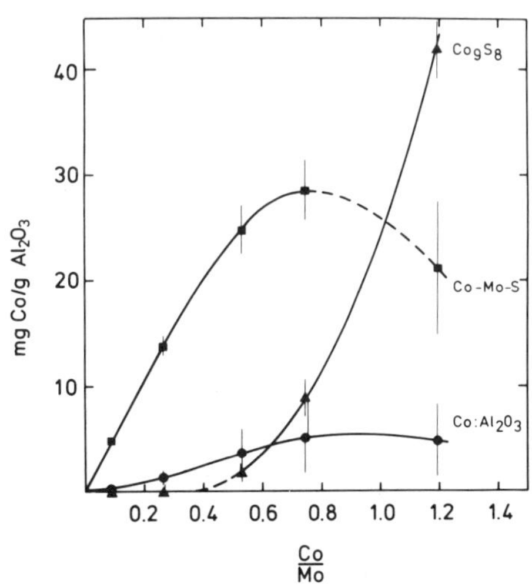

Figure 7. Distribution of cobalt in different phases in the sulfided catalysts as a function of the cobalt to molybdenum ratio. Adapted from [11].

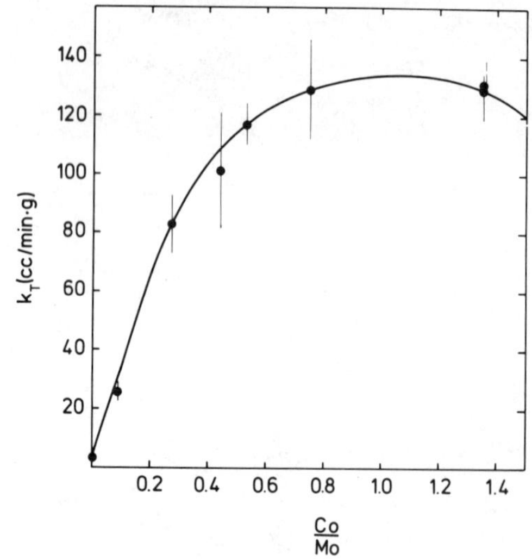

Figure 8. Catalytic activity of cobalt-molybdenum/Al_2O_3 HDS catalysts as a function of the cobalt to molybdenum ratio. Adapted from [11].

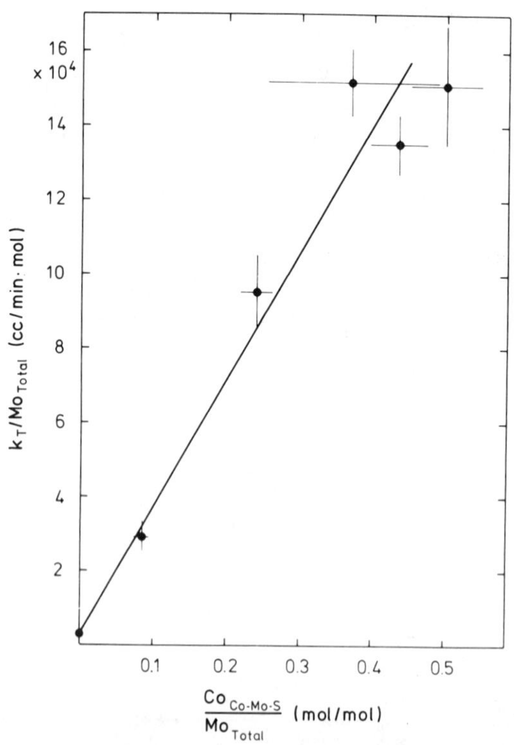

Figure 9. Activity parameter plotted as a function of the amount of cobalt in the Co-Mo-S phase. Both quantities are normalized to unit molybdenum content. Reproduced with permission from [11].

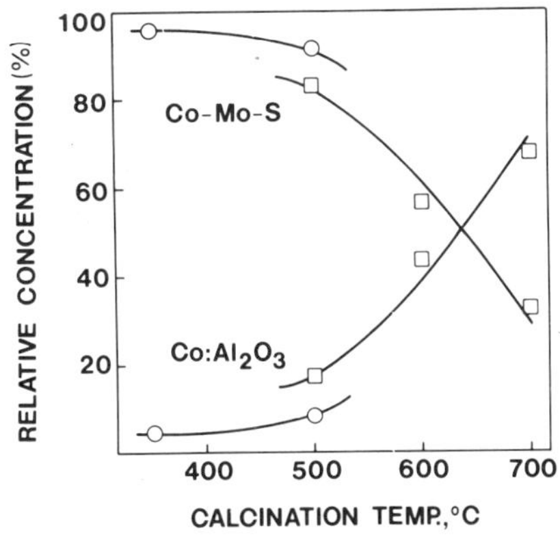

Figure 10. The effect of calcination temperature on the phase distribution of cobalt in sulfided catalysts. Circles and squares refer to catalysts with a cobalt to molybdenum atomic ratio of 0.27 to 0.50, respectively. Reproduced with permission from [21]

cobalt sites in this structure are responsible for the promotion of the HDS activity [11,12].

In the calcined state of the catalysts, the chemical state of cobalt also depends on the cobalt loading (Figure 5). Specifically, the spectra of the catalysts with high cobalt loadings show the presence of Co_3O_4. It is interesting that the amount of this phase as a function of the cobalt loading was found to follow the amount of Co_9S_8 in the sulfided state [13]. These results show that Co_3O_4 is transformed to Co_9S_8 upon sulfidation. Because Co_9S_8 has a very low catalytic activity compared to Co-Mo-S [19] one can conclude that it is important to prepare the calcined catalysts in such a way that the formation of Co_3O_4 is suppressed. A more detailed comparison of the spectra of the calcined and sulfided catalysts suggests that the precursor of the catalytically active Co-Mo-S phase is cobalt in octahedral sites in the alumina, possibly associated with molybdenum atoms on the surface of the alumina [20], whereas cobalt in tetrahedral sites in the alumina does not seem to be transformed to Co-Mo-S after typical sulfidation [13].

INFLUENCE OF CALCINATION AND SULFIDATION TEMPERATURES

The calcination temperature is known to have a significant influence on the catalytic activity of HDS catalysts. Specifically, a decrease in the activity has been observed for catalysts calcined at temperatures above 850K [21].

Mössbauer studies of catalysts calcined at different temperatures show that this decrease in activity is related to structural changes in the calcined catalysts [13,21]. The spectra of catalysts calcined at 1000K are similar to those of $CoAl_2O_4$. Thus, during the high-temperature calcination, the amount of cobalt atoms in tetrahedral sites has increased at the expense of cobalt in octahedral sites. The spectra obtained after subsequent sulfidation showed a corresponding decrease in the amount of cobalt in the Co-Mo-S phase. The amounts of cobalt in Co-Mo-S and in the oxidic phase in catalysts calcined at different temperatures and sulfided at 675K are shown in Figure 10. This study illustrates that cobalt in octahedral sites in the alumina is the most important precursor phase of the Co-Mo-S structure [21].

In another study the influence of the sulfidation temperature has been investigated [22]. Catalysts with cobalt to molybdenum ratios of 0.09 and 0.27 were calcined at 773K and subsequently sulfided at different temperatures. Mössbauer studies of these sulfided samples show that there is a tendency toward an increase in the amount of Co_9S_8 at the expense of cobalt in Co-Mo-S with increasing sulfidation temperature. Also after high temperature sulfiding the HDS activity is found to be correlated with the amount of cobalt in Co-Mo-S. It is interesting to note that the Co-Mo-S structure was found to be stable at temperatures up to about 1175K [22].

STRUCTURE OF THE Co-Mo-S PHASE

MoS_2 crystallizes in a layer structure consisting of S-Mo-S slabs made up of a molybdenum layer having a close-packed sulfur layer on each side. Each of the molybdenum atoms is surrounded by six sulfur atoms in a trigonal prismatic coordination. The S-Mo-S slabs are held together by relatively weak van der Waals forces. Figure 11 shows the MoS_2 structure and some of the possible locations of cobalt, corresponding to cobalt intercalated in the bulk, (a); cobalt "pseudo-intercalated" at edges, (b); cobalt substituting for molybdenum in the interior, (c); cobalt substituting for molybdenum at edges, (d); and cobalt at basal planes, (e). Several results suggest, as discussed by Topsøe et al. [10], that the intercalation positions (a) or (b) are not the sites of the cobalt atoms. For example, the Co-Mo-S phase is observed in catalysts for which infrared spectroscopy results show that molybdenum is present, mainly as single S-Mo-S slabs [23,24]. In such Co-Mo-S structures intercalation sites are not present because at least two S-Mo-S slabs must be stacked on top of each other in order to form the van der Waals gaps. Therefore, sites (c), (d), or (e) appear to be more likely sites for cobalt in the Co-Mo-S structure.

The Co-Mo-S structure seems to be predominantly a two-dimensional "single slab"-like structure in alumina-supported catalysts after typical sulfiding procedures. In contrast, three-dimensional "multiple-slab" Co-Mo-S structures are observed in, for example, unsupported cobalt-molybdenum catalysts and in alumina-supported catalysts after severe sulfiding [24]. The locations of the cobalt atoms in the single-slab and multiple-slab structures seem to be similar because the values of the isomer shift, the temperature dependence of the quadrupole splitting, and the Debye temperature are essentially the same in both cases [10]. Although the temperature dependencies of the quadrupole splitting seem to be the same for all the

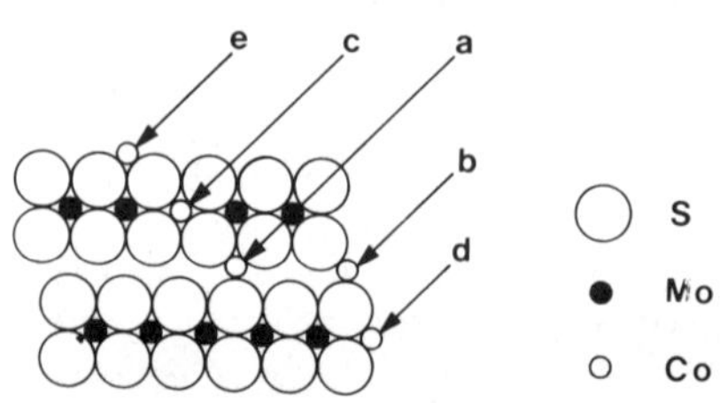

Figure 11. Possible locations for cobalt in MoS_2 (a) intercalated; (b) "pseudo-intercalated"; (c) substituted for molybdenum in the slab; (d) substitutional sites at the surface; (e) at basal plane. Reproduced from [12].

samples studied, the absolute value of the quadrupole splitting decreases slightly with increasing cobalt loadings [12]. This effect is also seen from a comparison of the quadrupole splittings of Co/MoS_2 and typical catalysts [10].

The electric field gradient at the nucleus, which determines the size of the quadrupole splitting, is in general given by the sum of two contributions, i.e. the lattice contribution, arising from the charges of neighboring ions, and the valence contributions, originating from the valence electrons of the Mössbauer atom itself. The lattice contribution is essentially independent of temperature, whereas the valence contribution decreases with increasing temperature because of the temperature dependence of the population of the electronic states. The similar temperature dependencies of the quadrupole splittings of the Co/MoS_2 sample and typical catalysts suggest that the Mössbauer atoms are present in sites with similar crystal field splittings. The temperature-independent difference in the quadrupole splittings of the two samples suggest that the lattice contributions are slightly different. At low cobalt loadings, the lattice contribution is expected to be determined by molybdenum and sulfur neighboring atoms. However, when the cobalt loading increases, the lattice contribution may also contain contributions from cobalt atoms in the neighborhood of the Mössbauer atoms. This may explain the gradual decrease in the quadrupole splitting with increasing cobalt loading [12]. This result shows that the Co-Mo-S phase is not a well-defined stoichiometric compound but rather represents a phase which may have a variable composition.

Recent investigations with high resolution electron microscopy, HREM, and analytical microscopy, AEM, electron have shown that the most likely cobalt site is the edge position (d) [25].

POSSIBLE HDS REACTION MECHANISM

The mechanism for the hydrodesulfurization reaction is generally believed to require one or more anion vacancies, where the sulfur-containing molecule can be adsorbed. In MoS_2 such an anion vacancy can be created according to the reaction

$$2Mo^{4+} + S^{2} + H_2 \rightarrow 2Mo^{3+} + \square + H_2S$$

resulting in the formation of molybdenum(III). In cobalt-molybdenum catalysts the formation of anion vacancies may also be associated with the presence of cobalt(III).

Mössbauer emission studies of supported cobalt-molybdenum catalysts [7] have indicated that the cobalt atoms in the Co-Mo-S phase are sensitive to changes in the gaseous environment. It was observed that alternating exposure to H_2/H_2S and H_2 gave rise to a reversible reduction-oxidation of the iron atoms produced by the decay of cobalt-57 in the Co-Mo-S phase. This behavior can be expressed as

$$2Fe^{3+} + S^{2-} + H_2 \rightleftarrows 2Fe^{2+} + \square + H_2S.$$

Figure 12 shows the room temperature spectra of two catalysts with cobalt/molybdenum ratios of 0.1 and 0.4 [26]. The spectra of the catalysts obtained after sulfidation at 725K show, in accordance with Figure 6, the presence of a iron(III) quadrupole doublet due to iron-57 atoms, produced by the decay of cobalt-57 in the Co-Mo-S phase, and a iron(II) quadrupole doublet due to cobalt atoms located in the alumina support. The spectra obtained after reduction in hydrogen at 725K show that a large fraction of the iron(III) atoms has been transformed to iron(II) [26]. This effect is

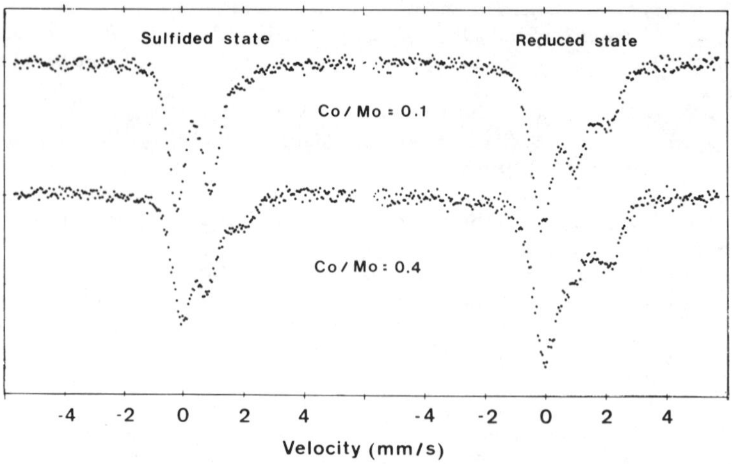

Figure 12. Mössbauer emission spectra of cobalt-molybdenum/Al_2O_3 catalysts in the sulfided and reduced state. Adapted from [26].

most pronounced for the catalysts with the high cobalt loading. Thus, the fraction of iron(II) atoms in the reduced state of the catalysts increases with increasing amounts of cobalt in the Co-Mo-S phase.

After exposure of the catalysts to hydrogen at 725K the average quadrupole splitting of the iron(III) component was found to increase. As discussed above, the quadrupole splitting of iron(III) in Co-Mo-S increases with decreasing cobalt concentration. Therefore, this result suggests that the promoter atoms present in regions with a high local concentration of cobalt are more easily reduced than the atoms present in regions with lower concentrations of cobalt. This interpretation is in accordance with the observation that the high-loading catalysts are more easily reduced than the low-loading catalysts.

The iron(II) component, which appears in the spectra after reduction in hydrogen, has Mössbauer parameters which are essentially identical for the two catalysts. The Mössbauer parameters of the iron(II) component are significantly different from those of the stoichiometric cobalt- and cobalt-molybdenum sulfides discussed above. Specifically, the isomer shift is larger than those typical of iron sulfides. This result indicates a decrease in the number of neighboring sulfur atoms upon exposure to hydrogen, i.e., the sulfur vacancies seem to be created during reduction. Spectra obtained with a catalyst alternatingly exposed to hydrogen/thiophene and hydrogen gave results similar to those shown in Figure 12 [7].

Although the Mössbauer results give information only on the valence state of iron atoms produced by the decay of cobalt-57 atoms in the catalysts, the chemical similarity of iron and cobalt, as well as the similarity of iron-molybdenum/Al_2O_3 and cobalt-molybdenum/Al_2O_3 catalyst [9], suggests that such valence changes also take place for the cobalt atoms. These Mössbauer results thus indicate that the promoter atoms may participate directly in the catalytic reaction.

In view of the present results on the state of cobalt and those obtained by other investigators on molybdenum, the following reaction sequence may take place. HSR is here assumed to represent the sulfur-containing compound

$$Mo^{4+} + Co^{2+} + \square + S^{2-} + HSR \rightarrow Mo^{3+} + Co^{3+} + SH^{-} + SR^{-}$$

$$Mo^{3+} + Co^{3+} + SH^{-} + SR^{-} + H_2 \rightarrow Mo^{3+} + Co^{3+} + 2SH^{-} + HR\uparrow$$

$$Mo^{3+} + Co^{3+} + 2SH^{-} \rightarrow Mo^{4+} + Co^{2+} + \square + S^{2-} + H_2S\uparrow$$

The results, which show that the HDS activity is proportional to the amount of cobalt in Co-Mo-S, are of course consistent with the active sites being associated with the cobalt atoms in Co-Mo-S.

CONCLUSIONS

The studies of HDS catalysts have illustrated that Mössbauer spectroscopy is a powerful technique for studies of poorly crystallized systems. The in situ studies have led to the identification of the catalytically active structure which was not known earlier. It has also been found that the catalytic activity is proportional to the amount of cobalt in the active Co-Mo-S phase. Thus studies of this parameter as a function of the preparation conditions are useful for determination of the optimum preparation conditions.

REFERENCES

1. F. E. Massoth, Adv. Catal. 27, 265 (1978).

2. B. Delmon, in "Proceedings of the Climax Third Intern. Conf. on Chemistry and Uses of Molybdenum", H. F. Barry and P. C. H. Mitchell, eds., Climax Molybdenum Co., Ann Arbor, Michigan, p. 73, (1979).

3. P. Grange, Catal. Rev.-Sci. Eng. 21, 135 (1980).

4. B. C. Gates, J. R. Katzer and G. C. A. Schuit, in "Chemistry of Catalytic Processes", McGraw-Hill, New York, Chapt. 5 (1979).

5. H. Topsøe, in "Surface Properties and Catalysis by Non-Metals: Oxides, Sulfides, and Other Transition Metal Compounds", p. 326, Reidel, Dordrecht (1983).

6. H. Topsøe and S. Mørup, in "Proceedings Intern. Conf. on Mössbauer Spectroscopy", A. Z. Hrynkiewicz and J. A. Sawicki, eds., vol. 1 p. 305, Poland, Cracow (1975).

7. B. S. Clausen, S. Mørup, H. Topsøe, and R. Candia, J. Physique Colloq. 37, C6-249 (1976).

8. H. Topsøe, B. S. Clausen, N. Burriesci, R.Candia and S. Mørup, in "Preparation of Catalysts II", B. Delmon, P. Grange, P. Jacobs, and G. Poncelet, eds., p. 479, Elsevier, Amsterdam/New York (1979).

9. S. Mørup, B. S. Clausen, and H. Topsøe, J. Physique Colloq. 40, C2-88 (1979).

10. H. Topsøe, B. S. Clausen, R. Candia, C. Wivel, and S. Mørup, J. Catal. 68, 433 (1981).

11. C. Wivel, R. Candia, B. S. Clausen, S. Mørup, and H. Topsøe, J. Catal. 68, 453 (1981).

12. H. Topsøe, B. S. Clausen, R. Candia, C. Wivel, and S. Mørup, Bull. Soc. Chim. Belg. 90, 1189 (1981).

13. C. Wivel, B. S. Clausen, R. Candia, S. Mørup, and H. Topsøe, J. Catal. 87, 497 (1984).

14. M. Breysse, B. A. Bennett, D. Chadwick, and M. Vrinat, Bull. Soc. Chim. Belg. 90, 1271 (1981).

15. H. Topsøe, J. A. Dumesic, and S. Mørup, in "Applications of Mössbauer Spectroscopy", R. L. Cohen, ed., vol. II, p. 55, Academic Press, New York, (1980).

16. S. Mørup, J. A. Dumesic, and H. Topsøe, in "Applications of Mössbauer Spectroscopy", R. L. Cohen, ed., vol. II, p. 1, Academic Press, New York (1980).

17. B. S. Clausen, H. Topsøe, R. Candia, J. Villadsen, B. Lengeler, J. Als-Nielsen, and F. Christensen, J. Phys. Chem. 85, 3868 (1981).

18. B. S. Clausen, B. Lengeler, R. Candia, J. Als-Nielsen, and H. Topsøe, Bull. Soc. Chim. Belg. 12, 1249 (1981).

19. R. Candia, B. S. Clausen, J. Bartholdy, N.-Y. Topsøe, B. Lengeler, and H. Topsøe, in "Proceedings 8th Intern. Congr. Catal.", Berlin, Verlag Chemie, Weinheim, vol. II, p. 375 (1984).

20. N.-Y. Topsøe and H. Topsøe, J. Catal. 77, 293 (1982).

21. R. Candia, N.-Y. Topsøe, B. S. Clausen, C. Wivel, R. Nevald, S. Mørup, and H. Topsøe, in "Proceedings of the Fourth Climax Intern. Conf. on Chemistry and Uses of Molybdenum", H. F. Barry and P. C. H. Mitchell, eds., p. 374, Climax Molybdenum Co., Ann Arbor, Michigan (1983).

22. R. Candia, H. Topsøe, and B. S. Clausen, in "Proceedings 9th Iberoamerican Symposium on Catalysis", Lisbon, Portugal, p. 211 (1984).

23. N.-Y. Topsøe, J. Catal. 64, 235 (1980).

24. N.-Y. Topsøe and H. Topsøe, Bull. Soc. Chim. Belg. 12, 1311 (1981).

25. O. Sørensen, B. S. Clausen, R. Candia, and H. Topsøe, Appl. Catal. 13, 363 (1985).

26. P. H. Christensen, S. Mørup, B. S. Clausen, and H. Topsøe, in "Proceedings of the Intern. Conf. on the Application of the Mössbauer Effect", Alma Ata, Gordon and Breach, 1983, p. 1747, in press.

APPLICATIONS OF THE MÖSSBAUER EFFECT IN ELECTROLYSIS

A. Vèrtes and I. Czakò-Nagy
Laboratory of Nuclear Chemistry
L. Eötvös University
Budapest, Hungary

M. L. Varsànyi

Department of Physical Chemistry and Radiology
L. Eötvös University
Budapest, Hungary

During electrolysis ions are removed from the electrolytic solution as the electrolytic processes occur at the electrode-electrolyte interfaces. The processes are heterogeneous reactions accompanied by the transfer of electric charge across the interface. As a consequence of these generally complicated multistep cathodic and anodic reactions, neutral atoms or molecules are formed on the electrode surfaces. Recently Mössbauer spectroscopy has been successfully applied to the study of these products not only under ex situ, but also under in situ conditions.

In order to obtain a better understanding of the applicability of this technique, we shall describe in this chapter a few specific examples in detail. Other examples will be reviewed only briefly.

I. MÖSSBAUER STUDIES OF ELECTRODEPOSITS

The importance of electrodeposits is well known for the protection of metals from corrosion and for changing the mechanical, magnetic, or electrical properties properties of metal surfaces.

Although surface electrodeposits are extensively used commercially, their chemical structures are often unknown. Mössbauer spectroscopy is an excellent technique for studying the chemical composition, phase, and magnetic properties of the deposits if they contain one or more Mössbauer active elements.

ELECTRODEPOSITS CONTAINING COBALT

Cobalt containing electrodeposits can be investigated by cobalt-57 emission Mössbauer spectroscopy. Cobalt-57 decays to iron-57 by electron capture and the lifetime of the Mössbauer nuclear excited state is 10^{-7} s. This lifetime is not long enough for the transformation of the chemical environment at the iron which has resulted from the nuclear decay, so the Mössbauer spectrum still remembers and reflects the chemical environment of the parent cobalt site. An example of this memory occurs for cobalt electrodeposited in gold [1]. These gold electrodeposits are important in the electronics industry in protecting substrates from corrosion. In many cases the electroplated gold is hardened by the codeposition of 0.3 to 1.6 atomic percent of cobalt. Such cobalt-hardened gold deposits have satisfactory contact resistances and exhibit serviceable wear resistance [2].

The plating baths utilized to form the gold deposits were proprietary solutions that are available and used commercially. Three different plating solutions were used. One was a mixed phosphate-citrate bath and the other two were all-citrate baths. Cobalt was added as the sulfate in one all-citrate bath and as a proprietary chelate in the other two baths. Cobalt-57 was added to 10ml of each bath as the chloride so as to achieve an activity of approximately one mCi. The deposit thickness in all cases was equivalent to 15(3)mg/cm^2 of gold and the substrate was a 15x15x0.5mm copper foil.

The iron-57 Mössbauer spectra of the gold electrodeposits recorded immediately after deposition exhibited a broad and complex shape with a linewidth of 1.3 to 1.6mm/s. The linewidths decreased dramatically with aging at room temperature indicating that many of the cobalt atoms in the initial deposit are metastable and subsequently rearrange to different crystallographic site. The spectrum obtained shortly after deposition probably represents a summation of changes that occurred while the spectrum was being recorded because it generally required 16 to 48 hours to obtain a satisfactory spectrum. The computer fit spectrum of one deposit obtained after aging at room temperature for 15 days, is shown in Figure 1.

The spectra are interpreted on the basis of the three components present in all electrodeposits studied. The main spectral component was a singlet with an isomer shift, δ, between -0.60 and -0.69mm/s and a linewidth, Γ, between 0.7 and 0.8mm/s. This line is assigned to cobalt in substitutional sites in the gold lattice because the normal value of δ for cobalt-57 in substitutional sites in gold is -0.62 to -0.63mm/s [3,4].

The second spectral component consists of a species with δ of -0.44 to -0.53;mm/s and a quadrupole splitting, ΔE_Q, of 0.91 to 1.06mm/s. On the basis of previous emission Mössbauer studies of different hydroxides, oxides, and oxyhydroxides of cobalt [5] this species was assigned to β-CoOOH (see Table I). This material, obtained in a pure state by anodic oxidation of

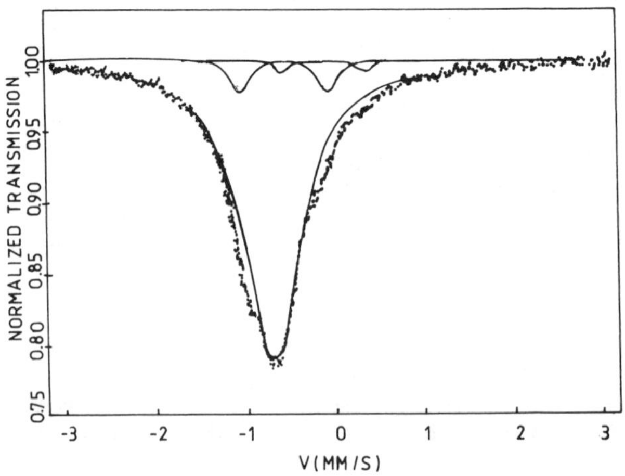

Figure 1. Room temperature Mössbauer spectrum of a specimen recorded 15 days after deposition.

Table I. Emission Mössbauer Parameters of Different Cobalt Hydroxides, Oxides and Oxyhydroxides[a]

Compound	δ	ΔE_Q	Remarks	References
α-$Co(OH)_2$	-1.34	2.32-2.80	ΔE_Q is dependent upon preparation	5
β-$Co(OH)_2$	-1.28	2.91		5
CoO	-1.21 -1.28 -0.55	 0.50	The amount of Fe(III) is a function of stoichiometry	5
Co_3O_4	-0.53 -0.53	 0.73	A site with cubic symmetry B site with trigonal symmetry	5
CoO_2	+0.15			4
β-CoOOH	-0.57	0.81		5

[a]All data in mm/s obtained at room temperature. δ is relative to α-iron.

electrochemically deposited $Co(OH)_2$, has a δ value of -0.48mm/s and a ΔE_Q of 0.81mm/s. The larger value for ΔE_Q for the species present in the electrode posit is presumably a result of the more asymmetric environment within the gold host lattice. The larger value of ΔE_Q thus would be consistent with a very small particle size for the β-CoOOH within the electrodeposit, in agreement with the observations of Okinaka and Nakahara [6].

The third component in the spectrum has δ values of -0.01 to -0.11mm/s and a wide range of ΔE_Q values from 0.60 to 1.15mm/s. This component is presumably a low-spin state of daughter iron(II) and iron(III) resulting from an environment containing the cyanide ion. It is worth noting that Okinaka, et al., [7] have reported isolating $Co_3[Co(CN)_6]_2$ and $K_3[Co(CN)_6]$ from cobalt-hardened gold electrodeposits.

No magnetic hyperfine splitting was observed in the Mössbauer spectra of the deposits taken at 78K. This absence of hyperfine splitting suggests that the cobalt in the atomic state is not sufficiently aggregated to yield a magnetic spectrum.

Cohen et al. have also studied the cobalt-hardened gold electrodeposits. They used both gold as a Mössbauer active element in absorption geometry [8] and cobalt-57 for emission Mössbauer spectroscopy [9]. The authors confirmed, in part, the results presented above [1]. They found more than 70 percent of the cobalt is substitutionally dissolved in the gold, and identified the $Co(CN)_6^{3-}$ complexes. However, these authors did not find CoOOH in their samples. Schubert has proposed the formation of CoOOH on the surface of cobalt-hardened gold electrodeposits on the basis of secondary ion mass spectroscopy analysis [10].

The presence of cobalt in electrodeposited zinc increases its corrosion resistance [11,12]. Zinc is frequently used for the electrogalvanization of steel. Thus those studies which are aimed at understanding the mechanism of the corrosion inhibition of zinc by cobalt are important. Emission Mössbauer measurements indicated that the codeposited cobalt was present as isolated atoms in the zinc matrix [13]. The cobalt content of the deposits was 0.01

to 2 percent. The electrodeposited zinc-cobalt alloys are thus supersaturated solutions of cobalt in zinc. The function of cobalt as a corrosion inhibitor results from its influence on the surface oxide film which it enters during active corrosion [14].

GUEST ATOM MÖSSBAUER SPECTROSCOPY

The emission Mössbauer spectroscopy can be used in certain cases where the material does not normally contain Mössbauer active elements. In such experiments, Mössbauer isotopes such as cobalt-57, are used as detectors of the local environment. For example, electrodeposited nickel can be investigated by the Mössbauer effect if the deposit is doped with cobalt-57. Nickel-61 is a Mössbauer active isotope but its Debye-Waller factor is too low to be useful because of the relatively high, 67.4keV, energy of the gamma rays. The chemical properties of cobalt are close to those of nickel and thus cobalt-57 emission spectra give reliable information about the structure of the host nickel. However, the concentration of cobalt-57 should not exceed 10ppm.

It is well known that nickel electrodeposits contain stresses ranging from highly compressive to highly tensile, depending upon the plating conditions and composition of the plating solution. The magnitude of the stress can be measured by the cobalt-57 emission Mössbauer spectroscopy if the crystal orientation for the samples is the same. The nickel in these deposits is ferromagnetic and consequently, the Mössbauer spectrum of the cobalt-57 shows a magnetic splitting in the nickel matrix.

The intensity ratios of the lines in a Mössbauer effect Zeeman split magnetic sextet are,

$$3(1+\cos^2\theta):4\sin^2\theta:(1+\cos^2\theta), \qquad (1)$$

where θ is the angle between the direction of magnetization and the γ-ray. Saturation magnetization of the sample causes the direction of magnetization to be the same over the entire sample. For example, if a sample is magnetized by a strong magnetic field applied perpendicular to the sample plane, and hence parallel to the γ-ray direction, θ is 0° at each iron site and the intensity ratio is 3:0:1. If the magnetization direction is perpendicular to the γ-ray, θ is 90°, and the intensity ratio is 3:4:1. In the absence of external magnetic field, a domain pattern is formed. The magnetization vector within the different domains may point in different directions so that θ may vary from domain to domain. In the case of randomly varying magnetization directions, ratio is averaged and the intensity ratio is 3:2:1. Even if only a small number of different magnetization directions are present, the intensity ratio is often close to 3:2:1. However, in some cases the direction of the domain magnetizations is not completely random and the intensity ratio is different from the random value of 3:2:1. These are the cases where the domain structure, and its change as a result of mechanical deformation can be studied by Mössbauer spectroscopy.

The formation of the domain structure is complex phenomenon whose details can be found in standard textbooks [15-17]. In summary, all the domain magnetizations point in special directions. These so-called easy directions are determined both by the magnetocrystalline anisotropy and by the mechanical stress resulting from magnetostriction. The distribution of the easy directions is determined by the requirement that the magnetostatic energy be minimized. The domain structure, and especially its size is also affected by the presence of domain walls, where the orientation of the spins deviate from the easy direction of magnetization. All these effects are coupled together in an actual sample and the domain pattern is determined by a local minimum in the sum of the magnetocrystalline, magnetoelastic,

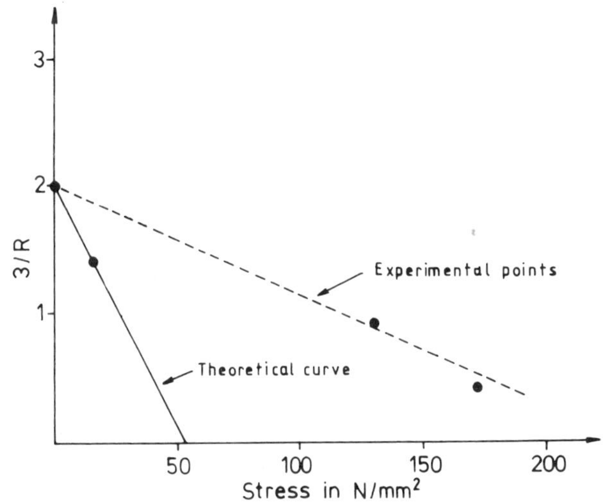

Figure 2. Comparison between the theoretical curve [19] and the experimental value for electrodeposits with a (200) preferred orientation. The effective magnetic field at the cobalt-57 nucleus was 26.7T.

elastic, magnetostatic, and domain wall energies. The relative intensities of the lines in a magnetic sextet provide information on the easy directions and their relative abundance. Therefore the theoretical determination of the easy direction of magnetization is important in the interpretation of magnetic Mössbauer spectra.

When the electrodeposition of nickel is carried out in a Watts bath at 30 to 50°C, samples are produced with (200) texture [18], i.e., the <200> direction is perpendicular to the sample plane in many of the crystallites. For the (200) nickel texture, a correlation can be derived between the intensity ratio of the Mössbauer lines of cobalt-57 and the stress of the electrodeposit [19]. This correlation is shown in Figure 2. In this figure R is the ratio $3I_2/I_1$, where I_1 and I_2 are the intensities of the first and second lines in a magnetic sextet Mössbauer spectrum. Unfortunately this technique cannot be used as a general tool for the measurement of the stress in electrodeposited nickel because the crystalline orientation perturbes the ratio of the Mössbauer effect lines and the crystalline texture is strongly affected by the temperature, pH, and current density during the electrodeposition and by the composition of the plating solution [20-22].

MÖSSBAUER SPECTROSCOPY OF TIN CONTAINING ELECTRODEPOSITS

Several tin containing electrodeposits have a high resistance to corrosion. For example a nickel-tin codeposit has an exceptionally high resistance to corrosion in various atmospheres and also in aqueous solutions [23-25]. Under similar conditions the pure metal components are often severely degraded.

For Mössbauer spectral measurements [26], several tin-nickel alloys with a range of compositions were plated from the fluoride bath described by Cuthbertson et al. [27]. The substrate was a 30x25x0.005cm copper foil. Different intermatallic compounds of nickel and tin, prepared metallurgically were used as standards in the investigation of the electrodeposited samples [26].

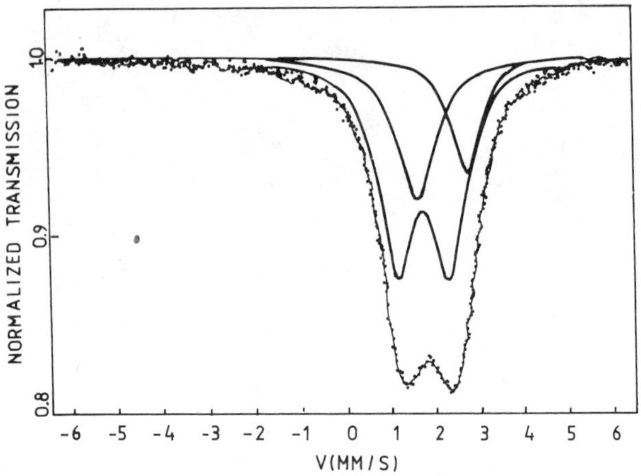

Figure 3. The room temperature Mössbauer spectrum of an electroplated tin-nickel alloy.

A typical Mössbauer spectrum of a tin-nickel alloy electrodeposit is shown in Figure 3. The spectra were interpreted in terms of three components. A doublet with an isomer shift of approximately 1.8mm/s was attributed to a nickel-tin intermetallic compound with the composition of NiSn. A singlet with an isomer shift of 2.5(6)mm/s was attributed to elemental tin and a singlet with an isomer shift of 1.5 to 1.8mm/s was attributed to a copper-tin alloy. The last singlet was also present when pure tin was electrodeposited onto copper as is illustrated in Figure 4a. Further evidence that the singlet at 1.5 to 1.8mm/s is due to a copper-tin alloy is found in the conversion electron spectrum shown in Figure 5, which also shows the absorption spectrum of the same sample. The component attributed to the copper-tin alloy is visible in the absorption spectrum but is not as apparent in the conversion electron spectrum. This indicates that the copper is not present in significant concentration near the surface of the deposit. It is well known that copper-tin alloys are formed at the interface when tin is electrodeposited on copper [28] and all the data are thus consistent with the presence of a copper-tin alloy. A significant increase in the amount of the copper-tin alloy was observed when a pure tin electrodeposited on copper was heated in air at 300°C for 3hr as shown in Figure 4b. Silver et al. have also published the Mössbauer parameters of several cast and electrodeposited nickel-tin alloys [29]. Their results agree well with those mentioned above.

Cobalt-tin electrodeposited alloys have been proposed as an alternative to chromium plating. These alloys are generally bright and coherent, with good corrosion resistance. Further, they are more ductile than an electrodeposited nickel-tin alloy, which is brittle due to the intermetallic compound NiSn. The structure of cobalt-tin alloys electrodeposited from a mildly alkaline sulfate bath have been studied [30,31] by using Mössbauer spectroscopy. The tin-119 Mössbauer spectra of alloys obtained from fresh plating baths containing 20 to 25 mole percent of cobalt, indicated that they consist of dissolved metallic tin and the cubic intermetallic compound CoSn. This compound does not appear in the phase diagram and can be obtained only by electrolysis. Upon exhaustation of the plating bath after eight depositions, the electrodeposited alloys become richer in cobalt and the formation of $\gamma' Co_3Sn_2$ is favored.

The cobalt-tin alloys were also studied by using cobalt-57 emission Mössbauer spectroscopy. The main conclusion from these studies was that in

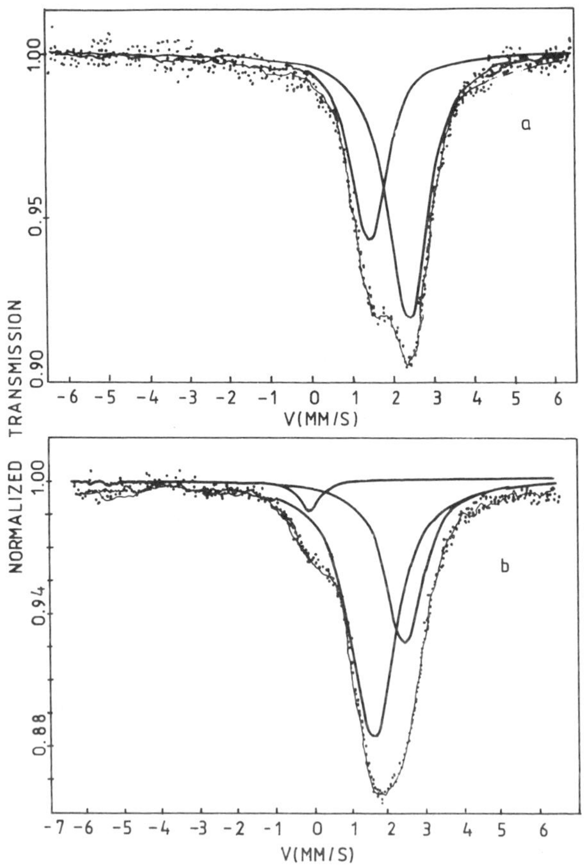

Figure 4. The Mössbauer spectra of a tin electrodeposit on a copper substrate before heating (a) and after heating at 300°C for 3hr. (b) The minor peak at zero mm/s results from a small amount of SnO_2 on the surface.

the early stages of deposition cobalt is alloyed in the form of an intermetallic compound. However, as the bath was exhausted cobalt particles were also formed as a separate phase.

Electrodeposited bronze finds many applications because of its excellent wear properties, attractive appearance, good corrosion resistance, excellent machinability, and solderability. Kudra and Galinker [32,33] suggested a non-toxic and inexpensive plating solution for bronze formation which used sodium tripolyphosphate as a complexing agent. In this plating solution, the current efficiency is 80 to 90 percent. In the absence of organic additives, the current density is less than 0.6 to 0.7A/dm^2, but in their presence, the current density can reach 1.2A/dm^2.

Bright electrodeposited copper-tin alloys containing 12 to 53 percent tin were prepared from a plating solution by using $Na_5P_3O_{10}$ as a complexing agent [34]. The metals were provided in the form of $[Cu(P_3O_{10})_2]^{8-}$ and $[Sn(P_3O_{10})_2]^{8-}$ with a copper to tin mass ratio of approximately 3.2. The structure of the resulting electrodeposited bronzes was highly dependent upon the potential applied during deposition. The Mössbauer spectra revealed that a tin(IV) compound and ε-Cu_3Sn were present in the alloys obtained in the

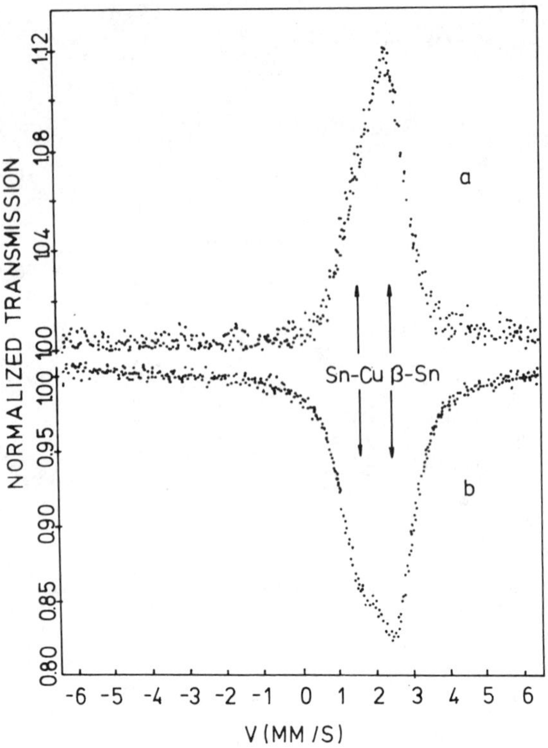

Figure 5. The conversion electron spectrum (a) and absorption spectrum (b) of a copper-tin alloy containing 86 atomic percent tin.

deposition range of -250 to -550mV versus the normal hydrogen electrode. At -770mV a tin(IV) species, a β-Cu-Sn alloy, η-Cu_6Sn, and β-tin were formed. The tin(IV) species has an isomer shift of approximately zero. However, the strong temperature dependence of its Debye-Waller factor suggests that it is not pure SnO_2 but another tin(IV) compound such as a phosphate or hydroxide or $CuSnO_3$.

INTERFACIAL PHENOMENA FOR ELECTRODEPOSITED TIN AND METAL SUBSTRATES

As was illustrated in Figures 3 and 4, an interphase can form at the interface of the electrodeposited tin layer and a copper substrate. This phenomenon can be considered as the electrochemical implantation of tin into the substrate. Recently, a systematic investigation of the electrochemical implantation of tin was carried out [35]. The authors used the eletroplating bath with the composition of 25g/dm^3 Sn-119I_4, 12g/dm^3 NaOH, 15g/dm^3 CH_3COONa and 1cm^3/dm^3 of 30 percent H_2O_2. The applied current density was 1A/dm^2 and the plating bath was kept at 75°C. The estimated rates of the electrochemical implantation were $5x10^{-4}$, 10^{-3}, $5x10^{-3}$, and $5x10^{-2}$ mg/cm^2min in nickel, iron, copper, and silver substrates respectively [35]. These values are the average rates for the first 30 seconds of the deposition. Nevertheless, in the case of the aluminum and platinum substrates no interfacial phase could be detected [36,37]. Hence, the rates of the electrochemical implantation of tin into aluminum and platinum are close to zero.

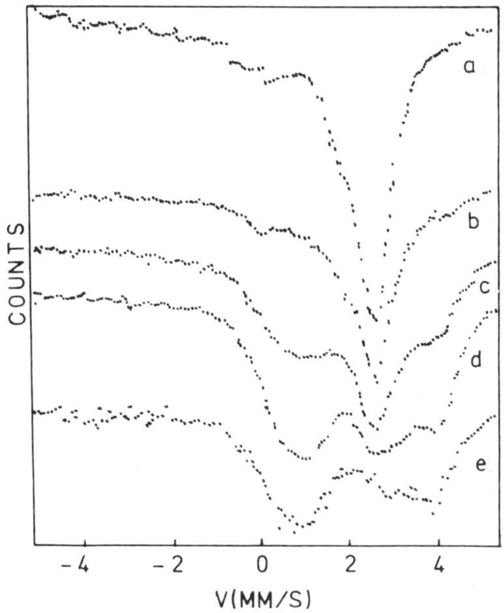

Figure 6. Mössbauer spectra of (a) Sn/Fe prior heating; (b), (c), and (d): Sn/Fe heat treated at 373K for 3h, at 423K for 7h and at 473K for 3h, respectively; and (e) $FeSn_2$. The activation energy of diffusion in the Sn/Fe interphase was 70kJ/mol.

A heat treatment of the electrodeposited samples results in an increase in the diffusion of both the electrodeposited atoms and substrate. Consequently, heat treatment increases the thickness of the interphase. Mössbauer spectroscopy is an excellent tool for the study of this phenomenon. The thickness of the interphase and its chemical composition can be followed simultaneously in the Mössbauer spectra. Some results for a tin deposit on an iron substrate [38] indicated that the interphase consisted mainly of Fe_3Sn_2 after the electrodeposition. Aging at higher temperature promoted the Fe_3Sn_2 to $FeSn_2$ transformation as is illustrated in Figure 6. The thickness, χ, of the interface changes with time according to a parabolic law,

$$\Delta\chi^2 = K\Delta t,$$

where K is proportional to the diffusion coefficient, and t is the time. When heat treatment results in a change in the diffusion coefficient of the interphase, the parabolic law can be used only during the initial period of annealing.

The thickness of the interphase can be measured by the Mössbauer spectroscopy and hence makes it possible to study the diffusion coefficient and to obtain the activation energy of diffusion [38]. The thickness of the interface, can be written as

$$\chi = \frac{d}{1 + R(S_\beta/S_a)} \frac{C_\beta}{C_a},$$

where d is the original thickness of the β-tin. S_β and S_a are the corresponding spectral areas found for β-tin and for the alloy interphase, C_β and

C_a are the concentrations of tin atoms in mol/cm³ in β-tin and in the alloy interphase, and R is the ratio of the recoil free fractions, f_a/f_β

STUDIES OF IRON-NICKEL ELECTRODEPOSITS

The use of iron-nickel alloys as a protective decorative coating is attractive because they have the same mechanical properties and brightness as nickel but are less expensive.

Iron-nickel alloys with 59.9 to 90.5 percent nickel were electrodeposited [39] onto a copper substrate from plating solutions containing $NiSO_4 \cdot 7H_2O$, $NiCl_2 \cdot 6H_2O$, H_3BO_3, $Na_2SO_4 \cdot 10H_2O$, $FeSO_4 \cdot 7H_2O$, a stabilizer, and a brightener. The depositions were carried out at a pH between 3.2 and 3.5 and at 333 to 338K with a current density of 5A/dm². The deposited layer was mechanically removed from the copper substrate for Mössbauer measurements in absorption geometry [39]. It was found that the average magnetic field of the electrodeposited alloys varies linearly with the nickel content in the studied concentration range (Figure 7) and is somewhat smaller than that found in a cast alloy of the same composition. The difference in the magnetic field was reported to be due to the absorbed hydrogen in the electrodeposited alloys.

II. METAL PASSIVATION AND THE ELECTROCHEMICAL PROCESSES ON AN ELECTRODE SURFACES

PASSIVITY OF A COBALT ELECTRODE

The passivation of a metal results from the formation of a thin, protective film on the metal surface. This film may consist of adsorbed oxygen, oxide, hydroxide, or a salt formed as a result of metal dissolution. Several studies of the surface film in nonaqueous solutions indicate that the passivation takes place by formation of metal oxide or hydroxide, because permanent passivity in non-aqueous media occurs only in the presence of water [40-43].

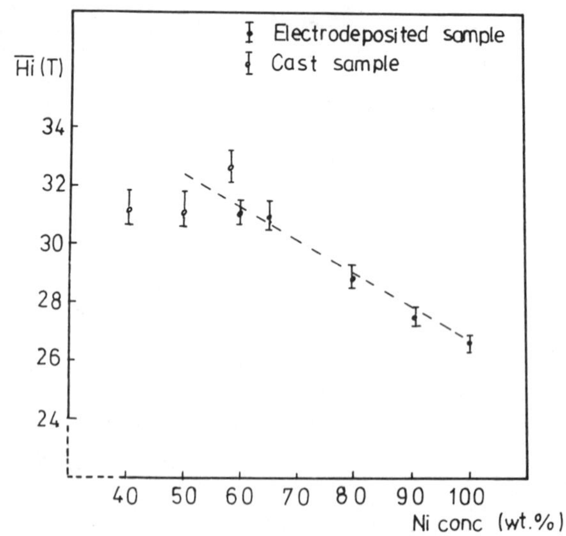

Figure 7. The dependence of the average magnetic field on the nickel concentration in some electrodeposited iron-nickel alloys.

Progress in research on passivity requires a knowledge of the chemical composition of the surface film. It is well known that electrochemical techniques provide limited information about the structural and chemical nature of the protective film. However, electron and X-ray diffraction as well as optical and microchemical methods can be used to clarify the structural nature of the film. Mössbauer spetroscopy complements these methods.

Studies of the passivity of cobalt by cobalt-57 emission Mössbauer spectroscopy have been carried out by Simmons et al. [5]. The anodic polarization of cobalt in a buffered borate solution, at a pH of 8.5, produced a classical potential vs. current dependence. The Mössbauer hyperfine parameters for specimens after polarization at -100mV were characteristic of α-$Co(OH)_2$. In contrast, the passive film obtained by polarization in the region of +200 and +500mV gave parameters characteristic of β-CoOOH.

The spectrum of the anodic film formed at +800mV is similar to that found for the surface anodically treated at +200 and +500mV, but with an additional line having an isomer shift of 0.14mm/s. This new spectral component in the transpassive film was attributed to the presence of tetravalent cobalt-57. The authors assumed that the outer layer of the transpassive film was CoO_2; a conclusion supported by further experiments. In these measurements CoO_2 was formed by anodic treatment of $Co(OH)_2$ in a borate solution (44). On the basis of the Pourbaix diagram, potentials of +600, +800, and +900mV, relative to a standard Calomel electrode, were chosen to oxidize the $Co(OH)_2$ to cobalt(IV). A doublet attributed to β-CoOOH and a singlet attributed to CoO_2 gave a satisfactory fit to the Mössbauer spectrum. A comparison of the Mössbauer hyperfine parameters for β-CoOOH and CoO_2, after polarization at +600, +800, and +900mV indicates that the anodic oxidation of β-CoOOH to CoO_2 is more complete at the higher anodic potentials. The line due to CoO_2 only appears at +900mV. The hyperfine parameters are given in Table 1.

OXIDE LAYER FORMATIONS ON IRON ELECTRODES

O'Grady carried out in situ absorption geometry Mössbauer experiments with the iron electrode [45]. The iron electrode served as the window of the electrochemical cell and was prepared by the electrode position of a 25 to 30nm thick layer of 90 percent enriched iron-57 onto a 25nm thick vacuum deposited gold film which was supported on 0.04mm thick mylar. A comparison of the measured Mössbauer parameters with those of the crystalline and amorphous iron oxides and hydroxides suggested that the passive film formed in a 0.15N boric acid-sodium borate buffer solution, with a pH of 8.4, is not highly structured but is amorphous and polymeric in nature. It probably consists of chains of iron ions bonded together by dioxy- and dihydroxy-bridging bonds. These chains are further linked together by water to form a continuous film. The temperature dependence of the Mössbauer spectra showed that drying resulted in the formation of superparamagnetic α-Fe_2O_3 on the surface of the iron electrode.

Lin et al. studied the effect of chloride ions on the chemical and crystalline structure of the passive film at the iron electrode [46]. They used an experimental technique similar to that of O'Grady [45]. They concluded that 0.013N chloride ions promoted formation of α-Fe_2O_3 in the passive film on the iron surface in 0.15N boric acid-sodium borate buffer solution. The passivation of the iron electrode has also been studied in a solution of 0.5M CH_3COONa in 95 percent CH_3COOH and 5 percent H_2O, in the presence of O_2 by Mössbauer spectroscopy [47]. This study found that a passive film, more than 20nm thick and consisting of $Fe(CH_3COO)_{3-x}(OH)_x$ where x is 1 or 2, was formed when the iron electrode was polarized from -300 to +200mV.

The structure of the passive film formed on the surface of an iron electrode in concentrated sulfuric acid at 50°C by Chibirova et al. [48]. They found that the passivation was due to the formation of a layer consisting of hydrated iron(II) or iron(III) sulfates, depending on the potential applied. At 400mV, vs. the normal hydrogen electrode, iron(II) was formed, at 900mV both iron(II) and iron(III) was formed, and at 1500mV only trivalent iron was formed.

Geronov et al. [49] investigated by in situ Mössbauer spectroscopy the surface of an iron electrode while cycling an alkaline solution over the surface. The surface products were determined quantitatively. On the basis of the Mössbauer spectra the authors concluded that β-FeOOH was invariably formed at the end of the second anodic reaction in 5N KOH. At the end of the charging there was always a certain amount of residual $Fe(OH)_2$. After three to five cycles in 5N KOH containing 9g/dm^3 of LiOH, β-FeOOH was converted into Fe_3O_4. At the same time a drop in the capacity of the electrode was observed. At the end of charging under these conditions a certain amount of Fe_3O_4 was always found.

In situ Mössbauer spectroscopy was applied to the study of the mechanism of the catalytic activity of the polymeric iron-phtalocyanine oxygen electrode [50]. The in situ experiment was carried out in a half-cell arrangement in which only one side of the electrode was in contact with the electrolyte solution. The electrode was prepared by the precipitation of the iron-phtalocyanine on activated carbon. A correlation between the amount of iron dissolved in the sulfuric acid electrolyte and the electrode potential was observed during the first 100 hours of operation. The experimental results showed that iron(II) was oxidized to iron(III) at the electrode and later was dissolved into the electrolyte [50].

In situ Mössbauer spectroscopy has been also utilized [51-52] on an operating fuel cell to examine the iron phtalocyanine, FePc, catalyst adsorbed on a carbon matrix under conditions very similar to those encountered in its technological applications. It has been found that the crystalline FePc plays an important role in the electroreduction of oxygen in alkaline solution. In contrast, FePc adsorbed at monolayer coverages on the carbon surface does not show higher activity for oxygen reduction, than that exhibited by the carbon alone.

PASSIVATION OF A TIN ELECTRODE

The study of tin passivity by Mössbauer spectroscopy is of interest from a practical standpoint, and the anodic behavior of tin both in acid and alkaline solutions, has been the subject of several investigations [53-61]. The general consensus arising from these investigations, based largely on galvanostatic and coulometric studies, is that, in hydroxide and carbonate solutions, SnO or $Sn(OH)_2$ is formed at the more negative potentials, and SnO_2 or $Sn(OH)_4$ is formed at the more positive potentials. The situation in a borate solution is much less clear. The nature of oxide is not known even after large number of investigations.

The aim of Mössbauer measurements was to determine the chemical nature of the passive film formed in borate solution under anodic polarization [62]. In order to carry out Mössbauer experiments in absorption geometry on a thin oxide layer on a tin-substrate, it is necessary that the ratio of the volume of oxide to the volume of the substrate be as large as possible. The strategy adopted was to electroplate a thin layer of tin enriched in tin-119 on both sides of a 0.0065cm thick aluminum foil. Aluminum was selected because of its low absorption cross section for the 23.87keV gamma ray and its inert behavior under polarization conditions in the borate electrolyte. Tin was electrodeposited at a current density of 0.2A/dm^2 for 30s while the plating bath was maintained at 70°C. The composition of the bath was $25g/dm^3$ of SnI_4 in which the tin-119 content was enriched to 89 percent, $12g/dm^3$ of NaOH, $1.5g/dm^3$ of CH_3COONa, and $1.1g/dm^3$ of 30 percent H_2O_2. Mössbauer spectra were recorded on samples passivated in the deaerated borate buffer at various potentials and for various periods of time. After passivation, the foils were immediately plunged into liquid nitrogen and the spectra were recorded while the samples were maintained in the liquid nitrogen.

The Mössbauer spectra of the samples, polarized over a potential range of -800 to + 1090mV, relative to the standard Calomel electrode showed that the predominant tin species in the passive film contained tin(IV). The potential range was selected on the basis of potentiodynamic curves recorded with a scanning rate of 0.6V/min. The average isomer shift for SnO_2 obtained for all the experiments was -0.067mm/s relative to $BaSnO_3$. The extreme values were -0.20 and +0.08mm/s. The spectrum of $Sn(OH)_4$ obtained by Cohen et al. [63] yields a value of 0.0mm/s. Thus, the isomer shift does not give useful information that can distinguish between the anhydrous and hydrated forms of SnO_2.

Recently, Kapusta and Hackerman [64] reported the stepwise oxidation of tin to tin(IV) in a borate buffer solution over the same potential range in which no tin(II) could be detected by Mössbauer spectroscopy [62]. The aim of recent electrochemical and in situ Mössbauer studies on tin was to clear up this contradiction [65]. The potentiodynamic curves were recorded at scanning rates of 9V/min and 3.6V/min. At a fixed scanning rate the anodic limit of polarization was changed. Any previously formed oxide on the electrode was removed by polarizing the specimen cathodically, in a borate solution at -1.6V for 10 minutes, prior to anodic polarization. Two potentiodynamic curves obtained with a scanning rate of 9V/min are shown in Figure 8. In one case (dashed line) the electrode was polarized up to -850mV, whilein the second case (solid line) the maximum value of the anodic potential was +500mV. The most important feature of these curves is the presence of the previously unobserved current peak at -1180mV. The current of the maxima at about -780 and -580mV are in agreement with all the previous studies. The current maximum at -1100mV, which can be observed only under conditions of high scanning rate and high current sensitivity, suggests that, at this

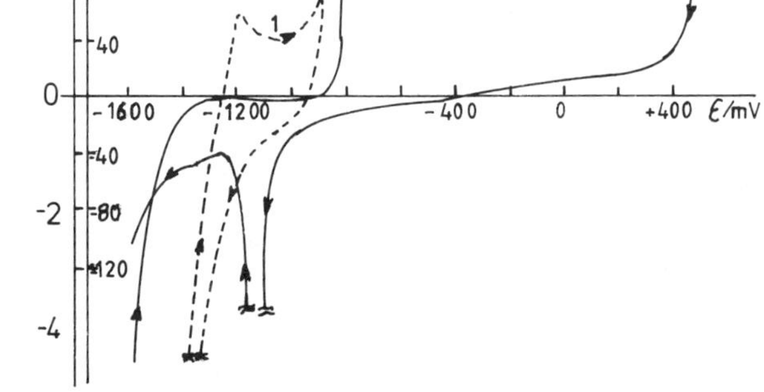

Figure 8. Potentiodynamic curves of a tin specimen.

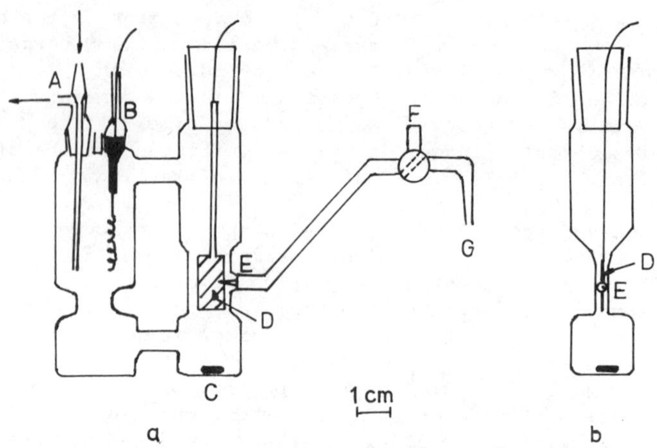

Figure 9. (a) An in situ Mössbauer cell showing (A) the gas inlet-outlet, (B) the platinum counter electrode, (C) a magnetic stirrer, (D) the specimen, (E) the Luggin capillary, (F) three-way stopcock, and (G) the reference electrode. (b) The specimen compartment.

potential, the formation of tin(II) oxide or hydroxide takes place. We have used in situ Mössbauer spectroscopy to gain further evidence for this assumption and to further characterize the surface film formed at the more negative potentials of -1200 to -780mV. These measurements were carried out in an electrochemical cell (Figure 9) designed specifically to study the composition of the passive film. The thickness of the glass is small enough to permit the transmission of the Mössbauer gamma rays. In Figure 10 the in situ Mössbaeur spectrum of the sample polarized at -900mV is illustrated.

The spectrum has a low velocity peak with an isomer shift of 0.03(2)mm/s which can be attributed to SnO_2 or $Sn(OH)_4$. The main peak of the spectrum, with an isomer shift of 2.50(3)mm/s is assigned to metallic β-tin. The spectrum has a poor statistics due to experimental limitations, but the small

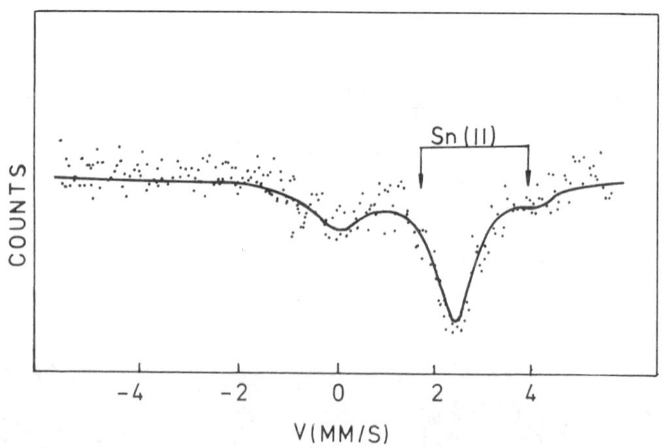

Figure 10. The in situ Mössbauer spectrum obtained during polarization at -900mV in a borate buffer solution.

absorption peak at ca. 4.17(5)mm/s is significant. The velocity of this peak is too high to result from a singlet, but attempts to fit the spectrum by using a doublet for the large velocity component were not successful. Among the oxygen containing tin compounds, only $Sn(OH)_2$ with an isomer shift of 2.78mm/s and a quadrupole splitting of 2.20mm/s [66] shows a line near the given velocity. Thus it can be assumed that this component belongs to a Sn(II) species, probably highly amorphous $Sn(OH)_2$, or a hydrated stannous oxide, which can have a somewhat higher quadrupole splitting than bulk $Sn(OH)_2$. Polarizations at +200mV, -750mV, and -780mV gave SnO_2 or $Sn(OH)_4$ and β-tin as components of the in situ Mössbauer experiments [62].

CONCLUSIONS

The reviewed results demonstrate that the Mössbauer technique is an unique tool for the study of electrolysis. The Mössbauer spectra of electro-formed layers give information about their chemical composition, and crystal-line and magnetic structure. By using in situ experiments one can learn the mechanism of the electrode processes occurring on the surface.

REFERENCES

1. H. Leidheiser, Jr., A. Vèrtes, M. L. Varsànyi, I. Nagy-Czakò, J. Electrochem. Soc. 126, 391 (1979).

2. M. Antler, in "Gold Plating Technology", F. H. Reid and W. Goldie, eds., Electrochemistry Publications, Ltd. Ayr, Scotland, pp. 259-294 (1974).

3. W. A. Styert and R. D. Taylor, Phys. Rev. 134, A716 (1964).

4. C. F. Steen, D. G. Howard, and R. H. Nussbaum, Solid State Commun. 9, 865 (1971).

5. G. W. Simmons, E. Kellerman, and H. Leidheiser, Jr., J. Electrochem. Soc. 123, 1276 (1976).

6. Y. Okinaka and S. Nakahara, J. Electrochem. Soc. 123, 1284 (1976).

7. Y. Okinaka, F.B. Koch, C. Wolowodiuk, and D. R. Blessington, J. Electrochem. Soc. 125, 348C (1978).

8. R. L. Cohen, K. W. West, M. Antler, J. Electrochem. Soc. 124, 342 (1977).

9. R. L. Cohen, F. B. Koch, L. N. Schoenberg, K. W. West, J. Electrochem. Soc. 126, 1608 (1979).

10. R. Schubert, J. Electrochem. Soc. 128, 126 (1981).

11. T. Adaniya, M. Ohmura, Metall. 31, 1100 (1977).

12. T. Adaniya, Steel Metal Indus. Iner. 55, 73 (1978).

13. H. Leidheiser, Jr., A Vèrtes, M. L. Varsànyi, J. Electrochem. Soc. 128, 1456 (1981).

14. H. Leidheiser, Jr., G. W. Simmons, S. Nagy, S. Music, J. Electrochem. Soc. 129, 1658 (1982).

15. R. Bocker and W. Döring, "Ferromagnetism", Springer, Berlin, 1939.

16. R. M. Bozorth, "Ferromagnetism", D. van Nostrand Co., Princeton, NJ, 1951.

17. S. Chikazumi, "Physics of Magnetism", J. Wiley and Sons, New York City, 1964.

18. F. Denise, H. Leidheiser, Jr., J. Electrochem. Soc. 100, 490 (1953).

19. L. Takàcs, A. Vèrtes, H. Leidheiser, Jr., Physica Status Solidi 74, K45 (1982).

20. A. Vèrtes, Zs. Kajcsos, Cs. Szeles, I. Czakò-Nagy, M. Lakatos-Varsànyi, H. Leidheiser, Jr., Proceedings of the Sixth International Conference on Positron Annihilation, Arlington, Texas, (USA) April 2-7, 1982.

21. A. Vèrtes, Zs. Kajcsos, I. Czakò-Nagy, M. Lakatos-Varsànyi, L. Csordàs, G. Brauer, H. Leidheiser, Jr., Nuclear Instruments and Methods 199, 353 (1982).

22. H. Leidheiser, Jr., A. Vèrtes, I.Czakò-Nagy, M. Lakatos-Varsànyi, L. Csordàs, Zs. Kajcsos, 33rd Meeting of the International Society of Electrochemistry, Lyon, France, September 6-10, 1982.

23. N. Parkinson, C. S. Britton, and R. Angles, Sheet Met. Ind. 28, 1 (1951).

24. M. Clarke and C. S. Britton, Corr. Sci. 3, 207 (1963).

25. M. Antler, M. Feder, C. F. Hornig, and J. Bohland, Plating Surf. Finish. 63, 30 (1976).

26. H. Leidheiser, Jr., I. Czakò-Nagy, M. L. Varsànyi, A. Vèrtes, J. Electrochem. Soc. 126, 204 (1979).

27. J. W. Cuthbertson, N. Parkinson, and H. P. Rooksby, J. Electrochem. Soc. 100, 107 (1953).

28. C. P. Thwaites, Trans. Inst. Metal Finish, 3, 203 (1959).

29. J. Silver, C. A. Mackay, F. D. Donaldson, J. Materials Science 11, 836 (1976).

30. J. Jaèn, M. L. Varsànyi, È. Kovàcs, I. Czakò-Nagy, A. Buzàs, A. Vèrtes, and L. Kiss, Electrochim. Acta 29, 1119 (1986).

31. A Vèrtes, J. Jaèn, M. L. Varsànyi, Radioanal. and Nucl. Chem. 90, 383 (1985).

32. O. K. Kudra, V. S. Galinker, and M. N. Nasonova, Ukr. Chim. Zhurn, 38, 875 (1972).

33. V. S. Galinker, O. K. Kudra, and M. N. Nasonova, Zashcita Metallov 8, 221 (1972).

34. J. Jaèn, A. Vèrtes, K. M. Tjutina, L. V. Kasmodamianskaya, L. Kiss, and L. V. Shepeleva, Electrochim. Acta 30, 535 (1985).

35. A. Vèrtes, S. Nagy, M. Lakatos-Varsànyi, International Conference on the Application of Mössbauer Spectroscopy, Jaipur, India, December 14-18, 1981.

36. R. L. Cohen, C. J. Raub, T. Muramaki, J. Electrochem. Soc. 125, 34 (1978).

37. H. Mehner and A. Vèrtes, Proceedings of the Int. Conf. on the Application of the Mössbauer Effect, pp. 780-781, Jaipur, India, December 14-18, 1981.

38. A. Vèrtes, S. Nagy, M. Z. Awad, Nuclear Instruments and Methods 199, 367 (1982).

39. J. Jaèn, A. Pashovsky, G. Raichevsky, I. Czakò-Nagy, L. Kiss, and A. Vèrtes, Radiochem. Radioanal. Letters 58, 111 (1983).

40. A. T. Tsinman and A. N. Katrevich, Zashchita Metallov 5, 517 (1969).

41. A. T. Tsinman, L. M. Pischik, V. M. Brusentsova, L. E. Zanarenkova, Zashchita Metallov 7, 700 (1971).

42. Ya. M. Kolotirkin, G. G. Kossii, Zashchita Metallov 1, 272 (1965).

43. L. Kiss, D. N. Lien, and M. L. Varsànyi, Coll. Czechosl. Chem. Com. 36, 914 (1971).

44. G. W. Simmons, A. Vèrtes, M. L. Varsànyi, and H. Leidheiser, Jr., Electrochem. Soc. 126, 188 (1979).

45. W. E. O'Grady, J. Electrochem. Soc. 127, 555 (1980).

46. M. C. Lin, R. G. Barnes, and T. E. Furtak, AIP Conference Proceedings No. 84, F. C. Schwerer, ed., American Institute of Physics, New York, 1982.

47. G. N. Markosyan, L. Sziràki, A. I. Molodov, L. M. Versànyi, V. V. Losev, J. Jaèn, A. Vèrtes, and L. Kiss, to be published in Elektrokhimiya (1986).

48. F. H. Chibirova, T. V. Revina, V. A. Makarov, and S. I. Reiman, Zaschcita Metallov 18, 905 (1982).

49. Y. Geronov, T. Tomov, S. Georgiev, J. Appl. Electrochem. 5, 351 (1975).

50. J. Blomquist, U. Helgeson, L. C. Moberg, L. Y. Johansson, and R. Larson, Electrochim. Acta 27, 1445 (1982) and 27, 1453 (1982).

51. D. Scherson, S. B. Yao, E. B. Yeager, J. Eldridge, M. E. Kordesch, and R. W. Hoffman, J. Phys. Chem. 87, 932 (1983).

52. D. Scherson, C. Fierro, E. B. Yeager, M. E. Kordesch, J. Eldridge, and R. W. Hoffman, International Soc. of Electrochemistry 34th Meeting, Erlangen, Germany, September 18-23, 1983.

53. R. Korr, J. Soc. Chem. Ind. 57, 405 (1938).

54. D. Bianchi, Chim. Ind. 29, 295 (1947).

55. S. E. S. El Wakkad, A. M. Shams El Din, and J. A. El Sayd, J. Chem. Soc. 3103 (1954).

56. M. Katoh, Bull. Chem. Soc. Japan 35, 1711 (1962).

57. S. N. Shah and D. E. Davies, Electrochim. Acta 8, 663 (1963).

58. A. M. Shams El Din and F. M. A. El Wakab, Electrochim. Acta 9, 883 (1963).

59. N. A. Hampson and N. E. Spencer Br., Corrosion J. 3, 1 (1986).

60. D. E. Davies and S. N. Shah, Electrochim. Acta 8, 703 (1963).

61. M. Pugh, L. M. Warner, and D. R. Gabe, Corrosion Science 7, 807 (1967).

62. A. Vèrtes, H. Leidheiser, Jr., M. L. Varsànyi, G. W. Simmons, and L. Kiss, J. Electrochem. Soc. 125, 1946 (1978).

63. R. L. Cohen, C. J. Raub, and T. Muramaki, J. Electrochem. Soc. 125, 34 (1978).

64. S. D. Kapusta and N. Hackerman, Electrochim. Acta 25, 1625 (1980).

65. M. L. Varsànyi, J. Jaèn, A. Vèrtes, and L. Kiss, to be published in Electrochim. Acta (1986).

66. W. D. Honnick and J. J. Zuckerman, Inorg. Chem. 15, 3034 (1976).

67. J. Jaèn, Ph.D. Dissertation, Eötvös University, Budapest, 1984.

68. H. Mehner, J. Juhàsz, M. Suba, and A. Vèrtes, Radiochem. Radioanal. Letters 56, 57 (1982).

ROLE OF TIN COMPLEXES IN BONDING SILVER TO GLASS: MÖSSBAUER STUDIES*

T. M. Thomas

Solar Energy Research Institute
1617 Cole Boulevard
Golden, CO 80401, USA

D. L. Williamson

Department of Physics
Colorado School of Mines
Golden, CO 80401, USA

D. P. Rainville

Department of Chemistry
University of Wisconsin-Riverfalls
Riverfalls, WI 54022, USA

ABSTRACT

The nature of the tin surface complex formed during the sensitization of glass in a silver mirror-making process has been studied by tin-119 Mössbauer spectroscopy. The tin is found to be in the 4+ oxidation state which eliminates proposed bonding mechanisms based upon simple acid/base or ligand exchange reactions, and confirms a mechanism based on oxidation of the tin(II) ions upon bonding to silicon atoms on the glass surface. Further, the isomer shift of the tin complex is consistent with tin bonding to several hydroxide groups, in agreement with the oxidation reaction model. Although Auger electron spectral surface analyses show that the tin complex covers less than a monolayer of the glass surface, useful Mössbauer data are obtained by both transmission and backscatter, conversion electron, Mössbauer techniques. This is possible in part because of careful sample preparation but also due to effective surface areas of the glass slides which are significantly larger than that corresponding to perfectly flat surfaces.

INTRODUCTION

Since the 1870's [1] mirror manufacturers have been applying a solution of $SnCl_2 \cdot 2H_2O$ and HCl to the surface of glass prior to silver deposition. This step was called sensitization of the glass surface. They found that after the glass had been contacted with this solution and rinsed with distilled water, the silver deposited would have a strong adherence to the glass. The mechanism of sensitization was still not understood as late as

*Work performed under DOE Contract No. DE-AC02-83CH-10093

the 1970's [1]. With the advent of surface sensitive analytical techniques, the chemistry of surface overlayers and thin films could be studied. Also, with the increased research into components for concentrating solar energy after 1974, many studies of the commercial mirroring process were commissioned. The combination of these events culminated in several proposed mechanisms for the sensitization reactions [2-5].

The proposed mechanisms for sensitization all suggest that the process can be broken into three parts. In the first step a tin ion or complex is bonded to the glass. In the second step the glass surface is rinsed with water which removes all traces of chlorine from the surface. In the last step silver is deposited onto the sensitized surface.

This work is part of a systematic research program on the complex set of reactions that make-up the mechanism of tin sensitization and the role of tin in bonding silver to glass surfaces [5]. Here the focus is on the bonding of tin to glass and not on subsequent silver bonding.

The proposed mechanisms for the bonding of tin chloride to the glass surface differ from model to model. One model [2] suggests that the tin deposits onto the surface as an ionic salt from the neutralization of the glass surface:

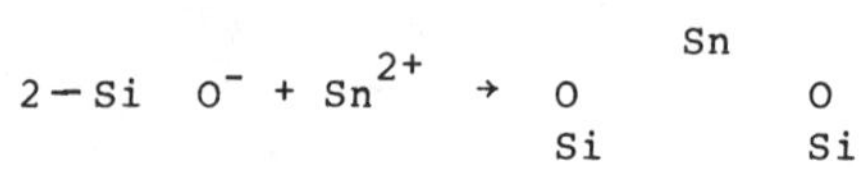

(1)

Another suggestion [3,4] is that a ligand exchange reaction takes place:

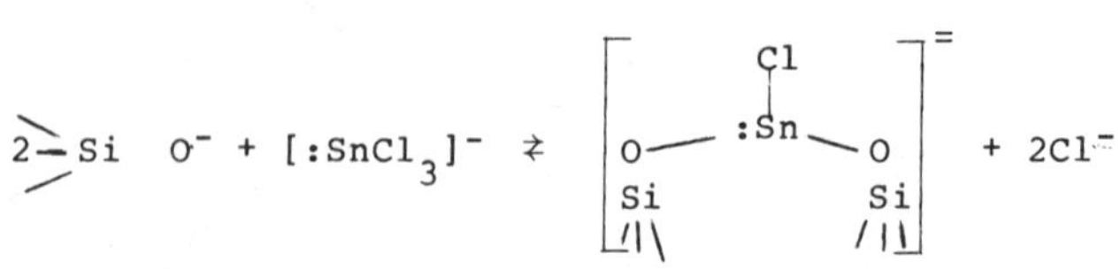

(2)

A more recent model [5] suggests that the tin undergoes an oxidation reaction upon bonding to the glass:

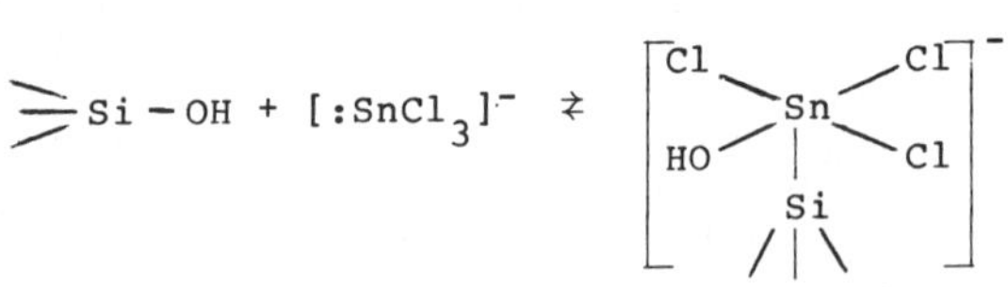

(3)

Thus oxidative addition, also called insertion, of the tin complex occurs which converts tin(II) to tin(IV).

Mössbauer spectroscopy using the tin-119 isotope can distinguish between these proposed complexes and oxidation states of the tin atom. Mössbauer isomer shifts and quadrupole splittings due to varying numbers and geometries of ligands around the stannous atom might differentiate the first two suggested models and a large isomer shift due to the change from the stannous to a stannic state would easily distinguish mechanism three from the first two.

This paper presents the results of detailed Mössbauer analyses and complementary surface analysis of sensitized glass surfaces. The proposed models will then be discussed in light of the current results.

EXPERIMENTAL

Six sensitized samples and one unsensitized sample were investigated by tin-119 Mössbauer spectroscopy. The nature of the samples is given Table I. A stack of twenty 22 x 22mm square micro cover glass slides manufactured by VWR each having a thickness of 140mm$\pm$20μm was used as a single sample for analysis. Both sides of each slide were sensitized. The thickness of the stack corresponds to twice the 1/e absorption length for the 23.8keV γ-rays. A glass powder, Micro-Grain Precision Abrasive Glass-beads, manufactured by MDC Industries, Inc., Philadelphia, PA was used as sample B. The average diameter of the beads was approximately 60μm. Since the powder sample was prepared to have the same mass per unit area as the twenty cover slips, the glass surface area exposed to the γ-ray beam for this sample was approximately 7 times that of the stack of twenty coverslips.

In the mirror preparation procedure used by commercial mirror manufacturers, the sensitization step is preceded by abrasion and polishing the glass surface. Here the glass cover slips were polished with ultrex grade CeO_2 powder on deionized water soaked cotton. The glass beads (sample B) could not be polished, so were carried through the remaining steps as received. The sample were then rinsed in deionized water to remove any residue. In the second sensitization step, the samples are dipped or sprayed with a proprietary tin solution. The exact composition of the commercial solutions is unknown, but is a mixture of hydrochloric acid, stannous chloride and usually contains a wetting agent to reduce surface tension to facilitate a uniform coverage of the glass surface by the solution. The composition of the sensitization solution used in this study was that suggested by Schweig [1]. The samples were soaked in this solution, followed by multiple rinses with copious amounts of deionized water. The samples were then air dried.

Table I. Mössbauer Absorbers

Sample	Preparation
A	Stack of 20 glass slides each sensitized with natural tin (8.6% tin-119)
B	Glass powder (~ 60μm diameter) sensitized with natural tin (without CeO_2 scrub)
C	Stack of 20 glass slides each sensitized with enriched tin (85% tin-119)
C'	One of the glass slides from C was used for CEMS
D	Stack of 20 glass slides each sensitized with enriched tin, then reimmersed in a solution containing only natural tin.
E	Stack of 20 glass slides each sensitized with enriched tin (exposed to solution 50% longer than C)
F	Stack of 20 glass slides as received

Tin-119 enriched (85%) metal shot was obtained from New England Nuclear, Boston, MA for the enriched sensitizing solutions. All stannous chloride salts used for this study were prepared from tin metal shot by the procedure of Pass and Sutcliffe [7]. Portions of the prepared salts were then pressed into KBr pellets and analyzed by transmission IR spectroscopy. These spectra were compared to the spectrum obtained from a KBr pellet of reagent grade $SnCl_2 \cdot 2H_2O$ obtained from Mallinckrodt to check the purity of the prepared salts.

The Auger electron spectra (AES) were obtained from a commercial 545C surface analysis system from Physical Electronics. AES analyses were used to check surface coverage, composition and film uniformity.

RESULTS AND DISCUSSION

A comparison of the Mössbauer spectra of samples B and C to SnO_2 is shown in Figure 1. These results definitely show that the tin compound has undergone oxidation to the tin(IV) state upon bonding to the glass surface. This confirms the observations of a previous photoelectron spectroscopy study [5] of sensitized glass surfaces that indicated the tin was bonded to at least 5 ligands, which could only take place if the tin was tin(IV). Samples A, C, C', D and E all yielded similar spectra and were fitted with a single Lorentzian resonance line by the method of least-squares. Spectral parameters for all the Lorentzian fits are listed in Table II. Sample B was fitted with two quadrupole pairs of lines as indicated in Figure 1. The high velocity quadrupole pair is due to tin(II) similar to SnO. Cohen et al. [8] show an almost identical spectrum to sample B in their Mössbauer studies of tin sensitized polymer films. They suggest that the 2+ state is due to

Table II. Tin-119 Mössbauer Spectral Parameters

Sample	T(K)	Site	δ(mm/s)	Γ(mm/s)	Δ(mm/s)	ε(%)
A	300	1	-0.27(5)	1.1(2)	0	0.31(5)
	76	1	-0.17(2)	1.07(5)	0	0.91(4)
B	300	1	-0.12(5)	0.8(1)	0.6(1)	0.72(7)
		2	3.1(2)	0.8(1)	1.3(2)	0.12(3)
	76	1	-0.03(2)	0.93(4)	0.54(4)	2.08(6)
		2	2.88(7)	1.0(1)	1.84(6)	0.35(3)
C	300	1	-0.25(5)	1.2(2)	0	0.45(4)
	76	1	-0.14(2)	1.17(5)	0	1.28(4)
C'	300	1	-0.1(1)	1.4(2)	0	1.4(1)
D	76	1	-0.15(2)	1.14(7)	0	1.07(4)
E	300	1	-0.12(6)	0.9(2)	0	0.46(6)
	76	1	-0.12(2)	1.19(6)	0	1.32(4)

T is the sample temperature. δ is the isomer shift relative to $CaSnO_3$ at 300K. Γ is the full linewidth at half maximum. ε is the size of the effect relative to the off-resonance count-rate. The uncertainty in the last significant figure is given in parentheses.

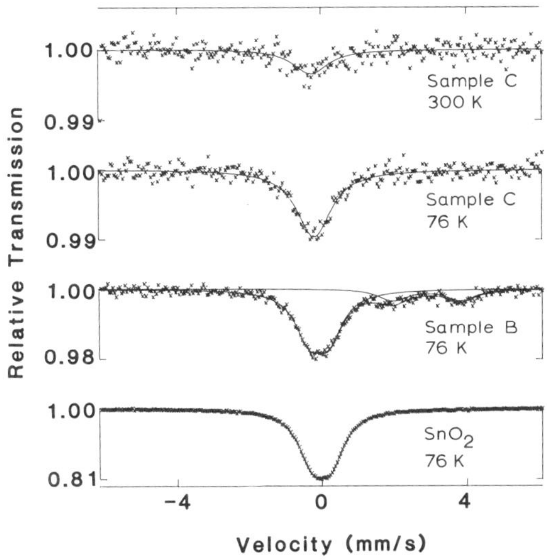

Figure 1. Mössbauer spectra of tin-sensitized glass surfaces and of SnO_2 powder. The solid lines are least squares fits of single Lorentzian lines for sample C, two quadrupole pairs for sample B and a single quadrupole pair for SnO_2.

residues left on the surface instead of a surface bonded species. They were able to centrifuge the 2+ containing residue from the sensitizing solution and obtain its Mössbauer spectrum. Thus the powder nature of sample B probably entrained some of the residue which was not able to be rinsed away after sensitization of the powder. The only sample which showed neither the 4+ nor the 2+ state of tin was sample F which was the unsensitized stack of 20 cover slips. No measurable Mössbauer resonance was observed from this sample.

Table III contains Mössbauer parameters for some tin compounds taken from the literature. By comparing the isomer shift of SnO_2 from Table III with those of site 1 in Table II, it is apparent that the tin complex on the glass surface after sensitization corresponds to tin(IV), similar to SnO_2, but it clearly has a more negative isomer shift. Averaging the 76K values of δ and subtracting the δ for SnO_2 at 76K gives a value of -0.15 ± 0.03mm/s. Such a negative shift corresponds to a lower electron desity at the nucleus and may be indicative of the presence of OH^- ligands. The mechanism proposed by Thomas et al. [5] included a tin complex having five or six hydroxide groups surrounding the tin atom. The data for $Sn(OH)_6^{2-}$ and SnO_3^{2-} ($CaSnO_3$ and $BaSnO_3$) from Table III are consistent with this hypothesis since hydroxide groups do cause a negative isomer shift relative to oxide groups.

The fact that here the isomer shift is more negative than that of $Sn(OH)_6^{2-}$ could be due to the complex being on the surface where the OH bond length could be increased, leading to a lower electron density relative to the bulk tin sites in $Sn(OH)_6^{2-}$.

Note from Table II that the single line fits all yield linewidth (Γ), which are significantly larger than 0.8 mm/s which is expected for a tin site with no electric field gradient at the nucleus. Fitting the 76K spectra from samples A, C, D, and E with a closely spaced quadrupole pair of lines yields $\Gamma = 0.9\pm0.1$mm/s and $\Delta = 0.4\pm0.2$mm/s, similar to $\Gamma(SnO_2)$ and $\Delta(SnO_2)$ in Table III. Thus, there is no detectable increase in the electric field gradient at

Table III. Mössbauer Parameters of Relevant Tin Compounds

Compound	T(K)	δ(mm/s)	Γ(mm/s)	Δ(mm/s)	Θ_m(K)	Ref.
SnO_2	RT	0.004(3)	0.97(2)	0.50(1)	313(6)	9
	76	0.024(1)	0.90(1)	0.51(1)	this work	
SnO	RT	2.678(3)	0.92(1)	1.36(1)	203(1)	9
	296	2.61(2)*	0.85(2)	1.31(1)	229(3)	10
	78	2.70(2)*	0.85(2)	1.33(2)	229(3)	10
$Sn(OH)_6^{2-}$	77	-0.07				11

*Relative to $BaSnO_3$ which is the same as $CaSnO_3$ within experimental error [12]. Θ_m is the Mössbauer-related Debye temperature. The remaining symbols are defined in Table II.

Table IV. Tin Surface Density From Mössbauer Measurements

Sample	Site	r	Θ_m(K)	f	d(atoms/cm^2)
A	1	2.13	223	0.33	1.3×10^{16}
B	1	2.00	234	0.37	4.6×10^{15}
	2	2.52	202	0.26	1.1×10^{15}
C	1	2.07	228	0.35	2.0×10^{15}
D	1	----	---	----	1.6×10^{15}*
E	1	2.72	195	0.24	2.1×10^{15}

*Assumes same f as sample C for d determination. In this table, r is the ratio of integrated resonance intensities of 76 and 300K spectra. Θ_m is the Mössbauer-related Debye temperature evaluated from r with the Debye model. f is recoil free fraction at 300K based on Θ_m. d is the surface density resonance intensity. Samples A, C, D, and E have been adjusted for the 40 surfaces in the γ-ray beam and sample B has been adjusted by an additional factor of 7 due to the extra surface area exposed to γ-ray beam (based on 60μm spherical particles).

the tin(IV) in the surface complex on the glass compared to tin(IV) in the distorted octahedron of SnO_2 (rutile). The Mössbauer-related Debye temperature (Θ_m) for the tin(IV) sites (Θ_m (average) = 220K from Table IV) is clearly less than that of tin(IV) in bulk SnO_2 (Θ_m = 313K from Table III). This is consistent with the surface state nature of the tin(IV) where a weaker bond is expected due to the reduced coordination.

An evaluation of the amount of tin on the surface of each sample was made as follows. From the ratio (r) of the integrated resonance intensities at 76 and 300K a recoilless fraction (f) was obtained on the basis of the

Debye model. By using f and other known tin-119 Mössbauer parameters [13] the tin atom area density (d) can be obtained from the integrated intensities. The results of the surface density calculation are presented in Table IV.

Due to the crudeness of the Debye model and the uncertainities in r and the other factors, such as γ-ray purity in the 23.8keV counting channel, we estimate d to be accurate to within a factor of two. Relative values of d from Table IV should be accurate to about ±10%.

When sample C was prepared with 85% enriched tin-119, a factor of ten increase in the Mössbauer effect was expected over sample A. However, Table II shows less than a factor of two improvement. By comparing the surface density of the unenriched sample A to the tin-119 enriched sample C from Table IV, it is concluded that sample A has about six times as many tin atoms per unit area as sample C. One of the cover slips from each of these samples was analyzed by AES. For sample A, the silicon to tin ratio indicated that 29% [14] of its surface sites were covered by tin complexes. For sample C, only 9% of its surface sites were covered by tin complexes. The relative values of d thus seem useful as a measure of relative surface coverage and thus sensitization efficiency.

The absolute surface coverage cannot be obtained from the Mössbauer values of d in the present work because the total surface area per sample was not measured (e.g., by the B.E.T. technique). For a glass surface which is perfectly flat, full monolayer coverage would correspond to about 1 x 10^{15} atoms/cm^2. Since all of the calculated d values are higher than this, the flat surface approximation is clearly not valid. However, if surface roughness is similar (due to uniform CeO_2 scrubbing) among the samples, ratios of d will reflect relative surface coverage.

The experiment with sample D was done to test the possible exchange of tin ions in the sensitizing solution with atoms already occupying tin surface sites. Assuming the same surface coverage before and after the reimmersion in the solution containing the natural tin, then the occurence of exchange would lead to a reduction in the Mössbauer signal by as much as a factor of ten. The data in Tables II and IV show that the signal was reduced by only a factor of 20% compared to sample C which would suggest that some exchange may have occured since the relative values are good to ±10%. However, the

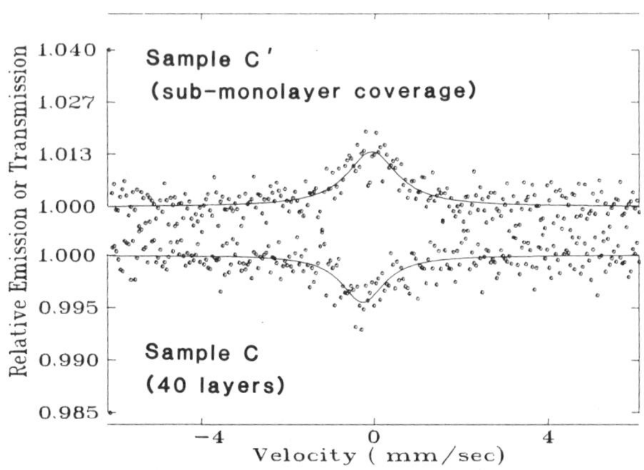

Figure 2. Comparison of room temperature Mössbauer spectra of tin-sensitized glass surfaces. The top spectrum is from CEMS and the bottom spectrum is from the standard transmission mode. The solid lines are least-squares fits of Lorentzian lines.

difference may also be due to a 20% lower average surface coverage in sample D compared to C.

The same value of the surface coverage for sample E compared to C shows that a longer soak in the sensitizing solution had no effect on coverage. The only preparatory differences between sample A and samples C, D, and E were possibly in the details of the CeO_2 scrubbing technique. This emphasizes the importance of the preparation steps prior to sensitization. The coverage difference between sample B and sample A may also result because sample B was not able to be scrubbed prior to sensitization.

Finally, we illustrate the possible use of CEMS as a surface sensitive technique for these types of samples. Figure 2 is a comparison of the conversion electron Mössbauer spectrum of sample C\` with the conventional Mössbauer spectrum from sample C. Note that 1/40 of the sample in CEMS gives very similar signal-to-noise compared to that of a full sample in the transmission mode. Note also the AES analysis of sample C\` showed that less than one tenth of a monolayer of tin covers the glass surface and yet is easily measurable by CEMS using a 2mCi source.

CONCLUSIONS

Only one type of tin oxidation state was identified on sensitized glass cover slips and it is similar to tin(IV) in SnO_2. The isomer shift of the tin complex is close to that of $Sn(OH)_6^{2-}$. These observations are consistent with the proposed sensitization mechanism by Thomas et al. [5] (Eq. 3) and precludes the others [2-4] (Eqs. 1 and 2) based upon simple ion exchange or ligand exchange reactions, which do not change the oxidation state of the tin.

The resonance of the tin(IV) state is slightly broadened, corresponding to a quadrupole splitting of ca. 0.5mm/s. This must be due to the tin atom being in a non-cubic environment. Samples B, C, D, and E all have relative surface coverages which are lower than sample A. This probably results from no CeO_2 scrubbing in the case of sample B and slightly different scrubbing procedures for samples C, D, and E. The calculated relative surface coverage results from Mössbauer data are consistent with those measured by AES. AES results indicate that tin sensitization of glass leaves a submonolayer tin coverage of the surface.

The similarity in coverage data between sample C and sample D suggests that after bonding to the glass, the tin complex has a very slow exchange rate with those in solution at room temperature.

When analyzing samples which have only surface bonded Mössbauer active species, three methods can be used to increase surface sensitivity: analyze a high surface area powder, analyze a multiple stack of very think sections, or use a surface sensitive technique such as CEMS.

REFERENCES

1. B. Schweig, \`Mirrors,\` Pelham Books, London, 1973.

2. C. H. De Minjer and P. F. J. van der Boom, J. Electrochem. Soc. 120. 1644 (1973).

3. L. R. Pederson and M. T. Thomas, Solar Energy Mater. 3, 151 (1980).

4. L. R. Pederson, Solar Energy Mater. 6, 221 (1982).

5. M. T. Thomas, J. R. Pitts, and A. W. Czanderna, Appl. Surf. Sci. 15, 75 (1983).

6. A. Sette-Camara and W. Keune, Corr. Sci. 15, 75 (1983).

7. G. Pass and H. Sutcliffe, `Practical Inorganic Chemistry,` Chapman and Hall Ltd., London, 1968, p. 6.

8. R. L. Cohen, J. F. D`Amico and K. W. West, J. Electrochem. Soc. 118, 2042 (1971).

9. G. S. Collins, T. Kachnowski, and N. Benczer-Koller, Phys. Rev. B19, 1369 (1979).

10. R. H. Herber, Phys. Rev. B27, 4013 (1983).

11. See, e.g., P. A. Flinn, in 'Mössbauer Isomer Shifts', G. K. Shenoy and F. E. Wagner, eds, North-Holland, New York, 1978, p. 593.

12. J. G. Stevens, Hyp. Int. 13, 221 (1983).

13. 'Mössbauer Effect Data Index`, J. G. Stevens and V. E. Stevens, eds., Plenum Publ. Co., New York, 1978, p. 103.

14. XPS, ISS and AES studies of sensitized quartz as well as glass substrates all seem to suggest that 30% coverage of the surface by tin is the upper limit. J. R. Pitts, Ph.D. Dissertation, University of Denver, Denver, CO, 1985.

USE OF MÖSSBAUER SPECTROSCOPY IN THE STUDY OF IV-VI SEMICONDUCTORS

Ida Ortalli

Faculty of Medicine
Physical Science Institute
University of Parma
Parma, Italy

Vincenzo Fano

Maspec Institute of CNR
Parma, Italy

INTRODUCTION

The IV-VI semiconductor group has received much attention over the last 15 years because it may be used in several different types of technological applications, such as optoelectronic devices like photovoltaic detectors and tunable infrared diode lasers, and in thermoelectric energy conversion. The technological properties of this group are directly related to both their fundamental properties and their methods of synthesis. For instance, some pseudobinary semiconductors, such as $Pb_{1-x}Sn_xTe$, have a composition-dependent energy gap which, according to the proposal of Dimmock [1], has a band model which can be made arbitrarily small by varying the stoichiometry, i.e., the lead to tin ratio, as shown in Figure 1. However, the commonly used single crystal growth methods from the melt induce a significant concentration gradient along the growth axis.

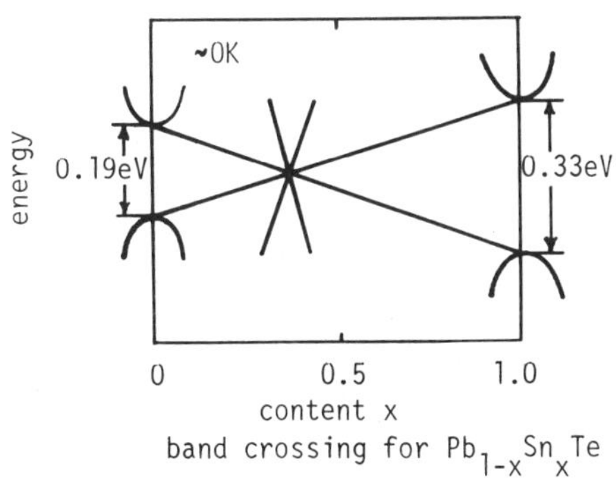

Figure 1. Schematical diagram of the band structure of $Pb_{1-x}Sn_xTe$ at the L point.

Although these materials have been widely studied for use in the above mentioned fields, some basic questions need to be answered about the phase transitions found in some binary and pseudo-binary systems, the nature of the chemical bonding, the stabilization forms of some doping elements, and the influence of the powder metallurgy operations on the physical properties in the syntering process.

For these reasons we have used Mössbauer spectroscopy to examine some binary and pseudobinary compositions. When compared with other microscopic methods, Mössbauer spectroscopy was found to provide a very high density of information, however, in some cases the information is difficult to extract in an unambiguous way. For solid state materials of which tin is a constituent, tin-119 Mössbauer effect spectroscopy has proven to be a fruitful technique for the characterization of the electronic environment around the probe atom and for testing the lattice dynamic properties.

EXPERIMENTAL: SYNTHESIS AND CHARACTERIZATION

In all experiments the starting elements were 99.999 percent pure and underwent further oxygen purification by melting in a hydrogen flux. In this way the interaction between the melt and the quartz tube during the synthesis is avoided. A direct synthesis of the compounds was then carried out. The samples used for powdering were cut from the central part of monocrystalline ingots. By the use of single crystals for powdering, we eliminate the uncertainty about the inhomogeneity of the starting materials. In fact, when the polycrystals are synthesized by usual methods, uncontrolled convective flows are present in the melt. These flows produce significant compositional fluctuations. Another cause of compositional variation is the contemporary formation of several centers of solidification which cause different segregation effects. These compositional variations can be readily observed by thermoelectric microprobe. The single crystals were grown from the melt by the vertical Bridgman technique with a growth velocity of 0.5cm/hr, and a thermal gradient of 20°C/cm [2].

For the ternary systems, $Pb_{1-x}Sn_xTe$, the compositions were determined by chemical analysis and scanning electron microscopy. In this case the uncertainty in composition is caused by the normal segregation along the growth axis. In the first approximation this segregation obeys the well known equation

$$C_s K_{eff} C_o (1-g)^{K_{eff}-1},$$

where C_o is the mean concentration in the starting melt and g is the fraction of solidified melt. The segregation coefficient, K_{eff}, is unknown. Nevertheless one can select samples with fixed stoichiometry, when the distribution curves are known by compositional analysis, in our case by scanning electron microscopy. Before the Mössbauer measurements all powders were annealed at 400°C in an inert atmosphere. In this way the concentration of some defects, which affect the Mössbauer peaks, decreases significantly. The evidence of this is shown by Fano, et al. [3] who examined SnTe doped with iron. The resonant peaks appear only in the annealed samples.

The electrical properties, such as the electrical conductivity and Hall effect are determined by simultaneously utilizing the cross modulation of the sample current, at a frequency of 84Hz, and magnetic field, at a frequency of 36Hz at 1500G.

STABILIZATION FORMS OF DOPING ELEMENTS

A factor which affects the efficiency and lifetime of thermoelectric generators is the diffusion of impurities from bridging materials. In fact the transition metals, such as iron, are used as a bridge in thermoelectric devices made of IV-VI semiconductor materials and may diffuse in the matrix at operating temperatures. The situation is complicated by transition-metal impurities in the lattice which can replace atoms, either at substitutional or interstitial sites, manifesting donor or acceptor behavior depending on the form of the stabilization in the matrix.

A Mössbauer effect study of iron ions diffused into the tin chalcogenide matrices, SnTe, SnSe, or SnS, in the form of cobalt-57, which is used as a source for the γ-nuclear resonance, has been carried out [4]. The description of data relative to these three compounds makes it easier to interpret the state of charge and the stabilization forms of iron in the semiconductors.

The same chalcogenide samples were oxidized in air. In this way an attempt was made to vary the environments of the iron by doping with oxygen, an element chemically analogous to sulfur, selenium and tellurium, and whose position in the lattice is established. In fact, for all semiconductors of the SnTe type, i.e., PbTe, $Pb_{1-x}Sn_xTe$, etc., it is generally assumed that oxygen has a behavior analogous to tellurium when present in microquantities, that is, it substitutes tellurium in the lattice. In the same way as tellurium, oxygen stabilizes the p-type carrier and does not give a localized level in the gap as shown by Koval'chilk et al. [5].

The Mössbauer spectrum obtained at room temperature for SnTe doped with cobalt-57 is shown in Figure 2a. Figure 2b and 2c show spectra of samples submitted to oxidization for five hours and nine hours respectively. In these cases the center of gravity of the spectra shifts towards smaller values. The spectrum shown in Figure 2a may be resolved into one quadrupole doublet, the spectrum 2b into two quadrupole doublets, and the spectrum 2c into three quadrupole doublets of nearly equal area (see reference 4). The associated isomer shifts range from 0.71 to 0.47mm/s, typical value for iron(III), and to 0.21mm/s, which can still be attributed to iron(III) in the interstitial positions. The quadrupole splitting may be attributed to the tellurium environment of iron-57 with a variable degree of oxygen substitution. This substitution increases in passing from the second doublet to doublet the third. The possibility that the spectra 2b and 2c refer to an environment of iron-57 in which the tellurium belonging to second neighbors and not that of first neighbors, may be replaced, can be excluded. This is the case because, a similar notable variation of the isomer shift with the unchanged values of the quadrupole splitting would not be expected. Therefore, passing to the substitution of one ion of tellurium with one of oxygen, the isomer shift shifts to the value of 0.24mm/s or $1.15 \cdot 10^{-7}$ eV which is nearly equal to that of the further shift which occurs on the substitution of a second tellurium ion with oxygen, i.e., 0.26mm/s or $1.25 \cdot 10^{-7}$ eV.

On passing from the first to the second to the third doublet, the percentage of 4s character at the nucleus increases from $3d^5s^{0.20}$ to $3d^5s^{0.36}$ to $3d^5s^{0.48}$ according to the Danon diagram [6]. The decrease in the value of the isomer shift, and the increase in the percentage of the s-electrons, means that the size of these ions decreases when passing from the tellurium to the oxygen. Also the energy of these orbitals is reduced and the electrons will be less apt to be transferred to the metal. This reduced electron density lowers the shielding of the s-electron cloud, displacing the isomer shift towards increasingly less positive values.

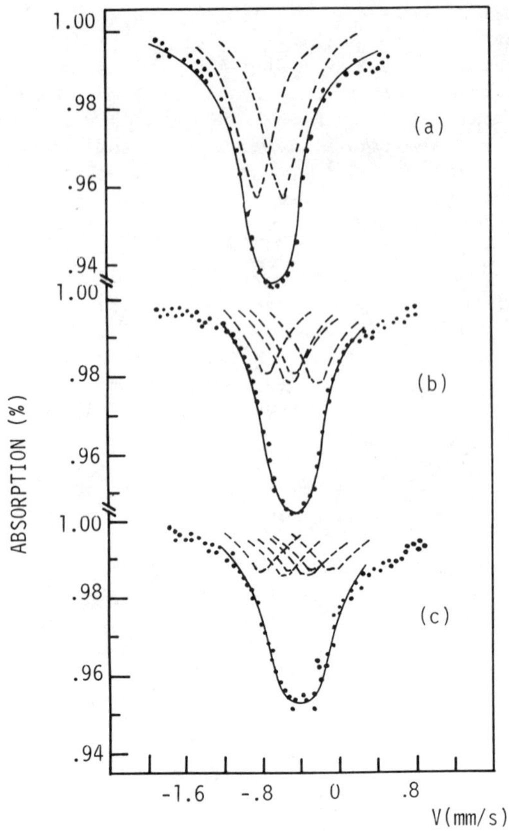

Figure 2. Emission Mössbauer spectra of SnTe containing cobalt-57 (a), and after 5h (b) and after 9h (c) of oxidation. The solid line indicates the fit by Lorentzian curves.

Finally, for even greater oxygen content, the spectrum is further complicated by the appearance of new peaks, with even smaller isomer shifts, as shown in Figure 3. However, their interpretation is somewhat difficult because the spectrum is complicated. Moreover, there is no exact knowledge of the types of defects present under these conditions of oxidation. It must be noted that, after 35 hours of oxidation with the introduction of oxygen, Figure 3c, when even the monocrystalline habit has been destroyed (the samples are pulverized) divalent iron appears with a quadrupole splitting of Q.S. ~ 2.5mm/s and an isomer shift of I.S. ~ 1.0mm/s.

Similar oxidation processes have also been carried out with the samples of SnS and SnSe. These too, after a few hours of oxidation, resulted in a second quadrupole doublet analogous to the second doublet in SnTe. However, the oxidation of the samples was not carried out as in the case of SnTe because they were easily shattered after further oxidation processes.

We can conclude that an association exists between iron and oxygen successively diffused into the matrix and that the doping elements, such as oxygen, partially substitute for the tellurium in the lattice. Thus, it can be assumed that the $FeTe_4$ cluster formed by iron-57, in an interstitial position, is partially transformed into $FeTe_xO_{1-x}$ with x values of 1, 2, 3 or 4, due to the oxidation process.

At this point a question arises as to whether the existence of these clusters reflects a random distribution of oxygen in the various crystallographic sites of the single crystal, or that the oxygen substitution of tellurium in the vicinity of the iron is due to the preference of iron to combine itself with the oxygen. Therefore, measurements were carried out on the variation of the Mössbauer peak intensities. These peak intensities correspond to the various clusters after annealing at 400°C in an inert atmosphere. The spectrum of SnS doped with iron-57 has been resolved into two quadrupole doublets (Q_1 and Q_2). The first has the same Mössbauer parameters as the non-oxidized samples. The second has values which can be ascribed to iron-57 where oxygen is present. The intensity of this doublet increases when the annealing time in inert atmosphere increases. In contrast, the intensity of the first doublet decreases when the annealing time increases. Therefore, random oxygen distribution can be excluded in the presence of iron microtraces. The large amount of iron having the neighboring tellurium partially substituted by oxygen proves the existence of an association between interstitial iron and substituted oxygen.

The sample of SnTe doped with iron-57 and oxidized for the same length of time behaves in the same way. In any case, the appearance of a more complex cluster is dependent upon a higher oxygen concentration within these materials.

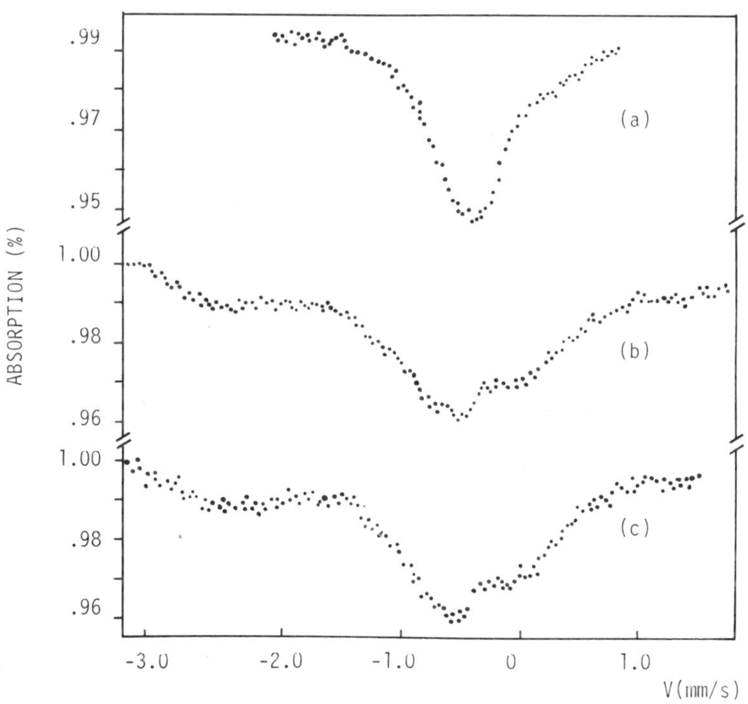

Figure 3. Emission Mössbauer spectra of SnTe containing cobalt-57 after (a) 15, (b) 25 and (c) 35 hours of oxidation.

SnS, SnSe, and SnTe

The properties of IV-VI binary semiconductors change with the variation of structure, which at room temperature changes from cubic in SnTe to orthorombic in SnSe and SnS. A detailed analysis of the chemical bond and structural properties of the IV-VI semiconductor group is complex due to the orbital hybridization and bond multiplicity. A complete overview of all these effects has not been considered to date, and we shall limit ourselves to describing qualitatively some results obtained by Mössbauer spectroscopy.

The parameters of the resonance absorption of the 23.8Kev radiation emitted by tin-119 in $BaSnO_3$ were measured for SnS, SnSe, and SnTe. The isomer shifts at 300K have been found [8] to be 3.33(3), 3.31(3), and 3.48(3)mm/s, respectively. One can relate the isomer shift to a more general electron configuration. Lees and Flinn [9] assume that the charge density is a function of both n_s and n_p, where n_x and n_p represent the number of 5s and 5p electrons, respectively, which enter the bond. For the tin(II) compounds, n is two and, for the isomer shift relative to the $BaSnO_3$ source at 300K, we obtain

$$\delta = 3.77 - 2.58n_p - 0.018n_p^2.$$

From this expression, the n_p values of 0.56 for SnS, 0.56 for SnSe, and 0.49 for SnTe were obtained at 300K. This confirms the existence of ionic-covalent bonds, in agreement with the data for the interatomic distances, which are intermediate between the values of the sum of the covalent radii and the sum of the ionic radii. For example, for SnTe the interatomic distance is 3.15Å, the sum of the covalent radii is 2.77Å, and the sum of the ionic radii is 3.33Å.

It is also possible to consider the effects of the crystal structure on the bonding type. In fact, the orthorombic SnS and SnSe can be considered as a distortion of the SnTe structure which is of the NaCl type at room temperature. In all three cases the tin atom has a coordination number of six. One can thus explain the presence of a quadrupole doublet only in SnS and SnSe. Moreover, from the data reported by Fano and Ortalli [8], it can be seen that the quadrupole splitting changes very little between 80 and 300K. One can then expect that, in this case, the quadrupole splitting will be determined only by the electric field gradient and does not depend on the 5p electrons.

Finally, in this class of semiconductors, it is possible to relate the energy gap with the isomer shift values as reported in Figure 4 where one can see a qualitatively linear behavior.

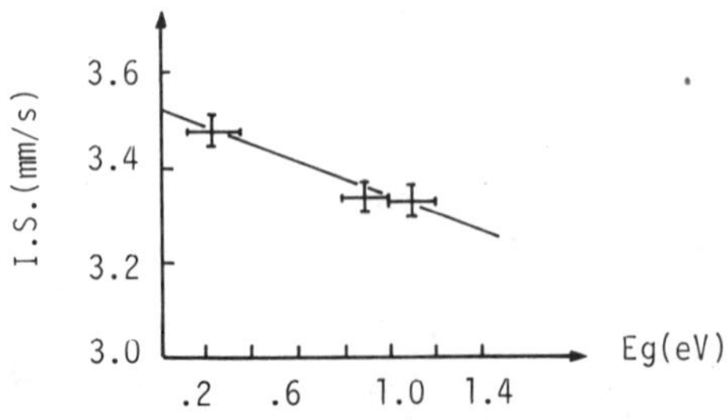

Figure 4. Relation between isomer shift and gap energy for SnTe, SnSe, and SnS.

$Pb_{1-x}Sn_xTe$ Systems

The technological properties of the $Pb_{1-x}Sn_xTe$ system, which forms a solid solution over the whole range of composition, are directly related to the fundamental properties of the alloy and to the lead to tin stoichiometric ratio. In fact, much evidence supports the band inversion model by Dimmock et al. [1], which explains the observed variation of the valence-conduction band gap, E_g, with x.

Mössbauer absorption spectra were recorded with a source of iodine-125 in copper for tellurium-125 studies [10] and with a $CaSnO_3$ [8] for the tin-119 studies. The tellurium-125 and tin-119 Mössbauer spectra of the samples consist of single lines. The linewidth at half maximum and the isomer shift are plotted as a function of x in Figure 5 for tellurium-125 and Figure 6 for tin-119, respectively. For tellurium-125 the linewidth values are the same, within the errors for the five compositions investigated, whereas the isomer shift values vary with a minimum in the composition range $0.3 < x < 0.4$. In contrast, for tin-119 the linewidth values indicate that, in the last two compositions, a quadrupole interaction at the tin atom is present [11]. The estimated quadrupole splitting values are equal to those observed in SnTe (0.38mm/s) as reported in reference 8. The isomer shift values are constant within the errors.

It is known that the valence and conduction band edge of $Pb_{1-x}Sn_xTe$ occurs at the L point in the Brillouin zone. Moreover, for lead telluride, the valence band edge is an L_6^+ state and the conduction band edge is an L_6^- state. These levels are inverted for tin telluride and the L_6^+ valence band states in PbTe are s-like at the lead atom and p-like at the tellurium atom, while L_6^- conduction band states are p-like at the lead atom and s-like at the tellurium atom, as shown by Tsang and Cohen [12]. In addition the change of the sign for the tin ions in the $L_6^+ \rightarrow L_6^-$ band crossover region has been predicted by S. H. Wengle [13]. These considerations suggest that the supposed charge transfer occurring when the lead to tin stoichiometric ratio changes can be related to isomer shift measurement. However, within the limits of experimental error, we found no variation of the isomer shift [10]. Consequently, one can state that the hypothesized charge transfer does not significantly involve the s-like electrons of tin.

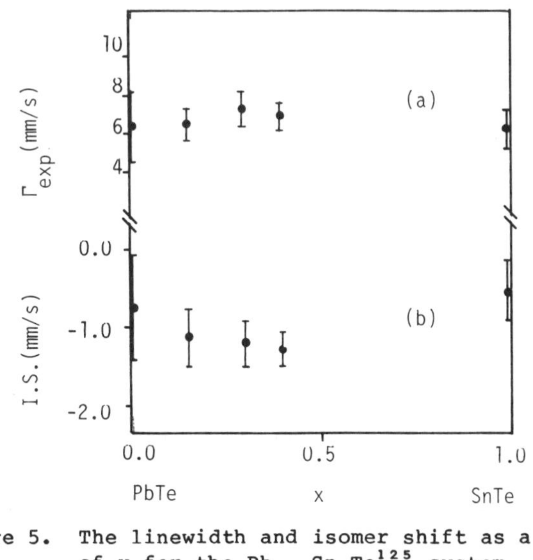

Figure 5. The linewidth and isomer shift as a function of x for the $Pb_{1-x}Sn_xTe^{125}$ system.

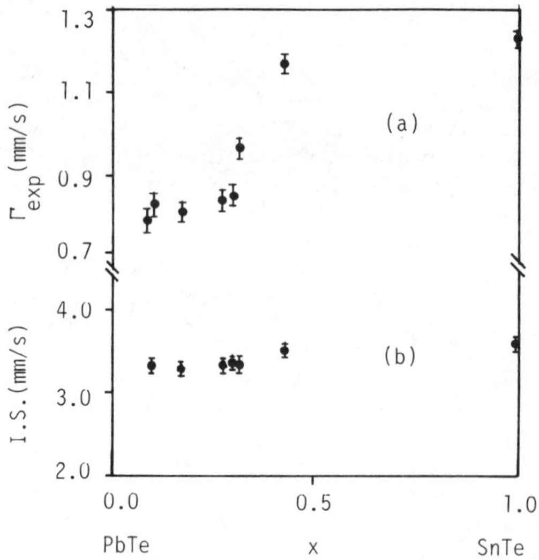

Figure 6. The linewidth and isomer shift as a function of x for $Pb_{1-x}Sn_x{}^{119}Te$ system.

On the other hand in the tellurium compounds the decrease of the isomer shift corresponds to the decrease of the electron density. From our data we can deduce that the isomer shift values of tellurium, for intermediate values of x, are smaller than those found for PbTe and SnTe by Stevens and Gettys [14], and their behavior is similar to that of the absolute value of the gap as a function of the composition as reported [10]. The minimum electron density of the tellurium nucleus occurs for an x value in the 0.3 to 0.4 range, when the energy gap is zero. Thus the minimum isomer shift can be attributed to the increased metallic bond associated with the smaller band gap value.

Finally in the IV-VI semiconductor group the tellurium $5s^2$ electrons are inactive and the chemical bonds are caused by the p electrons as has been demonstrated [15,16]. The existence of ionic-covalent bonds has been confirmed, and in particular, a $5s^2\ p^{0.46}$ electronic configuration for tellurium in tin-119 tellurium at 77K has been found [8,17].

PHASE TRANSITIONS

SnTe Compound

Some attempts have been made to clarify whether a phase transition below room temperature for SnTe, with the same number of free carriers, exists. In this case the Mössbauer studies have the advantage of probing the lattice microscopically and of giving information on changes in the lattice symmetry and the phonon spectrum at the phase transition. The first results (Figure 7) were interpreted in terms of a structural phase transition at 160K with the reasonable assumption that the transition from one cubic phase to another creates a small electric field gradient in tin atoms, thus broadening the line as reported in reference 8.

In addition, some Mössbauer experiments show [18,19] an anomalous behavior in the recoilless fraction, f_a, at very low temperatures. A phase transition at 22K was observed in different samples [18,49]. At this temperature, the f_a versus temperature curve shows a dip and the resistivity measurements show a slight jump at the same temperature (Figure 8).

This temperature dependence of f_a suggests a phase transition that might be associated with an anomaly of the Debye-Waller factor. Moreover, it is possible to suppose, as shown by Shirane and Yamade [20], that in SnTe there

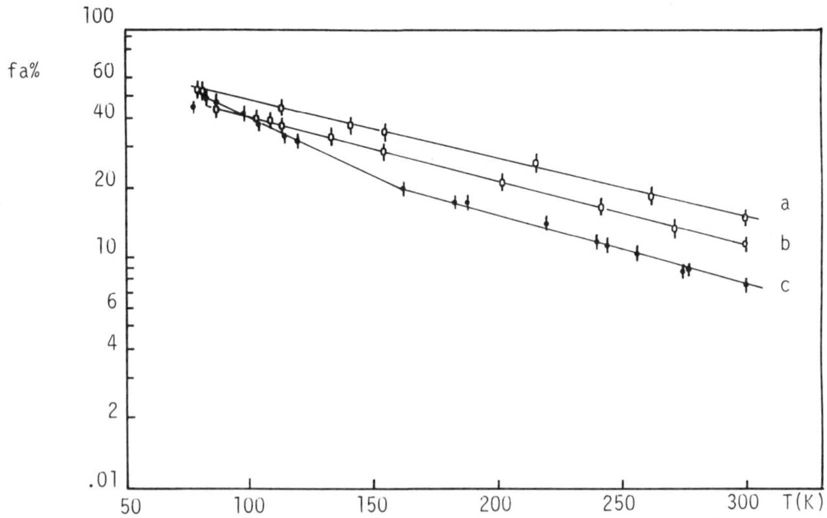

Figure 7. The temperature dependence of the $\log_{10} f_a$ for SnS (a), SnSe (b) and Sne (c).

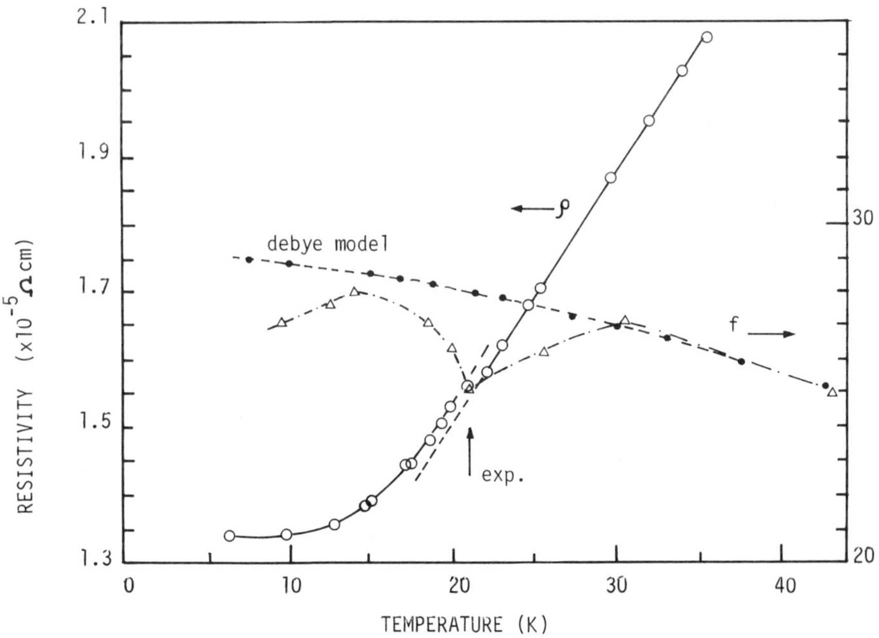

Figure 8. Experimental data for the Mössbauer area plus the theoretical curve in the Debye approximation and resistivity measurements for SnTe.

are lattice modes at the Brillouin zone boundary where the frequency approaches zero at the transition temperature. Usually the anomalous phonon mode is considered to originate from a softening of a transverse optical mode at the center of the Brillouin zone. However the observation of soft transverse acoustic modes near the boundary of the Brillouin zone has been reported by Canner et al. [21]. For the SnTe "paraelectric material" with lattice dynamic properties very close to $SrTiO_3$, one can suppose that the phase transition is antiferroelectric due to transverse acoustic modes boundary. The correlation between our anomaly of the phonon spectrum (soft mode) and the phase transition then gives an indication of an antiferroelectric phase transition at the Brillouin zone boundary.

Another anomaly at 90K was found by means of neutron diffraction as shown by Izumi et al. [22], and by electrical resistivity and specific heat measurements as shown by Hatta and Kobayashi [23]. These results were interpreted as due to a transverse optical phonon softening accompanying the cubic to rhombohedral phase transition. This anomaly is not confirmed by Mössbauer measurements and the phase transition at 160K was not confirmed by specific heat and resistivity measurements.

On the basis of these experimental results, some conclusions [24] can be reached. The phase transition at 22K could be identified as a first order rhombohedral to orthorombic phase transition. According to Birman [25], the three phase transitions can be classified as from cubic to tetragonal at 160K, a structural phase transition, from tetragonal to rhombohedral at 90K, a ferroelectric phase transition with a soft mode due to a transverse optical mode at the Brillouin zone center, and from rhomobohedral to orthorombic at 22K, an antiferroelectric phase transition, due to a transverse acoustic mode at the Brillouin zone boundary. But the first and second phase transitions

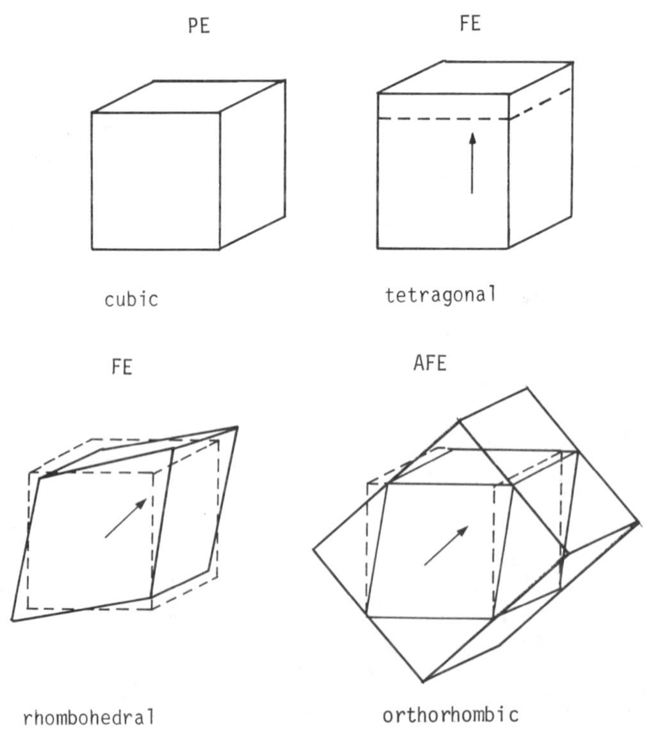

Figure 9. Distortion of the unit cell in its polymorphic forms.

can be considered to have a closer correlation. In fact the unit cell of tin telluride to 160K is cubic. At lower temperatures the structure is no longer cubic, and there is a variation of cell parameter indicating a transformation from cubic to tetragonal (Figure 9). At 90K the transformation is completed showing a rhombohedral structure. In the last phase transition a rhombohedral to orthorombic transformation occurs.

$Sn_{1-x}Ge_xTe$ System

In the $Sn_{1-x}Ge_xTe$ solid solutions, the phonon dispersion curves obtained by Lefkowitz et al. [26] from neutron diffraction studies for the composition of x of 0.935, show that the transverse optical mode frequency decreases at lower temperatures, whereas the short wavelength mode frequency is practically independent of temperature. Therefore, it seems that the phase transition in $Sn_{1-x}Ge_xTe$ is a displacive type ferroelectric transition and that this occurs as a consequence of the crystal instability caused by the large fluctuations introduced by the transverse optical mode for vanishing frequencies.

The correlation between the ferroelectric phase transition and the anomaly of the phonon spectrum (soft mode) is not clear, but usually suggests that this transition could be associated with an anomaly of the Debye-Waller factor and therefore with the temperature dependence of the recoil-free fraction. In particular the problem of the effect of the soft modes on the Mössbauer parameters has not yet been completely clarified for $Sn_{1-x}Ge_xTe$ at low germanium concentrations. In this regard, measurements on different absorbers, taken from the central part of the same crystal, were carried out. For example, the absorption peak area values of the $Sn_{0.9}Ge_{0.1}Te$ sample [27] obtained from different series of Mössbauer measurements show that no appreciable systematic anomaly was found in the 110 to 300K range (Figure 10). This is in disagreement with the conclusion reported by Chekin et al. [28] and Rigamonti and Petrin [29], but it is in agreement with those reported by Knauer [30], and this supports the debate about the observation, through the

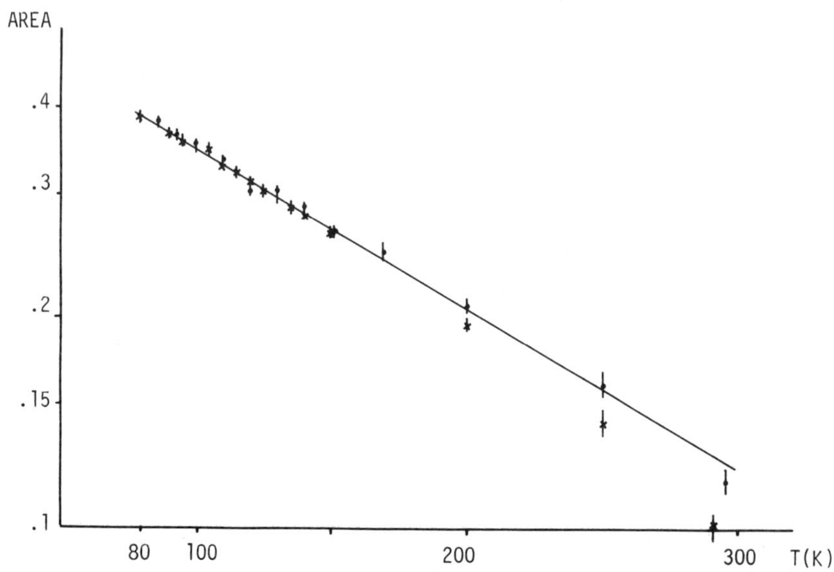

Figure 10. The area under the tin-119 Mössbauer resonance curves versus temperature for $Sn_{0.90}Ge_{0.10}Te$ single crystal samples.

f_a Mössbauer parameter, of ferroelectric phase transitions eventually assisted by soft mode phonons.

For sufficiently low concentrations of germanium, the behavior of the linewidth as a function of temperature as reported in [31] is similar to that of undoped SnTe, which presents a structural phase transition at 160K. However, the change in the linewidth slope disappears for higher x values, for instance, for the $Sn_{48.7}Ge_{1.6}Te_{50.3}$ sample. From the analysis of the germanium contents it follows that this behavior is related to the total number of cationic atoms, tin plus germanium at a stoichiometric ratio in the homogeneity region. For an x of 0.012, the Mössbauer parameters are approximately those of undoped SnTe, only because the germanium maintains the SnTe inside the homogeneity region. For high germanium concentrations the behavior of the systems is the same as the $Sn_{1-x}Ge_xTe$ pseudobinary alloy.

To conclude the studies performed to these compounds have revealed a new interesting application of Mössbauer spectroscopy by showing how it is possible to obtain the minimum value of doping element concentration at which the system become a true pseudobinary alloy.

CONCLUSION

It has been shown that Mössbauer spectroscopy is useful in a wide range of problems related to the technological properties of tellurium and/or tin based IV-VI semiconductor group. The evaluation of the electronic configuration gives a contribution to understanding the transport properties. Evidence of phase transitions in some systems is observed. The use of iron as a probe for testing the local lattice distortion of the reactivity of intentionally introduced impurities is demonstrated.

ACKNOWLEDGEMENT

The authors appreciate the financial support of the National Research Group on the Structure of Matter, C. N. R., and the Ministry of Education, MPI.

REFERENCES

1. J. O. Dimmock, J. Melngailis, and A. Y. Strauss, Phys. Rev. Lett. 16, 1193 (1966).

2. V. Fano, Prog. Crystal Growth Charach. 3, 287 (1981).

3. V. Fano, E. Maniezzi, I. Ortalli, and R. Pergolari, Materials Chemistry, 7, 509 (1982).

4. V. Fano and I. Ortalli, J. Phys. Chem. Solids, 37, 729 (1976).

5. T. L. Koval'chik and Yu. P. Maslakovets,Zh. Texh. Fiz. 26, 2417 (1956).

6. J. Dannon, Rev. Mod. Phys. 36, 459 (1964).

7. V. Fano and I. Ortalli, J. Mater. Science 12, 210 (1977).

8. V. Fano and I. Ortalli, J. Chem. Phys. 61, 5017 (1974).

9. J. K. Lees and P. A. Flinn, J. Chem. Phys. 48, 882 (1968).

10. A. Antonini, V. Fano, I. Ortalli, and A. Vera, Phys. Stat. Sol. (b) 114, 585 (1982).

11. A. Antonini, V. Fano, and I. Ortalli, Phys. Stat. Sol. (b) 120, K15 (1983).

12. Y. N. Tsang and M. L. Cohen, Solid State Com. 10, 871 (1972).

13. S. H. Wengle, Phys.Lett. 45A, 401 (1973).

14. J. S. Stevens and W. L. Gettys, Mössb. Eff. Data, Decemb. 14, 1981 (India).

15. T. C. Gibb, R. Greatrex, N. N. Greenwood, and A. C. Sarmo, J. Chem. Soc., A212 (1970).

16. H. Krebs, Acta Cryst. 9, 95 (1956).

17. S. L. Ruby and G. K. Shenoy, in "Mössbauer Isomer Shifts", G. K. Shenoy and F. E. Wagner, eds., North-Holland Publ. p. 627 (1978).

18. V. Fano, G. Fedeli, and I. Ortalli, Solid State Comm. 22, 467 (1977).

19. I. Ortalli, G. Fedeli, K. L. I. Kobayashi, and K. F. Komatsubara, Journal de Physique 40, C2-696 (1979).

20. G. Shirane and Y. Yamada, Phys. Rev. 117, 858 (1969).

21. J. P. Canner, C. M. Yagnik, R. Gerson, and J. James, Journ. Appl. Phys. 42, 4708 (1971).

22. M. Izumi, Y. Hamaguchi, K. F. Komatsubara, and Y. Kato, J. Phys. Soc. Japan 38, 443 (1975).

23. O. Hatta and K. L. I.Kobayashi, Solid State Communication 22, 775 (1977).

24. I. Ortalli, Ferroelectrics, 54, 325 (1984).

25. J. L. Birman, Journal de Physique 29, C4-151 (1968).

26. M. Lefkowitz, M. Shields, G. Dolling, W. J. C. Buyers, and R. A. Cowley, J. Phys. Soc. Japan Suppl. 28, 249 (1970).

27. V. Fano, E. Maniezzi, I. Ortalli, N. Burriesci, M. Petrera, Journal de Physique 40, C2-624 (1979).

28. V. V. Chekin, O. P. Balkashin, and G. D. Sultanov, Ukr. Fiz. Zh. 14, 102 (1969).

29. A. Rigamonti and G. Petrini, Phys. Status Solidi, 41, 581 (1970).

30. R. C. Knauer, Phys. Lett. 33A, 313 (1970).

31. V. Fano, E. Maniezzi, and I. Ortalli, Journal de Physique, 41, C1-325 (1980).

THE MÖSSBAUER CHARACTERIZATION OF MULTILAYERED FILMS WITH ARTIFICIAL SUPERSTRUCTURES

Teruya Shinjo and Toshio Takada

Institute for Chemical Research
Kyoto University
Uji, Kyoto-fu 611, Japan

INTRODUCTION

Artificially structured metallic superlattices have attracted much attention in recent years [1]. Artificial superstructures can be fabricated, not only from similar elements that form solid solutions but also by combining elements with considerably different atomic radii. A very interesting example of the latter case is the multilayer consisting of iron and magnesium, because these are mutually insoluble elements even in the liquid state. As will be described below, the present authors have confirmed that artificial periodicities with very short wavelengths can be constructed in iron-magnesium multilayers [2].

In Figure 1, two types of cross sections for multilayers are sketched schematically. In the case that the two elements have similar atomic radii and the original crystal structures are the same, it is likely that an epitaxial relation is held at each interface (Figure 1A). On the other hand, if the atomic sizes are quite different, epitaxial growth seems to be impossible. Figure 1B shows an example where no crystallographic coherency exists between the individual layers. In the early stage of metallic superlattice study, attempts were only made to prepare multilayers of the epitaxial type.

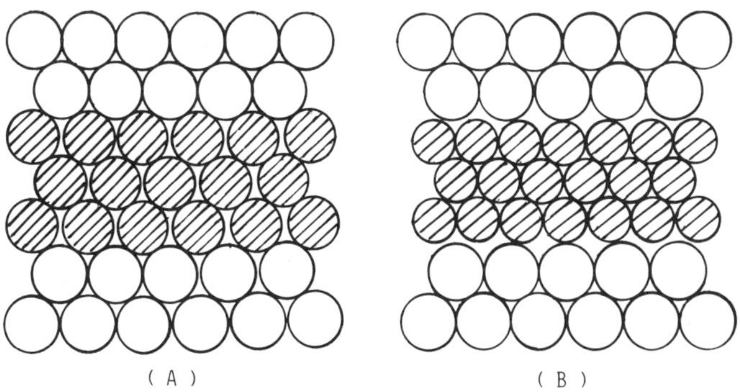

Figure 1. Two types of multilayers. (A) An epitaxial interface; the atomic sizes of the two elements are slightly different. (B) Nonepitaxial interface; the atomic sizes are considerably different.

Typical examples of the combination of similar elements are nickel-copper multilayers, whose magnetic properties have been rather intensively studied by several groups in the United States [3]. The present authors have prepared multilayers consisting of iron and vanadium, both of which are 3d metal elements with bcc structures and have investigated their magnetic properties by means of Mössbauer spectroscopy and polarized neutron diffraction [4].

Very recently nonepitaxial multilayers have been fabricated in several systems. Schuller has succeeded in preparing multilayers consisting of the eutectic combination, of bcc niobium and fcc copper, and studied its superconducting properties [5,6]. Other examples prepared by the present author besides iron-magnesium are iron-antimony [7], cobalt-antimony [8], and vanadium-silver multilayers [9]. These films are interesting as new materials which cannot exist naturally.

In this article, the results of Mössbauer spectroscopic studies on the two types of multilayers, iron-magnesium and iron-vanadium, are introduced and the utility of Mössbauer spectroscopic studies on the two types of multilayers is discussed.

IRON-MAGNESIUM MULTILAYERS

Vacuum deposition and sputtering techniques are the common preparative methods for metallic thin films. In the present study, multilayers have been prepared by alternate deposition in ultrahigh vacuum. Iron and magnesium are

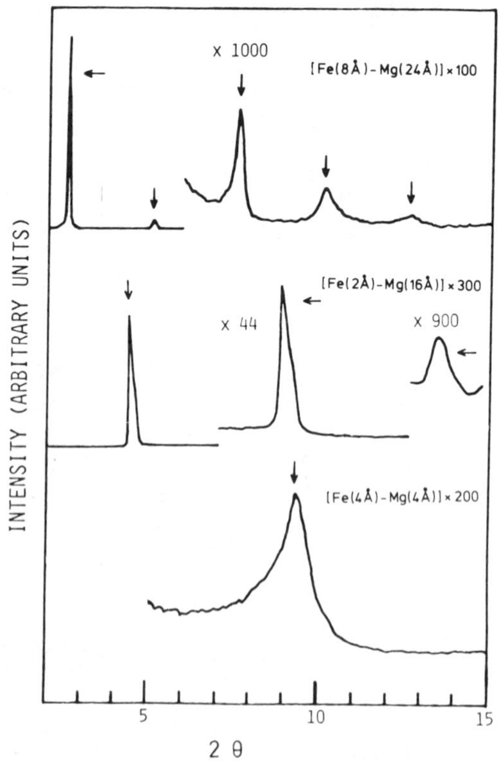

Figure 2. X-ray diffraction patterns in a small angle region for iron-magnesium multilayered films.

heated by electron beam guns in a vacuum system (10^{-10} torr range as the base and 10^{-9} torr during the deposition) and alternately deposited onto a cold substrate mylar film at about -50°C. Individual layer thicknesses are regulated with two shutters controlled by a programmable quartz oscillating thickness monitor. In each sample, the number of stacked bilayers is 100 ~ 300.

The established artificial periodicities are confirmed from X-ray diffraction. As shown in Figure 2, the Bragg reflections due to the superstructures are observed in a small angle region. For most of the prepared samples, the higher order peaks, up to the fifth order, are easily observed, which suggest the compositional modulation has nearly a square-wave mode. Even when the iron layer thickness is 2Å, which roughly corresponds to the monolayer thickness, the periodic structure is evidently established. The shortest wavelength so far constructed is iron(4Å)-magnesium(4Å), is also shown in the same figure. The wavelengths of the compositional modulation realized in the deposited films are estimated from the diffraction peak angles. For this evaluation, peak angles larger than 2° were used where the effect of refraction is negligible. In Figure 3, the observed wavelengths are compared with the "design" value, i.e., the sum of single iron and single magnesium layer thickness determined by the thickness monitor during the deposition. The agreement between "design" and "observed" periods is excellent. There is thus evidence that the thicknesses of the deposited layers are controlled on the atomic scale.

X-ray diffraction peaks corresponding to the bulk iron and magnesium structures are observed at higher angles, when the individual layers are at least 10Å thick. Although the substrate is cold and has no orientation, the deposited iron-magnesium films show a definite texture structure. The (110) reflection of bcc iron and (002)reflection of hcp magnesium are preferentially oriented in the film plane. The details of X-ray diffraction analysis will be published elsewhere [10].

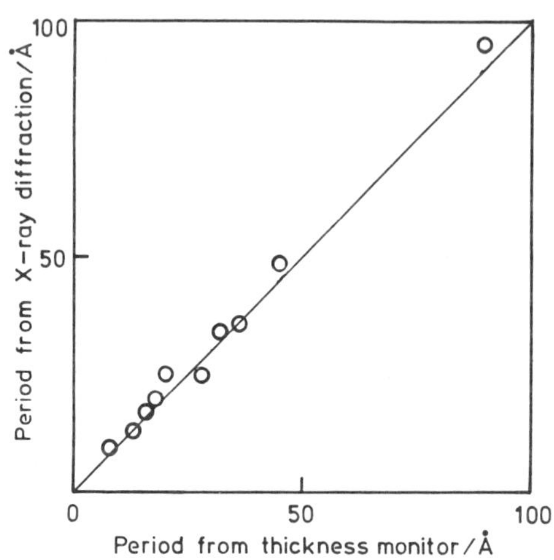

Figure 3. Comparison of "design" and "observed" wavelengths of compositional modulation in iron-magnesium multilayers. The period determined by the thickness monitor during the deposition is the "design" and that estimated from X-ray diffraction peak angles is the "observed" value.

Figure 4 shows the Mössbauer spectra for some iron-magnesium multilayers. In these samples, the magnesium layer thicknesses are not very small and therefore the magnetic interaction through the magnesium layer is negligible. The present result elucidates the thickness dependence of the magnetic properties in the iron layer sandwiched in between magnesium layers. At both 300K and 4.2K, the spectra for the 30Å iron film are nearly the same as the standard α-iron patterns. The 15Å iron film also shows a sextet, but the adsorption lines have asymmetric broadenings, suggesting the existence of reduced hyperfine fields at the interface sites. If the iron layer is thinner than 8Å, the spectra at 300K have no magnetic hyperfine structure and only quadrupole doublets are observed. At 4.2K, magnetic hyperfine splittings are observed for all samples, even when the iron layer thickness is 2Å. The average hyperfine field becomes smaller with the decrease of the thickness but the reduction compared with the bulk hyperfine field of 340kOe, is not drastic and is 250kOe for the 2Å sample. It is thus evident that all iron atoms have local magnetic moments and are participating in magnetically ordered states.

In order to synthesize an amorphous alloy film consisting of magnesium and iron, the co-deposition technique has to be used [11]. From the Mössbauer spectra for a 99.5 percent magnesium-0.5 percent iron film, isolated iron impurities in a magnesium matrix are confirmed to show a sharp singlet at any temperature down to 0.1K [12]. Because the spectra shown in Figure 4 have no singlet fraction, it is concluded that there is no diffusion or intermixing between iron and magnesium. In other words, the iron-magnesium multilayers have chemically "sharp" interfaces.

It is noteworthy that the intensity ratio of the six lines in case of the 2Å film is close to 3:0:1:1::0:3. When the iron layer thickness is larger than 15Å, the observed intensity ratio is 3:4:1:1:4:3, which is reasonable because the spontaneous magnetization in a thin film is usually oriented in the film plane by the shape anisotropy. However the relative

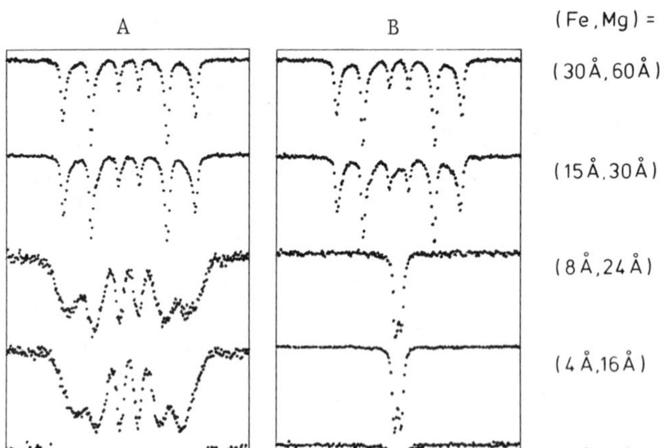

Figure 4. Iron-57 Mössbauer absorption spectra obtained at 4.2K (A) and 300K (B) for iron-magnesium multilayers with artificial superstructures.

intensities of 2 and 5 lines become weaker with the decrease of the iron layer thickness. Because the gamma-ray direction is perpendicular to the sample plane, this means that the direction of the spontaneous magnetization changes from in-plane to the surface normal. In the case of the 2Å iron film, the average angle between the magnetization and the film plane is estimated to be about 70°, from the relative intensities of 2 and 5 lines.

Some years ago, Gradmann et al. [13] claimed that the spontaneous magnetizations in very thin ferromagnetic films of iron-nickel alloy or cobalt with the thickness of a few Å are aligned along the surface normal. This claim was based on their magnetic torsion measurements [13]. They have attributed the phenomenon to the surface anisotropy which was originally proposed by Nèel [14]. Such a perpendicular magnetization seems to be a characteristic of very thin ferromagnetic films where the surface anisotropy plays a dominant role. On the other hand, one cannot account for this result by assuming the formation of iron clusters elongated along the surface normal. Such cluster formations apparently contradict the observed superstructure with the wavelength of iron(2Å)-magnesium(16Å). The 2Å iron layer is supposed to consist of fractional monolayers but the structure is not clear at the present time.

IRON-VANADIUM MULTILAYERS

The conditions required for preparing multilayers consisting of iron and vanadium is similar to that for iron-magnesium multilayers. Samples with various compositional ratios or iron and vanadium have been prepared and examined by X-ray diffraction. The shortest artificial period observed by X-ray diffraction is 12Å; iron(4Å)-vanadium(8Å) and iron(8Å)-vanadium(4Å). Both iron and vanadium have bcc structures and can form solid solutions in all compositional ratios, although the atomic sizes differ by about 6 percent. The structure of the prepared multilayers is confirmed to be bcc as expected; the (110) axis is preferentially oriented along the normal to the

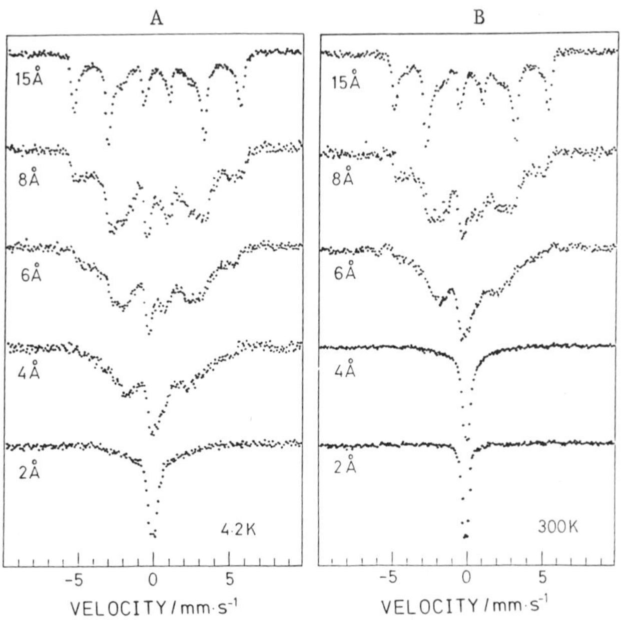

Figure 5. Iron-57 Mössbauer absorption spectral obtained at 4.2K and 300K for iron-vanadium multilayers with artificial superstructures.

Figure 6. (A) Observed hyperfine field distributions for iron-57 from Mössbauer and for vanadium-57 from NMR measurements on a iron(15Å)-vanadium(30Å) multilayered film. (B) Calculated hyperfine field distributions for ideal interface (atomically flat and no mixing as Figure 6C). (C) The profile of the compositional modulation at an ideal interface. The atom layers with $i \leq 4$ are pure iron and those with $i \geq 5$ are pure vanadium. Horizontal broken lines indicate the magnetic moment densities at respective atom layers. (D) Calculated hyperfine field distributions based on the compositional modulation at the interface shown in Figure 6E, which gives the best fit for the vanadium hyperfine field distribution in Figure 1A. (E) The profiles of compositional and also magnetic moment density modulation at the interface between iron and vanadium layers. The compositions for the atom layers, i = 3, 4, and 5, are 95 percent iron-5 percent vanadium, 50 percent iron-50 percent vanadium, and 5 percent iron-95% vanadium, respectively. The broken lines show the magnetic moments of iron and also vanadium at each atom layer.

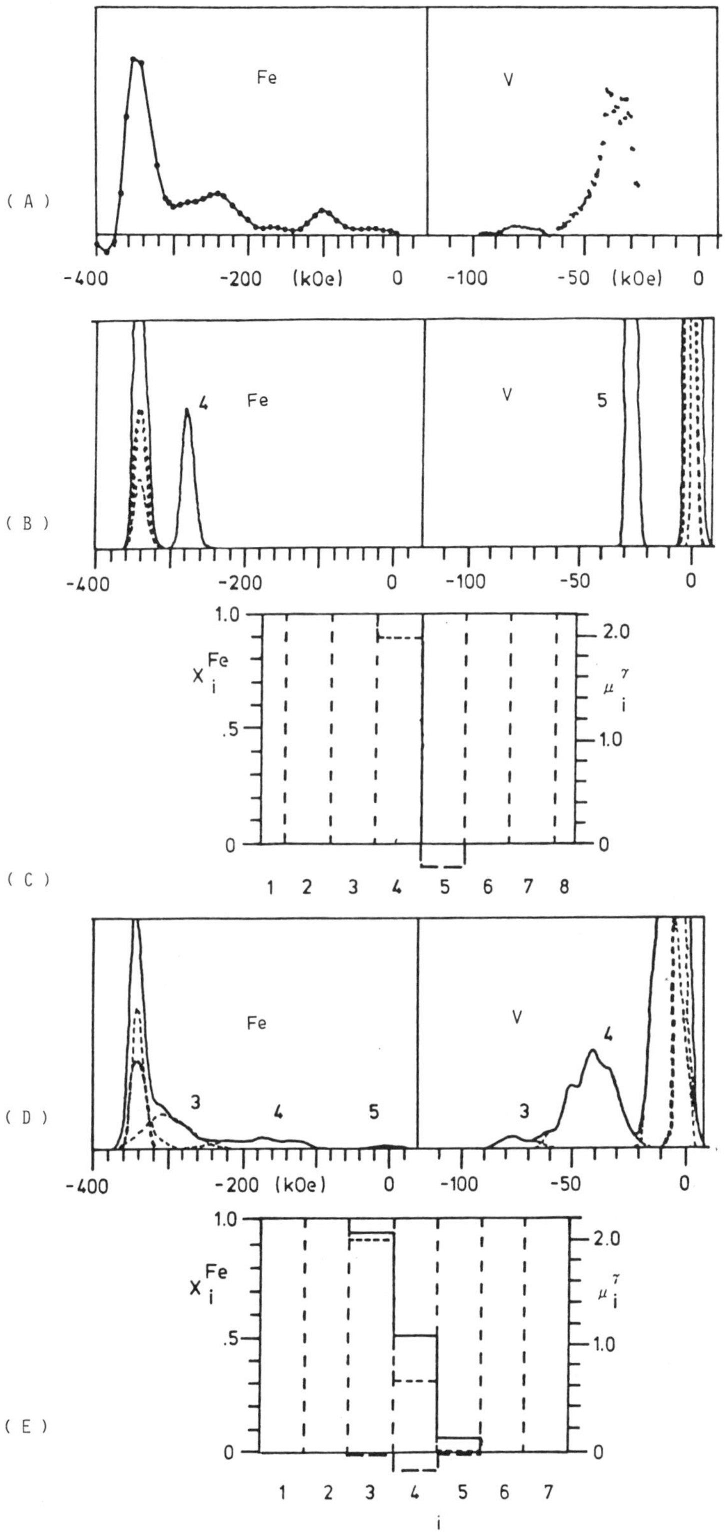
(A)
Fe
V
-400
-200
(kOe)
0
-100
-50
(kOe)
0
(B)
4
Fe
V
5
-400
-200
0
-100
-50
0
(C)
1.0
.5
0
X_i^{Fe}
2.0
1.0
0
μ_i^{γ}
1 2 3 4 5 6 7 8
(D)
Fe
V
3
4
5
3
4
-400
-200 (kOe)
0
-100
-50
0
(E)
1.0
.5
0
X_i^{Fe}
2.0
1.0
0
μ_i^{γ}
1 2 3 4 5 6 7
i

film plane. However, the crystallographic coherent length is fairly limited. The broadening of the X-ray diffraction peaks suggests the size of the crystallites is the order of 100Å [15].

The Mössbauer spectra at 4.2 and 300K as a function of the iron layer thickness are shown in Figure 5. The result at 300K is similar to the case of iron-magnesium multilayers. A nonmagnetic fraction appears in the 8Å spectrum and grows with the decrease of the thickness. The critical iron layer thickness required to stabilize the magnetization at 300K is around 8Å.

The degree of intermixing, or interdiffusion, at the interface can be estimated from the analysis of hyperfine field distribution. Not only iron-57 hyperfine fields, but also vanadium-51 hyperfine fields are available for the analysis of iron-vanadium multilayers. Vanadium-51 NMR measurements for the iron(15Å)-vanadium(30Å) film have been performed by Takanashi et al. [16}, and furnish complementary information to the iron-57 Mössbauer results. In the iron-vanadium alloy system, the dependences of the local magnetic moment and the hyperfine field at iron (or vanadium) on the local atomic configuration has been well understood and the parameters necessary to estimate the hyperfine field are already known [17]. Therefore, we can roughly calculate the hyperfine field from the coordination numbers of iron in the first and second nearest neighbors. If we assume the profile of the compositional modulation at the interface, the distribution of the hyperfine fields are estimated.

Figure 6A shows the observed hyperfine field distribution of vanadium at 1.3K together with the distribution curve for the iron hyperfine field deduced from the Mössbauer spectrum at 4.2K. The hyperfine field observed a the vanadium site in the absence of an external field shows the existence of a local magnetic moment induced by ferromagnetic iron atoms at the near neighbor sites. Therefore, only vanadium atoms in the vicinity of the interface contribute to the NMR signals. In the case of an ideal interface, namely atomically flat and with no intermixing, as shown in Figure 6C, only the top interface vanadium atom layer has a large hyperfine field but the field at the second layer should be negligibly small. Consequently, the predicted distribution curve for vanadium hyperfine fields is like a delta function (Figure 6B). In the iron side, the top interface atom layer has a unique, reduced hyperfine field, but the next layer has one similar to the α-iron. Then the distribution curve for the iron hyperfine fields should have one sharp satellite peak. The experimental results in Figure 6A do not agree with this model but show considerable broadening, suggesting an intermixing to be a certain extent.

The best fit for the distribution of vanadium hyperfine fields is obtained if the composition at each atom layer changes as shown in Figure 6E.. Namely the interface atom layer is assumed to have 50 percent iron-50 percent vanadium composition, and the mixing extending to one more atom distance is taken into account. The calculated curves for the hyperfine field distributions are shown in Figure 6D. This result indicates that significant layer intermixing occurs only at the interface atom layer and does not extend for more than three atom layers. It is to be noted that the short range order parameter, used in the fitting procedure, indicates that there is a tendency toward clustering rather than the homogeneous alloying of iron and vanadium in the interface atom layer. Therefore, the main reason or the mixing is not thermal diffusion but is supposed to be the microscopic surface roughness of the deposited film. For more details of the analysis, the original reports should be consulted [16,18,19].

In contrast to the previous results on iron-magnesium films, the Mössbauer spectra at 4.2K for extremely thin iron layers (less than 6Å) show nonmagnetic fractions and the relative amount increases with the decrease of the thickness. The intermixing at the interface seems to play an important role. According to the result of a theoretical study on iron-vanadium alloys [17], the iron magnetic moments depend on the microscopic environment, i.e., the number of vanadium neighbors, and the macroscopic composition ratio. For instance, in a vanadium-rich region, an iron atom loses the local magnetic moment if the number of vanadium nearest neighbors is more than four. Even when the intermixing is limited to within a few atomic layers, there should be a number of different interface sites. Then there is a large distribution of hyperfine field and eventually some of the iron atoms lose their local magnetic moment. In the case of 2Å film, almost all iron atoms are nonmagnetic. The observed hyperfine field is an applied field of 45kOe is actually the same as the external field. It is therefore evident that iron atoms in the nominally 2Å layer have no local moment. However, whether or not an ideal iron monolayer which is pure, atomically flat, and infinitively large, and sandwiched in between vanadium layers is magnetic is uncertain at this time.

DISCUSSION

Iron-57 Mössbauer spectroscopy has been applied to characterize iron-magnesium and iron-vanadium multilayers whose superstructures were confirmed by x-ray diffraction. The results on iron-magnesium films show that no mixing occurs between iron and magnesium. A novel magnetic property has been revealed in the iron monolayer sandwiched in between magnesium layers.

The compositional profile at the interface between iron and vanadium layers is estimated by combining the iron-57 hypefine field distribution observed in the Mösbauer spectra with the vanadium-51 distribution estimated by NMR technique. The intermixing is suggested to be confined to within three atoms layers. Because the substrates are cooled during the deposition, the thermal diffusion is inhibited but the size of the crystallite is very small. For the purpose of preparing single crystal multilayers, one has to choose an appropriate temperature to balance crystal growth and thermal diffusion.

Multilayered films with controlled layer thicknesses in the accuracy of atomic scale are interesting mainly from the following two viewpoints: (1) As a model system for a specific purpose in fundamental research. We can design the structure of our sample according to our purpose. The iron magnesium film is an example of a two dimensional ferromagnet. Dimensional cross-over in superconductivity has been studied extensively in the United States [20]. (2) In the fabrication of new materials. The multilayering technique is a promising method to synthesize new materials which can not exist in thermal equilibrium. In this sense, the multilayers of type B in Figure 1 are of great interest. The iron-magnesium multilayers are typical examples of this type. Concerning the stability, the difference between the types A and B is clearly revealed by heat treatment. In the case of iron-vanadium multilayers, the superstructures are smeared by heating and finally are totally homogenized. In contrast, iron-magnesium films show decomposition into the two elements. The success in preparing iron-magnesium multilayers suggests that we have a great number of possibilities in the choice of the constituents of superlattices. A lot of new materials can be produced, which have not yet been known. No doubt some of them will be useful for industrial applications. At least for the multilayers including iron, Mössbauer spectroscopy is a helpful tool for characterization.

ACKNOWLEDGMENTS

The authors thank N. Hosoito, K.Kawaguchi, R. Yamamoto, Y.Endoh, Y. Fujii, H.Yasuoka, K. Takanashi, N. Hamada, and K.Terakura for their cooperation. This work has been supported by the Grant-in-Aid for Special Distinguished Research from the Ministry of Education, Science, and Culture.

REFERENCES

1. I. K. Schuller and C. M. Falco, in "VLSI Electronics: Microstructure Sciences", Vol. 4, Academic Press, New York, p. 183 (1983).

2. T. Shinjo, K. Kawaguchi, R. Yamamoto, N. Hosoito, and T. Takada, Solid State Commun. 52, 257 (1984).

3. E. M. Gyorgy, D. B. McWhan, J. F. Dillon, Jr., L.R. Walker, and J. V. Waszczak, Phys. Rev. B25, 6739 (1982).

4. N. Hosoito, K. Kawaguchi, T. Shinjo, T. Takada, and Y. Endoh, J. Phys. Soc. Jpn. 53, 2659 (1984).

5. I. K. Schuller, Phys. Rev. Letters 44, 1597 (1980).

6. C. S. Chun, G.-G. Zheng, J. L. Vicent, and I. K. Schuller, Phys. Rev. B29, 4915 (1984).

7. T. Shinjo, N. Hosoito, K. Kawaguchi, T. Takada, Y. Endoh, Y. Ajiro and J. M. Friedt, J. Phys. Soc. Jpn. 52, 3154 (1983).

8. K. Takanashi, H. Yasuoka, K. Takahashi, N. Hosoito, T. Shinjo, and T. Takada, J. Phys. Soc. Jpn. 53, 2445 (1984). The report on the structure analysis is in preparation.

9. T. Yamada, N. Hosoito, T. Shinjo, and T. Takada, in preparation.

10. Y. Fujii, T. Ohnishi, T. Ishihada, Y. Yamada, K. Kawaguchi, N. Nakayama, and T. Shinjo, J. Phys. Jpn. 55 No. 1 (1986).

11. A. M. van der Kraan, and K. H. J.Buschow, Phys. Rev. B25, 3311 (1982).

12. T. Shinjo, K. Kawaguchi, and T. Kusuda, in preparation.

13. See review by U.Gradmann, J. Magn. Magn. Mater. 6, 173 (1974).

14. L. Nèel, Compt. Rend. 237, 1468 (1953).

15. Y. Endoh, K. Kawaguchi, N. Hosoito, T. Shinjo, T. Takada, Y. Fujii, and T. Ohnishi, J. Phys. Soc. Jpn. 53, 3481 (1984).

16. K. Takanashi, H. Yasuoka, K. Kawaguchi, N. Hosoito, and T. Shinjo, J. Phys. Soc. Jpn. 53, 4315 (1984).

17. H. Hamada, K. Terakura, and A. Yanase, J. Phys. F. Met. Phys. 14, 2371 (1984).

18. The theoretical backgrounds are described in the following paper: N. Hamada, K. Terakura, K. Takanashi, and H. Yasuoka, J. Phys. F. Met. Phys. 15, 835 (1985).

19. Independently of the present authors, Jaggi and Schwartz also prepared iron-vanadium multilayers and measured the Mössbauer spectra at room temperature. They analyzed the hyperfine field distributions with a thermal diffusion model and obtained the diffusion length to be about one atom distance. Concerning the degree of intermixing at the iron-vanadium interface, their result is in good agreement with the present one. N. K. Jaggi, L. H. Schwartz, H. K. Wong, and J. B. Ketterson, J. Magn. Magn. Mater. 49, 1 (1985).

20. S. T. Ruggiero, T. W. Barbee, Jr. and M. R. Beasley, Phys. Rev. B26, 4894 (1982).

AUTHOR INDEX

Page numbers are cited first, followed by the reference number(s).

Aaron, H. B., 266, 54
Aaronson, H. I., 266, 53,54
Abe, N., 169, 15
Abe, Y., 325, 14
Abouchacra, G., 119, 57
Ackermann, M., 59,69, 70,71; 369,11; 397, 16
Adaniya, T., 711, 11, 12
Adhikari, M. 170, 35
Adolph, J., 680, 29,39
Afanasev, A. M., 407, 12
Agarwal, J. C., 511, 28
Aharoni, A., 80, 8
Aikin, T. L. H., 521, 10
Ajiro, Y., 748, 7
Alben, R., 59, 66
Aldred, A. T., 265, 38
Aldridhe, L. P., 563, 8
Alexander, L. E., 446, 41
Allen, W. C., 465, 9
Allerton, S. E., 582, 11
Allia, P., 354, 29,33
Allmann, R., 593, 13
Alonso, J. A., 24, 14
Als-Nielsen, J., 696 17,18
Altounian, Z., 60, 78,82
Amelse, J. A., 632, 20; 633, 49; 634, 61; 679, 5
Amsel, G., 292, 42; 293, 43
Anantharaman, T. R., 369, 1; 370, 20
Anderegg, J. W., 326, 40
Anderson, J. R., 634, 69
Anderson, P. D., 353 14
Anderson, P. W., 407, 1
Anderson, R. B., 631, 3
Andreone, D., 354, 29
Andresen, A. F., 57, 12,16
Angles, R., 712, 23
Animesh, K., 169, 22
Annersten, H., 563, 16
Antler, M., 711, 2,8; 712, 25
Antonini, A., 737, 10, 11
Antonione, C., 353, 7, 16,17,25
Anttila, A., 291, 18; 293, 46
Aoki, K., 264, 6
Aouchacra, G., 120, 66
Araki, M., 633, 36
Arents, R. A., 632, 26
Arif, S. K., 445, 22
Arnold, G., 58, 28
Asano, A., 266, 52
Asano, K., 117, 13; 134, 4
Aschenbeck, D. P., 633 55; 634, 72; 647, 13
Asenov, S., 532, 13
Ashmead, D., 511, 31
Atkinson, R., 135, 12; 136, 30; 292, 36
Atsuko, I., 397, 15
Atusita, K., 420, 7
Aubertin, F., 59, 72
Auric, P., 446, 27
Aust, K. T., 60, 91
Avrami, M., 421, 36, 36,37,38
Awad, M. Z., 713, 38
Babeshkin, A. M., 563, 23
Babu, T. G. N., 369, 10
Bach, B., 608, 9
Bač̄k, P. J., 397, 21
Backer, I. W., 119, 55
Baer, W. H., 169, 27
Baerns, M., 632, 29
Baes, C. F., 582, 6
Bahadur, D., 420, 9
Bai-Shi, U., 445, 8
Bailey, M. G., 326, 39
Baily, S. W., 563, 15
Bainbridge, J., 135, 26; 293, 60
Baker, R. B., 511, 34
Baker-Sherman, J. S., 583, 32
Baldokhin, Y. V., 398, 34; 489, 7
Balint, A., 152, 13
Balkashin, O. P., 737, 28
Ballet, O., 563, 12
Ballivet-Tkatchenko, D., 679, 12,13,14
Bandyopadhyay, A. K., 446, 27
Bara, J. J., 407, 11
Bara Jozef, A., 169, 16
Barandiaran, J. M., 353, 23
Barbee, T. W., 749, 20
Bardani, L., 135, 11; 169, 20; 292, 39; 325, 10
Bardos, D. I., 265, 38
Barin, J., 236, 23
Barnavon, T., 292, 29; 293, 48,,47,49,52
Barnes, R. G., 236, 15; 713, 46
Baron, M., 293, 50

Barrett, P. H., 608, 12,13; 681, 43,46
Barros De, B. A. S., 136, 29
Bartholdy, J., 696, 19
Bartholomew, C. H., 634, 79
Bartlett, M. W., 607, 2,3
Bartley, W. J., 631, 8; 646, 2
Barton, L. S., 370, 32
Bartram, S. F., 354, 44
Bartunik, H. D., 608, 18
Basset, J. M., 680, 24
Battezzati, L., 353, 17,18; 354, 30
Bauminger, E. R., 563, 9
Baumvol, I. J. R., 119, 54,55; 136, 29; 291, 1,6,19; 293, 61,64; 294, 66,69
Baverstam, U., 134, 5; 135, 6
Bayreuther, G., 135, 15
Bean, C. P., 81, 36
Beasley, M. R., 749, 20
Beaubien, P. E., 325, 15
Beck, H., 59, 55
Beck, P. A., 407, 4
Becker, J. J., 398, 27
Bedworth, R., 236, 10
Beeby, J. L., 265, 38
Behar, M., 293, 61; 294, 66,69
Behrisch, R., 119, 60,61
Bein, Th, 679, 16; 680, 19,22,33,36
Bell, A. T., 631, 9
Belozerskii, G. N., 135, 7,8; 325, 7
Belugou, P., 510, 6
Belyustin, A. A., 445, 8
Benczer-Koller, N., 135, 15; 265, 25; 723, 9
Benecke, W., 292, 28
Benett, L. H., 398, 30
Bennemann, K. H., 681, 44
Bennet, L. H., 57, 17; 169, 6
Bennett, B. A., 696, 14
Bennett, C. O., 632, 24,25; 633, 42
Berezowski, R. M., 563, 6,7
Berkowitz, A. E., 80, 24; 354, 47
Berman, M., 510, 14
Bernal, J. D., 58, 33; 593, 9
Berry, F. J., 664, 1; 665; 6,8,12,13
Berti, M., 293, 45
Besnus, M. J., 354, 40
Bhanu, Prasad B., 369, 6,8; 370, 20
Bhasin, M. M., 631, 8; 646, 2
Bhatnagar, A. K., 369, 6,8,10; 370, 20,24, 31; 397, 8
Bhattacharya, D. L., 199, 36
Bianchi, D., 713, 54; 632, 24; 632, 25
Bill, E., 680, 21
Biloen, P., 631, 10
Bils, R., 582, 11
Binczycka, H., 119, 43,56
Birchall, T., 200, 47; 477, 9
Birchenall, C. E., 236, 2
Birman, J. L., 737, 25
Birringer, R., 81, 32
Bishay, A. M., 445, 22
Blaauw, C., 563, 11
Black, P. J., 216, 17
Blaes, N., 57, 19,24, 25; 61, 106
Blaine-Cecil, C., 532, 23
Blanchard, F. N., 546, 5
Blandin, A., 24, 13
Blas, R., 680, 21
Blasius, A., 57, 18, 22,24
Blasse, G., 446, 47
Blessington, D. R., 711, 7
Blokhin, V. F., 445, 24
Blomquist, J., 713, 50
Blume, M., 57, 27
Blyholder, G., 680, 25
Bocker, R., 711, 15
Bodlund-Ringstrom, B., 134, 5,6
Boekema, C., 446, 45
Bogacz, B. F., 169, 16; 407, 11
Bogner, R. L., 169, 8
Bohland, J., 712, 25
Bohm, C., 134, 5; 135, 6,7,8
Bohna, C., 325, 7
Bokemeyer, H., 118, 19
Boliang, Y., 60, 82
Bolster, R. N., 292, 23,24,25
Bommannavar, A. R., 511, 29
Bommel, H., 80, 9; 521, 521, 14
Bonchev, Z., 117, 10
Bonzel, H. P., 680, 26,34
Boon, J. A., 445, 16
Borcar, S., 632, 24,25
Borchardt, J., 353, 22
Borg, R. J., 119, 62
Borgesen, P., 119, 58,60
Borggard, O. K., 563, 21
Borran, M., 531, 5
Bortain, F., 531, 5
Boudart, M., 634, 79; 681, 41,42
Bowden, M. E., 563, 5,17
Bowen, H. K., 445, 7
Bowen, L. H., 583, 28
Boyle, J. F., 446, 43
Bozorth, R. M., 712, 16
Brand, R. A., 61, 107; 117, 15; 118, 16; 134, 3; 135, 12; 292, 40
Brauer, G., 712, 21
Brenner, A., 58, 29
Brett, N. H., 562, 1
Breysse, M., 696, 14
Briancon, C., 118, 17
Briggs, J., 420, 5
Britton, C. S., 712, 24
Broden, G., 680, 34
Brophy, G. R., 264, 2
Brouwer, G., 169, 7
Brown, D. K., 398, 29
Brown, L. J., 477, 7; 521, 11; 607, 1

Brown, W. F., 80, 5,6,7
Bruckner, R., 445, 25
Brusentsova, V. M., 713, 41
Brussler, M., 200, 43
Buchanan, D. N. E., 265, 27; 445, 18
Buckel, W., 58, 30
Budkowski, A., 118, 33
Budnick, J. I., 59, 66; 60, 97; 265, 29; 398, 33
Burch, T. J., 265, 29
Burger, K., 608, 11
Burriesci, N., 695, 8; 737, 27
Buschow, K. H. J., 60, 84,90; 748, 11
Bush, R. H., 198, 4; 216, 4
Bussemeier, B., 631, 5,7
Bussiere, P., 632, 22; 632, 33; 680, 24;
Butt, J. B., 632, 20,29; 633, 37,49; 634, 60,61; 679, 4,5
Butter, S. A., 679, 15
Buxh, R. H., 197, 2
Buyers, W. J. C., 737, 26
Buzas, A., 712, 30
Bykov, G. A., 264, 14

Cable, J. W., 265, 35
Cadeville, M. C., 24, 9
Cadman, P., 608, 9
Cadogan, J. M., 61, 107
Cahn, R. W., 353, 26
Cai, Z., 170, 33
Calas, G., 446, 28
Cambell, C., 169, 6
Cammack, P., 152, 13
Campany, R. G., 24, 4
Campbell, I. A., 24, 13; 265, 21,32,34,
Campbell, S. J., 57, 19; 59, 72; 397, 21
Candia, R., 664, 3; 665, 7; 695, 7,8, 10,11,12; 696, 13, 17,18,19,21,22,25,
Canner, J. P., 737, 21
Capes, C. E., 511, 32
Capps, W., 421, 28
Capps, W., 421, 29
Carbucicchio, M., 135, 11; 169, 20; 292, 39; 325, 10
Cargill, G. S., 59, 66; 370, 28
Carius, C., 593, 1
Carlson, L., 582, 16
Carruthers, T. G., 420, 5
Casalot, A., 407, 9
Cashion, J. D., 477, 7; 521, 1,2,3,4,10, 11; 532, 18, 563, 4, 607, 1; 608, 15
Cassidy, P. J., 521, 4
Celotti, G., 293, 54
Chabanel, M., 136, 27; 293, 56
Chadwick, D., 696, 14
Chakravorty, D., 420, 9; 446, 31
Chambaere, D., 446, 29,44,48; 521, 13
Chamberod, A., 607, 6
Chan, N. D., 679, 13
Chandra, D., 294, 80
Chang, A. L., 293, 50
Channing, D. A., 135, 26; 293, 60
Chao, C. C., 647, 14
Charles, S. W., 80, 15
Chassagne, A., 120, 66
Chattergee-Fischert, R., 169, 19
Chattopadhyay, S., 370, 16
Chaturvedi, U. K.,. 291, 10,11
Chauhan, P. K., 170, 34
Chekin, V. V., 737, 28
Chem, J., 170, 33
Chen, C. W., 326, 40
Chengyu, Wang, 665, 6
Cher, L., 445, 8
Chibirova, F. H., 713, 48
Chien, C. L., 58, 48, 51; 59, 53; 60, 85, 86,89,99,100; 370, 17,23,26; 397, 4, 11,19
Chien, H. S., 58, 38; 58, 38; 60, 91; 370, 17, 26; 397, 4,19
Chikazumi, S., 135, 16; 712, 17
Child, H. R., 265, 35
Childs, C. W., 583, 29,32
Chilton, J. P., 531, 4
Choplin, A., 680, 24
Chou, C. P., 59, 65; 353, 6
Choudhury, K., 170, 35
Chow. H. K., 169, 8; 266, 46
Chowdhury, A. R., 198, 11
Christ, B. W., 169, 14
Christensen, P. H., 34,35; 634, 62; 696, 26
Christenson, F., 696, 17
Chukhrov, F. V., 582, 17
Chun, C. S., 748, 6
Claes, J., 444, 3
Clark, P. E., 521, 3; 532, 11,15,17,18; 563, 4
Clarke, M., 712, 24
Clausen, B. S., 80, 3, 17; 81, 28,29,34, 35; 634, 62; 664, 3,4; 665, 7; 695, 7,8,9,10,11,12,13; 696, 17,18,19,21, 22,25,26
Clausen, C. A., 665, 11
Clayton, C. G., 510, 12; 291, 5, 292, 33,34
Clerc, A., 264, 5
Cocco, G., 353, 7, 25
Cochrane, R. W., 370, 28
Coey, J. M. D., 60, 82; 61, 110; 81, 26; 521, 15; 563, 10,12; 607, 6
Cohen, M. L., 737, 12
Cohen, R. L., 711, 8, 9; 712, 36; 714, 63; 723, 8
Collins, G. S., 198, 11; 723, 9
Collins, M. F., 265, 33,40
Collins, R. L., 169, 27,28
Colmenero, J., 353, 23
Comly, J. B., 265, 28

Conde, A., 370, 33
Constabaris, G., 80, 9; 521, 14
Constant, A., 264, 5; 521, 2,3,4; 608, 15
Cooper, A. R., 445, 15
Copperthwaite, R. G., 608, 20
Corby, R. N., 216, 17
Corliss, L. M., 57, 11,14,15
Cornils, B., 631, 5
Couch, D. E., 58, 29
Coudurier, G., 679, 12,13,14
Cowley, R. A., 737, 26
Cox, D. M., 681, 47
Crangle, J., 265, 37
Cranshaw, T. E., 24, 10; 168, 5; 266, 50,51
Crawford, J. E., 679, 8,10
Crites, J. W., 679, 8
Cromer, D. T., 608, 17
Crussard, C., 264, 5
Cser, L., 354, 36
Csordas, L., 712, 21, 22
Cui, F. Z., 293, 53
Cumbrera, F. L., 370, 33
Cusumano, J. A., 631, 6
Cuthbertson, J. W., 712, 27
Czako-Nagy, I., 712, 20,21,22,26,30; 713, 39
Czanderna, A. W., 723, 5
Czhzek, G., 61, 110
Cziraki, A., 216, 14

d'Amico, J. F., 723, 8
d'Artemare, E., 292, 42
Dalmon, J. A., 680, 24
Dande, Y. D., 170, 34
Danielewski, M., 236, 6
Daniels, F., 353, 22
Dannon, J., 736, 6
Das, D., 170, 35
Das, T. P., 265, 24
Dasgupta, D. R., 593, 9
Dash, J. G., 264, 15
Dastur, Y. N., 187, 3
David, D., 292, 42
Davies, D. E., 713, 57; 714, 60
Davies, H. A., 354, 40
Davies, M. A., 353, 5
Davies, N. C., 608, 9
Davis, L. A., 353, 6; 398, 32
Davis, L. E., 61, 111
Day, D. E., 445, 11
Daynyak, L. G., 563, 20
De Barros, B. A. S., 293 64
De Batist, R., 445, 4,5,6; 446, 29,35
De Boer, F. R., 24, 12
De Bruyn, P. L., 582, 7,8,9
De Chatel, P. F., 24, 12
De Grave, E., 445, 4; 446, 29,40,44,46, 48; 521, 13
De Launay, J., 266, 48
De Minjer, C. H., 722, 2
De Sitter, J., 446, 46
De Souza, J. P., 293, 64; 294, 66
Dean, J. A., 60, 88
Dearnaley, G., 291, 2, 3
Debusmann, K. 117, 15; 118, 16; 134, 3; 292, 40
Deckwer, W. D., 679, 2
Decristofaro, N., 265, 42; 398, 32
Dederichs, P. H., 57, 6; 199, 29
Degass, W. N., 665, 9
Dehe, G., 532, 19
Deiss, E., 593, 5
Dejonghe, P., 444, 3
Delbouille, A., 681, 41
Delcroix, P., 216, 13
Deleford, J., 199, 19
Delgass, W. N., 631, 11; 632, 15,21,23, 31; 633, 45,46; 634, 68,72,73; 647, 10,11,13; 679, 3
Della, Gatta G., 353, 16
Delmon, B., 695, 2
Delong, W. T., 169, 6
Demarthe, J. M., 531, 5
Demopoulos, G. P., 532, 14
Den Breeder, F. J. A., 58, 52; 397, 10
Denise, F., 712, 18
Denner, S. G., 169, 25
Derouane, E. G., 680, 20
Desitter, J., 446, 48
Deutchman, A. H., 291, 22
Dickerhoof, D. W., 510, 9; 532, 16
Diegle, R. B., 369, 5
Diehl, J. R., 679, 10
Dillon, J. F., 748, 3
Dimigen, H., 292, 26
Dimitrov, R., 532, 13
Dimmock, J. O., 736, 1
Distin, P. A. G. P., 532, 14
Dixon, J. B., 583, 21
Dluy, W., 631, 7
Dolenko, A. V., 151, 1
Dolling, G., 737, 26
Domke, M., 135, 9,10; 325, 3
Donald. I. W., 353, 5; 354, 40
Donaldson, F. D., 712, 29
Dongbai, L., 665, 6
Donner, J. T., 646, 7
Doppler, G., 680, 21
Doring, W., 711, 15
Dormann, J. L., 80, 2
Dos Santos, C. A., 136, 29; 291, 1; 293, 61,64; 294, 66,69
Doudart, M., 665, 9
Douglas, L. J., 679, 10
Douglas, R. W., 445, 14
Dousma, J., 582, 7,8, 9
Doyle, F. G., 292, 30
Drako, V. M., 293, 51
Dreshnan, A., 265, 25
Drigo, A. V., 293, 45, 54
Drijver, J. W., 24, 16; 353, 20
Drozdova, S. V., 236, 20
Drwiega, M., 119, 47, 56
Dry, M. E., 631, 4
Dubiel, S. M., 198, 9, 10; 236, 7,8; 294, 73,74,76,77,78,79;

354, 37,38
Dubois, J. M., 58, 41, 44; 61, 107,109; 294, 81; 353, 10; 354, 45; 398, 22; 632, 30, 33
Duff, K. J., 265, 24
Duke, B. J., 647, 9
Dulong, F., 532, 23
Dumesic, J. A., 80, 1, 21; 532, 12; 631, 1,2; 632, 28; 633, 35,50; 646, 4; 664, 2; 681; 41,42; 696, 15,16
Dumont, G., 444, 3
Dunmyre, G. R., 326, 26; 521, 7
Durand, J., 60, 97; 354, 46
Dutrizac, J. E., 531, 6; 532, 7,26
Duwez, P., 58, 34,35; 647, 14
Dvanarayanan, S., 398, 35
Dwyer, D. J., 633, 43

Eaton, H. C., 521, 9
Eckardt, D., 118, 19
Eckert, H., 421, 25
Edwards, S. F., 407, 1
Egachi, T., 198, 16
Egami, T., 60, 87; 353, 27,28
Egiazarov, B. G., 151, 1
Ehrenberg, J., 119, 60
Eibschutz, M., 58, 46; 370, 30; 464, 6
Eichinger, P., 292, 26
Eickel, K. H., 293, 58
Ekdahl, T., 135, 7,8; 325, 7
El Sayd, J. A., 713, 55
El-Raghy, S. M., 532, 8
El'ner, V. Ya, 198, 14
El Wakab, F. M. A., 713, 58
El Wakkad, S. E. S., 713, 55
Eldridge, J., 713, 51,52
Ellgen, P. C., 631, 8; 646, 2
Endo, K., 326, 25
Endoh, Y., 187, 8; 748, 4,7,15
Ensling, J., 325, 9
Enzo, S., 353, 7
Ergun, S., 510, 14
Ermilova, L. P., 582, 17
Errington, P. R., 187, 9
Eszterle, M., 634, 78
Evans, B. J., 446, 39; 532, 23
Evans, E. L., 608, 9
Evans, U. R., 327, 44
Ewan, G. T., 293, 54
Eymery, J. P., 199, 18,19

Fahnle, M., 370, 21
Falco, C. M., 748, 1
Fano, V., 736, 2,3,4, 7,8; 737, 10,11,18, 27,31
Farkas, J., 354, 34
Farmer, V. C., 583, 27
Fatseas, G. A., 532, 25
Fayeulle, S., 292, 29, 31; 293, 47,48,49, 52
Fedeli, G., 737, 18; 737, 19
Feder, M., 712, 25
Feitknecht, W., 593, 4
Feldkamp, L. A., 265, 30
Feller, H. G., 292, 28
Ferguson, M. M., 293, 54
Fey, M. V., 583, 21
Fierro, C., 713, 52
Filatova, T. A., 445, 24
Fine, M. E., 169, 13; 266, 45
Fink, J., 61, 110
Firrao, D., 294, 68
Firsov, V. M., 445, 23
Fischer, H., 58, 41; 61, 106; 398, 23
Fischer, P., 57, 12, 16
Fischer, W. R., 582, 15; 583, 34
Fisher, R. M., 187, 9
Fleisch, J., 325, 9
Fleisch, T., 633, 45,46
Flinn, P. A., 265, 26; 266, 46; 445, 17, 19; 723, 11; 736, 9
Flipse, C. F. J., 632, 14,16; 634, 63; 679, 7; 664, 5
Floyd, J. M., 477, 2
Fnidiki, A., 199, 18, 14
Fogarassy, B., 216, 14; 354, 40; 397, 5
Follstaedt, D. M., 292, 27,30
Ford, J. C., 398, 33
Forester, D. W., 80, 24
Fort, D., 680, 23
Fortunato, F. A., 634, 72; 647, 13
Fortune, J. M., 521, 15
Fox, K. E., 445, 21
Fox, R. A., 446, 43
Franck, J., 81, 42,43
Franco, M., 489, 2
Franke, H., 370, 36
Frattini, R., 294, 65, 67,68,71
Freeman, A. G., 607, 4
Freeman, A. J., 81, 30,31; 265, 23
Freyhardt, H. C., 59 73,74
Friedt, J. M., 24, 9; 532, 28; 748, 7
Fries, S. M., 57, 19; 60, 81
Frohning, C. D., 631, 5,7
Fryt, E., 236, 27
Fujii, Y., 748, 15
Fujinami, M., 325, 16; 326, 24,36,41
Fujita, F. E., 58, 32, 50; 265, 44; 293, 59
Fukamichi, K., 60, 76, 77,93,95
Fukushima, T., 634, 81; 646, 8; 647, 15
Fullmer, L. D., 521, 12
Fultz, B., 264, 9,10 13; 265, 41
Furtak, T. E., 713, 46
Furukawa, T., 420, 8; 445, 21
Fuyat, R. K., 546, 6
Fyfe, W. S., 445, 16
Fysh, S. A., 532, 11, 15,17,18,29; 563, 4

Gaa, R., 59, 71

Gabe, D. R., 714, 61
Gadiyar, H. S., 170, 34
Gadsden, J. A., 489, 9
Gafner, G., 680, 34
Galakhov, V. R., 397, 7
Galinker, V. S., 712, 32,33
Gambino, R. J., 60, 93
Ganiel, U., 464, 6
Garcia, E. A., 293, 61
Garten, R. L., 633, 56,58; 634, 65,66, 67,71; 646, 5
Garzanov, I. Y., 151, 2,3
Gates, B. C., 695, 4
Gavrilyuk, G. G., 265, 43
Gel'd, P. V., 198, 13, 14,15
Geller, S., 521, 12
Gemperle, R., 398, 36
Gendler, T. S., 563, 20
Gentry, S. J., 634, 75
Georges-Gilbert, H., 198, 6
Georgiev, S., 713, 49
Gerard, A., 464, 2,4
Gerard, P., 119, 51
Gerlach, G., 197, 3
Gerlach, R., 216, 3
Geronov, Y., 713, 49
Gerson, R., 737, 21
Gettys, W. L., 731, 14
Ghafari, M., 59, 54, 59,60,69; 60, 75; 61, 108
Gibb, T. C., 326, 35; 477, 8; 647, 9; 737, 15
Giberti, R. A., 511, 28
Gielen, P., 445, 26
Giles, P. M., 169, 14
Giordano, L., 293, 44; 294, 67
Girard, E., 292, 42
Gladkin, I., 354, 36
Glazer, A. A., 397, 7
Gleiter, H., 81, 32
Goens, D. N., 511, 30
Gohy, C., 464, 2,4
Goldanskii, V. I., 151, 1; 216, 11; 607, 5
Goldstein, J. I., 264, 11
Golumbic, N., 631, 3
Gonser, U., 57, 5,7, 19,22,24,25; 58, 31,32,41,42,43; 59, 54,55,59,68,69,70, 71,72; 60, 75,81; 61, 102,106,108; 197, 1; 198, 7,8; 199, 23,26,30; 216, 216, 6,7,8; 353, 4; 369, 11; 370, 18; 397, 16; 398, 26; 680, 21
Good, M. L., 646, 7; 665, 11
Goode, P. D., 291, 19
Goodenough, J. B., 532, 25
Gooding, E. J., 421, 27
Goodman, B. A., 583, 22
Gorbachev, V. V., 445, 24
Gorczyca, S., 294, 76
Gord, G. A., 199, 33
Gorlitz, C., 353, 4; 370, 18
Gorshkov, A. I., 582, 17
Gorskii, V. V., 199, 22
Gosselin, J. P., 445, 15
Goto, H., 199, 27,37
Gotz, G., 61, 110
Gotz, J., 420, 4
Govaert, A., 446, 44, 48; 521, 13
Gradmann, U., 118, 35; 135, 19
Graham, M. J., 325, 15
Granas, L., 398, 35
Granasy, L., 352, 1,2; 370, 15
Grandjean, F., 464, 2,4
Grange, P., 695, 3
Grant, R. W., 464, 5; 521, 12
Grant, W. A., 293, 54
Greatrex, R., 737, 15
Greenwood, N. N., 326, 35; 477, 8,9; 737, 15
Greer, A. L., 353, 19
Greneche, J. M., 353, 8; 398, 24
Gridnev, V. N., 265, 43
Gries, W., 119, 42
Griffin, R. A., 477, 5; 521, 6
Griger, A., 216 2,9, 10,15,16
Grimm, R., 325, 9
Grodzins, L., 445, 15
Groger. B., 398, 37
Groger, V., 200, 43
Groma, G., 200, 46
Gros, Y., 24, 17
Gruber, J., 325, 9
Gruffroy, D., 397, 3
Gruner, G., 265, 20
Grynkewich, G., 633, 49
Guczi, L., 633, 59; 634, 78; 646, 6
Guermazi, M., 119, 49
Guettinger, T. W., 510, 9; 532, 16
Guiraldenq, P., 292, 29; 293, 48,52
Gumanskij. G. A., 293, 51
Gunsser, W., 680, 19, 29,31
Guntherodt, H. J., 58, 40; 59, 55; 61, 107
Gutlich, P., 313, 2,3,4; 325, 6,9
Guttmann, H.-J., 313, 3,4
Guy, C. N., 407, 7
Gyorgy, E. M., 59, 67; 748, 3
Gzowski, O., 118, 24

Haacke, G., 293, 57
Haasen, P., 59, 62
Habashi, F., 531, 1
Habig, K. H., 169, 19
Hackerman, N., 714, 64
Haddon, J. R., 170, 37
Hafner. S. S., 446, 39
Haga, H., 264, 7
Hagane, T. T., 187, 1
Hagenmuller, P., 407, 9
Haggstrom, L., 61, 104; 398, 35
Halg. W., 57, 12,16
Hall, J. S., 521, 11
Hall, S. H., 563, 15
Hallam, G. C., 265, 37
Haly, J., 407, 9
Hamada, H., 748, 17
Hamada, N., 748, 18
Hamaguchi, Y., 737 22

Hampson, N. A., 714 59
Hancock, R. D., 608 20
Handa, A., 169, 21; 325, 11,12,13,14
Handrich, K., 370, 19
Haneda, K., 80, 4; 81, 25,27
Hang, H. G., 608, 16
Hangen, S., 236, 3
Hannaford, C. S., 521, 11
Hargitai, C., 354, 36
Harten, R. L., 631, 6
Hartley, N. E. W., 136, 28; 291, 12, 13,14; 292, 32; 293, 55
Hasegawa, R., 58, 48, 51; 59, 53,61; 353, 6; 397, 11
Hashimoto, K., 59. 57; 369, 4; 583, 24
Hasselbach, K. M., 61, 105
Hastings, J. M., 57, 11,14,15
Hatta, O, 737, 23
Hauffe, K., 235, 1
Haugh, J. J., 398, 31
Havlicek, S., 151, 4
Hayashi, H., 119, 59
Hayashi, M., 81, 33
Heberle, J., 264, 16
Hehemann, R. F., 264 3
Heidemann, A., 57, 20,21
Heidenreich, H., 509 2
Heiman, H., 60, 83
Hekimova, A., 532, 13
Helgeson, U., 713, 50
Heller-Kallai, L., 563, 2,9
Henry, G., 264, 5
Herber. G., 407, 6
Herber, R. H., 216, 11; 608, 10; 723, 10
Herman, H., 292, 33, 34
Hesse, J., 60, 96; 353, 9; 397, 17
Hidvegi, E., 216, 2, 10
Hien, P. J., 264, 14
Hill, C. G., 633, 35
Hilsch, R., 58, 30
Hinckley, C. D., 477, 5; 521, 6
Hines, W. A., 398, 33
Hirai, T., 420, 10,11, 14; 421, 16,17
Hiroyoshi, H., 60, 76, 77,95
Hirvonen, J. K., 291, 4,5,15,18; 292, 33, 34; 293, 46
Hisatake, K., 117, 13,14; 119, 52; 134, 4; 135, 13, 16,17
Hobbs, L. W., 197, 2
Hodgson, D., 200, 47
Hoffman, R. W., 713, 51,52
Hogg, C. S., 563, 19
Hohlwein, D., 607, 6
Hokenemser, C., 198, 11
Holden, T. M., 265, 28
Hollitt, M. J., 477, 7
Holloway, S., 680, 35
Homonnay, 216, 14,15
Honnick, W. D., 714, 66
Hoogendoorn, J. C., 631, 4
Hooley, J. G., 607, 2,3
Horn, G., 631, 7
Hornig, C. F., 712, 25
Hosoito, N., 744, 2,4, 7,8,9,15,16
Hosoyama, K., 60, 79
Howard, D. G., 607, 1; 711, 4
Hu, W. W., 292, 33,34
Huffman, G. P., 24, 2,6,7; 151, 7,8, 9,10,11; 168, 1,2, 4; 169, 12,14; 187, 4,7,9; 326, 26; 465, 10; 510, 16, 17; 521, 5,7; 532; 24; 608, 14
Huggins, F. E., 24, 2,6; 151, 7,8,9, 11; 168, 1,2,4,; 169, 12; 187, 4; 465, 7,10; 510, 16, 17; 521, 5,7; 532, 24; 608, 14
Hultgren, R., 353, 14
Hupkes, J., 446, 38
Hurst, N. W., 634, 75

Ibach, H., 135, 25
Ibe, G., 199, 30
Ichikawa, M., 634, 81; 646, 1,3,8; 647, 15
Ichinose, H., 198, 5
Ignatiev, A., 136, 31
Ignatova, T. S., 477, 3
Ilarraz, J., 353, 23
Inden, G., 236, 24
Ingalls, R., 446, 36
Injoue, A., 369, 4
Ino, H., 265, 44; 293; 59
Inouye, K., 326, 34
Inoyatov, A., 118, 17
Irminger, P. F., 511, 28
Isakov, L. V., 151, 1
Ishida, Y., 198, 5; 199, 27,31,37,38
Ishihada, T., 748, 10
Ishikawa, Y., 187, 8; 199, 21
Ismail, M., 117, 15; 134, 2,3
Isobe, T., 420, 15
Isozumi, Y., 118, 31
Ito, A., 199, 21
Itoh, J., 117, 14; 119; 52; 135, 13, 16,17
Ivanov, A. O., 421, 30
Izumi, M., 373, 22

Jaccarino, V., 265, 27; 354, 35
Jack, D. H., 293, 63
Jack, K. H., 293, 63
Jacobs, I. S., 80, 24; 81, 36; 168, 3; 510, 11; 679, 1,16; 680, 22,33,36
Jacques, A. G., 582, 1
Jadhao, V. G., 608, 7
Jaen, J., 712, 30,31, 34; 713, 39,47; 714, 65,67
Jaffrezic, H., 293, 49
Jagannathan, R., 369, 8,10; 370, 20,24; 397, 8
Jaggi, N. K., 24, 4; 151, 5,6; 169, 10, 11,17,18; 170, 30, 34; 632, 29; 749, 19
Jaghadeesh, M. S., 532, 22
Jahrb, N., 593, 13
Jain, K., 169, 23

Jambor, J. L., 532, 26
James, J., 737, 21
Jameson, R. F., 313, 5
Janot, C., 58, 49; 136; 27; 198, 6; 216, 13; 293, 56; 294, 72
Jepsen, H. P., 593, 13
Jiang, X.-Z., 632, 28
Johansson, L. Y., 713 50
Johnson, C. E., 266, 50
Johnson, D. P., 264, 15
Johnson, K. H., 681 45
Johnson, R. G., 532, 23
Johnson, W. A., 421, 35
Johnston, J. H., 446, 32; 582, 18; 583, 25,29
Jones, A., 634, 75
Jones, C. H. W., 521, 8
Jones, R. D., 169, 25
Jones, R. V., 265, 40
Jones, W., 117, 2; 199, 33
Jordanov, A., 117, 10
Joshi, G. M., 608, 7
Juhasz, J., 714, 68
Jujii, Y., 748, 10
Jung, H.-J., 632, 23

Kabacoff, L. T., 398, 33
Kachnowski, T., 723, 9
Kaeding, W. W., 679, 15
Kaindl, G., 57, 20; 135, 9
Kaiser, R., 80, 23
Kajcsos. Z., 712, 20,21
Kalawa, E., 118, 21
Kaldor, A., 681, 47
Kalvius, G. M., 120, 64
Kaneko, K., 326, 34
Kaneoko, K., 199, 31
Kaneyoshi, K., 292, 40
Kankeleit, E., 117, 9; 118, 19
Kant, R. A., 292, 34
Kaplow, R., 265, 42
Kapusta, S. D., 714, 64
Kardashuk, V. I., 151, 2
Karl, R. E., 24, 3
Kasindl, G., 608, 18
Kasmodamianskaya, L. V., 712, 34
Katada, M., 421, 21, 22; 608, 10
Katano, M., 118, 31
Kato, A., 326, 25
Kato, M., 198, 5; 198, 5; 199, 27
Kato, Y., 737, 22
Katoh, M., 713, 56
Katrevich, A. N., 713, 40
Katzer, J. R., 695, 4
Kawaguchi, K., 748, 2, 4,7,10,12,15,16
Kawai, M., 646, 3
Kazama, N., 60, 83
Keller, G., 593, 4
Kellerman, E., 170, 32; 325, 4; 711, 5
Kelley, K. K., 353, 14
Kemeny, T., 352, 1; 354, 32,40; 397, 5
Keramides, V. G., 120, 65
Ketterson, J. B., 749, 19
Keune, W., 117, 15; 118, 16; 134, 3; 135, 12,14,18,20 24; 292, 40; 723, 6
Khammouma, S., 681, 41
Kiado, A., 216, 1
Kiauko, W., 117, 15; 134, 3
Kim, C. S., 532, 21
Kim, J. I., 264, 12
Kim, K. J., 169, 9; 264, 8
King, D. L., 631, 6
King, H. W., 353, 15
Kingery, W. D., 445, 7
Kiraley, J., 352, 2; 370, 15
Kiraly, J., 634, 78
Kirkaidy, J. S., 236, 27
Kisdi-Koszo, I., 352, 1
Kiss, L., 352, 1; 354, 34; 521, 1; 712, 30,34; 713, 39,43, 47; 714, 62,65
Kitaguchi, M., 58, 50
Kittel, C., 632, 13
Klein, H. P., 59, 69; 61, 108; 81, 32; 407, 14
Klein, R., 680, 21
Klement, W., 58, 34
Kleppa, O. J., 563, 18
Klinger, R., 292, 28
Klug, H. P., 446, 41
Klyukin, A. B., 445, 24
Knacke, O., 236, 23
Knapp, J. A., 292, 27, 30
Knappwost, A., 680, 31
Knauer, R. C., 737, 30
Knight. R. J., 582, 13
Knoechel, D. J., 633, 48
Knudsen, J. E., 80, 13
Kobayashi, J., 326, 31
Kobayashi. K., 60, 80
Kobayashi. K. L. I., 737, 19,23
Kobs, K., 292, 26
Koch, C. J. W., 81, 42,43
Koch, F. G., 711, 7,9
Kojima, R., 326, 19
Kolotirkin, Y. M., 713, 42
Komatsu, T., 420, 3
Komatsubara, K. F., 737, 19,22
Kopcewicz, M., 58, 42, 43
Koptsik, S. V., 489, 7
Korber, F., 236, 26
Kordesch, M. E., 713, 51,52
Korecki, J., 80, 18; 118, 35; 135, 19; 407, 8
Korecz, L., 608, 11
Korr, R., 713, 53
Kosmack, D. A., 510, 16
Kossii, G. G., 713, 42
Kossowsky, R., 293, 50; 398, 31
Kosugi, N., 647, 15
Kotelnikov, V. A., 151, 2
Koustoudoand, E. P., 511, 26
Kovacs, E., 354, 34;

712, 30
Kovacs-Crselonyi, E., 200, 46
Koval'chik, T., 736, 5
Kowalski, J., 118, 26, 27,32; 119, 45, 46, 47; 326, 27,29
Kozlova, O. S., 398, 34
Kozubowski, J., 292, 34
Krakauer, H., 81, 30
Kraus, L., 398, 36
Krause, R. F., 398, 31
Krebs, H. U., 59, 73; 737, 16
Kreidl, N. J., 420, 2
Krishnan, A., 135, 15
Krolas, K., 24, 11
Kron, M., 264, 5
Kronmuller, H., 398, 37
Krop, K., 80, 18, 294; 73
Kroupa, F., 398, 36
Kubaschevski, O., 634, 64
Kudra, O. K., 712, 32, 33
Kudryavtseva, T. N., 477, 3
Kulikov, V. A., 151, 3
Kulkarni, V. N., 169, 23
Kun, Y., 291, 21
Kundig, W., 80, 9; 521, 14
Kunii, K., 489, 4
Kurakado, M., 118, 31
Kurkjian, C. R., 445, 12, 18; 446, 30
Kuroda, H., 647, 15
Kuspert, B., 680, 30
Kusuda, T., 748, 12
Kuzmin, R. N., 563, 20
Kuznetsov, B. N., 634, 83
Kyvelos, B., 135, 9, 10; 325, 3

Labar, C., 445, 26
Labushkin, V. G., 151, 2,3
Lahut, J. A., 80, 24
Lakatos-Varsanyi, M., 712, 20,21,22,35
Lakner, J., 200 46
Lam, Y. L., 633, 56; 634, 67
Lamb, A. B., 582, 1
Lamparter, P., 354, 41,43
Land, J. S., 511, 26
Lanford, W. A., 293, 43
Lang, N. D., 680, 35
Langer, S. H., 532, 12
Laras, W., 80, 18
Larkins, F. P., 521, 3
Larson, R., 713, 50
Lauer, H. V., 446, 446, 33,34
Laundy, B. J., 665 13
Laursen, T., 293, 54
Le Caer, G., 58, 41, 44; 61, 107,109; 294, 81; 353, 10; 354, 45; 398, 22; 632, 30, 32, 33
Lebsanft, E., 57, 13, 18
Lee, J. B., 532, 29
Lee, J. K., 266, 53
Lee, R. J., 510, 16
Lees, J. K., 736, 9
Lefevre, H. W., 236, 16
Lefkowitz, M., 737, 26
Legrand, B., 118, 17
Leider, H. R., 57, 26
Leidheiser, H., 170, 31,32; 325, 4; 327, 43; 711, 1,5,13,14; 712, 18,20,21,22, 26; 713, 44; 714, 62
Leiper, W., 563, 11
Lengeler, B., 696, 17, 18,19
Lentze, W., 583, 26
Leslie, W. C., 187, 3
Letocha, E., 119, 43
Leutenecker, R., 292, 26
Levin, E. M., 465, 8
Levin, K., 681, 44
Levinson, L. M., 80, 24; 168, 3; 510, 11
Levy, A. L., 445, 17
Lewis, D. G., 582, 18; 583, 22,33
Leyman, R., 446, 40
Li, H., 291, 21
Li, H. D., 293, 53
Liebermann, H. H., 353, 21
Liebsch, A., 681, 44
Lien, D. N., 713, 43
Lienard, A., 61, 110, 3
Liengme, B. V., 607, 3
Lilienthal, H., 58, 47
Liljequist, D., 117, 15; 118, 16; 134, 1,2,3,6,7,8,12,14; 325, 7
Lin, M. C., 236, 15; 713, 46
Lin, S. C. H., 58, 35, 36
Lindgreen, H. B., 563, 21
Lindhard, J., 135, 22
Lindquiest, R. H., 80, 9; 521, 14
Lines, M. E., 58, 46; 370, 30
Liou, S. H., 60, 99, 100
Lipka, J., 80, 11
Litvinov, P. I., 445, 23
Liwu, Lin, 665, 6
Lo, C., 679, 11
Lo Russo, S., 135, 21; 291, 10,11,16,17; 292, 35; 293, 44; 294, 65,67,70
Logan, J., 397, 20
Loktev, S. M., 632, 26
Long, G. J., 608, 19
Longworth, G., 58, 36; 119, 55; 135, 12; 136, 28,30; 292, 36 293, 55,62
Lorenzie, L., 511, 26
Losev, V. V., 713, 47
Lovas, A., 352, 1,2; 354, 34; 370, 15
Low, G. G., 265, 28, 33
Lowde, R. D., 265, 40
Luborsky, F. E., 59, 56; 60, 92; 353, 21; 369, 2,12
Lucci, A., 353, 16,17, 18,25
Lupis, C. H. P., 445, 17,19

Ma, R., 398, 25
Maccrone, R. K., 292, 34

Mackay, A. L., 593, 9
Mackay, C. A., 712, 29
Mackenzie, K. J. D., 562, 1; 563, 3,5,6, 7,17
Madare, M., 119, 51
Maddin, R., 28, 37,39
Madhusudahan, C. P., 646, 7
Madon, R. J., 632, 28
Madsen, M. B., 81, 42, 42,43
Maeda, K., 353, 27
Maeda, Y., 293, 59
Mager, S., 198, 17
Magrini, M., 294, 71; 464, 3; 477, 6
Maguire, B., 521, 1
Majumdar, C. K., 170, 35
Makarov, E. F., 151, 1,2; 398, 34; 713, 48
Makhnavetskii, A. S., 477, 3
Maksimov, Y. V., 632, 26
Maletta, M., 532, 28
Malysheva, T. V., 563, 22
Mangin, P., 58, 49
Maniezzi, E., 736, 3; 737, 27,31
Manifold, R., 236, 17
Manninger, I., 634, 78
Mansel, W., 57, 6; 199, 28,29,40,41
Mantl, S., 200, 42
Manzione, A. V., 510, 13
Mararov, E. F., 151, 3
Marchal, G., 58, 49
Marcus, H. L., 169, 13; 266, 45; 370, 34
Marest, G., 119, 46, 48,49,51; 293, 47, 49
Marino, F., 353, 25
Markosyan, G. N., 713, 47
Marquard, P., 81, 32
Marquez, R., 370, 33
Marroquim, R. P., 489, 2
Marschall, C.W., 264, 3
Marshall, M., 521, 3
Martin, P., 119, 51
Maslakovets, Y. P., 736, 5
Massenet, O., 118, 22
Massoth, F. E., 695, 1
Massumoto, T., 369, 4
Masuda, I., 489, 1,2
Masumoto, T., 58, 39, 50; 59, 57; 60, 79, 80,94; 369, 3; 370, 29
Matallk, Z. F., 59, 72
Matsumoto, H., 633, 40,42
Matsushima, Y., 326, 19
Matteazzi, P., 293, 62
Matusek, K., 634, 78
Maurer, M., 24, 9
Mazak, R. A., 169, 27, 28
Mazzoldi, P., 291, 16, 17; 292, 35
Mbadcam, K. J., 679, 8
McDonagh, C. M., 563, 10
McDonald, G. W., 532, 12
McDougall, G. J., 608, 20
McEnaney. B., 170, 36; 593, 2
McGill, I. R., 593, 2
McLliney, A. E., 511, 32
McMahon, K. C., 679, 18
McMillan, P. W., 420, 1
McMurdie, H. F., 465, 8
McNab, T. K., 446, 43
McNicol, B. D., 634, 75
McWhan, D. B., 748, 3
Meads, R. E., 563, 19
Mehl, K. F., 421, 35
Mehner, H., 713, 37; 714, 68
Meier, W. M., 679, 9
Meisel, W., 313, 2,3, 4,6; 325, 6; 532, 19,20
Meiser-Komor, P., 120, 64
Mello, I., 354, 44
Melngailis, J., 736, 1
Melson, G. A., 679, 10
Melson, G. M., 679, 8
Mendelsohn, L. I., 59 61
Merchant, H. D., 200, 47
Merest, G., 292, 31
Meriaudeau, P., 633, 44
Mesmer, R. E., 582, 6
Messinev, M. Yu, 489, 7
Messmer, R. P., 354, 31
Meussner, R. A., 236, 2
Meyer, A. J., 265, 32
Meyer, M., 59, 74
Meyers, R. A., 511, 26,27
Michaelsen, C., 59, 73,73
Miczko, M., 119, 56
Miedema, A. R., 24, 12; 61, 101; 120, 63
Milburn, R. M., 582, 3,4
Miles, K. E., 582, 5
Millan, M., 370, 33
Miller, A. J., 264, 2
Miller, C. E., 58, 38
Mimura, H., 264, 6
Minai, Y., 634, 81; 646, 8
Minkova, A., 117, 10; 118, 17
Misawa, T., 583, 24
Miskolczy, G. J., 80, 23
Mitchell, I. V., 293, 54
Mitrofanov, K. P., 477, 4
Miyamoto, Y., 421, 20
Miyasaka, H., 119, 52; 135, 13
Mizoguchi, T., 59, 63, 64
Moberg, L. C., 713, 50
Modzelewski, C. U., 398, 33
Mohs, E., 325, 6
Mokrushin, A. D., 445, 24
Molodov, A. I., 713, 47
Moncoffre, N., 292, 31; 293, 48,49
Monstry, F. M., 510, 21
Montano, P. A., 510, 7,8,18; 511, 29; 532, 22,27; 608, 12,13,40; 681, 43,

46
Montiel-Montoya, 60, 81
Morice, J. A., 326, 38
Moriya, T., 265, 44; 293, 59
Morley, J. G., 420, 6
Morris, J. W., 264, 9, 10,12,13; 265, 41
Morris, R. V., 446, 34
Morrish, A. H., 80, 4; 31, 25,27; 119, 50; 369, 9; 397, 1,2,9, 12,13,14
Morup, S., 80, 1,3,10, 11,12,13,14,16,17, 21; 81,28,29,34,35, 40,41,42,43,44; 446, 42; 563, 21; 583, 31; 631, 2; 634, 62; 664, 3,4, 7; 695, 6,7,8,10, 11,12; 696, 13,15, 16,21,26
Mossbauer, R. L., 608, 18
Motte, J. P., 136, 27; 293, 56
Moukarika, A., 563, 10,12
Moulin, J., 292, 42
Mozzanega, H., 679, 14
Mrowec, S., 236, 6,11, 21
Muir, D. M., 608, 21
Mulay, L. N., 582, 10; 632, 23; 679, 11
Muller, H. G., 200, 43
Muller-Warmuth, W., 421,
Murad, E., 546, 9; 582, 19; 583, 30; 593, 11,12
Murakami, Y., 199, 24, 25,26; 216, 7; 354, 42; 712, 36; 714, 63
Murday, J. S., 291, 20
Murgas, L., 216, 14
Murthy, M. K., 421, 31,32
Music, S., 711, 14
Musser, D., 397, 4
Muthukumarasamy, P., 370, 27
Mysen, B. O., 445, 20

Nadai, J. P., 292, 42
Nagarajan, T., 370, 27
Nagarathna, H. M., 532 22
Nagy, J. B., 680, 20
Nagy, S., 200, 46; 216, 1,2,10,14,15; 711, 14; 712, 35; 713, 38
Nagy-Czako, I., 711, 1
Nahon, V. N., 632, 22
Naka, M., 59, 57,59; 60, 75; 353, 4,13; 369, 4; 370, 18
Nakagawa, H., 326, 30
Nakahara, S., 711, 6
Nakamura, A., 421, 23
Nakamura, Y., 370, 35
Nakayama, N., 748, 10
Narayan, P. B. V., 326, 40
Narayanaswamy, A., 370, 27
Narita, T., 236, 4
Nasonova, M. N., 712, 32,33
Nasrallah, M. M., 445, 22
Nasu, S., 57, 7; 59, 68; 198, 7,8; 199; 23,24,25,26,30; 216, 6,7,8
Nathan, Y., 546, 7
Navarro, J. M. F., 445, 25
Navrotsky, A., 563, 18
Nazama, N. S., 370, 29
Nazarova, T. I., 477, 3
Neel, L., 80, 19; 407, 3; 748, 14
Nemoshkalenko, V. V., 199, 22; 265, 43
Nevald, R., 696, 21
Nicol, M. J., 608, 20
Nielsen, H. J. V., 59 58
Nielsen, P., 80, 17; 664, 4
Niemantsverdriet, J. W., 632, 14,15, 16,17,18,19,27; 633, 38,47,48,54; 634, 63,68,72,73, 74,77,80; 647, 10, 11,12,13; 664, 5; 679, 6,7
Nieminen, R. M., 117, 12
Niessen, A. K., 24, 12
Nikitina, A. P., 563, 23
Niningerm R. C., 81, 38,39; 489, 6
Nishida, K., 236, 4
Nishida, T., 420, 10, 11,12,13,14,15; 421, 16,17,18,19, 20,21,22,23,24
Nishio, M., 199, 25
Nold, E., 354, 41,43
Nomura, K., 325, 1,8; 326, 19,20,21,22, 23,30
Nonaka, T., 420, 12; 421, 24
Nonnekens, R. C. H., 634, 80
Norskov, J. K., 680, 35
Nose, M., 60, 80,94
Novikov, Y. N., 607, 5
Novomlinova, M. I., 477, 3
Nowik, I., 398, 29
Nozik, A. J., 293, 57
Nussbaum, R. H., 135, 12; 711, 4

O'Grady, W. E., 713, 45
O'Handley, R. C., 59, 61,65; 354, 39
O'Keefe, J., 421, 26
Obermyer, R. T., 679, 11
Odeurs, J., 199, 39; 200, 45
Oelsen, W., 236, 26
Ohhasi, K., 608, 8
Ohi, K., 198, 16
Ohlerich, U., 680, 36
Ohmura, M., 711, 11
Ohnishi, T., 748, 10, 10,15
Ohnuma, S., 60, 80, 95
Ohring, M., 118, 20
Ohyabu, M., 326, 32, 33
Ok, H. N., 397, 2,12; 532, 21
Okabe, I., 325, 12,14
Okamoto, G., 313, 1
Okinaka, Y., 711, 6,7
Okkerse, B., 58, 31
Olbrich, H., 354, 41
Olesch, M., 563, 16
Olivier, D., 680, 24
Ollis, D. F., 634, 65
Olson, D. H., 679, 9
Olson, F. A., 532, 9, 10
Onishchenko, E. V., 407, 12

Ono, K., 199, 21
Onodera, H., 60, 79
Ooka, T., 264, 4
Orowan, E., 56, 1
Orr, R. L., 353, 14
Ortalli, I., 736, 3, 4,7,8; 737, 10,11, 737, 18,19,24,27,31
Osamura, K., 354, 42
Oskovie-Tabrizki, M., 679, 11
Ostanevich, Yu M., 445, 8
Oswald, R., 118, 20
Ott, G. L., 633, 45, 46
Ottrey, A. L. 521, 10, 11
Ouchi, C., 187, 7
Ovcharov, V. P., 398, 34
Ovchinnchov, V. V, 198, 13
Owen, D. G., 326, 39
Owen, W. S., 265, 42
Ozawa, T., 264, 6; 353, 12

Palombarini, G., 169, 20; 325, 10
Panissod, P., 60, 97
Paoluzi, A., 398, 33
Papp. H., 632, 29
Pargamin, L., 445, 19
Parish, R. V., 608, 19
Parker, P. M., 61, 103
Parkinson, N., 712, 23,27
Parravano, G., 665, 9
Partyka, R. J., 291, 22
Pashovsky, A., 713, 39
Pass, G., 723, 7
Pasternak, M., 81, 30
Patil, M. D., 521, 9; 646, 7
Pattyn, H., 199, 39; 200, 45
Paul, D. I., 398, 29
Pauleve, J., 24, 17
Pauling, L., 200, 44
Pease, D. M., 398, 33
Pebler, J., 407, 13, 15
Pederson, L. R., 722, 3,4
Pekala, M., 407, 8,11
Pendlebury, R. E., 135, 26; 293, 60
Pense, A. W., 264, 1
Perez, A., 119, 46,48, 49,51,53,57
Perfil'ev, Y. D., 563, 23
Pergolari, R., 736, 3
Perrichon, V., 632, 22; 632, 33
Pervukhin, N. A., 198, 13,15
Pesic, B., 532, 9,10
Petiau, J., 446, 28
Petrera, M., 737, 27
Petrini, G., 737, 29
Petrovic, L. F., 511, 28
Petrovykh, N. V., 445, 23
Petruk, W., 531, 3
Petry, W., 199, 40,41; 200, 42,43
Pfannes, H.-D., 135, 12,14,18,24
Pfannes, H. L., 398, 23
Phalippou, J., 446, 27
Pick, M. A., 57, 14
Pickles, A. T., 265, 36
Picone, P. J., 119, 50
Picraux, S. T., 118, 39; 292, 27,30
Piecuch, M., 58, 49
Pijolat, M., 632, 33
Pilling, N., 236, 10
Ping, J., 398, 25
Pipkorn, D. N., 57, 26; 119, 62
Pisarevskii, A. M., 445, 8
Pischik, L. M., 713, 41
Pisharody, K. P. R., 608, 7
Pitsch, W., 236, 24; 293, 58
Pitts, J. R., 723, 5, 14
Plateau, J., 264, 5
Plattner, H., 293, 54
Podgurski, H. H., 24, 24, 7
Polaczek, A., 407, 8,11
Polanyi, M., 56, 2
Pollack, S. S., 510, 15
Polushkin, Y. A., 265, 43
Polyakova, N. P., 563, 22
Pond, R., 38, 37
Ponec, V., 633, 36,51; 634, 82
Pope, L. E., 292, 27, 30
Popplewell, J., 80, 15
Porter, C. R., 511, 30
Potapov, A. P., 397, 7
Potzel, W., 608, 18
Pouchard, M., 407, 9
Povitskii, V. A., 489, 7
Prasad, A., 420, 9
Prasad, B. B., 370, 24; 397, 8
Prasad, K. G., 199, 32
Preston, R. S., 57, 7, 22,24,25; 59, 55; 197, 3; 198, 7,8; 199, 30; 216, 3,5,8
Price, D. C., 397, 21; 531, 4
Principi, G., 135, 21; 293, 44,62; 294; 65,67,68,70,71; 464, 3; 477, 6
Pritchard, A. M., 170, 37
Prusakov, V. A., 607, 5
Przybyszewska, J., 236, 6
Psaras, D., 679, 18
Puddington, A. F., 511, 32
Puddington, I. E., 511, 32
Pugh, M., 714, 61

Radelaar, S., 24, 16; 353, 20
Radhakrishnan, T. S., 370, 27
Rahmel, A., 235, 1
Raichevsky, G., 713, 39
Rainier-Harbach, A., 354, 41
Raj, K., 265, 29
Ralek, M., 679, 2
Raman, T., 446, 31
Ramous, E., 293, 44, 62; 464, 3; 477, 6
Ranganathan, S., 353, 24
Rao, C. N. R., 608, 7
Rao, G. N., 446, 31

Rao, K. R. P. M., 24, 5; 151, 5,6; 169, 11,17,22; 170, 30, 34,35
Rao, V. S., 679, 8,11
Rasumov, O. N., 199, 22
Rathnam, N. R. M., 369, 6
Raub, C. J., 712, 36; 714, 63
Raupp, G. B., 631, 11; 632, 21; 679, 3
Rautavuoma, A. O. I., 633, 34
Ravi, N., 370, 20,31
Ray, R., 59, 65; 353, 6
Razamovskaya, I. V., 445, 24
Razavi, F., 60, 82
Razumov, O. N., 265, 43
Readman, P. W., 607, 6
Rebouillat, J. P., 61, 110
Redhead, P. A., 680, 37
Rees, L. V. C., 326, 38; 680, 28
Reichelt, J., 59, 73
Reid, A. F., 477, 9
Reidinger, F., 57, 11,14,15
Reilly, J. J., 57, 10,11,14,15
Reiman, S. I., 713, 48
Reintsoma, R. S., 200, 45
Renner, J., 582, 11
Renon, H., 531, 5
Renyuan, Tang, 665, 6
Revina, T. V., 713, 48
Reymond, J. P., 633, 44
Rickard, D. T., 326, 38
Ridout, M. S., 266, 50; 326, 37
Riedi, P. C., 265, 31
Rigamonti, A., 737, 29
Riggs, W. M., 61, 111
Rindone, G. E., 445, 11
Riontino, G., 353, 7, 16,17,18,25; 354, 29,30,33
Ritter, G., 407, 17
Riviere, J. P., 199, 18
Robbins, C. R., 465, 8
Robbrecht, G., 446, 44,48; 521, 13
Robelet, M., 292, 29; 293, 47,52
Robelet, R., 293, 48
Robert, M., 264, 5
Rodgers, M. D., 24, 8
Rogoff, M. H., 511, 33
Rojas, D., 680, 24
Romig, A. D., 264, 11
Ron, M., 197, 1
Rooksby, H. P., 712, 27
Rosenberg, M., 370, 36
Rosenzweig, A., 608, 17
Rosso, M., 294, 68
Rothbaum, H. P., 546, 3
Roy, S. K., 199, 36
Rozenson, I., 563, 2, 9
Ruane, M., 608, 21
Rubartsch, A., 60, 96; 353, 9; 17
Ruby, S. L., 265, 26; 353, 11; 532, 28; 737, 17
Rucklidge, J. C., 563, 13
Ruud, C. O., 169, 26
Ruggiero, S. T., 749, 20
Ruppersberg, H., 353, 4; 370, 18
Ruskov, T., 532, 13
Russell, K. C., 266, 53
Ryan, D. H., 60, 82
Ryssel, H., 292, 26
Ryzhenko, B. V., 198, 13,14,15

Sachtler, W. M. H., 631, 10; 633, 52
Saegusa, N., 397, 9,14
Sakamoto, I., 119, 59
Sakka, S., 420, 7
Sakurai, H., 264, 6
Salamon, M. B., 370, 32
Salmon, J. E., 313, 5
Salomon, D., 57, 20; 326, 28
Saltman, P., 582, 11
Sams, J. R., 607, 2,3
Samsonov, G. V., 236, 20
Sanderson, R. T., 680, 27
Sanek, J., 119, 45
Saneyoshi, K., 117, 13,14,15; 118, 16; 134, 3,4; 135, 16,17
Sano< H., 326, 25
Saporoschenko, M., 477, 5; 521, 6
Sareen, S. S., 511, 28
Sarmo, A. C., 737, 15
Sassa, K., 198, 5; 199 27,31,37,38
Satarova, L. M., 563, 22
Sato, N., 313 1
Sato, Turtelli R., 354 29,33
Sawatzky, G. A., 265, 22; 446, 38,45
Sawicka, B. D., 117, 1,4,6; 118, 26,27, 28; 119, 40,41,42, 43,44,45,46,47,48, 49,50,60,61; 199, 34,35; 326, 27,29
Sawicki, J. A., 117, 1,4,6,7; 118, 18, 21,23,24,25,26,27, 28,29,30,32,33,34, 36,37,38; 119, 41, 42,43,44,45,46,47, 48,49,56,57,58,60, 61; 120, 64; 199, 34,35; 292, 38; 325, 17; 326, 27,29
Schaafsma, A. S., 353, 20;354, 32
Schafer, W., 57, 18
Schafer-Stahl, H., 633, 39
Scharff, M., 135, 22
Schefer, J., 57, 16
Scherson, D., 713, 51,52
Scherzer, B. M. U., 119, 60
Scherzer, J., 680, 23
Schikorr, G., 593, 5
Schiott, H. E., 135, 22,23
Schlapbach. L., 57, 12,16; 60, 81

Schmidt, F., 679, 17; 680, 19,22,29,31, 32,33,36,38,39
Schmidt, K., 407, 15
Schmiester, G., 680, 19
Schmitt, H., 61, 110
Schneeweiss, O., 198, 17
Schneider, J., 398, 36
Schoefer, E. A., 169, 6
Schoenberg, L. N., 711, 9
Schreckenbach, K., 120, 64
Schreurs, J., 293, 50
Schroeder, K., 200, 42
Schroeer, D., 81, 38, 39; 489, 6
Schubert, R., 711, 10
Schuit. G. C. A., 695, 4
Schulkes, J. A., 446, 47
Schuller, I. K., 748, 1,5,6
Schulze, D. G., 546, 8
Schurer, P. J., 81, 27; 369, 9; 397, 1, 13
Schwartz, L. H., 24, 1; 169, 9,13,15; 264, 8; 266, 45,47; 294, 80; 370, 34; 632, 20,29; 633, 37,49; 634, 60,61; 679, 4,5; 749, 19
Schwartzendruber, L. J., 169, 6
Schweig, B., 722, 1
Schwertmann, U., 546, 9; 582, 14,15,16, 19; 583, 26,30,34; 593, 12
Schwoeble, A. J., 151, 10
Scotoni, I., 291, 16,17; 292, 35
Scott, M. G., 60, 98
Scroggie, B., 421, 32
Seehra, M. S., 532, 22
Seel, F., 680, 21
Seifert, F., 445, 20
Sekizawa. K., 646, 3
Selman, B., 632, 14
Selwood, P. W., 582, 10
Senateur, J. P., 632, 30
Senftle, F. E., 532, 23
Serughetti, J., 119, 57; 120, 66
Servidori, M., 293, 54
Sette-Camara, A., 723, 6
Shaafma, A. S., 397, 6
Shah, S. N., 713, 57; 714, 60
Shams El Din, A. M., 713, 55,58
Shanfield, Z., 608, 12,13; 681, 43,46
Sharp, J. H., 572, 1
Shartsis, L., 421, 28,29
Shchukin, N. F., 489, 7
Shelby, J. E., 421, 33,34
Shen, W. M., 633, 35,50
Shenoy, G. K., 532, 28; 737, 17
Shepeleva, L. V., 712, 34
Shibue, K., 354, 42
Shields, M., 737, 26
Shiga, M., 370, 35
Shigematsu, T., 135, 12,14,18,24
Shikakura, K., 646, 3
Shikazone, N., 370, 35
Shiley, R. H., 477, 5; 521, 6
Shimodaira, S., 583, 24
Shimony, U., 445, 15
Shingu, P. H., 199, 26; 216, 7
Shinjo, T., 80, 20; 748, 2,4,7,8,9,10, 12,15,16
Shiotsuki, T., 420, 13; 421, 18,19
Shirakawa, K., 60, 80,94
Shirane, G., 737, 20
Shirley, D. A., 264, 17; 407, 14
Shoesmith, D. W., 326, 39
Shtrikman, S., 464, 6
Shull, C. G., 265, 39
Shvorneva, L. I., 445, 24
Siddall, T. H., 582, 2
Sidorenko, F. A., 198, 12,13,14,15,
Sidorov, T. A., 477, 4
Siedel, B., 532, 19
Sigety, E. A., 445, 12
Silver, J., 712, 29 712 29
Silverman, M. P., 511, 133
Silverman, S. R., 546, 6
Simmons, G. W., 170, 31,32; 325, 4,43; 711, 5,14; 713, 44; 714, 62
Simozar, S., 24, 14
Sinfelt, J. H., 633, 53; 634, 71
Singer, I. L., 291, 7,8,9,20; 292, 23,24,25
Singhu, R. M., 608, 7; 420, 9
Skalski, S., 265, 29
Skoutarides, C., 293, 47
Skrzypek, S., 118, 21
Smaltzer, W. W., 236, 5,27
Smit, P. H., 60, 84
Smith, C. H., 398, 32; 521, 6
Smith, D. C., 170, 36; 593, 2
Smith, G. V., 477, 5
Smithells, C. J., 236, 28
Smykatz, W., 489, 8
Snijders, H., 353, 20 397, 6
Snow, R. B., 465, 9
Soga, N., 420, 3
Solomon, P. R., 510, 13
Somorjai, C. A., 632, 12; 633, 43
Sood, D. K., 169, 22,23
Sorensen, O., 696, 25
Soubeyroux, J. L., 563, 12
Souza De, J. P., 136, 29
Speich, G. R., 151, 10
Spencer, N. E., 714, 59
Spiering, H., 61, 105
Spijkerman, J. J., 117, 11; 187, 5; 292, 37
Spinner, S., 421, 28,29
Spiro, T. G., 582, 11

Spitler, C. A., 679, 10
Sproule, F. I., 325, 15
Stampfl, P. P., 593, 3
Stanek, J., 118, 27, 32,37,38; 119, 45, 46,47,56; 199, 34; 326, 29
Stanfield, R. M, 632, 23,31
Staniek, S., 135, 12, 14,18
Stanworth, J. E., 445, 9
Stapel, U., 680, 38
Stavan, M. J., 397, 13
Stearns, M. B., 236, 14,25; 264, 18; 265, 19,30
Steeb, S., 354, 41, 43
Steele, F. N., 445, 14
Steen, C. F., 711, 4
Stefaniay, V., 216, 9,15
Steiner, P., 57, 19
Stencel, J. M., 679, 8,10
Sterten, A., 236, 3
Stevens, J. G., 117, 8; 723, 12,13; 737, 14
Stevens, V. E., 117, 8; 723, 13
Steward, A. M., 370, 25
Stickels, C. A., 197, 2; 198, 4; 216, 4
Stoklosa, A., 236, 6
Stoop, F., 633, 57; 634, 76,80
Storch, H. H., 631, 3
Stout, R. D., 264, 1
Strafford, K. N., 236, 17
Strauss, A. Y., 736, 1
Stroink, G., 563, 11
Strom-Olsen, J. O., 60, 78,82
Stucki, G., 57, 12, 16
Stukan, R. A., 607, 5
Styert, W. A., 711, 3
Su, Z., 665, 6
Suba, M., 714, 68
Subiel, S. M., 294, 75
Subrahamanyam, J., 169, 22
Sucksmith, W., 265, 36
Sudakova, L. P., 477, 3
Sugawara, Y., 326, 19
Sugino, K., 264, 4
Suib, S. L., 679, 18
Sultanov, G. D., 737, 28
Sumitomo, Y., 265, 44; 293, 59
Sun, E., 397, 20
Sutcliffe, H., 723, 7
Suzdalev, I. P., 632; 26; 407, 16
Suzuki, K., 369, 3
Suzuki, R., 354, 42
Swanson, K. R., 117, 11; 187, 5
Swanson, M. L., 293, 54
Swartz, J. C., 398, 31
Swartzendruber, L. J., 57, 17; 398, 30
Sylva, R. N., 582, 12,13
Syn, C. K., 264, 9
Szabo, I., 216, 14
Szawlowska-Wallisch, K., 236, 18
Szekely, T., 326, 42
Szeles, Cs, 216, 1 712, 20
Sziraki, L., 713, 47

Taeymans, A., 444, 3
Takacs, L., 354, 34; 712, 19
Takada, T., 398, 28; 748, 2,4,7,8,9,15
Takahashi, T., 119, 59; 748, 8,16,18
Takashima, T., 420, 10,11,12,13,14,15; 421, 16-24
Takeda, M., 647, 9
Tamiya, M., 198, 16
Tammann, G., 236, 9
Tamura, I., 81, 33
Taneja, S. P., 521, 8
Tang, D., 521, 11
Tanji, Y., 266, 49
Tarasov, B. V., 477, 3
Tarnoczy, T., 326, 42
Tau, L. M., 632, 25
Taylor, C. A. J., 327, 44
Taylor, G. I., 56, 3
Taylor. H. F. W., 593, 10
Taylor, P., 326, 29
Taylor, R. D., 711, 3
Taylor, R. M., 582, 14; 583, 23; 593, 6,7,8,11
Taziani, A., 464, 3
Teichner, S. J., 633, 44
Tempersay, A. A., 236, 16
Terakura, K., 748, 17, 18
Terasaka, M., 325, 16
Terra, A., 510, 5
Terrel, J. H., 170, 29
Terzis, A., 582, 11
Teule-Gay, F., 632, 24
Tewari, A. K., 199, 36
Thevenard, P., 119, 49
Thiel, R., 582, 20
Tholance, J. L., 407, 2
Thomas, J. M., 117, 2; 325, 5
Thomas, M. T., 722, 3; 723, 5
Thompson, P., 57, 11, 14,15
Thorpe, R. K., 117, 2
Thorsen, G., 531, 2
Thrane, N., 80, 17; 664, 4
Thwaites, C. P., 712, 28
Tian, J., 291, 21
Tieman, J. W., 510, 20
Timmermans, W., 445, 4,5,6
Tittlebaum, M. E., 521, 9
Tiverois, B., 291, 17
Tiveron, B., 294, 65
Tiziani, A., 477, 6
Tjon, J. A., 57, 27
Tjutina, K. M., 712, 34
Tkatchendo, I., 679, 14
Tomilov, S. B., 445, 8
Tominaga, T., 634, 81; 646, 8
Tomov, T., 532, 13; 713, 49
Tooley, F. V., 445, 10
Topsoe, H., 80, 1,3, 10,11,14,17,21; 81, 28,29,34,35,40,41; 446, 42; 631, 1,2; 634, 62; 664, 2,3, 4; 665, 7; 681, 41,42; 695, 5,6,7, 8,9,11,12,13,15,16,

17,18,22; 696, 24, 25,26
Topsoe, N.-Y., 696, 19-21,23
Torgesen, D. R., 236, 5
Toriyama, T., 117, 13, 14; 119, 52; 134, 4; 135, 13,16,17
Tosello, C., 135, 21; 291, 16,17; 292, 35; 293, 44; 294, 65,67,70
Tosto, S., 135, 11; 291, 16; 292, 35,39
Tougaard, S., 136, 31
Tournier, R., 207, 2
Tousset, J., 292, 29; 293, 47,48,49,52
Towata, S., 198, 16
Trautwein, A. X., 680, 21
Travis, J. C., 634, 70
Tregeux, D., 292, 29; 293, 48,52
Trevor, D. Y., 681, 47
Tricker, M. J., 117, 2,3; 199, 33; 292, 41; 325, 5; 608, 9
Troiano, A. R., 264, 3
Tromp, K. F., 509, 4
Trueman, N. A., 546, 4
Trukhtanov, V. A., 151, 1
Tsang, Y. N., 737, 12
Tsinman, A. T., 713, 40,41
Tsuchida, N., 608, 21
Tsuei, C. C., 58, 36, 47; 647, 14
Tsujikawa, I., 608, 8
Turkdogan, E. T., 464, 1
Turlier, P., 632, 22
Turmezey, T., 216, 2,9,10,15
Turner, R. C., 582, 5
Turner, W. E. S., 421, 27
Twardowska, H., 477, 5; 521, 6
Twist, W., 477, 8
Tyliszczak, T., 118, 18,21,24-27,32-34; 119, 43,45-49,56; 326, 27,29
Uhlmann, D. R., 445, 7
Ujihira, Y., 169, 21; 325, 1,2,8,11,12, 13,14,16; 326, 19, 20,21,22,23,24,31, 32,33,36,41
Ullrich. B. M., 326, 18
Ulmo, J., 510, 6
Umeyama, S., 199, 38
Unmuth, E. E., 633, 37; 634, 60; 679, 4
Unruh, K. M., 60, 85, 86,89,100
Ura, M., 58, 50

Valkalahti, S., 117, 12
Van De Voorde, N., 444, 3; 446, 35
Van Der Baan, H. S., 632, 27; 679, 6
Van Der Boom, P. F. J. 722, 2
Van Der Kraan, A. M., 81, 37; 632, 14,15, 16,18,27,38; 634, 63,68,73,77,80; 647, 10,11; 664, 5; 665, 10; 679. 6,7; 748, 11
Van Der Woude, F., 24, 16; 60, 98; 61, 101; 120, 63; 265, 22; 353, 20; 354, 32; 397, 6; 446 38,45; 489, 5
Van Diepen, A. M., 24, 15; 58, 52; 397, 10
Van Dijk, W. L., 632, 27; 633, 41,47; 679, 6
Van Eenoo, M., 680, 20
Van Grondelle, J., 634, 77
Van Iseghem, P., 445, 4,5,6; 446, 29,35
Van Kaam, J. A. C., 634, 63
Van Loef, J. J., 632, 14,16; 634, 73; 647, 10; 664, 5; 679, 7
Van Nice, L. J., 511, 26
Van Wouve, D., 679, 1
Van't Blik, H. F. J., 633, 54; 634, 74; 647, 12
Vandenberghe, R. E., 397, 3; 446, 37,46
Vangeel, J., 444, 3; 445, 4
Vanleerberghe, R., 446, 37,40
Vannice, M. A., 632, 23; 633, 56; 634, 68
Varjoranta, T., 291, 18
Varret, F., 353, 8; 398, 24
Varsanyi, M. L., 711, 1,13; 712, 26,30, 31; 713, 43,44; 714, 62,65
Vasihnava, P. P., 680, 40
Vaughan, D. J., 326, 37
Venturello, G., 353, 7,16,17
Vera, A., 737, 10
Verbiest, E., 199, 39; 200, 45
Versanyi, L. M., 713, 47
Vertes, A., 200, 46; 216, 1,2,10,14,15; 326, 42; 354, 34; 608, 11; 711, 1,13; 712, 19,20,21,22, 26,30,31,34,35,37, 38,39,44,47; 714, 62,65,68
Vicent, J. L., 748, 6
Vigier, P., 370, 33 370 33
Villadsen, J., 81, 42,43; 696, 17
Vinai, V., 354, 29, 33
Vincze, I., 60, 98; 265, 20,21,32; 354, 32,40; 397, 5
Virgo, D., 445, 20
Vitek, V. 353, 27
Vogel. G., 57, 6; 199, 28,29,40,41; 200, 42,43
Vogler, G. L., 632, 28
Vol'pin, M. E., 607, 5
Volovic, T. J., 532, 12
Von Der Gathem, R., 511, 35
Von Heimendhal, M., 353, 24
Vosburgh, W. C., 582, 2,3

Vrinat, M., 696, 14
Vylov, Ts, 118, 17

Waard, H. De, 118, 36
Wadsworth, M. E., 532, 8
Wagner, C. P., 61, 111
Wagner, F. E., 57, 9,23; 120, 64
Wagner, H. A., 59, 73,74
Wagner, H. G., 57, 19; 58, 41-43,45; 59, 54,59,69,70,71,72; 60, 75,81; 61, 108; 369, 11; 397, 16
Wakita, N., 264, 6
Walker, J. C., 80, 22
Walker, L. R., 354, 35; 748, 3
WallenR. J., 118, 17
Walter, J. L., 352, 3; 354, 44,47; 370, 13
Walther, H., 680, 38
Walton, G. N., 170, 37
Wanatabe, S., 489, 3
Wang, C., 170, 33
Wang, C. S., 81, 30,31
Wang, Z., 170, 33
Wappling, R., 398, 35
Warner, L. M., 714, 61
Warren, G. W., 532, 8
Waseda, Y., 60, 87,91
Wasinski, H., 407, 15
Waszczak, J. V., 748, 3
Watanabe, H., 370, 29
Watkins, R. E. J., 119, 55
Watson, R. E., 57, 17; 265, 23
Wecker, J., 59, 73
Wegener, H., 216, 12
Wegener, H. H. F., 407, 17
Wehner, H. L., 407, 17
Weise, R. F., 266, 46
Weiser, G., 407, 15
Weiser, J. D., 546, 6
Weisman, I. D., 398, 30
Welker, H., 57, 4
Wellington, O. L., 608, 20
Welter, J. M., 58, 28
Wender, I., 511, 33
Wengle, S. H., 737, 13
Werber, T., 236, 7, 18,19,21,22
Wernick, J. H., 199, 20; 265, 27
Wertheim, G. K., 199, 20; 265, 27; 354, 35
West, K. W., 711, 8,9; 723, 8
West, P. J., 326, 28
Westenbarger, G. A., 264, 17
Weyer, G., 326, 28
Weyl, W. A., 445, 13
Wheelock, T. D., 510, 23
Whetten, R. L., 681, 47
White, M. S., 546, 1,2
White, W. B., 420, 8; 445, 21
Whitlow, W. H., 135, 26; 293, 60
Whittaker, E. J. W., 563, 14
Whittle, G. L., 370, 25
Whitton, J. L., 293, 54
Wickman, H. H., 407, 14
Wicks, F. J., 463, 14 14
Wiedersich, H., 57, 5
Wilk, W., 118, 34
Wilkinson, M. K., 265, 39
Willens, R. H., 58, 34
Willett, R., 593, 7
Williams, E. K., 58, 29
Williamson, D. L., 510, 9; 532, 16
Wilson, T. P., 631, 8; 646, 2
Window, G., 370, 22; 397, 18
Winterbottom, A. P., 325, 5
Wipf, H., 57, 20,21
Wirel, C., 664, 3
Wisall, R. H., 57, 10
Wivel. C., 81, 43,44; 665, 7; 695, 10,11, 12,13; 696, 21
Wohlfahrt, K., 118, 19
Wohlfarth, E. P., 60, 92; 407, 5,10
Wolf, b., 680, 21
Wolf, C., 509, 3
Wolf, G., 291, 17
Wolowodiuk, C., 711, 7
Wong, H. K., 749, 19
Wood, J. C., 293, 57
Wordel, R., 57, 23
Wormald, M. R., 510, 12
Wortmann, G., 57, 8,9, 20,21
Wright, P. A., 477, 1
Wrzeniewska, T., 294, 76
Wuckowska, L., 510, 10

Xu, Z., 398, 25

Yagnik, C. M., 169, 169, 27,28; 737, 21
Yamada, T., 748, 9
Yamada, Y., 737, 20; 748, 10
Yamamoto, H., 60, 70; 236, 13
Yamamoto, R., 478, 2
Yamauchi, H., 60, 79
Yamauchi, K., 59, 63
Yanase, A., 748, 17
Yang, T., 135, 15; 265, 25
Yano, S., 264, 6
Yao, S. B., 713, 51
Yashida, H., 199, 38; 748, 8,16,18
Yavtukhovich, N. A., 477, 3
Yeager, E. B., 713 51,52
Yelsukov, Y. P., 397, 7
Yermakov, Y. I., 634, 83
Yershova, Z. P., 563, 23
Yestropovb, K. S., 421, 30
Yokoyama, T., 647, 15
Yonekura, Y., 119, 52; 135, 13,17
Yoshida, K., 325, 16
Yoshinaga, M., 489, 3
Yost, F. G., 292, 27, 30
Young, D. J., 236, 5
Yurchikov, Y. Y., 397, 7
Zak, T., 198, 17
Zambo-Balla, K., 352, 2; 370, 15
Zanarenkova, L. E., 713, 41
Zarzycki, J., 446, 27
Zelger, M., 57, 23

Zemcik, T., 151, 4; 198, 17
Zentilli, M., 563, 11
Zhang, H., 170, 33
Zhang, Q. R., 59, 73
Zhang, X. Z., 293, 53
Zhsug, G. L., 291, 17
Zinn, W., 198, 9,10
Zoki-Awad, M., 200, 46
Zomorrodian, A., 136, 31
Zuckerman, J. J., 24, 3; 714, 66
Zukrowski, J., 80, 18; 294, 74,77,78; 354, 37
Zurek, Z., 237, 7,8, 19,22
Zussman, J., 563, 13
Zvyagin, B. B., 582, 17

SUBJECT INDEX

Abnormal magnetic anisotropy, 386
Absorber thickness, 241
Absorption area, 481
Absorption area determination, 524f
Absorption edge of chromium, 143
Acetic acid Fischer-Tropsch synthesis, 609
Acid mine drainage, 495
Acid soil conditions, 546
Acidophilic chemautotrophic bacteria, 508
Activated carbon, 595f
Activated carbon gold bonding, 606
Activated carbon-metal bonding, 596
Activated charcoal powder, 597
Activation energy diffusion, 705
Activation energy of crystallization, 343
Activation energy glass, 417
Active data acquisition, 102
Active catalytic phases, 615
Active potential, 298
Activity coefficient-steel, 448
Adherence to glass surface, 715
Adsorption of alkali metals, 668
Adsorption of $Fe(CO)_5$, 670
Akaganeite, 316, 567
Al-Cu alloys, 233
Al-Fe alloys, 189, 201f
Al-Fe-Si alloys, 201f
Alkali ion diffusion in glass, 414
Alkali metal vapor adsorption, 668
Alkaline earth glass modifiers, 428
Allowing elements, 271
Alloy atomic mobility, 342
Alloy charge density, 347
Alloy clustering, 346
Alloy crystallization temperature, 332, 340
Alloy crystallization enthalpy, 332
Alloy crystallization energy, 332
Alloy domain fluctuation, 337
Alloy effective atomic volume, 341
Alloy electrical resistivity, 333
Alloy electron distributions, 337
Alloy elemental distribution, 227
Alloy enthalpy of formation, 14
Alloy free volume, 347
Alloy heat of formation, 13
Alloy homogeneity, 330
Alloy hyperfine fields, 237
Alloy hyperfine field distribution, 624
Alloy-local magnetic properties, 347
Alloy magnetic behavior, 46
Alloy-mean field, 350
Alloy-mean isomer shift, 331
Alloy order parameter, 229
Alloy particle size, 625
Alloy particle surface enrichment, 74
Alloy-quenched in nuclei, 352
Alloy random distribution, 229
Alloy single crystals, 19
Alloy spin density, 347
Alloy-structural homogeneity, 351
Alloy-structural relaxation, 330, 346
Alloy surface layer composition, 226
Alloy thermal stability, 343
Alloy-topological order, 346
Alloying elements, 18
Alloys containing boron, 341
Alloys containing titanium, 330
AlOOH, 567
Alpha-iron, 128
$AlPO_4$, 536
Alumina structural promoter, 667
Alumina supported catalyst bonding, 660

Aluminized steel surfaces, 162
Aluminized steels, 163
Aluminized surface, 154
Aluminogoethite, 539
Aluminosilicate hearthbrick, 467
Aluminosilicate refractory linings, 468
Aluminum alloys, 201f
Aluminum-copper alloys, 233
Aluminum heat treatment, 208, 212
Aluminum hydrous layer silicates, 560
Aluminum-iron-alloys, 189
Aluminum-iron impurities, 201
Aluminum manufacturing, 189
Aluminum solutes in steel, 243
Aluminum substituted goethite, 567
Aluminum(III) in glass, 433, 437
Aluminum(III) in magnetite, 442
$Al_{0.26}Fe_{0.09}V_2O_5$, 399
$Al_{13}Fe_4$, 191f
Al_2O_3, 428, 432, 455, 467, 620, 683
Al_2S_3, 233f
Al_3Fe, 202f, 205f, 212
Al_4Cu_9, 233
Al_5Fe_2, 191f
Al_6Fe, 119f
$Al_6Si_2O_{13}$, 562
Amesite, 547f
Amicon Diafilters, 568
Ammonia in coal treatment, 502
Ammonia production, 652
Ammonium heptamolybdate, 683
Amomet, 32
Amorphous alloy-asymmetry parameters, 372
Amorphous alloy-crystallization, 343
Amorphous alloy films, 742
Amorphous alloy ribbons, 41, 383
Amorphous alloys, 329f, 355f, 360, 371f, 378
Amorphous alloys-ferromagnetic, 371f
Amorphous alloys-magnetic anisotropy, 371f
Amorphous alloys-magnetic relaxation, 371f
Amorphous alloys-magnetic texture, 371f
Amorphous alloys-quenched, 376
Amorphous alloys-short range order, 394
Amorphous alloys-statistical asymmetry, 378
Amorphous FeB phase, 162
Amorphous iron, 316
Amorphous metal-metal systems, 44
Amorphous metals, 25f, 31
Amorphous phase composition, 392
Amorphous phase on steel, 162
Amorphous phosphate, 546
Amorphous ribbon surface oxidation, 391
Amorphous ribbons, 41, 383
Amorphous surface layer, 162
Amorphous to crystalline transformation, 388
Analytical electron microscopy, 239, 693
Anchor onto silica support, 645
Ancient ceramic artifacts, 547
Angular resolved depth selective analysis, 122
Anisotropic fields in alloys, 331
Anisotropic hyperfine fields, 19
Anisotropy fields, 309
Anisotropy magnetic surface, 744
Annealing magnetic anisotropy, 387
Annealing of $Fe_{16}B_{27}Si_7C_4$, 387
Annealing treatment, 41
Anodic polarization, 707
Anodic protection, 166
Antiferromagnetic exchange, 47
Antiferromagnetic order, 309
Antiferromagnetism, 172, 483
Antigorite, 552f
Antimony-cobalt multilayers, 740
Antimony enriched surfaces, 662
Antimony-iron alloys, 12
Antimony-iron multilayers, 740
Antimony oxide, 662
Antimony pentachloride, 661
Antimony-tin catalysts, 660
Antimony-121, 660
Apatite, 534f
Applications to coal, 137
Applications to quality control, 137
Applications to steel, 137
Applied field CEMS, 99
Applied magnetic fields, 50, 54, 68, 100, 255, 309, 603
Argon arc furnace, 656
Argon implantation, 269
Argon ion sputtering, 55
Arrhenius activation energy, 260
Arrhenius equation, 260, 414
Artificial superstructures, 739f
Ash content in coal, 148f
Ash deposits, 153f
Ash determination in coal, 151
Ash fouling, 514
Ash precipitator, 513
Ash treatment, 495
Asperomagnetism, 46
ASTM coal test, 147
Asymmetric field distributions, 442
Asymmetric lineshapes, 567
Asymmetry in Mossbauer spectra, 371

Asymmetry parameter, 33, 53
Atmospheric oxidation of
steel, 301
Atmospheric pollution, 664
Atom mobility, 235
Atom packing-strain free, 105
Atomic mobility in alloys, 342
Atomic volume, 15, 18
AuCN, 607
AuFe, 112
Auger electron spectroscopy, 99
125, 271f, 311, 620
Auger electron surface
analyses, 715f
Auger electrons, 84, 125
Auger intensities, 301
Auger microprobe analysis, 297
AuMg alloys, 115
Austenite, 8, 10, 138f, 155f,
171, 176, 186, 237f, 253f,
606
Austenite chemical
composition, 256
Austenite determination-
steel, 139
Austenite diffusion, 261
Austenite in steel, 139, 141
Austenite meter, 8
Austenite nucleation
kinetics, 260
Austenite retained in
steel, 277
Austenite tempering, 259
Austenite to martensite
ratio, 155
Austenite volume fractions
in coal, 151
Austenitemeter, 144
Austenitic stainless steels, 108,
143, 270
Austenitic steel corrosion, 108
Average hyperfine field, 393f
Average magnetic moment, 678
Avrami exponent, 418
Avrami law, 344

Backscattered Mossbauer spectra,
8, 121, 141, 156, 238
Bacterial desulfurization
of coal, 502
Ball on disk wear tests, 271
Band inversion in semi-
conductors, 731
Barium(II) in glass, 431
Barrandite, 535
Basic oxygen furnace, 157f
Basic oxygen furnace slag, 138,
448
Basic oxygen process, 447
Basic steel making slag, 475
$BaSnO_3$, 719, 730
Bauxite, 523, 529
Beam current density, 278
Beaverite, 528
Bed moist brown coal, 513
Beneficiated ores, 7
Bernal model, 31
Bernal structure, 38
Berthierine, 547f
Bessel functions, 524
BET technique, 721
Beta-AlFeSi, 209f
Bidisperse iron phase, 668
Bifunctional zeolite
catalysts, 667f
Bilayer scale structure, 230
Bimetallic catalysts under
synthesis gas, 629
Bimetallic catalysts, 635f, 654
Bimetallic metal oxide
catalysts, 649f
Binding energy of impurities, 13
Binomial distribution, 107, 229
Biological leaching of coal, 491f
Bismuth, 16
Bituminous coal, 513
Black coal, 513, 606
Blast furnace, 153f, 447
Blocking distribution
function, 402
Blocking temperature, 399
Blocking temperature in coal, 516
Boiler ash, 154
Bond disorder, 362
Bonding of silver to glass, 715f
Bonding to alumina supports, 660
Borate glass, 409
Boride layer, 162
Boride layers on steel, 162
Borided steel surfaces, 161
Borided surface, 154
Bornite, 523f
Boron containing alloys, 341
Boron depletion, 41
Boron glass network former, 412
Boron-iron glasses, 352
Boron selective oxidation, 41f
Boronized steel, 316f
Borophosphate glass, 409
Borosilicate glass, 409, 424f
Brillouin curve, 361
Brillouin zone, 731
Brillouin zone boundary, 734
Bromine methanol-iron
extraction, 461
Bronze electrodeposits, 703
Brookite, 476
Brown coal, 606
Brown coal ash, 513
Brown coal-bed moist, 513
Brown coal utilization, 513
Brownmillerite, 516
Brucite, 592
Brucite layers, 585

Brucite trioctahedral layer, 592
Bulk compressive stress, 395f
Bulk crystallization, 41
Burger vectors, 32
Butadiene formation, 664

$CaCO_3$, 536
Calcination, 654
Calcination temperature, 692
Calciowustite, 459
Calciphos, 534, 545
Calcium aluminoferrite, 513
Calcium aluminosilicate, 516
Calcium chloride, 492
Calcium ferrite, 158
Calcium sulfate, 516
Calcium(II) in glass, 431
Calcium(II) in magnetite, 442
Canted spins, 309
CaO, 428
Carbide bulk structure, 616
Carbided steel, 271
Carbided surface, 154
Carbon active sites, 596
Carbon diffusion in cobalt, 616
Carbon diffusion in nickel, 616
Carbon effect on hyperfine field, 248f
Carbon iron surface dissociation, 615
Carbon monoxide conversion reaction, 617, 635
Carbon pore size distribution, 605
Carbon-saturated liquid-iron, 447
Carbon tetrachloride adsorption, 596
Carbonaceous material, 595f
Carbonitrided steel, 276f
Carbonitriding, 8
Carbonized iron foil, 101
Carbonized materials, 595f
Carburization Fischer-Tropsch synthesis, 616f
Carburization of catalysts, 667f
Carburization of iron catalysts, 611f
Carburization rate, 619
Carburized steel, 316f
Cascade mixing, 275
$CaSnO_3$, 98, 719, 731
$CaSO_4.2H_2O$, 520
Cassiterite, 137
Cast iron boilers, 167
Catalyst air oxidation, 642
Catalyst charge distribution, 650
Catalyst for hydrodesulfurization, 683f
Catalyst-gas adsorption, 625
Catalyst hyperfine field distribution, 624
Catalyst-isotope enrichment, 650
Catalyst metal loading, 635
Catalyst particle size, 625f
Catalyst passivation, 628
Catalyst reoxidation, 657
Catalyst rereduction, 628
Catalyst rereduction with hydrogen, 642
Catalyst surface structure, 643
Catalysts-bimetallic, 654
Catalytic competition model, 616f
Catalytic surface layers, 650f
Catalytically active phases, 615
Cathodic protection, 166
Cation deficient magnetite, 166
Cation ordered minerals, 561
Cationic oxidation states, 650
$Ca_2Al_2O_5$, 455
$Ca_2Fe_2O_5$, 158, 449
Ca_2SiO_4, 459
$Ca_3(PO_4)_2$, 536
Ca_3SiO_5, 459
$Ca_5(PO_4)_3F$, 534f
Cementite, 606
CEMS, 41
CEMS at high temperatures, 98
CEMS at ultra high vacuum, 99
CEMS in applied fields, 99
CEMS ion implantation studies, 103
CEMS measurements at low temperature, 92
CEMS probing depth, 85
CEMS sensitivity, 89
Cendrex, 148
CeO_2 powder, 717
CeO_2 scrubbing, 721
Ceramics, 63
Ceramics manufacture, 547
Cesium borosilicate glass, 415
Chalcocite, 528
Chalcogenide matrices, 727
Chalcogenide oxidation, 727
Chalcopyrite, 494f, 523f
Channel electron multipliers, 89 92, 98
Channeltron, 123
Characteristics of coal, 497
Charcoal-activated powder, 597
Charcoal-decolorizing, 597
Charge density in alloys, 347
Charge distribution in catalysts, 650f
Charge transfer, 672
Chemical composition of austenite, 256
Chemical ordering in alloys, 346
Chemisorbed hydrogen, 71
Chemisorption, 73
Chemisorption of carbon monoxide, 629
Chill casting, 203
Chloride brine on steel, 316

Chlorite minerals, 548, 561
Christmas Island fertilizer rock, 533f
Chrome magnesite, 468
Chromium content in steel, 151, 286
Chromium-effect on hyperfine field, 248f
Chromium in ferritic stainless steel, 288
Chromium in martensitic steel, 271
Chromium in stainless steel, 143
Chromium in steel, 140, 143, 157, 161, 166, 241, 271, 277, 279
Chromium-iron coordination, 283f
Chromium K-shell absorption edge, 143
Chromium nitrides, 277
Chromium plating, 702
Chromium random distribution in steel, 287f
Chrysotile, 552f
Citrapos, 534
Citrate sodium dithionite extraction, 539
Clay in coal, 147, 157, 469, 514,
Clay minerals, 63
Clay minerology, 548
Closure domains, 100f
Cluster critical flip frequency, 405
Cluster distribution in spin glasses, 399
Cluster magnetization in spin glasses, 399
Cluster size distribution, 405
Cluster superparamagnetism, 399 403
Cluster distribution function, 399
Clustering in alloys, 346
CO chemisorption, 629
Co precipitation techniques, 661
$Co(OH)_2$, 699
Cobalt oxyhydroxides, 698
Coal, 145, 154f, 165
Coal-agglomeration, 513
Coal ash content, 148f
Coal ash treatment, 495
Coal autoclaving, 520
Coal bacterial desulfurization, 502
Coal biological leaching of sulfur, 491f
Coal biological treatment, 508
Coal-bituminous, 513
Coal-black, 513, 606
Coal-brown, 606
Coal-caloric value, 501
Coal char, 603
Coal characterization, 154, 497
Coal-clay, 514
Coal-clay content, 496
Coal combustion, 153f
Coal crushing, 498
Coal density, 496
Coal deposits, 153f
Coal desulfurization, 491f
Coal desulfurization and grain size, 501
Coal desulfurization methods, 503
Coal distribution of hyperfine fields, 514
Coal drying, 513
Coal-encapsulated sulfur, 497
Coal f-factor, 498
Coal-float sink separation, 491f
Coal-fluidized bed combustion, 494
Coal-power plant use, 492
Coal-froth flotation, 491f
Coal gas scrubbers, 492
Coal gasification, 218, 609
Coal-grain size, 496, 501
Coal gravi-melt desulfurization process, 506
Coal gravity separation, 493
Coal-heterocyclic sulfur, 508
Coal-hydration, 520
Coal-hydrocycloning, 494, 501
Coal hydroliquification, 513
Coal-Lignite, 513
Coal-magnetic desulfurization, 491f, 506
Coal magnetic susceptibility, 496
Coal Mohr index, 496
Coal-neutron activation analysis, 498
Coal-North American, 514
Coal-organic sulfur, 496
Coal-particle size reduction, 502
Coal pelletizing, 520
Coal preparation plant, 147
Coal-pyrite, 514
Coal-pyrite clusters, 495
Coal-pyrite density, 498
Coal pyrolysis, 513
Coal-rapid analysis, 137
Coal-scanning electron micro-probe analysis 498,
Coal scrubbers, 492
Coal separation point, 493
Coal-solvent refining, 494, 504
Coal-sulfur agglomerates, 498
Coal-sulfur content, 492
Coal-sulfur determination, 497
Coal-sulfur hydrogenation, 496
coal-sulfur leaching curves, 493
Coal-sulfur oxides, 491f
Coal-superparamagnetic, 516
Coal testing laboratories, 147
Coal-thioether content, 496

Coal-thiophene content, 496
Coal-treatment for sulfur, 491f
Coal treatment processes, 491f
Coal treatment-zinc chloride, 501
Coal usage, 491
Coal-zinc chloride heating, 595
$CoAl_2O_4$, 687f
Coated steels, 154
Coating adhesion on steel, 295
Coatings on steel, 295f
Cobalt-antimony multilayers, 740
Cobalt electrode passivity, 706
Cobalt electrodeposits, 697f
Cobalt Fischer-Tropsch catalysts, 667f
Cobalt hardened gold deposits, 699
Cobalt hydroxides, 698
Cobalt impurities in iron, 15
Cobalt in alloys, 12
Cobalt in catalysts, 683f
Cobalt in electrodeposited zinc, 699
Cobalt in Fischer-Tropsch synthesis, 615
Cobalt loading in catalysts, 688f
Cobalt-molybdate catalysts, 664
Cobalt-molybdenum catalysts, 683f
Cobalt nitrate, 683
Cobalt oxides, 687, 698
Cobalt-small particles, 70
Cobalt substitution in gold, 698
Cobalt sulfate, 698
Cobalt to molybdenum ratio, 692
Cobalt-57, 684, 687f
Cobalt-57 doped ferrite, 158
Cobalt-57 implanted aluminum, 194
Cobalt-57 in chromium, 165
Cobalt-57 in pyrite, 156
Cobalt-57 in tantalium, 165
Cobalt-57 in vanadium, 165
Cobalt-57 pyrite source, 145
Coconut char, 603
Coevaporation, 105
Coke, 165f
Cold hammered austenite, 259
Cold hammered martensite, 259
Cold reduction steel, 182
Cold rolled steels, 171f
Cold worked iron, 341
Cold working, 143
Cold working martensite, 172
Collapsed spectrum, 35
Collective excitations in glass, 442
Collective magnetic excitations, 65f, 70f
Collective spin dynamics, 405
Collector plate mechanism, 261
Collimated deuteron incident beam, 274
Collimators, 95
Columnar crystals, 228
Combustion of coal, 153f
Commercial aluminum alloys, 194f
Commercial glasses, 426
Commercial mirror manufacturing, 717
Commercial steel, 296
$CoMoO_4$, 687
Compact hematite, 479
Composition dependent energy gap, 725f
Composition fluctuations in semiconductors 726f,
Compositional thin layer modulation, 741
Compression in steel, 165
Compressive stress, 395f, 700
Compressive stress-implanted layers, 271
Compressive stress in steel, 269
Compton scattering, 160
Computer simulations, 54
Concentrate smelt, 467
Conduction electron polarization, 16, 241
Conduction electrons, 249
Constant velocity scans, 463
Constrained fitting, 240
Contact interaction, 18
Contact magnetic field, 378
Controlled thermonuclear fusion, 108
Conversion catalysts, 637
Conversion electron Mossbauer scattering, 8, 83, 121, 153f, 162, 268, 273, 296, 715f
Conversion electron spectra of steel, 315f
Conversion electrons, 121
CoO, 699
CoOOH, 698f
Coordination number in glass, 435
CoO_2, 699, 707
Copper aluminum alloys, 233
Copper concentrates, 523f
Copper concentrates-activation, 530
Copper extraction-hydro-metallurigical, 523f
Copper extraction tech-nology, 523f
Copper flotation concentra-tion, 523f
Copper iron sulfides, 523f
Copper leach residues, 523f
Copper leached concentrates, 529
Copper-nickel multilayers, 740f
Copper-tin alloy, 702f
Coquimbite, 495
Cordierite, 562
Core polarization, 241
Correlated bond disorder, 362
Corrosion, 122, 168

Corrosion by sulfur, 217f
Corrosion in austenitic steels, 108
Corrosion in ferritic steels, 108
Corrosion in Zircaloys, 108
Corrosion inhibitors, 167
Corrosion kinetics, 218
Corrosion of iron, 567, 585f
Corrosion of steel, 269, 315
Corrosion phenomena, 153
Corrosion potential, 297
Corrosion product analysis, 315
Corrosion products, 295f
Corrosion resistance, 155, 162, 356
Corrosion resistance in glasses, 428
Corundum, 483
Corundum mullite, 468
Coulomb repulsion, 242
Covalence in glass, 435
Covellite, 530
$Co_3(Co(CN)_6)_2$, 699
Co_3O_4, 691, 699
Co_3Sn_2, 702
Co_9S_8, 688f
Crack propagation, 228
Crack resistant steels, 237
Cracking in steel, 154f
Crandallite, 533f
Cristobalite, 468, 536
Critical exponent, 366f
Cronstedtite, 547f
Cronstedtite electron hopping, 552
Crossed dendritic crystallites, 389
Cryogenic steels, 237
Cryogenic treatment, 259
Crystal field potential, 475
Crystal growth in alloys, 343
Crystalline silicates in slag, 452
Crystallization activation energy, 332, 343, 415
Crystallization behavior, 355, 367
Crystallization enthalpy-alloys, 332
Crystallization of amorphous alloys, 329f
Crystallization of glass, 409
Crystallization temperature, 41, 358, 360, 388
Crystallization temperature of alloys, 340
Crystallographic coherence length, 746
$Cr_{23}C_6$, 288
Cu-Al alloys, 233
Cubic alpha-AlFeSi, 209f
$CuFeS_2$, 523
Curie temperature, 358, 360
Current density of implantation, 282
Cutting steel, 164
Cu_2FeS_2, 528
Cu_3Sn, 703
Cu_5FeS_4, 523, 528
Cu_6Sn, 704
$CaSnO_3$, 704
Cu_9Al_4, 233
Cyanides, 8
Cylindrical mirror analyzer, 88
C_3 hydrocarbons, 658

Danon diagram, 727
DC casting, 204
Debye model, 254, 482
Debye temperature, 254, 489, 721
Debye temperature-surface, 610
Debye-Waller factor, 193
Decolorizing charcoal, 597
Decomposition of $Fe(CO)_5$, 669
Deconvolution by harmonic analysis, 240
Decreptation of iron ores, 479
Defect concentration in semi-conductors, 726f
Defect mobility, 26
Defects in hematite, 566
Deformation, 205
Dehydroxylation, 544, 555f
Dehydroxylation processes, 560
Demagnetizing field, 68, 101
Dendritic crystallites, 389
Dense random packing, 38
Dense random packing model, 55
Deposited thin films, 741f
Depth distribution, 132
Depth distribution of iron nitride, 122f
Depth profile, 133
Depth profiling, 297
Depth selective analysis, 85
Depth selective conversion electrons, 121
Depth selective Mossbauer spectroscopy, 121
Depth selectivity, 126
Desulfurization of coal, 491f
Desulfurization of coal-Meyers process, 504
Desulfurization of coal TWR process, 504
Detection-large solid angle, 91
Detection of 6.3kev electrons, 139
Determination of austenite in steel, 139
Determination of pyrite in coal, 145
Diafilters, 568f
Dicalcium ferrite, 449

Difference spectra, 239, 250, 259
Difference spectrum method, 240
Differential scanning calorimetry, 41, 329, 343, 409, 480
Differential thermal analysis, 409, 480, 534
Diffusion coefficient, 31, 705
Diffusion in alloys, 330
Diffusion in amorphous alloys, 330
Diffusion in austenite, 261
Diffusionless transformations, 260
Diffusive mobilities, 260
Dilatation anisotropy, 479
Dilute ore extraction, 595f
Dimensional changes in steel, 268
Dimensional crossover, 747
Dimeric species of iron, 568
Dipolar fields, 22
Dipolar hyperfine field, 19
Dipolar magnetic field, 19, 372
Dipole-dipole interactions, 404
Discovery of Mossbauer effect, 2
Dislocations, 25
Disorder, 26
Dispersed catalysts, 686
Dissociative adsorption of carbon monoxide, 668
Distribution function-correlation, 51
Distribution of hyperfine fields, 33, 331, 373,
Distribution of hyperfine parameters, 435
Distribution of magnetic moments, 378f
Distribution of quadrupole splittings, 35
Distribution of small particles in glass, 441
Distribution profiles-implantation, 278
Domain fluctuation in alloys, 337
Domain rotation, 100
Domain structure, 700
Donetz coal, 508
Doppler velocities, 156
Doppler velocity amplitude, 157
Dose retention in steel, 270
Double ion implantation, 270, 276f
DO_3 alloys, 191
DO_3 ordered structure, 229
DO_3 superstructure, 234
Dual phase steels, 143
Ductile steel, 271
Ductile to brittle transition, 237
Ductility of steel, 143

Easy axis of magnetization, 21
Edge dislocations, 26
Effective atomic volume in alloys, 341
Effective iron-57 density, 130
Effective mass, 26
Effective spectrometer luminosity, 130
Efficiency of fitting spectra, 376
EFG orientation, 203
EFG principal axes, 28
EFG random distribution, 35
EFG reorientation, 28
Elastic energy, 18
Elastic strain, 26
Electric arc furnace, 447
Electric field gradients, 309
Electric field gradient differences, 51
Electrical properties, 25
Electrical resistivity of alloys, 333
Electrochemical implantation, 704
Electrochemical potentials, 295
Electrochemical surface processes, 706
Electrocodeposition, 698
Electrode electrolyte interface, 697f
Electrode iron oxide layers, 707
Electrode potentials, 296
Electrodeposit composition, 697
Electrodeposited bronze, 703
Electrodeposited zinc, 699
Electrodeposition, 31
Electrodeposition of oxygen, 708
Electrodeposits, 697f
Electrogalvanization, 699
Electrolysis Mossbauer studies, 697f
Electrolyte in situ studies, 296
Electron charge distribution, 650f
Electron conversion coefficients, 84
Electron density, 15
Electron density correlation functions, 337
Electron escape simulations, 273
Electron fluctuation domains, 337
Electron hopping, 19, 441
Electron hopping in Cronstedtite, 552
Electron microprobe analysis of gold, 598
Electron microscopy, 106, 217, 389, 649, 693
Electron microscopy of fractures, 228
Electron microscopy of slags, 468

Electron spectral surface analyses, 715f
Electron spectrometer, 121f
Electron spectroscopy chemical analysis, 315
Electron spin polarization, 242
Electron spin resonance, 649
Electronegativity, 425
Electronegativity difference, 15
Electroplated gold 698,
Electropositive promoters, 668
Electrostatic electron analyzer, 123
Electrostatic flyash precipitator, 514
Element partition in steel, 448
Elemental distributions alloys, 227
Elementary cells, 38
Ellipsoidal mirror analyzer, 88
Elliptically polarized Mossbauer spectra, 382
Embrittlement of metals, 11
Embrittlement of steel, 319
Emission Mossbauer spectroscopy, 664, 683, 697f
Energetic promoters, 667f
Energy dispersive analysis, 468
Enriched antimony surfaces, 662
Enthalpy of formation of alloys, 14
Environmental pollutants, 315
Environmental pollution, 165
Epitaxial layer growth, 739f
Epitaxial multilayers, 740
Epsilon phase steels, 171
Escape depth of photo-electrons, 125
Eshka method, 497
Etching, 121
Ethanol Fischer-Tropsch synthesis, 609
Ethanol rate of formation, 644
Ethanol synthesis, 620, 635f
Ethylene Fischer-Tropsch synthesis, 609
Europium valence states, 108
Europium-151, 89, 95, 103, 108, 273, 316
Eutectic crystallization, 41f
Eutectic transformations in alloys, 332
EXAFS, 55, 687
EXAFS catalysts studies, 644
Exchange interaction, 242
Exchange interaction in spin glass, 404
Exchange polarization, 242
Exhaust emission control, 660
Extreme glancing angles, 122

f-factor, 161, 165, 482
f-factor determination, 524f
f-factor for $FeCl_3$, 597
f-factor for iron(III) sulfate, 604
f-factor in pyrite, 527
f-factor of coal, 498
f-factor temperature dependence, 735
f-factors in coal, 156
f-factors in minerals, 524f
Fatigue, 91
Fatigue behavior, 122
Fatigue failure in steel, 272
Fe-Al alloys, 189, 201f, 217f, 230
$(Fe,Al)PO_4$, 546
Fe-Al-Si alloys, 191
$(Fe,Co)_{80}P_{13}Al_{13}$, 372
Fe-Ge alloys, 243
$(Fe,Ni)_3B$, 367
Fe-Ni-C alloys, 252
Fe-Ni-Cr alloys, 252, 261
Fe-Ni-Mn alloys, 250, 252, 257, 261
Fe-Ni-Mo-B alloys, 346
Fe-Ni non-dilute alloys, 243
Fe-Ni-X alloys, 237
Fe-Si alloys, 217f, 243
$Fe(CO)_5$, 508
$Fe(CO)_5$ in zeolites, 669
$Fe(NO_3)_3.9H_2O$, 572
$Fe(OH)_2$, 167f, 708
$FeAlO_3$, 562
$FeAl_3$, 202f
$FeAl_6$, 191f, 195f, 203f
FeAu, 112
$FeCl_2.2H_2O$, 315
$FeCl_2.4H_2O$, 304, 315
$FeCl_3$ f-factor, 597
$FeCl_3$ intercalated in graphite, 597
$FeCl_3$ intercalation, 606
$FeCl_3.xH_2O$, 319
$FeCl_3.6H_2O$, 636
FeCo catalyst, 621
$FeCO_3$, 157
$FeHP_2O_7.1.5H_2O$, 308
$FeH_3(PO_4)_2.xH_2O$, 304, 308
FeIr catalyst, 621, 636
FeI_2 on silica, 629
Feldspar, 428
FeNi catalyst, 621
FeO, 426f, 435, 448, 452f
FeOOH, 77, 163, 166f, 304, 315f, 322, 479, 526, 538, 566f, 591, 708
FePd catalyst, 621f
FePt catalyst, 621, 636
FePt on silica, 692
FeRh catalysts, 620f, 636
FeRhTi catalysts, 636
Fermi contact interaction, 241
Ferrate ligands, 476

Ferrihydrite, 520, 566f, 586f
Ferrihydrite stabilization, 581
Ferrite, 8f, 138
Ferrite in austenitic steels, 155
Ferrite in steel, 143
Ferritemeter, 7
Ferritic stainless steel, 143, 271, 288
Ferritic steel, 155f, 165, 271
Ferritic steel castings, 160
Ferritic steel corrosion, 108
Ferritic steel welds, 160
Ferroelectric phase transitions, 735
Ferrofluids, 63, 70
Ferrobacillus ferrooxidans, 508
Ferromagnet-two dimensional, 747
Ferromagnetic amorphous alloys, 38, 371f
Ferromagnetic amorphous alloys texture, 378
Ferromagnetic amorphous metals, 33
Ferromagnetic exchange, 47
Ferromagnetic glasses, 364
Ferrous deficient magnetite, 166
Ferrous sulfide, 218
Fertilizer manufacture, 533f
FeRu catalysts, 636
FeS, 218, 232f, 321, 523f
FeS nucleation, 233
$FeSn_2$, 705
$FeSO_3$, 528
$FeSO_4.H_2O$, 315, 328, 528
$FeSO_4.4H_2O$, 315
$FeSO_4.7H_2O$, 706
FeS_2, 154, 230, 232, 469, 491f, 518, 523, 528
$FeTe_4$ clusters, 728
$FeTi_2O_5$, 107
$Fe_{0.35}V_2O_5$, 399
$Fe_{109}Ni_{45}Mo_{17}B_{29}$, 360
$Fe_{16}N$, 276f
$Fe_{16}N_2$, 290
$Fe_{16}N_2$ precipitation in steel, 273
$Fe_{16}B_{27}Si_{27}C_4$, 377
$Fe_{2.2}C$, 613
$Fe_2(B,C)$, 388
$Fe_2(SO_4)_3$, 595f, 604
Fe_2Al_5, 191f
Fe_2B, 162
$Fe_2B_{27}C_4$, 393
Fe_2C, 614
Fe_2N, 10, 128, 133f, 270, 275f
Fe_2O_3, 73, 75, 167, 282, 301, 410, 428f, 432, 455, 479f, 514, 518, 523f, 562, 566f, 652, 708
Fe_2O_3 Neel temperature, 481
$Fe_{20}Zr_{80}$, 47
$Fe_{23}C_6$, 288
$Fe_{24}Zr_{76}$, 53
$Fe_3(CO)_{12}$, 668
$Fe_3(PO_4)_2.8H_2O$, 299
Fe_3Al, 205f, 212
Fe_3B, 43, 391
Fe_3C, 277, 606, 613
Fe_3N, 277
Fe_3O_4, 76, 166, 168, 301, 438, 441, 518
Fe_3O_4 microcrystals, 69
Fe_3Si, 217f, 232
Fe_3Si metal consumption, 230
Fe_3Sn_2, 705
Fe_3S_4, 315, 321
$Fe_{39}Ni_{39}Mo_4B_{12}Si_6$, 360
$Fe_4(P_2O_7)_3$, 308
Fe_4Al_{13}, 191f, 195f
Fe_4N, 275f
$Fe_{40}Ni_{38}Mo_4B_{18}$, 359f
$Fe_{40}Ni_{40}B_{20}$, 360, 367
$Fe_{40}Ni_{40}P_{14}B_6$, 373f, 383
Fe_5C_2, 613, 618, 652, 667f
$Fe_5HO_8.4H_2O$, 566f
$Fe_{50}Zr_{50}$, 49
$Fe_{59}Ni_{23}Mo_4B_{14}$, 360
$Fe_{72}Ni_4Mo_6B_{16}Si_2$, 355f, 358f, 365, 368
$Fe_{75}B_{25}$, 43, 391
$Fe_{75}P_{15}C_{10}$, 386
$Fe_{75}Tm_5B_{20}$ amorphous alloys, 329f
$Fe_{75}V_4B_{21}$, 331
$Fe_{76}Co_4B_{20}$, 331
$Fe_{76}Ni_5B_{19}$, 331
$Fe_{77}B_{23}$, 43
$Fe_{77}Cr_5B_{18}$, 331
$Fe_{77}Mn_4B_{19}$, 331
$Fe_{78}Si_{10}B_{12}$, 373f
$Fe_{80}BM_{20}$, 39
$Fe_{80}B_{20}$, 33, 329f, 372
$Fe_{80}PM_{20}$, 39
$Fe_{81}B_{13.5}Si_{3.5}C_2$, 373f
$Fe_{81}B_{19}$, 331
$Fe_{82}B_{12}Si_6$, 391
$Fe_{82}P_{18}$, 373f
$Fe_{89}Zr_{11}$, 47
$Fe_{90}Zr_{10}$, 48
$Fe_{92}Zr_8$, 48
$Fe_{96}Si_4$, 393
Field distribution function, 38
Film layer thickness control, 747
Film plane texture, 741
Filter functions, 240
Fischer-Tropsch catalysts, 609f 650f
Fischer-Tropsch deactivation, 617
Fischer-Tropsch industrial synthesis, 617
Fischer-Tropsch mechanism, 610
Fischer-Tropsch synthesis, 667f
Fischer-Tropsch synthesis-acetic acid, 609
Fischer-Tropsch synthesis-ethanol, 609

Fischer-Tropsch synthesis-
ethylene, 609
Fischer-Tropsch synthesis-
propylene, 609
Fissuring in steel, 154f
Fitting method, 371
Fitting overlapping spectra, 376
Flame barrier, 220
FLK granules, 433
FLK incinerator, 425
FLK incinerator operation, 427
FLK slags, 426
Float sink separation of
coal, 491f
Flotation concentration, 523f
Fluctuating magnetic fields, 442
Flue gas composition, 519
Fluidized bed combustion, 494
Fluoride brine on steel, 316
Fluoroapatite, 534
Flux of vacancies, 235
Fly ash, 63
FLY incinerator, 411
Fly ash composition, 513
Fly ash electrostatic
preciptiator, 514
Force constant, 26
Fossil fuels, 8
Fourier bessel functions, 337
Fractional monolayers, 744
Fracture electron microscopy, 228
Fracture toughness, 155
Framboidal pyrite crystals, 495
Free volume in alloys, 347
Freezing temperature spin
glass, 399
Friedel's model, 16
Froth flotation of coal, 491f
Frozen solution studies, 605
Frustrated antiferromagnetism, 400
Fuel cell studies, 708
Fuel injectors, 269

Gallium-iron alloy, 16
Galvanized steel, 163, 316f
Galvanized steel surfaces, 162
Galvanized surface, 154
Gamma-AlFeSi, 210f
Gamma-iron, 606
Gamma-phase austenite, 171
Gamma-ray attenuation factor, 130
Garnets, 106
Gas adsorbant properties, 595f
Gas adsorption, 625
Gas stream desulfurization, 502
Gas turbines, 217
Gasoline production, 667f
Gaussian depth distribution, 133
Gaussian field distribution, 34
Gaussian filter function, 240
Gaussian implantation profiles,
273
Gaussian peaks, 394
Gaussian probability distribution,
54
Gaussian rolloff, 240
Gedanken experiment, 25
Geophysical prospecting, 137
Germanate glass, 409
Germanium impurities in iron, 15
Germanium solutes in steel, 243
Germanium-tellurium-tin
semiconductors, 735
Glancing angle X-ray
diffraction, 273
Glancing electron emission, 128,
133
Glass absorption length, 717
Glass-alkali ion diffusion, 414
Glass-alkali ion migration, 414
Glass beads, 717
Glass borate anomaly, 410
Glass ceramic matrix, 413
Glass ceramics, 409, 411
Glass composition and
crystallinity, 442
Glass composition-quenching
mode, 423
Glass coordination number, 435
Glass-corrosion resistance, 428
Glass-covalency, 435
Glass crystallization kinetics,
413
Glass crystallization activation
energy, 415
Glass-distribution of small
particles, 441
Glass-germanate anomaly, 410
Glass heat treatment, 412
Glass-isothermal annealing, 419
Glass-long range order, 409
Glass-melt temperature, 428
Glass-melt time, 428
Glass network, 423f
Glass network former, 425
Glass network modifier, 425
Glass nucleation, 409
Glass-partially crystallized, 437
Glass powder, 717
Glass-quenching and crystal-
linity, 442
Glass-quenching mode, 423
Glass-quenching technique, 457
Glass-residual phase, 414
Glass-short range order, 409
Glass-silver bonding, 715f
Glass structure, 411
Glass structure interstitial, 425
Glass structure substitutional,
425
Glass surface-adherence, 715
Glass surface oxidation, 718
Glass surface-polishing, 717
Glass surface sensitization, 715f

Glass surface silicon atoms, 715f
Glass surface tension, 717
Glass wetting agents, 717
Glasses-commercial, 426
Glasses-standard reference, 425
Glassy alloy crystallization, 343
Glassy alloys, 330
Glassy rad waste matrix, 475
Goethite, 479, 488, 526, 533f, 538, 566f,
Goethite hyperfine field, 540
Gold adsorption, 596
Gold alloy thin films, 112
Gold alloys-metastable, 114
Gold catalysts, 654
Gold diffusion in silicon, 114
Gold diluted in magnesium, 116
Gold electrodeposit, 698
Gold electron microprobe analysis, 598
Gold implanted in MgO, 115
Gold-iron alloys, 112
Gold-magnesium alloys, 115
Gold metal, 607
Gold metallic precipitates, 115
Gold metallization, 115
Gold on activated carbon, 606
Gold oxidation state, 607
Gold silicides, 114
Gold-silicon interface, 114
Gold-197, 83, 89, 95, 103, 316, 595f, 603, 606
Gold-197 CEMS, 111
Gold-199, 113
Goldanskii-Karyagin asymmetry, 607
Goldanskii-Karyagin effect, 51
Grain boundaries, 11, 26, 295
Grain size influence on Morin transitions, 484
Granulometric degradation, 479
Graphite, 111, 221
Graphite intercalation, 597, 606
Graphite-iron(III), 605
Graphite protective tiles, 111
Gravi-melt process, 506
Gravimetric thermal analysis, 480
Green rusts, 585f
Greigite, 321
Group VIII metal catalysts, 609f
Guest atom Mossbauer studies, 700
Gypsum, 520

Hadfield steel, 171
Hall effect in semiconductors, 727
Handrich model, 362
Hardhead slag, 467
Hardness, 173
Harlan coal, 148
Harmonic analysis and deconvolution, 240
Hazardous nuclear waste, 424
Hearthbrick, 467f
Heat of formation of alloys, 13
Heat treatment, 11, 43, 262
Heat treatment of electrodeposits, 705
Heat treatment of glass, 412
Heat treatment of steel, 161
Helium counter efficiency, 91
Helium flow counter, 85
Hematite, 8, 65, 167, 479, 518, 523, 529, 536, 542f, 558, 562, 566f
Hematite compact, 479
Hematite-Debye temperature, 489
Hematite defects, 566
Hematite-non decreptate, 480
Hesse method, 371, 375
Hesse method modified, 376
Hesse-Rubartsch method, 35, 47, 288, 331
Heterocyclic sulfur in coal, 508
Heterogeneous catalysts, 649f
Heterogeneous reactions, 697f
Heterogeneous scale structure, 230
Hexagonal alpha-AlFeSi, 209f
High alloy steels, 140
High cooling rates, 43
High dispersion iron catalysts, 610
High dose ion bombardment, 106
High manganese steels, 171f
High mechanical strength glass, 409
High metal loaded catalysts, 657
High metal surface areas, 667
High polish surface, 10
High process temperatures, 268
High strength steels, 237
High temperature CEMS, 98
High temperature corrosion, 111
High temperature incineration, 423f
High volatile bituminous coal, 147
Highly dispersed catalysts, 686
Highly dispersed iron oxide reduction, 627
History of Mossbauer spectroscopy, 2
Hollandite, 424
Homogeneous macroscopic statistics, 385
Hot dip galvanizing, 163
Hot rolled steels, 171f
Hydride formation, 27
Hydrocarbon formation, 615
Hydrocarbon product distribution, 658
Hydrocarbon synthesis, 635f
Hydrocycloning of coal, 494

Hydrodesulfurization, 683
Hydrodesulfurization
catalysts, 683
Hydrodesulfurization
mechanism, 693
Hydrogen diffusion, 28f
Hydrogen diffusion in metals, 27
Hydrogen effusion in metals, 27
Hydrogen embrittlement of
steel, 319
Hydrogen implanted steel, 108
Hydrogen in metals, 27
Hydrogen induced stresses, 27
Hydrogen mobility, 29
Hydrogen phase diagram, 28
Hydrogen relaxation effects, 29
Hydrogen solubility in metals, 27
Hydrogen spillover mechanism, 628
Hydrogen storage materials, 27
Hydrogenation of sulfur in
coal, 496
Hydroliquification of coal, 513
Hydrolysis-molecular weight
fractions, 568f
Hydrolysis of iron intermediate
polymers, 565f
Hydrometallurgical treatment,
523f
Hydrometallurgical copper
extraction, 523f
Hydrous aluminosilicates, 547f
Hydrous iron layer silicates,
547f
Hydrous iron oxides, 565f
Hydroxides of iron, 565f
Hyperfine field distributions,
12, 33, 49, 244f, 288, 331,
362, 744
Hyperfine field distributions
in alloys, 338
Hyperfine field linear response
model, 248
Hyperfine field mean, 39, 245f
Hyperfine field skewness, 245f
Hyperfine field varience, 245f
Hyperfine fields in alloys, 237
Hyperfine fields in zeolites, 677
Hyperfine fields-large, 605
Hyperfine parameter
distribution, 435
H_3BO_3, 410
H_3PO_3, 296

Ideal iron monolayer, 747
Implantaion distribution
profiles, 278
Implantation depth, 133
Implantation distribution
shape, 132
Implantation profiles, 273
Implantation under argon, 269
Implanted aluminum atoms, 191
Implanted cobalt 57, 189
Implanted ions in steel, 270
Implanted layer compressive
stress, 271
Implanted layer oxidation
potential, 104
Implanted layers, 287
Implanted steel surface
analysis, 273
Implanted surface analysis, 273
Impurity atom pairs, 26
Impurity binding energy, 13
Impurity defect complexes, 106
Impurity host interactions, 104
Impurity impurity interactions,
104
In situ catalytic cells, 650f
In situ catalytic studies, 635,
650f
In situ electrolytic cells, 697f
In situ fuel cell studies, 708
In situ Mossbauer cell, 684
In situ Mossbauer reactors, 618
In situ Mossbauer scattering, 311
In situ Mossbauer spectroscopy,
610
Incinerator operation, 427
Increased hyperfine field, 677
Indium-119, 98
Industrial aluminum, 189, 201f
Industrial clay minerals 547f,
Industrial Fischer-Tropsch
synthesis, 617
Industrial instrumentation, 7
Industrial nuclear waste, 424
Inelastically scattered
electrons, 125f
Inhomogeneities in alloys, 330
Inhomogeneous surface layer, 396
Initially formed hydrolysis
species, 565f
Injection moulding, 269
Inner sulfidation, 218
Insulators with iron 57, 106
Integral CEMS, 128
Integrated circuit fab-
rication, 103
Interaction potential, 12
Interaction with the silica
surface, 643
Intercalated species, 595
Intercalation in graphite, 597
Intercluster correlation in
spin glass, 399
Interdiffusion interface, 746
Interface compositional profile,
747
Interface defects, 26
Interface electric charge
transfer, 697f
Interface hyperfine fields, 742
Interface interdiffusion, 746

Interface intermixing, 746
Interface sites, 742
Interfacial phenomena, 704
Intermixing interface, 746
Internal conversion coefficient, 84, 160
Internal conversion electrons, 84
Internal stresses quenched, 385
International conferences, 4
Interstitial atom, 26
Interstitial content in steel, 282
Interstitial glass structure, 425
Interstitial iron in semi-conductors, 727
Interstitial migration, 279
Interstitial nitrogen in steel, 277
Interstitials, 193
Iodine adsorption, 596
Iodine-125, 731
Ion beam current density, 270
Ion channeling measurements, 106
Ion exchange, 722
Ion implantation, 83, 122, 267f
Ion implantation beam current, 278
Ion implantation CEMS, 103
Ion implantation current density, 282
Ion implantation surface modifications, 272
Ion implanted layers, 287
Ion implanted steel, 267f, 316
Ion penetration depth, 270
IrFe catalysts, 636
Iridium-193, 95
Iron aggregates, 193
Iron alloy Fischer-Tropsch catalysts, 609f
Iron-aluminum alloys, 189, 201f, 217f, 230
Iron-aluminum phosphate, 534f
Iron-aluminum silicon alloys, 191, 201f
Iron-antimony alloys, 12
Iron-antimony multilayers, 740
Iron atom diffusion, 235
Iron bearing clay in coal, 147
Iron bearing glass, 158
Iron bearing minerals, 8, 138, 147
Iron bearing minerals in coal, 151
Iron bearing silicates, 158
Iron bimetallic catalysts, 620
Iron-boron alloys, 371f
Iron-boron based glasses, 352
Iron-boron-silicon alloys, 371f
Iron carbide, 8, 10
Iron carbide bulk structure, 616
Iron carbides in catalysts, 611f
Iron-carbon bond strength, 672
Iron-carbon surface bonding, 604
Iron carbonyl decomposition, 668
Iron carbonyls multinuclear, 667f
Iron catalysts, 652
Iron catalysts carburization, 611f
Iron catalysts high dispersion, 610
Iron catalysts partially carburized, 612f
Iron catalysts-promoted, 612f
Iron chemical state in catalysts, 640
Iron chromium coordination, 283f
Iron cluster in zeolite cages, 670
Iron clusters, 668
Iron cold worked, 341
Iron coloidal solutions, 568
Iron containing catalysts, 609f
Iron containing minerals, 547f
Iron dimers, 104
Iron doped in tin-tellurium, 726f
Iron extraction with bromine-methanol, 461
Iron-FAU zeolites, 667f
Iron-film oxidation, 616
Iron Fischer-Tropsch catalysts, 609f, 667f
Iron foil thin films, 100
Iron fractional monolayers, 744
Iron frozen solution studies, 605
Iron-gallium alloy, 16
Iron-germanium alloys, 243
Iron-gold alloys, 112
Iron hydrolysis rate, 581
Iron hydrolysis species, 565f
Iron hydrous silicates, 548
Iron hydroxide, 539, 565f
Iron ideal monolayer, 747
Iron impurity in aluminum, 201
Iron impurity isomer shift, 105
Iron in argon matrices, 605
Iron in brown coal ash, 513
Iron in reduced Fe-LTA zeolites, 674
Iron in Xenon matrices, 605
Iron initial hydrolysis species, 577
Iron intercalated species, 595
Iron ion mobility, 233
Iron-iron distances in carbon, 605
Iron layer silicates, 547f
Iron layer thickness, 742
Iron-liquid carbon saturated, 447
Iron-magnesium multilayers, 740f
Iron-manganese alloys, 180, 257
Iron-manganese carbides, 180
Iron-manganese diffusion, 742
Iron manganese oxide, 179

Iron-MOR zeolites, 667f
Iron multilayers, 739f
Iron-nickel alloys, 22, 237
Iron-nickel-carbon alloys, 252
Iron-nickel-chromium alloys, 24
252, 261
Iron-nickel electrodeposits, 706
Iron-nickel-manganese alloys,
250f, 261
Iron-nickel non-dilute alloys,
243
Iron nitride depth distribution,
122f
Iron nitride layers, 8
Iron nitrides, 8
Iron on alumina, 654
Iron ore, 153f
Iron ore decreptation, 479
Iron oxide electrode layers, 707
Iron oxide mineral character-
istics, 566f
Iron oxide morphology, 566f
Iron oxide precipitation, 565f
Iron oxide recoil-free
fraction, 618
Iron oxide supported catalyst,
616
Iron oxide surface layer, 653
Iron oxides, 539, 565f
Iron oxidizing bacteria, 508
Iron-palladium alloys, 12
Iron particles-magnetic
properties, 73
Iron pentacarbonyl, 508, 669
Iron phosphate, 316
Iron-phosphorus alloys, 371f
Iron phtalocyanine electrode, 708
Iron-platinum alloys, 13
Iron precarbided, 616f
Iron rapid oxidation, 585f
Iron reduction on alumina
supported catalysts 626,
Iron-rhodium catalysts, 635
Iron-rhodium catalyst surface
structure, 643
Iron-rhodium ratio in
catalysts, 639
Iron-ruthenium alloys, 616f, 656
Iron-ruthenium catalysts, 654
Iron-ruthenium supported on
silica, 657
Iron-silicon alloys, 22, 217f,
243, 389
Iron-silicon precipitation, 392
Iron small particles, 70
Iron storage proteins, 63
Iron sulfides, 315, 321, 491f,
687
Iron surface dissociation, 615
Iron surface enrichment
catalysts, 645
Iron-titanium alloys, 12
Iron trimeric hydrolysis
species, 575
Iron-vanadium homogeneous
alloys, 746
Iron-vanadium multilayers, 740f
Iron-zero valent, 624f
Iron-zinc alloy, 16
Iron-ZSM zeolites, 667f
Iron-57 hyperfine fields, 241
Iron-57 implanted aluminum, 193
Iron-57 implanted in copper, 103
Iron-57 in insulators, 106
Iron-57 thin films, 99
Iron(II) carbonate, 586
Iron(II) chloride, 304, 315
Iron(II) controlled oxidation,
585f
Iron(II) glasses, 433
Iron(II) in silicate glass, 454,
460
Iron(II) ion exchange, 668
Iron(II) oxalate, 433
Iron(II) phosphate, 299f
Iron(II) reference glasses, 436
Iron(II) sulfate, 315, 323, 527
Iron(II) sulfide, 218
Iron(II) vacancy in LiF, 106
Iron(III) acid phosphate, 304
Iron(III) chloride, 318, 492
Iron(III) dimeric species, 576
Iron(III) glasses, 423f, 429
Iron(III) hydroxide, 528
Iron(III) in graphite, 605
Iron(III) in alumina supported
catalysts, 624f
Iron(III) nitrate, 568, 654
Iron(III) phosphate, 299f
Iron(III) sulfate, 504, 708
Iron(III) sulfate f-factor, 604
Iron(III) vacancies in MgO, 106
Iron(0) in catalysts, 624f, 654
Iron(0) in rhodium-iron
catalysts, 638
Irridation of metals, 31
Isolated iron dimers, 104
Isomer shift in alloys, 331
Isomer shift-iron impurities, 105
Isothermal annealing, 396
Isothermal resistivity
measurment, 333
Isotope enrichment in
catalysts, 650
Isotropic recoil-free
fraction, 381
Itinerant magnetic model, 47

Jarosite, 494f, 501

K-conversion electrons, 121, 125
130
Kandites, 547
Kaolinite, 547f

$KAu(CN)_2$, 595f, 606
Kerr effect, 385
Kinetics of austenite
formation, 260
Kinetics of corrosion, 218f
Kinetics of glass crystal-
lization, 413
Kinetics of sulfidation, 218f,
230
K_2O, 428
$K_3Co(CN)_6$, 699
$K_4Fe(CN)_6.3H_2O$, 685

L-conversion electrons, 130
Langevin function, 68
Large hyperfine fields, 605
Large solid angle detection, 91
Large surface areas, 595f
Laser treatment of surfaces, 162
Lattice damage, 104
Lattice defects, 26
Lattice vibrational anisotropy,
51
Lattice vibrations, 677
Lavas, 63
Layer composition, 226
Layer lattice silicates, 547f
Layer microstructure, 224
Lead telluride, 731
Lead-tin-tellurium semi-
conductors, 725f
Ledgemont process, 506
Lepidocrocite, 566f, 590
Lepidocrocite formation, 581
Lepidocrocite reducing
conditions, 578
LiF, 106
LiF iron(II) vacancy, 106
Ligand exchange reactions, 722
Lignite coal, 513
Lignites, 147
Line defects, 26
Line shape, 35
Line shape distortion, 102
Linear response model of hyper-
fine fields, 248
Liquid quench techniques, 44
Lithiophile elements, 635
Lithium(I) in magnetite, 442
Lithium(I) network modifier, 425
Lizardite, 552f
Li_2O, 2
Locking field, 255
Long range order in glass, 409
Lorentz field, 68
Lorentzian lineshape, 472, 550
Low alloy steel, 91, 140f, 270
Low carbon steel, 270
Low energy electron
diffraction, 99
Low level radioactive waste, 423f
Low silica slag, 450
Low spin iron(III), 687
Low temperature CEMS
measurements, 92
Lubricated sliding steel, 269
Lubricated sliding surfaces, 10
Lubrication conditions, 270
Luena coal, 501

Mackinawite, 320
Macroscopic homogeneous
statistics, 385
Maghemite, 304, 566f
Magic angle Mossbauer spectra, 338
Magnesia structural promoter, 667
Magnesioaluminoferrite, 513, 516
Magnesioferrite, 107, 518
Magnesium-gold alloys, 115
Magnesium-iron multilayers, 740f
Magnesium silicate, 561
Magnesium(II) in glass, 431
Magnesium(II) in magnetite, 442
Magnesium hydrous layer
silicates, 560
Magnetic anisotropy abnormal, 386
Magnetic anisotropy energy,
63, 71
Magnetic anisotropy from
annealing, 387
Magnetic anisotropy in micro-
crystals, 70
Magnetic anisotropy in amorphous
alloys, 371f
Magnetic anisotropy thinning, 385
Magnetic dead layer, 73
Magnetic desulfurization of
of coal, 491f, 506
Magnetic dimensional cross-
over, 747
Magnetic dipole interactions, 77
Magnetic domain pattern, 700
Magnetic domains, 383
Magnetic field distributions-
asymmetric, 442
Magnetic inhomogenieties, 46
Magnetic moments-linear
response, 241
Magnetic order and super-
conductivity, 47
Magnetic order in small
particles, 75
Magnetic permeability, 174
Magnetic properties of
alloys, 347
Magnetic relaxation, 567, 599f
Magnetic separation, 153f
Magnetic separation of slags, 474
Magnetic surface anisotropy, 744
Magnetic surface effects, 72f
Magnetic susceptibility, 171
Magnetic texture, 378, 382f
Magnetic texture in amorphous
alloys, 371f, 378

Magnetic texture in quenched alloys, 383
Magnetic texture-statistical description, 378
Magnetic torsion measurements, 743
Magnetic wedge, 92
Magnetite, 8, 19, 168, 304, 441, 474, 518, 566f, 585f,
Magnetization direction, 19
Magnetization measurement, 47
Magnetization of Fe-LTA, 674
Magnetization-thin film, 742
Magnetization vector, 100
Magnetocrystalline anisotropy, 32, 700
Magnetometric CEMS, 100
Magnetostatic energy, 700
Magnetostriction, 700
Magnex magnetic separation, 508
Manganese-effect on hyperfine field, 248f
Manganese in steel, 140, 171f, 241
Manganese phosphate, 316
Marcasite, 494f
Marcasite-weathered, 494f
Martensite, 8, 10, 155, 176, 186, 240, 244, 253f, 260, 275
Martensite by cold working, 172
Martensite in steel, 143
Martensite lath boundaries, 238
Martensitic stainless steel, 143
Martensitic steel, 271
Mass balance calculations, 462
Mass spectrometry-thermo-gravimetry, 535
Materials fatigue, 83
Mean depth, 132
Mean field, 364
Mean fields in alloys, 350
Mean hyperfine field, 39, 245f
Mechanical coal treatment, 491f
Mechanical polishing of steel, 282
Mechanical properties, 25
Mechanical stress, 205, 700
Mechanical surface properties, 122
Mechanically induced phase transitions, 156
Mechanism of hydrodesulfurization, 693
Meisel method, 526
Melanterite, 494f
Melglas, 32
Melt spinning, 43f
Melt spinning techniques, 31 329f
Meltspun amorphous metals, 53
Met glasses, 424
Metal adsorption by activated carbon, 595f
Metal bonding to activated carbon, 596
Metal consumption zone, 230
Metal dislocations, 25
Metal embrittlement, 11
Metal extraction, 595f
Metal formation in Fe-LTA zeolites, 674
Metal-hydrogen, 27
Metal loading on catalysts, 635
Metal matrix sources, 138
Metal mechanical properties, 25
Metal metaloid alloys, 38
Metal passivation, 706
Metal plastic deformation, 25
Metal sublimation, 41
Metal-support interaction, 678
Metallic glasses, 355, 360, 362, 366
Metallic gold, 607
Metallic gold precipitates, 115
Metallic iron in slag, 458
Metallic luster, 228
Metallic metal oxide catalysts, 649f
Metallic superlattices, 739f
Metallic surface implantation, 267f
Metallographic methods, 228
Metallographic studies, 262
Metallography, 140
Metalloid transition metal alloys, 38
Metallurgical damage, 11
Metallurgical phases, 160
Metallurgical problems, 153f
Metastable gold alloys, 114
Metastable phase formation, 103
Metavariscite, 545
Metering pumps, 269
Metglas, 359
Methane synthesis, 620
Methanol-rate of formation, 640
Methanol synthesis, 620, 635f
Meyer process, 491
Meyers desulfurization process, 504
$MgAl_2O_4$, 518
(MgFe)O, 107
$MgFe_2O_4$, 107f
MgO, 108, 428, 516
MgO implanted with gold, 115
MgO iron(III) vacancy, 106
Microcrystal magnetic order, 75
Microcrystalline Fe_3O_4, 69
Microcrystalline goethite, 571
Microcrystalline magnetic anisotropy, 70
Microcrystals, 38, 63, 686
Microelectronics industry, 25
Microprobe analysis, 209,211,215
Microstructural analysis, 275
Microstructure of layers, 224

Microstructure of surface layers, 224
Miedema-van der Woude model, 113
Miedema-Korlas model, 15f
Miedema model of alloys, 105
Miedema method, 14
Migration of interstitials, 279
Migration of protons in minerals, 560
Millisite, 533f
Mineral cation ordering, 555
Mineral dehydroxylation, 555f
Mineral electron microprobe analysis, 554
Mineral heating with oxidation, 555
Mineral heating with reduction, 558
Mineral high temperature reactions, 561
Mineral lattice oxidation-reduction, 558
Mineral quantitative analysis, 523f
Mineral recoil free fractions, 524f, 539
Mineral substitutional mismatch, 554
Mineral thermal reactions, 547f, 560
Mineral thermal treatment, 550
Minerals with cation order, 561
Mirror analyzer, 88
Mirror manufacturing, 715f
Mixed metal oxide catalysts, 649f
Mixed phosphates, 300
MnO_2, 467
Modifying catalytic properties, 635
Modulation of thin layer structure, 741f
Mohr index, 496
Molecular sieves, 667f
Molten salt bath, 8
Molybdenum-cobalt catalysts, 683f
Molybdenum in steel, 143, 161
Moments theorem, 263
Monitoring vibrations, 8
Monocalcium phosphate, 534
Monolayers-fractional iron, 744
Monolayers on glass surface, 715f
Monomers, 195
Monte Carlo method, 362
Monte Carlo simulations, 85, 130, 273
Montgomeryite, 535
Montmorillonite, 547
Morin temperature, 484
Morin transition broadening, 484
Morphology of surface scale, 222f
Morwell coal, 516
Mossbauer based rapid analysis, 137
Mossbauer difference spectra, 239
Mossbauer effect conferences, 4
Mossbauer Effect Data Center, 4
Mossbauer emission spectroscopy, 683f
Mossbauer Nobel prize, 3
Mossbauer spectral asymmetry, 394
Mossbauer spectroscopy history, 2
Most probable field, 364
MoS_2, 687f
Mullite, 468f, 475
Multilayered films, 739f
Multilayers antimony-cobalt, 740
Multilayers of iron-magnesium, 739f
Multilayers of iron-vanadium, 740f
Multilayers vanadium-silver, 740
Multinuclear iron carbonyls, 667f

$NaAu(CN)_2$, 606
NaI detector, 274
Naphtha reforming, 620
Natural gas steam reforming, 609
$Na_{88}Al_{88}Si_{104}O_{385}$, 670
Near surface magnetization, 83
Near surface oxidation, 217
Near surface sulfidation, 217
Near surface zone, 217
Neel temperature of Fe_2O_3, 481
Network former, 425
Network modifier, 425
Neutron activation analysis of coal, 498
Neutron activation analysis of zeolites, 670
Neutron scattering, 55
Nickel-copper multilayers, 740
Nickel diffusion in steel, 261
Nickel-effect on hyperfine field, 244, 248f
Nickel electrodeposits, 700f
Nickel in Fischer-Tropsch synthesis, 615
Nickel in steel, 140, 166
Nickel-iron alloys, 12, 237
Nickel small particles, 70
Nickel solutes in steel, 244
Nickel-tin codeposits, 702
Nickel-61, 700
$NiCl_2.6H_2O$, 706
$NiFe_2O_4$, 73
Niggli values, 469
Nine-nickel steel, 237, 253
NiSn, 702
$NiSO_4.7H_2O$, 706
Nitride depth distribution models, 133
Nitride formation in steel, 282
Nitride interstitials in steel, 277

Nitride layers on iron, 8
Nitride precipitates in
steel, 271, 277
Nitrided layers, 162
Nitrided martensite, 276f
Nitrided steel, 162, 268, 271,
316f
Nitride subsurface phase, 133
Nitrided surface, 154
Nitriding methods, 268f
Nitriding temperatures, 268f
Nitrogen implantation layers
in steel, 273
Nitrogen implantation
profiles, 278
Nitrogen implanted steel, 267f,
288
Nitrogen ion implantation, 122
Nitrogen microanalysis, 274
Ni_3Al, 22
Ni_3Fe, 20
Ni_3Sn_2, 12
Ni_3Sn_4, 12
$Ni_{74}Fe_{26}$, 21
$Ni_{76}Fe_{24}$, 21, 23
Noble metal heterogeneous
catalysts, 635f
Nominal ion dose in steel, 270
Non-decreptate hematite, 480
Non-destructive testing, 7, 274
Non-epitaxial multilayers, 740
Non-linear least squares
minimization, 376
Non-polarized Mossbauer
spectroscopy, 383
Non-sulfidized alloys, 222
Nondestructive testing, 121, 144,
153f
Nonresonant absorption, 148
Nonresonant background, 90
Nonstoichiometric magnetite, 167
North American coal, 514
Nuclear magnetic resonance, 12
Nuclear power plants, 295
Nuclear reaction analysis, 268f,
274
Nuclear reactors, 168
Nuclear waste, 424
Nucleation, 260
Nucleation growth factors, 260
Nucleation in alloys, 343
Nucleation of FeS, 233
Nucleation of glass, 409
Nuclei quenched in alloys, 352
Nukundamite, 530
Nulcear fission reactors, 108

Off resonance count rate, 147
Off resonance radiation, 137
Oleic Acid, 70f
Olivine, 106, 558, 562
On resonance radiation, 137
One dimensional defects, 26
Optical microscopy, 217, 221
Optical microscopy of slags, 468
Optoelectronic devices, 725
Order parameter in alloys, 229
Ore containing magnetite, 154
Ore floation, 154
Ore internal water vapor
pressure, 479
Ore selection, 154
Ore washing, 153
Ores, 137
Organic sulfur in coal, 496, 508
Orthophosphates, 305
Orthophosphoric acid, 296
Overlap corrections, 529
Oversaturation of vacancies, 218
Oxidation of chalcogenides, 727
Oxidation of green rusts, 585f
Oxidation of Hydroxycarbonate,
585f
Oxidation of iron, 669
Oxidation-parabolic law, 218f
Oxidation potential-implanted
layer, 104
Oxygen distribution in semi-
conductors, 729
Oxygen leaching process, 506
Oxygen steel making process, 447
Oxyhydroxides of cobalt, 698
Oxyhydroxides of iron, 565f

p-type carriers, 727
P(H) distribution, 363
P(H) distribution functions, 385
Palladium-iron alloys, 12
Parabolic diffusion law, 705
Parabolic rate law, 218f
Paraelectric SnTe, 734
Paramagnetic iron nitride, 126f
Paramagnetic nitride phase, 126
Partial catalyst reoxidation, 657
Partial differential quotients,
375f
Partially crystallized glass, 437
Particle size in catalysts, 625f
Partition of slag and liquid
steel, 447
Passivated bimetallic
catalysts, 629
Passivation of catalysts, 628
Passivation of tin electrodes,
708
Passive potential, 298
Passivity-cobalt electrode, 706
Patent literature, 8
Pauli paramagnetism, 47
PbTe, 727
Pelletized ore, 479
Penetration depth in steel, 270
Penetration depth in hearth-
brick, 468

Percent ash in coal, 148f
Percent resonance effect, 137, 145
Periclase, 516
Perovskite, 424
Perpendicular magnetization, 744
Perturbed angular correlations, 14, 100
Phase analysis, 25, 253
Phase distribution in steel, 154
Phase transfromations, 218
Phase transitions in semi-conductos, 732f
Phenolic resin, 269
Phonon dispersion curves, 735
Phonon spectral anomaly, 735
Phosphate glass, 409
Phosphate layers on steel, 295
Phosphate minerals, 533f
Phosphate ore-aluminum contamination, 534f
Phosphate ore-iron contamination, 534f
Phosphate ore quality, 533f
Phosphate ores-calcined, 533f
Phosphate rock, 533
Phosphates, 305
Phosphates-mixed, 300
Phosphatization of steel, 295f
Phosphoric acid, 295f
Phosphorus partition in steel, 448
Photoeffect cross section, 89
Photoelectric cross sections, 160
Photoelectron escape depth, 125
Photoelectron spectra, 125
Photoelectron spectroscopy, 55, 649, 661
Photoreduction, 668
Photoreduction of iron, 668
Photovoltaic detectors, 725
Physical metallurgy, 25
Pigmenting power, 567
Pigments, 567
Pittsburgh coals, 147
Pittsburgh seam coal, 157
PIXE, 111
Plasma induced effects, 83
Plasma induced erosion, 109
Plasma wall interaction, 83, 108
Plastic deformation, 25f, 91
Plated film thickness, 8
Plating baths, 698
Platinum-iron alloys, 13
Platinum-197, 111, 598
Point defects, 26f
Polarized gamma-ray Mossbauer spectroscopy, 382
Polarized gamma-ray studies, 100
Polarized neutron diffraction, 740
Polishing surfaces, 44
Polling, 448
Polymerization of iron, 568
Polymers of iron, 568
Polymorphous crystallization, 43
Polyphosphates, 305
Poorly ordered polymeric material, 571
Pore size distribution, 605
Portable device for steel analysis, 144
Porus scale structure, 230
Porus zeolite structures, 667f
Position sensitive detector, 164
Post implantation processes, 103
Potassium borate glass, 416f
Potassium ferricyanide, 596
Potentiodynamic curves, 709
Potentiometric analytical methods, 565f
Pourbaix diagram, 707
Powder metallurgy, 726
Power generation with coal, 513
Precipitator ash, 516
Preferential film plane, 741
Pressing deformation, 143
Principal axes of the EFG, 28
Probability distributions, 193
Probing depth of CEMS, 85
Product distribution, 620
Production line fabrication, 103
Promoted iron catalysts, 612f
Promoter charge transfer, 672
Promoter effect in alkali metal, 667f
Promoter effect in zeolite, 667f
Promoters in Fischer-Tropsch synthesis, 611f, 667f
Proportional counter, 85
Propylene Fischer-Tropsch synthesis, 609
Protective barriers for radioactivity, 424
Protective boiler scale, 168
Protective oxide layer, 162
Protoferrihydrite, 566f
Proton migration in minerals, 560
Pseudo dipolar hyperfine field, 254
Pseudobrookite, 476
PtFe catalysts, 636
PtIr catalysts, 620
PtRe catalysts, 620
Pulse height analysis, 138
Pyrite, 138, 145f, 154, 157, 165, 469, 491f, 518, 523f
Pyrite chemical decomposition, 503
Pyrite clusters in coal, 495
Pyrite content, 154
Pyrite f-factor, 527

Pyrite-framboidal crystals, 495
Pyrite in coal, 145, 156, 514
Pyrite-weathered, 494f
Pyrite density in coal, 498
Pyritemeter, 7, 157
Pyritic sulfur, 145f
Pyritic sulfur in coal, 151
Pyroaurite, 585f
Pyrophosphates, 305
Pyrophyllite, 547
Pyrrhotite, 491f, 506, 523f
P_2O_5, 448

Q-basic oxygen process, 447
Quadrupole splitting-distribution, 35
Quality control, 7, 148
Quality control applications, 137
Quality of spectral fits, 373
Quartz, 148
Quenched alloys-magnetic texture, 383
Quenched amorphous alloys, 376
Quenched internal stresses, 385
Quenched ribbon, 385
Quenched steel, 277
Quenching process, 16
Quenching technique-glass, 457
Quenching techniques, 26, 31, 194

Radiation damage, 83
Radiation damage in steel, 269
Radioactive waste, 409
Radioactive waste incineration, 423f
Radioactivity protective barriers, 424
Radiofrequency field collapse, 34
Radionuclei isolation, 425
Random spin glass anisotropy, 399
Random dense packed hard spheres, 31
Random distribution, 12
Random distribution alloys, 229
Random distribution of implanted ions, 287f
Random tetrahedral network, 105
Rapid austenite analysis, 140
Rapid melt quenching, 162
Rapid method simulation, 138
Rapid Mossbauer analysis, 137, 531
Rapid Mossbauer methods, 157
Rapid pyrite determination, 145
Rapid quenching of steel, 143
Rapid transmission method, 157
Rate of iron hydrolysis, 581
Rate of sulfidation, 230
Rayleigh scattering, 160
Recoil free absorption, 160
Recoil free fraction, 145, 151, 155, 245, 524f, 539, 720
Recoil free fraction-determination, 524f
Recoil free fraction in minerals, 539
Recoil free fractions in semiconductors, 732f
Recoil free fraction-isotropic, 381
Recoil free fraction of iron oxide, 618
Reduction highly dispersed iron oxides, 627
Reduction of Fe_2O_3 in hydrogen, 652
Reference electrodes, 296
Reference glasses, 425
Reforming naptha, 620
Refractory linings, 467
Relativistic effects, 25
Relaxation lineshape, 552
Rereduction of catalysts, 628
Rereduction with hydrogen, 642
Residual glass phase, 414
Residual stress in austenitic steels, 164
Residual stress in steel, 164
Residual surface stress, 164
Resistance to corrosion, 356
Resistance to stress cracking, 155
Resistivity in semiconductors, 733
Resistivity of amorphous alloys, 329, 333
Resonant effect, 138
Resonant gamma-ray attenuation, 130
Resonant signal-time averaged, 138
Resonant signal-velocity averaged, 138
Resonsant absorber thickness, 241
Retained austenite, 143, 155, 277
Retained dose, 133
Reverberatory furnaces, 467
$RhCl_3.3H_2O$, 636
RhFe catalysts, 636
RhFeTi catalysts, 636
Rhodium based catalysts, 610
Rhodium-iron catalysts, 610, 635
Rhodium-iron catalyst-surface structure, 643
Rigid band model, 39
RKKY interations, 400
Roller bearings, 91
Rolling deformation, 378
Romerite, 495
Routine monitoring, 153
Rozenite, 494f
Rubidium borosilicate glass, 415
RuFe catalysts, 636
Rust, 567

Rust on steel, 304
Rust transformers, 295, 302
Ruthenium dioxide, 660
Ruthenium-99, 660
Ruthenium(III) chloride, 654
Rutherford backscattering, 106, 112
Rutherford backscattering spectroscopy, 273
Rutile, 720
Rutile related structures, 662

Samarium-151, 108
Sampling depth, 56
Sanderson electronegativity, 668, 678
Sapphirine, 562
$SbCl_5$, 661
Scale morphology, 222
Scale thickness, 218
Scanning electron microscopy, 217f, 320, 320, 726
Scattering geometry, 137f, 156
Scherrer approximation, 441
Scored surfaces, 10
Scuffing, 10
Sculling, 448
Sea water corrosion, 166
Second order Doppler shift, 677
Secondary electron recording, 99
Secondary electrons, 84, 95
Secondary ion mass spectrometry, 273, 699
Secondary Mossbauer absorption, 130
Selective oxidation of boron, 41
Semiconducting devices, 114
Semiconductor composition fluctuations, 726f
Semiconductor defect concentration, 726f
Semiconductor-doping elements, 727f
Semiconductor energy gap, 725f
Semiconductor ferroelectric phases, 735
Semiconductor industry, 25
Semiconductor phase transitions, 732f
Semiconductor recoil free fractions, 732f
Semiconductor resistivity, 733
Semiconductor specific heats, 734
Semiconductors, 725f
Semiconductors-interstitial iron, 727
Sensitivity of CEMS, 89
Sensitization of glass surface, 715f
Septechlorites, 548
Serpentine, 547f
Serpentine structure, 562
Shape selectivity in catalysts, 667f
Shock wave produced imperfections, 484
Short chain hydrocarbons, 658
Short range order-amorphous alloys, 394
Short range order in glass, 409
Shot peening steel, 273
Siderite, 157
Siderite in coal, 496
Silica, 558
Silica supported iron-ruthenium, 657
Silica supported noble metals, 635
Silica supported rhodium-iron catalysts, 635, 638
Silicate glass in slag, 450
Silicate glass-iron(II), 460, 454
Silicate glasses, 409, 423f
Silicate ligands, 476
Silicate spinel, 562
Silicate tin glass, 475
Silicates, 547f
Silicon based glasses, 431
Silicon-effect on hyperfine field, 248f
Silicon glass network former, 412
Silicon-gold interface, 114
Silicon in steel, 241
Silicon on a glass surface, 715f
Silicon satellite, 251
Silicon solutes in steel, 243
Silicon surface barrier detector, 274
Silicon(IV) network former, 425
Sillimanite, 468
Silver-vanadium multilayers, 740
Simulated spectra, 36
Single crystal alloys, 19
Single domain particles, 63
Sintering process, 154
SiO_2, 410, 426f, 467, 480, 620
SiS_2, 234
SiS_2 disintegration, 232
Site disorder, 362
Size mismatch, 18
Skewness of hyperfine fields, 245f
Slag, 157
Slag-attack on firebrick, 467f
Slag basicity, 448
Slag-crystalline silicates, 452
Slag from basic oxygen furnace, 448
Slag from basic steel processes, 475
Slag in tin smelting, 467f
Slag layer, 447
Slag-low silica, 450

Slag-magnetic separation, 474
Slag-metallic iron, 458
Slag phase equilibrium diagrams, 458f
Slag polling, 448
Slag sculling, 448
Slag-silicate glass, 450
Slag studies, 154
Slag-tin hardhead, 467
Slag tin-119 spectra, 470
Slags, 138
Slater-Pauling diagram, 39
Sliding steel wear tests, 270
Slip velocity, 10
Slow activation catalytic model, 616
Slow release fertilizer, 546
Small angle X-ray scattering, 329, 351
Small particle magnetic order, 75
Smectites, 547
Smelt concentrate, 467
$Sn(OH)_2$, 708
$SnCl_2$, 461
$SnCl_2.2H_2O$, 715f, 718
$SnCl_4$, 661
SnI_4, 704
SnNi, 702
SnO, 467, 476, 708
SnO_2, 91, 98, 137, 467, 475f, 662, 718f,
SnS, 727f
SnSe, 727f
SnTe, 727f
SnTe-paraelectric, 734
Sodium oxalate, 316
Sodium promoters, 678
Soft ferromagnets, 48
Soft magnetic alloys, 32
Soils, 565f
Solar energy mirrors, 716
Solid breeder ceramics, 108
Solid breeder materials, 111
Solid solutions, 193
Solid state catalysts, 649f
Solid state topotactic transformations, 585
Solute atomic volume, 242
Solvent refining of coal, 504
Somorjai rule, 610
Sortex, 148
Spatial electron distributions in alloys, 337
Spatial magnetic texture, 381
Specific heats of semi-conductors, 734
Spectral absorption area, 481
Spectral count times, 102
Spectral statistical quality, 90
Spectrometer luminosity, 130
Spin density, 26
Spin density in alloys, 347
Spin dynamics in spin glass, 405
Spin glass alloys, 112
Spin glass anisotropy, 405
Spin glass cluster distribution, 399
Spin glass freezing temperature, 399
Spin glass intercluster cor-relation, 399
Spin glass order, 399
Spin glass phase, 400
Spin glass phase transformation, 399
Spin glass random anisotropy, 399
Spin glass spin dynamics, 405
Spin glass thermal excitation, 399
Spin glass transition temperature, 406
Spin melting, 105
Spin polarization, 242
Spin wave excitation, 365
Spin wave stiffness coef-ficients, 365
Spinel, 518, 558, 562
Splat quenched alloys, 193
Splat quenching, 43
Spontaneous spalling, 143
Sputtering, 105, 121, 275
Sputtering-argon ion, 55
Sputtering effects, 133f
Sputtering techniques, 301
SrO, 536
$SrTiO_3$, 734
Stabilizing anchor, 639
Stainless steel castings, 155
Stainless steel source, 139, 141, 144
Stainless steel welds, 155
Stainless steels, 109
Stationary source, 139
Statistical description of texture, 378
Statistical quality of spectra, 90
Steam reforming of natural gas, 609
Steel activation energies, 417
Steel-active potential, 298
Steel activity coefficients, 448
Steel-alloying, 25
Steel-alloying elements, 271
Steel-atmospheric oxidation, 301
Steel-boronized, 316f
Steel-carbided, 271, 276f
Steel-carbonitrided, 276f
Steel-carburized, 316f
Steel-chromium content, 279
Steel coating adhesion, 295
Steel coatings, 295f
Steel-compressive stress, 268

Steel-corrosion by HCl, 315f
Steel-corrosion by H_2S, 315f
Steel-corrosion by SO_2, 315f
Steel corrosion potential, 297
Steel corrosion products, 295f, 315
Steel corrosion reduction, 269
Steel-crack resistant, 237
Steel cracking, 155
Steel deformation, 178
Steel-dimensional change, 268
Steel ductility, 143, 271
Steel-element partition, 448
Steel fatigue failure, 272
Steel fatigue lifetime, 268, 273
Steel fatigue reduction, 273
Steel-ferritic, 288
Steel-galvanized, 316f
Steel grain boundaries, 295
Steel-heat treatment, 161
Steel-high strength, 237
Steel-hydrogen embrittlement, 319
Steel implantation profiles, 273
Steel in chloride brine, 316
Steel in fluoride brine, 316
Steel-in H_2S, 319
Steel-in SO_2 environemnt, 322
Steel industry, 153f
Steel-ion implanted, 267f, 316
Steel-ion penetration depth, 270
Steel-low alloy, 91, 270
Steel-low carbon, 270
Steel-lubricated sliding, 269
Steel-mechanical polishing, 282
Steel-mechanical properties, 185
Steel nitride formation, 282
Steel-nitrided, 271, 316f
Steel nitrides-highly dispersed, 276f
Steel-nitrogen implanted, 267f, 288
Steel-nitrogen implantation layers, 273
Steel-nominal ion dose, 270
Steel passive potential, 298
Steel passivity, 298
Steel-phosphate layers, 295
Steel phosphatization, 295f
Steel-phosphorus partition, 448
Steel-quenched, 277
Steel radiation damage, 269
Steel-rapid analysis, 137
Steel-rapid quenching, 143
Steel residual stress, 164
Steel-retained dose, 284
Steel rust, 304
Steel rust transformers, 295, 302
Steel shot peening, 273
Steel-slag partition, 447
Steel sliding wear tests, 270
Steel sputtering techniques, 301
Steel stress corrosion, 164
Steel stress cracking, 319
Steel substrate adhesion, 295
Steel surface-amorphous phase, 162
Steel surface analysis, 273
Steel surface characterization, 267f
Steel surface corrosion, 315f
Steel surface finish, 269
Steel surface irregularities, 156
Steel surface layer growth, 323
Steel surface layers, 316f
Steel surface nitrides, 278
Steel surface protection, 295f
Steel surface reactions, 295
Steel surface wear resistance, 161
Steel-tempered, 277
Steel-tempering, 237f
Steel-thermal processing, 143
Steel-tin coated, 316
Steel-toughness, 237
Steel transpassive potential, 298
Steel-volumetric wear rate, 269
Steel wear reduction, 269
Steel wear resistance, 268, 271, 290
Step function distributions, 133
Strain free atom packing, 105
Stress corrosion cracking, 155, 164
Stress corrosion in steel, 164
Stress cracking of steel, 319
Stress in steel, 165
Structural homogeneity of alloys, 329f, 351
Structural microanalysis, 83
Structural promoters, 667f
Structural relaxation in alloys, 330
Structural relaxation in amorphous alloys, 371f
Structure of SiO_2 supported catalysts, 630
Subbituminous coal, 147
Submonolayer tin coverage, 722
Substitutional cobalt in aluminum, 194f
Substitutional cobalt in gold, 698
Substitutional glass structure, 425
Substitutional impurity, 26
Substitutional iron atoms, 104
Substitutional size mismatch, 554
Substrate adhesion on steel, 295
Subsurface microanalysis, 83
Subsurface nitride phase, 133f
Sulfates in coal, 494f
Sulfidation, 217f
Sulfidation-parabolic law, 218f
Sulfidation rate, 230

Sulfidation surface enrichment, 232
Sulfidation temperature, 692
Sulfide formation, 664
Sulfided catalysts, 683f
Sulfidized alloys, 222
Sulfur content of coal, 154
Sulfur corrosion, 217f
Sulfur corrosion protection, 217f
Sulfur determination in coal, 497
Sulfur dioxide, 664
Sulfur leaching curves, 493
Sulfur pollution, 683
Sulfur vapor, 220
Sulfuric acid, 492
Superconducting transition temperatures, 47
Superconductivity and magnetic order, 47
Superconductivity-dimensional crossover, 747
Superconductors, 46
Superferromagnetism, 79
Superoperator formalism, 54
Superparamagnetic blocking temperature, 516
Superparamagnetic coal, 516
Superparamagnetic component, 674
Superparamagnetic Fe_2O_3, 282, 654
Superparamagnetic goethite, 567
Superparamagnetic iron, 606
Superparamagnetic phase, 182
Superparamagnetic relaxation, 64, 70f, 399
Superparamagnetic relaxation-clusters, 403
Superparamagnetism, 191, 455
Superparamagnetism in catalysts, 610
Superphosphate, 533
Superstructure Bragg reflections, 741f
Supported bimetallic catalysts, 620f, 650f
Supported catalyst structure, 630
Supported Fischer-Tropsch catalysts, 611f
Supported noble metal catalysts, 635
Supported rhodium-iron catalysts, 635
Surface analysis, 297
Surface anisotropy, 744
Surface area, 595f
Surface atom mobility, 628
Surface atoms, 72
Surface boriding, 161
Surface characterization of steels, 267f
Surface complexes in catalysis, 616
Surface conductivity, 221
Surface corrosion, 166
Surface coverage-tin, 721
Surface crystallization, 41, 391
Surface crystalline layers, 41
Surface Debye temperature, 610
Surface density with tin, 721
Surface effects in glass, 442
Surface electrochemical processes, 706
Surface electrodeposits, 697
Surface enriched iron catalysts, 645
Surface enrichment, 232
Surface failure, 10
Surface hardness, 175
Surface-high polish, 10
Surface impurity atoms, 103
Surface impurity defects, 103
Surface irregularities, 156
Surface layer composition, 226
Surface layer growth, 323
Surface layer heat treatment, 41
Surface layer inhomogeneous, 396
Surface layer-iron oxide, 653
Surface layer-microstructure, 224
Surface layer sensitivity, 133
Surface luster, 228
Surface morphology, 162
Surface nitrides in steel, 278
Surface nitriding, 161
Surface oxidation of amorphous ribbons, 391
Surface oxidation potential, 104
Surface phase analysis, 121
Surface phase composition, 224
Surface polishing, 44
Surface protection of steel, 295f
Surface reactions on steel, 295
Surface scale-morphology, 222
Surface scale thickness, 218
Surface scoring, 10
Surface steel corrosion, 162
Surface structure-catalyst, 643
Surface sulfidation, 217
Surface thin films, 716
Surface treated steels, 153, 162
Surface unevenness, 385
Surface wear protection, 161
Surface-bulk crystallization, 41
Surfactants, 167
SYNROC, 410
SYNROC corrosion resistance, 424
SYNROC glasses, 424
Synterized ore, 479
Synthesis gas, 609f, 619
Synthesis gas bimetallic catalysts, 629
Synthesis of ethanol, 635f
Synthesis of hydrocarbons, 635f
Synthesis of methanol, 635f
Synthetic corrosion products, 586
Synthetic fuel production, 165

Synthetic rock, 424
Szomolnokite, 494f, 528

Talc, 547
Tantalum-181, 28, 85, 316
Ta_2O_5, 297, 302
Tellurium-tin-germanium semiconductors, 735
Tellurium-tin-lead semiconductors, 725f
Tellurium-125, 95, 731
Temper embrittlement, 11
Temperature dependence of f-factor, 735
Temperature programmed reduction, 627
Tempered steel, 277
Tempering, 259
Tempering of austenite, 259
Tempering of steel, 237f
Tempering sequences, 253
Tensile force in steel, 165
Tensile properties of steels, 173
Tensile stress, 700
Tetrahedral random network, 105
Texture, 26, 33, 204
Texture in ferromagnetic alloys, 378
Texture in thin films, 741
Texture-statistical description, 378
Thermal analysis, 533f
Thermal excitations in spin glass, 399
Thermal processing, 143
Thermal reaction mechanism, 547f
Thermal scanning techniques, 165
Thermal treatment, 378
Thermal treatment of minerals, 550
Thermo-mechanical stablity, 259
Thermoelectric energy conversion, 725
Thermoelectric microprobe analysis, 726f
Thermogravimetric analysis, 220, 534
Thermogravimetric zeolite studies, 671
Thermonuclear devices, 83
Thickness distortion, 241
Thickness effects, 33, 529
Thickness monitor, 741
Thin ferromagnetic films, 743
Thin film alloys, 742
Thin film evaporation, 99
Thin film gold alloys, 112
Thin film interdiffusion, 122
Thin film interface sites, 742
Thin film interfaces, 83
Thin film iron foils, 100
Thin film magnetization, 100, 742
Thin film microanalysis, 83
Thin film overlays, 83
Thin film spontaneous magnetization, 743
Thin film texture, 741
Thin films on glass, 716f
Thiobacillus ferrooxidans, 508
Thiobacillus thiooxidans, 508
Thioether in coal, 496
Thiourea, 167
Thoria structural promoter, 667
Time averaged resonant signal, 138
Tin-antimony catalysts, 660
Tin chalcogenides, 727f
Tin chloride, 716f
Tin chloride wetting agents, 717
Tin coated steel, 316f
Tin electrode passivation, 708
Tin electrodeposits, 702
Tin extraction, 467
Tin-germanium -tellurium semiconductors, 735
Tin glass surface complex, 715f
Tin insertion reactions, 716f
Tin-lead-tellurium semiconductors, 725f
Tin metal, 468
Tin-nickel codeposits, 702
Tin oxidation, 716f
Tin oxidation state, 718
Tin-oxidative addition, 716f
Tin penetration depth, 468
Tin smelting furnaces, 467f
Tin surface complex, 715f
Tin-tellurium doped with iron, 726f
Tin tetrachloride, 661
Tin vapor transport, 467
Tin-119, 84f, 89, 91, 95, 98, 103, 111, 138, 273, 316, 467f, 470, 661, 702, 715f, 726, 730f
Tin-119 CEMS, 108
Tin-119 enriched shot, 718
Tin-119 in iron-palladium alloys, 13
Tin-119 in nickel-iron alloys, 11
Tin(II) chloride, 461
Tin(II) compounds, 730
Tin(II) glassy phases, 467
Tin(II) hydroxide, 710
Tin(II) oxide, 710
Tin(IV) oxide, 662
TiO_2, 106, 467, 542
Titanium containing alloys, 330, 344
Titanium containing glasses, 344
Titanium-iron alloys, 12
Titanium-iron-hydrogen alloys, 28
Tokamak, 108
Tool inserts, 269

Tool steel, 269
Topological order in alloys, 346
Topotactic transformations, 585
Torque curve, 385
Tramp element, 11
Transition metal-metalloid alloys, 38
Transition probability, 31
Transmission electron microscopy, 10, 208, 238, 257, 273, 275
Transpassive potential, 298
Transverse acoustic modes, 734
Transverse optical modes, 734
Tribological effects, 269
Triboxidation, 162
Trimetallic catalysts, 635f
Trioctahedral brucite layer, 592
Tritium breeding reaction, 111
Tunable infrared diode lasers, 725
Tungsten-182, 84, 95
Two-dimensional defects, 26
Two disk machine, 10
TWR desulfurization process, 504
TWR process, 491
$T_{80}M_{20}$, 31, 38

Ultra-high vacuum CEMS, 99
Ultrex powder, 717
Ulvospinel ferrian, 476
Unpromoted Fischer-Tropsch synthesis, 611f
Unsupported Fischer-Tropsch catalysts, 611f
Unweathered coal, 157
Uranium(IV) network modifier, 425

Vacancies, 26, 193
Vacancy oversaturation, 218
Vacuum deposition, 740
Valence conduction band gap, 731
Validity diagram, 36
van der Waals bonding, 31
Vanadium containing alloys, 344
Vanadium containing glasses, 344
Vanadium field distributions, 746
Vanadium hyperfine fields, 746
Vanadium in steel, 161
Vanadium-iron multilayers, 740f
Vanadium oxide bronzes, 399
Vanadium-silver multilayers, 740
Vanadium-51 NMR, 746
Vapor condensation, 31
Vapor deposition, 46
Vapor deposition techniques, 44
Varience of hyperfine field, 245f
Velocity averaged resonant signal, 138
Verway transition, 75
Verway transition temperature, 75
Vibration monitoring, 8
Vicker's hardness, 173
Victorian brown coal, 513
Vitrification of waste products, 424
Vitrovac, 32
Vivianite, 299f
Volcanic activity, 567
Volume of atomic solutes, 242
Volume percent austenite in steel, 142
Volumetric wear rate, 269

Waste product vitrification, 424
Water gas, 606
Water purification, 595
Water vapor-internal pressure in ores, 479
Wavellite, 535
Wear layers, 8
Wear reduction in steel, 269
Wear resistance, 122
Wear resistance in steel, 268, 290
Wear surfaces, 10
Wear tests, 271
Weathered coal, 166
Weathered iron containing material, 565f
Weathered pyrite, 494f
Weathering of iron, 565f
Weight function, 130
Welding characteristics, 155
Welding steel, 164
Wet agglomeration changes in coal, 520
Wet coal agglomeration, 513
Wetting agents-glass, 717
White layer problem, 10
Whitlockite, 536
Wigner-Seitz cell, 15
Window method, 371
Wohlfarth distribution function, 401
Work hardening, 186
Work hardening rate, 173
Wustite, 157, 452f, 459

X-ray diffraction, 10
X-ray diffraction analysis, 531
X-ray diffraction line broadening, 441
X-ray fluorescence, 140, 157f, 239, 448
X-ray microanalysis, 217, 221, 224, 232
X-ray photoelectron spectroscopy, 55, 271, 628, 636
X-ray reemission Mossbauer spectra, 173
X-ray scattering in amorphous alloys, 351
X-satellite, 250
X-type zeolite, 671

Y-type zeolite, 671
Yallourn coal, 516
Young's modulus, 479
Ytterbium-170, 95

Zeolite based iron catalysts, 667f
Zeolite Fe-LTA, 670
Zeolite metal interaction, 678
Zeolite Na-X, 670
Zeolite Na-Y, 670
Zeolite neutron activation analysis, 670
Zeolite porus structures, 667f
Zeolite thermogravimetric studies, 671
Zeolite X-type, 671
Zeolite Y-type, 671
Zero dimension defects, 26
Zero valent iron, 624f
Zero velocity methods, 156
Zig-zag domains, 386, 395
Zinc chloride coal treatment, 501
Zinc chloride glass, 409
Zinc corrosion inhibition, 699
Zinc electrogalvanization, 699
Zinc-iron alloy, 16
Zinc phosphate, 316
Zircaloy corrosion, 108
Zirconium tetrafluoride glass, 409
Zirconolite, 424
ZnO, 467
ZrO_2, 55
$Zr_{10}Fe_{90}$, 48
$Zr_{11}Fe_{89}$, 47
$Zr_{50}Fe_{50}$, 49
$Zr_{76}Fe_{24}$, 28, 53
Zr_8Fe_{92}, 48
$Zr_{80}Fe_{20}$, 47